CRC Handbook of tables for Organic Compound Identification

CRC HANDBOOK

of tables for

ORGANIC COMPOUND IDENTIFICATION

Third Edition

Compiled by
ZVI RAPPOPORT, Ph.D.
Hebrew University of Jerusalem, Israel

CRC Press, Inc.
Boca Raton, Florida

Direct all inquiries to CRC Press, Inc., 2000 Corporate Blvd., N.W., Boca Raton, Florida, 33431.

© 1967 by CRC Press, Inc.
Formerly the Chemical Rubber Company
2nd Printing, April 1972
3rd Printing, May 1973
4th Printing, April 1975
5th Printing, January 1976
6th Printing, December 1976
7th Printing, September 1977
8th Printing, June 1979
9th Printing, September 1980
10th Printing, July 1981
11th Printing, March 1983
12th Printing, December 1983
13th Printing, November 1984

International Standard Book Number (ISBN) 0-8493-0303-6
Former International Standard Book Number (ISBN) 0-87819-303-0

Library of Congress Card No. 63-19660
Printed in the United States.

First Edition
Tables for Identification of Organic Compounds

Compiled by
Max Frankel, Ph.D.
Saul Patai, Ph.D.

Assisted by
Albert Zikha, Ph.D.
Robert Farkas—Kadmon

Second Edition
Tables for Identification of Organic Compounds

Compiled by
Max Frankel, Ph.D.
Saul Patai, Ph.D.

Assisted by
Albert Zilkha, Ph.D.
Zvi Rappoport, Ph.D.
Robert Farkas—Kadmon

Third Edition
Handbook of Tables for Organic Compound Identification

Compiled by
Zvi Rappoport, Ph.D.

CONTROLLED RELEASE PESTICIDES FORMULATIONS
By **Nate F. Cardarelli, M.S.,** University of Akron.

DIFFUSE REFLECTANCE SPECTROSCOPY IN ENVIRONMENTAL PROBLEM–SOLVING
By **R. W. Frei, Ph.D.,** Sandox, Ltd. (Switzerland), and **J. D. MacNeil, M.Sc., Ph.D.,** Canada Department of Agriculture.

DRUGS AS TERATOGENS
By **James L. Schardein, B.A., M.S.,** Parke, Davis and Company.

FUNDAMENTAL MEASURES AND CONSTANTS FOR SCIENCE AND TECHNOLOGY
By **Frederick D. Rossini, Ph.D.,** Rice University, Houston.

IMMUNOASSAYS FOR DRUGS SUBJECT TO ABUSE
Edited by **S. J. Mulé, Ph.D.,** New York State Narcotic Addiction Control Commission, et al.

MASS SPECTROSCOPY OF PESTICIDES AND POLLUTANTS
By **Stephen Safe, B.Sc., M.Sc., D.Phil.,** and Otto **Hutzinger, Ing.Chem., M.Sc., Ph.D.,** National Research Council of Canada.

MERCURY IN THE ENVIRONMENT
By **Lars T. Friberg, M.D.,** National Institute of Public Health (Stockholm), and **Jaroslav J. Vostal, M.D., Ph.D.,** University of Rochester.

ORGANOPHOSPHORUS PESTICIDES: ORGANIC AND BIOLOGICAL CHEMISTRY
By **Morifusa Eto, Ph.D.,** Kyushu University, Japan.

RECENT DEVELOPMENTS IN SEPARATION SCIENCE
Edited by **Norman N. Li, Sc.D.,** Exxon Research and Engineering Co.

TRACE ELEMENT MEASUREMENTS AT THE COAL-FIRED STEAM PLANT
By **W. S. Lyon, Jr. B.S., M.S.,** Oak Ridge National Laboratory.

CRC CRITICAL REVIEWS^{TM} JOURNALS:

CRC CRITICAL REVIEWS^{TM} IN ANALYTICAL CHEMISTRY
Edited by **Bruce H. Campbell, Ph.D.,** J. T. Baker Chemical Co.

CRC CRITICAL REVIEWS^{TM} IN TOXICOLOGY
Edited by **Leon Golberg, M.D., B.Chir., D.Sc., D.Phil., F.R.C.Path.,** Chemical Industry Institute of Toxicology

Direct inquiries to CRC Press, Inc.

PREFACE

The present volume is a revised and enlarged third edition of the book formerly titled TABLES FOR IDENTIFICATION OF ORGANIC COMPOUNDS. Four new classes of compounds, i.e., sulfonyl chlorides, sulfonamides, thiols and thioethers were added, bringing the number of classes included in the book to twenty-six. The tables of alkanes, alkenes, alkynes, aromatic hydrocarbons, phenols, nitriles and sulfonic acids were all thoroughly revised and considerably enlarged. In all, the addition of 2400 compounds to the third edition, raised the total number of parent compounds in the book to over 8150.

Three tables containing the dissociation constants of more than 1050 phenols, organic acids and organic bases were added. Correlation charts for I.R., Far I.R. and N.M.R. were also included.

Explanatory sections entitled "Explanations and References" precede the tables. In these sections the formulas of the derivatives and the full reaction equations for their preparation by the most important methods, together with some essential details and references to their preparations, are given.

An index covering both the names and synonyms of the compounds was added at the end of the book. An index listing the names of all tables and major subjects was also added.

The main objective of this book is to assist chemists in the identification of organic compounds. The organization of the compounds in classes according to increasing boiling points should also assist in the search for standard vapor phase chromatography work.

For further information of techniques of organic analysis one or more of the following books should be consulted:

N. D. Cheronis, J. B. Entrikin and E. M. Hodnett, *Semimicro Qualitative Organic Analysis*, 3rd Ed., Interscience Publishers, New York, 1965.

L. Meites, *Handbook of Analytical Chemistry*, McGraw Hill Book Co., 1963.

F. Feigl, *Spot Tests in Organic Analysis*, 6th Ed., Elsevier Publishing Co., 1960.

A. I. Vogel, *A Textbook of Practical Organic Chemistry*, 3rd Ed., Longmans Green and Co., London, 1957.

R. L. Shriner, R. C. Fuson and D. Y. Curtin, *The Systematic Identification of Organic Compounds*, 4th Ed., John Wiley and Sons, New York, 1956.

F. Wild, *Characterization of Organic Compounds*, 2nd Ed., Cambridge University Press, 1958.

The publication of the third edition would not have been possible without the work of those involved in the preparation of the earlier editions. The editor is thankful for the part in the earlier work contributed by Professors M. Frankel, S. Patai and A. Zilkha, and the late Mr. R. Farkas-Kadmon. Thanks are due also to Mrs. E. Shohamy and Mr. A. Glazer for assisting in collection of new data, to Mrs. Y. Elmaleh for typing the index, and especially to Prof. S. Patai, who as consulting editor, read all the new material and gave many helpful suggestions. The cooperation of Dr. Robert C. Weast and Mrs. F. Thomas in the publication of the book is gratefully acknowledged.

Z. R.

Jerusalem
January 1967

Contents

	Table No.	Page No.
Abbreviations		IX
Explanations and References to the Tables		1
Alkanes and Cycloalkanes	I	2–9
Gases, Liquids, Solids		
Alkenes, Cycloalkenes, Dienes and Polyenes	II	10–26
Liquids, Solids		
Alkynes (Acetylenes)	III	27–31
Liquids, Solids		
Aromatic Hydrocarbons	IV	32–51
Liquids, Solids		
Halides	V	52–76

A) Alkyl and cycloalkyl halides
    1. Chlorides, Liquids and Solids
    2. Bromides, Liquids and Solids
    3. Iodides, Liquids and Solids
B) Dihalides and polyhalides (non-aromatic)
    1. Fluorides
    2. Chlorides, Liquids and Solids
    3. Bromides, Liquids and Solids
    4. Iodides, Liquids and Solids
C) Aryl halides
    1. Fluorides, Liquids
    2. Chlorides, Liquids and Solids
    3. Bromides, Liquids and Solids
    4. Iodides, Liquids and Solids

Alcohols	VI	77–90
Liquids, Solids		
Phenols	VII	91–128
Liquids, Solids		
Ethers	VIII	129–140
Liquids, Solids		
Aldehydes	IX	141–160
Liquids, Solids		
Ketones	X	141–143; 161–179
Quinones	XI	180–185
Carboxylic Acids	XII	186–189; 186–210
Liquids, Solids		
Acyl Halides	XIII	186–189; 211–222
Acid Fluorides		
Acyl Bromides, Liquids, Solids		
Acyl Chlorides, Liquids, Solids		
Acyl Iodides, Liquids, Solids		
Acid Anhydrides	XIV	223–227
Liquids, Solids		
Amides and Imides	XV	228–247
Liquids, Solids		
Esters	XVI	248–281
Liquids, Solids		

Contents (Continued)

	Table No.	Page No.
Amino Acids	XVII	282–290
Amines	XVIII	291–325
Primary and Secondary Amines		
Tertiary Amines		
Liquids, Solids		
Carbohydrates	XIX	326–333
Liquids, Solids		
Nitro Compounds	XX	334–343
Liquids, Solids		
Nitriles	XXI	344–368
Liquids, Solids		
Sulfonic Acids	XXII	369–380
Sulfonyl Chlorides	XXIII	369–370; 381–390
Sulfonamides and Sulfonanilides	XXIV	391–410
Thiols (Mercaptans)	XXV	411–417
Thioethers (Sulfides)	XXVI	418–427
Acid Dissociation Constants of Organic Acids in Aqueous Solution	XXVII	428–433
Acid Dissociation Constants of Phenols in Aqueous Solution	XXVIII	428; 434–435
Dissociation Constants of Organic Bases in Aqueous Solution	XXIX	428; 436–439
I.R.—Infra-red Correlations Charts		441–447
Far I.R.—Far Infra-red Vibrational Frequency Correlation Charts		448–449
NMR—Characteristic NMR Spectral Positions for Hydrogen in Organic Structures		450
Miscibility of Organic Solvent Pairs		451–453
Emergent Stem Correction for Liquid-in-Glass Thermometers		454
Correction of Boiling Points to Standard Pressure		455–456
Molecular Elevation of the Boiling Point		457
Molecular Depression of the Freezing Point		457
Carbohydrates		458–465
Fats and Oils		466–467
Waxes		468
Diamagnetic Susceptibilities of Organic Compounds		469–474
Four-Place Logarithms		475–476
Periodic Table of the Elements		477
Table of Atomic Weights		478
Index of Organic Compounds		479–560
Index listing names of tables and major subjects		561–563
Request for New Data		564

LIST OF ABBREVIATIONS USED

$[\alpha]_D^t$	specific rotation	fum.	fuming	pr.	prisms
a.	acid	glac.	glacial	purp.	purple
abs.	absolute	glit.	glittering	pyr.	pyridine
abt.	about	glyc.	glycerol	rac.	racemic (or: racemate)
ac. a.	acetic acid	gran.	granular		
ac. anh.	acetic anhydride	grn.	green	rect.	rectangular
acet.	acetone	h.	hot	recr.	recrystallization
add.	addition	htng.	heating	redsh.	reddish
al.	alcohol	hyd.	hydrate or hydrolyses	rhomb.	rhombic
alk.	alkali			r.	rapid
amor.	amorphous	hyg., hygr.	hygroscopic	r.h.	rapid heating
anh.	anhydrous	*i.*	inactive	s.	soluble
arom.	aromatic	i.	insoluble	sec.	secondary
aqu.	aqueous	ign.	ignites	scar.	scarcely
asym., as.	asymmetric	insol.	insoluble	s. h.	slow heating
bl.	blue	l., L.	levorotatory (or: L-configuration)	sh.	short
blk.	black			sl.	slightly
boil.	boiling			sld.	solid
b.p., B.P.	boiling point	leaf., lf.	leaflets	slend.	slender
br.	brown	lg.	large	sm.	small
bz.	benzene	lgr.	ligroin	soft.	softens
brt.	bright	liq.	liquid	sol., soln.	solution(s)
brnsh.	brownish	lng.	long	solv.	solvent(s)
c.	cold	lt.	light	st.	steel
ca.	about	lvs.	leaves	stab.	stable
caust.	caustic	*m-*	meta	subl.	sublimes
chl.	chloroform	me., meth.	methyl	*sym.*	symmetrical
cl.-bz.	chlorobenzene	micr.	microscopic	tab., tabl.	tablet(s), tables
col.	colorless	mixt.	mixture	*tert.*	tertiary
comp.	compound	ml.	milliliter	tetr.	tetragonal
conc.	concentrated	mod.	modification	tol.	toluene
cor.	corrected	monohyd.	monohydrate	trans.	transparent
cr., cryst.	crystals	monocl.	monoclinic	thk.	thick
d.	decomposes	m.p., M.P.	melting point	tricl.	triclinic
d., D.	dextrorotatory (or: D-configuration)	need., nd.	needles	trim.	trimeric
		o-	ortho	uns.	unsymmetrical
deriv.	derivative	ol.	olive	unst.	unstable
deliq.	deliquescent	or.	orange	vac.	vacuum, in vacuo
dil.	dilute	ord.	ordinary	v.	very
dist.	distillate	org.	organic	var.	variable
dk.	dark	orth.	orthorhombic	*vic-*	vicinal
dl., d,l; D,L.	racemic	oxid.	oxidation	visc.	viscous
efflor.	efflorescent	*p-*	para	volat.	volatile or volatilizes
et.	ethyl	pa.	pale	vlt.	violet
et. ac.	ethyl acetate	part.	partly	w.	water
eth.	ether	pet.	petroleum	wh.	white
exp.	explodes	pet. eth.	petroleum ether	yel.	yellow
f.	from	ph.	phenyl	yelsh, ylsh.	yellowish
fl.	flakes	ph. hydraz.	phenyl hydrazine	>	above, greater than
fluores.	fluorescent	$PhNO_2$	nitrobenzene	<	below, smaller than
f.p.	freezing point	pl.	plates	∞	miscible
frz.	freezes	powd.	powder	xyl.	xylene

EXPLANATIONS AND REFERENCES TO THE TABLES

The following section gives explanations and references for the preparation of the derivatives appearing in the Tables. Formulas of the derivatives as well as the main methods for their preparation are given. Usually, only the reagents and the solvents required for the preparation of a derivative are mentioned without specific details for the reaction conditions and the exact procedure. The aim of these notes is mainly to enable the worker to choose the method preferable in the conditions and the reagents available to him in his laboratory for the derivatization of his specific compound. However, THIS IS ONLY A REFERENCE SECTION AND NOT AN INSTRUCTION MANUAL AND THE QUOTED REFERENCES SHOULD BE CONSULTED FOR THE ACTUAL PREPARATION OF DERIVATIVES, ESPECIALLY REGARDING SAFETY HAZARDS INVOLVED IN THE WORK.

References are usually given for the preparation of all the derivatives having separate columns in the Tables, as well as for important ones listed in the "miscellaneous" section of the Tables. References to five different popular analytical textbooks are given, assuming that at least one of them, or another equivalent publication, would be available to the worker. These are:

N. D. Cheronis, J. B. Entrikin and E. M. Hodnett, *Semimicro Qualitative Organic Analysis*, 3rd edition, Interscience, New York, 1965, quoted in the text as "Cheronis."

R. P. Linstead and B. C. L. Weedon, *A Guide to Qualitative Organic Chemical Analysis*, Butterworth Scientific Publication, London, 1956, quoted in the text as "Linstead."

R. L. Shriner, R. C. Fuson and D. Y. Curtin, *The Systematic Identification of Organic Compounds*, 4th edition, John Wiley and Sons, New York, 1956, quoted in the text as "Shriner."

A. I. Vogel, *A Textbook of Practical Organic Chemistry*, 3rd edition, Longmans, Green and Co., London, 1957, quoted in the text as "Vogel."

F. Wild, *Characterization of Organic Compounds*, 2nd edition, Cambridge University Press, Cambridge, 1958, quoted in the text as "Wild."

In addition, leading references from the original literature are also given. Although the literature coverage is not complete (especially for the common derivatives) it was attempted to describe different methods, and to give as many references as possible to less common derivatives having limited scope. More references can be found in the textbooks mentioned above.

Derivatives appear either in a separate column or in the "miscellaneous" section in the Tables, where separate columns are usually given for derivatives which should be tried first, and for which enough data are available. Derivatives which should be tried as a second choice, or preferred derivatives for which not enough data are available appear in the "miscellaneous" section. The explanations and the references for the different derivatives are arranged usually in the same order as in the Tables. Occasionally, this order is changed in the explanatory notes in order to describe the derivatives in a logical order (e.g., in Table 17 the phenylurethane appears in a separate column, while the phenylhydantoin appears in the "miscellaneous" section; in the "explanations and references" section the phenylhydantoin appears directly after the phenylurethane).

Derivatives which are followed by an asterisk are those recommended for first trial. Other derivatives should be tried after these.

Although "Ar" usually stands for monovalent aromatic group, we used it a few times in the following sections as a polyvalent aromatic residue. This was done only for demonstration purposes.

1

EXPLANATIONS AND REFERENCES TO TABLE I

As a result of their inertness no general suitable derivative exists for alkanes and cycloalkanes. Characterization is based only on the physical constants given in the Table: melting and boiling points, index of refraction and density. Any laboratory text-book will give adequate directions for the determination of these constants.

WARNING: This is not an instruction manual. References should be consulted for the preparation of derivatives.

TABLE I. ALKANES AND CYCLOALKANES

TABLE I. ALKANES AND CYCLOALKANES
a) Gases and Liquids (Listed in order of increasing b.p.*)**

No.	Name	Boiling point, °C	Melting point, °C	n_D^{20}	D_4^{20}
1	Methane	−161.49	−182.48[T]		
2	Ethane	−88.63	−183.27[T]		
3	Propane	−42.07	−187.69[T]		0.5005[S]
4	Cyclopropane	−32.86	−127.42		0.720[−79]
5	2-Methylpropane (Isobutane)	−11.73	−159.6		0.5572[S]
6	n-Butane	−0.50	−138.35	1.3326[S]	0.5788[S]
7	2,2-Dimethylpropane (Neopentane)	9.503	−16.55	1.342[S]	0.5910[S]
8	Cyclobutane	13.08[741]; 12.5	−80.	1.37520[0]	0.7038[0]
9	1,1-Dimethylcyclopropane	20.63	−108.96	1.3668	0.6589
10	2-Methylbutane	27.852	−159.9	1.35373	0.61967
11	trans-1,2-Dimethylcyclopropane	29		1.3713	0.6769
12	Ethylcyclopropane	35.94	−149.41	1.3786	0.6839
13	n-Pentane	36.074	−129.721	1.35748	0.62624
14	Methylcyclobutane	36.3		1.3830	0.6933
15	cis-1,2-Dimethylcyclopropane	37	−140.9	1.3822	0.6928
16	Spiropentane	38.977	−107.06	1.41200	0.755
17	Cyclopentane	49.262	−93.879	1.40645	0.74538
18	2,2-Dimethylbutane	49.741	−99.87	1.36876	0.64916
19	1,1,2-Trimethylcyclopropane	56–7[750]		1.3848[19.5]	0.6822[19.5]
20	2,3-Dimethylbutane	57.988	−128.538	1.37495	0.66164
21	2-Methylpentane	60.271	−153.67	1.37145	0.65315
22	3-Methylpentane	63.282		1.37652	0.66431
23	1,2,3-Trimethylcyclopropane	65–7[755]		1.3945[18]	0.6946[18]
24	n-Hexane	68.74	−95.348	1.37486	0.65937
25	Ethylcyclobutane	70.64	−142.85	1.4020	0.7280
26	Methylcyclopentane	71.812	−142.455	1.4097	0.74864
27	2,2-Dimethylpentane	79.197	−123.811	1.38215	0.67385
28	2,4-Dimethylpentane	80.5	−119.242	1.38145	0.67270
29	Cyclohexane	80.738	6.554	1.42623	0.77855
30	2,2,3-Trimethylbutane	80.882	−24.912	1.38944	0.69011
31	3,3-Dimethylpentane	86.064	−134.46	1.39092	0.69327
32	1,1-Dimethylcyclopentane	87.846	−69.795	1.41356	0.75448
33	2,3-Dimethylpentane	89.784		1.39196	0.69508
34	2-Methylhexane	90.052	−118.276	1.38485	0.67859
35	trans-1,3-Dimethylcyclopentane	90.773	−133.702	1.40894	0.74479
36	cis-1,3-Dimethylcyclopentane	91.725	−133.975	1.41074	0.74880
37	3-Methylhexane	91.850		1.38864	0.68713
38	trans-1,2-Dimethylcyclopentane	91.869	−117.58	1.41200	0.75144
39	3-Ethylpentane	93.475	−118.604	1.39339	0.69816
40	Quadricyclane (Quadricyclo [2,2,1,0[2,6],0[3,5]] heptane)	98		1.4804	
41	n-Heptane	98.427	−90.61	1.38764	0.68376
42	2,2,4-Trimethylpentane	99.238	−107.38	1.39145	0.69192
43	cis-1,2-Dimethylcyclopentane	99.532	−53.892	1.42217	0.77262
44	Methylcyclohexane	100.934	−126.593	1.42312	0.76939
45	Ethylcyclopentane	103.466	−138.446	1.41981	0.76647
46	1,1,3-Trimethylcyclopentane	104.893	−142.44	1.41119	0.74825
47	2,2-Dimethylhexane	106.84	−121.18	1.39349	0.69528
48	2,5-Dimethylhexane	109.103	−91.20	1.39246	0.69354
49	1,trans-2,cis-4-Trimethylcyclopentane	109.29	−130.78	1.41060	0.74727
50	2,4-Dimethylhexane	109.429		1.39534	0.70036
51	2,2,3-Trimethylpentane	109.841	−112.27	1.40295	0.71602
52	1,trans-2,cis-3-Trimethylcyclopentane	110.2	−112.705	1.4138	0.7535
53	3,3-Dimethylhexane	111.969	−126.1	1.40009	0.7100
54	2,3,4-Trimethylpentane	113.467	−109.2Γ	1.40422	0.71906
55	1,1,2-Trimethylcyclopentane	113.729	−21.64	1.42229	0.77252
56	2,3,3-Trimethylpentane	114.76	−100.70	1.40750	0.72619
57	2,3-Dimethylhexane	115.607		1.40113	0.71214
58	3-Ethyl-2-methylpentane	115.65	−114.96	1.40401	0.71932

*Derivative data given in order: m.p., crystal color, solvent from which crystallized.

**T = triple point; S = at saturation pressure.

No.	Name	Boiling point, °C	Melting point, °C	n_D^{20}	D_4^{20}
59	1,cis-2,trans-4-Trimethylcyclopentane	116.731	−132.55	1.41855	0.76345
60	1,cis-2,trans-3-Trimethylcyclopentane	117.5	−112.	1.4218	0.7704
61	2-Methylheptane	117.647	−109.04	1.39494	0.69792
62	4-Methylheptane	117.709	−120.955	1.39792	0.70463
63	3,4-Dimethylhexane	117.725	....	1.40406	0.71923
64	1,cis-2,cis-4-Trimethylcyclopentane	118.	....	1.422	0.766
65	3-Ethyl-3-methylpentane	118.259	−90.87	1.40775	0.72742
66	3-Ethylhexane	118.534	....	1.40162	0.71358
67	3-Methylheptane	118.925	−120.5	1.39848	0.70582
68	Cycloheptane (Suberane)	118–20	−7.98	1.4449	0.8275⁰
69	trans-1,4-Dimethylcyclohexane	119.351	−36.962	1.42090	0.76255
70	1,1-Dimethylcyclohexane	119.543	−33.495	1.42900	0.78094
71	cis-1,3-Dimethylcyclohexane	120.088	−75.573	1.42294	0.76603
72	trans-1-Ethyl-3-methylcyclopentane	120.8	−108.	1.4186	0.7619
73	trans-1-Ethyl-2-methylcyclopentane	121.2	....	1.4219	0.7690
74	cis-1-Ethyl-3-methylcyclopentane	121.4	....	1.4203	0.7724
75	1-Ethyl-1-methylcyclopentane	121.522	−143.80	1.42718	0.78093
76	2,2,4,4-Tetramethylpentane	122.284	−66.54	1.40694	0.71947
77	1,cis-2,cis-3-Trimethylcyclopentane	123.0	−116.43	1.4262	0.7792
78	trans-1,2-Dimethylcyclohexane	123.419	−88.194	1.42695	0.77601
79	2,2,5-Trimethylhexane	124.084	−105.78	1.39972	0.70721
80	cis-1,4-Dimethylcyclohexane	124.321	−87.436	1.42966	0.78285
81	trans-1,3-Dimethylcyclohexane	124.45	−90.108	1.43085	0.78472
82	n-Octane	125.665	−56.795	1.39743	0.70252
83	Isopropylcyclopentane	126.419	−111.375	1.42582	0.77653
84	2,2,4-Trimethylhexane	126.54	−120.	1.4033	0.7156
85	cis-1-Ethyl-2-methylcyclopentane	128.050	−105.95	1.42933	0.78522
86	cis-1,2-Dimethylcyclohexane	129.728	−50.023	1.43596	0.79627
87	2,4,4-Trimethylhexane	130.38	−113.38	1.40745	0.72381
88	n-Propylcyclopentane	130.8	−118.7	1.4266	0.7761
89	2,3,5-Trimethylhexane	131.34	−127.8	1.4061	0.7219
90	Ethylcyclohexane	131.783	−111.323	1.43304	0.78792
91	2,2-Dimethylheptane	132.69	−113.0	1.4016	0.7105
92	2,2,3,4-Tetramethylpentane	133.016	−121.09	1.41472	0.73895
93	2,4-Dimethylheptane	133.5	....	1.4033	0.716
94	Methylcycloheptane	133–5	....	1.4410	0.8052
95	2,2,3-Trimethylhexane	133.6	....	1.4105	0.7292
96	4-Ethyl-2-methylhexane	133.8	....	1.4068	0.723
97	3-Ethyl-2,2-dimethylpentane	133.83	−99.2	1.4123	0.7348
98	4,4-Dimethylheptane	135.2	....	1.4076	0.725
99	2,6-Dimethylheptane	135.21	−102.9	1.4007	0.7089
100	2,5-Dimethylheptane	136.0	....	1.4038	0.715
101	3,5-Dimethylheptane	136.0	....	1.4067	0.723
102	Bicyclo[4.2.0]octane	136.0	....	1.4613	0.8573
103	cis-Bicyclo[3.3.0]octane	136–6.5	....	1.4595²⁵	0.8638²⁵
104	2,4-Dimethyl-3-ethylpentane	136.73	−122.2	1.4137	0.7379
105	1,1,3-Trimethylcyclohexane	137–8	....	1.4362	0.7868²⁵
106	3,3-Dimethylheptane	137.3	....	1.4085	0.725
107	2,2,5,5-Tetramethylhexane	137.5	....	1.40550	0.71875
108	2,3,3-Trimethylhexane	137.68	−116.80	1.4141	0.738
109	3-Ethyl-2-methylhexane	138	....	1.4120	0.731
110	trans-1,3,5-Trimethylcyclohexane	138.5–9⁷⁵⁴	....	1.42740ₕ	0.7720
111	2,3,4-Trimethylhexane	139.0	....	1.4144	0.7392
112	cis-1,3,5-Trimethylcyclohexane	140–0.5⁷⁵²	....	1.43010ₕ	0.7773
113	trans-1,2,4-Trimethylcyclohexane	140–1	....	1.43121ₕₑ	0.7813
114	2,2,3,3-Tetramethylpentane	140.274	−9.9	1.42360	0.75666
115	4-Ethyl-3-methylhexane	140.4	....	1.416	0.742
116	3,3,4-Trimethylhexane	140.46	−101.2	1.4178	0.7454
117	2,3-Dimethylheptane	140.5	....	1.4085	0.7260

*Derivative data given in order: m.p., crystal color, solvent from which crystallized.

No.	Name	Boiling point, °C	Melting point, °C	n_D^{20}	D_4^{20}
118	3,4-Dimethylheptane	140.6		1.4111	0.7314
119	3-Ethyl-3-methylhexane	140.6		1.4142	0.741
120	4-Ethylheptane	141.2		1.4096	0.730
121	2,3,3,4-Tetramethylpentane	141.551	−102.123	1.42222	0.75473
122	2,3-Dimethyl-3-ethylpentane	142.		1.419	0.754
123	trans-1,2,3-Trimethylcyclohexane	142–3.5[762]		1.43582[He]	0.7914
124	1-Isopropyl-3-methylcyclopentane (Pulegan)	142–4		1.4236	0.7730[22]
125	4-Methyloctane	142.48	−113.2	1.4061	0.7199
126	1-Isopropyl-2-methylcyclopentane	142.5[759]		1.4279	0.7792
127	3-Ethylheptane	143.0		1.4093	0.727
128	2-Methyloctane	143.26	−80.4	1.4031	0.7134
129	cis-1,2,3-Trimethylcyclohexane	144–6[753]		1.43682[He]	0.7930
130	3-Methyloctane	144.18	−107.6	1.4062	0.7207
131	2,4,6-Trimethylheptane	144.8		1.4071	0.7225
132	cis-1,2,4-Trimethylcyclohexane	146		1.43209	0.786
133	3,3-Diethylpentane	146.168	−33.11	1.42051	0.75359
134	2,2-Dimethyl-4-ethylhexane	147		1.4131	0.733
135	2,2,4-Trimethylheptane	147.7		1.4092	0.7275
136	2,2,4,5-Tetramethylhexane	147.8		1.41318	0.73546
137	2,2,5-Trimethylheptane	148		1.409	0.726
138	2,2,6-Trimethylheptane	148.2		1.4059	0.7195
139	2,2,3,5-Tetramethylhexane	148.4		1.4142	0.7378
140	Nopinane (7,7-Dimethylbicyclo[3.1.1]heptane)	149[747]		1.4641	0.8611[22/22]
141	trans-1-Ethyl-4-methylcyclohexane	149.05–.15	−80.8	1.4304	0.7798
142	Cyclooctane	150[750]	14 (4.3)	1.4586	0.8349
143	1-Ethyl-2-methylcyclohexane	150–2		1.432	0.784
144	n-Nonane	150.81	−53.519	1.40542	0.71763
145	1,3,3-Trimethylbicyclo[2.2.1]heptane (Fenchane)	151–2		1.44714	0.8345
146	trans-1-Ethyl-4-methylcyclohexane	151.69		1.4382	0.7972
147	cis-1,1,3,5-Tetramethylcyclohexane	152.4–.5		1.4319	0.7813
148	cis-1-Ethyl-4-methylcyclohexane	152.55–.60		1.4374	0.7969
149	2,5,5-Trimethylheptane	152.8		1.4136	0.7368
150	2,4,4-Trimethylheptane	153		1.412	0.733
151	2,3,3,5-Tetramethylhexane	153		1.4196	0.746
152	2,2,4,4-Tetramethylhexane	153.3		1.4208	0.7470
153	Isopropylcyclohexane	154.5	−89.8	1.44095	0.80232
154	1,1,3,3-Tetramethylcyclohexane	154.8–5.0		1.4374	0.7936
155	2,2,3,4-Tetramethylhexane	154.9		1.4226	0.7548
156	2,2-Dimethyloctane	155		1.4082	0.7245
157	3-Ethyl-2,2,4-trimethylpentane	155.3		1.4223	0.7571
158	3,3,5-Trimethylheptane	155.6		1.4170	0.7428
159	2,3,6-Trimethylheptane	155.7		1.4125	0.7345
160	2,4-Dimethyloctane	155.8–6.0		1.4090	0.7259[20/20]
161	d,l-cis-1-Ethyl-3-methylcyclohexane	155.97		1.4432	0.8094
162	d,l-2,5-Dimethyloctane	156–8		1.4160	0.7370
163	1,1,3,5-Tetramethylcyclohexane	156.4–.5		1.4370	0.7929
164	n-Butylcyclopentane	156.56	−107.985	1.4316	0.7846
165	n-Propylcyclohexane	156.724	−94.90	1.43705	0.79360
166	2,3,5-Trimethylheptane	157		1.416	0.741
167	2,5-Dimethyl-3-ethylhexane	157		1.416	0.741
168	2,4,5-Trimethylheptane	157		1.4160	0.741
169	2,4-Dimethyl-3-isopropylpentane	157		1.42463	0.75830
170	2,2,3-Trimethylheptane	158		1.417	0.7420
171	2,4-Dimethyl-4-ethylhexane	158		1.419	0.747
172	2,2-Dimethyl-3-ethylhexane	159		1.420	0.749
173	2,2,3,4,4-Pentamethylpentane	159.3		1.43069	0.76703
174	1,1,3,4-Tetramethylcyclohexane	159.5–6.1		1.4380	0.7976
175	5-Ethyl-2-methylheptane	159.7		1.4134	0.736
176	2,7-Dimethyloctane	159.9		1.4086	0.7242

*Derivative data given in order: m.p., crystal color, solvent from which crystallized.

5

No.	Name	Boiling point, °C	Melting point, °C	n_D^{20}	D_4^{20}
177	3,6-Dimethyloctane......................	160	....	1.4145[18]	0.7363
178	3,5-Dimethyloctane......................	160	....	1.413	0.736
179	4-Isopropylheptane......................	160	....	1.417	0.741
180	2,3,3-Trimethylheptane.................	160	....	1.4202	0.7488
181	4-Ethyl-2-methylheptane...............	160	....	1.413	0.736
182	2,6-Dimethyloctane.....................	160–0.5	....	1.4113	0.7285
183	2,2,3,3-Tetramethylhexane.............	160.3	....	1.42818	0.76446
184	trans-1-Isopropyl-4-methylcyclohexane (p-Menthane)..........................	161	....	1.4393	0.792
185	4,4-Dimethyloctane.....................	161	....	1.4144	0.7347
186	2,3,4,5-Tetramethylhexane.............	161	....	1.424	0.757
187	5-Ethyl-3-methylheptane...............	161	....	1.414	0.737
188	3,3-Dimethyloctane.....................	161.2	....	1.4165	0.7390
189	4,5-Dimethyloctane.....................	162	....	1.4173	0.7458
190	3,4-Diethylhexane......................	162	....	1.420	0.754
191	4-Propylheptane........................	162	....	1.4150	0.7364
192	1,1,4-Trimethylcycloheptane (Eucarvane)........	162–3[720]	....	1.4420	0.8011
193	trans-1,2,3,5-Tetramethylcyclohexane.............	162–4	....	1.44657[He]	0.8140
194	2,3,4,4-Tetramethylhexane.............	162.2	....	1.4270	0.7639
195	2,3,4-Trimethylheptane.................	163	....	1.421	0.751
196	3-Isopropyl-2-methylhexane...........	163	....	1.421	0.751
197	2,2,7-Trimethylbicyclo[2.2.1]heptane (α-Fenchane)	163.5–4.5[753]	....	1.4590	0.8579
198	3-Ethyl-3-methylheptane...............	163.8	....	1.4208	0.7501
199	2,4-Dimethyl-3-ethylhexane............	164	....	1.424	0.759
200	3,4,4-Trimethylheptane.................	164	....	1.424	0.757
201	3,3,4-Trimethylheptane.................	164	....	1.424	0.757
202	3,4,5-Trimethylheptane.................	164	....	1.424	0.759
203	2,3-Dimethyl-4-ethylhexane...........	164	....	1.424	0.759
204	1-Methyl-3-propylcyclohexane........	164–5 (171–3)	....	1.4377	0.7895[21]
205	2,3-Dimethyloctane.....................	164.4–4.6[764]	....	1.4152	0.7377
206	d,l-Pinane.............................	164.5–5.0	....	1.4609	0.8551
207	2,3,3,4-Tetramethylhexane.............	164.6	....	1.4297	0.7694
208	3,3-Dimethyl-4-ethylhexane...........	165	....	1.427	0.764
209	5-Methylnonane........................	165.1	−87.7	1.4122	0.7326
210	4-Methylnonane........................	165.7	−98.7	1.4123	0.7323
211	3-Ethyl-2-methylheptane...............	166	....	1.418	0.746
212	3,4-Dimethyloctane.....................	166	....	1.4182	0.746
213	d-α-Pinane.......................	166–6.5[762]	....	1.4630	0.8560
214	d,l-1-Isopropyl-3-methylcyclohexane (d,l-m-Menthane).....................	166–7	....	1.44[24]	0.7965[24]
215	2,2,3,3,4-Pentamethylpentane..........	166.1	....	1.43606	0.78009
216	trans-1,2,4,5-Tetramethylcyclohexane.........	166.2–8.0	....	1.44446[He]	0.8100
217	3,3-Diethylhexane......................	166.3	....	1.428	0.767
218	2-Methylnonane........................	166.8	−74.5	1.4099	0.7281
219	d-1-Isopropyl-3-methylcyclohexane (d-m-Menthane)......................	167	....	1.446[23]	0.8116[23]
220	3-Ethyl-4-methylheptane...............	167	....	1.422	0.753
221	4-Ethyl-3-methylheptane...............	167	....	1.422	0.753
222	4-Ethyl-4-methylheptane...............	167	....	1.421	0.752
223	l-β-Pinane........................	167.5–8[748]	....	1.4605	0.8567
224	3-Methylnonane........................	167.8	−84.8	1.4125	0.7334
225	3-Ethyloctane..........................	168	....	1.416	0.740
226	4-Ethyloctane..........................	168	....	1.416	0.740
227	3-Ethyl-2,2,3-trimethylpentane........	168	....	1.436	0.781
228	l-1-Isopropyl-3-methylcyclohexane (l-m-Menthane)	168	....	1.4358	0.7938
229	cis-1-Isopropyl-4-methylcyclohexane (cis-p-Menthane).....................	168.5	....	1.4515	0.816
230	cis-1,2,3,5-Tetramethylcyclohexane.............	168–70[762]	....	1.44847[He]	0.8166
231	2,3-Dimethyl-3-ethylhexane............	169	....	1.427	0.765

*Derivative data given in order: m.p., crystal color, solvent from which crystallized.

6

TABLE I. ALKANES AND CYCLOALKANES
a) Gases and Liquids (Listed in order of increasing b.p.*) (Continued)

No.	Name	Boiling point, °C		n_D^{20}	D_4^{20}
232	1-Isopropyl-4-methylcyclohexane (*p*-Menthane)	169–70	....	1.4375[21]	0.7929
233	**3,4-Dimethyl-3-ethylhexane**	170	....	1.431	0.772
234	**3,3,4,4-Tetramethylhexane**	170	....	1.4368	0.7824
235	**Cyclononane**	170–2	9.7	1.4328[16]	0.8534[15.2]
236	1-Isopropyl-2-methylcyclohexane (*o*-Menthane)	171	....	1.447[21]	0.8135[21]
237	*cis*-1,2,4,5-Tetramethylcyclohexane	171[755]	....	1.44647$_{He}$	0.8122
238	1-Methyl-1-propylcyclohexane	172	....	1.4440	0.8101
239	*n*-Decane	174.123	−29.661	1.41189	0.73005
240	1-Methyl-4-propylcyclohexane	174.3–7.1	....	1.4393	0.798
241	1-Methyl-2-propylcyclohexane	175.5–6.0[756]	....	1.4468[19]	0.8130[19]
242	*n*-Pentylcyclopentane	180; 60[10]	−83.	1.4358	0.7912
243	*n*-Butylcyclohexane	180.947	−74.725	1.44075	0.79918
244	*trans*-Decahydronaphthalene (*trans*-Decalin)	187.25	−30.40	1.4695	0.8699
245	Isoamylcyclohexane	193.	....	1.4423	0.8023
246	*cis*-Decahydronaphthalene (*cis*-Decalin)	195.69	−43.01	1.4810	0.8965
247	*n*-Undecane (*n*-Hendecane)	195.89	−25.594	1.41716	0.74017
248	**Cyclodecane**	201	9.6	1.4692	0.8577[20.4]
249	*n*-Pentylcyclohexane	202.8	−57.5	1.4437	0.8037
250	*n*-Hexylcyclopentane	203	....	1.4392	0.7965
251	9-Methyl-*trans*-decahydronaphthalene	205; 77[10]	....	1.4631	0.8620
252	1,10-Dimethyl-*trans*-decahydronaphthalene	213; 83[10]	....	1.4659	0.8633
253	9-Methyl-*cis*-decahydronaphthalene	215; 85[10]	....	1.4804	0.8910
254	*n*-Dodecane	216.278; 51.84[1]	−9.587 (−12)	1.42160	0.74869
255	1,10-Dimethyl-*cis*-decahydronaphthalene	220; 88.99[10]	....	1.4812	0.8896
256	*n*-Hexylcyclohexane	224.0; 92.0[10]	−43	1.4462	0.8076
257	*n*-Heptylcyclopentane	224	....	1.4421	0.8010
258	9-Ethyl-*trans*-decahydronaphthalene	225; 93.06[10]	....	1.466	0.8610
259	9-Ethyl-*cis*-decahydronaphthalene	233; 99[10]	....	1.480	0.8860
260	1-Methyl-*trans*-decahydronaphthalene	235; 101[10]	....	1.4270	....
261	*n*-Tridecane	235.44; 66.35[1]	−5.392	1.4256	0.7564
262	**Bicyclohexyl**	236.5–7.5; 100[10]	3.5–4.0	1.4795	0.8848
263	*n*-Octylcyclopentane	243	....	1.4446	0.8048
264	*n*-Heptylcyclohexane	244	....	1.4484	0.8109
265	*n*-Tetradecane	235.57; 80.13[1]	5.863	1.4289	0.7628
266	*n*-Nonylcyclopentane	262	....	1.4467	0.8081
267	*n*-Octylcyclohexane	264	....	1.4503	0.8138
268	*n*-Pentadecane	270.63; 93.26[1]	9.926	1.4319	0.7685
269	*n*-Decylcyclopentane	279.3	....	1.44862	0.81097
270	*n*-Nonylcyclohexane	282	....	1.4519	0.8163
271	*n*-Undecylcyclopentane (*n*-Hendecylcyclopentane)	296	....	1.4503	0.8135
272	*n*-Decylcyclohexane	299	....	1.45338	0.81858
273	**2-Methylheptadecane**	311; 178.5[15]	....	1.4394[14]	0.7838[15]
274	*n*-Dodecylcyclopentane	312	....	1.4518	0.8158
275	*n*-Undecylcyclohexane (*n*-Hendecylcyclohexane)	316	5.8	1.4547	0.8206
276	*n*-Tridecylcyclopentane	327	5	1.4531	0.8178
277	*n*-Dodecylcyclohexane	331	12.5	1.4559	0.8223
278	*n*-Tetradecylcyclopentane	341	9	1.4543	0.8196

*Derivative data given in order: m.p., crystal color, solvent from which crystallized.

TABLE I. ALKANES AND CYCLOALKANES

b) Solids (Listed in order of increasing m.p.*)**

No.	Name	Melting point, °C	Boiling point, °C	n_D	D_4
1	Pentadecylcyclopentane	17	355	1.4554^{20}	0.8213^{20}
2	n-Hexadecane (Cetane)	18.165	286.793; 105.20[1]	1.43453^{20}	0.7734^{20}
3	Tridecylcyclohexane	18.5	346	1.4570^{20}	0.8239^{20}
4	Hexadecylcyclopentane	21	....	1.4543^{25}	0.8194^{25}
5	n-Heptadecane	21.98	301.82; 117.26[1]	1.4348^{25}; 1.4369_U^{20}	0.7745^{25}; 0.7780_U^{20}
6	Tetradecylcyclohexane	24	....	1.4559^{25}	0.8221^{25}
7	Heptadecylcyclopentane	27	....	1.4532^{30}	0.8173^{30}
8	n-Octadecane	28.18	316.12; 128.28[1]	1.4191^{70}; 1.4390_U^{20}	0.7751^{30}; 0.7819_U^{20}
9	Pentadecylcyclohexane	29		1.4545^{30}	0.8201^{30}
10	Octadecylcyclopentane	30		1.4541^{30}	0.8186^{30}
11	n-Nonadecane	32.1	329.7; 138.8[1]	1.4211^{70}; 1.4409_U^{20}	0.7787^{30}; 0.7855_U^{20}
12	Hexadecylcyclohexane	33.6		1.4596_U^{20}	0.8279_U^{20}
13	Nonadecylcyclopentane	35		1.4588_U^{20}	0.8266_U^{20}
14	n-Eicosane	36.8	342.7; 148.9[1]	1.4230^{70}; 1.4426_U^{20}	0.7550^{70}; 0.7887_U^{20}
15	Heptadecylcyclohexane	37.8		1.4603_U^{20}	0.8290_U^{20}
16	Eicosylcyclopentane	38		1.4595_U^{20}	0.8276_U^{20}
17	n-Heneicosane	40.5	356.5; 152.94[1]	1.4247^{70}; 1.4441_U^{20}	0.7583^{70}; 0.7917_U^{20}
18	Octadecylcyclohexane	41.6		1.4610_U^{20}	0.8300_U^{20}
19	Heneicosylcyclopentane	42		1.4602_U^{20}	0.8286_U^{20}
20	n-Docosane	44.4 (47)	368.6; 161.88[1]	1.4260^{70}; 1.4455_U^{20}	0.7631^{70}
21	Docosylcyclopentane	45		1.4608_U^{20}	0.8295_U^{20}
22	Nonadecylcyclohexane	45.2		1.4616_U^{20}	0.8310_U^{20}
23	n-Tricosane	47.6	380.2; 170.48[1]	1.4276^{20}; 1.4468_U^{20}	0.7641^{70}; 0.7969_U^{20}
24	Eicosylcyclohexane	48.5		1.4622_U^{20}	0.8318_U^{20}
25	Tricosylcyclopentane	49		1.4614_U^{20}	0.8304_U^{20}
26	n-Tetracosane	50.9	391.3; 178.7[1]	1.4286^{70}; 1.4480_U^{20}	0.7657^{70}; 0.7991_U^{20}
27	Tetracosylcyclopentane	51		1.4619_U^{20}	0.8312_U^{20}
28	Heneicosylcyclohexane	51.5		1.4627_U^{20}	0.8326_U^{20}
29	n-Pentacosane	53.7	401.9; 186.55[1]	1.4302^{70}; 1.4491_U^{20}	0.7693^{70}; 0.8012_U^{20}
30	Pentacosylcyclopentane	54		1.4624_U^{20}	0.8319_U^{20}
31	Docosylcyclohexane	54.4		1.4632_U^{20}	0.8334_U^{20}
32	Hexacosylcyclopentane	56		1.4628_U^{20}	0.8326_U^{20}
33	Nortricyclene (Tricyclo[2.2.1.0^{2.6}]heptane)	56	106–7		
34	n-Hexacosane	56.4	412.2; 194.18[1]	1.4310^{70}; 1.4501_U^{20}	0.7704^{70}; 0.8032_U^{20}
35	Cyclohexadecane	57	$93–8^{0.8}$	....	....
36	Tricosylcyclohexane	57		1.4637_U^{20}	0.8341_U^{20}
37	Heptacosylcyclopentane	59		1.4633_U^{20}	0.8333_U^{20}
38	n-Heptacosane	59.0	422.1; 201.54[1]	1.4321^{70}; 1.4511_U^{20}	0.7732^{70}; 0.8050_U^{20}
39	Tetracosylcyclohexane	59.5		1.4641_U^{20}	0.8347_U^{20}
40	Cyclopentadecane	60–1		$1.4592^{61.5}$	$0.8634^{61.5}$
41	Octacosylcyclopentane	61		1.4637_U^{20}	0.8339_U^{20}
42	n-Octacosane	61.4	431.6; 208.58[1]	1.4330^{70}; 1.4520_U^{20}	0.7750^{70}; 0.8067_U^{20}
43	Pentacosylcyclohexane	61.9		1.4645_U^{20}	0.8353_U^{20}
44	Nonacosylcyclopentane	63		1.4640_U^{20}	0.8345_U^{20}
45	n-Nonacosane	63.7	440.8; $172.36^{0.1}$	0.7797^{65}; 1.4529_U^{20}	1.4361^{65}; 0.8083_U^{20}

*Derivative data given in order: m.p., crystal color, solvent from which crystallized.

**U = undercooled liquid.

TABLE I. ALKANES AND CYCLOALKANES
b) Solids (Listed in order of increasing m.p.*)** (Continued)

No.	Name	Melting point, °C	Boiling point, °C	n_D	D_4
46	Hexacosylcyclohexane	64	....	1.4649^{20}	0.8359^{20}
47	Triacontylcyclopentane	65	....	1.4644^{20}	0.8350^{20}
48	d,l-Isobornane (2,2,3-Trimethylbicyclo[2.2.1]heptane)	65–7	....	1.44186^{67}	0.82757^{67}
49	n-Triacontane	65.8	449.7; 222.00^{1}	1.4348^{70}; 1.4536_U^{20}	0.7797^{70}; 0.8097_U^{20}
50	Heptacosylcyclohexane	66.1	....	1.4653_U^{20}	0.8365_U^{20}
51	Hentriacontylcyclopentane..................	67	....	1.4648_U^{20}	0.8356_U^{20}
52	n-Hentriacontane	67.9	458; $184.17^{0.1}$	1.4543^{20}	0.8111^{20}
53	Octacosylcyclohexane	68	....	1.4656_U^{20}	0.8370_U^{20}
54	Dotriacontylcyclopentane....................	69	....	1.4651^{20}	0.8360_U^{20}
55	n-Dotriacontane (Bicetyl)...................	69.7	467; 234.8^{1}	1.4550^{20}	0.7791^{75}; 0.8124^{20}
56	Nonacosylcyclohexane	69.9	....	1.4659^{20}	0.8374^{20}
57	Tritriacontylcyclopentane	70	....	1.4654^{20}	0.8365_U^{20}
58	Tritriacontane.............................	71.4	....	1.4557_U^{20}	0.8136_U^{20}
59	Triacontylcyclohexane	71.6	....	1.4662^{20}	0.8379_U^{20}
60	Tetratriacontylcyclopentane	72	....	1.4657_U^{20}	0.8370_U^{20}
61	Tetratriacontane...........................	72.6; 73.1	285.4^{3}	1.4296^{90}; 1.4563_U^{20}	0.7728^{90}; 0.8148_U^{20}
62	28-Methylnonacosane.......................	73–4	$222^{0.3}$	....	....
63	Hentriacontylcyclohexane	73.3	....	1.4665^{20}	0.8383_U^{20}
64	Pentatriacontylcyclopentane	74	....	1.4660^{20}	0.8374_U^{20}
65	Pentatriacontane...........................	74.7	331	1.4568^{20}	0.8157^{20}
66	Dotriacontylcyclohexane	74.8	....	1.4668^{20}	0.8388_U^{20}
67	Hexatriacontylcyclopentane	75	....	1.4662^{20}	0.8378_U^{20}
68	Hexatriacontane...........................	76.2	....	1.4573^{20}	0.8169^{20}
69	Tritriacontylcyclohexane	76.3	....	1.4670^{20}	0.8391^{20}
70	Heptatriacontane	77.7	....	1.4578^{20}	0.8179_U^{20}
71	Tetratriacontylcyclohexane	77.7	....	1.4673_U^{20}	0.8395^{20}
72	Octatriacontane	79	....	1.4583_U^{20}	0.8188^{20}
73	Pentatriacontylcyclohexane	79.1	....	1.4675_U^{20}	0.8399^{20}
74	Nonatriacontane...........................	80.3	....	1.4588^{20}	0.8197_U^{20}
75	Hexatriacontylcyclohexane	80.4	....	1.4678_U^{20}	0.8402^{20}
76	Tetracontane..............................	81.5	....	1.4593^{20}	0.8205_U^{20}
77	Norbornane (Bicyclo[2.2.1]heptane)	86–7, subl.	....	....	....
78	2,2,3,3-Tetramethylbutane..................	100.69	106.47	1.4695^{20}	$0.8242^{23}_{\text{solid}}$
79	Bornane (Camphane)	158–9, subl.	....	....	....
80	Adamantane	268 (252–3)	....	1.568	1.07_{solid}

*Derivative data given in order: m.p., crystal color, solvent from which crystallized.
**U = undercooled liquid.

9

*Bromine addition compound.**

$$RCH{=}CHR' \ + \ Br_2 \ \rightarrow \ RCHBr{-}CHBrR'$$

Bromine addition
compound

From the alkene and bromine in carbon tetrachloride.
For directions and examples see: Cheronis, p. 576; Shriner, p. 106.
From the alkene and bromine in water.
See: Vogel, p. 241.
From the alkene and bromine in chloroform.
See: C. G. Schmitt and C. E. Boord, *J. Amer. Chem. Soc.,* **54,** 751 (1932).

*2,4-Dinitrobenzenesulfenyl chloride addition compound.**

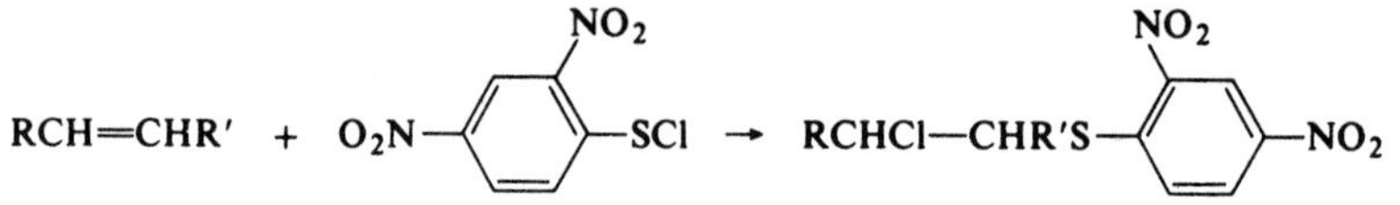

2,4-Dinitrobenzenesulfenyl
chloride adduct

From the alkene and 2,4-dinitrobenzenesulfenyl chloride in glacial acetic acid.
For directions and examples see: Cheronis, p. 577; N. Kharasch and C. M. Buess, *J. Amer. Chem. Soc.,* **71,** 2724 (1949); D. J. Cram, *J. Amer. Chem. Soc.,* **71,** 3883 (1949); N. Kharasch, C. M. Buess and S. I. Strashun, *J. Amer. Chem. Soc.,* **74,** 3422 (1952).
From the alkene and 2,4-dinitrobenzenesulfenyl chloride in benzene or in carbon tetrachloride.
See: N. Kharasch and C. M. Buess, *J. Amer. Chem. Soc.,* **71,** 2724 (1949).

*S-Alkylmercaptosuccinic açid.**

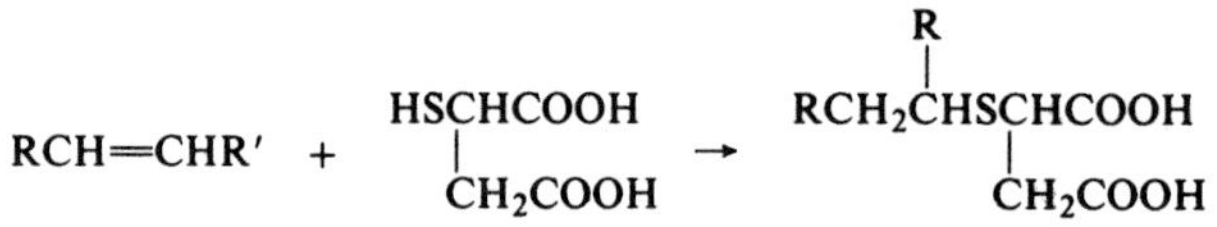

S-Alkylmercaptosuccinic
acid

From the alkene, mercaptosuccinic acid and benzoyl peroxide in methanol.
For directions and examples see: J. G. Hendrickson and L. F. Hatch, *J. Org. Chem.,* **25,** 1747 (1960).

Maleic anhydride adduct (from dienes).

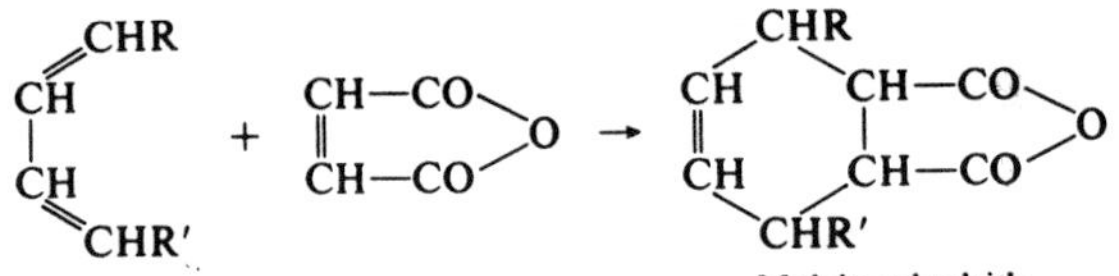

Maleic anhydride
adduct

From the diene and maleic anhydride in benzene.
For directions and examples see: Linstead, p. 51; Vogel, p. 943.
From the diene and maleic anhydride in xylene.
See: Linstead, p. 51.
For general references see: M. C. Kloetzel in *Organic Reactions*, Vol. 4 (Ed. R. Adams), John Wiley and Sons, New York, 1948, p. 1; H. L. Holmes in *Organic Reactions*, Vol. 4, (Ed. R. Adams), John Wiley and Sons, New York, 1948, p. 60; O. Diels and K. Alder, *Chem. Ber.,* **62,** 2081 (1929).

Nitrosochloride addition compound.

$$RCH{=}CHR' \ + \ NOCl \ \rightarrow \ RCH(NO)CHClR'$$

Nitrosochloride
adduct

*Derivatives recommended for first trial.
WARNING: This is not an instruction manual. References should be consulted for the preparation of derivatives.

From the alkene and nitrosyl chloride (prepared from sodium nitrite in concentrated hydrochloric acid) in ether-acetic acid mixture.

For directions and examples see: Linstead, p. 52; R. Perrot, *Compt. Rend.*, **203**, 329 (1936).

From the alkene and nitrosyl chloride (prepared from thionyl chloride and nitrogen trioxide) in ether.

See: M. Tuot, *Compt. Rend.*, **204**, 697 (1937).

---

*Derivatives recommended for first trial.

WARNING: This is not an instruction manual. References should be consulted for the preparation of derivatives.

TABLE II. ORGANIC DERIVATIVES OF ALKENES, CYCLOALKENES, DIENES AND POLYENES
a) Liquids 1) (Listed in order of increasing b.p.*)**

No.	Name	Boiling point, °C	Melting point, °C	n_D^{20}	D_4^{20}	x-Bromo-	B.P., °C	M.P., °C	n_D^{20}	D_4^{20}	Miscellaneous
							Bromine addition product				
1	Ethene (Ethylene)....	−103.71	−169.15[T]	....	0.384[0]	di-	131.36	9.85	1.53868	2.1792	....
2	Propene (Propylene)..	−47.70	−185.25[T]	....	0.5139	di-	141.99	−55.5	1.52004	1.93268	
3	Cyclopropene	−36[744]		....	....		....	....	....	....	....
4	Allene	−34.5	−136	....	....	tetra-		10.7	1.6200	2.703	
5	2-Methylpropene	−6.90	−140.35	1.3467	0.5942[S]	di-	149	....	1.5080	1.7595	2,4-Dinitrophenyl-sulfenyl chloride, 86–7
6	1-Butene	−6.26	−185.35	1.3465	0.5951	di-	166.3	....	1.5150	1.7951	2,4-Dinitrophenyl-sulfenyl chloride, 77.5–8.5
7	1,3-Butadiene	−4.41	−108.92	1.4292[.25]; 0.650[−6]	0.6255[−6.5]	tetra-	......	118, lgr.	....	....	
8	trans-2-Butene	0.88	−105.55	....	0.6042[S]	di-	161.0	....	1.5110	1.7852	
9	Cyclobutene	2.4	....		0.733[0]		....				
10	cis-2-Butene	3.72	−138.91	....	0.6306[1]	di-	161.0	....	1.5110	1.7852	
11	1,2-Butadiene (Methylallene)	10.85	−136.19	1.4208[3]	0.652[S]	tetra-	97.5[7]	−2	1.6070	2.5085	
12	3-Methyl-1-butene ...	20.06	−168.49	1.3643	0.6272	di-	61–2[12]	....	1.50932	1.6776	
13	1,4-Pentadiene	25.97	−148.28	1.38876	0.66706	tetra-	......	85.5–6.0, eth.	....	....	
14	1-Pentene	29.97	−165.22	1.37148	0.64050	di-	68[12]	....	1.5012[12]	1.592[19]	Mercaptosuccinic acid adduct, 107.3–.6
15	2-Methyl-1-butene ...	31.16	−137.56	1.3778	0.6504	di-	47.4–48[9]	....	1.5088	1.6711	Mercaptosuccinic acid adduct, 122.3–.6
16	3-Methylcyclobutene .	32	....	1.4005							
17	2-Methyl-1,3-butadiene (Isoprene).	34.07	−145.95	1.42194	0.68095	di- tetra-	90–6[12] 155–60[12]				Maleic anh. adduct, 63–4, lgr.
18	trans-2-Pentene	36.35	−140.24	1.3793	0.6482	di-	91.0[50]	....	1.5096	1.6809	
19	cis-2-Pentene	36.94	−151.39	1.3830	0.6556	di-	92.4[50]	....	1.5096	1.6817	
20	1-Methyl-1-cyclobutene	37.1	....	1.4088	0.7244						
21	2-Methyl-2-butene ...	38.57	−133.77	1.3874	0.6623	......					Nitrosochloride, 74; Mercaptosuccinic acid adduct, 153.7–4.0
22	3-Methyl-1,2-butadiene (1,1-Dimethyl-allene)	40	....	1.410	0.680	tetra-	150–2[17]	....	1.594[17]	2.305[17]	....
23	Cyclopentadiene	40.83[772]	−85	1.4398[19.5]	0.7983[19.5]	......		....	....	....	Dimer, 32; Maleic anh. adduct, 164–5; Benzoquinone adduct, 75–6
24	1,3-Pentadiene (Piperylene)	41.1	−88.9	1.4309	0.6803	tetra-	......	114.5, al.	....	....	Maleic anh. adduct, 61, pet. eth.; Oxid. by KMnO₄ → HCOOH + CH₃COOH
25	3,3-Dimethyl-1-butene	41.24	−115.2	1.3760	0.6529	di-	95.3–5.6[10]	....	1.5109	1.5615	....
26	1, trans-3-Pentadiene .	42.03	−87.47	1.43008	0.67603	tetra-	131[3]	115, al.	....	....	
27	1,cis-3-Pentadiene ...	44.07	−140.82	1.43634	0.69102	tetra-	131[3]	115, al.	....	....	

*Derivative data given in order: m.p., crystal color, solvent from which crystallized.

**T = triple point; S = at saturation pressure.

No.	Name	Boiling point, °C	Melting point, °C	n_D^{20}	D_4^{20}	Bromine addition product					Miscellaneous
						x-Bromo-	B.P., °C	M.P., °C	n_D^{20}	D_4^{20}	
28	Cyclopentene.......	44.24	−135.08	1.42246	0.77199	di-	71.5[12]	....	1.5510[19]	1.8713[19]	Mercaptosuccinic acid adduct, 142.8-3.1; Pseudo-nitrosite, 69–70; Perbenzoic acid oxid. → epoxy-cyclopentane, b.p., 102-3
29	1,2-Pentadiene (Ethylallene).......	44.86	−137.26	1.42091	0.69257	tetra-	94–6[0.1]	....	....	2.3469[18/18]	....
30	2,3-Pentadiene (1,3-Dimethylallene)	48.27	−125.26	1.42842	0.69502	....					
31	4-Methyl-1-pentene ..	53.88	−153.63	1.3828	0.6642	di-	87[21]	....	1.4980	1.5689	Mercaptosuccinic acid adduct, 102.6-.9
32	3-Methyl-1-pentene ..	54.14	−153.0	1.3842	0.6675	di-	99[30]		1.5060	1.6016	....
33	3-Methyl-1,4-penta-diene	55	....	1.405	0.695	....					
34	2,3-Dimethyl-1-butene	55.67	−157.27	1.3904	0.6779	di-	80[17]		1.5105	1.6033	
35	2-Methyl-1,4-penta-diene	56	....	1.405	0.694	......					
36	4-Methyl-cis-2-pentene	56.3	−134.43	1.3880	0.6690	di-	72–3[18]	....	1.5060	1.5983	
37	4-Methyl-trans-2-pentene	58.55	−140.81	1.3889	0.6686	di-	78[22]	....	1.5070	1.5996	....
38	1,5-Hexadiene (Biallyl)..........	59.46	−140.8	1.4042	0.6923	tetra-	....	52	....	....	Dil. HNO₃ → succinic ac., 185
39	2-Methyl-1-pentene ..	60.7	−135.72	1.3920	0.6817	di-	87–8[20]		1.5015	1.5581	
40	1-Hexene..........	63.49	−139.82	1.38788	0.67317	di-	89–90[18]		1.5024	1.5774	2,4-Dinitrophenyl-sulfenyl chloride, 61–2; Mercapto-succinic acid adduct, 94.5-5.7
41	2-Ethyl-1-butene.....	64–6	−131.53	1.3969	0.6894	di-	87[21]		1.5112	1.6045	....
42	trans-1,3-Hexadiene..	64.5–5.5	....	1.4060[19]	0.6925[19]	tetra-	....	19			
43	3-Methylcyclopentene	65.0		1.4207	0.7622	....					....
44	cis-3-Hexene........	66.44	−137.82	1.3947	0.6796	di-	80–1[13]		1.5045	1.6027	
45	3-Hexene (cis-trans mixture)	66.6–67	....	1.3942	0.6816	di-	80–1[13]		1.5045	1.6027	
46	trans-3-Hexene......	67.08	−113.43	1.3943	0.6772	di-	80–1[13]		1.5045	1.6027	....
47	2-Methyl-2-pentene ..	67.29	−135.7	1.4004	0.6863	di-	71–2[18]		1.5063	1.5849	Mercaptosuccinic acid adduct, 152.1-.6
48	3-Methyl-trans-2-pentene	67.63	−134.84	1.4016	0.6942	di-	72–4[15]	....	1.5085		....
49	trans-2-Hexene......	67.87	−132.97	1.3935	0.6784	di-	90[16]		1.5025	1.5812	
50	2-Hexene (cis-trans mixture)	67.9–8.1	....	1.3928	0.6813	di-	90[16]		1.5025	1.5812	....
51	2,3-Hexadiene......	68	....	1.395	0.680						
52	2,3-Dimethyl-1,3-butadiene	68.78	−76.01	1.4394	0.7267	di- tetra-	....	47, lgr. 138, bz.	....	....	Maleic anh. adduct, 78–9
53	cis-2-Hexene........	68.84	−141.14	1.3977	0.6869	di-	90[16]		1.5025	1.5812	
54	4-Methyl-1,2-penta-diene (1-Isopropyl-allene)	70	....	1.424	0.708	....					
55	3-Methyl-cis-2-pentene	70.45	−138.45	1.4045	0.6986	di-	72–4[15]	....	1.5085		....

*Derivative data given in order: m.p., crystal color, solvent from which crystallized.

TABLE II. ORGANIC DERIVATIVES OF ALKENES, CYCLOALKENES, DIENES AND POLYENES
a) Liquids 1) (Listed in order of increasing b.p.)* (Continued)

No.	Name	Boiling point, °C	Melting point, °C	n_D^{20}	D_4^{20}	Bromine addition product					Miscellaneous
						x-Bromo-	B.P., °C	M.P., °C	n_D^{20}	D_4^{20}	
56	2-Methyl-2,3-pentadiene (Trimethyl allene)	72	....	1.425	0.711	....	....	....	....	....	....
57	1,4-Hexadiene	72.3-2.5		1.4402[19]	0.7057[19]	tetra- 2 forms	....	a) 63-4; b) f.p. <-50			
58	4,4-Dimethyl-1-pentene	72.49	−136.6	1.3918	0.6827	di-	77-8[9]	....	1.4970	1.5129	Mercaptosuccinic acid adduct, 119.0-.5
59	1, (cis and/or trans)-3-hexadiene	73		1.438	0.705	....	....	....	....	....	....
60	2,3-Dimethyl-2-butene	73.21	−74.28	1.4122	0.7080	di-		173-4; 121			
61	2-Ethyl-1,3-butadiene	75		1.445	0.717	....	....	....	....	....	....
62	4-Methylcyclopentene	75.2	....	1.4306	0.7796						
63	1-Methylcyclopentene	75.8	−127	1.4330	0.7802						
64	2-Methyl-1,(cis and/or trans)-3-pentadiene	76	....	1.446	0.719						
65	1,2-Hexadiene (n-Propylallene)	76		1.4282	0.7149	tetra-	130[3]		1.5850	2.1873	
66	2-Methyl-1,(cis and/or trans)-3-pentadiene	76		1.446	0.719						
67	4-Methyl-1,3-pentadiene	76.3	....	1.451	0.719						
68	4,4-Dimethyl-trans-2-pentene	76.75	−115.24	1.3982	0.6889	di-	92.8-93[14]	....	1.5080	1.5538	
69	3,3-Dimethyl-1-pentene	77.54	−134.3	1.3984	0.6974	di-	95.3-6[10]	....	1.5109	1.5615	....
70	2,3,3-Trimethyl-1-butene	77.87	−109.85	1.4029	0.7050	di-	98-9[14]	38-9			
71	3-Methyl-1,3-pentadiene	78.0-.3	....	1.4494	0.7499						
72	trans-1,3,5-Hexatriene	78.5; 77-8.5		1.4884[13.5]	0.74229[15/15]	hexa-	....	78			
73	cis-1,3,5-Hexatriene	78.5		1.4577	0.7179						
74	3-Methyl-1,2-pentadiene	79; 70	....	1.425	0.715						
75	2,4-Hexadiene	79.4-81.6[765]	−79	1.4493	0.7152	2,5-di-tetra-	85[11]	182	1.534[19]	1.622[19]	Maleic anh. adduct, 95-6, lgr.; SO₂ adduct, 43-3.5
76	3-Methyl-1,5-hexadiene	80-1		1.4116	0.7103						
77	4,4-Dimethyl-1,2-pentadiene (tert-Butylallene)	80-3			0.7184						
78	1,3-Cyclohexadiene	80.31[757]	−104.8	1.4740	0.8413	di- ... tetra-2 forms	....	68, isomerizes → m. 108 1, trans-2,cis-3, trans-4, 92; 1,cis-2, trans-3, trans-4, 156	....	....	Maleic anh. adduct, 145-6, heptane; Benzoquinone adduct, 196-7, lgr.
79	4,4-Dimethyl-cis-2-pentene	80.42	−135.46	1.4024	0.6996	di-	92.8-3.0[14]	....	1.5080	1.5538	....

*Derivative data given in order: m.p., crystal color, solvent from which crystallized.

a) Liquids 1) (Listed in order of increasing b.p.)* (Continued)

No.	Name	Boiling point, °C	Melting point, °C	n_D^{20}	D_4^{20}	Bromine addition product					Miscellaneous
						x-Bromo-	B.P., °C	M.P., °C	n_D^{20}	D_4^{20}	
80	3,4-Dimethyl-1-pentene	81	....	1.3995	0.701	....	....	....	....	....	....
81	Cyclohexene	82.97	−103.51	1.44654	0.81096	di-	101–3[13]	....	1.5445[19]	1.7759[19]	Mercaptosuccinic acid adduct, 150.5–1.5; KMnO$_4$ oxid. → adipic ac., 154; 2,4-Dinitro-phenylsulfenyl chloride, 117–8; HBr → cyclo-hexyl bromide, b.p. 165
82	2,4-Dimethyl-2-pentene	83–4	−127.7	1.40165[22]	0.6958[22]	di-	88[17]	....	1.50920[22.5]	1.5431[22.5]	....
83	3-Methyl-1-hexene ...	84.0	....	1.397	0.695	di-	84.0–.2[6]	....	1.5028	1.5248	
84	2,3-Dimethyl-1-pentene	84.26	−134.8	1.4033	0.7051	di-	72.5–3.0[3]		1.5028	1.5245	
85	3-Ethyl-1-pentene	85.13	−127.4	1.3980	0.6962	di-	93.5[15]		1.5006	1.5251	
86	5-Methyl-1-hexene ...	85.31	....	1.3966	0.6920	di-	142.6–3.6[101]		1.4970	1.5072	....
87	5-Methyl-trans-2-hexene	86.0		1.400	0.700	di-	87–8[10]		1.4960	1.5027	
88	2-Methyl-3-hexene ...	86.0	....	1.399	0.694	di-	96[19]	....	1.5060	1.5310	
89	2,4-Dimethyl-2,3-pentadiene (Tetramethylallene)	86.5		1.40039	0.7006	....	....				
90	4-Methyl-1-hexene ...	86.73	−141.45	1.4000	0.6985	di-	94.7–5.7[11]		1.4980	1.5027	
91	3,4-Dimethyl-2-pentene	87.0	....	1.407	0.713	di-	65.5–6.0	....	1.5104	1.5400	
92	4-Methyl-cis-2-hexene	87.37	....	1.4024	0.6996	di-	91–2[11]		1.5045	1.5382	
93	4-Methyl-trans-2-hexene	87.6	−126.5	1.4023	0.6975	di-	91–2[11]		1.5045	1.5382	
94	3,3-Dimethylcyclo-pentene	88		1.423	0.771	....	....	....	....	....	
95	2-Ethyl-3-methyl-1-butene	89	....	1.410	0.715	di-	72.5–3.5[3]		1.5062	1.5261	
96	5-Methyl-cis-2-hexene	91		1.400	0.700	di-	89–90[11]	....	1.4990	1.5152	....
97	5-Methyl-1,4-hexadiene	91–2.5		1.4390	0.7258	1,2-di-	101–4[18]		1.5233[16]	1.566[16]	....
98	2-Methyl-1-hexene ...	92.0	−102.84	1.4034	0.7030	di-	100.5–1.5[23]	....	1.5000	1.5066	....
99	1,3-Dimethylcyclo-pentene	92	....	1.428	0.766	....	....				
100	2-Methyl-1,5-hexadiene	92.5[769]		1.423/6[17.3]	0.7289[18.5]	....	....				Nitrosochloride, 75–6
101	2,4-Dimethyl-1,3-pentadiene	93	−114	1.4412	0.7368	....	....				
102	1,4-Dimethylcyclo-pentene	93.2	....	1.4283	0.779	....	....				
103	3-Methyl-trans-3-hexene	93.5	....	1.4107	0.7099	....	....				
104	1-Heptene	93.64	−119.03	1.39980	0.69698	di-	106.2[13]	....	1.4990	1.5208	Mercaptosuccinic acid adduct, 103.4–.9

*Derivative data given in order: m.p., crystal color, solvent from which crystallized.

TABLE II. ORGANIC DERIVATIVES OF ALKENES, CYCLOALKENES, DIENES AND POLYENES
a) Liquids 1) (Listed in order of increasing b.p.)* (Continued)

No.	Name	Boiling point, °C	Melting point, °C	n_D^{20}	D_4^{20}	Bromine addition product					Miscellaneous
						x-Bromo-	B.P., °C	M.P., °C	n_D^{20}	D_4^{20}	
105	3-Methyl-*trans*-2-hexene	94.0	....	1.410	0.7120	di-	65.0–5.1[2]	....	1.5040	1.5240	....
106	2-Ethyl-1-pentene	94	....	1.405	0.708	di-	77–8[4]	....	1.4990	1.4929	
107	3-Methyl-*cis*-3-hexene	95.35	....	1.4123	0.7132	....	....				
108	2-Methyl-2-hexene	95.41	−130.35	1.4106	0.7082	di-	73.0–.1[8]	....	1.4990	1.5116	
109	*trans*-3-Heptene	95.67	−136.63	1.4043	0.6981	di-	105.5–6.5[23]	....	1.5010	1.5153	
110	*cis*-3-Heptene	95.75	....	1.4059	0.7030	di-	105.5–6.5[23]	....	1.5010	1.5153	
111	5-Methyl-1,2-hexadiene (Isobutylallene)	96	....	1.4282[19]	0.7225[19]	....	....				
112	3-Ethyl-2-pentene	96.01	....	1.4148	0.7204	di-	76.0–.4[3]	....	1.5090	1.5426	
113	2,3-Dimethyl-2-pentene	97.5	−118.3	1.4208	0.7277	di-	97–9[15]	....	1.517[22]	1.547[22]	
114	*trans*-2-Heptene	97.95	−109.48	1.4045	0.7012	di-	96.2[12]		1.5000	1.5129	
115	3-Ethylcyclopentene	98.1	....	1.4319	0.7830	....					
116	*cis*-2-Heptene	98.5	....	1.406	0.708	di-	96.2[12]		1.5000	1.5129	
117	5-Methyl-1,3-cyclohexadiene	100.5–1.5[762]	....	1.4662[22.5]	0.8252	....					
118	2,2-Dimethyl-*trans*-3-hexene	100.9	....	1.4063	0.7039	di-	96.5–7.0[8]		1.5032	1.4856	
119	1,4-Heptadiene	101	....	1.4202	0.7106	di-			1.5734	2.091	
120	2,4,4-Trimethyl-1-pentene	101.44	−93.48	1.4086	0.7150	....					
121	3,3-Dimethyl-1,5-hexadiene	101.6	....	1.4160	0.7249						
122	3,4-Dimethyl-1,5-hexadiene	101.8	....	1.4211	0.7304						
123	2,5-Dimethyl-3-hexene	102	....	1.406	0.710	di-	109[19]	....	1.5058	1.5034	....
124	5,5-Dimethyl-1-hexene	102.5	....	1.4049	0.709	....					
125	4-Methylcyclohexene	102.74	−115.5	1.4414	0.7947	di-	130[40]			1.650$^{15}_{15}$	
126	3-Methylcyclohexene	104.0	....	1.4444	0.8010	....					
127	2-Isopropyl-3-methyl-1-butene	104	....	1.4085	0.722						
128	3,4,4-Trimethyl-1-pentene	104	....	1.412	0.719						
129	3,5-Dimethyl-1-hexene	104	....	1.404	0.708						
130	3,3-Dimethyl-1-hexene	104	....	1.4070	0.7140						
131	5,5-Dimethyl-*trans*-2-hexene	104.1	....	1.4055	0.7066						
132	2,4,4-Trimethyl-2-pentene	104.91	−106.33	1.4160	0.7218						
133	3,3,4-Trimethyl-1-pentene	105	....	1.4144	0.729						
134	2,2-Dimethyl-*cis*-3-hexene	105.4	−137.4	1.4099	0.7128						
135	1,2-Heptadiene (*n*-Butylallene)	105.5–6.0	....	1.432[18]	0.7306[18]	2,3-di-	108–10[12];		1.5200[18]	1.5595[18]	
						tetra-	140[3]		1.5718	2.0675	
136	1,2-Dimethylcyclopentene	105.8	−90.4	1.4448	0.7976	....					
137	4-Ethylcyclopentene	106		1.440	0.798						
138	4,4-Dimethyl-2-hexene	106	....	1.413	0.722	di-	92–3[4]	......	1.5113	1.5148	
139	1-Ethylcyclopentene	106.3	−118.4	1.4410	0.7982	......	....				....
140	5,5-Dimethyl-*cis*-2-hexene	106.9	....	1.4113	0.7169	......	....				

*Derivative data given in order: m.p., crystal color, solvent from which crystallized.

No.	Name	Boiling point, °C	Melting point, °C	n_D^{20}	D_4^{20}	Bromine addition product x-Bromo-	B.P., °C	M.P., °C	n_D^{20}	D_4^{20}	Miscellaneous
141	2-Methyl-2,4-hexadiene	107	....	$1.4266^{24.5}$	0.7439	....	....	....	....	....	....
142	2-Methyl-1,3-cyclo-hexadiene	107–8	....	1.4662^{18}	0.8272_{18}^{18}	....	....	....	....	....	....
143	3-Methyl-2,4-hexadiene	107–8	....	1.46146^{15}	0.7625^{15}	....	....	....	....	....	....
144	4,4-Dimethyl-1-hexene	107.2	....	1.4102	0.7198	di-	224–5	....	1.5003	1.4689	....
145	3-Ethyl-4-methyl-1-pentene	107.5	....	1.4097	0.7200	....	....	....	....	....	....
146	2,4-Heptadiene	107.5–8.0	....	1.4578	0.7384	....	....	....	....	....	....
147	2,4-Dimethyl-trans-3-hexene	107.6	....	1.4126	0.7145	....	....	....	....	....	....
148	Quadricyclene (Quad-ricyclo [2,2,1,0²·⁶, 0³·⁵]heptane)......	108^{740} sl.d.	....	1.4804	....	....	....	....	....	....	....
149	2,3,4-Trimethyl-1-pentene	108	....	1.415	0.729	....	....	....	....	....	....
150	4-Methyl-1,3-hexadiene	108–10	....	1.4523	0.7558	....	....	....	....	....	....
151	2,3,3-Trimethyl-1-pentene	108.3	−69	1.4174	0.7352	....	....	....	....	....	....
152	4,5-Dimethyl-1-hexene	109	....	1.414	0.728	....	....	....	....	....	....
153	1,5,5-Trimethylcyclo-pentene (Iso-laurolene)	109^{754}	....	1.4324	0.7824	....	....	....	....	....	Reduces Tollen's reagent on warming
154	2,4-Dimethyl-cis-3-hexene	109	....	1.4140	0.7178	....	....	....	....	....	....
155	3,3-Dimethyl-2-ethyl-1-butene..........	110	....	1.4159	0.728	....	....	....	....	....	....
156	3-Ethyl-2-methyl-1-pentene	110	....	1.415	0.730	....	....	....	....	....	....
157	4,5-Dimethyl-2-hexene	110	....	1.413	0.725	....	....	....	....	....	....
158	1-Methylcyclohexene .	110.0	−121	1.4503	0.8102	di-	$100–2^{12}$	....	....	....	2,4-Dinitrophenyl-sulfenyl chloride, 139–40
159	2-Ethyl-4-methyl-1-pentene	110.3	....	1.4105	0.7195	....	....	....	....	....	....
160	3-Ethyl-1-hexene	110.3	....	1.407	0.715	....	....	....	....	....	....
161	2,3-Dimethyl-1-hexene	110.5	....	1.4113	0.7214	....	....	....	....	....	....
162	2,4-Dimethyl-2-hexene	110.6	....	1.4118	0.7213	....	....	....	....	....	....
163	3-Methyl-1-heptene .	111	....	1.406	0.711	....	....	....	....	....	....
164	2,4-Dimethyl-1-hexene	111.2	....	1.4110	0.720	....	....	....	....	....	....
165	2,5-Dimethyl-1-hexene	111.6	....	1.4105	0.7172	....	....	....	....	....	....
166	3-Ethyl-3-methyl-1-pentene	112	....	1.418	0.7305	....	....	....	....	....	....
167	3,4-Dimethyl-1-hexene	112	....	1.413	0.724	....	....	....	....	....	....
168	3,4,4-Trimethyl-2-pentene	112	....	1.4232	0.7395	....	....	....	....	....	....
169	3,5-Dimethyl-2-hexene	112	....	1.416	0.725	....	....	....	....	....	....
170	2-Methyl-3-heptene .	112	....	1.402	0.706	....	....	....	....	....	....
171	5-Methyl-3-heptene .	112	....	1.410	0.713	....	....	....	....	....	....
172	2,5-Dimethyl-2-hexene	112.2	....	1.4140	0.720	di-	88^{13}	....	1.4740	1.3980	....
173	3-Methyl-1,5-heptadiene	112.5	....	$1.4230^{22.5}$	....	....	....	....	....	....	....
174	2-Ethyl-3-methyl-1-pentene	112.5	....	1.4142	0.729	....	....	....	....	....	....

Derivative data given in order: m.p., crystal color, solvent from which crystallized.

a) Liquids 1) (Listed in order of increasing b.p.)* (Continued)

No.	Name	Boiling point, °C	Melting point, °C	n_D^{20}	D_4^{20}	x-Bromo-	B.P., °C	M.P., °C	n_D^{20}	D_4^{20}	Miscellaneous
						Bromine addition product					
175	4-Methyl-1-heptene ..	112.8	....	1.410	0.717	....					....
176	6-Methyl-3-heptene ..	113^{734}	....	1.4114	0.7256	di-	....	97			....
177	4-Ethyl-1-hexene	113	....	1.412	0.726	....					....
178	4-Ethyl-2-hexene	113	....	1.412	0.725	....					....
179	2-Isopropyl-1-pentene.	113	....	1.414	0.725	....					....
180	5-Methyl-1-heptene ..	113.3	....	1.4094	0.7164	....					....
181	4-Methyl-2-heptene ..	113.5–4.1	....	1.4096	0.7154	....					....
182	2,3-Dimethyl-3-hexene	114	....	1.416	0.728	....					....
183	4-Methyl-2-octene ...	114	....	1.4100^{25}	0.7188^{25}	....					....
184	2,4-Dimethyl-2,4-hexadiene	114–5	....	1.45457$^{16.5}$	0.7635$^{16.5}$	....					....
185	6-Methyl-2,4-heptadiene	114–6	....	1.4397^{25}	0.7041^{25}	....					....
186	3-Ethyl-4-methyl-trans-2-pentene.....	114.3	....	1.4210	0.7350	....					....
187	Cycloheptene (Suberene)........	114.38	−56	1.4580	0.8254	di-	unstable	....			Nitrosochloride, 118; Oxid. → pimelic ac., 105
188	1-Methyl-1,4-cyclohexadiene	114.5–4.8	< −70	1.4703	0.848	tetra-	....	171			....
189	3-Ethyl-4-methyl-cis-2-pentene	115	....	1.424	0.739	....					....
191	1,3,5-Cycloheptatriene (Tropilidene)	115.5	−79.49	1.5243	......	....					Maleic anh. adduct, 104.2–5.0, CCl₄
192	3,4-Dimethyl-2-hexene	116	....	1.418	0.737	....					....
193	3-Ethyl-3-hexene	116	....	1.418	0.729	....					....
194	6-Methyl-1,3-heptadiene	116–8	....	....	0.741^{22}	....					....
195	2,5-Dimethyl-1,3-hexadiene	116–8	> −80	1.45024	0.7412	....					....
196	2,3,4-Trimethyl-2-pentene	116.3	−113.3	1.4275	0.7434	....					....
197	4,4-Dimethylcyclohexene	116.98	−80.5	1.4420	0.7996	....					....
198	6-Methyl-2-heptene ..	117	....	1.412	0.718	....					....
199	2-n-Propyl-1-pentene .	117.7	....	1.4136	0.7240	....					....
200	5-Methyl-2-heptene ..	118	....	1.414	0.723	....					....
201	3,3-Dimethylcyclohexene	119	....	1.445	0.804	....					....
202	2-Methyl-1-heptene ..	119–22	−87.38	1.41195	0.72025	....					....
203	2,5-Dimethyl-1,5-hexadiene	119–23	solid at −80; liq. at −23	1.45054	0.7637	....					....
204	2-Ethyl-1-hexene	120	....	1.4157	0.7270	....					Mercaptosuccinic acid adduct, 101.9–2.7
205	d,l-1,2,3-Trimethyl-cyclopentene (Laurolene)........	120–1^{752}	....	1.4421	0.7950	....					....
206	4-Methyl-3-heptene ..	120.4	....	1.41712^{25}	0.7411^{25}	....					....
207	3-Ethyl-2-hexene	121	....	1.424	0.737	....					....
208	3-Methyl-3-heptene ..	121	....	1.418	0.728	....					....
209	1-Octene	121.28	−101.76	1.40870	0.71492	di-	240–2; 118.5^{15}	....	1.4970	1.4580	Mercaptosuccinic acid adduct, 96.1–.6

*Derivative data given in order: m.p., crystal color, solvent from which crystallized.

No.	Name	Boiling point, °C	Melting point, °C	n_D^{20}	D_4^{20}	Bromine addition product					Miscellaneous
						x-Bromo-	B.P., °C	M.P., °C	n_D^{20}	D_4^{20}	
210	trans-4-Octene	121.4[739]	f.p. −94	1.41157	0.71467	di-(meso)	103[8]	....	1.4967[24]	1.4525	....
211	1,3-Cycloheptadiene (Hydrotropilidene)..	121.52	−110.42	....	0.8929[0]₀	......					$H_2 \rightarrow$ Cyclo-heptane, b.p. 118-20
212	3-Methyl-2-heptene ..	121.6	....	1.4183	0.7296	......					....
213	cis-4-Octene	121.7[739]	f.p. −118	1.41361	0.72048	di-(d,l)	84.0–8.4[4.3]	....	1.4981	1.4569	....
214	2,3-Dimethyl-2-hexene	121.8	−115.1	1.4268	0.7408	......					....
215	3,4-Dimethyl-trans-3-hexene	122	....	1.430	0.747	di-	85-7[5]		1.5060	1.387	....
216	6-Methyl-1-heptene ..	122–4; 113–5	....	1.4070	0.7125						....
217	cis-3-Octene	122.3[741]	f.p. −126	1.41246	0.71888	....					....
218	trans-3-Octene	122.4[741]	f.p. −110.4	1.41241	0.71630	....					....
219	2-Methyl-2-heptene ..	123–5	....	1.4138	0.7241	....					....
220	trans-2-Octene	125.0	−87.7	1.4132	0.7199	....					Mercaptosuccinic acid adduct, 142.9-3.5
221	cis-2-Octene	125.64	−100.2	1.4150	0.7243	....					....
222	2-Methyl-1,3-hepta-diene	127–8[747]	....	1.4432	0.7432	....					....
223	1,4-Dimethylcyclo-hexene	128		1.446	0.802						Nitrosochloride, 83-4
224	1,5-Dimethylcyclo-hexene	128		1.448	0.8051						Nitrosochloride, 118-9
225	2,6-Dimethyl-2-heptene	128.9	....	1.412	0.722[15]₁₅	....					....
226	4-Vinylcyclohexene ..	129.5–30.5; 36[23]		1.4623	0.8320	α,β-di-		69.5–70, eth.			....
227	4-Methyl-2,4-hepta-diene	131–2		1.4621	0.7551	......					....
228	3,4-Dimethyl-2,4-hexadiene	132–4; 71–3[100]	....	1.4410	0.7832[19]						....
229	3-Methyl-2,4-hepta-diene	132–5	....	1.4649[15]	0.7667[15]						HBr → Dihydro-bromide, b.p. 109-11[16]
230	Bicyclo[4,2,0]oct-7-ene	132.5		1.4761	....	di-	74[0.5]				
231	4-Ethylcyclohexene ..	133	....	1.449	0.810						....
232	1,6-Dimethylcyclo-hexene	133		1.454	0.815	....					....
233	3,5-Dimethyl-2,4-heptadiene	133–44[740]	....	1.4487	0.7728	....					....
234	2,4-Octadiene	133.5–4.0		1.4542[25]	0.7427[25]						....
235	2,5-Dimethyl-2,4-hexadiene	133.6; 28[10]	14.6–.8	1.4796[19.5]	0.7646[18]	tetra-	......	101	......		Oxid. in air → poly-meric peroxide, 59
236	3-Ethylcyclohexene ..	134	....	1.451	0.814						....
237	1,2,3,3-Tetramethyl-cyclopentene (Campholene)......	134–5	....	1.44406	0.8035[15]						....
238	1-Ethylcyclohexene ..	136		1.4575	0.823	......					....
239	1,2-Dimethylcyclo-hexene	137		1.4588	0.8250	di-	......	142–3, acet.	....		Nitrosochloride, 58-60
240	1,3-Dimethylcyclo-hexene	137		1.445	0.802	....					....
241	Bicyclo[4,2,0]oct-2-ene	137–9	....	1.4810[30]	0.8948	......					....

*Derivative data given in order: m.p., crystal color, solvent from which crystallized.

19

No.	Name	Boiling point, °C	Melting point, °C	n_D^{20}	D_4^{20}	Bromine addition product					Miscellaneous
						x-Bromo-	B.P., °C	M.P., °C	n_D^{20}	D_4^{20}	
243	**1-Methylcycloheptene**	137.5–8.5	....	1.4581	0.8243^{22}	....	....	....	....	....	Nitrosochloride, 106; Nitrosate, 97–8
244	**2-Methyl-4-octene** ...	138^{739}	....	1.4181	0.7392						
245	**δ-Fenchene** (1,5,5-Tri-methylbicyclo[2,2,1] hept-2-ene)	139–40	....	1.44862	0.8433						Nitrosochloride, 131
246	**1,5,5-Trimethylcyclo-hexene** (α-Cyclo-geraniolene)	$139–41^{759}$	....	$1.44612^{21.5}$	0.7981^{23}						Nitrosochloride, 100–20, aq. me., al. Nitrosate, 102–4
247	**2,6-Dimethyl-2,4-heptadiene**	$139–43^{752}$	....	1.4587^{44}	0.74820						....
248	**1,4,4-Trimethylcyclo-hexene** (Pulenene)...	139.5–40.5	....	$1.444^{23.2}$	$0.8032^{18.8}$	di-	$120–0.5^{10}$	....	1.5247^{19}	1.5324^{19}	Nitrosochloride, 118–22, et. ac.
249	**1,5,6-Trimethylcyclo-hexene**	140	....	1.4572	0.831_{25}^{25}						....
250	**2,3-Dimethyl-2-nor-bornene** (Santene; 2,3-Dimethylbicyclo [2,2,1]-hept-2-ene) ..	140–1; 35^{15}	....	1.46699	0.8640						Dichloride, 88–9; Nitrosate, 216d.; Nitrosochloride, 109–10; Nitrosite, 3 forms: a) 122–4, bl.; b) 127–8, grn.; c) 104, col.
251	**2,6-Dimethyl-1,3-heptadiene** (Iso-geraniolene)	140–2; 31^7	....	1.4606^{22}	0.7923^{22}						....
252	**Cyclooctatetraene**....	140.56; 142–3; $42–2.5^{17}$	−4.68; −7	1.5290	0.9206						Maleic anh. adduct, 167–8; Benzo-quinone adduct, 141, al.; Acrylic ac. adduct, 112–3, lgr.; AgNO$_3$ adduct, 173–4
253	**7-Methyl-3-octene** ...	141^{746}	....	1.4168	0.7278	....	....	....	....	....	....
254	**2,6-Dimethyl-1,5-heptadiene** (Geraniolene)......	141–2; 165–70	−70	1.44361^{22}	0.7626^{22}	....	....	....	....	....	
255	**1,8-Nonadiene**.........	141–4	....	1.4302	0.7511						
256	**1,3,5-Trimethylcyclo-hexene** (Tetrahydro-mesitylene)	142.5–3.5	....	$1.449^{13.5}$	$0.8025^{14.3}$	....	....	....	....	....	Nitrosochloride, 134
257	**3-Methyl-2-octene** ...	$143–5^{734}$	....	1.4247	0.7409^{25}	....	....	....	....	....	....
258	**Cyclooctene**.........	$143.8–4.5^{773}$	....	1.4693		....	....	....	....	....	Br$_2$ → Bromocyclo-octene, b.p. 7–8^{23}; n_D^{20}: 1.5182; Di-chloride, b.p. 130.4–0.6^{25}; m.p. −5; n_D^{20}: 1.5061, D_4^{20}: 1.1620
259	**3,6-Dimethyl-2,4-heptadiene**	144–6	....	1.46335^{14}	0.7853^0	....	....	....	....	....	....
260	**4-Nonene**	144–6; $44–6^{12}$	....	1.4212^{18}	0.732^{18}	di-	$119–20^{12}$	....	1.4988^{17}	1.410^{17}	....
261	**1,4,5-Trimethylcyclo-hexene**	144–6	....	1.4482	0.805	....	....	....	....	....	....

*Derivative data given in order: m.p., crystal color, solvent from which crystallized.

No.	Name	Boiling point, °C	Melting point, °C	n_D^{20}	D_4^{20}	Bromine addition product					Miscellaneous
						x-Bromo-	B.P., °C	M.P., °C	n_D^{20}	D_4^{20}	
262	1-Vinylcyclohexene	145; 63–5[53]	....	1.4950[19]	....						50% $H_2SO_4 \rightarrow$ dimer, b.p. 118–9[5]
263	cis-1,4-Cyclooctadiene	145.1[758]	−53	....	0.8754	tetra-	....	139			
264	1,3,5-Cyclooctatriene	145–6; 76[90]	....	1.5035[25]	0.8971[25]	....					Maleic anh. adduct, 144–5; $AgNO_3$ adduct, 125–6, al.
265	4,4-Dimethyl-1,7-octadiene	145–8		1.4330	0.7647						
266	ξ-Fenchene (2,7,7-Tri-methylbicyclo[2,2,1]hept-2-ene)	146.2–6.8[752]	....	1.4865	0.8626						$[\alpha]_D^{20}$: −24.1
267	1,6,6-Trimethylcyclo-hexene	146.2–7.2[767]; 144–6		1.456[20.4]	0.8217[20.3]						Nitrosochloride, 133–4, et. ac.
268	1-Nonene	146.87	−81.37	1.41572	....	di-	141.5[20]		1.4942	1.3980	....
269	3-Nonene	147.4[750]		1.4173	0.7294	....					
270	1,5-Cyclooctadiene	148–9	....	1.4905	0.8818[24]						N-Bromosuccini-mide → bromo-cyclooctadiene, b.p. 64[1.9], n_D^{25}: 1.5410, D_4^{25}: 1.3420
271	4-Methyl-3,5-octa-diene	148–51		1.46285[25]	0.7640[25]	....					
272	7-Methyl-2,4-octa-diene	149		1.4543[18]	0.7521[18]	tetra-	184[18]				
273	1-Ethyl-4-methyl-cyclohexene	149; 153–4	....	1.453[16]	0.8169[16]						Nitrosochloride, 2 forms: a) 103–4, pr., eth.; b) 98–9, cr., eth.
274	1-Ethyl-3-methyl-cyclohexene	149–51	....	1.454	0.8296						
275	2-Nonene	149.4–9.9		1.420[21]	0.738[21]						
276	1,2,3-Trimethylcyclo-hexene	149.6–150[749]		1.463[12]	0.8347[12]						
277	1-Ethyl-5-methyl-cyclohexene	150		1.4527[25]	0.812[25]						
278	β-Fenchene (2,2-Di-methyl-5-methylene-bicyclo[2,2,1]heptane)	150.5–3.5		1.46511	0.8599	di-	....	81–2			$[\alpha]_D^{25}$: +62.5; Nitrosochloride, 120
279	2,6-Dimethyl-2,5-heptadiene	150.6–1.0	....	1.4490	....						
280	2,7-Nonadiene	150.6	−72.5	1.4358	0.7499						
281	Allylcyclohexane (3-Cyclohexylpropene)	151	....	1.4536[13]	0.8196[13]	di-	143–4[16]			1.537[0]	....
282	1-Ethylidene-4-methyl-cyclohexane	152–3		1.4571[21]	0.81[21]						Nitrosochloride, 2 forms: a) 117–8, least soluble; b) 113–4, more soluble
283	4,5-Dimethyl-2,6-octadiene	152.9–3.8		1.4375[25]	0.7611[25]						
284	1-Ethylidene-3-methylcyclohexane	153		1.458[4]	0.8135[19 19]						Nitrosochloride, 114, acet.
285	3,6-Dimethyl-2,6-octadiene	153–5	....	1.44453	0.7767						
286	2,6-Dimethyl-2,7-octadiene	155–6[720]	....	1.4385[18]	0.7605[18]	....					

*Derivative data given in order: m.p., crystal color, solvent from which crystallized.

TABLE II. ORGANIC DERIVATIVES OF ALKENES, CYCLOALKENES, DIENES AND POLYENES

a) Liquids 1) (Listed in order of increasing b.p.)* (Continued)

No.	Name	Boiling point, °C	Melting point, °C	n_D^{20}	D_4^{20}	Bromine addition product					Miscellaneous
						x-Bromo-	B.P., °C	M.P., °C	n_D^{20}	D_4^{20}	
287	α-Pinene	156.0–6.3	−50	1.4560	0.8600	di-	....	169–70, al.	....	....	Nitrosochloride, 109; Hydrobromide, 89; Nitrosobromide, 91–2d.; Acid KMnO₄ → pinonic ac., 103–5
288	1-Ethylidene-2-methylcyclohexane ..	158	....	1.47	0.823⁰						....
289	2,4-Dimethyl-2,4-octadiene	161–3⁷⁴⁶	....	1.4558⁹·⁸	0.7802⁹·⁸						
290	1,4,4-Trimethylcyclo-heptene (Eucarvene) .	161–5⁷²⁰	....	1.4561	0.8185						
291	β-Pinene (Nopinene; Pseudopinene)	163–4	....	1.4782	0.8694						$[\alpha]_D$: −22
292	2,7-Dimethyl-2,6-octadiene	163.5–4.5	....	1.44814	0.7849	tetra-	....	124–7	......		
293	3,7-Dimethyl-2,4-octadiene	164–7; 58¹²	....	1.456	0.7933						
294	l-4-Carene (3,7,7-Tri-methylbicyclo[2,2,1] hept-2-ene)	165.5–7.0⁷⁰⁷	....	1.474³⁰	0.8552³⁰₃₀						$[\alpha]_D^{30}$: +62.2
295	Myrcene (2-Methyl-6-methylene-2,7-octadiene)	166	....	1.4722	0.7982						Maleic anh. adduct, 33–4; 1,4-Naphthoquinone adduct, 81
296	2,6-Dimethyl-2,6-octadiene	168; 56¹⁴	....	1.45245¹⁵	0.775²¹						Methiodide, 130d.
297	l-3-Carene (3,7,7-Tri-methylbicyclo[2,2,1]-hept-3-ene)	168–9⁷⁰⁵; 123–4²⁰⁰	....	1.469³⁰	0.8586³⁰₃₀						$[\alpha]_D^{30}$: +7.69; Nitrosochloride, 100–1
298	3,8-o-Menthadiene (cis-3-Isopropenyl-4-methylcyclohexene)	169–70	....	1.4749	0.8507						....
299	5-Decene	170⁷⁵⁰	−112 to −111	1.4260	0.7474	di-	119⁹	....	1.4912	1.3484	....
300	p-8-Menthene (1-Iso-propenyl-4-methyl-cyclohexane).......	170	....	1.4523	0.8142						
301	d-m-8-Menthene (1-Isopropenyl-3-methylcyclohexane) .	170	....	1.4546	0.8179						$[\alpha]_D$: +9.73
302	l-m-8-Menthene (1-Isopropenyl-3-methylcyclohexane) .	170–1	....	1.4574	0.8189						$[\alpha]_D$: −8.06
303	6,8-o-Menthadiene (3-Isopropenyl-2-methylcyclohexene) .	170–1	....	1.4758	0.8481						
304	5,8-o-Menthadiene (4-Isopropenyl-3-methylcyclohexene) .	170–1	....	1.4778	0.8490¹⁷₁₇						
305	1-Decene	170.57	−66.31	1.42146	0.74081	di-	145–160¹⁸	....	1.4891²⁴	1.324²⁸	Mercaptosuccinic acid adduct, 93.5–.8
306	4-Decene	170.6	....	1.4243	0.7404	......	....	....	....	....	

*Derivative data given in order: m.p., crystal color, solvent from which crystallized.

22

a) Liquids 1) (Listed in order of increasing b.p.)* (Continued)

No.	Name	Boiling point, °C	Melting point, °C	n_D^{20}	D_4^{20}	Bromine addition product					Miscellaneous
						x-Bromo-	B.P., °C	M.P., °C	n_D^{20}	D_4^{20}	
307	*p*-Menthene (1-Iso-propylidene-4-methylcyclohexane) .	172–4	....	1.4568	0.819[21]	....	....		....	....	Nitrosochloride, 101–3
308	1-Isopropenyl-1,4-cyclohexadiene	172–6	....	1.5216	0.9068	*tetra-*	....	113	....	....	....
309	*o*-Menthene (1-Iso-propylidene-2-methylcyclohexane) .	173; 160–2	....	1.467	0.8345	....					....
310	*m*-Menthene (1-Iso-propylidene-3-methylcyclohexane) .	173–5	....	1.4670	0.8214						....
311	α-Terpinene (1-Iso-propyl-4-methyl-1,3-cyclohexadiene) .	173.5–4.8[755]	....	1.477	0.8375	....			....		Dihydrochloride, 53–4, me. al.; Di-hydrobromide, 58–9, me. al.; Di-hydroïodide, 76, me. al.; Nitrosite, 155; Maleic anh. adduct, 62; 66–7
312	2,4-*p*-Menthadiene(2-Isopropyl-5-methyl-1,3-cyclohexadiene) .	174–6	....	1.4845[27]	0.8441[27][27]	....			....		....
313	*l*-5-Isopropyl-2-methyl-1,3-cyclo-hexadiene	174–7	....	1.4732	0.8425	....			....		$[\alpha]_D^{20}$: −112.76; Nitrosite, α: 120–1; β: 105–6
314	1,5-*p*-Menthadiene (5-Isopropyl-2-methyl-1,3-cyclo-hexadiene)	175–6	....	1.4777	0.8463[25]	....			....		$[\alpha]_D$: +49.1; Nitro-site, α: 113–4; β: 105. Maleic anh. adduct, 126–7, pet. eth.
315	*d*-Silvestrene	175–8	....	1.4760	0.8479	*tetra-*	....	135	....	....	Dihydrochloride, 72; Nitrosochlo-ride, 106–7
316	*d*-Limonene	176–6.4	....	1.4743	0.8411	*tetra-*	....	104	....	....	Nitrosochloride, 100–4; 2,4-Dinitro-phenylsulfenyl chloride, 195–6
317	Isocarvestrene (5-Iso-propenyl-1-methyl-cyclohexene)	176–7[765]	....	1.4804	0.8496	*tetra-*	....	137–8, me. al.-chl.	....	....	Dihydrochloride, 71.5, me. al.
318	*l*-3-Isopropenyl-1-methylcyclohexene .	176–8		1.4761[18]	0.848[19]						$[\alpha]_D^{18}$: −68.2; Di-hydrochloride, 72
319	Dipentene (*d,l*-Limonene)	177.6–8.0		1.4727[20]	0.8402[20]	*tetra-*	....	125, eth.			Dihydrochloride, 50–1, al.
320	3,8-*m*-Menthadiene (1-Isopropenyl-5-methylcyclohexene) .	179[730]	....	1.4972							$[\alpha]_D$: +17.5
321	Menogerene (5-Iso-propylidene-2-methyl-1,3-cyclo-hexadiene)	180–1[769]	....	1.5005	0.8672	*di-*	....	115			
322	γ-Terpinene (1-Iso-propyl-4-methyl-1,4-cyclohexadiene)	183	....	1.4765[14.5]	0.849	*tetra-*	....	129–30, pet. eth.	....	....	$[\alpha]_D^{25}$: +36; Nitro-sochloride, 111; Nitrosate, 116d., ac. a.-me. al.
323	*d,l*-2,8-*m*-Menthadiene (1-Isopropenyl-3-methylcyclohexene) .	184–7	....	1.503	0.864[20][20]	....					....

*Derivative data given in order: m.p., crystal color, solvent from which crystallized.

TABLE II. ORGANIC DERIVATIVES OF ALKENES, CYCLOALKENES, DIENES AND POLYENES

a) Liquids 1) (Listed in order of increasing b.p.)* (Continued)

No.	Name	Boiling point, °C	Melting point, °C	n_D^{20}	D_4^{20}	Bromine addition product					Miscellaneous
						x-Bromo-	B.P., °C	M.P., °C	n_D^{20}	D_4^{20}	
324	**Menogene** (3-Iso-propylidene-6-methyl cyclohexene)	184.6[764.5]	. . . .	1.5026	0.8624	. . . .	. . . .	. . . .		. . . .	Nitrosite, 155, me. al. or chl.; Maleic anh. adduct, 205–8
325	**Terpinolene** (4-Iso-propylidene-1-methylcyclohexene) .	186	. . . .	1.4883	0.8633[15/15]	di- tetra- 2 forms	. . . .	69–70 a) 119, ac. a.; b) 122	. . . .		Maleic anh. adduct, 182
326	**2-Undecene** (2-Hendecene)	192–3; 78.5[14]	. . . .	1.43325	0.7735[15/15]	di-	145–6[9]	. . . .			
327	**5-Undecene** (5-Hendecene)	192.2	. . . .	1.4289	0.7511	. . . .	. . . .	. . . .			
328	**1-Undecene** (1-Undecene)	192.67	−49.19	1.42609	0.75032	di-	186[23]	. . . .	1.4916	1.3122	. . . .
329	*cis*-**Cyclodecene**	194–5[740]	. . . .	1.4854	0.8770	di-	. . . .	121	. . . .		$O_3 \rightarrow$ Sebacic acid, 134.5
330	**1-Dodecene**	213.36; 88.7[10]	−35.23	1.43002	0.75836	di-	. . . .	−15	. . . .		
331	**1-Tridecene**	232.78; 104.5[10]	−23.07	1.4336	0.7653	. . . .	. . . .	. . . .			
332	**1-Tetradecene**	251.1; 119.0[10]	−12.85	1.43631	0.7713	di-	. . . .	0	. . . .		Mercaptosuccinic acid adduct, 104.0–.8
333	**Cedrene**	262–3; 124–6[12]	. . . .	1.5001[19]	0.9359[15]	. . . .	. . . .	. . . .			
334	**1-Pentadecene**	268.17; 133.7[10]	−3.73	1.4389	0.77641	di-	204–5[17]	. . . .	1.4897	1.2235	. . . .
335	**1-Hexadecene**	284.4; 103.9[1]	4.12	1.44120	0.78112	di-	225–7[15]	13.5, al.	. . . .		1% Hot $KMnO_4 \rightarrow$ *n*-pentadecylic ac., 52.3; Mercapto-succinic acid adduct, 105.0–.8
336	**1-Heptadecene**	299.7; 116[1]	11.2	1.4432	0.7852	. . . .	. . . .	. . . .			. . . .
337	**2-Methyl-2-heptadecene** . . . . ·	314; 277[100]	−2.5	. . . .	0.7953	di-	267–8[28]	. . . .			. . . .
338	**1-Octadecene**	314.2; 128[1]	17.6	1.4449	0.7888	di-	. . . .	24, al.			

* Derivative data given in order: m.p., crystal color, solvent from which crystallized.

TABLE II. ORGANIC DERIVATIVES OF ALKENES, CYCLOALKENES, DIENES AND POLYENES
a) Liquids 2) (B.p. at reduced pressure only. Listed alphabetically)

No.	Name	Boiling point, °C	Melting point, °C	n_D^{20}	D_4^{20}	Miscellaneous
1	Bicyclo[12,2,2]octadeca-14,16,18-triene	163.5–4.5[5.5]	....	1.5204[25]	......	Maleic anh. adduct, 143–4
2	Bicyclo[4,2,0]oct-3-ene	81[140]	....	1.4832	......	....
3	Butylcyclooctatetraene	98[20]	....	1.5083[25]	0.8876[25]	....
4	*trans*-Cyclodecene	68–70[10]	....	1.4822	0.8672	$O_3 \rightarrow$ Sebacic acid, 134.5
5	1,5,9,13-Cyclohexadecatetraene	93–8[0.8]	....	1.5472		....
6	*trans*-Cyclononene	73–4[30]	....	1.4799	0.8615	Phenylazide adduct, 97.8–8.2
7	1,3-Cyclooctadecadiene	115[3]	....	1.4899	0.8814	....
8	1,3-Cyclooctadiene	48[25]	−57 to −55	1.4940[25]	0.8699[25]	....
9	1,3,6-Cyclooctatriene	68[60]	−62 to −56	1.4982	0.8940[25]	....
10	1,3-Cyclotetradecadiene	106–8[3]	....	1.4982	0.8723[25]	Nitrosochloride, 109–10; Nitrosate, 210d.
11	1,2-Dimethylcyclooctatetraene	107[96]	....	1.5219[25]	0.8950[25]	Maleic anh. adduct, 184.5–5.5, bz.-lgr.; AgNO₃ adduct, 142.5–4.5, al.
12	2,6-Dimethyl-2,5-octadiene	59.0–.5[12]	....	1.4500	0.733	....
13	Ethylcyclooctatetraene	81[37]	....	1.5187[25]	0.8996[25]	Maleic anh. adduct, 97–8.5, bz.-cyclohexane; AgNO₃ adduct, 124–5.5, al.
14	5-Methylcycloheptene	69–70[38]	....	1.42016[31]	0.76061[31]	....
15	Methylcyclooctatetraene	84.5[67]	....	1.5249[25]	0.8978[25]	....
16	7-Pentadecene	114[3.2]	....	1.4420	0.7765	....
17	Propylcyclooctatetraene	73[9]	....	1.5131[25]	0.8870[25]	....

*Derivative data given in order: m.p., crystal color, solvent from which crystallized.

TABLE II. ORGANIC DERIVATIVES OF ALKENES, CYCLOALKENES, DIENES AND POLYENES

b) Solids (Listed in order of increasing m.p.)*

No.	Name	Melting point, °C	Boiling point, °C	n_D^{20}	D_4^{20}	Bromine addition product					Miscellaneous
						x-Bromo-	B.p., °C	M.p., °C	n_D^{20}	D_4^{20}	
1	**1-Nonadecene**	23.4	328.0; 138.8[1]	1.4445	0.7886	....	......	....	.....	.....	....
2	Eicosene	28.6	341.2	1.4439[30]	0.7882[30]	....	......	....	.....	.....	....
3	**Bicyclo[4,7]pentadiene** (*endo*-4,7-Methylene-4,7,8,9-tetrahydroindene)	32	170	1.5070[25]	0.9766[33]	....	......	....	.....	.....	Phenylazide adduct, 128
4	**1-Heneicosene**	33.3	355	1.4494[S]	0.7977[S]	....	......	....	.....	.....	....
5	**1-Docosene**	37.8	....	1.4505[S]	0.8002[S]	....	......	....	.....	.....	....
6	**1-Tricosene**	41.6	379	1.4516[S]	0.8023[S]	....	......	....	.....	.....	....
7	**1-Tetracosene**	45.3	390	1.4527[S]	0.8045[S]	....	......	....	.....	.....	....
8	**1-Pentacosene**	48.7	401	1.4536[S]	0.8063[S]	....	......	....	.....	.....	....
9	*cis,cis,cis,*1,4,7-Cyclononatriene	49.5–50	....	....	....	....	......	....	.....	.....	AgNO₃ adduct, 243d.
10	*d,l*-Camphene (2,2-Dimethyl-3-methylenebicyclo[2,2,1]heptene)	50	159–60	....	....	*di-*	91–2	153.5[15]	.....	.....	
11	*l*-Camphene	51.3	159–60	....	....	....	......	....	.....	.....	2,4-Dinitrophenyl-sulfenyl chloride, 121–2; Hydrochloride, 125–7
12	**1-Heptacosene**	54.7	421	1.4552[S]	0.8097[S]	....	......	....	.....	.....	....
13	**1-Triacontene**	62.4	448	1.4573[S]	0.8141[S]	....	......	....	.....	.....	....
14	**1-Hentriacontene**	64.6	457	1.4580[S]	0.8153[S]	....	......	....	.....	.....	....
15	**1-Dotriacontene**	66.7	465	1.4585[S]	0.8165[S]	....	......	....	.....	.....	....
16	**1-Tritriacontene**	68.7	473	1.4591[S]	0.8176[S]	....	......	....	.....	.....	....
17	**1-Tetratriacontene**	70.5	481	1.4596[S]	0.8186[S]	....	......	....	.....	.....	....
18	**1-Pentatriacontene**	72.3	489	1.4601[S]	0.8196[S]	....	......	....	.....	.....	....
19	**1-Hexatriacontene**	73.9	496	1.4605[S]	0.8205[S]	....	......	....	.....	.....	....
20	**1-Heptatriacontene**	75.5	503	1.4610[S]	0.8214[S]	....	......	....	.....	.....	....
21	**1-Octatriacontene**	77	510	1.4614[S]	0.8223[S]	....	......	....	.....	.....	....
22	**1-Nonatriacontene**	78.4	517	1.4618[S]	0.8231[S]	....	......	....	.....	.....	....
23	**1-Tetracontene**	79.8	523	1.4622[S]	0.8238[S]	....	......	....	.....	.....	....
24	**Bicyclo(2,2,2)-oct-2-ene**	111–2	128–34	....	....	2,3-*trans*-di-	.....	55.0–5.5	.....	.....	

*Derivative data given in order: m.p., crystal color, solvent from which crystallized.

S = Supercooled liquid at 20°

EXPLANATIONS AND REFERENCES TO TABLE III

Hg salt (Mercuric acetylide). *

$$2\,RC{\equiv}CH \;+\; K_2HgI_4 \;+\; 2\,KOH \;\rightarrow\; (RC{\equiv}C)_2Hg \;+\; 4\,KI \;+\; 2\,H_2O$$
Mercuric salt

From the terminal alkyne and K_2HgI_4 (prepared from mercuric chloride, potassium iodide and potassium hydroxide).

For directions and examples see: Linstead, p. 52; J. R. Johnson and W. L. McEwen, *J. Amer. Chem. Soc,* **48,** 469 (1926).

Hydration to form carbonyl compound.

$$\text{Terminal acetylenes:}\quad RC{\equiv}CH \;+\; H_2O \;\xrightarrow[\text{HgSO}_4]{\text{H}_2\text{SO}_4}\; RCOCH_3$$
Methyl ketone

$$\text{Other acetylenes:}\quad RC{\equiv}CR' \;+\; H_2O \;\xrightarrow[\text{HgSO}_4]{\text{H}_2\text{SO}_4}\; RCOCH_2R' \;+\; RCH_2COR'$$
Mixture of ketones

From the alkyne in methanol and a catalyst composed of boron trifluoride etherate, red mercuric oxide and trichloroacetic acid.

For directions and examples see: J. G. Sharefkin and E. M. Boghosian, *Anal. Chem.,* **33,** 640 (1961).

From the alkyne, mercuric sulfate and sulfuric acid in 70% methanol, in 70% acetone or in 60% acetic acid.

See: Cheronis, p. 576; H. Erdmann and F. Kother, *Z. Anorg. Chem.,* **18,** 48 (1898); R. J. Thomas, K. N. Campbell and G. F. Hennion, *J. Amer. Chem. Soc.,* **60,** 718 (1938).

From the alkyne, mercuric oxide and sulfuric acid in alcohol.

See: J. R. Johnson, A. M. Schwartz and T. L. Jacobs, *J. Amer. Chem. Soc.,* **60,** 1882 (1938).

NOTE: For directions and examples for the preparation of the semicarbazones and the 2,4-dinitrophenyl-hydrazones of the formed carbonyl compounds see explanations and references to Table IX and X, p. 141, 142, 143.

---

*Derivatives recommended for first trial.

WARNING: This is not an instruction manual. References should be consulted for the preparation of derivatives.

TABLE III. ORGANIC DERIVATIVES OF ALKYNES (ACETYLENES)
a) Liquids (Listed in order of increasing b.p.)*

No.	Name	Boiling point, °C	Melting point, °C	n_D^{20}	D_4^{20}	Ketone	B.p., °C	2,4-Dinitrophenyl-hydrazone of ketone	Semicarbazone of ketone	Hg salt	Miscellaneous
						Hydration product (RC≡CR' → RCOCH₂R') and its derivatives					
1	**Ethyne** (Acetylene).....	−84.0 (sat. press.)	−80.8, subl.	....	0.6179^{-84}	(Acetalde-hyde)	(20.2)	(168)	(162–3)	....	....
2	**Propyne** (Methyl acetylene)	−23.22	−102.7	....	0.6174^{-23}	Acetone	56	128	190	204	
3	**1-Butyne** (Ethyl acetylene)	8.09; 8.3	−125.72	1.3962	0.6682^{8}	2-Butanone	82	116–7	135–6	162–3, al.	....
4	**1,3-Butadiyne** (Diacetylene)	10.3	−35 to −36	1.4120	0.7249	....			....	161.5–2.0	KOBr → 1,4-Di-bromo deriv., 49–50, bz.
5	**3-Methyl-1-butyne** (Isopropyl acetylene) ..	26.35; 28	−89.7	1.3723	0.666	....			....		Tetrabromo deriv., b.p. 275
6	**2-Butyne** (Dimethyl acetylene)	27.2–.6	−32.26	1.3921	0.6901	2-Butanone	82	116–7	135–6	....	Tetrabromo deriv., 243, eth.; 2,4-Di-nitrophenyl-sulfenyl chloride, 65–6
7	**3-Methyl-3-buten-1-yne** (Isopropenyl acetylene)	33	....	1.4158	0.6801^{11}	....			....		3,4-Dibromo deriv., b.p. $50.0–1.5^{10}$
8	**3,3-Dimethyl-1-butyne** (tert-Butyl acetylene) ..	38–9	−81.20	1.37725^{15}	0.6737^{15}	....			....	92.5–3.0	Cu salt, 140, red
9	**1-Pentyne** (Propyl acetylene)	40.18	−105.7	1.3852	0.6901	2-Pentanone	102.3	145	112; 106	118.4	Tetrabromo deriv., b.p. 275
10	**1-Penten-4-yne** (Allyl acetylene)	42–3	....	1.4125^{16}	0.738^{16}	Allyl methyl ketone	111–2	160			1,2-Dibromo deriv., b.p. $79.5–80.5^{10}$; 4,4,5,5-Tetrabromo deriv., b.p. $132–6^{10}$
11	*cis*-**3-Penten-1-yne** (*cis*-Propenyl acetylene)	44.6		1.4330		....			....	48	....
12	*trans*-**3-penten-1-yne** (*trans*-Propenyl acetylene)	52.2		1.4377	0.7270	....			....	155–7	1,2-Dibromo deriv., b.p., $60–2^{10}$; 3,4-Dibromo deriv., $66–76^{10}$
13	**2-Pentyne** (Ethyl methyl acetylene)	56.07	−109.3	1.4039	0.7107	2-Pentanone + 3-Pentanone	102.3 102	145 156	112; 106 138–9	....	KMnO₄ → formic ac. + propionic ac.
14	**1-Penten-3-yne** (Methyl vinyl acetylene).......	59.2–60.1	....	1.4496	0.740J	....			....		....
15	**4-Methyl-1-pentyne** (Isobutyl acetylene) ...	61.1–.2; 99	−105.1	1.3936_{α}^{15}	0.7092^{15}	....			....	100.0–.5	....
16	**3-Methyl-1-pentyne** (*sec*-Butyl acetylene) ..	$65–70^{770}$; 57.7	....	1.3916	0.7037	....			....	74–5	....
17	**1-Hexyne** (Butyl acetylene)	71.33	−131.9	1.3989	0.7155	2-Hexanone	128	106–7	125	96.2–.4	....
18	**4-Methyl-2-pentyne** (Isopropyl methyl acetylene)	72.0–.5	−110.37	1.4078^{19}	0.716^{19}	....			....		
19	**4,4-Dimethyl-1-pentyne** .	73–5	....	1.4028	0.7154					125–6.5	
20	**1,3-Pentadiyne**	75.0–.5; 55–6	−45 to −35	1.4431^{21}	0.7375^{21}	....			....		
21	**1,4-Hexadiyne**	78–83	....	....	0.825^{0}	....			....		
22	**3-Hexyne** (Diethyl acetylene)	81.5^{744}	−51	1.4112^{25}	0.7263^{25}	3-Hexanone	125	130	112	....	2,4-Dinitrophenyl-sulfenyl chloride, 65–6

*Derivative data given in order: m.p., crystal color, solvent from which crystallized.

28

No.	Name	Boiling point, °C	Melting point, °C	n_D^{20}	D_4^{20}	Hydration product (RC≡CR′ → RCOCH₂R′) and its derivatives — Ketone	B.p., °C	2,4-Dinitro-phenyl-hydrazone of ketone	Semicar-bazone of ketone	Hg salt	Miscellaneous
23	**4,4-Dimethyl-2-pentyne** (*tert*-Butyl methyl acetylene)	82.9–3.0	. . . .	1.4071	0.7176	. . . .	. . . .	. . . .	. . . .	. . . .	. . . .
24	**2-Hexyne** (Methyl propyl acetylene)	83.7–4.0	−92	1.4135	0.7317	2-Hexanone + 3-Hexanone	128 125	106–7 130	125 112	. . . .	. . . .
25	**1-Hexen-3-yne** (Ethyl vinyl acetylene).	85[758]	. . . .	1.4522 ·	0.7492	. . . .	. . . .	. . . .	. . . .	. . . .	1,2-Dibromo deriv., b.p. 87.0–.5[10]; 3,3,4,4-Tetra-bromo deriv., b.p. 140–50[10]
26	**1,5-Hexadien-3-yne** (Divinyl acetylene). . . .	85.0	−87.83	1.5045	0.7857	. . . .	. . . .	. . . .	. . . .	. . . .	. . . .
27	**1-Hexen-4-yne**	87[753]	. . . .	1.446[14]	0.767[14]	. . . .	. . . .	. . . .	. . . .	. . . .	1,2-Dibromo deriv., b.p. 93.0–.5[10] 1,2,4,5-Tetra-bromo deriv., b.p. 154.5–5.0[10]
28	**3-Ethyl-1-pentyne**	87.0–8.5	. . . .	1.4102	0.7246	. . . .	. . . .	. . . .	. . . .	. . . .	. . . .
29	**1,5-Hexadiyne** (Dipro-pargyl).	87.5–8.5[758]; 20[46]	−4.266	1.4381[23]	0.79943	. . . .	. . . .	. . . .	. . . .	. . . .	. . . .
30	**2-Hexen-4-yne**	88–9	. . . .	1.4918	0.7710	Allyl ethyl ketone	74.5–6.5[90]	95–106	. . . .	. . . .	. . . .
31	**2-Methyl-3-hexyne**	95.2	−116.7	1.4114	0.7263	. . . .	. . . .	. . . .	. . . .	. . . .	. . . .
32	**4-Methyl-2-hexyne**	95.94	f.p.: −107.63	1.4170	0.73855	. . . .	. . . .	. . . .	. . . .	. . . .	. . . .
33	**3-Ethyl-3-penten-1-yne**. .	96.5	. . . .	1.4338[25]	0.7886[25]	. . . .	. . . .	. . . .	. . . .	. . . .	. . . .
34	**5-Methyl-3-heptyne**	98–100[745]	. . . .	1.4102	0.7360	. . . .	. . . .	. . . .	. . . .	. . . .	. . . .
35	**1-Heptyne** (*n*-Pentyl acetylene)	99.74	−80.9	1.4087	0.7328	2-Heptanone	151.2	89	123; 127	61, me. al.	. . . .
36	**5-Methyl-2-hexyne**	102.46	−92.91	1.41762	0.73776	. . . .	. . . .	. . . .	. . . .	. . . .	. . . .
37	**8-Methyl-4-nonyne**	104.5	. . . .	1.4311	0.7681	. . . .	. . . .	. . . .	. . . .	. . . .	. . . .
38	**3-Heptyne** (Ethyl propyl acetylene)	105–6	. . . .	1.415	0.7337	4-Heptanone	144	75	132	. . . .	. . . .
39	**2-Heptyne** (*n*-Butyl methyl acetylene)	111.5–2.5	. . . .	1.4230	0.748	2-Heptanone + 3-Heptanone	151.2 148	89	123; 127 101	. . . .	. . . .
40	**2,2,5,5-Tetramethyl-3-hexyne** (Di-*tert*-butyl acetylene)	111.9[746]	19.4	1.4055	0.7120	. . . .	. . . .	. . . .	. . . .	. . . .	. . . .
41	**1,6-Heptadiyne**	112; 30[26]	−85	1.451[17]	0.8164[17]	. . . .	. . . .	. . . .	. . . .	. . . .	. . . .
42	**1-Octyne** (*n*-Hexyl acetylene)	126.2	−79.3	1.4159	0.7461	2-Octanone	173	64–5; 58	124–5	80.4–.7, me. al.	. . . .
43	**4-Octyne** (Dipropyl acetylene)	130.4–.6[745]	. . . .	1.4226	0.7484	. . . .	. . . .	. . . .	. . . .	. . . .	. . . .
44	**3-Octyne** (Butyl ethyl acetylene)	131.0–.5	. . . .	1.4261	0.748	. . . .	. . . .	. . . .	. . . .	. . . .	. . . .
45	**2-Octyne** (Hexyl methyl acetylene)	138.0–.4	. . . .	1.4285	0.761	2-Octanone + 3-Octanone	173 169–70[738]	64–5; 58 64–5	124–5 117.0–.5	. . . .	. . . .
46	**4-Nonyne** (*n*-Butyl propyl acetylene)	150–4[752]	. . . .	1.4296[25]	0.757[25]	4-Nonanone + 5-Nonanone	187–8 188.4	57–8	73–4 90	. . . .	. . . .

*Derivative data given in order: m.p., crystal color, solvent from which crystallized.

TABLE III. ORGANIC DERIVATIVES OF ALKYNES (ACETYLENES)
a) Liquids (Listed in order of increasing b.p.)* (Continued)

No.	Name	Boiling point, °C	Melting point, °C	n_D^{20}	D_4^{20}	Hydration product (RC≡CR' → RCOCH₂R') and its derivatives				Hg salt	Miscellaneous
						Ketone	B.p., °C	2,4-Dinitrophenylhydrazone of ketone	Semicarbazone of ketone		
47	**1-Nonyne** (Heptyl acetylene)	150.8	−50	1.4217	0.7568	2-Nonanone	195.3	55–6	118–20	67.8–8.5, me. al.	. . . .
48	**3-Nonyne** (Methyl pentyl acetylene)	153–5[745]; 92[97]	. . . .	1.4299	0.7616	3-Nonanone + 4-Nonanone	187[751] 187–8	55–6 57–8	111–2 73–4	. . . .	
49	**Cyclooctyne**	157.5–8.0[740]	. . . .	1.4850	0.868	. . . .	. . . .	. . . .	. . . .		
50	**2-Nonyne** (Methyl hexyl acetylene)	161	. . . .	1.4331	0.769	3-Nonanone + 2-Nonanone	187[751] 195.3	55–6 55–6	111–2 118–20	. . . .	
51	**1,8-Nonadiyne**	162; 55.0–.5[13]	−27.28	1.4490	0.8158	. . . .	. . . .	. . . .	. . . .		
52	**1-Decyne** (n-Octyl acetylene)	174.0	−44	1.4265	0.7655	2-Decanone	215.5	124	63; 81	80.0–.7	. . . .
53	**3-Decyne** (Ethyl hexyl acetylene)	175–6	. . . .	1.433[21]	·0.7765[21]	3-Decanone + 4-Decanone	211 206–7	. . . .	100–1 51–2		
54	**5-Decyne** (Dibutyl acetylene)	177; 100[80]	−73	1.4332	0.7688	5-Decanone	. . . .	60–1.5	57.5–8.0	. . . .	NaNH₂ at 210° → 1-Decyne, b.p. 174
55	**Cyclononyne**	177–8[740]	−36.4	1.4890	0.8972	Cyclononanone	m.p.: 34	146	184–5		
56	**2,7-Nonadiyne**	180	4.30	1.4674	0.8332	. . . .	. . . .	. . . .	. . . .		
57	**1-Undecyne** (1-Hendecyne; Nonyl acetylene).	195; 96.43[30]	−25	1.4306	0.7728	2-Undecanone	228	63	122.0–.5	79	. . . .
58	**Cyclodecyne**	203–4[740]; 78.5[12]	. . . .	1.4950	0.8975	Cyclodecanone	100–2[12]	. . . .	203–5		Ozonolysis → sebacic acid, 134.5
59	**6-Dodecyne** (Dipentyl acetylene)	209[745]; 90[8]	. . . .	1.4380[25]	0.7816[25]	. . . .	. . . .	. . . .	. . . .		NaNH₂ → 1-Dodecyne, b.p. 215
60	**1-Dodecyne** (Decyl acetylene)	215; 89.09[10]	19	1.4340	0.7788	. . . .	. . . .	. . . .	. . . .	84.2–.8	. . . .
61	**1-Tridecyne** (n-Undecyl acetylene)	234; 102.95[10]	−5	1.4371	0.7842	. . . .	. . . .	. . . .	. . . .		
62	**1-Tetradecyne** (Dodecyl acetylene)	252; 118.31[10]	0	1.4396	0.7888	. . . .	. . . .	. . . .	. . . .		
63	**1-Pentadecyne** (Tridecyl acetylene)	268; 129.79[10]	10	1.4419	0.7928	. . . .	. . . .	. . . .	. . . .		
64	**1-Hexadecyne** (Tetradecyl acetylene)	284; 103.3[1]	15	1.4440	0.7965	. . . .	. . . .	. . . .	. . . .		

*Derivative data given in order: m.p., crystal color, solvent from which crystallized.

TABLE III. ORGANIC DERIVATIVES OF ALKYNES (ACETYLENES)
b) Solids (Listed in order of increasing m.p.)*

No.	Name	Melting point, °C	Boiling point, °C	n_D^{20}	D_4^{20}	Miscellaneous
1	1-Heptadecyne	22	299	1.4437[25]	0.7961[25]	....
2	2,6-Octadiyne	27	62[19]	1.453[21]	0.828[30]	....
3	1-Octadecyne	27	313; 180[15]	1.4474	0.8025	....
4	2-Octadecyne	30	184[15]	....	0.8016	....
5	1-Nonadecyne	33	327; 144[1.5]	1.4488[S]	0.8050[S]	Hg salt, 96–7, n-BuOH
6	1-Eicosyne	35	340; 153[1.1]	1.4501[S]	0.8073[S]	....
7	2-Heneicosyne	35–6	180[2]	1.4499[40]	....	....
8	1-Heneicosyne	41	....	1.4513[S]	0.8094[S]	....
9	1-Docosyne	45	363	1.4524[S]	0.8114[S]	....
10	1-Tricosyne	49	374	1.4534[S]	0.8131[S]	....
11	1-Tetracosyne	52	385	1.4544[S]	0.8148[S]	....
12	1-Pentacosyne	55	395	1.4552[S]	0.8163[S]	....
13	1-Hexacosyne	57	405	1.456[S]	0.8177[S]	....
14	1-Heptacosyne	60	415	1.4568[S]	0.8190[S]	....
15	1-Octacosyne	62	426	1.4575[S]	0.8202[S]	....
16	1-Nonacosyne	65	432	1.4581[S]	0.8213[S]	....
17	1-Triacontyne	67	441	1.4587[S]	0.8224[S]	....
18	1-Hentriacontyne	69	449	1.4593[S]	0.8234[S]	....
19	1-Dotriacontyne	71	457	1.4598[S]	0.8243[S]	....
20	1-Tritriacontyne	73	464	1.4603[S]	0.825[S]	....
21	1-Tetratriacontyne	74	472	1.4608[S]	0.8260[S]	....
22	1-Pentatriacontyne	76	479	1.4612[S]	0.8268[S]	....
23	1-Hexatriacontyne	77	486	1.4617[S]	0.8275[S]	....
24	1-Heptatriacontyne	79	493	1.4621[S]	0.8282[S]	....
25	1-Octatriacontyne	80	499	1.4625[S]	0.8289[S]	....
26	1-Nonatriacontyne	82	505	1.4628[S]	0.8295[S]	....
27	1-Tetracontyne	83	512	1.4632[S]	0.8301[S]	....

Derivative data given in order: m.p., crystal color, solvent from which crystallized.

S = supercooled liquid at 20°

EXPLANATIONS AND REFERENCES TO TABLE IV

Nitro derivative. *

$$ArH \quad + \quad HNO_3 \quad \rightarrow \quad ArNO_2 \quad + \quad H_2O$$

Nitro
derivative

From the aromatic hydrocarbon with concentrated nitric and sulfuric acids.
For directions and examples see: Cheronis, p. 578–80; Linstead, p. 48, 49; Shriner, p. 249; Vogel, p. 520; Wild, p. 24.
From the aromatic hydrocarbon with fuming and concentrated nitric acids.
See: Shriner, p. 249; Wild, p. 24.
From the aromatic hydrocarbon with fuming nitric acid in acetic acid.
See: Vogel, p. 520.
From the aromatic hydrocarbon with nitric and sulfuric acids in chloroform.
See: Vogel, p. 580.

Acetamido and Benzamido derivatives. *

$$ArH \xrightarrow{HNO_3} ArNO_2 \xrightarrow{Sn/HCl} ArNH_2$$

$$\xrightarrow{(CH_3CO)_2O} ArNHCOCH_3 \quad + \quad CH_3COOH$$

Acetamido
derivative

$$\xrightarrow{C_6H_5COCl} ArNHCOC_6H_5 \quad + \quad HCl$$

Benzamido
derivative

Nitration of the aromatic hydrocarbon is followed by reduction with tin and hydrochloric acid. The resulting amine is acetylated with acetic anhydride or benzoylated with benzoyl chloride.
For directions and examples see: Cheronis, p. 581; V. L. Ipatieff and L. A. Schmerling, *J. Amer. Chem. Soc.*, **59**, 1056 (1937); **60**, 1476 (1938); **65**, 2470 (1943).

o-Aroylbenzoic acid (product with phthalic anhydride).

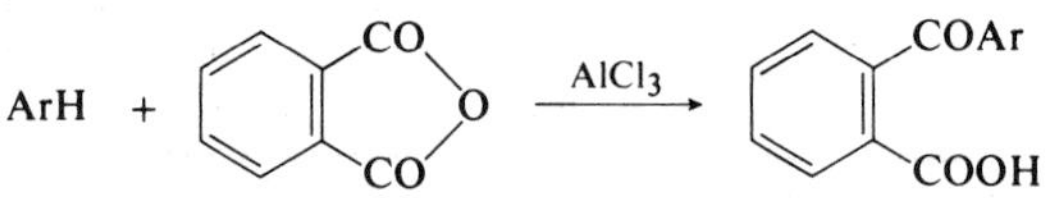

o-Aroylbenzoic acid

From the aromatic hydrocarbon, phthalic anhydride and aluminum chloride in carbon disulfide.
For directions and examples see: Cheronis, p. 548; Shriner, p. 250; Vogel, p. 519; Wild, p. 28; H. W. Underwood and W. L. Walsh, *J. Amer. Chem. Soc.*, **57**, 940 (1935).
From the aromatic hydrocarbon, phthalic anhydride and aluminum chloride without solvent.
See: G. F. Lewenz and K. T. Serijan, *J. Amer. Chem. Soc.*, **75**, 4087 (1953).

2,4-Dinitrobenzenesulfenyl chloride derivative (Aryl 2,4-dinitrophenyl sulfide).

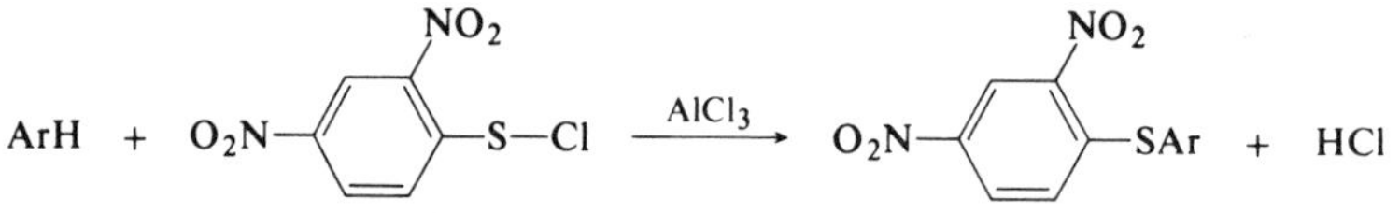

Aryl 2,4-dinitrophenyl
sulfide

From the aromatic hydrocarbon, 2,4-dinitrobenzenesulfenyl chloride and aluminum chloride in 1,2-dichloroethane.
For directions and explanations see: Cheronis, p. 585; C. M. Buess and N. Kharasch, *J. Amer. Chem. Soc.*, **72**, 3529 (1950).

Picrate.

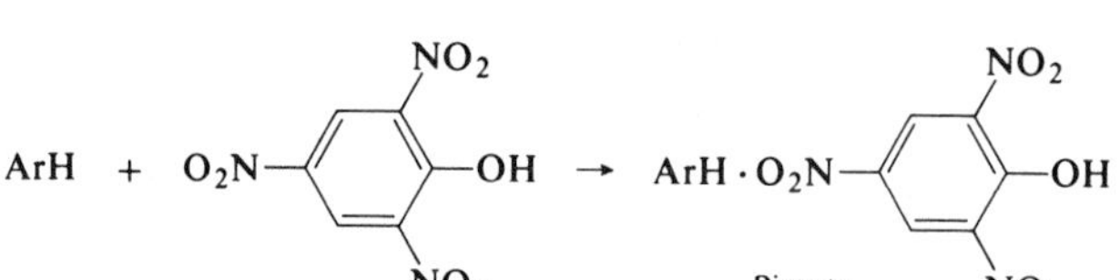

Picrate
(Molecular complex)

*Derivatives recommended for first trial.
WARNING: This is not an instruction manual. References should be consulted for the preparation of derivatives.

From the aromatic hydrocarbon and picric acid in alcohol.
For directions and examples see: Linstead, p. 50; Vogel, p. 518; Wild, pp. 29–30.
From excess of liquid aromatic hydrocarbon and picric acid without solvent.
See: Wild, pp. 28–9; Baril and Hauber, *J. Amer. Chem. Soc.*, **53,** 1087 (1931).
From the aromatic hydrocarbon in methanol or in dry benzene.
See: Cheronis, pp. 582–3.

Styphnate.

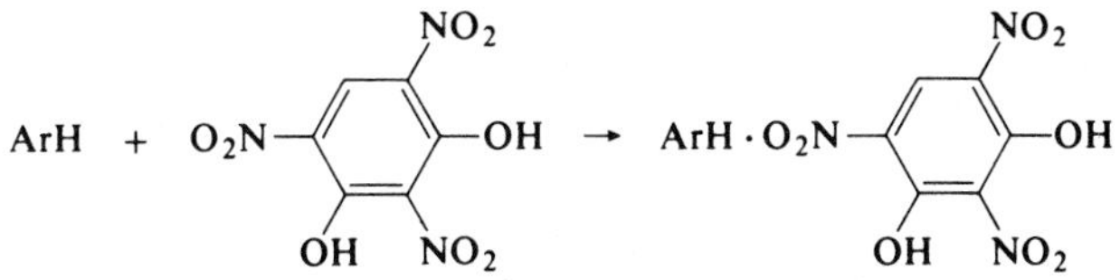

Styphnic acid

Styphnate (Molecular complex)

From the aromatic hydrocarbon and styphnic acid (2,4,6-trinitroresorcinol) in acetic acid.

For directions and examples see: Vogel, p. 519; W. J. Hickinbottom, *Reactions of Organic Compounds*, 2nd ed., Longmans, Green and Co., London, 1948, p. 76.

1,3,5-Trinitrobenzene derivative.

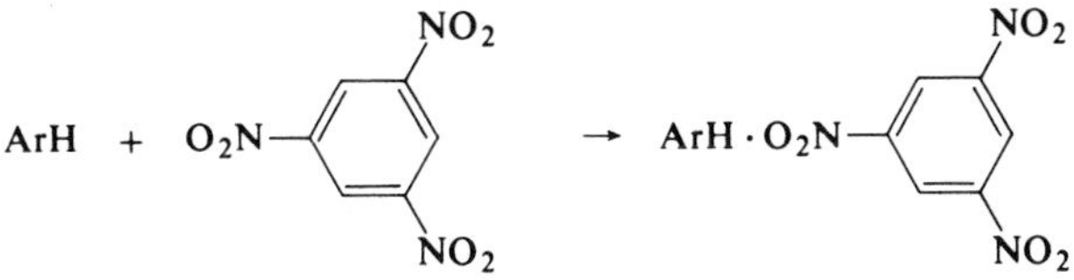

1,3,5-Trinitrobenzene molecular complex

From the aromatic hydrocarbon and 1,3,5-trinitrobenzene in alcohol, acetic acid, or benzene.
For directions and examples see: Vogel, p. 519.

*2,4,7-Trinitrofluorenone (TNF) derivative.**

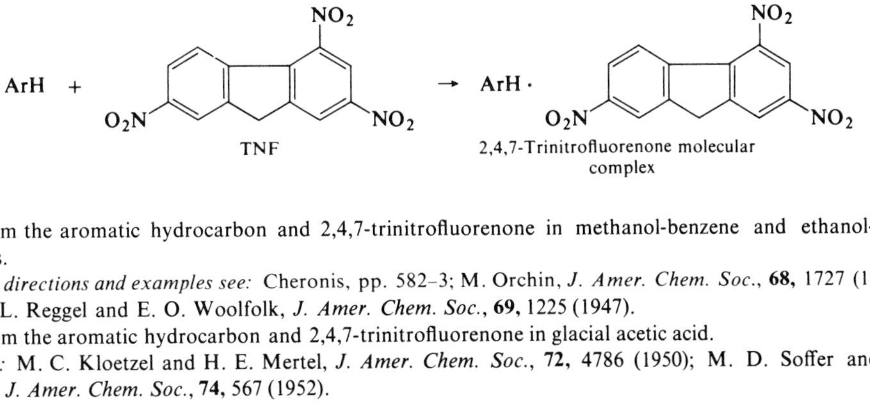

TNF

2,4,7-Trinitrofluorenone molecular complex

From the aromatic hydrocarbon and 2,4,7-trinitrofluorenone in methanol-benzene and ethanol-benzene mixtures.

For directions and examples see: Cheronis, pp. 582–3; M. Orchin, *J. Amer. Chem. Soc.*, **68,** 1727 (1946); M. Orchin, L. Reggel and E. O. Woolfolk, *J. Amer. Chem. Soc.*, **69,** 1225 (1947).

From the aromatic hydrocarbon and 2,4,7-trinitrofluorenone in glacial acetic acid.

See: M. C. Kloetzel and H. E. Mertel, *J. Amer. Chem. Soc.*, **72,** 4786 (1950); M. D. Soffer and R. A. Stewart, *J. Amer. Chem. Soc.*, **74,** 567 (1952).

From the aromatic hydrocarbon and 2,4,7-trinitrofluorenone without solvent.

See: D. E. Laskowski and W. C. McCrone, *Anal. Chem.*, **30,** 542 (1958).

Acids from side-chain oxidation.

$$\mathrm{ArR} \xrightarrow[\mathrm{NaOH}]{\mathrm{KMNO_4}} \mathrm{ArCOONa} \xrightarrow{\mathrm{H^+}} \mathrm{ArCOOH}$$

Aromatic acid

*Derivatives recommended for first trial.
WARNING: This is not an instruction manual. References should be consulted for the preparation of derivatives.

From the alkyl-substituted aromatic hydrocarbon with potassium permanganate in sodium hydroxide or sodium carbonate solution.

For directions and examples see: Cheronis, p. 585, 627; Linstead, p. 50; Shriner, p. 250; Vogel, p. 520; Wild, p. 26.

From the alkyl-substituted aromatic hydrocarbon with sodium bichromate and sulfuric acid.

See: Cheronis, p. 627; Shriner, p. 250; Wild, p. 26.

Sulfonamide. *

$$ArH \ + \ ClSO_3H \ \rightarrow \ \underset{\substack{\text{Sulfonyl} \\ \text{chloride}}}{ArSO_2Cl} \ \xrightarrow{NH_3} \ \underset{\text{Sulfonamide}}{ArSO_2NH_2}$$

From the aromatic hydrocarbon and chlorosulfonic acid in chloroform, followed by aqueous ammonia.

For directions and examples see: Linstead, p. 49; Wild, p. 27; E. H. Huntress and F. H. Carten, *J. Amer. Chem. Soc.*, **62**, 511 (1940); E. H. Huntress and J. S. Autenrieth, *J. Amer. Chem. Soc.*, **63**, 3446 (1941).

From the aromatic hydrocarbon with chlorosulfonic acid without solvent, followed by ammonolysis with dry ammonium carbonate.

See: Wild, p. 27.

---

*Derivatives recommended for first trial.
WARNING: This is not an instruction manual. References should be consulted for the preparation of derivatives.

TABLE IV. ORGANIC DERIVATIVES OF AROMATIC HYDROCARBONS
a) Liquids. (Listed in order of increasing b.p.)*

No.	Name	Boiling point, °C	Melting point, °C	n_D^{20}	D_4^{20}	Picrate	1,3,5-Trinitrobenzene derivative	Nitro derivative	Acetamido derivative	Phthalic anhydride derivative	2,4-Dinitrophenyl sulfenyl chloride derivative	Miscellaneous
1	Benzene	80.1	5.5	1.5011	0.87901	84	......	1,3-di: 89; 1,3,5-tri: 122	....	127	120	Sulfonamide, 156
2	Toluene	110.6	−95	1.49613	0.86694	88.2, pa. yel.	......	2,4-di: 70	2,4-di: 221	137	102–3	Oxid. → benzoic acid, 121; Sulfonamide, 137
3	Ethylbenzene	136.2	−93.9	1.49594	0.86690	96.6, pa. yel.	......	2,4,6-tri: 37	2,4-di: 223	122; 128	97	Oxid. → benzoic acid, 121; Sulfonamide, 109
4	1,4-Xylene	138.3	13.26	1.49581	0.86105	90	......	2,3,5-tri: 139	....	132; 148	134–5	Oxid. → terephthalic acid, >300, subl.; Sulfonamide, 147
5	1,3-Xylene	139.1	−47.89	1.49722	0.86417	91	......	2,4,6-tri: 183	....	126; 142	......	Oxid. → isophthalic acid, 348, h.w.; Sulfonamide, 137
6	1,2-Xylene	144.4	−25.18	1.50545	0.88020	88	......	4,5-di: 118	....	178	......	Oxid. → phthalic acid, 206–8; Sulfonamide, 144
7	Isopropylbenzene (Cumene)	152.4	−96.04	1.49146	0.86179	......	......	2,4,6-tri: 109	4-mono: 106; 2,4-di: 216	133	......	Oxid. → benzoic acid, 121; Sulfonamide, 106
8	n-Propylbenzene	159.2	−99.59	1.49202	0.86204	103	......	2,4-di: b.p. 150[1]	4-mono: 96; 2,4-di: 208	125	......	Oxid. → benzoic
9	1-Ethyl-3-methylbenzene (m-Ethyltoluene)	161.3	−96.55	1.49661	0.86455	......	......	....	....	......	......	Oxid. → isophthalic acid, 348, h.w.
10	1-Ethyl-4-methylbenzene (p-Ethyltoluene)	162.1	−62.35	1.49500	0.86118	......	......					Oxid. → terephthalic acid, >300, subl.
11	1,3,5-Trimethylbenzene (Mesitylene)	164.7	−44.72	1.49937	0.86518	97	......	2,4-di: 86; 2,4,6-tri: 235	....	212	......	Oxid. → trimesic acid, 380; Sulfonamide, 141
12	1-Ethyl-2-methylbenzene (o-Ethyltoluene)	165.2	−80.83	1.50456	0.88069							Oxid. → phthalic acid, 206–8
13	tert-Butylbenzene	169.1	−58.34	1.49266	0.86650	......	......	2,4-di: 62; 2,4,6-tri: 124	4-mono: 170; 2,4-di: 210	......	130–1	Oxid. → benzoic acid, 121
14	1,2,4-Trimethylbenzene (Pseudocumene)	169.4	−43.91	1.50484	0.87582	97	......	3,5,6-tri: 185	....			Oxid. → trimellitic acid, 225–35 d.
15	Isobutylbenzene	172.8	−51.53	1.48646	0.85321	......	......		4-mono: 127.0–7.5	......	99–100	Oxid. → benzoic acid, 121
16	sec-Butylbenzene	173.3	−75.57	1.49020	0.86207	......	......	2,4-di: b.p. 161–2[5]	4-mono: 126; 2,6-di: 192	......	88–9	Oxid. → benzoic
17	3-Isopropyl-1-methylbenzene (3-Isopropyltoluene; m-Cymene)	175.1	−63.75	1.4930	0.8610	......	....	....	....	......	......	....

*Derivative data given in order: m.p., crystal color, solvent from which crystallized.

TABLE IV. ORGANIC DERIVATIVES OF AROMATIC HYDROCARBONS
a) Liquids. (Listed in order of increasing b.p.)* (Continued)

No.	Name	Boiling point, °C	Melting point, °C	n_D^{20}	D_4^{20}	Picrate	1,3,5-Trinitrobenzene derivative	Nitro derivative	Acetamido derivative	Phthalic anhydride derivative	2,4-Dinitrophenyl sulfenyl chloride derivative	Miscellaneous
18	1,2,3-Trimethylbenzene (Hemimellitene)..........	176.08	−25.41	1.51393	0.89438	90.5	......	....	....	......	......	Oxid. → hemimellitic acid, 190–7 d.
19	trans-Propenylbenzene.......	176.5–7.5	−27.1 to −25.9	1.5463²⁵	0.902	......						....
20	Indane...................	177	−51.4	1.5381	0.9645							
21	4-Isopropyl-1-methylbenzene (4-Isopropyltoluene; 4-Cymene)..............	177.1	−67.94	1.4909	0.8537	......	......	2,6-di: 54; 2,3,6-tri: 118		123–4	......	Sulfonamide, 115
22	2-Isopropyl-1-methylbenzene (2-Isopropyltoluene; 2-Cymene)	178.35	−71.71	1.5006	0.8766						......	Br₂ → Tetrabromo, 59.5–60.5
23	1,3-Diethylbenzene..........	181.1	......	1.49552	0.86394	......	......	2,4,6-tri: 62		114		
24	1-Methyl-3-propylbenzene (m-Propyltoluene)	181.8	......	1.4936	0.8610							
25	Indene	182.4	−2	1.5764	0.9915	98, yel.	......				......	Acid → polymer
26	n-Butylbenzene	183.27	−88.15	1.48979	0.86013	......	......		4-mono: 105; 2,4-di: 214	97	72–3	....
27	1-Methyl-4-propylbenzene (p-Propyltoluene)	183.3	......	1.4919	0.8584							
28	1,2-Diethylbenzene...........	183.4	......	1.50346	0.87996							
29	1,4-Diethylbenzene...........	183.8	......	1.49483	0.86196							
30	1,3-Dimethyl-5-ethylbenzene..	183.8	−84.4	1.4981	0.8648			2,4,6-tri: 117.0–7.6	....			Br₂ → Tribromo, 89
31	1-Methyl-2-propylbenzene (o-Propyltoluene)..........	184.8	−60.2	1.4998	0.8744							
32	2,2-Dimethyl-1-phenylpropane (Neopentylbenzene)..........	186	......	1.4880	0.858							
33	1,4-Dimethyl-2-ethylbenzene..	186.9	......	1.5043	0.8772			3,5,6-tri: 127–8, al.	4-mono: 142; 2,4-di: 181	......		Sulfonamide, 107–8
34	2-Methylindane	187.0	......	1.5070	0.9034							
35	3-Methyl-2-phenylbutane	188	......	1.486	0.8701				4-mono: 147–8; 2,4-di: 193	....	......	4-Benzamido deriv., 141–2
36	1-Methylindane	188–90	......	1.5274	0.939							Heat with Pt. at 310–350 → Naphthalene, 80.3
37	1,3-Dimethyl-4-ethylbenzene..	188.4	−63.0	1.5038	0.8763			2,5,6-tri: 127.5–9.0	....			Br₂ → 2,5,6-Tribromo, 94–5; 81–2
38	3-tert-Butyl-1-methylbenzene (3-tert-Butyltoluene)	189.3	−41.39	1.4944	0.8657							
39	1,2-Dimethyl-4-ethylbenzene..	189.55	−67.1	1.5031	0.8745						......	Oxid. → trimellitic acid, 225–35 d.
40	1,3-Dimethyl-2-ethylbenzene..	190	....	1.5107	0.8904							Oxid. → hemimellitic acid, 190–7 d.

*Derivative data given in order: m.p., crystal color, solvent from which crystallized.

No.	Name	Boiling point, °C	Melting point, °C	n_D^{20}	D_4^{20}	Picrate	1,3,5-Trinitro-benzene derivative	Nitro derivative	Acetamido derivative	Phthalic anhydride derivative	2,4-Dinitro-phenyl sulfenyl chloride derivative	Miscellaneous
41	3-Phenylpentane	191	......	1.4877	0.8649	......	......	......	4-*mono*: 145–6; 2,4-*di*: 199–200	......	......	4-Benzamido deriv., 154
42	1-Ethyl-3-isopropylbenzene	192	......	1.4955	0.859	......	......	......	......			
43	2-Methyl-2-phenylbutane	192.38	......	1.4934	0.8737	......	......	......	4-*mono*: 142; 2,4-*di*: 181	......	......	4-Benzamido deriv., 112–3
44	4-*tert*-Butyl-1-methylbenzene (4-*tert*-Butyltoluene)	192.8	−52.49	1.4918	0.8612	......	......	2,6-*di*: 96	......			
45	1-Ethyl-2-isopropylbenzene	193	......	1.5080	0.888	......	......	......				
46	2-Phenylpentane	193	......	1.4876	0.8576	......	......	......	4-*mono*: 107; 2,4-*di*: 181–2	......	......	4-Benzamido, 127–8
47	1,2-Dimethyl-3-ethylbenzene	193.9	−49.5	1.5117	0.8921	......	......	......	......	......	......	Oxid. → hemimellitic acid, 190–7 d.
48	3-*sec*-Butyl-1-methylbenzene (3-*sec*-Butyltoluene)	194	......	1.490	0.858	......	......	......				
49	3-Isobutyl-1-methylbenzene (3-Isobutyltoluene)	194	......	1.4888	0.8536							
50	d-2-Methyl-1-phenylbutane	194	......	1.4880	0.8617							
51	1,3-Dimethyl-5-isopropyl-benzene	194.5; 191	......	1.4955	0.8591							Oxid. → trimesic acid, 380
52	2-Phenyl-*cis*-2-butene	194.5	......	1.5402^{25}	0.9191^{25}						81–2	......
53	4-Isobutyl-1-methylbenzene (*p*-Isobutyltoluene)	196	......	1.4874	0.8517							
54	2-*sec*-Butyl-1-methylbenzene (2-*sec*-Butyltoluene)	196	......	1.497	0.873							
55	2-Isobutyl-1-methylbenzene (*o*-Isobutyltoluene)	196	......	1.4935	0.8649							
56	1,4-Dimethyl-2-isopropyl-benzene	196.2	......	1.5010	0.8738							
57	1-Ethyl-4-isopropylbenzene	196.6	......	1.4923	0.8585							
58	d,l-2-Methyl-1-phenylbutane	197	......	1.486	0.859	......	......	......	4-*mono*: 115–6; 2,4-*di*: 193–4	......	......	4-Benzamido, 126
59	1,2,3,5-Tetramethylbenzene (Isodurene)	197.9	......	1.5125	0.8899	......	......	4,6-*di*: 181; 157	......	213		
60	3-Methyl-1-phenylbutane (Isopentylbenzene)	198.9; 196	......	1.4847	0.8558	......	......	......	4-*mono*: 114; 2,4-*di*: 215–6			
61	1,3-Dimethyl-2-isopropyl-benzene	199	......	1.509	0.890	......	......					
62	1,3-Dimethyl-4-isopropyl-benzene	199.1; 195	......	1.5018	0.869							
63	3-Methylindene	199.2–200; 198.5	......	1.55907^{27}	0.9640	76–8, or.-yel., al.	......	......				
64	4-*sec*-Butyl-1-methylbenzene (*p*-*sec*-Butyltoluene)	200	......	1.4932	0.8650	......	......	......				

*Derivative data given in order: m.p., crystal color, solvent from which crystallized.

No.	Name	Boiling point, °C	Melting point, °C	n_D^{20}	D_4^{20}	Picrate	1,3,5-Trinitro-benzene derivative	Nitro derivative	Acetamido derivative	Phthalic anhydride derivative	2,4-Dinitrophenyl sulfenyl chloride derivative	Miscellaneous
65	2-*tert*-Butyl-1-methylbenzene (2-*tert*-Butyltoluene)	200.5	....	1.5076	0.8897	....	....			....	....	
66	3,5-Diethyl-1-methylbenzene (3,5-Diethyltoluene)	200.7	−74.12	1.4969	0.8630	....	....	2,4,6-*tri*: 106–6.5				
67	2-Butyl-1-methylbenzene (2-Butyltoluene)	201; 208	....	1.4958	0.8721							
68	1-Ethyl-3-propylbenzene	201		1.4930	0.8607							
69	1,2-Dimethyl-4-isopropyl-benzene	201.8	....	1.4993	0.8699							
71	1,2-Dimethyl-3-isopropyl-benzene	202.6	....	1.508	0.888							
72	1-Ethyl-2-propylbenzene	203	....	1.4992	0.8744							
73	1,3-Di-isopropylbenzene	203.2	−63.1	1.4883	0.85593			4,6-*di*: 76.9–7.2, 2-ProH				
74	1,2-Diethyl-4-methylbenzene	203.6		1.5039	0.8762	....	....					
75	1,2-Di-isopropylbenzene	203.8	....	1.4960	0.8771		.,					
76	1,4-Dimethyl-2-propyl-benzene	204.3		1.4999	0.8717							
77	1,2,3,4-Tetramethylbenzene (Prehnitene)	205.0	−6.3	1.5201	0.9053	92–5	....	5,6-*di*: 176				
78	1-Ethyl-4-propylbenzene	205	....	1.4921	0.8594	....						
79	3-Butyl-1-methylbenzene (*m*-Butyltoluene)	205	....	1.491	0.859							
80	2,4-Diethyl-1-methylbenzene (2,4-Diethyltoluene)	205	....	1.5027	0.8748							
81	*n*-Pentylbenzene	205.4	−75	1.4878	0.8585				4-*mono*: 101–2; 2,4-*di*: 202	....	....	4-Benzamido, 128–9
82	3-Methyl-3-phenylpentane	206	....	1.4958	0.8755							
83	1,3-Dimethyl-5-*tert*-butyl-benzene	206–6.5	−21.5	1.4958	0.8645		....	2,4,6-*tri*: 107 (one form); 114 (another form)				
84	1,3-Dimethyl-4-propylbenzene	206.6	....	1.4998	0.8723	....						
85	1,2-Diethyl-3-methylbenzene	206.6		1.5105	0.8910							
86	4-Butyl-1-methylbenzene (4-Butyltoluene)	207		1.490	0.857	....						
87	2,5-Diethyl-1-methylbenzene (2,5-Diethyltoluene)	207.1	....	1.5034	0.8758							
88	1,2,3,4-Tetrahydronaphthalene (Tetralin)	207.6	−35.79	1.54135	0.9702	....	....	5,7-*di*: 95		153–5	....	$Cl_2 \rightarrow$ 5,6,7,8-Tetrachloro, 172
89	1,3-Diethyl-2-propylbenzene	207.6		1.5063	0.8856							
90	2,6-Diethyl-1-methylbenzene (2,6-Diethyltoluene)	208.8	....	1.5106	0.8907							
91	1,2-Dimethyl-4-propylbenzene	208.9	....	1.5000	0.8715							
92	1,3-Dimethyl-5-propylbenzene	209	....	1.4933	0.8610							
93	2-Methyl-3-phenylpentane	209	....	1.4912	0.8678							
94	4-*tert*-Butyl-1,3-dimethyl-benzene	210–4		1.5030^{37}	0.9372^{30}			2,5,6-*tri*: 112, al.				
95	1,4-Di-isopropylbenzene	210.4	−17.1	1.48983	0.85676	....						
96	1,2-Dimethyl-3-propylbenzene	210.7	....	1.5075	0.8864	....						

*Derivative data given in order: m.p., crystal color, solvent from which crystallized.

No.	Name	Boiling point, °C	Melting point, °C	n_D^{20}	D_4^{20}	Picrate	1,3,5-Trinitrobenzene derivative	Nitro derivative	Acetamido derivative	Phthalic anhydride derivative	2,4-Dinitrophenyl sulfenyl chloride derivative	Miscellaneous
97	1-*tert*-Butyl-4-ethylbenzene . . .	211	. . . . . .	1.4950	0.8635	. . . . . .	. . . . . .	2,6-*di*: 94–5, al.	. . . .	. . . . .	. . . . .	. . . .
98	d,l-3-Phenylhexane	211; 208.3	. . . . . .	1.4867	0.8596	. . . . . .	. . . . . .		2,4-*di*: 207–8	. . . . .	. . . . .	. . . .
99	2-Ethyl-1,3,5-trimethyl-benzene	212.4	−12.2	1.5074	0.883	. . . . . .	. . . . . .	4,6-*di*: 111, al.	. . . .	. . . . .	. . . . .	. . . .
100	3-Ethyl-4-isopropyl-1-methyl-benzene	213	. . . . . .	1.5006	0.8722	. . . . . .	. . . . . .		. . . .	. . . . .	. . . . .	. . . .
101	5-Ethyl-1,2,4-trimethylbenzene	213	−13.5	1.5075	0.833	. . . . . .	. . . . . .	3,6-*di*: 87–8, al.	. . . .	. . . . .	. . . . .	3,6-Dibromo, 60–1, acet.
102	6-Ethyl-1,2,4-trimethylbenzene	213	. . . . . .	1.5118	0.8897	. . . . . .	. . . . . .		. . . .	. . . . .	. . . . .	. . . .
103	2-Phenylhexane	214	. . . . . .	1.4882	0.8600	. . . . . .	. . . . . .		2,4-*di*: 178	. . . . .	. . . . .	. . . .
104	2-Methyl-1-phenylpentane	215	. . . . . .	1.4847	0.8624	. . . . . .	. . . . . .		. . . .	. . . . .	. . . . .	. . . .
105	4-Isopropyl-1-propylbenzene . .	215	. . . . . .	1.4972	0.8614	. . . . . .	. . . . . .		. . . .	. . . . .	. . . . .	. . . .
106	1,3-Dipropylbenzene	215–8	. . . . . .	1.5155[16]	0.9137[17]	. . . . . .	. . . . . .		. . . .	. . . . .	. . . . .	. . . .
107	5-Ethyl-1,2,3-trimethylbenzene	215.8	. . . . . .	1.5101	0.8863	. . . . . .	. . . . . .		. . . .	. . . . .	. . . . .	. . . .
108	3-Ethyl-1,2,4-trimethylbenzene	216.6	. . . . . .	1.5133	0.895	. . . . . .	. . . . . .	5,6-*di*: 79–80, al.	. . . .	. . . . .	. . . . .	. . . .
109	1,2,4-Triethylbenzene	217.7	. . . . . .	1.4982	0.8791	. . . . . .	. . . . . .		. . . .	. . . . .	. . . . .	. . . .
110	1,3,5-Triethylbenzene	218; 211.2	. . . . . .	1.4965	0.8568[25]	. . . . . .	. . . . . .	2,4,6-*tri*: 112.4–2.6	. . . .	129	. . . . .	2,4,6-Tribromo, 105
111	2-Methyl-1,2,3,4-tetrahydro-naphthalene (2-Methyl-tetralin)	218	. . . . . .	1.5311	0.952	. . . . . .	. . . . . .		. . . .	. . . . .	. . . . .	. . . .
112	1-Methyl-1,2,3,4-tetrahydro-naphthalene (1-Methyl-tetralin)	219	. . . . . .	1.5357	0.9580	. . . . . .	. . . . . .		. . . .	. . . . .	. . . . .	. . . .
113	4-Ethyl-1,2,3-trimethyl-benzene	220.4	. . . . . .	1.5180	0.9019	. . . . . .	. . . . . .		. . . .	. . . . .	. . . . .	. . . .
114	1,4-Dipropylbenzene	221	. . . . . .	1.4914	0.8564	. . . . . .	. . . . . .		. . . .	. . . . .	. . . . .	. . . .
115	3-Methyl-1-phenylpentane	221	. . . . . .	1.4876	0.8605	. . . . . .	. . . . . .		. . . .	. . . . .	. . . . .	. . . .
116	2-Propyl-1,3,5-trimethyl-benzene	221	. . . . . .	1.5033	0.8782	. . . . . .	. . . . . .		. . . .	. . . . .	. . . . .	. . . .
117	1,1-Dimethyl-1,2,3,4-tetra-hydronaphthalene (1,1-Dimethyltetralin)	221	. . . . . .	1.5292	0.950	. . . . . .	. . . . . .	Ar-x, x-*di*: 64.5	. . . .	. . . . .	. . . . .	. . . .
118	3-*tert*-Butyl-1-isopropyl-benzene	222	. . . . . .	1.4832	0.8512	. . . . . .	. . . . . .		. . . .	. . . . .	. . . . .	. . . .
119	1-Methyl-3-pentylbenzene (3-Pentyltoluene)	223	. . . . . .	1.4911	0.8593	. . . . . .	. . . . . .		. . . .	. . . . .	. . . . .	. . . .
120	4-*tert*-Butyl-1-isopropyl-benzene	224	. . . . . .	1.4872	0.8665	. . . . . .	. . . . . .		. . . .	. . . . .	. . . . .	. . . .
121	2-Methyl-2-phenylhexane	225	. . . . . .	1.4943	0.8737	. . . . . .	. . . . . .		. . . .	. . . . .	. . . . .	. . . .
122	2,4-Di-isopropyl-1-methyl-benzene (2,4-Di-isopropyl-toluene)	225	. . . . . .	1.4990	0.8664	. . . . . .	. . . . . .		. . . .	. . . . .	. . . . .	. . . .
123	3-Methyl-3-phenylhexane	226	. . . . . .	1.4980	0.8776	. . . . . .	. . . . . .		. . . .	. . . . .	. . . . .	. . . .
124	n-Hexylbenzene	226.1	−61.2	1.4864	0.8575	. . . . . .	. . . . . .		2,4-*di*: 205–6	. . . . .	. . . . .	. . . .
125	3-Phenylheptane	227	. . . . . .	1.4862	0.8607	. . . . . .	. . . . . .		. . . .	. . . . .	. . . . .	. . . .
126	2,6-Di-isopropyl-1-methyl-benzene (2,6-Di-isopropyl-toluene)	228	. . . . . .	1.5032	0.8768	. . . . . .	. . . . . .		. . . .	. . . . .	. . . . .	. . . .

*Derivative data given in order: m.p., crystal color, solvent from which crystallized.

No.	Name	Boiling point, °C	Melting point, °C	n_D^{20}	D_4^{20}	Picrate	1,3,5-Trinitrobenzene derivative	Nitro derivative	Acetamido derivative	Phthalic anhydride derivative	2,4-Dinitrophenyl sulfenyl chloride derivative	Miscellaneous
127	5-Propyl-1,2,4-trimethyl-benzene	228	......	1.5095	0.887	......	......	......	......	......	......	
128	6-Methyl-1,2,3,4-tetra-hydronaphthalene (6-Methyltetralin)	229	......	1.5357	0.9537	......	......	......	......	......	......	
129	2,2-Dimethyl-1,2,3,4-tetra-hydronaphthalene (2,2-Dimethyltetralin)	230	......	1.5200	0.935	......	......	......	......	......	......	
130	2-Phenylheptane	231	......	1.4863	0.8610							
131	5-Methyl-1,2,3,4-tetra-hydronaphthalene (5-Methyltetralin)	234.4	......	1.54395	0.9720							
132	2-Ethyl-1,2,3,4-tetrahydro-naphthalene (2-Ethyltetralin)	235	......	1.523	0.938							
133	Cyclohexylbenzene	235–6	7–8	1.5329	0.9502							
134	1-Ethyl-1,2,3,4-tetrahydro-naphthalene (1-Ethyltetralin)	236	......	1.5321	0.9535							
135	2,5-Dimethyl-1,2,3,4-tetra-hydronaphthalene (2,5-Dimethyltetralin)	236	......	1.526	0.946							
136	2,8-Dimethyl-1,2,3,4-tetra-hydronaphthalene (2,8-Dimethyltetralin)	236	......	1.526	0.941							
137	2,7-Dimethyl-1,2,3,4-tetra-hydronaphthalene (2,7-Dimethyltetralin)	237–8	......	1.526	0.941							
138	2,6-Dimethyl-1,2,3,4-tetra-hydronaphthalene (2,6-Dimethyltetralin)	238	......	1.526	0.941							Oxid. → trimellitic acid, 225–35 d.
139	1,4-Di-sec-butylbenzene	239	......	1.4892	0.8590							
140	1,5-Dimethyl-1,2,3,4-tetra-hydronaphthalene (1,5-Dimethyltetralin)	239	......	1.526	0.9410							
141	3-Ethyl-3-phenylhexane	239	......	1.4943	0.875							
142	6-Ethyl-1,2,3,4-tetrahydro-naphthalene (6-Ethyltetralin)	241	......	1.5331	0.9568							
143	2-Methyl-1-phenyl-1-butene	241–2	......	1.528^{18}	......							Nitrosit, 129–30
144	5-Ethyl-1,2,3,4-tetrahydro-naphthalene (5-Ethyltetralin)	242	......	1.540	0.973							
145	n-Heptylbenzene	244	......	1.4875	0.8595	......	......	......				
146	1-Methylnaphthalene	244.8	−30.57	1.6174	1.02025	142, or.-red, al.	153.5–4.5, al.	4-mono: 71; 4,5-di: 143	......	68	......	Styphnate, 135, al.
147	5,6-Dimethyl-1,2,3,4-tetra-hydronaphthalene (5,6-Dimethyltetralin)	252	......	1.552	0.975	......	......				......	Oxid. → melophanic acid, 238–42
148	6,7-Dimethyl-1,2,3,4-tetra-hydronaphthalene (6,7-Dimethyltetralin)	252	10	1.5360	0.954	......	......	5,8-di: 203			......	
149	5,7-Dimethyl-1,2,3,4-tetra-hydronaphthalene (5,7-Dimethyltetralin)	253.1	−6	1.5405	0.9583	......	......	......			......	Heating with S at 320° → 1,3-Dimethylnaphthalene, b.p. 263

*Derivative data given in order: m.p., crystal color, solvent from which crystallized.

TABLE IV. ORGANIC DERIVATIVES OF AROMATIC HYDROCARBONS

No.	Name	Boiling point, °C	Melting point, °C	n_D^{20}	D_4^{20}	Picrate	1,3,5-Trinitrobenzene derivative	Nitro derivative	Acetamido derivative	Phthalic anhydride derivative	2,4-Dinitrophenyl sulfenyl chloride derivative	Miscellaneous
150	**5,8-Dimethyl-1,2,3,4-tetrahydronaphthalene (5,8-Dimethyltetralin)**	254	......	1.547	0.967	......	......		......	......	......	Heating with S at 230° → 1,4-Dimethylnaphthalene, b.p. 268
151	**2-Ethylnaphthalene**	257.9	−7.5	1.59761	0.9922	77.0±7.5, al.	88-9, 2-PrOH	......		......	......	Styphnate, 88-90
152	**1-Ethylnaphthalene**	258.67	−13.88	1.6062	1.00816	98.5	111.5-12, al.	......		......	......	Styphnate, 111-3, al.
153	**1,7-Dimethylnaphthalene**	263	−13	1.60831	1.0115	121	137	......		......	......	Styphnate, 143
154	**1,6-Dimethylnaphthalene**	263	−14	1.6072	1.003	114-5, or., al.	139, yel.	......		......	......	Styphnate, 122
155	**1,3-Dimethylnaphthalene**	263	−4.0	1.6078	1.0063	118	135, yel., al.	......		......	......	Styphnate, 117-8, w.-me. al.; 2,4,7-Trinitrofluorenone deriv., 142-5, or.
156	*n*-**Octylbenzene** (1-Phenyloctane)	264.5	−36	1.4845	0.8562	......	......	2,4-*di*: 2	....	......	......	......
157	**1-Allylnaphthalene**	265-7	......	1.6140	1.0228	69	......				......	......
158	**1-Isopropylnaphthalene**	267.9	−16	1.5950	0.99565	85-6	......				......	Dimer, 198.5-9.5; Tetrabromo, 141-2
159	**1,4-Dimethylnaphthalene**	268; 262-4	7.66	1.6127	1.0166	144, or., me. al.	165-6, yel., me. al.	......		......	......	Styphnate, 126-7 or., me. al.
160	**1,1-Diphenylethane**	268-70	......	1.5761	1.0033	......	......				......	Oxid. → benzophenone, 49
161	**2-Isopropylnaphthalene**	a) 268.2; b) 262	......	1.5772; 1.5861	0.9795	93-5; 91-3	......				......	......
162	**2-Propylnaphthalene**	273.5; 277-9	......	1.5872	0.9770	93-4, or., al.	99				......	......
163	**1-Propylnaphthalene**	277; 272.5	−10	1.5952	0.9918	91-2	86-7, al.				......	......
164	**1,3,7-Trimethylnaphthalene**	280	13.5	1.5759	1.007	144, or., al.	......				......	Styphnate, 151.5, or., me. al.
165	**1-Isopropyl-7-methylnaphthalene** (Apocadalene)	282	......	1.5884	0.9833	102, or., al.	......				......	Styphnate, 166 (163-4), yel., al.
166	*n*-**Nonylbenzene** (1-Phenylnonane)	282	−24	1.4838	0.8558	......	......				......	4-Sulfonamide, 94.5-5.0; Maleic anhydride → 3-(4-Nonylbenzoyl)acrylic acid, 82-3
167	**2-Butylnaphthalene**	283-5; 292	−8.1	1.57774	0.9673	71-3, or.-yel., al.	......				......	......
168	**2-*tert*-Butylnaphthalene**	285-90	−4	1.5768	0.9687	102-3	......				......	......
169	**1-*tert*-Butylnaphthalene**	287-9	......	1.5726	0.9629	96, yel.	......				......	......
170	**1-Butylnaphthalene**	289.34	−19.76	1.5819	0.97673	104-5, or.-yel.	......				......	......

*Derivative data given in order: m.p., crystal color, solvent from which crystallized.

TABLE IV. ORGANIC DERIVATIVES OF AROMATIC HYDROCARBONS
a) Liquids. (Listed in order of increasing b.p.)* (Continued)

No.	Name	Boiling point, °C	Melting point, °C	n_D^{20}	D_4^{20}	Picrate	1,3,5-Trinitro-benzene derivative	Nitro derivative	Acetamido derivative	Phthalic anhydride derivative	2,4-Dinitro-phenyl sulfenyl chloride derivative	Miscellaneous
171	**4,5-Benzindane** (1,2-Cyclo-pentanonaphthalene)	294–5	......	1.6290	1.066	110	119–20	....	......	......	......	2,4,7-Trinitro-fluorenone deriv., 133
172	*n*-Decylbenzene (1-Phenyl-decane)	300	−14.38	1.48319	0.85553	......						....
173	**1-Pentylnaphthalene**	307	−22	1.5725	0.9656	......	75, yel.					
174	**2-Pentylnaphthalene**	310	−21	1.5694	0.9561	......	74, yel.					
175	*n*-Undecylbenzene (*n*-Hendecyl-benzene; 1-Phenylundecane)..	316	−5	1.4828	0.8553	......						4-Sulfonamide, 95.7–6.2
176	**1-Hexylnaphthalene**	322	−17.7	1.5647	0.9566	......	69–74	....				....
177	**2-Hexylnaphthalene**	324	−5.6	1.620	0.9479	......	67–8, yel.					....
178	*n*-Dodecylbenzene (1-Phenyl-dodecane)	331	3	1.4824	0.8551							4-Sulfonamide, 97.5
179	**1-Heptylnaphthalene**	340	......	1.5582	0.9491	......						....
180	**2-Heptylnaphthalene**	341	1	1.5556	0.9410	......						....
181	**Tridecylbenzene** (1-Phenyl-tridecane)	346	10	1.4821	0.8550							
182	**1-Octylnaphthalene**	356	−2.0	1.5532	0.9427							
183	**2-Octylnaphthalene**	357	2 forms: stable: −0.5; meta-stable: 13	1.5501	0.9356	......						
184	**1-Nonylnaphthalene**	372	......	1.5477	0.9371	......	......	....	......	......	......	....
185	**2-Nonylnaphthalene**	372	12	1.5454	0.9298	......	......	....	......	......	......	....
186	**1-Decylnaphthalene**	387	......	1.5435	0.9322	......	......	....	......	......	......	....

*Derivative data given in order: m.p., crystal color, solvent from which crystallized.

TABLE IV. ORGANIC DERIVATIVES OF AROMATIC HYDROCARBONS
b) Solids. (Listed in order of increasing m.p.)*

No.	Name	Melting point, °C	Boiling point, °C	Picrate	Styphnate	sym-Trinitrobenzene derivative	2,4,7-Trinitrofluorenone derivative	Nitro derivative	Phthalic anhydride derivative	2,4-Dinitrophenyl sulfenyl chloride derivative	Miscellaneous
1	1,2,6-Trimethylnaphthalene....	14	146^{10}	122–3, al.	150–1, al.	....	....	....	....	....	n_D^{20}: 1.6010
2	Diphenylmethane	26–7	264.7; 261–2; 120^{10}	....				2,4,2',4'-tetra: 172			n_D^{20}: 1.5770; D_{25}^{25}: 1.0056; CrO_3 → Benzophenone, 49
3	1,2,3-Trimethylnaphthalene....	27–8, al.	125–30^{12}	143, or., al.	143.5, yel.	154–6, al.	....	....			n_D^{25}: 1.5725
4	1,6,7-Trimethylnaphthalene....	28, me. al.	285; 138^{12}	125–6, or., me. al.	148–9, or., me. al.	142–3, yel., me. al.					
5	2-Isopropylazulene	31, bl.-vlt.	....	....	....	di: 113–4					
6	1,4-Dimethyl-7-isopropyl-azulene (δ-Guaiazulene)	31.5, bl.-vlt., al.	167–8^{12}	122–2.5, bl., al.	105–6, bl., me. al.	151–1.5					D_4^{19}: 0.9728 (super-cooled); 2,4,6-Trinitrotoluene deriv., 89
7	2,6-Dimethylphenanthrene.....	33–4, me. al.	....	135–6, yel., al.	148–50, yel., me. al.						
8	1,2,5-Trimethylnaphthalene....	33.5; 31–2, al.	147–8^{11}	138–40, al.	131, al.	159–60, al.					n_D^{20}: 1.6110; D_{20}^{20}: 1.0103; 2,4,6-Trinitrotoluene deriv., 90–0.5
9	1-Propylphenanthrene	34–5, me. al.	....	100–1, yel., me. al.							
10	5-Isopropylazulene	34.5				134; 122–3					
11	2-Propylphenanthrene	35–6, al.	$170^{0.2}$	91–2, al.							
12	2-Methylnaphthalene	37–8; 34.4	240–2; 110–2^{16}	116, al.	....	123, yel.	125–6, al.	1-mono: 81	....		CrO_3 → β-Naphthoic acid, 182
13	1-Ethyl-5-methylnaphthalene...	40, al.	133^{10}	97, or., al.	....						
14	9-Isopropylphenanthrene	41–2		109–10							
15	6-Isopropylazulene	43	....	....	....	124					
16	2-Ethyl-6-methylnaphthalene...	44–5	145–50^{11}	109, or.	119, yel.	116–7, yel.					2,4,6-Trinitrotoluene deriv., 62, yel.
17	2-Isopropylphenanthrene ,.....	44–5, al.	....	108, yel., me. al.							
18	6-Isopropyl-1-methyl-phenanthrene...............	45–6		143, or.							
19	2-Ethylazulene	45.5; 44–5, bl.	....	110–1	....	107					
20	2,5-Dimethylphenanthrene	46–7, al.	204–5^{15}	127–9, yel., al.	132–3, or., al.						
21	1,3,5-Trimethylnaphthalene....	47, me. al.	139.5^{10}	141–2, me. al.	138, yel.						
22	3-Ethyl-6-methylphenanthrene .	47–8	....	156–6.5, or.							
23	2-Methylazulene.............	47–8		130–1, bl., al.		140–1, dk. red, al.					
24	1,3,8-Trimethylnaphthalene....	48, me. al.		127.5, or., al.	140.5, al.	....					
25	4-Methylphenanthrene	49–50, 95% al.	....	140–1, al.	135, or., al.						

*Derivative data given in order: m.p., crystal color, solvent from which crystallized.

TABLE IV. ORGANIC DERIVATIVES OF AROMATIC HYDROCARBONS
b) Solids. (Listed in order of increasing m.p.)* (Continued)

No.	Name	Melting point, °C	Boiling point, °C	Picrate	Styphnate	*sym*-Tri-nitro-benzene derivative	2,4,7-Tri-nitrofluo-renone derivative	Nitro derivative	Phthalic anhy-dride deriva-tive	2,4-Di-nitro-phenyl sulfenyl chloride deriva-tive	Miscellaneous
26	1,4-Dimethylphenanthrene.....	50–1 (cor.), me. al.	....	143.5, or.-yel.	135.5–6.5, or.	....					
27	Bibenzyl (1,2-Diphenylethane) .	53	284	....	....	102	....	4,4'-*di*: 180; 2,2',4,4'-*tetra*: 169	.....	132–3	D_{50}^{50}: 0.9782; CrO$_3$ → Benzoic acid, 121
28	Methylenefluorene (Bi-phenyleneethylene)	53	....	152–3							
29	3,5-Dimethylphenanthrene.....	53–4, me. al.	....	139, or., me. al.	124–5, or.-yel.	....					
30	1,3-Dimethylazulene..........	54	....	164–6	....	164–6, al.					
31	7-Methyl-3,4-benzphenanthrene	54.0–4.5, al.		134.0–4.5, red, al.	....		178.5–8.8				
32	Pentamethylbenzene..........	54.3; 51	231.8	131	....	121	....	6-*mono*: 154	......		
33	1,2,4-Trimethylnaphthalene....	55–6; 50, me. al.	146[12]	148–8.5, or., me. al.	123.5, yel., me. al.	165–6.5, me. al.	....				
34	3,3'-Dimethylstilbene (*sym*-Di-*m*-tolylethylene)	55–6		97							
35	1,4,5,7-Tetramethylnaphthalene	56	162–5[11]	153			152.8–3.4				
36	1,2,4,8-Tetramethylnaphthalene	56–7	150[10]	145.5	....	167, red, me. al.					2,4,6-Trinitrotoluene deriv., 88, yel.
37	2,9-Dimethylphenanthrene.....	56–7, al.	....	138, yel., al.	....						
38	1,5-Dimethylphenanthrene.....	57–8, me. al.	....	134–5, or., me. al.							
39	2-Benzylnaphthalene	58	350	93–4, yel., al.	....		124.3–5.4				D^0: 1.176
40	1-Benzylnaphthalene	58–9	350	103–4, yel.	....						D^{17}: 1.166
41	1,2-Dimethylazulene..........	58–9, bl., al.	....	129–30, blk., al.	....	166–7, br.-blk., al.					
42	9-Propylphenanthrene	59	265–70[22]	99, yel., al.	....						
43	1,7-Dimethyl-4-isopropyl-naphthalene...............	60, al.-w.	....	92, or.-red, al.	120, yel.						
44	3-Methylphenanthrene	62–3	140–50[6]	137–8, yel., al.	....						
45	3,4-Dimethylphenanthrene.....	62–3, me. al.	....	129–30, or.-red, al.	142–3, or.-red, al.	....					
46	1-Ethylphenanthrene	62.5, al.	....	108–9, or., al.	144, yel., al.						
47	*sym*-Diphenylacetylene (Tolane)	62.5, al.	....	111, yel.	....	96, yel.					
48	9-Ethylphenanthrene	62.5–3.0; 66, bz.-pet. eth.	198–200	123–4, or.-red, al.	....					....	D_4^{78}: 1.0603; n_D^{78}: 1.6582

*Derivative data given in order: m.p., crystal color, solvent from which crystallized.

TABLE IV. ORGANIC DERIVATIVES OF AROMATIC HYDROCARBONS
b) Solids. (Listed in order of increasing m.p.)* (Continued)

No.	Name	Melting point, °C	Boiling point, °C	Picrate	Styphnate	sym-Trinitrobenzene derivative	2,4,7-Trinitrofluorenone derivative	Nitro derivative	Phthalic anhydride derivative	2,4-Dinitrophenyl sulfenyl chloride derivative	Miscellaneous
49	1,4,5-Trimethylnaphthalene....	63	145[12]	144–5, red, al.	129–30	....	....	....	....	....	....
50	4-Methylfluorene	63, al.	....	....	....	....	....	....	....	....	In H₂SO₄ sol. → grn.
51	1,4,6,7-Tetramethylnaphthalene	63–4, al.	....	148–9	....	....	172.4–3.4	....	....	....	....
52	1,2,3-Trimethylphenanthrene...	63.8–4.8	....	187–8, or.	....	200.7–1.5, yel., bz.-al.	....	....	....	....	....
53	1,8-Dimethylnaphthalene......	65; 63	140[18]	156; 148	160	....	....	....	....	....	....
54	8-Methyl-3,4-benzphenanthrene	65–6, al.	....	107–8, red, al.	....	....	....	....	....	....	....
55	2-Ethylphenanthrene	67–8, me. al.; 64–5	....	95.5–6.0, yel., al.; 92–3	....	....	180.7–0.9	....	....	....	....
56	3,4-Benzphenanthrene	68, al.	....	120.8–8.5, red, al.	....	....	170.8–1.1	....	....	....	....
57	1,3,7-Trimethylphenanthrene ..	68–9, me. al.	....	163–4, al.	160–1	....	....	....	....	....	....
58	4-Isopropyl-1-methyl-phenanthrene...............	68–8.5, me. al.	....	113.6–4, or., al.	....	....	....	....	....	....	....
59	4,8-Dimethylazulene.........	69–70, bl., al.	....	157–8, blk., al.	....	179–80, red-br., al.	....	....	....	....	....
60	Biphenyl	69.2; 71	254–5; 145[22]	....	....	....	....	4,4'-di: 237; 229; 2,2',4,4'-tetra: 150	225	142–3	n₄²⁰: 1.475; D₄²⁰: 0.866
61	2-Methyl-3,4-benzphenanthrene	70.4–1.0, al.	....	141.8–3.2, red, bz.-al.	....	145, bz.-pet. eth.	158–8.5, or.-red, aq. al.	....	....	....	....
62	3-Methylpyrene	71–2, al.	....	211–2, br.-red, bz.	....	....	....	....	....	....	Conc. H₂SO₄ sol. → yel. with grn. fluorescence; on heating → olive grn. with vlt. fluorescence
63	1,4,7-Trimethylphenanthrene...	72–3	....	141–2, me. al.	129–30	....	....	....	....	....	....
64	1,4-Dimethylanthracene.......	74, al.	....	140	....	....	....	....	....	....	....
66	4,9-Dimethyl-1,2-benz-anthracene................	75, me. al.	....	116, br., me. al.	....	124–5, red, me. al.	....	....	....	....	....
67	Benzalfluorene (ω-Phenyl-dibenzfulvene)	76, al.	....	115–6	....	....	....	....	....	....	Dibromide, 116 d.
68	1,3-Dimethylphenanthrene.....	76–7, ac. a.	....	153–5, or., al.	165–6	....	....	....	....	....	....
69	1-Methyl-3,4-benzphenanthrene	77.8	210[0.4]	112–3, red, al.	....	....	171.8–2.2	....	....	....	....
70	3-Isopropyl-1-methyl-phenanthrene.............	79	180[1.5]	150	155	....	....	....	....	....	....

*Derivative data given in order: m.p., crystal color, solvent from which crystallized.

No.	Name	Melting point, °C	Boiling point, °C	Picrate	Styphnate	sym-Trinitrobenzene derivative	2,4,7-Trinitrofluorenone derivative	Nitro derivative	Phthalic anhydride derivative	2,4-Dinitrophenyl sulfenyl chloride derivative	Miscellaneous
71	1,2'-Binaphthyl	79-80; 76	....	127.0-7.5 (cor.), or.	,.......	....	145.0-6.9	....	......	....	....
72	2,3-Dimethylphenanthrene	79-80	.....	146-7, or.-red, al.	147-8, or., al.	....	....				
73	1,2,4,5-Tetramethylbenzene (Durene)	79.2; 80	196-8	....				3,6-di: 205	263		
74	1-Ethyl-2-methylphenanthrene	80, me. al.	....	134-5, me. al.		152.5-3.0					
75	1,5-Dimethylnaphthalene	80.0-0.5, 85% al.	....	140							
76	6-Methyl-3,4-benzphenanthrene	80-1, al.	206-8²	118.0-8.5, red, me. al.			144.2-4.5				
77	Naphthalene	80.3	218	149	....	153	153-4	1-mono: 61; 57	172	173-4	....
78	1,3,6,8-Tetramethylnaphthalene	81	115-6²	151-2		175-6					
79	1-Ethyl-7-methylphenanthrene (Homopimanthrene)	81, al.	....	115-6, yel., me. al.							Quinoxaline deriv., 154, ac. a.
80	9-Methylanthracene	81.5	196-7¹²	137 d., red-br.							D_4^{99}: 1.065, n_D^{99}: 1.6959; Irradiation in acetone → dimer, 228.0-8.5; Photo-oxide, ca. 80, exp.
81	1-Isopropyl-7-methyl-phenanthrene	82-3	170¹	119-20, or.	148-9, yel.		163-4, yel.				....
82	6-Methylazulene	83, bl.-vlt.	....	137; 125	....		141				
83	1,3-Dimethylanthracene	83	....	136							bl. fluorescence
84	2,2'-Dimethylstilbene (sym-Di-o-tolylethylene)	83	176-80¹⁰	102							
85	1-Methylanthracene	85-6, me. al.	199-200	115.6-6.2, red, me. al.	176.4-7.0, red, me. al.			219.0-9.8, red, bz.			D_4^{99}: 1.0471; n_D^{99}: 1.6802; Irradiation → dimer, 246
86	1,7-Dimethylphenanthrene	86	....	132, me. al.	159, me. al.						
87	1,6-Diphenylnaphthalene	86-7	....	106-8							
88	1,6-Dimethylphenanthrene	87-8, me. al.	....	134, yel., al.							
89	1,9-Dimethylphenanthrene	88, al.	....	163.5, or.-yel., me. al.	181						
90	9-Methylphenanthrene	90-1, aq. al.	....	152-3, al.							
91	1,2,10-Trimethylanthracene	90.6-1.4, yel.	....	138.5-9.5, al.			169.6-170.2, al.				
92	7-Ethyl-1-methylphenanthrene	91.0-1.5; 84-5	....	....	141-2	135-6					
93	Triphenylmethane	92	358	....				4,4',4''-tri: 206	......	......	Br₂ → bromo deriv., 152
95	5-Isopropylnaphthanthracene	92, ac, a.	....	157	....						

*Derivative data given in order: m.p., crystal color, solvent from which crystallized.

TABLE IV. ORGANIC DERIVATIVES OF AROMATIC HYDROCARBONS
b) Solids. (Listed in order of increasing m.p.)* (Continued)

No.	Name	Melting point, °C	Boiling point, °C	Picrate	Styphnate	sym-Tri-nitro-benzene derivative	2,4,7-Tri-nitrofluo-renone derivative	Nitro derivative	Phthalic anhy-dride deriva-tive	2,4-Di-nitro-phenyl sulfenyl chloride deriva-tive	Miscellaneous
96	3,9-Dimethyl-1,2-benz-anthracene...............	93	....	137–8, red, al.	....	145, red, bz.-lgr.	....				
97	5,6-Benzindane (2,3-Cyclo-pentenonaphthalene).......	94	....	120–1	....						
98	12-Isopropylnaphthanthracene .	94–5, ac. a.	....	157–8	....						
99	Acenaphthene	96.2, yel., al.	278	162, or.-red, al.	....	168, yel. al.	175–6	5-mono: 101	198	187–9 d.	n_D^{99}: 1.6066; D_4^{99}: 1.0242; 1,2-Di-bromide, 121–3
100	2,7-Dimethylnaphthalene......	96–7, al.	262	136.0–6.5, yel., me. al.	158.0–9.5, yel., me. al.						
101	7-Isopropyl-1-methylfluorene...	96.5–7.0, al.	....	....	....			di: ca. 245			
102	Azulene	98.5–99	270 d.; 115–35[10]	120 d.	....	167					Heat at 270° → Naphthalene, 80.3; 2,4,6-Trinitrotolu-ene deriv., 95.5–100
103	Retene (7-Isopropyl-1-methyl-phenanthrene)	100.5–101, al.	390; 158–6.5[0.2]	124–5	141–2	139					D: 1.035
104	Phenanthrene	101, al.; 96.3	340; 332	144; 132.8	....	158; 145	197			250–1	n_D: 1.5973; D: 1.182
105	2,7-Dimethylphenanthrene.....	101–2, me. al.	....	152–3, or., al.	....						
106	2,3,6-Trimethylnaphthalene....	102; 92–3	286; 146–8[14]	130, yel., me. al.	165, yel., me. al.						
107	2-Phenylnaphthalene	102–3	....	....	....			169.5–70.5			
108	1,2,3,4-Tetrahydroanthracene ..	103–5	....	....	....			182.4			
109	2,3-Dimethylnaphthalene (Guaiene)	104.0–4.5 (subl.), al.	265–6	123–4	....						D_4^{20}: 1.008
110	Ethylidenefluorene	104, ac. a.	....	155–6	....						Dibromide, 93.5
111	1,7-Dimethylfluorene (Gibberene)	107.0–7.5	....	85–6, or.-red	....	98, yel., al.					In H_2SO_4 sol. → bl.
112	1,1'-Dinaphthylmethane.......	110, al.	>360; 270[14]	142	....	141.5	216				
113	Fluoranthrene	110	....	185–6	....		216				
114	2,6-Dimethylnaphthalene......	111	261–2	143	....		156				
115	2,4-Dimethylphenanthrene.....	111, al.	....	138–9; 142, me. al.	....						
116	Fluorene	113.5; 116–7	293–5	87; 77	....	105	179	2-mono: 156; 2,7-di: 199	228	....	CrO_3 → Fluorenone, 84
117	4,10-Dimethyl-1,2-benzan-thracene...................	114, pa. yel., me. al.	....	162, blk.	....						
118	4H-Cyclopenta(def)phenan-threne (Phenanthrindene)	116, al.	353	166	....						Benzylidene deriv., 108
119	1,3,8-Trimethylphenanthrene...	116	....	174–5	....	188	199				
120	11-Methylnaphthanthracene ...	117–8	....	159–60, dk. red	....	170, or.	238.2–8.6				

*Derivative data given in order: m.p., crystal color, solvent from which crystallized.

TABLE IV. ORGANIC DERIVATIVES OF AROMATIC HYDROCARBONS
b) Solids. (Listed in order of increasing m.p.)* (Continued)

No.	Name	Melting point, °C	Boiling point, °C	Picrate	Styphnate	*sym*-Tri-nitro-benzene derivative	2,4,7-Tri-nitrofluo-renone derivative	Nitro derivative	Phthalic anhy-dride deriva-tive	2,4-Di-nitro-phenyl sulfenyl chloride deriva-tive	Miscellaneous
121	5-Methylchrysene	117.2–7.8, bz.-al.	. . . .	142.6–3.0, or.-red, al.	. . . .	172.6–3.6, bz.-al.	. . . .	. . . . . .	. . . . . .	. . . . .	
122	1,2,5,6-Tetramethylnaphthalene	118	150–60[10]	156–7, red.	166, bz.	180.0–0.5, bz.	. . . .				
123	Cyclohept(fg)acenaphthene (Acepleiadene)	118–20 (subl.), red, al.	. . . .	150 d.							Maleic anh. add. comp., 248–50, bz.
124	1,2,7-Trimethylphenanthrene. . .	120–1, al.	. . . .	148–9	169–70						
125	1′,10-Dimethyl-1,2-dibenzan-thracene.	122–3, al.		147–8, red, al.							In H_2SO_4 sol. → red
126	9,10-Dimethyl-1,2-benzan-thracene.	122–3	. . . .	112–3, blk., al. *di*: 102–6, red, al.							Highly carcinogenic
127	Benz(bc)aceanthrylene	122–3, al.	. . . .	141.5–2.0, dk. red., al.	. . . .	162.5–3.0, or., al.	. . . .				
128	1-Methylphenanthrene	123, aq. al.	. . . .	139, yel., al.	152–3, yel.	. . . .					
129	1,6,7-Trimethylphenanthrene. . .	123–4, al.	. . . .	165–6, or.	111–2						
130	1,1′-Diacenaphthene	(a) 124, al.; (b) 169, pet. eth.	. . . .	*di*: 270, red	. . . .						
131	*trans*-Stilbene	124, al.	305[720]	94–5	. . . .	115–20					
132	3,4-Benzfluorene	124–5, al.	. . . .	130–1, red, al.	. . . .		191.8				
133	9-Isopropylnaphthanthracene . .	125, al.	. . . .	152							
134	6-Methylnaphthanthracene	126.2–7.2, al.	. . . .	149–50, red-br., al.			221.4–1.8; 163–4, al.				
135	5,8-Dimethyl-1,2-benzan-thracene.	131, bz.-al.	. . . .	175, red, al.							
136	8-Isopropylnaphthanthracene . .	132–3	. . . .	118							
137	1,4,5,8-Tetramethylnaphthalene	132–3	. . . .	154.6–5.4	143.4–4.2		158–9				
138	12-Methylnaphthanthracene . . .	138, yel.	. . . .	115–6, red	. . . .		234.5–5.0; 209.5–9.7, red, al.				
139	2-Methyl-1′,2′-benzpyrene	138–9, pa. yel., me. al.; after fusing, 140.0–0.2	. . . .	184–5, br., bz.-lgr.	. . . .	211.5–2.0, red, bz.-lgr.	. . . .				
140	1,5-Dimethylanthracene	139–40, pa. yel.	. . . .	166–7, scar., al.	. . . .						
141	7-Methylnaphthanthracene	140	. . . .	174, dk. red	. . . .		236.1–6.5				

*Derivative data given in order: m.p., crystal color, solvent from which crystallized.

No.	Name	Melting point, °C	Boiling point, °C	Picrate	Styphnate	sym-Tri-nitro-benzene derivative	2,4,7-Tri-nitrofluo-renone derivative	Nitro derivative	Phthalic anhy-dride deriva-tive	2,4-Di-nitro-phenyl sulfenyl chloride deriva-tive	Miscellaneous
142	3,6-Dimethylphenanthrene.....	141, al.	....	172–3, or.-yel., me. al.	....	....	....	....	....	....	....
143	5-Methyl-3,4-benzphenanthrene	141.4–1.9	....	....	....	....	130.6–1.4, red	....	....	....	....
144	1,4-Dimethylchrysene........	142	....	141, red, al.	....	....	....	....	....	....	....
145	1,2-Dimethylphenanthrene.....	142–3, al.	....	148, or., al.	153, yel., al.	....	....	....	....	....	....
146	8,10-Dimethyl-1,2-benzan-thracene..................	146, bz.-al.	....	166, red, al.	....	....	....	....	....	....	....
147	1,2,8-Trimethylphenanthrene...	146–7, al.	210–20[15]	164–5, or., al.	....	193.0–3.5, al.	....	....	....	....	....
148	3-Methyl-1′,2′-benzpyrene.....	146.7–8.1, yel., al.-eth.	....	179.5–80.0, br.-red, bz.-lgr.	....	....	210.5–11.0, bz.-lgr.	....	....	....	....
149	9-Methyl-1′,2′-benzpyrene.....	146.8–8.0, yel., hexane	....	....	....	....	218.5–9.5, red, bz.-lgr.	....	....	....	....
150	9-Phenylfluorene............	147–8, al.	....	....	....	....	....	di: ca. 240; tetra: ca. 235 d.	....	....	Bromide, di: 181–2; tri: 167–71
151	2-Methylnaphthanthracene	149–50, al.	....	180	....	....	218.7–9.2	....	....	....	....
152	Pyrene....................	149–50, pa. yel.	385	222, red, al.; 220; 227	....	....	242–3	....	....	....	....
153	9-Methylnaphthanthracene	150.5–1.5, al.	....	157–8	....	....	225.1–5.4	....	....	....	....
154	4-Methylchrysene............	151.0–1.5, bz-al.	....	two forms: 135.0–5.5, red, bz.-lgr.; 137.5–8.0, or., bz.-lgr.	....	....	....	....	....	....	....
155	trans-trans-1,4-Diphenyl-1,3-butadiene (trans-trans-Distyryl)	152.5	350	152–3	....	....	....	....	....	....	Maleic anh. add. comp., 198–200
156	Cinnamalfluorene............	155, pa. yel., ac. a.	....	di: 178–9	....	....	....	....	....	....	Tetrabromide, ca. 160 d.
157	5-Methylnaphthanthracene	155.9–6.9, bz.-pet. eth.	....	153	....	....	235.4–5.6	....	....	....	....
158	1,2-Benzanthracene	159–60	....	133	....	....	160	....	....	....	....
159	8-Methylnaphthanthracene	160.0–0.6	....	166	....	....	243.2–3.6	....	....	....	....
160	1,1′-Binaphthyl	160.5	240–4[12]	145	....	....	....	....	....	....	....
161	Di-1-naphthastilbene (sym.-Di-1-naphthylethylene)......	161, pa. yel., al.	....	tri: 210	....	....	....	....	....	....	....

*Derivative data given in order: m.p., crystal color, solvent from which crystallized.

TABLE IV. ORGANIC DERIVATIVES OF AROMATIC HYDROCARBONS
b) Solids. (Listed in order of increasing m.p.)* (Continued)

No.	Name	Melting point, °C	Boiling point, °C	Picrate	Styphnate	sym-Trinitrobenzene derivative	2,4,7-Trinitrofluorenone derivative	Nitro derivative	Phthalic anhydride derivative	2,4-Dinitrophenyl sulfenyl chloride derivative	Miscellaneous
162	6-Methylchrysene	161.0–1.4, et. ac.-al.		170.0–0.6, or., bz.-al.		189.8–190.6, yel., bz.-al.					
163	3-Methylnaphthanthracene	163.0–3.9, al.		146.0–6.8			239.0–9.6				
164	2,6-Dimethyl-1,2-benzanthracene	164, ac. a.		199–200							
165	Cyclopentadienophenanthrene	164–5		146–7		172–3					In H_2SO_4 sol. → bl.
166	10,11-Benzfluoranthene	165; 166,		194–5		220.0–0.5					
167	Hexamethylbenzene	165	264	170		174					
168	3-Methylchrysene	170.0–0.5, bz.-pet-eth.		164.0–4.5, grn., al.							
169	Cholanthrene	170–1; 173 (subl.), pa. yel., bz.-al.		167–8, vlt.-blk., bz.			245–6				
170	6-Methyl-1',2'-benzpyrene	171.0–1.5, yel., bz.-lgr.		181.5–2.5, br., bz.-lgr.	209–10, red, bz.-lgr.						
171	6,7-Dimethyl-1,2-benzanthracene	174, et. ac.		170							
172	1,2-Benzpyrene	176.5–7.5, pa. yel., bz.-me. al.	310–2^{10}	197–8, vlt.-blk.							
173	5,10-Dimethyl-1,2-benzanthracene	177, bz.-al.		174, red-blk., bz.							
174	4,5-Benzpyrene	178–9, bz.		229–30, red, bz.							
175	9,10-Dimethylanthracene	180–1, al.		176–7 d.							
176	10-Methylnaphthanthracene	183.0–3.6, yel., al.		159.0–9.4							2,4,6-Trinitrotoluene deriv., 224.8–5.0
177	5,6-Dimethyl-1,2-benzanthracene	187–8, al.		191–3, red, al.							
178	2,2'-Binaphthyl	188; 181	452	184				171			Lt. bl. fluorescence; $KMnO_4$ → Phthalic acid, 206–8
179	1,2-Benzfluorene (Chrysofluorene)	189–90, ac. a.; 183–4	413; 398–400	di: 127.5; 124–6		144–5	213.5–5.5				
180	1,8-Dimethylphenanthrene	191–2, bz.		151–2, yel.			193–4				
181	8-Methyl-1',2'-benzpyrene	192–3, yel., bz.		205, dk. br., bz.		233 d., red, bz.					
182	Bifluorenylidene (Diphenyleneethylene)	194–5 (cor.), red		177–8				two di-forms: 171, dk. red; 170, or. red			Dibromide, 312, red, bz.
183	1,2,7,8-Dibenzanthracene	196, bz.		212, brt. red							Bl.-grn. fluorescence in sol.

*Derivative data given in order: m.p., crystal color, solvent from which crystallized.

No.	Name	Melting point, °C	Boiling point, °C	Picrate	Styphnate	sym-Trinitrobenzene derivative	2,4,7-Trinitrofluorenone derivative	Nitro derivative	Phthalic anhydride derivative	2,4-Dinitrophenyl sulfenyl chloride derivative	Miscellaneous
184	4-Methylnaphthanthracene	197.4–8.0, al.	....	139–40, lt. red	....	....	228.2–8.8	....	......	......	
185	1,2,3,4-Dibenzanthracene	200–2, pa. yel., ac. a.	....	207, red	....	....	....	....	......	.....	In conc. H_2SO_4 sol. → pa. vlt.-red
186	Di-2-fluorenylmethane	201–2, al.	....	....	....	....	....	di: 256–7	......	.....	$Na_2Cr_2O_7$ → Di-2-fluorenyl ketone, 297–8, yel., ac. a.
187	2,3-Benzfluorene	208–9	....	....	....	....	221.2–2.0	....	....	....	....
188	5-Methyl-1',2'-benzpyrene.....	215.7–6.2, yel., eth.-al.	....	207–8, vlt.-blk., bz.-lgr.	....	230–1, red, bz.-lgr.	....	....	....	....	....
189	Anthracene	216.2	340 (cor.); 226.5[53]	138	....	164	194	....	....	....	D_4^{27}: 1.25; Dibromide, 122; CrO_3 → Anthraquinone, 273
190	11,12-Benzfluoranthene	217	480	170–1	....	182	236–7	....	....	....	....
191	4-Methyl-1',2'-benzpyrene.....	217.5–8.0, yel.	....	203–4, vlt.-br., bz.	....	....	....	....	....	....	....
192	2,8-Dimethylchrysene.........	218, bz.	....	171–2, bz.	204, or., bz.	195, yel., bz.	....	....	....	....	....
193	2-Methylchrysene.............	224.5–5.5, bz.-al.	....	143–6, yel., al.	....	....	....	....	....	....	....
194	6,12-Dimethylchrysene........	237	....	....	207 d., bz.	222	....	....	....	....	....
195	1,2-Benzphenanthrene (Chrysene)	254, bz.	448	273	....	186	248–9	....	214	......	Dibromide, 275
196	Di-2-naphthastilbene (sym-Di-2-naphthylethylene)......	254–5, bz.	....	215	....	....	....	....	....	....	....
197	1-Methylchrysene.............	254–5 (cor.), (vacuum), bz.	....	....	....	174–6, yel., bz.	....	....	....	....	....
198	2,3,6,7-Dibenzphenanthrene ...	257, grn.-yel.	....	184, or.-red	....	....	....	....	....	....	Bl. fluorescence in sol.; intense yel.-grn. in u.v.
199	2,3,5,6-Dibenzphenanthrene ...	261, grn.-yel., ac. a.	....	di: 213, or.-red	....	....	....	....	....	....	Bl. fluorescence in bz. sol.; grn. fluorescence in u.v.
200	1,2,5,6-Dibenzanthracene	262, met., ac. a.	....	di: 214, or.	....	....	....	....	....	....	....
201	Perylene	273–4	....	....	....	....	270–1	....	....	....	....
202	Picene (1,2,7,8-Dibenzphenanthrene)	365–6, xyl.	518–20	....	....	....	257–8	....	....	....	Dibromide, 295; 2,7-Dianthraquinone add. comp., 299–300
203	1,2,3,4,5,6,7,8-Tetrabenzanthracene.................	428–9	....	....	....	....	318–9, red	....	....	....	....
204	Coronene (Hexabenzobenzene).	438–40 (cor.), yel., bz.	525	>250 d., red, bz.	....	>280 d., or., bz.	....	....	....	....	....

*Derivative data given in order: m.p., crystal color, solvent from which crystallized.

S-Alkylthiuronium picrate (S-Alkylisothiourea picrate)**

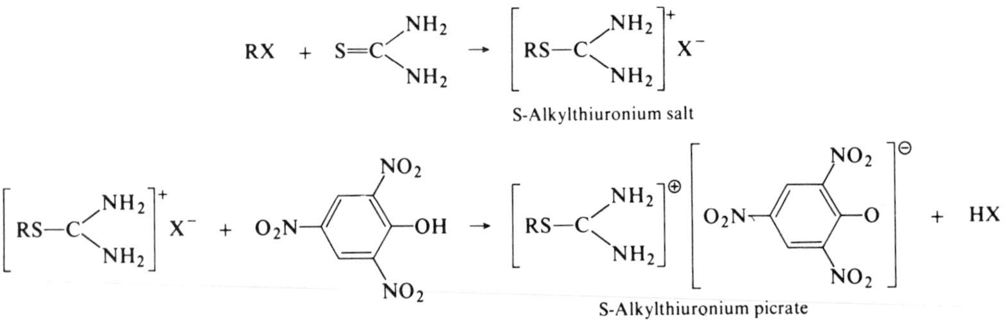

$$RX \; + \; S\!\!=\!\!C\!\!\begin{array}{c} NH_2 \\ NH_2 \end{array} \; \rightarrow \; \left[RS\!\!-\!\!C\!\!\begin{array}{c} NH_2 \\ NH_2 \end{array} \right]^{+} X^{-}$$

S-Alkylthiuronium salt

$$\left[RS\!\!-\!\!C\!\!\begin{array}{c} NH_2 \\ NH_2 \end{array} \right]^{+} X^{-} \; + \; O_2N\!\!-\!\!\langle \rangle\!\!-\!\!OH \; \rightarrow \; \left[RS\!\!-\!\!C\!\!\begin{array}{c} NH_2 \\ NH_2 \end{array} \right]^{\oplus} \left[O_2N\!\!-\!\!\langle \rangle\!\!-\!\!O \right]^{\ominus} \; + \; HX$$

S-Alkylthiuronium picrate

From the alkyl halide with thiourea in 95% ethanol, followed by addition of picric acid in ethanol.

For directions and examples see: Linstead, pp. 82–3; Shriner, p. 245; Vogel, pp. 291–2; Wild, p. 43; E. L. Brown and N. Campbell, *J. Chem. Soc.*, 1699 (1937); W. J. Levy and N. Campbell, *J. Chem. Soc.*, 1442 (1939).

From the alkyl halide with thiourea in ethylene glycol, followed by addition of picric acid in ethanol.

See: Cheronis, p. 550; H. M. Crosby and J. B. Entrikin, *J. Chem. Ed.*, **41**, 360 (1964).

1-Naphthylamide (α-Naphthalide).*

$$RX \; + \; Mg \; \rightarrow \; RMgX$$

$$RMgX \; + \; 1\text{-}C_{10}H_7N\!\!=\!\!C\!\!=\!\!O \; \rightarrow \; 1\text{-}C_{10}H_7N\!\!=\!\!\overset{\overset{R}{|}}{C}OMgX \; \xrightarrow{\;H_2O\;} \; 1\text{-}C_{10}H_7NHCOR \; + \; Mg(OH)X$$

1-Naphthylamide

From the Grignard reagent (prepared from the alkyl halide and magnesium in dry ether) with 1-naphthylisocyanate in ether.

For directions and examples see: Cheronis, pp. 551, 553; Linstead, p. 83; Shriner, p. 244; Vogel, pp. 290–1; Wild, pp. 35–37; H. Gilman and M. Furry, *J. Amer. Chem. Soc.*, **50**, 1214 (1928); H. W. Underwood and J. C. Gale, *J. Amer. Chem. Soc.*, **56**, 2117 (1934).

Anilide.*

$$RX \; + \; Mg \; \rightarrow \; RMgX$$

$$RMgX \; + \; C_6H_5N\!\!=\!\!C\!\!=\!\!O \; \rightarrow \; C_6H_5N\!\!=\!\!\overset{\overset{R}{|}}{C}OMgX \; \xrightarrow{\;H_2O\;} \; C_6H_5NHCOR \; + \; Mg(OH)X$$

Anilide

From the Grignard reagent (prepared from the alkyl halide and magnesium in dry ether) with phenylisocyanate in ether.

For directions and examples see: Cheronis, pp. 551, 554; Linstead, p. 83; Shriner, p. 244; Vogel, pp. 290–1; Wild, pp. 35–7; A. M. Schwartz and J. R. Johnson, *J. Amer. Chem. Soc.*, **53**, 1063 (1931); H. W. Underwood and J. C. Gale, *J. Amer. Chem. Soc.*, **56**, 2117 (1934).

Alkylmercuric halide.*

$$RX \; + \; Mg \; \rightarrow \; RMgX$$

$$RMgX \; + \; HgX_2 \; \rightarrow \; RHgX \; + \; MgX_2$$

Alkyl mercuric halide

From the Grignard reagent (prepared from the alkyl halide and magnesium in dry ether) with the mercuric salt of the same halogen in ether.

For directions and examples see: Cheronis, pp. 551, 554; Shriner, p. 244; Vogel, p. 291; Wild, p. 38; C. S. Marvel, C. Gauerke and E. L. Hill, *J. Amer. Chem. Soc.*, **47**, 3009 (1925); E. L. Hill, *J. Amer. Chem. Soc.*, **50**, 167 (1928); K. H. Slotta and K. R. Jacobi, *J. prakt. Chem.*, **120**, 249 (1929).

---

*Derivatives recommended for first trial.

WARNING: This is not an instruction manual. References should be consulted for the preparation of derivatives.

Picrate of alkyl 2-naphthyl ether (Molecular complex).*

Alkyl 2-naphthyl
ether

Picrate of alkyl 2-naphthyl ether
(Molecular complex)

The alkyl 2-naphthyl ether is obtained from the alkyl halide with 2-naphthol in ethanolic sodium or potassium hydroxide.

For directions and examples see: Cheronis, p. 551; Linstead, p. 83; Shriner, p. 244; Vogel, p. 292; Wild, p. 44.

The picrate is obtained from the alkyl 2-naphthyl ether and picric acid in chloroform or ethanol.

See: Cheronis, p. 551; Linstead, p. 83; Vogel, p. 292; Wild, p. 44; O. L. Baril and G. A. Megrdichian, *J. Amer. Chem. Soc.,* **58**, 1415 (1936); V. H. Dermer and O. C. Dermer, *J. Org. Chem.,* **3**, 289 (1938).

Alkyl 2,4-dinitrophenyl thioether (Alkyl 2,4-dinitrophenyl sulfide).

Alkyl 2,4-dinitrophenyl
thioether

From the alkyl bromide or iodide with 2,4-dinitrothiophenol in butyl carbitol (2-(2-butoxyethoxy)ethanol) and aqueous potassium hydroxide.

For directions and examples see: Cheronis, p. 557; R. W. Bost, P. K. Starnes and E. L. Wood, *J. Amer. Chem. Soc.,* **73**, 1968 (1951).

From the alkyl chloride with 2,4-dinitrothiophenol in butyl carbitol (2-(2-butoxyethoxy)ethanol) with potassium iodide and aqueous potassium hydroxide.

See: Cheronis, p. 557; R. W. Bost, P. K. Starnes and E. L. Wood, *J. Amer. Chem. Soc.,* **73**, 1968 (1951).

Alkyl 2,4-dinitrophenyl sulfone.

Alkyl 2,4-dinitrophenyl
sulfone

From the alkyl 2,4-dinitrophenyl thioether (prepared from the alkyl halide as above) in glacial acetic acid, with aqueous potassium permanganate.

For directions and examples see: R. W. Bost, J. O. Turner and R. D. Norton, *J. Amer. Chem. Soc.,* **54**, 1985 (1932).

*Derivatives recommended for first trial.

WARNING: This is not an instruction manual. References should be consulted for the preparation of derivatives.

6-Nitro-2-mercaptobenzothiazole derivative.

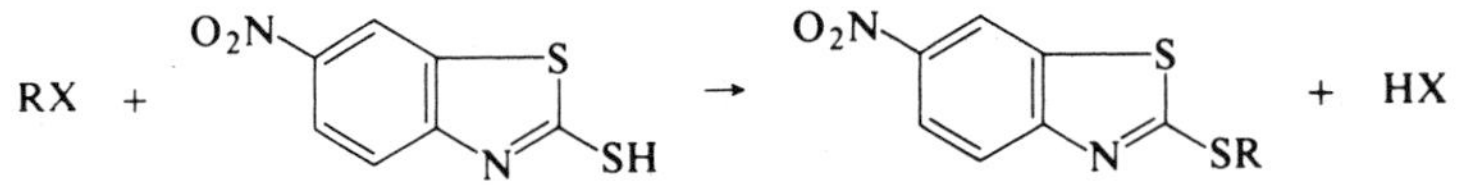

$$RX \ + \ \text{(6-nitro-2-mercaptobenzothiazole, -SH)} \ \rightarrow \ \text{(Alkyl 6-nitrobenzothiazolyl sulfide, -SR)} \ + \ HX$$

Alkyl 6-nitrobenzothiazolyl
sulfide

From the alkyl halide (especially a dihalide) and 6-nitro-2-mercaptobenzothiazole in butyl carbitol (2-(2-butoxyethoxy)ethanol) and aqueous sodium hydroxide.

For directions and examples see: Cheronis, p. 557; H. B. Cutter and H. R. Golden, *J. Amer. Chem. Soc.*, **69**, 831 (1947); H. B. Cutter and A. Kreuchunas, *Anal. Chem.*, **25**, 198 (1953).

Substituted N-alkylphthalimides.

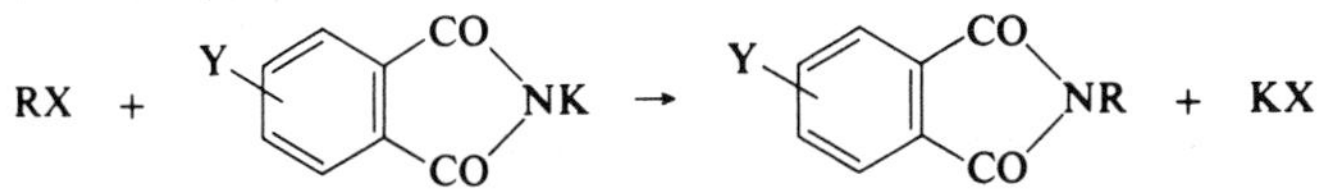

$$RX \ + \ \text{(Substituted potassium phthalimide, NK)} \ \rightarrow \ \text{(Substituted N-alkylphthalimide, NR)} \ + \ KX$$

Substituted potassium
phthalimide

Substituted
N-alkylphthalimide

From the alkyl halide with the potassium salt of the substituted phthalimide.
For directions and examples see: Wild, p. 41.

From the alkyl halide with the potassium salt of the substituted phthalimide or with the substituted phthalimide and potassium carbonate in dimethylformamide.
See: J. H. Billman and R. V. Cash, *J. Amer. Chem. Soc.*, **75**, 2499 (1953).

From the alkyl halide with the substituted phthalimide and potassium hydroxide in methanol-dioxane mixture.
See: C. H. Allen and R. V. V. Nicholls, *J. Amer. Chem. Soc.*, **56**, 1409 (1934).

Nitro derivative. *

$$ArX \ \xrightarrow{HNO_3} \ Ar(NO_2)X$$

Nitroaryl
halide

From the aromatic halide with fuming and concentrated nitric acids.
For directions and examples see: Wild, p. 450.

From the aromatic halide with 100% nitric acid.
See: Cheronis, pp. 559–561, 563.

From the aromatic halide with nitric and sulfuric acids.
See: Cheronis, pp. 560, 563; Vogel, p. 543.

Sulfonamide. *

$$ArX \ \xrightarrow{ClSO_3H} \ Ar(X)SO_2Cl$$

Sulfonyl
chloride

$$Ar(X)SO_2Cl \ \xrightarrow{NH_3} \ Ar(X)SO_2NH_2$$

Sulfonamide

The sulfonyl chloride is prepared from the aromatic halide and chlorosulfonic acid in chloroform or without solvent. The sulfonamide is obtained from the sulfonyl chloride with concentrated ammonia or dry ammonium carbonate.

For directions and examples see: Cheronis, pp. 564, 638, 639; E. H. Huntress and F. H. Carten, *J. Amer. Chem. Soc.*, **62**, 511 (1940).

NOTE: For additional information regarding directions and examples for the derivatization of aromatic halides (through the above reactions or additional ones, e.g., side-chain oxidation) see explanations and references to Table IV, pp. 32, 33, 34.

---

*Derivatives recommended for first trial.
WARNING: This is not an instruction manual. References should be consulted for the preparation of derivatives.

TABLE V. ORGANIC DERIVATIVES OF HALIDES
A) Alkyl and cycloalkyl halides 1. Chlorides
a) Liquids (Listed in order of increasing atmospheric b.p.)*

No.	Name	Boiling point, °C	n_D^{20}	D_4^{20}	S-Alkyl thiuronium picrate	1-Naphthyl amide	Anilide	Alkyl mercuric halide	Picrate of 2-naphthyl ether	2,4-Dinitrophenyl thioether	2,4-Dinitrophenyl sulfone	Miscellaneous
1	Methyl chloride	−24	......	......	224	160	114	167	......	128	185; 189	Methyl-2-naphthyl ether, 70
2	Vinyl chloride	−14	......	......	104		......	......	......	......	......	Polymerizes to solid on irradiation
3	Ethyl chloride	13	......	0.917_6^6	188	126	104	192	104	115	156; 160	......
4	Isopropyl chloride	36.5	1.378	0.859	196;148	......	103	......	95			
5	1-Chloropropene	37	......	......	......		114					
6	Allyl chloride	44.5	1.416	0.940	155	......	114	......	99	......	......	3-Nitrophthalimide deriv., 100–1
7	n-Propyl chloride	46.5	1.388	0.889	181; 176	121	92	147	8ǀ	84	126	......
8	tert-Butyl chloride	51	1.386	0.846	160–1	147	128	122–3	......	......	......	......
9	Chloroprene	59	1.458	0.9583								Heating with maleic anh. and boiling the adduct in water → 4-chloro-1,2,3,6-tetrahydro-phthalic acid, 173–5
10	sec-Butyl chloride	68	1.397	0.874	190; 166	129	108	30.5	86	66	120	......
11	Isobutyl chloride	69	1.398	0.881	174	125	109	......	85	76	105	......
12	Methallyl chloride (3-Chloro-2-methyl-1-propene)	72	1.4340	0.9475	......	......	......	......	......	......	......	Phthalimide deriv., 89–90
13	n-Butyl chloride	78	1.402	0.886	180; 177	112	63	128	67	66	92	......
14	Neopentyl chloride	85	......	0.879	......	......	130–1	117–8	......			......
15	tert-Amyl chloride	86	1.405	0.865	......	138	92	......				......
16	3-Chloro-1-pentene	93–4	1.4254	0.8978	......	......	......	......	......	......	......	Phthalimide deriv., 78–9
17	DL-3-Chloro-2-methyl-1-butene	94	1.4304	0.9088								$Br_2 \rightarrow$ dibromo deriv., 197–8
18	Trimethylvinyl chloride (3-Chloro-2-methyl-2-butene)	94	1.4320	0.925 (0.905)								Br_2 in ether → dibromo deriv., 197
19	DL-2-Chloropentane	97	1.4079	0.8695	......	102–3	94–6					
20	3-Pentyl chloride (3-Chloropentane)	97	1.4082	0.8723	......	117–8	127; 122					
21	Isoamyl chloride	100	1.409	0.872	179; 173	111	108	86	94	80	124	......
22	n-Amyl chloride (n-Pentyl chloride)	106	1.412	0.882	154	112	96	110	67	80	83	......
23	1-Chloro-2-pentene	109–10	1.435^{21}	$0.908_4^{21.5}$	......	......	......	......	......	......	......	Phthalimide deriv., 69–70
24	2-Chloro-2-methylpentane	110–3	1.4126	0.863	......	116–8	71–4					
25	3-Chloro-2,2-dimethyl-butane (Pinacolyl chloride)	112	1.4181	0.8767	......	......	......	89–90				
26	Cyclopentyl chloride	114–5	1.4510	1.005	......	......	......	108				
27	4-Chloro-2,2-dimethyl-butane	115	1.4160	0.8670	......	......	138–9	133				
28	3-Chloro-3-methylpentane	115–7	1.421	0.89	......	......	87–8					
29	2-Chloro-2,3-dimethyl-butane	117–9	......	0.8769^{22}	......	......	......	......				Carbonation of Grignard and conversion of acid to amide, 125–7; $Br_2 \rightarrow$ 2,3-dibromo deriv., 166–8 (173–4)
30	3-Hexyl chloride (3-Chlorohexane)	123	1.4163	0.870_{20}^{20}	......	......	......	......				Grignard reagent $+O_2 \rightarrow$ 3-hexanol $\xrightarrow{CrO_3/H_2SO_4}$ 3-hexanone, 2,4-Dinitrophenylhydrazone, 147–8; Semicarbazone, 110–11

*Derivative data given in order: m.p., crystal color, solvent from which crystallized.

55

TABLE V. ORGANIC DERIVATIVES OF HALIDES
A) Alkyl and cycloalkyl halides 1. Chlorides
a) Liquids (Listed in order of increasing atmospheric b.p.)* (Continued)

No.	Name	Boiling point, °C	n_D^{20}	D_4^{20}	S-Alkyl thiuronium picrate	1-Naphthyl amide	Anilide	Alkyl mercuric halide	Picrate of 2-naphthyl ether	2,4-Dinitrophenyl thioether	2,4-Dinitrophenyl sulfone	Miscellaneous
31	2-Hexyl chloride (2-Chlorohexane)	123–4	1.4142^{21}	0.8694_4^{21}	......	......	91–2	......	......	......	......	......
32	1-Chloro-2-ethylbutane	125–7	1.4230	0.8914	......	......	81–2; 83–4					
33	3-Chloro-2,2,3-trimethyl-butane	133	......	......	......	......	......	......				Carbonation of Grignard → acid, 80
34	n-Hexyl chloride	133	1.420	0.878	157	106	69	125	......	74	97	
35	Cyclohexyl chloride	143	1.462	0.989	......	188	146	......	......	......	......	6-Nitro-2-mercapto-benzothiazole deriv., 100–1; 2-sulfone, 189
36	5-Chloro-2,3-dimethyl-pentane	152	1.4299	0.8825	......	......	80–1					
37	n-Heptyl chloride	159	1.426	0.877	142	95	57	120	......	82	101	
38	Benzyl chloride	179	1.539	1.100	188	166	117	104	123	130	178; 182	Quaternary salt with dimethyl aniline, 110
39	n-Octyl chloride	180; 184	1.431	0.875	134	91	57	115	......	78	98	......
40	β-Phenylethyl chloride	190	......	......			97	......	84		......	
41	4-Methylbenzyl chloride	192	1.5380	1.0512	......		......				......	Phthalimide, 120; 117; Carbonation of Grignard → 4-tolylacetic acid, 92
42	α-Phenylethyl chloride	195	......	......			133					
43	3-Methylbenzyl chloride	195–6	1.5327^{25}	1.064_{20}^{20}	......							Phthalimide deriv., 117–8; Carbonation of Grignard → 3-tolylacetic acid, 61
44	2-Methylbenzyl chloride	197–9										Phthalimide deriv., 148–9; Heating with pyridine → alkyl pyridinium chloride, 183
45	β-Chlorostyrene	197–9	1.571^{25}	1.109								Br₂ in chl. → dibromo deriv., 32; Oxid. → benzoic acid, 122
46	n-Nonyl chloride	202	1.434	0.870	131					86	92	......
47	2-Chlorobenzyl chloride	213–4										5-Nitro deriv., 66; Carbonation of Grignard → 2-chlorophenylacetic acid, 94–5
48	4-Chlorobenzyl chloride	214; 222	......	......	......	......	166	......				Oxid. → 4-chlorobenzoic acid, 242
49	3-Chlorobenzyl chloride	216	......	1.2695_4^{15}								Oxid. → 3-chlorobenzoic acid, 158; 155; Heating with 2,4-dichlorophenol in toluene → 2-(3-chlorobenzyl)-4,6-dichlorophenol, 59–60
50	n-Decyl chloride	223	1.437	0.868	137							
51	4-Isopropylbenzyl chloride	226–9	......									Carbonation of Grignard → acid, 52
52	n-Undecyl chloride (n-Hendecyl chloride)	241	1.440	0.868	139	......						
53	n-Dodecyl chloride (Lauryl chloride)	243–4	1.4425	0.8673	......	......	......	114	......			Refluxed with pyridine → alkyl pyridinium chloride, 92
54	Cetyl chloride (Hexadecyl chloride)	286 d.	......	......	155	......	......	102	......			3-Nitrophthalimide deriv., 101; Alkyl saccharin deriv., 98

*Derivative data given in order: m.p., crystal color, solvent from which crystallized.

TABLE V. ORGANIC DERIVATIVES OF HALIDES
A) Alkyl and cycloalkyl halides 1. Chlorides
b) Solids (Listed in order of increasing m.p.)*

No.	Name	Melting point, °C	Boiling point, °C	S-Alkyl thiuronium picrate	1-Naphthyl amide	Anilide	Alkyl mercuric halide	Picrate of 2-naphthyl-ether	2,4-Dinitrophenyl thioether	2,4-Dinitrophenyl sulfone	Miscellaneous
1	**1,3-Bis(chloromethyl)benzene** (m-Xylylene dichloride)	32–4	250–5	......	......	......	......	......	......	......	Diphthalimide deriv., 237
2	**4-Bromobenzyl chloride**	36–8; 50	236	219	......	......	......	......	......	......	Oxid. → 4-bromobenzoic acid, 251
3	**2,4,6-Trimethylbenzyl chloride**	37	130^{22}	......	......	......	......	......	......	......	Hydrolysis → 2,4,6-trimethyl-benzyl alcohol, 88–9; Phthalimide deriv., 209–10
4	**2,6-Dichlorobenzyl chloride**	39–40	......	......	......	......	......	......	......	......	Carbonation of Grignard → 2,6-dichlorophenylacetic acid, 157–8
5	**1-Chloro-2,2,3,3-tetramethyl-butane**	52–3	......	......	......	......	170–1	......	......	......	Grignard treated with O_2 at −5 → carbinol, 149–50
6	**1,2-Bis(chloromethyl)benzene** (o-Xylylene dichloride)	54–5	239–41	......	......	......	......	......	......	......	Oxid. → phthalic acid, 200–6
7	**4-Nitrobenzyl chloride**	71	......	......	......	......	......	......	......	......	Oxid. → 4-nitrobenzoic acid, 241
8	**1,4-Bis(chloromethyl)benzene** (p-Xylylene dichloride)	98–100	240–5	......	......	......	......	......	......	......	Heating with benzyl alcohol +KOH → dibenzyl ether, 67; Boiling with $Pb(NO_3)_2$ → terephthaldehyde, 115
9	**Triphenylmethyl chloride** (Trityl chloride)	113	......	......	......	......	......	......	......	......	Boiling with H_2O → triphenyl carbinol, 162

*Derivative data given in order: m.p., crystal color, solvent from which crystallized.

TABLE V. ORGANIC DERIVATIVES OF HALIDES
A) Alkyl and cycloalkyl halides 2. Bromides
a) Liquids (Listed in order of increasing atmospheric b.p.)*

No.	Name	Boiling point, °C	n_D^{20}	D_4^{20}	S-Alkyl thiuronium picrate	1-Naphthyl amide	Anilide	Alkyl mercuric halide	Picrate of 2-naphthyl ether	2,4-Dinitrophenyl thioether	2,4-Dinitrophenyl sulfone	Miscellaneous
1	Methyl bromide	3.5	......	......	224	160	114	172; 160	....	128	185; 189	....
2	Vinyl bromide	16	......	......	....	......	104	....	.....	....	....	....
3	Ethyl bromide	38	1.425	1.460	188	126	104	193; 198	104	115	156; 160	....
4	1-Bromopropene	60	1.452	1.4133	....	......	114	....	.....	....	....	....
5	Isopropyl bromide	60	1.425	1.314	196;148	......	103	93	92	95	140	....
6	Allyl bromide	71	1.46545	1.398	155	......	114	....	99	71	....	....
7	n-Propyl bromide	71	1.4341	1.353	181; 177	121	92	138	75	84	126	....
8	tert-Butyl bromide	72–3	......	1.211	....	147	128	.....	.....	....	....	....
9	Isobutyl bromide	91	1.435	1.253	174;167	125	109	55	84	76	105	....
10	sec-Butyl bromide	91	1.437	1.256	190; 166	129	108	39	85	66	120	....
11	n-Butyl bromide	101	1.440	1.274	180; 177	112	63	136; 129	67	66	92	....
12	tert-Amyl bromide	108	1.442	1.198_4^{18}	....	138	92	.....	.....	....	....	....
13	Neopentyl bromide	109	......	1.225	....	......	126	.....	.....	....	....	....
14	DL-2-Pentyl bromide	117; 113	1.442	1.212	....	102–3	93	.....	.....	....	....	....
15	3-Pentyl bromide	118	1.443	1.211	....	......	124	.....	.....	....	....	....
16	Isoamyl bromide	120–1	1.442	1.213	179; 173	111	108	80	94	80	124	....
17	n-Amyl bromide (n-Pentyl bromide)	129	1.445	1.219	154	112	96	127; 122	67	80	83	....
18	Cyclopentyl bromide	137	1.489	1.387	....	......	....	.....	.....	....	....	....
19	2-Hexyl bromide (2-Bromohexane)	146	1.4832^{25}	1.1658	....	......	91–2	.....	.....	....	....	....
20	n-Hexyl bromide	155; 157	1.448	1.175	157	106	69	127; 119	....	74	97	....
21	Cyclohexyl bromide	165	1.495	1.336	....	188	146	153	.....	....	....	....
22	n-Heptyl bromide	180; 174	1.451	1.140	142	95	57	118	....	82	101	....
23	Benzyl bromide	198	......	1.438	188	166	117	119	123	130	178; 182	....
24	n-Octyl bromide	201; 204	1.453	1.112	134	91	57	109	....	78	98	....
25	α-Phenylethyl bromide	205	......	......	....	......	133	.....	.....	....	....	....
26	β-Phenylethyl bromide	218	1.556	1.359	....	......	97	169	84	....	....	....
27	n-Nonyl bromide	220	1.454	1.090	131	......	....	109	....	86	92	....
28	β-Bromostyrene	221	......	......	....	217	115	91	.....	....	....	....
29	n-Dodecyl bromide (Lauryl bromide)	130^6	1.458	1.038	....	......	....	108	.....	....	....	....
30	n-Tetradecyl bromide	179^{20}	1.460	1.017	....	......	....	.....	.....	94	104	m.p. 5

*Derivative data given in order: m.p., crystal color, solvent from which crystallized.

TABLE V. ORGANIC DERIVATIVES OF HALIDES
A) Alkyl and cycloalkyl halides 2. Bromides
b) Solids (Listed in order of increasing m.p.)*

No.	Name	Melting point, °C	Boiling point, °C	S-Alkyl thiuronium picrate	1-Naphthyl amide	Anilide	Alkyl mercuric halide	Picrate of 2-naphthyl ether	2,4-Dinitrophenyl thio-ether	2,4-Dinitrophenyl sulfone	Miscellaneous
1	*n*-Hexadecyl bromide (Cetyl bromide)	14	201^9	155; 137	.....	.....	101–2	.....	95	105	n_D^{20}: 1.462; D_4^{20}: 1.001
2	**2-Bromobenzyl bromide**	31	.....	222	.....	.....	.....	.....	.....	.....	CrO$_3$ → 2-bromobenzoic acid, 150
3	**3-Bromobenzyl bromide**	41	.....	205	.....	.....	.....	.....	.....	.....	CrO$_3$ → 3-bromobenzoic acid, 155
4	**2-Nitrobenzyl bromide**	46–7	.....	....	.....	.....	.....	.....	.....	.....	Oxid. → 2-nitrobenzoic acid, 146–8
5	**4-Chlorobenzyl bromide**	51	.....	194	.....	.....	.....	.....	.....	.....	CrO$_3$ → 4-chlorobenzoic acid, 242
6	**3-Nitrobenzyl bromide**	58–9	.....	....	.....	.....	.....	.....	.....	.....	Oxid. → 3-nitrobenzoic acid, 141
7	**4-Bromobenzyl bromide**	62	.....	219	.....	.....	.....	.....	.....	.....	CrO$_3$ → 4-bromobenzoic acid, 251
8	**4-Nitrobenzyl bromide**	99	.....	....	.....	.....	.....	.....	.....	.....	Oxid. → 4-nitrobenzoic acid, 240

*Derivative data given in order: m.p., crystal color, solvent from which crystallized.

TABLE V. ORGANIC DERIVATIVES OF HALIDES
A) Alkyl and cycloalkyl halides 3. Iodides
a) Liquids (Listed in order of increasing atmospheric b.p.)*

No.	Name	Boiling point, °C	n_D^{20}	D_4^{20}	S-Alkyl thiuronium picrate	1-Naphthyl amide	Anilide	Alkyl mercuric halide	Picrate of 2-naphthyl ether	2,4-Dinitro-phenyl thioether	2,4-Dinitro-phenyl sulfone	Miscellaneous
1	Methyl iodide	43	1.532	2.282	224	160	114	152; 145	117	128	185; 189	. . . .
2	Vinyl iodide	56	. . . .	. . . .	. . . .	. . . .	104	. . . .	. . . .	. . . .	. . . .	. . . .
3	Ethyl iodide.	72	1.514	1.940	188	126	104	186; 182	104	115	156; 160	. . . .
4	Isopropyl iodide	90	1.499	1.703	196; 148	. . . .	103	. . . .	92	95	140	. . . .
5	n-Propyl iodide.	102–3	1.505	1.743	181; 176	121	92	113	75	84	126	. . . .
6	Allyl iodide	103	1.578	1.777	155	121	114	112	99	. . . .	. . . .	. . . .
7	tert-Butyl iodide.	103; 98	. . . .	. . . .	188	147	128	. . . .	. . . .	. . . .	. . . .	. . . .
8	sec-Butyl iodide.	120	1.499	1.592	190; 166	129	108	. . . .	85	66	120	. . . .
9	Isobutyl iodide	120	1.496	1.602	174; 167	125	109	72	84	76	105	. . . .
10	tert-Amyl iodide.	128	. . . .	1.479	. . . .	138	92	. . . .	. . . .	. . . .	. . . .	. . . .
11	n-Butyl iodide.	130	1.499	1.616	180; 177	112; 110	63	117	67	66	92	. . . .
12	2-Pentyl iodide	142	1.496	1.510	. . . .	. . . .	93	.	. . . .	. . . .	. . . .	. . . .
13	3-Pentyl iodide	142	1.497	1.511	. . . .	. . . .	124	. . . .	. . . .	. . . .	. . . .	. . . .
14	Isoamyl iodide	148	1.493	1.503	179; 173	111	108	122	94	80	124	. . . .
15	n-Amyl iodide (n-Pentyl iodide).	155	1.496	1.512	154	112	96	110	67	80	83	. . . .
16	Cyclopentyl iodide	166–7	1.5447	1.7096	. . . .	. . . .	. . . . . .	. . . .	. . . .	. . . .	. . . .	. . . .
17	Cyclohexyl iodide.	179, sl. d.	. . . .	1.626_{15}^{15}	. . . .	188	146	. . . .	. . . .	. . . .	. . . .	. . . .
18	n-Hexyl iodide	179	1.493	1.437	157	106	69	110	. . . .	74	97	. . . .
19	n-Heptyl iodide.	204	1.490	1.373	142	95	57	103	. . . .	82	101	. . . .
20	n-Nonyl iodide	220	. . . .	. . . .	131	. . . .	. . . . . .	. . . .	. . . .	86	92	. . . .
21	n-Octyl iodide	225–6	1.489	1.330	134	. . . .	. . . . . .	. . . .	. . . .	78	98	. . . .

*Derivative data given in order: m.p., crystal color, solvent from which crystallized.

TABLE V. ORGANIC DERIVATIVES OF HALIDES
A) Alkyl and cycloalkyl halides 3. Iodides
b) Solids (Listed in order of increasing m.p.)*

No.	Name	Melting point, °C	Boiling point, °C	S-Alkyl thiuronium picrate	1-Naph-thyl amide	Anilide	Alkyl mercuric halide	Picrate of 2-naphthyl ether	2,4-Dinitro-phenyl thioether	2,4-Dinitro-phenyl sulfone	Miscellaneous
1	*n*-**Hexadecyl iodide** (Cetyl iodide).........	22	......	155; 137	......	......	82	....	95	105	....
2	**Benzyl iodide**...........................	24	......	188	166	117	....	123	130	178; 182	....

*Derivative data given in order: m.p., crystal color, solvent from which crystallized.

61

TABLE V. ORGANIC DERIVATIVES OF HALIDES
B) Dihalides and polyhalides (non-aromatic)
1. Fluorides (Listed in order of increasing atmospheric b.p.)*

No.	Name	Boiling point, °C	n_D^{20}	D_4^{20}	Miscellaneous
1	Perfluorocyclopentane	22	....	1.648_4^{25}	m.p. 10
2	1,3-Difluoropropane	41–2	1.3190^{26}	1.0057_4^{25}	....
3	Perfluorocyclohexane	50	....	....	m.p. 49
4	Perfluoro-n-hexane	57	1.2515^{22}	1.6995_4^{25}	....
5	Perfluoro-2-methylpentane	58	1.2564^{22}	1.7326	....
6	Perfluoro-n-heptane	84	1.2770	1.801_4^{25}	....
7	Perfluoro-n-nonane	127	1.2865^{25}	1.860_4^{25}	....
8	Perfluoro-n-decane	150	1.2890^{25}	1.873_4^{25}	m.p. 36
9	Perfluoro-n-undecane (Perfluoro-n-hendecane)	161	1.2960^{25}	1.919_4^{25}	m.p. 57

*Derivative data given in order: m.p., crystal color, solvent from which crystallized.

TABLE V. ORGANIC DERIVATIVES OF HALIDES

B) Dihalides and polyhalides (non-aromatic)
2. Chlorides a) Liquids (Listed in order of increasing atmospheric b.p.)*

No.	Name	Boiling point, °C	n_D^{20}	D_4^{20}	Miscellaneous
1	**Dichloromethane** (Methylene chloride)	41	1.4237	1.336	6-Nitro-2-mercaptobenzothiazole deriv., 232–3; Di-(2-naphthyl) ether, 133; S-Alkyl *bis*-(thiuronium picrate), 267
2	*trans*-**1,2-Dichloroethylene**	48	1.452	1.2569	$Br_2 \rightarrow$ dibromo deriv., 190–5
3	**1,1-Dichloroethane**	57	1.4164	1.175	1,1-Di-(1-naphthyl) ether, 117
4	*cis*-**1,2-Dichloroethylene**	60	1.4428[25]	1.282	$Br_2 \rightarrow$ dibromo deriv., 190–5
5	**Chloroform**	61	1.446	1.489	Gives carbylamine test with primary amines
6	**2,2-Dichloropropane**	70	1.4117	1.093	. . . .
7	**1,1,1-Trichloroethane**	74	1.4380	1.349	. . . .
8	**Carbon tetrachloride**	77	1.4630	1.595	. . . .
9	**Ethylene dichloride** (1,2-Dichloroethane)	84	1.4443	1.256	1,2-Di-(2-naphthyl) ether, 217
10	**1,1,2-Trichloroethylene**	87	1.4773	1.464	HgO + NaOEt + KCN in al. shaken 1 hr. at 40–60 → mercury *bis*-(trichloroethylenide), $Hg(-CCl{=}CCl_2)_2$, 83, eth.
11	**1,2-Dichloropropane**	96	1.4388	1.155	1,2-Di-(2-naphthyl) ether, 152; 1,2-Diphenyl ether, 32
12	**1-Bromo-2-chloroethane**	106–7	. . . . . .	. . . . . .	6-Nitro-2-mercaptobenzothiazole deriv., 202–3; Di-(2-naphthyl) ether, 217
13	**1,1,2-Trichloroethane**	114	1.4707	1.443	. . . .
14	**1,1,2,2-Tetrachloroethylene** (Perchloroethylene)	121	1.5055	1.623	With paraformaldehyde + conc. $H_2SO_4 \rightarrow \alpha,\alpha$-dichloro-$\beta$-hydroxypropionic acid, 88–9
15	**1,2-Dichlorobutane**	123–4	1.440	. . . . . .	6-Nitro-2-mercaptobenzothiazole deriv., 164–5
16	**1,3-Dichloropropane**	125	1.449	1.189_{18}^{18}; 1.177_4^{25}	1,3-Di-(1-naphthyl) ether, 103–4; 1,3-Di-(2-naphthyl) ether, 148–9; 1,3-Diphenyl ether, 60
17	**1,3-Dichloro-2-methylpropane**	135–6	1.4627[19]	1.131_{20}^{20}	. . . .
18	**1-Bromo-3-chloropropane**	143–4	1.4861	1.594	1,3-Di-(1-naphthyl) ether, 103–4; 1,3-Di-(2-naphthyl) ether, 148–9; 1,3-Diphenyl ether, 60
19	**1,1,2,2-Tetrachloroethane**	146	1.4942	1.600	
20	**1,2,3-Trichloropropane**	158	1.4585	1.417	. . . .
21	**Pentachloroethane**	161	1.504	1.681	. . . .
22	**Benzalchloride**	207; 214	1.5515	1.295[16]	Oxid. → benzoic acid, 122; Hydrolysis → benzaldehyde, 2,4-Dinitrophenylhydrazone 237, Semicarbazone, 222
23	**Benzotrichloride**	221	. . . . . .	1.374[17]	Hydrolysis → benzoic acid, 122
24	**2-Chlorobenzalchloride** (2-Chlorobenzylidene chloride) . .	228–9	1.5670[16]	1.399_{15}^{15}	Oxid. → 2-chlorobenzoic acid, 141; Hydrolysis → 2-chlorobenzaldehyde, 2,4-dinitrophenylhydrazone, 213; 209
25	**3-Chlorobenzalchoride** (3-Chlorobenzylidene chloride) . . .	237–40	. . . . . .	. . . . . .	Oxid. → 3-chlorobenzoic acid, 158; Hydrolysis → 3-chlorobenzaldehyde, 2,4-dinitrophenylhydrazone, 256; 248
26	**4-Chlorobenzalchloride** (4-Chlorobenzylidene chloride) . .	237	. . . . . .	. . . . . .	Oxid. → 4-chlorobenzoic acid, 240; Hydrolysis → 4-chlorobenzaldehyde, 47, 2,4-dinitrophenylhydrazone, , 265

*Derivative data given in order: m.p., crystal color, solvent from which crystallized.

TABLE V. ORGANIC DERIVATIVES OF HALIDES
B) Dihalides and polyhalides (non-aromatic) 2. Chlorides
b) Solids (Listed in order of increasing m.p.)*

No.	Name	Melting point, °C	Boiling point, -°C	Miscellaneous
1	**3,4-Dichlorobenzotrichloride**	26	. . . .	Hydrolysis → 3,4-Dichlorobenzoic acid, 202
2	**DDT** (2,2-Bis-(4-chlorophenyl)-1,1,1-trichloroethane). . .	108	260	Heating with Cl_2 + trace PCl_3 in CCl_4 → 1,1,1,2-tetrachloro deriv., 91–2; Nitration → nitro deriv., 148; $AlCl_3$ + benzene → 1,1,2,2-tetraphenylethane, 211
3	**γ-Benzene hexachloride** (Gammexane, 666).	112	. . . .	. . . .
4	**α-Benzene hexachloride**	157	288	Heat above m.p. → HCl + 1,2,4-trichlorobenzene, 17, b.p. 213, mononitro deriv., 56, dinitro, 103
5	**Hexachloroethane**.	187 subl.	185	. . . .
6	**β-Benzene hexachloride**	310	. . . .	Unreactive to boiling pyridine; Unattacked by boiling HNO_3 or H_2SO_4

*Derivative data given in order: m.p., crystal color, solvent from which crystallized.

TABLE V. ORGANIC DERIVATIVES OF HALIDES
B) Dihalides and polyhalides (non-aromatic)
3. Bromides a) Liquids (Listed in order of increasing atmospheric b.p.)*

No.	Name	Boiling point, °C	n_D^{20}	D_4^{20}	Miscellaneous
1	Dibromomethane (Methylene bromide)	98–9	1.538	2.496	6-Nitro-2-mercaptobenzothiazole deriv., 232–3; S-Alkyl *bis*-(thiuronium picrate), 267
2	1,1-Dibromoethane	112	1.5128	2.055	6-Nitro-2-mercaptobenzothiazole deriv., 145–6; 1,1-Di-(1-naphthyl) ether, 117
3	1,2-Dibromoethane	132	1.5379	2.179	m.p. 10; 6-Nitro-2-mercaptobenzothiazole deriv., 201–2; 1,2-Di-(2-naphthyl) ether, 217
4	DL-1,2-Dibromopropane	141–2	1.5203	1.933	6-Nitro-2-mercaptobenzothiazole, 194–5; 1,2-Di-(2-naphthyl) ether, 152; 1,2-Diphenyl ether, 32
5	1,2-Dibromo-2-methylpropane....	149	1.512	1.783	....
6	1,2-Dibromo-1-butene..........	150	......	1.887	....
7	Bromoform	150–1	1.598	2.890⁰	m.p. 8
8	1,3-Dibromopropene...........	156	1.538²⁵	2.097⁰	....
9	1,1-Dibromo-2-methylpropene	156–7	1.530	1.866²⁰₂₀	....
10	2,3-Dibromobutane............	157	1.515	1.792	....
11	1,2-Dibromobutane............	166	......	1.820	6-Nitro-2-mercaptobenzothiazole deriv., 164–5
12	1,3-Dibromopropane...........	167–8	1.523	1.982	1,3-Di-(1-naphthyl) ether, 103–4; 1,3-Di-(2-naphthyl) ether, 148–9; 1,3-Diphenyl ether, 60
13	1,3-Dibromo-2-butene..........	168–9	1.548	1.877	....
14	1,3-Dibromobutane............	174	1.507	1.820⁰	....
15	1,1,2-Tribromoethane..........	189	1.5933	2.6211	....
16	1,4-Dibromobutane............	197–8	......	1.847⁰	*p*-toluidine (3 moles) $\xrightarrow{\text{heat}}$ N-4-tolylpyrrolidine, 42, dil. al.
17	1,2,3-Tribromopropane	220	1.582	2.402	`
18	1,5-Dibromopentane	221	1.514¹⁵	1.694²⁵₄	6-Nitro-2-mercaptobenzothiazole deriv., 132–3; S-Alkyl *bis*-(thiuronium picrate), 247
19	1,1,2,2-Tetrabromoethane	243–4	1.638	2.967	....
20	1,8-Dibromooctane............	270–2	1.501¹⁵	1.468¹⁵	m.p. 15–6; S-Alkyl *bis*-(thiuronium picrate), 214
21	1,9-Dibromononane	285–8	......	1.415¹⁵	m.p. −2.5; S-Alkyl *bis*-(thiuronium picrate), 193

*Derivative data given in order: m.p., crystal color, solvent from which crystallized.

TABLE V. ORGANIC DERIVATIVES OF HALIDES
B) Dihalides and polyhalides (non-aromatic)
3. Bromides b) Solids (Listed in order of increasing m.p.)*

No.	Name	Melting point, °C	Boiling point, °C	Miscellaneous
1	**1,7-Dibromoheptane**	42	263	S-Alkyl *bis*-(thiuronium picrate), 208
2	**Carbon tetrabromide**	92	190	

*Derivative data given in order: m.p., crystal color, solvent from which crystallized.

TABLE V. ORGANIC DERIVATIVES OF HALIDES
B) Dihalides and polyhalides (non-aromatic)
4. Iodides a) Liquids (Listed in order of increasing atmospheric b.p.)*

No.	Name	Boiling point, °C	n_D^{20}	D_4^{20}	Miscellaneous
1	**Di-iodomethane** (Methylene iodide)	181	1.7425	3.325	6-Nitro-2-mercaptobenzothiazole deriv., 232–3; Di-(2-naphthyl) ether, 133; S-Alkyl *bis*-(thiuronium picrate), 267
2	**1,3-Di-iodopropane**	224	1.6423	2.5755	1,3-Di-(1-naphthyl) ether, 103–4; 1,3-Di-(2-naphthyl) ether, 148–9; 1,3-Diphenyl ether, 60

*Derivative data given in order: m.p., crystal color, solvent from which crystallized.

TABLE V. ORGANIC DERIVATIVES OF HALIDES

B) Dihalides and polyhalides (non-aromatic)
4. Iodides b) Solids (Listed in order of increasing m.p.)*

No.	Name	Melting point, °C	Boiling point, °C	Miscellaneous
1	**1,2-Di-iodoethane**	81	. . . . . .	6-Nitro-2-mercaptobenzothiazole deriv., 202–3; 1,2-Di-(2-naphthyl) ether, 217
2	**Iodoform**	119	. . . . . .	Comp. with quinoline, 65

*Derivative data given in order: m.p., crystal color, solvent from which crystallized.

TABLE V. ORGANIC DERIVATIVES OF HALIDES
C) Aryl halides 1. Fluorides (Listed in order of increasing atmospheric b.p.)*

No.	Name	Boiling point, °C	Melting point, °C	n_D^{20}	D_4^{20}	Nitro derivative M.P.	Nitro derivative Position of nitro groups	Sulfonamide M.P.	Sulfonamide Position of sulfonamide group	Miscellaneous
1	1,3-Difluorobenzene	82	......	1.4404^{18}	1.1473_4^{25}	74	1,3	......	....	....
2	Fluorobenzene	87	......	1.466	1.024	......	....	125	4	....
3	1,4-Difluorobenzene	88	......	1.4423^{18}	1.1632_4^{25}	......	....	......	....	Boiling with NaOH → 4-fluorophenol, 48, b.p. 186–8, n_D^{56}: 1.5010, D^{56}: 1.1889
4	1,2-Difluorobenzene	92	−34	1.4451^{18}	1.1496^{25}	......	....	......	....	....
5	2-Fluorotoluene	114	......	......	......	......	....	105	5	Oxid. → 2-fluorobenzoic acid, 127
6	3-Fluorotoluene	116	......	......	......	......	....	174	6	Oxid. → 3-fluorobenzoic acid, 124
7	4-Fluorotoluene	117	......	1.496	0.998	......	....	141	2	Oxid. → 4-fluorobenzoic acid, 182
8	1-Fluoronaphthalene	214	......	1.594	1.134	......	....	......	....	Picrate, 113
9	2-Fluoronaphthalene	......	60	......	......	......	....	......	....	Picrate, 101

*Derivative data given in order: m.p., crystal color, solvent from which crystallized.

TABLE V. ORGANIC DERIVATIVES OF HALIDES

C) Aryl halides 2. Chlorides a) Liquids (Listed in order of increasing atmospheric b.p.)*

No.	Name	Boiling point, °C	Melting point, °C	n_D^{20}	D_4^{20}	Nitro derivative M.P.	Position of nitro groups	Sulfonamide M.P.	Position of sulfonamide group	Miscellaneous
1	Chlorobenzene	132	......	1.525	1.107	52	2,4	144	4	2,4-Dinitrobenzenesulfenyl chloride adduct, 123–4
2	2-Chlorotoluene	159	......	1.524	1.082	63	3,5	128	5	Oxid. → 2-chlorobenzoic acid, 141
3	3-Chlorotoluene	162	......	1.521	1.072	91	4,6	185	6	Oxid. → 3-chlorobenzoic acid, 158
4	4-Chlorotoluene	162	7	1.521	1.071	38	2	143	2	Oxid. → 4-chlorobenzoic acid, 240
5	1,3-Dichlorobenzene	173	......	1.546	1.288	103	4,6	182	6	....
6	1-Chloro-2-ethylbenzene	178; 180	......	1.5218	1.057	......	....	......	....	Oxid. → 2-chlorobenzoic acid, 141
7	1,2-Dichlorobenzene	179	......	1.552	1.305	110	4,5	135; 140	4	....
8	1-Chloro-3-ethylbenzene	184	......	1.5199	1.053	......	....	......	....	Oxid. → 3-chlorobenzoic acid, 158
9	1-Chloro-4-ethylbenzene	184; 180–1	......	1.5175	1.045	......	....	......	....	Oxid. → 4-chlorobenzoic acid, 240
10	2-Chloro-1,4-dimethyl-benzene	184–5	2	......	1.059^{20}_{20}	77; 101	5; 5,6	155	5′	Sulfonyl chloride, 50
11	1-Chloro-2-vinylbenzene (o-Chlorostyrene)	189	......	1.5649	1.100	......	....	......	....	Polymerizes on heating with benzoyl peroxide
12	1-Chloro-2,3-dimethyl-benzene	190	......	......	....	......	....	......	....	Oxid. → 3-chloro-2-methylbenzoic acid, 159
13	1-Chloro-2-isopropyl-benzene	191	......	1.5168	1.0341	......	....	......	....	Oxid. → 2-chlorobenzoic acid, 141
14	1-Chloro-2,4-dimethyl-benzene	192; 187	......	1.5230^{25}	1.0598^{20}_{20}	42	6	195	6	Oxid. $\xrightarrow{\text{CrO}_3/\text{H}_2\text{SO}_4}$ 4-chloro-3-methylbenzoic acid, 209–10; Oxid. $\xrightarrow{\text{aq. KMnO}_4}$ 4-chloroisophthalic acid, 294–5
15	1-Chloro-4-vinylbenzene (p-Chlorostyrene)	192	......	1.5660	1.0868	......	....	......	....	Polymerizes on heating with peroxide
16	1-Chloro-3,4-dimethyl-benzene	194–5	–6	......	1.069^{15}_{15}	63	5	207	5	$\text{Cl}_2 \xrightarrow{\text{Fe}}$ 1,2-dichloro-4,5-dimethyl-benzene, 76
17	1-Chloro-4-isopropyl-benzene (p-Chloro-cumene)	198	......	1.5117	1.0208	......	....	91	....	Oxid. → 4-chlorobenzoic acid, 240
18	2,6-Dichlorotoluene	199	......	1.5510	1.2686	50; 121	3; 3,5	......	....	Chlorosulfonic acid in chl. → 3-sulfonyl chloride, 54–6; 60; Oxid. → 2,6-dichlorobenzoic acid, 139
19	2,5-Dichlorotoluene	199	4–5	......	1.2535^{20}_{20}	50–1; 100–1	4; 4,6	......	....	Oxid. $\xrightarrow{\text{dil. HNO}_3}$ 2,5-dichlorobenzoic acid, 154
20	2,4-Dichlorotoluene	200	......	1.549	1.249	104	3,5	......	....	Oxid. → 2,4-dichlorobenzoic acid, 164
21	3,5-Dichlorotoluene	201	......	......	....	61–2; 99–100	2; 2,6	168–9	2	Oxid. → 3,5-dichlorobenzoic acid, 188
22	2-Chloro-1,3,5-trimethyl-benzene	204–6	......	1.5212^{30}	1.0337^{30}	178	4,6	165–6	....	Oxid. $\xrightarrow{\text{aq. KMnO}_4}$ 2-chlorobenzene tricarboxylic acid, 285 (anh.), 278 (hyd.)
23	2,3-Dichlorotoluene	207	......	1.5511	....	51; 71–2	4; 4,6	......	....	Oxid. $\xrightarrow{\text{alk. KMnO}_4}$ 2,3-dichloro-benzoic acid, 163
24	3,4-Dichlorotoluene	209	......	1.5471	1.2526	63; 91–2	6; 2,6	......	....	Oxid. → 3,4-dichlorobenzoic acid, 206
25	1,2,4-Trichlorobenzene	213	17	......	....	56; 103	5; 3,5	>200	....	Sulfonyl chloride, 31–4
26	2-Chloro-4-isopropyl-1-methylbenzene (2-Chloro-p-cymene)	217	......	1.5178^{17}	1.015^{17}_{4}	109–10	5,6	......	....	Boiling with dil. HNO_3 → 3-chloro-4-methylbenzoic acid, 196
27	3-Chloro-4-isopropyl-1-methylbenzene (3-Chloro-p-cymene)	217	......	1.5179^{18}	1.018^{18}_{4}	102–3; 106	2,6	......	....	....
28	1-Chloronaphthalene	259	......	1.633	1.191	180	4,5	......	....	Picrate, 137
29	3-Chlorobiphenyl	284–5	16	......	....	202–3	4,4′	......	....	Oxid. → 3-chlorobenzoic acid, 158; 155

*Derivative data given in order: m.p., crystal color, solvent from which crystallized.

TABLE V. ORGANIC DERIVATIVES OF HALIDES
C) Aryl halides 2. Chlorides b) Solids (Listed in order of increasing m.p.)*

No.	Name	Melting point, °C	Boiling point, °C	Nitro derivative M.P.	Position of nitro groups	Sulfonamide M.P.	Position of sulfonamide group	Miscellaneous
1	2-Chlorobenzotrichloride	29	283	....	....	....	....	Hydrolysis → 2-chlorobenzoic acid, 142
2	2,4,6-Trichlorotoluene..............	33–4; 38	....	54;50; 178–80	3; 3,5	....	....	Oxid. → 2,4,6-trichlorobenzoic acid, 160–1
3	2-Chlorobiphenyl	34	273	....	....	....	....	Oxid. → 2-chlorobenzoic acid, 141
4	1,2-Dichloronaphthalene	35	296	169	di	....	....	n_D^{48}: 1.6337; D_4^{48}: 1.3147; Oxid. $\xrightarrow{CrO_3/ac.\,a.}$ 5,6-dichloro-1,4-naphthoquinone, 181
5	2,3,4-Trichlorotoluene..............	41	231[761]	60; 140–1	5 or 6; 5,6	....	....	Oxid. → 2,3,4-trichlorobenzoic acid, 186–7
6	1,2,3,4-Tetrachlorobenzene..........	44–5	254	63–5; 151	5; 5,6	....	....	
7	3,4,5-Trichlorotoluene..............	44–5	245[768]	81–2; 163–4	2; 2,6	....	....	Oxid. → 3,4,5-trichlorobenzoic acid, 203; $Cl_2 \xrightarrow{Al/Hg}$ 2,3,4,5-tetrachlorotoluene, 97–8
8	2,3,5-Trichlorotoluene..............	45–6	232	58–9; 149–50	4 or 6; 4,6	....	....	Oxid. $\xrightarrow{dil.\,HNO_3}$ 2,3,5-trichlorobenzoic acid, 162
9	1,6-Dichloronaphthalene	48	....	119	4	216	4	....
10	1,2,3,5-Tetrachlorobenzene..........	50–1	246	40–1; 161–2	4; 4,6	....	....	....
11	1,2,3-Trichlorobenzene	52–3	218–9	56; 92–3	4; 4,6	226–30	4	Sulfonyl chloride, 65
12	1,4-Dichlorobenzene	53	173	54	2	180; 186	2	....
13	4,4′-Dichlorodiphenylmethane	55	337	198–9	3,3′	....	....	Oxid. $\xrightarrow{CrO_3/ac.\,a.}$ 4,4′-dichlorobenzophenone, 145
14	2-Chloronaphthalene	56; 61	265	175	1,8	126	....	Picrate, 81
15	2,2′-Dichlorobiphenyl	60	....	203–5	5,5′	....	....	....
16	1,3-Dichloronaphthalene	61	291[775]	150 and 158	di	....	....	Oxid. $\xrightarrow{dil.\,HNO_3}$ phthalic acid, 200–6
17	1,3,5-Trichlorobenzene	63	208	68	2	210–2	....	Sulfonyl chloride, 35–40
18	1,7-Dichloronaphthalene	63–4	286	138–9	....	226	4	Sulfonyl chloride, 118
19	1-Bromo-4-chlorobenzene	67	197	72	2	....	....	....
20	1,4-Dichloronaphthalene	68	286[740]	92	8	244	6	Oxid. $\xrightarrow{CrO_3/ac.\,a.}$ 5,8-dichloro-1,4-naphthoquinone, 173–4; Boiling HNO_3 (D = 1.3) → 3,6-dichlorophthalic acid, 194; 185
21	2,5-Dichloro-1,4-dimethylbenzene (2,5-Dichloro-p-xylene)	68	224	....	....	....	....	Oxid. → 2,5-dichloroterephthalic acid, 306
22	4-Chlorobiphenyl	77	293	....	....	....	....	Oxid. → 4-chlorobenzoic acid, 240
23	2,4,5-Trichlorotoluene..............	82	230[715]	89–90; 226–7	3; 3,6	....	....	Oxid. → 2,4,5-trichlorobenzoic acid, 168
24	Pentachlorobenzene................	86; 84	276	143; 146	6	....	....	....
25	1,8-Dichloronaphthalene	89	....	....	....	228	4	Sulfonyl chloride, 141
26	1,5-Dichloronaphthalene	107	....	142	8	204	3	Oxid. $\xrightarrow{CrO_3/ac.\,a.}$ 3-chlorophthalic acid, 185–7; Picrate, 87
27	2,7-Dichloronaphthalene	114–5	....	141–2	mono	218	3	Oxid. $\xrightarrow{dil.\,HNO_3}$ 4-chlorophthalic acid, 157
28	2,6-Dichloronaphthalene	135–6	285	....	....	269	4	Sulfonyl chloride, 136; Cl_2 in chl. → 1,2,6-trichloronaphthalene, 92; Oxid. $\xrightarrow{CrO_3/ac.\,a.}$ 1,4-naphthoquinone deriv., 148–9
29	1,2,4,5-Tetrachlorobenzene..........	140	245	99; 232	3; 3,6	....	....	Chlorosulfonic acid → hexachlorobenzene, 229
30	4,4′-Dichlorobiphenyl	149	....	....	....	....	....	Oxid. → 4-chlorobenzoic acid, 240

*Derivative data given in order: m.p., crystal color, solvent from which crystallized.

No.	Name	Melting point, °C	Boiling point, °C	Nitro derivative		Sulfonamide		Miscellaneous
				M.P.	Position of nitro groups	M.P.	Position of sulfona-mide group	
31	5,6,7,8-Tetrachlorotetralin	174	180[26]	. . . .	. . . .	. . . .	. . . .	Br_2 in CS_2 → 1,2-dibromo-5,6,7,8-tetra-chloronaphthalene, 142
32	1,2,3,4-Tetrachlorotetralin	182; 187	. . . .	. . . .	. . . .	. . . .	. . . .	Oxid. $\xrightarrow{CrO_3/ac. a.}$ 2,4-dichloro-1-naphthol, 106–7; Boiling HNO_3 → phthalic acid, 200–6
33	Octachloronaphthalene	198; 200	442	. . . .	. . . .	. . . .	. . . .	Oxid. $\xrightarrow{fuming\ HNO_3}$ hexachloro-1,4-naphthoquinone, 222; $SbCl_5$ in CCl_4 → cherry-red color
34	9,10-Dichloroanthracene (meso-Di-chloroanthracene)	209–10, yel.	. . . .	. . . .	. . . .	279	2	Sulfonyl chloride, 221–5; Oxid. → 9,10-anthraquinone, 286; Maleic anh. adduct, 258–9
35	Hexachlorobenzene	229, subl.; 226	309	. . . .	. . . .	. . . .	. . . .	Boiling with fuming HNO_3 + conc. H_2SO_4 → tetrachloro-1,4-benzoquinone (chloroanil), 290

*Derivative data given in order: m.p., crystal color, solvent from which crystallized.

TABLE V. ORGANIC DERIVATIVES OF HALIDES
C) Aryl halides 3. Bromides a) Liquids (Listed in order of increasing atmospheric b.p.)*

No.	Name	Boiling point, °C	Melting point, °C	n_D^{20}	D_4^{20}	Nitro derivative M.P.	Nitro derivative Position of nitro groups	Sulfonamide M.P.	Sulfonamide Position of sulfonamide group	Miscellaneous
1	Bromobenzene.............	156	......	1.560	1.494	70–2	2,4	166; 161	4	1-Naphthylamide, 161; 2,4-Dinitrobenzenesulfenyl chloride adduct, 140–1
2	2-Bromotoluene............	182	......	......	1.425	82	3,5	146	5	Oxid. → 2-bromobenzoic acid, 150
3	3-Bromotoluene............	184	......	......	1.410	103	4,6	168	6	Oxid. → 3-bromobenzoic acid, 155
4	1-Bromo-2-ethylbenzene......	199	......	1.5486	1.355	......	....	......	....	Oxid. → 2-bromobenzoic acid, 150
5	1-Bromo-4-ethylbenzene......	205	......	1.5448	1.342	......	....	......	....	Oxid. → 4-bromobenzoic acid, 251
6	1-Bromo-2-vinylbenzene (o-Bromostyrene)	210	......	1.5927	1.4160	......	....	......	....	Polymerizes on heating with benzoyl peroxide
7	1-Bromo-2-isopropylbenzene ..	210	......	1.5408	1.3020	......	....	......	....	Oxid. → 2-bromobenzoic acid, 150
8	1-Bromo-4-vinylbenzene (p-Bromostyrene)	212	......	1.5947	1.398	......	....	......	....	Polymerizes on heating with benzoyl peroxide; Oxid. → 4-bromobenzoic acid, 251
9	1-Bromo-2,3-dimethylbenzene .	217	......	......	......	......	....	......	....	Oxid. → 3-bromophthalic acid, 188
10	1,3-Dibromobenzene........	219	......	1.606	1.952	61	4	190	6	....
11	1-Bromo-4-isopropylbenzene ..	219	......	1.5361	1.2854	......	....	......	....	Oxid. → 4-bromobenzoic acid, 251
12	1,2-Dibromobenzene........	219	......	1.609	1.956	114	4,5	176	4	....
13	2-Bromocymene	234	......	......	1.267	97	....	......	....	Anilide, 143
14	2,5-Dibromotoluene	236	......	......	1.811	......	....	......	....	Oxid. → 2,5-dibromobenzoic acid, 157
15	3,4-Dibromotoluene	240	......	......	1.81	......	....	......	....	Oxid. → 3,4-dibromobenzoic acid, 235
16	1-Bromonaphthalene........	281	......	1.658	1.484	85	4	191–3	4	Picrate, 134; Carbonation of Grignard → 1-naphthoic acid, 162
17	2-Bromobiphenyl...........	297	......	......	......	......	....	......	....	Oxid. → 2-bromobenzoic acid, 150

*Derivative data given in order: m.p., crystal color, solvent from which crystallized.

TABLE V. ORGANIC DERIVATIVES OF HALIDES
C) Aryl halides 3. Bromides b) Solids (Listed in order of increasing m.p.)*

No.	Name	Melting point, °C	Boiling point, °C	Nitro derivative		Sulfonamide		Miscellaneous
				M.P.	Position of nitro groups	M.P.	Position of sulfona-mide group	
1	**4-Bromotoluene**	28–9	184	. . . . . .	. . . .	. . . . . .	. . . .	Oxid. → 4-bromobenzoic acid, 251
2	**2-Bromonaphthalene**	59	281	. . . . . .	. . . .	208	8	Picrate, 86; 79; 2,4,7-Trinitrofluorenone adduct, 138–40
3	**1,2-Dibromonaphthalene**	67	. . . . . .	. . . . . .	. . . .	. . . . . .	. . . .	Oxid. → 3,4-dibromophthalic acid, 196
4	**1,4-Dibromonaphthalene**	82	. . . . . .	. . . . . .	. . . .	. . . . . .	. . . .	Oxid. → 3,6-dibromophthalic acid, 135
5	**1,4-Dibromobenzene**	89	219	84	2,5	195	2	. . . .
6	**4-Bromobiphenyl**	89	310	. . . . . .	. . . .	. . . . . .	. . . .	Oxid. → 4-bromobenzoic acid, 251
7	**1,3,5-Tribromobenzene**	120	271	. . . . . .	. . . .	222	2	. . . .
8	**4,4′-Dibromobiphenyl**	164	. . . . . .	. . . . . .	. . . .	. . . . . .	. . . .	Oxid. → 4-bromobenzoic acid, 251
9	**1,2,4,5-Tetrabromobenzene**	180	. . . . . .	168	3	. . . . . .	. . . .	. . . .

*Derivative data given in order: m.p., crystal color, solvent from which crystallized.

74

TABLE V. ORGANIC DERIVATIVES OF HALIDES
C) Aryl halides 4. Iodides a) Liquids (Listed in order of increasing atmospheric b.p.)*

No.	Name	Boiling point, °C	Melting point, °C	n_D^{20}	D_4^{20}	Nitro derivative M.P.	Nitro derivative Position of nitro groups	Sulfonamide M.P.	Sulfonamide Position of sulfona-mide group	Miscellaneous
1	**Iodobenzene**.....................	188–7	......	1.620	1.831	171	4	......	....	Br$_2$ → 1-Bromo-4-iodobenzene, 91
2	**3-Iodotoluene**	204	......	.....	1.698	108	4,6	......	....	Oxid. → 3-iodobenzoic acid, 187
3	**2-Iodotoluene**	211	......	.....	1.698	103	6	......	....	Oxid. → 2-iodobenzoic acid, 162
4	**1-Iodo-4-isopropylbenzene**	236–8	......	.....	.....	......	....	......	....	Cl$_2$ in chl. → dichloride (ArICl$_2$), 110
5	**1-Iodonaphthalene**	305	......	.....	.....	......	....	......	....	Picrate, 128

*Derivative data given in order: m.p., crystal color, solvent from which crystallized.

TABLE V. ORGANIC DERIVATIVES OF HALIDES
C) Aryl halides 4. Iodides b) Solids (Listed in order of increasing m.p.)*

No.	Name	Melting point, °C	Boiling point, °C	Nitro derivative		Sulfonamide		Miscellaneous
				M.P.	Position of nitro groups	M.P.	Position of sulfona-mide group	
1	4-Iodotoluene	35	211	. . . . . .	. . . .	. . . . . .	. . . .	HNO₃ at 200° → 4-iodobenzoic acid, 270
2	1-Iodo-2,4,5-trimethylbenzene	37	256–8	. . . . . .	. . . .	. . . . . .	. . . .	Cl₂ in chl. → dichloride (ArICl₂), 66
3	1,3-Di-iodobenzene	40	285	. . . . . .	. . . .	. . . . . .	. . . .	. . . .
4	2-Iodonaphthalene	55	309	. . . . . .	. . . .	. . . . . .	. . . .	Picrate, 95
5	4-Iodobiphenyl	114	320 d.	. . . . . .	. . . .	. . . . . .	. . . .	Cl₂ in chl. → dichloride (ArICl₂), 102
6	1,4-Di-iodobenzene	129	289	171	2,5	. . . . . .	. . . .	. . . .

*Derivative data given in order: m.p., crystal color, solvent from which crystallized.

76

EXPLANATIONS AND REFERENCES TO TABLE VI

Phenylurethane.

$$ROH \ + \ C_6H_5N{=}C{=}O \ \rightarrow \ C_6H_5NHCOOR$$

Phenylurethane
(Phenylcarbamate)

From the dry alcohol with phenylisocyanate without solvent.

For directions and examples see: Linstead, pp. 34–35; Shriner, p. 211; Vogel, p. 264; B. T. Dewey and N. F. Witt, *Ind. Eng. Chem., Anal. Ed.*, **12**, 459 (1940); **14**, 648 (1942).

From the dry alcohol with phenylisocyanate in petrol ether.

See: Wild, pp. 55–57.

1-Naphthylurethane (α-Naphthylcarbamate).*

$$ROH \ + \ 1\text{-}C_{10}H_7N{=}C{=}O \ \rightarrow \ 1\text{-}C_{10}H_7NHCOOR$$

1-Naphthylurethane
(α-Naphthylcarbamate)

From the dry alcohol with 1-naphthylisocyanate without solvent.

For directions and examples see: Cheronis, pp. 475–479; Linstead, pp. 34–35; Shriner, p. 211; Vogel, p. 264; V. T. Bickel and H. E. French, *J. Amer. Chem. Soc.*, **48**, 747 (1926); H. E. French and A. F. Wirtel, *J Amer. Chem. Soc.*, **48**, 1736 (1926).

From the dry alcohol with 1-naphthylisocyanate in petrol ether.

See: Cheronis, pp. 476–477; Wild, pp. 55–57.

p-Xenylurethane (4-Biphenylylurethane).

$$ROH \ + \ p\text{-}C_6H_5C_6H_4N{=}C{=}O \ \rightarrow \ p\text{-}C_6H_5C_6H_4NHCOOR$$

p-Xenylurethane
(4-Biphenylylcarbamate)

From the alcohol with *p*-xenylisocyanate in toluene.

For directions and examples see: G. T. Morgan and A. E. J. Pettet, *J. Chem. Soc.*, 1124 (1931); B. Witten and E. E. Reid, *J. Amer. Chem. Soc.*, **69**, 2470 (1947).

From the alcohol with *p*-xenylisocyanate in a benzene—petrol ether mixture.

See: M. J. van Gelderen, *Rec. Trav. chim.*, **52**, 969 (1933).

p-Nitrobenzoate.*

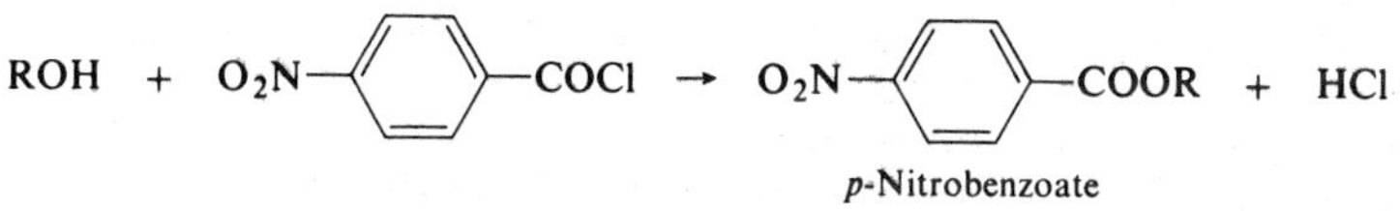

p-Nitrobenzoate

From the alcohol in excess and *p*-nitrobenzoyl chloride.

For directions and examples see: Cheronis, pp. 467–469, 471; Shriner, p. 212; Vogel, p. 263; Wild, p. 52; M. D. Armstrong and J. E. Copenhaver, *J. Amer. Chem. Soc.*, **65**, 2252 (1943).

From an aqueous solution of the alcohol with *p*-nitrobenzoyl chloride in a ligroin-benzene mixture.

See: Cheronis, pp. 468–469, 472.

From the alcohol with *p*-nitrobenzoyl chloride in pyridine.

See: Shriner, p. 212; Wild, p. 52; L. F. King, *J. Amer. Chem. Soc.*, **61**, 2383 (1939).

From the alcohol with *p*-nitrobenzoyl chloride in aqueous sodium hydroxide.

See: Wild, p. 52.

From the alcohol with *p*-nitrobenzoyl chloride in an aqueous solution of sodium acetate and potassium hydroxide at low temperature.

See: Wild, p. 53; F. A. Menalda, *Rec. Trav. chim.*, **49**, 967 (1930); H. Henstock, *J. Chem. Soc.*, 216 (1933).

3,5-Dinitrobenzoate.*

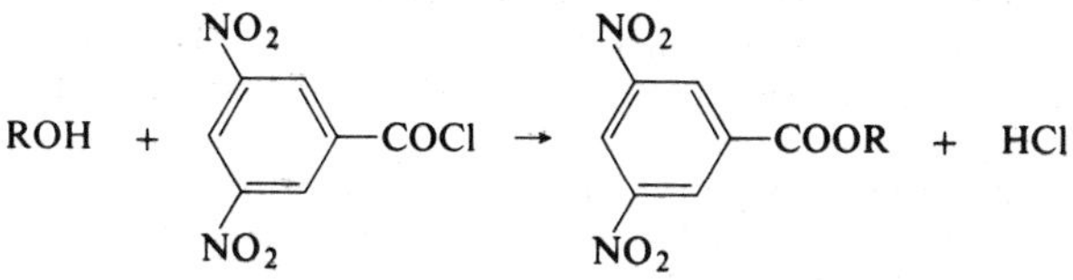

3,5-Dinitrobenzoate

*Derivatives recommended for first trial.

WARNING: This is not an instruction manual. References should be consulted for the preparation of derivatives.

From the alcohol in excess with 3,5-dinitrobenzoyl chloride.

For directions and examples see: Cheronis, pp. 467–470; Shriner, pp. 212–213; Vogel, p. 262; G. B. Malone and E. E. Reid, *J. Amer. Chem. Soc.*, **51**, 3424 (1929).

From an aqueous solution of the alcohol with 3,5-dinitrobenzoyl chloride in a ligroin-benzene mixture.

See: Cheronis, pp. 468–469, 472.

From the alcohol with 3,5-dinitrobenzoyl chloride and pyridine in benzene.

See: Linstead, p. 34; Wild, p. 53; T. Reichstein, *Helv. chim. Acta*, **9**, 799 (1926); W. M. D. Bryant, *J. Amer. Chem. Soc.*, **54**, 3758 (1932).

From the alcohol with 3,5-dinitrobenzoyl chloride and a catalytic amount of pyridine in isopropyl or *n*-butyl ether.

See: Cheronis, pp. 469, 471.

From the alcohol with 3,5-dinitrobenzoyl chloride in pyridine.

See: Cheronis, p. 469; Shriner, pp. 212–213; Vogel, pp. 262–263.

From the alcohol with 3,5-dinitrobenzoyl chloride in aqueous potassium hydroxide.

See: Linstead, p. 34.

From the alcohol in aqueous sodium hydroxide and potassium acetate with 3,5-dinitrobenzoyl chloride in a benzene-ligroin mixture.

See: Wild, pp. 53–54; W. N. Lipscomb and R. H. Baker, *J. Amer. Chem. Soc.*, **64**, 179 (1942).

*Hydrogen phthalate.**

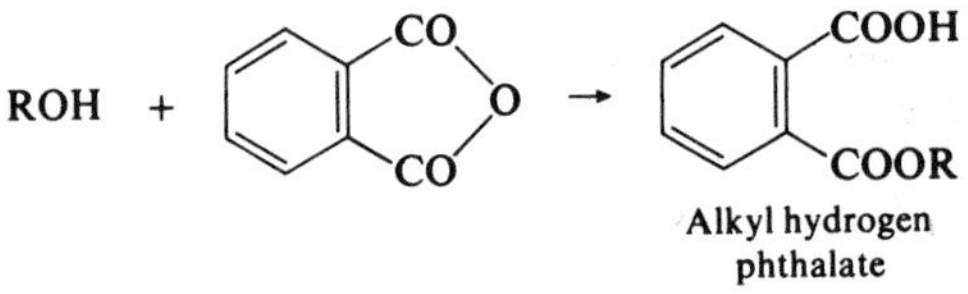

Alkyl hydrogen phthalate

$$R_3COH + C_2H_5MgBr \rightarrow R_3COMgBr + C_2H_6$$

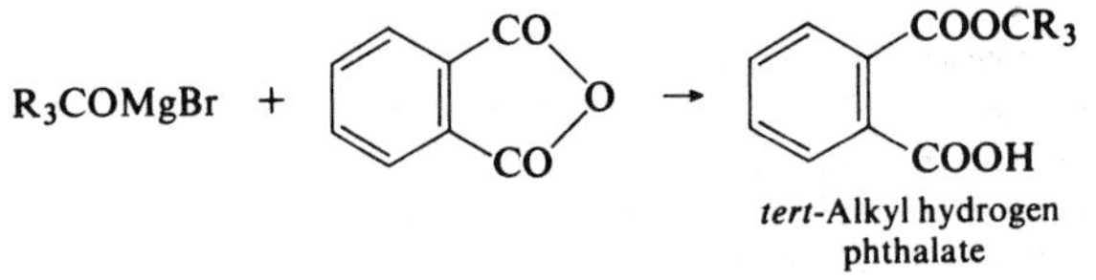

tert-Alkyl hydrogen phthalate

From the alcohol with phthalic anhydride.

For directions and examples see: E. E. Reid, *J. Amer. Chem. Soc.*, **39**, 1250 (1917); J. F. Goggans and J. E. Copenhaver, *J. Amer. Chem. Soc.*, **61**, 2909 (1939).

From the alkoxymagnesium halide derived from a tertiary alcohol (prepared from the alcohol with ethylmagnesium bromide) with phthalic anhydride in ether or an ether-dioxan mixture.

See: W. A. Fessler and R. L. Shriner, *J. Amer. Chem. Soc.*, **58**, 1384 (1936).

*Hydrogen 3-nitrophthalate.**

Main product   Trace

Hydrogen 3-nitrophthalates
(3-Nitrophthalic acid monoalkyl esters)

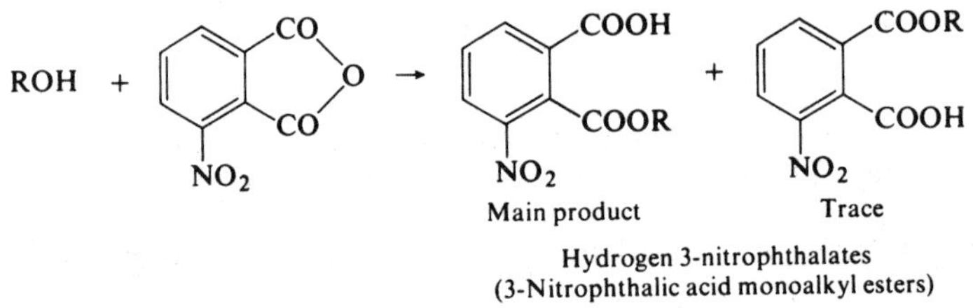

From the alcohol with 3-nitrophthalic anhydride.

For directions and examples see: Cheronis, pp. 473–474; Linstead, p. 35; Shriner, p. 213; Vogel, p. 265; Wild, p. 59; G. M. Dickinson, L. H. Crosson and J. E. Copenhaver, *J. Amer. Chem. Soc.*, **59**, 1094 (1937); B. H. Nicolet and J. Sacks, *J. Amer. Chem. Soc.*, **47**, 2348 (1925); A. J. Veraguth and H. Diehl, *J. Amer. Chem. Soc.*, **62**, 233 (1940).

From a high boiling alcohol with 3-nitrophthalic anhydride in toluene.

See: Linstead, p. 35; Shriner, p. 213; Vogel, p. 265; Wild, p. 59; G. M. Dickinson, L. H. Crosson and J. E. Copenhaver, *J. Amer. Chem. Soc.*, **59**, 1094 (1937).

---

*Derivatives recommended for first trial.

WARNING: This is not an instruction manual. References should be consulted for the preparation of derivatives.

Pseudosaccharin ether (Pseudosaccharin derivative).

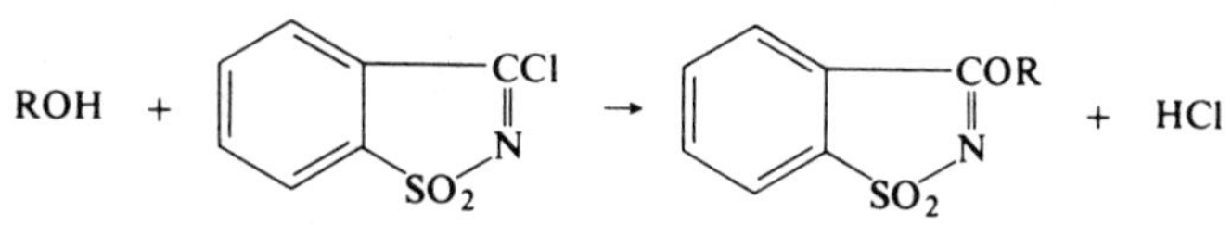

$$ROH \quad + \qquad \text{(Pseudosaccharin chloride)} \qquad \rightarrow \qquad \text{(Pseudosaccharin ether)} \qquad + \quad HCl$$

From the alcohol with pseudosaccharin chloride without solvent.

For directions and examples see: Vogel, p. 266; Wild, pp. 60–61; J. R. Meadoe and E. E. Reid, *J. Amer. Chem. Soc.*, **65**, 457 (1943).

From the alcohol with pseudosaccharin chloride in chloroform.

See: H. Bohme and H. Opper, *Z. Anal. Chem.*, **139**, 255 (1953).

From the alcohol with pseudosaccharin chloride and a catalytic amount of pyridine in chloroform.

See: Cheronis, p. 482; H. Bohme and H. Opper, *Z. Anal. Chem.*, **139**, 255 (1953).

Allophanate

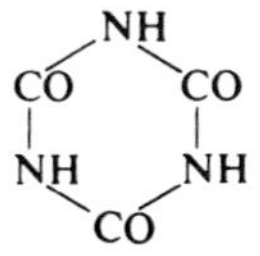

$$\text{Cyanuric acid} \qquad \rightarrow \quad HN{=}C{=}O \quad \rightarrow \quad H_2NCON{=}C{=}O$$
Cyanuric acid   Cyanic acid   Cyanic acid dimer

$$ROH \quad + \quad H_2NCON{=}C{=}O \quad \rightarrow \quad H_2NCONHCOOR$$
Allophanate

From the alcohol with cyanic acid (prepared from the depolymerization of cyanuric acid).

For directions and examples see: Linstead, p. 36; A. Behál, *Compt. rend.*, **168**, 945 (1919); M. A. Spielman, J. D. Barnes and W. J. Close, *J. Amer. Chem. Soc.*, **72**, 2520 (1950); H. W. Blohm and E. I. Becker, *J. Amer. Chem. Soc.*, **72**, 5342 (1950); *Chem. Revs.*, **51**, 471 (1952).

From the alcohol with sodium cyanate and dry hydrochloric acid in dioxane.

See: E. S. Lane, *J. Chem. Soc.*, 2764 (1951).

*Benzoate.**

$$ROH \quad + \quad C_6H_5COCl \quad \rightarrow \quad C_6H_5COOR \quad + \quad HCl$$
Benzoate

Especially for polyhydric alcohols.

From the alcohol with benzoyl chloride.

For directions and examples see: Shriner, p. 212.

From the alcohol with benzoyl chloride in anhydrous pyridine.

See: Cheronis, pp. 481–482; Shriner, p. 212; Vogel, pp. 243, 447.

From the alcohol with benzoyl chloride in aqueous sodium hydroxide.

See: Vogel, p. 447.

Acetate.

$$ROH \quad + \quad (CH_3CO)_2O \quad \rightarrow \quad CH_3COOR \quad + \quad CH_3COOH$$

$$R_3COH \quad + \quad CH_3COCl \quad \rightarrow \quad CH_3COOCR_3 \quad + \quad HCl$$
Acetate

Especially for polyhydric alcohols.

From the alcohol with acetic anhydride and sodium acetate.

For directions and examples see: Linstead, pp. 35, 39; Shriner, p. 212.

From the alcohol with acetic anhydride in pyridine.

See: Shriner, p. 212.

From the alcohol (especially a tertiary alcohol) with acetyl chloride in the presence of magnesium.

See: A. Spassow, *Chem. Ber.*, **70B**, 1926 (1937).

NOTE: For additional information regarding directions and examples for the preparation of derivatives of polyhydric alcohols see explanations and references to Table XIX, p. 326.

---

*Derivatives recommended for first trial.

WARNING: This is not an instruction manual. References should be consulted for the preparation of derivatives.

TABLE VI. ORGANIC DERIVATIVES OF ALCOHOLS
a) Liquids (Listed in order of increasing atmospheric b.p.)*

No.	Name	Boiling point, °C	Melting point, °C	n_D^{20}	D_4^{20}	Phenyl-urethane	1-Naph-thyl-urethane	4-Nitro-benzoate	3,5-Dinitro-benzoate	Hydrogen 3-nitro-phthalate	Hydro-gen phthalate	Miscellaneous
1	**Methanol** (Methyl alcohol).............	64.65	f.p.: −97	1.3306^{15}	0.7915	47, al.	124, lgr.	96, dil. al.	108 (cor.), al.	153 (cor.)	82.5 (cor.)	Pseudosaccharin ether, 182 (cor.)
2	**Ethanol** (Ethyl alcohol)..	78.32	f.p.: −117.3	1.3610	0.7894	52	79, lgr.	57, al.	93, al.	158 (cor.), w.	48	Pseudosaccharin ether, 219 (cor.)
3	**2-Propanol** (Isopropyl alcohol).............	82.4	−89.5	1.37927	0.78507	75–6, lt. pet.	106	110.5, lt. pet.; 108	123, pet. eth.	154 (cor.), w.	....	Pseudosaccharin ether, 137 (cor.)
4	**_d,l_-3-Buten-2-ol** (Methyl vinyl carbinol)........	94–6	....	....	....	....	....	....	....	43–4	....	Allophanate, 152; Constant boil. mixt. with 21.76% w., b.p.: 80
5	**2-Propen-1-ol** (Allyl alcohol).............	97.1	....	1.41345	0.8540	70	108	28	49–50	124	....	....
6	**1-Propanol** (_n_-Propyl alcohol).............	97.1	....	1.38499	0.80359	57, pet.	80; 76	35, pet.	74, pet. eth.	145.5 (cor.), w.	54.1–4 (cor.), pet. eth.-bz. (9:1)	Pseudosaccharin ether, 124.5 (cor.)
7	**2-Butanol** (_d,l-sec_-Butyl alcohol; Ethyl methyl carbinol).............	99.5	....	1.39495^{25}	0.80692	64.5, pet.	97	25–6, dil. al.	76	131 (cor.)	59–60	Pseudosaccharin ether, 65.5 (cor.)
8	**2-Methyl-2-butanol** (_tert_-Amyl alcohol)........	102.3	−8.55	1.4052	0.80889	42, pet. eth.	72	85	116; 117–8	....	....	....
9	**2-Fluoroethanol**	105	....	1.3633^{25}	....	....	128					
10	**2-Methyl-1-propanol** (Iso-butyl alcohol)........	108.1	....	1.3939^{25}	0.80196	86, lgr.	104	69	87	180.5 (cor.)	65, pet. eth.	Pseudosaccharin ether, 100 (cor.)
11	**3-Buten-1-ol**	122.5–3.5^{755}	....	....	....	23.4–4.5	....					
12	**_d,l_-3-Methyl-2-butanol** (_sec_-Isoamyl alcohol; _d,l_-Isopropyl methyl carbinol).............	114; _d_: 110–2	....	1.3973	0.8180	68	109		127	39; _d_: 34; _l_: 34	_d_: $[\alpha]_D^{20}$: +5.34, in al.	
13	**3-Pentanol** (_sym-sec_-Amyl alcohol; Diethyl carbinol)	116.1	....	1.4103	0.82037	48–9	95, lgr.	17	101; 99; 97	121	....	....
14	**1-Butanol** (_n_-Butyl alcohol).............	117.6; 116	−90.2	1.3974^{25}	0.80960	61	71	70; 64; 35–6	64; 62.5	147 (cor.)	73.1–.5 (cor.)	Pseudosaccharin ether, 96 (cor.)
15	**_d,l_-2-Pentanol** (_sec_-Amyl alcohol).............	119.85	....	1.4060	0.80919	....	74.5; 76; _d_: 88–91	17	62	102–3	60–1; _d_: 34; _l_: 34	Pseudosaccharin ether, 38 (cor.)
16	**3,3-Dimethyl-2-butanol** (_d,l_-Pinacolyl alcohol; _tert_-Butyl methyl carbinol).............	120.4	5.3	1.4148	0.8185	77–8, pet. eth.	....		107, yel.-wh., pet. eth.	....	85–6, lt. pet.	....
17	**2,3-Dimethyl-2-butanol** (Dimethyl isopropyl carbinol).............	120.5	−14	1.4140	0.8208	65–6, pet. eth.	101		111, yel., bz.-pet. eth.	....	....	....
18	**3-Methyl-3-pentanol**	123	−22	1.4166^{25}	0.82334_4^{25}	43.5	83.5	....	96.5, yel., pet. eth.; 62.5	....	....	Allophanate, 152 (cor.)

*Derivative data given in order: m.p., crystal color, solvent from which crystallized.

TABLE VI. ORGANIC DERIVATIVES OF ALCOHOLS
a) Liquids (Listed in order of increasing atmospheric b.p.)* (Continued)

No.	Name	Boiling point, °C	Melting point, °C	n_D^{20}	D_4^{20}	Phenyl-urethane	1-Naph-thyl-urethane	4-Nitro-benzoate	3,5-Dinitro-benzoate	Hydrogen 3-nitro-phthalate	Hydro-gen phthalate	Miscellaneous
19	2-Methyl-2-pentanol (Dimethyl *n*-propyl carbinol)	123; 121	−103; −108	1.4113	0.81341	. . . .	. . . .	. . . .	72	. . . .	. . . . . .	Benzoate, 182–3, al.; Allophanate, 128
20	2-Methoxyethanol (Methyl cellosolve; Ethylene glycol mono-methyl ether)	124.5	. . . .	1.40238	0.9647	. . . .	112.5–3.0	50.5, dil. al.	. . . .	129, dil. al.	. . . . . .	Diphenyl ure-thane, 51
21	2-Methyl-3-pentanol (Ethyl isopropyl carbinol)	127.5	. . . .	1.4168	0.82487	50	. . . .	. . . .	85, yel. pet. eth.	150.7	70; 69–71, ra-cemic	. . . .
22	1-Chloro-2-propanol	127	. . . .	. . . .	. . . .	. . . .	. . . .	. . . .	77	. . . .	. . . . . .	. . . .
23	2-Methyl-1-butanol (Active amyl alcohol; *d-sec*-Butyl carbinol) . . .	128.9	. . . .	1.4107	0.8193	31	82, lgr.	. . . .	70	157–8, w.	. . . . . .	$[\alpha]_D^{20}$: −5.756
24	2-Chloroethanol (Ethylene chlorohydrin)	131	. . . .	. . . .	. . . .	51	101	. . . .	. . . .	98	. . . . . .	. . . .
25	*d,l*-4-Methyl-2-pentanol (Isobutyl methyl car-binol)	132	. . . .	1.4011	0.80713	143, et. ac.	88	26	65, yel., pet. eth.	. . . .	. . . . . .	. . . .
26	3-Methyl-1-butanol (*prim*-Isoamyl alcohol)	132	−117	1.40851[15]	0.80918	56–7, lgr.	68	21	61	166.3 (cor.), 30% al.; 165–6, w.	. . . . . .	Pseudosaccharin ether, 64 (cor.)
27	*d,l*-2-Chloro-1-propanol . .	133.4	. . . .	1.436	1.103	. . . .	. . . .	. . . .	76	. . . .	. . . . . .	Alkali + heat → propylene oxide, b.p. 35
28	3-Methyl-2-pentanol (*sec*-Butyl methyl carbinol) .	134.2[749]	. . . .	. . . .	. . . .	. . . .	72	. . . .	43.5, yel., pet. eth.; 41	. . . .	. . . . . .	. . . .
29	2-Ethoxyethanol (Ethyl-ene glycol monoethyl ether)	135	. . . .	1.40797	0.9297	. . . .	67.3–.5	. . . .	75, al.	118–8.6 (anh.); 94.2–4 (mono-hyd.), w.-al.	. . . . . .	Diphenyl ure-thane, 43
30	3-Hexanol (Ethyl *n*-propyl carbinol)	136	. . . .	1.4159	0.81851	. . . .	. . . .	. . . .	97, yel.-wh., pet. eth.	. . . .	76–7, pet. eth.	. . . .
31	2,2-Dimethyl-1-butanol (*tert*-Amyl carbinol) . . .	136.7	. . . .	1.4208	0.82834	65–6	80–1, lgr.	. . . .	. . . .	51, yel., pet. eth.	. . . . . .	Pseudosaccharin ether, 68–9, lt. pet.
32	1-Pentanol (*n*-Amyl alcohol)	138 (cor.)	−78.5	1.40994	0.81479	46	68	11	46.4	136 (cor.)	75.5	Pseudosaccharin ether, 62 (cor.)
33	*d,l*-2-Hexanol (*n*-Butyl methyl carbinol)	138–9[745]	. . . .	1.4126[25]	0.80977$_4^{25}$	. . . .	60.5	40	38.5	. . . .	*d*: 29	3,5-Dinitrophenyl urethane, 40
34	2,4-Dimethyl-3-pentanol .	140	. . . .	1.42259	0.8288	95, eth.-pet. eth.; 96–9	95; 99	155	. . . .	150–1	. . . . . .	Camphor-like odor
35	Cyclopentanol	140.85	. . . .	1.4530	0.94688	132.5, al.	118	. . . .	. . . .	. . . .	. . . . . .	. . . .
36	2-Isopropoxyethanol (Ethylene glycol mono-isopropyl ether)	141.5[736]	. . . .	1.40954	0.9030	. . . .	. . . .	. . . .	. . . .	. . . .	. . . . . .	Triphenylmethyl ether, 71.0–.5, me. al.

*Derivative data given in order: m.p., crystal color, solvent from which crystallized.

TABLE VI. ORGANIC DERIVATIVES OF ALCOHOLS.
a) Liquids (Listed in order of increasing atmospheric b.p.)* (Continued)

No.	Name	Boiling point, °C	Melting point, °C	n_D^{20}	D_4^{20}	Phenyl-urethane	1-Naph-thyl-urethane	4-Nitro-benzoate	3,5-Dinitro-benzoate	Hydrogen 3-nitro-phthalate	Hydro-gen phthalate	Miscellaneous
37	3-Ethyl-3-pentanol (Triethyl carbinol).....	142	....	1.4305	0.83889	....	....			....	....	Camphor-like odor; Allophanate, 152 (cor.)
38	2,3-Dimethyl-1-butanol ..	145	....	1.4195	$0.8297_4^{20.5}$	28–9	....		51.5, pa. yel., pet. eth.	....		....
39	3-Hydroxy-2-butanone (d,l-Acetoin; Acetyl methyl carbinol)......	145	−72	1.4178	0.9861_4^{30}	....	....				....	Semicarbazone, 185, al.; 202; 2,4-Dinitrophenylhydrazone, 318, or., PhNO₂-tol.
40	1-Hydroxy-2-propanone (Acetol; Acetyl carbinol)	146	−17	1.4295	1.0824_{20}^{20}	....	....				....	Semicarbazone, 196, al.; 2,4-Dinitrophenylhydrazone, 128.5 (cor.), or., al.
41	2-Methyl-1-pentanol (2-Methyl-n-amyl alcohol)	148.0	....	1.4190	0.8208	....	75–6		50.5, yel., pet. eth.	145; 141, bz.	....	....
42	2-Ethylbutanol	148.9	....	1.4224	0.83345	....			51.5, pet. eth.		....	....
43	2-Bromoethanol (Ethylene bromohydrin)........	149d.	....	....	....	....	86					
44	2-n-Propoxyethanol (Ethylene glycol mono-n-propyl ether).......	150.0⁷³⁶	....	1.41328	0.9112	....						
45	Trichloroethanol........	151	19	....	....	87	120	71, al.	142.3		....	Urethane, 64–5
46	3-Methyl-1-pentanol	151–2; 153.7–4.1	....	1.4188	0.8242	....	58; d,l: 40–1; d: 38–40; l: 37–8		38, yel., pet. eth.		....	
47	4-Methyl-1-pentanol (Isoamyl carbinol; Isohexyl alcohol)......	152–3	....	1.4153	0.8131	48 (cor.)	....		72, pet. eth.; 69.8 (cor.)	138.5–40, bz.-pet. eth.	....	
48	d,l-4-Heptanol (Di-n-propyl carbinol)......	156	−41.5	1.4205	0.8183	....	78–80	35	64	....	60	....
49	1-Hexanol (n-Hexyl alcohol).............	157.5	−51.6; −46.1	1.41778	0.81893	42	59; 62	5	58.4 (cor.); 60–1	124 (cor.); 123	25	Pseudosaccharin ether, 60 (cor.)
50	d,l-2-Heptanol (n-Amyl methyl carbinol; sec-Heptyl alcohol)	158.7	....	1.4210	0.8167	....	54		49.4	....	57.5; d,l: 76.5	....
51	2-Isobutoxyethanol (Ethylene glycol mono-isobutyl ether)	159.3⁷⁴⁶	....	1.41428	0.8900	....						
52	2-sec-Butoxyethanol (Ethylene glycol mono-sec-butyl ether).......	159.3⁷⁴⁶	....	1.41606	0.8966	....						
53	2,4-Dimethyl-1-pentanol .	159.8	....	1.427	0.793	....	....		154–5, bz.-pet. eth.	....		p-Xenylurethane, 74–5, pet.

*Derivative data given in order: m.p., crystal color, solvent from which crystallized.

TABLE VI. ORGANIC DERIVATIVES OF ALCOHOLS
a) Liquids (Listed in order of increasing atmospheric b.p.)* (Continued)

No.	Name	Boiling point, °C	Melting point, °C	n_D^{20}	D_4^{20}	Phenyl-urethane	1-Naph-thyl-urethane	4-Nitro-benzoate	3,5-Dinitro-benzoate	Hydrogen 3-nitro-phthalate	Hydro-gen phthalate	Miscellaneous
54	**3-Chloro-1-propanol** (3-Chloropropyl alcohol)	161.2	....	....	....	38	76	....	77	....	....	....
55	**2-Methyl-1-hexanol**	164–5	....	1.4250	0.8270	....	....	....	....	131–2, wh., pet.	....	p-Xenylurethane, 88.0–.5, pet.
56	**2-Ethyl-1-pentanol** (2-Ethyl-n-amyl alcohol) ..	164–6	....	....	....	....	....	....	....	127–8, bz.-pet.	....	p-Xenylurethane, 77–77.5, pet.
57	**d,l-4-Methyl-1-hexanol** ..	165; 173	....	1.4219	0.8239	....	50	....	....	149	....	Odor of Amyl alcohol
58	**d,l-cis-2-Methylcyclo-hexanol** (cis-Hexa-hydro-o-cresol)	165.3	−9.3	1.4640	0.9340	90–1; 93–4	....	51–2; 55–6	98–9	....	103–4; 104–5	....
59	**4-Hydroxy-4-methyl-2-pentanone** (Diacetone alcohol)	166	....	....	0.9306^{25}	....	....	48	55	....	....	Oxime, 57.5–8.5, lgr.-eth.
60	**d,l-trans-2-Methyl-cyclohexanol** (trans-Hexahydro-o-cresol)	167.4	21	1.4611	0.9235	105; mixt. cis + trans: 90–105	....	65; mixt. cis + trans: 35–6	114–5; mixt. cis + trans: 85–90	....	124–5; mixt. cis + trans: 95–6	....
61	**2-n-Butoxyethanol** (Ethylene glycol mono-n-butyl ether)	170–6^{743}	....	1.4177^{26}	0.9188	62	....	....	....	120.0–.6	....	4-Nitrophenyl-urethane, 58.7–9.1, CCl$_4$
62	**2-Aminoethyl alcohol** (Ethanolamine)	171				....	....	....	....	....	....	N-1-Naphthyl-urea, 186 (cor.); Picrate, 160
63	**2,6-Dimethyl-4-heptanol** (Di-isobutyl carbinol) ..	171.4–3.4	....	1.4242	0.8129^{20}_{20}	61–2, lgr.-al.	....	....	....	....	118	p-Xenylurethane, 118; Allophanate, 156
64	**Furfuryl alcohol** (2-Furyl carbinol)	172; 170	....	1.4863	1.1351^{20}_{20}	45	129–30, lgr.; 133	76	80–1	....	85	Urethane, 50; Pseudosaccharin ether, 55
65	**d,l-cis-3-Methylcyclo-hexanol** (cis-Hexa-hydro-m-cresol)	173–4	....	1.4572	0.919	87–8	128–9	65	91–2	....	82–3	....
66	**d,l-cis-4-Methylcyclo-hexanol** (cis-Hexahydro-p-cresol)	173–4^{750}	....	1.4549	0.914	118–9	....	94	134	....	72–3	....
67	**d,l-trans-4-Methylcyclo-hexanol** (trans-Hexa-hydro-p-cresol)	173–4.5^{745}	....	1.4534	0.913	124–5; mixt. cis + trans: 112–5	....	67	139–40; mixt. cis + trans: 125–30	....	119–25, ac. a.	....
68	**d,l-trans-3-Methylcyclo-hexanol** (trans-Hexa-hydro-m-cresol)	174–5	....	1.4550	0.9145	93–4; mixt. cis + trans: 75–85	122	58	97–8; mixt. cis + trans: 80–5	....	93–4	....
69	**2,6-Dimethylcyclohexanol**	174–5^{748}	....	1.4619	0.9115; 0.9235	158	....	....	....	....	....	....
70	*cis*-**2,5-Dimethylcyclo-hexanol**	175	....	1.4522^{17}	0.9096^{17}_{17}	....	....	....	....	....	....	Allophanate, 157–8
71	*trans*-**2,4-Dimethylcyclo-hexanol**	175	....	1.4560	0.900	96	....	....	....	....	....	Acetate, b.p.: 198^{765}
72	**1,3-Dichloro-2-propanol**	176	....	....	....	73	115	....	....	....	....	....

*Derivative data given in order: m.p., crystal color, solvent from which crystallized.

TABLE VI. ORGANIC DERIVATIVES OF ALCOHOLS

TABLE VI. ORGANIC DERIVATIVES OF ALCOHOLS
a) Liquids (Listed in order of increasing atmospheric b.p.)* (Continued)

No.	Name	Boiling point, °C	Melting point, °C	n_D^{20}	D_4^{20}	Phenyl-urethane	1-Naph-thyl-urethane	4-Nitro-benzoate	3,5-Dinitro-benzoate	Hydrogen 3-nitro-phthalate	Hydrogen phthalate	Miscellaneous
73	3-Bromo-1-propanol (Triethylene bromo-hydrin)	176d.					73					
74	cis-2,4-Dimethylcyclo-hexanol	176		1.4582	0.907							
75	1-Heptanol (n-Heptyl alcohol)	176.8	−34.6; −33.8	1.4245	0.82242	60; 65	62	10	46; 47	127 (cor.)	16.5–17.5	Pseudosaccharin ether, 55 (cor.)
76	trans-2,5-Dimethylcyclo-hexanol	177		1.4545^{17}	0.9079_{17}^{17}							Allophanate, 125
77	2,2-Dimethylcyclohexanol	177	8	1.4648	0.9225	85						
78	Tetrahydrofurfuryl alcohol	$177-8^{743}$		1.45167	1.0544	61, pet. eth.		46–8	83–4			Diphenylurethane, 81, me. al.
79	2-Methyl-1,2-propanediol (Isobutylene glycol)	178		1.4358^{17}	0.999_4^{14}	bis: 140.5						
80	d,l-2-Octanol	179		1.4265	0.8205	oil	63–4; 62.5	28	32		55; d,l: 75	
81	2,2-Dibromoethanol	179–81										Urethane, 90–1
82	1,3,5-Trimethylcyclo-hexanol	181		$1.454^{16.3}$	$0.8876_4^{16.8}$							
83	2,3Butanediol (2,3-Butylene glycol)	meso: 181.7^{742}; d,l: 176.7^{742}	meso: 34.4; d,l: 7.6	meso: 1.43637	meso: 1.0433	meso: bis: 201						Dibenzoate: d,l: 53–4; meso: 75.5–6.2
84	Cyclohexyl carbinol (Hexahydrobenzyl alcohol)	182		1.4649	0.9280							Acetate, b.p.: $199-201^{740}$
85	2,3-Dichloropropanol	182				73	93	37–8				2-Naphthylure-thane, 99
86	4-Methyl-1-heptanol	182.7								133		Pseudosaccharin ether, 34 (cor.)
87	2-Ethyl-1-hexanol	184.6		1.4328	0.8328	33–4	60–1			108		p-Xenylurethane, 80, pet.; Pseudo-saccharin ether, 53.5 (cor.)
88	3,3-Dimethylcyclohexanol	185^{754}	11–2	1.4606^{15}	0.9128_4^{14}			83				Acetate, b.p.: $194-5^{750}$; o-Nitrobenzoate, 62
89	cis-3,5-Dimethylcyclo-hexanol	187		1.454^{21}	0.9109_4^{21}							Acetate, b.p.: 201–2
90	trans-3,5-Dimethyl-cyclohexanol	d,l: 187		1.4579	d: 0.9146; l: 0.9166							Acetate, d,l; b.p.: 196; d, $[\alpha]_D^{26}$: +4.55; l, $[\alpha]_D^{17}$: −7.74
91	d,l-1,2-Propanediol (α-Propylene glycol)	187.4		1.43162^{25}	1.0354_4^{23}	bis: 153; 143–4						Monostearate, 59.5; Distearate, 72.3
92	3,4-Dimethylcyclohexanol	189		1.458^{16}	0.9073_4^{16}	119						
93	2,4,5-Trimethylcyclo-hexanol	cis: 191–3; trans: 196				cis: 83.5, al.; trans: 95, al.					81–3.5, eth.-lgr.	
94	1,3,3-Trimethylcyclo-hexen-6-ol	193 (cor.)			0.9310_4^{13}							Acetate, b.p.: 206–7
95	2,3,6-Trimethylcyclo-hexanol	$193-5^{747}$			0.9119_4^{17}							

*Derivative data given in order: m.p., crystal color, solvent from which crystallized.

TABLE VI. ORGANIC DERIVATIVES OF ALCOHOLS
a) Liquids (Listed in order of increasing atmospheric b.p.)* (Continued)

No.	Name	Boiling point, °C	Melting point, °C	n_D^{20}	D_4^{20}	Phenyl-urethane	1-Naphthyl-urethane	4-Nitro-benzoate	3,5-Dinitro-benzoate	Hydrogen 3-nitro-phthalate	Hydrogen phthalate	Miscellaneous
96	2-(2-Methoxyethoxy)-ethanol (Diethylene glycol monomethyl ether)	194	......	1.4244	1.035_{20}^{20}	......	......	92		91.4–2.2 (anh.); 87–90 (mono-hyd.), w.-al.	......	4-Nitrophenyl-urethane, 73.5; 76
97	5-Nonanol	194^{743}	......	1.4289^{18}	......							
98	1-Octanol (n-Octyl alcohol)	195	−16; −16.7	1.4274^{25}	0.8249	74;72	67	12	61–2	128 (cor.)	45 22	Allophanate, 158 Pseudosaccharin ether, 46 (cor.)
99	2-Methyl-2,4-pentanediol	196; 198	......	$1.42976^{16.7}$	0.9240_4^{17}							Odor of pinacol; Heating with 2% HBr → diene, b.p.: 75.5–76.0
100	2-(2-Ethoxyethoxy)-ethanol (Diethylene glycol monoethyl ether)	196^{763}	......	1.4298	1.023_{20}^{20}	......		oil	oil	oil	......	4-Nitrophenyl-urethane, 65.8–6.3
101	Glycol (1,2-Ethanediol)	197.85	−12.6	1.43192	1.11361	di: 157	di: 176	140; 141	di: 169	......		bis-4-Nitro-phenylurethane, 135.5
102	d,l-2-Nonanol	198.2	......	1.4290^{25}	0.81910_4^{25}		55.5, lt. pet.		42.8 (cor.)	......	42–4, d,l: 58–9	......
103	l-Linalool (l-Linalyl alcohol)	199	......	1.46238	0.8622	65–6	53	70	......			$[\alpha]_D$: −3 to −17
104	Benzyl alcohol	205.5	−15.3	1.53955	1.04540	77; 75.5–76, pet. eth.	134	85	113	176	106; 104	Pseudosaccharin ether, 130 (cor.)
105	d,l-1,3-Butanediol (d,l-1,3-Butylene glycol)	207.5; 204	......	$1.44252^{19.5}$	1.0053	122–3; d: 115–6	184					Diphenyl-urethane, l: 127–8
106	d,l-2-Decanol (Methyl n-octyl carbinol)	211	......	d: 1.4344	d: 0.8250		69, lt. pet.				48–9; d: 38–9	......
107	1-Nonanol (n-Nonyl alcohol)	213.5	......	1.43105	0.8271	60; 69; 62–4	65.5	60; 66	52.2	125 (cor.)	42.5	Pseudosaccharin ether, 49 (cor.)
108	1,3-Propanediol (Tri-methylene glycol)	214.7; 210–2	−30	1.43983	1.0538	di: 137	di: 164	di: 119	di: 178			Dibenzoate, 57; 59
109	3-Methylbenzyl alcohol (3-Tolyl carbinol)	217	......	......	0.9157^{17}	......	116					
110	α,4-Dimethylbenzyl alcohol (α-Methyl-4-tolyl carbinol)	219	......	......	$0.9668_4^{15.5}$	96, pet. eth.						
111	1-Phenyl-n-propyl alcohol (d,l-Phenylethyl carbinol)	219	......	1.5257	1.0056_{20}^{20}	......	102	59–60; 56.5–7.8				
112	2,3-Dibromo-1-propanol	219d.	......	......	......	84		59–60				3,5-Dinitro-phenylurethane, 71
113	2-Phenethyl alcohol (2-Phenylethanol)	219.8	−25.8	1.5240	1.0235_4^{25}	78; 79–80, al.	119	61.5–2 (cor.); 62–3, al.	108	123	188–9	......
114	d,l-α-Terpineol	221	35; d,l: 37–8	1.4834	0.9337	112–3, me. al.; d,l: 110	152; 147	139, me. al.	78–9, lgr.	......	117–8, ac. a.	Commercial liquid, lilac-like odor

*Derivative data given in order: m.p., crystal color, solvent from which crystallized.

TABLE VI. ORGANIC DERIVATIVES OF ALCOHOLS
a) Liquids (Listed in order of increasing atmospheric b.p.)* (Continued)

No.	Name	Boiling point, °C	Melting point, °C	n_D^{20}	D_4^{20}	Phenyl-urethane	1-Naph-thyl-urethane	4-Nitro-benzoate	3,5-Dinitro-benzoate	Hydrogen 3-nitro-phthalate	Hydro-gen phthalate	Miscellaneous
115	Citronellol	222; 118[17]										Oxid. → Adipic acid, 89; Rose-like odor
116	α-Isopropylbenzyl alcohol (d,l-Isopropyl phenyl carbinol)	222–4		1.51932[18.7]	0.9790_{20}^{20}		116–7					Oxid. → Iso-propylphenyl ketone, b.p.: 222
117	d,l-2-Undecanol (d,l-2-Hendecanol; Methyl n-nonyl carbinol)	228–9			0.8263[18]						49–50	
118	2-(2-n-Butoxyethoxy)-ethanol (Diethylene glycol mono-n-butyl ether)	228–30		1.4341	0.957_{20}^{20}							4-Nitrophenyl-urethane, 54.5–5.3
119	1,4-Butanediol (Tetra-methylene glycol)	230; 235	19.0–.5	1.4467	1.0171	di: 183–3.5, chl.; 180	di: 199, xyl.	di: 175, ac. a.				Dibenzoate, 81–2, eth.
120	Geraniol	230		1.4766	0.8894		47–8	35	62–3	117	47, lgr.	
121	1-Decanol (n-Decyl alcohol)	231	5.99; 6.4	1.43682	0.8292	59.6, bz., then al.	73	30.2, al.	57.7	122.8 (cor.)	38 (cor.)	Pseudosaccharin ether, 47.5 (cor.)
122	2-Phenoxyethanol (Ethylene glycol monophenyl ether)	237; 245		1.534	1.102[22]					112–3		Benzoate, 64; 4-Toluene-sulfonate, 80, al.
123	3-Phenylpropanol (Hydrocinnamyl alcohol)	237.4; 235		1.53565	1.0079	45; 47–8, al.		45–6; 47	92	117		4-Nitrophenyl-urethane, 104, pet. eth.
124	1,5-Pentanediol (Penta-methylene glycol)	238–9		1.4499	0.9939_{20}^{20}	di: 174–5 (cor.), abs. al.	di: 147	di: 104–5, bz.-al.				
125	1-Undecanol (1-Hende-canol; n-Undecyl alcohol)	243	15.85; 14.3			62, al.; 52		99.5, al.	55	123.3 (cor.)	43.8–4.1	Allophanate, 156; Pseudosaccharin ether, 58–5 (cor.)
126	Diethylene glycol (β,β'-Dihydroxydiethyl ether)	244.5	f.p.: −10.45	1.4475	1.1212_{15}^{15}		149	151 (cor.); 149, ac. a.				
127	2-Methoxybenzyl alcohol (Saligenin-2-methyl ether)	247		1.549[17]	1.0495_{15}^{15}		135–6					Allophanate, 180; Benzoate, 59, lgr.
128	2-Benzyloxyethanol (Ethylene glycol mono-benzyl ether)	265.0		1.5225	1.0700_{20}^{20}							Triphenylmethyl ether, 76–7, eth.
129	n-Hexyl phenyl carbinol	275		1.501	0.946	75						
130	Triethylene glycol (Ethylene glycol di-(β-hydroxyethyl)ether)	285; 165[14]	−9.4	1.4578[15]	1.1274_4^{15}							bis-Triphenyl-methyl ether, 142–2.5, acet.
131	Glycerol (1,2,3-Trihydroxypropane)	290d.	17.9	1.4729	1.26134	tri: 180	tri: 191–2, al.	tri: 188				4-Nitrophenyl-urethane, 216; Tribenzoate, 71–2; 75–6
132	3,4-Dimethoxybenzyl alcohol (Veratryl alcohol)	296–7[732]		1.555[17]	1.179_{17}^{17}	118						Acetate, b.p.: 170[12], n_D^{17}: 1.5245; Benzoate, 36–7

*Derivative data given in order: m.p., crystal color, solvent from which crystallized.

TABLE VI. ORGANIC DERIVATIVES OF ALCOHOLS
a) Liquids (Listed in order of increasing atmospheric b.p.)* (Continued)

No.	Name	Boiling point, °C	Melting point, °C	n_D^{20}	D_4^{20}	Phenyl-urethane	1-Naph-thyl-urethane	4-Nitro-benzoate	3,5-Dinitro-benzoate	Hydrogen 3-nitro-phthalate	Hydro-gen phthalate	Miscellaneous
133	**4-Methoxyphenyl methyl carbinol** (4-Anisyl methyl carbinol).......	310d. (cor.)	....	1.557	1.086_4^{16}	82–3	....	....	....	....	....	Odor of anise; Oxid. → 4-Methoxy-acetophenone, 38
134	*cis*-**Octa-9-decen-1-ol** (Oleyl alcohol; *cis*-Octadecenyl alcohol)...	333–5	....	1.4607	0.8489	oil	β: 44–5, al.	....	....	....	....	Allphanate, 135, chl.; 129, chl.; 4-Nitrophenyl-urethane, 85–91

*Derivative data given in order: m.p., crystal color, solvent from which crystallized.

TABLE VI. ORGANIC DERIVATIVES OF ALCOHOLS
b) Solids (Listed in order of increasing m.p.)*

No.	Name	Melting point, °C	Boiling point, °C	Phenyl-urethane	1-Naph-thyl-urethane	4-Nitro-benzoate	3,5-Di-nitro-benzoate	Hydrogen 3-nitro-phthalate	Hydrogen-phthalate	Pseudo-sac-charine ether	Miscellaneous
1	1-Phenylethyl alcohol (d,l-Methyl phenyl carbinol)	20; 20.1	202	92; 91–2, lgr.	106	43 (cor.), al.	95; 93	......	108; ac. a.	......	n_D^{20}: 1.5275; D_4^{20}: 1.0129
2	trans-2-Methylcyclohexanol	21	167.4	d,l: 105	d,l: 155	....	d,l: 115				Oxalate: d, l: 61
3	1-Dodecanol (n-Dodecyl alcohol; Lauryl alcohol).	24; 26	259	74	80	45; 42	60	124 (cor.)	50.3 (cor.)	54 (cor.)	....
4	1,1-Dimethyl-2-phenylethanol (Benzyl dimethyl carbinol)	24	216	....							n_D^{16}: 1.5174
5	4-Methoxybenzyl alcohol (4-Anisyl carbinol)	24–5	259	92 (cor.)	....	....	....				Benzoate, 38; Me. eth., b.p.: 225–6; n_D^{25}: 1.5422; D_{15}^{15}: 1.1129
6	Cyclohexanol	25.1	161.1	82	129	50	112–3, al.	160	99	......	n_D^{25}: 1.46477; D_4^{20}: 0.94155
7	tert-Butyl alcohol (Trimethyl carbinol)	25.5	82.5	136, eth.	101	116, al.	142, pet. eth.				n_D^{20}: 1.38779; D_4^{20}: 0.78670
8	3-Nitrobenzyl alcohol	27	175–80³	....							Benzoate, 71–2; Ox. → 3-nitro-benzoic acid, 140
9	2,3,3-Trimethylcyclohexanol	28	197	....							..
10	Diethanolamine (β,β'-Dihydroxy-diethylamine)	28	270; 217–8¹⁵⁰	....							Picrate, 109–10
11	2,4-Hexadien-1-ol	30.5–1.5	76¹²	78–9			85				
12	1-Tridecanol	α: 30.6; β: 28.3	155–6¹⁵			37.4 (cor.)		124 (cor.)	52.5	66 (cor.)	D_4^{31}: 0.8223.
13	Cinnamyl alcohol	33	257	90.0–1.5	114	78; 76.5	121				
14	2-Methylbenzyl alcohol (2-Tolyl carbinol)	36; 35	219	79 (cor.)	....						
15	trans-Octa-9-decen-1-ol (Elaidyl alcohol)	36.7; 35; 34	333; 216¹⁸	56–7	71	....					
16	d,l-Fenchyl alcohol	38–9; d,α: 45; l,α: 47; l,β: 3–4	201.5; d,α: 201–2	104; d,α: 82; l,α: 82	149	109; α: 108–9; β: 94–5; l,α: 109; l,β: 83	104	95	169; d,α: 145; l,α: 146; l,β: 153		
17	1-Tetradecanol (Myristyl alcohol)	39; 37.7	170–3²⁰	74; 71	82	51.2 (cor.)	67	123.5 (cor.)	60 (cor.)	62 (cor.)	....
18	1,2,2-Trimethylcyclohexanol	41 (+½ H_2O)	81.4–8²⁰	....							$n_D^{18.4}$: 1.469; $D_4^{18.4}$: 0.9274
19	Pinacol (Tetramethylethylene glycol)	43; 45–6	172	....							Diacetate, 65; Hydrate, 46
20	l-Menthol	44; 42.5; 35; 33; 31; d: 38–40; d,l: 34	216	111–2, al.; d,l: 103–4	119; 126	61–2	153	......	112; 129–31, ac. a.	......	Allophanate, 215
21	1-Pentadecanol	α: 44; β: 38.9	....	72, bz.	72	45.8 (cor.)	....	122.5 (cor.)	60.4	72 (cor.)	....
22	1,3,5-Trimethyl-1-cyclohexen-3-ol	46	87–90¹⁷	....							$n_D^{19.3}$: 1.4735; $D_4^{20.2}$: 0.9132
23	d,l-α-Propylbenzyl alcohol	d: 49; l: 49	168–70¹⁰⁰	....	99	58			91; 53–4		
24	1,1,1-Trichloroisopropanol	50–1	161.8⁷⁷³	....							Camphor-like odor
25	1-Hexadecanol (Cetyl alcohol) ...	50; 49.27	190¹⁸	73	82	58.4 (cor.); 52	66	122 (cor.)	66.8	69.5 (cor.)	....
26	2,2,6-Trimethylcyclohexanol	51, al.	186–7⁷⁵³	....							n_D^{20}: 1.4600; D_4^{20}: 0.9128

*Derivative data given in order: m.p., crystal color, solvent from which crystallized.

TABLE VI. ORGANIC DERIVATIVES OF ALCOHOLS
b) Solids (Listed in order of increasing m.p.)* (Continued)

No.	Name	Melting point, °C	Boiling point, °C	Phenyl-urethane	1-Naph-thyl-urethane	4-Nitro-benzoate	3,5-Di-nitro-benzoate	Hydrogen 3-nitro-phthalate	Hydrogen-phthalate	Pseudo-sac-charine ether	Miscellaneous
27	3,3,5-Trimethylcyclohexanol	trans: 52; 37	cis: 201–3[750]; trans: 196.5[770]	....						......	Acetate, cis: b.p.: 209–10; trans: b.p.: 209–10
28	2,2-Dimethyl-1-propanol (tert-Butyl carbinol; Neopentyl alcohol)	52–3	113	144, lgr.	100	....			71		....
29	4-Methylbenzhydrol (Phenyl 4-tolyl carbinol)	53; 58; 42	....	....							Ox. → Phenyl 4-tolyl ketone, 60
30	1-Heptadecanol	α: 54	310	....	88.5	53.8 (cor.)	121.5 (cor.)	121.0–.8	66.6–.7 (cor.)	76 (cor.)	....
31	Piperonyl alcohol	58	....	102.5	....	....					Benzoate, 66; Allo-phanate, 176.5
32	4-Methylbenzyl alcohol (4-Tolyl carbinol)	59–60	217	79	....	....	117–8				
33	1-Octadecanol (Stearyl alcohol)	59.5	210.5[15]	79–80	....	64.3 (cor.)	66	119	72.5 (cor.)	74.5 (cor.)	....
34	1-Nonadecanol	62	....	....		58.9	....		71	80.5	
35	Eicosanol	65	220[3]	....		69.4			77	......	Acetate, 40
36	1-(1-Naphthyl) ethanol (d,l-Methyl 1-naphthyl carbinol)	66	....						131–2, bz.		
37	1,2-Diphenylethanol (d,l-Benzyl phenyl carbinol)	67	167[10]	....					131 (cor.), eth.-lt. pet.	......	
38	4,4'-Dimethylbenzhydrol (Di-4-tolyl carbinol)	68, al.	....	....						......	Carbinyl bromide, 48.5–9, lgr.
39	Benzhydrol (Diphenyl carbinol)	68, lgr.	288; 180[20]	139–40, bz.	135–6; 139	131–2	141	......	164–5	......	Acetate, 41–2; Benzoate. 88–9, al.
40	1-Glyceryl phenyl ether	69, eth.	....								Conc. H_2SO_4 → pa. red
41	Erythritol (d,l-1,2,3,4-Tetra-hydroxybutane)	72	....								Tetraacetate, 53
42	Dihydroxyacetone (1,3-Di-hydroxy-2-propanone)	72	....								Diacetate, 48; Di-benzoate, 120.5; 2,4-Dinitrophenyl-hydrazone, 277–8
43	2-Nitrobenzyl alcohol	74	270; 165[20]	....							Benzoate, 101–2
44	1,10-Decanediol (Decamethylene glycol)	75.5; 72	....								1,10-Dibromide, 27.4; b.p.: 162–5.5[10]
45	2,2,2-Tribromoethanol	80	92–3[10]	....							Urethane, 86–7
46	10-Nonadecanol (Myricyl alcohol)	85	....	96	54						....
47	Phenacyl alcohol (α-Hydroxy-acetophenone; Benzoyl carbinol)	86	118–20[11]	....		128.6					Benzoate, 118.5; 3-Nitrobenzoate, 104.5
48	n-Triacontanol (1-Hydroxytria-contane)	86.5, bz.	....								Acetate, 69, pet. eth.
49	2-Hydroxybenzyl alcohol (Saligenin)	86–7	....								2-Benzoate, 66
50	d-Sorbitol	89–93; 112 (anh.)	....								Hexaacetate, 99; Hexabenzoate, 216–7, et. ac.
51	3-Nitrophenacyl alcohol (3-Nitro-benzoyl carbinol)	92.5–3.0, pa. yel.	....								Acetate, 53, eth.-lgr.; Semicarba-zone, 214, al.

* Derivative data given in order: m.p., crystal color, solvent from which crystallized.

No.	Name	Melting point, °C	Boiling point, °C	Phenyl-urethane	1-Naph-thyl-urethane	4-Nitro-benzoate	3,5-Di-nitro-benzoate	Hydrogen 3-nitro-phthal-ate	Hydrogen-phthalate	Pseudo-sac-charine ether	Miscellaneous
52	**4-Nitrobenzyl alcohol**	93	185[12]	. . . . . .	. . . . . .	. . . . . .	. . . . . .	. . . . . .	. . . . . .	. . . . .	Acetate, 78; Benzoate, 94–5
53	**4,4′,4″-Trimethyltriphenyl carbinol** (Tris-p-tolyl carbinol)	96	. . . .	. . . . . .	. . . . . .	. . . . . .	. . . . . .	. . . . . .	. . . . . .	. . . . .	Et. eth., 111, pet. eth.; $H_2SO_4 \rightarrow$ gr.-red
54	*meso*-**Erythritol**	121; 120 (cor.)	330	. . . . . .	. . . . . .	. . . . . .	. . . . . .	. . . . . .	. . . . . .	. . . . .	Tetraacetate, 85; 89; Dibenzylidene, 201–2
55	**4-Nitrophenacyl alcohol** (4-Nitro-benzoyl carbinol)	121		. . . . . .	. . . . . .	. . . . . .	. . . . . .	. . . . . .	. . . . . .	. . . . .	Acetate, 124, et. ac.-lgr.; Phenyl-hydrazone, 178, bz.
56	*d,l*-**Benzoin** (Benzoyl phenyl carbinol)	137; 133	344	165	140	123	. . . . . .	. . . . . .	. . . . . .	. . . . .	4-Nitrophenyl-urethane, 183; Acetate, 83
57	**Cinchol** (β-Sitosterol)	137	. . . .	. . . . . .	. . . . . .	. . . . . .	202–4	. . . . . .	. . . . . .	. . . . .	Acetate, 134; Benzoate, 145
58	**Furoin** (Furoyl furyl carbinol) . . .	138–9 (cor.); 135	. . . .	. . . . . .	. . . . . .	. . . . . .	. . . . . .	. . . . . .	. . . . . .	. . . . .	Acetate, 76–7; Benzoate, 92–3
59	**Cholesterol** (anh.) (*l*-Cholesterol) .	148.5	360d.	168	176	185; 190–3	. . . . . .	. . . . . .	161	. . . . .	4-Nitrophenyl-urethane, 205; Benzoate, 151–2
60	**Triphenylmethanol** (Triphenyl carbinol)	161–2, bz.	380	. . . . . .	. . . . . .	. . . . . .	. . . . . .	. . . . . .	. . . . . .	. . . . .	Acetate, 87–8
61	**Ergosterol**	165	. . . .	185	202	. . . . . .	202, chl.	. . . . . .	. . . . . .	. . . . .	Acetate, 176; 180, eth.; Benzoate, 168
62	*d*-**Mannitol**	166; *l*: 163–4; *d,l-α*: 168	. . . .	303	. . . . . .	. . . . . .	. . . . . .	. . . . . .	. . . . . .	. . . . .	Hexaacetate, 126, eth.; Benzoate, 149–50; 147–8, al.
63	**1,1,1-Tribromo-*tert*-butyl alcohol** (Brometone)	167–76, w.-al.		. . . . . .	. . . . . .	. . . . . .	. . . . . .	. . . . . .	. . . . . .	. . . . .	Acetate, 43.4, al.; Benzoate, 27, al.
64	**Dulcitol** (1,2,3,4,5,6-Hexanehexol)	188.5	. . . .	. . . . . .	. . . . . .	. . . . . .	. . . . . .	. . . . . .	. . . . . .	. . . . .	Hexaacetate, 171; 168–9, abs. al.; Hexabenzoate, 189–91, eth.-chl.
65	**4,4′,4″-Triaminotriphenyl carbinol** (Pararosaniline base)	205		. . . . . .	. . . . . .	. . . . . .	. . . . . .	. . . . . .	. . . . . .	. . . . .	4,4′,4″-Triacetyl, 192, acet.-eth.; Me. eth., 105, eth.; 135, bz.
66	*d*-**Borneol** ("Borneo camphor") . .	208; 205; *d,l*: 210.5	212	138–9	132; *iso*: 130; 127	153; *d,l*: 134; 137	154–5	. . . . . .	161.4, ac. a.; 165 (cor.)	. . . . .	. . . .
67	*meso*-**Inositol** (1,2,3,4,5,6-Hexa-hydroxycyclohexane)	225 (cor.); 218	. . . .	. . . . . .	. . . . . .	. . . . . .	86, al.	. . . . . .	. . . . . .	. . . . .	Hexaacetate, 212, subl., tol.; Hexa-benzoate, 258, al.
68	*d*-**Quercitol** (Pentahydrocyclo-hexane)	232; 234	. . . .	. . . . . .	. . . . . .	. . . . . .	. . . . . .	. . . . . .	. . . . . .	. . . . .	Pentabenzoate, 155
69	**Pentaerythritol** (2,2-Bishydroxy-methyl-1,3-propanediol)	262; 253	. . . .	. . . . . .	. . . . . .	. . . . . .	. . . . . .	. . . . . .	. . . . . .	. . . . .	Tetraacetate, 84, wh., al.; Tetra-benzoate, 99–101, al.

*Derivative data given in order: m.p., crystal color, solvent from which crystallized.

EXPLANATIONS AND REFERENCES TO TABLE VII

Phenylurethane.

$$ArOH \ + \ C_6H_5N{=}C{=}O \ \rightarrow \ C_6H_5NHCOOAr$$

Phenylurethane
(Phenylcarbamate)

From the phenol with phenylisocyanate without solvent.

For directions and examples see: Linstead, pp. 34–35; J. B. McKinley, J. E. Nickels and S. S. Sidue, *Ind. Eng. Chem., Anal. Ed.*, **16**, 304 (1944).

From the phenol with phenylisocyanate in kerosene.

See: Cheronis, p. 489.

From the phenol with phenylisocyanate and a catalytic amount of pyridine.

See: Shriner, p. 265.

From the phenol with phenylisocyanate in toluene.

See: O. L. Brady and J. Harris, *J. Chem. Soc.*, **127**, 2175 (1925).

*1-Naphthylurethane (α-Naphthylcarbamate).**

$$ArOH \ + \ 1\text{-}C_{10}H_7N{=}C{=}O \ \rightarrow \ 1\text{-}C_{10}H_7NHCOOAr$$

1-Naphthylurethane
(α-Naphthylcarbamate)

From the phenol with 1-naphthylisocyanate without solvent.

For directions and examples see: Cheronis, p. 488; Linstead, pp. 34–35; Vogel, p. 683; Wild, p. 68.

From the phenol with 1-naphthylisocyanate and a catalytic amount of pyridine, triethylamine or trimethylamine in ether.

See: Shriner, p. 211; Vogel, p. 683; Wild, p. 68; H. E. French and A. F. Wirtel, *J. Amer. Chem. Soc.*, **48**, 1736 (1926).

From the phenol with 1-naphthylisocyanate and a catalytic amount of a tertiary aliphatic amine in petrol ether.

See: Cheronis, p. 488; Vogel, p. 684.

*p-Nitrobenzoate.**

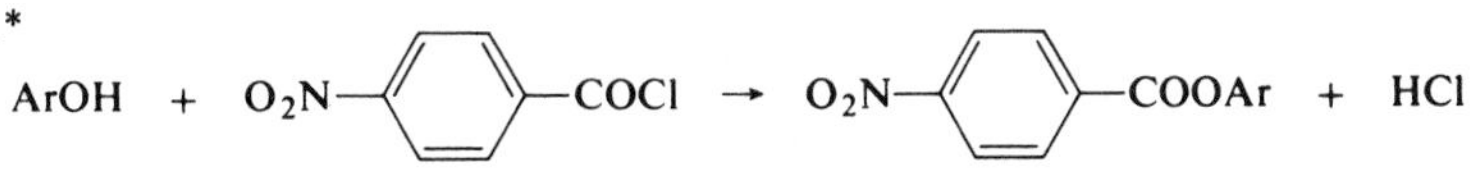

$$ArOH \ + \ O_2N{-}C_6H_4{-}COCl \ \rightarrow \ O_2N{-}C_6H_4{-}COOAr \ + \ HCl$$

p-Nitrobenzoate

From the phenol with *p*-nitrobenzoyl chloride in pyridine.

For directions and examples see: Vogel, p. 682; Wild, p. 52.

From the phenol with *p*-nitrobenzoyl chloride without solvent.

See: Wild, p. 52.

*3,5-Dinitrobenzoate.**

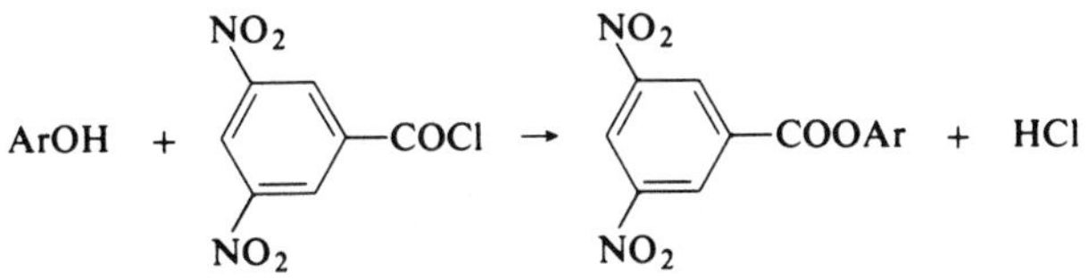

$$ArOH \ + \ (NO_2)_2C_6H_3{-}COCl \ \rightarrow \ (NO_2)_2C_6H_3{-}COOAr \ + \ HCl$$

3,5-Dinitrobenzoate

From the phenol with 3,5-dinitrobenzoyl chloride in pyridine.

For directions and examples see: Cheronis, p. 486; Vogel, p. 682; Wild, p. 65; R. C. Brown and R. E. Kremers, *J. Amer. Pharm. Ass.*, **11**, 607 (1922); M. Phillips and G. L. Keenan, *J. Amer. Chem. Soc.*, **53**, 1924 (1931).

From the phenol with 3,5-dinitrobenzoyl chloride without solvent.

See: Shriner, pp. 212–213.

Benzoate.

$$ArOH \ + \ C_6H_5COCl \ \rightarrow \ C_6H_5COOAr \ + \ HCl$$

Benzoate

From the phenol with benzoyl chloride in aqueous sodium hydroxide.

For directions and examples see: Cheronis, pp. 481–482, 487; Linstead, p. 20; Wild, p. 65.

*Derivatives recommended for first trial.

WARNING: This is not an instruction manual. References should be consulted for the preparation of derivatives.

From the phenol with benzoyl chloride in pyridine.
See: Linstead, p. 19.

Acetate.

$$ArOH \;+\; (CH_3CO)_2O \;\rightarrow\; CH_3COOAr \;+\; CH_3COOH$$

Acetate

From the phenol with acetic anhydride in aqueous sodium hydroxide.
For directions and examples see: Linstead, p. 21; Vogel, p. 682; Wild, p. 64; F. D. Chattaway, *J. Chem. Soc.*, 2495 (1931).

From the phenol with acetic anhydride and sodium acetate.
See: Linstead, p. 21; Wild, p. 64.

From the phenol with acetic anhydride and a catalytic amount of sulfuric acid.
See: Cheronis, p. 487.

p-Phenylazobenzoate.

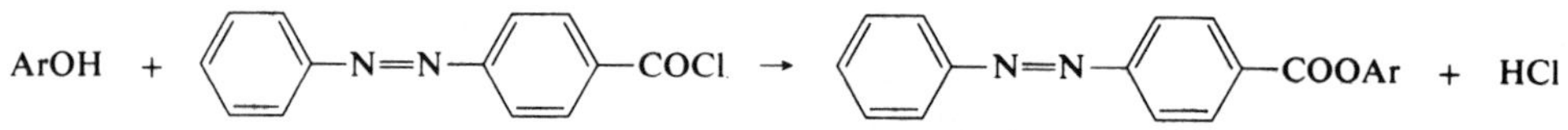

$$ArOH \;+\; \langle \rangle-N=N-\langle \rangle-COCl \;\rightarrow\; \langle \rangle-N=N-\langle \rangle-COOAr \;+\; HCl$$

p-Phenylazobenzoate

From the phenol with *p*-phenylazobenzoyl chloride in pyridine.
For directions and examples see: Cheronis, p. 486; E. O. Woolfolk and J. M. Taylor, *J. Org. Chem.*, **22**, 827 (1957).

p-Toluenesulfonate.

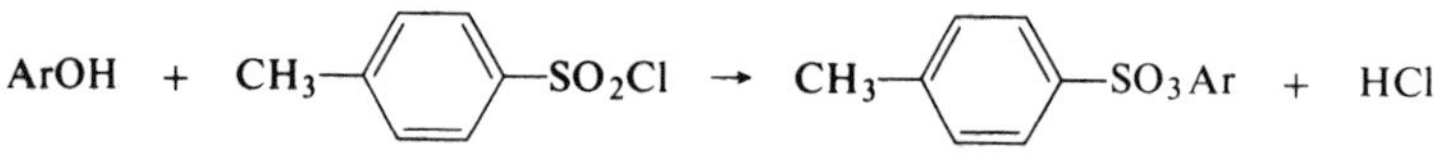

$$ArOH \;+\; CH_3-\langle \rangle-SO_2Cl \;\rightarrow\; CH_3-\langle \rangle-SO_3Ar \;+\; HCl$$

p-Toluenesulfonate

From the phenol with *p*-toluenesulfonyl chloride in pyridine.
For directions and examples see: Linstead, p. 20; Vogel, p. 684; Wild, p. 66.

From the phenol with *p*-toluenesulfonyl chloride in aqueous sodium hydroxide.
See: Linstead, p. 20.

From the phenol with *p*-toluenesulfonyl chloride and sodium hydroxide in aqueous acetone.
See: Wild, p. 66.

Bromo derivative. *

$$ArOH \;+\; Br_2 \;\rightarrow\; Ar(OH)Br \;+\; HBr$$

Mono-
bromophenol

$$ArOH \;+\; nBr_2 \;\rightarrow\; Ar(OH)Br_n \;+\; nHBr$$

Poly-
bromophenol

From the phenol in aqueous methanol, in ethanol, in acetone or in dioxane with bromine in aqueous potassium bromide.
For directions and examples see: Cheronis, p. 490; Shriner, p. 264.

From the phenol in aqueous hydrochloric acid with bromine in water.
See: Linstead, p. 19.

From the phenol in glacial acetic acid and bromine.
See: Wild, p. 73.

From the phenol with bromine in carbon disulfide.
See: Vogel, p. 679.

For a discussion on the effect of solvents in the bromination of phenols.
See: N. D. Cheronis, *Micro and Semimicro Methods (Technique of Organic Chemistry)*, Vol. 6, Interscience, New York, 1954, pp. 286–287.

*Derivatives recommended for first trial.
WARNING: This is not an instruction manual. References should be consulted for the preparation of derivatives.

*Aryloxyacetic acid.**

$$ArONa \; + \; ClCH_2COOH \; \rightarrow \; ArOCH_2COOH \; + \; NaCl$$
Aryloxyacetic acid

From the phenol in aqueous sodium hydroxide with aqueous chloroacetic acid.

For directions and examples see: Cheronis, pp. 489–490; Linstead, p. 20; Shriner, p. 264; Vogel, p. 683; Wild, p. 72; C. F. Koelsch, *J. Amer. Chem. Soc.*, **53**, 304 (1931); N. V. Hayes and G. E. K. Branch, *J. Amer. Chem. Soc.*, **65**, 1555 (1943).

Aryl 2,4-dinitrophenyl ether.

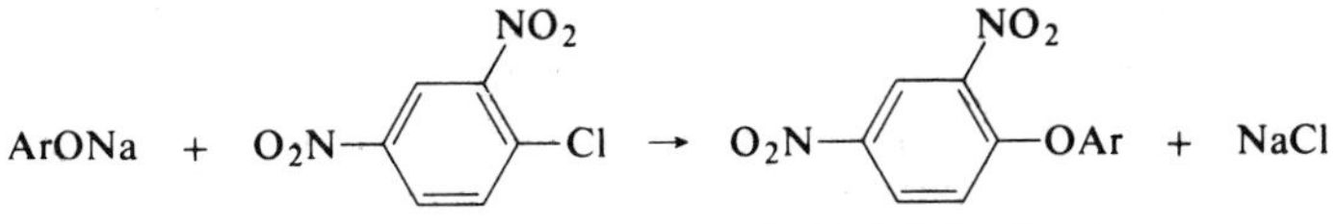

Aryl 2,4-dinitrophenyl ether

From the phenol in aqueous sodium hydroxide with 2,4-dinitrochlorobenzene in alcohol.

For directions and examples see: Cheronis, pp. 490–491; Vogel, p. 684; Wild, p. 71; R. W. Bost and F. Nicholson, *J. Amer. Chem. Soc.*, **57**, 2368 (1935).

From the phenol with 2,4-dinitrochlorobenzene and aqueous potassium hydroxide.

See: Linstead, pp. 20–21.

NOTE: For additional information regarding directions and examples for the preparation of derivatives of phenols which are similar to those of alcohols (e.g., 1-naphthylurethanes, 3,5-dinitrobenzoates, etc.) see explanations and references to Table VI, p. 77, 78, 79.

*Derivatives recommended for first trial.

WARNING: This is not an instruction manual. References should be consulted for the preparation of derivatives.

TABLE VII. ORGANIC DERIVATIVES OF PHENOLS
a) Liquids. (Listed in order of increasing atmospheric b.p.)*

No.	Name	Boiling point, °C	Melting point, °C	n_D^{20}	D_4^{20}	Phenyl-urethane	α-Naph-thyl-urethane	p-Nitro-benzoate	3,5-Di-nitro-benzoate	Bromo derivative	p-Toluene sulfonate	Miscellaneous
1	2-Chlorophenol	175.6	7	1.5473^{40}	1.2410^{18}_{15}	121	120	115	143	mono: 48–9; di: 76	74	Aryloxyacetic acid, 145; 2,4-Dinitrophenyl ether, 99; 4,6-Bis(dimethylaminomethyl) deriv., 62–3; p-Phenylazobenzoate, 120–1
2	2-Bromophenol	195	5	......	1.4924	......	129	......	......	95	78	p-Phenylazobenzoate, 126–7
3	2-Chloro-4-methylphenol (2-Chloro-p-cresol)	195–6	.....	1.5200^{27}	1.1785^{27}							Benzoate, 71–2; Aryloxyacetic acid, 108; Acetate, b.p. 238
4	2-Hydroxybenzaldehyde (Salicylaldehyde)	197 (cor.)	1.6	1.574	1.1690^{20}_{20}	133	......	128			63–4	Aryloxyacetic acid, 132; Acetate, 39
5	3-Methylphenol (m-Cresol)	203	12	1.540	1.03401	125; 121–2, al.-lgr.	127–8	90	165.4 (cor.), al.	tri: 84	51	Aryloxyacetic acid, 103; 2,4-Dinitrophenyl ether, 74; Benzoate, 55
6	2-Ethylphenol	207	.....	......	1.0371^{0}	143.4; 141	......	56–7	108			Aryloxyacetic acid, 141; Benzoate, 38–9, al.
7	2-Isopropylphenol	212	16	1.5315	1.012	......						Aryloxyacetic acid, 132–3; Methyl urethane, 96–7; 4,6-Dinitro deriv., 53
8	2-Bromo-4-ethylphenol	213–4	.....	......	......						121	
9	3-Ethylphenol	217	–4	......	1.0250^{0}	137; 138.8	......	68				Aryloxyacetic acid, 77; Benzoate, 52, 95% al.
10	2-Allylphenol	220	–6	1.5181	1.0255^{15}_{15}	116; 106.0–6.5	......			6-mono: 50	......	Aryloxyacetic acid, 148.5–50
11	2-Chloro-4,6-dimethylphenol	221–3	.....	......		129–30	......	94–5				....
12	Methyl salicylate (Methyl 2-hydroxybenzoate)	224	–8	1.5369	1.184	117, bz.	......	128				Acetate, 52; Benzoate, 92, al.
13	2-Propylphenol	224.6–6.6; 220.0–0.5	.....	1.5280	1.000^{15}_{15}	111, formic a.	......	......	96			Aryloxyacetic acid, 99–100
14	2-sec-Butylphenol	227–8; 116^{21}	12–3	1.5288	0.9876	86, lgr.	......					Aryloxyacetic acid, 109.5–11'0
15	4-Allylphenol	230–1^{750}	16	1.5441^{18}	1.033^{18}	......			103.0–3.5		63–4	Urethane, 122–3; Acetate, b.p. 238–9
16	Ethyl salicylate (Ethyl 2-hydroxybenzoate)	234	1.3	1.5226	1.131	98–100	......	107–8, yel., bz.				Benzoate, 79–80; 87, al.; 3,5-Dinitro deriv., 92–3
17	2-n-Butylphenol	234–7; 113–5^{14}	....	$1.5180^{25.5}$	0.975				97			Aryloxyacetic acid, 104–5; Acetate, 105.5
18	4-Isobutylphenol	236; 235–9	.....	1.5319^{25}	0.9796^{20}_{20}	......						Aryloxyacetic acid, 124–5
19	Carvacrol (5-Isopropyl-2-methylphenol)	237.8	1	1.524	0.9760	134–5; 138	116, lgr.	51	83; 76–7	46	......	Aryloxyacetic acid, 151; p-Xenylurethane, 116, al.
20	Isopropyl salicylate (Isopropyl 2-hydroxybenzoate)	240–2	.....	1.50650	1.0729	......						3,5-Dinitro deriv., 101–2, al.

*Derivative data given in order: m.p., crystal color, solvent from which crystallized.

TABLE VII. ORGANIC DERIVATIVES OF PHENOLS
a) Liquids. (Listed in order of increasing atmospheric b.p.)* (Continued)

No.	Name	Boiling point, °C	Melting point, °C	n_D^{20}	D_4^{20}	Phenyl-urethane	α-Naph-thyl-urethane	p-Nitro-benzoate	3,5-Di-nitro-benzoate	Bromo derivative	p-Toluene sulfonate	Miscellaneous
21	3-Methoxyphenol (Resorcinol monomethyl ether)	243; 244	−17.5	......	....	....	128−9	......	....	tri: 104	......	Aryloxyacetic acid, 118; 111−3, w.; 2,4-Dinitrophenyl ether, 87−8
22	2-Allyl-6-methoxyphenol	250−1; 115[9]	.....	1.5393	....	....	....	96−7	....			....
23	4-Allyl-2-methoxyphenol (Eugenol)	254.8; 127[15]	−9.1; 16−9	1.5410	1.0664	95.5	122, lgr.	81	130.8 (cor.), al.	tetra: 118	85	Acetate, 29, al.; Benzoate, 70, al.; Aryloxyacetic acid, 81; 100; 2,4-Dinitrophenyl ether, 115; N,N-Diphenyl-urethane, 108
24	Isobutyl salicylate (Isobutyl 2-hydroxybenzoate)..	260−2	.....	1.50872	1.0639	....	....	....	....		....	3,5-Dinitro deriv., 72−3, al.
25	d,l-1,2,3,4-Tetrahydro-2-naphthol	264[716]	.....	1.5523[17]	1.0715[17]	99	....				....	Acetate, b.p. 169[34]; Benzoate, b.p. 254−5[40]
26	2-Methoxy-4-propenylphenol (Isoeugenol)	267.5	.....	cis: 1.5700; trans: 1.5782	cis: 1.0851; trans: 1.0852	cis: 118; trans: 152	149−50, lgr.	109	154−8 (cor.), n-BuOH	....		Acetate, 79−80; bz.-lgr.; Benzoate, cis, 68; trans, 103−4, al.; Aryloxyacetic acid, 116 (94); 2,4-Dinitrophenyl ether, 128
27	n-Butyl salicylate (n-Butyl 2-hydroxybenzoate)	270−2; 259−60	.−5.9	1.51148	1.0728	....	....				....	3,5-Dinitro deriv., 60−1, al.; p-Nitrobenzyl ether, 92
28	Isoamyl salicylate (Isopentyl salicylate)	276−8	.....	1.50799	1.0535	....					....	3,5-Dinitro deriv., 61−2, al.
29	3-Acetoxyphenol (Resorcinol monoacetate)	283	.....	1.5328	....	....						Saponification → resorcinol, 110 + ac. a.; Diacetate, b.p. 130−1[7]

*Derivative data given in order: m.p., crystal color, solvent from which crystallized.

TABLE VII. ORGANIC DERIVATIVES OF PHENOLS
b) Solids (Listed in order of increasing m.p.)*

No.	Name	Melting point, °C	Boiling point, °C	Phenyl ure-thane	α-Naph-thyl-ure-thane	p-Nitro-ben-zoate	3,5-Di-nitro-benzoate	Bromo derivative	p-Toluene sul-fonate	Acetate	Benzoate	Miscellaneous
1	**2-Benzylphenol (2-Hy-droxydiphenylmethane) (Labile form)**	21; stable: 52; 54	312	117–8	......	......	......				......	Me. eth., 30; Ure-thane, 110–1
2	**4-n-Propylphenol**	22	232	129	......		123			b.p. 245–6[745]	38	$n_D^{2?}$: 1.5220
3	**4-n-Butylphenol**	22	248; 138–9[18]	115, al.	......	67–8, yel., al.	92			b.p. 138–41[15]	27	$n_D^{25.5}$: 1.5165; D_4^{20}: 0.978; Aryloxy-acetic acid, 81
4	**4-n-Amylphenol (4-n-Pentylphenol)**	23	248–53							51.0–0.5, al.		n_D^{23}: 1.5272; D_{20}^{20}: 0.9621; Aryloxy-acetic acid, 90
5	**3-n-Propylphenol**	26	228	......		75 (117–8)					......	n_D^{20}: 1.5223; D_4^{20}: 0.9887; Aryloxy-acetic acid, 96–7, lgr.
6	**2-Chloro-3,4-dimethyl-phenol**	27, pet. eth.	......	......	......	......	......			87, aq. al.		....
7	**2,4-Dichloro-3-methyl-phenol (2,4-Dichloro-m-cresol)**	27	240.5–2.5	......					92.0–2.5, al.		90.5	Benzenesulfonate, 70, al.
8	**2,4-Dimethylphenol (m-4-Xylenol)**	27.8	211.5 (cor.)	103; 112, CCl₄	135	102; 105	164.6 (cor.), 95% al.	......			37–8, ac. a.	n_D^{14}: 1.5420; D_4^{14}: 1.0276; p-Xenyl-urethane, 184; Aryloxyacetic acid, 141; p-Phenylazo-benzoate, 110–3; 2,4-Dinitrophenyl ether, 102–3
9	**2-Ethoxyphenol**	28	217	......							31	Allophanate, 212
10	**2-Acetylphenol (2-Hydroxyacetophenone)**..	28	215	......						89, al.	87–8, al.	n_D^{20}: 1.5590; D_4^{20}: 1.131; Semicarba-zone, 210; Oxime, 118; Phenylhydra-zone, 110
11	**2-Methylphenol (o-Cresol)**	31	191–2	141; 143	141–2, lgr.	94	138.4 (cor.), al.	di: 56	54–5, pyr.		......	Aryloxyacetic acid, 152; p-Phenylazo-benzoate, 110–11.5; 2,4-Dinitro-phenyl ether, 90; N,N-Diphenylure-thane, 73
12	**2-Methoxyphenol (Guaiacol)**	32; 28.2	205	136, al.	118	93	141.2 (cor.), al.	4,5,6-tri: 116, al.	85, lgr.	......	57–8	n_D^{20}: 1.5441; $D_{(vac)}^{20.4}$: 1.1287; Aryloxy-acetic acid, 116; 2,4-Dinitrophenyl ether, 97
13	**5-Fluoro-2-nitrophenol**....	32, lgr.	....	......						......	110–11	Me. eth., 52, lgr.
14	**3-Bromophenol**	33	236	......	108	......			52.4	b.p. 149[40]	86	Aryloxyacetic acid, 108; p-Phenylazo-benzoate, 125–6
15	**3-Chlorophenol**	33	214	......	158	99	156				71	Aryloxyacetic acid, 110; 2,4-Dinitro-phenyl ether, 75; p-Phenylazoben-zoate, 127–8
16	**2-Bromo-4-chlorophenol** ..	33–4	123[10]	......	......	......	......		......		99–100	Aryloxyacetic acid, 139–40

*Derivative data given in order: m.p., crystal color, solvent from which crystallized.

TABLE VII. ORGANIC DERIVATIVES OF PHENOLS
b) Solids (Listed in order of increasing m.p.)* (Continued)

No.	Name	Melting point, °C	Boiling point, °C	Phenyl ure-thane	α-Naph-thyl-ure-thane	p-Nitro-ben-zoate	3,5-Di nitro-benzoate	Bromo derivative	p-Toluene sul-fonate	Acetate	Benzoate	Miscellaneous
17	**4-Methylphenol (p-Cresol)**	36	202	115	146	98	188.6 (cor.), al.	di: 49; tetra: 198–9, al.	69–70, al.	......	70	p-Phenylazobenzoate, 134.5–6.5; N,N-Diphenylurethane, 94; p-Xenylurethane, 198
18	**2,4-Dibromophenol**	36; 40	238–9	......	......	183.5	....	6-mono: 95–6	120	36	97.5	Aryloxyacetic acid, 153; 2,4-Dinitrophenyl ether, 135; Me. eth., 61.3; b.p. 272; Et. eth., 53.5
19	**4-Methyl-2-nitrophenol** (2-Nitro-p-cresol)	36.5, yel., w.-al.	125^{22}	......	......	......	192	....			100–1	Me. eth., 8.5, pa. yel., b.p. 274; Et. eth., b.p. 275–85 d.
20	**4-Chlorophenol**	37; 43	217	148.5	166	171	186	2-mono: 33–4; 2,6-di: 90	71	7–8	88	Aryloxyacetic acid, 156; 2,4-Dinitrophenyl ether, 126; p-Phenylazobenzoate, 153–4; N,N-Diphenylurethane, 97
21	**2,4-Diethylphenol**	37–8	219	170–1	......	......	......				......	Aryloxyacetic acid, 67–8
22	**3-Fluoro-2-nitrophenol**	39, lgr.	....	......	......	......	......				114	Me. eth., 43.5, lgr.
23	**2,6-Dichloro-4-methyl-phenol** (2,6-Dichloro-p-cresol)	39; 42	120–5^{14}	......					48		9	NH$_4$ salt, 125; Me. eth., b.p. 234
24	**3-Ethoxyphenol** (Resorcinol monoethyl ether).............	40	246–7, pa. yel.; 254–8	......		183.5					97–5	2,4-Dinitrophenyl ether, 114–5; Picrate, 105–6, red, chl.
25	**3-Iodophenol**	40		138	......	133	183	......	60–1	38, pet. eth.	72–3, pet. eth.	Aryloxyacetic acid, 115
26	**2-Benzoylphenol** (2-Hydroxybenzophenone)	41		......	......	124	....				......	p-Nitrobenzyl ether, 124–5, acet.; Phenylhydrazone, 155; Semicarbazone, 250–1; Oxime, 141–3
27	**3-Methyl-2-nitrophenol** (2-Nitro-m-cresol)	41, yel., pet. eth.	....							59, al.	79, al.	Me. eth., 54, yel., al.
28	**Phenol**	41.8; 42	182; 183	126, bz.	132–3, lgr.	127, bz.	145.8 (cor.), al.	tri: 95	95–6, al.	......	69	Aryloxybenzoic acid, 99; p-Phenylazobenzoate, 148–50; p-Xenylurethane, 173; N,N-Diphenylurethane, 105
29	**3-Fluoro-4-nitrophenol**	42, w. or lgr.	173^{12}	......	......	......	......			......	118	Me. eth., 56.5
30	**Phenyl salicylate** (Salol; Phenyl 2-hydroxy-benzoate)	42; 38.8; 28.5 (three forms)	....	111–2, bz., 242	......	111	....			99.5	81	N,N-Diphenyl-urethane, 144

* Derivative data given in order: m.p., crystal color, solvent from which crystallized.

No.	Name	Melting point, °C	Boiling point, °C	Phenyl urethane	α-Naphthyl urethane	p-Nitrobenzoate	3,5-Dinitrobenzoate	Bromo derivative	p-Toluene sulfonate	Acetate	Benzoate	Miscellaneous
31	**2-Iodophenol**	43	186–7[160]	122	......	......	......	......	......	98–101	34, pet. eth.	D^{80}: 1.8757; Aryloxyacetic acid, 135; p-Phenylazobenzoate, 126–8; 2,4-Dinitrophenyl ether, 95
32	**5-Bromo-2-nitrophenol**	44, yel.	......							74.5	......	Me. eth., 85.5; Et. eth., 79.5–80.5, al.
33	**2,4-Dichlorophenol**	45	210	......	......	......	142–3	68	125	......	97	Aryloxyacetic acid, 141, 135; β-Naphthylurethane, 166; 2,4-Dinitrophenyl ether, 119
34	**2-Nitrophenol**	45, yel., al.	216	......	113	141	155	di: 117	83	40–1, lgr.	59	D_4^{40}: 1.2942; Aryloxyacetic acid, 158; 2,4-Dinitrophenyl ether, 142; N,N-Diphenylurethane, 114; p-Phenylazobenzoate, 136.5–7.0
35	**5-Chloro-2-hydroxybiphenyl**	46		......		......					88	....
36	**2-Chloro-5-methylphenol** (6-Chloro-m-cresol)	46	196						96	b.p. 122–3[11]	31; 40	Benzenesulfonate, 99
37	**4-Bromo-5-isopropyl-2-methylphenol**	46, lgr.				53–4					......	Me. eth., b.p. 147–50[15]
38	**2-tert-Butyl-5-methylphenol** (6-tert-Butyl-m-cresol)	46–7	127[11]	......						b.p. 139[17]	......	Me. eth., 22, b.p. 225–7
39	**3-Chlorocatechol** (3-Chloro-1,2-dihydroxybenzene)	46–8; 48–50	110–11[11]	......							110–11	2-Me. eth., 31.5–3.0
40	**4-Ethylphenol** (4-Hydroxyethylbenzene)	47	219	120	128	80–1	132–3	......			59–60, al.	n_D^{25}: 1.5239; D_{20}^{20}: 1.0123; Aryloxyacetic acid, 97; p-Phenylazobenzoate, 117–8
41	**3-Methyl-2,4,6-trichlorophenol** (2,4,6-Trichloro-m-cresol)	47, w.	265	......				92–3		35; 32, eth.	53	....
42	**4-Chloro-2-methylphenol** (4-Chloro-o-cresol)	48	222–5							......	71	Aryloxyacetic acid, 115–7; Me. eth., b.p. 213–5
43	**2,6-Dibromo-4-methylphenol** (2,6-Dibromo-p-cresol)	49; 54	......			141.2				67	94–5	....
44	**Thymol** (2-Isopropyl-5-methylphenol)	49.7; 51.5, acet.	233.5	107, aq. al.	160, lgr.	70	103.2, al.	55	71	b.p. 242–3	33	Aryloxyacetic acid, 149; 2,4-Dinitrophenyl ether, 67; p-Xenylurethane, 194; p-Phenylazobenzoate, 85–6

*Derivative data given in order: m.p., crystal color, solvent from which crystallized.

TABLE VII. ORGANIC DERIVATIVES OF PHENOLS
TABLE VII. ORGANIC DERIVATIVES OF PHENOLS
b) Solids (Listed in order of increasing m.p.)* (Continued)

No.	Name	Melting point, °C	Boiling point, °C	Phenyl urethane	α-Naphthylurethane	p-Nitrobenzoate	3,5-Dinitrobenzoate	Bromo derivative	p-Toluene sulfonate	Acetate	Benzoate	Miscellaneous
45	**2-Hydroxy-4-methoxy-acetophenone** (Peonol)...	52–3, al.	....	....	....	....	....	....	....	46.5	....	$n_D^{81.2}$: 1.54322; $D^{81.2}$: 1.310; m-Nitro-benzoate, 109; p-Nitrophenyl-hydrazone, 238–9, ac. a.
46	**2-Benzylphenol** (2-Hydroxydiphenyl-methane) (stable form)..	52; 54; labile: 21–2	312	117.5–8.0, lgr.	....	....	....	....	....	....	....	Benzyl eth., 38, me. al.
47	**2-Chloro-3-methylphenol** (2-Chloro-m-cresol)	55–6; 49–50	194	....	....	....	....	....	96	....	55–6	Benzenesulfonate, 58
48	**4-Methoxyphenol** (Hydroquinone monomethyl ether).............	56; 55	243–4	....	....	....	....	....	....	32	87, al.	Aryloxyacetic acid, 110–2
49	**3-Methyl-6-nitrophenol** (6-Nitro-m-cresol)......	56, yel., bz.	....	....	....	....	....	....	....	48, al.	77	Me. eth., 62; Et. eth., 55, pet. eth.; 50–1
50	**3-Bromo-4-methylphenol** (3-Bromo-p-cresol).....	56	245	....	....	....	....	....	....	....	75	Me. eth., b.p. 103–5^{10}
51	**2-Phenylphenol** (2-(Hydroxybiphenyl).....	56; 67.5 (cor.)	275	....	....	....	....	64–6, dil. al.	62.5–3, pet. eth.	....	75–6	p-Phenylazobenzoate, 141–4; 2,4-Dinitrophenyl ether, 113–4
52	**2,6-Dibromophenol**......	56–7	162^{21}	....	....	....	93.3	....	....	46	68	Me. eth., 13; b.p. 143–5^{34}; Et. eth., 40.6
53	**4-n-Caproylresorcinol**....	56–7, tol.-pet. eth.	343–5 d.	....	....	89–91, pa. yel., al.	....	....	....	....	....	....
54	**2-Bromo-4-methylphenol** (2-Bromo-p-cresol).....	56–7	213–4; 102–4^{20}	....	....	....	....	121	b.p. 120^3	....	....	$D_{24.5}^{24.5}$: 1.547
55	**5-Bromo-3-methylphenol** (5-Bromo-m-cresol).....	56–7; 54	....	....	....	....	....	....	83	....	....	Me. eth., b.p. 139–40^{20}; Picrate, 130
56	**2-Cyclohexylphenol**......	56–7	....	....	....	....	....	....	....	....	....	4,6-Dinitro deriv., 106, al.; 2,4-Dinitrophenyl ether, 76–7
57	**2,3-Dichlorophenol**......	56–7	206	....	....	....	....	di: 90	....	....	....	Me. eth., 31
58	**4,6-Dibromo-2-methylphenol** (4,6-Dibromo-o-cresol)...............	57	....	....	....	136–7	....	....	....	....	62	....
59	**3-Hydroxy-6-nitrobiphenyl**..............	57–8, bz.	....	....	....	135	171	....	....	....	....	2.4-Dinitrophenyl ether, 131; 4-Nitro deriv., 176–8
60	**2,3,6-Trichlorophenol**....	58, pet. eth.	....	....	....	....	....	....	....	90, al.	....	
61	**2,6-Dichloro-3-methylphenol** (2,6-Dichloro-m-cresol)...............	58	234	....	....	....	....	....	100–1, al.	78.0–8.5, al.	....	
62	**2,5-Dichlorophenol**......	58–9, pet. eth.	212	....	....	....	....	....	....	....	69	
63	**4-Propylcatechol** (1,2-Dihydroxy-4-propyl-benzene)...............	60, bz.	175–80^{30}	....	....	....	....	....	....	....	....	3-Me. eth., b.p. 240–2; Di-me. eth., b.p. 247
64	**1,2-Dihydroxynaphthalene** (1,2-Naphthalenediol)...	60 (hyd.); 103–4 (anh.)	....	....	....	....	....	....	....	....	di: 106	....
65	**4-Isopropylphenol**.......	61	223–5	....	....	....	....	....	....	....	71–2	....

*Derivative data given in order: m.p., crystal color, solvent from which crystallized.

TABLE VII. ORGANIC DERIVATIVES OF PHENOLS
b) Solids (Listed in order of increasing m.p.)* (Continued)

No.	Name	Melting point, °C	Boiling point, °C	Phenyl urethane	α-Naphthyl urethane	p-Nitrobenzoate	3,5-Dinitrobenzoate	Bromo derivative	p-Toluene sulfonate	Acetate	Benzoate	Miscellaneous
66	5,6,7,8-Tetrahydro-2-naphthol	61.5–2.5	275[705]	......	......	113; 106.5	....	......	......	b.p. 158[14]	96	N,N-Diphenylurethane, 114; Cinnamate, 77.5
67	2,3,5-Trichlorophenol	62 (hyg.)	....	......	......	......	....	......	......	......	103, lgr.	Me. eth., 84, al.
68	2-Methyl-3,5,6-trichlorophenol (3,5,6-Trichloro-o-cresol)	62, ac. a.	....	......	......	......	....	......	......	......	110, al.	....
69	3,4-Dimethylphenol (o-4-Xylenol)	62–5, w.	225[757]	120, dil. al.	141–2, lgr.	......	181.6 (cor.), al.	tri: 171	......	22	58.5	Aryloxyacetic acid, 162.5; 2,4-Dinitrophenyl ether, 105–6; p-Xenylurethane, 183, al.; Picrate, 83.8, yel.; Me. eth., 204–5; p-Phenylazobenzoate, 104–7
70	2,6-Dinitrophenol	63–4	....	......	......	......	....	......	135	......	......	Picrate, 122; Me. eth., 118
71	4-Bromo-3-methylphenol (4-Bromo-m-cresol)	63.5, pet. eth.	118–23[7]	......	......	......	....	......	84–5	......	83.5	Benzenesulfonate, 79–80; Me. eth., b.p. 108.5[12]
72	4-Bromo-2-methylphenol (4-Bromo-o-cresol)	64	235 (subl.)	......	......	......	....	......	......	b.p. 132[12]	67–8	Et. eth., b.p. 238–40
73	1-Acetyl-2-naphthol	64, lgr.	....	......	......	......	....	......	......	......	85–6, pyr.	
74	4-Bromophenol	64; 66.4	238	140	169	180	191	tri: 171	......	22	58.5	Aryloxyacetic acid, 157; 2,4-Dinitrophenyl ether, 141; N,N-Diphenylurethane, 99; p-Phenylazobenzoate, 167.5–8.5
75	2-Methyl-1-naphthol	64–5	....	......	......	......	....	......	......	81–2	94–5	Picrate, 133–4
76	4-Methylcatechol (3,4-Dihydroxytoluene)	65, bz.	252	di: 166	......	......	....	......	......	di: b.p. 260–4	di: 58	Diaryloxyacetic acid, 58; Di-p-xenylurethane, 193
77	3-Bromo-2-nitrophenol	65–7, pet. eth. (anh.); 35 (hyd.), w.	....	......	......	......	....	......	136.5–7.5, al.	......	133	Me. eth., 73
78	4-Chloro-3-methylphenol (4-Chloro-m-cresol)	66; 55	235	......	153–4	......	....	......	98	......	86	2,4-Dinitrophenyl ether, 112; Benzenesulfonate, 66
79	4-Hydroxy-3-nitrobiphenyl	66	....	......	......	......	....	......	......	85–6, lgr.	111	Me. eth., 91–2
80	4-Methyl-2,3,5-trichlorophenol (2,3,5-Trichloro-p-cresol)	66–7, ac. a.	....	......	......	......	....	......	......	37–8, w.-ac. a.	89, w.-al.	....
81	2,6-Dichlorophenol	67	219–20; 80–5[4]	......	......	......	....	......	......	......	74.0–4.5	Me. eth., b.p. 105–6[20]
82	3,5-Dimethylphenol (m-5-Xylenol)	68	219.5 (subl.)	148; 151	......	109	195.4 (cor.), al.	tri: 166	83, ac. a.	b.p. 130[26]	24	Aryloxyacetic acid, 111; 81; 2,4-Dinitrophenyl ether, 100; p-Phenylazobenzoate, 104.5–6.5; p-Xenylurethane, 150
83	2-Bromo-6-nitrophenol	68, yel.	....	......	......	......	....	......	......	39.5–40	......	Me. eth., 67

*Derivative data given in order: m.p., crystal color, solvent from which crystallized.

TABLE VII. ORGANIC DERIVATIVES OF PHENOLS
b) Solids (Listed in order of increasing m.p.)* (Continued)

No.	Name	Melting point, °C	Boiling point, °C	Phenyl urethane	α-Naphthyl urethane	p-Nitrobenzoate	3,5-Dinitrobenzoate	Bromo derivative	p-Toluene sulfonate	Acetate	Benzoate	Miscellaneous
84	3-Methylcatechol (2,3-Dihydroxytoluene)	68, bz.	....	....	....	....	....	....	....	....	....	2-Me. eth., 39, b.p. 204; Di-Me. eth., b.p. 202–3
85	3,5-Dichlorophenol......	68	233	....	....	....	....	tri: 189	116	38	55	....
86	2,6-Di-iodophenol.......	68	....	....	....	....	....	....	....	107, ac. a.	....	Me. eth., 35, ac. a.; Et. eth., 41–2, ac. a.
87	2,4,5-Trichlorophenol	68, pet. eth.	....	....	....	....	....	....	....	....	92–3, al.	Aryloxyacetic acid, 157; Me. eth., 77–5, al.
88	5,6,7,8-Tetrahydro-1-naphthol	68.5–9.0; 74–5	264.5–5.0	....	....	....	....	....	....	73–5	46	Me. eth., b.p. 124[10]; Et. eth., b.p. 259[705]
89	2-Bromo-4,6-dichlorophenol	68–9	268 d.	....	....	....	....	....	82–3	....	....	2,4-Dinitrophenyl ether, 140–1
90	2-Bromo-4-methyl-6-nitrophenol (2-Bromo-6-nitro-p-cresol)	69, yel.	....	....	....	....	....	....	128	110–11	....	....
91	2,4,6-Trichlorophenol	69.5, ac. a.; 68	245	....	....	105–6	....	....	b.p. 261–2	....	75.5, al.	Aryloxyacetic acid, 182–6; 2,4-Dinitrophenyl ether, 136; N,N-Diphenylurethane, 143; Me. eth., 61–2, al.; Et. eth., 43–4
92	2,4,6-Trimethylphenol	70; 69	220	141–2, lgr.	....	....	....	di: 158	....	....	62, pet. eth.	Aryloxyacetic acid, 142
93	2,3,4,6-Tetrachlorophenol .	70	150[15]	....	....	....	....	....	....	65–6	108	Me. eth., 64–5; Et. eth., 55; Benzenesulfonate, 127
94	1-Chloro-2-naphthol......	70, lgr.; 72	....	....	....	....	....	....	....	42–3	99–100	Me. eth., 70–1; Et. eth., 58
95	Methyl 3-hydroxybenzoate	70	280	115–6, bz.	....	....	....	....	....	....	....	....
96	2-Methyl-6-nitrophenol (6-Nitro-o-cresol)......	70, w.-al.	....	....	....	....	....	66, al.	....	....	42, al.	Me. eth., 30, pet. eth.; Et. eth., b.p. 249–50, yel.
97	3-Propylcatechol (1,2-Dihydroxy-3-propylbenzene)	70–2, pet. eth.	....	....	....	....	....	....	....	....	....	3-Me. eth., b.p. 144–6[25]; Di-Me. eth., b.p. 134–7[22]; Aryloxyacetic acid, 132; p-Xenylurethane, 196
98	2,4,5-Trimethylphenol (Pseudocumenol).......	71	232	110	....	....	....	35	....	34.0–4.5, pet. eth.	63, al.	Aryloxyacetic acid, 132; p-Xenylurethane, 196
99	4,6-Dichloro-3-methylphenol (4,6-Dichloro-m-cresol)..............	72	235–6; 110[18]	....	....	....	....	....	104–5, al.	....	57.5, al.	n_D^{20}: 1.5722; Benzenesulfonate, 86, al.
100	2,4-Di-iodophenol.......	72, w.	....	....	....	....	....	165–7 (cor.)	....	70–1, w.-al.	98	Me. eth., 68, w.-al.; Et. eth., 46, me. al.; 51
101	2-Chloro-4,5-dimethylphenol	72	....	....	....	....	....	....	....	....	43	....
102	3,5-Dimethylcatechol (4,5-Dihydroxy-m-xylene) ...	73–4, w.	....	....	....	....	....	....	di: 161, ac. a.	....	....	4-Me. eth., b.p. 227–8

*Derivative data given in order: m.p., crystal color, solvent from which crystallized.

TABLE VII. ORGANIC DERIVATIVES OF PHENOLS
b) Solids (Listed in order of increasing m.p.)* (Continued)

No.	Name	Melting point, °C	Boiling point, °C	Phenyl urethane	α-Naphthyl urethane	p-Nitrobenzoate	3,5-Dinitrobenzoate	Bromo derivative	p-Toluene sulfonate	Acetate	Benzoate	Miscellaneous
103	5-Chloro-2-methylphenol (5-Chloro-o-cresol)	73–4	225	......	......	......	......	*tri*: 190	110	......	53–4	Me. eth., b.p. 206–8; Et. eth., b.p. 210–20
104	2,5-Dibromophenol	73–4							110			
105	3-Iodo-2-nitrophenol	73.5, w.								102.5		Me. eth., 83–4
106	Ethyl 3-hydroxybenzoate	73.8, bz.	295; 282							35	58, al.	
107	1,3-Dibromo-2-naphthol	75								102		
108	2,5-Dimethylphenol (p-2-Xylenol)	75, al.-eth.	212	160–1, bz.; 162; 166	172–3, lgr.	87	137.2 (cor.)	*tri*: 178			61	Aryloxyacetic acid, 118; p-Xenylurethane, 162; p-Phenylazobenzoate, 95.5–7.5; Picrate, 81–2, or., al.
109	2,3-Dimethylphenol (o-3-Xylenol)	75, w.-al.	193.5									Aryloxyacetic acid, 187; p-Phenylazobenzoate, 134–6; Me. eth., 29, b.p. 199; Et. eth., 10, b.p. 212.5
110	8-Hydroxyquinoline (Oxine)	75–6, dil. al.	266.6^{752}			174–5			115	b.p. 280	118–20, al.	Picrate, 203–4; Methiodide, 143 d. (hyd.)
111	4-(Dimethylamino)phenol	76							130	78		
112	3-Chloro-4-hydroxybiphenyl	76–7; 80									110–11; 95–7	2,4-Dinitrophenyl ether, 109–11
113	3-Bromo-2,6-dichlorophenol	76.5, lgr.	264–70^{751}								102	Me. eth., b.p. 260–5^{750}
114	2-Methyl-4,5,6-trichlorophenol (4,5,6-Trichloro-o-cresol)	77, pet. eth.								45, w.-me. al.		Me. eth., 51.5, al.
115	2-Bromo-4-methyl-3-nitrophenol (2-Bromo-3-nitro-p-cresol)	77, yel., w.								81		Me. eth., 74
116	3-(Diethylamino)phenol	78	276–80								22–3	Methylurethane, 85–6; Methiodide, 140–3
117	3-Phenylphenol (3-Hydroxybiphenyl)	78; 75	>300						52.5		60–1, al.	Et. eth., 34
118	4-Bromo-2,6-dinitrophenol	78, yel.							136, al.	110.5, bz.	154, al.	Me. eth., 88; Et. eth., 66
119	4-Chloro-2-iodophenol	78		128						57	88; 84	
120	2,4-Diaminophenol (4-Hydroxy-m-phenylenediamine)	78–80; unstable										Me. eth., 68; Et. eth., 67.8; 4,N-Acetyl, 249; N,N'-Diacetyl, 220–2; O,N,N'-Triacetyl, 180–2; N,N'-Dibenzoyl, 253–4; Picrate, 120 d., yel.
121	4-Methyl-3-nitrophenol (3-Nitro-p-cresol)	79, yel., eth.								91		Me. eth., 17, b.p. 266–7
122	2,3,4,6-Tetramethylphenol (Isodurenol)	79–81	230–50	178–9, wh., w.-al.							71–2, wh., w.-al.	
123	2-Bromo-4,5-dimethylphenol	80, pet. eth.									51	Me. eth., b.p. 86^3; o-Nitrobenzoate, 151–2

*Derivative data given in order: m.p., crystal color, solvent from which crystallized.

TABLE VII. ORGANIC DERIVATIVES OF PHENOLS
b) Solids (Listed in order of increasing m.p.)* (Continued)

No.	Name	Melting point, °C	Boiling point, °C	Phenyl ure- thane	α-Naph- thyl- ure- thane	p-Nitro- ben- zoate	3,5-Di- nitro- benzoate	Bromo derivative	p- Toluene sul- fonate	Acetate	Benzoate	Miscellaneous
124	5-Bromo-2-methylphenol (5-Bromo-o-cresol)	80	....	......	......	......	......	......	......	......	41	....
125	4-Iodo-2-nitrophenol	80–1	....	......	......	......	......	......	......	......	102–3	Me. eth., 98; Et. eth., 80
126	4-Hydroxy-3-methoxy- benzaldehyde (Vanillin)..	81	285	116–7	......	......	......	......	115	102	78	Aryloxyacetic acid, 187; 2,4-Dinitro- phenyl ether, 131; 2,4-Dinitrophenyl- hydrazone, 271 d.
127	3,5-Dibromophenol	81	....	......	......	......	......	......	......	53	77	Me. eth., 140
128	4-Chloro-2,6-dinitrophenol	81	....	......	......	......	......	......	......	110–11	......	Me. eth., 66; Et. eth., 54–5
129	4-Methyl-2-naphthol	81–2	....	......	......	......	......	......	......	117–8	......	....
130	5-Chloro-2,3-dimethyl- phenol	81–2	....	......	......	......	......	......	......	88, al.	......	....
131	3-Methyl-2,4,6-tribromo- phenol (2,4,6-Tribromo- m-cresol)...............	81.5–2.0, al.	....	......	......	......	......	......	113–4, al.	68, al.	84–5, al.	p-Phenylazoben- zoate, 130–2
132	2-Hydroxy-1-naphthalde- hyde................	82	192^{27}	......	......	......	......	......	......	87, al.; tri: 124, al.	......	Me. eth., 84; Et. eth., 115; Oxime, 157; Picrate, 120
133	3,4-Di-iodophenol.......	83	....	......	......	......	......	......	......	......	123	....
134	2-Hydroxyazobenzene (2- Benzeneazophenol)	83	....	......	......	......	......	......	......	–20	93, pet. eth.	Me. eth., 41; Et. eth., 44; Cu deriv., 225–6, al.
135	5-Propylresorcinol (1,3- Dihydroxy-5-propyl- benzene; Divarinol).....	83–4, bz., (anh.); 51, w.	....	......	......	......	......	......	......	di: 12–5	......	Di-Me. eth., b.p. 147^{29}
136	2,3,4-Trichlorophenol	83.5, lgr.	....	......	......	......	......	......	......	141, al.	......	Me. eth., 69.5, al.
137	4-Benzylphenol (4-Hy- droxydiphenylmethane) .	84	321; 308	......	......	......	......	......	......	mono: 87, pet.	......	2,4-Dinitrophenyl ether, 75.6; Benzyl eth., 49.5, al.
138	4-Chloro-2,3-dimethyl- phenol	84	....	......	......	......	......	......	......	102	......	....
139	2,4-Dimethyl-1-naphthol ..	84–5	169–70^{10}	174–5	......	......	......	......	......	......	......	Me. eth., b.p. 150–1^{12}; Picrate, 143–4, dk. red
140	4-Methyl-1-naphthol	84–5	....	......	......	......	......	......	......	......	81	....
141	1-Bromo-2-naphthol......	85	....	......	......	......	......	......	......	56	......	Me. eth., 85
142	3-Bromo-2-naphthol......	85	....	......	......	......	......	......	......	94	......	Me. eth., 77–8
143	2-Nitroresorcinol (1,3- Dihydroxy-2-nitro- benzene)	85, or.-red, w.-al.	....	......	......	......	......	......	......	di: 63	......	Di-Me. eth., 131, yel., al.; Di-Et. eth., 106–7
144	3-(Dimethylamino)phenol (3-Hydroxydimethyl- aniline)	85	265–8; 138^{10}	......	......	......	......	......	......	36.5; b.p. 160^5	94	Picrate, 162; Methylurethane, 87; Ethylurethane, 150; 99–100; Me. eth., b.p. 237; Et. eth., b.p. 247
145	3-Hydroxy-2-nitrobiphenyl	85–6, bz.; 81–2, yel., al.	....	......	......	......	......	......	......	61.5–2.5	131.0– 2.5	Benzenesulfonate, 130–1; 4-Nitro deriv., 126–7; 6-Nitro deriv., 214; 4,6-Dinitro deriv., 168–70
146	2-Hydroxybenzyl alcohol (Saligenin)	86–7, w.	....	......	......	......	......	......	......	di: 51, 70% al.	......	Aryloxyacetic acid, 120, w.

* Derivative data given in order: m.p., crystal color, solvent from which crystallized.

No.	Name	Melting point, °C	Boiling point, °C	Phenylurethane	α-Naphthylurethane	p-Nitrobenzoate	3,5-Dinitrobenzoate	Bromo derivative	p-Toluenesulfonate	Acetate	Benzoate	Miscellaneous
147	**4,6-Dinitro-2-methylphenol** (4,6-Dinitro-o-cresol) ...	86.5	......	......	......	......	....	....	......	95–6	135; 132	....
148	**3-Nitrocatechol** (1,2-Dihydroxy-3-nitrobenzene)	86.5, yel., pet. eth.								di: b.p. 103–4[1]	2-mono: 66	1-Me. eth., 62, yel.; 2-Me. eth., 102–3; Di-Me. eth., 64–5, al.
149	**4-Chloro-2-nitrophenol** ...	86–7	......	......	......	......	....	....		47–8	......	Me. eth., 98; Et. eth., 61–2
150	**2,4,5-Tribromophenol**	87, CH₂Cl₂-pet. eth.									99	Me. eth., 105; o-Br-p-toluenesulfonate, 107–8, al.
151	**4-Bromocatechol** (4-Bromo-1,2-dihydroxybenzene) ..	87									di: 111	1-Me. eth., 65
152	**4-(Methylamino)phenol** ...	87, bz.								135, bz.-lgr.	43, pet. eth.	173–4, 50% al.	Aryloxyacetic acid, 213–4; Me. eth., 37
153	**2-Methyl-3,4,5-tribromophenol** (3,4,5-Tribromo-o-cresol)	89, pet.									....	....
154	**2-Amino-6-methylphenol** (6-Amino-o-cresol)	89, w.							89; 90	78–9	......	N-Acetyl, 100–1
155	**N-Benzylidene-2-aminophenol**	89								93–6	......	Et. eth., b.p. 215–6[20]
156	**3-Nitro-2,4,6-tribromophenol**	89–90, lgr.								146–7, al.			Me. eth., 82, al.; Et. eth., 79, eth.
157	**2-Amino-6-chloro-4-methylphenol** (2-Amino-6-chloro-p-cresol)	89–90											N,O-Diacetyl, 162–3
158	**1,3-Dimethyl-2-naphthol** ..	89–90		197						85–6			Picrate, 132–3
159	**2-Propylhydroquinone** (1,4-Dihydroxy-2-propylbenzene)	90, bz.											Di-Me. eth., b.p. 240–6
160	**4-Chlorocatechol** (4-Chloro-1,2-dihydroxybenzene)	90–1	136[8.5]								di: b.p. 145–7[7]	di: 96–7, eth.	Di-Me. eth., b.p. 242.4
161	**3-Chloro-4,6-dimethylphenol**	90–1											
162	**2-Methyl-3,4,6-tribromophenol** (3,4,6-Tribromo-o-cresol)	91, pet.									76–7, ac, a.	85, 133, bz.	Me. eth., 71, al.
163	**5-Chloro-2,4-dinitrophenol**	92									69	......	Me. eth., 105; Et. eth., 112; 6-Nitro deriv., 113–5
164	**4-Chloro-2,6-dibromophenol**	92								107–8	......	......	2,4-Dinitrophenyl ether, 145–6; Me. eth., 74
165	**4-Bromo-2-nitrophenol** ...	92; 89									75		Me. eth., 88; Et. eth., 47; Benzenesulfonate, 83–4
166	**4-tert-Amylphenol**	92–3	260–5	108	......	......	....	....		55	61	....
167	**4-Hydroxyphenylethyl alcohol** (Tyrosol)	93, chl.	310; 156[15]								β-mono: 59, eth.-lgr.; di: b.p. 187[18]	di: 111, al.	4-Et. eth., 40

*Derivative data given in order: m.p., crystal color, solvent from which crystallized.

TABLE VII. ORGANIC DERIVATIVES OF PHENOLS

TABLE VII. ORGANIC DERIVATIVES OF PHENOLS
b) Solids (Listed in order of increasing m.p.)* (Continued)

No.	Name	Melting point, °C	Boiling point, °C	Phenyl ure-thane	α-Naph-thyl-ure-thane	p-Nitro-ben-zoate	3,5-Di-nitro-benzoate	Bromo derivative	p-Toluene sul-fonate	Acetate	Benzoate	Miscellaneous
168	4-Iodophenol	93–4, w.	....	148, bz.	......	......	....		99, me. al.	32	119	Aryloxyacetic acid, 156; N,N-Di-phenylurethane, 127; 2,4-Dinitro-phenyl ether, 156
169	1-Naphthol (α-Naphthol)	94	278–80	177–8, al.	152, lgr.	143; 140	217.4 (cor.), yel., al.	2,4-di: 105	89	48–9, al.	56, al.	Aryloxyacetic acid, 193.5; p-Xenyl-urethane, 190; 2,4-Dinitrophenyl ether, 128; p-Phenylazo-benzoate, 118–9
170	2-Iodo-4-nitrophenol	94; 86–7	......	......	......	......	......			68	....	Me. eth., 97; Et. eth., 96
171	2,4,6-Tribromophenol (Bromol)	95, HCOOH; 94, ac. a. (+1 ac. a.)	....	....	153	153	174	tetra: 120	113, al.	82; 87, ac. a.	81, al.	Aryloxyacetic acid, 200; N,N-Di-phenylurethane, 153; 2,4-Dinitro-phenyl ether, 137–8; p-Phenylazo-benzoate, 116–9
172	2,2'-Stilbenediol (2,2'-Dihydroxystilbene)	α: 95, al.; β: 197, al.								α: di: 107–8		β: Di-Me. eth., 136
173	6-Chloro-1-naphthol	95	......	......						47	....	Picrate, 165
174	3-Bromo-4-hydroxybi-phenyl	95	......	......						74–5	93–4	Benzenesulfonate, 102–3
175	1,3,6-Trihydroxynaph-thalene (1,3,6-Naph-thalenetriol)	95								tri: 112–3		Me. eth., 103–4
176	2,3,4-Tribromophenol	95, w.-HCOOH	......							95		Me. eth., 106
177	2,3,5-Trimethylphenol	95–6	233	174, pet. eth.	......					241	50, pet. eth.	....
178	2-Naphthyl salicylate (Betol)	95.5, stable; 93.5, labile	....	268, yel., ac. a.	......	......	......			136, al.		
179	3-Amino-2,4,6-trichloro-phenol	95.5–6.0, lgr.	......	......								N-Acetyl, 185.0–6.5, tol.
180	3-Acetylphenol (3-Hydroxyacetophenone)	96	296; 153[5]	......						44.0–4.5	52–3	D^{109}: 1.099; n_α^{109}: 1.5348; Me. eth., b.p. 240; Semi-carbazone, 194–6
181	5-Iodo-2-nitrophenol	96, yel., pet. eth.	......							95	122	Me. eth., 92; Et. eth., 86–7
182	2-Methoxy-5-propenyl-phenol	96; 92	147[19]	......						101		Et. eth., 49–50
183	2-Methyl-4-nitrophenol (4-Nitro-o-cresol)	96, yel., bz.; 30–40 (+1 H_2O), w.	186–90	......				107, al.	....		128	Me. eth., 64, al.; Et. eth., 71, w.-al.
184	2-(Methylamino)phenol	96–7; 86–7, pet. eth.	....								157–9	N-Benzoyl, 160–1; Me. eth., 33.0–3.5
185	3-Nitrophenol	97	194[70]	129	167	174	159	di: 91	112–3	55–6	95	D^{100}: 1.2797; Aryloxybenzoic acid, 156; 2,4-Dinitrophenyl ether, 136; p-Phenylazo-benzoate, 160.5–2.5

*Derivative data given in order: m.p., crystal color, solvent from which crystallized.

No.	Name	Melting point, °C	Boiling point, °C	Phenyl urethane	α-Naphthyl urethane	p-Nitrobenzoate	3,5-Dinitrobenzoate	Bromo derivative	p-Toluene sulfonate	Acetate	Benzoate	Miscellaneous
186	2,3-Dibromo-5,6-dimethyl-phenol	97, aq. al.								78, pet. eth.	153, al.	
187	5-Methyl-1-naphthol	97–8									77–8	
188	3-Acetylcatechol (2,3-Dihydroxyacetophenone)	97–8, yel., w.								di: 109, bz.		1-Me. eth., 152–3; Di-Me. eth., 48; Oxime, 96–7; Semicarbazone, 166–7
189	2-Cyanophenol	98									106, pet. eth.	$D^{99.6}_4$: 1.1052; $n^{99.6}_\alpha$: 1.53716
190	3-Chloro-4,5-dimethyl-phenol	98									42, al.	
191	2,5-Di-iodophenol	99, pet. eth.								70, ac. a.		
192	4-tert-Butylphenol	100	237	148.5	110			50; di: 64–7	109–10		81–2	Aryloxyacetic acid, 86.5; 2,4-Dinitrophenyl ether, 108–10; Benzenesulfonate, 70–1
193	3,4,5-Trichlorophenol	101, lgr.; 91	271–7[746]								120, al.	Me. eth., 130, b.p. 256–61; m-Nitrobenzenesulfonate, 176
194	3,5-Dibromo-2-methyl-phenol (3,5-Dibromo-o-cresol)	101; 98–101									91–3	
195	4,6-Dinitro-3-methylphenol (4,6-Dinitro-m-cresol)	101; 73–4							110–11		95	Me. eth., 115; Et. eth., 97
196	2-Acetyl-1-naphthol	102, pa. grn., al; 98, br.-yel., bz.	325, sl. d.							107.5, al.	128, al.	
197	2-Methyl-3,4,5-trinitro-phenol (3,4,5-Trinitro-o-cresol)	102, or.-yel., acet.										Me. eth., 111–2, w.-al.
198	4-Methyl-2,3,6-tribromo-phenol (2,3,6-Tribromo-p-cresol)	102, pet.								77, lgr.		
199	1-Nitro-2-naphthol	103, yel., al.								61, pet. eth.		m-Nitrobenzenesulfonate, 176, ac. a.; Et. eth., 104–5, yel., al.
200	3,3'-Dihydroxydiphenyl-methane	103, yel., al.								di: 57.5–8.5, lgr.		
201	4-Chloro-2-naphthol	104								56		
202	3,5-Di-iodophenol	104, w.								79, me. al.	93	Me. eth., 85, pet. eth.; Et. eth., 30, me. al.
203	2-Bromo-4-methyl-5-nitro-phenol (2-Bromo-5-nitro-p-cresol)	104, yel.								121		Me. eth., 94, yel., eth.
204	3-Hydroxybenzaldehyde (3-Formylphenol)	104; 108 (cor.)	240	158–60						b.p. 203	38; 48.5–9.0	Aryloxyacetic acid, 148; 2,4-Dinitrophenylhydrazone, 257d.; Semicarbazone, 198
205	5-Methyl-2-naphthol	104–5									107–8	Picrate, 156–7

*Derivative data given in order: m.p., crystal color, solvent from which crystallized.

No.	Name	Melting point, °C	Boiling point, °C	Phenyl urethane	α-Naphthyl urethane	p-Nitrobenzoate	3,5-Dinitrobenzoate	Bromo derivative	p-Toluene sulfonate	Acetate	Benzoate	Miscellaneous
206	3-Hydroxy-4-nitrobiphenyl	104–5, al.	....	.....	......	157	199	....	......	....	....	....
207	Catechol (Pyrocatechol; 1,2-Dihydroxybenzene)..	105	245.6	di: 169	175	mono: 159; di: 169, al.	di: 152	tetra: 192–3, wh.-vlt.	....	mono: 57–8; di: 65	mono: 181; 131; di: 84, al.-eth.	Aryloxyacetic acid, 136–8; Monobenzenesulfonate monoacetate, 86, me. al.; Dibenzenesulfonate, 155–6, acet.
208	2,4-Dibromo-1-naphthol ..	105; 111	....	......	......	....	....	....	....	92–3	......	Me. eth., 54–5, al.; sym.-TNB add. comp., 97; Picrate, 97, yel.
209	2,4-Dichloro-5-nitrophenol	105–6	....	......	......	....	....	....	....	....	111–2	m-Nitrobenzoate, 154; o-Nitro-p-toluenesulfonate, 143
210	Chlorohydroquinone	106	....	......	......	....	....	....	....	mono: 62; di: 72; 99	......	....
211	4,6-Dinitro-2-iodophenol ..	106–7, w.	....	......	......	....	....	149	....	113	......	....
212	1,6-Dibromo-2-naphthol ..	106–7 (+1 ac. a., 84)	....	......	......	....	....	....	....	125	......	Me. eth., 102; Et. eth., 94
213	2-Amino-5-chloro-4-methylphenol (2-Amino-5-chloro-p-cresol)	106–7	....	......	......	....	....	....	....	....	......	N-Acetyl, 115, aq. al.; Me. eth., 106, lgr.
214	4-Chlororesorcinol (4-Chloro-1,3-dihydroxybenzene)	106–7; 89	259; 147[18]	......	......	....	....	....	di: 46–7	di: 66	......	3-Me. eth., 79–80; Di-Me. eth., b.p. 135–7[17]
215	2-Hydroxypyridine (α-Pyridone)	106–7, bz.	280–1	......	......	120	....	....	158–9; 53	b.p. 150–60[0.09]	42	Picrate, 176–7; Benzyl eth., 42
216	2-Phenoxyphenol (2-Hydroxydiphenyl ether) .	106–7	151–5[11]	......	......	....	....	....	....	b.p. 358–60	48.5	Me. eth., 79, lgr.
217	5-Methylresorcinol (Orcinol; 3,5-Dihydroxytoluene)..............	106.5–8; 56–8 (+1 H₂O)	287–90	154	160, lgr.	214	190	tri: 104	....	di: 25	di: 88, al.	Aryloxyacetic acid, 217; p-Xenylurethane, 196; 2,4-Dinitrophenyl ether, 153–4
218	6-Amino-4-chloro-2-methylphenol (6-Amino-4-chloro-o-cresol)	107	....	......	......	....	....	....	....	2,3-di: 196	....	....
219	2,4-Dichloro-1-naphthol ..	107–8	....	......	......	....	....	....	....	74–6	......	Me. eth., 58
220	1,2-Dihydroxynaphthalene (1,2-Naphthalenediol; β-Naphthohydroquinone).............	108; 105.5	....	......	......	....	....	....	....	di: 104–6; 109, ac. a.	di: 106	Diaryloxyacetic acid, 104–6; 1-Me. eth., 90.5; Di-Me. eth., 31
221	1,2,3,4-Tetrahydroanthranol...............	108, lgr.	....	......	......	....	....	....	....	109, al.	142	Me. eth., b.p. 197[14]
222	4-Methyl-5,6,7,8-tetrahydro-2-naphthol	108	....	......	......	114–6	....	....	....	....	89	....
223	4-Acetylphenol (4-Hydroxyacetophenone)	109	....	......	......	....	....	....	....	54	......	Semicarbazone, 199; 2,4-Dinitrophenylhydrazone, 261.5 (cor.), br., al.
224	3-Methyl-2,4,6-trinitrophenol (2,4,6-Trinitro-m-cresol).............	109–10, yel., al.	....	......	......	....	....	....	....	135, pa. yel., bz.	140	Di-Me. eth., 155, al.; Di-Et. eth., 36–7, w.-al.

*Derivative data given in order: m.p., crystal color, solvent from which crystallized.

No.	Name	Melting point, °C	Boiling point, °C	Phenyl ure-thane	α-Naph-thyl-ure-thane	p-Nitro-ben-zoate	3,5-Di-nitro-benzoate	Bromo derivative	p-Toluene sul-fonate	Acetate	Benzoate	Miscellaneous
225	**2-Iodo-6-nitrophenol**	109–10	....	,......	......	......	....	....		96–7	....	Me. eth., 96–7
226	**Resorcinol** (1,3-Dihydroxy-benzene)	110 (stable); 108–8.5 (labile)	280.8 (cor.); 275.9	*di*: 164, chl.	......	*di*: 182; 175	*di*: 201	*tri*: 112	*di*: 80–1, acet.-dil. al.	57–8, dil. al.	*mono*: 135–6; *di*: 117, dil. al.	Aryloxyacetic acid, 175; 195; 2,4-Dinitrophenyl ether, 194; N,N-Diphenylurethane, 194; Me. eth., 43–4, me. al.
227	**2,2'-Biphenol** (2,2'-Dihydroxybiphenyl)	110; 109; tol.	326	*di*: 145, dil. al.				*di*: 188	190	*di*: 95, xyl.	*di*: 101, al.	Di-Me. eth., 155, al., Di-Et. eth., 36–7, w.-al.
228	**Bromohydroquinone**	110	....	......	......	......	....	*di*: 186	......	*di*: 72	....	....
229	**4-Hydroxy-2'-nitro-biphenyl**	110–1; 116	....	......	......	......	....	....	122		156–7	Aryloxyacetic acid, 160–1; Benzene-sulfonate, 106, yel., al.; Me. eth., 60.0–0.5; Et. eth., 51; 5-nitro deriv., 151–2
230	**7-Methyl-1-naphthol**	110–11								39–41	....	Picrate, 164–5
231	**1-Methyl-2-naphthol**	111								66	116–7	Picrate, 163–4
232	**2-Chloro-4-nitrophenol** ...	111, w.								63	....	Me. eth., 98; Et. eth., 82; 142
233	**2-Amino-6-nitrophenol**....	111–2, aq. al.								2-*mono*: 102–3 (hyd.); 122 (anh.)	....	Me. eth., 198
234	**2,4,6-Tribromoresorcinol** (1,3-Dihydroxy-2,4,6-tri-bromobenzene)	112, w.								*mono*: 114, CS₂; *di*: 108	*mono*: 120, chl.-pet.-eth.	Me. eth., 104; Di-Me. eth., 68–9, w.-al.
235	**4',5-Dimethyl-2-hydroxy-azobenzene** (2-p-Toluene-azo-p-cresol)...........	112–3, tol.								91, yel., ac. a.	95, yel., al.	Et. eth., 43, ac. a.; Propionate, 62, lgr.
236	**4-Hydroxyphenanthrene** (4-Phenanthrol)	112–3.5, pet. eth.								58–9, al.	....	Me. eth., 68, me. al.
237	**2,2'-Dihydroxy-3,3'-dimethylbiphenyl**	113, pet. eth.									*di*: 147, me. al.	....
238	**2,6-Dinitro-4-iodophenol** ..	113, w.							138		175	....
239	**2-Chloro-4,6-dinitrophenol**	113; 110							155			....
240	**4,4'-Dihydroxy-2,2'-dimethylbiphenyl**	114								*di*: 75	*di*: 127	....
241	**2-Bromo-4-nitrophenol** ...	114								62; 86	131–2	Me. eth., 106; Et. eth., 98, yel.
242	**4-Nitrophenol**	114	....	156	150–1	159	186; 188 (+1 ac. a.)	2,6-*di*: 142	97	81–2, w.-al.	142.5	Aryloxyacetic acid, 187; 2,4-Dinitro-phenyl ether, 120; N,N-Diphenylure-thane, 112; p-Phenylazoben-zoate, 203–6
243	**2,4-Dinitrophenol**........	114	....	......	......	139	....	6-*mono*: 118	121	72	132	2,4-Dinitrophenyl ether, 248
244	**2,3,5-Tri-iodophenol**	114, bz.-lgr.	....	......	......	......	....	....		123	....	Et. eth., 121

*Derivative data given in order: m.p., crystal color, solvent from which crystallized.

TABLE VII. ORGANIC DERIVATIVES OF PHENOLS

b) Solids (Listed in order of increasing m.p.)* (Continued)

No.	Name	Melting point, °C	Boiling point, °C	Phenyl ure-thane	α-Naph-thyl-ure-thane	p-Nitro-ben-zoate	3,5-Di-nitro-benzoate	Bromo derivative	p-Toluene sul-fonate	Acetate	Benzoate	Miscellaneous
245	4-Hydroxy-3-methoxy-benzyl alcohol	115								4-mono: 51; di: 48, bz.-lgr.	4-mono: 90, et. ac.-al.; di: 121	4-Et. eth., 56–7
246	1-Hydroxyfluorenone	115, yel.								130–1, aq. al.	128–9	Me. eth., 141.5–2.5; Oxime, 169–70; Phenylhydrazone, 173–4
247	Ethyl 4-hydroxybenzoate	115; 116	297–8									94, eth.	
248	4-Chloro-3,5-dimethyl-phenol	115–6									47–8		
249	3,5-Dimethyl-2,4-dinitro-phenol	115–6; 106								171	148	156	
250	4-Bromo-1,5-dihydroxy-naphthalene (4-Bromo-1,5-naphthalenediol)	116									di: 138		Di-Me. eth., 115
251	3-Benzoylphenol (3-Hy-	116, al.											Oxime, anti: 76, bz.; syn: 126 (on heating anti)
252	4-Hydroxybenzaldehyde	116–7										90	Aryloxyacetic acid, 198; 2,4-Dinitro-phenylhydrazone, 270–1
253	3-Bromo-1,2-dihydroxy-naphthalene (3-Bromo-1,2-naphthalenediol)	117									di: 160		Naphthalene add. comp., 100
255	Phloroglucinol (1,3,5-Trihydroxybenzene)	117 (+2H$_2$O); 217–9 (anh.)				283		tri: 151		tri: 104	tri: 185	Picrate, 101–3
256	2,4'-Dihydroxydiphenyl-methane	117–8, w.-al.									di: 70, ac. a.	di: 108	Di-Me. eth., 26; Di-Et. eth., 60, w.-al.
257	2,3,5,6-Tetramethylphenol (Durenol)	118, wh., pet.	249					4-mono: 118, or., dil. al.					
258	4-Amino-2-methyl-6-nitro-phenol (4-Amino-6-nitro-o-cresol)	118, al.									4-mono: 217, yel., al.		
259	2,4-Dibromo-6-nitrophenol	118, ac. a.								140		89	Me. eth., 76–7, yel.; Et. eth., 46
260	2-Methyl-5-nitrophenol (5-Nitro-o-cresol)	118, yel., lgr.								123–4, al.	74		Me. eth., 74, al.; Et. eth., 61, al.
261	2-Bromo-4,6-dinitrophenol	118–9, yel.								157	104.5	94, aq. al.	Me. eth., 48, yel.
262	2-Bromo-5-nitrophenol	118.5–21, pet. eth.								131.5–2.5, al.			Me. eth., 104, al.
263	2-Amino-4-methyl-6-nitro-phenol (2-Amino-6-nitro-p-cresol)	119; 110, br., al.											N-Acetyl, 143, yel., al.
264	3-Amino-2,4,6-tribromo-phenol	119, pet. eth.								146–7, al.			O,N,N-Triacetyl, 136

*Derivative data given in order: m.p., crystal color, solvent from which crystallized.

No.	Name	Melting point, °C	Boiling point, °C	Phenyl urethane	α-Naphthyl urethane	p-Nitrobenzoate	3,5-Dinitrobenzoate	Bromo derivative	p-Toluene sulfonate	Acetate	Benzoate	Miscellaneous
265	2-Methylresorcinol (1,3-Dihydroxy-2-methylbenzene; 2,6-Dihydroxytoluene)	119–20, bz.	271 (cor.)	......	......	......	......	......	......	......	di: 105–6, me. al.	Di-Me. eth., 39
266	4-Acetylcatechol (3,4-Dihydroxyacetophenone)	119.2–.7; 116							di: 144–5, et. ac.	4-mono: 58; di: 91; 88	di: 118, al.	Oxime, 184 d., et. ac.; 1-Me. eth., 91; 2-Me. eth., 115; Di-Me. eth., 51
267	2-Chloro-5-nitrophenol	119.5, w.		......	......	......	......	......	......	82	127–8	Me. eth., 83; Et. eth., 64.5, al.
268	5-Methyl-1,2,3-trihydroxybenzene (3,4,5-Trihydroxytoluene; 5-Methylpyrogallol)	120 (subl.), bz.		......	......	......	......	......	......	tri: 99	......	3,5-Di-Me. eth., 36, al.
269	9-Hydroxyanthracene (9-Anthrol; Anthranol)	120, yel.		......	......	......	......	......	......	126–31	286–8	....
270	6-Bromo-2-methyl-4-nitrophenol (6-Bromo-4-nitro-o-cresol)	120 d., yel., lgr.								137		
271	2,2'-Dihydroxy-4,4'-dimethylbiphenyl	120		......						......	di: 148, al.-acet.	....
272	4-Nitro-2-naphthol	120, yel., pet. eth.		......	......	......	......	......	122, pa. yel., al.	......	......	m-Nitrobenzenesulfonate, 149, ac. a.; Me. eth., 100–3, br., bz.-al.
273	4-Chloro-1-naphthol	120–1; 116–7, al. or chl.		......	......	......	......	......	......	44	......	Picrate, 171; Carbonate, 228
274	2-Chloro-3-nitrophenol	120.5, w.		......	......	......	......	......	......	51.5	94	Me. eth., 94; Et. eth., 51, me. al.
275	4-Chloro-3,5-dibromophenol	121		......	......	......	......	......	......		132	Me. eth., 82.5
276	3,4-Dimethyl-1-naphthol	121.5–3.0, lgr.	205–10¹⁵	......	......	......	222–4	......	......	89.5–91	......	Turns red in air; Picrate, 168
277	3-Aminophenol (3-Hydroxyaniline)	122		......	......	143	179	......	......	......	......	N-Acetyl, 148; N-p-Toluenesulfonyl, 157; N-Phenylthiourea, 156
278	2-Amino-3-chlorophenol	122		......	......	......	......	......	......	......	......	N-Acetyl, 123; N-Benzoyl, 123; HNO₃ salt, 137
279	4-Bromo-2-naphthol	122		......	......	......	......	......	......	61	......	Me. eth., 64
280	4-Nitroresorcinol (1,3-Dihydroxy-4-nitrobenzene)	122, yel., CCl₄	178–9¹¹	......	......	......	......	......	......	di: 90–1, al.	1-mono: 124, ac. a.; 3-mono: 189, ac. a.; di: 110	Di-Et. eth., 85
281	2,4,6-Trinitrophenol (Picric acid)	122 (subl. on slow htng.; exp. on rapid htng.)		......	......	......	......	......	......	76	......	Naphthalene add. comp., 149–51

*Derivative data given in order: m.p., crystal color, solvent from which crystallized.

TABLE VII. ORGANIC DERIVATIVES OF PHENOLS
b) Solids (Listed in order of increasing m.p.)* (Continued)

No.	Name	Melting point, °C	Boiling point, °C	Phenyl urethane	α-Naphthyl urethane	p-Nitrobenzoate	3,5-Dinitrobenzoate	Bromo derivative	p-Toluene sulfonate	Acetate	Benzoate	Miscellaneous
282	3-Hydroxyphenanthrene (3-Phenanthrol)	122–3, al.								114–5, aq. al.		Me. eth., 63, me. al.; Et. eth., 46, me. al.; Picrate, 124–5, red, al.
283	2,4-Dichloro-6-nitrophenol	122–3								77		3-Nitrobenzoate, 149–50
284	2-Naphthol (β-Naphthol)	123	286	155–6, al.	156–7, lgr.	169	210.2 (cor.), al.	84	125, al.	71–2; 70	106–7	Aryloxyacetic acid, 95; 2,4-Dinitrophenyl ether, 95; N,N-Diphenylurethane, 141; p-Phenylazobenzoate, 190–3
285	2,3,4,5-Tetrabromophenol	123, aq. al.						225–6		110.5, aq. ac. a.	133, al.	
286	2,4-Dimethyl-2-methoxyphenol	123							137–8	114		Et. eth., 91
287	2-Nitro-4,5,6-tribromophenol	123, yel., bz.										Me. eth., 109, al.; Et. eth., 74
288	1,4-Dichloro-2-naphthol	123–4								90–1		
289	4-Butyl-2-methylphenol (4-Butyl-o-cresol)	124	127–9[15]							268–70		
290	3,3'-Biphenol (3,3'-Dihydroxybiphenyl)	124, w.								di: 82.5, dil. al.	di: 92	Di-Me. eth., 36, w.-al., b.p. 328
291	1,3-Dihydroxynaphthalene (1,3-Naphthalenediol)	124, w.								di: 56, w.-ac. a.		sym-Trinitrobenzene add. comp., 174.5, red
292	3-Iodo-4-nitrophenol	124, yel., pet. eth.								73–7	119	Me. eth., 69–70
293	2-Methylhydroquinone (2,5-Dihydroxytoluene)	124–5, bz.						84	125	mono: 92, pet. eth.; di: 49; 45, w.	di: 119–20	Aryloxyacetic acid, 153; Di-Et. eth., 24–5, b.p. 247–9
294	4-Hydroxybenzyl alcohol (α,4-Dihydroxytoluene)	124.5–5.5								α-mono: 84; di: 75	α-mono: 88–9	
295	4,6-Dimethylresorcinol (4,6-Dihydroxy-m-xylene)	124.5–5.0, w. (+1 w.)	276–9							di: 45, al.; b.p. 285–7		Di-Me. eth., 76; Di-Et. eth., 75, al.
296	4-Benzyl-1-naphthol	125–6								87–8	103	
297	Pentamethylphenol	126	267	215						273	127	Me. eth., 163–4
298	3,5-Dinitrophenol	126; 122								126–7		Me. eth., 105–6; Et. eth., 97
299	7-Chloro-2-naphthol	126.5								104.5		
300	4-Chloro-3-nitrophenol	126.5, w.								83.5	96–7	Me. eth., 98, al.; Et. eth., 47.5, al.
301	2-Nitro-3,4,6-tribromophenol	127, pa. yel., w.-HCOOH								118		Me. eth., 72, w.-HCOOH
302	2-Nitro-1-naphthol	127–8, yel., al.								118		Me. eth., 80; Et. eth., 84, yel., lgr.
303	1,3-Dibromo-2,4-dihydroxynaphthalene (1,3-Dibromo-2,4-naphthalenediol)	128–9, ac. a.						186, CS₂		4-mono: 148; di: 125, al.		

*Derivative data given in order: m.p., crystal color, solvent from which crystallized.

TABLE VII. ORGANIC DERIVATIVES OF PHENOLS

b) Solids (Listed in order of increasing m.p.)* (Continued)

No.	Name	Melting point, °C	Boiling point, °C	Phenyl ure-thane	α-Naph-thyl-ure-thane	p-Nitro-ben-zoate	3,5-Di-nitro-benzoate	Bromo derivative	p-Toluene sul-fonate	Acetate	Benzoate	Miscellaneous
304	4-Iodo-2-naphthol	128.5								59		Me. eth., 67
305	4-Bromo-1-naphthol	129								51		
306	3-Methyl-4-nitrophenol (4-Nitro-m-cresol)	129, w.								34, al.	74	Me. eth., 55, lgr.; Et. eth., 45, al.
307	3-Hydroxypyridine (β-Pyridone)	129								b.p. 210; 92⁹	50.0-0.5	Hydrochloride, 105–7; Picrate, 200–1
308	4-Hydroxy-3-nitroazo-benzene	129, lgr.								120.5, ac. a.	132, bz.	Me. eth., 107
309	3-Amino-2-methylphenol (3-Amino-o-cresol)	129, w.								108		Turns br. in air
310	6-Bromo-2-naphthol	129–30								103		Me. eth., 108
311	8-Nitro-1-naphthol	130										m-Nitrobenzene-sulfonate, 166, ac. a.
312	4-Amino-2-nitrophenol	131, dk. red										Me. eth., 243; Et. eth., 170, N-Acetyl, 157–8, yel.
313	Methyl 4-hydroxybenzoate	131		134–5, bz.						85	135	
314	5-Methyl-1,2,4-trihydroxy-benzene (2,4,5-Trihy-droxytoluene)	131–2, bz.								tri: 114–5, al.		4-Me. eth., 124, w.; 4-Et. eth., 131 (subl.), bz., Tri-me. eth., 55, w.-me. al.
315	5-Chloro-1-naphthol	131–2, w. or CS₂								53		Picrate, 160
316	4-Cyclohexylphenol	132		145.5		137, al.	168 (cor.)			35	118.5, me. al.	
317	Pyrogallol (1,2,3-Tri-hydroxybenzene)	133	309	tri: 173		tri: 230	tri: 205	di: 158		di: 110–11; tri: 165, 173	mono: 140; di: 108; tri: 90, al.	Tris-N,N-Diphenyl-urethane, 212
318	2-Amino-4,6-dimethyl-phenol	134–5, al.										N-Acetyl, 96, aq. al.; O,N-Dibenzoyl, 153.5; Me. eth., b.p. 239–40
319	4-Amino-5-isopropyl-2-methyl-6-nitrophenol (4-Amino-5-isopropyl-6-nitro-o-cresol)	134–5, yel., al.									1-mono: 280–3; di: 222–5, bz.	
320	4,6-Dibromo-2-naphthol	134–5								128	128–9	Me. eth., 103; Et. eth., 98
321	2,3-Dichloro-1,4-dihy-droxynaphthalene	135; 155–6, aq. al.								di: 239–40	di: 252	Di-Me. eth., 107–8; Dipropionate, 166–7; Dibutyrate, 128
322	4-Benzoylphenol (4-Hy-droxybenzophenone)	135								81, me. al.	115; 94–5	2,4-Dinitrophenyl-hydrazone, 242.4 (cor.); Semicar-bazone, 194

*Derivative data given in order: m.p., crystal color, solvent from which crystallized.

TABLE VII. ORGANIC DERIVATIVES OF PHENOLS
b) Solids (Listed in order of increasing m.p.)* (Continued)

No.	Name	Melting point, °C	Boiling point, °C	Phenyl ure-thane	α-Naph-thyl-ure-thane	p-Nitro-ben-zoate	3,5-Di-nitro-benzoate	Bromo derivative	p-Toluene sul-fonate	Acetate	Benzoate	Miscellaneous
323	2-Amino-4-methylphenol (2-Amino-p-cresol)	135, w.										N-Benzoyl, 191, al.; O,N-Dibenzoyl, 190–1, al.; O,N-Diacetyl, 128–9; Me. eth., 93–4, pet. eth.
324	3-Amino-4,6-dichloro-phenol	135–6, w.							113–4			N-Acetyl, 233–6; Me. eth., 51
325	2,6-Dinitrohydroquinone	135–6 (anh.)								4-mono: 95–6; di: 135–6	4-mono: 150–1	4-Me. eth., 102; Di-Me. eth., 95–6; 112
326	1,4-Dimethyl-2-naphthol	135–6								77–8	124–5	
327	Trichlorohydroquinone	136, w.								di: 153 (subl.)		Di-Et. eth., 68.5, al.
328	5-Iodo-3-nitrophenol	136, w.								110	100.5	Me. eth., 84
329	Trichlorophloroglucinol	136, al.								tri: 167–8, ac. a.		Di-Me. eth., 93–5; Tri-Me. eth., 130–1, al.
330	1,6-Dihydroxynaphthalene (1,6-Naphthalenediol)	137–8, bz.								di: 73, al.	di: 103–4	Di-Me. eth., 60–1, pet. eth.; Di-Et. eth., 83, lgr.; 2-Naphthylamine add. comp., 110.5
331	2,4-Dinitro-1-naphthol	138									174	Me. eth., 97; Et. eth., 92
332	4-Amino-2,6-dimethyl-phenol	138 d., bz.										N-Acetyl, 136–7, aq. al.; O,N-Di-acetyl, 160; Me. eth., 66, w.
333	2-Amino-3-bromophenol	138							120–1, al.			Me. eth., 65; Hydrochloride, 225
334	2-Amino-4-chlorophenol	138 (unstable)							115	73–4		N-Acetyl, 185; O,N-Dibenzoyl, 157–8; Me. eth., 82–3
335	1,4,6-Trihydroxynaphtha-lene (1,4,6-Naphthalene-triol)	138–40								94.5		
336	2,6-Dibromo-3,4,5-tri-hydroxybenzoic acid (Dibromogallic acid)	139 d. (+1H₂O)								tri: 168	tri: 95–6	Triacetate of Me. eth., 150–2
337	2-Hydroxy-2'-nitro-biphenyl	139–40				116	180	3,5-di: 149	100		116	3-Nitrobenzenesul-fonate, 161; 4-Nitrobenzenesul-fonate, 147; 2,4-Dinitrophenyl ether, 118; Me. eth., 82; 3,5-Di-nitro deriv., 123
338	4-Phenyl-1-naphthol	140									73–4	
339	1,3-Dihydroxy-2-methyl-naphthalene (2-Methyl-1,3-naphthalenediol)	140								118		
340	1,2,4-Trihydroxybenzene	140–5, eth.								tri: 96–7, wh., abs. al.	tri: 120, al.	Tri-Et. eth., 34, al.; Picrate, 96
341	2-Amino-5-iodophenol	141										N-p-Toluenesul-fonyl, 175–6

*Derivative data given in order: m.p.; crystal color, solvent from which crystallized.

No.	Name	Melting point, °C	Boiling point, °C	Phenyl ure-thane	α-Naph-thyl-ure-thane	p-Nitro-ben-zoate	3,5-Di-nitro-benzoate	Bromo derivative	p-Toluene sul-fonate	Acetate	Benzoate	Miscellaneous
342	4-Hydroxy-2-nitro-biphenyl	141–3								169	105–6	Me. eth., 72
343	1,8-Dihydroxynaphthalene (1,8-Naphthalenediol)	142, ac. a.; 140, w.								di: 155, ac. anh.	di: 174–5	Di-Me. eth., 50, pet. eth.; Picrate, 135–7; 2-Naphthyl-amine add. comp., 124
344	2-Amino-4-nitrophenol	142–3 (anh.); 80–90 (hyd.)				122			122, yel., al.			N-Acetyl, 279; N-Benzoyl, ca. 200 d.; Me. eth., 124–5; Et. eth., 97–8
345	3,4-Dihydroxy-phenanthrene (Morphol)	143, pet. eth.								di: 159, eth.		Me. eth., 62–3; Di-Me. eth., 45, me. al.
346	3-Chloro-1-naphthol	143; 134–5, lgr.								69, lgr.	118–9	Me. eth., b.p. 162–4[18]
347	2-Isopropyl-5-methyl-hydroquinone (Thymo-quinol)	143; 139.5	290	232–3, al.	147–8, al.					di: 73–5	141–2, al.	
348	2,6-Dibromo-4-nitrophenol	144							128–9	181		
349	3,4,5-Tribromocatechol	144 (+1H$_2$O)								di: 120		Di-Me. eth., 86–7
350	3-Amino-4-methylphenol (3-Amino-p-cresol)	144										N-Acetyl, 178; O,N-Diacetyl, 128–9; N-Benzenesul-fonyl, 183; Me. eth., 47, w.
351	8-Nitro-2-naphthol	144–5, yel.								101–2, w.-al.		m-Nitrobenzene-sulfonate, 144–6, ac. a.; Me. eth., 69; Et. eth., 72–3, yel., pet. eth.
352	4-Benzoylresorcinol (2,4-Dihydroxybenzo-phenone)	144–6								di: 78		
353	1-Hydroxyacenaphthene (1-Acenaphthenol)	144.5-5.5 (cor.); 148		137						b.p. 166–8[5]		
354	4-Amino-2,3,6-trinitro-phenol	145 d., ac. a.										N-Acetyl, 178–9; N-Benzoyl, 205, ac. a.; Me. eth., 138–9, al.
355	4-Benzoylcatechol (3,4-Dihydroxybenzo-phenone)	145; 134									di: 95	Me. eth., 131–2; Di-Me. eth., 103–4
356	3-Bromo-5-nitrophenol	145								99		Me. eth., 88
357	3,4-Dihydroxybiphenyl (4-Phenylcatechol)	145								di: 77.5–8.0		
358	2,4,6-Tri-iodoresorcinol	145, CS$_2$								di: 170		
359	2,4'-Dihydroxychalcone	145								di: 94–5	di: 120	4'-Me. eth., 148 d., al.
360	2,6-Diamino-4-methyl-phenol (4-Hydroxy-3,5-toluenediamine)	146										N,N'-Diacetyl, 225–7, dil. al., 1,2,6-Triacetyl, 228, al.
361	5-Aminoresorcinol (3,5-Dihydroxyaniline)	146–52										N-Benzoyl, 139; Di-Me. eth., 46; Picrate, 167–70 d., yel., aq. al.

*Derivative data given in order: m.p., crystal color, solvent from which crystallized.

No.	Name	Melting point, °C	Boiling point, °C	Phenyl ure-thane	α-Naph-thyl-ure-thane	p-Nitro-ben-zoate	3,5-Di-nitro-benzoate	Bromo derivative	p-Toluene sul-fonate	Acetate	Benzoate	Miscellaneous
362	4-Acetylresorcinol (Resacetophenone; 2,4-Dihydroxyacetophenone)	147	....	.....	.....	.....	....	.....	.....	3-mono: 119–20; 4-mono: 74; di: 38	3-mono: 67; 4-mono: 106–7	Semicarbazone, 216; 2,4-Dinitro-phenylhydrazone, 218; Oxime, 199; Phenylhydrazone, 159
363	3-Chloro-5-nitrophenol ...	147	....	.....	.....	.....	....	.....	.....	84	78	Me. eth., 101; Et. eth., 47, al.
364	5-Nitro-2-naphthol.......	147, yel., w.								....	....	m-Nitrobenzene-sulfonate, 106, ac. a.; Et. eth., 115, yel., al.
365	2-Methyl-3-nitrophenol (3-Nitro-o-cresol)	147, yel., w.							94, al.	....	97	Me. eth., 52–3, w.-me. al.
366	2-Chloro-4-hydroxybenzal-dehyde..............	147–8	....	.....	.....	.....	....	.....	.....	52	97	Semicarbazone, 214; Oxime, 194; p-Nitrophenyl-hydrazone, 288 d.
367	1,5-Dibromo-4,8-di-hydroxynaphthalene.....	147.5	....	.....	.....	.....	....	.....	.....	di: 131	....	....
368	5-Acetylresorcinol (3,5-Dihydroxyacetophenone)	148								di: 91–2	....	Semicarbazone, 205–6; p-Nitro-phenylhydrazone, 236–7
369	2,4-Dinitroresorcinol	148							mono: 126–7	di: 119–20	di: 182–4	Aryloxyacetic acid, 155; Me. eth., 108; Di-Et. eth., 57
370	4-Propionylphenol (4-Hydroxypropiophenone)	148	.....	.....	.....	.....	....	.....	.....	62, lgr.	107.5	2,4-Dinitrophenyl-hydrazone, 229
371	9,10-Dihydroxyphenan-threne (9,10-Phenan-threnediol)	148	....	.....	.....	.....	....	.....	.....	mono: 168–70, di: 202	di: 216–7	....
372	3,4,5-Trihydroxyphenan-threne (3,4,5-Phenan-threnetriol)	148, w.	....	.....	.....	.....	....	.....	.....	tri: 138, bz.-pet. eth.	....	Tri-me. eth., 90, me. al.
373	4-Hydroxypyridine (γ-Pyridone)	148.5 (anh.); 65 (+1 H₂O)	>350	.....	.....	.....	....	.....	.....	140–50	81	Picrate, 240; Methiodide, 108–9
374	3-Nitrosalicylic acid.....	148.5–9.0								....	.....	Amide, 155; 145
375	3-Acetamidophenol (3-Hydroxyacetanilide)	148–9, w.								99.5–100.5		....
376	2-Amino-3,6-dimethyl-phenol	149–50, bz.								.....	.....	N-Benzoyl, 210–11; O,N-Dibenzoyl, 178–9
377	2,6-Dimethylhydroquinone	149–51, xyl.								.....	.....	4-Me. eth., 77, pet. eth.
378	2,4′-Dihydroxybenzo-phenone	150	.....	.....	.....	.....	....	.....	.....	84.5	....	....
379	5-Amino-2-bromophenol ..	150 d.							135–6			....
380	2-Amino-5-bromophenol ..	150	.....	.....	.....	.....	....	.....	.....			O,N-Dibenzoyl, 133–4; H₂SO₄ salt, 239–40
381	1-Hydroxyanthracene (1-Anthrol)	150–3, br., al.								128–30	....	Me. eth., 70; Et. eth., 69
382	2-Amino-4-chloro-6-nitro-phenol	152	.....	.....	.....	.....	....	.....	.....			N-Acetyl, 150–60, yel.

*Derivative data given in order: m.p., crystal color, solvent from which crystallized.

TABLE VII. ORGANIC DERIVATIVES OF PHENOLS
b) Solids (Listed in order of increasing m.p.)* (Continued)

No.	Name	Melting point, °C	Boiling point, °C	Phenyl ure-thane	α-Naph-thyl-ure-thane	p-Nitro-ben-zoate	3,5-Di-nitro-benzoate	Bromo derivative	p-Toluene sul-fonate	Acetate	Benzoate	Miscellaneous
383	4-Hydroxyazobenzene (4-Benzeneazophenol)	152, al.								89; 84-5, yel., al.	136-8, yel., al.	Hydrochloride, 169; $HNO_3 \rightarrow$ 2,4-Dinitrophenol, 113; Propionate, 75, red
384	Tribromophloroglucinol. . .	152-3 (anh.), w.							mono: 169, bz.-pet. eth.; tri: 181-3, al.			Me. eth., 123, bz.; Tri-Me. eth., 145, al.; Di-Et. eth., 63-5, w.-ac. a.; Tri-Et. eth., 102-4, ac. a.
385	4-Amino-2-chlorophenol . .	153							116-7, aq. al.			N-Acetyl, 144; O,N-Diacetyl, 124; Me. eth., 62
386	2,2'-Dihydroxy-5,5'-dimethylbiphenyl	153-4, w.								di: 88		Diformate, 61, 70% al.
387	2-Amino-5-chlorophenol . .	154, aq. al.										O,N-Dibenzoyl, 140, al.; Me. eth., 46; Hydrochloride, 226 d.
388	1,2,4-Trihydroxynaph-thalene (1,2,4-Naph-thalenetriol)	154								tri: 134-5		. . . .
389	6-Hydroxybiphenyl-2-carboxylic acid	154 (anh.)							177, ac. a.	88-9, al.	121; 150	Benzenesulfonate, 104-5; Butyrate, 59-60
390	4-Amino-3-nitrophenol. . . .	154, dk. red							134, al.	185		N-Acetyl, 218, yel., w.; O,N-Diacetyl, 146; O-m-nitro-benzenesulfonate, 184
391	2-Azoxyphenol (2,2'-Di-hydroxyazoxybenzene) . .	154-5								di: 150, bz.	di: 108, al.	Di-Me. eth., 81; Di-Et. eth., 102
392	3,5-Stilbenediol (3,5-Di-hydroxystilbene; Pino-sylvyn)	155.5-6.0, ac. a.								di: 100-1, me. al.	di: 150-1, ac. a.-me. al.	Me. eth., 122-3, ac. a.; Di-Me. eth., 56-7, me. al.
393	1-Hydroxyphenanthrene (1-Phenanthrol)	156, eth.								135-6, al.		Me. eth., 105, me. al.; Picrate, 182, or.-red, me. al.
394	4-Iodo-3-nitrophenol	156, yel., al.								107.5		Me. eth., 62; Et. eth., 63.5
395	1,4,5-Trichloro-2-naphthol	157-8, ac. a.								129		. . . .
396	9-Hydroxyphenanthrene (9-Phenanthrol)	158, bz; 153								77-8, al.	99-100	Me. eth., 95-6, me. al.; Picrate, 185, red; Propionate, 95, ac. a.
397	5-Nitroresorcinol	158, w.								di: 105		Me. eth., 141-2, yel.; Di-Me. eth., 89, eth.-et. ac.; Et. eth., 80, w.
398	4,4'-Dihydroxydiphenyl-methane.	158, w.								di: 69-70, al.	di: 156, al.	Di-Me. eth., 52; Di-Et. eth., 38-9
399	2,4,6-Tri-iodophenol	158-9, w.-al.				181				156, bz.	137	Me. eth., 98-9, bz.; Et. eth., 83, eth.

*Derivative data given in order: m.p., crystal color, solvent from which crystallized.

TABLE VII. ORGANIC DERIVATIVES OF PHENOLS

b) Solids (Listed in order of increasing m.p.)* (Continued)

No.	Name	Melting point, °C	Boiling point, °C	Phenylurethane	α-Naphthylurethane	p-Nitrobenzoate	3,5-Dinitrobenzoate	Bromo derivative	p-Toluene sulfonate	Acetate	Benzoate	Miscellaneous
400	**Salicylic acid (2-Hydroxybenzoic acid)**	158.3	211[20]	......	......	205, me. al.	....		......	135, wh.	132	Aryloxyacetic acid, 191; Amide, 142; Anilide, 136; p-Toluidide, 156
401	**4-Amino-2,3,6-trichlorophenol**	159, al.		....	....					153, yel.	......	N-2,4-Dinitrophenyl, 211, or.-red
402	**4-Amino-3-chlorophenol** ..	160									160	N-Acetyl, 121
403	**6-Amino-2,4-dinitro-3-methylphenol (6-Amino-2,4-dinitro-m-cresol)**	160, al.										N-Acetyl, 225, yel.
404	**2,2′-Dihydroxychalcone**...	160–1		....	....					di: 85.5–6.0	di: 114	2-Me. eth., 112, al.; 2-Et. eth., 61, al.
405	**4-Amino-6-nitroresorcinol** .	160–1 d., eth.-lgr.										N-Acetyl, 261, yel., ac. a.; Di-Me. eth., 136–7, red. aq. al.
406	**4,4′-Dihydroxy-3,3′-dimethylbiphenyl**	160–1								di: 131, al.; 135.5	185, ac. a.	....
407	**2,3-Dihydroxynaphthalene (2,3-Naphthalenediol)**...	160–1, w.		....							di: 235	Me. eth., 108; Di-Me. eth., 116.5, lgr., Et. eth., 109–10; Di-Et. eth., 96–7; 2-Naphthylamine add. comp., 168
408	**1,2-Dihydroxyanthracene (1,2-Anthracenediol)**	160–2		....						di: 157.0–7.5, al.-ac. a.		....
409	**6-Retenol (6-Hydroxyretene)**	161, xyl.		....	....				110.5–11	134–5	......	Picrate, 152–2.5, red, al.
410	**5-Amino-2-methylphenol (5-Amino-o-cresol)**	161, w.							111–2, aq. al.			N-Acetyl, 225; N,N-Diacetyl, 132–3; O,N-Dibenzoyl, 162; Me. eth., 58
411	**4,4′-Dihydroxytriphenylmethane**	161, w.-al.								di: 109–10	......	Di-Me. eth., 100–1, chl.-me. al.
412	**3-Hydroxyphthalic acid** ...	161 d.								114–6	148	....
413	**1,2,3,4-Tetrahydroxybenzene (Apionol; Phenetrol)**	161, pink, et. ac.								tetra: 142	tetra: 191–2	1,4-Di-Me. eth., 106; Tetra-Me. eth., 89
414	**5-Amino-2-nitrophenol**	162, or.-yel., w.										N-Acetyl, 221, yel., ac. a.; O,N-Diacetyl, 149, w.
415	**1,3,4-Trichloro-2-naphthol**	162								133.5–4.0, ac. a.	....	
416	**2-Amino-5-methylphenol (6-Amino-m-cresol)**	162 d.		....								N-Benzoyl, 169, acet.-pet. eth.; O,N-Dibenzoyl, 162–3, me. al.; Me. eth., 171, 50% al.
417	**1-Amino-2,4-dihydroxynaphthalene**	162 (at 130 → vlt.)		....	....							O,O,N-Triacetyl, 155–6, bz.

*Derivative data given in order: m.p., crystal color, solvent from which crystallized.

TABLE VII. ORGANIC DERIVATIVES OF PHENOLS
b) Solids (Listed in order of increasing m.p.)* (Continued)

No.	Name	Melting point, °C	Boiling point, °C	Phenyl ure-thane	α-Naph-thyl-ure-thane	p-Nitro-ben-zoate	3,5-Di-nitro-benzoate	Bromo derivative	p-Toluene sul-fonate	Acetate	Benzoate	Miscellaneous
418	2,4′-Biphenol (2,4′-Di-hydroxybiphenyl)	162–3	342	......						di: 94, al.	......	Di-Me. eth., 70
419	2,5-Dimethylresorcinol (2,6-Dihydroxy-p-xylene)	163, bz.	277–80	......						di: 69, al.	......	Me. eth., 118–21, bz.
420	2-Amino-3,5-dimethyl-phenol	163, bz.		....						186–7	......	O,N-Diacetyl, 87–8; N-Benzoyl, 211–2, al.; O,N-Di-benzoyl, 148–9, me. al.; Hydro-chloride, 270–80, dil. HCl
421	4-Nitro-1-naphthol	164, w.		....							......	m-Nitrobenzene-sulfonate, 135; Me. eth., 81, yel., bz.; 85–6; Et. eth., 120, al.; 116–7
422	3,5-Dinitrocatechol (1,2-Dihydroxy-3,5-dinitro-benzene)	164		....						120; di: 112–4; 124, al.	......	1-Me. eth., 123; 2-Me. eth., 80; Di-Me. eth., 102, al.; 1-Et. eth., 155, al.; Di-Et. eth., 78; 94–5
423	2,2′-Dihydroxy-6,6′-dimethylbiphenyl	164		....						di: 87, al.	di: 136, al.	....
424	2-Amino-3,4-dihydroxy-naphthalene...........	164		....						di: >200 d.	......	N-Acetyl, 170 d.; 3-O,N-Diacetyl, 195 d., acet.
425	2,6-Dibromohydroquinone.	164		....						di: 117	......	
426	2-Benzoylphloroglucinol (2,4,6-Trihydroxybenzo-phenone).............	165		....							tri: 125–6	2,4-Di-Me. eth., 83; 2,6-Di-Me. eth., 178–9
427	5-Amino-3-nitrophenol....	165, yel.		....							......	N-Acetyl, 260–70, et. ac.
428	2-Methoxy-4-nitrophenol	165 d., yel., chl.									156–8	188, yel., al.	Me. eth., 105–6
429	3-Amino-4,5-dimethyl-phenol	165, eth., br.; (173–5, subl.)		....							......	N-Acetyl, 191 (sinters, 184); O,N-Diacetyl, 157, al., O,N,N-Triacetyl, 100–1, al.; N-Benzoyl, 195–6
430	1,2,3,5-Tetrahydroxy-benzene	165									tri: 74	......	1,3-Di-Me. eth., 159; 83; 1,2,3-Tri-Et. eth., 105
431	4-Phenylphenol (4-Hydroxybiphenyl)......	165	305–8	167.5	....				177, ac. a.; 179, al.-acet.	88–9; 87–8, al.	150–1, al.	p-Phenylazo-benzoate, 213.5–4.0; 2,4-Dinitro-phenyl ether, 118
432	4-Amino-2-bromophenol	165; 155		....							......	N-Benzoyl, 145; O,N-Dibenzoyl, 192
433	Benzeneazocatechol (3,4-Dihydroxyazobenzene)..	165 d., dk. red, al.		....							......	3-Me. eth., 70–1, red, lgr.; Di-Me. eth., 53–4, red, lgr.; 44–5
434	4-Amino-2-chloro-3-nitrophenol	165.5 d., bz.		....							......	N-Acetyl, 184–5; Et. eth., 74, or.

*Derivative data given in order: m.p., crystal color, solvent from which crystallized.

TABLE VII. ORGANIC DERIVATIVES OF PHENOLS
b) Solids (Listed in order of increasing m.p.)* (Continued)

No.	Name	Melting point, °C	Boiling point, °C	Phenyl urethane	α-Naphthyl urethane	p-Nitrobenzoate	3,5-Dinitrobenzoate	Bromo derivative	p-Toluene sulfonate	Acetate	Benzoate	Miscellaneous
435	4-Amino-2,5-dinitrophenol	166–7, al.										N-Acetyl, 144–5; Me. eth., 153, bz.-lgr., red; Et. eth., 139
436	4-Amino-2,6-dinitro-3-methylphenol (4-Amino-2,6-dinitro-m-cresol)	167, red, 50% al.										N-Acetyl, 231 d., ac. a.; Et. eth., 96–7; Hydrochloride, 200 d., dil, HCl
437	2-Benzamidophenol (2-Hydroxybenzanilide	167 d.							109–10, al.	134–40, al.		Me. eth., 60
439	2,3,4-Trichloro-1-naphthol	168, lgr.								123–4		
440	2-Hydroxyphenanthrene (2-Phenanthrol)	168, al.								142–3	139–40, al.	Et. eth., 112, al.; Picrate, 156, red
441	1,5-Di-(2-hydroxyphenyl)-1,4-pentadiene-3-one	168							di: 128, al.		135, al.	Di-Me. eth., 125, al.; Di-Et. eth., 89, al.
442	3,5-Diaminophenol	168–70										3,5-Diacetyl, 195
443	4-Acetamidophenol (4'-Hydroxyacetanilide)	169								150–1		
444	2-Amino-4,6-dinitrophenol (Picramic acid)	169, dk. red, al.										N-Acetyl, 201; N-Benzoyl, 230; 220; N-p-Toluenesulfonyl, 191
445	4-Amino-2-chloro-5-nitrophenol	169.5 d.										N-Acetyl, 166, yel.; Et. eth., 128.5, or.
446	3,3'-Dihydroxybenzophenone	170								89–90	101–2	
447	1,4-Dihydroxy-2-methylnaphthalene (2-Methyl-1,4-naphthohydroquinone)	α: 170; β: 60, ac. a.							di: 113, al.		181	
448	4-Amino-2,6-dinitrophenol (Isopicramic acid)	170, br., w.										N-Acetyl, 182; N-Benzoyl, 263; 250; Me. eth., 212; Et. eth., 172
449	Benzeneazoresorcinol (2,4-Dihydroxyazobenzene)	170, dk. red							di: 104, or.-red, al.			Di-Me. eth., 92, red, al.; Di-Et. eth., 70, yel.-red, al.
450	3,4-Diaminophenol	170–2; 167–8; unstable										N,N'-Diacetyl, 205–7; 3,4-Dibenzoyl, 203–5; 1,3,4-Tribenzoyl, 225; Et. eth., 71–2, b.p. 294–6
451	5-Nitro-1-naphthol	171, dk. yel.-red, w.								114, w.-al.	109, me. al.	Me. eth., 96–7, yel., pet. eth.
452	Hydroquinone (1,4-Dihydroxybenzene)	171; 172	286	di: 224; 205–7		di: 258, al.	di: 317	di: 186	mono: 98–9, bz.; di: 159, 25% al.	di: 123, w.	di: 199; 204 (cor.), tol.	Aryloxyacetic acid, 250; 2,4-Dinitrophenyl ether, 243–6; N,N-Diphenylurethane, 250

*Derivative data given in order: m.p., crystal color, solvent from which crystallized.

TABLE VII. ORGANIC DERIVATIVES OF PHENOLS
b) Solids (Listed in order of increasing m.p.)* (Continued)

No.	Name	Melting point, °C	Boiling point, °C	Phenyl ure-thane	α-Naph-thyl ure-thane	p-Nitro-ben-zoate	3,5-Di-nitro-benzoate	Bromo derivative	p-Toluene sul-fonate	Acetate	Benzoate	Miscellaneous
453	2-Azophenol (2,2'-Dihydroxyazobenzene)..	172, yel., bz.								di: 150, or.-red	......	Di-Me. eth., 153, or.; Di-Et. eth., 131, red
454	2,3,4-Trihydroxyaceto-phenone (Gallaceto-phenone).............	173; 192.0–2.5								2,4-di: 107–8; 3,4-di: 78–81; tri: 85	......	Semicarbazone, 225 d.; Oxime, 162–3; Picrate, 133
455	3,5-Dinitrosalicylic acid...	173–4 (anh.)								......	163	Amide, 181; Anilide, 180 (subl.); 194
456	2-Aminophenol (2-Hydroxyaniline)	174								......	175	N-Acetyl, 209; 201; N-Benzenesulfonyl, 141; N-p-Toluenesulfonyl, 146
457	3-Benzamidophenol (3-Hydroxybenzanilide)	174, tol.								......	153	Et. eth., 103
458	4-Amino-2-methylphenol (4-Amino-o-cresol)	175, bz.							109–10, bz.-lgr.		......	N-Acetyl, 179; O,N-Dibenzoyl, 194, ac. a.; Me. eth., 59.0–9.5; 92–3, aq. al.
459	1,6,7-Trihydroxynaphtha-lene (1,6,7-Naphthalene-triol)	175								tri: 143–4	......	....
460	4-Nitrocatechol	176, yel., w.								di: 98	di: 156, al.	Di-Me. eth., 96, w.-al.; Di-Et. eth., 73–5, pa. yel.
461	1,4-Dihydroxynaphthalene (1,4-Naphthalenediol; 1,4-Naphthohydro-quinone).............	176; 192								di: 128–30, al.	di: 169, ac. a.	Me. eth., 131; Di-Me. eth., 85, CS$_2$
462	1,7-Dihydroxynaphthalene (1,7-Naphthalenediol)...	178, bz.	....	203–4	......	182–3	....			di: 108, bz.	di: 101.5; 113–5	Di-Et. eth., 67
463	1,2-Dihydroxyphenan-threne (1,2-Phenanthrene-diol)................	178, w.-al.								di: 147, me. al.	......	Di-Me. eth., 100–2, lgr.
464	4-Amino-6-isopropyl-3-methylphenol (4-Amino-6-isopropyl-m-cresol) ...	178–9, bz.									......	N-Benzoyl, 178–9; O,N-Dibenzoyl, 166–7, aq. al.; Hydrochloride, 255
465	1,8,9-Trihydroxyanthra-cene (1,8,9-Anthracene-triol; Anthralin).......	178–80; 176–7								tri: 209–10		....
466	4-Amino-3-methylphenol (4-Amino-m-cresol).....	179, 50% al.									92, pet. eth.	N-Acetyl, 138; Hydrochloride, 215; Me. eth., 29–30
467	2,5-Dihydroxyphenan-threne (2,5-Phenanthrene-diol)................	180, xyl.								di: 144, w.-ac. a.	......	Di-Me. eth., 117, ac. a.
468	9,10-Dihydroxyanthracene (9,10-Anthradiol)......	180, yel.								di: 260, ac. a.	di: 292, yel. xyl.-chl.	Di-Me. eth., 202, bz.; Di-Et. eth., 148, bl. fluor., al.

*Derivative data given in order: m.p., crystal color, solvent from which crystallized.

TABLE VII. ORGANIC DERIVATIVES OF PHENOLS

b) Solids (Listed in order of increasing m.p.)* (Continued)

No.	Name	Melting point, °C	Boiling point, °C	Phenyl urethane	α-Naphthyl urethane	p-Nitrobenzoate	3,5-Dinitrobenzoate	Bromo derivative	p-Toluene sulfonate	Acetate	Benzoate	Miscellaneous
469	1,2,5,8-Tetrahydroxynaphthalene (1,2,5,8-Naphthalenetetrol)	180 d.								*tetra*: 202		
470	4-Hydroxy-3-iodo-5-methoxybenzaldehyde (5-Iodovanillin)	180								105–6	135.5–6.5	Oxime, 178–9; Semicarbazone, 187–8 d.
471	6-Nitro-1-naphthol	181–2								121		
472	4-Amino-3,5-dimethylphenol	181.2, chl.										N-Acetyl, 178–80; N-Benzyl, 104–5; Me. eth., 43, pet. eth.
473	2-Amino-4-nitroresorcinol	182, br., aq. al.										N-Acetyl, 213, yel.
474	3-Azoxyphenol (3,3′-Dihydroxyazoxybenzene)	183								*di*: 102, 50% al.	*di*: 75, ac. a.	Di-Me. eth., 49–50; 47.6
475	N-Benzylidene-4-aminophenol	183								92	144	Me. eth., 62; Et. eth., 76 (71); Hydrochloride, 167–77
476	4-Aminophenol (4-Hydroxyaniline)	184 (subl.)				178.5 (cor.)				168		O,N-Diacetyl, 150; 2,4-Dinitrobenzoyl, 204.5 (cor.); N-p-Toluenesulfonyl, 253
477	1,9-Dihydroxyphenanthrene (1,9-Phenanthrenediol)	184–5, bz.								*di*: 154–5, w.-al.		9-Me. eth., 131–2, bz.-pet. eth.; Di-Me. eth., 113–4, me. al.
478	4,6-Diacetylresorcinol (Resodiacetophenone)	185, al.								*di*: 120	*mono*: 214–5; *di*: 118	Di-Me. eth., 171.5; Dioxime, 242; Phenylhydrazone, 233
479	4-Amino-2-naphthol	185							137			O,N-Dibenzoyl, 309
480	7-Hydroxy-4-methylcoumarin (4-Methylumbelliferone)	185–6, al.		155–6		143				150, al.	159–60, al.	Me. eth., 159; m-Nitrobenzoate, 210–11; Picrate, 108
481	3-Amino-4-nitrophenol	185–6, or., w.										N-Benzoyl, 166; Me. eth., 131; Et. eth., 105–6
482	1,2,6-Trihydroxynaphthalene (1,2,6-Naphthalenetriol)	188								*tri*: 262		
483	α,2,4-Trihydroxyacetophenone (ω-Hydroxyresacetophenone)	189								*tri*: 129	α-mono: 200	α-Me. eth., 136; p-Nitrophenylhydrazone, 205 d.; Oxime, 105–7
484	2,7-Dihydroxynaphthalene (2,7-Naphthalenediol)	190; 185–6, w.						150, chl.		*mono*: 171–2, me. al.; *di*: 136, w.	*mono*: 199, me. al.; *di*: 139, al.	Aryloxyacetic acid, 149; N,N-Diphenylurethane, 176; Me. eth., 117; Di-Me. eth., 139; Di-Et. eth., 104, al.
485	1,4,5,8-Tetrahydroxynaphthalene (1,4,5,8-Naphthalenetetrol)	190								*tetra*: 277–9		

*Derivative data given in order: m.p., crystal color, solvent from which crystallized.

TABLE VII. ORGANIC DERIVATIVES OF PHENOLS

No.	Name	Melting point, °C	Boiling point, °C	Phenyl ure-thane	α-Naph-thyl-ure-thane	p-Nitro-ben-zoate	3,5-Di-nitro-benzoate	Bromo derivative	p-Toluene sul-fonate	Acetate	Benzoate	Miscellaneous
486	1,5-Diacetyl-2,3,4-tri-hydroxybenzene (4,6-Diacetylpyrogallol; Gallodiacetophenone)...	190–1, w.								mono: 207–9	tri: 189	Tri-Me. eth., 73–4
487	3,9-Dihydroxyacridine....	190–2, lt. br. → dk. in air, al.										9-Me.-3-Et. eth., 144, yel., w.; In al. → grn. fluor.
488	Pentachlorophenol.......	190.2							145	149–50	164–5; 159	Aryloxyacetic acid, 196
489	5-Amino-2-naphthol......	191										O,N-Diacetyl, 187; O,N-Dibenzoyl, 223
490	1,4-Dihydroxynaphthalene (1,4-Naphthalenediol)...	192								di: 128–30	di: 169	....
491	5-Amino-1-naphthol......	192										O,N-Dibenzoyl, 276
492	Tetrabromocatechol......	192–3								di: 215–6	di: 197–8, bz.-lgr.	Me. eth., 162–3; Di-Me. eth., 151–2; 118–20
493	6-Hydroxyquinoline......	193, al.							98	36–8	230–1, ac. a.	Methiodide, 236 d.; Picrate, 235–6
494	3-Methyl-2,4,5,6-tetra-bromophenol (2,4,5,6-Tetrabromo-m-cresol)...	194, ac. a.								165–6	153–4	Me. eth., 145–6, al.; Et. eth., 108, eth.
495	1,5-Dichloro-4,8-dihy-droxynaphthalene.......	194, ac. a.								4-mono: 148–60, ac. a.; di: 154; 143, acet.	4-mono: 157–8, ac. a.; di: 179; acet.	....
496	1,6-Dinitro-2-naphthol....	195 d.							181			Me. eth., 204; 198; Et. eth., 144, yel.; Alc. NH₃ at 160° → 1,6-di-nitro-2-naphthyl-amine, 248; Oxid. → 4-nitro-phthalic acid, 165
497	5,6-Dihydroxyace-naphthene (5,6-Acenaph-thenediol)............	196–9, bz.								di: 194–5		....
498	1,2,7-Trihydroxynaphtha-lene (1,2,7-Naphthalene-triol)...............	197								181–2		
499	4-Methyl-2,3,5,6-tetra-bromophenol (2,3,5,6-Tetrabromo-p-cresol)...	199								156		....
500	2-Hydroxyquinoline......	199										2-Ph. eth., 69
501	2-Amino-4-methyl-5-nitro-phenol (2-Amino-5-nitro-p-cresol)............	199–200 d.										N-Acetyl, 242, ac. a.; Me. eth., 132, yel.
502	6-Amino-1-naphthol......	199.5										O,N-Diacetyl, 130; N-Benzoyl, 203; O,N-Dibenzoyl, 230
503	Hexahydroxybenzene	200 d.								hexa: 203	hexa: 313	....

*Derivative data given in order: m.p., crystal color, solvent from which crystallized.

TABLE VII. ORGANIC DERIVATIVES OF PHENOLS
b) Solids (Listed in order of increasing m.p.)* (Continued)

No.	Name	Melting point, °C	Boiling point, °C	Phenyl urethane	α-Naphthyl urethane	p-Nitrobenzoate	3,5-Dinitrobenzoate	Bromo derivative	p-Toluene sulfonate	Acetate	Benzoate	Miscellaneous
504	Methylenedi-2-naphthol (2,2'-Dihydroxy-1,1'-dinaphthylmethane)	200, ac. a.								di: 214, al.		Di-Me. eth., 144–7, al.; Picrate, 178–9, red-br.
505	3-Hydroxybenzoic acid	200								131		Aryloxyacetic acid, 206; Amide, 170; p-Bromophenacyl ester, 176
506	Methyl gallate (Methyl 3,4,5-trihydroxybenzoate)	200–1								tri: 120; 120–2, al.	tri: 139, al.	
507	3-Hydroxyquinoline	200–1, bz.							90			Picrate, 240–5
508	7-Amino-2-naphthol	201; 208									177	N-Acetyl, 232; O,N-Diacetyl, 156; N-Benzoyl, 243–6; O,N-Dibenzoyl, 181
509	4-Amino-3-methyl-2-nitrophenol (4-Amino-2-nitro-m-cresol)	201, al.										O,N-Diacetyl, 127–8
511	2-Acetamidophenol (2-Hydroxyacetanilide)	201–3; 209, aq. al.								122	140	O,N-Diacetyl, 77
512	2-Acetylhydroquinone (2,5-Dihydroxyacetophenone)	202								di: 68	di: 113	p-Nitrophenylhydrazone, 215–6; Oxime, 149–50
513	2,8-Dihydroxyphenanthrene (2,8-Phenanthrenediol)	202, w.-al.								di: 125, al.		
514	4-Hydroxy-4'-nitrobiphenyl	203; 200–1							159, bz.	138–9, lgr.	208–10, ac. a.	Me. eth., 111, yel., al.; 3,5-Dinitro deriv., 197–8
515	3,5-Dimethoxy-4-hydroxybenzoic acid	205								191	229–32	
516	4-Hydroxycoumarin (Benzotetronic acid)	206; 232–3								103		Me. eth., 124, w.
517	2,3,5-Trihydroxyacetophenone	206–7, yel., ac. a.								tri: 106–7, lgr.	tri: 106–7	p-Nitrophenylhydrazone, 241–2 d., w.-al.
518	2,4,5-Trihydroxyacetophenone	206–7, red								2,4-di: 165–6, bz.		
519	6-Chloro-3-hydroxy-4-methylbenzoic acid	206–8									146	Amide, 239–40; Anilide, 222
520	8-Amino-2-naphthol	207 d.										O,N-Diacetyl, 178; O,N-Dibenzoyl, 208
521	3-Azophenol (3,3'-Dihydroxyazobenzene)	207, yel.								di: 144, yel., al.	di: 188, w.-al.	Di-Me. eth., 73–4; Di-Et. eth., 91, yel.
522	2-Amino-5-nitrophenol	207–8, or.; 201–2, w.							188, yel., al.			N-Acetyl, 271–2; O,N-Diacetyl, 187; Me. eth., 139
523	2-Methyl-3,4,5,6-tetrabromophenol (3,4,5,6-Tetrabromo-o-cresol)	208								154		
524	2,6-Dichloro-3,4,5-tribromophenol	209									202	Me. eth., 143–4

Derivative data given in order: m.p., crystal color, solvent from which crystallized.

TABLE VII. ORGANIC DERIVATIVES OF PHENOLS
b) Solids (Listed in order of increasing m.p.)* (Continued)

No.	Name	Melting point, °C	Boiling point, °C	Phenyl ure-thane	α-Naph-thyl-ure-thane	p-Nitro-ben-zoate	3,5-Di-nitro-benzoate	Bromo derivative	p-Toluene sul-fonate	Acetate	Benzoate	Miscellaneous
525	4-Hydroxy-3-methoxy-benzoic acid (Vanillic acid)	210 (subl.)	....	......	......	140–1 d.	....			110, w.	178, aq. al.	Hydrazide, 207; Phenylhydrazide, 150–2.5
526	4,4'-Dihydroxybenzo-phenone	210	....	......	......	....	....			....	156; 152	Me. eth., 151–2; Di-Me. eth., 146; 2,4-Dinitrophenylhy-drazone, 190–2
527	cis-1,2-Dihydroxyace-naphthene (cis-Acenaph-thylene glycol)	212–3, w.								mono: 122–3, al.; di: 130, me. al.		....
528	4-Hydroxy-4'-nitroazo-benzene	212–3; 219.0–9.5	....	......	......	....	....			147, or.	195, or.	Me. eth., 157.5–8.0
529	6-Amino-2-naphthol	213 d.								....	......	O,N-Diacetyl, 220
530	2,4-Dihydroxybenzoic acid	213; 216 d.	....	......	......	....	....			....	......	Amide, 222; Anilide, 126–7; p-Nitrophenyl ester, 189
531	Methylphloroglucinol (2,4,6-Trihydroxy-toluene)	214–6, et. ac. ...	....	......	......	....	....			tri: 76, lgr.	......	2-Me. eth., 91 (+1H₂O); 117–9 (anh.); 4-Me. eth., 124; Tri-Me. eth., 10–13, b.p. 140–2[18]
532	4-Hydroxybenzoic acid	215; 210	....	......	......	....	....			187	221–3	Aryloxyacetic acid, 278; Anilide, 198; p-Toluidide, 204
533	4,6-Dinitroresorcinol	215, yel.			di: 178		....		mono: 135	di: 139	di: 343–4	Me. eth., 113; Di-Me. eth., 157; Et. eth., 77; Di-Et. eth., 133
534	4-Chloro-2,3,5,6-tetra-bromophenol	215	....	......	......	....	....			......	203	Me. eth., 161
535	2,6-Dihydroxynaphthalene (2,6-Naphthalenediol)	215; 218	....	......	......	....	....			di: 175	di: 215	Di-Me. eth., 50, bz.; Di-Et. eth., 162, al.
536	1,2,4,5-Tetrahydroxy-benzene	215–20	....	......	......	....	....			tetra: 226–7	......	2-Me. eth., triacetyl, 142; Tetra-Me. eth., 103
537	4-Azophenol (4,4'-Di-hydroxyazobenzene)	α: 216, grn. (anh.); β: 216, dk. red. (anh.)								di: 198–9, yel., ac. a.	di: 210.5–1.5; 249–51, red-yel., bz.	Di-Me. eth., 160.5–2.5, me. al.; Di-Et. eth., 157–9, yel., al.; 160
538	2-Amino-3-nitrophenol	216–7, red	....	......	......	....	....		136	....	......	N-Acetyl, 172; Me. eth., 75–6
539	4-Benzamidophenol (4-Hy-droxybenzanilide)	216–7; 227	....	......	......	....	....			171	235	Me. eth., 153–4, al.; Et. eth., 173, aq. al.; Benzyl eth., 226–7
540	2,5-Dimethylhydro-quinone (Hydro-phlorone)	217	....	......	......	....	....			mono: 117; di: 135	mono: 162–3, pet. eth.; di: 159, me. al.	Me. eth., 90, lgr.

*Derivative data given in order: m.p., crystal color, solvent from which crystallized.

TABLE VII. ORGANIC DERIVATIVES OF PHENOLS
b) Solids (Listed in order of increasing m.p.)* (Continued)

No.	Name	Melting point, °C	Boiling point, °C	Phenyl urethane	α-Naphthyl urethane	p-Nitrobenzoate	3,5-Dinitrobenzoate	Bromo derivative	p-Toluene sulfonate	Acetate	Benzoate	Miscellaneous
541	Phloroglucinol (1,3,5-Trihydroxybenzene)	217–9, rapid htng.; 200–9, slow htng.	....	tri: 190–1	.....	283	tri: 162	tri: 151	......	tri: 104–6, al.	tri: 173–4, al.	Monobenzenesulfonate, 163–4 (anh.); Monobenzenesulfonate, diacetate, 95–6, bz.; Dibenzenesulfonate, 120–1, bz.; Dibenzenesulfonate monoacetate, 81, me. al.
542	2-Amino-3,5-dinitrophenol	218, yel., al.	....	......	......	......	......	186	......	......	......	O,N-Di-p-Toluenesulfonyl, 188; N-Acetyl, 171; Me. eth., 181
543	2,2'-Dihydroxy-1,1'-binaphthyl	218	....	......	......	......	......	......	......	di: 109, al.	mono: 204; di: 160	Dipicrate, 175–6; Di-Me. eth., 190, Di-Et. eth., 90
544	7-Hydroxybenzo [a]pyrene	218–9, yel.	....	......	......	......	......	......	......	194–5, yel.	191–2, yel.	Me. eth., 183–4, pyr.; p-Nitrobenzyl eth., 252–3
545	Acetylphloroglucinol (2,4,6-Trihydroxyacetophenone; Phloracetophenone)	219, w.; 222	....	......	......	......	......	......	......	tri: 103	2-mono: 168; 4-mono: 210–11; tri: 117–8	2-Me. eth., 205–7; 4-Me. eth., 139–40; Tri-Me. eth., 103; Tri-Et. eth., 75, aq. al.
546	1,1'-Dihydroxy-2,2'-binaphthyl	220	....	......	......	......	......	......	......	di: 169	......	Di-Me. eth., 122, lgr.
547	3,5-Dihydroxypyrene	220 d., ac. a.	....	......	......	......	......	......	......	di: 155	......	Di-Me. eth., 177–8
548	3,6-Dihydroxyphenanthrene (3,6-Phenanthrenediol)	221, w.-al.	....	......	......	......	......	......	......	di: 124.5, al.	......	3-Me. eth., 135–6, w.-me. al.; Di-Me. eth., 104–5, w.-me. al.
549	2,3-Dimethylhydroquinone (3,6-Dihydroxy-o-xylene)	221, sl. d., w.	....	......	......	......	......	......	......	mono: 174–5, acet.-pet. eth.; di: 182, acet.-pet. eth.	......	Di-Et. eth., 68–9
550	3-Hydroxy-2-naphthoic acid	222	....	......	......	......	......	......	......	184–6	......	Amide, 218; Anilide, 244; p-Toluidide, 222
551	4-Amino-6-chloro-3-methylphenol (4-Amino-6-chloro-m-cresol)	223–5 d.	....	......	......	......	......	......	......	......	......	O,N-Diacetyl, 162; O,N-Dibenzoyl, 220
552	4-Azoxyphenol (4,4'-Dihydroxyazoxybenzene) ..	224 d.	....	......	......	......	......	......	......	di: 163, or., al.	mono: 200, al.; 212, bz.; di: 187–90	Di-Me. eth., 118–9, yel.; Di-Et. eth., 137–8

*Derivative data given in order: m.p., crystal color, solvent from which crystallized.

No.	Name	Melting point, °C	Boiling point, °C	Phenyl ure-thane	α-Naph-thyl-ure-thane	p-Nitro-ben-zoate	3,5-Di-nitro-benzoate	Bromo derivative	p-Toluene sul-fonate	Acetate	Benzoate	Miscellaneous
553	5-Hydroxyquinoline	224							85			Hydrochloride, 240; Methiodide, 224; Picrate, 187
554	2-Amino-4-chloro-5-nitrophenol	225 d., yel.										N-Acetyl, 193; Me. eth., 132
555	1,2,3,4-Tetrahydroxy-naphthalene (1,2,3,4-Naphthalenetetrol)	225								tetra: 220		
556	1,8-Dihydroxyanthracene (1,8-Anthracenediol; Chrysazol)	225, yel., al.-w.								di: 184, et. ac.		Di-Me. eth., 198, al.; Di-Et. eth., 139, al.
557	4-Amino-6-nitrocatechol	228, yel.										O,O,N-Triacetyl, 207, ac. a.
558	5-Hydroxycoumarin	229								88–9; 84		Me. eth., 85–7
559	2-Hydroxy-5-nitro-benzoic acid	229–30										Amide, 225; Ani-lide, 224
560	Pentabromophenol	229.5; 225–6								197; 171		Me. eth., 173–4; Et. eth., 136
561	2,6-Dihydroxyphenan-threne (2,6-Phenanthrene-diol)	234, w.-al.								di: 122–3, al.	di: 252–3	Di-Me. eth., 87, me. al.
562	3-Amino-2-naphthol	235										O,N-Diacetyl, 188; O,N-Dibenzoyl, 184
563	5-Hydroxy-1-naphthoic acid	235								202	241	
564	2,3-Dihydroxyacridine (2,3-Acridinediol)	235 d.										Di-Me. eth., 107, yel.-wh., al.
565	Tetrachlorohydroquinone	236–7								di: 245	di: 233	
566	7-Hydroxyquinoline	238–40, al., (br. at 200)							116		88–9, al.	Methiodide, 251 d., al.; Picrate, 244–5
567	1,7-Dihydroxyanthrone (Euxanthone)	240								7-mono: 160, al.; di: 185, bz.	di: 221–2; 214	7-Me. eth., 130–5
568	1,6-Dihydroxypyrene (1,6-Pyrenediol)	240 d.; sinters at 175										Sol. red with grn. fluor.; In alkali red soln. with bl. fluor.; Zn dust → pyrene, 149–50
569	4-Amino-2,5-dimethyl-phenol	242 d., al.										N-Acetyl, 177–9, al.; Et. eth., 69.5
570	3,8-Dihydroxyphenan-threne (3,8-Phenanthrene-diol)	247, lt. red, w.-al.								di: 184, al.		Di-Me. eth., 117, me. al.
571	3-Hydroxy-1-naphthoic acid	248								169–70	222–3	Amide, 209–11; Anilide, 112–3
572	2,5-Dihydroxypyridine (2,5-Pyridinediol)	248								5-mono: 156		Hydrochloride, 106 (hyd.); 154 (anh.)
573	1,4-Dihydroxyisoquino-line (1,4-Isoquinoline-diol)	>250, ac. a. turns red at 200								4-mono: 207–8		4-Me. eth., 171, acet.-pet. eth.
574	2,3-Dihydroxyquinoline (3-Hydroxycarbostyril; 2,3-Quinolinediol)	257–8; >300								3-mono: 211	3-mono: 286–7; di: 45–6, pet. eth.	

*Derivative data given in order: m.p., crystal color, solvent from which crystallized.

TABLE VII. ORGANIC DERIVATIVES OF PHENOLS

b) Solids (Listed in order of increasing m.p.)* (Continued)

No.	Name	Melting point, °C	Boiling point, °C	Phenyl ure-thane	α-Naph-thyl-ure-thane	p-Nitro-ben-zoate	3,5-Di-nitro-benzoate	Bromo derivative	p-Toluene sul-fonate	Acetate	Benzoate	Miscellaneous
575	1,5-Dihydroxynaphthalene (1,5-Naphthalenediol)...	265; 258	....	....	....	....	....	....	....	di: 159–61, dil. al.	di: 235; 242, pyr.	Di-Me. eth., 183–4; Di-Et. eth., 130, w.-al.; 2-Naph-thylamine add. comp., 229.5
576	2,7-Dihydroxyphenan-threne (2,7-Phenanthrene-diol).............	265, w.-al.	....	....	....	....	....	....	....	di: 181.5, al.	......	Di-Me. eth., 169–70, me. al.
577	Phenolphthalein	265 (cor.); 261	....	di: 135, bz.	......	....	....	....	....	di: 143, al.	di: 169, bz.-lgr.	....
578	1,5-Dihydroxyanthracene (Rufol; 1,5-Anthracene-diol).............	265 d., yel.	....	....	....	....	....	....	....	di: 198, et. ac.	......	Di-Me. eth., 224, me. al.; Di-Et. eth., 179, al.
579	4,4′-Biphenol (4,4′-Di-hydroxybiphenyl)	274–5, al.	....	....	....	....	....	....	di: 189–90, bz.	di: 161, dil. al.; 164 (cor.)	di: 241, ac. a.	Aryloxyacetic acid, 274; Di-Me. eth., 173 (subl.); Di-Et. eth., 176
580	1-Amino-2-naphthol	276 d.	....	....	....	....	....	....	....			O,N-Diacetyl, 206; O,N-Dibenzoyl, 235
581	2,7-Dihydroxyanthracene (2,7-Anthracenediol)....	280–5 d., bz., turns dk. at 250	....	....	....	....	....	....	....	282		Di-Me. eth., 216–7, ac. a.; Di-Et. eth., 192–3
582	2,3-Dihydroxyanthracene (2,3-Anthracenediol)....	282 d., yel.	....	....	....	....	....	....	....	di: 175	......	Di-Me. eth., 204, al.
583	4,4′-Dihydroxystilbene (4,4′-Stilbenediol)	284, ac. a.	....	....	....	....	....	....	....	di: 213	......	Di-Me. eth., 214–5; Di-Et. eth., 208
584	3,5′-Dimethoxy-5,7,4′-trihydroxyflavonol (Syringetin)	288–9, pa. yel., ac. a.	....	....	....	....	....	....	....	tetra: 224–6	......	4′-Benzyl eth., 240–1
585	2,7-Dihydroxy-4-methyl-quinoline	290–300, w.-al. (+1H₂O); turns br. at 280	....	....	....	....	....	....	....	7-mono: 250–4, al.	7-mono: 288, al.	....
586	2,6-Dihydroxyanthracene (Flavol; 2,6-Anthracene-diol)...............	295–300 d., al.; turns dk. at 270	....	....	....	....	....	....	....	di: 260–1, ac. a.	......	Di-Me. eth., 255–6, ac. a.; Di-Et. eth., 230–1
587	Bi-α-naphthol (4,4′-Di-hydroxy-1,1′-binaphthyl)	300; 250	....	....	....	....	....	....	....	di: 217	......	Di-Me. eth., 252; Di-Et. eth., 211
588	2,8-Dihydroxyacridine (2,8-Acridinediol)	>300, turns red at 275	....	....	....	....	....	....	....			Di-Me. eth., 138–9; Di-Et. eth., 142–3
589	2,6-Dibromo-1,5-dihy-hydroxynaphthalene.....	>300; 224 d. turns dk. at 200	....	....	....	....	....	....	....	1-mono: 273; di: 228	di: 262, pyr.	1-Acetate, 5-ben-zoate, 164; Di-Me. eth., 161; Di-Et. eth., 148
590	3,7-Dihydroxyacridine (3,7-Acridinediol)	324, pa. yel., w.-al.	....	....	....	....	....	....	....			In al. sol. → grn. fluor.
591	3,8-Dihydroxypyrene (3,8-Pyrenediol)	330, Tri-cl-bz.-ph. hydraz.; turns dk. at 280	....	....	....	....	....	....	....	di: 224, ac. a.	......	Di-Me. eth., 244, cl-bz.

*Derivative data given in order: m.p., crystal color, solvent from which crystallized.

TABLE VII. ORGANIC DERIVATIVES OF PHENOLS
b) Solids (Listed in order of increasing m.p.)* (Continued)

No.	Name	Melting point, °C	Boiling point, °C	Phenyl ure-thane	α-Naph-thyl-ure-thane	p-Nitro-ben-zoate	3,5-Di-nitro-benzoate	Bromo derivative	p-Toluene sul-fonate	Acetate	Benzoate	Miscellaneous
592	**1,3-Dihydroxyacridone** ...	370	....	......	....	....	....	....	......	*mono*: 200, yel.	*mono*: 295–7	Me. eth. (i) 203, dk. br.; (ii) 252, yel.; Di-Me. eth., 286–7 d.; Anil., 269–70; Zn dust → acri-dine, 111

*Derivative data given in order: m.p., crystal color, solvent from which crystallized.

Cleavage to alkyl bromide or alkyl iodide.

$$ROR' \;+\; 2HBr \;\rightarrow\; RBr \;+\; R'Br \;+\; H_2O$$

Mixture of alkyl
bromides

$$ROR' \;+\; 2HI \;\rightarrow\; RI \;+\; R'I \;+\; H_2O$$

Mixture of
alkyl iodides

$$ArOR \;+\; HI \;\rightarrow\; ArOH \;+\; RI$$

Phenol   Alkyl
iodide

From the ether with concentrated hydrochloric acid.

For directions and examples see: Cheronis, p. 543; Linstead, pp. 46–7; Shriner, p. 116; Vogel, p. 316.

NOTE: For directions and examples for preparation of derivatives of alkyl iodides and alkyl bromides formed on cleavage of ethers see explanations and references to Table V, pp. 52, 53, 54.

*Alkyl 3,5-dinitrobenzoate.**

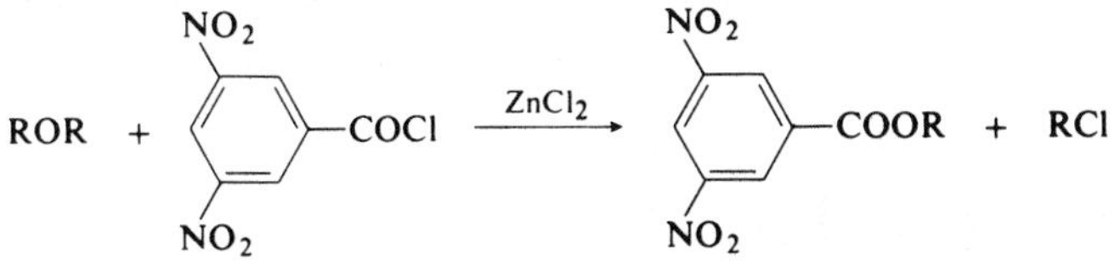

Alkyl 3,5-
dinitrobenzoate

From a symmetrical aliphatic ether with freshly fused zinc chloride and 3,5-dinitrobenzoyl chloride.

For directions and examples see: Cheronis, pp. 542, 543; Linstead, p. 46; Shriner, p. 239; Vogel, p. 316; Wild, p. 96; H. W. Underwood, O. L. Baril and G. C. Toone, *J. Amer. Chem. Soc.,* **52,** 4087 (1930).

Bromo derivative.

$$ArOR \;\xrightarrow{\;Br_2\;}\; Ar(OR)Br$$

Bromoaryl
ether

$$ArOR \;\xrightarrow{\;nBr_2\;}\; Ar(OR)Br_n$$

Polybromoaryl
ether

From alkyl aryl or diaryl ether with bromine in glacial acetic acid or chloroform.

For directions and examples see: Cheronis, p. 545; Shriner, p. 240.

From the aromatic ether with bromine in alcohol, acetic acid, ether, chloroform or petrol ether.

See: Wild, pp. 98–9; 101; H. W. Underwood, O. L. Baril and G. C. Toone, *J. Amer. Chem. Soc.,* **52,** 4087 (1930).

*Sulfonamide.**

$$ArOR \;\xrightarrow{\;ClSO_2OH\;}\; Ar(OR)SO_2Cl \;\xrightarrow{\;(NH_4)_2CO_3\;}\; Ar(OR)SO_2NH_2$$

Sulfonyl   Sulfonamide
chloride

The sulfonyl chloride is prepared from the aromatic ether with chlorosulfonic acid in chloroform or without solvent. The sulfonamide is obtained from the sulfonyl chloride with ammonium carbonate and/or aqueous ammonia.

For directions and examples see: Cheronis, pp. 545, 546; Linstead, pp. 47, 50; Shriner, p. 241; Vogel, p. 672; Wild, pp. 27, 101; E. H. Huntress and F. H. Carten, *J. Amer. Chem. Soc.,* **62,** 511, 603 (1940).

---

*Derivatives recommended for first trial.

WARNING: This is not an instruction manual. References should be consulted for the preparation of derivatives.

Picric acid and 1,3,5-trinitrobenzene addition complexes.

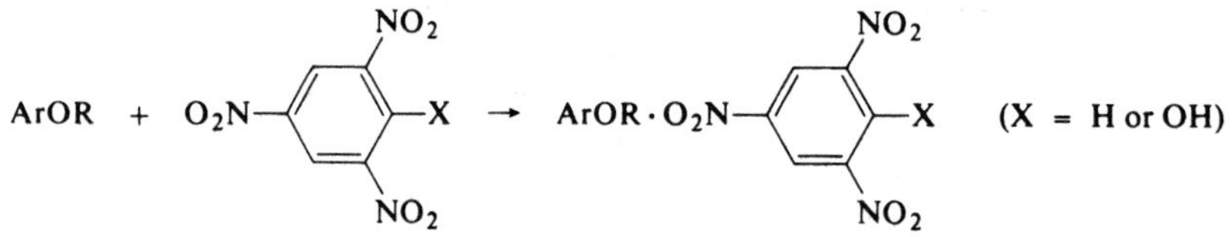

$$\text{ArOR} + O_2N\!-\!\!\!\overset{NO_2}{\underset{NO_2}{\bigcirc}}\!\!\!-X \rightarrow \text{ArOR} \cdot O_2N\!-\!\!\!\overset{NO_2}{\underset{NO_2}{\bigcirc}}\!\!\!-X \quad (X = H \text{ or } OH)$$

Picric acid or
1,3,5-Trinitrobenzene
molecular complex

From the aromatic ether with the aromatic polynitro compound in chloroform.

For directions and examples see: Cheronis, pp. 545, 547; Linstead, pp. 47, 50; Shriner, p. 241; Vogel, p. 672; Wild, p. 100; O. L. Baril and G. A. Megrdichian, *J. Amer. Chem. Soc.*, **58**, 1415 (1936); E. K. Andersen, *Acta Chem. Scand.*, **8**, 157 (1954).

From the aromatic ether with picric acid in ethanol.

See: V. H. Dermer and O. C. Dermer, *J. Org. Chem.*, **3**, 289 (1938).

NOTE: For additional information regarding directions and examples for the derivatization of aromatic ethers (nitration, side-chain oxidation, etc.) see explanations and references to Table IV, pp. 32, 33, 34.

*Derivatives recommended for first trial.
WARNING: This is not an instruction manual. References should be consulted for the preparation of derivatives.

TABLE VIII. ORGANIC DERIVATIVES OF ETHERS

TABLE VIII. ORGANIC DERIVATIVES OF ETHERS
a) Liquids (Listed in order of increasing atmospheric b.p.)*

No.	Name	Boiling point, °C	Melting point, °C	n_D^{20}	D_4^{20}	Picrate	Sulfon-amide	Nitro derivative	Bromo derivative	1,3,5-Tri-nitro-benzene addition com-pound	3,5-Dinitro-benzoate	Miscellaneous
1	Ethylene oxide (Epoxyethane)	10.7	−111.7	1.3614^4	0.89713_4^0	....	....	....	....	....	....	HBr → Ethylene bromohydrin, b.p.: 149
2	Ethyl methyl ether ...	10.8; 10	....	....	0.7260_4^0	....	....	....	....	....	....	....
3	Furan	31.27	−85.6	1.42157	0.9366	....	....	....	....	....	....	Maleic anhydride → 3,6-Endoxo-Δ^4-tetrahydrophthalic anhydride, 125d.; 118d., abs. eth.
4	Diethyl ether (Ethyl ether).............	34.60	−116.3, stab.; −123.3, unst.	1.3526	0.71352	....	....	....	....	....	93	Iodide, b.p.: 72
5	Propylene oxide (1,2-Epoxypropane)	35	....	1.466	0.830	....	....	....	....	....	....	Heating with dil. H_2SO_4 → d,l-pro-pylene glycol, b.p.: 187.4
6	Ethyl vinyl ether	35.75	−115.8	1.3768	0.7589	....	....	....	....	....	....	Dil. acid → al. + acetaldehyde
7	Methyl n-propyl ether	39	....	1.3579	0.7356_4^{13}	....	....	....	....	....	....	
8	Allyl methyl ether....	46	....	....	....	....	....	....	....	β,γ-di: b.p.: 185, D_4^{20}: 1.8329	....	
9	Ethyl isopropyl ether .	53–4	....	....	0.7211 (0.745$_4^0$)	....	....	....	....	....	....	Heating with 1% H_2SO_4 (sealed tube) → al. + iso-propyl alcohol
10	tert-Butyl methyl ether	55.2	....	1.3689	0.7405	....	....	....	....	....	....	Constant boil. mixt. with w., b.p.: 52.6, with 4% w.
11	2,3-Epoxybutane	cis: 58–9^{745}; trans: 53–4^{741}	....	....	cis: 0.8226_4^{25}; trans: 0.8010_4^{25}	....	....	....	....	....	....	Normal crude mixt. is 65% trans + 35% cis
12	Chloromethyl methyl ether	59	....	1.3974	1.015	163	....	....	....	....	....	
13	α-Butylene oxide (1,2-Epoxybutane)......	61–2	....	1.385^{17}	0.837^{17}	....	....	....	....	....	....	
14	Ethyl n-propyl ether ..	63.6	< −79	1.36948	0.7386	....	....	....	....	....	....	Constant boil. mixt. with al., b.p.: 61.2, with 25% al.
15	2-Methylfuran (Sylvan)	64	....	1.434	0.913	....	....	....	....	....	....	
16	Tetrahydrofuran.....	65	....	1.407	0.889	....	....	....	....	....	....	
17	Allyl ethyl ether	66–7^{742}	....	1.3881	0.7651	....	....	....	....	β,γ-di: b.p.: 193–5	....	Heating with 2% H_2SO_4 → al. + allyl alcohol
18	Di-isopropyl ether (Isopropyl ether) ...	67.5	−60	1.3688	0.726	....	....	....	....	....	123; 120–1, CCl_4	....

*Derivative data given in order: m.p., crystal color, solvent from which crystallized.

TABLE VIII. ORGANIC DERIVATIVES OF ETHERS
a) Liquids (Listed in order of increasing atmospheric b.p.)* (Continued)

No.	Name	Boiling point, °C	Melting point, °C	n_D^{20}	D_4^{20}	Picrate	Sulfon-amide	Nitro derivative	Bromo derivative	1,3,5-Tri-nitro-benzene addition com-pound	3,5-Dinitro-benzoate	Miscellaneous
19	*n*-Butyl methyl ether ..	70	−115.5	1.3728; 1.3736	0.7455; 0.774	....	....	....	....	....	....	Oxid. by alkaline $KMnO_4$ at 35–40° → ac. a. + methoxyacetic acid
20	*tert*-Butyl ethyl ether	73.1 (cor.)	....	1.3760	0.7404	....					....	Constant boil. mixt. with w., b.p.: 65.2, with 6% w.
21	Tetrahydrosylvan	79	....	1.407	0.855							....
22	Chloromethyl ethyl ether	80; 83d.	....	1.40398	1.014	....						....
23	Ethyl isobutyl ether ..	81.1 (cor.)	....	1.3739^{25}	0.7323_4^{25}							
24	*sec*-Butyl ethyl ether..	81.2 (cor.)	....	1.3802	0.7503							
25	Isopropyl *n*-propyl ether	83	....	1.376	0.7370							
26	Ethylene glycol di-methyl ether	84.7	....	1.37965	0.8665							
27	Dihydropyran	86	....	1.440	0.923							....
28	*tert*-Amyl methyl ether	86.3	....	1.3885	0.7703	....					....	Constant boil. mixt. with w., b.p.: 73.8, with 9% w.; Constant boil. mixt. with me. al., b.p.: 62.3, with 50% me. al.
29	Tetrahydropyran	88	....	1.421	0.881	....						....
30	Di-*n*-propyl ether (*n*-Propyl ether)	90.1	−122	1.38829	0.74698	....					74	Constant boil. mixt. with w., b.p.: 75.4; Constant boil. mixt. with *n*-propyl alcohol, b.p.: 85.8
31	*n*-Butyl ethyl ether ...	92.3 (cor.)	−124	1.3820	0.7505							....
32	2,5-Dimethylfuran ...	94	....	$1.4363^{21.6}$	$0.888_4^{20.1}$	....		....	*penta*: 180, chl.	....		Maleic anhydride → 3,6-Endoxo-3,6-dimethyl-Δ⁴-tetra-hydrophthalic anhydride, 78, eth.
33	*α*-Chloroethyl ethyl ether	98	....	1.404	0.966							....
34	*n*-Amyl methyl ether .	99–100	....	1.3873	0.761							
35	*tert*-Amyl ethyl ether .	101	....	1.3912	0.7657	....					....	Constant boil. mixt. with 13% w., b.p.: 81.2
36	1,4-Dioxane	101.4	11.8	1.4232	1.03361	....		....	65–6	....	....	Iodine derivative, 84–5; Constant boil. mixt. with 48 mole % dioxane, b.p.: 82.8

*Derivative data given in order: m.p., crystal color, solvent from which crystallized.

TABLE VIII. ORGANIC DERIVATIVES OF ETHERS
a) Liquids (Listed in order of increasing atmospheric b.p.)* (Continued)

No.	Name	Boiling point, °C	Melting point, °C	n_D^{20}	D_4^{20}	Picrate	Sulfon-amide	Nitro derivative	Bromo derivative	1,3,5-Tri-nitro-benzene addition com-pound	3,5-Dinitro-benzoate	Miscellaneous
37	Ethylene glycol mono-ethyl monomethyl ether (1-Ethoxy-2-methoxyethane)	102	....	1.38677	0.8529	......	......				....	....
38	Cyclopentyl methyl ether	105	....	1.4206	0.862	......	......					
39	β-Chloroethyl ethyl ether	107	....	1.411	0.989	......	......					
40	n-Butyl isopropyl ether	108[738]	....	1.3889$_{5461}^{24.9}$	0.7594[15]							Boil. HI → n-butyl iodide + isopropyl iodide
41	α-Epichlorohydrin ...	115–7	....	1.438	1.181							
42	α,α'-Dichloroethyl ether	116; 114		1.4183[24]	1.138$_4^{12}$							
43	n-Amyl ethyl ether ...	118	....	1.3927	0.762							
44	Di-sec-butyl ether	121	....	1.3928[25]	0.760				b.p.: 90–1; n_D^{25}: 1.250		75.5	....
45	Cyclopentyl ethyl ether	122	....	1.423	0.853	......	......					
46	Di-isobutyl ether (Iso-butyl ether).......	123	....	....	0.7616$_{15}^{15}$						87; 84.5–5.5	
47	Ethylene glycol mono-methyl mono-n-propyl ether (1-Methoxy-2-n-propoxyethane)	124.5	....	1.39467	0.8472	......	......					
48	n-Hexyl methyl ether .	126	....	1.3972	0.772							
49	2-Methoxy-1-propanol	130[758]	....	....	....						97	α-Naphthylure-thane, 60
50	Cyclohexyl methyl ether	134	....	1.435	0.875	......	......					
51	Ethylene glycol mono-ethyl ether ("Cello-solve"; 2-Ethoxy-ethanol)	134.8; 135.1	....	1.40797	0.9297; 0.9311						75	Miscible with w., with al. and with eth.; Diphenyl-urethane, 43; Xan-thate, 202.5 (cor.), acet.-abs. eth. Phenylurethane, 226
52	3-Ethoxy-2-methyl-2-butanol	141	....	....	....						....	
53	n-Hexyl ethyl ether ..	142	....	1.4008	0.772							
54	Di-n-butyl ether (Butyl ether).......	142.4; 144	−98	1.3989	0.76829						62–3; 64	....
55	Cyclohexyl ethyl ether	149	....	1.435	0.864	......						
56	Anisole (Methoxyben-zene; Methyl phenyl ether).............	153.8; 155 (43[10])	−37.5	1.52211	0.99393	79–81, unst. in. air.	113; 110–1, al.	2,4-di: 86.9, al.; 95.5	2,4-di: 61, al.	....	87	
57	3-Methoxy-2-methyl-1-propanol.........	155	....	1.4140[27]	....						64	
58	Diethylene glycol dimethyl ether......	162.0	−75	1.4099	0.9440$_{20}^{20}$							

* Derivative data given in order: m.p., crystal color, solvent from which crystallized.

No.	Name	Boiling point, °C	Melting point, °C	n_D^{20}	D_4^{20}	Picrate	Sulfonamide	Nitro derivative	Bromo derivative	1,3,5-Trinitrobenzene addition compound	3,5-Dinitrobenzoate	Miscellaneous
59	Furfuryl alcohol (2-Furancarbinol).....	170	....	1.4868	1.1351	....	....	....	....	....	80–1, pyr.	p-Nitrobenzoate, 76; N-Phenylurethane, 45
60	Ethylene glycol mono-n-butyl ether (n-Butyl "cellosolve")......	171^{743}	....	1.4177	▸0.9188	....	....	....	172	....	oil	3-Nitrophthalate, 120
61	Benzyl methyl ether ..	170–1 (cor.)	....	1.5008	0.9649	115–6	....	....	....	....	....	....
62	2-Methoxytoluene (2-Methyl anisole; Methyl 2-tolyl ether; 2-Cresyl methyl ether).............	171	....	1.505	0.9853	116; 113–4, pa. yel.	137, al.	3,5-di: 69, pa. yel., me. al.	5-mono: 63–4, al.	....	....	Oxid. → o-Methoxybenzoic acid, 101
63	Phenyl ethyl ether (Phenetole)........	172	−33	1.5080; 1.5074	0.9666	92	150	p-mono: 58	....			....
64	Di-isoamyl ether (Isoamyl ether)........	172.5	....	1.409	0.778	....	....	....	....	....	60–1	Constant boil. mixt. with w., b.p.: 97.2
65	4-Methoxytoluene (4-Methylanisole; Methyl 4-tolyl ether; 4-Cresyl methyl ether).............	173; 176	....	1.512	0.970	88–9, yel.-or.	182, al.					Oxid. → p-Anisic acid, 184–6; 184, w.
66	3-Methoxytoluene (3-Methylanisole; Methyl 3-tolyl ether; 3-Cresyl methyl ether).............	173; 177 (cor.)	....	1.513	0.972	113–4, yel.-or.	129–30, al.	2-mono: 54–5, pet. eth.; 2,4,6-tri: 92, al.	....			Oxid. → m-Methoxybenzoic acid, 110
67	Tetrahydrofurfuryl alcohol............	117	....	1.45167	1.0544	....	....	....	....	....	83–4	p-Nitrobenzoate, 46–8; N-Phenylurethane, 61, pet. eth.
68	Phenyl isopropyl ether	178	....	1.4992	0.975	....	....	....	....	....	....	conc. H_2SO_4 + ac. a. → o-Isopropylphenol, b.p.: 213–4, m.p.: 130
69	β,β'-Dichloroethyl ether	178	....	1.4568	1.220							
70	2-Ethoxytoluene (2-Cresyl ethyl ether; Ethyl 2-tolyl ether)..	184	....	1.505	0.953	117.5–8.5, pa. yel.	148–9, al.	di: 51	....		....	Oxid. → o-Ethoxybenzoic acid, 25; 19.0–.5, well dried
71	Benzyl ethyl ether (Homophenetole) ..	184–6 (cor.)	....	1.4958	0.9478	....	....	....	....			Refluxed in bz. + P_2O_5 → Ethylene + Diphenylmethane, 25.1
72	Di-n-amyl ether (n-Amyl ether)	187.5	−69.3	1.416	0.78298	....	....	....	....	....	42–3	....
73	Diethylene glycol diethyl ether	188	....	1.411	0.906							
74	Phenyl n-propyl ether (n-Propoxybenzene)	188; 189.3 (cor.)	....	1.5014; 1.5011	0.9494$_{20}^{20}$	....	116–7, al.	....				

*Derivative data given in order: m.p., crystal color, solvent from which crystallized.

No.	Name	Boiling point, °C	Melting point, °C	n_D^{20}	D_4^{20}	Picrate	Sulfon-amide	Nitro derivative	Bromo derivative	1,3,5-Tri-nitro-benzene addition compound	3,5-Dinitro-benzoate	Miscellaneous
75	4-Ethoxytoluene (4-Cresyl ethyl ether; Ethyl 4-tolyl ether)..	190.5	....	1.505	0.949	110–1, or.-yel.	138.0–.5, al.	....	....	....	....	Oxid. → p-Ethoxy-benzoic acid, 198; 195.0–.5, al.
76	3-Ethoxytoluene (3-Cresyl ethyl ether; Ethyl 3-tolyl ether)..	190.5		1.506	0.949	114–5, or.-yel.	110–1, al.					Oxid. → m-Ethoxy-benzoic acid, 137
77	Diethylene glycol monomethyl ether ...	194	....	1.4244	1.035_{20}^{20}	....		....			....	p-Nitrophenyl-urethane, 73.5
78	3-Chloroanisole	194	....	....	....	....	131	....	....	....	....	....
79	2-Chloroanisole	195		1.5433^{25}	1.1865_4^{25}	....	....	95				
80	Diethylene glycol monoethyl ether	196		1.4298	1.023_{20}^{20}						oil	p-Nitrophenyl-urethane, 66
81	4-Chloroanisole	200		....	$1.1851_4^{12.8}$	....	151	2-mono: 95				
82	n-Butyl phenyl ether.. (n-Butoxybenzene)..	206	—	1.5049	....	110–2, pa. yel., chl.	103–4, al.					
83	2-Chlorophenetole ...	208	17	....	....	....	133	82				
84	2-Bromoanisole	210	....	....	....	....	140	106				
85	Benzyl isobutyl ether .	210–2 (cor.)	....	1.4826	0.9233	....						
86	p-Bromoanisole	215; 216	....	....	1.494_4^9	....	148	88				
87	Methyl thymyl ether..	216	....	....	0.954_4^9	....		tri: 92				
88	Triethylene glycol dimethyl ether......	216	−47	1.4233	0.9871_{20}^{20}							
89	1,3-Dimethoxybenzene (Resorcinol dimethyl ether)............	217 (cor.)	−58; −52	1.4233	1.0552_{25}^{25}	56–8, or.-yel., unst. in air	166–7, al.	2,4-di: 72, pa. yel., al.; 4,6-di: 157, al.; 2,4,6-tri: 123–4, pa. red, al.	4,6-di: 140, al.	....		
90	2-Bromophenetole ...	218	....	....	....	....	135	98	....	....		
91	Benzyl n-butyl ether..	219–21 (cor.)	....	1.4833	0.9227	....						
92	Creosol (4-Methyl-catechol 2-methyl ether)............	221	5.5	1.5353^{25}	1.0919_4^{25}	112, yel.	....					
93	n-Butyl 2-tolyl ether (2-Butoxytoluene)..	223	—	....	0.9437_0^0		95–6, al.					
94	2-Methoxyaniline (2-Anisidine)	225	5	....	1.0978_{15}^{15}							N-Formyl, 83.5; N-Acetyl, 87–8
95	Di-n-hexyl ether (n-Hexyl ether)	$228–9^{761}$	....	....	0.7936						54.5–5.5	....
96	Safrole (4-Allyl-1,2-methylenedioxy-benzene)	233	11	1.5383	1.100	104.0–5.5, or.-red	....		tri: 108; penta: 169–70, bz.	51	....	
97	4-Bromophenetole ...	233; 229	12	....	....	....	145	47	....	....		....

*Derivative data given in order: m.p., crystal color, solvent from which crystallized.

No.	Name	Boiling point, °C	Melting point, °C	n_D^{20}	D_4^{20}	Picrate	Sulfonamide	Nitro derivative	Bromo derivative	1,3,5-Trinitrobenzene addition compound	3,5-Dinitrobenzoate	Miscellaneous
98	**Resorcinol diethyl ether**	235	12.4	....	....	109	184	....	*tri*: 69	....	....	....
99	**Eugenol methyl ether** (4-Allyl-1,2-dimethoxybenzene)	244	....	1.5360	1.0336	114–5, red-br., chl.	7	....	*tri*: 78, abs. al.			
100	**2-Iodophenetole**	246	....	....	1.800	84	....	....	*tri*: 110	....	....,	
101	*trans(β)*-**Isosafrole** (1,2-Methylenedioxy-4-propylbenzene)	248	6.8	1.5782	1.122	74–5, dk. red, chl.	....		*di*: 52–3, eth.; *tri*: 109–10, pet. eth.	85–6, brt. scar.	....	
102	**3-Methoxyaniline** (3-Anisidine)	251	....		....	169d., yel.	....	....	....	....	....	N-Formyl, 57; N-Acetyl, 81; N-*p*-Toluenesulfonyl, 68
103	**Di-*n*-heptyl ether** (*n*-Heptyl ether)	263; 260	....	1.427	0.8056_{20}^{20}	....			....	....	47	
104	**Isoeugenol methyl ether**	264	16–7	1.5692	1.0528	42–5, dk. red, chl.	....	....	*di*: 101.0-.5, abs. eth.	69–70, brt. scar.	....	Oxid. → Veratric acid, 181
105	**Tetraethylene glycol dimethyl ether**	266; 275	....	1.432	1.009	....						
106	**Methyl 1-naphthyl ether** (1-Methoxynaphthalene)	271 (cor.)	< −10	1.6940^{25}	1.09159	129.5–30.5; yel.-or., chl.	156–7, al.	*2-mono*: 80; *4-mono*: 85, yel., al.	*4-mono*: b.p.: 181–2; *5-mono*: 67.5–8.0; *x-mono*: 46, al.; *2,4-di*: 54–5, al.	139–40; 137–8, yel.	....	
107	**2-Nitroanisole**	277	10	1.562	1.254	....	....	....	....	....	....	Reduct. → *o*-anisidine, b.p.: 225
108	**Ethyl 1-naphthyl ether** (1-Ethoxynaphthalene)	280.5 (cor.)	5.5	1.5973^{25}	1.074	118.5–9.0 (cor.)	164–5, al.	*2-mono*: 84; *4-mono*: 116–7, al.	*4-mono*: 48, al.	125.5, yel.	....	
109	**Dibenzyl ether** (Benzyl ether)	290–300d.	3.6	....	1.0428	77–8, or.-yel., chl.	....	....	*di*: 107–8, al.	....	112	
110	**Isoamyl 1-naphthyl ether**	317.5 (cor.)	< −10	$1.57049^{14.2}$	$1.00689_4^{14.2}$	96.0–7.0	....					

*Derivative data given in order: m.p., crystal color, solvent from which crystallized.

TABLE VIII. ORGANIC DERIVATIVES OF ETHERS
b) Solids (Listed in order of increasing m.p.)*

No.	Name	Melting point, °C	Boiling point, °C	n_D^{20}	D_4^{20}	Picrate	Sulfon-amide	Nitro derivative	Bromo derivative	1,3,5-Tri-nitro-benzene addition com-pound	3,5-Dinitro-ben-zoate	Miscellaneous
1	**4-Chlorophenetole** (1-Chloro-4-ethoxybenzene)	21	212	1.5227^{19}	1.1231_{20}^{20}	....	134	....	2,6-*di*: 54	......	......	....
2	**Veratrole** (1,2-Dimethoxy-benzene)	22.5	207; 205	1.5287^{21}	1.080	56–7, red	135–6, al.	95	4,5-*di*: 92–3	......	......	....
3	**Anethole** (1-Methoxy-4-propenylbenzene)	22.5, al.	235	1.558	0.989_4^{28}	69–70d., or.-red, al.	....		*di*: 67, eth.; 65; 62–4; *tri*: 108, pet. eth.	......	......	....
4	**4-Methoxybenzyl alcohol** (Anisyl alcohol)	24	151^{27}; 138^{14}	....	....	....	....	....	....	......	......	Phenylurethane, 93; *p*-Nitroben-zoate, 94
5	***n*-Amyl 2-naphthyl ether** ..	24.5	327.5 (cor.)	1.5587^{30}	....	66.5–7.0, or., h. al.	159	*di*: 135	*di*: 58	......	......	....
6	**4-Iodophenetole** (1-Ethoxy-4-iodobenzene)........	27	252	....	....	....	....	96	....	......	......	....
7	**Diphenyl ether** (Phenyl ether)................	28	259	1.5826^{24}	1.073	110	*di*: 159, al.	4,4'-*di*: 144.4, al.; 2,4,2',4'-*tetra*: 195–7, pa. yel., ac. a.	4,4'-*di*: 54–5, al.	......	......	....
8	**Isoamyl 2-naphthyl ether** ..	28.0–.5	321.0 (cor.)	....	....	93.5–4.0, al.	....	....	....	......	......	....
9	**2-Methoxyphenol** (Guaia-col; Catechol mono-methyl ether)	28.2	205	1.5441	$1.1287_{vac.}^{20.4}$	86–7	....	....	4,5,6-*tri*: 116, al.	......	141.2 (cor.), al.	....
10	**2-Methoxybiphenyl** (2-Biphenyl methyl ether) ..	29, pet. eth.	274					5-*mono*: 95–6, pa. yel., me. al.		......	......	....
11	**β-Bromoethyl phenyl ether**.	32	......	....	....	....	....		56	......	......	....
12	**Isobutyl 2-naphthyl ether** ..	33	304	....	....	84–5				......	......	....
13	**Didodecyl ether** (Dilauryl ether)................	33; 28–30	$190–5^1$							......	......	....
14	**2-Ethoxybiphenyl** (2-Bi-phenyl ethyl ether)......	34	132^6							......	......	....
15	***sec*-Butyl 2-naphthyl ether**	34	298.5 (cor.)	....	....	86				......	......	....
16	**3-Ethoxybiphenyl** (3-Bi-phenyl ethyl ether)......	35	158^8	....	....	....				......	......	....
17	**3-Nitrophenetole**	35	284							......	......	Reduct. → *m*-Phenetidine, (Picrate, 158)
18	***n*-Butyl 2-naphthyl ether** ..	35.5	309 (cor.)	....	....	67				......	......	....
19	**Ethyl 2-naphthyl ether** (Neonerolin)...........	36; 37	282 (cor.)	1.5932^{47}	1.064	101	161–3, al.	....	1-*mono*: 66, pet. eth.; 1,6-*di*: 94, pet. eth.	......	......	....
20	**3-Aminodiphenyl ether**	37, lgr.	315	....	....	....	....	....	....	......	......	N-Hydrochloride, 141, 139; N-Acetyl, 83, lgr.

*Derivative data given in order: m.p., crystal color, solvent from which crystallized.

No.	Name	Melting point, °C	Boiling point, °C	n_D^{20}	D_4^{20}	Picrate	Sulfonamide	Nitro-derivative	Bromo derivative	1,3,5-Trinitrobenzene addition compound	3,5-Dinitrobenzoate	Miscellaneous
21	3-Nitroanisole	39	258	....	1.373^{18}	....	....	....	....	......	......	Reduct. → m-Anisidine, (Picrate, 169d.)
22	Isopropyl 2-naphthyl ether	40	285			95						
23	2-Naphthyl n-propyl ether	40	297			81						
24	Catechol diethyl ether (1,2-Diethoxybenzene)	43, dil. al.	217	....		69–71, red-br., unst. in air	3,4-di: 162–3, al.	tri: 122				
25	bz-Tetrahydro-6-methoxyquinoline	43	130	1.5718^{50}								
26	2,4,6-Trichlorophenetole	44	246					di: 100				
27	8-Methoxyquinoline	45	175^{29}			162						
28	2-Aminodiphenyl ether	47	173^{14}									Acetate, 81
29	Pyrogallol trimethyl ether (1,2,3-Trimethoxybenzene)	47, dil. al.	241			78.5–80.0, yel.	2,3,4-tri: 123–4	5-mono: 106, ac. a.	mono: oil; di: oil; 4,5,6-tri: 73–4	81, pa. yel.		
30	4-Iodoanisole	52	139									
31	Phloroglucinol trimethyl ether (1,3,5-Trimethoxybenzene)	52–3, al.	255.5 (cor.)						2-mono: 96–7, dil. al.; 2,4-di: 129–30, al.; 2,4,6-tri: 145, al.			
32	Hydroquinone monomethyl ether	52.5	$244^{754.2}$									
33	4-Nitroanisole	54	274	1.5707^{60}	1.233_4^{20}							Reduct. → p-Anisidine, 57
34	Hydroquinone dimethyl ether (1,4-Dimethoxybenzene)	56, 75% al.	213 (cor.)			47–8, or.-red, unst. in air	148, al.	2-mono: 72, yel.; 2,3-di: 177; 2,5-di: 202	di: 142, ac. a.	di: 86.5, red		
35	4-Methoxyaniline (4-Anisidine)	57	246	......	1.071_4^{55}							N-Formyl, 81; N-Acetyl, 130-2, w.; N-Benzoyl, 216-7
36	4-Cyclohexylphenyl methyl ether	59	116^4									
37	2,4,6-Trichloroanisole	60						di: 95				
38	1,3-Diphenoxypropane (Trimethylene glycol diphenyl ether)	61, al.	338–40 (cor.)				4,4'-di: 245–55, al.					
39	Catechol dibenzyl ether (1,2-Dibenzyloxybenzene)	63–4, wh., me. al.	......					4-mono: 98, pa. yel., al.				
40	1-Phenyl-2-phenoxymethanol	64										p-Nitrobenzoate, 84
41	α-Glyceryl phenyl ether	70	187^{15}									
42	2,4,6-Tribromophenetole	72						79				

*Derivative data given in order: m.p., crystal color, solvent from which crystallized.

No.	Name	Melting point, °C	Boiling point, °C	n_D^{20}	D_4^{20}	Picrate	Sulfon-amide	Nitro-derivative	Bromo derivative	1,3,5-Tri-nitro-benzene addition com-pound	3,5-Dinitro-ben-zoate	Miscellaneous
43	Hydroquinone diethyl ether (1,4-Diethoxybenzene) ..	72	......	....	......	....	......	*mono*: 49, yel.; 2,3-*di*: 130, yel., al.; 2,5-*di*: 176, yel., al.	....	86.5	......	....
44	Methyl 2-naphthyl ether (Nerolin).............	73 (cor.), eth.	273	......	......	116.5–7.0, dk. yel., 113.0–3.5	150–1, al.	1-*mono*: 128; 1,6,8-*tri*: 215 d.	*x-mono*: 62–3, pet. eth.; 1-*mono*: 83–4; 3-*mono*: 77–8; 6-*mono*: 108	93.5, yel.	......	....
45	2,4,5-Tribromophenetol ...	73	......	....	......	....	......	79	......			
46	4-Ethoxybiphenyl (4-Biphenyl ethyl ether)....	76; 74	188[13]	......	......							
47	Benzyl 1-naphthyl ether ...	77 (cor.)	200[12]	......	......	85–100 (cor.)						
48	2-Phenyl-2-phenoxy-ethanol	81										*p*-Nitrobenzoate, 87
49	4-Aminodiphenyl ether....	83.5; 84, w.	189[14]	......	......							N-Acetyl, 127; N-HCl, 122
50	Biphenylene oxide (Di-benzofuran)	86, wh., al.	288 (cor.)	......	......	94	......	3-*mono*: 181–2; *di*: 245, ac. a.	....	96, yel.	......	....
51	4-Methoxybiphenyl (4-Biphenyl methyl ether; 4-Phenylanisole)	90; 89					3-*mono*: 91–2, al.; 3,5-*di*: 137–8, yel., al.; 3,4'-*di*: 171	3-*mono*: 79; 4'-*mono*: 144, pet.; 3,4'-*di*: 134; 3,5-*di*: 87, pet.	....			....
52	1,2-Diphenoxyethane (Ethylene glycol diphenyl ether)...............	98, al.	......	....	......	....	4,4'-*di*: 228–9, al.	2',4'-*di*: 215.2 (cor.), pa. yel., acet.	*di-p*: 134–5, al.	......	......	....
53	Benzyl 2-naphthyl ether (2-Benzyloxynaphthalene)..	101.5 (cor.), al.	d.	......	......	123.0 (cor.)						
54	Anisoin (4,4'-Dimethoxy-benzoin)	113	......	....	......	....	......	....	....	....	......	Methyl ether, 52–3, yel., CCl₄; Ethyl ether, 103–4, al.-w.
55	Hydroquinone dibenzyl ether (1,4-Dibenzyloxy-benzene)	128–9, al.	......	....	......	....	......	83	....	....	......	....

*Derivative data given in order: m.p., crystal color, solvent from which crystallized.

TABLE VIII. ORGANIC DERIVATIVES OF ETHERS
b) Solids (Listed in order of increasing m.p.)* (Continued)

No.	Name	Melting point, °C	Boiling point, °C	n_D^{20}	D_4^{20}	Picrate	Sulfon-amide	Nitro-derivative	Bromo derivative	1,3,5-Tri-nitro-benzene addition com-pound	3,5-Dinitro-ben-zoate	Miscellaneous
56	**Anisil** (4,4'-Dimethoxy-benzil)...............	133, yel.	......	....	....	....	....	....	....	......	......	Monoxime, 133; *syn*-Dioxime, 217; *anti*-Di-oxime, 195, bz.; Disemicarba-zone, 254–5, dil. ac. a.; Dihydra-zone, 118
57	**Antiarol** (5-Hydroxy-1,2,3-trimethoxybenzene).....	148, w.	......	....	....	....	....	....	....	......	......	Acetyl, 74, al.; Methyl ether, 47, b.p.: 271
58	**Anhalamine** (6,7-Di-methoxy-8-hydroxy-1,2,3,4-tetrahydroiso-quinoline)............	187–8, al.	......	....	....	234–6	....	....	....	......	......	Dibenzoyl, 128–9; N-Benzoyl, 167–8
59	**7-Methoxyquinoline**......	210	287[758]	....	....	229	....	....	....	......	......	....

*Derivative data given in order: m.p., crystal color, solvent from which crystallized.

*Phenylhydrazone.**

$$RCHO + H_2NNHC_6H_5 \rightarrow RCH{=}NNHC_6H_5 + H_2O$$

$$RR'CO + H_2NNHC_6H_5 \rightarrow RR'C{=}NNHC_6H_5 + H_2O$$

Phenylhydrazone

From the carbonyl compound with phenylhydrazine in methanol or ethanol.
For directions and examples see: Cheronis, pp. 497–8; Shriner, p. 131.
From the carbonyl compound with phenylhydrazine in aqueous acetic acid.
See: Wild, p. 111.
From the carbonyl compound with phenylhydrazine in methanol in the presence of acetic acid.
See: Cheronis, p. 511.
From the carbonyl compound in alcohol with phenylhydrazine hydrochloride and sodium acetate in water.
See: Vogel, p. 721.

*p-Nitrophenylhydrazone.**

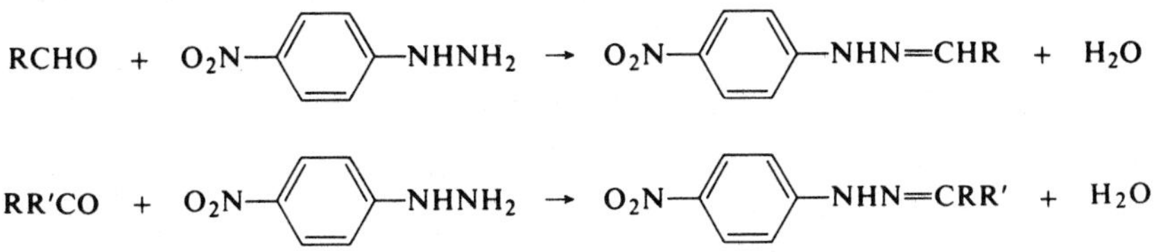

p-Nitrophenylhydrazone

From the carbonyl compound with *p*-nitrophenylhydrazine and a catalytic amount of acetic acid in alcohol.
For directions and examples see: Shriner, pp. 131, 219; Vogel, p. 722; Wild, p. 112.
From the carbonyl compound in alcohol or water with *p*-nitrophenylhydrazine in aqueous acetic-hydrochloric acids.
See: G. Petit, *Bull. Soc. Chim. France*, 141 (1948).

*2,4-Dinitrophenylhydrazone (DNP-derivative).**

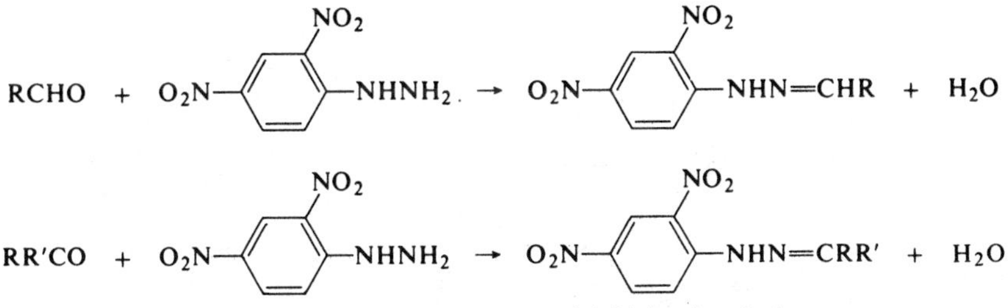

DNP · 2,4-Dinitrophenylhydrazone

From the carbonyl compound with 2,4-dinitrophenylhydrazine and sulfuric acid in methanol or ethanol.
For directions and examples see: Linstead, p. 26; Shriner, p. 219; Vogel, p. 344; Wild, pp. 114–5; O. L. Brady and G. V. Elsmie, *Analyst*, **51**, 77 (1926); O. L. Brady, *J. Chem. Soc.*, 756 (1931); H. H. Strain, *J. Amer. Chem. Soc.*, **57**, 758 (1935); O. L. Brady and S. G. Jarret, *J. Chem. Soc.*, 1021 (1950).
From the carbonyl compound with 2,4-dinitrophenylhydrazine and 1% hydrochloric acid in methanol or ethanol.
See: Cheronis, pp. 499–501, 511; Vogel, p. 722; Wild, pp. 112–4; C. F. H. Allen, *J. Amer. Chem. Soc.*, **52**, 2955 (1930); C. F. H. Allen and J. H. Richmond, *J. Org. Chem.*, **2**, 222 (1937).
From the carbonyl compound with 2,4-dinitrophenylhydrazine and acetic acid in diglyme (diethylene glycol dimethyl ether).
See: H. J. Shine, *J. Org. Chem.*, **24**, 1790 (1959).
From the carbonyl compound in 95% ethanol with 2,4-dinitrophenylhydrazine and concentrated hydrochloric acid in diglyme (diethylene glycol dimethyl ether).
See: Cheronis, p. 501; H. J. Shine, *J. Org. Chem.*, **24**, 252 (1959); *J. Chem. Ed.*, **36**, 575 (1959).

*Derivatives recommended for first trial.
WARNING: This is not an instruction manual. References should be consulted for the preparation of derivatives.

From the carbonyl compound in ethanol with 2,4-dinitrophenylhydrazine in 85% phosphoric acid.
See: Vogel, p. 344; G. D. Johnson, *J. Amer. Chem. Soc.*, **73**, 5888 (1951); **75**, 2720 (1953).
From the carbonyl compound with 2,4-dinitrophenylhydrazine and sulfuric acid in isopropyl alcohol.
See: N. R. Campbell, *Analyst*, **61**, 391 (1936).
From the carbonyl compound with 2,4-dinitrophenylhydrazine in pyridine.
See: E. A. Braude and C. J. Timmons, *J. Chem. Soc.*, 3131 (1953).

*Semicarbazone.**

$$RCHO + H_2NNHCONH_2 \rightarrow RCH{=}NNHCONH_2 + H_2O$$

$$RR'CO + H_2NNHCONH_2 \rightarrow RR'C{=}NNHCONH_2 + H_2O$$

$$RR'CO + (CH_3)_2C{=}NNHCONH_2 \rightarrow RR'C{=}NNHCONH_2 + (CH_3)_2CO$$

Semicarbazone

From the carbonyl compound with aqueous semicarbazide hydrochloride and sodium acetate.
For directions and examples see: Cheronis, pp. 503–504, 512; Shriner, p. 218; Vogel, p. 344; Wild, p. 121;
A. Michael, *J. Amer. Chem. Soc.*, **41**, 417 (1919).
From the carbonyl compound in ethanol with aqueous semicarbazide hydrochloride and sodium acetate.
See: Linstead, p. 27; Shriner, p. 218; Wild, p. 122; R. L. Shriner and T. A. Turner, *J. Amer. Chem. Soc.*, **52**,
1267 (1930).
From the carbonyl compound and acetone semicarbazone in acetic acid.
See: B. Angla, *Ann. Chim. Anal. Chim. Appl.*, **22**, 10 (1940).

*Thiosemicarbazone.**

$$RCHO + H_2NNHCSNH_2 \rightarrow RCH{=}NNHCSNH_2 + H_2O$$

$$RR'CO + H_2NNHCSNH_2 \rightarrow RR'C{=}NNHCSNH_2 + H_2O$$

Thiosemicarbazone

From the carbonyl compound with thiosemicarbazide and sodium acetate in water, alcohol or acetic acid.
For directions and examples see: Cheronis, pp. 503, 512; Wild, p. 128; F. J. Wilson and R. Burns, *J.
Chem. Soc.*, **121**, 873 (1922); W. Baird, R. Burns and F. J. Wilson, *J. Chem. Soc.*, 2527 (1927); M. Busch, *J.
prakt. Chem.*, **124**, 301 (1930); P. P. T. Sah and T. C. Daniels, *Rec. Trav. Chim.*, **69**, 1545 (1950).

*Phenylsemicarbazone.**

$$RCHO + H_2NNHCONHC_6H_5 \rightarrow RCH{=}NNHCONHC_6H_5 + H_2O$$

$$RR'CO + H_2NNHCONHC_6H_5 \rightarrow RR'C{=}NNHCONHC_6H_5 + H_2O$$

Phenylsemicarbazone

From the carbonyl compound with phenylsemicarbazide in alcohol or acetic acid.
For directions and examples see: P. P. T. Sah and T. -S. Ma, *J. Chinese Chem. Soc.*, **2**, 32 (1934), *C.A.*,
28, 3713 (1934).

*Oxime.**

$$RCHO + NH_2OH \rightarrow RCH{=}NOH + H_2O$$

$$RR'CO + NH_2OH \rightarrow RR'C{=}NOH + H_2O$$

Oxime

From the carbonyl compound with hydroxylamine hydrochloride and pyridine in ethanol or without
solvent.
For directions and examples see: Cheronis, p. 513; Shriner, p. 254; Vogel, p. 345; J. B. Buck and W. S. Ide,
J. Amer. Chem. Soc., **53**, 1536 (1931); W. E. Bachmann and C. H. Boatner, *J. Amer. Chem. Soc.*, **58**, 2097
(1936); W. E. Bachmann and M. X. Barton, *J. Org. Chem.*, **3**, 300 (1938).
For a modification of the above method in aqueous alcohol.
See: W. M. D. Bryant and D. M. Smith, *J. Amer. Chem. Soc.*, **57**, 57 (1935).
From the carbonyl compound with hydroxylamine hydrochloride and sodium hydroxide in methanol or
aqueous ethanol.

*Derivatives recommended for first trial.
WARNING: This is not an instruction manual. References should be consulted for the preparation of derivatives.

See: Cheronis, p. 513; Shriner, p. 255; Vogel, p. 721; Wild, p. 121.

From the carbonyl compound with hydroxylamine hydrochloride and potassium hydroxide in 95% ethanol. *See:* Shriner, p. 255.

From the carbonyl compound with hydroxylamine hydrochloride and sodium or potassium acetate in water or aqueous ethanol.

See: Linstead, p. 27; Vogel, pp. 343, 345; J. S. Buck and W. S. Ide, *J. Amer. Chem. Soc.*, **53**, 1536 (1931).

From the carbonyl compound with hydroxylamine hydrochloride and sodium carbonate or bicarbonate in water or aqueous ethanol.

See: Wild, p. 120.

Dimethone derivative (Methone derivative).*

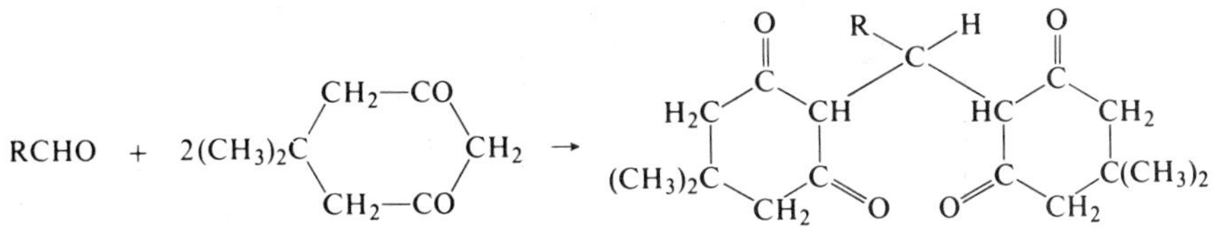

$$RCHO + 2(CH_3)_2C\underset{CH_2-CO}{\overset{CH_2-CO}{\big<}}CH_2 \rightarrow$$

Methone (Dimedone)   Dimethone derivative

This derivative is specific for aldehydes only.

From the aldehyde and methone (dimedone; 5,5-dimethyl-1,3-cyclohexanedione; dimethyl dihydroresorcinol) in aqueous ethanol or methanol.

For directions and examples see: Cheronis, p. 505; Linstead, p. 27; Shriner, p. 220; Vogel, p. 333; Wild, pp. 136-7; D. Vorlander, *Z. Anal. Chem.*, **77**, 241 (1929); *Z. Angew. Chem.*, **42**, 46 (1929); W. Weinberger, *Ind. Eng. Chem., Anal. Ed.*, **3**, 365 (1931).

From the aldehyde with methone and a catalytic amount of piperidine in aqueous ethanol.

See: E. C. Horning and M. G. Horning, *J. Org. Chem.*, **11**, 95 (1946).

Anhydride of dimethone derivative (substituted octahydroxanthene).*

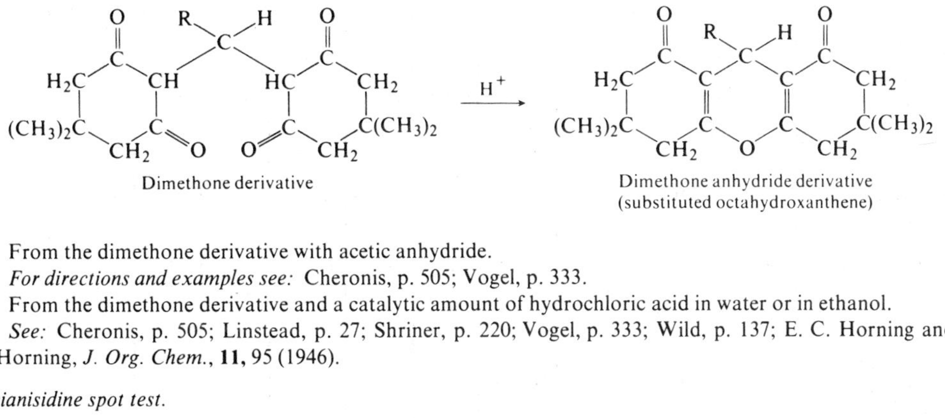

Dimethone derivative $\xrightarrow{H^+}$ Dimethone anhydride derivative (substituted octahydroxanthene)

From the dimethone derivative with acetic anhydride.

For directions and examples see: Cheronis, p. 505; Vogel, p. 333.

From the dimethone derivative and a catalytic amount of hydrochloric acid in water or in ethanol.

See: Cheronis, p. 505; Linstead, p. 27; Shriner, p. 220; Vogel, p. 333; Wild, p. 137; E. C. Horning and M. G. Horning, *J. Org. Chem.*, **11**, 95 (1946).

o-Dianisidine spot test.

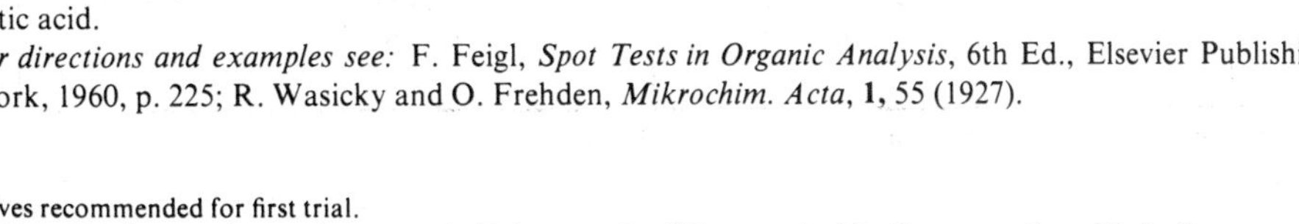

$$RCHO + H_2N-\langle\rangle-\langle\rangle-NH_2 \rightarrow RCH=N-\langle\rangle-\langle\rangle-N=CHR$$

Schiff base (colored)

This test is usually applicable to aldehydes only.

From the aldehyde and a saturated solution of *o*-dianisidine (4,4'-diamino-3,3'-dimethoxybiphenyl) in glacial acetic acid.

For directions and examples see: F. Feigl, *Spot Tests in Organic Analysis*, 6th Ed., Elsevier Publishing Co., New York, 1960, p. 225; R. Wasicky and O. Frehden, *Mikrochim. Acta*, **1**, 55 (1927).

---

*Derivatives recommended for first trial.

WARNING: This is not an instruction manual. References should be consulted for the preparation of derivatives.

TABLE IX. ORGANIC DERIVATIVES OF ALDEHYDES
a) Liquids 1) Listed in order of increasing atmospheric b.p.*

No.	Name	Boiling point, °C	Melting point, °C	n_D^{20}	Semi-carba-zone	2,4-Di-nitro-phenyl-hydra-zone	p-Nitro-phenyl-hydra-zone	Phenyl-hydra-zone	Oxime	Dimeth-one deriv. (Dime-done deriv.)	Dimeth-one anhy-dride	Miscel-laneous	o-Dianisidine spot test Cold	Hot	Limit, γ
1	Formaldehyde (Methanal)	−21	−91	....	169	167, yel., al.	181–2, yel., bz.	145	oil	189, al.; 191.4	......	....	pa. yel.	or. br.	50
2	Trifluoroacetalde-hyde	−20	......	....	......	151	......	......	......	......	......	....	....	....	....
3	Acetaldehyde (Ethanal).......	20.2	−123.5	1.3392[18]; 1.3316	162–3	stable: 168, al.; un-stable: 157; mix-ture: 148	128.5	57; 99	47	139, al.	175–6, al.	Thio-semicar-bazone, 146	or.	dk. br.	30
4	Propionaldehyde (Propanal).......	48–9	−81	1.364	89, bz.-lgr.; 154, w.	148, or.; 150, red; 155	125, yel., 50% al.	oil	40	154–6, al.	143	Picrate, 156–7	dk. ol. gn.	red	20
5	Glyoxal ....'.	50	15	....	270	328	311	180	178	mono: 186; di: 228	mono: 224	Phenyl-osazone, 169–70	.....	....	......
6	Acrolein (Acralde-hyde)	52.4	−87.7	1.4025	171, w.	165	150–1	50–1, hot lgr., pyra-zoline	......	192, 50% al.	163, al.	....	red br.	vlt. br.	0.1
7	Propynal (Propargyl aldehyde).......	55	......	......	......	......	......	......	......	......	......	Cu deriv., 160	.....	....	....
8	2,2,2-Trifluoro-propionaldehyde ..	56[745]	......	....	...,....	151	......	......	......	......	......	....	....	....	....
9	Isobutyraldehyde...	64	−65.9	1.3730	125.6	187, or.-yel., al.; 182	130–1, or.-yel., al.	oil	oil	154	144	....	....	....	....
10	2-Methyl-2-propenal (Methacrolein) ...	73.5	......	1.4191	198	206	......	74, py-razo-line	......	......	......	....	....	....	....
11	n-Butyraldehyde (Butanal).......	74.7	−97.1	1.38433	95.5, lgr.; 106	123, al.	87, yel., al.; 93–5, red	93–5	b.p. 152[715]	134; 142	141	....	....	....	....
12	Trimethylacetalde-hyde (Pivaldehyde)	75	3; 6	1.3791	190	210, yel.	......	......	41						
13	Chloroacetaldehyde	85–6	......	....	134–5d; 148, al.	......	......	......	oil						
14	2-Chloropropion-aldehyde........	86	......	1.431[17]	......	......	......	......	......			Hydrate, b.p. 80.5–81			
15	Dichloroacetalde-hyde	89.5–90.5	......	....	155–6, using only 1 equiv-alent of re-agent	......	......	......	b.p.: 67–9[17], using only 1 equiv-alent of re-agent						

*Derivative data given in order: m.p., crystal color, solvent from which crystallized.

TABLE IX. ORGANIC DERIVATIVES OF ALDEHYDES
a) Liquids 1) Listed in order of increasing atmospheric b.p.* (Continued)

No.	Name	Boiling point, °C	Melting point, °C	n_D^{20}	Semi-carba-zone	2,4-Di-nitro-phenyl-hydra-zone	p-Nitro-phenyl-hydra-zone	Phenyl-hydra-zone	Oxime	Dimeth-one deriv. (Dime-done deriv.)	Dimeth-one anhy-dride	Miscel-laneous	o-Dianisidine spot test		
													Cold	Hot	Limit, γ
16	Methoxyacetaldehyde	92	......	1.3950	....	124–5	115	......	......	......	......	......	.....	...	.....
17	3-Methylbutanal (Isovaleraldehyde)	92.5	−51	1.39225	107	123, yel.-or., al.	110–1, al.	oil	48.5	154–5, al.	173 (cor.)	Thio-semicar-bazone, 52–3	.....	...	.....
18	2-Methyl-1-butanal (α-Methylbutyr-aldehyde)	92–3	20, tri-	1.3942	103–5, bz.-pet. eth.	120						....			
19	Trichloroethanal (Chloral; Tri-chloroacetalde-hyde)	98	−57.5	1.45572	90d.	131	131, yel.	......	56	......	......	Hydrate, 51.7	.....		
20	Pentanal (Valeral-dehyde)	103.4	−91.5	1.3947	....	98, yel., al.; 107	......	.....	52, pet. eth.	104.5	113	Thio-semicar-bazone, 65	.....		
21	tert-Butylacetalde-hyde	103	......	1.4150	....	147	......	......	......	......		....			
22	2-Butenal (Croton-aldehyde)	104	−69	1.4362²⁰·⁵	199	190, crim., bz.-lt. pet.	184–5	56	119	183	163, sint.; 167	Phenyl-semicar-bazone, 126–7	dk. red	dk. br.-red	2
23	Dimethylethylacet-aldehyde	104	......	....	....	....	......	......	......	......	......	....	.....	...	.....
24	Ethoxyacetaldehyde	106	......	1.3956	....	116–7, me. al.	113–4, al.								
25	2-Isopropylacrolein	107–9	.....	1.4223	....	165									
26	2-Butynal	105–110⁷⁵⁵	.....	1.446¹⁹	....	136	......	......	......	......	......	....	.....	...	.....
27	Methylisopropyl-acetaldehyde	114	.....	1.3998²⁵	....	124									
28	2-Bromoisobutyr-aldehyde	115	.....	1.4518²⁵	....	....						Decom-poses in w.			
29	Diethylacetaldehyde (2-Ethylbutyralde-hyde)	116; 117	......	1.4025	99, bz.-lt. pet.	95, pa.-or., lt. pet.; 129–30, al.	......	......		102, me. al.	......				
30	Methyl-n-propyl-acetaldehyde	116⁷³⁷	.....	....	102	103	......	......	......	......	......	....	.....	...	.....
31	2-Methyl-2-butenal	116–9	.....	....	216	....									
32	n-Propoxyacetalde-hyde	119⁷⁴⁸			....	86									
33	Isobutylacetalde-hyde (Isocapro-aldehyde)	121⁷⁴³	.....	....	127	99				b.p. 103³⁵					
34	Paraldehyde (Acet-aldehyde trimer)	124.4⁷⁵²	12.6	1.4049	....	....	......	......	......	......		Dilute acid → Acetal-dehyde, b.p. 20.2	dk. ol. grn.	dk. red br.	4
35	2-Pentenal	125	.....	....	180	....	123	......	......	......	......	....	.....	...	.....
36	3-Methoxyisobutyr-aldehyde	129	.....	1.4030²⁷	....	102									

*Derivative data given in order: m.p., crystal color, solvent from which crystallized.

145

TABLE IX. ORGANIC DERIVATIVES OF ALDEHYDES
a) Liquids 1) Listed in order of increasing atmospheric b.p.* (Continued)

No.	Name	Boiling point, °C	Melting point, °C	n_D^{20}	Semicarbazone	2,4-Dinitrophenylhydrazone	p-Nitrophenylhydrazone	Phenylhydrazone	Oxime	Dimethone deriv. (Dimedone deriv.)	Dimethone anhydride	Miscellaneous	o-Dianisidine spot test Cold	Hot	Limit, γ
37	3-Chloropropionaldehyde	130–1	......	1.475^{15}	....	....	......	....	....	......	......	Trimer, 35.5, dil. HCl-abs. al.; b.p. 170–5^{12-5}	.....	....	......
38	Hexanal (Caproaldehyde)	131	......	1.4068	106, bz.-pet. eth.	104, or.-yel.	......	......	51, pet. eth.-me. al.	108.5, dil. al.	......	Phenyl-semicarbazone, 135–6			
39	Ethylisopropylacetaldehyde	133.5	......	1.4086^{25}	....	121									
40	3,3-Dimethylpentanal	134	......	1.4292	....	102									
41	3-Methyl-2-butenal (3-Methylcrotonaldehyde)	135	......	1.4526	223	182									
42	Cyclopentanecarboxaldehyde	136	......	....	124	....									
43	2-Methylpenten-2-al-1 (3-Ethyl-2-methylacrolein)	136.8		1.4488	207	159, red, al.	......	58–60	48–48.8						
44	Tetrahydrofurfural	$142–3^{779}$	......	1.4473; 1.43658	166	134	......					Conc. HCl → brt. red col.; α-Benzyl-α-phenylhydrazone, 67, me. al.	.....		
45	5-Methylhexanal	144^{750}	......	1.4114	117	117									
46	3-Furaldehyde	144^{732}	......	1.4945	211	....	......	149.5	......				......		
47	1-Cyclopentenylformaldehyde	146	......	1.4828^{21}	208	....	188	......							
49	2-Chloro-2-butenal (2-Chlorocrotonaldehyde)	147–50	......	1.478^{23}	....	....						Cyanohydrin, b.p. $137–8^{26}$, n_D^{21}: 1.4762			
50	2-Hexenal	150	......	1.4470^{13}	176	....	139	......	......						
51	3-Hexenal	150	......	....	147	....									
52	Heptanal (Enanthaldehyde)	155	–45	1.4125	109, al.	108, yel., al.	73	......	57	135	112	....	red-br.	red	9
53	Ethylisobutylacetaldehyde	155	......	....	98	....									
54	Di-n-propylacetaldehyde	161	......	1.4142^{15}	101	....	......	.....	b.p. 126^{47}				....	....	......

*Derivative data given in order: m.p., crystal color, solvent from which crystallized.

TABLE IX. ORGANIC DERIVATIVES OF ALDEHYDES
a) Liquids 1) Listed in order of increasing atmospheric b.p.* (Continued)

No.	Name	Boiling point, °C	Melting point, °C	n_D^{20}	Semi-carba-zone	2,4-Di-nitro-phenyl-hydra-zone	p-Nitro-phenyl-hydra-zone	Phenyl-hydra-zone	Oxime	Dimeth-one deriv. (Dime-done deriv.)	Dimeth-one anhy-dride	Miscel-laneous	o-Dianisidine spot test Cold	Hot	Limit, γ
55	2-Furancarbox-aldehyde (Fur-fural)	161.7	−36.5	1.52608	202	212–4, yel.; 230 (cor.); red; mix-ture: 185	54	97	α: 75–6, pet. eth.; β: 91–2, al.	160d.	162–5	Phenyl-semicar-bazone, 180–1	dk. red-vlt.	dk. bl.-vlt.	0.02
56	Hexahydro-benzaldehyde.....	162	......	1.4495[19]	173; 176, w.	172	......	......	90–1, pet. eth.	......	......	Oxime-HCl, 107–8d.	......	......	......
57	2-Ethylhexanal-1 (n-Butylethyl-acetaldehyde)	163	......	1.4150	254d.	114–5, dil. al.; 120–1, yel., al.						....			
58	2,2,3-Trichloro-n-butyraldehyde (n-Butylchloral; Crotonchloral) ...	164.5–5.5	......	1.47554	....	....	......	......	65			$NH_3 \rightarrow$ Butyl-chloral ammo-nia, 62			
59	Butanedial (Succinaldehyde) .	169–70d	......	1.4254	....	280	......	......	di: 172	......	......	Polymer, 65	......		
60	Octanal (n-Octalde-hyde; Capryl-aldehyde)........	171	......	1.42167	98, dil. me. al.; 101	106, yel., al.; 96	80, brt. yel.	......	60, me. al.	90, dil. al.	101	Thio-semicar-bazone, 94–94.5	......		
61	2-Ethyl-3-n-propyl-acrolein	173	......	1.4518[22]	150–1; 153	124–5; 122									
62	3-Fluorobenz-aldehyde........	173	......	......	....	....	202	114	63						
63	2,2,2-Tribromo-ethanal (Bromal)..	174, yel.			......	......	......	......	115			Mono-hydrate, 53.5	no reac.	dk. grn.	40
64	4-Fluorobenz-aldehyde........	174.5[752]	......		......	....	212	147	syn: 116–7; anti: 86	......					
65	2-Fluorobenz-aldehyde........	175	−44.5	....	....	....	205	90	63	......					
66	Benzaldehyde......	179	−26; f.p.: −55.6	1.5446	222; 233–5, r. htng.	237, or, al.	190, red, al.; 234–6; 262	158; 154–5	α: 35 (sta-ble); β: 130, eth.	193	200	Phenyl-semicar-bazone, 180–1	or.	red-or.	3
67	Nonanal (Pelargon-aldehyde)........	185	......	1.4273	100; 84, me. al.	100 (cor.), yel., al.	......	......	64, pet. eth.	86	......	Phenyl-semicar-bazone, 131–2	......		
68	5-Methylfurfural ...	187	......	1.5147[25]	211*	212 (cor.)	130, red	147–8	syn: 112; anti: 51–2	......	......	......			
69	Glutaraldehyde	187–9d.	......	1.4330[25]	....	....	169	......	di: 175; 178, w.						
70	Phenylethanal (Phenylacetal-dehyde)	194	33	1.53191	153, dil. al.; 156	121, grn.-yel., al.; 110	......	58, lgr.; 62–3	97–8, eth.; 100	165	126	....	dk. br.-red	dk. br.	polym.

*Derivative data given in order: m.p., crystal color, solvent from which crystallized.

TABLE IX. ORGANIC DERIVATIVES OF ALDEHYDES
a) Liquids 1) Listed in order of increasing atmospheric b.p.* (Continued)

No.	Name	Boiling point, °C	Melting point, °C	n_D^{20}	Semi-carba-zone	2,4-Di-nitro-phenyl-hydra-zone	p-Nitro-phenyl-hydra-zone	Phenyl-hydra-zone	Oxime	Dimeth-one deriv. (Dime-done deriv.)	Dimeth-one anhy-dride	Miscel-laneous	o-Dianisidine spot test		
													Cold	Hot	Limit, γ
71	2-Hydroxybenz-aldehyde (Salicyl-aldehyde)	197 (cor.)	−7; f.p.: 1.6	1.574	231	248, red, abs. al.; 252d., lt. red, ac. a.	227, red-br., al.	142	57; 63	. . . . .	208, 70% al.	p-Nitro-benzo-ate, 128	or.	or.	5
72	2-Thiophenecarbox-aldehyde	198	. . . . .	1.5950[16]	. . . . .	242	. . . . .	119; 139	. . . . .	. . . . .					
73	3-Methylbenz-aldehyde (3-Tolu-aldehyde)	199		1.5413[21]	204; 223–4	212; 194	157	91, lgr.; 84	60, lgr.	172	206	. . . . .	dk. or.-red	ch. red	5
74	2-Methylbenz-aldehyde (2-Tolu-aldehyde)	200		1.5481	209, al.; 212; 218	193–4, red, ac. a.	222, red, al.	101; 105–6; 111	49	167	215	dk. or.-red		ch. red	5
75	4-Methylbenz-aldehyde (4-Tolu-aldehyde)	204–5		1.5454	234, al.; 215	232.5–4.5 (cor.), or.-yel., al., PhNO₂	200.5 (cor.), dk. red, ac. a.	112–3, al.; 121	79–80; 110	. . . . .			dk. or.-red	ch. red	5
76	d-Citronellal (d-Rhodinal)	207		1.4485	83–4, chl., ppt. by lgr; 91–2	78, yel., al.	. . . . .	. . . . .	oil	77–9, dil. al.	173	. . . . .	dk. grn.	brt. red	10
77	Decanal (Capralde-hyde)	207–9		1.4287	102	104, yel.	. . . . .	. . . . .	69, dil. me. al.	91.7, dil. al.	. . . . .	Thio-semicar-bazone, 99–100	pa. ol.	dk. br.	200
78	2-Chlorobenz-aldehyde	213–4	11	1.56708	146, yel.; 225, pyr.; 229–30, me. al.	213.6 (cor.); 209, or. red, xyl.	237–8, red, al.; 241, br.-red; 249, or.	86	α: 75–6, al.; β: 101–3	205d., al.	224–6 (cor.), al.	. . . . .	. . . . .	. . . .	. . . . .
79	Phenoxyethanal (Phenoxyacetalde-hyde; Glycolalde-hyde phenyl ether)	215d.		1.5380[21]	145	. . . . .	. . . . .	86, pa. yel., al.	95, pet. eth.	. . . . .					
80	3,5-Dimethylbenz-aldehyde	220–2	9	1.5385	201–2	. . . . .	. . . . .	. . . . .	. . . . .	. . . . .		Oxid. → acid, 170, al.	. . . . .		
81	3-Phenylpropion-aldehyde (Hydro-cinnamaldehyde)	224			127, al.	149, yel., al.	122–3, yel., dil. al.	. . . . .	93–4.5, dil. al.; 97 (cor.)	. . . . .					
82	Citral a. (Geranial)	228d.		1.48752	164, me. al.	108–10, red-or., al.; 116	. . . . .	. . . . .	143–5	. . . . .	208	p-Nitro-benzo-ate	dk. red	red blk	0.1
83	Citral b. (Neral)	228d.		1.4900	HCl: 171; mix-ture: 132, NaOAc	96, red-or., al.	. . . . .	. . . . .	. . . . .	. . . . .			dk. red	red blk.	0.1

*Derivative data given in order: m.p., crystal color, solvent from which crystallized.

TABLE IX. ORGANIC DERIVATIVES OF ALDEHYDES
a) Liquids 1) Listed in order of increasing atmospheric b.p.* (Continued)

No.	Name	Boiling point, °C	Melting point, °C	n_D^{20}	Semi-carba-zone	2,4-Di-nitro-phenyl-hydra-zone	p-Nitro-phenyl-hydra-zone	Phenyl-hydra-zone	Oxime	Dimeth-one deriv. (Dime-done deriv.)	Dimeth-one anhy-dride	Miscel-laneous	o-Dianisidine spot test Cold	Hot	Limit, γ
84	2,6-Dimethylbenz-aldehyde.........	228^{742}	11	....	158	....	....	....	....	....	....	....	....	....	....
85	3-Methoxybenz-aldehyde (3-Anis-aldehyde)........	230	3–4	1.5538	233d.	....	171	76	39–40 pet. eth.; 112	....	....	Phenyl-thio-semicar-bazone, 153	dk. or	red br.	0.4
86	3-Bromobenz-aldehyde.........	234–6	......	....	205	....	220	141	72	....	....	....	....	....	....
87	4-Isopropylbenz-aldehyde (Cumal-dehyde).........	236	......	1.5301	211, me. al.	241, red, bz.; 243, red, ac. a.; 244–5, al.-chl.	190, al.	129, al.	α: 52, al.; β: 111	170–1, al.	172–3	....	dk. red	ol. yel.	3
88	3-Ethoxybenz-aldehyde.........	245.5	......	1.5408	....	....	....	....	....	....	....	....	....	....	....
89	4-Methoxybenz-aldehyde (4-Anisal-dehyde).........	248	2.5	1.5731	210; 203	253–4d., red, ac. a.; 250, red, xyl.	160, red-vlt.	120–1, wh., dil. al.	α': 64–5, bz. α: 4–5 (from α' on fu-sion); β: 133, bz.	144–5 (cor.), al.	243 (cor.), al.	....	dk. or.	red-br.	0.4
90	3-Phenylpropenal (Cinnamaldehyde)	252d.	–7.5	1.61949	215–6, w.	255d., red, ac. a.	195, red, al.	168, yel., dil. al.	α: 64–5, lgr.; β: 138.5, bz.	213 (cor.), al.; 161, al.	175, al.	....	dk. ch. red	ch. red	0.05
91	4-Ethoxybenz-aldehyde.........	255; 249	13–4	....	202d., al.; 208	....	....	....	syn: 157; anti: 118	....	....	....	....	....	....
92	3,4-Diethoxybenz-aldehyde (Proto-catechualdehyde diethyl ether).....	277–80	......	....	....	....	....	....	98	....	....	Oxid. → acid, 165	....	....	....
93	Diphenylacetalde-hyde...........	315–6d	......	....	162	....	....	....	α: 120; β: 106	....	....	Oxid. → Benzo-phenone, 48	....	....	....

*Derivative data given in order: m.p., crystal color, solvent from which crystallized.

TABLE IX. ORGANIC DERIVATIVES OF ALDEHYDES
a) Liquids 2) Reduced pressure b.p. only (listed in order of increasing semicarbazone m.p.)*

No.	Name	Boiling point, °C	Melting point, °C	n_D^{20}	Semicarbazone	2,4-Dinitrophenylhydrazone	p-Nitrophenylhydrazone	Phenylhydrazone	Oxime	Dimethone deriv. (Dimedone deriv.)	Dimethone anhydride	Miscellaneous	o-Dianisidine spot test Cold	Hot	Limit, γ
1	7-Methyloctanal	94[120]	......	......	80	100									
2	3-(2-Furyl)propionaldehyde	70[14]	......	1.4470	80	......									
3	2-Methyloctanal	83[20]	......	......	80										
4	2,3-Dichloro-n-butyraldehyde	58–60[20]	......	1.4618[21]	96–7					oil					
5	Octanal (n-Octaldehyde)	81[32]	......	1.4217	98; 101	106, yel.	80	......		59–60	89.8				
6	Undecanal (Hendecanal)	120[20]	−4	1.4324[23]	103, me. al.	104, yel.	......	......	72, wh., me. al.			Timer, 47–8			
7	Tridecanal (n-Tridecylaldehyde)	136[8]	15	......	106, al.	108	......	......	80.5, dil. al.			Trimer, 61.5, eth.			
8	2-Hydroxypropionaldehyde	114[9]	......		114, w.	......	127								
9	2-n-Amylcinnamaldehyde (2-n-Pentylcinnamaldehyde; Jasminaldehyde)	161–3[19]	......	1.5381	118	164, red, al.	......	......	74, al.-w.						
10	2-Methyl-3-phenylpropionaldehyde	90[6]	......		123										
11	2-Hydroxy-2-methyl hexanal (n-Butylmethylglycolaldehyde)	87–8[35]	......		143										
12	Phenoxyacetaldehyde	83[5]	......	1.5360	146	138	......	......	95						
13	2-Ethyl-2-hexenal	73[20]	......		152	125									
14	2-Ethyl-3-hexenal	84[52]	......		156										
15	Cyclohexylacetaldehyde	58[10]	......	1.4509[25]	159	125	......								
16	2-Nonenal	126[21]	......	1.4426	165	126	113								
17	2-Heptenal	85[14]	......	1.4314	169	......	116								
18	2,3,6-Trimethylbenzaldehyde	114[10]	......		169	......	......	......	126						
19	3,5-Dimethylhexahydrobenzaldehyde	71[14]	......		171										
20	2-Hydroxy-2-phenylpropionaldehyde (Methylphenylglycolaldehyde)	101[4]	......		182–3										
21	2,4,6-Trimethylbenzaldehyde	98[6]; 128[15]	......	1.5524	188										
22	2-Hydroxy-2-phenylbutyraldehyde (Ethylphenylglycolaldehyde)	110–11[5]	......		188										
23	2-Hydroxybutyraldehyde (Aldol)	83[20]	......		194	......	109–11, red-yel., dil. al.	......	*syn*: 112; *anti*: 51–2	146–8, 30%, me. al.	126	4-Bromophenylhydrazone, 127–8			
24	1,2,3,4-Tetrahydro-2-naphthaldehyde	92[0.5]	......		197	......	......	......							
25	2-(1-Naphthyl)propionaldehyde	132[2]	......	......	204										

*Derivative data given in order: m.p., crystal color, solvent from which crystallized.

TABLE IX. ORGANIC DERIVATIVES OF ALDEHYDES
a) Liquids 2) Reduced pressure b.p. only (listed in order of increasing semicarbazone m.p.)* (Continued)

No.	Name	Boiling point, °C	Melting point, °C	n_D^{20}	Semi-carba-zone	2,4-Di-nitro-phenyl-hydra-zone	p-Nitro-phenyl-hydrazone	Phenyl-hydra-zone	Oxime	Di-meth-one deriv. (Dime-done deriv.)	Di-meth-one anhy-dride	Miscellaneous	o-Dianisidine spot test		
													Cold	Hot	Limit, γ
26	**1,6-Hexanedial** (Adipic dialdehyde)	94[12]; 70[3]	......	1.4350	*di*: 206	......	....	......	*di*: 185–6, w.	......	......	....	...	...	...
27	**2-Methylcinnamalde-hyde**	124[14]	......	1.6057[17]	208, al.-w.	......	....	......		......	......	....			
28	**Phenylglyoxal**	108[15]	......	....	*mono*: 208–9d., yel., al. *bis*: 229d.	......	309	......		......	......	91 (mono-hyd.) w.; 2-Thio-semicarba-zone, 170, yel. al.	...	...	...
30	**Cyclohexenecarbox-aldehyde**...........	70[13]	......	1.4921[17]	213	......	....	......	99	......	......	....	...	...	...
31	**2-Phenoxybenzaldehyde**	153[1]	......	....	215	......	....	......		......	......	....			

*Derivative data given in order: m.p., crystal color, solvent from which crystallized.

TABLE IX. ORGANIC DERIVATIVES OF ALDEHYDES
a) Liquids 3) Miscellaneous; reduced pressure b.p. only (listed alphabetically)*

No.	Name	Boiling point, °C	Melting point, °C	n_D^{20}	Semicarbazone	2,4-Dinitrophenylhydrazone	p-Nitrophenylhydrazone	Phenylhydrazone	Oxime	Dimethone deriv. (Dimedone deriv.)	Dimethone anhydride	Miscellaneous	o-Dianisidine spot test Cold	Hot	Limit, γ
1	2-Chloroacrolein.......	29–31[17]	.....	1.463	.....	.....	.....	.....	.....	.....	.....	Diethylacetal, b.p. 158-60	....	...	...
2	4-Chloro-n-butyraldehyde	50-1[13]	.....	1.44662[8.5]	.....	134–5	110	.....	74.5						
3	d,l-2,3-Dichloropropionaldehyde.............	48[14]	.....	1.4762	.....	.....	.....	.....	.....	.....	.....	Dimethylacetal, b.p. 78-82[13], n_D^{18}: 1.144	....	...	...
4	2-Heptynal	54[13]	.....	1.4521[17]	.....	74	.....	.....	.....						
5	2,4-Hexadienal (Sorbaldehyde)...............	65[11]	.....	1.5372[22]	.....	.....	.....	102	160						
6	4-Hydroxy-n-butyraldehyde.............	68[8]	.....	1.4403	.....	118									
7	3-Hydroxy-2-isopropylpropionaldehyde	84[10]	.....	.....	.....	126									
8	3-Hydroxy-3-methyl-n-butyraldehyde........	67[13]	.....	.....	.....	.....	142								
9	4-Methoxy-2-methyl-n-butyraldehyde	66[55]	.....	1.4280[25]	.....	88									
10	5-Methyl-2-thiophenecarboxaldehyde........	114[25]	.....	1.5782[29]	.....	.....	.....	126							
11	3-Methyl-2-thiophenecarboxyaldehyde.......	114[25]	.....	1.5833[25]	.....	.....	.....	149							
12	4-Octenal.............	84[13]	.....	1.4463[25]	.....	108									
13	Phenylpropargyl aldehyde	116–7[17]	.....	1.6032[25]	.....	.....	.....	.....	108, lgr.						
14	2-Phenylpropionaldehyde .	76[4]	.....	.....	.....	135									
15	3-Pyridinecarboxaldehyde (Nicotinaldehyde)......	99[26]	.....	.....	.....	.....	.....	.....	158						
16	2,2,4-Trichloro-n-butyraldehyde.............	.....	f.p.-78	.....	.....							HNO₃ → acid, 73-5			
17	2,4,6-Trihydroxybenzaldehyde (Phloroglucinaldehyde)	.....	d.	.....					195d., (hyd.), w.			2,4,6-Triacetate, 156-7, al.; 2-Benzoate, 198-200, chl.			
18	4-Vinylbenzaldehyde (4-Formylstyrene).......	93[14]	.....	1.5960[25]	.....	.....	.....	131	.....				....	...	...

*Derivative data given in order: m.p., crystal color, solvent from which crystallized.

TABLE IX. ORGANIC DERIVATIVES OF ALDEHYDES
b) Solids (Listed in order of increasing m.p.)*

No.	Name	Melting point, °C	Boiling point, °C	Semicarbazone	2,4-Dinitrophenylhydrazone	p-Nitrophenylhydrazone	Phenylhydrazone	Oxime	Dimethone deriv. (Dimedone deriv.)	Dimethone anhydride	Miscellaneous	o-Dianisidine spot test		
												Cold	Hot	Limit, γ
1	3-Chlorobenzaldehyde	17–8	213–4	228, pyr.; 230, me. al.	248, dk. red, xyl.; 256, or.-yel.	216, dil. al.	134–5, abs. al.	α, anti: 70–1, al.; β, syn: 118	. . . . . .	. . . .	n$_D^{20.2}$: 1.55908	. . . .	. . . .	. . . .
2	2,3,5,6-Tetramethyl-benzaldehyde	20	135[11]	270d.	. . . .	. . . .	. . . . . .	125	. . . . . .					
3	2-Ethoxybenzaldehyde (Salicylaldehyde ethyl ether)	20–2; 6–7	247–9	219, al.	. . . .	. . . .	. . . . . .	57–9, pet. eth.		. . . .	Diacetate, 88–9, ac. anh.			
4	Tetradecanal (Myrist-aldehyde)	23–3.5	166[24]	106.5, dil. al.	108	95, brt. yel.	. . . . . .	82.5–3.5, dil. al.			Trimer, 65			
5	Pentadecanal	24–5	160[14]	106.5, al.	106–7, yel., pyr.-al.	94–5, yel., al.	. . . . . .	86, dil. al.			Trimer, 69–70			
6	Hexadecanal (Palmit-aldehyde)	34	. . . . . .	107, dil. al.; 108–9	108	96.5, yel., eth.	. . . . . .	88, yel.	. . . . . .		Trimer, 73; Thiosemi-carbazone, 106–9			
7	1-Naphthaldehyde	34	292; 162[18]	221	. . . .	224	80	98; 90	. . . . . .					
8	Phenylacetaldehyde	34	195	156; 163	121	151	mono: 63; 58; di: 101–2	99; 103	165		. . . . .			
9	4-Methoxy-1-naphthal-dehyde	34, wh.	200[11]	. . . . .	. . . .	. . . .	113	. . . . . .			Azine, 185, yel., al.			
10	(5-Hydroxymethyl)furfural	35–6	115–20[0.5]	195d.; al, recr. tol.-lgr.	184, red	185, dk. red, al.	140–1, tol.	77–8; 108	. . . . . .					
11	Heptadecanal (Margaric aldehyde)	35–6; 63	. . . . . .	. . . . . .	. . . . .	. . . . . .	. . . . . .	89.5, et. ac.		. . . .	Trimer, 77–8, lt. pet.			
12	3,4-Methylenedioxy-benzaldehyde (Piperonal)	37	263	230; 234; 237	265d., xyl.; 266d., red, ac. a.	199–200, red	102–3, yel., al.; 99; 106	syn: 146, me. al.; anti: 112, w.	177–8; 193, yel., al.	220 (cor.)	. . . .	brt. red	dk. red	4
13	2-Iodobenzaldehyde	37	129[14]	206	. . . . .	. . . .	79	108	. . . . . .	. . . .				
14	Octadecanal (Stearaldehyde)	38	. . . . . .	108–9	101; 110	101, yel., me. al.	. . . . . .	89	. . . . . .	. . . .	Thiosemicar-bazone, 111			
15	2-Methoxybenzaldehyde (o-Anisaldehyde; Salicylaldehyde methyl ether)	38–9	243–6 (cor.)	215d., al.	253.5 (cor.), red, xyl.	204–5, br. red	. . . . . .	92, dil. al.	. . . . . .	. . . .	n$_D^{20}$: 1.5598	. . . .		
16	2-Aminobenzaldehyde	40	. . . . . .	. . . . .	. . . . .	220	221	135	. . . . . .	. . . .				
17	4-Diethylaminobenz-aldehyde	41	172[7]	241d., al.	. . . . .	. . . .	103, yel.-br.	93						
18	Dodecanal (Lauralde-hyde)	2 forms: a) 42–3; b) 11	238	103; 106	106, yel.	90	. . . . . .	76–7, eth.; 77–8, me. al.	. . . . . .		Thiosemicar-bazone, 100	. . . .	. . . .	. . . .

*Derivative data given in order: m.p., crystal color, solvent from which crystallized.

TABLE IX. ORGANIC DERIVATIVES OF ALDEHYDES
b) Solids (Listed in order of increasing m.p.)* (Continued)

No.	Name	Melting point, °C	Boiling point, °C	Semicarbazone	2,4-Dinitrophenylhydrazone	p-Nitrophenylhydrazone	Phenylhydrazone	Oxime	Dimethone deriv. (Dimedone deriv.)	Dimethone anhydride	Miscellaneous	o-Dianisidine spot test		
												Cold	Hot	Limit, γ
19	3,4-Dichlorobenzaldehyde	43–4	247–8	....	....	276–7, or.	......	syn: 120, al.; on fusing → anti: 114–5; 118–9	....	....		.....	.....	.....
20	3-Phenylcinnamaldehyde	44	210[14]	214–5	196	....	173, yel.	....						
21	2,4,5-Trimethylbenzaldehyde	44	120[10]	243	....	....	127	....						
22	2-Nitrobenzaldehyde	44	....	256	265	263	156	anti: 102; syn: 154, bz.	....	....		grn. br.	red br.	5
23	3,4-Dimethoxybenzaldehyde (Veratraldehyde)	44; 58	285	177	261–3 (cor.), or., PhNO₂; 264–5	....	121, al.	94–5, lgr.	173					
24	4-Chlorobenzaldehyde	48	214.5–6.5	230, pyr., 233, me. al.	254 (cor.), or.	237, dk. br., al.	127–7.5, lt. yel., dil. al.	α: 110; β: 146				.....	.....	.....
25	2-Pyrrolecarboxaldehyde	50	217–9	183.5, w.	....	182–3, red, xyl.	139, lgr.	164, bz.	....		n_D^{16}: 1.5939	.....	.....	.....
26	Benzylglycolaldehyde	52	121[4]	137	....	....	....	....	....		Benzoate, 70			
27	Furfural diacetate	52, eth.	220	....	....									
28	4-Chloro-2-hydroxybenzaldehyde	52.5	....	212, pa. yel., ac. a.	....	257, or., ac. a.	......	155, col., al.						
29	Quinoline-4-carboxaldehyde	51–3, tol. (anh.); 84–4.5 (mono-hyd.)	123[4]	....	....	261–2, yel., al.	......	181–2, me. al.	....		Picrate, 179	.....	.....	.....
30	2-Ethyl-4-hydroxybenzaldehyde	53	145[1]	....	....	....	....					.....	.....	.....
31	2-(2-Furyl)-acrolein	54, lgr.	95[9]	219.5	....	....	132, pet. eth.	110–1				.....	.....	.....
32	2,3-Dimethoxybenzaldehyde (o-Veratraldehyde)	54	137[12]	231d.	....	....	138	99, al.-w.						
33	9-Hydroxynonanal	54	120[0.1]	....	....	....	....							
34	2,3-Diphenylpropionaldehyde	54	170[11]	125	....	....	....							
35	3-Benzyloxybenzaldehyde	54	218[20]	....	....	....	....							

*Derivative data given in order: m.p., crystal color, solvent from which crystallized.

TABLE IX. ORGANIC DERIVATIVES OF ALDEHYDES

b) Solids (Listed in order of increasing m.p.)* (Continued)

No.	Name	Melting point, °C	Boiling point, °C	Semicarbazone	2,4-Dinitrophenylhydrazone	p-Nitrophenylhydrazone	Phenylhydrazone	Oxime	Dimethone deriv. (Dimedone deriv.)	Dimethone anhydride	Miscellaneous	o-Dianisidine spot test Cold	Hot	Limit, γ
36	3-Chloro-2-hydroxybenzaldehyde (3-Chlorosalicylaldehyde)	54.5–5.5		240–3, 50% ac. a.				167–8, dil. al.			5-Nitro deriv., 129, yel., dil. al.			
37	Octatrienal	55												
38	Isoquinoline-1-carboxaldehyde	55.5		197			171–2							
39	Phthalaldehyde	56					di: 191					brt. yel. ppt.	brt. yel. ppt.	
40	2,3,5-Trichlorobenzaldehyde	56, col., dil. al.									Oxid. $\xrightarrow{KMnO_4}$ acid, 162–3			
41	2-Hydroxy-5-methylbenzaldehyde (5-Methylsalicylaldehyde)	56, dil. al.	217–8				149, yel., al.	105, w.			Diacetate, 94, al.			
42	3-Iodobenzaldehyde	57		226		212	155	62						
43	4-Bromobenzaldehyde	57		228; 229	128; 257	207–8	113	syn: 157; anti: 111						
44	3-Nitrobenzaldehyde	58		246	293d.	247	120; 124	120; 122						
45	2,5-Dichlorobenzaldehyde	58	231–3				104–5, al.	127.5–8, dil. al.						
46	2,4,6-Trichlorobenzaldehyde	58–9												
47	2-Phenanthraldehyde	59; 59.5		282				175						
48	Paraisobutyraldehyde (2,4,6-Tri-isopropyl-1,3,5-trioxan)	59–60	195 (cor.), sl. de-polym.								See: Isobutyraldehyde, b.p. 64			
49	1-Hydroxy-2-naphthaldehyde	59–60, gm.-yel., al.-w.						145, bz.						
50	2-Naphthaldehyde	60, w.	150^{15}	245, al.	270	230	205–6 d., al.; 217–8	156, dil. al.						
51	4-Phenylbenzaldehyde	60		243d., al.	239d., scar., xyl.		189d.	149–50						
52	3-Methoxy-1-naphthaldehyde	60, pet. eth.		200, al.-w.		197, red, ac. a.		102, al.-w.						
53	4-Ethoxy-3-methoxybenzaldehyde	64												
54	3,5-Dichlorobenzaldehyde	65	$235–40^{748}$				106.5, yel., pet. eth.	112						
55	2,3-Dichlorobenzaldehyde	65–7												
56	5-Methoxy-1-naphthaldehyde	66, yel., pet. eth.		246, ac. a.-w.		246, red, ac. a.-w.		104, w.						

*Derivative data given in order: m.p., crystal color, solvent from which crystallized.

TABLE IX. ORGANIC DERIVATIVES OF ALDEHYDES
b) Solids (Listed in order of increasing m.p.)* (Continued)

No.	Name	Melting point, °C	Boiling point, °C	Semi-carba-zone	2,4-Di-nitro-phenyl-hydra-zone	p-Nitro-phenyl-hydrazone	Phenyl-hydra-zone	Oxime	Dimeth-one deriv. (Dime-done deriv.)	Di-meth-one anhy-dride	Miscellaneous	o-Dianisidine spot test Cold	Hot	Limit, γ
57	Dibenzofuran-2-carboxaldehyde	68	....	....	....	....	162	....	.....	.....	....	.....	.....	.....
58	2,6-Dichlorobenzalde-hyde	70–1	....	....	....	....	....	....	.....	.....	o-Nitro-phenylhydra-zone, 154; p-Bromo-phenylhydra-zone, 142	.....	.....	.....
59	2,4-Dimethoxybenz-aldehyde (β-Resorcylaldehyde dimethyl ether)	71, dil. al.; 69	165¹⁰	....	....	....	....	106, w.	....	.....	5-Nitro deriv., 188–9, me. al.	.....	.....	.....
60	Quinoline-2-carbox-aldehyde	71, pet. eth. (anh.); 51 (mono-hyd.), w.	....	....	250, yel.; 225, subl.	....	204, yel., al.	188	....	.....	....	.....	.....	.....
61	4-Ethoxy-1-naphthal-dehyde	72								Hydrazone, 160–182, dk. red; Azine, 209, yel., PhNO₂			
62	4-Aminobenzaldehyde	72	....	153	....	....	156	124	....	.....	....	or. br.	red br.	0.4
63	2,4-Dichlorobenzalde-hyde	72; 74.5						136–7			Oxime HCl, 133.5	.....		
64	4-Dimethylamino-benzaldehyde	74	....	222	325	182	148	185			....			
65	4-Methylthiazole-5-carboxaldehyde	75; 72.5	118²¹	....			159; 161	....						
66	Quinoline-6-carbox-aldehyde	75–6 (anh.); 55 (hyd.)		239, yel., al.			185, red, al.	191, yel., al.			Methiodide, 218, yel., al.	.....		
67	3,4,5-Trimethoxy-benzaldehyde (Gallaldehyde tri-methyl ether)	78; 75	163–5¹⁰	219–20	....	201–2	....	83–4						
68	4-Iodobenzaldehyde	78	....	224	257	201	121	....						
69	3-Phenanthraldehyde	80	....	275	....	....	....	145	....					
70	4-Hydroxy-3-methoxy-benzaldehyde (Vanillin)	80–1, w.	285d.	230; 240d.	271d. (cor.), red, ac. ac.; 268	227, ac. a.; 223	105, bz.	117, w.; 122	196–8 (cor.), al.	228	2,4-Dinitro-phenyl ether, 131	brt. or. red	ch. red	3
71	2-Hydroxy-1-naphthal-dehyde	82, al.	192²⁷	240, yel., me. al.	....	....	....	157			Picrate, 120	or.	brk. red	10
72	Stilbene-2-carbox-aldehyde	83	....											
73	2-Methoxy-1-naphthal-dehyde	84, al.	200–1¹¹	....							Azine, 255–6, yel., PhNO₂	.....	.....	.....

*Derivative data given in order: m.p., crystal color, solvent from which crystallized.

No.	Name	Melting point, °C	Boiling point, °C	Semi-carbazone	2,4-Di-nitro-phenyl-hydrazone	p-Nitro-phenyl-hydrazone	Phenyl-hydra-zone	Oxime	Dimeth-one deriv. (Dime-done deriv.)	Di-meth-one anhy-dride	Miscellaneous	o-Dianisidine spot test		
												Cold	Hot	Limit, γ
74	2,3,6-Trichlorobenz-aldehyde	86–7	......	......	......	......	......	......	......		Ac. anh. + ac. a. + NaOAc → 2,3,6-tri-chlorocin-namic acid, 189, ac. a.	......	......	....
75	Isophthalaldehyde	89	......	......	......	......	242	180	......					
76	3,4,5-Trichlorobenz-aldehyde	90–1, al.	......	252–4, al.	......	342d., or., PhNO₂	147	......			2-Nitro deriv., 118.5–9.0; Oxid. alk. KMnO₄ → 3,4,5-tri-chloroben-zoic acid, 210			
77	Phenylglyoxal hydrate	91	......	α: 217d.	......	309	di: 152	α: 129; di: 168	......					
78	2,3,4-Trichlorobenz-aldehyde	91	......	......	......	......	......	......	......		Ac. anh. + ac. a. + NaOAc → 2,3,4-tri-chlorocin-namic acid, 185			
79	Quinoline-8-carbox-aldehyde	94–5, dil. al.	......	238–9, bz.	......	......	176, yel., al.	115, dil. al.	......		......			
80	2-Phenylcinnamalde-hyde	94, al.; 95	195–200[17]	188–9, al.; 195	......	......	125–6, yel., ac. a.; 141	165–6, al.	......					
81	3,5-Dichloro-2-hy-droxybenzaldehyde (3,5-Dichlorosalicyl-aldehyde)	95–6	......	227d., ac. a.	......	......	153, pa. yel., al.	195–6, al.-w. (4:1)			Ac. anh. + ac. a. + NaOAc → dichloro-coumarin, 160, bz.	......	......	....
82	Hydroxyacetaldehyde (Glycolaldehyde)	96–7	......	......	......	......	162				Phenylosa-zone, 178–9, yel., eth.; p-Nitrophenyl-osazone, 311			
83	3-Chloro-n-butyr-aldehyde (Trimer)	96–7	28–33[13]	......	......	......	......	......			......			
84	2,2-Dimethyl-3-hy-droxypropionaldehyde	97	85[15]											
85	2,3,4,6-Tetrachloro-benzaldehyde	97–8	......											
86	Benzaldehyde-2-carboxylic acid (2-Formylbenzoic acid; Phthalaldehydic acid)	98–9 (hyd.); 240–50 (anh.)	......	202				120, w.						
87	3-Hydroxy-2-naphthal-dehyde	99–100	......	>270, me. al.	......	......	246–8	207d.	......					

*Derivative data given in order: m.p., crystal color, solvent from which crystallized.

TABLE IX. ORGANIC DERIVATIVES OF ALDEHYDES

TABLE IX. ORGANIC DERIVATIVES OF ALDEHYDES
b) Solids (Listed in order of increasing m.p.)* (Continued)

No.	Name	Melting point, °C	Boiling point, °C	Semicarbazone	2,4-Dinitrophenylhydrazone	p-Nitrophenylhydrazone	Phenylhydrazone	Oxime	Dimethone deriv. (Dimedone deriv.)	Dimethone anhydride	Miscellaneous	o-Dianisidine spot test Cold	Hot	Limit, γ
88	5-Chloro-2-hydroxybenzaldehyde	99–100	105[12]	286–7, ac. a.	....	....	150–2	128, w.; 123–4				......	......	......
89	9-Phenanthraldehyde..	101	......	223	....	265	......	157				......	......	......
90	3-Hydroxybenzaldehyde	104; 108 (cor.), w.	240	198; 199	259, scar., xyl.; 260d., red, al.	221–2, dil. ac. a.	130–1.5, tol.; 147, recr. bz.	90; 88				dk. br. red	dk. ch. red	4
91	9-Anthraldehyde	105	......	219, yel., al.	....	....	207, or., al.	187				......	......	......
92	4-Nitrobenzaldehyde..	106	......	221; 211	320	249	159; 153	anti: 133; 129; syn: 182–4			......	or. br.	red br.	1
93	2,3,4,5-Tetrachlorobenzaldehyde	106–6.5	......	......	....	....	......	......				......	......	......
94	2,3-Dihydroxybenzaldehyde	108	......	226d.	....	....	167	......				......	......	......
95	2-Chloro-5-hydroxybenzaldehyde	110.5–1.5, ac. a.	......	236, pa. yel.	259, ...	....	250–1, red, dil. al.	......	146–7, abs. al.			......	......	......
96	1-Phenanthraldehyde..	111.5	......	......	....	....	......	189				......	......	......
98	2,4,5-Trichlorobenzaldehyde	112–3, al.	......	......	....	....	......	......			Ac. anh. + ac. a. + NaOAc → 2,4,5-trichlorocinnamic acid, 200–1	......	......	......
99	3-Hydroxy-2,4,6-trichlorobenzaldehyde..	113–6.5, 50% ac. a.	......	......	....	272–3d., yel. or.	......	170–2, dil. al.				......	......	......
100	2-Ethoxy-1-naphthaldehyde	115, al.	......	214–5, yel., al.	....	....	91	......			Azine, 184, yel., PhNO₂-al.	......	......	......
101	Metaldehyde	115; 246; (polymers)	......	......	....	....	......	......			Dil. a. → acetaldehyde, b.p. 20.2	......	......	......
102	4-Hydroxybenzaldehyde	116–7, w.	......	224; 280d.	280d., purp., ac. a.; 260 (monohyd.), red, w.	266	177–8, al.; 184, slow htng.	72; 112 (anh.)	188–90 (cor.); 184	246	....	dk. or. red	ch. red	5
103	Terephthalaldehyde...	116; 118	245	......	....	di: 281	di: 278d.; 154	di: 200				......	......	......
104	2,4,6-Trimethoxybenzaldehyde	118	......	......	....	....	......	201–3, me. al.				......	......	......

*Derivative data given in order: m.p., crystal color, solvent from which crystallized.

TABLE IX. ORGANIC DERIVATIVES OF ALDEHYDES
b) Solids (Listed in order of increasing m.p.)* (Continued)

No.	Name	Melting point, °C	Boiling point, °C	Semicarbazone	2,4-Dinitrophenylhydrazone	p-Nitrophenylhydrazone	Phenylhydrazone	Oxime	Dimethone deriv. (Dimedone deriv.)	Dimethone anhydride	Miscellaneous	o-Dianisidine spot test		
												Cold	Hot	Limit, γ
105	1-Bromo-2-naphthaldehyde	118									Oxid. → acid, 186			
106	4-Chloro-3-hydroxybenzaldehyde	121		238–9, pa. yel.		226–7, vlt.-red, dil. al.		126 (anh.); 106–10d. (mono-hyd.)						
107	Pyrene-3-carboxaldehyde	126												
108	1,2,3,4-Tetrahydrophenanthrene-9-carboxaldehyde	129												
109	4,6-Dichloro-3-hydroxybenzaldehyde	129–30									2-Nitro deriv., 157			
110	2,4-Dihydroxybenzaldehyde (β-Resorcylaldehyde)	135–6, yel., w.		260d.	286d., brt. red, AmOH		156–60d.	191, w.						
111	3-Chloro-4-hydroxybenzaldehyde	139 (cor.)		210d., yel., v. dil. ac. a.				144–5, w.						
112	2-Chloro-3-hydroxybenzaldehyde	139–9.5		236–7, pa. yel.		244–5, or. red, al.		149, dil. al.						
113	2,6-Dichloro-3-hydroxybenzaldehyde	140–2						174–5, dil. al.						
114	2,4-Dichloro-3-hydroxybenzaldehyde	141, ac. a.				277–8, or. red		188, al.						
115	d,l-Glyceraldehyde (dimer)	142, 40% me. al.		160d.	166–7 (cor.), 50% me. al.			117–8	197 (cor.), 50% al.; 203	172, 50% al.				
116	2-Chloro-4-hydroxybenzaldehyde	147–8, w.		214, yel., al.		284d., dk. red, al.		194, al.						
117	3,4-Dihydroxybenzaldehyde (Protocatechualdehyde)	153–4, w.		230d.	275d., dk. red, me. al.		175–6d., w.; 121–8	157, xyl.	145d., al.		Dibenzoate, 96–7, al.			
118	3,5-Dihydroxybenzaldehyde (α-Resorcylic aldehyde)	156–7		223–4										
119	3,5-Dichloro-4-hydroxybenzaldehyde	158–9 (cor.); 156, dil. al.		236–7d. (cor.), grn.-yel., ac. a.				185, dil. al.						
120	Hydroxypyruvic aldehyde	160						135						
121	Diphenylglycolaldehyde	163		242				124						
122	Benzaldehyde-3-carboxylic acid (3-Formylbenzoic acid)	175, w.		265			164	188d.						

*Derivative data given in order: m.p., crystal color, solvent from which crystallized.

TABLE IX. ORGANIC DERIVATIVES OF ALDEHYDES

b) Solids (Listed in order of increasing m.p.)* (Continued)

No.	Name	Melting point, °C	Boiling point, °C	Semicarbazone	2,4-Dinitrophenylhydrazone	p-Nitrophenylhydrazone	Phenylhydrazone	Oxime	Dimethone deriv. (Dimedone deriv.)	Dimethone anhydride	Miscellaneous	o-Dianisidine spot test		
												Cold	Hot	Limit, γ
123	2-Hydroxybenzaldehyde-3-carboxylic acid (3-Formylsalicylic acid)......	179	......	......	......	......	188, al.	193, yel., w.	......	.....	......	.....	.....	.....
124	4-Hydroxy-1-naphthaldehyde	181, yel., w.	......	224	......	......	......	......			Hydrazone, 220–36, dk. red; Azine, 236, yel., PhNO$_2$	.....	.....	
125	Indole-3-carboxaldehyde	195; 198	......	......	......	......	198	......						
126	Pentachlorobenzaldehyde	202.5	......	......	......	......	152.5 (cor.), yel., al.	201 (cor.), bz.	......					
127	3,4-Benzypyrene-5-carboxaldehyde	203	......	......	......	......	......							
128	3,4,5-Trihydroxybenzaldehyde (Gallaldehyde)	212d., (monohyd.)	.			226; 234–6d.	......	195–200d.						
129	4-Hydroxybenzaldehyde-3-carboxylic acid (5-Formylsalicylic acid)......	248–9				......	219, al.	179	......					
130	Benzaldehyde-4-carboxylic acid (4-Formylbenzoic acid).	256, w., subl.	......			......	226	208–10						

*Derivative data given in order: m.p., crystal color, solvent from which crystallized.

TABLE X. ORGANIC DERIVATIVES OF KETONES
a) Liquids 1) (Listed in order of increasing atmospheric b.p.)*

No.	Name	Boiling point, °C	Melting point, °C	n_D^{20}	D_4^{20}	Semi-carbazone	2,4-Di-nitrophenyl-hydrazone	p-Nitro-phenyl-hydrazone	Phenyl-hydrazone	Oxime	Miscellaneous
1	Acetone (2-Propanone)	56	−95	1.3592	....	190, w.	126; 128, yel., al.	148-9, yel., al.	42	59	Thiosemicar-bazone, 179
2	3-Buten-2-one (Methyl vinyl ketone)	81	......	1.4095[22]	....	141; 140	....	....			....
3	2-Butanone (Ethyl methyl ketone)	80; 82	−86.4	1.3791	0.804	146	116-7; 115, yel., al.	128-9, yel., w.-al.	oil	b.p. 152	Phenylsemicar-bazone, 168
4	3-Butyn-2-one (Ethynyl methyl ketone)	86	......	......	....	....	181	143	....		
5	2,3-Butanedione (Biacetyl)	88, gr.-yel.	f.p. −2.4	1.3927	....	mono: 235 (cor.), w.; di: 278-9, ac. a.	di: 314-5 (cor.), red-or., PhNO₂	mono: 230, or.-yel.	mono: 134, yel., dil. al.; di: 243d., yel., bz.	mono: 76; di: 245-6 (cor.); 234-5, subl., dil. al.	
6	2-Methyl-3-butanone (Isopropyl methyl ketone)	94.3	......	1.3879	0.8046	113-4; 112-3, al.	120; 117, or.-yel., al.-chl.	108-9, or.-yel., al.	oil	oil	....
7	2-Methyl-1-buten-3-one (Iso-propenyl methyl ketone)	97[7.44]	......	1.4232; 1.4235	....	173	181	....			
8	Cyclobutanone	100	......	1.4189[25]	....	....	146				
9	3-Pentanone (Diethyl ketone)	102	−39.8	1.3922	....	138-9	156, pa.-or., al.	144, or.-yel., 50°, al.	oil	b.p. 165	....
10	2-Pentanone (Methyl n-propyl ketone)	102.3	......	1.3902; 1.39012	0.80639	112; 106	143-4, yel.-or., al.	117	oil	b.p. 167	....
11	1-Penten-3-one (Ethyl vinyl ketone)	102[7.10]	......	1.4192	....	129					
12	3,3-Dimethyl-2-butanone (tert-Butyl methyl ketone; Pinacolone)	106	−49.8	1.3960; 1.3956	0.8114	157-8	125, or.-yel., al.; fusion → 131		oil	75; 79	
13	1-Methoxy-2-propanone (Methoxymethyl methyl ketone)	115[7.56]	......	1.3981	....	163; 159	111; 109	....			
14	1-Methoxy-3-butanone (1-Methoxyethyl methyl ketone)	116[7.39]	......	1.3936	....	141					
15	4-Methyl-2-pentanone (Isobutyl methyl ketone)	116.8	......	1.3956	0.8008	132; 135	95, or.-red, al.			b.p. 176	
16	3-Methyl-2-pentanone (sec-Butyl methyl ketone)	118	......	1.3990	....	94-5, pet. eth.	71.2			oil, b.p. 89[20]	
17	1-Chloro-2-propanone (Chloro-acetone)	119	......	......	....	150; 164d.	125			b.p. 171[7.30]	
18	2-Methyl-1-penten-3-one (Ethyl isopropenyl ketone)	119[7.51]	......	1.4270[24]	....	161					
19	1,1-Dichloro-2-propanone (1,1-Dichloroacetone)	120	......	......	1.305[18_15]	163					
20	2,4-Dimethyl-3-pentanone (Di-isopropyl ketone)	124	......	1.4001	0.8108	160 (cor.); 149	88; 85-6, or.; 94-8				
21	Methyl neopentyl ketone	125; 122	......	1.4018[25]	....	....	100				
22	3-Hexanone (Ethyl n-propyl ketone)	125	......	1.4007	0.81491[22]	113	130			b.p. 86[17]	
23	2,2-Dimethyl-3-pentanone (tert-Butyl ethyl ketone)	125[7.29]	......	1.4052	....	....	144				
24	2-Hexanone (n-Butyl methyl ketone)	128	......	1.40069	0.81127	125 (cor.); 121, rapid htng.	106, red-or., al.; 110	88	oil	49	Thiosemicar-bazone, 110

*Derivative data given in order: m.p., crystal color, solvent from which crystallized.

TABLE X. ORGANIC DERIVATIVES OF KETONES
a) Liquids 1) (Listed in order of increasing atmospheric b.p.)* (Continued)

No.	Name	Boiling point, °C	Melting point, °C	n_D^{20}	D_4^{20}	Semicarbazone	2,4-Dinitrophenylhydrazone	p-Nitrophenylhydrazone	Phenylhydrazone	Oxime	Miscellaneous
25	4-Methyl-3-penten-2-one (Isopropylideneacetone; Mesityl oxide)	130	.....	1.44397	0.86532	α: 164; β: 133-4, bz.	200, red, al.; 203, red, ac. a.	132-4, or.-yel., al.	142	β: 48-9, me. al.	.....
26	3,3-Dimethyl-2-pentanone (tert-Amyl methyl ketone)	130[733]	.....	1.4100	.....	.....	112	.....	.....	.....	.....
27	Cyclopentanone	130.7	−51.3	1.4366; 1.4370	0.94869	210; 203; 216-7, rapid htng.	146, or., ac. a.; 142, or.-yel., al.	154	55, lt. pet.	56.5, pet. eth	.....
28	5-Hexen-2-one (Allylacetone)	132	.....	1.4174[25]	.....	102	108	.....	.....	.....	.....
29	1-Methoxy-2-butanone (Ethyl methoxymethyl ketone)	133[757]	.....	1.4063	.....	.....	198	.....	.....	.....	.....
30	2,2,4-Trimethyl-3-pentanone (tert-Butyl isopropyl ketone)	135	.....	1.4065	.....	132	.....	.....	.....	144	.....
31	5-Methyl-3-hexanone (Ethyl isobutyl ketone)	135[735]	.....	1.407	.....	152	.....	.....	.....	.....	.....
32	2-Methyl-1-penten-4-one	135-45d.	.....	.....	.....	192	.....	.....	.....	.....	.....
33	1-Bromo-2-propanone (Bromoacetone)	136	.....	.....	.....	135d.	.....	.....	.....	36	.....
34	2-Methyl-3-hexanone (Isopropyl n-propyl ketone)	136	.....	1.4075	.....	119	.....	.....	.....	.....	.....
35	4-Methyl-3-hexanone (sec-Butyl ethyl ketone)	136	.....	1.402	.....	137	78	.....	.....	.....	.....
36	2-Methoxy-3-pentanone (Ethyl 1-methoxyethyl ketone)	136[750]	.....	1.4019	.....	120	.....	.....	.....	.....	.....
37	Cyclobutyl methyl ketone	136	.....	1.4283[28]	.....	149	.....	.....	.....	.....	.....
38	3-Methyl-2-hexanone	137	.....	.....	.....	70	.....	.....	.....	.....	.....
39	3-Methyl-1-hexen-5-one	138	.....	1.4197[25]	.....	112	.....	.....	.....	.....	.....
40	1-Chloro-2-butanone (Chloromethyl ethyl ketone)	138	.....	1.4372	.....	.....	.....	.....	.....	.....	.....
41	3-Methyl-1-penten-4-one	138	.....	.....	.....	201	.....	.....	.....	75-6	.....
42	3,4-Dimethyl-2-pentanone	138	.....	1.4094	.....	113	.....	.....	.....	.....	.....
43	4-Hexen-3-one	139	.....	1.4388	.....	157	.....	.....	.....	.....	.....
44	2,4-Pentanedione (Acetylacetone)	139	−30	1.4465[25.6]	0.976	mono: 122; di: 209	209, yel., al.	.....	.....	di: 149, al.	.....
45	2-Methylcyclopentanone	139	.....	1.4364	.....	184; 182	.....	.....	.....	b.p. 103[22]	.....
46	3-Ethyl-2-pentanone	139[746]	.....	1.4073	.....	99	.....	.....	.....	.....	.....
47	3-Hydroxy-3-methyl-2-butanone (Acetyl diethyl carbinol)	140	.....	.....	.....	165	.....	.....	.....	87	.....
48	4-Methyl-2-hexanone	142; 139	.....	1.4057[25]	.....	120; 128	.....	.....	.....	.....	.....
49	1-Propoxy-2-propanone (Isopropoxymethyl methyl ketone)	142	.....	1.4004	.....	.....	142; 144	.....	.....	.....	.....
50	d-3-Methylcyclopentanone	143	.....	1.4340[19]	.....	184-5	.....	.....	.....	α: 91-2; β: 67-9	$[\alpha]_D^{12}$: + 132.9
51	3,4-Dimethyl-4-penten-2-one	144	.....	.....	.....	114	.....	.....	.....	.....	.....
52	4-Heptanone (Di-n-propyl ketone)	144	−34.0	1.4069	0.8175	132, pet. eth.	75, yel.-or., al.	.....	.....	b.p. 193	.....
53	2,4-Dimethyl-3-hexanone (sec-Butyl isopropyl ketone)	145	.....	1.4059; 1.4080	.....	71	.....	.....	.....	.....	.....
54	3-Hydroxy-2-butanone (d, l-Acetoin)	145; 148	−72; 15	1.4178	0.9861[30]	185, al.; 202	di: 318, or., PhNO₂-tol.	.....	.....	.....	Phenylosazone, 243d., yel., bz.
55	d,l-3-Methylcyclopentanone	145[755]	.....	1.4329	.....	185	.....	.....	.....	.....	.....
56	1-Methoxy-3-methyl-2-butanone (Isopropyl methoxymethyl ketone)	145[748]	.....	1.4078	.....	163	.....	.....	.....	.....	.....

*Derivative data given in order: m.p., crystal color, solvent from which crystallized.

TABLE X. ORGANIC DERIVATIVES OF KETONES
a) Liquids 1) (Listed in order of increasing atmospheric b.p.)* (Continued)

No.	Name	Boiling point, °C	Melting point, °C	n_D^{20}	D_4^{20}	Semi-carbazone	2,4-Di-nitrophenyl-hydrazone	p-Nitro-phenyl-hydrazone	Phenyl-hydrazone	Oxime	Miscellaneous
57	2,2-Dimethyl-3-hexanone (tert-Butyl n-propyl ketone)	145[738]	......	1.4107	....	....	124; 116	....			
58	1-Hydroxy-2-propanone (Acetol)	146	−17	1.4295	....	196, al.	128.5 (cor.), or., al.	173	....		
59	4-Chloro-3-methyl-2-butanone (α-Chloroisopropyl methyl ketone)	146	......	1.4390	....	....	116				
60	1-Hepten-4-one	146–7				110, w.-al.			....	b.p. 92–3[13]	
61	3,4-Dimethyl-3-penten-2-one	147	......	1.4506[14]	....	200					
62	1-Ethoxy-2-butanone (Ethoxy-methyl ethyl ketone)	147[752]	......	1.4068							
63	2,5-Dimethyl-3,4-hexanedione (Di-isobutyryl)	148	......	1.42057						mono: 125; di: 172	
64	3-Heptanone (n-Butyl ethyl ketone)	148	......	1.4092	....	101; 103; 152	....				
65	5-Methyl-4-hexen-3-one	148	......	1.4496[15]		163					
66	5-Methyl-5-hexen-2-one (Meth-allylacetone)	149	......	1.4285[25]		137					
67	2-Methyl-4-heptanone (Isobutyl n-propyl ketone)	150[750]				124					
68	4,4-Dimethyl-3-hexanone (tert-Amyl ethyl ketone)	150–2	......			98					
69	2-Heptanone (n-Amyl methyl ketone)	151.2	−35.5	1.40069	....	123, al.; 127	89, yel.-or., al.; 74	....	207		
70	3-Ethyl-5-hexen-2-one	152	......	1.4260[25]	....		53				
71	5-Hepten-2-one (Crotylacetone)	153		1.4280[25]	0.8446	105; 97					
72	1-Methoxy-2-pentanone (Meth-oxymethyl n-propyl ketone)	153[745]	......	1.4119							
73	3,3-Dimethylcyclopentanone	153[748]	......	....		178					
74	2,2,4,4-Tetramethyl-3-pentanone (Di-tert-butyl ketone)	154	......	1.4392; 1.4194							
75	1-Bromo-2-butanone (Bromo-methyl ethyl ketone)	155; 50[12]	......	1.4670							
76	Cyclopentyl methyl ketone	155	......	....		143					
77	2,2,5,5-Tetramethylcyclo-pentanone	155	......	1.4280							
78	1-Methoxy-3-hexanone (1-Meth-oxymethyl n-propyl ketone)	155[746]		1.4091		169; 170					
79	Cyclohexanone	156	−16.4	1.4507		166–7	160; 162, yel. al.	146–7, 90% al.	81–2, 50% al.	91, lgr.	
80	4-Methyl-6-hepten-3-one	156					80				
81	2-Hepten-4-one	156–7				147, w.-me. al.					
82	2,3-Hexanedione	158								di: 175	
83	3,4-Dimethyl-2-hexanone	158; 155				120; 118; 126					
84	3,4-Dimethyl-3-hexen-2-one	158	......	1.4476[15]		142					
85	2,2,4-Trimethyl-3-hexanone (tert-Butyl isobutyl ketone)	158	......	....		145					
86	2-Ethyl-1-hexen-3-one	158[742]	......	1.4408[18]		119					
87	3,3-Dimethyl-1-methoxy-2-butanone (tert-Butyl methoxy-methyl ketone)	159[743]		1.4193							
88	1-Isopropoxy-3-methyl-2-butanone	160	......	....		....	88			....	
89	2-Methyl-3-cyclopentenone	161	......	1.4771	....	220	....			127	....

*Derivative data given in order: m.p., crystal color, solvent from which crystallized.

TABLE X. ORGANIC DERIVATIVES OF KETONES

a) Liquids 1) (Listed in order of increasing atmospheric b.p.)* (Continued)

No.	Name	Boiling point, °C	Melting point, °C	n_D^{20}	D_4^{20}	Semi-carbazone	2,4-Dinitrophenyl-hydrazone	p-Nitro-phenyl-hydrazone	Phenyl-hydrazone	Oxime	Miscellaneous
90	1-Ethylcyclopentanone	161[755]				189					
91	3-Methyl-2-heptanone	162		1.415		82					
92	4,5-Dimethyl-5-hexen-3-one	162[750]				110					
93	3,5-Dimethyl-4-heptanone (Di-sec-butyl ketone)	162; 170–3				83–4					
94	6-Methyl-3-heptanone (Ethyl isoamyl ketone)	163; 160				132					
95	2-Methoxy-3-methyl-2-pentanone (sec-Butyl methoxymethyl ketone)	164[757]		1.4162							
96	1-Methoxy-4-methyl-2-pentanone (Isobutyl methoxymethyl ketone)	164[751]		1.4140							
97	2-Methylcyclohexanone	165.1; 166	−14.0	1.4885	0.92500	191; 197d., al., rapid htng.	135.5–7.0, al.; 137 (cor.)	132	b.p. 220[35-40]	43, eth.	
98	4-Hydroxy-4-methyl-2-pentanone (Diacetone alcohol)	166; 164					202–3			58	3,5-Dinitro-benzoate, 55
99	4,5-Dimethyl-4-hexen-3-one	166[750]				209					
100	2,2-Dimethyl-3-heptanone (n-Butyl tert-butyl ketone)	166[745]		1.4167		145					
101	2,6-Dimethyl-4-heptanone (Di-isobutyl ketone; Isovalerone)	168.0		1.4173[25]		122; 126	66, or.-red; 92				
102	2-Methyl-4-octanone (n-Butyl isobutyl ketone)	168				132					
103	2,5-Dimethyl-4-heptanone (sec-Butyl isobutyl ketone)	169; 167				133					
104	d-3-Methylcyclohexanone	169		1.4456[21]		180, me. al.				43	$[\alpha]_D^{15}$: +13.38
105	d, l-3-Methylcyclohexanone	168; 169.6	−73.5	1.4430; 1.4463	0.91535	179, me. al.; 191.4d., rapid htng.	155, yel.	119	94, w.-al.		
106	1-Methoxy-2-hexanone (n-Butyl methoxymethyl ketone)	169[744]		1.4173							
107	4-Octanone (n-Butyl n-propyl ketone)	170				96					
108	Methyl acetoacetate	170	−40.6	1.41964	1.0765	152					
109	2,2-Dimethylcyclohexanone	170; 171		1.4482		201; 193	140–2				
110	6-Methyl-2-heptanone (Isohexyl methyl ketone)	171		1.4146		154	77				
111	trans-2,4-Dimethylcyclohexanone	171		1.4429[16]		136					
112	4-Methylcyclohexanone	171.25	−40.6	1.4445	0.91562	199, me. al.; 203.5d., rapid htng.		128.5, yel., al.	109–10, al.	37–9	
113	d,l-2,5-Dimethylcyclohexanone	171–3		1.4446		α: 122; β: 173				111, al.	
114	2-Octanone (Hexyl methyl ketone)	173	−21.5	1.41518; 1.4154	0.81853	122–3 (cor.), pet. eth.-al.	58, or., al.	92–3, yel., al.			
115	5-Ethyl-3-heptanone	173				134					
116	d-2,5-Dimethylcyclohexanone	173–4				176–7				97–8	$[\alpha]_D^{20}$: +11.6
117	3-Ethyl-2-methylcyclopentanone	174				170, al.					

*Derivative data given in order: m.p., crystal color, solvent from which crystallized.

TABLE X. ORGANIC DERIVATIVES OF KETONES

a) Liquids 1) (Listed in order of increasing atmospheric b.p.)* (Continued)

No.	Name	Boiling point, °C	Melting point, °C	n_D^{20}	D_4^{20}	Semi-carbazone	2,4-Di-nitrophenyl-hydrazone	p-Nitro-phenyl-hydrazone	Phenyl-hydrazone	Oxime	Miscellaneous
118	2,6-Dimethylcyclohexanone	174		1.4500; 1.4470							
119	2-Isopropylcyclopentanone	174		1.4395[29]		202					
120	Acetoxyacetone	174–5; 74[18]		1.4150	1.0749	145, me. al.		144, yel., bz.	60d., eth.	b.p. 144[20]	
121	2,4-Heptanedione (n-Butyryl-acetone)	174–5									Cu salt, 165; 161, pa. bl.
122	3-Methyl-3-hepten-2-one	175				164					
123	cis-2,4-Dimethylcyclohexanone	176		1.4430[25]		200; 190				98–9	
124	4-Ethyl-4-hydroxy-3-hexanone	178[742]				177					
125	2,3-Dimethylcyclohexanone	178–9		1.4505		203–4					
126	2,2,6-Trimethylcyclohexanone	179[767]		1.4480		209	141				
127	3,3-Dimethylcyclohexanone	179[748]		1.4482[17]		219					
128	5-Ethyl-4-hepten-3-one	179[740]				105					
129	3-Ethyl-4-methylcyclopentanone	180				208–9, w.-al.				oil, b.p. 117[11]	
130	Cyclohexyl methyl ketone	180		1.4514		177		154		60	
131	trans-3,5-Dimethylcyclohexanone	d, l: 180–1		1.4475[21]	d,l: 0.897; d: 0.9083; l: 0.9074	d,l: 193–4; d: 193–4; l: 189				d,l: oil, b.p. 116–8[14]	d: $[\alpha]_D^{20}$: +4.65 l: $[\alpha]_D^{20}$: −7.91
132	5-Hydroxy-4-octanone (Butyroin)	180–90					99				
133	Ethyl acetoacetate	181				133; 129d.	93				
134	Cycloheptanone	181; 182				163	148	137		23	
136	cis-3,5-Dimethylcyclohexanone	182–3		1.4407	0.890	202–3				74	
137	2-Propylcyclopentanone	183		1.4429		214d., al.				oil, b.p. 109–11[9]	
138	2,2,6,6-Tetramethylcyclohexanone	184[772]	15	1.4473							
139	3-Methyl-2,4-hexanedione	184; 183									Cu salt, 177
140	5-Nonanone (Di-n-butyl ketone)	186–7	f.p.-5.9	1.421[15]	0.8222	90, al.					
141	3,4-Dimethylcyclohexanone	187		1.4520; 1.4507	0.906	189					
142	3-Nonanone (Ethyl n-hexyl ketone)	187[751]				112					
143	2,5-Dimethylcyclohexen-3-one	189–90		1.4753[22]		165, me. al.				92–3, me. al.; 169, al.	
144	Methyl 2-pyridyl ketone (2-Acetyl-pyridine)	190								121	
145	3-Propylcyclopentanone	190–1		1.4456[12]		178–9, me. al.				oil, b.p. 121–2[12]	
146	3-Ethylcyclohexanone	192		1.4537; 1.4511		182; 175					
147	1,5-Dimethylcyclohexen-4-one	192–3; 194								102	
148	2,5-Hexanedione (Acetonyl-acetone)	194	−9	1.428; 1.449	0.97370	mono: 185d.; di: 224	di: 257, pyr.	di: 210–2, red., al.	di: 120, dil. al.	mono: b.p. 130[11]; di: 137	

*Derivative data given in order: m.p., crystal color, solvent from which crystallized.

TABLE X. ORGANIC DERIVATIVES OF KETONES
a) Liquids 1) (Listed in order of increasing atmospheric b.p.)* (Continued)

No.	Name	Boiling point, °C	Melting point, °C	n_D^{20}	D_4^{20}	Semi-carbazone	2,4-Di-nitrophenyl-hydrazone	p-Nitro-phenyl-hydrazone	Phenyl-hydrazone	Oxime	Miscellaneous
149	d-Fenchone	195–6	......	1.46355^{18}	0.947^{19}	184; 172	140	....	b.p. 202–3[18]	d or l: α+: 165; β+: 123; d,l: α+: 159; β+: 129	Hydrazone, 56–7
150	2-Nonanone (n-Heptyl methyl ketone)	195.3	−8	1.42072	0.82133; 0.82217	118–9, al.	....				....
151	1-Acetyl-4-methylcyclohexanone	195–7	......	1.4509^{18}	....	α: 159, me. al.; β: 175	....			57–9	....
152	Methyl levulinate	196.0	......	1.42333	1.04945	143	142, al.	....	96; 105, al.	....	....
153	4-Fluoroacetophenone	196	......	1.5081^{25}	....	219					
154	d,l-2-Ethyl-5-methylcyclo-hexanone	197	......	1.4485	....	178–81				80, w.-al.	
155	1-Acetyl-2-methylcyclohexanone	197–200	....	....	....	172–3	....				
156	2-n-Propylcyclohexanone	198–9[748]	......	1.4558^{13}	....	133d., w.-al.				67–8	
157	1-Acetylcyclohexene	200	......	1.4892	....	220	....			59	
158	3-(Trifluoromethyl)acetophenone	202	......	....	....						
159	β-Thujone	202; 76[10]	......	....	....	174	114				
160	Acetophenone (Methyl phenyl ketone)	205; 202	20	1.541; 1.5339	1.02810	198–9 (cor.), 50% al.; 203	238–40, al.; 249–50, or.-red., ac. a.	184–5, or.-red	105, wh., al., → dk.	60	....
161	2-Ethyl-4-methylcyclohexanone	205[747]	......	1.4452	....	....					
162	Ethyl levulinate	206	......	1.42288	1.01114	148	102	....	104		....
163	4-Decanone	206–7	......	....	....	51–2					
164	1,5-Dimethylcyclohexen-3-one	208–9	......	1.4819^{22}	....	179–80, yel., al.; 168–71	....		76–8	....	Thiosemicar-bazone, 195d.
165	l-Menthone	209; 207	−6.6	1.4505	....	189; 187; 184	146, or., al.		53	59, eth.	....
166	2-Decanone (Methyl n-octyl ketone)	211; 215.5	14; f.p. 3.1	1.42523	0.82370	124; 126, pet. eth.	....				
167	Methyl 4-pyridyl ketone	212	....	....	....	....				142	
168	4-n-Propylcyclohexanone	212[740]	......	1.4514^{25}	....	180					
169	2-Acetylthiophene (Methyl 2-thienyl ketone)	213	10.5	1.5666	....	190, bz.	....	181	96	81	
170	2-Methylacetophenone (Methyl 2-tolyl ketone)	214; 216	......	1.5320	....	205, al.; 210	159, yel., al; 161			61	
171	6-Bromo-2-hexanone	214[720]	......	1.4713	....	....	81	....			
172	1,5,5-Trimethylcyclohexen-3-one (Isophorone)	215	......	$1.4789^{21.5}$	$0.9255^{20.5}_4$	199.5d., al.; 191	....		68, dil. al.	79.5, pet. eth.; 76	....
173	n-Propyl 2-pyridyl ketone	217–8	....	....	....	....			82	48	Picrate, 75
174	Propiophenone (Ethyl phenyl ketone)	218; 220	20; 18.6	1.5270	1.0105	173–4 (cor.), al.; 182, rapid htng.	190–1, red, bz.; 189			54; 53, pet. eth.	....
175	Methyl 3-pyridyl ketone	220; 218	....	....	....	....			137	113	....
176	3-Methylacetophenone (Methyl 3-tolyl ketone)	220	......	1.5306	1.007	198; 203	207	....		55; 57, al.	....

*Derivative data given in order: m.p., crystal color, solvent from which crystallized.

TABLE X. ORGANIC DERIVATIVES OF KETONES

TABLE X. ORGANIC DERIVATIVES OF KETONES
a) Liquids 1) (Listed in order of increasing atmospheric b.p.)* (Continued)

No.	Name	Boiling point, °C	Melting point, °C	n_D^{20}	D_4^{20}	Semi-carbazone	2,4-Di-nitrophenyl-hydrazone	p-Nitro-phenyl-hydrazone	Phenyl-hydrazone	Oxime	Miscellaneous
177	**Isobutyrophenone** (Isopropyl phenyl ketone)...............	222; 217	......	1.5190	$0.9863_4^{16.9}$	181, al.	163, or.-red, dil. ac. a.	....	73	94, lt. pet.	....
178	**Acetyl benzoyl ketone**..........	222, yel.	......	1.537^{10}	....	*di*: 229–32	....	*di*: 256	α: 143	α, mono: 166; β, mono: 114; *di*: 240	....
179	**5-Ethyl-1-methylcyclohexene-3-one**....................	223–7	......	......	....	162–8d., al.	....	....	....	....	Thiosemicar-bazone, 150–1
180	**4[8]-*p*-Menthen-3-one** (Pulegone).	224; 221–2	......	1.48705^{18}	....	174; 175–6	142	....	....	119	....
181	**Pivalophenone** (*tert*-Butyl phenyl ketone)	224^{750}	......	1.5082; 1.5102	....	150	194–5	....	....	167	....
182	**1-Phenyl-2-butanone** (Benzyl ethyl ketone)................	226	......	......	....	135; 146	....	....	....	....	....
183	**6-Undecanone** (6-Hendecanone; Di-*n*-amyl ketone)	228 (cor.)	15	1.42875	0.82471	oil	....	....	....	oil	....
184	**3-Chloroacetophenone**	228	......	......	....	232	....	176	....	88	....
185	**Ethyl 1-thienyl ketone**	228	......	......	....	167	....	....	....	55–6	....
186	**2-Undecanone** (2-Hendecanone; Methyl *n*-nonyl ketone)........	228	12.1; 12.7	1.42899	0.82564	122.0–.5	63, al.	90–1, yel., al.	....	44–5	....
187	**2,4-Dimethylacetophenone**.......	228; 234–5	......	1.5381; 1.5340	....	185–7	....	....	....	63–4, pet. eth.	....
188	**2-Chloroacetophenone**	229	......	1.685^{25}	....	160	....	....	....	113	....
189	**1-Phenoxy-2-propanone** (Phenoxy-acetone)....................	229–30; 120^{19}	......	1.5228	1.0903	173; 176 (cor.), 50% al.	....	....	....	....	....
190	*n*-**Propyl 4-pyridyl ketone**.......	229–31	......	......	....	....	....	....	....	....	Picrate, 96
191	*n*-**Butyrophenone** (Phenyl *n*-propyl ketone).............	230; 218–21	11.5–13.0	1.5196; 1.5203	....	187–8, al.; 191	190, or.-red, dil. ac. a.	....	....	50, abs. eth.	$[α]_D^{20}$: +62.9
192	*d*-**Carvone**	230	......	1.49952	0.9608	162–3; 142–3; *d, l*: 154–6	191, red, ac. a.	174–5, red-br.	....	*d*, α: 72–3, al.: *d*, β: 56–7; *l*, α, (−): 72; *l*, β, (+): 57–8; *d, l*: 93–4	....
193	**2,5-Dimethylacetophenone**.......	230	......	1.5291; 1.5306	....	168–9	....	....	....	....	....
194	**4-Chloroacetophenone**	232; 236	12	......	....	204; 160; 146	231	239	114	95	....
196	**4-Phenyl-2-butanone** (Methyl-β-phenylethyl ketone)	235	......	......	....	142	....	....	....	87	....
197	**Isovalerophenone** (Isobutyl phenyl ketone)	236	......	......	....	210	....	....	....	76; 64.5	....
198	**3,5-Dimethylacetophenone**.......	236–7	......	1.5276^{25}	....	....	179–80, yel., ac. a.	....	....	114, me. al.	....
199	**2-Methoxyacetophenone** (2-Acetylanisole)	239; 245	......	1.5395	1.089	183	....	....	114, al.	83; 96.0–.5, pet.	....
200	**3-Methoxyacetophenone** (3-Acetylanisole)...............	240; 252	......	$1.5583^{15.4}$	$1.0993_4^{15.4}$	196	....	....	....	....	....
201	**5-Phenyl-3-pentanone**..........	244	......	1.5125	....	80	....	....	....	....	....
202	**5-Isopropyl-2-methylacetophenone** (2-Acetyl-*p*-cymene)...........	245	......	1.51849	0.9654_{20}^{20}	147	140–2, clearing at 160	....	....	91–2.5	....

*Derivative data given in order: m.p., crystal color, solvent from which crystallized.

TABLE X. ORGANIC DERIVATIVES OF KETONES
a) Liquids 1) (Listed in order of increasing atmospheric b.p.)* (Continued)

No.	Name	Boiling point, °C	Melting point, °C	n_D^{20}	D_4^{20}	Semi-carbazone	2,4-Di-nitrophenyl-hydrazone	p-Nitro-phenyl-hydrazone	Phenyl-hydrazone	Oxime	Miscellaneous
203	α-Bromopropiophenone	245–50	......	1.5686^{25}	....	....	....	....	....	85	....
204	3,4-Dimethylacetophenone.......	246–7; 251	......	1.5400	....	233–4	....	....	....	85	....
205	2,4,5-Trimethylacetophenone	246–7				204				85–6	....
206	n-Propyl 3-pyridyl ketone	246–52	......	1.5128	....	169–70			182		Picrate, 104
207	n-Valerophenone (n-Butyl phenyl ketone)	248.5; 242	......	1.5150	$0,988_{20}^{20}$	160, w.-al.	166, brt. red, ac. a.	161.5–2.5, or.-red, al.	162	52.0–.5, pet. eth.	....
208	2-Aminoacetophenone	250–2d.	20	....	....	290d., al.	....		108, al.	109, subl., w.	
209	2,5-Dichloroacetophenone	251	14							130	
210	4-Isopropylacetophenone	252–4									
211	1,1,1-Tribromoacetone..........	255d.	......								
212	Ethyl benzoylacetate	265				125					
213	n-Enanthophenone (n-Hexyl phenyl ketone)	283.3	16.4	1.50760_{He}^{16}	0.95155	119, dil. al.		127–8		55	
214	3-Phenylcyclohexanone	$287–8^{736}$				167, al.				128–9, al.	
215	1-Acetylnaphthalene (Methyl 1-naphthyl ketone)...........	302	......	1.629		288.5–9.5; 232–3			146	140; 137.5	
216	Ethyl 1-naphthyl ketone........	305–7	......	1.6109						58	Picrate, 77–8, al.; 79
217	2-Benzoylpyridine (Phenyl 2-pyridyl ketone)............	317	......	1.6056	....	....	199		136, yel., al.	150; 165	Picrate, 130, al.
218	2,4,5-Trimethylbenzophenone	328	......	....	1.0332_4^{18}	....				....	....
219	2,4,4'-Trimethylbenzophenone....	340								132, al.	....
220	α-Methylstyryl phenyl ketone (Dypnone)	340–5 sl. d.	......		1.108_0^{20}	151, bz.				syn: 134, al.; anti: 78	....
221	1,5-Diphenyl-3-pentanone (Di-benzylacetone)	352; 348	13–4	....	....	....				95–6	....

*Derivative data given in order: m.p., crystal color, solvent from which crystallized.

TABLE X. ORGANIC DERIVATIVES OF KETONES

TABLE X. ORGANIC DERIVATIVES OF KETONES
a) Liquids 2) (Reduced pressure b.p. only) (Listed in order of increasing semicarbazone m.p.)*

No.	Name	Semi-carbazone	Boiling point, °C	n_D^{20}	D_4^{20}	Miscellaneous
1	**1-Chloro-2-methyl-3-pentanone**	70	64^9	....	....	....
2	**4-Methyl-2-octanone**	70	94^{40}	....	....	....
3	**1-Hepten-5-one**	82–3, w.-al.	$46–7^{12}$	$1.4254^{18.3}$	$0.8487^{18.3}$	....
4	**3-Dodecanone**	89, w.-al.	134^{18}	....	....	m.p. 19
5	**1,3-Diethoxy-2-propanone** (sym-Diethoxyacetone)	91	105^{35}	1.4202	....	....
6	**1-Ethoxy-2-propanone** (Ethoxyacetone)	96	36^{28}	1.4000	....	....
7	**3-Cyclopentyl-2-butanone** (α-Cyclopentyl-α-methylacetone)	98	79^{17}	1.4470	....	....
8	**7-Methyl-1-octen-5-one**	101–2	$62–3^{14}$	$1.4288^{12.5}$	....	....
9	**Cyclohexyl methoxymethyl ketone**	102	111^{21}	1.4552^{25}	....	....
10	**1-Phenoxy-2-butanone** (Ethyl phenoxymethyl ketone)	102	100^5	1.5201	....	....
11	**1-Naphthoxyacetone**	103	$205–8^{14}$	....	....	....
12	**1-Hepten-6-one**	108, w.-al.	$41–3^{10}$	1.4350^{18}	0.8673^{18}	....
13	**1-Phenoxy-2-pentanone** (Phenoxymethyl n-propyl ketone)	108	112^4	1.5148	....	....
14	**3-Octyn-2-one**	109	76^{15}	1.4446^{25}	....	2,4-Dinitrophenylhydrazone, 88
15	**3-Hepten-6-one**	109–10, wh.	$61–2^{20}$	1.4290^{21}	0.8618^{21}	....
16	**3-Methyl-4-phenyl-2-butanone**	112; 114	130^{17}	1.5090^{19}	....	....
17	**3-Methyl-3-hepten-5-one**	114	$82–6^{42}$	1.4488^{25}	....	....
18	**1-Chloro-2-ethyl-3-hexanone**	115	92^{12}	....	....	....
19	**1,3-Dimethoxy-2-propanone** (sym-Dimethoxyacetone)	120	78^{18}	1.4174	....	....
21	trans-**3-Hepten-2-one**	125; 128	60^{16}	1.4421; 1.4430	0.8445	....
22	**5-Hydroxy-5-methyl-3-heptanone**	125	86^{14}	1.4386^{14}	....	....
23	α-**Ethoxyacetophenone**	128	122^{15}	1.5250	....	....
24	**3-Propylpropiophenone** (Ethyl 3-propylphenyl ketone)	128	145^{20}	....	....	....
25	α-**Methoxyacetophenone**	129	126^{19}	....	....	....
26	**1-Phenyl-4-hexen-1-one**	130	97^1	1.5270^{25}	....	....
27	**5-Phenyl-2-pentanone**	130	122^6	....	....	....
28	**2-Methyl-3-octen-6-one**	131–2	$73–7^{19}$	1.44533	....	....
29	**1-Phenyl-1-hexen-5-one**	132, et. ac.	$153–5^{10}$	1.5458^{25}	....	....
30	**3-Phenyl-1-hexen-5-one**	132	$153–5^{10}$	1.5193^{25}	....	2,4-Dinitrophenylhydrazone, 103
31	**4-Phenyl-2-pentanone**	137	115^{13}	1.5124	....	....
32	**3-Propyl-3-hexen-2-one**	142	72^9	....	....	....
33	**3-Acetylfuran**	150	84^{21}	....	....	....
34	**3-Hydroxy-3-methyl-2-pentanone**	150	73^{50}	1.4200	....	....
35	**1-Cyclopentyl-2-propanone** (1-Cyclopentylacetone)	150	67^{12}	....	....	....
37	cis-**3-Hepten-2-one**	152	70^{15}	1.4505^{22}	0.8555^{22}	....
38	**3-Hydroxy-3-methyl-2-heptanone**	152	84^{19}	....	....	....
39	d-**2-Ethyl-5-methylcyclohexanone**	152–4	$83–4^{18}$	....	0.9016_4^{15}	$[\alpha]_D$: +8.5
40	**5-Hydroxy-2-pentanone**	155	86^{10}	1.4350^{25}	....	....
41	**3-Phenyl-2-butanone**	158	107^{22}	1.5092	....	....
42	**3-Phenyl-3-hexen-5-one**	158	138^{14}	....	....	....
43	**Dicyclopentyl ketone**	162	112^{12}	....	....	....
44	**2-Ethylcyclohexanone**	162; 163	76^{20}	1.4522	....	2,4-Dinitrophenylhydrazone, 162
45	**2,3-Dimethyl-2-hepten-6-one**	163	76^{13}	....	....	....
46	**3-n-Propylcyclohexanone**	169	$42^{0.7}$	1.4530	....	....
48	**2-Cyclohexenone**	172; 168	68^{22}	1.4879	....	2,4-Dinitrophenylhydrazone, 163; 117
49	**2-Chloropropiophenone** (2-Chlorophenyl ethyl ketone)	173	106^{12}	....	....	....
50	d,l-**1-Acetyl-3-methylcyclohexanone**	174–5	$99–100^{38}$	....	....	....
51	**3-n-Propyl-2-cyclohexenone**	175	$60^{0.4}$	1.4876^{25}	....	2,4-Dinitrophenylhydrazone, 156
52	**2-Bromoacetophenone**	177	112^{10}	....	....	2,4-Dinitrophenylhydrazone, 189

*Derivative data given in order: m.p., crystal color, solvent from which crystallized.

169

No.	Name	Semi-carbazone	Boiling point, °C	n_D^{20}	D_4^{20}	Miscellaneous
53	2-Methyl-3-butenyl phenyl ketone	177	$100^{2.1}$	1.5223^{25}		
54	4-Methoxycyclohexanone	178	85^{14}	1.4560		2,4-Dinitrophenylhydrazone, 150
55	5,5-Dimethyl-3-hexen-2-one	178	79^{40}	1.4430		
56	Ethyl 2-methylstyryl ketone	178	152^{14}			
57	2-Bromopropiophenone (2-Bromophenyl ethyl ketone)	179	118^{11}			
58	3-Isopropyl-2-cyclohexenone	179	$60^{0.3}$	1.4842		2,4-Dinitrophenylhydrazone, 155
59	2-Ethylacetophenone	180	118^{29}	1.5249		
60	3-Pyridylacetone	185	123^{1}			
61	l-6-Isopropyl-3-cyclohexenone	185	$98–100^{10}$	1.484		$[\alpha]_D^{18}$: −64.5; 2,4-Dinitrophenyl-hydrazone, 137–8; p-Nitrophenylhydrazone, 168–9
62	4-n-Butylacetophenone	185	141^{14}			
63	3-Phenyl-2-hexen-5-one	185	138^{14}			
64	3-Ethyl-2-cyclohexenone	186	$57^{0.9}$	1.4913		
65	3-Methyl-3-phenyl-2-butanone	186	77^{15}	1.5083		
66	2-Methyl-3-octen-5-one	187–8	$68–78^{24}$	1.4748		
67	4-Isopropylcyclohexanone	188; 188–9	91^{13}	1.4560		p-Nitrophenylhydrazone, 123–4
68	4,4,6-Trimethyl-2-cyclohexen-1-one	188; 185–7	$73–5^{13}$			
69	2-Acetyl-5-methylfuran	191	73^{8}			
71	3-Phenyl-2-pentanone		110^{18}	1.5051		
72	4-Phenylhexahydroacetophenone (Methyl 4-phenylcyclohexyl ketone)	191	$121^{1–2}$			
73	2,6,6-Trimethylcycloheptanone	191–2	85.5^{12}	1.4568^{18}	0.9095^{18}	
74	Acetyl phenyl carbinol	194	137			2,4-Dinitrophenylhydrazone, 170; 126; Oxime, 113
75	2-Tetralone	194	131^{11}	1.5555^{25}		m.p. 18; Oxime, 88
76	3-Isopropylcyclohexanone	195	51^{1}	1.4540		
77	1-Propionylcyclohexene	195; 189	102^{14}			Oxime, 78
78	α-Thienylacetone (1-(α-Thienyl)-2-propanone)	195	106^{12}	1.5366^{14}		
79	5,5-Dimethyl cyclohexen-3-one	195	88.5^{32}			$H_2SO_4 \rightarrow$ red $\rightarrow$ vlt. $\rightarrow$ col.
80	2-Acetylbiphenyl	197	105^{1}			
81	3-Methyl-2-cyclohexen-1-one	199; 201	78^{14}	1.4945		2,4-Dinitrophenylhydrazone, 176; 178
82	1-Acetyl-2,2-dimethylcyclohexene	201	118^{49}	1.4810^{25}		
83	1,1-Dimethyl-2-tetralone	204	$96^{0.5}$	1.538		
84	2-Methyl-1-tetralone	205; 195	138^{16}	1.5447		
85	6-Propionyltetralin	209	163^{11}	1.5508^{29}		
86	3-Methyl-2-n-propylcyclopentanone	210	58^{2}	1.4778		
87	4-Methyl-1-tetralone	211	111^{1}			
88	1-Acetylcyclopentene	211	74^{12}			
89	2-Acetyl-5-methylthiophene	217	83^{2}	1.5622		
90	1-Tetralone	217, yel., al.	170^{49}			Oxime, 102, prisms; 88–9, needles, me. al.
91	Neopentyl phenyl ketone	218	116^{11}	1.5078		Oxime, 114
92	3-Methyl-2,5-hexanedione	220	71^{10}	1.4260		
93	cis-1-Decalone	220d.	116^{18}	1.4939		
94	3-Acetylbiphenyl	223	138^{1}	1.6140^{25}		
95	1,2-Dimethyl cyclohexen-3-one	225d.	$118–9^{12}$			
96	3-Chloroacetophenone	232	113^{11}	1.5494		
97	3-Bromoacetophenone	233	132^{17}	1.5755		
99	6-Acetyltetralin	234	156^{10}	1.5591^{25}		
100	2,3-Dimethyl-2-cyclopentenone	250	92^{25}	1.4830		

*Derivative data given in order: m.p., crystal color, solvent from which crystallized.

TABLE X. ORGANIC DERIVATIVES OF KETONES

a) Liquids 2) (Reduced pressure b.p. only) (Listed in order of increasing semicarbazone m.p.)* (Continued)

No.	Name	Semi-carbazone	Boiling point, °C	n_D^{20}	D_4^{20}	Miscellaneous
101	1-Ethyl-2-methyl-3-cyclohexenone	250	105^{19}	. . . .	. . . .	. . . .
102	3-Acetylthianaphthene	250	137^3	. . . .	. . . .	. . . .
103	2-Oxopropionaldehyde (Methylglyoxal; Pyruvic aldehyde).	*di*: 254	52^{12}	. . . .	. . . .	*bis*-2,4-Dinitrophenylhydrazone, 299–300, red, PhNO$_2$; Dioxime, 157, al.

*Derivative data given in order: m.p., crystal color, solvent from which crystallized.

TABLE X. ORGANIC DERIVATIVES OF KETONES
b) Solids (Listed in order of increasing m.p.)*

No.	Name	Melting point, °C	Boiling point, °C	Semi-carbazone	2,4-Di-nitrophenyl-hydrazone	p-Nitro-phenyl-hydrazone	Phenyl-hydrazone	Oxime	Miscellaneous
1	**2-Aminoacetophenone**	20	250–2d.	290d., al.	. . . .	. . . .	108, al.	109, subl., w.	. . . .
2	**2-Dodecanone** (*n*-Decyl methyl ketone)	20.5	246–7	122–3, dil. al.	. . . .	. . . .	. . . .	. . . .	n_D^{30}: 1.42855; D_4^{30}: 0.81982
3	*n*-Amyl phenyl ketone (*n*-Pentyl phenyl ketone)	24.7 .	265.2	131.5, 50% al.; 133 (cor.)	168 (cor.), ac. a.	. . . .	. . . .		n_D^{25}: 1.50272
4	**3,5-Dichloroacetophenone**	26	134–6[17]	. . . .	. . . .	. . . .	. . . .	138	. . . .
5	**Benzyl methyl ketone** (Phenyl-acetone)	27	216.5 (cor.)	199–9.5, al.; 188	156	145	86–7, lgr.; 83	68–70	n_D^{20}: 1.5168
6	**4-Methylacetophenone**	28	226	204–5, al.; 197	260.4 (cor.), scar., ac. a.	198	97, al.	87–8, pet. eth.	n_D^{20}: 1.5348; D_4^{20}: 1.003
7	**2-Hydroxyacetophenone**	28	215	210	210–12; 212–3	. . . .	110	118	. . . .
8	**Phorone** (Di-isopropylidene-acetone)	28, yel.-gr.	198	221, w.; 186	118	. . . .	. . . .	48	Tetrabromide, 88–9, al.
9	**2-Tridecanone** (Methyl *n*-undecyl ketone; *n*-Hendecyl methyl ketone)	28	263	123, al.; 126; 117	69, or.-yel.	101–2	. . . .	56–7, al.-pet. eth.	n_D^{30}: 1.43175
10	**Furfuralacetophenone**	29, yel.-red	317d.	. . . .	169, scar.	. . . .	. . . .	. . . . . .	. . . .
11	**Ethyl 2-furyl ketone**	30; 28	183	189	. . . .	. . . .	. . . .	. . . . . .	. . . .
12	**4-Cyclohexylcyclohexanone**	30–31; 29.2	100[0.1]	216	137	. . . .	. . . .	101–2; 104–5	. . . .
13	**2-(1′-Hydroxycyclopentyl)-cyclo-pentanone**	31	99[3]	. . . .	. . . .	. . . .	. . . .	78	. . . .
14	**7-Tridecanone** (Di-*n*-hexyl ketone)	31–2	255; 264	. . . .	. . . .	96	. . . .	. . . . . .	. . . .
15	**2-Acetylfuran** (2-Furyl methyl ketone)	32; 33	169–73	150, me. al.; 148	220	185–6, red, w.-al.	86, yel., w.-al.	104, eth.-pet. eth.; 92	n_D^{20}: 1.5017; D^{20}: 1.098
16	**Levulinic acid** (3-Acetyl propionic acid)	33	245–6	. . . .	206 (cor.), or.-yel., ac. a.	174–5	108, bz.	45–6	. . . .
17	**2-Tetradecanone** (*n*-Dodecyl methyl ketone)	33–4, dil. al.	. . . .	115–6, al.	. . . .	. . . .	. . . .	. . . . . .	. . . .
18	**Methyl 1-naphthyl ketone** (1-Acetylnaphthalene)	34	302	222.5–4	. . . .	. . . .	146	136	n_D^{20}: 1.6257; Picrate, 116, yel.
19	**1,3-Diphenyl-2-propanone** (1,3-Diphenylacetone; Dibenzyl ketone)	34; 35; 30	330	145–6, abs. al.; 125–6, dil. al.	100	. . . .	121, al.; 128–9	125; 123	. . . .
20	**2,2,6,6-Tetramethyl-4-piperidone**.	34.9, red, dry eth.; 58 (mono-hyd.), eth.	205–6	219–20, al.	. . . .	. . . .	. . . .	153, al.	. . . .
21	**4-Chloropropiophenone**.	36; 35	118[2]	177; 175–6	. . . .	. . . .	. . . .	62–3	. . . .
22	**2-Phenylcyclopentanone**	37	126–7[10]	218d.	. . . .	. . . .	. . . .	. . . . . .	. . . .
23	**Furfuralacetone**	38; 39	116[10]	. . . .	241 (cor.)	. . . .	131–2, al.	. . . . . .	. . . .
24	**4-Methoxyacetophenone**	38; 37	258	197–8, dil. al.	220 (cor.), red; 231.8 (cor.)	195–5.5, or., al.	142, yel., al.	86–7, wh., pet. eth.	. . . .
25	**1-Phenyl-1-hepten-3-one** (*n*-Butyl styryl ketone)	38–9; 40	159–67[11]	. . . .	. . . .	. . . .	98, yel.	. . . .	Dimer, 175–6

*Derivative data given in order: m.p., crystal color, solvent from which crystallized.

TABLE X. ORGANIC DERIVATIVES OF KETONES

b) Solids (Listed in order of increasing m.p.)* (Continued)

No.	Name	Melting point, °C	Boiling point, °C	Semi-carbazone	2,4-Di-nitrophenyl-hydrazone	p-Nitro-phenyl-hydrazone	Phenyl-hydrazone	Oxime	Miscellaneous
26	4,4-Dimethylcyclohexanone	38–41	73^{14}	204	....	....	106–7	....	n_D^{24}: 1.4537; D_4^{20}: 0.932
27	3,4-Methylenedioxypropio-phenone (Propiopiperone)	39	....	187–8	....	....	97	104	....
28	2-Hydroxybenzophenone	39, w.-al.; 153	250^{560}	....	....	....	155, al.	syn: 141, bz.; anti: 143, bz.	
29	2-Methoxybenzophenone (2-Anisyl phenyl ketone).........	39	....	....	....	....		145–8, dil. al.; fusion → 130	....
30	3-Bromopropiophenone.........	40	....	183	....	....			
31	Benzalacetone	41	212	187	227; 223	166	157	115	Thiosemicarbazone, 148
32	1-Indanone (α-Hydrindone)	42; 38	$241–2^{739}$	233; 239	258	234–5, ac. a.	130–1; 134–5, me. al.; 124–8	146; 144, bz.-pet. eth.	
33	2,4-Dichloroacetophenone	42	235–40	208	....	....	....	152	n_D^{20}: 1.5640
34	3-Aminopropiophenone.........	42, yel.	$168–9^{15}$	196–7	....	....	....	112; 112–3	N-p-Toluenesulfonyl deriv., 97, al.
35	2-Bromobenzophenone	42	345	....	....	....		133	....
36	8-Pentadecanone (Di-n-heptyl ketone)...................	42	178	....	....	....		120	....
37	4-Phenyl-3-buten-2-one (Methyl styryl ketone)...............	42	262	186, al.; 198; 142	227, red, ac. a.; 223, or.-red, al.	165–7, red, al.	156–7, yel., al.; 159	115–6, 60% al.; 87	$H_2SO_4 →$ or.-red
38	3-Chlorobenzyl phenyl ketone....	43	....	....	....	....		102	....
39	8-Acetylguanine...............	43.5	$116^{0.7}$	....	253	....			
40	2-Benzoylfuran (2-Furyl phenyl ketone)...................	44	150^3	....	....	....		122	
41	1,3-Dichloro-2-propanone (1,3-Dichloroacetone).........	45	175; 173	....	....	....			n_D^{46}: 1.47144; D_4^{46}: 1.3826
42	2-Acetylquinoline	46	....	....	....	....	54		
43	3-Chloropropiophenone.........	46	....	179–80	....	....			
44	4-Bromopropiophenone.........	46	140^2	171	....	....		90–1	
45	Phenyl n-undecyl ketone (n-Hendecyl phenyl ketone; Laurophenone)..............	47; 44	....	94.5–95	....	....		63–63.5	
46	2-Aminopropiophenone.........	47, pet. eth.	....	190d., al.	....	....		88–9, w.	Hydrochloride, 184–5; N-Acetyl, 70–1
47	Benzophenone	48; 49	306 (cor.)	164–5, al.; 167	238–9, or.-yel., ac. a.; 229	154–5, yel., al.; 144, red, al.	137–8, al.	142–3, me. al.; 144; 140	D_{50}^{50}: 1.0976
48	2,4,6-Heptanetrione (Diacetyl-acetone)	49	121^{10}	mono: 203	....	....	di: 142	2,6-di: 68.5	n_D^{20}: 1.4930; D_{40}^{40}: 1.0681
49	5-Phenoxy-2-pentanone (Methyl 3-phenoxypropyl ketone)......	50	121^2	108–10	110	....			
50	9-Heptadecanone (Di-n-octyl ketone)...................	50–50.5	250–3	....	....	54		111–2	
51	Phenacyl bromide (ω-Bromo-acetophenone)............	50; 51	....	146	212–3, yel., or.	....		89.5; 97	
52	4-Bromoacetophenone.........	51; 50.5	225	208	230; 237 (cor.)	....	126	128; 129	
53	3-Bromophenacyl bromide	51, lgr.	174^{14}	163–4d., al.	....	....			
54	Ethynyl phenyl ketone	51	....	214	....	....			

*Derivative data given in order: m.p., crystal color, solvent from which crystallized.

TABLE X. ORGANIC DERIVATIVES OF KETONES
b) Solids (Listed in order of increasing m.p.)* (Continued)

No.	Name	Melting point, °C	Boiling point, °C	Semi-carbazone	2,4-Di-nitrophenyl-hydrazone	p-Nitro-phenyl-hydrazone	Phenyl-hydrazone	Oxime	Miscellaneous
55	3,4-Dimethoxyacetophenone (Acetoveratrone)............	51, w.-al.	286–8	218, w.-al.	206–7	227	131	140, w.-al.	....
56	Diphenylacetoin................	52	....	169	....	....	....	....	4-Nitrobenzoate, 84
57	Methyl 2-naphthyl ketone (2-Acetylnaphthalene)...........	53–4, al.; 56	301	234–5; 237	262d., red, ac. a.	....	176–7; 171	149; 145–6	Picrate, 82
58	Chloromethyl 2-phenylethyl ketone...................	54	244.5	156	147; 146	....	....	89	....
59	2-Furyl-2-thenoylmethane.......	55.5	195[6]	....	....	....	....	....	Cu salt, 274
60	Phenyl 2-thienyl ketone (2-Benzoylthiophene)	56	300	195–7	....	....	....	93	....
61	2-Nonadecanone (n-Heptadecyl methyl ketone)	56	....	124–5; 125.5–6	....	....	....	77	....
62	4-n-Caproylresorcinol (2,4-Di-hydroxy-n-caproylbenzene)....	56–7	243–5d.	....	....	....	....	190–1d., 50% al.	....
63	2,2′-Dichlorobenzoin...........	57	....	....	....	....	....	....	....
64	3-Propionylphenanthrene.......	57	....	....	....	....	....	53.6–4.7	Picrate, 113
65	9-Propionylphenanthrene	57	....	....	....	....	....	....	Picrate, 107
66	Benzalacetophenone (Phenyl styryl ketone; Chalcone).......	58, pa. yel., al.; 57–8	345	α: 168; β: 170, yel.; γ: 179–80d.	244d., or.-red, ac. a.; 245 (cor.)	....	120	140; 68	Picrate, 93–7; Dimers, 124; 178; 195; 225–6
67	2-Indanone..................	58; 58–9	....	218; 212–15 (dec.)	....	232 (dec.)	....	155	....
68	Phenacyl chloride (ω-Chloro-acetophenone)..............	59	244	156; 149	213–4d.; 214–5d.	....	....	89	....
69	3-Phenyl-2,4-pentanedione	60	134[20]	....	....	....	....	....	Cu salt, 224
70	Mesitylacetone	60	130[10]	205; 197	....	....	....	....	....
71	Ethyl 2-naphthyl ketone	60	312–4	202	....	....	....	133, w.-al.	....
72	Difurfuralacetone	60, yel.	....	....	....	....	121–2	....	....
73	Desoxybenzoin (Benzyl phenyl ketone)....................	60, al.	321 (cor.)	148, dil. al.	204 (cor.), or.	163, red-br.; 160	116, yel., al.	98, al.	....
74	4-Methylbenzophenone (Phenyl 4-tolyl ketone)...............	60; 55	326 (cor.)	121–2, al.	202.4 (cor.); 199–200, or.	....	109, wh., ac. a.	154, less soluble in w.-ac. a.; 115, more soluble in w.-ac. a.	....
75	1-Phenyl-1,3-butanedione (Benzoylacetone; Methyl phenacyl ketone).............	61	261–2	....	151, pa. yel., al.	100–1, me. al.	150–3	....	....
76	1,1-Diphenylacetone	61, al.	....	170	....	....	131, bz.	165	....
77	2,4′-Dibromobenzophenone	62, pet. eth.	381–4 sl. d.	....	....	....	....	141–2	....
78	5-Isopropyl-1,3-cyclohexanedione	62	....	....	....	....	....	....	....
79	Benzyl 3-chlorophenyl ketone....	62	....	....	....	....	....	120	....
80	4-Methoxybenzophenone (4-Anisyl phenyl ketone).........	62–4, al.	354	....	180, dk. or.	198–9, al.	132; 90	α: 140–1; 137–8; β: 115–6	....
81	2-Phenylcyclohexanone.........	63, al.; 60	160[15]	190	139	....	....	169	....
82	4-Methoxybenzil	63	....	....	....	....	....	124	....
83	1-Acetyl-2-hydroxynaphthalene ..	64	....	....	....	....	....	....	Benzoate, 85–6, pyr.
84	11-Heneicosanone (Di-n-decyl ketone)...................	64, w.-al.	....	....	....	....	....	27.5	....
85	Benzoylformic acid (Phenylgly-oxylic acid)	66	147–51[12]	....	196–7d. (cor.), yel.	....	....	α: 127, eth.; β: 145d., w.	Thiosemicarbazone, 188–9 (cor.), yel.
86	2,2′-Dimethylbenzophenone (Di-2-tolyl ketone)..............	67	....	....	190	....	....	105	....

*Derivative data given in order: m.p., crystal color, solvent from which crystallized.

TABLE X. ORGANIC DERIVATIVES OF KETONES
b) Solids (Listed in order of increasing m.p.)* (Continued)

No.	Name	Melting point, °C	Boiling point, °C	Semi-carbazone	2,4-Di-nitrophenyl-hydrazone	p-Nitro-phenyl-hydrazone	Phenyl-hydrazone	Oxime	Miscellaneous
87	1,5-Diphenyl-1,5-pentanedione (1,3-Dibenzoylpropane).......	67.5; 62–3	....	....	....	....	....	*di*: 165–6d.	....
88	3,5-Dibromoacetophenone	68, al.; 65	198[15]	268d., w.-ac. a.	....	....	109–10, yel., al.	....	....
89	Cinnamalacetone.............	68, eth.	....	186, yel., al.	222–3, vlt.-red, ac. a.; 218–20, br.-red, chl.-me. al.	....	180, yel., al.	153, al.; 152	....
90	Benzoyl-2-furoylmethane	68	169[3]	....	....	....	....	....	Cu salt, 248
91	12-Tricosanone (Di-*n*-undecyl ketone; Di-*n*-hendecyl ketone; Laurone)	69.5	....	179	....	....	....	39–40, al.	....
92	Diphenyl triketone............	70; 69, yel., lgr.	....	....	....	....	....	....	Heating with excess phenylhydrazine → 4-Benzene-azo-1,3,5-tri-phenylpyrazole, 156–7, yel.-red, al.
93	2-Chlorobenzyl phenyl ketone....	71	....	....	....	....	....	86	....
94	Dihydroxyacetone.............	72	....	....	277–8	160	....	84	Diacetate, 48; Dibenzoate, 120.5; Dimer, 78–81
95	3-Acetylphenanthrene..........	72	....	230	....	....	193–4	....	Picrate, 125–6
96	4-Methoxybenzalacetone (Anisal-acetone)...................	73, me. al.	....	....	229 (cor.), red, ac. a.	....	119–20	....	....
97	Benzylacetophenone............	73	....	144	....	....	....	87	....
98	Phenoxymethyl phenyl ketone (α-Phenoxyacetophenone).....	74	187[8]	187	....	....	....	....	....
99	9-Acetylphenanthrene..........	74	170[1]	201	....	....	....	154–5	Picrate, 107–8
100	1-Naphthyl phenyl ketone	75.5–6.0	225[15]	385	α: 246–7, red; β: 243–4, or.; mixt.: 220	....	....	161	....
101	9-Acetylfluorene	75.5	....	....	....	....	139	....	....
102	2-Benzofuryl methyl ketone	76; 72	136[11]	207	....	....	154	....	....
103	Benzyl 4-methoxyphenyl ketone..	77	....	....	....	....	....	118	....
104	4-Methoxybenzalacetophenone (Anisalacetophenone; 4-Methoxychalcone)	77, yel., al.	....	α: 168, al.; β: 190, al.	....	....	....	....	Picrate, 87, or.
105	2-Naphthoxyacetone...........	78; 77	....	203	....	....	154	123	....
106	4-Phenylcyclohexanone.........	78	....	212; 229d., al.	....	....	....	110	....
107	Benzoyl-2-thenoylmethane	78	201[4]	....	....	....	....	....	Cu salt, 278
108	4-Chlorobenzophenone	78	....	....	185	....	106	β: 105–6	....
109	1,4-Cyclohexanedione.........	79	132[20]	*mono*: 221–2; *di*: 231	240	....	....	188	....
110	Di-*n*-tridecyl ketone (Myristone).	79	....	....	....	....	....	57	....
111	Phenacyl phenyl ketone (Di-benzoylmethane; ω-Benzoyl-acetophenone)...............	keto: 81; enol: 78, al.	....	205	....	....	*mono*: 105	*mono*: 165	Dibromo deriv., 94–5, eth.
112	3-Nitroacetophenone..........	81	....	257	228	....	128; 135	132	....
113	4-Bromobenzophenone	82	....	350	230	....	126	α: 116–7; β: 110–1	....
114	1-Benzoylpropionic acid	82–3d.	....	....	....	....	100–4, br., bz.	....	....

*Derivative data given in order: m.p., crystal color, solvent from which crystallized.

175

TABLE X. ORGANIC DERIVATIVES OF KETONES

b) Solids (Listed in order of increasing m.p.)* (Continued)

No.	Name	Melting point, °C	Boiling point, °C	Semi-carbazone	2,4-Di-nitrophenyl-hydrazone	p-Nitro-phenyl-hydrazone	Phenyl-hydrazone	Oxime	Miscellaneous
115	Fluorenone	83, yel., bz.	341.5	....	283–4	269	151–2, yel., al.	195–6 (cor.)	....
116	4-Iodoacetophenone	85, eth.	153[18]	....	....	....	....	....	....
117	Di-n-Heptadecyl ketone	88.5, lgr.; 89	....	....	....	....	....	67; 62–3	....
118	4,4'-Dimethylbenzophenone (Di-4-tolyl ketone)	95, al.	335	143–4	218–9	....	100, yel., al.	163, al.	Hydrazone, 108–10
119	Benzil (Dibenzoyl)	95, yel.	347	*mono*: 174–5 d.; *di*: 243–4d., al.	*di*: 189, yel., al.; 185	*mono*: 192–3, dk. or., ac. a.; *di*: 290, yel., pyr.-eth.	*mono*: 134, yel., al.; *di*: 235, chl. rapid htng.	α: *mono*: 137–8; 140, bz.; *di*: 237	....
120	β,β-Diphenylpropiophenone	96; 92	....	....	....	....	....	133	....
121	3-Hydroxyacetophenone	96; 95	296	189–91	261	....	....	....	....
122	4-Chlorophenacyl bromide	96–8	....	....	....	....	....	....	Acetate, 72
123	2,3-Dihydroxyacetophenone (3-Acetylcatechol)	97–8, dk. yel., bz.-lgr.	....	166–7	....	....	....	96–7	Diacetate, 109, bz.; Di-Me. eth., b.p. 143–4[14] 143–4[14]
124	2-Naphthylglyoxal	98, (hyd.), w.; 109	183[20]	....	....	....	....	....	Diacetate, 150, bz.-pet. eth.; osazone, 184
125	3-Aminoacetophenone	98–9, pa. yel., al.	....	196d., w.	....	....	....	192–4, w.-al.	Hydrazone, 98, al.
126	2-Benzoylacrylic acid	99 (anh.); 64 (mono-hyd.)	....	190	....	....	197	168d.	Hydrazone, 185–6, al.
127	Di-2-thenoylmethane	100	....	....	....	....	....	....	Cu salt, 263
128	Di-1-naphthyl ketone	100	....	....	....	....	....	200	Picrate, 121.5–2.0
129	1,3-Dibenzoylbenzene	101–2, al.	....	....	....	....	*mono*: 201; *di*: 70–3	....	....
130	2-Acetyl-1-hydroxynaphthalene	102, gr.-yel., al.; 98, yel., bz.	325 sl. d.	245–50, pa. yel.	....	....	136–7, wh., dil. al.	168–9	Acetate, 107.5; Benzoate, 128, al.
131	Cinnamacetophenone	102, yel., al.	....	....	222, red, ac. a.; 218–9d.	α: 135, al.	....	....	....
132	1,3-Cyclohexanedione	104	....	....	....	....	....	156	....
133	4-Tolil (4,4'-Dimethylbenzil)	104–5	....	....	....	....	....	225	....
134	2-Propionylphenanthrene	105; 104	....	....	....	....	....	....	Picrate, 107
135	2-Aminobenzophenone	105–6, pa. yel., al.	....	....	....	....	....	156, alk.-stab; 127, a.-stab.	....
136	Benzoylnitromethane	106	....	....	....	....	105–5.5, yel., al.	96, w.	....
137	4-Aminoacetophenone	106, w.	294	250, yel. al.	266–7; 263	....	....	147–8, al.	4-Toluenesulfonamide, 203
138	Benzalacenaphthenone	107, yel., w.-al.	....	....	....	....	....	48	....
139	Benzyl 4-chlorophenyl ketone	108	....	....	....	....	....	123	....
140	4-Bromophenacyl bromide	108–9	....	....	....	....	....	115	Acetate, 72
141	4-Hydroxyacetophenone (4-Acetylphenol)	109	....	199	210, br., al.	....	151, wh. → yel.	145; 143–4, bz.	Acetate, 54; Benzoate, 134–5, al.
142	trans-1,2-Dibenzoylethylene	110	....	....	....	....	....	211	....
143	Piperonalacetone	110–1, pa. yel.	....	α: 217, al.; β: 168, bz.	....	....	163	186, al.	....

*Derivative data given in order: m.p., crystal color, solvent from which crystallized.

TABLE X. ORGANIC DERIVATIVES OF KETONES
b) Solids (Listed in order of increasing m.p.)* (Continued)

No.	Name	Melting point, °C	Boiling point, °C	Semi-carbazone	2,4-Di-nitrophenyl-hydrazone	p-Nitro-phenyl-hydrazone	Phenyl-hydrazone	Oxime	Miscellaneous
144	2-Acetyldibenzothiophene.......	112	....	235	....	....	....	....	....
145	1,5-Diphenyl-3-pentadienone (Dibenzalacetone)...........	112; 111	....	187–90, al.	180, red, ac. a.	173, yel., bz.	152–3; 147–8, yel., al.	142–4	Picrate, 114
146	4,4'-Dimethoxybenzoin (Anisoin)	113, dil. al.	....	185 (cor.), w.-al.	....	....	....	....	Acetate, 94–5, al.-pet. eth.
147	1,4-Diacetylbenzene	114	130³	....	....	....	....	240	....
148	3,4-Dihydroxyacetophenone (4-Acetylcatechol)...........	116, w.	127–33¹¹	....	....	....	....	184d., et. ac.	3,4-Diacetate, 91
149	3-Hydroxybenzophenone	116, al.	....	....	....	....	....	syn: 76; anti: 126, bz.	....
150	α-Hydroxyacetophenone (Phenacyl alcohol; Benzoyl carbinol)	117–8	118¹¹	146–6.5, al.	....	....	112, lgr.-eth.	70	Acetate, 49, eth.; Benzoate, 118, dil. al.
151	7-Acenaphthenone.............	121 (cor.), al.	....	....	....	....	90, dk., al.	175, al.; 183–4, bz.; di: 222d.	Picrate, 113
152	4-Phenylacetophenone (4-Acetyl-biphenyl; Methyl 4-xenyl ketone)....................	121, al.	325–7	....	241.5–2, red-or.	....	....	186–7, al.	....
153	Piperonalacetophenone (3,4-Methylenedioxychalcone).....	122, yel., al.	....	α: 203–5, abs. al.	....	....	....	....	Dibromide, 152, bz.-lgr. (1:1); Picrate, 126–8
154	9-Hydroxyxanthene (Xanthydrol)	122–4d.	....	....	....	....	....	....	Heat → Dixanthyl ether, 219; Dixanthyl, 204, lgr.
155	4-Aminobenzophenone	124, w.-al.	....	....	....	....	....	168, al.; 127, w.-al.	Hydrazone, 139–40, yel., al.
156	4-Phenylphenacyl bromide	124–5	....	....	....	....	....	....	Acetate, 111
157	2-Naphthoin...................	126	....	....	....	....	....	172	....
158	4-Phenylphenacyl chloride	126–7	....	....	....	....	....	....	Acetate, 111
159	Vanillalacetone (4-Hydroxy-3-methoxystyryl methyl ketone)..	129, yel., al.	....	....	230 (cor.), red	....	127–8, yel.	....	....
160	Dianisalacetone...............	129, yel., bz.-pet. eth.	....	....	82–3	....	....	147–8	....
161	4,4'-Dimethoxybenzil (Anisil) ...	133, yel.	....	di: 254–5, dil. ac. a.	....	....	....	mono: 133; syn, di: 217; anti, di: 195, bz.	Dihydrazone, 118
162	d,l-Benzoin...................	133, al.; 129	344	α: 205–6, w.	245, yel., al.; 234	....	α: 158–9, bz.-lgr.; β: 106	α: 151–2, bz.; β: 99	Acetate, 83, al.; Benzoate, 124–5, 75% al.
163	cis-1,2-Dibenzoylethylene.......	134	....	....	....	....	....	di: 210–1d.	....
164	3,4,5-Tribromoacetophenone	134–5, al.	....	265d., ac. a.	....	....	129–34d., yel., pet. eth.	....	....
165	4-Hydroxybenzophenone (4-Benzoylphenol).............	134–5, w.	....	194, bz.	242.4 (cor.), or.	....	144, pet. eth.	81; Heat → 152	Acetate, 81, al.; Benzoate, 94–5; 115
166	1-Furoin.....................	135; 138–9 (cor.)	....	....	216–7, or.-red, al.	....	79–81, lgr.-bz.	α: 161, al.; β: 102, pa. yel., eth.	Acetate, 76–7; Benzoate, 92–3
167	Mesityl phenyl ketone	137	....	....	232	....	....	....	....
168	4-Chlorobenzyl phenyl ketone....	138	....	....	....	....	....	96	....

* Derivative data given in order: m.p., crystal color, solvent from which crystallized.

TABLE X. ORGANIC DERIVATIVES OF KETONES
b) Solids (Listed in order of increasing m.p.)* (Continued)

No.	Name	Melting point, °C	Boiling point, °C	Semi-carbazone	2,4-Di-nitrophenyl-hydrazone	p-Nitro-phenyl-hydrazone	Phenyl-hydrazone	Oxime	Miscellaneous
169	α-Anisal-α'-cinnamalacetone....	139, red, al.	....	....	....	....	Pyrazo-line, 155–6, yel., al.	....	Dibromide, 139–40, CS₂; Tetrabromide, 155–6, CS₂
170	4-Aminopropiophenone.........	140, w.	....	....	....	....	....	153, al. 181–2d.	....
171	3,3'-Dibromobenzophenone.....	141	....	....	....	....	....		....
172	2-Acetylphenanthrene..........	143	....	260	....	....	187–8		....
173	Dicinnamalacetone	144, yel., abs. al.	....	....	195.7 (cor.), dk. red	....	166, yel., al.	....	Light → d.
174	1,2-Dibenzoylethane	147	....	....	di: 265	....	mono: 116; di: 179	204	....
175	2,4-Dihydroxyacetophenone (Resacetophenone)..........	147; 144	....	218; 214–20d.	206–8	....	156–8	202–3	2,4-Diacetate, 38; 2,4-Dibenzoate, 81
176	3,5-Dihydroxyacetophenone.....	147–8, w.	....	205–6, al.	....	236–7, red, w.-ac. a.	....	....	3,5-Diacetate, 91–2, lgr.
177	4,4'-Dichlorobenzophenone	147–8	....	....	238–40	....	....	135	....
178	1,2-Dibenzoylbenzene	148	....	....	....	....	....	mono: 150	....
179	9-Benzoylanthracene (9-Anthra-phenone)	148, yel., bz.	....	....	....	....	....	....	H₂SO₄ → brief blue color
180	4-Hydroxypropiophenone.......	148	....	....	229	....	....	....	....
181	5,5-Dimethyl-1,3-cyclo-hexanedione (Methone; Dimedone)...............	148–9	....	....	....	....	....	mono: 115; di: 176	....
182	2,3,4,2'-Tetrahydroxybenzo-phenone	149 (anh.); 100 (hyd.), yel., w.	....	....	....	....	....	....	2,3,4,2'-Tetraacetate, 118, a
183	Anthrone	154, ac. a.	....	....	....	....	....	....	Al. sol. → bl. fluorescence
184	1,4-Dibenzoylbenzene	161, al.	....	....	....	....	....	mono: 212–3; di: 235	....
185	Furil (2,2'-Bifuroyl)	165, yel., bz.	....	....	....	....	mono: 82–3, yel.; di: 184, yel., lgr.	α, mono: 106; β, mono: 97–8; α, di: 100; 166; β, di: 188–90	....
186	Benzanthrone.................	170, yel., al.	....	....	....	....	....	....	H₂SO₄ → or.-red sol., gr. fluorescence
187	Quinhydrone	171, dk. gr., subl.	....	....	....	....	152	161	....
188	2,3,4-Trihydroxyacetophenone (Gallacetophenone)	172, w.	....	225, rapid htng.	....	....	162–3		2,3,4-Triacetate, 85; Picrate, 133
189	Xanthone (Diphenylene ketone oxide).....................	174	350	....	....	....	152	161	....
190	4,4'-Bis(dimethylamino)benzo-phenone..................	174	....	....	....	....	174–5	233	Hydrazone, 150; Picrate, 156
191	4,4'-Dibromobenzophenone	177	395	....	....	....	....	150–2	Hydrazone, 92–4
192	d,l-Camphor	d, l: 178; d: 179	209; 205	247–8d. (cor.); 236–7, al.	d, l: 164; d: 177, or., al.	217	233	d: 118–9	d: [α]²⁰_D: + 44; Hydrazone, 55
193	3-Acetylindazole...............	182, yel., ac. a.	....	....	>320, red, PhNO₂	....	....	222, bz.-al.	N-Acetyl deriv., 123, ac. a.
194	3,4,5-Trihydroxyacetophenone...	187–8, w.	....	216–7, al.	....	260d., red, w.-ac. a.	....	....	3,4,5-Triacetate, 111–2, lgr.

*Derivative data given in order: m.p., crystal color, solvent from which crystallized.

TABLE X. ORGANIC DERIVATIVES OF KETONES

b) Solids (Listed in order of increasing m.p.)* (Continued)

No.	Name	Melting point, °C	Boiling point, °C	Semi-carbazone	2,4-Di-nitrophenyl-hydrazone	p-Nitro-phenyl-hydrazone	Phenyl-hydrazone	Oxime	Miscellaneous
195	3-Acetylindole	190–1, bz.						144–7, w.	N-Acetyl deriv., 151, subl., pet. eth.-bz.; Picrate, 183, yel., pet. eth.; Diacetyl deriv., 165–6, bz.
196	2,5,2′,6′-Tetrahydroxybenzo-phenone	200–2d., w.							2,5,2′,6′-Tetraacetate, 118–9, al.
197	2,5-Dihydroxyacetophenone (Quinacetophenone)	202, yel.-gr., w.				215–6		149–50, tol.	2,5-Diacetate, 68, ac. a.
198	2,4,5-Trihydroxyacetophenone	206–7, red							2,4,5-Triacetate, 165–6, bz.
199	2,3,5-Trihydroxyacetophenone	206–7, yel., ac. a.				241–2d., w.-al.			2,3,5-Triacetate, 106–7, lgr.
200	2,4,6-Trihydroxyacetophenone (Phloracetophenone)	219, anh., w.							2,4,6-Triacetate, 103; 2,4,6-Tribenzoate, 117–8
201	2,3-(3′,4′,5′-Triphenylcyclo-pentadieno)-indone	222					246–7		
202	3,4,3′4′-Tetrahydroxybenzo-phenone	227–8, w.						145, al.	
203	Ninhydrin (Triketohydrindene hydrate)	243; 241d.					di: 207–8, or.-red, al.	201	Warm aqueous sol. of α-amino acids → intense bl., vlt., etc.; At 125–30, loses w., turns red.
204	Phenacridone	304–5, yel., ac. a.							N-Benzyl deriv., 188–9, yel., al.

*Derivative data given in order: m.p., crystal color, solvent from which crystallized.

The derivatives of quinones are usually similar to those of ketones. Therefore, for directions and examples for their preparation see explanations and references to Tables 9 and 10, pages 141, 142, 143.

The derivatives listed as *phenylhydrazones*, *p-nitrophenylhydrazones* and *2,4-dinitrophenylhydrazones* of quinones in most cases are not real substituted phenylhydrazones, but either addition compounds or hydroxy derivatives of variously substituted aromatic azo compounds.

Semicarbazone.*

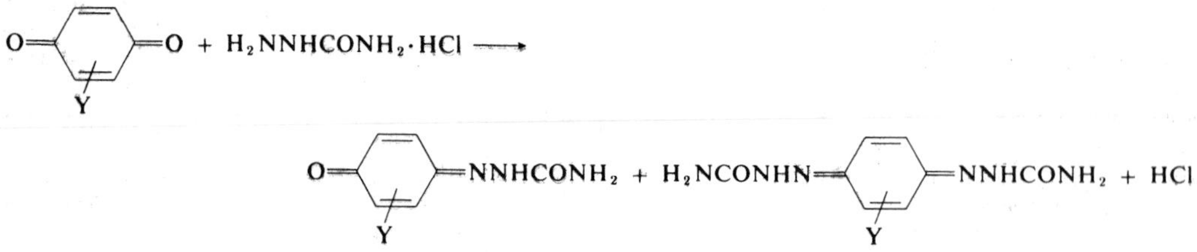

$$O=\!\!\!\!\raisebox{-0.5em}{\text{(ring, Y)}}\!\!\!\!=O \; + \; H_2NNHCONH_2 \cdot HCl \longrightarrow$$

$$O=\!\!\!\!\raisebox{-0.5em}{\text{(ring, Y)}}\!\!\!\!=NNHCONH_2 \; + \; H_2NCONHN=\!\!\!\!\raisebox{-0.5em}{\text{(ring, Y)}}\!\!\!\!=NNHCONH_2 \; + \; HCl$$

Monosemicarbazone   Disemicarbazone

From the quinone with semicarbazide hydrochloride in water.
For directions and examples see: Linstead, p. 30; Vogel, p. 749.

Oxime.*

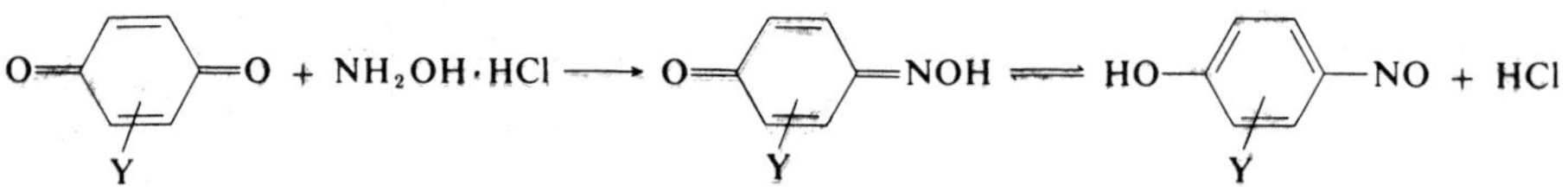

$$O=\!\!\!\!\raisebox{-0.5em}{\text{(ring, Y)}}\!\!\!\!=O \; + \; NH_2OH \cdot HCl \longrightarrow O=\!\!\!\!\raisebox{-0.5em}{\text{(ring, Y)}}\!\!\!\!=NOH \rightleftharpoons HO\!\!\!\!\raisebox{-0.5em}{\text{(ring, Y)}}\!\!\!\!-NO \; + \; HCl$$

Monooxime   Substituted *p*-nitrosophenol

From the quinone with hydroxylamine hydrochloride in water.
For directions and examples see: Linstead, p. 30.

---

*Derivatives recommended for first trial.
WARNING: This is not an instruction manual. References should be consulted for the preparation of derivatives.

TABLE XI. ORGANIC DERIVATIVES OF QUINONES
(Listed in order of increasing m.p.)*

No.	Name	Melting point, °C	Boiling point, °C	Semi-carbazone	2,4-Di-nitrophenyl-hydrazone	p-Nitro-phenyl-hydrazone	Phenyl-hydrazone	Oxime	Miscellaneous
1	2,3,5-Trimethyl-1,4-benzoquinone	32, yel., eth.	....	252–3 d., yel.	....	....	....	1-*mono*: 181–2, yel., al.; 4-*mono*: 134, yel.	....
2	2-Isopropyl-5-methyl-1,4-benzoquinone (Thymoquinone) .	45.5, or.-yel.	232	*mono*: 201–2d., yel., al.; *di*: 237, yel.	*mono*: 179–80, dk. red, al.	....	93	1-*mono*: 160–2, pa. yel., chl.	....
3	2,3-Dimethyl-1,4-benzoquinone (*o*-Xylo-*p*-quinone)...........	55 (subl.), yel.	....	....	....	....	....	*mono*: 166, yel.	....
4	2-Bromo-1,4-benzoquinone......	56	....	....	....	....	....	1-*mono*: 184, or.; 4-*mono*: 196	....
5	2-Chloro-1,4-benzoquinone......	57, yel.-red	....	4-*mono*: 185d.	....	....	....	1-*mono*: 184, pa. grn.-yel.; 4-*mono*: 148d.	....
6	2-Methyl-1,4-benzoquinone (*p*-Toluquinone)	67.6–68.4	....	4-*mono*: 178–9, yel., al.; *di*: 240d., or.-red	*di*: 269, PhNO₂	....	130	*mono*: 134–5d., w.; *di*: 220d., yel.-wh.	....
7	2,6-Dimethyl-1,4-benzoquinone (*m*-Xylo-*p*-quinone)	68–71, yel.	....	....	....	....	....	1-*mono*: 175, yel., al.-w.; 4-*mono*: 170–1, yel., bz.	....
8	4-Chloro-1,2-benzoquinone......	78, pa. yel.-red, hexane	....	....	....	....	....	*di*: 128, br.	....
9	2,5-Dimethyl-1,4-naphthoquinone	94, yel., me. al.	....	....	....	....	*mono*: 226, red, ac. a.	....	....
10	Dunnione	98–9, or.-red, w.	....	*mono*: 232–3, w.-me. al.	266–8, or.	....	....	....	[α]$_D^{18}$: +310° in chl.
11	2-Methyl-1,4-naphthoquinone ...	106, yel., me. al.	....	....	4-*mono*: 299d.	....	....	*mono*: 160; *di*: 166–8	Benzoyl chloride + Zn dust → 1,4-dibenzyloxy-2-methylnaphthalene, 180.0–0.5, col., al.
12	Quinone (1,4-Benzoquinone)	116, yel.; 113	....	*mono*: 166, yel.; 165–6d., red; 178; *di*: 243d., red.	*mono*: 185.6, br., al.; *di*: 231	....	152	240	Picrate, 79; Quinhydrone, 171
13	2-Chloro-1,4-naphthoquinone	117, yel., w.	....	....	....	....	....	4-*mono*: 200 d.; 198, pa. yel., bz.	....
14	1,4-Naphthoquinone (α-Naphtho-quinone)...................	117–8, yel., al.	....	*mono*: 247, grn.-yel., ac. a.	*mono*: 278, yel., pyr.	*mono*: 277–9, or.-red, PhNO₂	*mono*: 205–6d., dk. vlt., bz.	....	....

*Derivative data given in order: m.p., crystal color, solvent from which crystallized.

TABLE XI. ORGANIC DERIVATIVES OF QUINONES
(Listed in order of increasing m.p.)* (Continued)

No.	Name	Melting point, °C	Boiling point, °C	Semi-carbazone	2,4-Dinitro-phenyl-hydrazone	p-Nitro-phenyl-hydrazone	Phenyl-hydrazone	Oxime	Miscellaneous
15	2,5-Dimethyl-1,4-benzoquinone (p-Xylo-p-quinone)	125, yel., al.	....	....	....	....	122–4, yel.; 154–5, or.	mono: 168, yel., w.; di: 272, yel., al., 254	Monobenzoylphenyl-hydrazone, 122–4, yel., lgr.
16	2,6-Dibromo-1,4-benzoquinone	131, yel., al.	....	4-mono: 225d., yel., al.	....	....	....	4-mono: 170d., br.	....
17	2,6-Diphenyl-1,4-benzoquinone	137–8, red	....	....	....	....	....	242–4d.	....
18	2,5-Dihydroxy-3-n-dodecyl-1,4-benzoquinone (Embelin)	143, or.-red, eth.-bz.	....	di: 236	....	....	di: 189–90	tetra: 175	Diacetate, 54; Dibenzoate, 97–8
19	1,2-Naphthoquinone (β-Naphtho-quinone)	145–7d., red, eth.; 120	....	1-mono: 184d., yel., al.	....	1-mono: 250–1; 2-mono: 236, dk. red, ac. a.	2-mono: 138, dk. red, al.	1-mono: 109.5; 2-mono: 162–4d., di: 169, yel., al.	....
20	3,7-Dimethyl-1,2-naphthoquinone	151–2, red, al.	....	....	....	....	....	2-mono: 222d., or.	
21	5-Hydroxy-1,4-naphthoquinone (Juglone)	153–4, or., bz.	....	....	....	....	....	mono: 167.0–7.5, red, ac. a.; di: 225 (exp.), dk. br., ac. a.	Acetate, 154–5
22	3-Chloro-1,2-naphthoquinone	172, red, chl.	....	....	....	....	....	1-mono: 167–8, or.	
23	1,8-Dihydroxy-2-methyl-9,10-anthraquinone (2-Methylchry-sazin)	175	....	....	....	....	....	....	Diacetate, 205
24	2-Methyl-9,10-anthraquinone	177–9; 177, pa. yel.	....	....	....	....	....		Diacetate, 217, pa. yel. ac. a.
25	3,5-Dihydroxy-2-methyl-1,4-naphthoquinone (Droserone)	181, pa. yel., al.	....	....	....	....	....	di: 151	Diacetate, 119, me. al.
26	2,3,4-Trihydroxy-9,10-phenan-thraquinone	185d., br.-red	....	mono: 270d., br.-red, al.	....	....	....	....	Phenazine, 255d., br., al.
27	6-Bromo-1,4-dihydroxy-9,10-anthraquinone (6-Bromo-quinizarin)	185.5, red-br., bz.	....	....	....	....	....		Di-Me. eth., 176.5, or.-yel., al.; Diacetate, 220.5, yel., al.
28	1,2-Anthraquinone	185–90d., or.-br.	....	....	....	....	....	1-mono: 188d., or.-br.; 2-mono: 200d., or.-br.	....
29	4-Chloro-1,2-naphthoquinone	188, red-br., bz.	....	....	....	....	....	2-mono: 157, pa. yel.	....
30	7-Isopropyl-1-methyl-9,10-phenanthraquinone (Retene-quinone)	197, or., al.	....	mono: 200, yel., pyr.	....	mono: 222–3, red, ac. a.	mono: 160, or., bz.-al.	mono: 128.5; 130–1 (cor.), yel., al.	....

*Derivative data given in order: m.p., crystal color, solvent from which crystallized.

TABLE XI. ORGANIC DERIVATIVES OF QUINONES
(Listed in order of increasing m.p.)* (Continued)

No.	Name	Melting point, °C	Boiling point, °C	Semi-carbazone	2,4-Dinitro-phenyl-hydrazone	p-Nitro-phenyl-hydrazone	Phenyl-hydrazone	Oxime	Miscellaneous
31	Camphorquinone	199, yel., dil. al.	....	3-*mono* (α): 236d., al.	*mono*: 36; *di*: 190	239	183–90, pa. yel., al.; 170	*mono*: α: 153; β: 114–5; *di*: (four forms): α: 201d.; β: 248d.; γ: 136; δ: 194d.	3-*p*-Bromophenylhydrazone, 215–6, ac. a.; Hydrazone, 182
32	1-Hydroxy-9,10-anthraquinone	200, or.-red, al.; 193	subl.	....	....	....	....	....	Acetate, 176–9, yel., al.
33	1,4-Dihydroxy-9,10-anthraquinone (Quinizarin)	200–2 (cor.), red, ac. a.; 195	....	....	....	....	....	....	Diacetate, 207.8, yel., pyr.; 200–1, yel., ac. anh.
34	9,10-Phenanthraquinone	206–7.5, or.-yel.; 208, or.	>360; subl. → or.-red	*mono*: 220d.	*mono*: 312–3d., dk. red	*mono*: 245, red, xyl.	*mono*: 165, dk. red, al.	*mono*: 158, grn.-yel., al.; *di*: 202d.	Conc. H₂SO₄ → dull grn.
35	1,4-Anthraquinone	218d., yel.; turns dark at 200–10	....	....	....	....	....	233, br.; turns dark at 205	....
36	4-Chloro-1,2,3-trihydroxy-9,10-anthraquinone	233, yel.	....	....	....	....	....	....	Triacetate, 187, yel.; Conc. H₂SO₄ → red
37	2-Bromo-9,10-phenanthraquinone	233–4, red-yel., ac. a.	....	....	....	....	*mono*: 163–4, or.	....	....
38	1-Bromo-9,10-phenanthraquinone	233–4, yel.	....	....	....	....	*mono*: 213d.	....	....
39	Chrysoquinone (Chrysene-quinone)	239.5, red	....	....	....	....	*mono*: 161, or.	....	Conc. H₂SO₄ → bl.
40	1,2,8-Trihydroxy-9,10-anthraquinone (2-Hydroxychrysazin)	239–40, red, ac. a.	....	....	....	....	....	....	Triacetate, 224, yel.; 2-Me. eth., 220, or., chl.-me. al.; 2,8-Di-Me. eth., 193, br.-yel., chl.-me. al.; 1,2,8-Tri-Me. eth., 157, yel., me. al.
41	1-Amino-9,10-anthraquinone	251, or.-red	....	....	....	....	....	....	N-Acetyl, 218, or.-red; N-Benzoyl, 255, grn.; N-*p*-Toluenesulfonyl, 228–9
42	3-Amino-9,10-phenanthraquinone	254, dk. red-br., al.	....	....	....	....	....	*mono*: 247d., red-br.	....
43	1,2,4-Trihydroxy-9,10-anthraquinone	259, dk. red, abs. al.	....	....	....	....	....	....	2-Me. eth., 232–3, red, bz.; 2,4-Di-Me. eth. 186–9, or.; 2-Acetate, 179–80, or., al.; Triacetate, 198–200 (sinters 193), pa. yel.
44	Acenaphthenequinone	261 (cor.), yel., ac. a.	....	*mono*: 192–3, ac. a.; *di*: 271, al.	....	*mono*: 247, or.-red, ac. a.	*mono*: 179, or.-red, al., *di*: 219, dk. yel., al.	*mono*: 230, dil. al.; 220d.	....

* Derivative data given in order: m.p., crystal color, solvent from which crystallized.

No.	Name	Melting point, °C	Boiling point, °C	Semi-carbazone	2,4-Dinitro-phenyl-hydrazone	p-Nitro-phenyl-hydrazone	Phenyl-hydrazone	Oxime	Miscellaneous
45	**1,3-Dihydroxy-4-methyl-9,10-anthraquinone** (4-Methylpur-puroxanthin)	265–6, or., bz.	....	....	....	....	....	....	Di-Me. eth., 162, yel., chl.; Diacetate, 181–2, yel., ac. a.
46	**2-Bromo-1,4-dihydroxy-9,10-anthraquinone** (2-Bromo-quinizarin)	265–8, br.-red	....	....	....	....	....	....	Diacetate, 226–9, yel.
47	**3-Bromo-9,10-phenanthraquinone**	268, dk.-yel., ac. a.	....	*mono:* 242d.	....	....	*mono:* 177, red	*mono:* 198; *di:* 212d., grn.	....
48	**Aceanthrenequinone** (3,4-Benzacenaphthenequinone)....	270, red, bz.	....	....	....	....	*mono:* 203, or., bz.	*mono:* 251d., yel., ac. a.	subl.
49	**1,2,5-Trihydroxy-9,10-anthra-quinone** (Hydroxyanthrarufin)..	273–4, red, ac. a.	....	....	....	....	....	....	2-Me. eth., 229, yel., al.; 1,2-Di-Me. eth., 231, or., al.; 1,2,5-Tri-Me. eth., 203–4, yel., al.; Triacetate, 228–9, yel., al.
50	**1,5-Dihydroxy-9,10-anthra-quinone** (Anthrarufin)........	280 (subl.), pa. yel.	379–81	....	....	....	....	....	Diacetate, 245d., pa. yel., ac. a.
51	**Chloranilic acid** (2,5-Dichloro-3,6-dihydroxy-1,4-benzo-quinone)...................	283–4, red, w. (+2H₂O)	....	....	....	....	....	....	Di-Me. eth., 141–2, red; Di-Et. eth., 107, red; Di-acetate, 182.5, yel.
52	**9,10-Anthraquinone**...........	286	382; 376.8 (cor.)	....	....	....	....	*mono:* 224, pa. yel., (rapid htng.)	Diacetate, 260, col., ac. a.
53	**Chloranil** (2,3,5,6-Tetrachloro-1,4-benzoquinone)	290, yel., ac. a., (slow htng., sealed tube)	subl.	....	....	....	....	....	SO₂—Tetrachlorohydro-quinone; In w.-alkali sol. → alkali salts of chloranilic acid
54	**Alizarin** (1,2-Dihydroxy-9,10-anthraquinone)	290, or., al.	430	....	....	....	....	....	2-Benzoate, 214–6, al.; 2-p-Bromobenzoate, 195, yel.; Diacetate, 184, yel., al.; Di-Me. eth., 215
55	**2-Amino-9,10-phenanthraquinone**	>300, dk. vlt., w.; sinters at 205–10	....	....	....	....	....	....	N-Acetyl, 324, dk. red-vlt., PhNO₂; N-Benzoyl, 297–8, br.-red, PhNO₂
56	**3-Aminoalizarin** (3-Amino-1,2-dihydroxy-9,10-anthraquinone)	>300, dk. red, ac. a.	....	....	....	....	....	....	N-Acetyl, 238–40, yel.-br.; Monobenzoyl, 275, dk. yel.; Dibenzoyl, 252, yel.
57	**1,4-Diamino-5,8-dihydroxy-9,10-anthraquinone** (5,8-Diamino-quinizarin).................	>300, br.-vlt., PhNO₂	....	....	....	....	....	....	N,N′-Dibenzoyl, 284–5, br.-vlt., xyl.; N,N′-Di-phenyl, 258–60, dk. bl., bz.-lgr.; 1,4-Di-Me. eth., 250d., vlt.-blk., ac. a.
58	**Dianthraquinone** (9,9′-Di-anthranyl-10,10′-quinone).....	>300, yel.	....	....	....	....	....	....	Conc. H₂SO₄ → vlt.-red; CrO₃ → Anthraquinone; Zn dust + ac. a. → di-anthranol, 230 (enol), 250 (keto)

*Derivative data given in order: m.p., crystal color, solvent from which crystallized.

No.	Name	Melting point, °C	Boiling point, °C	Semi-carbazone	2,4-Dinitro-phenyl-hydrazone	p-Nitro-phenyl-hydrazone	Phenyl-hydrazone	Oxime	Miscellaneous
59	**2-Amino-9,10-anthraquinone**	306; 303–6, red, al.	....	....	....	....	....	....	N-Acetyl, 262, yel., N,N-Diacetyl, 258, yel., ac. a.; N-Benzoyl, 227–8, yel., ac. a.
60	**2-Hydroxy-9,10-anthraquinone** ..	305, yel., al.	306	....	....	....	....	....	Acetate, 159–60, al.; Benzoate, 202–4, ac. a.
61	**1,2,3-Trihydroxy-9,10-anthra-quinone** (Anthragallol)	313–4d., br.-or. (290, subl.)	....	....	....	....	....	....	Triacetate, 181–2, yel., al.; 188–9, pyr.; 2,3-Di-p-toluenesulfonate, 196–8, yel., pyr.; Tri-Me. eth., 167–9, grn.-yel., bz.-pet.; Triacetate, 181–2, yel., ac. a.; Tribenzoate, 213–5, pa. yel., al.-bz.
62	**1,2,6-Trihydroxy-9,10-anthra-quinone** (Flavopurpurin)	>330, (>160, subl.)	....	....	....	....	....	....	2,6-Di-Me. eth., 239, yel.; 1,2,6-Tri-Me. eth., 225–6, yel.; Diacetate, 238; Tri-acetate, 202–3
63	**1,2,7-Trihydroxy-9,10-anthra-quinone** (Anthrapurpurin)	369, or., al.	....	....	....	....	....	....	2-Acetate, 296–8, yel., al.; 2,7-Diacetate, 192–3, yel., al.-ac. a.; Triacetate, 223, pa. yel., ac. a.; Tri-Me. eth., 201, yel., al.

*Derivative data given in order: m.p., crystal color, solvent from which crystallized.

EXPLANATIONS AND REFERENCES TO TABLES XII, XIII AND XIV

The derivatives of three classes of compounds (carboxylic acids, acyl halides and acid anhydrides) are essentially the same as those of carboxylic acids, and are prepared either directly from the acid or *via* the acyl halide. All of them appear therefore under the same title.

Hydrolysis of acid halide or acid anhydride to the corresponding carboxylic acid.

$$RCOCl + H_2O \longrightarrow RCOOH + HCl$$
$$\text{Acid}$$

$$(RCO)_2O + H_2O \xrightarrow{NaOH} 2\,RCOONa \xrightarrow{H^+} RCOOH$$
$$\text{Acid}$$

From the acyl halide in water.
For directions and examples see: Wild, p. 180.
From the acyl halide with aqueous sodium hydroxide.
See: Vogel, p. 369; Wild, p. 180.
From the acid anhydride with water.
See: Vogel, p. 376; Wild, p. 184; A. C. D. Rivett and N. V. Sidgwick, *J. Chem. Soc.*, **97**, 1677 (1910).
From the acid anhydride with aqueous sodium hydroxide.
See: Linstead, pp. 16–7; Wild, p. 184.

*Amide.**

$$RCOOH \xrightarrow{SOCl_2} RCOCl \xrightarrow{NH_3} RCONH_2 + NH_4Cl$$
$$\text{Acid}$$
$$\text{chloride}$$

$$RCOOH + NH_3 \longrightarrow RCONH_2$$

$$(RCO)_2O + NH_3 \longrightarrow RCONH_2$$
$$\text{Amide}$$

Acid chloride is prepared from the acid and thionyl chloride. Amide is formed on addition of aqueous ammonia.
For directions and examples see: Cheronis, p. 440; Shriner, p. 200; Vogel, p. 361; Wild, p. 181.
From the acid chloride in benzene with aqueous ammonia.
See: D. Swern, J. M. Stutzman and E. T. Roe, *J. Amer. Chem. Soc.*, **71**, 3017 (1942).
By passing gaseous ammonia through a benzene or ether solution of the acyl chloride.
See: Linstead, p. 14; Wild, p. 182.
From the neat acid with gaseous ammonia.
See: J. A. Mitchell and E. E. Reid, *J. Amer. Chem. Soc.*, **53**, 1879 (1931).
From the acid anhydride with aqueous ammonia.
See: Wild, p. 184, 185.

*Anilide.**

$$RCOOH \xrightarrow{SOCl_2} RCOCl \xrightarrow{C_6H_5NH_2} RCONHC_6H_5 + C_6H_5NH_3{}^+Cl^-$$

$$RCOOH + C_6H_5NH_2 \longrightarrow RCONHC_6H_5 + H_2O$$

$$(RCO)_2O + C_6H_5NH_2 \longrightarrow RCONHC_6H_5 + C_6H_5NH_3{}^+RCOO^-$$
$$\text{Anilide}$$

From the acid chloride (prepared from the acid and thionyl chloride) and aniline in benzene or in ether.
For directions and examples see: Cheronis, p. 445; Linstead, p. 14; Shriner, pp. 98, 200–1; Vogel, pp. 361, 369, 458; Wild, p. 182; P. W. Robertson, *J. Chem. Soc.*, **115**, 1210 (1919).
From the acid chloride with aniline in aqueous sodium hydroxide.
See: Wild, pp. 181, 219.
From the acid and aniline at high temperatures.
See: Vogel, p. 362.

---

*Derivatives recommended for first trial.
WARNING: This is not an instruction manual. References should be consulted for the preparation of derivatives.

From the sodium salt of the acid with aniline and concentrated hydrochloric acid.
See: Shriner, p. 201; Wild, p. 154.
From the acid anhydride with aniline without solvent.
See: Linstead, p. 17; Vogel, p. 377; Wild, p. 185.
From the acid anhydride with aniline in benzene.
See: Linstead, p. 15; Wild, p. 185.

*p-Toluidide.**

$$RCOOH \xrightarrow{SOCl_2} RCOCl \xrightarrow{p\text{-}CH_3C_6H_4NH_2} RCONHC_6H_4CH_3\text{-}p \ + \ p\text{-}CH_3C_6H_4NH_3{}^+Cl^-$$

$$RCOOH \ + \ p\text{-}CH_3C_6H_4NH_2 \longrightarrow RCONHC_6H_4CH_3\text{-}p \ + \ H_2O$$

$$(RCO)_2O \ + \ p\text{-}CH_3C_6H_4NH_2 \longrightarrow RCONHC_6H_4CH_3\text{-}p \ + \ p\text{-}CH_3C_6H_4NH_3{}^+RCOO^-$$

From the acid chloride with *p*-toluidine in ether or benzene.
For directions and examples see: Cheronis, pp. 441, 444, 458; Linstead, p. 14; Shriner, pp. 200–1; Vogel, p. 361.
From the acid and *p*-toluidine at high temperatures.
See: Cheronis, pp. 441, 442–3; Vogel, p. 362.
From the sodium salt of the acid, *p*-toluidine and concentrated hydrochloric acid.
See: Shriner, p. 201; Wild, p. 154.
From the acid anhydride with *p*-toluidine without solvent.
See: Cheronis, p. 459; Linstead, p. 17.
From the acid anhydride with *p*-toluidine in benzene.
See: Wild, p. 185.

*1- and 2-Naphthylamide.**

$$RCOOH \xrightarrow{SOCl_2} RCOCl \xrightarrow{1\text{- or }2\text{-}C_{10}H_7NH_2} RCONHC_{10}H_7 \ + \ C_{10}H_7NH_3{}^+Cl^-$$

Naphthylamide
(1- or 2-)

From the acid chloride with the naphthylamine.
For directions and examples see: Cheronis, p. 446; P. W. Robertson, *J. Chem. Soc.,* **115,** 1210 (1919).

*p-Nitrobenzyl ester.**

$$RCOONa \ + \ p\text{-}NO_2C_6H_4CH_2X \rightarrow RCOOCH_2C_6H_4NO_2\text{-}p \ + \ NaX \quad (X = Cl, Br, I)$$

p-Nitrobenzyl ester

From an aqueous solution of the sodium salt of the acid, with the *p*-nitrobenzyl halide in ethanol.
For directions and examples see: Cheronis, pp. 447, 448; Shriner, p. 200; Vogel, p. 362; Wild, pp. 144–5.
From an aqueous solution of the sodium salt of the acid with *p*-nitrobenzyl bromide in acetone.
See: F. F. Blicke and F. D. Smith, *J. Amer. Chem. Soc.,* **51,** 1947 (1929).
From the sodium or the potassium salt of the acid and *p*-nitrobenzyl bromide in 1:2 water-ethanol.
See: E. E. Reid, *J. Amer. Chem. Soc.,* **39,** 124 (1917).
From the sodium or the potassium salt of the acid and *p*-nitrobenzyl chloride or iodide in 1:2 water-ethanol.
See: J. A. Lyman and E. E. Reid, *J. Amer. Chem. Soc.,* **39,** 701 (1917).

*p-Bromophenacyl ester.**

$$RCOONa \ + \ p\text{-}BrC_6H_4COCH_2Br \longrightarrow RCOOCH_2COC_6H_4Br\text{-}p \ + \ NaCl$$

p-Bromophenacyl
bromide

$$RCOOH \ + \ p\text{-}BrC_6H_4COCHN_2 \xrightarrow{CuCl_2} RCOOCH_2COC_6H_4Br\text{-}p \ + \ N_2$$

p-Bromodiazoaceto-
phenone

p-Bromophenacyl ester

From the sodium salt of the acid and *p*-bromophenacyl bromide in aqueous ethanol.

*Derivatives recommended for first trial.
WARNING: This is not an instruction manual. References should be consulted for the preparation of derivatives.

For directions and examples see: Cheronis, pp. 447, 448; Linstead, p. 14; Shriner, p. 200; Vogel, p. 362; Wild, p. 146.

From the sodium salt of the acid (neutralization with sodium carbonate) with *p*-bromophenacyl halide in 1:2 water-ethanol.

See: W. L. Judefind and E. E. Reid, *J. Amer. Chem. Soc.*, **41**, 1043 (1920).

From the sodium salt of the acid (neutralization with sodium hydroxide) with *p*-bromophenacyl bromide in 95% ethanol.

See: R. M. Hann, E. E. Reid and G. S. Jamieson, *J. Amer. Chem. Soc.*, **52**, 818 (1930); C. G. Moses and E. E. Reid, *J. Amer. Chem. Soc.*, **54**, 2101 (1930).

From the acid and *p*-bromodiazoacetophenone in dioxane in the presence of catalytic amounts of cupric chloride.

See: J. L. E. Erickson, J. M. Dechary and M. R. Kesling, *J. Amer. Chem. Soc.*, **73**, 5301 (1951).

p-Phenylphenacyl ester.*

$$RCOONa + p\text{-}C_6H_5C_6H_4COCH_2Br \longrightarrow RCOOCH_2COC_6H_4C_6H_5\text{-}p + NaBr$$

p-Phenylphenacyl bromide

$$RCOOH + p\text{-}C_6H_5C_6H_4COCHN_2 \xrightarrow{CuCl_2} RCOOCH_2COC_6H_4C_6H_5\text{-}p + N_2$$

p-Phenyldiazoaceto-
phenone

p-Phenylphenacyl ester

From the sodium salt of the acid (neutralization with sodium carbonate) and *p*-phenylphenacyl bromide in aqueous alcohol.

For directions and examples see: Linstead, p. 14; Vogel, p. 363; N. L. Drake and J. Bronitsky, *J. Amer. Chem. Soc.*, **52**, 3715 (1930).

From the sodium salt of the acid (neutralization with sodium hydroxide) and *p*-phenylphenacyl bromide in aqueous alcohol.

See: Shriner, p. 200; N. L. Drake and J. P. Sweeney, *J. Amer. Chem. Soc.*, **54**, 2059 (1932).

For dibasic acids: from the acid, ethylamine and *p*-phenylphenacyl bromide in aqueous ethanol.

See: Wild, p. 147; N. L. Drake and J. P. Sweeney, *J. Amer. Chem. Soc.*, **54**, 2059 (1932).

From the acid and *p*-phenyldiazoacetophenone in dioxane in the presence of catalytic amounts of cupric chloride.

See: J. L. E. Erickson, J. M. Dechary and M. R. Kesling, *J. Amer. Chem. Soc.*, **73**, 5301 (1951).

Methyl ester.

$$RCOOH + CH_3OH \xrightarrow{H_2SO_4} RCOOCH_3 + H_2O$$

$$RCOOH + CH_2N_2 \longrightarrow RCOOCH_3 + N_2$$

Methyl ester

From the acid with methanol and a catalytic amount of sulfuric acid.

For directions and examples see: Linstead, p. 16; Vogel, p. 383.

From the acid and diazomethane in ether.

See: B. Eistert, in *Newer Methods of Preparative Organic Chemistry*, Interscience, New York, 1948, p. 513.

Ethyl ester.

$$RCOOH + C_2H_5OH \longrightarrow RCOOC_2H_5 + H_2O$$

$$RCOOAg + C_2H_5I \longrightarrow RCOOC_2H_5 + AgI$$

$$RCOOH \xrightarrow{SOCl_2} RCOCl \xrightarrow{C_2H_5OH} RCOOC_2H_5 + HCl$$

Ethyl ester

From the acid and ethanol in the presence of a catalytic amount of sulfuric acid.

For directions and examples see: Vogel, pp. 383, 385, 386, 387.

From the silver salt of the acid with ethyl iodide.

See: Vogel, p. 388.

---

*Derivatives recommended for first trial.
WARNING: This is not an instruction manual: References should be consulted for the preparation of derivatives.

From the acid chloride and ethanol.

See: Vogel, p. 389.

NOTE: The same methods can be used for the formation of other esters.

S-Benzylthiuronium salt. *

$$RCOONa \; + \; [C_6H_5CH_2SC(NH_2)_2]^+Cl^- \; \rightarrow \; [C_6H_5CH_2SC(NH_2)_2]^+RCOO^- \; + \; NaCl$$

S-Benzylthiuronium chloride   S-Benzylthiuronium salt

From the sodium or the potassium salt of the acid and S-benzylthiuronium chloride in water.

For directions and examples see: Linstead, p. 15; Vogel, p. 36; Wild, p. 149; S. Veibel and H. Lillelund, *Bull. Soc. Chim.* [5], **5**, 1153 (1938), S. Veibel and K. Ottung, *Bull. Soc. Chim.* **6**, 1434 (1939).

From the sodium or the potassium salt of the acid in water or in aqueous ethanol with an ethanolic solution of S-benzylthiuronium chloride.

See: Cheronis, p. 449; Shriner, p. 202; J. J. Donleavy, *J. Amer. Chem. Soc.*, **58**, 1004 (1936).

Phenylhydrazide.

$$RCOOH \; + \; H_2NNHC_6H_5 \; \rightarrow \; RCONHNHC_6H_5 \; + \; H_2O$$

Phenylhydrazide

From the acid with phenylhydrazine without solvent.

For directions and examples see: Shriner, p. 201; Wild, p. 152; G. H. Stempel and G. S. Schaffel, *J. Amer. Chem. Soc.*, **64**, 470 (1942).

From the acid with phenylhydrazine in benzene.

See: Shriner, p. 201; Wild, p. 152.

---

*Derivatives recommended for first trial.

WARNING: This is not an instruction manual. References should be consulted for the preparation of derivatives.

TABLE XII. ORGANIC DERIVATIVES OF CARBOXYLIC ACIDS
a) Liquids 1) (Listed in order of increasing atmospheric b.p.)*

No.	Name	Boiling point, °C	Melting point, °C	n_D^{20}	D_4^{20}	p-Toluidide	Anilide	p-Bromophenacyl ester	Amide	Methyl ester	Ethyl ester	Miscellaneous
1	**Thioacetic acid**	93	. . . .	. . . .		1.074_4^{10}	130	76	. . . .	108	. . . . . .	. . . . . .	
2	**Formic acid**	100.7	8.4	1.37137	1.22026	53	50	140; 135	. . . . . .	. . . . . .	. . . . . .	p-Nitrobenzyl ester, 31
3	**Acetic acid** (Ethanoic acid) . . .	118.2	16.6	1.36976; 1.3721	1.04926	153; 147	114	86.0	82	. . . . . .	. . . . . .	p-Nitrobenzyl ester, 78
4	**Difluoroacetic acid**	134–5	. . . . . .	. . . . . .	. . . .	. . . . . .	. . . . . .	. . . . . .	52			. . . .
5	**Acrylic acid**	141; 140	13	1.4224	1.0621_4^{16}	141	104–5, w.	. . . . . .	84–5, pet. eth.			. . . .
6	**Propionic acid** (Propanoic acid)	141	−20.8	1.3868	0.99336	126; 123	106	63.4	81	. . . . . .	. . . . . .	p-Nitrobenzyl ester, 31
7	**Propiolic acid**	144d.	18	. . . .	1.139_{15}^{15}	. . . . . .	87	. . . . . .	61–2			. . . .
8	**Isobutyric acid** (Isobutanoic acid)	154.7	−46.1	1.3920	0.94791	108.5–9.5	105	76.8	128; 129			. . . .
9	**Methacrylic acid**	161	16	1.429	1.015	. . . . . .	. . . . . .	. . . . . .	102–6			p-Bromoanilide, 116
10	**n-Butyric acid** (n-Butanoic acid)	162.5; 164	−5.5; −8	1.3983; 1.3979	0.95790	75	96; 97	63	115–6			p-Nitrobenzyl ester, 35
11	**Pyruvic acid** (α-Oxopropionic acid)	165d.; 80^{25}	13.6	1.4138	1.2668_4^{15}	109; 130	104, subl.	. . . . . .	124–5; 145			2,4-Dinitrophenylhydrazone, 218, yel., al.
12	**Vinylacetic acid** (3-Butenoic acid)	169; 163	−35	1.4221	1.0094	. . . . . .	58	. . . . . .	73			. . . .
13	**Isocrotonic acid** (cis-(β)-Crotonic acid; cis-2-Butenoic acid)	169	15	1.4456	1.0265	132	101–2	81	101–2			
14	**d,l-2-Methylbutanoic acid** (Ethylmethylacetic acid)	176–7; 174	. . . . . .	1.4052	0.938_{20}^{20}	92.5–3.0	110	55	112			
15	**Isovaleric acid** (3-Methylbutanoic acid)	176.5	−30.0	1.4043	0.92623	106–7	109.5 (cor.)	68.0	135; 137			
16	**n-Amylpropiolic acid** (1-Heptyne-1-carboxylic acid) .	180–220d.	f.p.: 2–5	. . . .		68, bz.	. . . . . .		91			Nitrile, b.p.: 194–6; o-Toluidide, 60, pet. eth.
17	**3,3-Dimethylbutanoic acid** (tert-Butylacetic acid)	184; 96^{26}	6.7	1.4096	0.9124	134	132, et. ac.-pet. eth.	. . . . . .	132	b.p.: 126		
18	**d,l-α-Chloropropionic acid** . . .	186	. . . . . .	. . . . . .	. . . . . .	124	92	. . . . . .	80			
19	**Cyclopropanecarboxylic acid** .	186; 182–4	17; 18–9	1.43901	1.0885	. . . . . .	125			. . . . .	b.p.: 134; n_D^{20}: 1.41902; D_4^{15}: 0.96078	. . . .
20	**n-Pentanoic acid** (n-Valeric acid)	186.4	−34.5	1.4086	0.93922	74	63	75	106			. . . .
21	**2,2-Dimethylbutanoic acid** (Dimethylethylacetic acid) .	187; 190	−15.0	1.4141; 1.4145	0.9276	83.0–.5	92; 90–1	. . . . . .	103			p-Phenylphenacyl ester, 86
22	**Allylacetic acid** (4-Pentenoic acid)	188–9	. . . . . .	1.4341^7; 1.4283	0.9843_4^{18}	. . . . . .	. . . . . .	. . . . . .	94, b.p.: 230		b.p.: 144–6	. . . .
23	**Cyclopropylacetic acid**	190^{750}	. . . . . .	1.4320^{25}	. . . .							p-Phenylphenacyl ester, 83
24	**d,l-2,3-Dimethylbutanoic acid** (Isopropylmethylacetic acid)	191.7	−1.5	1.4146	0.9275	112.6	78.4	. . . . . .	132			p-Phenylphenacyl ester, 74
25	**Dichloroacetic acid**	194	5–6	1.4659	1.5634	153	118	99	98, subl.	. . . . . .	. . . . . .	

*Derivative data given in order: m.p., crystal color, solvent from which crystallized.

TABLE XII. ORGANIC DERIVATIVES OF CARBOXYLIC ACIDS
a) Liquids 1) (Listed in order of increasing atmospheric b.p.)* (Continued)

No.	Name	Boiling point, °C	Melting point, °C	n_D^{20}	D_4^{20}	p-Toluidide	Anilide	p-Bromophenacyl ester	Amide	Methyl ester	Ethyl ester	Miscellaneous
26	Cyclobutanecarboxylic acid ..	195		1.4403^{25}	1.0599	....	......	......	......	152–3	b.p.: 136.0 –.5	b.p.: 159–62	....
27	2-Ethylbutanoic acid (Diethylacetic acid)	195	−31.8	1.4132	0.9239	116.2	127.5	......	112; 107	......	....	
28	d,l-2-Methylpentanoic acid (Methyl-n-propylacetic acid)	195–6	......	1.4136	0.9230	81	95	......	79.6	......	....	
29	d,l-3-Methylpentanoic acid ...	197.5	−41.6	1.4159	0.9262	74.8	87; 88	......	124.9	......	....	
30	4-Methylpentanoic acid (Isocaproic acid; Isobutylacetic acid)	199.1^{752}	−33	1.4144	0.9225	63.0	112.0; 110.5; 111.5	77.3	120–1			
31	Methoxyacetic acid (Glycolic acid methyl ether)	204; 203	......	1.41677	1.1768	......	58, pet. eth.	......	96.5–7.0; 92–4			
32	2-Ethyl-2-methylbutanoic acid (Diethylmethylacetic acid) ..	204	......	1.4256	....	......	......	......	78			
33	Hexanoic acid (n-Caproic acid)	205.35	−3.9; f.p.: −1.5–2	1.41635	0.93568	74–5	94–5	72.0	100; 101		b.p.: 166–7	
34	Ethoxyacetic acid (Glycolic acid ethyl ether)	206–7	......	1.41937	1.1021	32, eth.	95; 92	104.8	80–2			
35	5-Methylhexanoic acid......	207^{752}		1.4220	....	....	75		103			
36	2-Ethylpentanoic acid (Ethyl-n-propylacetic acid)..	209				129	94		104–5			
37	2-Methylhexanoic acid (n-Butylmethylacetic acid).....	209.6	......	1.4189^{25}	....	85	98		73; 70–2.5			
38	α-Chloroisovaleric acid	20–2									b.p.: 178–9	Nitrile, b.p.: 154–5; Chloride, b.p.: 149
39	α-Bromobutanoic acid	217d.				92	98	......	112; 108			
40	4-Methylhexanoic acid.......	$217-8^{754}$	....	1.4211	0.9194	......	76.5		98	......	....	
41	2,2-Dimethylhexanoic acid ...	218	....	....	....	......	......	......	89			
42	4-Ethyl-4-methylbutanoic acid (active-Amylacetic acid).	221			0.9149				b.p.: 158–64	b.p.: 173–9		$[\alpha]_D^{15}$: +7.6 in me. al.
43	2-Chloro-n-valeric acid (2-Chloropentanoic acid)	222	....	....	....	......	......	......	b.p.: 160	b.p.: 185–6		Nitrile, b.p.: 160
44	n-Heptanoic acid (n-Heptoic acid)	223.0	−7.46	1.4234	0.91808	81	70; 65	72.0	96; 96.5	......	....	
45	2-Ethylhexanoic acid (α-Ethylcaproic acid)	228	....	....	....	......	......	......	102			p-Phenylphenacyl ester, 53–4; 49.5–50
46	Cyclohexylacetic acid	237	....	....	....	......	......	......	172	......	....	
47	n-Caprylic acid (n-Octanoic acid)	237; 239.3	16.3	1.4268	0.90884	70	57	67.4	110; 106	......	b.p.: $207-8^{753}$	
48	Pelargonic acid (n-Nonanoic acid)	254.4	12.3	1.43446_{He}^{15}, yel.	0.90552	84	57	68.5	99			
49	d-Citronellic acid (2,6-Dimethyl-1-octene-8-carboxylic acid)	257			0.9308	......	......		84–5	......	b.p.: $113-5^{12}$; $[\alpha]_D^{16}$: +0.3	$[\alpha]_D$: +21; Nitrile, b.p.: 230, D^{20}: 0.8645
50	2-Phenylpropionic acid	265	....	....	....	......	......	......	92	......	....	
51	4-Acetylbutanoic acid (γ-Acetobutyric acid)	275d.; 195–200^{65}	13–4	....	....	123, w.	......	......	114, chl.			Semicarbazone, 175d. (+1 H_2O), w.; Oxime, 104–5. bz.

*Derivative data given in order: m.p., crystal color, solvent from which crystallized.

TABLE XII. ORGANIC DERIVATIVES OF CARBOXYLIC ACIDS
a) Liquids 2) (Reduced pressure b.p. only) (Listed in order of increasing amide m.p.)*

No.	Name	Amide	Boiling point, °C	Melting point, °C	n_D^{20}	D_4^{20}	p-Toluidide	Anilide	p-Bromophenacyl ester	Methyl ester	Ethyl ester	Miscellaneous
1	3-Ethoxypropionic acid	51	120[17]	......	1.4216							
2	3-Heptynoic acid	67	102[2]	14	1.4635[25]							
3	2-Heptynoic acid	68-9, al.	135[20]	......	1.4619	0.978				b.p.: 91-3[19]; n_D^{20}: 1.4455; D_4^{20}: 0.937		
4	trans-Oleic acid	75-6	216[5]; 250 (super-heated steam)	α: 13.36; β: 16.25	1.4597		42.5	41	40; 46			
5	2-Fluoropropionic acid	76	60[8]	......								
6	2-Azidoisovaleric acid (2-Triazoisovaleric acid)	78-9, bz.	82[0.1]	......	......	1.0638_{33}^{33}	......				b.p.: 82[16]; D_{20}^{20}: 1.0295	......
7	d,l-Lactic acid	78.5-9.0 (cor.), bz.-al. (3:1)	122[15]	18	......		107	58.5-9.0, w.	112.8	144-8	154	......
8	d,l-2-Azidopropionic acid (d,l-2-Triazopropionic acid)	80, bz.	121.5[20]								b.p.: 70[16]; n_D^{25}: 1.428-57; D_{23}^{23}: 1.065	Explodes on heating
9	2-Ethyl-3-Hexenoic acid	80	132[19]	......	......							
10	4-Phenoxybutanoic acid	80	197[18]	......	......							
11	2-Methoxypropionic acid	81	89[10]	......	......							
12	l-Citronellic acid (2,6-Dimethyl-1-octen-8-carboxylic acid)	84-5	117.9[0.6]	......	1.4563[24]	0.9274_{25}^{25}	93-4	76		b.p.: 86[1.1]	......	$[\alpha]_D^{24}$: -6.6
13	2-Ethyl-4-methylpentanoic acid (Ethylisobutylacetic acid)	89	115[20]									
14	2-Octynoic acid	90	133[10]	......	1.4595		60					
15	6-Methyloctanoic acid	91	149[23]	......	1.4337							
17	3-Methylhexanoic acid	98	112[16]	......	1.4222							
18	2,3-Dimethylpentanoic acid	102	92[15]	......								
19	2-Cyanopropionic acid	105; 81	142-5[11]	......							b.p.: 192-3	
20	7-Methyloctanoic acid	106	105[2]	......								
21	1-Chlorocyclohexane carboxylic acid	110, me.al.-w.	138-40[13]									Ethylamide, 53
22	2-Cyanobutanoic acid	113	153-5[15]	......							b.p.: 207-9	
23	Methylneopentylacetic acid	123	108[14]	......								
24	2-Isopropylbutanoic acid (2-Ethyl-3-methylbutanoic acid)	135	105[15]	......								
25	5-Cyclopentylpentanoic acid	136	123[4.5]	......								
26	2-Methylcyclopentanecarboxylic acid	148	107[9]	......	1.4504[22]							
27	3,4,4-Trimethylpentanoic acid	167	98[4]	......	1.4320[21]							
28	cis-4-Methylcyclohexanecarboxylic acid	175	130[13]	......								
29	Cyclopentanecarboxylic acid	179	123[27]	......								

*Derivative data given in order: m.p., crystal color, solvent from which crystallized.

TABLE XII. ORGANIC DERIVATIVES OF CARBOXYLIC ACIDS
b) Solids (Listed in order of increasing m.p.)*

No.	Name	Melting point, °C	Boiling point, °C	p-Toluidide	Anilide	p-Nitro-benzyl ester	p-Bromo-phenacyl ester	Amide	Methyl ester	Ethyl ester	Miscellaneous
1	β-Bromobutyric acid (β-Bromo-butanoic acid)............	20	122^{16}	....	....	......	....	92–3	....	b.p.: 183–4	....
2	trans-β-Ethyl-α-methylacrylic acid	24	112^{12} cis: 94^{10}	....	....	......	91; cis: 46	80	....	....	n$_D^{20}$: 1.4578; cis: n$_D^{25}$: 1.4485
3	d,l-α-Azidobutyric acid (d,l-α-Triazobutyric acid)	24	81$^{0.2}$	....	....	......	....	38–9, bz.-pet. eth.	....	b.p.: 64^7, D$_{20}^{20}$: 1.038	....
4	Undecylenic acid (10-Undecen-1-oic acid; 10-Hendecen-1-oic acid)...................	24.5	275	....	....	......	....	87	....	....	Cu salt, 232–4; Pb salt, 80
5	2-Ethoxybenzoic acid (Salicylic acid ethyl ether)............	24.5–5.5	300d.	....	....	......	....	132	....	....	....
6	d,l-α-Bromopropionic acid	25.7	203.5	125	99; 100	......	....	123	b.p.: 145–50	b.p.: 159–60d.	....
7	n-Undecylic acid (n-Undecanoic acid; n-Hendecanoic acid)....	28.5; α: 13.4; β: 16.3	280; 284	80	71	......	68.2	103	....	....	....
8	α-Chloroisobutyric acid (α-Chloroisobutanoic acid)	31	118^{50}	....	69–70, al.	......	....	....	b.p.: 133–5	b.p.: 148–9	....
9	α-Azidoisobutyric acid (α-Azidoisobutanoic acid; α-Triazoisobutyric acid)	31	75$^{0.2}$	....	....	......	....	93–4	....	b.p.: 71^{16}, D$_{20}^{20}$: 1.0344	D$_{33}^{33}$: 1.1433
10	Cyclohexanecarboxylic acid (Hexahydrobenzoic acid)	31; 30–1	233	....	146 (cor.)	......	....	185–6	....	....	....
11	α-Ketobutyric acid (2-Oxo-butanoic acid)............	31	78^{25}	....	....	......	....	117	....	....	n$_D^{20}$: 1.3975; p-Nitrophenyl-hydrazone, 194
12	Fluoroacetic acid	31–2	167–9	....	....	......	....	108	....	....	....
13	Capric acid (n-Decanoic acid).................	31.5	268–70	78	70	......	67, al.; 66	108; 100.1; 98; 99	b.p.: 224	b.p.: 243–5	n$_D^{40}$: 1.42855
14	Bromochloroacetic acid	31.5	215 sl. d.	....	....	......	....	126; 117	....	b.p.: 174d.	Phenyl ester, 46.5, b.p.: 266
15	2-Hexenoic acid	32	....	....	110	......	....	....	....	....	....
16	n-Butylmethylglycolic acid.....	33	....	....	....	......	....	58	....	....	....
17	cis-13-Docosenoic acid (Erucic acid).....................	33–4	264^{15}	75–8	55	......	62.5; 61.0	84	....	....	D: 0.860^{55}
18	Levulinic acid (γ-Ketovaleric acid; β-Acetylpropionic acid) .	33–5 deliq.	245–6	108–9, w.	102, w.	61	84	107–8d.	....	....	Oxime, 96
19	Pivalic acid (Trimethylacetic acid)	35.5	163–4	119–20	132–3; 128 (cor.)	......	75–6	155–7; 153–4, et. ac.-pet. eth.	....	....	....
20	d,l-α-Methylhydrocinnamic acid (α-Benzylpropionic acid).	36.5	272	130; d: 115–6	....	......	....	107–8; d: 113–4	....	....	....
21	1-Cyclohexenylcarboxylic acid .	38	107^3	....	....	......	....	128	....	....	....
22	n-Hexylmethylglycolic acid	40	....	....	....	......	....	59	....	....	....
23	5-Acetyl-n-valeric acid (δ-Acetylpentanoic acid).......	40–2; 31–2	250–3^{280}	....	....	......	....	....	....	....	Semicarbazone, 144–6, ac. a.
24	d, l-α-Campholytic acid (1,5,5-Trimethylcyclopenten-4-carboxylic acid)..........	40.5	162–5^{45}; l: 240–3	....	....	......	....	103, w.	b.p.: 200	....	Nitrile, b.p.: 200–5
25	β-Chloropropionic acid	41, w.; 39, lgr.	204	....	....	......	....	....	b.p.: 155–7, D^0: 1.198	b.p.: 162, D$_4^{20}$: 1.1086, n$_D^{20}$: 1.42537	Nitrile, b.p.: 175–6, D$^{18.5}$: 1.1443
26	d,l-α-Ethylphenylacetic acid (d,l-α-Phenylbutanoic acid) ..	42	270	....	....	......	....	85–7; 83	....	....	....

*Derivative data given in order: m.p., crystal color, solvent from which crystallized.

TABLE XII. ORGANIC DERIVATIVES OF CARBOXYLIC ACIDS

b) Solids (Listed in order of increasing m.p.)* (Continued)

No.	Name	Melting point, °C	Boiling point, °C	p-Toluidide	Anilide	p-Nitrobenzyl ester	p-Bromophenacyl ester	Amide	Methyl ester	Ethyl ester	Miscellaneous
27	α-Ethylpimelic acid	43	223[17]	......	145	......	......	......	....	....	....
28	Tridecylic acid (n-Tridecanonic acid)	43; 41.6	312; 177[10]	88	80	......	75.0	100	....	....	....
29	Lauric acid (n-Dodecanoic acid)	44; 42	299	87	78	......	76	100; 99	....	....	....
30	d,l-α-Bromoisovaleric acid	44	230d.	124	116	......	......	133, bz.	b.p.: 174	b.p.: 186	....
31	Elaidic acid (trans-Oleic acid)	44–5; 51	234[15]	......	......	......	65	93–4; 89–90	....	....	p-Phenylphenacyl ester, 73.5; p-Chlorophenacyl ester, 56
32	4-Cyanobutanoic acid (γ-Cyanobutyric acid)	45	....	......	....	......	......	69–70, sealed tube	....	b.p.: 245	....
33	Dimethylneopentylacetic acid	45	230[732]	......	....	......	......	71	....	....	....
34	trans-2-Methyl-2-butenoic acid (Angelic acid)	45–6	185 (cor.)	......	126, bz.	......	......	127–8	....	....	Isobutyl ester, b.p.: 177; 2-Naphthylamide, 135, bz.; Heating 2 hours in sealed tube → tiglic acid, 64–5
35	Dibromoacetic acid	48	232–5	......	....	......	......	156	....	....	....
36	α-Bromoisobutyric acid (α-Bromoisobutanoic acid)	48–9	198–200	92.5, al.	83, al.-w.	......	......	148	....	....	o-Toluidide, 63
37	tert-Butylpropiolic acid	48–9	110[10]	......	....	......	......	....	b.p.: 66[13], D^0: 0.9209	b.p.: 75[15], D^0: 0.9209	....
38	Hydrocinnamic acid (β-Phenylpropionic acid)	48.7; 40	279–80 (cor.)	135	98; 96	36.3	104	105; 82	....	....	p-Phenylphenacyl ester, 95
39	β-Cyanopropionic acid	48–50	....	......	....	......	......	97, sealed tube	b.p.: 215, n_D^{20}: 1.42427, D^{20}: 1.0792	b.p.: 220[754] D^{20}: 1.0353	....
40	Benzylpyruvic acid	49–50 ($+\frac{1}{2}$ H_2O), w.	....	......	....	......	......	180	....	....	Semicarbazone, 175d.; Oxime, 65; Phenylhydrazone, 144–5
41	Bromoacetic acid	50	208	......	131	88	......	91	b.p.: 144d.	b.p.: 168–9	....
42	2-Pentynoic acid (Ethylpropiolic acid)	50	100[10]	......	....	......	......	146	....	....	....
43	γ-Phenylbutyric acid	52	290	......	....	......	......	84	....	....	....
44	n-Pentadecylic acid (n-Pentadecanoic acid)	52.3	212[16]	......	78	39.5–40 (cor.)	77.2	102.5	....	....	....
45	β-Campholenic acid	53.5	245	......	....	......	......	86	....	222–5	Nitrile, 225, D^{20}: 0.9093
46	Myristic acid (Tetradecanoic acid)	53.9	202[16]	93	84	......	81	103	....	....	....
47	Trichloroacetic acid	57–8	197.5	113	97; 94	80	......	141	b.p.: 153.8	b.p.: 168	Phenylhydrazide, 123
48	α-Acetoxypropionic acid (O-Acetyllactic acid)	57–60; 39–40	167–70[78]	......	....	......	......	....	....	....	Nitrile, b.p.: 172–3
49	β-Acetylglutaric acid	58	....	......	....	......	......	mono: 141–2, al.-eth.	89	....	....
50	sec-n-Amylmalonic acid (2-Ethylbutane-1,1-dicarboxylic acid; sec-n-Pentylmalonic acid)	58, bz.	....	......	di: 219–20	......	......	....	....	di: b.p.: 243–5	....

*Derivative data given in order: m.p., crystal color, solvent from which crystallized.

TABLE XII. ORGANIC DERIVATIVES OF CARBOXYLIC ACIDS

b) Solids (Listed in order of increasing m.p.)* (Continued)

No.	Name	Melting point, °C	Boiling point, °C	p-Toluidide	Anilide	p-Nitro-benzyl ester	p-Bromo-phenacyl ester	Amide	Methyl ester	Ethyl ester	Miscellaneous
51	*trans*-Brassidic acid	59.7	256[10]	....	78	......	94.2	94	....	....	....
52	**5-Phenylpentanoic acid**	60	....		90			109			
53	**β-Cyclohexylacrylic acid**	60	154[11]					159			
54	**β-Chloroisocrotonic acid**	61	195 (subl.)		108			110	b.p.: 142	b.p.: 161	
55	**Margaric acid** (*n*-Heptadecanoic acid).............	61.2	231[16]	....	....	48.5–9.0 (cor.)	82.6	108; 106			
56	**Chloroacetic acid**	α: 61.3; β: 56.2; γ: 52.5	189	162	136–7	......	104	121	b.p.: 130	b.p.: 145–6	....
57	**β-Bromopropionic acid**	62.5	....					111	b.p.: 80[27], D^{17}: 1.4897	b.p.: 70[12], D^{15}: 1.2609	2-Naphthylamide, 174
58	**Palmitic acid** (Hexadecanoic acid).................	62.7	222[16]	98	90.6, al.	42.5	86; 82	106–7; 105.3, al.	....	....	....
59	*cis*-2-Methyl-2-butenoic acid (Tiglic acid)...............	64.5–5.0	198.5 (cor.)	70.0–1.5	77, pet. eth.	64	68	75–6, bz.			
60	**Cyanoacetic acid**............	66	....		198–9	......	......	119–20	b.p.: 200	b.p.: 207	Nitrile, 29–30, b.p.: 218–9
61	**Benzoylformic acid**	66						91			Nitrile, 32–3, b.p.: 206–8; 2,4-Dinitrophenyl-hydrazone, 196–7d. (cor.); Phenylhydrazone, 64
62	**Acetoxyacetic acid**	66–8, bz.	145[12]	....	89–90, w.					b.p.: 179, D^{17}: 1.0993	....
63	**2,3-Dibromopropionic acid**	67, stab.; 51, unst.	160[20]	....	....			130			
64	**3,3-Dimethylacrylic acid**	67	106[20]					108			
65	**2-Furylacetic acid**............	67	....		85						
66	**3,3,4,4-Tetramethylpentanoic acid**	67	....					138			
67	**4-Ketocyclohexanecarboxylic acid** (4-Oxocyclohexane-carboxylic acid)............	67–8, bz.-pet. eth.							b.p.: 140[20]	b.p.: 158[40]	Semicarbazone, 200d.; Oxime, 147, eth.
68	**d,l-2-Phenyllactic acid**	68 (+½ H₂O), w.; 94 (anh.); *d*: 116–7, w.; *l*: 115–6, bz.	....					101–2, di-chloro-ethylene; *l*: 62.5–3.5, bz.	....		*d,l*-Et. eth., 60–2, lgr.; *d*-Quinine salt, 216d., al.; *l*-1-Menthyl ester, 55.5–6.0
69	*d*-Chaulmoogric acid (*d*-ω-Cyclopentyltridecanoic acid)	68.5	247–8[20]	100	89	......	......	106, al.	22	b.p.: 230[20], D_4^{15}: 0.9064	$[\alpha]_D$: +62 in chl.
70	**Stearic acid** (Octadecanoic acid).....................	70–1; 69.6	....	102	95.5, al.	......	92	109; 108.4, al.	....		
71	*trans*-Crotonic acid (*trans*-2-Butenoic acid).............	72, w.	189 (cor.)	132, bz.	118, w.; 115	67.4	95–6	159–60, bz.	b.p.: 121	b.p.: 138, D_4^{20}: 0.9175, n_D^{20}: 1.42524	....

*Derivative data given in order: m.p., crystal color, solvent from which crystallized.

No.	Name	Melting point, °C	Boiling point, °C	p-Toluidide	Anilide	p-Nitro-benzyl ester	p-Bromo-phenacyl ester	Amide	Methyl ester	Ethyl ester	Miscellaneous
72	γ-Bromocrotonic acid.........	74, lgr.	....	....	....	......	......	101	b.p.: 87[15], D$_4^{19}$: 1.490, n$_D^{19}$: 1.498	b.p.: 97–8[15], D$_4^{16}$: 1.402, n$_D^{16}$: 1.490	....
73	Caproylacetic acid..........	74d.	....	....	....	......	......	100	b.p.: 118[19], D$_4^0$: 0.9916	b.p.: 127[19], D^0: 0.9721	Nitrile, b.p.: 127–8[14], D[15]: 0.9914
74	3-Ketocyclohexanecarboxylic acid (3-Oxocyclohexane-carboxylic acid)............	75–6, bz.	195–7[20]	....	....	......	......			b.p.: 138[18]	Semicarbazone, 183–4, al.; Oxime, 170d., w.; Phenylhydra-zone, 125, yel.
75	2-Thienylacetic acid.........	76						148			
76	sec-Butylmalonic acid (Iso-pentane-1,1-dicarboxylic acid)...................	76	....	....	....	......	......	di: 242	di: b.p.: 217–8[748]	di: b.p.: 245–50[762]	....
77	Phenylacetic acid...........	76.5, subl.	256.5 (cor.)	135–6	117–8	65	89	156	....	....	
78	Eicosanoic acid (Arachidic acid).....................	77; 75	204[1]	96	92	......	89	108–9			
79	Glycolic acid (Hydroxyacetic acid).....................	78–9; 80	....	143, w.	97, w.	106.8	138	120, al.-et. ac.		....	On prolonged heating at 100° → anh., 128–30
80	α-Hydroxyisobutyric acid (2-Hydroxyisobutanoic acid; Dimethylglycolic acid)......	79	212	132–3, w.	136, w.	80.5	98, acet.	....			
81	α-Methylcinnamic acid......	81; 74						128			
82	2-Ketocyclohexanecarboxylic acid (2-Oxocyclohexane-carboxylic acid)...........	81–2, eth.								b.p.: 107–8[12]; 159–60[100]	Alcoholic sol. + FeCl$_3$ → blue color
83	n-Docosanoic acid (Behenic acid).....................	81–2	....	....	101–2	......	......	111	54	50	....
84	β-Iodopropionic acid........	82; 85						101; 142			
85	α-Benzoylpropionic acid.....	82–3, bz.-pet. eth.			137–8, al.			145–6			Phenylhydrazone, 100–4, br., bz.
86	Iodoacetic acid.............	83						95			
87	γ-Chlorocrotonic acid.......	83	117–8[13]					130–2, w.		b.p.: 191–3[750]	Nitrile, b.p.: 73[15], D^0: 1.1495
88	Lignoceric acid (Tetracosanoic acid).....................	84	....	....	......	......	91				p-Chlorophenacyl ester, 100
89	β-Methyladipic acid........	85; 91	223[18]		200						
90	α-Benzoylbutyric acid (α-Benzoylbutanoic acid).......	85–7						148–9		b.p.: 168–71[19]	Nitrile, b.p.: 134–5[3]
91	d,l-α-Bromophenylacetic acid..	86; 84	....	....	......	......	......	148; 144			Nitrile, 29, b.p.: 242d.
92	(4-Methoxyphenyl)acetic acid..	87; 84						189			
93	Dineopentylacetic acid.......	88						140			
94	Dibenzylacetic acid..........	89	....	175, abs. al.	155, abs. al.	......	......	128–9, bz.			
95	(2-Tolyl)acetic acid..........	90; 88						161			
96	α-Thienylglyoxylic acid......	91						88			
97	Δ⁵-Campholytic acid (4,5,5-Trimethylcyclopentene-1-carboxylic acid)...........	91						90, lgr.			
98	Citraconic acid (Methylmaleic acid)....................	92d.; 92–3; 91d., eth.-lgr.	....	mono: 170–1, yel., eth.	mono: 153; di: 175.5, al.	di: 70.6	....	di: 185–7d.	di: b.p.: 210–1	di: b.p.: 231	....

*Derivative data given in order: m.p., crystal color, solvent from which crystallized.

TABLE XII. ORGANIC DERIVATIVES OF CARBOXYLIC ACIDS

b) Solids (Listed in order of increasing m.p.)* (Continued)

No.	Name	Melting point, °C	Boiling point, °C	p-Toluidide	Anilide	p-Nitrobenzyl ester	p-Bromophenacyl ester	Amide	Methyl ester	Ethyl ester	Miscellaneous
99	2-Bromobenzoylformic acid	93–101	......	....	......	....	......	136–7, w.	....	....	Nitrile, 62–4, yel.; Oxime, 162–4d.
100	β-Chlorocrotonic acid........	94	206–11 sl. d.	....	123–4	....	......	100–1	b.p.: 64–7[14], D_4^{22}: 1.555, n_D^{20}: 1.463	b.p.: 180, 184; D_4^{20}: 1.1062, n_D^{20}: 1.459	1-Naphthyl-amide, 169–70
101	Phenyl-n-propylglycolic acid ...	94	......	....	......	....	......	132	....	....	
102	(2-Chlorophenyl)acetic acid....	95, w.	......	170	138.5	....	......	175, w.	....	....	Nitrile, 25, b.p.: 251; o-Toluidide, 174
103	(4-Tolyl)acetic acid	95; 91	159[15]	....	......	....	......	185	....	....	
104	o-Chlorohydrocinnamic acid ...	96.5, w.	......	....	......	....	......	119, bz.	b.p.: 255	....	Nitrile, b.p.: 267–8
105	2-Hydroxy-3-phenylpropionic acid	97; 96	......	....	......	....	......	112	....	....	
106	1,2,3,4-Tetrahydro-2-naphthoic acid	97	......	....	......	....	......	139	....	....	
107	Glutaric acid (1,3-Propane-dicarboxylic acid)..........	98	302–4	di: 218	di: 223–4	di: 69	di: 136.8	di: 175–6	....	....	
108	3-Phenoxypropionic acid	98	......	....	......	....	......	119	....	....	
109	α-Crotonic acid...............	99	212	....	......	....	......	112	161	176	Nitrile, b.p.: 136
110	Phenoxyacetic acid..........	98–9; 99–100	285d.	....	99, al.	....	148.5	101.5	....	....	
111	2-Benzofurylacetic acid	99	......	....	......	....	......	164	....	....	
112	1-Naphthylglycolic acid	99; 124–5	......	....	......	....	......	135	....	....	
113	Citric acid	100 (+1 H₂O); 153 (anh.)	......	tri: 189, al.	tri: 192	tri: 102	tri: 148	tri: 210–5d., w.	....	....	
114	2-Methoxybenzoic acid (o-Anisic acid; Salicylic acid methyl ether).............	100–1	200	....	......	....	113	129	....	....	
115	l-Malic acid (Hydroxysuccinic acid)..................	100–1	......	di: 206–7	di: 197	mono: 87.2; di: 124.5	di: 179	di: 156–7; d,l: 162–3	....	....	
116	Oxalic acid	101 (+2 H₂O) (rapid htng.); 189.5 (anh.); subl. at 150–60	......	mono: 169; di: 268	mono: 148–9; di: 254; 246, bz.	di: 204	......	mono: 219; di: 419d.	....	....	
117	Acetylpyruvic acid (2,4-Diketo-n-valeric acid; 2,4-Dioxo-pentanoic acid).............	101, bz.	......	....	......	....	......	131–2d., al.	63–4	....	....
118	n-Butylmalonic acid (Pentane-1,1-dicarboxylic acid)	101	......	....	di: 193	....	......	di: 200	....	di: b.p.: 235–40	Mononitrile, 122.5–6.5, w., subl.
119	α-Cyanohydrocinnamic acid (Benzylcyanoacetic acid).....	101–2	......	....	......	....	......	130	....	b.p.: 176–85[21]	....
120	2-Chloro-6-methylbenzoic acid	102, w.	......	....	......	....	......	167	....	....	Nitrile, 82–3, pet. eth.
121	Aleuritic acid (9,10,16-Trihydroxypalmitic acid)	102, w.	......	....	......	....	......	....	63–4	....	Hydrazide, 139–40; Azide, 50d., al.

*Derivative data given in order: m.p., crystal color, solvent from which crystallized.

TABLE XII. ORGANIC DERIVATIVES OF CARBOXYLIC ACIDS

b) Solids (Listed in order of increasing m.p.)* (Continued)

No.	Name	Melting point, °C	Boiling point, °C	p-Toluidide	Anilide	p-Nitrobenzyl ester	p-Bromophenacyl ester	Amide	Methyl ester	Ethyl ester	Miscellaneous
122	(2-Bromophenyl)acetic acid	103–4; 109	......	......	......	......	......	186–7	......	......	Nitrile, b.p.: 145–7[14]
123	Benzylacetic acid	103–4d.	......	......	107–8	......	......	113	......	......	Nitrile, 80–1; 1-Menthyl ester, 41
124	Pimelic acid	104–5, subl.	223[15]	di: 206, al.	mono: 108–9; di: 155–6, me. al.-w.	......	di: 136.6	di: 175	......	......	......
125	2-Toluic acid (2-Methylbenzoic acid)	104–5; 107–8	259[751]	144	125	90.7	57	142.8 (cor.)	......	......	......
126	Allylmalonic acid	105, eth.	......	......	......	46, al.	......	......	......	di: b.p.: 222–3	Mononitrile, b.p.: 223; Dinitrile, b.p.: 217–8
127	(4-Chlorophenyl)acetic acid	105–6; 104	......	190	164–5, al.	......	......	175, al.	......	32; b.p.: 260	o-Toluidide, 190, bz.: Nitrile, 30, b.p.: 265–7
128	Δ²-Cyclogeranic acid (1,5,5-Trimethylcyclohexene-6-carboxylic acid)	106, lgr.	138[11]	......	......	......	......	120–1, bz.-pet. eth.	......	b.p.: 101–2[10]	......
129	d-Campholic acid (d-1,2,2,3-Tetramethylcyclopentane-1-carboxylic acid)	106; d,l: 109	255	......	91	......	......	80; d,l: 90	b.p.: 208	b.p.: 220	Nitrile, 73; Anhydride, 56; d,l: 66
130	Atropic acid (1-Phenylacrylic acid)	106–7, w.	......	......	134	......	......	121–2, w.	......	b.p.: 124[16], n_D^{16}: 1.52605	......
131	l-Campholic acid	106–7	250	......	......	......	......	78–9	b.p.: 211	b.p.: 228	Anhydride, 57–8
132	Azelaic acid (Heptane-1,7-dicarboxylic acid)	106.5	>360 sl. d.; 237[15]	di: 201–2	mono: 107–8, dil. al.; di: 186–7	di: 43.8	di: 130.6	mono: 93–5; di: 175	......	......	......
133	Methylneopentylglycolic acid	109	......	......	......	......	......	116	......	......	......
134	trans-4-Methylcyclohexane-carboxylic acid	111	......	......	......	......	......	226	......	......	......
135	cis-α-Chloroallocinnamic acid	111	......	......	138–9, al.-w.	......	......	134, bz.	b.p.: 153–4[28]	b.p.: 157–8[10], D_4^{25}: 1.1569 n_D^{25}: 1.5525	......
136	Ethylmalonic acid	111	......	......	150	75	......	di: 214	......	......	......
137	3-Toluic acid (3-Methylbenzoic acid)	111–3; 110–1	263, subl.	118	126	86.6	108	94; 97	......	......	......
138	O-Benzoyllactic acid (Lactic acid benzoate)	112	......	......	......	......	......	124	......	b.p.: 288	Nitrile, b.p.: 269–70; 1-Naphthylamide, 155; 2-Naphthylamide, 177
139	2,4,6-Triethylbenzoic acid	113	......	......	......	......	......	156	......	......	......
140	2-Phenylbenzoic acid	113	......	......	......	......	......	177	......	......	......
141	(1-Naphthyl)glyoxylic acid	113	......	......	......	......	......	151	......	......	......
142	Bromomalonic acid	113d.	......	di: 217, ac. a.	......	......	......	di: 181, al.	di: b.p.: 215–25	di: b.p.: 233–5d.	Dinitrile, 65–6
143	(4-Bromophenyl)acetic acid	114, subl.	......	......	......	......	......	192–4	......	30	Nitrile, 46–7
144	2-Acetylbenzoic acid (Acetophenone-o-carboxylic acid)	114–5	......	......	......	......	......	116.5, w.	......	......	Oxime, 159; 2,4-Dinitrophenylhydrazone, 186

*Derivative data given in order: m.p., crystal color, solvent from which crystallized.

No.	Name	Melting point, °C	Boiling point, °C	p-Toluidide	Anilide	p-Nitro-benzyl ester	p-Bromo-phenacyl ester	Amide	Methyl ester	Ethyl ester	Miscellaneous
145	Pyrotartaric acid (Methyl-succinic acid)	115	....	164	mono: 159, et. ac.; 123, chl.; di: 200	....	....	di: 225	....	....	....
146	2-Phenoxypropionic acid	115–6; 112–3	....	115	117; 118–9	....	....	132–3; 130	....		
147	3-Benzoylpropionic acid	116	....	....	150; 145	....	....	145–6, w.	....	18–9	Semicarbazone, 181d.
148	Benzylmalonic acid	117d.; 121	....	....	di: 217	di: 119.5	....	di: 225	....	....	Dinitrile, 91; 79
149	d,l-Tropic acid (3-Hydroxy-2-phenylpropionic acid)	117–8; d: 130	....	....	....	....	....	169			
150	Cuminic acid (4-Isopropyl-benzoic acid)	117–8, al.	....	....	....	....	....	133	....	b.p.: 263–4	Nitrile, b.p.: 243–4[734]
151	d,l-Mandelic acid (α-Hydroxy-phenylacetic acid)	118; d: 133; l: 134	....	172, al.	151–2, al.	123–4	....	133–4 (cor.)			
152	l-Arabonic acid	118–9, al.	....	200	204	....	....	136, me. al.			o-Toluidide, 172
153	α-Chlorodiphenylacetic acid ...	118–9d., bz.-lgr.	....	....	88	....	....	115	....	43–4	Anhydride, 219
154	d,l-Citramalic acid (d,l-2-Hydroxy-2-methylsuccinic acid).	119	....	....	....	....	....	mono: 140–1	di: b.p. 112[15]	....	Me. eth., 90–2; Et. eth., 81–3
155	Anilinomalonic acid	119d, al.-lgr.	....	....	mono: 157d., w.; di: 162; 246–7, ac. a.	....	....	di: 156	di: 68	di: 45, al.	....
156	4-Chloromandelic acid	119–22; 112–3	....	....	....	....	....	122–3			Nitrile, 43; Me. eth., 85–8, bz.-pet. eth.
157	(3-Nitrophenyl)acetic acid	120	....	....	....	:	....	110			
158	cis-2-Bromoallocinnamic acid ..	120	....	....	....	....	....	129	b.p.: 111[0.6], D_4^{20}: 1.4726	b.p.: 173–4[30], D_4^{25}: 1.3713	....
159	3-Furoic acid (3-Furancar-boxylic acid)	121	....	....	....	....	....	169			
160	1-Cyclopentenylcarboxylic acid.	121	....	122	126	....	....	....			
161	Cetylmalonic acid (Heptade-cane-1,1-dicarboxylic acid) ...	121.5–2.0, ac. a.	....	....	....	....	....	mono: 130–50d., lgr.-al.	di: 44, eth.	di: 22	....
162	cis-1,3-Cyclopentanedicar-boxylic acid	122; trans: 161	....	....	....	....	....	226			
163	d,l-trans-Camphenic acid (d,l-trans-Camphene-camphoric acid)	122–3, ac. a.	....	....	di: 165, ac. a.	....	....	di: 231–2, ac. a.			
164	Benzoic acid	122.4; at 100, subl.	249	158	160, boil. 50% al.	89	119.0	130	b.p.: 199.6	b.p.: 212.6	....
165	3,3,3-Trichlorolactic acid	124	....	....	164	....	....	96			
166	3-Nitrosalicylic acid	125 (hyd.)	....	....	....	....	....	145			
167	Diethylmalonic acid	125	....	....	....	di: 91	....	mono: 146; di: 224			
168	1,14-Tetradecanedicarboxylic acid	126	....	....	163	....	....				

*Derivative data given in order: m.p., crystal color, solvent from which crystallized.

TABLE XII. ORGANIC DERIVATIVES OF CARBOXYLIC ACIDS
b) Solids (Listed in order of increasing m.p.)* (Continued)

No.	Name	Melting point, °C	Boiling point, °C	p-Toluidide	Anilide	p-Nitrobenzyl ester	p-Bromo-phenacyl ester	Amide	Methyl ester	Ethyl ester	Miscellaneous
169	2,4-Dimethylbenzoic acid	127 (anh.); 90 (hyd.)			141			179–81			
170	cis-2-Chloroallocinnamic acid	127						112			
171	2-Benzoylbenzoic acid (Benzo-phenone-2-carboxylic acid)	128; 91 (+1 H₂O), w.			195	100.4		165 (cor.)	52		
172	1,10-Decanedicarboxylic acid	128						185			
173	2-Thenoic acid (2-Thiophene-carboxylic acid)	129						180			
174	4-Bromopyromucic acid (4-Bromofuran-2-carboxylic acid)	129						155–6		29; b.p.: 235–6	
175	Maleic acid (cis-Butenedioic acid)	130 (+30% Fumaric a.); 137 (pure)		di: 142, eth.	mono: 198; 187, yel., al.; di: 187, al.	di: 91 (cor.)	168–70; 190	mono: 172–3, w.; 153 sl. d.; di: 260; 181, me. al.			Heated at 160 → anh., 60, b.p.: 202
176	Tribromoacetic acid	131; 135	245d.					122			
177	(1-Naphthyl)acetic acid (1-Naphthaleneacetic acid)	131; 135			155, al.; 159.6			180–1, al.			
178	trans-α-Bromocinnamic acid	131–2						119	23	b.p.: 294–6	
179	2,5-Dimethylbenzoic acid	132			140			186			
180	cis-β-Chloroallocinnamic acid	132		142	134.5			76	34	b.p.: 265 sl. d.	
181	trans-Cinnamic acid	133	300	168	153; 109	116.8	145.6	147–8			Nitrile, 20–1, b.p.: 255–6
182	Chloromalonic acid	133			di: 118, w.			di: 170	di: b.p.: 206–8⁷⁷²	di: b.p.: 222	Di-p-bromo-anilide, 239
183	Sebacic acid (Decanedioic acid; Octane-1,8-dicarboxylic acid)	33, subl.	243¹⁵	di: 201	mono: 122–3; di: 201–2	di: 73.5; 72.6	di: 147	mono: 170; di: 210; 208			Phenylhydrazide, 194
184	2-Furoic acid (2-Furancar-boxylic acid; Pyromucic acid)	133–4; 132	230–2	170.5, al.	123.5, al.	133.5	138.5	142–3		34, b.p.: 195	
185	Malonic acid (Propanedioic acid)	134.8–.9		mono: 156d.; di: 252–3, al.	mono: 132; di: 230; 227–8, al.	di: 85.5		mono: 106–10; di: 170, w.-al.		di: b.p.: 199	Phenylhydrazide, 194
186	O-Acetylsalicylic acid (Aspirin)	135, rapid htng.	140d.		136	90.5		138			Phenacyl ester, 105
187	β-Campholytic acid (1,5,5-Tri-methylcyclopentene-2-carboxylic acid)	135	247–9; 255–6	114	104, al.-w.			130	b.p.: 203–4	b.p.: 214	
188	2-Anilinoisovaleric acid	135, w.						102–3		b.p.: 275–80	
189	Acetone-1,3-dicarboxylic acid	135d., et. ac.			di: 155, al.						Oxime, 53–4; w.; Acids or alkalis → acetone, b.p. 56 + CO₂
190	d,l-cis-Camphenic acid (d,l,cis-Camphenecamphoric acid)	135–7			di: 212			di: 225			
191	Phenylpropiolic acid	136–7, subl.		142	128; 126; 125	83		99–100; 109			Melts under w. at 80

*Derivative data given in order: m.p., crystal color, solvent from which crystallized.

No.	Name	Melting point, °C	Boiling point, °C	p-Toluidide	Anilide	p-Nitrobenzyl ester	p-Bromophenacyl ester	Amide	Methyl ester	Ethyl ester	Miscellaneous
192	*trans*-Glutaconic acid	136–8	....	....	*mono*: 167; *di*: 228	.....	.....	....	....	....	Acetic anhydride → anh., 88
193	3-Ethoxybenzoic acid	137	....	....	....	.....	.....	139.0–.5	....	....	....
194	Methylmalonic acid	137; 138d.	....	*mono*: 145d.; *di*: 228; 214	....	.....	.....	217; 206	....	....	....
195	*trans*-α-Chlorocinnamic acid	137–8	....	116, al.	118, al.	.....	.....	121–2	33	b.p.: 209^{75}, D$_4^{25}$: 1.1719, n$_D^{25}$: 1.5705	....
196	3-Thenoic acid (3-Thiophenecarboxylic acid)	138	....	....	....	.....	130	180	....	....	....
197	*cis*-Cyclobutane-1,2-dicarboxylic acid	138	....	....	....	.....	.....	*di*: 228, w.	*di*: b.p.: 225	*di*: b.p.: 238–42^{720}	Anhydride, 75; 71–3
198	2-Pyridinecarboxylic acid (Picolinic acid)	138	....	....	....	.....	.....	107	....	....	....
199	(3-Chloromethyl)benzoic acid	138, w.	....	....	....	.....	.....	124	....	b.p. 168–9^{25}	Nitrile, 67, al., b.p.: 258–60
200	5-Chloro-2-nitrobenzoic acid	139, w.	....	....	164, eth.	.....	.....	154, eth.	48.5, me. al.	....	Methylamide, 134, al.-w.; Dimethylamide, 104.5
201	Anhydrocamphoronic acid	139	....	....	202–3	.....	.....	α: 138; β: 45	....	....	....
202	Butane-1,1,4-tricarboxylic acid	139–40, bz.-et. ac.	....	....	*mono*: 177	.....	.....	....	....	*tri*: b.p.: 175–6^{18}, D^{15}: 1.0726	....
203	*meso*-Tartaric acid	140	....	....	*mono*: 193–4, pa. yel., w.	93	.....	*di*: 187; 189–90, dil. me. al.	....	....	Diphenylhydrazide, 245
204	3-Nitrobenzoic acid	140	....	162	154	141	132	143	78.5; 70	47; 40–1	....
205	3-Bromo-4-toluic acid (2-Bromo-4-methylbenzoic acid)	140	....	....	....	.....	.....	137, subl.	....	....	Nitrile, 47
206	2-Chloro-4-nitrobenzoic acid	140–2	....	....	168	.....	.....	172, al.	73–5	....	....
207	(2-Nitrophenyl)acetic acid	141; 138	....	....	....	.....	.....	161	....	....	....
208	Furanacrylic acid (β-(2-Furyl)acrylic acid)	141	286	....	....	.....	.....	168–9	....	....	....
209	2-Anilinobutyric acid	141	....	....	92, al.	.....	.....	123, w.	....	26, b.p.: 278	Nitrile, 39
210	(2-Naphthyl)acetic acid (2-Naphthaleneacetic acid)	141–2; 143	....	....	....	.....	.....	200; 205	....	....	....
211	4-Chloro-2-nitrobenzoic acid	142, w.	....	....	....	.....	.....	172, al.	41–3	....	Nitrile, 98
212	*trans*-β-Chlorocinnamic acid	142	....	122–5	128	.....	.....	118	29	b.p.: 293	....
213	2-Chlorobenzoic acid	142; 140	....	131	114; 118, pet. eth.	106	106	142; 202	b.p.: 234	b.p.: 243	....
214	(2-Bromophenoxy)acetic acid	142.5, al.	....	....	....	.....	.....	151, al.	....	b.p.: 160–70^{16}	....

Derivative data given in order: m.p., crystal color, solvent from which crystallized.

TABLE XII. ORGANIC DERIVATIVES OF CARBOXYLIC ACIDS

b) Solids (Listed in order of increasing m.p.)* (Continued)

No.	Name	Melting point, °C	Boiling point, °C	p-Toluidide	Anilide	p-Nitrobenzyl ester	p-Bromophenacyl ester	Amide	Methyl ester	Ethyl ester	Miscellaneous
215	**Suberic acid** (Octanedioic acid; Hexane-1,6-dicarboxylic acid)	144; 139–41	....	*di*: 218; 219	*mono*: 128–9; *di*: 186–7	*di*: 85	*di*: 144.2	*mono*: 125–7; *di*: 216–7	....	*di*: b.p.: 282–6	....
216	**Asaronic acid** (2,4,5-Tri-methoxybenzoic acid)	144, bz.-pet. eth.	*ca.* 300	....	154.5	....	....	184.5	97.5, yel.	72, yel.	Nitrile, 112–4, al.
217	**(2-Chlorophenoxy)acetic acid** ..	145–6, w.	....	....	121	....	....	149.5	b.p.: 186–8	32	....
218	**2-Nitrobenzoic acid**	146	....	....	155	112	107	176	b.p.: 275; 269, D_4^{20}: 1.286	30, b.p.: 148–50[10]	....
219	**Phthalonic acid**	146	....	....	*mono*: 176; *di*: 208	....	....	α: 179d.; β: 155d.	....	....	Phenylhydrazone, 171–2
220	**(2-Hydroxyphenyl)acetic acid** ..	147; 149; 141	....	....	....	....	....	118	....	....	....
221	**2-Anilinovaleric acid** (2-Anilinopentanoic acid)	147–8, al.-w.	....	....	....	....	....	99, eth.-pet. eth.	....	....	Nitrile, 51, pet. eth.
222	**Diphenylacetic acid**	148	....	172–3	180	....	....	167.5–8.0	....	....	....
223	**Diglycolic acid**	148; 142	....	*mono*: 148, w.	*mono*: 118; *di*: 152, eth.-al. (2:1)	....	....	*mono*: 135	....	....	....
224	**Oxanilic acid**.............	148–9	....	....	*di*: 154	....	....	228	....	....	....
225	**(4-Hydroxyphenyl)acetic acid** ..	148–50; 148, w.	....	....	....	....	....	175	....	....	Benzoate of amide, 167–9
226	**2-Bromobenzoic acid**	150	....	....	141	110	102	155	b.p.: 243–4	b.p.: 254–5	....
227	**Benzilic acid** (α-Hydroxy-diphenylacetic acid)........	150	....	189–90	174–5	99.5	152	153, chl.; 154–5	74–5	34	Acetate, 98, ac. a.
228	**Citric acid** (2-Hydroxypropane-1,2,3-tricarboxylic acid)	153 (slow htng.)	....	*tri*: 189, al.	*tri*: 192, al.-w.	*tri*: 102	*tri*: 148	*tri*: 210–5d.	....	....	Triphenyl ester, 124
229	**(4-Nitrophenyl)acetic acid**	153	....	....	198	....	207	198	....	....	....
230	**2,5-Dichlorobenzoic acid**	153	....	....	....	....	....	155	....	....	....
231	**Phenylmalonic acid**..........	153	....	....	....	....	....	233	....	....	....
232	**Adipic acid** (Butane-1,4-dicarboxylic acid)..........	153–4 (cor.)	216[15]	241	*mono*: 151–3, w.; *di*: 240–1, al.	106	154.5; 152.6	*mono*: 125–30, w.; *di*: 220	3, b.p.: 162[10]	*di*: 8, b.p.: 112[10], f.p.: 0	Diphenyl ester, 105–6, al.-w.
233	**(4-Bromophenoxy)acetic acid** ..	153–4, al.	....	....	....	....	....	....	....	54, al.	Phenyl ester, 73, al.
234	**Phenylpyruvic acid**	154	....	....	....	....	....	....	....	....	Oxime, 159
235	**3-Bromobenzoic acid**	155	....	....	136	105	120	155	31–2	b.p.: 254–5	....
236	**2,4,6-Trimethylbenzoic acid**....	155; 153	....	....	....	....	....	188	....	....	....
237	**2-Chloro-4-methylbenzoic acid** .	155–6	....	....	....	....	....	182	....	....	Nitrile, 61–2, subl.
238	**(4-Chlorophenoxy)acetic acid** ..	155–6, w.; 158	....	....	125	....	136	133	b.p.: 177–80	49	....
239	**3-(1-Naphthyl)propionic acid** ..	156	....	....	....	....	....	104	....	....	....
240	**Tartronic acid** (Hydroxy-malonic acid)	156–8d.	....	....	....	....	....	*di*: 198, dil. al.; 195–6d.	....	....	....
241	**Benzoylpyruvic acid**	156–8d. (+1 H₂O), al.-w.	....	....	....	....	....	138d.	59; 62	46	1-Oxime, 98–100d. (+1 H₂O)

*Derivative data given in order: m.p., crystal color, solvent from which crystallized.

No.	Name	Melting point, °C	Boiling point, °C	p-Toluidide	Anilide	p-Nitrobenzyl ester	p-Bromophenacyl ester	Amide	Methyl ester	Ethyl ester	Miscellaneous
242	Cyclobutane-1,1-dicarboxylic acid	157, w.; 158	......	......	di: 214–5	......	......	di: 275–7	......	b.p.: 218[762]; di: b.p.: 222–6	Dihydrazide, 109–10, al.-w.
243	3-Chlorobenzoic acid	158; 155	......	......	122–5, al.	107	116	134	21, b.p.: 231	b.p.: 245	....
244	Salicylic acid (2-Hydroxy-benzoic acid)	158.3; subl. at 76	......	156	136	97–8	140	142; 139	−8.6, b.p.: 223.3	1.3, b.p.: 234; 231.5	
245	1-Naphthoic acid	161–2 (cor.)	......	......	162–3; 164	......	135.5	202; 205	......	......	
246	2-Iodobenzoic acid	162	......	......	141	111	143	110	184	......	
247	2-Anilinopropionic acid	162, w.	......	......	127, al.	......	......	144	......	b.p.: 272	Nitrile, 92, al.; N-Acetyl, 143, w.
248	4-Dibenzothienylacetic acid	162	......	......	......	......	......	206	......	......	
249	Alloxanic acid	162–3d., eth.	......	......	......	......	......	191, w.	171, et. ac.	115, acet.-chl.	Phenylamide, 99, eth.
250	5-Chloro-3-nitrosalicylic acid (5-Chloro-2-hydroxy-3-nitro-benzoic acid)	163	......	......	......	......	......	199	91, al.	......	....
251	2-Benzofurylacetic acid	163	......	......	......	......	......	210	......	......	
252	Cholanic acid	164, ac. a.	......	......	......	......	......	175	......	93–4, 80% al.	$[\alpha]_D^{14}$: +21.74 in chl.; Propyl ester 56–7; Butyl ester, 53
253	4-Nitrophthalic acid	165	......	mono: 172	192	......	......	200d.	......	......	....
254	Itaconic acid (Methylene-succinic acid)	165	......	......	mono: 151.5, eth.	di: 90.6	di: 117.4	di: 191.2–.8, al.	......	......	
255	5-Bromosalicylic acid (5-Bromo-2-hydroxybenzoic acid)	165	......	......	222	......	......	232	61, b.p.: 264–6	50	....
256	6-Chloro-3-nitrobenzoic acid	165, w.	......	......	......	......	......	178, w.	73, me. al.	28–9	Nitrile, 105–6
257	3,4-Dimethylbenzoic acid	166; 164	......	......	104; 108	......	......	130	......	......	
258	Tricarballylic acid (Propane-1,2,3-tricarboxylic acid)	166	......	......	tri: 252, PhNO₂	......	tri: 138.2	tri: 205–7d.	......	......	
259	Mesitylenic acid (3,5-Dimethyl-benzoic acid)	166	......	......	......	......	......	133	......	......	
260	d,l-Phenylsuccinic acid	167–8; 84 (anh.); d,l: 173–4	......	mono (α): 175; mono (β): 168–9	mono (α): 175; mono (β): 171; di: 222	......	......	mono (α): 158–9; mono (β): 145; di: 211			
261	Mesitylacetic acid	168	......	......	......	......	......	210	......	......	....
262	3-Chloro-4-hydroxybenzoic acid	169–70, w.; 164–5	......	......	......	......	......	180–2	106–7	77–8	Nitrile, 155
263	d-Tartaric acid	169–71	......	......	mono: 180d.; 194 (cor.), ac. a.; di: 263–4d., al.	di: 163	di: 204	mono: 171–2; di: 196d., al.	......	......	Phenylhydrazide, 240

*Derivative data given in order: m.p., crystal color, solvent from which crystallized.

No.	Name	Melting point, °C	Boiling point, °C	p-Toluidide	Anilide	p-Nitrobenzyl ester	p-Bromophenacyl ester	Amide	Methyl ester	Ethyl ester	Miscellaneous
264	Azobenzene-3-carboxylic acid	170–1, or., al.						198–9, or., al.	58, me. al.		
265	2,2-Diphenylpropionic acid	171; 175						149			
266	8-Chloro-1-naphthoic acid	171–2, al.-w.						207.5, red, al.		50	
267	4-Bromo-2,5-dimethylbenzoic acid	171.5–2.5, lgr.						209–10			Nitrile, 103–4
268	5-Chlorosalicylic acid (5-Chloro-2-hydroxybenzoic acid)	172, w.						226–7	50, b.p.: 249d.	25	Nitrile, 165–7; Phenyl ester, 81–3; Me. eth., 81–2; Et. ester, 118
269	4-Chloro-2-methylbenzoic acid	172						183		b.p.: 258	Nitrile, 67
270	2,4-Dibromobenzoic acid	174						198			
271	3-Aldehydeobenzoic acid (3-Formylbenzoic acid)	175						190d.	53	b.p.: 278	Nitrile, 79–81, eth.; Semicarbazone, 265; Phenylhydrazone, 164
272	3-Thianaphthenecarboxylic acid	175			173			198			
273	Apiolic acid (2,5-Dimethoxy-3,4-methylenedioxybenzoic acid)	175, w.							71–2, w.		Nitrile, 135.5, al.-w.
274	Allomucic acid (2,3,4,5-Tetrahydroxyadipic acid)	176d., w.						175–6, w.; di: 209d., w.		di: 139–41, al.	Polyphenylhydrazide, 218d., al.
275	8-Bromo-1-naphthoic acid	178, bz.			151, al.			179–80	33, pet. eth.	52, pet. eth.	
276	3-Phenanthrylacetic acid	178						176			
277	Acetylenedicarboxylic acid	179						di: 294d.			
278	4-Toluic acid (4-Methylbenzoic acid)	179–80, subl.; 182	275 (cor.)	160; 165	144–5; 148; 140	104.5	153	160; 158			
279	6-Bromo-3-nitrobenzoic acid	180			166, al.			197–8	82	66	Nitrile, 117, subl.
280	5-Bromo-2,4-dimethylbenzoic acid	180–1						197.5–8.5			Nitrile, 88–9
281	Veratric acid (3,4-Dimethoxybenzoic acid)	181 (anh.)			154			164			
282	N-Benzoylanthranilic acid (2-Benzamidobenzoic acid)	181			279			218–9	100	98	Nitrile, 156
283	4-Chloro-3-nitrobenzoic acid	181–2			131			156, al.	83, me. al.	59, yel.	Nitrile, 100–1
284	3,5-Dinitrosalicylic acid (3,5-Dinitro-2-hydroxybenzoic acid)	182 (anh.); 174 (+1H$_2$O)						181			
285	4-Fluorobenzoic acid	182; 182.6						154; 154.5			Nitrile, 35
286	4-Chloropicolinic acid (4-Chloropyridine-2-carboxylic acid)	182d.						158; 152–4	57–8		Phenyl ester, 68, pet. eth.
287	2,4-Dinitrobenzoic acid	183				142	158	203			
288	2-Naphthoic acid	184; 185.5		192, al.	171, bz.; 173			192–3, al.; 195			
289	3-Bromosalicylic acid (3-Bromo-2-hydroxybenzoic acid)	184						165			Nitrile, 49–50
290	2-Anilinoisobutyric acid (2-Anilinoisobutanoic acid)	184–5, w.			155, al.			136			Nitrile, 93–4, al.
291	4-Anisic acid (4-Methoxybenzoic acid)	184–6; 184.2 (cor.)	275–80	186	169–71	132	152	167; 162–3, w.			

*Derivative data given in order: m.p., crystal color, solvent from which crystallized.

TABLE XII. ORGANIC DERIVATIVES OF CARBOXYLIC ACIDS

No.	Name	Melting point, °C	Boiling point, °C	p-Toluidide	Anilide	p-Nitro-benzyl ester	p-Bromo-phenacyl ester	Amide	Methyl ester	Ethyl ester	Miscellaneous
292	(2-Carboxyphenyl)acetic acid (Homophthalic acid)	185; 180						228			
293	Acetylanthranilic acid	185, ac. a.			167–8, al.			177, al.			Nitrile, 133, w.; N-Methylamide, 172, al.
294	Succinic acid (Butanedioic acid; Ethane-1,2-dicarboxylic acid).	185; 182.8	235d.	mono: 179–80; di: 254.5–5.5; 260	mono: 148.5; di: 230, al.	di: 88	di: 211.0	mono: 157; di: 260d., w.			
295	5-Bromopyromucic acid (5-Bromofuran-2-carboxylic acid)	186, w.						144–5		b.p.: 234	Hydrazide, 135.5–6.0; Azide, 66–7
296	Hippuric acid	187			208	136	151	183			
297	3-Iodobenzoic acid	187				121	128	186			
298	5-Bromo-2-toluic acid (4-Bromo-2-methylbenzoic acid).	187, subl.						180			Nitrile, 70
299	2-Cyanobenzoic acid (Phthalic acid mononitrile)	187; 192						173	151	70	
300	3-Nitroanisic acid (4-Methoxy-3-nitrobenzoic acid)	187			163						
301	Fluorene-2-acetic acid	187						266			
302	Coumarin-3-carboxylic acid	187d., w.			250			236	116–7	94	Nitrile, 182
303	d-Camphoric acid (1,2,2-Trimethylcyclopentane-1,3-dicarboxylic acid).	187.5–8.0; l: 187; d,l: 202; 208		(α): 212–4; (β): 190–6	mono (α): 204; (β): 209–10; mono (β): 196; di: 226; di, l: 226	65.5		mono (α-amide-β-acid): 176; mono (β-amide-α-acid): 182–3; di: 192–3	(α): 77; (β): 86; di: b.p.: 263–4	(α): 47–8; (β): 57; di: b.p.: 285–6	
304	3-Bromophthalic acid	188								mono: 127–8	Anhydride, 132–4
305	4-Bromo-3,5-dinitrobenzoic acid	188, al.						188, pa. yel., al.-w.	125, me. al.-w.	118, al.-w.	
306	Butane-1,2,3,4-tetracarboxylic acid (low melting form)	189, w.			187 (rapid htng.), al.-w.			di: 181d., dil. H₂SO₄ tetra: 310d., w.	tetra: 75–6, w.	di: 168, w.	Heating → monoanh. of high melting form
307	Anthroxanic acid	190; 196d.						211–2, w.	70	64–5	
308	α-Chrysenic acid (o-2-Naphthylbenzoic acid)	190						169–70	63		
309	l-Ascorbic acid	190d.; d,l: 168–9									$[\alpha]_D^{23}$: +48 in me. al.; Diphenyl-hydrazone, 178d., red; Di-p-nitrophenyl-hydrazone, 262d., al.; Di-2,4-dinitro-phenylhydrazone, 282d., br.-red

*Derivative data given in order: m.p., crystal color, solvent from which crystallized.

No.	Name	Melting point, °C	Boiling point, °C	p-Toluidide	Anilide	p-Nitrobenzyl ester	p-Bromophenacyl ester	Amide	Methyl ester	Ethyl ester	Miscellaneous
310	**Chlorofumaric acid**..........	191–2, ac. a.	....	....	186, al.	*di:* 138.5	....	*di:* b.p.: 224	*di:* b.p.: 250 sl. d.		Ethyl ester amide, 102
311	*N*-Methylacetylanthranilic acid.	192–3	....	....	....	......	155	....	....	....	*N*-Methylamide, 171–2; *N*-Ethyl-amide, 140
312	**Coumarilic acid** (Coumarone-2-carboxylic acid)..........	192–3, w.	310–5 sl. d.	....	159	......	....	159	....	27	Nitrile, 36; Phenyl ester, 101
313	**Dimethylmalonic acid**........	193, subl.	....	....	....	83.6	....	*di:* 269	....	....	
314	*trans*-**Aconitic acid**	194–5d. (cor.); *cis*: 125 → *trans* on heating	....	....	*di:* 189; *cis*, *mono*: 170d., al.	......	*tri:* 186	*tri:* 250 → br.; 260 → sinters	....	....	Heat → Itaconic acid, 165
315	**Benzylidenemalonic acid**	195–6d.	....	....	....	......	....	*di:* 189–90	*di:* 44	85; *di:* 32	Mononitrile, 183; Dinitrile, 87
316	**4-Ethoxybenzoic acid**	198; 195–6	....	....	169; 170; 172	......	....	202			
317	*trans*-**3-Nitrocinnamic acid**	199; *cis*: 138	....	....	....	174	178; 173	196			
318	**Chrysodiphenic acid** (2-Phenyl-naphthalene-1,2'-dicarboxylic acid).....................	199	....	....	....	......	....	1-*mono*: 275; 2'-*mono*: 220	1-*mono*: 171.5, me. al.; 2'-*mono*: 124; *di*: 90	....	
319	**3,4-Dihydroxybenzoic acid** (Protocatechuic acid)........	199–200d.	....	....	166	188	....	212	134.5, w.	....	
320	**3-Hydroxybenzoic acid**	200, subl.	....	163, dil. al.	156–7, w.; 155	106–8	176; 176.1–.4	170; 167, w.			
321	**Phthalic acid** (Benzene-1,2-dicarboxylic acid)..........	200–6; 191 (sealed tube); 230 (rapid htng.)	....	*mono*: 150 (slow htng.); 160–5 (rapid htng.); *di*: 201	*mono*: 170; *di*: 253–5	*di:* 155.5	*di:* 152.8	*mono*: 149; *di*: 220			
322	**3,4-Dichlorobenzoic acid**	201–2; 208–9	....	....	....	......	....	133			
323	**(4-Chloromethyl)benzoic acid** ..	203	....	....	....	......	....	173			Nitrile, 79–80, al., b.p.: 263
324	**5-Bromo-2-nitro-4-toluic acid** (2-Bromo-4-methyl-5-nitro-benzoic acid).............	203; 200	....	....	....	......	....	191		61	Nitrile, 132
325	**4-Bromo-3-nitrobenzoic acid** ...	203–4	....	....	156, or.-yel., al.	......	....	156	104	74	Nitrile, 120
326	*d,l*-**Tartaric acid**	203–4 (+1 H₂O); 205–6 (anh.)	....	....	*di:* 235–6	*di:* 147.6	....	*di:* 226, w.-me. al.			
327	*cis*-**Apocamphoric acid**	204, w.; *trans*: 190–1, w.	....	....	*mono:* 212	......	....	....			Anhydride, 178, al.
328	**3,5-Dinitrobenzoic acid**	204–5	....	....	234	157	159	183	....	....	

*Derivative data given in order: m.p., crystal color, solvent from which crystallized.

No.	Name	Melting point, °C	Boiling point, °C	p-Toluidide	Anilide	p-Nitrobenzyl ester	p-Bromophenacyl ester	Amide	Methyl ester	Ethyl ester	Miscellaneous
329	Mesaconic acid (Methylfumaric acid)	204.5 (cor.), subl.	....	*mono* (α): 196; *di*: 212, al.	*mono* (α): 202; *mono* (β): 163; *di*: 185.7	*di*: 134 (cor.)	....	*mono* (α): 222; *mono* (β): 174; *di*: 176.5	....	....	....
330	5-Bromo-3-nitro-4-toluic acid (6-Bromo-4-methyl-2-nitro-benzoic acid)	206	....	....	....	....	....	171	....	....	Nitrile, 130, subl.
331	Anthracene-9-carboxylic acid (*meso*-Anthroic acid)	207, pa. yel., al.							111, yel.	....	Nitrile, 170–2, lgr.
332	Vanillic acid	207; 210		....	140d.				....	44, b.p.: 293	....
333	*trans*-2-Coumaric acid (*trans*-2-Hydroxycinnamic acid)	207–8d., subl., w.	....	....	152.5			209d.	....	....	Acetate, 154–5; 146, bz.
334	Oxamic acid	210		....	148–9		....	419d.	....	....	....
335	Pentamethylbenzoic acid	210						206			
336	4-Coumaric acid (4-Hydroxy-cinnamic acid)	210–3; 206 (anh.)						194	137; 126		Acetate, 200–5
337	*trans*-2-Chlorocinnamic acid	212, yel., al.			176			168	10.5, b.p.: 278–9	b.p.: 162^{12}	Nitrile, 40
338	2,4-Dihydroxybenzoic acid (β-Resorcylic acid)	213d. (rapid htng.); 216d.; 217		....	126–7	188–9	....	222	....	....	Loses H_2O of crystallization at 100. Easy loss of CO_2 gives m.p. varying from 194 to 236.
339	2-Bromo-3,5-dinitrobenzoic acid	213		....	....	....	....	216, pa. yel., al.-w.	109, me. al.-w.	74, al.	....
340	4-Dibenzofurylacetic acid	214		....	....	....	....	212	....	....	....
341	Mucic acid	214d.; (varies with htng. rate), 223–255		....	....	310	225	*mono*: 192d.; *di*: 220	....	....	....
342	3-Chloroanisic acid (3-Chloro-4-methoxybenzoic acid)	214–5		....	....	....	....	193	94–5		
343	4-Hydroxybenzoic acid	215; 213–4; 210	....	203–4, al.	196–7, yel., w.	180–2	191.5 (cor.); 184	162 (+1 H_2O), w.			
344	Piperic acid	216		....	....	145		....			
345	3-Chloro-2-naphthoic acid	216–7, me. al.-w.		....	....	....	....	237	58, me. al.	50	
346	3-Nitrophthalic acid	218		*di*: 226	*di*: 234	189	....	*di*: 201d.			
347	Acenaphthene-5-carboxylic acid	219, bz.		....	....	....		198	....	....	Nitrile, 110–1
348	4-Cyanobenzoic acid (Terephthalic acid mono-nitrile)	219; 214		....	179	189	....	223	62	54	....
349	4-Phenylbenzoic acid	221		....	....	....		223	....	....	
350	3-Hydroxy-2-naphthoic acid	222–3 (cor.)		221–3	243–4, ac. a.; 249 (cor.)	....		217–8 (cor.), yel., al.	....	....	

*Derivative data given in order: m.p., crystal color, solvent from which crystallized.

TABLE XII. ORGANIC DERIVATIVES OF CARBOXYLIC ACIDS

b) Solids (Listed in order of increasing m.p.)* (Continued)

No.	Name	Melting point, °C	Boiling point, °C	p-Toluidide	Anilide	p-Nitrobenzyl ester	p-Bromophenacyl ester	Amide	Methyl ester	Ethyl ester	Miscellaneous
351	4-Hydroxy-2-naphthoic acid	225–6	....	206	....	....	....	217–8	....	....	
352	5-Bromo-3-nitro-2-toluic acid (4-Bromo-2-methyl-6-nitrobenzoic acid)	226						235			Nitrile, 106–7, subl.
353	9-Fluorenecarboxylic acid	227; 230; 225						251			
354	Biphenyl-2,2'-dicarboxylic acid (2,2'-Diphenic acid)	227; 233; 229			mono: 176; di: 229–30, al.	di: 187; 182.6		mono: 193; 190–1; di: 212, w.			
355	Methyliminodiacetic acid	227d.						mono: 169; di: 169			
356	2,4,6-Trinitrobenzoic acid	228						264d.			
357	Piperonylic acid	229; 228						169, al.			
358	5-Nitrosalicylic acid	229–30			224			225			
359	4-Chloro-3-hydroxy-2-naphthoic acid	231, yel.						225	116, yel.		
360	1-Phenanthroic acid	232						284			
361	3-Pyridylacrylic acid	233						148			
362	4-Bromo-3-hydroxy-2-naphthoic acid	233–5d., yel., al.-ac. a.			161–2						Acetate, 183
363	4-Chloro-1-hydroxy-2-naphthoic acid	234, al.		143–4	180–1				120–1	92–3	o-Toluidide, 148–9; m-Toluidide, 188–9
364	3-Chloro-2-nitrobenzoic acid	235, w.			186						
365	Benzophenone-2,4-dicarboxylic acid	235, w.						di: >288	di: 107		
366	5-Bromo-2-hydroxy-3-toluic acid (5-Bromo-2-hydroxy-3-methylbenzoic acid)	236			125, al.-w.			75–8	109	75	
367	2-Thianaphthenecarboxylic acid	236						177			
368	Butane-1,2,3,4-tetracarboxylic acid (high melting form)	236–7 (slow htng.)			di: 168, acet. (slow htng.)			di: 169d.	tetra: 63–4, lgr.		Di-imide, 320d., w.
369	7-Bromo-1-naphthoic acid	237 (cor.), 60% al.			202, al.-w.			247, 50% al.	55 (cor.), 60% me. al.	46	
370	3-Pyridinecarboxylic acid	237–8; 235; 232		150	85; 132, bz.-lgr.; 265, w.			128; 122			
371	trans-2-Nitrocinnamic acid	240; cis: 146–7				132	141	185			
372	2-Chloroquinoline-3-carboxylic acid	240						200–1			
373	4-Nitrobenzoic acid	241		204; 192	211; 204	168	137	201; 198	96	57	
374	Azobenzene-4-carboxylic acid	241, red, al.						224–5, red	123–4, or., me. al.	86–7, or.-red, al.	Nitrile, 120–1, br., bz.; Propyl ester, 64, red, lgr.
375	4-Chlorobenzoic acid	243; 240			194, al.	129.5, al.	126	179; 170	44	b.p.: 238	
376	7-Chloro-1-naphthoic acid	243, 60% al.			185, al.-w.			237, 50% al.	54, 60% al.		
377	3-Chlorocinchonic acid (2-Chloroquinoline-4-carboxylic acid)	244, al.			202, al.			334–5, al.-w.; 276–8 (after fusion)	89–90, acet.	64.5	

* Derivative data given in order: m.p., crystal color, solvent from which crystallized.

No.	Name	Melting point, °C	Boiling point, °C	p-Toluidide	Anilide	p-Nitrobenzyl ester	p-Bromophenacyl ester	Amide	Methyl ester	Ethyl ester	Miscellaneous
378	5-Chloro-1-naphthoic acid	245; 241–2 subl.	. . . . . .	. . . . . .	. . . . .	. . . . .	. . . . .	239	. . . .	42	Nitrile, 145
379	Azobenzene-2,2′-dicarboxylic acid	245, dk. yel., al.	. . . . . .	. . . . . .	. . . . .	. . . . .	. . . . .	*mono:* 215d., red-br., et. ac.; *di:* 294d., red-br., ac. a.	*di:* 101, red, me. al.	*di:* 85, pa. red, al.	. . . .
380	Anthracene-1-carboxylic acid (α-Anthroic acid)	245, yel., al.; 252	. . . . . .	. . . . . .	. . . . .	. . . . .	. . . . .	260, al.	108, ac. a.	. . . .	Nitrile, 126, yel.; Phenyl ester, 207–9, yel.
381	2-Amino-9,10-anthraquinone-1-carboxylic acid	250–2, or.-red	. . . . . .	. . . . . .	. . . . .	. . . . .	. . . . .	300, or., PhNO₂	. . . .	. . . .	. . . .
382	4-Bromobenzoic acid	251–3	. . . . . .	. . . . . .	197	180	. . . . .	189–90, w.	. . . .	. . . .	p-Phenylphenacyl ester, 193
383	9-Phenanthroic acid	251; 253	. . . . . .	. . . . . .	. . . . .	. . . . .	. . . . .	233	. . . .	. . . .	. . . .
384	4-Bromocinnamic acid	251–3	. . . . . .	. . . . . .	183	. . . . .	. . . . .	. . . . .	80	. . . .	. . . .
385	Cinchoninic acid (Quinoline-4-carboxylic acid)	253–4 (+1 or 2H₂O)	. . . . . .	. . . . . .	. . . . .	. . . . .	. . . . .	181	24	13, b.p.: 173[15]	Nitrile, 102
386	Gallic acid (3,4,5-Trihydroxy-benzoic acid).	253–4d.; 222–40d.	. . . . . .	. . . . . .	207	141	134	189	. . . .	. . . .	. . . .
387	1-Acenaphthoic acid	256	. . . . . .	. . . . . .	. . . . .	. . . . .	. . . . .	228	. . . .	. . . .	. . . .
388	2-Phenanthroic acid . . . . .:	260	. . . . . .	. . . . . .	. . . . .	. . . . .	. . . . .	243	. . . .	. . . .	. . . .
389	Cinchomeronic acid (Pyridine-3,4-dicarboxylic acid)	260d., w.	. . . . . .	. . . . . .	*di:* 199–206	. . . . .	. . . . .	3-*mono:* 200d.; 4-*mono:* 170d., w.; *di:* 163–5d.	3-*mono:* 182; 4-*mono:* 154–72; *di:* 141	4-*mono:* 131–3, bz.; *di:* b.p.: 172[21]	Imide, 229–30, subl.
390	5-Bromo-1-naphthoic acid	261; 256	. . . . . .	. . . . . .	. . . . .	. . . . .	. . . . .	241	. . . .	48–9	Nitrile, 147, subl.
391	Chelidonic acid (γ-Pyrone-2,6-dicarboxylic acid).	262	. . . . . .	. . . . . .	. . . . .	. . . . .	. . . . .	245	*di:* 122.5	227; *di:* 63	p-Phenylphenacyl ester, 195–8d.
392	4-Iodobenzoic acid	270; 265	. . . . . .	. . . . . .	210	141	146	217	. . . .	. . . .	. . . .
393	5-Chloro-2-naphthoic acid	270, al.	. . . . . .	. . . . . .	202.5	. . . . .	. . . . .	186–7	81	45	Nitrile, 144
394	3-Phenanthroic acid	270	. . . . . .	. . . . . .	. . . . .	. . . . .	. . . . .	234	. . . .	. . . .	. . . .
395	Quinoline-3-carboxylic acid	272	. . . . . .	. . . . . .	. . . . .	. . . . .	. . . . .	198	. . . .	. . . .	. . . .
396	1-Chloroanthraquinone-2-carboxylic acid	272, pa. yel., al.	. . . . . .	. . . . . .	248–9, pa. yel., bz.-ac. a.	. . . . .	. . . . .	317, yel.	161.5, yel., acet.; 155	142, yel., al.	Benzyl ester, 135–6, yel., al.
397	Anthracene-2-carboxylic acid (β-Anthroic acid)	281, yel., al.	. . . . . .	. . . . . .	. . . . .	. . . . .	. . . . .	293–5, yel., al.	. . . .	134–5	. . . .
398	*trans*-4-Nitrocinnamic acid	285	. . . . . .	. . . . . .	. . . . .	186 150.8	191	204; 217	. . . .	. . . .	. . . .
399	Fumaric acid (*trans*-Butanedioic acid)	286–7; (sealed tube); >200, subl.; 293–5	. . . . . .	. . . . . .	*mono:* 233.0–4.5; *di:* 313–4, ac. a.	. . . . .	. . . . .	270; 300–2 subl.; *di:* 266d.	*di:* 102, b.p.: 192	*mono:* 66; *di:* b.p.: 218	At 230 → maleic anh., 56
400	Muconic acid.	289d. (slow htng.); 306 (rapid htng.)	. . . . . .	. . . . . .	. . . . .	. . . . .	. . . . .	*di:* 240d.	. . . .	. . . .	*trans-trans*: 296–8; *cis-cis*: 195

*Derivative data given in order: m.p., crystal color, solvent from which crystallized.

TABLE XII. ORGANIC DERIVATIVES OF CARBOXYLIC ACIDS

b) Solids (Listed in order of increasing m.p.)* (Continued)

No.	Name	Melting point, °C	Boiling point, °C	p-Toluidide	Anilide	p-Nitrobenzyl ester	p-Bromophenacyl ester	Amide	Methyl ester	Ethyl ester	Miscellaneous
401	**9,10-Anthraquinone-2-carboxylic acid**	290–2, yel., ac. a.	....	....	258–60	......	....	280, ac. a.-bz.	170	147	....
402	**9,10-Anthraquinone-1-carboxylic acid**	293–4, pa. yel., ac. a.	....	....	288–9, pa. pa. yel., PhNO$_2$	......	....	280, pa. yel., al.	189, pa. yel., me. al.	169, yel., al.	Nitrile, 247, yel., ac. a.
403	**Bromoterephthalic acid** (2-Bromobenzene-1,4-dicarboxylic acid)	299	....	....	....	......	....	di: 270	1-mono: 145; 4-mono: 164; di: 54	....	....
404	**Terephthalic acid** (Benzene-1,4-dicarboxylic acid)	300, subl. without melting	....	....	di: 334–7, PhNO$_2$	di: 263.5	di: 225	di: >225	....		
405	**Chloroterephthalic acid** (2-Chlorobenzene-1,4-dicarboxylic acid)	>300, w.	....	....	....	......	....	di: >300	di: 60	....	....
406	**4-Pyridinecarboxylic acid** (Isonicotinic acid)	324	....	....	....	......	....	156			
407	**9,10-Anthraquinone-2,3-dicarboxylic acid**	240–2, yel., ac. a.	....	....	....	......	....	mono: >340, br., ac. a.	....	....	Anhydride, 290
408	**Isophthalic acid** (Benzene-1,3-dicarboxylic acid)	348, subl.	....	....	....	202.5	179.1	mono: 280; di: 280	....		Ba salt (+6 H$_2$O) very soluble—differentiates from Terephthalic acid.
409	**Benzophenone-4,4′-dicarboxylic acid**	subl. <360	....	....	....	......	....	di: >300	di: 224; 231	....	Dinitrile, 204–5
410	**Trimesic acid** (Benzene-1,3,5-tricarboxylic acid)	380 (cor.)	....	....	tri: 118–20d., ac. a.	......	tri: 197 (sealed tube)	365d. (cor.)	tri: 143–4, me. al.	tri: 132–3, al.; 133 after sintering at 127	....

* Derivative data given in order: m.p., crystal color, solvent from which crystallized.

TABLE XIII. ORGANIC DERIVATIVES OF ACYL HALIDES
I. Acyl Fluorides (Listed in order of increasing b.p.) *

No.	Name	Boiling point, °C	Melting point, °C	n_D	Density g/ml	Acid B.p., °C	Acid M.p., °C	Amide	Anilide	p-Toluidide	2-Naphthyl amide	Miscellaneous
1	Acetyl fluoride	20–1	. . . . . .	. . . .	1.002[15]	118	. . . .	82	114	147	134	. . . .
2	Propionyl fluoride	44–6	. . . . . .	. . . .	. . . .	141	. . . .	81	106	126	. . . .	. . . .
3	Fluoroacetyl fluoride	50.5–51	. . . . . .	. . . .	. . . .	167–9	31–2	108	. . . .	. . . .	. . . .	. . . .
4	Trichloroacetyl fluoride	66–8	. . . . . .	. . . .	. . . .	197	57–8	141	97	113	. . . .	. . . .
5	n-Butyryl fluoride	67	. . . . . .	. . . .	. . . .	162.5	. . . .	115	96	75	125	. . . .
6	Chloroacetyl fluoride	73–5	. . . . . .	. . . .	. . . .	. . . . . .	63	120	137	162	117–8	. . . .
7	Phthaloyl difluoride	224–6	42–3	. . . .	. . . .	. . . . . .	206	220 (di)	253 (di)	201 (di)	. . . .	. . . .
8	Phenylacetyl fluoride	88–9[17]	. . . . . .	. . . .	. . . .	. . . . . .	76	156	118	136	159	. . . .

ˣ Derivative data given in order: m.p., crystal color, solvent from which crystallized.

TABLE XIII. ORGANIC DERIVATIVES OF ACYL HALIDES
II. Acyl Chlorides a) Liquids 1) (Listed in order of increasing atmospheric b.p.)*

No.	Name	Boiling point, °C	Melting point, °C	n_D	Density g/ml	Acid B.p., °C	Acid M.p., °C	Amide	Anilide	p-Toluidide	2-Naphthyl amide	Miscellaneous
1	Acetyl chloride	51–2	. . . . .	1.3897^{20}	1.105^{20}_{4}	118	. . . . . .	82	114	147	134	. . . .
2	Oxalyl chloride	64	–12	1.434^{13}	$1.488^{13.4}_{4}$	. . . . . .	101 (dih-yd.)	419d.	246	268	. . . . .	. . . .
3	Fluoroacetyl chloride	72–3	. . . . .	. . . .	. . . .	167–9	31–2	108	. . . .			
4	Acrylyl chloride	76	. . . . .	1.4343^{20}	1.114^{20}_{4}	140	. . . . . .	85	105	141		
5	Propionyl chloride	80	. . . . .	1.4051^{20}	1.065^{20}_{4}	141	. . . . . .	81	106	126		
6	Isobutyryl chloride	92	. . . . .	1.4079^{20}	1.017^{20}_{4}	154.5	. . . . . .	128	105	107		
7	Methacrylyl chloride	95–6	. . . . .	1.4435	. . . .	160.5	15–6	102–6				
8	Vinylacetyl chloride	98				163	. . . . . .	72–3	58	. . . . .		
9	n-Butyryl chloride	101–2	. . . . .	1.4121^{20}	1.028^{20}_{4}	162.5	. . . . . .	115	96	75		
10	Pivalyl chloride	105–6				. . . . .	35	154	129	120		
11	Dichloroacetyl chloride	108				194	. . . . . .	98	118	153		
12	Chloroacetyl chloride	108–10	. . . . .	1.454^{20}	1.3997^{18}_{4}	189	63	120	137	162		
13	DL-α-Chloropropionyl chloride	110–11	. . . . .	1.440^{20}	1.285^{20}_{4}	185–6	. . . . . .	80	92	124		
14	Methoxyacetyl chloride	113				204	. . . . . .	97	58			
15	DL-Ethylmethylacetyl chloride	115–6				176		112	110	93		
16	Isovaleryl chloride	115	. . . . .	$1.4136^{24.3}$	$0.985^{24.3}_{4}$	176	. . . . . .	135	109–10	107	138.5	
17	Acetylglycyl chloride	115–8					206	137	. . . .		. . . . .	Hydrazide, 115; Me. ester, 58–9
18	Trichloroacetyl chloride	118	. . . . .	1.470^{20}	1.620^{20}_{4}	197	57–8	141	97	113		
19	Cyclopropane carbonyl chloride	120			1.152^{20}_{0}	186	18	125	. . . .			
20	Ethoxyacetyl chloride	123–4				207	. . . . . .	80–2				
21	trans-Crotonyl chloride	126	. . . . .	1.46^{18}	1.08^{20}_{4}	. . . . . .	72	161	118	132		
22	n-Pentanoyl chloride (n-Valeryl chloride)	126	. . . . .	1.420^{20}	1.0004^{20}_{4}	186	. . . . . .	106	63	74	112	
23	Allylacetyl chloride	128			1.074^{16}	188–9	. . . . . .	94				
24	Bromoacetyl chloride	133–5			1.908^{0}	208	50	91	131	. . . . . .	134	
25	Cyclobutane carbonyl chloride	137; 142–3				195	. . . . . .	153				
26	β-Methoxypropionyl chloride	138		1.424		107^{10}	. . . . . .	50.5				
27	Diethylacetyl chloride	140		1.4234		190		107				
28	β-Chloropropionyl chloride .	144		1.455	1.331^{13}	. . . . . .	42		119	121	. . . . . .	
29	Isocaproyl chloride (4-Methylpentanoyl chloride)	147		. . . .	0.9725^{20}_{4}	199	. . . . . .	121	112	63	. . . . . .	
30	α-Acetoxypropionyl chloride (O-Acetyl lactoyl chloride) .	150 part. d.		1.4241^{17}	1.192	. . . . . .	57–60; 40					
31	DL-α-Bromobutyryl chloride	150–2				217d.	–4	112	98	92		
32	n-Hexanoyl chloride (n-Caproyl chloride)	153	. . . . .	1.426^{20}	0.975^{20}_{4}	205	. . . . . .	100	95	75	107	
34	Furoyl chloride	173–4				. . . . . .	133–4	142–3	124	107		
35	n-Heptanoyl chloride (Enanthoyl chloride)	175	. . . . .	1.4345^{15}	0.963^{20}_{4}	223		96	70 (65)	81	101	
36	Hexahydrobenzoyl chloride (Cyclohexane carboxylic acid chloride)	183–4	. . . . .	1.4766^{15}; 1.4711^{20}	1.096^{15}_{4}	. . . . . .	29–30	185	143–4			
37	3-Fluorobenzoyl chloride . . .	189; 204					124	130	. . . . . .			
38	Succinyl dichloride	190d.	20	1.473^{15}	1.395^{15}_{4}	. . . . . .	186	260 (di)	230 (di)	255 (di)		
39	4-Fluorobenzoyl chloride . . .	193	9	. . . .	. . . .		183	154.5	. . . .			

*Derivative data given in order: m.p., crystal color, solvent from which crystallized.

TABLE XIII. ORGANIC DERIVATIVES OF ACYL HALIDES
II. Acyl Chlorides a) Liquids 1) (Listed in order of increasing atmospheric b.p.)* (Continued)

No.	Name	Boiling point, °C	Melting point, °C	n_D	Density g/ml	Acid B.p., °C	Acid M.p., °C	Amide	Anilide	p-Toluidide	2-Naphthyl amide	Miscellaneous
40	n-Octanoyl chloride (n-Capryloyl chloride)	196	. . . . . .	. . . .	0.949_4^{20}	239	16	110; 106	57	70	103	. . . .
41	Benzoyl chloride	197	−1	1.558^{15}	1.212_4^{20}	. . . . . .	122	130	163	158	. . . . . .	. . . .
42	Diethyl malonyl dichloride . .	197	. . . . . .	1.5537^{20}	1.2187_{15}^{15}	. . . . . .	125	224 (di)	. . . .	. . . .	. . . . . .	. . . .
43	2-Fluorobenzoyl chloride . . .	206	4	. . . .	. . . .	. . . . . .	126.5	116	. . . .	. . . .	. . . . . .	. . . .
44	Phenylacetyl chloride	210	. . . . . .	1.533^{20}	1.1685_4^{20}	. . . . . .	76	156	118	136	159	. . . .
45	n-Nonanoyl chloride (Pelargonyl chloride)	215	. . . . . .	. . . .	0.946_4^{15}	255	12	99	57	84	103	. . . .
46	Glutaryl dichloride	218	. . . . . .	1.473^{20}	1.324_4^{20}	302	97–8	175–6 (di)	224	. . . .	. . . . . .	. . . .
47	4-Chlorobenzoyl chloride . . .	222	16	1.579^{20}	1.362_4^{20}	. . . . . .	240	179; 170	194	. . . .	. . . . . .	. . . .
48	Hydrocinnamoyl chloride (β-Phenylpropionyl chloride)	225 d.	. . . . . .	. . . .	1.135_{21}^{21}	. . . . . .	48	105	98, pet. eth.	135	. . . . . .	. . . .
49	3-Chlorobenzoyl chloride . . .	225	. . . . . .	. . . .	. . . .	. . . . . .	158	134	122	. . . .	. . . . . .	. . . .
50	Phenoxyacetyl chloride	225–6	. . . . . .	. . . .	. . . .	. . . . . .	98–9	101.5	101	. . . .	. . . . . .	. . . .
51	4-Methylbenzoyl chloride (4-Toluyl chloride)	225–6	−3.9	1.545^{20}	1.1686_4^{20}	. . . . . .	179–80	160	145	160; 165	. . . . . .	. . . .
52	n-Decanoyl chloride	232	. . . . . .	. . . .	. . . .	268–70	31	108; 98	70	78	104	. . . .
53	2-Chlorobenzoyl chloride . . .	233	. . . . . .	. . . .	. . . .	. . . . . .	142	142	118	131	. . . . . .	. . . .
54	3-Methoxybenzoyl chloride .	242–4	. . . . . .	. . . .	. . . .	. . . . . .	110; 105	. . . .	. . . .	. . . .	. . . . . .	Benzylamine salt of acid, 112
55	3-Bromobenzoyl chloride . . .	243; 239	. . . . . .	. . . .	. . . .	. . . . . .	155	155	136	. . . .	. . . . . .	Hydrazide, 151
56	2-Bromobenzoyl chloride . . .	245	11	. . . .	. . . .	. . . . . .	150	155–6	141	. . . .	. . . . . .	Hydrazide, 153
57	2-Methoxybenzoyl chloride .	254	. . . . . .	. . . .	. . . .	. . . . . .	. . . . . .	129	131	. . . .	. . . . . .	Phenyl ester, 59
58	4-Methoxybenzoyl chloride (Anisoyl chloride)	262–3 sl. d.	22	1.58^{20}	1.261_4^{20}	. . . . . .	184	162–3	169; 163	186	. . . . . .	Anisidide, 202
59	Phthaloyl dichloride	276	15–6	1.569^{20}	1.406_4^{20}	. . . . . .	200–6	220 (di)	253–5 (di)	201 (di)	. . . . . .	. . . .
60	3-Nitrobenzoyl chloride	278	35	. . . .	. . . .	. . . . . .	140	143	154	162	. . . . . .	. . . .
61	1-Naphthoyl chloride	297.5	20	. . . .	. . . .	. . . . . .	161	202	. . . .	. . . .	. . . . . .	Piperidide, 85–7

*Derivative data given in order: m.p., crystal color, solvent from which crystallized.

TABLE XIII. ORGANIC DERIVATIVES OF ACYL HALIDES
II. Acyl Chlorides a) Liquids
2) (Reduced pressure b.p. only) (Listed in order of increasing m.p. of the corresponding amides)*

No.	Name	Boiling point, °C	Melting point, °C	n_D	Density g/ml	Acid B.p., °C	Acid M.p., °C	Amide	Anilide	p-Toluidide	2-Naphthyl amide	Miscellaneous
1	β-Ethoxypropionyl chloride	78[52]				120[17]		51				
2	Azidoacetyl chloride	50[20]					16	58				
3	Oleyl chloride	163[2]					16	75–6	41	42.5		
4	γ-Phenoxybutyryl chloride	155[20]					64–5; 60	80				
5	γ-Phenylbutyryl chloride	140–2[12]				290	52	84.5				
6	ω-Undecenoyl chloride (ω-Undecylenoyl chloride; ω-Hendecenoyl chloride)	128[14]				275	24.5	87				
7	Benzoylformyl chloride	125[9]					64–6	91				2,4-Dinitrophenyl-hydrazone of acid, 196–7, yel.
8	Iodoacetyl chloride	49–52[15]			2.25[25]		83	95	143–4			
9	β-Iodopropionyl chloride	81[15]					82; 85	101; 142				
10	Palmitoyl chloride	194[17]	11–2				63	106–7	90	98	109	
11	Myristoyl chloride	174[16]	1–3				54; 58	107	84	93	108	
12	Phenylpropiolyl chloride	115–6[17]					136–7	108–9; 99–100	126	142		
13	Dodecanoyl chloride (Lauroyl chloride)	145[18]	−17	1.446[20]		299	44	110; 102	78	87	106	
14	α-Phenoxybutyryl chloride	128–31[38]				258	82–3; 99	111; 123	93–4			Phenyl ester, 48–9
15	Cyanoacetyl chloride	57[0.5]					66	119–20	198–9			
16	Nicotinyl chloride	90[15]				235		122	85			
17	Dibenzylacetyl chloride	202[18]					89	129	155			
18	α-Phenoxypropionyl chloride	115[20]					115–6; 112–3	132	117	115	117	
19	DL-α-Bromoisovaleryl chloride	59[15]					44	133	116	124	145	
20	3-Ethoxybenzoyl chloride	135–40[16]	27–8				137	139				
21	α-Bromoisobutyryl chloride	52[30]		1.475[23]			48–9	148	83	92.5	135	
22	4-Isopropylbenzoyl chloride	121[10]				256–8		153				
23	Benzilic acid chloride (α-Hydroxydiphenylacetyl chloride)	193–5[27]					150	154–5	175	190		Me. ester, 74–5
24	1-Naphthoxyacetyl chloride	194[10]					190	155	144			4-Phenetidide, 145–6
25	Hexahydrophenylacetyl chloride (Cyclohexylacetyl chloride)	98–100[23]				244–6	33	171–2				
26	Azelayl dichloride	166[18]; 140[0.4]					106.5	175 (di)	186–7 (di)	191 (di)		
27	2-Nitrobenzoyl chloride	148[9]	20				146	176	155			
28	Mesaconyl dichloride (Methylfumaryl dichloride)	64–5[14]					240.5	177 (di)	186 (di)	212		
29	1-Naphthylacetyl chloride	188[23]					131	180–1; 154				
30	2,4,6-Trimethylbenzoyl chloride	155–6[18]		1.5263[25]	1.0967[25]$_4$		152	188				
31	3-Formylbenzoyl chloride	130[20]					175	190				Me. ester, 53; Semicarbazone of acid, 265
32	4-Ethoxybenzoyl chloride	160[20]					198	202	170			
33	Sebacoyl dichloride	182[16]		1.4684	1.1212[20]$_4$		134.5	210 (di)	198 (di)			

*Derivative data given in order: m.p., crystal color, solvent from which crystallized.

TABLE XIII. ORGANIC DERIVATIVES OF ACYL HALIDES
II. Acyl Chlorides a) Liquids
2) (Reduced pressure b.p. only) (Listed in order of increasing m.p. of the corresponding amides)* (Continued)

No.	Name	Boiling point, °C	Melting point, °C	n_D	Density g/ml	Acid B.p., °C	Acid M.p., °C	Amide	Anilide	p-Toluidide	2-Naphthyl amide	Miscellaneous
34	Adipyl dichloride	130–2[18]	. . . . . .	. . . .	. . . .	. . . . . .	153	220 (*di*)	240–1	241	. . . .	Di-N-methyl-amide, 152–3
35	Benzylmalonyl dichloride. . .	141[15]	. . . . . .	. . . .	. . . .	. . . . . .	117d.	225 (*di*)	217 (*di*)	. . . . . .	. . . .	. . . .
36	Phenylmalonyl dichloride . .	122[15]	. . . . . .	. . . .	. . . .	. . . . . .	152–3	233	. . . .	. . . . . .	. . . .	Di-Me. ester, 51
37	*trans*-Aconityl trichloride (1,2,3-Propylenetricar-boxylic acid trichloride) . .	155–7[20]	. . . . . .	. . . .	. . . .	. . . . . .	194–5	260 (*tri*) (sin-ters)	. . . .	. . . . . .	. . . .	. . . .
38	Fumaryl dichloride	63[13]	. . . . . .	1.5004[18]	1.408$_4^{20}$	. . . . . .	300–2	266 (*di*)	314 (*di*)	. . . . . .	. . . .	. . . .

*Derivative data given in order: m.p., crystal color, solvent from which crystallized.

TABLE XIII. ORGANIC DERIVATIVES OF ACYL HALIDES
II. Acyl Chlorides b) Solids (Listed in order of increasing m.p.)*

No.	Name	Boiling point, °C	Melting point, °C	n_D	Density g/ml	Acid B.p., °C	Acid M.p., °C	Amide	Anilide	p-Toluidide	2-Naphthyl amide	Miscellaneous
1	Salicyloyl chloride	92^{15}	19–20	....	....	......	158	142	136	156	189	....
2	Stearoyl chloride	$202–3^6$	23	....	....	......	70–1	109	95	102	112	....
3	2-Iodobenzoyl chloride.....		35–40; 30–1	....	....	......	162	184	....	....	....	....
4	trans-Cinnamoyl chloride ..	258	35–6	1.6202^{37}	1.1632^{37}	......	133	147–8	151	168	....	....
5	4-Bromobenzoyl chloride ...	245–7	42	....	....	......	251–3	189–90	197	....	....	Hydrazide, 164
6	Isophthaloyl dichloride.....	276	43–4	1.570^{47}	1.388_4^{47}	......	345–7	280 (di)	....	....	....	Di-Me. ester, 68; Dihydrazide, 220
7	2-Naphthoyl chloride	304–6	43	....	....	......	184–5	192	....	....	....	Piperidide, 88–90
8	2,4-Dinitrobenzoyl chloride .		46	....	....	......	183	203	....	....	....	Me. ester, 70
9	4-Formylbenzoyl chloride ..	258	48	....	....	......	256	....	....	....	....	Me. ester, 63; Phenylhydrazone of acid, 226
10	4-Nitrophenylacetyl chloride		48	....	....	......	153	198	198	....	....	....
11	2-Naphthoxyacetyl chloride.		54	....	....	......	156	147	145	....	....	4-Phenetidide, 164–5; Et. ester, 48–9
12	Diphenylacetyl chloride		56–7	....	....	......	148	168	180	173	191–2	....
13	trans-2-Nitrocinnamoyl chloride		64.5	....	....	......	240	185	....	....	....	Me. ester, 73
14	3,5-Dinitrobenzoyl chloride .		68–9; 74	....	....	......	204–5	183	234	....	....	Me. ester, 108; Et. ester, 93
15	4-Nitrobenzoyl chloride	$150–2^{15}$	75	....	....	......	241	201	211	204	....	....
16	3-Nitrophthaloyl dichloride .		77	....	....	......	218	201 (di)	234 (di)	226 (di)	....	....
17	Benzylidene malonyl dichloride		77	....	....	......	195–6	189 (di)	....	....	....	Di-Me. ester, 45
18	Fluorene-9-carboxylic acid chloride		77	....	....	......	230–2	251	....	....	....	Me. ester, 63
19	Terephthaloyl dichloride ...		83–4	....	....	......	....	>250 (di)	334–7 (di)	....	....	Di-1-naphthyl-amide, 334
20	4-Iodobenzoyl chloride.....		83; 77–8				270; 265	218	210	....	....	....
21	Diphenylcarbamyl chloride (Diphenylaminoformyl chloride)		86					189	....	....	....	Me. ester, 86; Et. ester, 72, lgr.
22	2,2'-Diphenic acid dichloride		94				228–9	212	....	....	....	....
23	α,α-Diphenylpropionyl chloride.........		95–6				173–4	149	....	....	....	Benzyl ester, 71–2
24	Phenanthrene-2-carboxylic acid chloride		101				259–60	242–3	217–8	....	....	....
25	Phenanthrene-9-carboxylic acid chloride		102				252	232; 226	218	....	....	....
26	Diphenyl-4-carbonyl chloride (4-Phenylbenzoyl chloride)		114–5				228	223	....	....	....	Me. ester, 117–8; Et. ester, 46
27	Phenanthrene-3-carboxylic acid chloride		116–7				269	233; 227	216–7	....	....	....
28	trans-4-Nitrocinnamoyl chloride		124				286	217	....	....	....	Me. ester, 161
29	Fluorenone-4-carboxylic acid chloride		128, yel.				227, yel.	230; 225	....	....	....	Oxime of acid, 263; Me. ester, 132; Et. ester, 103
30	Fluorenone-1-carboxylic acid chloride		140, yel.	....	....	....	....	229–30	....	....	....	Oxime of acid, 230; Me. ester, 86–9; Et. ester, 84–6

*Derivative data given in order: m.p., crystal color, solvent from which crystallized.

No.	Name	Boiling point, °C	Melting point, °C	n_D	Density g/ml	Acid B.p., °C	Acid M.p., °C	Amide	Anilide	p-Toluidide	2-Naphthyl amide	Miscellaneous
31	Azobenzene-4,4'-dicar-boxylic acid dichloride	. . . .	144–5, red	. . . .	. . . .	. . . . . .	330d.	. . . .	. . . .	. . . .	. . . .	Di-Me. ester, 242; Di-Et. ester, 146
32	9, 10-Anthraquinone-2-car-boxylic acid chloride	. . . .	147	. . . .	. . . .	. . . . . .	290, yel.	280	258–60	. . . .	. . . .	Me. ester, 170
33	Di-(1-naphthyl) acetyl chloride	. . . .	167–9	. . . .	. . . .	. . . . . .	228.5	. . . .	. . . .	. . . .	. . . .	. . . .
34	9, 10-Anthraquinone-2,6-di-carboxylic acid dichloride .	. . . .	197–8	. . . .	. . . .	. . . . . .	>400	>370	. . . .	. . . .	. . . .	. . . .
35	9,10-Anthraquinone-1,4-di-carboxylic acid dichloride .	. . . .	203–5	. . . .	. . . .	. . . . . .	>300	. . . .	. . . .	. . . .	. . . .	. . . .
36	9,10-Anthraquinone-1,5-di-carboxylic acid dichloride .	. . . .	260–3	. . . .	. . . .	. . . . . .	>390	. . . .	. . . .	. . . .	. . . .	Di-Me. ester, 236; Di-Et. ester, 155, yel.

*Derivative data given in order: m.p., crystal color, solvent from which crystallized.

TABLE XIII. ORGANIC DERIVATIVES OF ACYL HALIDES
III. Acyl Bromides a) Liquids 1) (Listed in order of increasing atmospheric b.p.)*

No.	Name	Boiling point, °C	Melting point, °C	n_D	Density g/ml	Acid B.p., °C	Acid M.p., °C	Amide	Anilide	p-Toluidide	2-Naphthyl amide	Miscellaneous
1	Oxalyl dibromide	64	......	....	....	......	101 (hyd.)	419d. (*di*)	254 (*di*)	268 (*di*)	....	....
2	Acetyl bromide	81	......	1.4538^{16}	1.6625^{16}_4	118	....	82	114	147	132	....
3	Propionyl bromide	103	......	....	....	141	....	81	106	126	....	....
4	Chloroacetyl bromide......	127	......	....	....	......	63	120	137	162	117–8	....
5	*n*-Butyryl bromide	128	......	....	....	162.5	....	115	96	75	125	....
6	Isovaleryl bromide	138–40	......	....	....	176	....	135	....	107	138.5	....
7	Trichloroacetyl bromide ...	143	......	1.90^{15}_{18}	....	197	57–8	162.5	95–7	113	....	4-Nitroanilide, 146–7
8	Bromoacetyl bromide......	150	......	....	2.425	208	50	91	131	....	134	....
9	DL-α-Bromopropionyl bromide...............	154–5	......	....	2.061^{16}_4	204	25.7	123	99; 110	....	....	....
10	α-Bromoisobutyryl bromide	162–4	......	....	....	198–200	48–9	148	83	92.5	135	....
11	DL-α-Bromobutyryl bromide...............	172–4	......	....	....	127^{25}	–4	112; 108	....	....	....	....
12	*n*-Hexanoyl bromide (*n*-Caproyl bromide)	175–6	......	....	....	205	....	100	95	75	107	....
13	DL-α-Bromoisovaleryl bromide...............	184–94	......	....	....	......	44	133	116	124	145	....
14	Benzoyl bromide	218–9	......	....	1.570^{15}	......	122	130	163	158	....	....

*Derivative data given in order: m.p., crystal color, solvent from which crystallized.

218

TABLE XIII. ORGANIC DERIVATIVES OF ACYL HALIDES

III. Acyl Bromides a) Liquids
2) (Reduced pressure b.p. only) (Listed in order of increasing m.p. of the corresponding amides)*

No.	Name	Boiling point, °C	Melting point, °C	n_D	Density g/ml	Acid B.p., °C	Acid M.p., °C	Amide	Anilide	p-Toluidide	2-Naphthyl amide	Miscellaneous
1	3-Methylbenzoyl bromide ...	137[52]	......	....	....	......	111	94; 97	126	118	....	....
2	n-Pentanoyl bromide (n-Valeryl bromide)	64[66]	......	....	....	186.5	....	106	63	74	....	....
3	3-Chlorobenzoyl bromide...	145[40]	......	....	....	......	158; 155	134	122	....	....	....
4	2-Chlorobenzoyl bromide...	144[37]	......	....	....	......	140	142	114; 118	131	....	....
5	2-Methylbenzoyl bromide ..	135[37]	......	....	....	......	104–5	143	125	144	....	....
6	2-Bromobenzoyl bromide...	167[18]	......	....	....	......	150	155	141	....	....	....
7	Phenylacetyl bromide......	150–5[50]	......	....	....	......	76	156	117–8	135–6	159	....
8	4-Methylbenzoyl bromide ..	147[42]	......	....	....	......	179–80	160	145; 148	160; 165	....	....
9	4-Methoxybenzoyl bromide.	185[27]	......	....	....	......	184–6	167; 163	169–71	186	....	....
10	4-Chlorobenzoyl bromide...	142[27]	......	....	....	......	240	179; 170	194	....	....	....
11	4-Bromobenzoyl bromide...	136[18]	......	....	....	......	251	189	197	....	....	....
12	Succinyl dibromide........	105–6[13]	......	....	....	......	186	260 (di)	230 (di)	255 (di)	....	....

*Derivative data given in order: m.p., crystal color, solvent from which crystallized.

TABLE XIII. ORGANIC DERIVATIVES OF ACYL HALIDES
III. Acyl Bromides b) Solids (Listed in order of increasing m.p.)*

No.	Name	Boiling point, °C	Melting point, °C	n_D	Density g/ml	Acid B.p., °C	Acid M.p., °C	Amide	Anilide	p-Toluidide	2-Naphthyl amide	Miscellaneous
1	3-Nitrobenzoyl bromide....	....	43	....	....	......	140	143	154	162	....	....
2	trans-Cinnamoyl bromide ..	....	48	....	....	......	133	148	151; 153	168	....	....
3	4-Iodobenzoyl bromide.....	....	55	....	....	......	270; 265	217	210	....	....	....
4	3,5-Dinitrobenzoyl bromide.	....	60	....	....	......	204–5	183	234	....	....	....
5	4-Nitrobenzoyl bromide....	....	64	....	....	......	241	201; 198	211	203	....	....
6	Phthaloyl dibromide.......	....	80	....	....	......	206	220 (di)	253 (di)	201 (di)	....	....

*Derivative data given in order: m.p., crystal color, solvent from which crystallized.

220

TABLE XIII. ORGANIC DERIVATIVES OF ACYL HALIDES
IV. Acyl Iodides. Liquids 1) (Listed in order of increasing atmospheric b.p.)*

No.	Name	Boiling point, °C	Melting point, °C	n_D	Density g/ml	Acid B.p., °C	Acid M.p., °C	Amide	Anilide	p-Toluidide	2-Naphthyl amide	Miscellaneous
1	Acetyl iodide	108	......	....	1.98¹⁷	118	....	82	114	147	134	....
2	Propionyl iodide	127	......	....	....	141	....	81	106; 103	126	....	....
3	n-Butyryl iodide	146–8	......	....	....	162.5	....	115	96	75	125	.......,.......

*Derivative data given in order: m.p., crystal color, solvent from which crystallized.

TABLE XIII. ORGANIC DERIVATIVES OF ACYL HALIDES
IV. Acyl Iodides. Liquids
2) (Reduced pressure b.p. only) (Listed in order of increasing m.p. of the corresponding amides)*

No.	Name	Boiling point, °C	Melting point, °C	n_D	Density g/ml	Acid B.p., °C	Acid M.p., °C	Amide	Anilide	p-Toluidide	2-Naphthyl amide	Miscellaneous
1	Dichloroacetyl iodide	55[15]	......	1.5754	....	194	....	98	118	153	....	....
2	Chloroacetyl iodide	37[4]	......	1.5903	....	......	63	120	137; 134	162	117–8	....
3	Benzoyl iodide	109[10]	......	....	....	......	122	130	163	158	....	....
4	Trichloroacetyl iodide	74[30]	......	1.5711	....	197	57–8	141	97; 94	113	....	....

*Derivative data given in order: m.p., crystal color, solvent from which crystallized.

TABLE XIV. ORGANIC DERIVATIVES OF ACID ANHYDRIDES
a) Liquids 1) (Listed in order of increasing atmospheric b.p.)*

No.	Acid anhydride	Boiling point, °C	Melting point, °C	n_D	Density g/ml	Acid B.P.	Acid M.P.	Amide	Anilide	p-Toluidide	2-Naphthyl-amide	Miscellaneous
1	Trifluoroacetic	39	.....	1.269^{25}	1.490^{25}_4	72	.....	75	88	....	....	....
2	Perfluoropropionic	72	.....	1.273^{25}	1.571^{25}_4	96	.....	95	....	....	....	....
3	Perfluoro-n-butyric	108	.....	1.285^{20}	1.665^{20}	120	.....	105	93	....	....	....
4	Acetic	140	−73	1.3904^{20}	1.0811^{20}	118	16	82	114	153	....	....
5	n-Propionic	167	−45	1.404^{20}	1.017^{15}	141	.....	81	106	126 (124)	....	....
6	Perfluoro-n-caproic (Perfluoro-n-hexanoic)	176	.....	1.295^{20}	1.769^{25}	157	.....	117	....	....	....	....
7	Isobutyric	182	.....	....	0.957^{17}	154	.....	128	105	107	....	....
8	Pivalic (Trimethylacetic)	190	.....	....	....	164	35	154	129	120	....	....
9	n-Butyric	198	.....	....	0.978^{15}	162	.....	115	96	75 (73)	125	....
10	Citraconic (Methylmaleic)	214	7–8	$1.471^{21.5}$	1.238^{25}_4	.....	92d.	185–7 (di)	175 (di)	....	....	....
11	Isovaleric	215	.....	....	....	176	.....	135 (137)	....	107	138	....
12	Dichloroacetic	216d.	.....	•.....	....	194	.....	98	118	153	....	....
13	Valeric (Pentanoic)	218	.....	....	0.922^{17}_4	186	.....	106	63	74	112	....
14	Crotonic	248	.....	1.4745^{20}	1.0397^{20}_4	189	72	161 (158)	118 (115)	132	....	....
15	Caproic (n-Hexanoic)	254–7 (245)	.....	1.4297^{20}	0.92^{20}	205	.....	100	95 (92)	75 (73)	107	....
16	n-Heptanoic	258	.....	1.4335^{15}	0.9175^{20}_4	223	.....	96	70 (65)	81	101	....
17	α-Methylglutaric	272–5	.....	....	....	.....	79	....	175–6 (di); mono: (2 forms) 114 or 100	174–5 (di); mono: (2 forms) 126 or 98–9	227–8 (di), 115–9 (mono)	....
18	Caprylic (n-Octanoic)	280–5	−1	1.436^{17}	0.9065^{17}_4	239	16	110 (106)	57 (55)	70	103	....
19	cis-Hexahydroisophthalic	304	.....	....	....	.....	187–9	....	298–9 (di)	....	....	....

*Derivative data given in order: m.p., crystal color, solvent from which crystallized.

223

TABLE XIV. ORGANIC DERIVATIVES OF ACID ANHYDRIDES

a) Liquids 2) (Reduced pressure b.p. only) (Listed in order of increasing m.p. of the corresponding amide derivative)*

No.	Acid anhydride	Boiling point, °C	Melting point, °C	n_D	Density g/ml	Acid B.P.	Acid M.P.	Amide	Anilide	p-Toluidide	2-Naphthyl-amide	Miscellaneous
1	**DL-α-Bromobutyric**.........	148–52^{10}	......	....	....	127^{25}	−4	112	....	....	....	4-Nitrophenyl ester, 48–9; 2-Naphthyl ester, 54
2	**DL-α-Bromopropionic**.......	120^5 (123–4^{10})	......	....	....	204	26	123	....	....	....	....

* Derivative data given in order: m.p., crystal color, solvent from which crystallized.

No.	Acid anhydride	Melting point, °C	Boiling point, °C	Acid B.P.	Acid M.P.	Amide	Anilide	p-Toluidide	2-Naphthyl-amide	Miscellaneous
1	Oleic	22	.....	.....	16	76	41	43	169	....
2	Capric (n-Decanoic)	24	.....	268–70	31	108; 98	70	78	104	D_D^{70}: 0.8596; n_D^{70}: 1.4234
3	β-Ethyl-β-methylglutaric	25	185[20]	.....	87	....	105 (mono)	....		1-Naphthylamide, 126
4	Hexahydrobenzoic	25	280–3	232	29–30	185–6				
5	α,α-Dimethylsuccinic (unsym.-Dimethylsuccinic)	29	220	.....	141	....				α-Me. ester, 41; β-Me. ester, 52; Anil, 87
6	cis-Hexahydrophthalic	32	145[18]	.....	192	....				Conc. HCl at 180 → trans form, 221
7	DL-Methylsuccinic	37	244–8		115 (112)	225 (di)	123, chl. (mono); 159, et, ac, (mono)	164 (mono)	155 (mono)	Anil, 109–10
8	n-Undecanoic (n-Hendecanoic)	37	.....	284	30	103; 99	71	80	....	
9	2-Methylbenzoic (o-Toluic)	39	.....		104–5	143	125	144		
10	β-Methylglutaric	41	276–8	.....	87	....	200 (di); 121 (117) (mono)	mono: 135	143 (mono)	
11	Bromoacetic	41–2	.....	208	50	91	131	....	134	
12	Lauric (n-Dodecanoic)	42	.....	299	44 (42)	110; 100	78	87	106	
13	Benzoic	42	360	.....	122	130	163	158		
14	trans-α,β-Dimethylsuccinic	43			198 (208)	238 (di); 165–7 (mono)	....			Imide, 78
15	Iodoacetic	46	.....	.....	83	95	143–4	....		
16	Chloroacetic	46	.....	189	63	121	134	162	117–8	
17	n-Tridecanoic	50	.....	312	44	100	80	88		
18	DL-Phenylsuccinic	54	204–6[22]	.....	168	209–10 (di); 158–9 (α-); 144–5 (β-)	222 (di); 175 (α-); 170–1 (β-)	175 (α-); 168–9 (β-)		Imide, 90
19	Myristic (n-Tetradecanoic)	54	.....	202[16]	54	107; 103	84	93	108	D_4^{70}: 0.8502, n_D^{70}: 1.4335
20	Glutaric	56	.....	200[20]	97	di: 175–6	224	218		
21	Maleic	56; 52–4	198	.....	130	181 (172) (mono); 266 (di)	173–5 (mono); 187 (di)	di: 142		
22	Suberic (dimer) (Octanedioic)	56–7			144 (141)	127 (mono); 217 (di)	128 (mono); 186 (di)	di: 218		
23	α-Bromoisobutyric	63–5	198–200	48–9		148	83	92	135	1-Naphthylamide, 116
24	Palmitic (n-Hexadecanoic)	64	.....	222[16]	63	106–7	90	98	109	D_4^{70}: 0.847; n_D^{70}: 1.4357
25	Margaric (n-Heptadecanoic)	67	.....	231[16]	61	108	....	....		
26	Itaconic (Methylenesuccinic)	67–8	.....	.....	165	di: 192	190; 185			
27	Sebacic (dimer) (Decanedioic)	68	.....	243[15]	133	210 (di); 170 (mono)	201 (di), 122 (mono)	201		
28	Stearic (n-Octadecanoic)	70	.....	.....	70	109	95	102	112	
29	3-Methylbenzoic (m-Toluic)	71	.....	111–3		94	126	118	....	Hydrazide, 97
30	Phenylacetic	72	.....		76–7	156	118	136	159	
31	2-Bromobenzoic	75–6	.....		150	155–6	....			Hydrazide, 153
32	Arachidic (n-Eicosanoic)	77–8	.....		77	108–9	92	96	112	
33	2-Chlorobenzoic	79	.....		142; 140	142	114; 118	131		
34	cis-α-Methylglutaconic	85	.....		118	....	148 (mono)			
35	cis-β-Methylglutaconic	86	.....		147–9	....	143 (mono)			Mono-Et. ester, 73
36	cis-α,β-Dimethylsuccinic	87	.....		129; 122	148–9 (mono); 244 (di)	222 (di)			Imide, 111; 101; Anil, 146
37	3,5-Dichlorophthalic	89	.....		164	....				Imide, 208; N-Phenyl-imide, 150
38	4-Methylphthalic	92	295	.....	152	188 (di)	....			Imide, 196
39	3-Chlorobenzoic	95	.....		158; 155	134	122			
40	4-Methylbenzoic (p-Toluic)	95	.....		179–80	160	145	160		
41	Diphenylacetic	98	.....		148	167–8	180	....		Me. ester, 60

*Derivative data given in order: m.p., crystal color, solvent from which crystallized.

No.	Acid anhydride	Melting point, °C	Boiling point, °C	Acid B.P.	Acid M.P.	Amide	Anilide	p-Toluidide	2-Naphthyl-amide	Miscellaneous
42	**4-Chlorophthalic**	99	. . . . . .	. . . . . .	157	. . . .	. . . .	. . . .	. . . .	Imide, 210–11, Di-Me. ester, 37; Anil, 174
43	**Anisic (4-Methoxybenzoic)**	99	. . . . . .	. . . . . .	184–6	167; 163	169–71	186	. . . .	
44	**DL-Benzylsuccinic**	102	. . . . . .	. . . . . .	161	. . . .	. . . .	. . . .	. . . .	Imide, 97–8; Dihydrazide, 146
45	**β-Phenylglutaric**	105	. . . . . .	. . . . . .	140	. . . .	171 (168) (*mono*)	154 (*mono*)	. . . .	Imide, 174; Di-Me. ester, 86–7
46	**4-Ethoxybenzoic**	108	. . . . . .	. . . . . .	198	202	170	. . . .	. . . .	Hydrazide, 124
47	**α-Benzylcinnamic**	108–9	. . . . . .	. . . . . .	158	. . . .	. . . .	. . . .	. . . .	Me. ester, b.p. 278; Et. ester, 38–9
48	**3,5-Dinitrobenzoic**	109	. . . . . .	. . . . . .	204–5	183	234	. . . .	. . . .	. . . .
49	**4-Bromophthalic**	113; 109	. . . . . .	. . . . . .	173–5; 166	. . . .	. . . .	. . . .	. . . .	Di-Me. ester, 40
50	**3-Methylphthalic**	114–5; 110	. . . . . .	. . . . . .	157	. . . .	. . . .	. . . .	. . . .	Imide, 189–90
51	**4-Nitrophthalic**	119	. . . . . .	. . . . . .	165	200d.	192	*mono:* 172	. . . .	
52	**Succinic**	120	261	. . . . . .	186	157 (*mono*); 260 (*di*)	148 (*mono*); 230 (*di*)	180 (*mono*); 255 (*di*)	. . . .	
53	**Nicotinic (3-Pyridine-carboxylic)**	123	. . . . . .	. . . . . .	237–8	128	85	150	. . . .	
54	**3-Chlorophthalic**	124–5	. . . . . .	. . . . . .	186	. . . .	. . . .	. . . .	. . . .	Imide, 118–20 (sealed tube)
55	**4-Iodophthalic**	125–6	. . . . . .	. . . . . .	182; 185	. . . .	. . . .	. . . .	. . . .	Imide, 224–4
56	**3,4-Dimethylphthalic**	126	. . . . . .	. . . . . .	201	. . . .	. . . .	. . . .	. . . .	Imide, 240–1; Methyl-imide, 98–9
57	**3-Bromophthalic**	132–4	. . . . . .	. . . . . .	188; 178	. . . .	. . . .	. . . .	. . . .	Mono-Et. ester, 127–8
58	**Phthalic**	132	. . . . . .	. . . . . .	206; 200	149 (*mono*); 220 (*di*)	170 (*mono*); 253–5 (*di*)	*mono:* 160	. . . .	
59	**3-Iodobenzoic**	134	. . . . . .	. . . . . .	187	187	. . . .	. . . .	. . . .	Me. ester, 54–5
60	**2-Naphthoic (β-Naphthoic)**	135	. . . . . .	. . . . . .	184	192–3	171–2	192	. . . .	
61	**2-Nitrobenzoic**	135	. . . . . .	. . . . . .	146	176	155	. . . .	. . . .	
62	**Cinnamic**	136	. . . . . .	. . . . . .	133	148	151; 153	168	. . . .	
63	*trans*-**DL-Hexahydrophthalic acid**	140	. . . . . .	. . . . . .	221	*mono:* 196	. . . .	. . . .	. . . .	Mono-Me. ester, 96; Di-Me. ester, 33
64	**Homophthalic (2-Carboxy-phenylacetic)**	141	. . . . . .	. . . . . .	180–1	230 (2-), 185 (α-)	231	. . . .	. . . .	α-Me. ester, 96–8; 2-Me. ester, 143–5; α-Et. ester, 107–8; Imide, 233
65	**1-Naphthoic (α-Naphthoic)**	146	. . . . . .	. . . . . .	162	202	163	. . . .	. . . .	Piperidide, 85–7
66	**3-Bromobenzoic**	148–9	. . . . . .	. . . . . .	155	155	. . . .	. . . .	. . . .	Hydrazide, 151
67	**3-Iodophthalic**	159–61	. . . . . .	. . . . . .	206	. . . .	. . . .	. . . .	. . . .	Di-Me, ester, 89; Di-Et. ester, 70; Imide, 238
68	**2,4-Dinitrobenzoic**	160	. . . . . .	. . . . . .	183	203	. . . .	. . . .	. . . .	
69	**3-Nitrobenzoic**	160; 163	. . . . . .	. . . . . .	140	143	154	162	. . . .	
70	**3-Nitrophthalic**	162	. . . . . .	. . . . . .	218	201 (*di*)	234 (*di*)	226 (*di*)	. . . .	
71	**3,5-Dinitrophthalic**	163–4; 161	. . . . . .	. . . . . .	226	. . . .	. . . .	. . . .	. . . .	Et. ester, 187; Di-Et. ester, 73
72	**4,5-Dichlorophthalic**	188	313	. . . . . .	200; 188	. . . .	. . . .	. . . .	. . . .	Mono-Et. ester, 133–4
73	**4-Nitrobenzoic**	189	. . . . . .	. . . . . .	241	201; 198	211; 204	204; 192	. . . .	
74	**4-Chlorobenzoic**	194	. . . . . .	. . . . . .	240	179; 170	194	. . . .	. . . .	
75	**3,6-Dichlorophthalic**	194–5	339	. . . . . .	194–5	. . . .	. . . .	. . . .	. . . .	Mono-Et. ester, 130–1; Di-Et. ester, 60; Imide, 242; Anil, 191
76	**3,4-Di-iodophthalic**	198	. . . . . .	. . . . . .	212–3	. . . .	. . . .	. . . .	. . . .	Anil, 270
77	**β-Phenylglutaconic**	206	. . . . . .	. . . . . .	154–5	138 (*mono*)	174 (*mono*)	184 (*mono*)	. . . .	Mono-Et. ester, 78; Imide, 256–7
78	**4,5-Dimethylphthalic**	208	. . . . . .	. . . . . .	123;196	. . . .	. . . .	. . . .	. . . .	Methylimide, 150; Ethylimide, 89

*Derivative data given in order: m.p., crystal color, solvent from which crystallized.

TABLE XIV. ORGANIC DERIVATIVES OF ACID ANHYDRIDES

b) Solids (Listed in order of increasing m.p.)* (Continued)

No.	Acid anhydride	Melting point, °C	Boiling point, °C	Acid B.P.	Acid M.P.	Amide	Anilide	p-Toluidide	2-Naphthyl-amide	Miscellaneous
79	**2,2'-Diphenic**...............	217	......	......	229	191 (*mono*); 212 (*di*)	176 (*mono*); 230 (*di*)	....	....	H$_2$SO$_4$ at 100-120 → Fluorenone-4-carboxylic acid, 227
80	**4-Bromobenzoic**	218	......	......	251	189	197	....	....	....
81	D-**Camphoric**...............	221	......	......	188	177 (*mono*); 193 (*di*)	209 (204) (*mono*); 226 (*di*)	α: 212–4; β: 190–6	....	....
82	**4-Iodobenzoic**	228	......	......	270; 267	217–8	....	....	....	Me. ester, 114
83	**Tetrachlorophthalic**	256; 249	......	......	250 d.	....	....	....	....	Mono-Me. ester, 142; Mono-Et. ester, 94–5; Imide, 338–9; 2-Naphthylimide, 287; Anil, 268–9
84	**1,8-Naphthalenedicarboxylic** ...	274	......	......	274	....	250–82 (*di*)	....	....	Heat with aq. NH$_3$ → 1,8-Naphthalimide, 300; N-Phenylimide, 202; Di-Me. ester, 102–3
85	**Tetrabromophthalic**	280	......	......	266	....	....	....	....	Mono-Me. ester, 267; Imide, > 380, yel.; Anil, 279–80; p-Tolil, 280; 2-Naphthylimide, 306–8
86	**Tetraiodophthalic**	318; 325, yel.	......	......	324–7	....	....	....	....	Mono-Me. ester, 298

*Derivative data given in order: m.p., crystal color, solvent from which crystallized.

Hydrolysis of amide or imide to the corresponding carboxylic acid and amine.

$$RCONHR' + H_2O + HCl \rightarrow RCOOH + NH_3R'^+Cl^-$$

Acid $\qquad\qquad \downarrow OH^-$

$$NH_2R'$$

Amine

$$RCONHR' + NaOH \rightarrow RCOONa + NH_2R'$$

$\downarrow H^+$ $\qquad\qquad$ Amine

$$RCOOH$$

Acid

From the amide with aqueous hydrochloric acid.

For directions and examples see: Cheronis, pp. 607, 608; Vogel, pp. 404, 808.

From the amide with 85% or 100% phosphoric acid.

See: Cheronis, p. 609; G. Berger and S. C. J. Olivier, *Rec. Trav. chim.*, **46**, 600 (1927); W. M. Dehn and K. E. Jackson, *J. Amer. Chem. Soc.*, **55**, 4284 (1933).

From the amide with 70% sulfuric acid.

See: Wild, p. 193.

From the amide with aqueous sodium hydroxide.

See: Cheronis, p. 609; Vogel, pp. 404, 799; Wild, p. 193.

NOTE: For directions and examples for preparation of derivatives of carboxylic acids formed on hydrolysis of amides and imides see explanations and references to Tables XII, XIII and XIV, p. 186, 187, 188, 189.

For directions and examples for preparation of derivatives of amines formed on hydrolysis of amides and imides see explanations and references to Table XVIII, p. 291, 292, 293, 294.

N-Xanthylamides.*

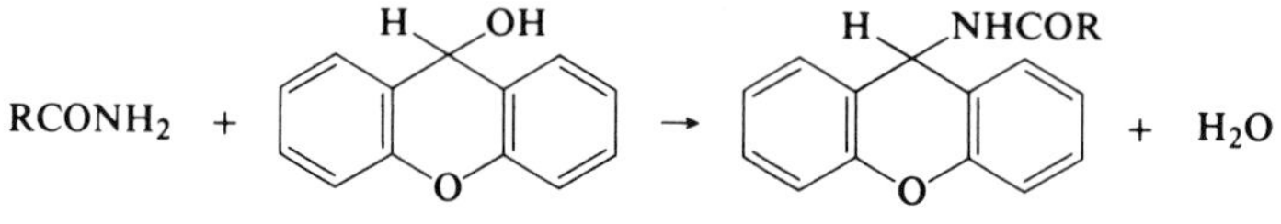

N-Xanthylamide

From the amide with xanthydrol in glacial acetic acid.

For directions and examples see: Cheronis, p. 610; Linstead, p. 66; Shriner, p. 222; Vogel, p. 405; Wild, p. 195; R. F. Phillips and B. M. Pitt, *J. Amer. Chem. Soc.*, **65**, 1355 (1943); W. Andriani, *Rec. Trav. chim.*, **35**, 180 (1916).

From the amide with xanthydrol in ethanol-water-acetic acid mixture.

See: Shriner, p. 222; Wild, p. 195; R. F. Phillips and B. M. Pitt, *J. Amer. Chem. Soc.*, **65**, 1355 (1943).

Hg salt (Hg derivative).*

$$RCONH_2 + HgO \rightarrow (RCONH)_2Hg + H_2O$$

Mercuric salt

From the amide with mercuric oxide in methanol or ethanol.

For directions and examples see: Cheronis, pp. 610, 611; Wild, p. 197; J. W. Williams, W. T. Rainey and R. S. Leopold, *J. Amer. Chem. Soc.*, **64**, 1738 (1942).

From the amide with mercuric oxide in water.

See: Vogel, p. 405.

From the amide with yellow mercuric oxide without solvent.

See: Cheronis, p. 611; Wild, p. 196; J. W. Williams, W. T. Rainey and R. S. Leopold, *J. Amer. Chem. Soc.*, **64**, 1738 (1942).

Oxalate.

$$RCONH_2 + (COOH)_2 \rightarrow RCONH_2 \cdot (COOH)_2$$

Amide oxalate

From the amide with anhydrous oxalic acid in the presence of ethyl acetate.

---

*Derivatives recommended for first trial.

WARNING: This is not an instruction manual. References should be consulted for the preparation of derivatives.

For directions and examples see: Cheronis, p. 611; Wild, p. 196; C. A. MacKenzie and W. T. Rawles, *Ind. Eng. Chem., Anal. Ed.,* **12,** 737 (1940).

N-Acylphthalimide (Phthalimide derivative).

$$RCONH_2 \; + \; \underset{\text{Phthaloyl chloride}}{\text{COCl} \atop \text{COCl}} \; \rightarrow \; \underset{\text{N-Acylphthalimide}}{\text{CO} \atop \text{CO}} NCOR \; + \; 2\,HCl$$

From the amide with phthaloyl chloride in toluene or without solvent.

For directions and examples see: Cheronis, p. 611; T. W. Evans and W. M. Dehn, *J. Amer. Chem. Soc.,* **51,** 3651 (1929).

*Derivatives recommended for first trial.
WARNING: This is not an instruction manual. References should be consulted for the preparation of derivatives.

TABLE XV. ORGANIC DERIVATIVES OF AMIDES AND IMIDES
a) Liquids (Listed in order of increasing b.p.)*

No.	Name	Boiling point, °C	Melting point, °C	n_D^{20}	D_4^{20}	Derived acid				Derived amine				Miscellaneous
						M.P., °C	B.P., °C	p-Nitro benzyl ester	p-Bromo-phenacyl ester	M.P., °C	B.P., °C	Acet-amide	Benz-amide	
1	N,N-Dimethylformamide	153; 76[39]	−61	1.42938[22.4]	0.9484[22.4]	8.4	100.7	31	140; 135	....	7	.....	41	....
2	N,N-Diethylformamide	176–8; 68[15]	.....	....	0.908[16]	8.4	100.7	31	140; 135	....	56	.....	42	....
3	N-Methylformamide	180–5; 131[90]	−3.8; −5.4	1.4310[25]	1.011[19]	8.4	100.7	31	140; 135	....	−6	28	80	....
4	Formamide	193; 195d.	2.55	....	....	8.4	100.7	31	140; 135	NH₃		.....		Xanthyl deriv., 184; Oxalate, 107.4–7.7
5	N-Ethylformamide	197–9	.....	....	0.952[21]	8.4	100.7	31	140; 135	....	16.5; 19	.....	71	....
6	N-Formylpiperidine (Form-N-piperidide)	222	.....	....		8.4	100.7	31	140; 135	....	106	.....	48	
7	N-Acetylpiperidine (Aceto-N-piperidide)	226	.....	....		166	118.2	78	86.0	....	106	.....	48	
8	N-Methylformanilide	243–4; 249–51; 128–9[15]	12.5	....	1.0928[23]	8.4	100.7	31	140; 135	....	196	102	63	....
9	N-Ethylformanilide	258[728]; 123[11]	.....	....	1.0549	8.4	100.7	31	140; 135	....	205	54	60	....
10	N-Propylformanilide	267[731] (cor.)	.....	....	1.044[16]	8.4	100.7	31	140; 135	....	222	47	....	....
11	N-Isobutylformanilide	274[731] (cor.)	.....	....		8.4	100.7	31	140; 135	....	227	.....		....
12	N-Isoamylformanilide	285–6[728]	.....	....	1.004[16]	8.4	100.7	31	140; 135	....	254.5 (cor.)			....

*Derivative data given in order: m.p., crystal color, solvent from which crystallized.

TABLE XV. ORGANIC DERIVATIVES OF AMIDES AND IMIDES
b) Solids (Listed in order of increasing m.p.)*

No.	Name	Melting point, °C	Xanthyl-amide	Derived acid				Derived amine				Miscellaneous
				M.P., °C	B.P., °C	p-Nitro-benzyl ester	p-Bromo-phenacyl ester	M.P., °C	B.P., °C	Acet-amide	Benz-amide	
1	n-Butyranilide..............	35	......	− 5.5; − 8	162.5;164	....	63	.....	184	114	160	....
2	Oleanilide..................	41	......	α: 13.36; β: 16.25	250 (super-heated steam); 216[5]	....	40; 46	.....	184	114	160	....
3	N-Benzpiperidide (N-Benzoyl piperidine)..............	48	......	122.4	249	89	119.0	.....	106		48	....
4	Ethyl urethane (Ethyl carbamate)...............	49; 48	169	....	....	....	....	NH₃				
5	Formanilide	50; 47	......	8.4	100.7	31	140; 135	.....	184	114	160	B.p. 271: Benzyl chloride → benzyl form-anilide, 48
6	Malonic acid monoamide (Malonamic acid)	50	......	134.8–.9	....	di: 85.5	....	NH₃				....
7	N-Propylacetanilide.........	50	......	16.6	118.2	78	86.0	.....		47		....
8	Difluoroacetamide	52	......	....	134–5	....	....	NH₃		....		....
9	N-Benzyl-n-caproamide......	53	......	− 3.9	203.35	....	72	.....	184–5	60	105	....
10	Phenyl urethane (Phenyl carbamate)..............	53	......	....	....	....	....	.....	184	114	160	....
11	Methyl urethane (Methyl carbamate)..............	54; 52	193	....	....	....	....	NH₃				....
12	N-Ethylacetanilide..........	54	......	16.6	118.2	78	86.0	.....	205	54	60	B.p. 249
13	n-Butyl urethane (n-Butyl carbamate)...............	54	......	....	....	....	....	NH₃		....		....
14	N-Benzylisovaleramide	54	......	− 30	176.5	68.0	135; 137	.....	184–5	60	105	....
15	Acetoacetamide	54	......	....	....	....	....	NH₃		....		....
16	Isobutyl urethane (Isobutyl carbamate)..............	55	148	....	....	....	....	NH₃				....
17	N-Methyl-2-acetotoluidide	56	......	16.6	118.2	78	86.0	.....	208	56	66	....
18	Carprylanilide	57;55	......	16.3	237; 239.3	....	67.4	.....	184	114	160	....
19	Pelargonanilide	57	......	12.3	254.4	....	68.5	.....	184	114	160	....
20	N-Benzylacetanilide.........	58	......	16.6	118.2	78	86.0	37	298	58	107	....
21	Methoxyacetanilide.........	58	......	....	204; 203	....	....	.....	184	114	160	....
22	d,l-Lactanilide	59	......	18	122[15]	....	112.8	.....	184	114	160	....
23	N-Ethylbenzanilide	60	......	122.4	249	89	119.0	.....	205	54	60	....
24	n-Propyl urethane (n-Propyl carbamate)...............	60	......	....	....	....	....	NH₃		....		....
25	N-Benzylformamide.........	60	......	8.4	100.7	31	140; 135	.....	184–5	60	105	....
26	N-Benzylacetamide	61	......	16.6	118.2	78	86.0	.....	184–5	60	105	....
27	Propiolamide	61–2	......	18	144d.	....	....	NH₃		....		....
28	n-Valeranilide (n-Pentananilide)..........	63	......	− 34.5	186.4	....	75	.....	184	114	160	....
29	Isoamyl urethane (Isoamyl carbamate)...............	64	145	....	....	....	....	NH₃		....		....
30	Erucanilide	65; 55	......	33–4	264[15]	....	62.5	.....	184	114	160	....
31	3-Acetotoluidide (N-Acetyl-m-toluidine)	66; 65	......	16.6	118.2	78	86.0	.....	203	65	125	....
32	N-Methyl-3-acetotoluidide	66	......	16.6	118.2	78	86.0	.....	206–7	66	125	....
33	Ethyl oxanilate..............	66–7	......	....	....	....	....	.....	184	114	160	....
34	Benzindole	68	......	122.4	249	89	119	52	253	157–8	68	....
35	Heptananilide..............	70; 65	......	− 7.46	223	....	72	.....	184	114	160	....
36	Capranilide (n-Decananilide) ..	70	......	31.3	268.7	....	60	.....	184	114	160	....
37	n-Undecananilide (n-Hendecananilide)........	71	......	28.5; α: 13.4; β: 16.3	280; 284	....	68.2	.....	184	114	160	....

*Derivative data given in order: m.p., crystal color, solvent from which crystallized.

TABLE XV. ORGANIC DERIVATIVES OF AMIDES AND IMIDES

b) Solids (Listed in order of increasing m.p.)* (Continued)

No.	Name	Melting point, °C	Xanthyl-amide	Derived acid				Derived amine				Miscellaneous
				M.P., °C	B.P., °C	p-Nitro-benzyl ester	p-Bromo-phenacyl ester	M.P., °C	B.P., °C	Acet-amide	Benz-amide	
38	2-Methylhexanamide	72	......	......	209.6			NH$_3$	......	......		
39	N,N-Diphenylformamide	73	......	8.4	100.7	31	140; 135	53–4	......	101	180	......
40	Vinylacetanilide	73	......	−35	169; 163				184	114	160	
41	Oleamide	76	......	α: 13.36 β: 16.25	250 (super-heated steam); 216[5]	......	40–6	NH$_3$		......	......	
42	Thioacetanilide	76	......	......	93	......			184	114	150	......
43	Tiglamide	76	.....	64.5–5.0	198.5 (cor.)	64	68	NH$_3$				
44	Tiglanilide	77	.....	64.5–5.0	198.5 (cor.)	64	68		184	114	160	
45	4-Methylhexananilide	77	......	......	217–8[254]				184	114	160	
46	Lauranilide (Dodecan-anilide)	78	......	44; 42	299	......	76		184	114	160	
47	d,l-2,3-Dimethylbutananilide	78	......	−1.5	191.7	......			184	114	160	
48	Pentadecananilide	78	......	52.3	212[16]	39.5–40 (cor.)	77.2		184	114	160	
49	d,l-Lactamide	78.5–9.0; 76		18	122[15]	......	112.8	NH$_3$		......	......	
50	Transbrassidanilide	79	......	59.7	256[10]	......	94.2		184	114	160	
51	2-Acetophenetidide (N-Acetyl-o-phenetidine)	79	......	16.6	118.2	78	86.0		229	79	104	
52	d,l-2-Methylpentanamide	79.6; 80	......	......	195–6			NH$_3$				
53	Tridecananilide	80	......	43; 41.6	312; 177[10]	......	75.0		184	114	160	
54	d,l-α-Chloropropionamide	80	......		186			NH$_3$				
55	Propionamide	81; 77	214; 211	−20.8	141	31	63.4	NH$_3$		......	......	Mercury deriv., 201; Oxalate, 80.8–1.0
56	Acetamide	82	245; 238–40	16.6	118.2	78	86.0	NH$_3$		......	......	Mercury deriv., 196–7; Oxalate : 127.3; Phthali-mide, 135–6
57	Ethoxyacetamide	82	......		206–7	......	104.8	NH$_3$		......		
58	N-Methyl-4-acetotoluidide	83	......	16.6	118.2	78	86.0		210	83	......	
59	α-Bromoisobutyranilide	83	......	48–9	198–200				184	114	160	
60	γ-Phenylbutyramide	84	......	52	290			NH$_3$				
61	Myristanilide	84	......	53.9	212[16]	......	81		184	114	160	
62	Acrylamide	85	......	13	141; 140			NH$_3$				
63	Allylurea	85							58			
64	Acetoacetanilide	85							184	114	160	
65	d,l-α-Ethylphenylacetamide (α-Phenylbutyramide)	86; 83		42	270			NH$_3$				
66	4-n-Propylacetanilide	87	......	16.6	118.2	78	86.0		225	87	115	
67	α-Undecylenamide (α-Hendecyleneamide)	87		24.5	275			NH$_3$				
68	Propiolanilide	87	......	18	144d.	......	......		184	114	160	
69	3-Bromoacetanilide	87	......	16.6	118.2	78	86.0	18	251	87	120; 136	
70	d,l-3-Methylpentananilide	87	......	−41.6	197.5	......			184	114	160	
71	n-Butyl oxamate	88	......	189.5 (anh.); 101(+ 1H$_2$O)	......	di: 204		NH$_3$		......		
72	2-Chloroacetanilide	88	......	16.6	118.2	78	86.0		209; 207	87	99	
73	D-Chaulmoogranilide	89	......	68.5	247–8[20]	......			184	114	160	......
74	Palmitanilide	90	......	62.7	222[16]	42.5	86;82		184	114	160	
75	N-Phenylmaleimide	91	......	130	......	di: 91 (cor.)	168–70; 190		184	114	160	

*Derivative data given in order: m.p., crystal color, solvent from which crystallized.

No.	Name	Melting point, °C	Xan-thyl-amide	Derived acid				Derived amine				Miscellaneous
				M.P., °C	B.P., °C	p-Nitro-benzyl ester	p-Bromo-phenacyl ester	M.P., °C	B.P., °C	Acet-amide	Benz-amide	
76	Bromoacetamide	91	......	50	208	88	....	NH₃		......	......	
77	α-Phenylpropionamide.......	92	158	....	265	....	....	NH₃		......	......	
78	Isopropyl urethane (Isopropyl carbamate).......	92	......	....	....	....	....	33		......	......	
79	Arachidanilide	92	......	77; 75	204¹	....	89	......	184	114	160	
80	d,l-α-Chloropropionanilide	92	......	....	186	....	....	......	184	114	160	
81	2,2-Dimethylbutananilide	92	......	−15.0	187; 190	....	....	......	184	114	160	
82	2-Nitroacetanilide	92; 94	......	16.6	118.2	78	86.0	71	......	92; 94	98; 110	
83	Maleimide.................	93	......	130	....	di: 91	168–70; 190	NH₃				
						(cor.)						
84	Elaidamide	93–4	......	44–5; 51	234¹⁵	....	65	.. NH₃		......	......	
85	Transbrassidamide..........	94	......	59.7	256¹⁰	....	94.2	NH₃		......	......	
86	2-Ethylpentananilide	94	......	....	209	....	....	......	184	114	160	
87	N-Methyl-N-(1-naphthyl) acetamide	94	......	16.6	118.2	78	86.0	......	294	94–5	121	
88	n-Caproanilide (n-Hexananilide)	95; 92	......	−3.9	205.35	....	72	......	184	114	160	
89	N-Butyranilide	95	......	−5.5; −8	162.5; 164	35	63	......	184	114	160	
90	N-Benzylpalmitamide	95	......	62.7	222¹⁶	42.5	86; 82	......	184–5	60	105	
91	Azelaic acid monoamide	95	......	106.5	>360 sl. d.; 237¹⁵	di: 43.8	di: 130.6	NH₃		......	......	
92	d,l-2-Methylpentananilide	95	......	83	195–6	....	....	......	184	114	160	
93	Iodoacetamide	95	......	83	....	....	....	NH₃		......	......	
94	Stearanilide	95	......	70.1; 69.6	....	....	92	......	184	114	160	
95	Heptanamide	96	154–5	−7.46	223.0	....	72	NH₃		......	......	
96	Trichlorolactamide..........	96	......	124	....	....	....	NH₃		......	......	
97	Semicarbazide	96	......	....	....	....	....	1.4	113.5	mono: 67; di (sym): 138	mono: 172–5; di (sym): 241 (cor.)	Gives semicar-bazones with aldehydes and ketones
98	3-Toluamide	97; 94	......	111–3; 110–1	236, subl.	86.6	108	......	......	......	......	Mercury deriv., 200
99	Trichloroacetanilide..........	97; 94	......	57–8	197.5	80	....	......	184	114	160	
100	3-Acetophenetidide..........	97	......	16.6	118.2	78	86.0	......	248	37	103	
101	N-Benzylstearamide.........	97	......	70.1; 69.6	....	....	92	......	184–5	60	105	
102	Glycollanilide	97	......	78–9; 80	....	106.8	138	......	184	114	160	
103	Methoxyacetamide..........	97	......	....	204; 203	....	....	NH₃		......	......	
104	Hydrocinnamanilide (β-Phenylpropionanilide)	98; 96	......	48.7; 40	279–80 (cor.)	36.3	104	......	184	114	160	
105	α-Hydroxyisobutyramide	98	......	79	212	80.5	98	NH₃		......	......	
106	Dichloroacetamide..........	98, subl.	......	5–6	194	....	99	NH₃		......	......	
107	2-Methylhexananilide	98	......	....	209.6	....	....	......	184	114	160	
108	4-Methylhexanamide.........	98	......	....	217–8⁷⁴⁵	....	....	NH₃		......	......	
109	4-Methyl-2-nitroacetanilide ...	99; 94	......	16.6	118.2	78	86.0	117	......	99	148	
110	Capramide (n-Decanamide) ...	99	148	31.3	268.7	....	67.0	NH₃		......	......	
111	2-Bromoacetanilide	99	......	16.6	118	78	86.0	32	250	99	116	
112	Phenoxyacetanilide	99	......	98–9; 99–100	285d.	....	148.5	......	184	114	160	
113	Pelargonamide	99	147.5–8.5	12.3	254.4	....	68.5	NH₃		......	......	
114	α-Bromopropionanilide	99; 110	......	25.7	203.5	....	....	......	184	114	160	
115	n-Caproamide (n-Hexaneamide)..........	100; 101	160	−3.9	203.35	....	72	NH₃		......	......	Oxalate, 71.1–.3
116	Tridecanamide	100	......	43; 41.6	312; 177¹⁰	....	75	NH₃		......	......	
117	Phenylpropiolamide..........	100; 109	......	136–7, subl.	....	83	....	NH₃		......	......	

*Derivative data given in order: m.p., crystal color, solvent from which crystallized.

TABLE XV. ORGANIC DERIVATIVES OF AMIDES AND IMIDES

b) Solids (Listed in order of increasing m.p.)* (Continued)

No.	Name	Melting point, °C	Xan-thyl-amide	Derived acid				Derived amine				Miscellaneous
				M.P., °C	B.P., °C	p-Nitro-benzyl ester	p-Bromo-phenacyl ester	M.P., °C	B.P., °C	Acet-amide	Benz-amide	
118	N,N-Diphenylacetamide	101	......	16.6	118.2	78	86.0	53–4	......	101	180	....
119	β-Iodopropionamide.........	101 ...		82; 85	......	......	......	NH₃		......	......	
120	N,N'-Diacetyltrimethylene-diamine	101	......	16.6	118.2	78	86.0	......	136	mono: 126; di: 101	mono: 140; di: 147	
121	Methylurea.................	101	230	....	....	....	....		–6	28	80	Picrate, 127d.
122	Phenoxyacetamide..........	101	......	98–9; 99–100	285d.	......	148.5	NH₃		......	......	
123	N-Methylacetanilide	102	......	16.6	118.2	78	86.0	......	196	102	63	B.p. 237; Conc. HNO₃ + H₂SO₄ → p-Nitro deriv., 153
124	Levulinanilide	102	......	33–5, deliq.	245–6	61	84		184	114	160	....
125	Pentadecanamide	102	......	52.3	212[16]	39.5–40 (cor.)	77.2	NH₃		......	......	
126	Isocrotonamide.............	102	......	15	169	......	81	NH₃		......	......	
127	Isocrotonanilide	102	......	15	169	......	81		184	114	160	
128	n-Undecanamide (n-Hendecanamide)	103; 99	......	28.5; α:13.4; β:16.3	280–4	......	68.2	NH₃		......	......	
129	3-Benzophenetidide (N-Benzoyl-m-phenetidine)...	103	......	122.4	249	89	119	......	248	97	103	
130	N,N-Dicyclohexylacetamide...	103	......	16.6	118.2	78	86.0	20	225 sl. d.	103	153	
131	3,4-Dimethylbenzanilide	104	......	166; 164	....	....	....		184	114	160	
132	2-Benzophenetidide (N-Benzoyl-o-phenetidine)...	104	......	122.4	249	89	119		229	79	104	
133	Pyruvanilide	104, subl.	......	13.6	165d; 80[25]				184	114	160	
134	N-Cyclohexacetamide......	104	......	16.6	118.2	78	86.0		134	104	149	
135	2-n-Propylacetanilide........	104–5	......	16.6	118.2	78	86.0		222–4	104–5	119	
136	2-Ethylpentanamide	105; 103	......		209	......		NH₃				
137	Isobutyranilide	105	......	–46.1	154.7		76.8		184	114	160	
138	4-n-Butylacetanilide........	105	......	16.6	118.2	78	86.0		261	105	126	
139	Acrylanilide	105	......	13	141; 140	......			184	114	160	
140	β-Phenylpropionamide (Hydrocinnamamide)	105	189	48.7; 40	279–80 (cor.)	36.3	104	NH₃		......	......	
141	Propionanilide	106; 103	......	–20.8	141	31	63.4		184	114	160	Conc. HNO₃ + H₂SO₄ → p-Nitro deriv., 182
142	Palmitamide.................	106	142	62.7	222[16]	42.5	86; 82	NH₃		......	......	
143	N-Benzylbenzamide..........	106	......	122.4	249	89	119	......	184–5	60	105	
144	D-Chaulmoogramide	106	......	68.5	247–8[20]	......		NH₃		......	......	
145	sym-Dimethylurea	106	......	......	......	......	......		–6	28	80	
146	n-Valeramide (n-Pentanamide).	106	167	–34.5	186.4	......	75	NH₃		......	......	Oxalate, 61.1–.4
147	Myristamide................	107; 103	......	53.9	212[16]	......	81	NH₃		......	......	
148	N-Benzylbenzanilide.........	107	......	122.4	249	89	119	37	298	58	107	
149	Margaramide (Heptadecanamide)........	108	......	61.2	231[16]	48.5–9.0 (cor.)	82.6	NH₃		......	......	
150	Thioacetamide	108	......	93	....	......	......	NH₃				
151	Fluoroacetamide............	108	......	31–2	167–9	......	......	NH₃				
152	Levulinamide	108d.	......	33–5, deliq.	245–6	61	84	NH₃				
153	Azelaic acid monoanilide	108	......	106.5	> 360 sl. d.; 237[15]	di: 143.8	di: 130.6	......	184	114	160	
154	Arachidamide	108–9	......	77; 75	204[1]	......	89	NH₃		......	......	

*Derivative data given in order: m.p., crystal color, solvent from which crystallized.

TABLE XV. ORGANIC DERIVATIVES OF AMIDES AND IMIDES
b) Solids (Listed in order of increasing m.p.)* (Continued)

No.	Name	Melting point, °C	Xan-thyl-amide	Derived acid				Derived amine				Miscellaneous
				M.P., °C	B.P., °C	p-Nitro-benzyl ester	p-Bromo-phenacyl ester	M.P., °C	B.P., °C	Acet-amide	Benz-amide	
155	Anthranilamide..............	109	......	147d.	....	......	......	NH₃	......	......	......	....
156	2-Iodoacetanilide	109	......	16.6	118.2	78	86	61; 58	......	109	139	....
157	Pimelic acid monoanilide	109	......	104–5, subl.	223¹⁵	....	di: 136.6	......	184	114	160	
158	Stearamide	109	139–41	70.1; 69.6	....	....	92	NH₃	......	......	......	
159	α-Methylhydrocinnamamide...	109	......	....	....	....	......	NH₃				
160	Lauramide (Dodecanamide)...	170; 102	......	44; 42	299	....	76	NH₃				
161	n-Caprylamide (n-Octanamide)	110; 108	148	16.3	237; 239.3	....	67.4	NH₃	......	......	......	
162	Isovaleranilide.............	110	......	−30.0	176.5	....	68.0	......	184	114	160	
163	d,l-2-Methylbutananilide......	110	......	....	176–7; 174	....	55	......	184	114	160	
164	3-Nitrophenylacetamide.......	110	......	120	....	....	......	NH₃				
165	2-Ethylacetanilide	111	......	16.6	118.2	78	86.0	......	210–1	111	147	
166	β-Bromopropionamide........	111	......	62.5	....	....	......	NH₃				
167	3-Aminobenzamide...........	111	......	174d.	....	....	......	NH₃				
168	2-Ethylbutanamide...........	112; 107	......	−31.8	195	....	......	NH₃				
169	4-Methylpentananilide (Isocaproanilide)...........	112; 110	......	−33	199.1⁷⁵²	....	77.3	......	184	174	160	
170	d,l-2-Methylbutanamide	112	......	....	176–7; 174	....	55	NH₃				
171	2-Acetotoluidide............	112	......	16.6	118.2	78	86.0	200	......	110–1¹	146	KMnO₄ → Acetylanthra-nilic acid, 185
172	d-Hydnocarpamide (α-Cyclo-pentylundecylamide).......	112–3	......	60.5	....	....	......	NH₃				
173	4-Aminobenzamide..........	114	......	118d.	....	....	......	NH₃				
174	Acetanilide	114 (cor.)	......	16.6	118.2	78	86.0	......	184	114	160	B.p.304; Conc. HNO₃ + H₂SO₄ → p-nitro deriv., 210
175	N-Benzylcrotonamide	114	......	72	189 (cor.)	67.4	95–6	......	184–5	60	105	
176	Ethyl oxamate	114	......	189.5 (anh.); 101(+ 2H₂O) (rapid htng.)	....	di: 204	......	16.5; 19	......	......	71	
177	2-Chlorobenzanilide.........	114; 118	......	142; 140	....	106	106	......	184	114	160	....
178	Dihydroacetic acid monoanilide	115	......	109	270	....	......	......	184	114	160	
179	n-Butyramide	115	185–7	−5.5; −8	162.5; 164	....	63	NH₃	......	......	......	Mercury deriv., 222–4; Oxalate, 65.9–6.2
180	Methacrylamide.............	116	......	16	161	....	......	NH₃				
181	α-Bromoisovaleranilide	116	......	44	230d.	....	......	......	184	114	160	
182	trans-α-Crotonanilide	118; 115	......	72, w.	189 (cor.)	67.4	95–6	......	184	114	160	....
183	Phenylacetanilide............	118	......	76.5, subl.	256.5 (cor.)	65	89	......	184	114	160	Conc. HNO₃ → 2,4-dinitro-phenylacetic acid, 189
184	Dichloroacetanilide	118	......	5–6	194	....	99	......	184	114	160	....
185	N-Ethyl-4-nitroacetanilide.....	118	......	16.6	118.2	78	86.0	96	......	119	98	....
186	Trimesic acid trianilide	118–20d.	......	380 (cor.)	....	....	tri: 197 (sealed tube)	......	184	114	160	....
187	3-Iodoacetanilide	119	......	16.6	118.2	78	86.0	33; 27	......	119	157	....
188	Cyanoacetamide.............	120	222–3	66	....	....	......	NH₃	......	......	......	

*Derivative data given in order: m.p., crystal color, solvent from which crystallized.

235

TABLE XV. ORGANIC DERIVATIVES OF AMIDES AND IMIDES
b) Solids (Listed in order of increasing m. p.)* (Continued)

No.	Name	Melting point, °C	Xanthylamide	Derived acid				Derived amine				Miscellaneous
				M.P., °C	B.P., °C	p-Nitrobenzyl ester	p-Bromophenacyl ester	M.P., °C	B.P., °C	Acetamide	Benzamide	
189	Glycolamide	120	.....	78–9; 80	....	106.8	138	NH_3		.....	.....	
190	Isocaproamide (Isohexanamide)	120–1	159–60	−33	199.1[752]	....	77.3	NH_3				
191	Chloroacetamide	121	208–9	α:61.3; β:56.2; γ:52.5	189	....	104			.....		
192	2-Chlorophenoxyacetanilide	121	.....	145–6, w.	.....	.....	.....		184	114	160	
193	Tribromoacetamide	122	.....	131; 135	245d.	.....	.....	NH_3				
194	Sebacic acid monoanilide	122	.....	33, subl.	243[15]	di: 73.5; 72.6	di: 147		184	114	160	
195	3-Chlorobenzanilide	122–5	.....	158; 155	....	107	116		184	114	160	
196	α-Bromopropionamide	123	.....	25.7	203.5	.....	.....	NH_3				
197	Furanilide	123.5	.....	133–4; 132	230–2	133.5	138.5		184	114	160	
198	Pyruvamide	124	.....	13.6	165d.; 80[25]	.....	.....	NH_3				
199	3-Methylpentanamide	125	.....	−41.6	197.5	.....	.....	NH_3				
200	4-Chlorophenoxyacetanilide	125	.....	155–6, w.; 158	.....	.....	136		184	114	160	
201	3-Benzotoluidide (N-Benzoyl-m-toluidine)	125	.....	122.4	249	89	119.0		203	65	125	
202	2-Toluanilide	125	.....	104–5; 107–8	259[751]	90.7	57		184	114	160	
203	Acetyl-β-phenylhydrazine	125–6	.....	16.6	118.2	78	86.0	19; 23	243	128; di: 107	168; di: 177	
204	Adipic acid monoamide	125–30	.....	153–4 (cor.)	216[15]	106	154.5; 152.6	NH_3				
205	Succinimide	126	245–7	185; 182.8	235d.	di: 88	di: 211	NH_3				
206	3-Toluanilide	126	.....	111–3; 110–1	263, subl.	86.6	108		184	114	160	
207	Angelanilide	126	.....	45–6	185 (cor.)	.....	.....		184	114	160	
208	β-Resorcylanilide (2,4-Dihydroxybenzanilide)	126–7	.....	213d. (rapid htng.) 216; 217	.....	188–9	.....		184	114	160	
209	2-Ethylbutananilide	127; 127.5	.....	−31.8	195	.....	.....		184	114	160	
210	Suberic acid monoamide	127	.....	144; 139–41	....	di: 85	di: 144.2	NH_3				
211	4-Methoxyacetanilide	127	.....	16.6	118.2	78	86.0	58	240	130; 127	154; 157	
212	Angelamide	127–8	.....	45–6	185 (cor.)	.....	.....	NH_3				
213	Phenylpropiolanilide	128; 126	.....	136–7, subl.	....	83	.....		184	114	160	
214	Nicotinamide	128; 129; 129–31	.....	237–8, subl.	.....	.....	.....	NH_3				B.p. 150–60[5.10⁻⁴]; Chloroaurate, 205; N-Ethyl, 188–9; N-isopropyl, 184–6
215	Pivalanilide	128 (cor.); 132–3	.....	35.5	163–4	.....	75–6		184	114	160	
216	Suberic acid monoanilide	128–9	.....	144; 139–41	.....	di: 85	di: 144.2	.....	184	114	160	
217	Isobutyramide	129; 127	210–1	−46.1	154.7	.....	76.8	NH_3				
218	2-Methoxybenzamide	129	.....	100–1	200	.....	113	NH_3				Mercury deriv., 241
219	2-Bromo-4-nitroacetanilide	129	.....	16.6	118.2	78	86.0	105		129	160	
220	Benzamide	130; 129	222.5–3.5; 224	122.4	249	89	119	NH_3				Mercury deriv., 222; Phthalimide, 168

*Derivative data given in order: m.p., crystal color, solvent from which crystallized.

No.	Name	Melting point, °C	Xanthylamide	Derived acid				Derived amine				Miscellaneous
				M.P., °C	B.P., °C	p-Nitrobenzyl ester	p-Bromophenacyl ester	M.P., °C	B.P., °C	Acetamide	Benzamide	
221	3,4-Dimethylbenzamide	130	. . . . . .	166; 164	. . . .	. . . .	. . . .	NH$_3$			. . . . . .	. . . . . .
222	α,β-Dibromopropionamide	130; 133	. . . . . .	67 (stab.); 50 (un-stab.)	160^{20}	. . . .	. . . .	NH$_3$			. . . . . .	. . . . . .
223	Anthranilanilide	131	. . . . . .	147	. . . .	. . . .	. . . .		184	114	160	. . . .
224	Bromoacetanilide	131	. . . . . .	50	208	88	. . . .		184	114	160	. . . .
225	2-Methoxybenzanilide	131	. . . . . .	100–1	200	. . . .	113		184	114	160	. . . .
226	Nicotinanilide	132, bz.; 85 (+ 2H$_2$O), w.; 265 (anh.)	. . . . . .	237–8, subl.	. . . .	. . . .	. . . .		184	114	160	. . . .
227	2,2-Dimethylbutanamide	132; 103	. . . . . .	−15.0	187; 190	. . . .	. . . .		184	114	160	. . . .
228	d,l-2,3-Dimethylbutanamide . . .	132	. . . . . .	−1.5	191.7	. . . .	. . . .	NH$_3$			. . . . . .	. . . . . .
229	3,3-Dimethylbutananilide	132	. . . . . .	6.7	184; 96^{26}	. . . .	. . . .		184	114	160	. . . .
230	3,3-Dimethylbutanamide	132	. . . . . .	6.7	184; 96^{26}	. . . .	. . . .	NH$_3$			. . . . . .	. . . . . .
231	2,5-Dichloroacetanilide	132	. . . . . .	16.6	118.2	78	86.0	50	. . . . . .	132	120	. . . .
232	Malonic acid monoanilide	132	. . . . . .	134.8–.9	. . . .	di: 85.5	. . . .		184	114	160	. . . .
233	Urea (Carbamide)	132.8	265; 274	. . . .	. . . .	. . . .	. . . .	NH$_3$			. . . . . .	Phthalimide, 188–90; Picrate, 148
234	2,4-Dimethylacetanilide	133; 130	. . . . . .	16.6	118.2	78	86.0	. . . . . .	217	133; 130	192	. . . .
235	Mesityleneamide (3,5-Dimethylbenzamide)	133	. . . . . .	16.6	. . . .	. . . .	. . . .	NH$_3$			. . . . . .	. . . . . .
236	4-Isopropylbenzamide	133	. . . . . .	177, al.	. . . .	. . . .	. . . .	NH$_3$			. . . . . .	. . . . . .
237	α-Bromoisovaleramide	133	. . . . . .	44	230d.	. . . .	. . . .	NH$_3$			. . . . . .	. . . . . .
238	4-Chlorophenoxyacetamide	133	. . . . . .	155–6, w.; 158	. . . .	. . . .	136	NH$_3$			. . . . . .	. . . . . .
239	d,l-Mandelamide	133–4 (cor.)	. . . . . .	118	. . . .	123–4	. . . .	NH$_3$			. . . . . .	. . . . . .
240	N-(2-Naphthyl)acetamide	134	. . . . . .	16.6	118.2	78	86.0	112	. . . . . .	132	162	Br$_2 \rightarrow$ 1-Bromo deriv., 140
241	3-Chlorobenzamide	134	. . . . . .	158; 155	. . . .	107	116	NH$_3$			. . . . . .	Mercury deriv., 245
242	Phenacetin (4-Aceto-phenetidide)	134	. . . . . .	16.6	118.2	78	86.0	2–3	248; 254	137	173	10% HNO$_3 \rightarrow$ 3-Nitro deriv., 103
243	Chloroacetanilide	134; 136–7	. . . . . .	α:61.3; β:56.2; γ:52.5	189	. . . .	104	. . . . . .	184	114	160	. . . .
244	2,3-Dimethylacetanilide	135	. . . . . .	16.6	118.2	78	86.0	. . . . . .	221–2	135	189	. . . .
245	Isovaleramide	135; 136	182–3	−30.0	176.5	. . . .	68.0	NH$_3$			. . . . . .	. . . . . .
246	Acetyl salicylanilide	136	. . . . . .	135 (rapid htng.)	140d.	90.5	. . . .		184	114	160	. . . .
247	3-Bromobenzanilide	136	. . . . . .	155	. . . .	105	120		184	114	160	. . . .
248	Salicylanilide	136	. . . . . .	158.3 (subl. at 76)	. . . .	97–8	140		184	114	160	. . . .
249	α-Hydroxyisobutyranilide	136	. . . . . .	79	212	80.5	98		184	114	160	. . . .
250	N,N-Diacetyltetramethyl-enediamine	137	. . . . . .	16.6	118.2	78	86.0	27	159	di: 137	di: 177	. . . .
251	Benzo-2-iodoanilide (N-Ben-zoyl-o-iodoaniline)	139	. . . . . .	122.4 (subl. at 100)	249	89	119	61; 58	. . . . . .	109	139	. . . .
252	3-Ethoxybenzamide	139	. . . . . .	137	. . . .	. . . .	. . . .	NH$_3$			. . . . . .	. . . . . .
253	2,5-Dimethylacetanilide	139	. . . . . .	16.6	118.2	78	86.0	15.5	213–5	139	140	. . . .
254	3-Aminobenzanilide	140	. . . . . .	174d.	. . . .	. . . .	. . . .		184	114	160	. . . .

*Derivative data given in order: m.p., crystal color, solvent from which crystallized.

No.	Name	Melting point, °C	Xanthyl-amide	Derived acid M.P., °C	Derived acid B.P., °C	p-Nitro-benzyl ester	p-Bromo-phenacyl ester	Derived amine M.P., °C	Derived amine B.P., °C	Acet-amide	Benz-amide	Miscellaneous
255	2-Toluamide	140; 143	199–200.5	104–5; 107–8	259[7.51]	90.7	57	NH₃		......	......	Mercury deriv., 196
256	Trichloroacetamide	141	......	57–8	197.5	80	......	NH₃		......	......	....
257	2-Iodobenzanilide	141	......	162	......	111	143	......	184	114	160	....
258	2-Bromobenzanilide	141	......	150	......	110	102	......	184	114	160	....
259	3-Tolylurea	142		......	......	......	......	......	203	65	125	
260	2-Chlorobenzamide	142	......	142; 140	......	106	106	NH₃		......		
261	Salicylamide	142	......	158.3 (subl. at 76)	......	97–8	140	NH₃		......		Mercury deriv., 190
262	Furamide	142–3	210	133–4; 132	230–2	133.5	138.5	NH₃		......		....
263	3-Nitrobenzamide	143; 142	......	140	......	141	132	NH₃		......		
264	Iodoacetanilide	143–4	......	83	......	......	......	......	184	114	160	
265	3,5-Dimethylacetanilide	144; 140	......	16.6	118.2	78	86	......	220	144; 140	136	
266	2-Benzotoluidide (N-Benzoyl-o-toluidine)	144	......	122.4	249	89	119	......	200	110–11	146	KMnO₄ → Benzoylanthra-nilic acid, 177
267	3-Nitrosalicylamide	145 (hyd.)	......	125	......	......	......	NH₃		......		....
268	2,4-Dichloroacetanilide	145	......	16.6	118.2	78	86.0	63	......	145	177	
269	α-d,l-Phenylsuccinamide (β form)	145	......	167–8; 84 (anh.)	......	......	......	NH₃		......		
270	4-Toluanilide	145; 148	......	179–80; 182 subl.	275 (cor.)	104.5	153	......	184	114	160	
271	Diethylmalonic acid monoamide	146	......	125	......	di: 91	......	NH₃		......		
272	Cyclohexancarboxanilide (Hexahydrobenzanilide)	146; 131	......	30–1	233	......	......	......	184	114	160	
273	Phenylurea	147	225	......	......	......	......	......	184	114	160	
274	N,N'-Dibenzoyltrimethylene-diamine	147	......	122.4	249	89	119	......	136	mono: 126; di: 101	mono: 140; di: 147	....
275	Benzo-2-ethylanilide (N-Benzoyl-o-ethylaniline)	147	......	122.4	249	89	119	......	210–11	111	147	....
276	2,4,6-Trichloroacetanilide	148	......	16.6	118.2	78	86.0	78	......	148	172	
277	α-Bromoisobutyramide	148	......	48–9	198–200	......	......	NH₃		......		
278	Cinnamamide	148; 142	......	133	300	116.8	145.6	NH₃		......		
279	Succinic acid monoanilide	148.5	......	185; 182.8	235 d.	di: 88	di: 211	......	184	114	160	
280	N-Cyclohexybenzamide	149	......	122.4	249	89	119	......	134	104	149	
281	Benzylurea	149	......	......	......	......	......	......	184–5	60	105	
282	Phthalic acid monoamide (Phthalamic acid)	149	......	200–6; 197 (sealed tube); 230 (rapid htng.)	......	di: 155.5	di: 152.8	NH₃		......		
283	β-Benzoylpropionanilide	150; 145	......	116	......	......	......	......	184	114	160	....
284	Ethylmalonic acid monoanilide	150	......	111	......	75	......	......	184	114	160	
285	2-Chlorophenoxyacetamide	150	......	145–6, w.	......	......	......	NH₃		......		
286	Adipic acid monoanilide	151–3	......	153–4 (cor.)	216[15]	106	154.5; 152.6	......	184	114	160	....
287	Cinnamanilide	151; 153	......	133	300	116.8	145.6	......	184	114	160	Acid KMnO₄ → Benzoic acid, 122.4

*Derivative data given in order: m.p., crystal color, solvent from which crystallized.

TABLE XV. ORGANIC DERIVATIVES OF AMIDES AND IMIDES
b) Solids (Listed in order of increasing m.p.)* (Continued)

No.	Name	Melting point, °C	Xanthyl-amide	Derived acid				Derived amine				Miscellaneous
				M.P., °C	B.P., °C	p-Nitro-benzyl ester	p-Bromo-phenacyl ester	M.P., °C	B.P., °C	Acet-amide	Benz-amide	
288	4-Fluoroacetanilide	152	......	16.6	118.2	78	86.0	−1	186	152	185	....
289	d,l-Mandelanilide	152	......	118	....	123–4	....	......	184	114	160	....
290	4-Acetotoluidide (N-Acetyl-p-toluidine)	153; 147	......	16.6	118.2	78	86.0	45	200	147	158	KMnO₄ → 4-Acetamido-benzoic acid, 256; Br₂ → 3-Bromo deriv, 117
291	N-Methyl-4-nitroacetanilide	153	......	16.6	118.2	78	86.0	152	......	153	112	....
292	Veratranilide	154	......	181	....	....	....	......	184	114	160	
293	4-Fluorobenzamide	154	......	182; 182.6	....	....	....	NH₃		......	......	
294	3-Nitrobenzanilide	154	......	140	....	141	132	......	184	114	160	
295	Benzilamide	154–5; 155	......	150	....	99.5	152	NH₃		......	......	
296	3-Nitrosalicylamide	155; 145	......	125 (+ H₂O)	....	....	....	NH₃				
297	2,5-Dichlorobenzamide	155	......	153	....	....	....	NH₃				
298	2-Bromobenzamide	155	......	150	....	110	102	NH₃		......	......	Mercury deriv., 242
299	Dibenzylacetanilide	155	......	89	....	....	....	......	184	114	160	....
300	2-Nitrobenzanilide	155	......	146	....	112	107	......	184	114	160	....
301	3-Bromobenzamide	155	......	155	....	105	120	NH₃		......	......	Mercury deriv., 235
302	N-(1-Naphthyl) acetanilide	155; 159.6	......	131; 135	....	....	....	......	184	114	160	
303	Phthalonamide (β form)	155d.	......	146	....	....	....	NH₃				
304	3-Nitroacetanilide	155	......	16.6	118.2	78	86.0	......	114	mono: 155; di: 76	mono: 155; di: 150	....
305	Pimelic acid dianilide	155–6	......	104–5, subl.	223¹⁵	....	di: 136.6	......	184	114	160	
306	Pivalamide	155–7; 153–4	......	35.5	163–4	....	75–6	NH₃		......	......	
307	N-Phenylsuccinimide	156	......	185; 182.8	235d.	di: 88	di: 211	......	184	114	160	
308	Dibromoacetamide	156	......	48	232–5	....	....	NH₃		......	......	
309	L-Malamide	156.5–8.0	......	100–1	....	mono: 87.2; di: 124.5	di: 179	NH₃				
310	Phenylacetamide	156; 157	196	76.5, subl.	256.5 (cor.)	65	89	NH₃		......	......	
311	3-Hydroxybenzanilide	157; 155	......	200, subl.	....	106–8	176; 176.1–.4	......	184	114	160	
312	Succinic acid monoamide (Succinamic acid)	157	......	185; 182.8	235d.	di: 88	di: 211	NH₃		......	......	
313	N-Acetylindole	157–8	......	16.6	118.2	78	86	52	253	157–8	68	
314	d,l-Phenylsuccinamide (α form)	158	......	167–8	....	....	....	NH₃		......	......	
315	4-Benzotoluidide (N-Benzoyl-p-toluidine)	158	......	122.4	249	89	119	45	200	147	158	CrO₃ → 4-Benzamido benzoic acid, 278
316	N-(1-Napthyl) acetamide	159	......	16.6	118.2	78	86.0	50	......	159	160	Br₂ → 4-Bromo deriv., 193; Fuming HNO₃ →dinitro deriv., 250

*Derivative data given in order: m.p., crystal color, solvent from which crystallized.

TABLE XV. ORGANIC DERIVATIVES OF AMIDES AND IMIDES
b) Solids (Listed in order of increasing m.p.)* (Continued)

No.	Name	Melting point, °C	Xan-thyl-amide	Derived acid M.P., °C	B.P., °C	p-Nitro-benzyl ester	p-Bromo-phenacyl ester	Derived amine M.P., °C	B.P., °C	Acet-amide	Benz-amide	Miscellaneous
317	4-Toluamide	159; 160	224–5	179–80; 182, subl.	275 (cor.)	104.5	153	NH₃				Mercury deriv., 260
318	Crotonamide	161; 158	......	72, w.	189 (cor.)	67.4	95–6	NH₃		......	......	....
319	N-(1-Naphthyl)benzamide	161	......?	122.4	249	89	119	50	......	159	160	....
320	N-(2-Naphthyl)benzamide	162	......	122.4	249	89	119	112	162	132	162	....
321	4-Aminoacetanilide	162	......	16.6	118.2	......	86.0	140; 147	267	mono: 162–3; di: 304	mono: 128; di: 300	Azo-β-naphthol deriv., 261
322	4-Hydroxybenzamide	162 (hyd.)	......	215; 213–4; 210	....	180–2	191.5 (cor.); 184	NH₃		......		....
323	Benzanilide	163; 160	......	122.4	249	89	119	......	184	114	160	Br₂→4-Bromo deriv., 204
324	1-Naphthanilide	163; 164		161–2 (cor.)	....	....	135.5	......	184	114	160	
325	Trichlorolactanilide	164		124				......	184	114	160	
326	Veratramide	164	......	181	....	....	....	NH₃		......		
327	2-Benzoylbenzamide	165	......	128; 91 (hyd.)		100.4	......	NH₃				
328	L-Glutamamide	165	......	211 (rapid htng.); 197 (slow htng.)				NH₃				
329	Protocatechuanilide	166	......	199–200d.	....	....	188	......	184	114	160	....
330	4-Anisamide	167; 163	......	184–6; 184.2 (cor.)	275–80	132	152	NH₃		......	......	Mercury deriv., 222
331	4-Bromoacetanilide	167; 168	......	16.6	118.2	78	86.0	66	245	168	204	....
332	Acetylanthranilanilide	167	......	185, ac. a.				......	184	114	160	
333	Diphenylacetamide	168	......	148	....	....	....	NH₃		......		
334	α-Benzoyl-β-phenylhydrazine	168	......	122.4	249	89	119	19; 23	243	128	168	
335	2-Furanacrylamide (β-(2-Furyl)-acrylic acid)	168–9	......	141	286			NH₃				
336	Piperonylamide	169	......	229; 228	....	....	....	NH₃				
337	d,l-Tropamide	169	......	117–8	....	....	....	NH₃				
338	Methyliminodiacetic acid monoamide	169	......	227d.	....	....	....	NH₃				
339	Methyliminodiacetic acid diamide	169	......	227d.	....	....	....	NH₃				
340	4-Hydroxyacetanilide	169	......	16.6	118.2	78	86.0	184; 186	......	mono: 168; di: 150	N-mono: 216–7; N,O-di: 234	
341	4-Anisanilide	169–71	......	184–6; 184.2 (cor.)	275–80	132	152	......	184	114	160	....
342	3-Hydroxybenzamide	170; 167	......	200, subl.	....	106–8	176; 176.7–.4	NH₃		......		
343	Sebacic acid monoamide	170	......	33, subl.	243¹⁵	di: 73.5; 72.6	di: 147	NH₃		......		
344	cis-Aconitanilide	170d.	......	125	....	....	tri: 186	......	184	114	160	....

*Derivative data given in order: m.p., crystal color, solvent from which crystallized.

TABLE XV. ORGANIC DERIVATIVES OF AMIDES AND IMIDES
b) Solids (Listed in order of increasing m.p.)* (Continued)

No.	Name	Melting point, °C	Xan-thyl-amide	Derived acid M.P., °C	B.P., °C	p-Nitro-benzyl ester	p-Bromo-phenacyl ester	Derived amine M.P., °C	B.P., °C	Acet-amide	Benz-amide	Miscellaneous
345	Malonic acid diamide (Malondiamide)...........	170	270	134.8–.9	....	di: 85.5	....	NH$_3$		....	....	....
346	D,L-Phenylsuccinic acid monoanilide (β form).......	170	......	167–8; 84 (anh.)	....				184	114	160	
347	4-Ethoxybenzanilide	170; 172	......	198; 195–6	....				184	114	160	
348	2-Naphthylanilide...........	171; 173	......	184; 185.5					184	114	160	
349	N,N'-Diacetylethylenediamine .	172	......	16.6	118.2	78	86.0	8.5	116	di: 172	di: 244	
350	Benzo-2,4,6-trichloroanilide (N-Benzoyl-2,4,6-trichloroaniline)...........	172	......	122.4	249	89	119	78	......	148	172	
351	Azelaic acid diamide	172	......	106.5	>360 sl. d.; 237^{15}	di: 43.8	di: 130.6	NH$_3$		....	....	
352	Maleic acid monoamide (Maleamic acid)...........	172–3; 153 sl.d.	......	137–8; 130 (+3% fumaric acid)	....	di: 91 (cor.)	168–70; 190	NH$_3$		....		
353	4-Phenetylurea (Dulcin)	173	....	....	....	....	....	2–3	248; 254	137	173	
354	4-Benzophenetidide .... (N-Benzoyl-p-phenetidine) ...	173	......	122.4	249	89	119	2–3	248; 254	137	173	
355	Benzylanilide	174–5	......	150	....	99.5	152	......	184	114	160	
356	Pimelic acid diamide	175	......	104–5, subl.	223^{15}	....	di: 136.6	NH$_3$		....		
357	d,l-Phenylsuccinic acid monoanilide (α form).......	175	......	167–8; 84 (anh.)	....				184	114	160	
358	4-Hydroxyphenylacetamide....	175	......	148–50; 148, w.	....			NH$_3$				
359	Citraconic acid dianilide	175.5	......	92–3; 22d.	....	70.6	....		184	114	160	
360	Glutaric acid diamide........	175–6	......	98	302–4	di: 69	di: 139.8	NH$_3$				
361	2-(4-Toluyl) benzamide	175–6	......	139–140	....			NH$_3$				
362	2-Nitrobenzamide...........	176; 175	......	146		112	107	NH$_3$		....		
363	Phthalonic acid monoanilide ...	176	......	146	....				184	114	160	
364	N,N-Dibenzoyltetra-methylenediamine...........	177	......	122.4	249	89	119	27	159	di: 137	di: 177	
365	Acetylanthranilamide.........	177	......	185, ac. a.	....			NH$_3$				
366	2,6-Dimethylacetanilide.......	177	......	16.6	118.2	78	86.0	......	215; 218	177	168	
367	D-Camphoric acid monoamide .	177	..X...	187.5–8.0	....	65.5	....	NH$_3$		....		
368	Mesaconic acid diamide.......	177	......	204.5 (cor.), subl.	....	di: 134 (cor.)	....	NH$_3$		....		
370	4-Cyanobenzanilide	179	......	219; 214	....	189	....		184	114	160	....
371	N-Benzylphthalamide	179	......	200–6; 191 (sealed tube); 230 rapid htng.)	....	di: 155.5	di: 152.8	......	184–5	60	105	....
372	4-Chloroacetanilide	179	......	16.6	118.2	78	86.0	72	......	179	192	....
373	α-Phthalonamide	179d.	......	146	....	....	....	NH$_3$		....		

*Derivative data given in order: m.p., crystal color, solvent from which crystallized.

TABLE XV. ORGANIC DERIVATIVES OF AMIDES AND IMIDES
b) Solids (Listed in order of increasing m.p.)* (Continued)

No.	Name	Melting point, °C	Xan-thyl-amide	Derived acid				Derived amine				Miscellaneous
				M.P., °C	B.P., °C	p-Nitro-benzyl ester	p-Bromo-phenacyl ester	M.P., °C	B.P., °C	Acet-amide	Benz-amide	
374	2,4-Dimethylbenzamide	179–81	......	127; 90 (+H₂O)	....	....	....	NH₃				....
375	D-Tartaric acid monoanilide	180d.	......	169–71	....	di: 163	di: 204	......	184	114	160	....
376	α-Acetyl-β-methylurea	180	......	16.6	118.2	78	86.0		−6	28	80	
377	Diphenylacetanilide	180	......	148	....	....	....		184	114	160	
378	3,5-Dinitrosalicylamide	181	......	182	....	....	....	NH₃				....
379	1-Naphthylacetamide	181	......	131; 135	....	....	....	NH₃				
380	Maleic acid diamide	181; 180	......	137	....	di: 91	168–70; 190	NH₃				
381	unsym-Dimethyl urea	182	225	....	....	....	....		7	......	41	
382	Thiourea	182	......	....	....	....	....	NH₃				
383	Hippuramide	183	......	187	....	136	151	NH₃				
384	4-Tolylurea	183	......	....	....	....	....	45	200	147	158	
385	3,5-Dinitrobenzamide	183	......	204–5	....	157	159	NH₃				
386	2-Iodobenzamide	184	......	162	....	111	143	NH₃				
387	4-Iodoacetanilide	184	......	16.6	118.2	78	86.0	67–8	......	184	222	
388	Benzo-4-fluoroanilide (N-Benzoyl-p-fluoroaniline)	185	......	122.4	249	89	119	−1	186	152	185	
389	N,N′-Diacetyl-o-phenylene-diamine	185	......	16.6	118.2	78	86.0	102		di: 185	di: 301	
390	2-Nitrocinnamamide	185	......	240	....	132	141	NH₃				
391	Mesaconic acid dianilide	185.7	......	204.5 (cor.)	....	di: 134 (cor.)	....		184	114	160	
392	Citraconic acid diamide	185–7	......	92–3; 92d.	....	di: 70.6	....	NH₃				
393	Chlorofumaranilide	186	......	191–2	....	di: 138.5	....		184	114	160	
394	3-Iodobenzamide	186	......	187	....	121	128	NH₃				
395	Suberic acid dianilide	186–7	......	144; 139–41	....	di: 85	di: 144.2	......	184	114	160	
396	Maleic acid dianilide	187	......	137	....	di: 91	168–70; 190 (cor.)	......	184	114	160	
397	Maleic acid monoanilide	187	......	137	....	di: 91	168–70; 190 (cor.)	......	184	114	160	....
398	Hippuric acid	187.5	......	122.4	249	89	119	Gly-cine: 228–30; 262d.	......	206	187.5	Acetamide, 136; Benzamide, 151
399	unsym-Diphenylurea	189	180	....	....	....	....	53–4	......	101	180	
400	Aconitic acid dianilide	189	......	194–5d. (cor.)	....	....	tri: 186		184	114	160	
401	meso-Tartaric acid diamide	189–90; 187	......	140	....	93	....	NH₃				
402	4-Bromobenzamide	189–90	......	251–3	....	180	....	NH₃		....	....	Mercury deriv., 266
403	Itaconic acid dianilide	190; 185	......	165	....	di: 90.6	di: 117.4	......	184	114	160	....
404	Picramide	190	......	122.5	....	....	....	NH₃				Acetamide, 230; Benzamide, 196
405	N,N′-Diacetyl-m-phenylene-diamine	191	......	16.6	118.2	78	86.0	63	......	mono: 87–9; di: 191	mono: 125 di: 240	....
406	Itaconic acid diamide	191.2–.8	......	165	....	di: 90.6	di: 117.4	NH₃		....	....	
407	Mucic acid monoamide	192d.	......	214d.; varies with htng. rate; 223–255	....	310	225	NH₃		....	....	

*Derivative data given in order: m.p., crystal color, solvent from which crystallized.

No.	Name	Melting point, °C	Xanthyl-amide	Derived acid				Derived amine				Miscellaneous
				M.P., °C	B.P., °C	p-Nitro-benzyl ester	p-Bromo-phenacyl ester	M.P., °C	B.P., °C	Acet-amide	Benz-amide	
408	4-Nitrophthalanilide.........	192	......	165	....	....	....		184	114	160	....
409	2-Tolylurea................	192	228	....	....	....	....		200	110–11	146	....
410	Benzo-4-chloroaniline (N-Benzoyl-p-chloroaniline) .	192	......	122.4	249	89	119	......	72	179; 172	192	....
411	Biuret	192d.	......	....				NH₃				
412	2-Naphthamide.............	192; 195	......	184; 185.5	....	....	....	NH₃				
413	D-Camphoric acid diamide	193	......	187.5–8.0	....	65.5	....	NH₃				
414	4-Chlorobenzanilide..........	194	......	243; 240	....	129.5	126	......	184	114	160	
415	4-Coumaramide	194	......	210–3; 206 (anh.)	....	....	....	NH₃				
416	2-Benzoylbenzanilide	195	......	128; 91 (+1 H₂O)	....	100.4	....		184	114	160	
417	3-Nitrocinnamamide	196	......	199	....	174	178; 173	NH₃	......	......		
418	D-Tartaric acid diamide......	196d.	......	169–71	....	di: 163	di: 204	NH₃				
419	2-Methyl-4-nitroacetanilide ...	196; 202	......	16,6	118.2	78	86.0	130	......	202		
420	L-Malanilide (Hydroxysuccin-anilide)	197	......	100–1	....	mono: 87.2; di: 124.5	di: 179	......	184	114	160	
421	4-Hydroxybenzanilide	198; 196–7	......	215; 213–4	....	180–2	191.5 (cor.)	......	184	114	160	
422	4-Nitrophenylacetamide.......	198	......	153	....	......	207	NH₃	......	......		
423	4-Nitrophenylacetanilide	198	......	153	....	......	207	147–8	......	215	di: 193.; 203	
424	Cyanoacetanilide	198–9	......	66	....	....	....		184	114	160	
425	Citric acid trianilide........	199; 192 (+1 H₂O)	......	100; 153 (anh.)	....	tri: 102	tri: 148		184	114	160	
426	4-Nitrophthalamide	200d.	......	165				NH₃	......	......		
427	cis-Aconitic acid dianilide	200	......	125	....		tri: 186		184	114	160	
428	Methylsuccinic acid dianilide ..	200	......	115					184	114	160	
429	4-Nitrobenzamide............	200	232	241	....	168	137	NH₃	......	......		
430	2-Naphthylacetamide........	200; 205	......	141–2; 143	....			NH₃				
431	Isatin.....................	200–1		....	....	....	......		......	......		Acetamide, 147
432	3-Nitrophthalic acid diamide ...	201d.	......	218		189	......	NH₃				
433	Sebacic acid dianilide........	201	......	33, subl.	243¹⁵	di: 73.5; 72.6	di: 147	......	184	114	160	
434	1-Naphthoamide.............	202	......	161–2 (cor.)		......	135.5	NH₃				
435	4-Ethoxybenzamide	202	......	198; 195–6	....			NH₃				
436	2,4-Dinitrobenzamide	203	......	183		142	158	NH₃				
437	D-Camphoric acid monoanilide	204	......	187.5–8.0		65.5			184	114	160	
438	Benzo-4-bromoaniline (N-Benzoyl-p-bromoaniline) .	204	......	122.4	249	89	119	66	245	168	204	....
439	4-Nitrocinnamamide	204; 217	......	285	....	186	191	NH₃				
440	N-Phenylphthalimide........	205	......	200–6; 191 (sealed tube); 230 (rapid htng.)	....	di: 155.5	di: 152.8		184	114	160	

*Derivative data given in order: m.p., crystal color, solvent from which crystallized.

TABLE XV. ORGANIC DERIVATIVES OF AMIDES AND IMIDES
b) Solids (Listed in order of increasing m.p.)* (Continued)

No.	Name	Melting point, °C	Xan-thyl-amide	Derived acid				Derived amine				Miscellaneous
				M.P., °C	B.P., °C	p-Nitro-benzyl ester	p-Bromo-phenacyl ester	M.P., °C	B.P., °C	Acet-amide	Benz-amide	
441	Carballylic acid triamide	207d.; 205–7d.	......	166	....	....	tri: 138.2	NH₃	......	......		....
442	Gallanilide	207	......	253–4d.; 222–40d.	....	141	134	......	184	114	160	
443	Phthalonic acid dianilide	208	......	146	....	....	....	......	184	114	160	
444	Hippuranilide	208	......	187	....	136	151	......	184	114	160	
445	2-Coumaramide	209d.	......	207–8, subl.	....	152.5	....	NH₃				
446	Sebacic acid diamide	210; 208	......	133, subl.	243¹⁵	di: 73.5; 72.6	di: 147	NH₃				
447	4-Iodobenzanilide	210	......	270; 265	....	141	146	......	184	114	160	
448	Citric acid triamide	210–5	......	153 (slow htng.)	....	tri: 102	tri: 148	NH₃				
449	4-Nitrobenzanilide	211; 204	......	241	....	168	137	......	184	114	160	
450	d,l-Phenylsuccinic acid diamide	211	......	167–8; 84 (anh.)	....	....	....	NH₃				
451	Protocatechuamide	212	......	199–200d.	....	188	....	NH₃				
452	2,2'-Diphenic acid diamide	212	......	227; 233	....	di: 187; 182.6	....	NH₃				
453	4-Nitroacetanilide	213–4	......	16.6	118.2	78	86.0	147–8	......	215	di: 199; 203	....
454	Ethylmalonic acid diamide	214	......	111	....	75	....	NH₃				
455	3-Nitrophthalimide	216	......	218	....	189	....	NH₃				
456	Suberic acid diamide	216–7	......	144; 139–41	....	di: 85	di: 144.2	NH₃				
457	Methylmalonic acid diamide	217; 206	......	137; 138d.	....	....	....	NH₃				
458	4-Iodobenzamide	217	......	270; 265	....	141	146	NH₃				
459	Benzylmalonic acid dianilide	217	......	117d; 121	....	di: 119.5	....	......	184	114	160	
460	4-Hydroxy-2-naphthamide	217–8	......	225–6	....	....	....	NH₃				
461	Acetylurea	218	......	16.6	118.2	78	86.0	NH₃				
462	3-Hydroxy-2-naphthamide	218	......	222–3 (cor.)	....	....	....	NH₃				
463	sym.-Di-3-tolylurea	218	......	....	....	....	....	......	203	65	125	
464	Phthalic acid diamide	220	......	200–6; 191 (sealed tube); 230 (rapid htng.)	....	di: 155.5	di: 152.8	NH₃				
465	Mucic acid diamide	220	......	214d.; varies with htng. rate; 223–255	....	310	225	NH₃				....
466	Saccharin	220	198–9	206	....	....	....	NH₃	......	......		....
467	Adipic acid diamide	220; 229	......	153–4 (cor.)	216¹⁵	106	152.6; 154.5	NH₃	......	......		....
468	d,l-Phenylsuccinic acid dianilide	222	......	167–8; 84 (anh.)	....	136	....		184	114	160	

*Derivative data given in order: m.p., crystal color, solvent from which crystallized.

TABLE XV. ORGANIC DERIVATIVES OF AMIDES AND IMIDES

b) Solids (Listed in order of increasing m.p.)* (Continued)

No.	Name	Melting point, °C	Xanthylamide	Derived acid				Derived amine				Miscellaneous
				M.P., °C	B.P., °C	p-Nitrobenzyl ester	p-Bromophenacyl ester	M.P., °C	B.P., °C	Acetamide	Benzamide	
469	β-Resorcylamide	222	......	213d. (rapid htng.) 216d; 217	....	188–9	....	NH₃	......	......	....	....
470	4-Cyanobenzamide	223	......	219; 214	....	189	....	NH₃	......	....	....	
471	Diethylmalonic acid diamide	224	......	125	302–4	di: 91	....	NH₃				
472	Glutaric acid dianilide	224	......	98	....	di: 69	di: 139.8	......	184	114	160	
473	5-Nitrosalicylanilide	224	......	229–30	....	....	....	......	184	114	160	
474	5-Nitrosalicylamide	225	......	229–30	....	....	....	NH₃				
475	Benzylmalonic acid diamide	225	......	117d.; 121	....	di: 119.5	....	NH₃				
476	Methylsuccinic acid diamide	225	......	115	....	....	....	NH₃				
477	Malonic acid dianilide	225; 227; 230	......	134.8–.9	....	di: 85.5	....	184	114	160		
478	d,l-Tartaric acid monoamide	226	......	203–4 (+1 H₂O); 205–6 (anh.)	....	di: 147.6	....	NH₃	......	....	....	
479	D-Camphoric acid dianilide	226	......	187.5–8.0	....	65.5	....	184	114	160		....
480	Succinic acid dianilide	230; 226	......	185; 182.8	235d.	di: 88	di: 211	......	184	114	160	Heat above m.p. → succinanil, 155–6 + aniline, b.p. 184
481	2,2'-Diphenanilide	230	......	227; 229; 233	....	di: 187; 182.6	....	184	114	160		....
482	Phthalimide	233.5 (cor.); 238	176–7	200–6; 191 (sealed tube); 230 (rapid htng.)	249	89	119	122; 119	......	232	198; 204	....
483	3,5-Dinitrobenzanilide	234	......	204–5	....	157	159	......	184	114	160	....
484	3-Nitrophthalic acid dianilide	234	......	218	....	189		......	184	114	160	
485	Terephthalic acid monoanilide	234–7	......	300, subl. without melting		di: 263.5	di: 225		184	114	160	
486	d,l-Tartaric acid monoanilide	236	......	203–4 (+1 H₂O); 205–6 (anh.)	....	....			184	114	160	....
487	N-Benzylacrylamide	237	......	13	141; 140	....		......	184–5	60	105	
488	Carbanilide (sym-Diphenylurea)	238; 240	......						184	114	160	
489	Muconic acid diamide	240d.	......	289d. (slow htng.); 306 (rapid htng.)	....	....	....	NH₃	......	....	....	
490	Adipic acid dianilide	241	......	153–4 (cor.)	216¹⁵	106	152.6; 154.5	......	184	114	160	....

*Derivative data given in order: m.p., crystal color, solvent from which crystallized.

TABLE XV. ORGANIC DERIVATIVES OF AMIDES AND IMIDES

No.	Name	Melting point, °C	Xan-thyl-amide	Derived acid				Derived amine				Miscellaneous
				M.P., °C	B.P., °C	p-Nitro-benzyl ester	p-Bromo-phenacyl ester	M.P., °C	B.P., °C	Acet-amide	Benz-amide	
491	N,N'-Dibenzoylethylene-diamine	244	......	122.4	249	89	119	8.5	116	di: 172	di: 244	....
492	Gallamide	245	......	253–4d.; 222–40d.	....	141	134	NH₃		....	....	....
493	3-Hydroxy-2-naphthanilide	249 (cor.); 244	......	222–3 (cor.)	....	....	....		184	114	160	
494	sym-Di-2-tolylurea	250	......	....	....	....	....		200	110–1	146	
495	Carballylic acid trianilide	252	......	166	....	....	tri: 138.2		184	114	160	....
496	Phthalic acid dianilide	253–5	......	200–6; 191 (sealed tube); 230 (rapid htng.)	....	di: 155.5	di: 152.8	......	184	114	160	
497	Oxalic acid dianilide	254	......	189.5 (anh.); 101 (+2 H₂O); (rapid htng.)	....	di: 204	....	......	184	114	160	....
498	Succinic acid diamide	255; 260d.	275	185; 182.8	235d.	di: 88	di: 211	NH₃		....	....	
499	2,4,6-Trinitrobenzamide	264d.	......	228	....	....	....	NH₃				
500	D-Tartaric acid dianilide	264d.	......	169–71	....	di: 163	di: 204	......	184	114	160	
501	Fumaric acid diamide	266d.	......	293–5; 286–7 (sealed tube); 200, subl.	....	150.8	....	NH₃				
502	sym-Di-4-tolylurea	268	......	....	....	....	....	45	200	147	158	....
503	Isophthalic acid monoamide	280	......	348, subl.	....	202.5	179.1	NH₃				
504	Isophthalic acid diamide	280	......	348, subl.	....	202.5	179.1	NH₃				
505	Acetylenedicarboxylic acid diamide	294d.	......	179	....	....	....	NH₃				
506	Mucic acid monoanilide	310	......	214; varies with htng. rate; 223–255	....	310	225	......	184	114	160	
507	N,N'-Diacetyl p-phenylenedi-amine	310	......	16.6	118.2	78	86.0	140; 147	267	mono: 162–3; di: 304	mono: 128; di: 300	....
508	Fumaric acid dianilide	314	......	286–7 (sealed tube); 293–5; 200, subl.	....	150.8	....		184	114	160	....
509	N,N'-Diacetylbenzidine	317	......	16.6	118.2	78	86.0	127	......	mono: 199; di: 317	mono: 203–5; di: 352	....

*Derivative data given in order: m.p., crystal color, solvent from which crystallized.

246

No.	Name	Melting point, °C	Xan-thyl-amide	Derived acid				Derived amine				Miscellaneous
				M.P., °C	B.P., °C	p-Nitro-benzyl ester	p-Bromo-phenacyl ester	M.P., °C	B.P., °C	Acet-amide	Benz-amide	
510	Terephthalic acid dianilide	334–7	......	300, subl. without melting	....	di: 263.5	di: 225	......	....	....	....	....
511	N,N'-Dibenzoylbenzidine	352	......	122.4	249	89	119	127	....	*mono*: 199; di: 317	*mono*: 203–5; di: 352	....
512	Trimesic acid triamide	365d.	......	380 (cor.)	....	....	tri: 197 (sealed tube)	NH₃		....	....	
513	Oxalic acid diamide	419d. (sealed tube)	......	189.5 (anh.); 101 (+2 H₂O); (rapid htng.)	....	di: 204	....	NH₃		....	....	

*Derivative data given in order: m.p., crystal color, solvent from which crystallized.

Hydrolysis to the corresponding acid and alcohol. *

$$RCOOR' + KOH \rightarrow R'OH + RCOOK$$

Alcohol $\quad\downarrow_{H^+}$

RCOOH

Acid

From the ester with aqueous sodium or potassium hydroxide.
For directions and examples see: Cheronis, p. 539; Linstead, p. 42; Vogel, pp. 390, 391, 786; Wild, p. 187.
From the ester and potassium hydroxide in anhydrous or aqueous diethylene glycol.
See: Cheronis, p. 538; C. E. Redemann and H. J. Lucas, *Ind. Eng. Chem., Anal. Ed.*, **9**, 514 (1937).
From the ester with sodium methoxide in methanol or with sodium ethoxide in ethanol.
See: Linstead, p. 40; Vogel, p. 391.
From an α-hydroxy ester in water without catalyst.
See: A. Findlay and E. M. H. Hickmans, *J. Chem. Soc.*, **95**, 1004 (1909).

Saponification equivalent.

The saponification equivalent (S. E.) measures the number of equivalents of base required for complete hydrolysis of an ester. It is defined as:

$$S.E. = \frac{\text{Milligrams ester taken for hydrolysis}}{\text{Milliequivalents of KOH required for complete hydrolysis}}$$

For directions and examples see: Cheronis, pp. 975–6; Shriner, p. 235; Vogel, p. 392.
NOTE: For directions and examples for preparation of derivatives of the carboxylic acids formed on hydrolysis of esters see explanations and references to Tables XII, XIII and XIV, pp. 186, 187, 188, 189.
For directions and examples for preparation of derivatives of alcohols formed on hydrolysis of esters see explanations and references to Table VI, pp. 77, 78, 79.

Amide.

$$RCOOR' + NH_3 \rightarrow RCONH_2 + R'OH$$

From the ester with aqueous or alcoholic ammonia.
For directions and examples see: Linstead, p. 42; Wild, p. 188.

Anilide and p-Toluidide.

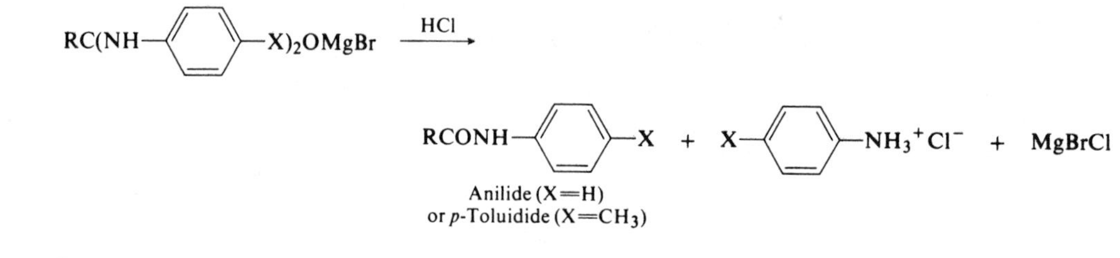

$$X\text{—}C_6H_4\text{—}NH_2 + C_2H_5MgBr \rightarrow X\text{—}C_6H_4\text{—}NHMgBr + C_2H_6$$

Aniline (X=H)
or *p*-Toluidine (X=CH_3)

$$2\ X\text{—}C_6H_4\text{—}NHMgBr + RCOOR' \rightarrow RC(NH\text{—}C_6H_4\text{—}X)_2OMgBr + R'OMgBr$$

$$RC(NH\text{—}C_6H_4\text{—}X)_2OMgBr \xrightarrow{HCl}$$

$$RCONH\text{—}C_6H_4\text{—}X + X\text{—}C_6H_4\text{—}NH_3^+Cl^- + MgBrCl$$

Anilide (X=H)
or *p*-Toluidide (X=CH_3)

From the ester and the N-magnesium bromide derivative of aniline or *p*-toluidine (prepared from the amine and ethylmagnesium bromide) in anhydrous ether.
For directions and examples see: Cheronis, p. 537; Linstead, pp. 42–3; Vogel, p. 394; Wild, p. 190; C. F. Koelsch and D. Tenenbaum, *J. Amer. Chem. Soc.*, **55**, 3049 (1933); D. V. N. Hardy, *J. Chem. Soc.*, 398 (1936).

*Derivatives recommended for first trial.
WARNING: This is not an instruction manual. References should be consulted for the preparation of derivatives.

Hydrazide.

$$RCOOR' + H_2NNH_2 \rightarrow RCONHNH_2 + R'OH$$
Hydrazide

From the ester and 90% hydrazine hydrate.
For directions and examples see: Linstead, p. 42; Wild, p. 191.
From the ester and 85% hydrazine hydrate in alcohol.
See: Cheronis, p. 535; Shriner, p. 237; Vogel, p. 395; P. P. T. Sah, *Rec. Trav. chim.*, **59**, 1036 (1940).

N-(β-Aminoethyl)morpholide. *

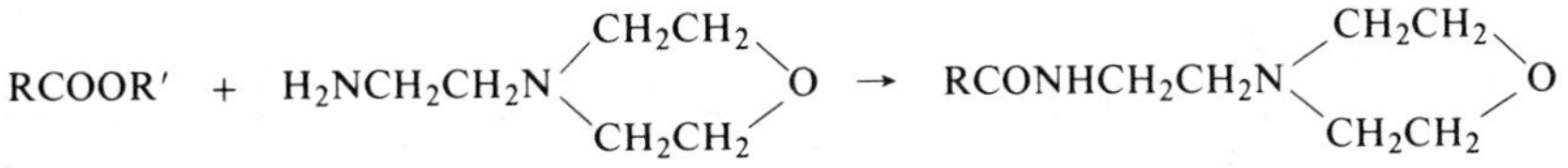

$$RCOOR' + H_2NCH_2CH_2N\begin{matrix}CH_2CH_2\\ \\CH_2CH_2\end{matrix}O \rightarrow RCONHCH_2CH_2N\begin{matrix}CH_2CH_2\\ \\CH_2CH_2\end{matrix}O$$
N-(β-Aminoethyl)morpholide

From the ester and N-(β-aminoethyl)morpholine in ethylene glycol or without solvent.
For directions and examples see: Cheronis, p. 536; R. W. Bost and L. V. Mullen, *J. Amer. Chem. Soc.*, **73**, 1967 (1951).

3,5-Dinitrobenzoate. *

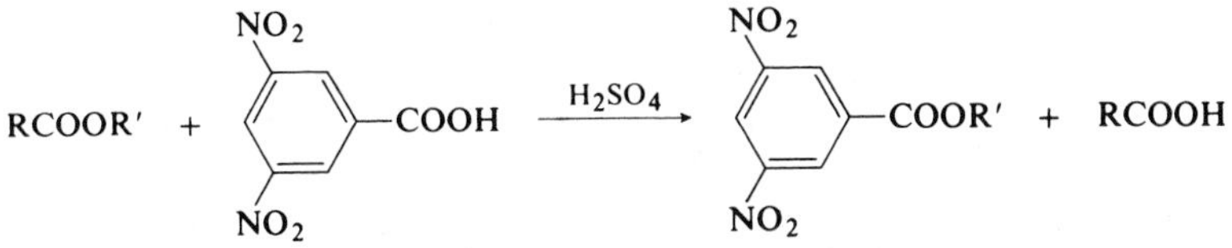

$$RCOOR' + \begin{matrix}NO_2\\ \\ \\NO_2\end{matrix}COOH \xrightarrow{H_2SO_4} \begin{matrix}NO_2\\ \\ \\NO_2\end{matrix}COOR' + RCOOH$$
3,5-Dinitrobenzoate

From the ester with 3,5-dinitrobenzoic acid with a catalytic amount of sulfuric acid.
For directions and examples see: Linstead, p. 43; Shriner, p. 238; Vogel, p. 393; W. B. Renfrow and A. Chaney, *J. Amer. Chem. Soc.*, **68**, 150 (1946).
From the ester with 3,5-dinitrobenzoyl chloride and pyridine.
See: Cheronis, p. 538.

*Derivatives recommended for first trial.
WARNING: This is not an instruction manual. References should be consulted for the preparation of derivatives.

TABLE XVI.

TABLE XVI. ORGANIC DERIVATIVES OF ESTERS
Including esters of inorganic acids
a) Liquids. 1) (Listed in order of increasing b.p.).*

No.	Name	Boiling point, °C	Melting point, °C	n_D^{20}	D_4^{20}	Saponification Equivalent	Acid M.P., °C	Acid B.P., °C	Alcohol M.P., °C	Alcohol B.P., °C	Amide	p-Toluidide	3,5-Di-nitro-benzoate	Miscellaneous
1	Ethyl nitrite	17	. . . . . .	. . . .	0.900_4^{15}	75	. . . .	. . . . . .	−117.3	78.32	. . . .	. . . . .	93, al.	. . . .
2	Methyl formate . .	31.50	−99	1.34648^{15} He (yel)	0.97421	60	8.4	100.7	−97	64.65	2.55	53	108 (cor.), al.	. . . .
3	Ethyl formate	54.15	−79.4	1.35975	0.92247	74	8.4	100.7	−117.3	78.32	2.55	53	93, al.	. . . .
4	Methyl acetate . . .	57.1	−98.7	1.36170; 1.3639	0.9274; 0.93347	74	16.6	118.2	−97	64.65	82	153; 147	108 (cor.), al.	. . . .
5	Ethyl trifluoro-acetate	60.5	. . . . . .	1.3093^{15}	. . . .	142	−15.25	72.4	−117.3	78.32	. . . .	. . . . .	93, al.	. . . .
6	Methyl nitrate . . .	65	. . . . . .	. . . .	1.217_4^{15}	77	. . . .	. . . . . .	−97	64.65	. . . .	. . . . .	108 (cor.), al.	. . . .
7	Isopropyl formate	71; 68	. . . . . .	. . . .	0.8728	88	8.4	100.7	−89.5	82.4	2.55	53	123, pet. eth.	. . . .
8	Butyl nitrite	75	. . . . . .	. . . .	0.911	103	. . . .	. . . . .	−90.2	117.6; 116	. . . .	. . . . .	64; 62.5	
9	Methyl chloro-formate	75	. . . . . .	1.38675	1.2231	94.5			−97	64.65			108 (cor.), al.	
10	Ethyl acetate	77.15	−83.6	1.372	0.90055	88	16.6	118.2	−117.3	78.32	82	153; 147	93, al.	. . . .
11	Methyl propionate	79.65	−87.5	1.3779	0.9151	88	−20.8	141	−97	64.65	81; 81.3; 79	126; 123	108 (cor.), al.	. . . .
12	Methyl acrylate . .	80.3	. . . . . .	1.3984	$0.961^{19.2}$	86	13	140; 141	−97	64.65	84–5, pet. eth.	141	108 (cor.), al.	Polymerizes on standing
13	n-Propyl formate .	80.85; 81	−92.9	1.37789	0.9071; 0.918	88	8.4	100.7	. . . .	97.1	2.55	53	74, pet. eth.	. . . .
14	tert-Butyl formate	83	. . . . . .	. . . .	. . . .	102	8.4	100.7	25.5	82.5	2.55	53	142, pet. eth.	. . . .
15	Allyl formate	83.6	. . . . . .	. . . .	0.946	86	8.4	100.7	. . . .	97.1	2.55	53	49–50	. . . .
16	Ethyl nitrate	87	−112	. . . .	1.106	91	. . . .	. . . . . .	−117.3	78.32	. . . .	. . . . .	93, al.	. . . .
17	Isopropyl acetate .	88.9; 91	−73.4	1.3740^{25}	0.872	102	16.6	118.2	−89.5	82.4	82	153; 147	123, pet. eth.	. . . .
18	Dimethyl carbon-ate (Methyl car-bonate)	90.5	. . . . . .	1.3687	1.0702; 1.0694	90	. . . .	. . . . . .	−97	64.65	. . . .	. . . . .	108 (cor.), al.	. . . .
19	Methyl iso-butyrate	92.6; 92.26	−87.7; −84.7	1.3840	0.8906	102	−46.1	154.7	−97	64.65	128; 129	108.5–9.5	108 (cor.), al.	. . . .
20	Ethyl chloro-formate	93	. . . . .	1.3974	1.13519	108.5	. . . .	. . . . . .	−117.3	78.32	. . . .	. . . . .	93, al.	. . . .
21	sec-Butyl formate	97	. . . . . .	1.384	0.884	102	8.4	100.7	. . . .	99.5	2.55	53	76	. . . .
22	tert-Butyl acetate .	97.8	. . . . . .	1.386	0.867	116	16.6	118.2	25.5	82.5	82	153; 147	142, pet. eth.	. . . .
23	Isobutyl formate .	98.4	−95.8	1.38568	0.88535; 0.8755	102	8.4	100.7	. . . .	108.1	2.55	53	87	
24	Isoamyl nitrite . . .	99	. . . . . .	1.38708^{21}	0.880^{15}	117	. . . .	. . . . . .	−117	132	. . . .	. . . . .	61	
25	Ethyl difluoro-acetate	99	. . . . . .	1.3463	. . . .	124	. . . .	134–5	−117.3	78.32	52	. . . . .	93, al.	. . . .

*Derivative data given in order: m.p., crystal color, solvent from which crystallized.

TABLE XVI. ORGANIC DERIVATIVES OF ESTERS
Including esters of inorganic acids
a) Liquids. 1) (Listed in order of increasing b.p.)* (Continued)

No.	Name	Boiling point, °C	Melting point, °C	n_D^{20}	D_4^{20}	Saponification					Amide	p-Toluidide	3,5-Di-nitro-benzoate	Miscellaneous
						Equivalent	Acid		Alcohol					
							M.P., °C	B.P., °C	M.P., °C	B.P., °C				
26	Methyl meth-acrylate	99; 100–1	−50	1.413	0.936	100	16	161	−97	64.65	102–6	......	108 (cor.), al.	Polymerizes on standing or heating; 4-Bromo-anilide, 116
27	Ethyl propionate .	99.1	−73.85	1.3853	0.8889	102	−20.8	141	−117.3	78.32	81; 81.3; 79	126; 123	93, al.	N-(β-Amino-ethyl)mor-pholide, 85
28	Ethyl acrylate ...	101	......	1.4059$^{19.4}$	0.9136^{15}	100	13	141	−117.3	78.32	84–5, pet. eth.	141	93, al.	Polymerizes on standing or heating
29	Methyl pivalate (Methyl tri-methylacetate)..	101	......	1.4228	0.891^0	116	35.5	163–4	−97	64.65	155–7; 153–4, et. ac.-pet. eth.	119–20	108 (cor.), al.	....
30	n-Propyl acetate .	101.55	−95	1.38468	0.8834	102	16.6	118.2	......	97.1	82	153; 147	74, pet. eth.	....
31	Methyl n-butyrate	102.3	−84.8	1.3879	0.8982	102	−5.5; −8	162.5; 164	−97	64.65	115–6	75	108 (cor.), al.	....
32	Allyl acetate.....	104	......	1.40488	0.9276	100	16.6	118.2	......	97.1	82	153; 147	49–50	....
33	Trimethyl ortho-formate (Methyl orthoformate)..	105; 102	......	1.3793	0.9676	106	8.4	100.7	−97	64.65	2.55	53	108 (cor.), al.	....
34	Methyl iso-crotonate	106.2–108.2 (cor.)	......	....	....	100	15	169	−97	64.65	101–2	132	108 (cor.), al.	....
35	n-Butyl formate ..	106.6	−91.9	1.38940	0.8885	102	8.4	100.7	−90.2	117.6; 116	2.55	53	64; 62.5	....
36	Ethyl isobutyrate	109.8; 111	−88.2	1.3903	0.86930	116	−46.1	154.7	−117.3	78.32	128; 129	108.5–9.5	93, al.	Hydrazide, 104, eth.-al.
37	n-Propyl nitrate .	110	......	1.3979	1.063	105	......	......	......	97.1	......	......	74, pet. eth.	....
38	Chloromethyl acetate	111	......	....	1.094^{15}	108.5	16.6	118.2	......	......	82	153; 147	......	
39	Isopropyl propi-onate.........	111.3	......	....	0.8931^0	116	−20.8	141	−89.5	82.4	81; 81.3; 79	126; 123	123, pet. eth.	....
40	sec-Butyl acetate .	112.0	......	1.3865^{25}	0.872; 0.8648^{25}	116	16.6	118.2	......	99.5	82	153; 147	76	
41	n-Propyl chloro-formate.......	113; 115	......	1.40350	1.0901	122.5	......	......	......	97.1	......	......	74, pet. eth.	
42	Methyl isovalerate	116.7	......	1.3900^{25}	0.8808	116	−30.0	176.5	−97	64.65	135; 137	106–7	108 (cor.), al.	....
43	Isobutyl acetate ..	117.2; 118	......	1.39008	0.8747	116	16.6	118.2	......	108.1	82	153; 147	87	
44	Ethyl pivalate (Ethyl trimethyl-acetate).......	118.15	......	1.39061	0.85467	130	35.5	163–4	−117.3	78.32	155–7; 153–4, et. ac.-pet. eth.	119–20	93, al.	....
45	Ethyl meth-acrylate	118.5^{753}	......	1.41472	0.91063	114	16.	161	−117.3	78.32	102–6	......	93, al.	Polymerizes on heating; 4-Bromo-anilide, 116

*Derivative data given in order: m.p., crystal color, solvent from which crystallized.

TABLE XVI. ORGANIC DERIVATIVES OF ESTERS
Including esters of inorganic acids
a) Liquids. 1) (Listed in order of increasing b.p.)* (Continued)

No.	Name	Boiling point, °C	Melting point, °C	n_D^{20}	D_4^{20}	Saponification Equivalent	Acid M.P., °C	Acid B.P., °C	Alcohol M.P., °C	Alcohol B.P., °C	Amide	p-Toluidide	3,5-Dinitrobenzoate	Miscellaneous
46	Methyl crotonate	118.8–119.3	......	....	0.9806^4	100	72	189 (cor.)	−97	64.65	118, w.; 115	132, bz.	108 (cor.), al.	....
47	Isopropyl iso-butyrate	120.76	.....	....	$0.84708^{21.3}$	130	−46.1	154.7	....	82.4	128; 129	108.5–9.5 pet. eth.	123, al.	Hydrazide, 104, eth.-al.
48	Ethyl n-butyrate	121.6	−100.8	1.40002	0.87917	116	−5.5; −8	162.5; 164	−117.3	78.32	115–6	75	93, al.	....
49	n-Propyl pro-pionate	122.2; 123.4	−75.9	1.39325	0.8809	116	−20.8	141	......	97.1	81; 81.3; 79	126; 123	74, pet. eth.	....
50	tert-Amyl acetate (Dimethylethyl-carbinyl acetate)	124	......	1.392	0.8738^{19}	130	16.6	118.2	−8.55	102.3	82	153; 147	116; 117–8	....
51	Allyl propionate	124	......	....	....	114	−20.8	141	......	97.1	81; 81.3; 79	126; 123	49–50	....
52	Isoamyl formate	124.2	−93.5	1.39756	0.8820	116	8.4	100.7	−117	132	2.55	53	61	
53	Ethyl isocrotonate	125.5–126^{749}	......	1.42423	0.91820	114	15	169	−117.3	78.32	101–2	132	93, al.	
54	n-Butyl acetate	126.1	−73.5	1.39614	0.881	116	16.6	118.2	−90.2	117.6; 116	82	153; 147	64; 62.5	....
55	Diethyl carbonate (Ethyl carbonate)	126.5	−43.0	1.3852	0.9752	118	......	......	−117.3	78.32	......	......	93, al.	
56	tert-Butyl iso-butyrate	126.7	......	1.3921	....	144	−46.1	154.7	25.5	82.5	128; 129	108.5–9.5 pet. eth.	142, al.	Hydrazide, 104, eth.-al.
57	Methyl n-valerate (Methyl n-pentanoate)	127.7	−91.0	1.397	0.885	116	−34.5	186.35	−97	64.65	106	74	108 (cor.), al.	....
58	Isopropyl n-butyrate	128	......	....	0.8652^{13}	130	−5.5; −8	162.5; 164	−89.5	82.4	115–6	75	123, pet. eth.	....
59	Isobutyl chloro-formate	130; 128.8	......	$1.40711^{17.9}_{He}$	....	136.5	......	......	......	108.1	......	......	87	....
60	Methyl methoxy-acetate	130	......	1.39636	1.0511	104	......	203	−97	64.65	96.5–7.0; 92–4	......	108 (cor.), al.	....
61	Methyl chloro-acetate	130; 132	......	1.4221	1.238	108.5	α: 61.3; β: 56.2; γ: 52.5	189	−97	64.65	121	162	108 (cor.), al.	....
62	Ethyl methoxy-acetate	132	......	....	1.0118^{15}	118	......	203	−117.3	78.32	96.5–7.0; 92–4	......	93, al.	
63	n-Amyl formate	132.1; 130	−73.5	1.39916	0.8853	116	8.4	100.7	−78.5	138 (cor.)	2.55	53	46.4	
64	sec-Amyl(3) acetate (Diethyl-carbinyl acetate)	133	......	1.4005	....	130	16.6	118.2	......	116.1	82	153; 147	101; 99; 97	....
65	sec-Amyl(2) acetate (Methyl n-propylcarbinyl acetate)	133.5	......	1.3960	0.8692^{18}	130	16.6	118.2	......	119.85	82	153; 147	62	....
66	Allyl isobutyrate	134	......	....	....	128	−46.1	154.7	......	97.1	128; 129	108.5–9.5	49–50	Hydrazide, 104, eth.-al.
67	Ethyl isovalerate	134.7	−99.3	1.4009	0.86565	130	−30.0	176.5	−117.3	78.32	135; 137	106–7	93, al.	....
68	n-Propyl iso-butyrate	135	......	1.3959	0.8843^0	130	−46.1	154.7	......	97.1	128; 129	108.5–9.5	74, pet. eth.	Hydrazide, 104, eth.-al.

*Derivative data given in order: m.p., crystal color, solvent from which crystallized.

TABLE XVI. ORGANIC DERIVATIVES OF ESTERS
Including esters of inorganic acids
a) Liquids. 1) (Listed in order of increasing b.p.)* (Continued)

No.	Name	Boiling point, °C	Melting point, °C	n_D^{20}	D_4^{20}	Saponification Equivalent	Acid M.P., °C	Acid B.P., °C	Alcohol M.P., °C	Alcohol B.P., °C	Amide	p-Toluidide	3,5-Dinitro-benzoate	Miscellaneous
69	n-Butyl nitrate ...	136	.....	....	1.048⁰	119	....	....	−90.2	117.6; 116	....	.....	64; 62.5	....
70	Methyl pyruvate .	136.8–138; 134–7	.....	....	1.154⁰	102	13.6	165d.	−97	64.65	124–5; 145	109; 130	108 (cor.), al.	2,4-Dinitro-phenylhyd-razone, 186.5–7.5 (cor.), yel., diox.-me. al.
71	Methyl α-hy-droxyisobutyrate	137	.....	....	....	118	79	212	−97	64.65	....	132–3, w.	108 (cor.), al.	....
72	Isobutyl pro-pionate	138; 137	−71.4	1.3975	0.8876⁰	130	−20.8	141	....	108.1	81; 81.3; 79	123–6	87	....
73	Ethyl crotonate ..	138; 136.7⁷⁴⁹	.....	1.42524	0.91752	114	72, w.	189 (cor.)	−117.3	78.32	159–60, bz.	132, bz.	93, al.	....
74	Allyl n-butyrate ..	142	.....	....	....	128	−5.5; −8	162.5; 164	....	97.1	115–6	75	49–50	....
75	Isoamyl acetate (3-Methylbutyl acetate).......	142	.....	1.40034	0.8674	130	16.6	118.2	−117	132	82	153; 147	61	....
76	Isopropyl iso-valerate.......	142	.....	1.3938²⁵	0.8538¹⁷	144	−30	176.5	−89.5	82.4	135; 137	106–7	123, pet. eth.	....
77	n-Propyl n-buty-rate	143.8	−95.2	1.4005	0.872	130	−5.5; −8	162.5; 164	....	97.1	115–6	75	74, pet. eth.	....
78	β-Methoxyethyl acetate (Ethylene glycol mono-methyl ether acetate; Methyl cellosolve acetate).......	144	.....	....	1.088	118	16.6	118.2	....	124.5	82	153; 147	.....	....
79	Methyl bromo-acetate	144d.	.....	....	1.657	153	50	208	−97	64.65	91	.....	108 (cor.), al.	....
80	n-Butyl chloro-formate.......	145; 139	.....	1.417⁸·⁴	1.079	136.5	....	....	−90.2	117.6; 116	....	.....	64; 62.5	....
81	β-Chloroethyl acetate	145	.....	1.4234	1.178	122.5	16.6	118.2	....	131	82	153; 147	.....	....
82	Methyl d,l-lactate	145; 144.8	.....	1.4144	1.0931	104	16.8; 18	122¹⁵	−97	64.65	78.5–9.0 (cor.), bz.-al. (3:1)	107	108 (cor.), al.	....
83	Ethyl chloro-acetate	145	−26	1.42274	1.158	122.5	α: 61.3; β: 56.2; γ: 52.5	189	−117.3	78.32	121	162	93, al.	....
84	tert-Butyl n-butyrate	145–6.6	.....	1.4001¹⁷·⁵	....	144	−5.5; −8	162.5; 164	25.5	82.5	115–6	75	142, pet. eth.	....
85	Triethyl ortho-formate (Ethyl-orthoformate) ..	145.5	.....	1.3922	0.8909	148	8.4	100.7	−117.3	78.32	2.55	53	93, al.	....
86	Ethyl n-valerate (Ethyl n-pentanoate)	145.5	−91.2	1.40094	0.8739	130	−34.5	186.35	−117.3	78.32	106	74	93, al.	....
87	Ethyl α-chloro-propionate	146	.....	....	1.087	136.5	....	186	−117.3	78.32	80	124	93, al.	Phenylhydra-zide, 95

*Derivative data given in order: m.p., crystal color, solvent from which crystallized.

253

TABLE XVI. ORGANIC DERIVATIVES OF ESTERS
Including esters of inorganic acids
a) Liquids. 1) (Listed in order of increasing b.p.)* (Continued)

No.	Name	Boiling point, °C	Melting point, °C	n_D^{20}	D_4^{20}	Equiv-alent	Acid M.P. °C	Acid B.P. °C	Alcohol M.P. °C	Alcohol B.P. °C	Amide	p-Tolui-dide	3,5-Di-nitro-benzoate	Miscellaneous
88	n-Butyl propionate	146.8	−89.6	1.4038^{15}	1.401	130	−20.8	141	−90.2	117.6; 116	81; 81.3; 79	126; 123	64; 62.5	....
89	Benzyl chloro-acetate	147	......	1.5246^{18}	1.2223^{4}	170.5	α: 61.3; β: 56.2; γ: 52.5	189	−15.3	205.5	121	162	113	
90	Di-isopropyl carbonate (Isopropyl carbonate)	147.2 (cor.)	......	1.3932	0.9162	146	....	......	−89.5	82.4	....	......	123, pet. eth.	
91	Methyl isobutyl-carbinyl acetate	148	......	....	0.8805^{0}	130	16.6	118.2	−97	64.65	82	153; 147	108 (cor.), al.	
92	Methyl ethoxy-acetate	148	......	....	1.0112^{15}	118	....	206–7	−97	64.65	80–2	32, eth.	108 (cor.), al.	....
93	Isobutyl iso-butyrate	148.6	−80.65	1.3999	0.87496^{0}	144	−46.1	154.7	....	108.1	128; 129	108.5–9.5	87	Hydrazide, 104, eth.-al.
94	n-Amyl acetate (n-Pentyl acetate)	149.25	−70.8	1.4031	0.8756	130	16.6	118.2	−78.5	138 (cor.)	82	153; 147	46.4	....
95	Ethyl α-hydroxy-isobutyrate	150	......	......	....	132	79	212	−117.3	78.32	....	132–3, w.	93, al.	....
96	Methyl glycolate	151.2	......	......	1.1677^{18}	90	78–9; 80	......	−97	64.65	120, al.-et. ac.	143, w.	108 (cor.), al.	
97	Methyl n-caproate (Methyl n-hexanoate)	151.25	−71.0	1.405	0.88464	130	−3.9	205.35	−97	64.65	100–1	74–5	123, pet. eth.	
98	Isopropyl n-valerate (Isopropyl n-pentanoate)	153.5	......	1.4009	0.8579	144	−34.5	186.35	−89.5	82.4	106	74	61	....
99	Isoamyl chloro-formate	154	......	1.41916^{15}_{He}	1.032^{15}	150.5	....	......	−117	132	....	......	93, al.	
100	Ethyl d,l-lactate	154.5	......	1.410	1.030	118	16.8; 18	122^{15}	−117.3	78.32	78.5–9.0 (cor.), bz.-al., (3:1)	107	93, al.	
101	Ethyl pyruvate	155	......	$0.0596^{15.6}$	$1.408^{15.6}$	116	13.6	165d.	−117.3	78.32	124–5; 145	109; 130	93, al.	Phenylhydrazone, 118, dil. al.; 4-Nitrophenylhydrazone, 185–7; 2,4-Dinitrophenylhydrazone, 154.5–155 (cor.), diox.-al.
102	n-Propyl iso-valerate	155.5	......	$1.40413^{17.8}$	$0.8643^{17.8}$	144	−30	176.5	....	97.1	135; 137	106–7	74, pet. eth.	....
103	n-Hexyl formate	155.51	−62.65	1.40898^{15} He (yel)	0.88133	130	8.4	100.7	−51.6; −46.1	157.5	2.55	53	58.4 (cor.)	....

*Derivative data given in order: m.p., crystal color, solvent from which crystallized.

TABLE XVI. ORGANIC DERIVATIVES OF ESTERS
Including esters of inorganic acids
a) Liquids. 1) (Listed in order of increasing b.p.)* (Continued)

No.	Name	Boiling point, °C	Melting point, °C	n_D^{20}	D_4^{20}	Saponification Equiv-alent	Acid M.P., °C	Acid B.P., °C	Alcohol M.P., °C	Alcohol B.P., °C	Amide	p-Toluidide	3,5-Dinitrobenzoate	Miscellaneous
104	β-Ethoxyethyl acetate (Ethylene glycol monoethyl ether acetate; Ethyl cellosolve acetate)	156.2; 158	......	1.40292	0.9701	132	16.6	118.2	....	135	82	153; 147	75, al.	....
105	Isobutyl n-butyrate	157	......	1.40295[18.4]	0.8620	144	-5.5; -8	162.5; 164		108.1	115-6	75	87	....
106	Ethyl dichloroacetate	158	......	1.43860	1.2821	157	5-6	194	-117.3	78.32	98 (subl.)	153	93, al.	
107	Ethyl bromoacetate	159	......	1.451	1.506	167	50	208	-117.3	78.32	91	......	93, al.	
108	Ethyl glycolate	160	......	....	1.0869[15]	104	78-9; 80	....	-117.3	78.32	120, al.-et. ac.	143, w.	93, al.	
109	Isoamyl propionate	160.2	......	1.4065	0.870	144	-20.8	141	-117	132	81; 81.3; 79	126; 123	61	
110	Ethyl α-bromopropionate	162	......		1.524	181	25.7	203.5	-117.3	78.32	123	125	93, al.	
111	Cyclohexyl formate	162.5[750]	......	1.443	1.010	128	8.4	100.7	25.1	161.1	2.55	53	112-3, al.	
112	β-Bromoethyl acetate	163	......		1.524	167	16.6	118.2	....	149d.	82	153; 147	......	
113	Tetraethyl silicate (Ethyl orthosilicate)	165.5; 168.5	......	1.38619	0.93975	52	....	....	-117.3	78.32	....	......	93, al.	Gives SiO₂ on hydrolysis
114	n-Propyl n-valerate (n-Propyl n-pentanoate)	166.2; 167	-70.7	1.4065	0.8699	144	-34.5	186.35	....	97.1	106	74	74, pet. eth.	....
115	n-Butyl n-butyrate	166.6	-91.5	1.406	0.869	144	-5.5; -8	162.5; 164	-90.2	117.6; 116	115-6	75	64; 62.5	....
116	Ethyl n-caproate (Ethyl n-hexanoate)	167.7	-67.5	1.40727	0.8710	144	-3.9	205.35	-117.3	78.32	100-1	74-5	93, al.	....
117	Ethyl trichloroacetate	168	......	1.450	1.380	191.5	57-8	197.5	-117.3	78.32	141	113	93, al.	Phenylhydrazide, 123
118	Isopropyl d,l-lactate	168; 166-8	......	1.4082[25]	0.998	132	16.8; 18	122[15]	-89.5	82.4	78.5-9.0 (cor.), bz.-al. (3:1)	107	123, pet. eth.	....
119	Di-n-propyl carbonate (n-Propyl carbonate)	168.5 (cor.)	......	1.4014	0.9411	146	....	....	....	97.1	....	......	74, pet. eth.	
120	n-Amyl propionate (n-Pentyl propionate)	168.7	-73.1	1.4096[15]	0.8761[15]	144	-20.8	141	-78.5	138 (cor.)	81; 81.3; 79	126; 123	46.4	....
121	Ethylidene diacetate	169	......		1.061[12]	73	16.6	118.2	-12.6	197.85	82	153; 147	169	....
122	Isoamyl isobutyrate	169	......		0.8760[0]	158	-46.1	154.7	-117	132	128; 129	108.5-9.5	61	Hydrazide, 104, eth.-al.
123	Methyl acetoacetate	170	......	1.41964	1.0765[0]	116	....	....	-97	64.65	....	......	108 (cor.), al.	Semicarbazone, 152.5, me. al.
124	Isobutyl isovalerate	171	......	1.40569	0.8534	158	-30	176.5	....	108.1	135; 137	106-7	87	....

*Derivative data given in order: m.p., crystal color, solvent from which crystallized.

TABLE XVI. ORGANIC DERIVATIVES OF ESTERS
Including esters of inorganic acids
a) Liquids. 1) (Listed in order of increasing b.p.)* (Continued)

No.	Name	Boiling point, °C	Melting point, °C	n_D^{20}	D_4^{20}	Saponification Equivalent	Acid M.P., °C	Acid B.P., °C	Alcohol M.P., °C	Alcohol B.P., °C	Amide	p-Toluidide	3,5-Dinitrobenzoate	Miscellaneous
125	Methyl enanthate (Methyl n-heptanoate)	173.8	−55.8	1.412	0.88011	144	−74.7	223	−97	64.65	96; 96.5	81	108 (cor.), al.	....
126	Ethylene glycol diformate	174	.....	....	1.193^0	59	8.4	100.7	−78.5	138 (cor.)	2.55	53	46.4	....
127	sec-Butyl n-valerate (sec-Butyl n-pentanoate)	174.5	.....	1.4081	0.8605^{20}	158	−34.5	108.35	.....	99.5	106	74	76	
128	Cyclohexyl acetate	175	.....	1.442	0.970	142	16.6	118.2	25.1	161.1	82	153; 147	112-3, al.	....
129	n-Butyl chloroacetate	175	.....	....	1.081	150.5	α: 61.3; β: 56.2; γ: 52.5	189	−90.2	117.6; 116	121	162	64; 62.5	....
130	Furfuryl acetate	175–7	.....	1.4627	1.118	140	16.6	118.2	....	172; 170	82	153; 147	80–1	....
131	Methyl methylacetoacetate	177.4	.....	1.418	1.030	130	....	....	−97	64.65	....	.....	108 (cor.), al.	Semicarbazone, 138, al.
132	n-Heptyl formate	178.12	.....	1.41505^{15} He (yel)	0.87841	144	8.4	100.7	−34.6; −33.8	176.8	2.55	53	46; 47	....
133	n-Hexyl acetate	178.1	−80.9; −60.9	1.41122^{15} He (yel)	0.87336	144	16.6	118.2	−51.6; −46.1	157.5	82	153; 147	58.4 (cor.)	....
134	Isoamyl n-butyrate	178.6	.....	1.411	0.864	158	−5.5; −8	162.5; 164	−117	132	115–6	75	61	....
135	Ethyl β-bromopropionate	179	.....	....	1.425	181	62.5	.....	−117.3	78.32	111	.....	93, al.	2-Naphthylamide, 174
136	Ethyl acetylglycolate	179	.....	....	1.0993^{17}	73	....	....	−117.3	78.32	....	.....	93, al.	Hydrolysis → glycolic a. + ac.a. + al.
137	Isobutyl n-valerate (Isobutyl n-pentanoate)	179	.....	1.4099	0.8625	158	−34.5	186.35	.....	108.1	106	74	87	....
138	β-Hydroxyethyl formate (Ethylene glycol monoformate)	180	.....	....	1.1989^{15}	90	8.4	100.7	−78.5	138 (cor.)	2.55	53	46.4	....
139	Ethyl methylacetoacetate	180.8 (cor.); 187	.....	1.419	1.0191	144	....	....	−117.3	78.32	73, eth.	.....	93, al.	Anilide, 138-40; Semicarbazone, 194; 2,4-Dinitrophenyl hydrazone, 56-7
140	Ethyl acetoacetate	181	.....	1.41976	1.025	130	....	....	−117.3	78.32	....	.....	93, al.	Semicarbazone, 129; 133, eth.
141	Methyl pyromucate (Methyl 2-furoate)	181.3	.....	1.4860	1.180	126	133–4; 132	230–2	−97	64.65	142–3	170.5, al.	108 (cor.), al.	2,4-Dinitrophenylhydrazone, 93; 96, yel., al.

*Derivative data given in order: m.p., crystal color, solvent from which crystallized.

TABLE XVI. ORGANIC DERIVATIVES OF ESTERS

TABLE XVI. ORGANIC DERIVATIVES OF ESTERS
Including esters of inorganic acids
a) Liquids. 1) (Listed in order of increasing b.p.)* (Continued)

No.	Name	Boiling point, °C	Melting point, °C	n_D^{20}	D_4^{20}	Saponification Equivalent	Acid M.P., °C	Acid B.P., °C	Alcohol M.P., °C	Alcohol B.P., °C	Amide	p-Toluidide	3,5-Dinitro-benzoate	Miscellaneous
142	Dimethyl malonate (Methyl malonate)......	181.5	−62	1.41398	1.1539	66	134.8–.9	......	−97	64.65	mono: 106–110; di: 170, w.-al.	mono: 156d.; di: 252–3, al.	108 (cor.), al.	Phenylhydrazide, 194
143	Ethyl β-methoxyethyl carbonate .	182.6	......	1.4036^{25}	1.0424^{25}	148	......	......	−117.3	78.32	......	......	93, al.	....
144	Methyl cyclohexanecarboxylate (Methyl hexahydrobenzoate)......	183	......	1.45372^{15}	0.9954^{15}	142	30–1	233	−97	64.65	185–6	......	108 (cor.), al.	....
145	Diethyl oxalate (Ethyl oxalate)..	185.19	−41.5; −40.6	1.41043	1.0785	73	189.5 (anh.); 101 (+ 2H₂O)	......	−117.3	78.32	mono: 219; di: 419d.	mono: 169; di: 268	93, al.	N-(β-Aminoethyl)-morpholide, 170
146	n-Amyl n-butyrate (n-Pentyl n-butyrate)......	186.4	−72.3	1.412	0.866	158	−5.5; −8	162.5; 164	−78.5	138 (cor.)	115–6	75	46.4	....
147	n-Butyl n-valerate (n-Butyl n-pentanoate)	186.9	−92.8	1.4123	0.8678	158	−34.5	186.35	−90.2	117.6; 116	106	74	64; 62.5	....
148	β-Hydroxyethyl acetate (Ethylene glycol monoacetate)........	187–9	......	......	......	104	16.6	118.2	−78.5	138 (cor.)	82	153; 147	46.4	
149	n-Propyl n-caproate (n-Propyl n-hexanoate).....	187.15; 186	−74.0	1.417	0.86719	158	−3.9	205.35	......	97.1	100–1	74–5	74, pet. eth.	....
150	Dimethyl sulfate (Methyl sulfate).	188	−27	1.3874	1.3348^{15}	63	......	......	−97	64.65	......	......	108 (cor.), al.	
151	n-Butyl lactate...	188	−43	1.4216	0.984_{20}^{20}	146	16.8; 18	122^{15}	−90.2	117.6; 116	78.5–9.0 (cor.), bz.-al. (3:1)	107	64; 62.5	....
152	Ethyl enanthate (Ethyl n-heptanoate)......	188.6	−66.3	1.413	0.86856	158	−7.47	223	−117.3	78.32	96; 96.5	81	93, al.	....
153	Methyl ethylacetoacetate	189.7 (cor.)	......		0.989	144	......	......	−97	64.65	95–6, bz.	......	108 (cor.), al.	
154	Di-isobutyl carbonate (Isobutyl carbonate).....	189.8 (cor.)	......	1.4072	0.9138	174	......	......		108.1	......	......	87	
155	n-Hexyl propionate	190	−57.5	1.41621^{15} He (yel)	0.86980	158	−20.8	141	−51.6; −46.1	157.5	81; 81.3; 79	126; 123	58.4 (cor.)	
156	Ethylene glycol diacetate.......	190.2	−31	1.4150	1.1040	73	16.6	118.2	−78.5	138 (cor.)	82	153; 147	46.4	
157	Isoamyl isovalerate........	190.4	......	$1.41300^{18.7}$	0.870	172	−30	176.5	−117	132	135; 137	106–7	61	
158	n-Heptyl acetate .	192.45	......	1.41653^{15} He (yel)	0.87070^{15}	158	16.6	118.2	−34.6; −33.8	176.8	82	153; 147	46; 47	....
159	Cyclohexyl propionate	193^{750}	......	......	0.9718^{0}	156	−20.8	141	25.1	161.1	81; 81.3; 79	126; 123	112–3, al.	....

* Derivative data given in order: m.p., crystal color, solvent from which crystallized.

TABLE XVI. ORGANIC DERIVATIVES OF ESTERS

TABLE XVI. ORGANIC DERIVATIVES OF ESTERS
Including esters of inorganic acids
a) Liquids. 1) (Listed in order of increasing b.p.)* (Continued)

No.	Name	Boiling point, °C	Melting point, °C	n_D^{20}	D_4^{20}	Saponification Equiv-alent	Acid M.P., °C	Acid B.P., °C	Alcohol M.P., °C	Alcohol B.P., °C	Amide	p-Tolui-dide	3,5-Di-nitro-benzoate	Miscellaneous
160	Di-isopropyl oxalate (Iso-propyl oxalate) .	193–4	......	1.4100	1.0097	87	189.5 (anh.); 101 (+ 2H₂O)	......	−89.5	82.4	*mono*: 219; *di*: 419d.	*mono*: 169; *di*: 268	123, pet. eth.	....
161	Ethyl β-ethoxy-ethyl carbonate .	194.5	......	1.5064²⁵	1.0115²⁵	162	......	......	−117.3	78.32	......	......	93, al.	
162	sec-Octyl acetate .	194.5	......	1.4141	0.8606¹⁹	172	16.6	118.2		179	82	153; 147	32	....
163	Methyl n-capryl-ate (Methyl n-octanoate).....	194.6	−41	1.417	0.878	158	16.3	237; 239.3	−97	64.65	57	70	108 (cor.), al.	
164	α-Tetrahydro-furfuryl acetate .	195; 194	......	1.4350	1.0624²⁵	144	16.6	118.2	......	177–8⁷⁴³	82	153; 147	83–4	....
165	Methyl levulinate .	196; 191	......	1.42333	1.04945	130	33–5	245–6	−97	64.65	107–8, al.	108–9, w.	108 (cor.), al.	Semicarba-zone, 142–3; Phenyl-hydrazone, 94–6; 2,4-Dinitro-phenylhyd-razone, 141.5–2.5 (cor.), diox.-al.; Oxime, 96
166	Ethyl cyclo-hexanecar-boxylate (Ethyl hexahydroben-zoate)........	196	......	1.45012¹⁵	0.9672¹⁵	156	30–1	233	−117.3	78.32	185–6	......	93, al.	....
167	Diethyl methyl-malonate	196	......	....	1.019¹⁵	87	137; 138d.	......	−117.3	78.32	217; 206	*mono*: 145d.; *di*: 228; 214	93, al.	....
168	Phenyl acetate ...	196.7	......	1.503	1.078	136	16.6	118.2	41.8; 42	182	82	153; 147	145.8 (cor.), al.	....
169	Ethyl ethylaceto-acetate	198	......	1.422	0.9856	158	......	......	−117.3	78.32	95–6, bz.	......	93, al.	Ketone cleavage → 2-penta-none, b.p. 102
170	n-Octyl formate ..	198.8	−39.1	1.42082¹⁵ He (yel)	0.87435	158	8.4	100.7	−16; −16.7	195	2.55	53	61–2	....
171	2-Ethyl-1-hexyl acetate	199	......	....	0.8733²⁰₂₀	172	16.6	118.2	......	184.6	82	153; 147	......	....
172	Methyl benzoate .	199.2	−12.4	1.5164	1.0888	136	122.4	249	−97	64.65	130	158	108 (cor.), al.	
173	Diethyl malonate (Ethyl malonate)......	199.3	−51.5	1.41618	1.05513	80	134.8–.9	......	−117.3	78.32	*mono*: 106–10; *di*: 170, w.-al.	*mono*: 156d.; *di*: 252–3, al.	93, al.	Phenylhy-drazide, 194
174	Methyl cyano-acetate	200	−22.5	....	1.0962²⁵	99	66	......	−97	64.65	119–20	......	108 (cor.), al.	....

* Derivative data given in order: m.p., crystal color, solvent from which crystallized.

TABLE XVI. ORGANIC DERIVATIVES OF ESTERS
Including esters of inorganic acids
a) Liquids. 1) (Listed in order of increasing b.p.)* (Continued)

No.	Name	Boiling point, °C	Melting point, °C	n_D^{20}	D_4^{20}	Saponification Equiv-alent	Acid M.P., °C	Acid B.P., °C	Alcohol M.P., °C	Alcohol B.P., °C	Amide	p-Tolui-dide	3,5-Di-nitro-benzoate	Miscellaneous
175	Dimethyl mesaconate.....	203	......	1.45119	1.0914	79	204.5 (cor.)	......	−97	64.65	mono (α): 222; (β): 174; di: 176.5	mono (α): 196; di: 212, al.	108 (cor.), al.	Dihydrazide, 215d., dil. al.
176	Benzyl formate...	203	......	1.516	1.080	136	8.4	100.7	−15.3	205.5	2.55	53	113	....
177	Vinyl benzoate...	203	......	....	1.065	148	122.4	249	......	......	130	158	......	
178	Cyclohexyl iso-butyrate	204⁷⁵⁰	......	....	0.9489⁰	170	−46.1	154.7	25.1	161.1	128; 129	108.5–9.5, al.	112–3, al.	Hydrazide, 104, al.-eth.
179	Dimethyl maleate	204.4	7.6	1.44156	1.14513¹⁵	72	137	......	−97	64.65	mono: 172–3, w.; 153, sl. d.; di: 181, me. al.	di:, 142, eth.	108 (cor.), al.	....
180	α-Tetrahydro-furfuryl pro-pionate	204–7	......	....	1.044	158	−20.8	141	......	177–8⁷⁴³	81; 81.3; 79	126; 123	83–4	....
181	Ethyl levulinate ..	205.8	......	1.42288	1.01114	144	33–5	245–6	−117.3	78.32	107–8d.	108–9, w.	93, al.	Semicarba-zone, 147–8; Phenylhyd-razone, 103–4; 2,4-Dinitro-phenylhyd-razone, 101–2, diox.-al.; Oxime, 96
182	Ethyl allylaceto-acetate	206 sl. d.; 211–2 sl. d.	......	1.43¹⁷·⁶	0.9898	170	......	......	−117.3	78.32	......	......	93, al.	Semicarba-zone, 125, w.
183	n-Amyl n-valerate (n-Pentyl n-pentanoate)	207.4	......	1.4181¹⁵	0.8825⁰	172	−34.5	186.35	−78.5	138 (cor.)	106	74	46.4	....
184	Di-n-butyl car-bonate (n-Butyl carbonate).....	207.5 (cor.)	......	1.4117	0.9238	174	......	......	−90.2	117.6; 116	......	......	64; 62.5	....
185	n-Butyl n-caproate (n-Butyl n-hexanoate) ...	207.74	−63.1; −64.3	1.41877¹⁵ He (yel)	0.86530	172	−3.9	205.35	−90.2	117.6; 116	100–1	74–5	64; 62.5	....
186	n-Hexyl n-butyrate	207.88	−78	1.41875¹⁵ He (yel)	0.86519	172	−5.5; −8	162.5; 164	−51.6; −46.1	157.5	115–6	75	58.4 (cor.)	....
187	2-Tolyl acetate ("o-Cresyl acetate")	208	......	....	1.048	150	16.6	118.2	31	191–2	82	153; 147	134.8 (cor.), al.	
188	Diethyl sulfate (Ethyl sulfate) ..	208	......	1.4010¹⁸	1.172²⁵	77	......	......	−117.3	78.32	......	......	93, al.	
189	n-Propyl n-enan-thate (n-Propyl n-heptanoate) ..	208	−64.8	1.41835¹⁵	0.86556	172	−7.47	223	......	97.1	96; 96.5	81	74, pet. eth.	
190	Ethyl n-caprylate (Ethyl n-octanoate)	208.35	α: −43.1; β: −59.2	1.41775	0.8667	172	16.3	237; 239.3	−117.3	78.32	110; 106	70	93, al.	N-(β-Amino-ethyl) mor-pholide, 59

*Derivative data given in order: m.p., crystal color, solvent from which crystallized.

TABLE XVI.

TABLE XVI. ORGANIC DERIVATIVES OF ESTERS
Including esters of inorganic acids
a) Liquids. 1) (Listed in order of increasing b.p.)* (Continued)

No.	Name	Boiling point, °C	Melting point, °C	n_D^{20}	D_4^{20}	Saponification Equivalent	Acid M.P., °C	Acid B.P., °C	Alcohol M.P., °C	Alcohol B.P., °C	Amide	p-Toluidide	3,5-Dinitrobenzoate	Miscellaneous
191	Isobutyl enanthate (Isobutyl n-heptanoate)	209	......	....	0.8593	186	−7.47	223	......	108.1	96; 96.5	81	87	....
192	Isopropyl levulinate	209.3	......	1.42088	0.98724	158	33.5	245–6	......	82.4	107–8d.	108–9, w.	123, pet. eth.	Semicarbazone, 141–2; Phenylhydrazone, 108–9; 2,4-Dinitrophenylhydrazone, 88–9 (cor.); 90.9
193	Trimethylene glycol diacetate (1,3-Diacetoxypropane)	210	......	....	1.069	80	16.6	118.2	−30	214.7; 210–2	82	153; 147	178	....
194	n-Octyl acetate	210	......	....	0.8847⁰	172	16.6	118.2	−16; −16.7	195	82	153; 147	61–2	....
195	n-Heptyl propionate	210	−50.9	1.42605¹⁵ He (yel)	0.86786	172	−20.8	141	−34.6; −33.8	176.8	81; 81.3; 79	126; 123	46; 47	....
196	Dimethyl citraconate	210.5	......	1.44856	1.11531	79	92; 91d. eth.-lgr.	206–11, sl. d.	−97	64.65	100–1	......	108 (cor.), al.	Dihydrazide, 177, w.; 1-Naphthylamide, 169–70
197	n-Propyl pyromucate (n-Propyl 2-furoate)	211	......	1.4737²⁵·⁹	1.0745²⁵·⁹	154	133–4; 132	230–2	......	97.1	142–3	170.5, al.	74, pet. eth.	....
198	Ethylene glycol dipropionate	211	......		1.0544¹⁵₁₅	87	−20.8	141	−78.5	138 (cor.)	81; 81.3; 79	126; 123	46.4	....
199	Cyclohexyl n-butyrate	212⁷⁵⁰	......	....	0.9572⁰	170	−5.5; −8	162.5; 164	25.1	161.1	115–6	75	112–3, al.	....
200	3-Tolyl acetate ("m-Cresyl acetate")	212	12	1.4978	1.049	150	16.6	118.2	11.95	202.7	82	153; 147	164.5 (cor.), al.	....
201	Ethyl acetopyruvate	213–5	......	1.4757¹⁷	1.1251	158	101, bz.	......	−117.3	78.32	131–2d., al.	......	93, al.	
202	Ethyl benzoate	212.4; 213.2	−34.7	1.50570	1.04684	150	122.4	249	−117.3	78.32	130	158	93, al.	N-(β-Aminoethyl) morpholide, 123.4
203	4-Tolyl acetate ("p-Cresyl acetate")	212.5	......	1.500	1.051	150	16.6	118.2	36	202	82	153; 147	188.6 (cor.), al.	....
204	Di-n-propyl oxalate (n-Propyl oxalate)	213.9	−51.7	1.4168	1.0169	87	189.5 (anh.); 101 (+2 H₂O)	......	......	97.1	mono: 219; di: 419d.	mono: 169; di: 268	74, pet. eth.	NH₄OH → Di-n-propyl oxamate, 90–2, me. al.
205	Methyl pelargonate	214; 213–4	......	....	0.892	172	12.3	254.4	−97	64.65	99	84	108 (cor.), al.	
206	Dimethyl glutarate	214⁷⁵¹	−34.7	1.42415	1.0874	80	98	302–4	−97	64.65	di: 175	di: 218	108 (cor.), al.	....

*Derivative data given in order: m.p., crystal color, solvent from which crystallized.

TABLE XVI. ORGANIC DERIVATIVES OF ESTERS
Including esters of inorganic acids
a) Liquids. 1) (Listed in order of increasing b.p.)* (Continued)

No.	Name	Boiling point, °C	Melting point, °C	n_D^{20}	D_4^{20}	Saponification Equivalent	Acid M.P., °C	Acid B.P., °C	Alcohol M.P., °C	Alcohol B.P., °C	Amide	p-Toluidide	3,5-Dinitrobenzoate	Miscellaneous
207	2,6-Dimethyl-phenyl acetate (vic-m-Xylenyl acetate)	214–6	......	....	....	164	16.6	118.2	49	203	82	153; 147	158.8 (cor.), al.	....
208	Methyl 2-toluate (Methyl 2-methyl-benzoate)	215	......	....	1.068	150	104–5; 107–8	259^{751}	−97	64.65	142.8 (cor.)	144	108 (cor.), al.	....
209	Benzyl acetate	217	......	1.5200	1.055	150	16.6	118.2	−15.3	205.5	82	153; 147	113	N-(β-Amino-ethyl) morpholide, 95.2
210	Diethyl succinate	217.25	−20.8	1.41975	1.0398	87	185; 182.8	235d.	−117.3	78.32	mono: 157; di: 260d., w.	mono: 179–80; di: 254.5–5.5; 260	93, al.	N-(β-Amino-ethyl) morpholide, 174
211	Diethylene glycol monoethyl ether acetate	218	......	....	1.013	176	16.6	118.2	....	198	82	153; 147	oil	....
212	Diethyl fumarate	218.4; 213–4	0.55	1.44103	1.052	86	286–7 (sealed tube); 233–5; 200, subl.	.....	−117.3	78.32	mono: 270; 300–2, subl.; di: 266d.	mono: 233.0–4.5; di: 313–4 ac. a.	93, al.	....
213	Isopropyl benzoate	218.5	......	1.4890^{25}	1.013	164	122.4	249	−89.5	82.4	130	158	123, pet. eth.	....
214	Methyl phenyl-acetate	220	......	1.507	1.068	150	76.5, subl.	256.5 (cor.)	−97	64.65	156	135–6	108 (cor.), al.	....
215	l-Linalyl acetate	220	......	1.4460	0.8951	196	16.6	118.2	....	199	82	153; 147	......	$[\alpha]_D$: −3 to −17; 4-Nitrobenzoate, 70
216	Methyl 3-toluate (Methyl 3-methylbenzoate)	221; 215	......	....	1.061	150	111–3; 110–1	263, subl.	−97	64.65	94; 97	118	108 (cor.), al.	....
217	n-Propyl levulinate	221.2	......	1.42576	0.98955	158	33–5	245–6	....	97.1	107–8d.	108–9, w.	74, pet. eth.	Semicarbazone, 129–30; Hydrazone, 88–90; Phenylhydrazone, 67–8 (cor.), al.; Oxime, 96
218	Diethyl d,l-tartronate	222–5d.	......	....	1.152^{15}	88	156–8d.	.....	−117.3	78.32	di: 198, dil. al.; 195–6d.	......	93, al.	....

*Derivative data given in order: m.p., crystal color, solvent from which crystallized.

TABLE XVI. ORGANIC DERIVATIVES OF ESTERS
Including esters of inorganic acids
a) Liquids. 1) (Listed in order of increasing b.p.)* (Continued)

No.	Name	Boiling point, °C	Melting point, °C	n_D^{20}	D_4^{20}	Saponification					Amide	p-Toluidide	3,5-Dinitrobenzoate	Miscellaneous
						Equiv-alent	Acid M.P., °C	Acid B.P., °C	Alcohol M.P., °C	Alcohol B.P., °C				
219	Diethyl maleate	222.7	−17	1.44156	1.066	86	137	......	−117.3	78.32	mono: 172–3, w.; 152–3 sl. d.; di: 181, me. al.	di: 142, eth.	93, al.	....
220	Methyl salicylate	224	......	1.5369	1.184	152	158.3	......	−97	64.65	142; 139	156	108 (cor.), al.	....
221	Ethyl n-butoxy-ethyl carbonate	224	......	1.4143²⁵	0.9756²⁵	190	....	......	−117.3	78.32	......		93, al.	
222	sec-Butyl levulinate	225.8	......	1.42495	0.96698	172	33–5	245–6	......	99.5	107–8d.	108–9, w.	76	Oxime, 96
223	n-Heptyl n-butyrate	225.87	−57.5	1.42279¹⁵, He (yel)	0.86371	186	−5.5; −8	162.5; 164	−34.6; −33.8	176.8	115–6	75	46; 47	....
224	2,4-Dimethyl-phenyl acetate (unsym-m-Xylenyl acetate)	226 (cor.)	......	1.4990¹⁵	1.0298¹⁵·⁵	164	16.6	118.2	27	211.5 (cor.)	82	153; 147	164.6 (cor.), 95% al.	....
225	Methyl n-caprate (Methyl n-decanoate)	226	......	1.426	0.873	186	31.5	268–70	−97	64.65	108; 100.1	78	108 (cor.), al.	....
226	n-Amyl n-caproate (n-Pentyl n-hexanoate)	226.16	−50; −47	1.42280¹⁵ He (yel)	0.86349	186	−3.9	205.35	−78.5	138 (cor.)	100–1	74–5	46.4	....
227	n-Butyl n-enanthate (n-Butyl n-heptanoate)	226.2	......	1.42280¹⁵	0.86382	186	−7.47	223	−90.2	117.6; 116	96; 96.5	81	64; 62.5	....
228	n-Hexyl n-valerate (n-Hexyl n-pentanoate)	226.3	−63.1	1.42286¹⁵	0.86345	186	−34.5	186.35	−51.6; −46.1	157.5	106	74	58.4 (cor.)	....
229	n-Propyl n-caprylate (n-Propyl n-octanoate)	226.43	−46.2	1.42351¹⁵ He (yel)	0.86591	186	16.3	237; 239.3	......	97.1	110; 106	57	74, pet. eth.	....
230	Ethyl 2-toluate (Ethyl 2-methyl-benzoate)	227	......	1.507²¹·⁶	1.0325₄²¹·⁵	164	104–5; 107–8	259⁷⁵¹	−117.3	78.32	142.8 (cor.)	144	93, al.	....
231	Ethyl pelargonate	227	α: −55.0; β: −36.7	1.42200	0.8657	186	12.3	254.4	−117.3	78.32	99	84	93, al.	N-(β-Amino-ethyl) mor-pholide, 61.3
232	l-Menthyl acetate	227	......	....	0.9185	198	16.6	118.2	43	216	82	153; 147	153	....
233	Ethyl phenyl-acetate	227.5	......	1.49921¹⁸·⁵	1.0333	164	76.5, subl.	256.5 (cor.)	−117.3	78.32	156	135–6	93, al.	N-(β-Amino-ethyl) mor-pholide, 88.9
234	n-Octyl propionate	227.9	−41.6	1.42185¹⁵ He (yel)	0.86633	186	−20.8	141	−16; −16.7	195	81; 81.3; 79	126; 123	61–2	....
235	Diethyl itaconate	228	......	1.4377	1.0467	93	165	......	−117.3	78.32	di: 191.2–.8, al.	......	93, al.	....

*Derivative data given in order: m.p., crystal color, solvent from which crystallized.

TABLE XVI. ORGANIC DERIVATIVES OF ESTERS
Including esters of inorganic acids
a) Liquids. 1) (Listed in order of increasing b.p.)* (Continued)

No.	Name	Boiling point, °C	Melting point, °C	n_D^{20}	D_4^{20}	Saponification Equivalent	Acid M.P., °C	Acid B.P., °C	Alcohol M.P., °C	Alcohol B.P., °C	Amide	p-Toluidide	3,5-Dinitrobenzoate	Miscellaneous
236	Diethyl mesaconate........	229	......	1.4488	1.0453	93	204.5 (cor.), subl.	......	−117.3	78.32	mono (α): 222; (β): 174; di: 176.5	mono (α): 196; di: 212, al.	93, al.	....
237	Di-isobutyl oxalate (Isobutyl oxalate).......	229	......	1.4180	0.97373	101	189.5 (anh.); 101 (+2 H₂O)	......	....	108.1	mono: 219; di: 419d.	mono: 169; di: 268	87	....
238	Allyl benzoate ...	230	......	1.0674¼	1.0578¹⁵₁₅	162	122.4	249	....	97.1	130	158	49–50	....
239	Isobutyl levulinate	230.9	......	1.42677	0.96770	172	33–5	245–6	....	108.1	107–8d.	108–9, w.	87	Semicarbazone, 112–3; Phenylhydrazone, 84–6; 2,4-Dinitrophenylhydrazone, 55.6; Oxime, 96
240	n-Propyl benzoate	231	......	1.500	1.023	164	122.4	249	....	97.1	130	158	74, pet. eth.	....
241	Diethyl citraconate........	231	......	1.4442	1.0491	93	92d., eth.-lgr.	206–11 sl. d.	−117.3	78.32	100–1	......	93, al.	1-Naphthylamide, 169–70
242	Methyl 3-chlorobenzoate.......	231	......	....	....	170.5	158; 155	......	−97	64.65	134	......	108 (cor.), al.	....
243	β-Phenylethyl acetate........	232; 224	......	1.5108	1.057²²·⁵	164	16.6	118.2	−25.8	219.8	82	153; 147	108	....
244	Ethyl β-(2-furyl) acrylate.......	232	14	1.5286	1.0891¹⁵	166	141	286	−117.3	78.32	168–9, w.	......	93, al.	....
245	Di-(β-methoxyethyl) carbonate.	232	......	1.4193²⁵	1.0936²⁵	178	....	......	....	124.5	......	......	......	....
246	Di-isoamyl carbonate........	233 (cor.)	......	1.4174	0.9067	202	....	......	−117	132	......	......	61	....
247	Diethyl glutarate .	233.66	−23.8	1.02229	1.42395	94	98	302–4	−117.3	78.32	di: 175–6	di: 218	93, al.	N-(β-Aminoethyl) morpholide, 152.7
248	Ethyl 3-toluate...	234	......	1.505²¹·⁴	1.0265²¹·⁴	164	111–3; 110–1	263, subl.	−117.3	78.32	94; 97	118	93, al.	....
249	Ethyl salicylate ..	234	......	1.52542	1.1396	166	158.3; subl. at 76	......	−117.3	78.32	142; 139	156	93, al.	....
250	Methyl 2-chlorobenzoate.......	234	......	....	....	170.5	142; 144	......	−97	64.65	142; 202	131	108, (cor.), al.	....
251	Ethyl 4-toluate...	234.5	......	1.5089¹⁸·²	1.0269¹⁸·²	164	179–80, subl.	275 (cor.)	−117.3	78.32	160; 158	160; 165	93, al.	....
252	Diethyl bromomalonate	235	......	....	1.426¹⁵₁₅	239	113d.	......	−117.3	78.32	di: 181, al.	di: 217 ac. a.	93, al.	....

* Derivative data given in order: m.p., crystal color, solvent from which crystallized.

TABLE XVI. ORGANIC DERIVATIVES OF ESTERS
Including esters of inorganic acids
a) Liquids. 1) (Listed in order of increasing b.p.)* (Continued)

No.	Name	Boiling point, °C	Melting point, °C	n_D^{20}	D_4^{20}	Saponification Equiv-alent	Acid M.P., °C	Acid B.P., °C	Alcohol M.P., °C	Alcohol B.P., °C	Amide	p-Tolui-dide	3,5-Di-nitro-benzoate	Miscellaneous
253	Ethylene glycol di-n-butyrate ...	235–7⁷⁴⁹	......	1.42619_He	1.0005	101	−5.5; −8	162.5; 164	−78.5	138 (cor.)	115–6	75	46.4	....
254	Diethyl butyl-malonate	235–40		....	1.425	101	....	......	−117.3	78.32	di: 200	......	93, al.	....
255	2,4,6-Trimethyl-phenyl acetate (Mesityl acetate)	236		....		178	16.6	118.2	70; 69	220	82	153; 147	......	
256	2,5-Dimethyl-phenyl acetate (p-Xylenyl acetate).......	237⁷⁶⁸		....	1.0264¹⁵	164	16.6	118.2	74.5	212	82	153; 147	137.2 (cor.), al.	....
257	n-Butyl levulinate	237.8	......	1.42905	0.97353	172	33–5	245–6	−90.2	117.6; 116	107–8d.	108–9, w.	64; 62.5	Semicarba-zone, 102–3; Phenyl-hydrazone, 79–81; 2,4-Dinitro-phenylhydra-zone, 65.8; Oxime, 96
258	Benzyl n-butyrate	238–40	......		1.033¹⁶	178	−5.5; −8	162.5; 164	−15.3	205.5	115–6	75	113	....
259	Methyl hydro-cinnamate (Methyl β-phenylpro-pionate).......	239	......	....	1.0455⁰	164	48.7; 40	279–80 (cor.)	−97	64.65	105; 82	135	108 (cor.), al.	
260	n-Propyl salicyl-ate...........	239; 249–51	......	1.51610	1.0979	180	158.3	......		97.1	142; 139	156	74, pet. eth.	....
261	Guaiacol acetate (2-Methoxy-phenyl acetate) .	240	......	1.5101²⁵	1.1285²⁵	166	16.6	118.2	32; 28.2	205	82	153; 147	141.2 (cor.), al.	
262	Isopropyl salicyl-ate...,......	240–2; 237	......	1.50650	1.0729	180	158.3; subl. at 76	......	−89.5	82.4	142; 139	156	123, pet. eth.	....
263	Dimethyl l-malate	242	......	1.4425	1.2334	81	100–1	......	−97	64.65	di: 156–7	di: 206–7	108 (cor.), al.	$[\alpha]_D^{20}$: −6.85
264	Geranyl acetate ..	242	......	1.4660	0.9174¹⁵	196	16.6	118.2	....	230	82	153; 147	62–3	....
265	Isobutyl benzoate.	242.2 (cor.)	......	....	0.999	178	122.4	249	....	108.1	130	158	87	....
266	Di-n-butyl oxalate (n-Butyl oxalate)	243; 245.5	−29.6	1.4240	0.98732	101	189.5 (anh.); 101 (+ 2H₂O)	......	−90.2	117.6; 116	mono: 219; di: 419d.	mono: 169; di: 268	64; 62.5	....
267	Methyl 2-bromo-benzoate.......	244	......			215	150		−97	64.65	155	......	108 (cor.), al.	....
268	n-Octyl n-butyrate	244.1	−55.6	1.42674¹⁵ He (yel)	0.86288	200	−5.5; −8	162.5; 164	−16; −16.7	195	115–6	75	61–2	....
269	Ethyl n-caprate (Ethyl n-decanoate).....	244.9	β: −20; γ: −30.6	1.42575	0.8650	200	31.5	268–70	−117.3	78.32	108; 100.1	78	93, al.	N-(β-Amino-ethyl) mor-pholide, 60.1

*Derivative data given in order: m.p., crystal color, solvent from which crystallized.

TABLE XVI. ORGANIC DERIVATIVES OF ESTERS
Including esters of inorganic acids
a) Liquids. 1) (Listed in order of increasing b.p.)* (Continued)

No.	Name	Boiling point, °C	Melting point, °C	n_D^{20}	D_4^{20}	Saponification Equivalent	Acid M.P., °C	Acid B.P., °C	Alcohol M.P., °C	Alcohol B.P., °C	Amide	p-Toluidide	3,5-Dinitrobenzoate	Miscellaneous
270	Diethyl adipate	245; 133.8[15]	−21	1.42765	1.0090	101	153–4 (cor.)	216[15]	−117.3	78.32	*mono*: 125–30, w.; *di*: 220	241	93, al.	N-(β-Amino-ethyl) mor-pholide, 165
271	Methyl phenoxy-acetate	245	......	....	1.150[17.5]	166	98–9	285d.	−97	64.65	101.5	......	108 (cor.), al.	....
272	Thymyl acetate	245	......	....	1.009[0]	192	16.6	118.2	51.5	233.5	82	153; 147	103.2 (cor.), al.	....
273	Carvacryl acetate	245 (cor.)	......	1.49128[28]	0.98959[25]	192	16.6	118.2	1	237.5	82	153; 147	83	....
274	Diethylene glycol diacetate (β,β'-Diacetoxy di-ethyl ether)	245–51; 148[26]	......	1.4348	1.1078[15/15]	95	16.6	118.2	−10.45	244.5	82	153; 147	149, ac. a.	....
275	n-Butyl n-caprylate (n-Butyl n-octa-noate)	245.02	−41.9; −43	1.42647[15] He (yel)	0.86278	200	16.3	237	−90.2	117.6; 116	110; 106	70	64; 62.5	....
276	n-Heptyl n-valerate (n-Heptyl n-pentanoate)	245.2	−46.4	1.42536[15] He (yel)	0.86225	200	−34.5	186.35	−34.6; −33.8	176.8	106	74	46; 47	....
277	n-Amyl n-enan-thate (n-Pentyl n-heptanoate)	245.4	−49.5	1.42627[15] He (yel)	0.86232	200	−7.47	223	−78.5	138 (cor.)	96; 96.5	81	46.4	....
278	n-Hexyl n-Caproate (n-Hexyl n-decanoate)	245.4	−55.25	1.42637[15] He (yel)	0.86216	200	−3.9	205.35	−51.6; −46.1	157.5	100–1	74–5	58.4 (cor.)	....
279	Di-(β-ethoxy-ethyl) carbonate	245.5	......	1.4239[25]	1.0635[25]	206	....	......	....	135	......	......	75, al.	....
280	Diethylene glycol monobutyl ether acetate	246	......	....	0.983	204	16.6	118.2	....	228–30	82	153; 147	......	....
281	Isobutyl phenyl-acetate	247	......	....	0.999[18]	192	76.5 (cor.)	256.5 (cor.)	....	108.1	156	135–6	87	....
282	Ethyl hydrocinna-mate (Ethyl β-phenylpro-pionate)	247.2	......	....	1.0147	178	40; 48.7	279–80 (cor.)	−117.3	78.32	105; 82	135	93, al.	....
283	Di-n-propyl succinate	248; 246	−10.4	1.4252	1.011	101	185; 182.8	235d.	....	97.1	*mono*: 157; *di*: 260d., w.	*mono*: 179–80; *di*: 254.5-5.5; 260	74, pet. eth.	....
284	Methyl 2-meth-oxybenzoate	248	......	1.534[19.5]	1.1571[19]	166	100–1	200	−97	64.65	129	......	108 (cor.), al.	....
285	Methyl undecylen-ate (Methyl hendecylenate)	248	−27.5	1.43928	0.889[15]	198	24.5	275	−97	64.65	87	......	108 (cor.), al.	....

*Derivative data given in order: m.p., crystal color, solvent from which crystallized.

TABLE XVI. ORGANIC DERIVATIVES OF ESTERS
Including esters of inorganic acids
a) Liquids. 1) (Listed in order of increasing b.p.)* (Continued)

No.	Name	Boiling point, °C	Melting point, °C	n_D^{20}	D_4^{20}	Saponification					Amide	p-Toluidide	3,5-Di-nitro-benzoate	Miscellaneous
						Equiv-alent	Acid		Alcohol					
							M.P., °C	B.P., °C	M.P., °C	B.P., °C				
286	Isoamyl levulinate	248.8	......	1.43102	0.96136	186	33–5	245–6	−117	132	107–8d.	108–9, w.	61	Semicarba-zone, 91–2; Phenylhyd-razone, 70–2; 2,4-Dinitro-phenylhyd-razone, 50.5; Oxime, 96
287	Diethyl acetone-dicarboxylate . . .	250d.	......	....	1.113	101	135	......	−117.3	78.32	....	......	93, al.	Dianilide, 155, bz.; Semicarba-zone, 94–5; Cu(OAc)$_2$ → Cu enolate, 142–3, grn. bz.
288	n-Butyl benzoate	250.3	−22.4	....	1.000	178	122.4	249	−90.2	117.6; 116	130	158	64; 62.5	....
289	Ethyl phenoxy-acetate	251	......	....	1.104$^{17.5}$	180	98–9	285d.	−117.3	78.32	101.5	......	93, al.	....
290	Methyl 3-meth-oxybenzoate....	252	......	1.52236	1.131	166	109–10	......	−97	64.65	....	......	108 (cor.), al.	α-Phenyl-ethylamide, 128.6–9.0; Benzyl-amide, 111.8–2.8
291	Diethyl l-malate .	253	......	1.4362	1.1290	95	100–1	......	−117.3	78.32	di: 156–7.	di: 206–7	93, al.	$[\alpha]_D^{20}$: −10.18
292	n-Amyl levulinate (n-Pentyl levulinate)	253.4	......	1.43192	0.96136	186	33–5	245–6	−78.5	138 (cor.)	107–8d.	108–9, w.	46.4	2,4-Dinitro-phenylhyd-razone, 84.2; Oxime, 96
293	n-Butyl phenyl-acetate	254	......	1.489	0.994	192	76.5, subl.	256.5 (cor.)	−90.2	117.6; 116	156	135–6	64; 62.5	....
294	β-Methoxyethyl benzoate (Methyl "cellosolve" benzoate)	255	......	1.5040^{25}	1.0891^{25}	180	122.4	249	......	124.5	130	158	......	4-Nitro-benzoate, 50.5, dil. al.
296	Diethyl pimelate .	255	−23.8	1.42985	0.9929	108	105	223^{15}	−117.3	78.32	di: 175	di: 206, al.	93, al.	N-(β-Amino-ethyl) mor-pholide, 137.9
297	Ethyl benzoyl-formate........	256–7	......	1.5190^{25}	1.222^{25}	178	66	......	−117.3	78.32	91	......	93, al.	Phenylhyd-razone, 64; 2,4-Di-nitro-phenylhyd-razone, 196–7d. (cor.)

*Derivative data given in order: m.p., crystal color, solvent from which crystallized.

TABLE XVI. ORGANIC DERIVATIVES OF ESTERS
Including esters of inorganic acids
a) Liquids. 1) (Listed in order of increasing b.p.)* (Continued)

No.	Name	Boiling point, °C	Melting point, °C	n_D^{20}	D_4^{20}	Saponification					Amide	p-Toluidide	3,5-Di-nitro-benzoate	Miscellaneous
						Equiv-alent	Acid M.P., °C	Acid B.P., °C	Alcohol M.P., °C	Alcohol B.P., °C				
298	Glyceryl triacetate	258	.....	....	1.161^{15}	72.7	16.6	118.2	17.9	290d.	82	153; 147	.....	Tris-4-nitro-benzoate, 188
299	Ethyl 3-methoxy-benzoate	260	.....	1.5161	1.0993	180	109–10	.....	−117.3	78.32	....	.....	93, al.	Benzyl-amide, 111.8–2.8; α-Phenyl-ethylamide, 128.6–9.0
300	β-Ethoxyethyl benzoate ("Cello-solve" benzoate)	$260–1^{738.5}$	.....	1.4969^{25}	1.0585^{25}_{25}	194	122.4	249	....	135	130	158	75, al.	....
301	Isobutyl salicylate	260–2	.....	1.50872	1.0639	194	158.3, subl. at 76	.....	....	108.1	142; 139	156	87	....
302	n-Amyl n-capryl-ate (n-Pentyl n-octanoate)	260.21	−34.8	1.43019^{15} He (yel)	0.86132	214	16.3	237; 239.3	−78.5	138 (cor.)	110; 106	70	46.4	....
303	n-Hexyl n-enan-thate (n-Hexyl n-heptanoate)	260.9	−47.9	1.42939^{15} He (yel)	0.86114	214	−7.47	223	−51.6; −46.1	157.5	96; 96.5	81	58.4 (cor.)	....
304	n-Heptyl n-caproate (n-Heptyl n-hexanoate)	260.97	−34.4	1.42934^{15} He (yel)	0.86115	214	−3.9	205.35	−34.6; −33.8	176.8	100–1	74–5	46; 47	....
305	Ethyl 2-methoxy-benzoate	261	.....	1.5224	1.1124	261	100–1	200	−117.3	78.32	129	.....	93, al.	....
306	n-Octyl n-valerate (n-Octyl n-pentanoate)	261.1	−42.5	1.42743^{15} He (yel)	0.86148	214	−34.5	186.35	−16; −16.7	195	106	74	61–2	....
307	Isoamyl benzoate .	262.3	.....	1.4950	1.004	192	122.4	249	−117	132	130	158	61	....
308	Dimethyl d-cam-phorate	263	.....	$1.46334^{16.9}$	1.0747	114	187.5–8.0	.....	−97	64.65	mono: (α-amide β-acid), 176; (β-amide-α-acid), 182–3; di: 192–3	α: 212–4; β: 190–6	108 (cor.), al.	
309	Ethyl undecylenate (Ethyl hendecylenate)	264	.....	1.4449^{23}	0.88271^{15}_{15}	212	24.5	275	−117.3	78.32	87	.....	93, al.	....
310	Isobutyl succinate	265	.....	1.427	0.974	115	185; 182.8	235d.	....	108.1	mono: 157; di: 260d., w.	mono: 179–80; di: 254.5 −5.5; 260	87	

*Derivative data given in order: m.p., crystal color, solvent from which crystallized.

TABLE XVI. ORGANIC DERIVATIVES OF ESTERS

TABLE XVI. ORGANIC DERIVATIVES OF ESTERS
Including esters of inorganic acids
a) Liquids. 1) (Listed in order of increasing b.p.)* (Continued)

No.	Name	Boiling point, °C	Melting point, °C	n_D^{20}	D_4^{20}	Saponification Equiv-alent	Acid M.P., °C	Acid B.P., °C	Alcohol M.P., °C	Alcohol B.P., °C	Amide	p-Tolui-dide	3,5-Di-nitro-benzoate	Miscellaneous
311	Ethyl benzoyl-acetate	265 sl. d.; 270 d.	......	1.5498	1.116	192	....	....	−117.3	78.32	....	......	93, al.	Ketone cleavage → aceto-phenone, m.p. 19.65, b.p. 202
312	Di-isoamyl oxa-late (Isoamyl oxalate)	267–8; 262	......	1.427	0.961	115	189.5 (anh.); 101 (+ 2H₂O)	......	−117	132	mono: 219; di: 419d.	mono: 169; di: 268	61	....
313	Dimethyl suberate	268	−5	1.43326	1.0198	101	144; 139–41	......	−97	64.65	mono: 125–7; di: 216–7	di: 218; 219	108 (cor.), al.	....
314	Methyl laurate	268	......	1.432	0.870	214	44; 42	299	−97	64.65	100; 99	87	108 (cor.), al.	β-Naphthol → β-naphthyl methyl ether, 72
315	Ethyl 4-methoxy-benzoate (Ethyl anisate)	269	7	1.5254	1.1038	180	184.6; 184.2 (cor.)	275–80	−117.3	78.32	167; 162–3, w.	186	93, al.	N-(β-Amino-ethyl) mor-pholide, 130.6
316	Ethyl laurate	269	−1.7	1.4321	0.8671$^{19}_{19}$	228	44; 42	299	−117.3	78.32	100; 99	87	93, al.	....
317	Trimethyl aconitate	270	......	....	....	72	194–5 (cor.) d.	......	−97	64.65	tri: turns br. at 250; sinters at 260	......	108 (cor.), al.	....
318	n-Butyl salicylate	270–2; 268	......	1.51148	1.0728	194	158.3, subl. at 76	......	−90.2	117.6; 116	142; 139	156	64; 62.5	....
319	Ethyl cinnamate	271	6.70	1.55982	1.0490	176	133	300	−117.3	78.32	147–8	168	93, al.	N-(β-Amino-ethyl) mor-pholide, 121.9
320	Di-n-butyl suc-cinate	274.5	−29.3	1.4298	0.9760	115	185; 182.8	235d.	−90.2	117.6; 116	mono: 157; di: 260d., w.	mono: 179–80; di: 254.5 −5.5; 260	64; 62.5	....
321	Ethyl 2-nitro-benzoate	275	30	....	....	195	146	.....	−117.3	78.32	176		93, al.	....
322	Triethyl aconitate	275d.	......	1.45562	1.1064	86	194–5 (cor.)	.....	−117.3	78.32	tri: turns br. at 250; sinters at 260	......	93, al.	....
323	Di-isopropyl d-tartarate	275	......	....	1.1274	117	169–71	......	−89.5	82.4	mono: 171–2; di: 196d., al.	......	123, pet. eth.	$[\alpha]_D^{20}$: +14.886; Phenylhydrazide, 240
324	Ethyl 4-ethoxy-benzoate	275	......		1.076^{21}	194	198; 195–6	......	−117.3	78.32	202	......	93, al.	Hydrazide, 126–7, al.

*Derivative data given in order: m.p., crystal color, solvent from which crystallized.

TABLE XVI. ORGANIC DERIVATIVES OF ESTERS
Including esters of inorganic acids
a) Liquids. 1) (Listed in order of increasing b.p.)* (Continued)

No.	Name	Boiling point, °C	Melting point, °C	n_D^{20}	D_4^{20}	Equiv-alent	Acid M.P., °C	Acid B.P., °C	Alcohol M.P., °C	Alcohol B.P., °C	Amide	p-Tolui-dide	3,5-Di-nitro-benzoate	Miscellaneous
325	n-Octyl n-caproate (n-Octyl n-hexanoate)	275.2	−28.4	1.43256[15] He (yel)	0.86032	228	−3.9	205.35	−16; −16.7	195	100–1	74–5	61–2	....
326	Isoamyl salicylate	276–8	......	1.50799	1.0535	208	158.3, subl. at 76	......	−117	132	142; 139	156	61	....
327	n-Heptyl n-enanthate (n-Heptyl n-heptanoate)	277.2	−33.3	1.43183[15] He (yel)	0.86039	228	−7.47	223.0	−34.6; −33.8	176.8	96; 96.5	81	46; 47	....
328	n-Hexyl n-caprylate (n-Hexyl n-octanoate)	277.44	−30.6	1.43230[15] He (yel)	0.86033	228	16.3	237; 239.3	−51.6; −46.1	157.5	110; 106	70	58.4 (cor.)	..........,
329	Resorcinol di-acetate	278 sl. d.	......	....	1.179	97	16.6	118.2	110 (stab.); 108–8.5 (labile)	280.8 (cor.)	82	153; 147	di: 201	....
330	Diethyl suberate	282	5.9	1.43236	1.9807	115	144; 139–41	......	−117.3	78.32	mono: 125–7; di: 216–7	di: 218–9	93, al.	N-(β-Amino-ethyl) mor-pholide, 157.2
331	Resorcinol mono-acetate	283	......	....	....	152	16.6	118.2	110 (stab.); 108–8.5 (labile)	280.8 (cor.)	82	153; 147	di: 201	....
332	Dimethyl phthalate (Methyl phthal-ate)	283.8	......	1.5138	1.191	97	200–6; 191 (sealed tube)	......	−97	64.65	mono: 149; di: 220	mono: 150 (slow htng.); 160–5 (rapid htng.)	108 (cor.), al.	....
333	Diethyl iso-phthalate	286[733]	11.5	....	....	111	348 (subl.)	......	−117.3	78.32	mono: 280; di: 280	......	93, al.	....
334	Diethyl d-cam-phorate	286	......	1.45354[26.2]	1.0298	128	187.5–8.0	......	−117.3	78.32	mono: (α-amide-β-acid): 176; (β-amide-α-acid): 182–3; di: 192–3	α: 212–4; β: 190–6	93, al.	....
335	Glyceryl tri-propionate	289	......	....	1.083[19]	86.7	−20.8	141	17.9	290d.	81; 81.3; 79	126; 123	......	Tris-4-nitro-benzoate, 188

*Derivative data given in order: m.p., crystal color, solvent from which crystallized.

TABLE XVI. ORGANIC DERIVATIVES OF ESTERS

TABLE XVI. ORGANIC DERIVATIVES OF ESTERS
Including esters of inorganic acids
a) Liquids. 1) (Listed in order of increasing b.p.)* (Continued)

No.	Name	Boiling point, °C	Melting point, °C	n_D^{20}	D_4^{20}	Saponification Equivalent	Acid M.P., °C	Acid B.P., °C	Alcohol M.P., °C	Alcohol B.P., °C	Amide	p-Toluidide	3,5-Di-nitro-benzoate	Miscellaneous
336	Diethyl phthalate (Ethyl phthal-ate)..........	289.5; 298	......	1.5019	1.1175	111	200–6; 191 (sealed tube)	......	−117.3	78.32	mono: 149; di: 220	mono: 150 (slow htng.); 160–5 (rapid htng.)	93, al.	....
337	n-Heptyl n-caprylate (n-Heptyl n-octanoate).....	290.6	−10.2	1.43492[15] He (yel)	0.85958	242	16.3	237; 239.3	−34.6; −33.8	176.8	110; 106	70	46; 47	....
338	n-Octyl n-enan-thate (n-Octyl n-heptanoate)..	290.8	−21.5	1.43488[15] He (yel)	0.85961	242	−7.47	223.0	−16; −16.7	195	96; 96.5	81	61–2	....
339	Diethyl azelate...	291	−18.5	1.43509	0.97294	122	106.5	>360 sl. d.; 237[15]	−117.3	78.32	mono: 93–5; di: 175	di: 201–2	93, al.	N-(β-Amino-ethyl) mor-pholide, 141.3
340	Triethyl citrate ..	294	......	1.44554	1.1369	92	153 (anh.); 100 (+1 H₂O)	......	−117.3	78.32	tri: 210–5d., w.	tri: 189, al.	93, al.	....
341	Ethyl myristate ..	295	α: 11.9; β: 12.3	1.4362	0.8573[25]	256	53.9	202[16]	−117.3	78.32	103	93	93, al.	N-(β-Amino-ethyl) mor-pholide, 76
342	Di-n-propyl d-tartarate.......	297	......	....	1.1390	117	169–71	......	......	97.1	mono: 171–2; di: 196d., al.	......	74, pet. eth.	$[\alpha]_D^{20}$: +12.00; Phenylhyd-razide, 240
343	Isoamyl succinate	297	......	1.434	0.958	129	185; 182.8	235d.	−117	132	mono: 157; di: 260d., w.	mono: 179–80; di: 254.5–5.5; 260	61	....
344	Di-(β-n-butoxy-ethyl)carbonate .	297–8	......	1.4279[25]	0.9766[65]	262	......	......	......	......	170–6[743]	......	......	....
345	Diethyl benzyl-malonate	300	......	....	1.077[15]	125	117d.	......	−117.3	78.32	225	......	93, al.	....
346	α-Tetrahydro-furfuryl ben-zoate..........	300–2[750]	......	....	1.137[20]₀	206	122.4	249	......	177–8[743]	130	158	83–4	....
347	Di-isopropyl phthalate (Iso-propyl phthalate)	302	......	....	1.065[19]	115	200–6; 191 (sealed tube)	......	−89.5	82.4	mono: 149; di: 220	mono: 150 (slow htng.); 160–5 (rapid htng.)	123, pet. eth.	....
348	n-Octyl n-caprylate (n-Octyl n-octanoate).....	306.8	−15.1	1.43698[15] He (yel)	0.85919	256	16.3	237; 239.3	......	......	110; 106	70	61–2	....

*Derivative data given in order: m.p., crystal color, solvent from which crystallized.

TABLE XVI. ORGANIC DERIVATIVES OF ESTERS
Including esters of inorganic acids
a) Liquids. 1) (Listed in order of increasing b.p.)* (Continued)

No.	Name	Boiling point, °C	Melting point, °C	n_D^{20}	D_4^{20}	Equiv-alent	Saponification				Amide	p-Tolui-dide	3,5-Di-nitro-benzoate	Miscellaneous
							Acid		Alcohol					
							M.P., °C	B.P., °C	M.P., °C	B.P., °C				
349	Diethyl sebacate	307	1.3	1.43657	0.9631	129	33, subl.	243[15]	−117.3	78.32	mono: 170; di: 210; 208	di: 201	93, al.	....
350	2-Tolyl benzoate ("o-Cresyl" benzoate)	307	......	....	1.114[19]	212	122.4	249	31	191–2	130	158	138.4 (cor.), al.	....
351	Ethyl 1-naph-thoate	309	......	....	1.1274[15/15]	200	161–2 (cor.)	......	−117.3	78.32	202; 205	......	93, al.	....
352	Glyceryl tri-butyrate	318	......	....	1.033[17]	100.7	−5.5; −8	162.5; 164	17.9	290d.	115–6	75	......	Tris-4-nitro-benzoate, 188
353	Di-n-butyl d,l-tartarate (Di-n-butyl racemate)	320	......	....	1.0879[18]	131	205–6 (anh.); 203–4 (+1 H$_2$O)	......	−90.2	117.6; 116	di: 226, w.-me. al.	......	64; 62.5	Dianilide, 235–6
354	Benzyl salicylate	320	......	....	....	228	158.3, subl. at 76	......	−15.3	205.5	142; 139	156	113	....
355	Di-n-butyl phthalate (n-Butyl phthalate)	340.7	......	1.4900	1.047[20/20]	139	200–6; 191 (sealed tube)	......	−90.2	117.6; 116	mono: 149; di: 220	mono: 150 (slow htng.); 160–5 (rapid htng.)	64; 62.5	N-(β-Amino-ethyl) mor-pholide, 124
356	Di-n-butyl sebacate	345	......	....	0.9329[15]	157	133, subl.	243[15]	−90.2	117.6; 119	mono: 126.5; di: 210; 208	di: 201	64; 62.5	Phenylhyd-razide, 194
357	Di-isoamyl phthal-ate (Isoamyl phthalate)	349	......	....	1.024[17]	153	200–6; 191 (sealed tube)	......	−117	132	mono: 149; di: 220	mono: 150 (slow htng.); 160–5 (rapid htng.)	61	....
358	Tricresyl phos-phate	400d.; 275–80[20]	−30	1.5568	1.197[25]	122.7	......	......	36	202.32	......	......	188.6 (cor.), al.	....

*Derivative data given in order: m.p., crystal color, solvent from which crystallized.

TABLE XVI. ORGANIC DERIVATIVES OF ESTERS
Reduced pressure b.p. only
a) Liquids. 2) (Listed in order of increasing m.p. of the corresponding amide)*

No.	Name	Boiling point, °C	Melting point, °C	n_D^{20}	D_4^{20}	Saponification Equivalent	Acid M.P., °C	Acid B.P., °C	Alcohol M.P., °C	Alcohol B.P., °C	Amide	p-Toluidide	3,5-Dinitro-benzoate	Miscellaneous
1	**3,5-Dimethylphenyl acetate** (*Sym-m-*Xylenyl acetate)	130[26]; 120[11]	.....	....	....	164	16.6	118.2	63.2: 68	220.2	82	153; 147	195.4, al.	....
2	**Dimethyl azelate**	156[20]; 146.2[10]	.....	1.43607	1.0069	108	106	>360 sl. d.	−97	64.65	*mono*: 93–5; *di*: 172	*di*: 201–2; 198	108 (cor.), al.	....
3	**Dimethyl adipate**	107.6[11]	8.5	1.4277	1.0625	87	153–4 (cor.)	216[15]	−97	64.65	*mono*: 125–30, w.; *di*: 220	241	108 (cor.), al.	....
4	**β-n-Butoxyethyl benzoate** (Butyl "cellosolve" benzoate)	156.5–7.0[14.5]; 131.6–2.6[3.0]	.....	1.4925[25]	1.0277$^{25}_{25}$	222	122.4	249	.....	170–6[7.13]	130	158	....	4-Nitrophenylurethane, 58.7–9.1, CCl$_4$
5	**Ethyl furoylacetate** ...	170[20]; 143[10]	.....	1.5055[16]	1.165$^{17}_{17}$	.....	......	......	−117.3	78.32	159, al.	....	93, al.	Oxime, 131–2, dil., al.
6	**Methyl furoylacetate** ..	144–5[20]; 96–8[1]	.....	....	....	.....	......	......	−97	64.65	159, al.	....	108 (cor.), al.	Semicarbazone, 141–2, bz.-al. (3:1); Oxime, 124–5, bz.
7	**Di-n-propyl adipate** ...	155[16]	−20	1.4314	0.9790	115	153–4	.....	.....	97.1	*mono*: 161; *di*: 220	*di*: 241	74, pet. eth.	....
8	**Dimethyl pimelate**	119.3–9.6[10]	−20.6	1.42888	1.0383	94	105	.....	−97	64.65	*di*: 175	*di*: 206, al.	108 (cor.), al.	....
9	**Di-n-propyl maleate** ..	114–7[6]	.....	1.444[18.3]	1.026	100	137	.....	.....	97.1	*di*: 181, me. al.	*mono*: 195d., chl.; *di*: 142, eth.	74, pet. eth.	....

*Derivative data given in order: m.p., crystal color, solvent from which crystallized.

TABLE XVI. ORGANIC DERIVATIVES OF ESTERS
b) Solids (Listed in order of increasing m.p.)*

No.	Name	Melting point, °C	Boiling point, °C	Saponification Equivalent	Acid M.P., °C	Acid B.P., °C	Alcohol M.P., °C	Alcohol B.P., °C	Amide	p-Toluidide	3,5-Dinitro-benzoate	Miscellaneous
1	Dimethyl succinate.....	18.2	196	73	185; 182.8	235d.	−97	64.65	*mono*: 157; *di*: 260d.	*mono*: 179–80; *di*: 254.5–5.5; 260	108 (cor.), al.	n_D^{20}: 1.41965 D_4^{20}: 1.1192
2	Methyl myristate	18.5	323	242	53.9	202[16]	−97	64.65	103	93	108 (cor.), al.	n_D^{15}: 1.428
3	Ethyl piperonylate	18.5	286	194	....	228; 229	−117.3	78.32	169, al.	....	93, al.	....
4	Diethyl *d*-tartarate.....	18.6	280	103	169–71	......	−117.3	78.32	*mono*: 171–2; *di*: 196d., al.	....	93, al.	n_D^{20}: 1.44677; D_4^{20}: 1.2028; $[\alpha]_{Hg}^{20}$(grn): +7.87; Phenylhydrazide, 204
5	Phenyl propionate	20	211	150	−20.8	141	41.8; 42	182; 183	81	123; 126	145.8 (cor.), al.	D_{25}^{25}: 1.0467; Tribromo, 95
6	Ethyl margarate.......	20.6 (β)	......	298	61.2	231[16]	−117.3	78.32	108; 106	....	93, al.	....
7	Methyl 3-chlorobenzoate	21	231	170.5	158; 155	......	−97	64.65	134		108 (cor.), al.	....
8	Benzyl benzoate	21	323–4 (cor.)	212	122.4	......	−15.3	205.5	130	158	113	n_D^{21}: 1.5681; D^{19}: 1.1224
9	Di-*n*-butyl *d*-tartarate ..	22	......	131	169–71	......	−90.2	117.6; 116	*mono*: 171–2; *di*: 196d., al.	....	64; 62.5	D_4^{18}: 1.0886; $[\alpha]_D^{14}$: + 10.09; Phenylhydrazide, 240
10	3,4-Dimethylphenyl acetate..............	22	235	164	16.6	118.2	62.5	225	82	153; 147	181.6, rods, al.	....
11	Isobutyl stearate.......	(a) 22.5; (b) 28.9; (two forms)	......	340	70–1	......	....	108.1	109; 108.4, al.	102	87	
12	Isoamyl stearate.......	23	......	354	70–1	......	−117	132	109; 108.4, al.	102	61	....
13	Cetyl acetate (*n*-Hexadecyl acetate)...	α: 18.5; β: 24.2; 22	......	284	16.6	118.2	49.27	190[18]	82	153; 147	66	
14	Ethyl palmitate........	α: 19.4; β: 24.2	185	284	62.7	222[16]	−117.3	78.32	106–7; 105.3, al.	98	93, al.	....
15	Methyl anthranilate (Methyl 2-aminobenzoate)	24.4	299.8	151	146	......	−117.3	78.32	109	151	93, al.	N-(β-Aminoethyl) morpholide, 126; Picrate, 106
16	Di-*n*-propyl *d,l*-tartarate	25	286[765]	117	203–4 (+1 H₂O); 205–6 (anh.)	......	....	97.1	226, w.-me. al.	....	74, pet. eth.	D_4^{20}: 1.1256
17	Dimethyl sebacate	26.6; 27–8	......	115	133, subl.	243[15]	−97	64.65	*mono*: 170; *di*: 210; 208	*di*: 201	108 (cor.), al.	n_D^{28}: 1.43549; D_4^{28}: 0.98818; Phenylhydrazide, 194

* Derivative data given in order: m.p., crystal color, solvent from which crystallized.

No.	Name	Melting point, °C	Boiling point, °C	Saponification					Amide	p-Toluidide	3,5-Di-nitro-benzoate	Miscellaneous
				Equiv-alent	Acid		Alcohol					
					M.P., °C	B.P., °C	M.P., °C	B.P., °C				
18	Methyl β-(2-furyl) acrylate............	27	227	152	141	286	−97	64.65	168–9	....	108 (cor.), al.	....
19	n-Butyl stearate	27.5; 28	....	340	70–1	......	−90.2	117.6; 116	109; 108.4, al.	102	64; 62.5	....
21	d-Bornyl acetate	29	226; 221	196	16.6	118.2	204.5–5.5	212	82	153; 147	154	$n_D^{22.6}$: 1.4633; D_4^{15}: 0.991 (undercooled)
22	Methyl margarate	29	....	284	61.2	231[16]	−97	64.65	108; 106	....	108 (cor.), al.	....
23	Eugenyl acetate	30	282	206	16.6	118.2	−9.1	253	82	153; 147	130.8 (cor.), 95% al.	n_D^{20}: 1.52069; D_{15}^{15}: 1.087
24	Methyl palmitate	30	184[12]	270	62.7	222[16]	−97	64.65	106–7; 105.3, al.	98	108 (cor.), al.	n_D^{45}: 1.4317
25	n-Amyl stearate (n-Pentyl stearate)	30	....	354	70–1	......	−78.5	138 (cor.)	109; 108.4, al.	102	136 (cor.)	....
26	Ethyl 2-nitrobenzoate...	30	275	195	146	......	−177.3	78.32	176	....	93, al.	....
27	n-Octadecyl acetate	α: 29.97; β: 31.95	....	312	16.6	118.2	α: 57.95; 59.5	210.5[15]	82	153; 147	66	β-Form exist below 0°; α-Form: seeding with crys. or cooling soln. of the compound.
28	Ethyl 2-naphthoate.....	32	304	200	184; 185.5	......	−117.3	78.32	192–3; 195, al.	192, al.	93, al.	n_D^{20}: 1.596; D_4^{20}: 1.117
29	Methyl-3-bromo-benzoate	32	....	229	155	......	−97	64.65	155	....	108 (cor.), al.	....
30	Methyl 4-toluate.......	33	222.5; 217	150	179–80, subl.	275 (cor.)	−97	64.65	160; 158	160; 165	108 (cor.), al.	....
31	Thymyl benzoate	33	....	255	122.4	......	51.5	233.5	130	158	103.2; al.	....
32	Di-(β-ethoxyethyl) phthalate............	33	....	155	200–6; 191 (sealed tube)	......	....	134.8	mono: 149; di: 220	mono: 150 (slow htng.); 160–5 (rapid htng.) di: 201	75, al.	....
33	Ethyl stearate	α: 30.9; β: 33.5	199[10]	312	70–1	......	−117.3	78.32	109; 108.4, al.	102	93, al.	α → β Slowly on rubbing; N-(β-Aminoethyl) morpholide, 58
34	Di-isopropyl d,l-tartarate	34	275[765]	117	203–4 (+1H₂O); 205–6 (anh.)	......	−89.5	82.4	226, w.-me., al.	....	123, pet. eth.	D_4^{20}: 1.1166
35	Ethyl pyromucate (Ethyl furoate)............	34	197	140	133–4; 132	230–2	−117.3	78.32	142–3	170.5, al.	93, al.	n_D: 1.4797; $D_4^{20.8}$: 1.1174 (undercooled)
36	Diethyl 4-nitrophthalate	34	....	133.5	165	......	−117.3	78.32	200 d	mono: 172	93, al.	....
37	Ethyl benzilate	34	....	256	150	......	−117.3	78.32	153, chl.	189–90	93, al.	....

*Derivative data given in order: m.p., crystal color, solvent from which crystallized.

TABLE XVI. ORGANIC DERIVATIVES OF ESTERS
b) Solids (Listed in order of increasing m.p.)* (Continued)

No.	Name	Melting point, °C	Boiling point, °C	Saponification Equiv-alent	Acid M.P., °C	Acid B.P., °C	Alcohol M.P., °C	Alcohol B.P., °C	Amide	p-Toluidide	3,5-Di-nitro-benzoate	Miscellaneous
38	**2,4,5-Trimethylphenyl acetate** (Pseudocumenyl acetate)	34–4.5	245–6	178	16.6	118.2	71	232	82	153; 147	. . . .	. . . .
39	**Methyl cinnamate**	36	261	162	133	300	−97	64.65	147–8	168	108 (cor.), al.	. . . .
40	**Ethyl *d,l*-mandelate**	37	254	180	118	. . . . . .	−117.3	78.32	133–4	172, al.	93, al.	. . . .
41	**Dimethyl itaconate**	38	208	79	165	. . . . . .	−97	64.65	*di*: 191.2–.8, al.	. . . .	108 (cor.), al.	n_D^{20}: 1.44413; D_4^{18}: 1.12410
42	**Methyl sebacate**	38	288d.	216	133, subl.	243[15]	−97	64.65	*mono*: 170; *di*: 210, 208	*di*: 201	108 (cor.), al.	Phenylhydrazide, 194
43	**Methyl stearate**	38.8	214–5[15]	298	70–1	. . . . . .	−97	64.65	109; 108.4, al.	102	108 (cor.), al.	. . . .
44	**Benzyl cinnamate**	39	. . . .	238	133	300	−15.3	205.5	147	168	113	. . . .
45	**Methyl dibenzylacetate**	41	. . . .	254	89	. . . . . .	−97	64.65	128–9, bz.	175, abs. al.	108 (cor.), al.	. . . .
46	**Phenyl salicylate** (Salol)	42	. . . .	214	158.3, subl. at 76	. . . . . .	41.8; 42	182; 183	142; 139	156	145.8 (cor.), al.	Tribromo deriv. of phenol, 95
47	**Benzyl succinate**	42	. . . .	208	185; 182.8	235d.	−15.3	205.5	*mono*: 157; *di*: 260d.	*mono*: 179–80; *di*: 254.5–5.5; 260	113	. . . .
48	**Dibenzyl phthalate**	43 ‧	. . . .	173	200.6; 191 (sealed tube)	. . . . . .	. . . .	. . . . . .	*mono*: 149; *di*: 220	*mono*: 150 (slow htng.); 160–5 (rapid htng.); *di*: 201	113	. . . .
49	**Diethyl terephthalate** . . .	44	302	111	300, subl. without melting	. . . . . .	−117.3	78.32	>225	. . . .	93, al.	Dianilide, 334–7, PhNO₂
50	**Cinnamyl cinnamate** . . .	44	. . . .	264	133	300	33	257	147–8	168	121	. . . .
51	**Ethyl 2-nitrocinnamate** .	44	. . . .	221	240	. . . . . .	−117.3	78.32	185	. . . .	93, al.	. . . .
52	**Methyl 2-chloro-cinnamate**	44	. . . .	196.5	212, yel., al.	. . . . . .	−97	64.65	168	. . . .	108 (cor.), al.	. . . .
53	**Diethyl 3-nitrophthalate**	46	. . . .	133.5	218	. . . . . .	−117.3	78.32	*di*: 201d.	*di*: 226	93, al	N-(β-Aminoethyl) morpholide, 131.9
54	**Ethyl 3-nitrobenzoate** . . .	47	296	195	140	. . . . . .	−117.3	78.32	*di*: 187; 189–90, dil. me. al.	. . . .	93, al.	. . . .
55	**Dicyclohexyl oxalate** . . .	47; 42	. . . .	127	189.5 (anh.); 101 (+ 2H₂O)	. . . . . .	25.15	161.1	*mono*: 219; *di*: 419d.	*mono*: 168; *di*: 268	112–3, al.	. . . .
56	**2-Phenylethyl cinnamate**	47–8	. . . .	252	133	300	−25.8	219.8	147–8	168	108	. . . .

*Derivative data given in order: m.p., crystal color, solvent from which crystallized.

No.	Name	Melting point, °C	Boiling point, °C	Saponification Equiv-alent	Acid M.P., °C	Acid B.P., °C	Alcohol M.P., °C	Alcohol B.P., °C	Amide	p-Toluidide	3,5-Di-nitro-benzoate	Miscellaneous
58	1-Naphthyl acetate.....	48; 49	....	186	16.6	118.2	94	278–80	82	153; 147	217.4, yel., al.	....
59	Methyl 4-methoxy-benzoate (Methyl anisate).............	49; 45	255	166	184–6	275–80	−97	64.65	167; 162–5, w.	186	108 (cor.), al.	....
60	Phenacyl acetate (Benzoylcarbinyl acetate; ω-Acetoxy-acetophenone).......	49	....	178	16.6	118.2	86	118–20[11]	82	153; 147	....	....
61	Di-3-tolyl carbonate (Di-"m-cresyl" car-bonate).............	49	....	242	....	......		12	203	....	165.4 (cor.), al.	....
62	Triphenyl phosphate....	49	260[20]	108.7	....	......	41.8; 42	182; 183	....	....	145.8 (cor.), al.	D_4^{20}: 1.185; Tri-bromo deriv. of phenol, 95
63	Dibenzyl d-tartarate....	50	....	165	169–71	......	−15.3	205.5	mono: 171–2; di: 196d., al.	....	113	Phenylhydrazide, 240
64	Dibenzyl succinate	51–2	....	149	185; 182.8	235d.	−15.3	205.5	mono: 157; di: 260d.	mono: 179–80; di: 254.5–5.5; 260	113	....
65	Methyl piperonylate....	51–2	270–1[777]	180	229; 228	......	−97	64.65	169, al.	....	108 (cor.), al.	....
66	Cetyl palmitate (n-Hexadecyl palmitate)	51.6	....	480	62.7	222[16]	49.27	190[18]	106–7; 105.3, al.	98	66	....
68	Furfuryl diacetate......	52	220	99	16.6	118.2	....	......	82	153; 147	....	Hydrolysis → furfural, b.p., 161.7; Semicarbazone, 202
69	Ethylene glycol dilaurate	52	....	213	44; 42	299	−12.6	197.85	100; 99	87	169	....
70	Phenyl stearate........	52	....	360	70–1	......	41.8; 42	182; 183	109; 108.4, al.	102	145.8 (cor.), al.	Tribromo deriv. of phenol, 95
71	Methyl d,l-mandelate...	53.3	250 sl. d.	166	118	......	−97	64.65	133–4	172, al.	108 (cor.), al.	....
72	Dimethyl tartronate	53.4	....	74	156–8d.	......	−97	64.65	di: 198, dil. al.	....	108 (cor.), al.	....
73	Dimethyl oxalate	54	163.5[262]	59	189.5 (anh.); 101 (+ 2H$_2$O)	......	−97	64.65	mono: 219; di: 419d.	mono: 168; di: 268	108 (cor.), al.	....
74	Tetraethyl pyromellitate	54	....	91.5	275	......	−117.3	78.32	....	....	93, al.	....
75	Diethyl meso-tartrate...	55	....	103	140	......	−117.3	78.32	di: 187; 189–90, dil. me. al.	....	93, al.	Bis-phenylhydrazide, 245

*Derivative data given in order: m.p., crystal color, solvent from which crystallized.

TABLE XVI. ORGANIC DERIVATIVES OF ESTERS
b) Solids (Listed in order of increasing m.p.)* (Continued)

No.	Name	Melting point, °C	Boiling point, °C	Saponification Equivalent	Acid M.P., °C	Acid B.P., °C	Alcohol M.P., °C	Alcohol B.P., °C	Amide	p-Toluidide	3,5-Dinitrobenzoate	Miscellaneous
76	3-Tolyl benzoate ("m-Cresyl" benzoate)	55	314	212	122.4	249	12	203	130	158	165.4 (cor.), al.	....
77	Ethyl 4-nitrobenzoate...	56	186.3	195	241	......	−117.3	78.32	201; 198	204; 192	93, al.	N-(β-Aminoethyl) morpholide, 186.3
78	1-Naphthyl benzoate ...	56	......	248	122.4	249	94	278–80	130	158	217.4, yel., al.	....
79	Cetyl stearate (n-Hexadecyl stearate)...	56.6	......	508	70–1	......	49.27	190[18]	109; 108.4, al.	102	66	....
80	Di-isobutyl d,l-tartarate.	58	311	131	203–4 (+ 1H₂O); 205–6 (anh.)	......	....	108.1	di: 226, w.-me. al.	....	87	
81	Ethyl diphenylacetate...	58	......	240	148	......	−117.3	78.32	167.5-8.0	172–3	93, al.	....
82	Ethyl 2-benzoylbenzoate	58	......	254	128 (anh.); 91 (+ 1H₂O); w.		−117.3	78.32	165 (cor.); 162		93, al.	N-(β-Aminoethyl) morpholide, 150.2
83	Diethyl naphthalate	58–60	......	136	274	......	−117.3	78.32	....		93, al.	....
84	Methyl diphenylacetate .	60	......	226	148	......	−97	64.65	167.5-8.0	172–3	108 (cor.), al.	
85	Di-2-tolyl carbonate (Di-"o-cresyl" carbonate) .	60	......	242	....	......	30.75	190.8	....	....	138.4 (cor.), al.	
86	Methyl 2-(4-toluyl)-benzoate	61		254	139–46		−97	64.65	175–6, al.		108 (cor.), al.	
87	Dimethyl d-tartarate ...	61.5		89	169–71		−97	64.65	mono: 171–2; di: 196d., al.	....	108 (cor.), al.	Two other forms, m.p. 48 and 50; Phenylhydrazide, 240
88	Ethylene glycol dimyristate	63.0		241	53.9	202[16]	−12.6	197.85	103	93	169	....
89	Dimethyl 4-nitrophthalate............	66		119.5	165		−97	64.65	200d.	mono: 172	108 (cor.), al.	....
90	Dicyclohexyl phthalate .	66		165	200–6; 191 (sealed tube)		25.15	161.1	mono: 149; di: 220	mono: 150 (slow htng.); 160–5 (rapid htng.); di: 201	112–3, al.	n_D^{20}: 1.451; D_4^{20}: 1.383
92	Ethyl oxanilate........	66–7		193	148–9		−117.3	78.32	228	....	93, al.	....
93	Dimethyl isophthalate ..	67–8		97	348, subl.		−97	64.65	mono: 280; di: 280	....	108 (cor.), al.	....
94	Ethyl 2-(4-toluyl)benzoate	68; 69	314; 299	268	139–40		−117.3	78.32	175–6, w.	....	93, al.	....
95	Phenyl benzoate	69; 71	314	198	122.4	249	41.8; 42	182; 183	130	158	145.8 (cor.), al.	AlCl₃ → 4-Hydroxy benzophenone; Tribromo deriv. of phenol, 95

* Derivative data given in order: m.p., crystal color, solvent from which crystallized.

TABLE XVI. ORGANIC DERIVATIVES OF ESTERS
b) Solids (Listed in order of increasing m.p.)* (Continued)

| No. | Name | Melting point, °C | Boiling point, °C | Saponification | | | | | Amide | p-Toluidide | 3,5-Di-nitro-benzoate | Miscellaneous |
				Equiv-alent	Acid M.P., °C	Acid B.P., °C	Alcohol M.P., °C	Alcohol B.P., °C				
96	Dimethyl 3-nitro-phthalate...........	69	....	133.5	218	....	−97	64.65	di: 201d.	di: 226	108 (cor.), al.	....
97	Methyl 3-hydroxy-benzoate	70	....	152	200, subl.	....	−97	64.65	170; 167	163, dil. al.	108 (cor.), al.	....
98	Ethylene glycol di-palmitate...........	70.5; 69	....	269	62.7	222[16]	−12.6	197.85	106–7; 105.3, al.	98	169	
99	2-Naphthyl acetate,....	71	....	186	16.6	118.2	123	285–6	82	153; 147	210.2, al.	....
100	Glyceryl tristearate	71	....	297	70–1	....	17.9	290d.	109; 108.4, al.	102	....	n[80]: 1.4399; D[80]: 0.862
101	4-Tolyl benzoate ("p-Cresyl" benzoate)....	71	316	212	122.4	249	36	202	130	158	188.6 (cor.), al.	....
102	Glyceryl tribenzoate....	72, lgr.; 76, al.	....	135	122.4	249	17.9	290d.	130	158	....	
103	Phenyl cinnamate......	72	....	238	133	300	41.8; 42	182; 183	147–8	168	145.8 (cor.), al.	Tribromo deriv. of phenol, 95
104	Ethylene glycol diben-zoate	73	....	135	122.4	249	−12.6	197.85	130	158	di: 169	....
105	Methyl 2-nitro-cinnamate	73	....	207	240	....	−97	64.65	185	....	108 (cor.), al.	
106	Di-isobutyl d-tartarate..	73–4; 70	....	131	169–71	....	....	108.1	mono: 171–2; di: 169d., al.	....	87	Phenylhydrazide, 240
107	Ethyl 3-hydroxy-benzoate	73.8	282	166	200, subl.	....	−117.3	78.32	170; 167	163, dil. al.	93, al.	....
108	Diphenyl phthalate (Phenyl phthalate)....	74–5; 70	....	159	200–6; 191 (sealed tube)	....	41.8; 42	182; 183	mono: 149; di: 220	mono: 150 (slow htng.); 160–5 (rapid htng.); di: 201	145.8 (cor.), al.	Tribromo deriv. of phenol, 95
109	Methyl 3-hydroxy-2-naphthoate	75	....	202	222–3 (cor.)	....	−97	64.65	217–8 (cor.), yel., al.	221–3	108 (cor.), al.	
110	Methyl benzilate.......	75	....	242	150	....	−97	64.65	154–5, chl.	189–90	108 (cor.), al.	
111	Ethylene glycol di-n-stearate	76; 73	....	297	70; 69.6	....	−12.6	197.85	109; 108.4, al.	102	169	....
112	Trimethyl citrate	76; 78.9	283–7d.	78	100; 153 (anh.)	....	−97	64.65	tri: 210–5d., w.	tri: 189, al.	108 (cor.), al.	
113	Methyl 2-naphthoate ...	77	290	186	184; 185.5	....	−97	64.65	192–3, al.	192, al.	108 (cor.), al.	....
114	Methyl α-phenyl-n-butyrate...........	77–8	....	178	42	270	−97	64.65	85–7	....	108 (cor.), al.	....

*Derivative data given in order: m.p., crystal color, solvent from which crystallized.

TABLE XVI. ORGANIC DERIVATIVES OF ESTERS

b) Solids (Listed in order of increasing m.p.)* (Continued)

No.	Name	Melting point, °C	Boiling point, °C	Saponification					Amide	p-Toluidide	3,5-Di-nitro-benzoate	Miscellaneous
				Equiv-alent	Acid M.P., °C	Acid B.P., °C	Alcohol M.P., °C	Alcohol B.P., °C				
115	Diphenyl carbonate	78	306	214	....	......	41.8; 42	182; 183	....	....	145.8 (cor.), al.	Phenylhydrazine → N,N'-diphenylcar-bazide; Tribromo deriv. of phenol, 95
116	Methyl 3-nitrobenzoate .	78	279	181	140	......	−97	64.65	143	162	108 (cor.), al.	....
117	Ethyl 3-nitrocinnamate .	79	....	221	199	......	−117.3	78.32	196	....	93, al.	....
118	Isoeugenyl acetate	79	283	206	16.6	118.2	....	267.5	82	153; 147	158.4 (cor.), BuOH	Dibromide, 132–3
119	Methyl 2-benzoyl-benzoate	79–80; 52	352	240	128 (anh.); 91 (+1 H₂O), w.	......	−97	64.65	165 (cor.); 162	....	108 (cor.), al.	....
120	Benzyl oxalate	80	....	135	189.5 (anh.); 101 (+2 H₂O)	......	−15.3	205.5	mono: 219; di: 419d.	mono: 169; di: 268	113	....
121	Methyl 4-bromobenzoate	81	....	215	251–3	......	−97	64.65	189–90, w.	....	108 (cor.), al.	....
122	Benzoin acetate	83	....	254	16.6	118.2	133	344	82	153; 147	....	....
123	Pentaerythritol tetra-acetate.............	84	....	76	16.6	118.2	262; 253	......	82	153; 147	....	....
124	Pyrocatechol dibenzoate (Catechol dibenzoate) .	84	....	159	122.4	249	105	245.6	130	158	di: 152	....
125	Ethyl 3-hydroxy-2-naphthoate	85	291	216	222–3 (cor.)	......	−117.3	78.32	217–8 (cor.), yel., al.	221–3	93, al.	....
126	Diguaiacol carbonate (Di-(2-methoxyphenyl) carbonate; "Guaiacol carbonate")	87	....	274	....	......	32	205	....	....	141.2 (cor.), al.	Monobromo deriv., 178
127	Dimethyl d,l-tartarate ..	90 (stab.); 84 (meta-stab.)	282	89	203–4 (+1 H₂O); 205–6 (anh.)	......	−97	64.65	226, w.-me. al.	....	108 (cor.), al.	....
128	Di-2-tolyl oxalate (Di-"o-cresyl" oxalate) ...	91	....	135	189.5 (anh.); 101 (+2 H₂O)	......	30.75	190.8	mono: 219; di: 419d.	mono: 169; di: 268	138.4 (cor.), al.	....
129	Ethyl 3,5-dinitroben-zoate	94; 93	....	240	204–5	......	−117.3	78.32	183	....	93, al.	N-(β-Aminoethyl) morpholide, 189.5
130	2-Naphthyl salicylate...	95.5; 93.5	....	264	158.3 subl. at 76	......	123	285–6	142; 139	156	210.2, al.	....
131	Methyl 4-nitrobenzoate .	96	....	181	241	......	−97	64.65	201; 198	204; 192	108 (cor.), al.	....
132	n-Propyl 4-hydroxy-benzoate	96	....	180	215; 210	......	....	97.1	162 (+1 H₂O), w.	203–4, al.	123, pet. eth.	....
133	1,2,4-Triacetoxybenzene ("Hydroquinone tri-acetate")...........	96–7	....	76	16.6	118.2	140.5	......	82	153; 147	....	....

*Derivative data given in order: m.p., crystal color, solvent from which crystallized.

No.	Name	Melting point, °C	Boiling point, °C	Saponification					Amide	p-Toluidide	3,5-Di-nitro-benzoate	Miscellaneous
				Equiv-alent	Acid		Alcohol					
					M.P., °C	B.P., °C	M.P., °C	B.P., °C				
134	Ethyl 3,5-dinitro-salicylate	99	. . . . . .	256	182; 173 (+1 H$_2$O)	. . . . . .	−117.3	78.32	181	. . . .	93, al.	. . . .
135	Dimethyl fumarate	101.7	193.25	72	286–7 (sealed tube), subl. 200	. . . . . .	−97	64.65	*mono:* 270; 300–2, subl. *di:* 266d.	. . . .	108 (cor.), al.	$n_D^{110.5}$: 1.40625; $D^{110.5}$: 1.0397
136	Ethyl 5-nitrosalicylate . .	102	. . . . . .	211	229–30	. . . . . .	−117.3	78.32	225	. . . .	93, al.	. . . .
137	Dimethyl naphthalate . . .	104	. . . . . .	122	274	. . . . . .	−97	64.65	. . . .	. . . .	108 (cor.), al.	. . . .
138	Di-3-tolyl oxalate (Di-"*m*-cresyl" oxalate) . . .	105	. . . . . .	135	189.5 (anh.); 101 (+2 H$_2$O)	. . . . . .	12	203	*mono:* 219; *di:* 419d.	*mono:* 169; *di:* 268	164.5 (cor.), al.	. . . .
139	Phloroglucinol tri-acetate (1,3,5-Tri-acetoxybenzene).	105–6	. . . . . .	76	16.6	118.2	200–9 (slow htng.); 217–9 (rapid htng.)	. . . . . .	82	153; 147	*tri:* 162	. . . .
140	Diphenyl adipate	106	. . . . . .	149	153–4 (cor.)	216[15]	. . . .	. . . . . .	*mono:* 125–30, w.; *di:* 220	241	145.8 (cor.), al.	Tribromo deriv. of phenol, 95
141	Acetylsalicylaldehyde diacetate ("Salicyl-aldehyde triacetate"). .	107; 103	. . . . . .	89	16.6	118.2	. . . .	. . . . . .	82	153; 147	. . . .	Hydrolysis → sali-cylaldehyde, b.p. 197 (cor.)
142	2-Naphthyl benzoate . . .	107	. . . . . .	248	122.4	249	123	285–6	130	158	210.2, al.	. . . .
143	Methyl 3,5-dinitro-benzoate	108	. . . . . .	226	204–5	. . . . . .	−97	64.65	183	. . . .	108 (cor.), al.	. . . .
144	Dimethyl mesotartrate . .	111	. . . . . .	89	140	. . . . . .	−97	64.65	*di:* 187; 189–90, dil. me. al.	. . . .	108 (cor.), al.	Bis-phenylhydrazide, 245
145	Di-4-tolyl carbonate (Di-"*p*-cresyl" carbonate)	114	. . . . . .	242	. . . .	. . . . . .	36	202	. . . .	. . . .	188.6 (cor.), al.	. . . .
146	Ethyl oxamate	114–5	. . . . . .	117	210	. . . . . .	−117.3	78.32	419d.	. . . .	93, al.	. . . .
147	Cholesteryl acetate	114	. . . . . .	416	16.6	118.2	148.5	. . . . . .	82	153; 147	. . . .	. . . .
148	Ethyl 4-hydroxybenzoate	116	. . . . . .	166	215; 210	. . . . . .	−117.3	78.32	162 (+1 H$_2$O), w.	203–4, al.	93, al.	N-(β-Aminoethyl) morpholide, 184
149	Resorcinol dibenzoate . .	117	. . . . . .	159	122.4	249	110 (stab.); 108–8.5 (labile)	280.8 (cor.)	130	158	*bis:* 201	. . . .
150	Ethyl 3-nitrosalicylate . .	118	. . . . . .	211	125 (+1 H$_2$O)	. . . . . .	−117.3	78.32	145	. . . .	93, al.	. . . .
151	Methyl 5-nitrosalicylate	119	. . . . . .	197	229–30	. . . . . .	−97	64.65	225	. . . .	108 (cor.), al.	. . . .

*Derivative data given in order: m.p., crystal color, solvent from which crystallized.

TABLE XVI. ORGANIC DERIVATIVES OF ESTERS

No.	Name	Melting point, °C	Boiling point, °C	Saponification					Amide	p-Toluidide	3,5-Di-nitro-benzoate	Miscellaneous
				Equiv-alent	Acid		Alcohol					
					M.P., °C	B.P., °C	M.P., °C	B.P., °C				
152	Diphenyl succinate	121	. . . .	135	185; 182.8	235d.	41.8; 42	182; 183	*mono*: 157; *di*: 260d.	*mono*: 179–80; *di*: 254.5–5.5; 260	148.5 (cor.), al.	Tribromo deriv. of phenol, 95
153	Di-4-tolyl succinate (Di-"p-cresyl" succinate) . .	121	. . . .	149	185; 182.8	235d.	36	202	*mono*: 157; *di*: 260d.	*mono*: 179–80; *di*: 254.5–5.5; 260	188.6 (cor.), al.	. . . .
154	Hydroquinone diacetate .	124	. . . .	97	16.6	118.2	171	286	82	153; 147	*bis*: 317	. . . .
155	Methyl 3-nitro-cinnamate	124	. . . .	207	199	. . . . . .	−97	64.65	196	. . . .	108 (cor.), al.	. . . .
156	Methyl 3,5-dinitro-salicylate	127	. . . .	242	182	. . . . . .	−97	64.65	181; 173 (+1 H₂O)	. . . .	108 (cor.), al.	. . . .
157	Methyl 4-hydroxy-benzoate	131	. . . .	152	215; 210	. . . . . .	−97	64.65	162 (+1 H₂O), w.	203–4, al.	108 (cor.), al.	. . . .
158	Methyl 3-nitro-salicylate	132	. . . .	197	125 (+1 H₂O)	. . . . . .	−97	64.65	145	. . . .	108 (cor.), al.	. . . .
159	Triethyl trimesate	133	. . . .	98	380 (cor.)	. . . . . .	−117.3	78.32	*tri*: 365d. (cor.)	. . . .	93, al.	. . . .
160	Ethyl 4-nitro-cinnamate	137; 142	. . . .	221	285	. . . . . .	−117.3	78.32	204; 217	. . . .	93, al.	. . . .
161	Dimethyl terephthalate .	141	. . . .	97	300 (subl. without melting)	. . . . . .	−97	64.65	*di*: >225	. . . .	108 (cor.), al.	. . . .
162	Tetramethyl pyromellitate	142	. . . .	77.5	275	. . . . . .	−97	64.65	. . . .	. . . .	108 (cor.), al.	. . . .
163	Trimethyl trimesate	144	. . . .	84	380 (cor.)	. . . . . .	−97	64.65	*tri*: 365d. (cor.)	. . . .	108 (cor.), al.	. . . .
164	Di-4-tolyl oxalate (Di-"p-cresyl" oxalate) . . .	148–9	. . . .	135	189.5 (anh.); 101 (+2 H₂O)	. . . . . .	36	202	*mono*: 219; *di*: 419d.	*mono*: 169; *di*: 268	188.6 (cor.), al.	. . . .
165	Methyl 4-nitro-cinnamate	161	. . . .	207	285	. . . . . .	−97	64.65	204; 217	. . . .	108 (cor.), al.	. . . .
166	Diethyl mucate	163–4	. . . .	133	214d.; 223–55	. . . . . .	−117.3	78.32	*mono*: 192d.; *di*: 220	. . . .	93, al.	. . . .
167	Pyrogallol triacetate . . .	165; 172	. . . .	84	16.6	118.2	133	309	82	153; 147	*tri*: 205	. . . .
168	Dimethyl mucate	165–7d.	. . . .	119	214d.; 223–55	. . . . . .	−97	64.65	*mono*: 192d.; *di*: 220	. . . .	108 (cor.), al.	. . . .
169	Methyl gallate	200–1	. . . .	184	253–4d.; 222–40d.	. . . . . .	−97	64.65	189	. . . .	108 (cor.), al.	. . . .
170	Hydroquinone di-benzoate	204 (cor.); 199	. . . .	159	122.4	249	171; 172	286	130	158	*bis*: 317	. . . .

*Derivative data given in order: m.p., crystal color, solvent from which crystallized.

EXPLANATIONS AND REFERENCES TO TABLE XVII

*p-Toluenesulfonamide (p-Toluenesulfonyl derivative).**

$$H_2NCHRCOOH \ + \ CH_3-\!\!\langle\ \rangle\!\!-SO_2Cl \ \rightarrow \ CH_3-\!\!\langle\ \rangle\!\!-SO_2NHCHRCOOH$$

p-Toluenesulfonamide

From the amino acid in aqueous sodium hydroxide with *p*-toluenesulfonyl chloride in ether.

For directions and examples see: Cheronis, pp. 453–454; Linstead, pp. 77–78; Shriner, p. 231; Vogel, p. 437; Wild, pp. 169–170; E. Fischer and P. Bergell, *Chem. Ber.*, **35**, 3779, 3784 (1902); E. W. McChesney and W. K. Swann, *J. Amer. Chem. Soc.*, **59**, 1116 (1937).

From the amino acid in aqueous sodium hydroxide with *p*-toluenesulfonyl chloride.

See: J. I. Harris and T. S. Work, *Biochem. J.*, **46**, 582 (1950).

From the amino acid with *p*-toluenesulfonyl chloride and triethylamine in aqueous tetrahydrofuran.

See: D. Theodoropoulos and L. C. Craig, *J. Org. Chem.*, **21**, 1376 (1956).

For an extensive list of references for the preparation of the *p*-toluenesulfonyl derivatives of amino acids *see*: J. P. Greenstein and M. Winitz, *Chemistry of the Amino Acids*, Vol. 2, John Wiley and Sons, New York, 1961, pp. 886–889.

*3,5-Dinitrobenzamide (3,5-Dinitrobenzoyl derivative).**

$$H_2NCHRCOOH \ + \ \begin{matrix} NO_2 \\ \langle\ \rangle \\ NO_2 \end{matrix}\!-COCl \ \rightarrow \ \begin{matrix} NO_2 \\ \langle\ \rangle \\ NO_2 \end{matrix}\!-CONHCHRCOOH \ + \ HCl$$

3,5-Dinitrobenzamide

From the amino acid in aqueous sodium hydroxide with 3,5-dinitrobenzoyl chloride.

For directions and examples see: Cheronis, p. 453; Vogel, p. 436; Wild, p. 168; B. C. Saunders, *Biochem. J.*, **28**, 580 (1934); *J. Chem. Soc.*, 1397 (1938); B. C. Saunders, G. J. Stacey and I. G. E. Wilding, *Biochem. J.*, **36**, 368 (1942); B. W. Town, *Biochem. J.*, **35**, 578 (1941).

Benzamide (Benzoyl derivative).

$$H_2NCHRCOOH \ + \ C_6H_5COCl \ \rightarrow \ C_6H_5CONHCHRCOOH \ + \ HCl$$

Benzamide

From the amino acid in aqueous sodium carbonate or bicarbonate and benzoyl chloride.

For directions and examples see: Linstead, p. 77; Vogel, p. 436; Wild, p. 167.

From the amino acid in aqueous sodium hydroxide with benzoyl chloride.

See: Cheronis, p. 453; E. Fischer and P. Bergell, *Chem. Ber.*, **35**, 3779, 3784 (1902); **39**, 597 (1906).

Acetamide (Acetyl derivative).

$$H_2NCHRCOOH \ + \ (CH_3CO)_2O \ \rightarrow \ CH_3CONHCHRCOOH \ + \ CH_3COOH$$

Acetamide

From the amino acid with acetic anhydride in water.

For directions and examples see: Linstead, p. 77; Shriner, p. 226; Wild, p. 167; R. M. Herbst and D. Shemin in *Organic Syntheses*, Coll. Vol. 2, (Ed. A. H. Blatt), John Wiley and Sons, New York, 1943, p. 11.

From the amino acid in aqueous sodium hydroxide with acetic anhydride.

See: Cheronis, p. 454; M. Bergmann and L. Zervas, *Biochem. Z.*, **203**, 288 (1928).

Carbobenzoxamide (Carbobenzoxy derivative; Benzyloxycarbonyl derivative).

$$H_2NCHRCOOH \ + \ C_6H_5CH_2OCOCl \ \rightarrow \ C_6H_5CH_2OCONHCHRCOOH \ + \ HCl$$

Carbobenzoxamide

From the amino acid in aqueous sodium hydroxide with carbobenzoxy chloride (benzyl chloroformate).

For directions and examples see: M. Bergmann and L. Zervas, *Chem. Ber.*, **65**, 1192 (1932); M. Winitz, L. Bloch-Frankenthal, N. Izumiya, S. M. Birnbaum, C. G. Baker and J. P. Greenstein, *J. Amer. Chem. Soc.*, **78**, 2423 (1956).

---

*Derivatives recommended for first trial.
WARNING: This is not an instruction manual. References should be consulted for the preparation of derivatives.

For an extensive list of references for the preparation of the carbobenzoxy derivatives of amino acids *see:* J. P. Greenstein and M. Winitz, *Chemistry of the Amino Acids*, Vol. 2, John Wiley and Sons, New York, 1961, pp. 887–895.

Phenylurethane (Phenylurea derivative).*

$$H_2NCHRCOOH \quad + \quad C_6H_5N=C=O \quad \rightarrow \quad C_6H_5NHCONHCHRCOOH$$

Phenylurethane

From the amino acid with aqueous potassium hydroxide and phenylisocyanate.
For directions and examples see: Linstead, p. 78; Shriner, p. 211; Wild, p. 171.

Phenylhydantoin.*

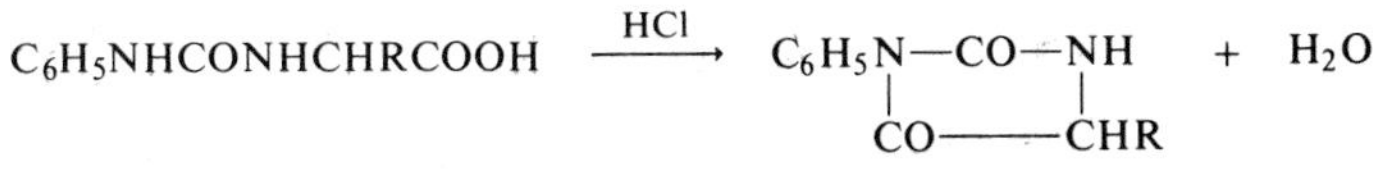

$$C_6H_5NHCONHCHRCOOH \quad \xrightarrow{HCl} \quad \begin{array}{c} C_6H_5N-CO-NH \\ | \qquad\qquad | \\ CO-\!\!-\!\!-CHR \end{array} \quad + \quad H_2O$$

Phenylhydantoin

From the phenylurethane (obtained as described above) in aqueous hydrochloric acid.
For directions and examples see: Cheronis, p. 456.

Rf Values.

For the determination of the Rf values of amino acids in aqueous phenol, in collidine-lutidine and in butanol-acetic acid mixtures *see:* R. J. Block, R. LeStrange and G. Zweig, *Paper Chromatography*, Academic Press, New York, 1952, pp. 51–66; E. Lederer and M. Lederer, *Chromatography*, Elsevier Publishing Co., New York, 1957, pp. 306–311.

NOTE: For additional information regarding directions and examples for the preparation of derivatives of amino acids which are similar to those of amines (e.g., phenylurethane, acetamide, etc.) see explanations and references to Table XVIII, p. 291, 292, 293, 294.

---

*Derivatives recommended for first trial.
WARNING: This is not an instruction manual. References should be consulted for the preparation of derivatives.

TABLE XVII. ORGANIC DERIVATIVES OF AMINO ACIDS
(Listed in order of increasing decomposition temperatures)* **

No.	Name	Decomp. temp., °C	Specific rotation				Rf values***			p-Toluene-sulfonyl	Phenyl urea	Benzoyl	3,5-Dinitro-benzoyl	Picrate	Miscellaneous
			$[\alpha]_D$	T, °C	Conc. (C) and solvent		Aqueous phenol	Collidine-Lutidine	Butanol-Ac. acid						
1	3-Aminohydro-cinnamic acid	84–5	....	....	....	....	....	....	....	....	....	....	....	Acetyl, 162; Hydrochloride, 191
2	N-Methyl-β-alanine	99–100	....	....	....	....	....	....	....	....	....	....	....	Hydrochloride, 105, aq. al.
3	Aminomalonic acid........	109	....	....	....	....	....	....	....	....	61, pet. eth.	....	....	Formyl, 48; Heat → glycine, decom. temp., 228–30; 262
4	2-Aminophenyl-acetic acid ...	119	....	....	....	....	....	....	....	....	179, al.	....	....	Acetyl, 158; Formyl, 110
5	2-Amino-1-Naphthoic acid	126	....	....	....	....	....	....	....	....	....	....	....	N-Acetyl, 195–6, al.
6	N-Phenyl-glycine	127	....	....	....	....	....	....	....	195	63	....	....	Acetyl; 124
7	4-Aminohydro-cinnamic acid .	132	....	....	....	....	....	....	....	....	194–5	....	....	Acetyl; 143 (anh.); 124 (hyd.)
8	L-Ornithine ...	140	11.5	25	c = 6.5	0.79	0.11	0.15	....	190	240 (mono); 189 (di)	....	208	α,δ-Dicarbo-benzyloxy, 112–4
9	Anthranilic acid (2-Amino-benzoic acid)	147	....	....	....	0.85	....	....	217	181	182	278	104	....
10	3-Aminophenyl-acetic acid ...	151	....	....	....	....	....	....	....	....	....	....	....	Amide, 164–6, al; Chloro-acetyl, 187–8, al.
11	5-Amino-pentanoic acid (ω-Amino-n-valeric acid) ..	157	....	....	....	....	....	....	....	....	105; 94	....	....	....
12	2-Aminocin-namic acid ...	158, yel.	....	....	....	....	....	....	....	....	191–3, al.	....	....	Monoacetyl, 250–1, al.; Diacetyl, 158, lgr.
13	DL-β-Amino-n-valeric acid DL-3-Amino-pentanoic acid)	160–5	....	....	....	....	....	....	....	....	145–6	....	....	N-β-Naphthal-enesulfonyl, 134
14	2-Amino-4-toluic acid....	165	....	....	....	....	....	....	....	....	....	......	....	Acetyl, 279–81
15	3-Aminoben-zoic acid.....	174	....	....	....	0.86	....	....	....	270; 264	248	270	....	....
16	trans-4-Amino-cinnamic acid	175–6	....	....	....	....	....	....	....	....	274	....	....	Acetyl, 259–60
17	ω-Aminotri-decylic acid ..	177	....	....	....	....	....	....	....	....	111; 105	....	....	Benzenesul-fonyl, 120, al.
18	3-Amino-4-toluic acid (Homoanthra-nilic acid)	177	....	....	....	....	....	....	....	....	....	....	....	Amide, 146; Acetyl, 184; Formyl, 186, al.

*Derivative data given in order: m.p., crystal color, solvent from which crystallized.

**Decomposition points of amino acids are only a first doubtful identification as they depend on velocity of heating and other conditions.

***Aqueous phenol: 80% w/w; for basic amino acids 3 vol.—% of a conc. aq. NH₃ has to be present (in a separate vessel) in the chamber, otherwise low values are obtained; Collidine-lutidine mixture: 2,6-lutidine (100 ml.), 2,4,6-collidine (100 ml.), water (100 ml.), diethylamine (3 ml.); Butanol-acetic acid-water: 40 ml.: 10 ml.: 10 ml.

No.	Name	Decomp. temp., °C	Specific rotation			Rf values***			p-Toluene-sulfonyl	Phenyl urea	Benzoyl	3,5-Dinitro-benzoyl	Picrate	Miscellaneous
			$[\alpha]_D$	T, °C	Conc. (C) and solvent	Aqueous phenol	Collidine-Lutidine	Butanol-Ac. acid						
19	**4-Amino-1-naphthoic acid**	177	......	....	....	......	......	......	......		......	......	......	Acetyl, 189; Amide, 175
20	*trans*-3-Amino-cinnamic acid	181, yel.									229			Acetyl, 237, al.
21	**3-Amino-1-naphthoic acid**	181	......	....	....	......	......	......	......				......	Acetyl, 254–5
22	**N-Ethylglycine.**	182	......	....	....	......	......	.,..						Hydrochloride, *ca.* 180
23	**L-Canavanine** .	184	7.9	20	....	0.51	......	......	......		*tri*: 186	......	163–4	....
24	**L-Glutamine** ..	185	......	....	....	0.6	......	......	......		......	......	......	Carbo-benzyloxy, 137
25	**Hippuric acid** ..	187											......	4-Nitrobenzyl ester, 136; 4-Bromo-phenacyl ester, 151
26	**4-Aminobenzoic acid**	188	......	....	....	0.81	......	......	......	300	278	290	......	
27	**DL-Canaline** ..	190–5							......		158–60 (*di*)	......	......	γ-Carboben-zyloxy, 208–10
28	**Betaine** (Tri-methyl-glycine)	193						0.43	......		......		......	Formyl, 183
29	**DL-β-Amino-butyric acid** ..	194				0.78					154			
30	**DL-Glutamic acid**	199	......	....	....	0.31	0.20	0.3	117	......	153; 155			Acetyl, 187
31	**4-Aminophenyl-acetic acid** ...	199–200	......	....	....						205–6, al.			Acetyl, 168–70; Chloroacetyl, 158–60
32	**β-Alanine**	200				0.66	0.22	0.37	....	168	120	202	......	Carboben-zyloxy, 106
33	**L-Isoserine**	200	......	....	....	0.39	......	......	......		107–9			
34	**ω-Amino-n-caproic acid** (6-Amino-hexanoic acid)	202	......	....	....	......	......	......	......		45–7			
35	**γ-Aminobutyric acid**	203				0.75								Hydrochloride, 135; Heat → pyrrolidone, 24, pet. eth.
36	**DL-Proline** ...	203 (mono-hyd.); 191	......	....	....	0.88	0.28	0.43	......	170	......	217	135–7	....
37	**γ-Amino-n-caproic acid** (4-Amino-hexanoic acid)	205–7									150–2	......	......	Hydrochloride, 120–1
38	**6-Amino-1-naphthoic acid**	205–6	......	....	....	......	......	......	......		......	......	......	Acetyl, 170–2; 252–3
39	**1-Amino-2-naphthoic acid**	205 r.h.	......	....	....	......	......	......	......		......	......	......	Heating with conc. HCl → 1-naphthyl-amine, 50

*Derivative data given in order: m.p., crystal color, solvent from which crystallized.

**Decomposition points of amino acids are only a first doubtful identification as they depend on velocity of heating and other conditions.

***Aqueous phenol: 80% w/w; for basic amino acids 3 vol.—% of a conc. aq. NH₃ has to be present (in a separate vessel) in the chamber, otherwise low values are obtained; Collidine-lutidine mixture: 2,6-lutidine (100 ml.), 2,4,6-collidine (100 ml.), water (100 ml.), diethylamine (3 ml.); Butanol-acetic acid-water: 40 ml.: 10 ml.: 10 ml.

TABLE XVII. ORGANIC DERIVATIVES OF AMINO ACIDS
(Listed in order of increasing decomposition temperatures)* ** (Continued)

No.	Name	Decomp. temp., °C	$[\alpha]_D$	T, °C	Conc. (C) and solvent	Aqueous phenol	Collidine-Lutidine	Butanol-Ac. acid	p-Toluene-sulfonyl	Phenyl urea	Benzoyl	3,5-Dinitro-benzoyl	Picrate	Miscellaneous
40	L-Arginine	207	11.8	20	c = 0.87 in 0.5 N-NaOH	0.89	0.17	0.2	....	....	298 (mono); 235 (di)	150	217 (mono); 190 (di)	Monocarbo-benzyloxy, 175
			12.5	20	c = 3.48 in water									
41	L-Glutamic acid	211 r.h.; 197 s.h.	31.2	22.4	c = 1.0 in 6N-HCl	0.31	0.2	0.3	131	....	....	....	....	Acetyl, 199; Carboben-zyloxy, 120
			11.5	18	c = 1.47 in water									
42	5-Amino-1-naphthoic acid	212	....		....	,,,,,	....	....				....	....	Acetyl, 296, al.
43	Sarcosine	212	....	...	....	0.78	....	....	102		104	153	....	Acetyl, 135
44	L-3,5-Di-iodo-tyrosine	213	2.9	20	c = 5.08 in 1.1 N-HCl	0.61	0.55							
45	L-α-Asparagine	213–5	....			0.47								Carboben-zyloxy, 164
46	DL-γ-Amino-n-valeric acid (DL-4-Amino-pentanoic acid)	214 (cor.)									132	....	....	....
47	L-Canaline....	214	....			0.77	....				99 (di)	....	192–3	....
48	β-Aminoiso-valeric acid ...	217	....											Hydrochloride, 120
49	β-Hydroxy-valine	218								182	153 (mono)			2-Naphthalene-sulfonyl, 261
50	L-Proline	220–2	−93.0	20	c = 2.42 in 0.6 N-KOH	0.88	0.28	0.43	130–3				154	Phenylhy-dantoin, 144
			−52.6	20	c = 0.57 in 0.5 N-HCl									
51	L-Citrulline ...	222	3.7	20	c = 2 in water	0.63	0.23	0.25	....		....	....	206	....
52	7-Amino-1-naphthoic acid	223–4	....											Acetyl, 229
53	L-Lysine......	224	14.6	20	c = 6.5 in water	0.81	0.11	0.14	....	184	235 (mono); 149 (di)	....	266	Monohydro-chloride, 235; Dihydrochlo-ride, 193; α,ε-Dicarbobenzyloxy, 150
			25.9	22.9	c = 2.0 in 6 N-HCl									
54	β-L-Asparagine	227	....	...	....	0.40	0.21	0.19	175	164	189	196	180	Carboben-zyloxy, 165
55	3,5-Diamino-benzoic acid ..	228	....										....	3,5-Diacetyl, 184; Et. ester, 84
56	L-Serine......	228	14.45	25	c = 9.34 in 1 N-HCl	0.36	0.28	0.27						Carboben-zyloxy, 121
			−6.83	20	c = 10.4 in water									
57	Glycine.......	228–30; 262	....	...	....	0.41	0.24	0.26	147	197	187.5	179	202	Carboben-zyloxy, 120
58	DL-Thyroxine .	230 s.h.; 250 r.h.	....	...	....	0.86	0.76				210–5	....	....	N-Chloro-acetyl, 201; Me. ester, 156
59	DL-β-Hydroxy-norvaline.....	230	....							156	170	....	....	Phenylhy-dantoin, 154–5

* Derivative data given in order: m.p., crystal color, solvent from which crystallized.
** Decomposition points of amino acids are only a first doubtful identification as they depend on velocity of heating and other conditions.
*** Aqueous phenol: 80% w/w; for basic amino acids 3 vol.—% of a conc. aq. NH_3 has to be present (in a separate vessel) in the chamber, otherwise low values are obtained; Collidine-lutidine mixture: 2,6-lutidine (100 ml.), 2,4,6-collidine (100 ml.), water (100 ml.), diethylamine (3 ml.); Butanol-acetic acid-water: 40 ml.: 10 ml.: 10 ml.

TABLE XVII. ORGANIC DERIVATIVES OF AMINO ACIDS
(Listed in order of increasing decomposition temperatures)* ** (Continued)

No.	Name	Decomp. temp., °C	Specific rotation			Rf values***			p-Toluene-sulfonyl	Phenyl urea	Benzoyl	3,5-Dinitro-benzoyl	Picrate	Miscellaneous
			$[\alpha]_D$	T, °C	Conc. (C) and solvent	Aqueous phenol	Collidine-Lutidine	Butanol-Ac. acid						
60	DL-β-Amino-hydrocinnamic acid........	231	......	....	....	......	....	....	60	....	194–6	....	....	Hydrochloride, 218; Formyl, 128–9; Acetyl, 161, al.
61	DL-Homo-aspartic acid (α-Amino-α-methyl succinic acid)....	232												Amide, 266
62	D-β-Amino-hydrocinnamic acid........	234–5	7.0	20	water	......	....	....						N-Formyl, 142–3
63	L-β-Amino-hydrocinnamic acid........	234–5	−7.5	25	water									N-Formyl, 142–3
64	3-Amino-salicylic acid .	235	......	....	....						189		....	Acetyl, 215; N-Benzenesulfonyl, 194
65	DL-Allo-threonine	237	......	....	....	0.50	0.34	....			176 (mono); 174 (di)	....		
66	DL-Arginine ..	238	......	....	....	0.89	0.17	0.2	....	....	230 (di, anh.); 176 (di, hyd.)	200 (mono); 196 (di)	....	
67	4-Dimethyl-aminobenzoic acid........	242	......	....	....						....		....	Amide, 206; Anilide, 182–3
68	β-Amino-β-phenyliso-butyric acid ..	243									205	....	....	
69	DL-Isoserine ..	246				0.39	....			184	151			Phenylure-thane, 183–4
70	DL-Serine	246	......			0.36	0.28	0.27	213	169	149–50	....		Carboben-zyloxy, 125
71	4-Hydroxy-phenylglycine .	248									117			Monoacetyl, 203; Diacetyl, 174–5; Amide, 135
72	Isoaspartic acid (α-Amino-α-methylmalonic acid).......	250	......	....	....									Diamide, 200–1; Heat → ala-nine, 295 d.
73	DL-Threonine .	251	......	....	....	0.50	0.34	0.35			177–8	145	....	Phenylhydan-toin, 164
74	DL-α-Amino-phenylacetic acid........	256 (subl.)									175, al.			Acetyl, 198.5; Formyl, 180
75	L-Cystine.....	260	−214.4 / −70.0	24.4 / 18.5	c = 1.0 in 1.02 N-HCl / c = 0.4 in 0.2 N-NaOH	......	....	0.1	204–5	160	181 (di)	180	....	Dicarboben-zyloxy, 123
76	DL-α-Amino-hydratropic acid (DL-α-Phenyl-alanine)......	260 (subl.)	......	....	....						....		....	N-Carbo-ethoxy, 191, al.

*Derivative data given in order: m.p., crystal color, solvent from which crystallized.

**Decomposition points of amino acids are only a first doubtful identification as they depend on velocity of heating and other conditions.

***Aqueous phenol: 80% w/w; for basic amino acids 3 vol.—% of a conc. aq. NH₃ has to be present (in a separate vessel) in the chamber, otherwise low values are obtained; Collidine-lutidine mixture: 2,6-lutidine (100 ml.), 2,4,6-collidine (100 ml.), water (100 ml.), diethylamine (3 ml.); Butanol-acetic acid-water: 40 ml.: 10 ml.: 10 ml.

TABLE XVII. ORGANIC DERIVATIVES OF AMINO ACIDS
(Listed in order of increasing decomposition temperatures)* ** (Continued)

No.	Name	Decomp. temp., °C	Specific rotation			Rf values***			p-Toluene-sulfonyl	Phenyl urea	Benzoyl	3,5-Dinitro-benzoyl	Picrate	Miscellaneous
			$[\alpha]_D$	T, °C	Conc. (C) and solvent	Aqueous phenol	Collidine-Lutidine	Butanol-Ac. acid						
77	Glycylglycine	260	......	......		0.39	......	0.22	178	176	208	210	......	Carbobenzyloxy, 178
78	DL-β-Phenyl-alanine	264	......	......		0.85	0.48	0.68	134–5	182	188	......	173	Carbobenzyloxy, 103
79	DL-Aspartic acid	270 (>300)	......	......		0.19	0.21	0.24	......		119 (hyd.); 176 (anh.)	......	......	Carbobenzyloxy, 116
80	DL-α-Amino-octanoic acid	270	......	......		0.89	0.55	......	......		128			
81	L-Aspartic acid	270	24.6	24	c = 2.0 in 6 N-HCl	0.19	0.21	0.24	140	162	185	......	......	Carbobenzyloxy, 116
82	DL-α-Amino-nonanoic acid	273	......								128			
83	L-Hydroxy-proline	274	−47.3; −75.2	20; 22.5	c = 1.31 in 1 N-HCl; c = 1.0 in water	0.63	0.28	......	153	175	100 (mono); 92 (di)			
84	DL-Trypto-phane	275	......	......		0.75	0.50	0.50	176	......	......	240	......	N-Benzenesulfonyl, 185; Carbobenzyloxy, 169–70
85	α-Aminoiso-butyric acid	280	......	......		0.74	0.32	......	......		198			1-Naphthyl-urea, 198
86	DL-α-Amino-heptanoic acid	281	......	......							135			
87	DL-Methionine	281	......	......		0.81	0.42	0.55	105	......	145			Acetyl, 114; Formyl, 100; Carbobenzyloxy, 112
88	5-Aminosalicy-lic acid	283	......	......							252	......	......	Monoacetyl, 218; Diacetyl, 184
89	DL-Lanthionine	283	......	......		0.26	0.11				195–8 (di)	......	......	......
90	L-Methionine	283	−8.11	25	c = 0.8 in water	0.81	0.42	0.55	......	......	94–5 (hyd.); 150 (anh.)			Acetyl, 98–9; 1-Naphthyl-urea, 188
91	L-(+)-Iso-leucine		40.6; 11.29	20; 20	c = 5.1 in 6.1 N-HCl; c = 3.1 in water	0.84	0.45	0.72	130–2	121	117	......	......	Formyl, 156
92	L-Phenylala-nine	283; 320	−35.1	20	c = 1.94 in water	0.85	0.48	0.68	164	181	146	93	......	Carbobenzyloxy, 126–8
93	L-Hexahydro-tyrosine	285	......											N-4-Nitro-benzoyl, 225
94	D-(−)-Iso-leucine	285	−40.86	20	c = 4.53 in 6.1 N-HCl	0.84	0.45	0.72	130–2	121	117	......	......	Formyl, 156
95	L-Histidine	288; 253	13.0; −39.0	22.7; 25	c = 1.5 in 6 N-HCl; c = 1.13 in water	0.69	0.27	0.2	202–4	......	230 (mono)	189	86	Carbobenzyloxy, 209
96	L-Cysteic acid	289; 260	8.66	20	c = 1.85 g. in 25 ml water	......	......	......	......	......	......	......	......	Diphenylacyl ester, 203

*Derivative data given in order: m.p., crystal color, solvent from which crystallized.

**Decomposition points of amino acids are only a first doubtful identification as they depend on velocity of heating and other conditions.

***Aqueous phenol: 80% w/w; for basic amino acids 3 vol.—% of a conc. aq. NH_3 has to be present (in a separate vessel) in the chamber, otherwise low values are obtained; Collidine-lutidine mixture: 2,6-lutidine (100 ml.), 2,4,6-collidine (100 ml.), water (100 ml.), diethylamine (3 ml.); Butanol-acetic acid-water: 40 ml.: 10 ml.: 10 ml.

TABLE XVII. ORGANIC DERIVATIVES OF AMINO ACIDS
(Listed in order of increasing decomposition temperatures)* ** (Continued)

No.	Name	Decomp. temp., °C	Specific rotation			Rf values***			p-Toluene-sulfonyl	Phenyl urea	Benzoyl	3,5-Dinitro-benzoyl	Picrate	Miscellaneous
			$[\alpha]_D$	T, °C	Conc. (C) and solvent	Aqueous phenol	Collidine-Lutidine	Butanol-Ac. acid						
97	L-Tryptophane	290; 252	−31.5 / 6.17	22.7 / 20	c = 1.0 in water / c = 2.42 in 0.5 N-NaOH	0.75	0.50	0.5	176	166	183	233	......	Carbobenzyloxy, 126
98	DL-N-Methyl-α-alanine	292	......	....	....	......	....	......	......	......	......	......	......	Methylamide, 43; Hydrochloride, 110
99	DL-Isoleucine .	292; 275 s.h.	......	....	....	0.84	0.45	0.72	141	120	118	......	......	Formyl, 121
100	L-(+)-α-Aminobutyric acid........	292; 303 s.h.	18.65	19	c = 4.8 in 6 N-HCl	0.71	0.31	0.45			121			Formyl, 126; 1-Naphthyl-urea, 195
101	DL-Leucine ...	293, s.h.; 332	......	....	....	0.84	0.45	0.73	......	165	137–41	187	......	....
102	DL-α-Amino-α-methyl-valeric acid (DL-α-Amino-α-methyl-pentanoic acid)........	295, s.h.												1-Naphthyl-urea, 196
103	DL-Alanine ...	295	......	....	....	0.60	0.28	0.38	139	......	166	177	......	Carbobenzyloxy, 114–5
104	D-α-Amino-hydratropic acid........	295 (subl.)	70.0	18	water	......	....	......						Formyl, 194–5
105	L-Alanine.....	297	14.7 / 2.7	15 / 22	c = 5.8 in 0.97 N-HCl / c = 10.3 in water	0.60	0.28	0.38	133	......	151	......	......	Carbobenzyloxy, 84; 1-Naphthyl-urea, 220–2
106	DL-Norleucine.	297; 327	......	....	....	0.84	0.45	0.74	124	......	......	......	......	Formyl, 114
107	DL-Valine	298, s.h.; 282	......	....	....	0.78	0.36	0.6	110	164	132	158	......	......
108	L-Djenkolic acid........	300–50	−60.5 / −47.5	20.5 / 25	1% in 1N-HCl / 2% in 1N-HCl	0.40	0.13	......			166 (mono); 88 (di)			
109	D-N-Methyl-α-alanine	300	5.6	20	water	......	....							Hydrochloride, 165–6
110	Taurine.......	300–5	......	....	....	0.42								N-Me., 241; N-Et., 147
111	L-(+)-Norleucine ...	301	21.3 / 6.26	20 / 20	c = 4.25 in 6 N-HCl / c = 0.7 in water	0.84	0.45	0.74	......	......	53			Formyl, 115–6; 2-Naphthalenesulfonyl, 149
112	DL-Norvaline .	303, s.h.	......	....	....	0.80	0.39	0.65	......	117	......	......	......	Formyl, 132; Phenylhydantoin, 103
113	Creatine (α-Methylguanidoacetic acid)	303	......	....	....								218–20	Diacetyl, 165
114	4-Aminocyclohexanecarboxylic acid ..	303–4	......	....	....									Heat → lactam, 191–2

*Derivative data given in order: m.p., crystal color, solvent from which crystallized.

**Decomposition points of amino acids are only a first doubtful identification as they depend on velocity of heating and other conditions.

***Aqueous phenol: 80% w/w; for basic amino acids 3 vol. —% of a conc aq. NH₃ has to be present (in a separate vessel) in the chamber, otherwise low values are obtained; Collidine-lutidine mixture: 2,6-lutidine (100 ml.), 2,4,6-collidine (100 ml.), water (100 ml.), diethylamine (3 ml.); Butanol-acetic acid-water: 40 ml.: 10 ml.: 10 ml.

TABLE XVII. ORGANIC DERIVATIVES OF AMINO ACIDS
(Listed in order of increasing decomposition temperatures)* ** (Continued)

No.	Name	Decomp. temp., °C	$[\alpha]_D$	T, °C	Conc. (C) and solvent	Aqueous phenol	Collidine-Lutidine	Butanol-Ac. acid	p-Toluene-sulfonyl	Phenyl urea	Benzoyl	3,5-Dinitro-benzoyl	Picrate	Miscellaneous
115	DL-α-Amino-butyric acid ..	304	......	...	......	0.71	0.31	0.45	......	170	147	......	......	1-Naphthyl-urea, 194; Phthalyl, 96
116	meso-Lan-thionine	304	......	...	......	0.26	0.11				198–200	......	......	Dicarbo-benzyloxy, 139–40
117	Creatinine	305; 260	......	...	......								220	....
118	L-α-Amino-phenylacetic acid........	305–10	−111.0; −157	20; 20	water dil. HCl									Acetyl, 191; Formyl, 190
119	meso-2,3-Di-aminosuccinic acid........	306	......	?...								(di) 212 hyd.	......	Diacetyl, 235
120	L-(+)-Norva-line	307	23.0	20	c = 10 in 10% HCl	0.80	0.39	......			64	......	......	Acetyl, 137
121	L-Tyrosine	314–8 r.h. 290–5 s.h.	−8.64; −13.2	20; 18	c = 4.4 in 6.3 N-HCl; c = 0.91 in 3 N-NaOH	0.51	0.51	0.45	N-: 188	104	N-: 166–7; 211–2 (di)	......	......	N-Acetyl, 148; Diacetyl, 172; N-Carboben-zyloxy, 101
122	L-(+)-Valine ..	315	28.8	20	c = 3.4 in 6 N-HCl	0.78	0.36	0.60	147	......	127	157–8	......	Carboben-zyloxy, 64–5
123	L-Leucine.....	337	15.1; −10.8	25.9; 24.7	c = 2.0 in 6 N-HCl; c = 2.0 in water	0.84	0.45	0.73	124	115	118	187	....	
124	DL-Tyrosine ..	340 r.h.; 295 s.h.	......			0.51	0.51	0.45	224–6	......	N-: 197	252–4 (di)		....
125	1-Aminocyclo-hexanecar-boxylic acid ..	350; 320 s.h.	......							......			209–10	Hydrochloride, 310
126	DL-2,3-Di-aminosuccinic acid........	......									164 (di) hyd.			Diacetyl, 235
127	DL-Ornithine..	......				0.79	0.11	0.15	188 (mono)	192	4-N-: 285–8; 188 (di)	......	208 (di)	....
128	L-(+)-Alloiso-leucine	......	38.1	20	c = 3.97 in 6 N-HCl					151	......			Formyl, 126; 1-Naphthylurea, 166; Benzene-sulfonyl, 147
129	DL-Lysine	......	......	...	......	0.81	0.11	0.14	......	196	249 (mono); 146 (di)	......	225 (mono)	Monohydro-chloride, 260–3; Dihy-drochloride, 187–9
130	L-Cysteine	......	......	...	......	0.57							......	S-Benzyl, 216; Oxid. → L-cystine, 260 d.; Hydrochlo-ride, 175–8, $[\alpha]_D^{25}$: 9.5 (c = 2 in water)

*Derivative data given in order: m.p., crystal color, solvent from which crystallized.

**Decomposition points of amino acids are only a first doubtful identification as they depend on velocity of heating and other conditions.

***Aqueous phenol: 80% w/w: for basic amino acids 3 vol. — % of a conc. aq. NH₃ has to be present (in a separate vessel) in the chamber, otherwise low values are obtained: Collidine-lutidine mixture: 2,6-lutidine (100 ml.). 2,4,6-collidine (100 ml.), water (100 ml.), diethylamine (3 ml.): Butanol-acetic acid-water: 40 ml.: 10 ml.: 10 ml.

EXPLANATIONS AND REFERENCES TO TABLE XVIII

Acetamide (Acetyl derivative). *

$$RNH_2 \ + \ (CH_3CO)_2O \ \rightarrow \ CH_3CONHR \ + \ CH_3COOH$$

$$RR'NH \ + \ (CH_3CO)_2O \ \rightarrow \ CH_3CONRR' \ + \ CH_3COOH$$

Acetamide

For primary and secondary amines only.

From the amine with acetic anhydride without solvent.

For directions and examples see: Cheronis, pp. 591–593; Linstead, p. 60; Vogel, p. 652; Wild, p. 218; J. J. Sudborough, *J. Chem. Soc.*, **79**, 533 (1901); L. C. Raiford, R. Taft and H. P. Lankelma, *J. Amer. Chem. Soc.*, **46**, 2051 (1924).

From the amine with acetic anhydride in aqueous sodium hydroxide.

See: F. D. Chattaway, *J. Chem. Soc.*, 2495 (1931).

From the amine with acetic anhydride in pyridine.

See: Cheronis, pp. 590, 592–593.

From the amine hydrochloride with acetic anhydride in aqueous sodium acetate.

See: Linstead, p. 60; Vogel, p. 652.

From the amine with acetic anhydride in water.

See: Shriner, p. 226.

From the amine with acetic anhydride in acetic acid.

See: Wild, p. 218.

Benzamide (Benzoyl derivative). *

$$RNH_2 \ + \ C_6H_5COCl \ \rightarrow \ C_6H_5CONHR \ + \ HCl$$

$$RR'NH \ + \ C_6H_5COCl \ \rightarrow \ C_6H_5CONRR' \ + \ HCl$$

Benzamide

For primary and secondary amines only.

From the amine with benzoyl chloride in aqueous sodium hydroxide.

For directions and examples see: Cheronis, pp. 591, 593–594; Linstead, p. 60; Vogel, p. 652; Wild, p. 219.

From the amine with benzoyl chloride in a pyridine-benzene mixture.

See: Shriner, p. 226.

From the amine with benzoyl chloride in benzene.

See: Shriner, p. 227.

From the amine with benzoyl chloride in an aqueous sodium hydroxide-chloroform mixture.

See: Shriner, p. 226.

From the amine with benzoyl chloride.

See: Vogel, p. 653.

Benzenesulfonamide (Benzenesulfonyl chloride derivative). *

$$RNH_2 \ + \ C_6H_5SO_2Cl \ \rightarrow \ C_6H_5SO_2NHR \ + \ HCl$$

$$RR'NH \ + \ C_6H_5SO_2Cl \ \rightarrow \ C_6H_5SO_2NRR' \ + \ HCl$$

Benzene-
sulfonamide

For primary and secondary amines only.

From the amine with benzenesulfonyl chloride in aqueous sodium hydroxide.

For directions and examples see: Cheronis, pp. 595–596; Shriner, pp. 103–104; Vogel, p. 653; O. Hinsberg, *Chem. Ber.*, **23**, 2962 (1890); **38**, 906 (1905).

From the amine in aqueous sodium hydroxide with benzenesulfonyl chloride in methanol.

See: Cheronis, p. 596.

From the amine with benzenesulfonyl chloride in benzene.

See: Wild, p. 221.

From the amine in aqueous sodium hydroxide with benzenesulfonyl chloride in acetone.

See: Wild, p. 221.

From the amine with benzenesulfonyl chloride in aqueous pyridine.

See: Wild, p. 221.

From the amine with benzenesulfonyl chloride in pyridine.

See: Vogel, p. 653.

---

*Derivatives recommended for first trial.

WARNING: This is not an instruction manual. References should be consulted for the preparation of derivatives.

From the amine with benzenesulfonyl chloride in aqueous sodium hydroxide.
See: Wild, p. 221.

p-Toluenesulfonamide (p-Toluenesulfonyl chloride derivative).*

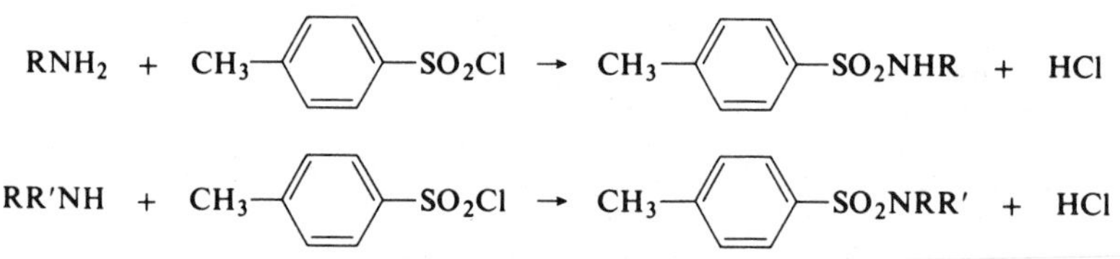

$$RNH_2 \ + \ CH_3-\!\!\!\langle\ \rangle\!\!\!-SO_2Cl \ \rightarrow \ CH_3-\!\!\!\langle\ \rangle\!\!\!-SO_2NHR \ + \ HCl$$

$$RR'NH \ + \ CH_3-\!\!\!\langle\ \rangle\!\!\!-SO_2Cl \ \rightarrow \ CH_3-\!\!\!\langle\ \rangle\!\!\!-SO_2NRR' \ + \ HCl$$

p-Toluenesulfonamide

For primary and secondary amines only.
From the amine with *p*-toluenesulfonyl chloride in aqueous sodium hydroxide.
For directions and examples see: Cheronis, p. 595; Linstead, p. 60; Shriner, pp. 103–104; Vogel, p. 653.
From the amine with *p*-toluenesulfonyl chloride in benzene.
See: Wild, p. 221.
From the amine with *p*-toluenesulfonyl chloride in acetone.
See: Wild, p. 221.
From the amine with *p*-toluenesulfonyl chloride in pyridine.
See: F. Bell, *J. Chem. Soc.*, 2787 (1929).
From the amine with *p*-toluenesulfonyl chloride in aqueous pyridine.
See: Wild, p. 221.

Phenylthiourea.*

$$RNH_2 \ + \ C_6H_5N{=}C{=}S \ \rightarrow \ C_6H_5NHCSNHR$$

$$RR'NH \ + \ C_6H_5N{=}C{=}S \ \rightarrow \ C_6H_5NHCSNRR'$$

Phenylthiourea

For primary and secondary amines only.
From the amine with phenylisothiocyanate in alcohol.
For directions and examples see: Cheronis, pp. 599–600; Shriner, p. 227; Vogel, p. 422; Wild, p. 227; T. Otterbacher and F. C. Whitmore, *J. Amer. Chem. Soc.*, **51**, 1909 (1929).
From the amine with phenylisothiocyanate without solvent.
See: Vogel, p. 422; Wild, p. 227; N. A. Lange, H. L. Ebert and L. K. Youse, *J. Amer. Chem. Soc.*, **51**, 1911 (1929).

1-Naphthylthiourea.*

$$RNH_2 \ + \ 1\text{-}C_{10}H_7N{=}C{=}S \ \rightarrow \ 1\text{-}C_{10}H_7NHCSNHR$$

$$RR'NH \ + \ 1\text{-}C_{10}H_7N{=}C{=}S \ \rightarrow \ 1\text{-}C_{10}H_7NHCSNRR'$$

1-Naphthylthiourea

For primary and secondary amines only.
From the amine with 1-naphthylisothiocyanate in ethanol.
For directions and examples see: Cheronis, pp. 600–601; Vogel, p. 422; Wild, p. 227; C. M. Suter and E. W. Moffet, *J. Amer. Chem. Soc.*, **55**, 2497 (1933).
From the amine with 1-naphthylisothiocyanate without solvent.
See: Vogel, p. 422; Wild, p. 227; C. M. Suter and E. W. Moffett, *J. Amer. Chem. Soc.*, **55**, 2497 (1933).

Phenylurea.

$$RNH_2 \ + \ C_6H_5N{=}C{=}O \ \rightarrow \ C_6H_5NHCONHR$$

$$RR'NH \ + \ C_6H_5N{=}C{=}O \ \rightarrow \ C_6H_5NHCONRR'$$

Phenylurea

For primary and secondary amines only.
From the amine with phenylisocyanate in petrol ether.

---

*Derivatives recommended for first trial.
WARNING: This is not an instruction manual. References should be consulted for the preparation of derivatives.

For directions and examples see: Linstead, p. 61; Wild, pp. 223–224.

From the amine with phenylisocyanate without solvent.

See: N. A. Lange, H. L. Ebert and L. K. Youse, *J. Amer. Chem. Soc.*, **51**, 1911 (1929).

1-Naphthylurea.*

$$RNH_2 \quad + \quad 1\text{-}C_{10}H_7N{=}C{=}O \quad \rightarrow \quad 1\text{-}C_{10}H_7NHCONHR$$

$$RR'NH \quad + \quad 1\text{-}C_{10}H_7N{=}C{=}O \quad \rightarrow \quad 1\text{-}C_{10}H_7NHCONR'R'$$

1-Naphthylurea

For primary and secondary amines only.

From the amine with 1-naphthylisocyanate without solvent.

For directions and examples see: Cheronis, p. 599; H. E. French and A. F. Wirtel, *J. Amer. Chem. Soc.*, **48**, 1736 (1926).

From the amine with 1-naphthylisocyanate in petrol ether.

See: Linstead, p. 61; Wild, pp. 223–224.

From the amine with 1-naphthylisocyanate with water.

See: Wild, p. 224.

Picrate.*

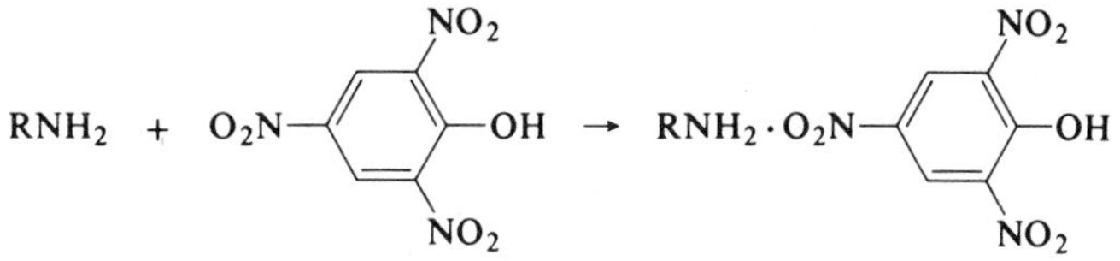

Picrate
(Molecular complex)

For primary, secondary and tertiary amines.

From the amine with picric acid in methanol or in ethanol.

For directions and examples see: Cheronis, pp. 603–604; Linstead, p. 50, 61; Shriner, p. 229; Vogel, pp. 422–423; Wild, p. 212.

From the amine with picric acid in water.

See: Vogel, p. 422.

From the amine with picric acid in acetone or benzene.

See: Wild, p. 212.

From the amine with picric acid.

See: Shriner, p. 229.

3,5-Dinitrobenzoic acid salt.*

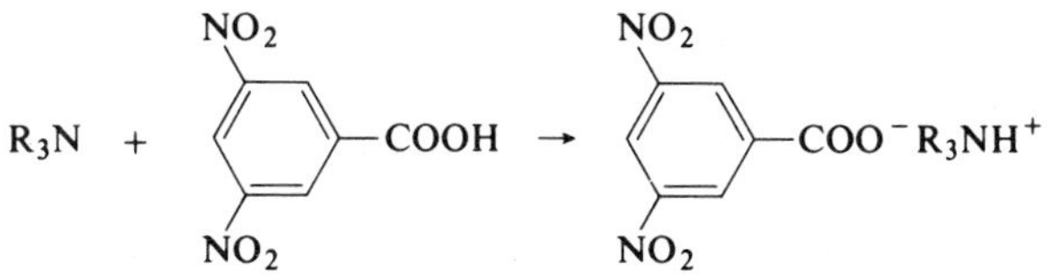

Trialkylammonium
3,5-dinitrobenzoate

Especially for tertiary amines.

From the amine with 3,5-dinitrobenzoic acid in methanol or ethanol.

For directions and examples see: Cheronis, pp. 602–603; Wild, p. 215; C. A. Buehler, E. J. Currier and R. Lawrence, *Ind. Eng. Chem., Anal. Ed.*, **5**, 277 (1933); C. A. Buehler and J. D. Calfree, *Ind. Eng. Chem., Anal. Ed.*, **6**, 351 (1934).

β-Resorcylic acid salt.

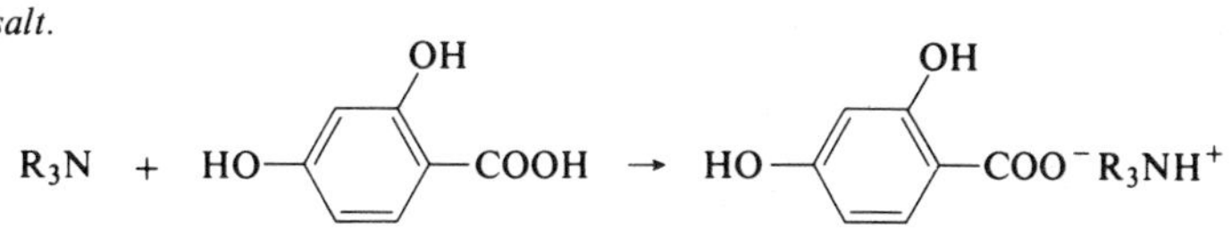

β-Resorcylic acid   Trialkylammonium resorcylate

*Derivatives recommended for first trial.
WARNING: This is not an instruction manual. References should be consulted for the preparation of derivatives.

Especially for tertiary amines.

From the amine with β-resorcylic acid in ether.

For directions and examples see: Cheronis, p. 603; K. W. Wilson, F. E. Anderson and R. W. Donohoe, *Anal. Chem.*, **23**, 1032 (1951).

Chloroplatinic acid salt. *

$$R_3N \;+\; H_2PtCl_6 \;\rightarrow\; [R_3NH]_2^+PtCl_6^=$$
$$\text{Chloroplatinate}$$

Especially for tertiary amines.

From the amine in aqueous hydrochloric acid with aqueous chloroplatinic acid.

For directions and examples see: Shriner, p. 230; Wild, p. 213.

Hydrochloride.

$$RNH_2 \;+\; HCl \;\rightarrow\; [RNH_3^+]Cl^-$$
$$\text{Hydrochloride}$$

For primary, secondary and tertiary amines.

From the amine with gaseous hydrogen chloride in ether, benzene or chloroform.

For directions and examples see: Cheronis, p. 601; Shriner, p. 224; Wild, p. 211.

From the amine with dilute aqueous hydrochloric acid.

See: Wild, p. 211.

Methiodide. *

$$R_3N \;+\; CH_3I \;\rightarrow\; [R_3NCH_3^+]I^-$$
$$\text{Methiodide}$$

Especially for tertiary amines.

From the amine with methyl iodide without solvent.

For directions and examples see: Linstead, p. 61; Shriner, p. 228; Vogel, p. 660; Wild, p. 232.

From the amine with methyl iodide in isopropyl ether.

See: Cheronis, p. 604.

From the amine with methyl iodide in ether or benzene.

See: Vogel, p. 660.

Metho-p-toluenesulfonate (Methyl p-toluenesulfonate). *

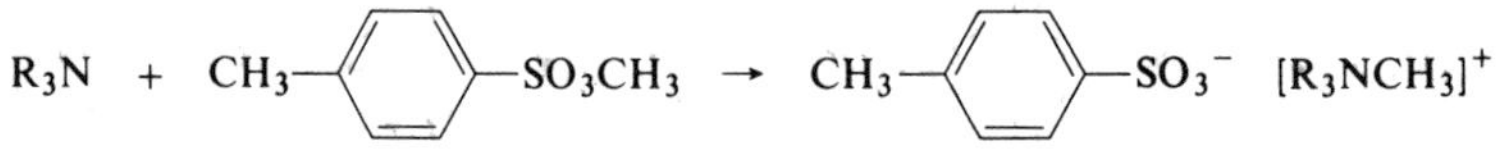

$$R_3N \;+\; CH_3-\!\!\langle\!\!\langle\bigcirc\!\!\rangle\!\!\rangle\!-SO_3CH_3 \;\rightarrow\; CH_3-\!\!\langle\!\!\langle\bigcirc\!\!\rangle\!\!\rangle\!-SO_3^- \; [R_3NCH_3]^+$$
$$\text{Metho-}p\text{-toluenesulfonate}$$

Especially for tertiary amines.

From the amine with methyl *p*-toluenesulfonate in isopropyl ether.

For directions and examples see: Cheronis, p. 604.

From the amine with methyl *p*-toluenesulfonate in benzene.

See: Linstead, p. 62; Shriner, p. 229; Vogel, p. 660; Wild, p. 233; C. S. Marvell, E. W. Scott and K. L. Amstutz, *J. Amer. Chem. Soc.*, **51**, 3638 (1929).

*Derivatives recommended for first trial.
WARNING: This is not an instruction manual. References should be consulted for the preparation of derivatives.

TABLE XVIII. ORGANIC DERIVATIVES OF AMINES
1. Primary and secondary amines a) Liquids 1) (Listed in order of increasing atmospheric b.p.)*

No.	Name	Boiling point, °C	Melting point, °C	n_D	Density g/ml	Acetamide	Benzamide	Benzene sulfon- amide	p- Toluene sulfon- amide	Phenyl thiourea	Picrate	Miscellaneous
1	Methylamine	−6	......	....	0.699_4^{-11}	28	80	30	75	113	207; 215	....
2	Dimethylamine	7	......	....	0.6804_4^0	....	41	47	79	135	158	....
3	Ethylamine	16.5; 19	......	....	0.7057_4^0	....	71	58	63	106; 135	165	....
4	Isopropylamine	33	......	1.377^{15}	0.691_4^{18}	....	....	26	......	101	......	1-Naphthylurea, 200; 1-Naphthylthiourea, 143
5	Ethyl methyl amine	36	......	....	....	....	....	......	......	......	196	Chloroplatinate, 207; Hydrochloride, 126–30
6	tert-Butylamine	46	......	1.3794^{18}	0.7004^{15}	101–2	134	......	......	120	198	Hydrochloride, 270–80
7	n-Propylamine	49	......	1.3901^{17}	0.714_4^{25}	....	84	36	52	63	135	1-Naphthylurea, 196
8	Isopropyl methyl amine	50	......	....	0.7026_4^{19}	....	....	......	......	120	135	Phenylurea, 131
9	Cyclopropylamine	50	......	1.421^{20}	0.824_4^{20}	....	99	120 (di)	......	......	149	Hydrochloride, 100
10	Ethyleneimine	56	......	....	0.832^{24}	....	....	......	52	......	142	Oxalate, 115
11	Diethylamine	56	......	1.3873^{18}	0.7108_4^{18}	....	42	42	60	34	155	1-Naphthylthiourea, 108
12	Allylamine	58	......	1.4194^{22}	0.7436_{20}^{20}	....	....	39	64	98	140	....
13	DL-sec-n-Butylamine (2-Aminobutane)	63	......	1.395^{17}	0.718^{20}	....	76	70	55	101	139–40	....
14	unsym-Dimethylhydra- zine	63	......	1.4075^{22}	0.7914^{22}	....	....	......	......	......	......	Hydrochloride, 81–2; Oxalate, 142; Sulfate, 105
15	Trimethyleneimine (Azetidine)	63	......	1.4287^{24}	0.8436^{20}	....	....	......	......	......	166–7	Chloroplatinate, 203; Chloroaurate, 192
16	Isobutylamine	69	......	1.3988^{17}	0.724_4^{25}	....	57	53	78	82	150	....
17	n-Butylamine	77	......	1.401	0.7401_4^{20}	....	....	......	......	65	151	1-Naphthylurea, 149; 1-Naphthylthiourea, 108-9; Hydrochloride, 195
18	2-Amino-2-methylbutane	78	......	....	0.756_4^0	....	....	......	......	......	183	....
19	DL-sec-Butyl methyl amine	78–9	......	....	0.740^{15}	....	....	......	......	......	78	Chloroplatinate, 151
20	Ethyl propyl amine	80–1	......	1.3966	0.773^{24}	....	....	......	......	......	......	Hydrochloride, 225; Chloroplatinate, 198; Chloroaurate, 86
21	Sym-Dimethylhydrazine	81	......	1.4209^{20}	0.8274_4^{20}	....	....	......	......	......	147–50	Oxalate, 119; Hydro- chloride, 168
22	Cyclobutylamine	82	......	1.4363^{19}	0.8328_4^{20}	....	....	......	......	......	......	Chloroplatinate, 210–5
23	Di-isopropylamine	84	......	....	0.722^{22}	....	....	......	......	......	140	N-Nitroso, 48; Chloroplatinate, 186–9; Hydrochlo- ride, 216–7
24	Pyrrolidine	89	......	1.4270^{15}	0.852^{22}	....	....	......	123	......	112, yel.; 163– 4, red	....
25	n-Butyl methyl amine	90–1	......	1.4018^{18}	0.7367_4^{15}	....	....	......	......	......	111	Hydrochloride, 170; Chloroplatinate, 205
26	5-Amino-1-pentene	91–4	......	....	....	....	....	......	......	......	......	Chloroplatinate, 166; Chloroaurate, 195
27	DL-2-Amino-n-pentane (sec-n-Amylamine)	92	......	....	0.7384_0^{20}	....	....	......	......	......	......	Hydrochloride, 168; Oxalate, 226; 131 Chloroaurate, 82–3
28	Isoamylamine	96	......	1.4096^{18}	0.751^{18}	....	....	......	......	102	138	1-Naphthylurea, 132

*Derivative data given in order: m.p., crystal color, solvent from which crystallized.

TABLE XVIII. ORGANIC DERIVATIVES OF AMINES
1. Primary and secondary amines a) Liquids 1) (Listed in order of increasing atmospheric b.p.)* (Continued)

No.	Name	Boiling point, °C	Melting point, °C	n_D	Density g/ml	Acetamide	Benzamide	Benzene sulfonamide	p-Toluene sulfonamide	Phenyl thiourea	Picrate	Miscellaneous
29	D-2-Methyl-*n*-butyl-amine (*active*-Amyl amine).............	96	.....	....	0.7505^{25}_4	.....	...	...	...	.....	.....	$[\alpha]^{25}_D$: −5.86; Hydrochloride, 176; Chloroplatinate, 240
30	2-Methylpyrrolidine....	97–8			0.84^{20}_{20}						88.5–9.5	Oxalate, 178–9; Chloroplatinate, 172–3 (anh.); 206–7, rapid htng; Picrate of N-methyl deriv., 235
31	3-Methylpyrrolidine....	103–5	......	1.4480^{20}	0.8654^{0}_4					......	106	Chloroplatinate, 194
32	*n*-Amylamine	104			0.7614^{20}_4					69	139	2-Naphthylthiourea, 114
33	2-(Methylamino)-*n*-pentane	105			0.947^{20}						77–8	Chloroplatinate, 138
34	Piperidine	106	.....	1.4530^{20}	0.8606^{20}_4	....	48	93–4	96	101	152	....
35	2-Aminodiethylether ...	108–9		1.4101^{20}	0.8512^{20}_4						122	Picrolonate, 204
36	Di-*n*-propylamine......	109–10		1.4046^{20}	0.7384^{20}_4			51		69	75	1-Naphthylurea, 93
37	2,5-Dimethylpyrrolidine	110–3	.....	1.4357^{15}	0.8185^{12}_4						117–8	Hydrochloride, 188–90; Chloroplatinate, 225
38	2,4-Dimethylpyrrolidine	115–7	.....	1.4325^{20}	0.8297^{20}_4	....				......	116–7	Chloroplatinate, 210
39	1,2-Ethylenediamine....	116	8.5	1.454^{26}	0.898^{25}_4	172 (*di*)	244 (*di*)	168 (*di*)	360 (*di*)	102	233 (*di*)	....
40	L-2-Methylpiperidine...	117	50–1								116–7	Hydrochloride, 190; Chloroplatinate, 194
41	DL-2-Methylpiperidine .	118–9	.....	1.4464^{24}	0.8436^{24}_4	....	45		55		164	Hydrochloride, 207; Oxalate, 125
42	DL-1,2-Diaminopropane	119–20	.....		0.878^{15}	139 (*di*)	192 (*di*)				135 (*di*)	
43	Isohexylamine	125			0.758^{25}_4						123–5	Hydrochloride, 220; Oxalate, 166; Chloroplatinate, 200
44	DL-3-Methylpiperidine .	126		1.446^{24}	0.845^{24}_4						138 (*di*)	Hydrochloride, 172; N-2,4-Dinitrophenyl deriv., 67
45	2,6-Dimethylpiperidine .	127–8	.....	1.4366^{25}	0.816^{25}_4	....	111	50		......	162–4	Chloroplatinate, 212
46	*n*-Hexylamine.........	130	−19		0.763^{25}_4	....	40	96		77	126	....
47	3-Amino-*n*-hexane	130										Hydrochloride, 227; Chloroplatinate, 190–200
48	Morpholine..........	130					75	118	147	136	146	
49	Cyclohexylamine	134		1.4372^{20}	0.8191^{20}_4	101	149	89		148		
50	Trimethylenediamine (1,3-Diaminopropane; 1,3-Propylenediamine)	136			0.884^{25}_4	1,3-*di*: 126; 107	1,3-*di*: 140; 147	96	148		250	Chloroplatinate, 240
51	2,2,6-Trimethyl-piperidine	138–9									195–6	Hydrochloride, 236; Chloroaurate, 128
52	Di-isobutylamine	139	−77	1.4093^{20}	0.745^{20}_4	86	....	55; 57		113	121	....
53	4-Amino-*n*-heptane	139–40			0.767^{20}_4							Hydrochloride, 246–7; Chloroplatinate, 235
54	1,3-Diaminobutane.....	141–2									240–5	Hydrochloride, 171–2
55	2-Amino-*n*-heptane	142		1.4199^{19}	0.7665^{19}_4							Hydrochloride, 133; Oxalate, 204–5; Chloroaurate, 63–4
56	*unsym*-Diethylethylene-diamine	145			0.827^{19}_{19}					115 (*mono*); 211 (*di*)	......	Chloroplatinate, 211
57	Furfurylamine (α-Furylmethylamine) ...	145–6					.....			......	150	Oxalate, 145; With CO_2 from air → comp., 75; Hydrochloride, 110

* Derivative data given in order: m.p., crystal color, solvent from which crystallized.

TABLE XVIII. ORGANIC DERIVATIVES OF AMINES

1. Primary and secondary amines a) Liquids 1) (Listed in order of increasing atmospheric b.p.)* (Continued)

No.	Name	Boiling point, °C	Melting point, °C	n_D	Density g/ml	Acetamide	Benzamide	Benzene sulfon-amide	p-Toluene sulfon-amide	Phenyl thiourea	Picrate	Miscellaneous
58	Cyclohexyl methyl amine............	145–7	......	....	....	....	85–6	......	......	......	170	Hydrochloride, 193
59	2-Ethylpiperidine (α-Ethylpiperidine)......	146–7	......	....	0.8651_0^0	....	....	64–5	......	......	133	Hydrochloride, 181; Chloroplatinate, 202
60	2,2,4-Trimethyl-piperidine	148	......	....	0.832^{15}	....	....	......	......	......	......	Chloroplatinate, 215; Methiodide, 266
61	Sym-Diethylethylene-diamine............	149–50	......	....	....	....	....	......	......	......	......	Hydrochloride, 260; Chloroplatinate, 223; Chloroaurate, 220
62	4-Ethylpiperidine (γ-Ethylpiperidine)...	151	......	1.4503^{25}	0.876^0	....	....	74–5	......	......	......	Chloroplatinate, 173; Chloroaurate, 105
63	3-Ethylpiperidine (β-Ethylpiperidine)...	153	......	....	0.871_4^{16}	....	....	......	......	......	63	Hydrochloride, 141; Chloroplatinate, 183
64	n-Heptylamine........	155	−23	1.41954_α^{26}	0.777^{20}	....	....	......	......	75	121	1-Naphthylthiourea, 68–9
65	Di-n-butylamine......	159	......	....	....	....	....	......	......	86	59	1-Naphthylthiourea, 123
66	Tetramethylenediamine (1,4-Diaminobutane; Putrescine).........	159	27	....	0.877_4^{25}	137 (di)	177 (di)	......	224 (di)	......	249–50 (di)	....
67	DL-2-Hydroxy-n-propylamine (Isopro-panolamine)........	163	......	....	0.973^{18}	....	....	......	......	......	142	Chloroplatinate, 195; Hydrochloride, 73
68	Hexahydrobenzylamine.	163.5	......	1.4646^{18}	0.87_4^{20}	....	98; 107	......	......	......	184–6	Hydrochloride, 254
69	Cyclohexyl ethyl amine .	164	......	....	0.868_0^0	....	....	......	......	......	133	Hydrochloride, 184
70	2-Ethylcyclohexylamine	170–1	......	1.4682^{20}	0.8744_4^{20}	....	....	121–2	......	......	190	Chloroplatinate, 239
71	2-Aminoethyl alcohol (Ethanolamine)......	171	......	1.4539^{20}	1.022_4^{20}	....	....	......	......	......	160	1-Naphthylurea, 186
72	3-Amino-2-hydroxy-pentane	172	......	....	0.906^{18}	....	....	......	......	......	......	N-Chloroacetyl deriv., 52–60; Monooxalate, 166; Dioxalate, 204
73	2-Aminopropyl alcohol ..	173–6	......	....	....	....	....	......	......	......	......	Hydrochloride, 86; Chloroplatinate, 198–9
74	2-Amino-3-hydroxy-pentane	174	......	1.4458	0.9289^{24}	....	....	......	......	......	......	Chloroplatinate, 154; Picrolonate, 215
75	2-Fluoroaniline........	176	−35; −29	....	....	80	113	......	......	......	......	Picrate of N,N-dimethyl deriv., 131
76	Pentamethylenediamine (1,5-Diaminopentane; Cadaverine).........	178–80	......	....	0.9174_4^0	....	135 (di)	119	......	148	237	....
77	n-Octylamine	180	......	....	0.777^{20}	....	....	......	......	......	112	1-Naphthylthiourea, 72
78	5-Methyl-2-pyrazoline ..	180	......	....	....	....	156	......	......	......	126	Phenylurea, 127
79	Benzyl methyl amine ...	181	......	....	0.945_{15}^{18}	....	....	......	95	......	117–8	Chloroplatinate, 197
80	Aniline	184	......	1.5863^{20}	1.022_4^{20}	114	160	112	103	54	180	....
81	Benzylamine..........	184–5	......	1.5401	0.9826_4^{19}	65	105	88	116; 185	156	194	
82	4-Fluoroaniline........	186	−1	1.5195^{20}	1.1725_4^{20}	152	185	......	......	......	......	N-4-Nitrobenzoyl deriv., 181
83	DL-α-Phenylethylamine (α-Aminoethyl-benzene)	187; 185	......	....	0.9395^{15}	57	120	......	......	......	......	Hydrochloride, 158
84	1,2-Diaminocyclohexane	187	......	....	....	260 (di)	....	......	......	......	210–5 (di)	Hydrochloride, 280
85	3-Fluoroaniline........	187–8, yel.	......	....	1.160^{16}	84	....	......	......	......	......	....
86	Di-isoamylamine	187–8	−44	1.4229^{21}	0.7672_4^{21}	....	....	......	......	72	94.5	1-Naphthylurea, 95; Methiodide, 221
87	3-Aminopropyl alcohol ..	188	......	1.457^{26}	0.982_4^{26}	....	....	......	......	......	222	Chloroplatinate, 199

*Derivative data given in order: m.p., crystal color, solvent from which crystallized.

TABLE XVIII. ORGANIC DERIVATIVES OF AMINES
TABLE XVIII. ORGANIC DERIVATIVES OF AMINES
1. Primary and secondary amines a) Liquids 1) (Listed in order of increasing atmospheric b.p.)* (Continued)

No.	Name	Boiling point, °C	Melting point, °C	n_D	Density g/ml	Acetamide	Benzamide	Benzene sulfonamide	p-Toluene sulfonamide	Phenyl thiourea	Picrate	Miscellaneous
88	1,2,3-Triaminopropane	190	......	....	....	200–2 (*tri*)	217–8 (*tri*)	......	..,....	......	>269	....
89	4-Amino-2,6-dimethyl-piperidine	195	......	....	....	....	....	......	......	......	220	Chloroplatinate, >250
90	N-Methylaniline	196	......	1.573^{16}	0.989^{20}_{4}	102	63	79	94	87	145	....
91	1-Phenylisopropylamine	196–7	......	1.5181^{25}	0.9424^{20}_{0}	....	159	......	......	......	......	Oxalate, 131; Hydrochloride, 236
92	β-Phenylethylamine (β-Aminoethylbenzene)	198	......	....	0.958^{24}_{4}	51	116	69	......	135	174; 167	....
93	Benzyl ethyl amine	199	......	....	0.935^{17}_{15}	....	....	......	50	......	118	Hydrochloride, 184
94	2-Methylaniline (o-Toluidine)	200	......	1.5688^{20}	1.0053^{20}_{20}	110–1	146	124	185–6	136	213 '	....
95	n-Nonylamine	201	......	....	....	34–5	49	......	......	......	111	....
96	3-Methylaniline (m-Toluidine)	203	......	1.5686^{20}	0.990^{25}_{25}	65	125	95	171–2	......	200	....
97	Di-n-amylamine	205								72	......	2-Naphthylthiourea, 126; Hydrochloride, 275
98	DL-2-Phenylisopropyl-amine	205	......			64 initially; 93 on standing	....				143	Hydrochloride, 145–7
99	N-Ethylaniline	205	......	1.5559^{20}	0.9625^{20}_{4}	54	60	......	87	89	132; 138	....
100	4-Methylpyrazole	207	......	1.4920^{20}_{He}	1.015^{20}_{4}	....	....	......	......	......	142	1-o-Nitrobenzoyl deriv., 107
101	3-Methylbenzylamine (m-Xylylamine)	207			0.9654^{20}_{0}	235–40	150				198; 156	Hydrochloride, 208; Chloroplatinate, 214
102	2-Chloroaniline	209; 207	......	1.5895^{20}	1.2125^{20}_{4}	87	99	129	193; 105	156	134	....
103	N,2-Dimethyl-aniline (N-Methyl-o-toluidine)	208	......		0.973^{15}	56	66				90	
104	2-Methylbenzylamine (o-Xylylamine)	208	−20	1.5436^{19}	0.977^{19}_{0}	69	88				215	Chloroplatinate, 220–3
105	4-Methylbenzylamine (p-Xylylamine)	208	13	1.5364^{20}	0.952^{20}_{0}	107–8	137				204	
106	3-Methylpyrazole	208	......	1.497^{16}_{He}	1.020^{16}_{4}	29–30	....	......	......	......	144	N-o-Nitrobenzoyl deriv., 120
107	1-Phenylpropylamine	208	......	1.5173^{25}	0.9347^{25}_{0}	....	115–6	81			......	Hydrochloride, 190
108	2-Phenylpropylamine	210					85				182	Hydrochloride, 123–4
109	N,4-Dimethyl-aniline (N-Methyl-p-toluidine)	210				83			67		131	N-Nitroso deriv., 52; Hydrochloride, 119.5
110	2-Ethylaniline	210–11	46.6	1.5584^{22}	0.9810^{20}	111	147				194–5	....
111	L-Menthylamine	212; 207	......			145	156			135	215	$[\alpha]^{19}_{D}$: −34.2
112	2,5-Dimethylaniline (p-2-Xylidine)	213–5, pa. yel.	14.2	1.5591^{21}	0.9735^{20}	139	140	138	232–3; 119	148	171	....
113	1-Phenylisobutylamine	214	......	1.5123^{20}	0.920^{20}_{0}						166–8	Oxalate, 120–2; Hydrochloride, 275–7
114	3-Methylpyridazine	215	......		1.0486^{26}_{26}						143–4	....
115	2,6-Dimethylaniline (m-2-Xylidine)	215; 218	11.2	1.5610^{20}	0.9842^{20}	177	168		212	204	180	....
116	4-Ethylaniline	216; 214	−6	1.5550^{20}	0.9690^{20}	94	151		104			....
117	2,4-Dimethylaniline (m-4-Xylidine)	217		1.561^{20}	0.9783^{20}_{4}	133; 130	192	130	181	152	209	....
118	2-Chloro-N-methyl-aniline	218	......	....	1.1735^{11}						133	....
119	2,4-Dimethylbenzyl-amine	218–9	......	....	....						223	Hydrochloride, 212; Chloroplatinate, 226

* Derivative data given in order: m.p., crystal color, solvent from which crystallized.

TABLE XVIII. ORGANIC DERIVATIVES OF AMINES

1. Primary and secondary amines a) Liquids 1) (Listed in order of increasing atmospheric b.p.)* (Continued)

No.	Name	Boiling point, °C	Melting point, °C	n_D	Density g/ml	Acetamide	Benzamide	Benzene sulfonamide	p-Toluene sulfonamide	Phenyl thiourea	Picrate	Miscellaneous
120	2-Amino-N,N-dimethyl-aniline	219	.....	....	....	72	51	.....	.....	.....	138–40	......
121	3,5-Dimethylaniline (m-5-Xylidine)	220	9.8	1.5581^{20}	0.9706^{20}	144; 140	144–5	.....	.....	153	200	N-Formyl deriv., 76
122	N-Ethyl-3-methylaniline (N-Ethyl-m-toluidine)	221; 215	.....	....	....	....	72	.....	.....	.....	.....	Hydrochloride, 159; Chloroplatinate, 182
123	3,5-Dimethylbenzyl-amine	221	.....	1.5305^{20}	0.950_0^{20}	....	....	.....	.....	.....	225	Hydrochloride, 245; Chloroplatinate, 204
124	2,3-Dimethylaniline (o-3-Xylidine)	221–2	3.5	1.5684^{20}	0.9931^{20}	135	189	.....	.....	.....	221	Hydrochloride, 254; N-Formyl deriv., 102
125	1-Aminoindane (1-Hydrindamine)	222	.....	....	....	....	142–3	.....	.....	.....	207	......
126	3-Phenylpropylamine	222	.....	....	0.976_4^{25}	....	57–8	.....	.....	.....	152–3	Hydrochloride, 218
127	2-Methyl-4,5,6,7-tetra-hydroindole	222	.....	....	0.987_4^{10}	....	....	86–91	.....	.....	141	Methiodide, 195; Chloroplatinate, 187
128	N-n-Propylaniline	222	.....	....	0.949^{18}	47	....	54	.....	104	.....	
129	2-n-Propylaniline	222–4	.....	....	....	104–5	119	.....	.....	.....	151	Hydrochloride, 173
130	2-Chloro-4-methylaniline	223	.....	....	....	113	137	.....	.....	.....	.....	
131	α-Amino-n-butylbenzene	223	.....	....	0.9367_0^{20}	....	128	.....	.....	.....	.....	Chloroplatinate, 184; Hydrochloride, 288
132	trans-9-Aminodecalin	223	−25	1.492_{He}^{20}	0.939_4^{20}	183	148–9	.....	.....	.....	.....	N-Formyl deriv., 172
133	γ-Amino-n-butylbenzene	223	.....	1.5152^{20}	0.9289_4^{15}	....	108, lgr.	.....	.....	.....	.....	Hydrochloride, 144; Chloroplatinate, 220
134	2-Methoxyaniline (o-Anisidine)	225	5–6	....	....	85; 88	60; 84	89	127	136	200	N-Formyl deriv., 84
135	4-Isopropylaniline (p-Cumidine)	225	.....	....	0.953_4^{20}	102	162	.....	.....	.....	.....	
136	4-n-Propylaniline	225	.....	....	....	93–4	115	.....	.....	.....	.....	Hydrochloride, 203–4
137	N-Isobutylaniline	227	.....	....	0.940_4^{18}	....	....	.....	122–3	.....	.....	
138	α-Methyl-α-phenyl-hydrazine	227	.....	1.5824^{20}	....	92	153	132	.....	.....	.....	
139	4-tert-Butylaniline	228	17	....	....	173	140	.....	179–80	.....	.....	N-Formyl deriv., 59; Hydrochloride, 270–4
140	cis-9-Aminodecalin	228	−13.5	1.498_{He}^{21}	0.951_4^{21}	127	147	.....	.....	.....	.....	N-Formyl deriv., 165–6
141	2-Ethoxyaniline (o-Phenetidine)	229	.....	....	....	79	104	102	164	137	.....	
142	2,4,6-Trimethylaniline (Mesidine)	229; 232	.....	....	....	216	204	137	167	193	189–91	
143	2-Aminoindane (2-Hydrindamine)	230	.....	....	....	127	155	.....	.....	.....	239	Hydrochloride, 241
144	3-Chloroaniline	230; 236	.....	1.5931^{20}	1.2225_{15}^{15}	72; 78	119–20	121	138; 210	124; 116	177	......
145	2,2'-Diaminodiethyl-sulfide	231–3	.....	....	....	....	....	.....	.....	.....	212	Dihydrochloride, 131
146	1,2,3,4-Tetrahydro-isoquinoline	233	.....	1.5798^{23}	1.064_4^{23}	46	129	154	.....	.....	200; 195	
147	2-tert-Butylaniline	233–5	.....	1.5453^{20}	0.977^{15}	159–61	....	.....	.....	.....	.....	
148	3-Amino-4-(dimethyl-amino)toluene	234	.....	....	....	....	....	.....	.....	.....	151	Hydrochloride, 192–3
149	4-Isobutylaniline	235, pa. yel.	.....	....	....	127	....	.....	136–7	.....	.....	
150	4-Aminoindane (4-Hydrindamine)	236	−3	....	....	126	136	.....	.....	.....	.....	
151	2-Aminoundecane (2-Aminohendecane; sec-n-Undecylamine)	237	.....	....	....	58	....	.....	.....	.....	111	Hydrochloride, 84
152	unsym-Ethylphenyl-hydrazine	237	.....	....	1.018^{15}	....	....	.....	.....	.....	.....	Hydrochloride, 137; Reduces warm Fehling

*Derivative data given in order: m.p., crystal color, solvent from which crystallized.

TABLE XVIII. ORGANIC DERIVATIVES OF AMINES
1. Primary and secondary amines a) Liquids 1) (Listed in order of increasing atmospheric b.p.)* (Continued)

No.	Name	Boiling point, °C	Melting point, °C	n_D	Density g/ml	Acetamide	Benzamide	Benzene sulfonamide	p-Toluene sulfonamide	Phenyl thiourea	Picrate	Miscellaneous
153	*sym*-Ethylphenylhydrazine	238–9	.....	1.55^{15}	1.004^{15}_{15}	....	100	.....	.....	.....	.....	Hydrochloride, 164; Oxalate, 167
154	2-Bromo-4-methylaniline (3-Bromo-*p*-toluidine)	240	26	....	1.51^{20}	118	149			.....		Hydrochloride, 221
155	1-Aminoundecane (1-Aminohendecane; *n*-Undecylamine; *n*-Hendecylamine)	240	15–6			48	60					Hydrochloride, 190
156	4-Chloro-N-methylaniline	240	.....		$1.169^{11.5}$	92–4	....				153	Nitrosamine, 51
157	4-Chloro-2-methylaniline (5-Chloro-*o*-toluidine)	241	29			140	....					
158	2-Amino-*p*-cymene (*p*-Cymidine)	241	.....	1.543^{19}	0.994^{20}	71	102	.....				Hydrochloride, 207
159	N-Butylaniline	241	.....	1.5381^{20}	0.9358^{20}_{4}	....	56	.....	56	.....		
160	Phenylhydrazine	243	19; 23	1.6081^{20}	1.0978^{20}_{4}	128; 107 (*di*)	168; 177 (*di*)	148	151	172	.....	
161	α,α'-Diamino-*m*-xylene (*m*-Xylylenediamine)	245–8		....		*di*: 134–5, bz.	N,N'-*di*: 172				185–90	Dihydrobromide, 266
162	2-Chloro-6-methoxyaniline (3-Chloro-*o*-anisidine)	246 sl. d.				123	135					
163	3-Ethoxyaniline (*m*-Phenetidine)	248				97	103	.....	157	138	158	....
164	4-Ethoxyaniline (*p*-Phenetidine)	248; 254	2–3	....	1.065^{16}_{4}	137	173	143	106	136	69	....
165	1,2,3,4-Tetrahydroquinoline	250	20	1.593^{24}	1.054^{24}_{4}	....	75	67				
166	2-Aminoacetophenone	250–2d.	20			76–7	98	.....	148	.....	.....	Semicarbazone, 290; Oxime, 109; Hydrochloride, 168 d.
167	3-Bromoaniline	251	18	1.626^{20}	1.579^{20}_{4}	87	120; 136	.....		143	180	
168	3-Methoxyaniline (*m*-Anisidine)	251				81	....		68		169	Hydrochloride, 167–8
169	3-Bromo-2-methylaniline (6-Bromo-*o*-toluidine)	254				163	176–7					
170	4-Amino-1,2,3,5-tetramethylbenzene (Isoduridine)	255	23–4			215–7					200	
171	Dicyclohexylamine	255 sl. d.	abt. 20	1.488^{18}	0.925^{18}	103	153	.....			173	
172	6-Methyl-1,2,3,4-tetrahydroisoquinoline	256			1.0235^{8}_{4}	....					205	Hydrochloride, 195–7; Methiodide, 144–5; N-Nitroso deriv., 98
173	4-Amino-N,N-diethylaniline	261				104	172					....
174	4-*n*-Butylaniline	261			0.945^{20}_{4}	105	126				.....	Chloroplatinate, 200–2
175	1-Amino-5,6,7,8-tetrahydronaphthalene	261–3	.....	1.5896^{23}	1.0625^{16}	158					.....	Hydrochloride, 259–61
176	7-Methyl-1,2,3,4-tetrahydroquinoline	264				....	70–2				153–4	Hydrochloride, 175
177	4-Methylindole	267	5		1.062^{20}_{4}	....					194–5	....
178	2-Amino-4-chloro-N,N-dimethylaniline	267–8				90					191	....
179	DL-3-Aminopropyleneglycol (2,3-Dihydroxypropylamine)	268 part. d.	.....	1.49^{10}	1.175^{20}_{4}	....	O,N-*di*: 109; O,O,N-*tri*: 113				.....	Chloroplatinate, 185; Picrolonate, 220; O,N-*di*-4-Nitrobenzoyl deriv., 139

*Derivative data given in order: m.p., crystal color, solvent from which crystallized.

No.	Name	Boiling point, °C	Melting point, °C	n_D	Density g/ml	Acetamide	Benzamide	Benzene sulfonamide	p-Toluene sulfonamide	Phenyl thiourea	Picrate	Miscellaneous
180	Diethanolamine (Di-(2-hydroxyethyl)-amine)	270	28	1.4776^{20}	1.0966_4^{20}	....	....	......	......	......	110	Nitrate, 69; Chloroplatinate, 160
181	3-(Dimethylamino) aniline	272	......	....	0.995^{25}	87; 69 (di)	163–4	......	......	......	187	N-Chloroacetyl deriv., 102
182	1-(N,N-Diethylamino) naphthalene	290	......	1.5961^{20}	1.015_{20}^{20}	....	....	......	......	......	152–4	1,3,5-Trinitrobenzene add. comp., 95, scar.
183	1-(Methylamino) naphthalene	294	......	....	....	94–5	121	......	164	......	......	....
184	Dibenzylamine	300	......	1.5743^{22}	1.0256_4^{22}	....	112	68	159	......	......	Hydrochloride, 256
185	α-Aminodiphenylmethane (Benzhydrylamine)	303–4	......	$1.5963^{21.5}$	$1.0635_0^{21.5}$	146–7	172; 167	......	......	......	205–6	N-Formyl deriv., 132
186	1,2-Diphenylethylamine	313	......	....	1.031^{15}	....	....	......	......	......	212–3	Oxalate, 158; Chloroplatinate, 188
187	2,3-Diphenylpropylamine	315–7	......	....	....	85 (di)	....	......	......	......	......	Hydrochloride, 188–90; Chloroaurate, 144–5
188	2-(Methylamino) naphthalene	317; 309	......	....	....	51	84	107	78	......	145	Hydrochloride, 182–3
189	2-(N,N-Diethylamino) naphthalene	320–2	......	....	....	....	....	......	......	......	......	1,3,5-Trinitrobenzene add. comp., 116, blk.; Hydrochloride, 177; Chloroplatinate, 95

*Derivative data given in order: m.p., crystal color, solvent from which crystallized.

TABLE XVIII. ORGANIC DERIVATIVES OF AMINES

1. Primary and secondary amines a) Liquids 2) (b.p. at reduced pressure only)
(Listed in order of increasing m.p. of the corresponding acetyl derivative)*

No.	Name	Boiling point, °C	Melting point, °C	n_D	Density g/ml	Acetamide	Benzamide	Benzene sulfonamide	p-Toluene sulfonamide	Phenyl thiourea	Picrate	Miscellaneous
1	3-Aminostyrene	$112\text{--}5^{12}$	......	....	1.0216^{20}_{20}	74–5	90, bz.-lgr.; 126, al.-w.	......	......	......	......	Polymerizes readily
2	2-Bromo-4-ethoxyaniline (3-Bromo-p-phenetidine)................	160^{23}	......	....	....	97	....	......	......	......	......	....
3	3-Aminothiophenol......	$180\text{--}90^{16}$	......	....	....	N,S-*di*: 97 *di*)	....	......	......	......	......	Hydrochloride, 232
4	2-*n*-Butylaniline	$122\text{--}5^{12}$	......	....	0.953^{20}_{4}	100	116–7	......	......	......	......	Hydrochloride, 137
5	2-(2-Aminophenyl)ethyl alcohol................	$147\text{--}8^{3.5}$	......	1.5849^{19}	....	103.5	....	......	......	......	......	Hydrochloride, 126
6	6-Amino-3,4'-dimethylbiphenyl.............	$165\text{--}7^{4}$	......	....	....	104	....	......	......	......	......	Hydrochloride, 216–26
7	2,2-Diphenylpropylamine	$179\text{--}82^{22}$	......	....	1.027^{18}	106–7	82–3	......	......	......	......	Hydrochloride, 261
8	2-Chloro-4-methoxyaniline (3-Chloro-p-anisidine)	156^{31}	......	....	....	114	....	......	......	......	......	Hydrochloride, 228
9	2-Aminostyrene	$97\text{--}8^{8}$	......	1.6130^{21}	1.015^{21}_{21}	129	....	......	......	......	......	Polymerizes readily
10	2-Aminothiophene	$77\text{--}9^{11}$	......	....	....	161–2	172–3	......	......	......	......	Oxidizes rapidly; N-2-toluenesulfonyl deriv., 183–4
11	4,4'-Diamino-2,3'-dimethylbiphenyl	244^{12}	......	....	....	N,N'-*di*: 253; *tetra*: 191	N,N'-*di*: 245	......	......	......	......	....

*Derivative data given in order: m.p., crystal color, solvent from which crystallized.

TABLE XVIII. ORGANIC DERIVATIVES OF AMINES
1. Primary and secondary amines c) Solids (Listed in order of increasing m.p.)*

No.	Name	Melting point, °C	Boiling point, °C	Acetamide	Benzamide	Benzene sulfonamide	p-Toluene sulfonamide	Phenyl thiourea	Picrate	Miscellaneous
1	Di-n-heptylamine	1	271	....	....	....	....	....	117–20	....
2	4-Aminostyrene	23.5	98⁴	142	160–1	....	....	....	....	D^{21}_{21}: 1.012; $n^{21.5}_D$: 1.625
3	3-Bromo-4-methylaniline (2-Bromo-p-toluidine)	25–6	254–7	117–8	132	....	....	....	....	....
4	2-Amino-3-methylpyridine (2-Amino-β-picoline)	26	224	64	220	....	....	....	229	....
5	2-Aminothiophenol	26	234	135 (N,S-di)	154 (N,S-di)	....	....	....	....	....
6	n-Dodecylamine	27–8	247–9	....	....	....	73	....	....	Hydrochloride, 98; Chloroplatinate, 215
7	DL-2,6-Dimethyl-1,2,3,4-tetrahydroquinoline	31–2	267	....	103–5	....	....	....	....	Hydrochloride, 180–3
8	5-Bromo-2-methylaniline (4-Bromo-o-toluidine)	32	253 part. d.	165	....	....	....	....	....	....
9	1-Amino-2-methylnaphthalene (2-Methyl-1-naphthylamine)	32	....	188	180	....	....	....	....	....
10	2-Bromoaniline	32	250	99	116	....	....	146; 161	129	....
11	4,4'-Dimethyldibenzylamine (p-Dixylylamine)	32.5	220³⁰	....	....	....	....	....	153	Nitrosamine, 52; Hydrochloride, 272
12	2-Aminodibenzyl	33	173–83¹¹	117	166	....	....	....	167–8	Hydrochloride, 198
13	3-Iodoaniline	33; 27	145–6¹⁵	119	157	....	128	....	....	....
14	unsym.-Diphenylhydrazine	34	220⁴⁰⁻⁵⁰	184	192	....	....	....	....	....
15	α,α'-Diamino-1,4-dimethylbenzene (p-Xylylenediamine)	35	....	194 (tetra)	193 (N,N'-di)	....	....	....	232	....
16	3-Bromo-5-methylaniline (5-Bromo-m-toluidine)	36	255–60	171–2	....	....	....	....	....	....
17	4-Amino-N-methylaniline	36	258	63	165	....	....	....	206	....
18	Pentadecylamine	36; 33	300	72	....	....	....	....	....	Hydrochloride, 199
19	N-Benzylaniline	37	298	58	107	119	148–9	103	48	....
20	3-Iodo-4-methylaniline (2-Iodo-p-toluidine)	37–8	....	130	....	....	....	....	....	Oxalate, 103
21	5-Aminoindane (5-Hydrindamine)	37–8	247–9⁷⁴⁵	106	137	....	....	....	....	....
22	2-Amino-5,6,7,8-tetrahydronaphthalene (5,6,7,8-Tetrahydro-2-naphthylamine)	38	275–7⁷¹³	107	167	....	....	....	204	....
23	2-Amino-5-bromonaphthalene (5-Bromo-2-naphthylamine)	38	207–10¹⁶	165	109	....	....	....	216	N-Benzal, 63
24	2,2-Diphenylethylamine	38	180³³	88	144–5	....	....	....	212–3	Phenylurethane, 191–2; Hydrochloride, 256–7
25	2,3-Dimethyl-1,2,3,4-tetrahydroquinoline	38–9	....	....	92	....	....	....	178	....
26	2-Iodo-4-methylaniline (3-Iodo-p-toluidine)	40	....	133	161	....	....	....	....	Oxalate, 120; Hydrochloride, 188
27	2-Chloro-4,6-dimethylaniline (5-Chloro-m-4-xylidine)	40	....	205–6	148	....	....	....	....	2-Naphthylthiourea, 154
28	2-Amino-6-methylpyridine (6-Amino-α-picoline)	41	208–9	90	90	....	....	....	202	Hydrochloride, 155; Chloroplatinate, 218
29	4-Amino-N,N-dimethylaniline	41; 53	262	132–3	228	....	....	....	188	....
30	1,6-Diaminohexane (Hexamethylenediamine)	42	204–5	125–7 (di)	155 (di)	154 (di)	....	....	220	....
31	1,3-Diaminoisopropyl alcohol	42	235	....	....	....	....	....	230	Hydrochloride, 185; Chloroplatinate, 240
32	4-Amino-3-methylbiphenyl	43	190¹⁵	165; 158	189	....	....	....	....	N-Benzal, 108
33	2,4'-Diaminobiphenyl	45	363	202 (di)	278 (di)	....	....	....	....	....
34	4-Methylaniline (p-Toluidine)	45	200	147	158	120	118	141	182	....
35	4-Aminothiophenol	46	140–5¹⁶	154 (N-); 144 (132) (N,S-di)	180 (N-)	....	....	....	....	Oxidized readily → 4,4'-diaminodiphenyl sulfide, 104–5

*Derivative data given in order: m.p., crystal color, solvent from which crystallized.

TABLE XVIII. ORGANIC DERIVATIVES OF AMINES
1. Primary and secondary amines c) Solids (Listed in order of increasing m.p.)* (Continued)

No.	Name	Melting point, °C	Boiling point, °C	Acetamide	Benzamide	Benzene sulfon-amide	p-Toluene sulfon-amide	Phenyl thiourea	Picrate	Miscellaneous
36	4-Aminobenzyl cyanide	46	....	97 (mono) 152–3 (di)	176–7	......		......	185	N-Formyl, 135
37	2-Aminopropiophenone...........	47	93$^{0.8}$	71	130	......				Oxime, 88–9; Semicarbazone, 190; N-Propionyl, 51
38	3-Bromo-4-ethoxyaniline (2-Bromo-p-phenetidine)................	47	189^{20}	114	....				178–9	....
39	4-Amino-2-thiocresol	47	....	95 (N-); 125 (N,S-di)						S-Me., 47
40	2-Iodo-5-methylaniline (4-Iodo-m-toluidine	48; 38	....	151; 146						N-Formyl, 129
41	1-Amino-4-fluoronaphthalene (4-Fluoro-1-naphthylamine)	48	162^{16}	....	197	......				Hydrochloride, 280
42	4-Aminodibenzyl	48	....	....	170–1					Hydrochloride, 210; Chloroplatinate, 286–9
43	3,4-Dimethylaniline (o-4-Xylidine; 4-Amino-o-xylene)............	49	224	99	....	118				N-Formyl, 52; Hydrochloride, 256
44	2-Ethoxy-6-nitroaniline	49, yel.	....	64	....					N-Me., 59
45	Heptadecylamine	49	335–40	62	91					....
46	2-Aminobiphenyl...............	49	299	121	102	......				N-Formyl, 75; N-Propionyl, 65
47	4-Amino-2-methyldiphenylamine ...	49–50	196^{4}	139–40	....					Hydrochloride, 185–7
48	2-Bromo-4,6-dimethylaniline (5-Bromo-m-4-xylidine)........	49–50	....	196–7	186	....			122	....
49	2,5-Dichloroaniline..............	50	251	132	120	......			86	Hydrochloride, 191–2
50	1-Aminooxindole...............	50	....	186–7	189	......				
51	1-Aminonaphthalene (α-naphthylamine)	50	....	159	160	167	157; 147	165	163; 181	
52	1-Amino-1,2,3-triazole	51			151				130	Hydrochloride, 114
53	2-Amino-1-methylnaphthalene (1-Methyl-2-naphthylamine)	51	......	188–9	222				..	
54	1-Amino-3-methylnaphthalene (3-Methyl-1-naphthylamine)	51–2	....	175–6	188–9	....				
55	1-Amino-4-methylnaphthalene (4-Methyl-1-naphthylamine)	51–2	176^{12}	166–7	238–9	....				Hydrochloride, 233–4
56	4-Chloro-2-methoxyaniline (5-Chloro-o-anisidine).........	52; 46	260	150	....				200	N-Formyl, 177–8; Hydrochloride, 238
57	Indole.......................	52	253	157–8	68	254			......	N-Nitroso, 171
58	2-Aminodiphenylmethane.........	52	190^{22}	135	116	....				Hydrochloride, 137
59	2,2′-Ditolylamine (Di-o-tolylamine)	52–3	318	....	114–5					
60	4-Aminobiphenyl...............	53	302	171; 120 (di)	230	......	255; 160		......	N-Formyl, 172
61	1,4-Bis-(methylamino)benzene (sym-Dimethyl-p-phenylenediamine)...................	53	150^{17}	....	....				186	N,N′-Dinitroso, 148
62	Diphenylamine	53–4	....	101	180	124	141	152	182	....
63	DL-α-Aminobenzyl cyanide	55	....	....	159–60				160–1	
64	4-Methoxy-3-nitroaniline (2-Nitro-p-anisidine)	57, or.		153						N-Chloroacetyl, 150; N,N-Di-Me., 46, red
65	5-Bromo-2-ethoxyaniline (4-Bromo-o-phenetidine).......	57; 53	....	133					135–7	....
66	2-Amino-5-bromobiphenyl	57	......	130	162	......				
67	4-Methoxyaniline (p-Anisidine) ...	58	240	130; 127	154; 157	95	114	157; 171		
68	1-Amino-7-methylnaphthalene (7-Methyl-1-naphthylamine)	58–9	162^{10}	182–3	204	......				
69	4-Bromo-2-methylaniline (5-Bromo-o-toluidine)	59	240	156–7	115	......				

*Derivative data given in order: m.p., crystal color, solvent from which crystallized.

TABLE XVIII. ORGANIC DERIVATIVES OF AMINES
1. Primary and secondary amines c) Solids (Listed in order of increasing m.p.)* (Continued)

No.	Name	Melting point, °C	Boiling point, °C	Acetamide	Benzamide	Benzene sulfonamide	p-Toluene sulfonamide	Phenyl thiourea	Picrate	Miscellaneous
70	2-Amino-1-chloronaphthalene (1-Chloro-2-naphthylamine)	59	....	147	98	131	....	....	......	N-Formyl, 136
71	2-Aminoazobenzene	59	....	126	122					
72	1-Amino-2-chloronaphthalene (2-Chloro-1-naphthylamine)	59–60	....	191 (mono); 88 (di)	....					
73	5-Methylindole	60	267						151	
74	2-Aminopyridine	60; 58	204	71	165 (di)				216–7	
75	3-(3-Indolyl)-propylamine	60–4	....						146–9 156	Hydrochloride, 170
76	2-Methylindole	61	271–2						139	1,3,5-Trinitrobenzene add. comp., 152; N-Formyl, 75
77	2-Iodoaniline	61; 58	....	109	139	....	....	......	112	Hydrochloride, 153–4
78	1-Naphthyl phenyl amine (N-Phenyl-1-naphthylamine)	62	226[2]	115	152					
79	3-Chloro-4-methoxyaniline (2-Chloro-p-anisidine)	62	....	94	....				186	
80	1-Amino-3-chloronaphthalene (3-Chloro-1-naphthylamine)	62	....	197	162					Hydrochloride, 219
81	2-Amino-4,4'-dimethylbiphenyl	62–3	....	118–9	95–6					
82	8-Amino-6-methylquinoline	62–4	....	91–2						1,3,5-Trinitrobenzene add. comp., 139
83	2-Amino-1-bromonaphthalene (1-Bromo-2-naphthylamine)	63	....	mono: 140; di: 105	....					N-Propionyl, 139; N-Benzal, 93–4; 1,3,5-Trinitrobenzene add. comp., 192
84	1,3-Diaminobenzene (m-Phenylenediamine)	63	282–4	191 (di) 87–9 (mono)	240 (di) 125 (mono)	194	172	......	184	....
85	2,4-Dichloroaniline	63	245	145	117	128	126	......	106	....
86	2-Amino-5-methylnaphthalene (5-Methyl-2-naphthylamine)	63–4	....	123–4	155–6					
87	3-Bromo-4-methoxyaniline (2-Bromo-p-anisidine)	64	....	111	....					Hydrochloride, 255
88	2,5-Diaminotoluene	64	273–4	220 (di)	307	2-mono: 147	2-mono: 150			
89	3-Aminopyridine	64	250–2	mono: 133; di: 88	119					
90	9-Aminofluorene	64	....	262	260–1					Hydrochloride, 255
91	4-Methylphenylhydrazine (p-Tolylhydrazine)	65	240–4d.	121	1-N-mono: 68–70; 2-N-mono: 146	....				
92	4-Aminobenzyl alcohol	65	....	188 (O,N-di)	4-mono: 150					Hydrochloride, 217
93	2-Bromo-6-methoxyaniline (3-Bromo-o-anisidine)	65	....		90					Hydrochloride, 225
94	1,3-Diamino-2,6-dimethylbenzene (2,4-Diamino-m-xylene)	65–6	....	>260 (di)	232 (227) (di)					N,N'-Diformyl, 220
95	Aminoacetamide (Glycineamide)	65–7	....							Hydrochloride, 186–9; Chloroaurate, 197–8; Hot H_2O → glycine + NH_3
96	2-Aminocyclohexanol	66	219							Hydrochloride, 175; N-Phenyl, 150
97	2-Amino-5-methylbenzophenone	66, yel.	....	159	118	....	....	......	145	....
98	4-Bromoaniline	66	245	168	204	134	......	148	180	
99	1,8-Diaminonaphthalene	66	....	311–2 (di)	....		207 (di)	......		....

*Derivative data given in order: m.p., crystal color, solvent from which crystallized.

305

No.	Name	Melting point, °C	Boiling point, °C	Acetamide	Benzamide	Benzene sulfonamide	p-Toluene sulfonamide	Phenyl thiourea	Picrate	Miscellaneous
100	3-Amino-5-bromopyridine	66–7	150¹²	76–8 (hyd.); 127 (anh.)	....	....	....	....	212–3	Chloroaurate, 185–7
101	1-Amino-8-methylnaphthalene (8-Methyl-1-naphthylamine)	67–8	....	183–4	195–6					
102	4-Iodoaniline	67–8	....	184	222			153	......	N-4-Nitrobenzoyl, 269; N-Benzal, 86
103	1-Amino-2,4,5-trimethylbenzene (Pseudocumidin)	68	....	162	167	136	....			
104	2,2′-Diaminodibenzyl	68		249 (di)	255 (di)	....	....		225–30	
105	2-Amino-4-methylnaphthalene (4-Methyl-2-naphthylamine)	68		172–3	194–5	....	....			
106	3-Nitro-N-methylaniline	68		95	105	83				
107	1-Amino-3-bromonaphthalene (3-Bromo-1-naphthylamine)	70	....	174	166	....				Hydrochloride, 247
108	1-Amino-5-methyl-1,2,3-triazole	70			158, 138 (di)	....				N-Benzal, 67–8; Hydrochloride, 138
109	8-Aminoquinoline	70; 65, yel.	....	103	98		154–6	......	......	Hydrochloride, 208–9
110	2-Nitroaniline	71, golden-yel.		92; 94	98; 110	104	142		73	....
111	3,4-Diaminotriphenylmethane	71–2	....	226 (di)	243 (di)	....				
112	4-Chloroaniline	72	....	179; 172	192	122	95; 119	152		
113	8-Amino-6-Chloroquinoline	73	....	....	....	....				Hydrochloride, 208; Chloroplatinate, 212; Methiodide, 178
114	4-Aminophenylurethane (N-carbethoxy-1,4-diaminobenzene)	73–4	....	202; 181	230					Hydrochloride, 242
115	4-Aminodiphenylamine	75 (anh.)	....	158	203	....				
116	3,5-Dimethylindole	75	278	....	....	....			180	....
117	Duridine (3-Amino-1,2,4,5-tetramethylbenzene)	75	261	207	....	....				Hydrochloride, 260
118	1-Amino-3,4,5-trimethylbenzene	75	240	163–4	....					N-Formyl, 98
119	2-Amino-1,4-dimethylnaphthalene (1,4-Dimethyl-2-naphthylamine)	75	333	219–20	....	....				
120	4-Nitromesidine (2-Amino-4-nitromesitylene)	75	....	191	169	163				
121	2-Methoxy-6-nitroaniline (3-Nitro-o-anisidine)	76, yel.		158–9	....					N-Me., 58, red
122	2-Bromo-1,4-diaminobenzene (2-Bromo-p-phenylenediamine)	76		200(di)	235(di)					
123	4,6-Dimethyl-2-nitroaniline (5-Nitro-m-4-xylidine)	76; 70		176; 173	185	....				
124	1-Amino-5-methylnaphthalene (5-Methyl-1-naphthylamine)	77–8		194–5	173–4	....			210 d.	
125	2,4,6-Trichloroaniline	78	263	204; 206	174	152–4	....		83	....
126	4-Methyl-3-nitroaniline (2-Nitro-p-toluidine)	78	....	148	172	160	164	171	......	....
127	2-Dibromoaniline	79	....	146	134	....	134	......	124	....
128	1-Naphthyl 4-tolyl amine	79	....	124	140	....				....
129	2-Aminodiphenylamine	79–80		121 (2-N-)	136 (2-N-)					....
130	2,4-Diaminophenol (4-Hydroxy-m-phenylenediamine)	79–80		220–2 (2,4-N-) 180–2 (tri)	253 (di)	....			120	....
131	4-Aminopyrazole	80–2	....	....	173 (di)	....			193–4	....

*Derivative data given in order: m.p., crystal color, solvent from which crystallized.

TABLE XVIII. ORGANIC DERIVATIVES OF AMINES

1. Primary and secondary amines c) Solids (Listed in order of increasing m.p.)* (Continued)

No.	Name	Melting point, °C	Boiling point, °C	Acetamide	Benzamide	Benzene sulfon-amide	p-Toluene sulfon-amide	Phenyl thiourea	Picrate	Miscellaneous
132	**4-Bromo-3-methylaniline** (6-Bromo-*m*-toluidine)..........	81	240	103–4	....	....	......	......	.....	N,N-Di-Me., 55
133	**2,2′-Diaminodibenzyl sulfide**......	81	....	209 (*di*)	....	....			203–4	N,N′-Diformyl, 163
134	**2,2′-Diaminobiphenyl**	81	162⁴	*mono*: 89; *di*: 161	159 (2-N-) 190 (*di*)	....				N,N′-Diformyl, 137
135	**3-Aminoacenaphthene**...........	81	....	192–3, al.	209, al.				221	N-Formyl, 151; Alc, FeCl₃ → bl.-vlt. col.
136	**2-Aminobenzyl alcohol**..........	82	270–80 part. d.	N-*mono*: 114	O-: 198–9	....		.....	110	Hydrochloride, 108
137	**1-Amino-1,3,4-triazole**	82–3	....	....	....				194–5	Hydrochloride, 153; N-Formyl, 117; Chloro-platinate, 230
138	**4-Chloro-3-methylaniline** (6-Chloro-*m*-toluidine)	83	....	91	119	....		......		2-Naphthylthiourea, 158
139	**2,6-Dibromoaniline**.............	83–4	262–4	210	....	....		123–4		....
140	**5-Chloro-2-methoxyaniline** (4-Chloro-*o*-anisidine)	84	....	104	77–8	....		194		
141	**4-Aminotriphenylmethane**	84, bz.	248¹²	168	198	....				N,N-Di-Me., 132
142	**1-Amino-3-iodonaphthalene** (3-Iodo-1-naphthylamine)	84	....	207	174					
143	**4-Aminobutyrophenone**	84		142	....					Hydrochloride, 178
144	**4,4′-Diaminodiphenyl disulfide**	85, yel.; 106	....	205 (*di*)	....					N,N′-Dicarbethoxy, 136–7
145	**7-Methylindole**	85	266	....	84				176	
146	**2-Amino-4-bromobenzaldehyde**	85	....							Oxime, 194; Phenylhydra-zone, 215
147	**2,2′-Diaminodiphenyl sulfide**.......	85–6	....	160 (*di*)	162–3 (*di*)	....		.....		....
148	**4-Aminoveratrol (3,4-Dimethoxy-aniline)**......................	85–6	174–6	133	177					Chloroplatinate, 227
149	**2-Aminophenanthrene**...........	85, pa. yel.		225	216					
150	**4-Aminobenzonitrile**	86		205	170				150	N-Formyl, 188–9; N-Propionyl, 169
151	**4-Iodo-2-methylaniline (5-Iodo-*o*-toluidine)**....................	87; 92	....	170; 162	184	....			189	N-Benzal, 55; Phenyl-urethane, 232
152	**4-Aminoacenaphthene**...........	87		175–6	196				190–200	......................·
153	**3-Aminoacetanilide**.............	87–9		191	....		241			
154	**3-Aminophenanthrene**...........	87.5		200–1	213					
155	**4-Amino-3-methyl-1-phenylpyrazole**	88	312	94–5 (hyd.); 120(anh.)	181	....			138	N-Formyl, 112 (anh.); 81 (hyd.); Chloroplatinate, 226
156	**2-Hydroxy-3-methylaniline** (3-Amino-*o*-cresol).............	89	....	N-*mono*: 78–9	....					N-Acetyl of Me. eth., 100
157	**3,4-Diaminotoluene**.............	89–90	265	210 (*di*); 95 (3-N); 131 (4-N)	263–4 (*di*)	178–9 (*di*)	4-*mono*: 140			
158	**2-Nitrophenylhydrazine**	90, red	....	140–1; *di*: 57–8	166	....		......		N-Formyl, 177
159	**2,2′-Diaminodibenzyl disulfide**	90–1	....	202–5 (*di*)	....					N,N′-Dipropionyl, 190–1
160	**4-Chloro-1,3-diaminobenzene**	91	....	242 (*di*)	178 (*di*)	......	215			....
161	**1-Amino-2,6-dimethylnaphthalene** (2,6-Dimethyl-1-naphthylamine)..	91		211	219–20					
162	**1-Amino-4-mercaptonaphthalene** (4-Amino-1-thionaphthol)	91–3	....	N-*mono*: 173	....					S-Me., 54
163	**2-Methyl-3-nitroaniline** (6-Nitro-*o*-toluidine)............	92; 97	....	158	168					
164	**2-Methyl-6-nitroaniline** (3-Nitro-*o*-toluidine)...........	92	305d.	158	167				.....	1-Naphthylthiourea, 171

*Derivative data given in order: m.p., crystal color, solvent from which crystallized.

No.	Name	Melting point, °C	Boiling point, °C	Acetamide	Benzamide	Benzene sulfon-amide	p-Toluene sulfon-amide	Phenyl thiourea	Picrate	Miscellaneous
165	3-Bromo-2-hydroxy-5-methylaniline (3-Amino-5-bromo-p-cresol)	93	....	N-mono: 129; di: 169	N-mono: 185; di: 166	N-mono: 157; di: 230	....	....	....	....
166	4-Chloro-2,6-dibromoaniline.......	93	....	226	194	....	....	....	....	....
167	2,2'-Diaminodiphenyl disulfide	93	....	156 (di)	....	....	....	....	141 (di)	....
168	1,18-Diaminooctadecane..........	93	....	....	150 (di)	....	....	....	....	Hydrochloride, >225
169	4,4'-Diaminodiphenylmethane	93	232[9]	236 (di); (228)	....	....	....	....	....	N,N'-Dibenzal, 130
170	3-Nitrophenylhydrazine	93, yel.	....	145; 150 (di)	151; 153 (di)	....	....	....	....	....
171	7-Aminoquinoline	93–4 (anh.); 73 (hyd.)	....	167	189	....	....	....	....	Chloroplatinate, 225
172	2-Aminodibenzfuran	94	....	178; 83 (di)	201	....	....	....	....	....
173	3-Aminoquinoline	94; 84	....	172; 167	....	....	....	....	210	....
174	7-Amino-2,4-dimethylquinoline	94–100	>300	212	....	....	....	....	215–7	....
175	Skatole (3-Methylindole)	95	267	68	....	....	....	....	170–1	Hydrochloride, 167–8; N-Propionyl, 45
176	2-Chloro-4,6-dibromoaniline.......	95	....	227	192	....	....	....	....	....
177	4-Hydroxybenzylamine	95	....	....	....	....	....	....	....	Hydrochloride, 195; N-Acetyl of Me. ether, 96
178	1-Amino-4,5-dimethyl-1,2,3-triazole......................	95	....	....	....	....	....	....	124–5	Hydrochloride, 131; Chloroplatinate, 215
179	1-Amino-8-hydroxynaphthalene (8-Hydroxy-1-naphthylamine; 8-Amino-1-naphthol)	95–7	....	N-mono: 181; N,O-di: 118	N-mono: 193; N,O-di: 206	....	N-mono: 189	....	163–4	N-Formyl, 140–50
180	5-Amino-2-methylpyridine (5-Amino-α-picoline)..........	96	....	126	111	....	....	....	201	Dihydrochloride, 215–8
181	4,4'-Diamino-3,3'-dimethyl-diphenyl sulfide	96	....	di: 220	di: 233	....	....	....	186 (di)	Dihydrochloride, 248–9
182	6,6'-Diamino-3,3'-dimethyl-diphenylmethane	96	....	226 (di); 152 (tetra)	....	....	....	....	199	Dihydrochloride, 248–9
183	N-Ethyl-4-nitroaniline...........	96	....	119	98	....	107	....	....	....
184	2,4-Di-iodoaniline	96	....	141; 171	181	....	....	....	....	....
185	1-Amino-8-nitronaphthalene (8-Nitro-1-naphthylamine).......	97, red	....	191	....	194	....	....	....	....
186	3-Aminobenzyl alcohol	97	....	N-mono: 106–7	N-mono: 115; N,O-di: 113–4	....	....	....	....	....
187	5-Bromo-2-methoxyaniline (4-Bromo-o-anisidine)..........	97–8	....	160	108	....	....	....	....	....
188	2-Amino-4-methylpyridine (2-Amino-γ-picoline)..........	98	....	102–3	114; 182–3 (di)	....	....	....	227	....
189	2-Aminophenacyl alcohol	98	....	N-mono: 141	N,O-di: 167	....	....	....	....	Phenylhydrazone, 198
190	1-Amino-4-chloronaphthalene (4-Chloro-1-naphthylamine)	98	....	186	....	....	....	....	....	....
191	1,2-Diaminonaphthalene	98	....	234 (di)	291 (di)	1-mono: 215	....	....	....	....
192	4-Amino-4'-methylbiphenyl	99	190[18]	221	....	....	....	....	....	Hydrochloride, 280–3
193	3-Aminoacetophenone...........	99	....	128–9	....	....	130	....	....	Semicarbazone, 196
194	2,4-Diaminotoluene.............	99	....	N,N'-di: 224	224 (di)	2-mono: 138; 2,4-di: 192	4-mono: 160; 2,4-di: 192–3	....	....	....

*Derivative data given in order; m.p., crystal color, solvent from which crystallized.

TABLE XVIII. ORGANIC DERIVATIVES OF AMINES

TABLE XVIII. ORGANIC DERIVATIVES OF AMINES
1. Primary and secondary amines c) Solids (Listed in order of increasing m.p.)* (Continued)

No.	Name	Melting point, °C	Boiling point, °C	Acetamide	Benzamide	Benzene sulfonamide	p-Toluene sulfonamide	Phenyl thiourea	Picrate	Miscellaneous
195	4-Amino-3,2'-dimethylazobenzene	100, yel.	....	185 (*mono*); 65 (75) (*di*), lgr.	....	....	....	......	......	N-Chloroacetyl, 171–2
196	1,2-Diaminobenzene (*o*-Phenylenediamine)	102	256–8	185 (*di*)	301 (*di*)	185	260 (*di*)	......	208	....
197	1-Amino-4-bromonaphthalene (4-Bromo-1-naphthylamine)	102; 95	....	193	....					N-Formyl, 172; 1,3,5-Trinitrobenzene add. comp., 196
198	2-Naphthyl 4-tolyl amine	103	....	85	139	....	....	......	......	....
199	3,4-Diaminobiphenyl	103	....	3-*mono*: 211; 4-*mono*: 155; 3,4-*di*: 163	3-*mono*: 186; 4-*mono*: 221; 3,4-*di*: 248	....				
200	6,6'-Diamino-3,3'-dimethyldiphenyl sulfide	103–4	....	165 (*di*)	185 (*di*)				179 (*di*)	
201	2-Amino-8-nitronaphthalene (8-Nitro-2-naphthylamine)	104, red	....	196	162	....	....			
202	Piperazine	104	140	*mono*: 52; *di*: 144	*mono*: 75; *di*: 196	282 (*di*)	173	......	280	....
203	4,4'-Diaminodibenzyl sulfide	104–5	....	188 (*di*)	224 (*di*)					
204	3-Amino-4,4'-dimethylbiphenyl	104–5	....	156–7	160–1					Hydrochloride, abt. 230
205	1,3-Diamino-4,6-dimethylbenzene (4,6-Diamino-*m*-xylene)	105	....	1-*mono*: 165; *di*: 295	258–9 (*di*)	......	221 (*di*)			N,N'-Diformyl, 182–3
206	2-Bromo-4-nitroaniline	105, yel.	....	129	160	....				N-Me., 118
207	Triphenylmethylamine	105	....	207–8	160–2					N-Benzyl, 110
208	4-Aminophenanthrene	105	....	190	224				216	
209	2-Aminobenzophenone	105–6, pa. yel.	....	72; 89	80	....				Oxime (alkali-stable), 156; (acid stable), 127; N-Propionyl, 78
210	3-Amino-6-phenylpyridine	105–6	....	148–9	201	....				
211	3-Amino-4-methylpyridine (3-Amino-γ-picoline)	106	260	84	81	....'...	....	......	179–80	Chloroplatinate, 227; Hydrochloride, 180
212	4-Bromophenylhydrazine	106	....	....	....					Acetophenone deriv., 112
213	4-Aminoacetophenone	106	294	167	205	128	203			Semicarbazone, 250; Oxime, 148
214	2,4-Diaminopyridine	107	....	....	191–2 (*di*)	....				Chloroplatinate, 224
215	9-Phenanthrylmethylamine	107	....	182–5	167	....			241	N-Benzal, 104
216	2-Methyl-5-nitroaniline (4-Nitro-*o*-toluidine)	107	....	151	....	172				N-Formyl, 178–9; N-4-Nitrobenzoyl, 214
217	2,5-Diaminopyridine	107–10	....	290 (*di*)	230 (*di*)	....				
218	2-Naphthyl phenyl amine	108	....	93	148; 136	....				
219	2-(4-Aminophenyl) ethyl alcohol	108	....	105	O-*mono*: 59–60; N,O-*di*: 136	93				Hydrochloride, 171
220	5-Aminoacenaphthene	108	....	238 (*mono*); 122 (*di*)	210; 199				190–200	N-Formyl, 172; FeCl$_3$ → bl. col.
221	4,4'-Diamino-2,2'-dimethylbiphenyl (*m*-Tolidine)	108–9	....	281 (275) (*di*)	....				225	N,N'-Dibenzal, 172–3
222	4-Aminoantipyrine (4-Amino-2,3-dimethyl-1-phenylpyrazolone-5)	109, yel.	....	199	....				144	
223	3-Amino-4-methylbenzophenone	109	....	108	....					Hydrobromide, 130, dil. HBr
224	2-Aminobenzamide	109–11	....	177	214–5	....				
225	1,4-Diamino-2-iodobenzene (2-Iodo-*p*-phenylenediamine)	110.5	....	211	254	....				

*Derivative data given in order: m.p., crystal color, solvent from which crystallized.

No.	Name	Melting point, °C	Boiling point, °C	Acetamide	Benzamide	Benzene sulfon-amide	p-Toluene sulfon-amide	Phenyl thiourea	Picrate	Miscellaneous
226	5-Aminoquinoline	110	310	178	....	....	203–4	......	......	
227	4-Amino-2-nitrostilbene	110–1, dk. red	....	192–3	....	....	....	......	......	Hydrochloride, 223
228	3-Aminocamphor	110–5	....	121	141	....	....	......	191	Oxime, 145; N-Formyl, 87
229	4-Bromo-2-nitroaniline	111, or.	....	104	137–8	....	....	......	......	N-Me., 102; N-Et., 91
230	3'-Amino-4-methylbenzophenone	111	....	139						Oxime, 146; Hydrochloride, 198
231	4-Amino-3-methylbenzophenone	112, pa. yel.	....	175	158					N-Propionyl, 128
232	2-Aminonaphthalene (β-Naphthylamine)	112		132	162	102	133	129	195	....
233	β-Aminopropiophenone (1-Amino-ethyl phenyl ketone)	112–4 (unst.)	....	90–1	104–5	......	....	......	164–5	Hydrochloride, 187; Chloroplatinate, 205
234	2-Hydroxybenzylaniline	113	....	93						Hydrochloride, 131; Chloroplatinate, 184
235	4-Ethoxy-2-nitroaniline (3-Nitro-p-phenetidine)	113; 108, red	....	104	....	72	94	......	......	
236	5-Bromo-2-hydroxy-3-methyl-aniline	113	....	N-mono: 119; di: 200	N-mono: 195	....	....			
237	3-Nitroaniline	114	....	mono: 155; di: 76	155; 150 (di)	136	138	160	143	....
238	6-Aminoquinoline	114 (anh.)	....	mono: 138; di: 75	169		193	......		Methiodide, 199
239	cis-2,5-Dimethylpiperazine	114	162	....	152 (di)	....	146–7 (di)	......		N,N'-Dinitroso, 95
240	3-Amino-2-phenylquinoline	115–6	223³	mono: 124; di: 173	179–80	......			194–5	Methiodide, 238; Ethiodide, 202
241	5-Amino-3-methyl-1-phenylpyrazole	116	333	110	....				160–2	Hydrochloride, 199–200
242	4-Aminotriphenylcarbinol	116	....	176						N,N-Dimethyl, 92–3
243	5-Bromo-2-hydroxy-4-methyl-aniline	116	....	N-mono: 199; di: 188	N-mono: 223	....	....			
244	4-Chloro-2-nitroaniline	116–7, yel.	....	104	....		110			
245	4-Methyl-2-nitroaniline (3-Nitro-p-toluidine)	117	....	99	148	102	146	......		Hydrochloride, 170–1
246	5-Amino-2-methylquinoline (5-Aminoquinaldine)	117–8 (anh.), grnsh.	....	205						N-Cinnamoyl, 257
247	trans-2,5-Dimethylpiperazine	118	162	....	228–9 (di)	....	225 (di)	......		N,N'-Dinitroso, 174
248	2,3,4,6-Tetrabromoaniline	118	....	228–9						1,3,5-Trinitrobenzene add. comp., 108
249	2-Methoxy-5-nitroaniline (4-Nitro-o-anisidine)	118, or.-red	....	175–6	160–1		128			N-Me., 87
250	4-Hydroxypyrazole	118	....	....	109 (di)				129	....
251	2-Amino-5,4'-dimethylazobenzene	118–9, or.-red	....	157	135					N-Carbethoxy, 94
252	1-Amino-5-nitronaphthalene (5-Nitro-1-naphthylamine)	119, red	....	220	....	183				N-Formyl, 199
253	5-Hydroxy-2,4,6-tribromoaniline	119	....	O,N,N-tri: 136	....		146–7			
254	1,4-Diaminonaphthalene	120	....	303 (di)	mono: 186; di: 280	....	mono: 187–8			
255	1,9-Diaminofluorene	120	....	293 (di)	abt. 310 (di)				205	....
256	2-(ω-Aminoethyl)-indole (2-(2-Indolyl)-ethylamine)	120	....	....	173–4	....				N-Benzal, 122
257	2,2'-Diamino-4,4'-dimethylbiphenyl	120	....	189 (di)	170 (di)	......	....	......		N,N'-Diformyl, 185

*Derivative data given in order: m.p., crystal color, solvent from which crystallized.

TABLE XVIII. ORGANIC DERIVATIVES OF AMINES

1. Primary and secondary amines c) Solids (Listed in order of increasing m.p.)* (Continued)

No.	Name	Melting point, °C	Boiling point, °C	Acetamide	Benzamide	Benzene sulfon-amide	p-Toluene sulfon-amide	Phenyl thiourea	Picrate	Miscellaneous
258	2,6-Diaminopyridine	121	. . . .	203 (di)	176 (di)	. . . .	. . . .	. . . .	240	. . . .
259	cis-4,4'-Diaminostilbene	121, pa. yel.	. . . .	172 (di)	253 (di)	. . . .	. . . .	. . . .	. . . . . .	. . . .
260	4-Aminobenzhydrol	121	. . . .	. . . .	145	. . . .	. . . .	. . . .	. . . . . .	Hydrochloride, 270–3
261	3-Amino-6-hydroxyacetophenone . . .	121; 110 yel.	. . . .	N-mono: 165; di: 174	. . . .	. . . .	. . . .	. . . .	. . . . . .	Oxime, 201; Et. eth., 60
262	2-Aminotriphenylcarbinol	121	. . . .	192	. . . .	. . . .	. . . .	. . . .	122–3	. . . .
263	3-Hydroxyaniline (3-Amino-phenol)	122	. . . .	mono: 148; di: 101	N-mono: 174 198; 204	. . . .	157	156	. . . . . .	. . . .
264	2,4,6-Tribromoaniline	122; 119	. . . .	232	. . . .	. . . .	. . . .	. . . .	. . . . . .	N-Formyl, 222
265	4-Amino-2-hydroxyacetophenone . . .	122–3	. . . .	N-mono: 91	. . . .	. . . .	. . . .	. . . .	. . . . . .	N,N-Di-Me., 120
266	1-Aminoisoquinoline	122–3	. . . .	. . . .	. . . .	. . . .	. . . .	. . . .	290–1	Hydrochloride, 233; Chloroplatinate, >300
267	3,4,5-Tribromoaniline	123	. . . .	255–6	210	. . . .	. . . .	. . . .	. . . . . .	. . . .
268	4,4'-Diamino-2,2'-dimethyl-diphenylmethane	123	. . . .	228 (di)	. . . .	. . . .	. . . .	. . . .	216	. . . .
269	cis-2,2'-Diaminostilbene	123; 107, red	. . . .	214–5 (di)	. . . .	. . . .	. . . .	. . . .	155–6	Hydrochloride, 230
270	2,4-Dimethyl-5-nitroaniline	123	. . . .	159	200	149	192	. . . .	. . . . . .	. . . .
271	2-Amino-4-nitrobenzaldehyde	124	. . . .	. . . .	. . . .	. . . .	. . . .	. . . .	. . . . . .	Oxime, 193; Semicarba-zone, 390; Anil, 147, red
272	4-Aminobenzophenone	124	. . . .	153	152	. . . .	. . . .	. . . .	. . . . . .	N-Propionyl, 139
273	7-Amino-8-hydroxyquinoline	124, br.	. . . .	N-mono: 177	. . . .	. . . .	. . . .	. . . .	205	. . . .
274	2-Amino-5-nitrobiphenyl	125, yel.	. . . .	133	. . . .	. . . .	169	. . . .	. . . . . .	. . . .
275	3-Hydroxy-4-methoxyaniline (4-Aminoguaiacol)	125–7	. . . .	N-mono: 116–9	N,O-di: 162–4	. . . .	. . . .	. . . .	. . . . . .	. . . .
276	2-Amino-1-nitronaphthalene (1-Nitro-2-naphthylamine)	126, or.-yel.	. . . .	123	168	156	160	. . . .	. . . . . .	. . . .
277	4-Aminoazobenzene	126	. . . .	146	211	. . . .	. . . .	. . . .	. . . . . .	N-propionyl, 170
278	Hydrazobenzene	126–7	. . . .	mono: 159; di: 105	mono: 126, bz.; di: 162	. . . .	. . . .	. . . .	. . . . . .	. . . .
279	5-Chloro-2-nitroaniline	126.5, gold-yel.	. . . .	121	. . . .	. . . .	. . . .	. . . .	. . . . . .	N-Me., 107; N,N-Di-Me., 49
280	Benzidine	127	. . . .	317 (di); 199 (mono)	352 (di); 203–5 (mono)	232 (di)	243 (di)	. . . .	. . . . . .	. . . .
281	2-Aminopyrimidine	127–8	. . . .	. . . .	. . . .	. . . .	. . . .	. . . .	237–8	Chloroplatinate, 216; Hydrochloride, 196
282	2-Amino-6-bromonaphthalene (6-Bromo-2-naphthylamine)	128	. . . .	192	218	. . . .	. . . .	. . . .	. . . . . .	. . . .
283	5-Bromo-2-hydroxyaniline (2-Amino-4-bromophenol)	128; 88	. . . .	177–9	. . . .	. . . .	. . . .	. . . .	. . . . . .	Me. eth., 97–8
284	5-Aminoisoquinoline	128	. . . .	. . . .	. . . .	. . . .	. . . .	. . . .	>200	Methiodide, 228; Ethiodide, 216
285	2-Aminovanillin (2-Amino-4-hydroxy-3-methoxybenzalde-hyde)	128–9	. . . .	97	. . . .	. . . .	. . . .	. . . .	. . . . . .	Oxime, 151–2; Phenyl-hydrazone, 165
286	2-Amino-3,7-dimethylnaphthalene (3,7-Dimethyl-2-naphthylamine) .	129; 134	. . . .	231	. . . .	. . . .	. . . .	. . . .	. . . . . .	Hydrochloride, 275
287	4-Methoxy-2-nitroaniline (3-Nitro-p-anisidine)	129; 123, dk. red	. . . .	117, yel.	140	. . . .	. . . .	. . . .	. . . . . .	N-4-Nitrobenzoyl, 204
288	2-Aminotriphenylmethane	129	. . . .	154–5	. . . .	. . . .	. . . .	. . . .	. . . . . .	N-Me., 130–2

*Derivative data given in order: m.p., crystal color, solvent from which crystallized.

TABLE XVIII. ORGANIC DERIVATIVES OF AMINES
1. Primary and secondary amines c) Solids (Listed in order of increasing m.p.)* (Continued)

No.	Name	Melting point, °C	Boiling point, °C	Acetamide	Benzamide	Benzene sulfonamide	p-Toluene sulfonamide	Phenyl thiourea	Picrate	Miscellaneous
289	**2-Aminoquinoline**	129							255–6	Methiodide, 247; Ethiodide, 232; 1,3,5-Trinitrobenzene add. comp., 186, red.
290	**4,4'-Diamino-3,3'-dimethylbiphenyl** (o-Tolidine)	129		mono: 103 (hyd.); di: 315; tetra: 211	198 (mono); 265 (di)					N,N'-Diformyl, 254; 1-Naphthylthiourea, 167; 3-Nitrophthalimide, 185
291	**2,4-Diaminodiphenylamine**	130		188 (di)	2-mono: 213					
292	**3-Aminocoumarin**	130, yel.		201–2	173					
293	**2-Methyl-4-nitroaniline** (5-Nitro-o-toluidine)	130		202		158	174			1-Naphthylthiourea, 165
294	**2-Amino-4-chloropyridine**	130–1		115–6	mono: 120; di: 165				243	
295	**4-Hydroxy-3-nitroaniline**	131; 127, red		N-mono: 157–8						N-Me., 113; Et. eth., 40
296	**4-Bromo-3-nitroaniline**	131–2		146						N,N'-Di-Me., 72
297	**2-Aminobenzothiazole**	132; 129		186	186				256	
298	**2-Amino-4-methylquinoline**	133	320						abt. 250	N-Phenyl, 129; Chloroplatinate, 230
299	**2,2'-Diaminobenzophenone**	133, pa. yel.		168 (154) (di)					164	
300	**2,2'-Diaminoazobenzene**	134, red		271 (di)						
301	**3,5-Dimethyl-2-hydroxyaniline**	134–5		N-mono: 96	154 (O,N-di)					N-Formyl, 68
302	**2-Hydroxy-5-methylaniline** (4-Hydroxy-m-toluidine)	135		N-mono: 160; N,O-di: 145	N-mono: 191; N,O-di: 190					N-Propionyl, 95–6; N,O-Dipropionyl, 91–2
303	**3-Methyl-4-nitroaniline** (6-Nitro-m-toluidine)	135		102						2-Naphthylthiourea, 159
304	**2-Amino-3-methylnaphthalene** (3-Methyl-2-naphthylamine)	135		181–2	190					
305	**2-Aminoacenaphthene**	135							260, yel., eth.	Hydrochloride, 270
306	**DL-2,2'-Diamino-6,6'-dimethylbiphenyl**	136		205 (di)	182 (di)		162–3 (di)			
307	**1,4-Diamino-2-nitrobenzene** (2-Nitro-p-phenylenediamine)	137, blk.		1-mono: 162; 4-mono: 189; di: 186	4-mono: 236					1,4-Di-4-nitrobenzoyl, >305
308	**9-Aminophenanthrene**	137–8; 104		207–8	199				190	
309	**4,4'-Diamino-3,3'-dimethoxybiphenyl** (Dianisidine)	137–8		242 (di)	236 (di)				225 (di)	
310	**3,5-Dimethyl-4-hydroxyaniline** (5-Amino-2-hydroxy-m-xylene)	137–8		160 (di)						Me. eth., 66
311	**2,6-Dinitroaniline**	138		197						
312	**2-(4-Aminophenyl)-quinoline**	138		mono: 189; di: 154	234					Methiodide, 220; N-Me., 82; N-Formyl, 160
313	**2-Amino-3,6-dimethylnaphthalene** (3,6-Dimethyl-2-naphthylamine)	139		207						Hydrochloride, 283
314	**2-Methoxy-4-nitroaniline** (5-Nitro-o-anisidine)	139–40, pa. yel.		153–4	150	181	175; 170			
315	**4-Aminopropiophenone**	140		161	190					Oxime, 153
316	**4-Amino-3-nitrobenzophenone**	140; 135, yel.			154–5					N,N-Di-Me., 116; N-Et., 100

*Derivative data given in order: m.p., crystal color, solvent from which crystallized.

No.	Name	Melting point, °C	Boiling point, °C	Acetamide	Benzamide	Benzene sulfonamide	*p*-Toluene sulfonamide	Phenyl thiourea	Picrate	Miscellaneous
317	2-Amino-4-methyldiphenylamine . . .	140	. . . .	. . . .	161	. . . .	. . . .	. . . . . .	. . . . .	Hydrochloride, 200
318	1,4-Diaminobenzene (*p*-Phenylene-diamine)	140; 147	267	*mono*: 162–3; *di*: 304	*mono*: 128; *di*: 300	247 (*di*)	266 (*di*)	. . . . . .	. . . . .	3,5-Dinitrobenzoate, 178
319	2-Hydroxy-5-nitroaniline	142–3 (anh.); 80–90 (hyd.), or.	. . . .	220	>200	. . . .	122 (O-)	. . . . . .	. . . . .	Me. eth., 118; Et. eth., 99
320	2-Amino-4-nitrostilbene	142–3, red	. . . .	N-*mono*: 220, yel., al.	. . . .	. . . .	. . . .	. . . . . .	. . . . .	Hydrochloride, 219
321	5-Amino-8-hydroxyquinoline	143	. . . .	N-*mono*: 221–2, al.; N,O-*di*: 206–7, al.	205 (O, N-*di*)	. . . .	. . . .	. . . . . .	. . . . .	Me. eth., 156, yel.
322	1-Amino-2-nitronaphthalene (2-Nitro-1-naphthylamine)	144, red-yel.	. . . .	199	175	. . . .	. . . .	. . . . . .	. . . . .	N-Et., 77, red
323	2-Amino-5-nitronaphthalene (5-Nitro-2-naphthylamine)	144, red	. . . .	186	182	. . . .	. . . .	. . . . . .	. . . . .	
324	5-Hydroxy-2-methylaniline (4-Hydroxy-*o*-toluidine)	144	. . . .	N-*mono*: 178; N,O-*di*: 128	. . . .	. . . .	183 (N-)	. . . . . .	. . . . .	
325	2,5-Dimethyl-4-nitroaniline (5-Nitro-*p*-2-xylidine)	144–5	. . . .	168–9	. . . .	162	185	. . . . . .	. . . . .	
326	4-Amino-4′-bromobiphenyl	145	. . . .	247	. . . .	. . . .	174	. . . . . .	. . . . .	
327	1-Aminophenanthrene	146	. . . .	220	. . . .	. . . .	. . . .	. . . . . .	204	
328	3,5-Dihydroxyaniline (5-Amino-resorcinol)	146–52	. . . .	119–21 (*tri*)	. . . .	. . . .	. . . .	. . . . . .	. . . . .	Di-Me. eth., 46; Picrate of Di-Me. eth., 167–70
329	4-Amino-2-chlorobenzaldehyde	147, yel.	. . . .	152	. . . .	. . . .	. . . .	. . . . . .	. . . . .	N,N′-Di-Me., 82; N-Et., 101
330	4-Nitroaniline	147–8, yel.	. . . .	215	199, 203 (*di*)	139	191	. . . . . .	100	1-Naphthylthiourea, 187
331	5-Amino-3-methyl-1,2,4-triazole . . .	148	. . . .	>270	285–90	. . . .	. . . .	. . . . . .	225	
332	2,2′-Dihydroxyhydrazobenzene	148	. . . .	. . . .	186 (*di*)	. . . .	. . . .	. . . . . .	. . . . .	Di-Me. eth., 102
333	7-Amino-2-methylquinoline	148 (anh.)	. . . .	. . . .	172–3	. . . .	. . . .	. . . . . .	213–4	
334	4,4′-Diamino-2,2′-dimethyl azoxybenzene	148, gold-yel.	. . . .	281 (*di*)	290	. . . .	. . . .	. . . . . .	. . . . .	
335	4-Bromo-3-hydroxyaniline	150	. . . .	210–2	. . . .	. . . .	135–6 (O-)	. . . . . .	. . . . .	
336	3-Amino-1-phenyl-1,2,4-triazole . . .	150	. . . .	3-*mono*: 168; 3,3-*di*: 118	. . . .	. . . .	. . . .	. . . . . .	220	Hydrochloride, 187
337	4-Aminochalcone (4-Aminobenzal-acetophenone).	151, golden	. . . .	179	. . . .	. . . .	. . . .	. . . . . .	. . . . .	Oxime, 139
338	4-Aminopyrimidine	151–2	. . . .	202	. . . .	. . . .	. . . .	. . . . . .	226	N-Me., 74–5; N-Phenyl, 142–3
339	Pentamethylaniline	151–2	277–8	213	. . . .	. . . .	. . . .	. . . . . .	. . . . .	N-Formyl, 217; N-Me., 60; N,N-Di-Me., 53–4
340	N-Methyl-4-nitroaniline	152	. . . .	153	112	121	. . . .	. . . . . .	. . . . .	. . . .
341	5-Bromo-2-nitroaniline	152, red-yel.	. . . .	139	. . . .	. . . .	. . . .	. . . . . .	. . . . .	N-Me., 115; N-Et., 90
342	3-Chloro-4-hydroxyaniline	153	. . . .	N-*mono*: 144; *di*: 124	. . . .	. . . .	116–7 (O-)	. . . . . .	. . . . .	Me. eth., 62

*Derivative data given in order: m.p., crystal color, solvent from which crystallized.

No.	Name	Melting point, °C	Boiling point, °C	Acetamide	Benzamide	Benzene sulfon-amide	p-Toluene sulfon-amide	Phenyl thiourea	Picrate	Miscellaneous
343	4-Chloro-2-hydroxyaniline.......	154	....	140 (di)	....	....	....	....	....	Hydrochloride, 226-7; Benzyl eth., 46-7; Me. eth., 52
344	4-Aminoquinoline (γ-Amino-quinoline)	154 (anh.)	....	178	....	....	....	....	274	Methiodide, 224; Ethio-dide, 232
345	3-Aminobenzopyrazole	154		177-8 (di)	182 (di)	....	....	....	....	
346	2,5-Dihydroxy-4-nitroaniline (2-Amino-5-nitrohydroquinone)..	154, red	....	mono: 226; di: 183-4		....	....	....	....	Di-Me. eth., 158
347	4-Hydroxy-2-nitroaniline	154, red		N-mono: 218; N,O-di: 146		....	....	....	....	Me. eth., 129; 123
348	3-Aminotriphenylcarbinol	155	....	164	....	....	....	....	....	O,N,N-Trimethyl, 81
349	L-2,2'-Diamino-6,6-dimethyl-biphenyl	156	....	205 (di)	172 (di)	....	....	....	....	
350	3,3'-Diaminoazobenzene..........	156; 140, or.-yel.	....	272 (di)	286 (di)	....	....	....	....	
351	5-Amino-1-phenyl-1,2,4-triazole ...	157		....	....	....	....	....	175	Chloroplatinate, 197
352	4-Nitrophenylhydrazine	157, or.-red		205	193	....	....	....	119-20	
353	4-Aminopyridine (γ-Amino-pyridine)	158	....	150 (anh.)	202	....	....	....	215-6	
354	4,4'-Diamino-3,3'-dimethyl-diphenylmethane	158-9	....	224 (di); 119 (tetra)	215 (di)	....	....	....	192-3	
355	2-Amino-8-nitroquinoline	159		211	166	....	....	....	257	
356	3-Amino-1,2,4-triazole	159		....	....	....	....	....	231	Hydrochloride, 153
357	2-Amino-4'-nitrobiphenyl	159, or.-red		199	....	....	163	....	....	
358	3-Amino-2-methylquinoline	159-60	270	165	161	....	....	....	235	N-Formyl, 163
359	1-Amino-4-hydroxy-3-nitro-naphthalene (4-Hydroxy-3-nitro-1-naphthylamine)	160, ma-roon	....	250; 238	330	....	....	....	....	
360	2,4-Dihydroxy-5-nitroaniline	160-1, red		N-mono: 261; tri: 176		....	....	....	....	Di-Me. eth., 136-7
361	1,3-Diamino-4-nitrobenzene (4-Nitro-m-phenylenediamine) ..	161; 157, yel.-red	....	1-mono: 200; 1,3-di: 246	222 (di)	....	169 (di)	....	....	N,N,N',N'-Tetramethyl, 81
362	3-Hydroxy-4-methylaniline (2-Hydroxy-p-toluidine)	161		N-mono: 225; di: 132-3		....	111-2 (O-)	....	....	N-Chloroacetyl, 154-5
363	2-Hydroxy-4-methylaniline (3-Hydroxy-p-toluidine)	162	....	N-mono: 171	N-mono: 169; N,O-di: 162	....	....	....	....	
364	4-Aminoacetanilide..............	162	....	304	....	....	....	....	....	
365	3-Hydroxy-4-nitroaniline	162; 158, or.-yel.		N-mono: 221; N,O-di: 149	....	....	....	....	....	Me. eth., 169; 161
366	2,4-Dimethyl-6-hydroxyaniline	163	....	mono: 186-7; di: 87-8	N-mono: 211; N,O-di: 148-9	....	....	....	....	Me. eth., 150
367	2-Aminofluorenone	163, vlt.-red		227	....	....	....	....	....	Hydrazone, 209; N-Car-bethoxy, 167-8
368	1-(2-Aminoethyl)-4-hydroxybenzene (4-(2-Aminoethyl)phenol; Tyramine)	164	....	....	N-mono: 162; di: 172	....	....	....	206	....

*Derivative data given in order: m.p., crystal color, solvent from which crystallized.

TABLE XVIII. ORGANIC DERIVATIVES OF AMINES

1. Primary and secondary amines c) Solids (Listed in order of increasing m.p.)* (Continued)

No.	Name	Melting point, °C	Boiling point, °C	Acetamide	Benzamide	Benzene sulfonamide	p-Toluene sulfonamide	Phenyl thiourea	Picrate	Miscellaneous
369	4-Aminophenacyl alcohol	165, yel.	....	N-*mono*: 176–7; O-*mono*: 130; N,O-*di*: 162	188 (O-)	....	....	....	......	Phenylhydrazone, 199
370	3-Bromo-4-hydroxyaniline	165; 155, pa. br.	....	N-*mono*: 157	N-*mono*: 184–5; *di*: 192	....	....	....		....
371	2,7-Diaminonaphthalene	166	....	261 (*di*)	267 (*di*)	....			210 (*di*)	....
372	4-Amino-4′-iodobiphenyl	166–7; 159, yel.	....			....		....	......	N-Benzal, 209; N-Piperonylidine, 150–1; Hydrochloride, 295
373	4-Amino-3-nitrobiphenyl	167, red	....	132	143	....		....	......	N-Me., 112
374	1,2-Diamino-4-hydroxybenzene (3,4-Diaminophenol)	167–8	....	1,2-*di*: 205–7	1,2-*di*: 203; *tri*: 225	....				....
375	4,4′-Diaminotriphenylcarbinol	168 s.h.; 175 r.h.	....	4,4′-*di*: 267	....					Me. eth., 162
376	4-Amino-2-phenylquinoline	168	....	108, 117 (*di*)	182	....				Methiodide, 274; Ethiodide, 244; N-Formyl, 275
377	2,6-Dinitro-4-methylaniline	168; 172	....	195	186	....				....
378	4-Amino-2-methylquinoline	168	333	....	....				197–9	N-Phenyl, 150; Chloroplatinate, 223
379	6-Aminocoumarin	168–70, yel.	....	216–7	173	159	....			N-Formyl, 175–6
380	Picramic acid (3,5-Dinitro-2-hydroxyaniline)	169, red	....	N-*mono*: 201; O-: 193	N-*mono*: 300; O-: 218	....	191	....		
381	3,3′-Diaminobenzophenone	173, yel.	....	226–7 (*di*)	....	....				Oxime, 177–8
382	4,5-Dimethyl-2-hydroxyaniline	173–5	....	N-*mono*: 191; N,O-*di*: 157	N-*mono*: 195–6; N,O-*di*: 152–3	....				....
383	2-Hydroxyaniline	174	....	*mono*: 209; 201; *di*: 124	N-*mono*: 165; O-: 185	141	146	......	......	
384	4-Hydroxy-3-methylaniline	175	....	N-*mono*: 179	N,O-*di*: 194	....	109–10 (O-)			
385	6-Amino-5,7-dimethylquinoline	175	>300	212	....			....	182	....
386	4,4′-Diaminodiphenyl sulfone	175–6	....	286 (*di*)						N,N′-Di-Me., 179–80; N,N-Tetramethyl, 260
387	*trans*-2,2′-Diaminostilbene	176; 168, gold-yel.	....	304 (*di*)					209	Hydrochloride, 267
388	6-Amino-5-nitroquinoline	178; 174, yel.	....				168		270	....
389	6-Aminothymol	178–9	....	N-*mono*: 74; *tri*: 91	N-*mono*: 178–9; N,O-*di*: 166–7	....				Oxid. → thymoquinone, 45.5
390	4-Hydroxy-2-methylaniline (5-Hydroxy-o-toluidine)	179	....	N-*mono*: 130	92 (O-)	....				Hydrochloride, 215
391	2,4-Dinitroaniline	180; 188	....	120	202; 220	....	219	......	......	....
392	1-Amino-4-chloroanthraquinone	180, red	....	203–4						N,N-Di-Me., 172
393	2,6-Dimethyl-4-hydroxyaniline	181	....	178–80						N-Benzal, 104–5; N-Me., 43; N-Et., 161–2
394	4-Amino-2,6-dimethylpyrimidine	183	....	....					214	N-Phenyl, 104
395	4-Aminobenzamide	183, yel.	....	275	....	....				N-Chloroacetyl, 241–3

*Derivative data given in order: m.p., crystal color, solvent from which crystallized.

No.	Name	Melting point, °C	Boiling point, °C	Acetamide	Benzamide	Benzene sulfonamide	p-Toluene sulfonamide	Phenyl thiourea	Picrate	Miscellaneous
396	4-Hydroxyaniline (4-Amino-phenol)	184; 186	....	150 (di); 168 (mono)	N-mono: 216–7; N,O-di: 234	125	N-mono: 252–4; O-: 142	150	......	3,5-Dinitrobenzoate, 178
397	1-Amino-3-hydroxynaphthalene (3-Hydroxy-1-naphthylamine) ...	185	....	N-mono: 179	309 (O,N-di)	....	137 (O-)		......	
398	5-Hydroxy-2-nitroaniline	185–6, or.	....	N-mono: 266	....				.....	Me. eth., 131, br.; Et. eth., 105, yel.
399	6,6′-Diamino-3,3′-dimethyl triphenylmethane.............	185–6	430, part. d.	217 (di)	196 (di)	....				
400	4-Amino-2,6-dimethylpyridine	186	246	113	....				194–5	Chloroplatinate, 250
401	4-Amino-4′-hydroxyazobenzene....	186	....	N-mono: 203; di: 236–7						N,N-Di-Me., 203
402	4-Amino-4′-methylbenzophenone ...	186–7	....	155	....				.....	Phenylhydrazone, 163
403	6-Amino-2-methylquinoline	187–8	....	168–9	....				.....	N-Cinnamoyl, 257
404	4-Amino-2,6-diethyl-5-methyl-pyrimidine (Cyanethine)	189	280 d.	59	....					
405	2,4,6-Trinitroaniline (Picramide)...	190	....	230	196	211				
406	1-Amino-6-hydroxynaphthalene (6-Hydroxy-1-naphthylamine) ...	190; 185	....	N-mono: 218; N,O-di: 187	N-mono: 152; N,O-di: 223	....			170	
407	4,4′-Diaminoazoxybenzene	190, yel.	....	275	....				.....	Sn + HCl → 1,4-Diamino-benzene, 147
408	4-Amino-3-nitrobenzaldehyde	191, yel.	....	155	....					Phenylhydrazone, 202; Oxime, 207; 4-Nitro-phenylhydrazone, 270–2
409	2-Amino-1,5-dinitronaphthalene (1,5-Dinitro-2-naphthylamine) ...	191	....	201	....		182			
410	DL-2,2′-Diamino-1,1′-dinaphthyl ..	193	....	235–6 (di)	235 (di)	....			185	
411	8-Amino-6-nitroquinoline	194, red	....	224	....				.....	Chloroplatinate, 180; Methiodide, 176
412	1-Amino-4-nitronaphthalene (4-Nitro-1-naphthylamine)	195	....	190	224	173; 158	185		......	
413	2-Amino-4,6-dimethylpyrimidine ...	197	....	....	....				230	Hydrochloride, 181; Chloroplatinate, 225; N-Me., 98
414	4,6-Diamino-2-methylquinoline	197	....	6-mono: 250	....					6-N-Cinnamoyl, 253–4; 4-N-Et., 195; 6-N-Et., 232
415	1,2-Diamino-4-nitrobenzene (4-Nitro-o-phenylenediamine)......	198, red	....	1-mono: 205; 2-mono: 195	235 (di)					N,N′-Di-Me., 172
416	2,4-Dinitrophenylhydrazine	199–200	....	197–8	206–7					
417	2-Hydroxy-5-methyl-4-nitroaniline .	200, yel.	....	N-mono: 242						Me. eth., 132; N-Acetyl of Me. eth., 156
418	4-Amino-3-nitropyridine..........	200, yel.	....						197–8	Hydrochloride, 258–9; Chloroplatinate, 256
419	4-Amino-4′-nitrobiphenyl	200, red	....	264; 240, yel.	....	174, yel.				
420	2-Amino-5-nitrobenzaldehyde	200, yel.	....	160–1	....		181–2		.....	Oxime, 203; N,N-Di-Me. 105, yel.
421	2-Amino-7-hydroxynaphthalene (7-Hydroxy-2-naphthylamine) ...	201	....	N-mono: 232; N,O-di: 156	N-mono: 243–6; N,O-di: 181	....				
422	2-Amino-1-bromo-3-methylanthra-quinone	202; 204	....	....	di: 243–4, pa. yel., al.	....				

*Derivative data given in order: m.p., crystal color, solvent from which crystallized.

No.	Name	Melting point, °C	Boiling point, °C	Acetamide	Benzamide	Benzene sulfonamide	p-Toluene sulfonamide	Phenyl thiourea	Picrate	Miscellaneous
423	4,4′-Diamino-1,1′-dinaphthyl	202	....	363 (di)	320 (di)	....	....	....	147	....
424	4,4′,4″-Triaminotriphenylmethane	203	....	201 (tri)	....	....	....	....		1,3,5-Trinitrobenzene add. comp., 140, blk.
425	Isatin	204	....	141		....	....	....		Oxime, 201–2
426	3-Amino-5-phenylacridine	204	....	256	246	....	....	....		....
427	2-Amino-1,4-naphthoquinone	204–5, or.-red	....	202	....	....	....	....		N-Phenyl, 191
428	1-Amino-2-methylanthraquinone	205, red	....	mono: 176–7; di: 203–6	....	....	218	....		....
429	4,4′,4″-Triaminotriphenylcarbinol (Pararosaniline)	205	....	192 (tri)	....	....	....	....		Me. eth., 135
430	1-Amino-7-hydroxynaphthalene (7-Hydroxy-1-naphthylamine)	205–7	....	N-mono: 165	N-mono: 208–9; N,O-di: 208	....				
431	4,6-Diaminoisophthaladehyde	208	....	270 (mono), 280 (di)	....	....				Dioxime, 220; Diphenylhydrazone, 337
432	N-4-Hydroxybenzylaniline	208	....	....	....	....				Me. eth., 65; Et. eth., 65
433	6-Aminobenzopyrazole	210	....	248 (6-N-), 184–5 (di)	....	....				Dihydrochloride, 230
434	4,4′-Diamino-2,5,2′,5′-tetramethyltriphenylmethane	210	....	217 (di)	250 (di)	....				
435	3-Indolylpyruvic acid	211, grey	....	....	....	....				4-Nitrophenylhydrazone, 153–4; Oxime, abt. 175
436	2-Amino-6-hydroxynaphthalene (6-Hydroxy-2-naphthylamine)	212–3	....	....	N,O-di: 228–30	....				Me. eth., 78; Et. eth., 91
437	5-Amino-1,4-dihydroxyanthraquinone (5-Aminoquinizarin)	212–3, br.-red	....	....	....	....				N-Phenyl, 223; Di-Me. eth., 242–3
438	2-Amino-4,5-dimethylpyrimidine	214–5	....	....	....	....			250	Chloroplatinate, 227
439	5-Bromo-4-hydroxy-2-methylaniline	215; 205	....	171–2 (di)	229 (di)	....				
440	4-Amino-4′-nitroazobenzene	216; 205	....	245	....	....				N-Me., 206–7, bl.
441	2-Hydroxy-6-nitroaniline	216, red	....	N-mono: 172	....	....	136 (O-)			Me. eth., 76
442	3,4-Diaminopyridine	218–9	....	....	222–3 (di)	....			235–7	Chloroplatinate, 231
443	1-Amino-5-chloroanthraquinone	219, red	....	219	218	....				
444	3,6-Dimethylcarbazole	219	....	129	....	....			192	N-Nitroso, 106
445	3-Aminothioxanthone	221–2, yel.-br.	....	236–7	....	....				Hydrochloride, 230
446	2-Amino-10-hydroxyphenanthrene	221	....	182 (O, N-di)	225 (O, N-di)	....				....
447	5-Amino-4-nitroacenaphthene	222, red	....	252	233	....				N-Formyl, 227
448	5-Chloro-4-hydroxy-2-methylaniline	223–5; 204–5	....	162 (di)	220 (di)	....				
449	2,6-Dimethylcarbazole	224	....	....	....	....			162	N-Nitroso, 113
450	2-Amino-1,8-dinitronaphthalene (1,8-Dinitro-2-naphthylamine)	226	....	238	....	....	221			
451	trans-4,4′-Diaminostilbene	231, yel.	....	353 (di)	352 (di)	....				
452	2-Amino-3-hydroxynaphthalene (3-Hydroxy-2-naphthylamine)	234	....	188 (O,N-di)	N-mono: 233–5	....				
453	1-Amino-2,4-dinitronaphthalene (2,4-Dinitro-1-naphthylamine)	242	....	259	252	....	166			
454	2,5-Dimethyl-4-hydroxyaniline	242	....	177–9	....	....				Et. eth., 69–70
455	1-Amino-3-bromoanthraquinone	243, red	....	214	....	....	227			
456	4,4′-Diaminobenzophenone	244	....	237 (di)	....	....				Phenylhydrazone, 240
457	Carbazole	246	....	69	98	....			185	
458	1-Amino-2-hydroxyanthraquinone	250, br.	....	N-mono: 170	....	....				Et. eth., 182, red

*Derivative data given in order: m.p., crystal color, solvent from which crystallized.

TABLE XVIII. ORGANIC DERIVATIVES OF AMINES
1. Primary and secondary amines c) Solids (Listed in order of increasing m.p.)* (Continued)

No.	Name	Melting point, °C	Boiling point, °C	Acetamide	Benzamide	Benzene sulfonamide	p-Toluene sulfonamide	Phenyl thiourea	Picrate	Miscellaneous
459	1-Aminoanthraquinone	252; 243	....	218	255	......	228–9	......	......	....
460	3-Aminocarbazole	254	....	3-mono: 217; di: 200; tri: 175	3-mono: 250	......				
461	2,7-Diaminocarbazole	260	....	320 (di)	....					N,N'-Dibenzal, 290
462	1,8-Diaminoanthraquinone	262, red	....	284 (di)	324 (di)	......				....
463	1,4-Diaminoanthraquinone	268, vlt.	....	271 (di)	284 (di), 280 (mono)	......				....
464	2-Amino-3-nitrofluorenone	269, vlt.	....	245–6	....					N-Carbethoxy, 204
465	1,1'-Diamino-2,2'-dinaphthyl	281	....	230 (di)	278 (di)	......				
466	1,7-Diaminoanthraquinone	290, red	....	283 (di)	325 (di)	......				
467	2,7-Diaminofluorenone	290, vlt.	....	222 (di)	....				230 (di)	Oxime, 255; Phenylhydrazone, 230; 4-Nitrophenylhydrazone, 280
468	1,6-Diaminoanthraquinone	292, red	....	295 (di)	275 (di)	......				
469	1-Amino-5-nitroanthraquinone	293, red	....	275	237	......				N-Me., 250–2, vlt.-blk; N-Et., 238
470	1-Amino-4-nitroanthraquinone	296, yel.-red	....	256–8	....					N-Me., 250
471	5,8-Diaminoquinizarin (1,4-Diamino-5,8-dihydroxyanthraquinone)	>300, br.-vlt.	....	5,8-di: 284	....	5,8-di: 275				
472	2-Aminoanthraquinone	305–8; 302	....	mono: 262; di: 258	228	271	304	......		
473	2-Amino-3-bromoanthraquinone	307, or.-yel.	....	259; 217	279	......				N-Benzal, 174
474	2-Aminoquinizarin (2-Amino-1,4-dihydroxyanthraquinone)	313–4, grn.-yel.	....	....						N-Phenyl, 255–6; N-4-Tolyl, 220
475	1,5-Diaminoanthraquinone	319, red	....	317 (di)	>350 (di)	......				....
476	2,7-Diaminoanthraquinone	>330, or.	....	>350	300 (di)	......				....
477	2,5-Dianilino-1,4-benzoquinone	345, red-br.	....							Anil, 203; Dianil, 240, red
478	2,8-Diaminoacridone	>350	....	>350 (di)	>250 (di)	......				N,N'-Dibenzal, 370

*Derivative data given in order: m.p., crystal color, solvent from which crystallized.

TABLE XVIII. ORGANIC DERIVATIVES OF AMINES
2. Tertiary amines a) Liquids 1) (Listed in order of increasing atmospheric b.p.)*

No.	Name	Boiling point, °C	Melting point, °C	n_D	Density g/ml	Methyl p-toluene sulfonate	Methio-dide	Picrate	Chloro-platinate	Miscellaneous
1	Trimethylamine............	3	......	......	0.6709_4^0	....	230	216; 225	....	p-Toluenesulfonate salt, 162
2	Dimethyl ethyl amine...........	37.5	......	......	....	....	193		....	Hydrochloride, 221
3	N-Methylpyrrolidine..........	78–80	......	......	....	....	221		233	....
4	Triethylamine	89	......	1.400^{20}	0.7255_4^{25}	....		173	....	2,4-Dinitrobenzoate salt, 81–3; β-Resorcylic acid salt, 120
5	1,2-Dimethylpyrrolidine........	96	......	1.4252^{20}	0.7994_4^{20}	....	235		223	
6	1,3-Dimethylpyrrolidine........	96–7	......	......	0.792_4^{15}	....		dimorphous, 181 or 110–5	58–9	HgCl₂ add. comp., 200
7	Pyridine	116	......	1.5092^{21}	0.978_4^{25}	139	117	167	241; 262–4	p-Toluenesulfonate salt, 160; Ethiodide, 91
8	1,2,5-Trimethylpyrrolidine.......	116	......	$1.4335_\alpha^{9,2}$	0.815_4^9	....	310	163		
9	2-Dimethylaminodiethyl ether	121	......	1.406^{20}	0.806^{20}	....	160–5	119–21		
10	Dimethylaminoacetone..........	123	......	......	....	....		176, s. h.		Oxime, 99
11	1-Methylpyrazole	127	......	1.4787_{He}^{14}	0.993_4^{14}	....	190	148	196–8	....
12	N-Ethylpiperidine	128	......	1.4416_α^{20}	0.8237_4^{20}	....		167.5	202	
13	2-Methylpyridine (α-Picoline)	129	......	1.503^{17}	0.9497_4^{15}	150	230	169	216; 195	p-Toluenesulfonate salt, 161; Ethiodide, 123
14	β-Dimethylaminoethyl alcohol (2-Dimethylaminoethyl alcohol).	135	......	1.43^{20}	0.8866_4^{20}	....		96–7		
15	1,3-Dimethylpyrazole	136	......	1.467_α^{15}	0.9628_4^{15}	....	256	138		
16	2-Methylpyrazine	136–7	......	......	1.029_4^{20}	....	129–30	133		
17	4-Methylpyrimidine	141–2	......	......	1.031_{16}^{16}	....		131–4		HgCl₂ double salt, 198
18	2,6-Dimethylpyridine (2,6-Lutidine)	142–3	......	......	....	....	233	168	208	
19	3-Methylpyridine (β-Picoline)	143	......	1.504^{24}	0.9515_4^{25}	....		150	202	Styphnate, 154; Oxid. → nicotinic acid, 228
20	4-Methylpyridine (γ-Picoline)	143	......	1.506^{19}	0.957_4^{15}	....		167	231	β-Resorcylic acid salt, 125
21	4-Chloropyridine..............	147–8	......	......	....	....			202	....
22	2-Ethylpyridine	149	......	......	0.9371^{17}	....		187–9	165–7	
23	3-Chloropyridine..............	149	......	......	....	....		135	168	
24	Tri-n-propylamine..............	156.5	......	1.4176^{20}	0.753_4^{20}	....	207–8	116	....	Ethiodide, 238
25	2,4-Dimethylpyridine (2,4-Lutidine)	157; 159	......	1.503^{14}	0.9273_4^{25}	....		183	216	β-Resorcylic acid salt, 143
26	2,5-Dimethylpyridine (2,5-Lutidine)	160	......	......	....	....		169	192–4	....
27	1,3,4-Trimethylpyrazole	160	......	1.4866_{He}^{18}	0.956_4^{18}	....		164		
28	3-Ethylpyridine	162–4	......	......	0.954^0	....		128–30	208–9; 196	
29	β-Diethylaminoethyl alcohol (2-Diethylaminoethyl alcohol).....	163	......	1.440^{25}	0.8601_{25}^{25}	....				4-Nitrophenylurethane, 60
30	Tropidine (2-Tropene)	163	......	1.4884_α^{19}	0.953_4^{20}	....	abt. 300	285	217	
31	2,3-Dimethylpyridine (2,3-Lutidine)	164	......	......	....	....		188	195	
32	3,4-Dimethylpyridine (3,4-Lutidine)	164						163	205	
33	4-Ethylpyridine	164–5	......	......	0.9417^{20}	....		168	213	
34	2,4,5-Triethylpyridine (2,4,5-Collidine)	165–8	......	......	....	....		128–31	205	
35	1,4-Bis-dimethylaminobutane	167	......	......	....	....		199		
36	Tropane	167	......	......	0.931_4^{20}	....	>300	281	230	
37	1-Diethylaminoisopropyl alcohol ..	167–72	......	......	0.8511_0^{20}	....		89	....	
38	2-Chloropyridine.............	170; 166	......	......	1.205^{15}	120				
39	3-Bromopyridine	170	......	1.5694^{20}	1.645_4^0	156	165	....	175	
40	3,5-Dimethylpyridine (3,5-Lutidine)	170–1	......	......	....	....		245	255	

*Derivative data given in order: m.p., crystal color, solvent from which crystallized.

No.	Name	Boiling point, °C	Melting point, °C	n_D	Density g/ml	Methyl p-toluene sulfonate	Methio-dide	Picrate	Chloro-platinate	Miscellaneous
41	2,4,6-Trimethylpyridine (2,4,6-Collidine)	172	......	....	0.917^{20}	....	....	156	223	....
42	1,4,5-Triethylpyrazole	176–7	......	1.4848^{18}_{He}	0.9685^{18}_4	....	....	175–6	....	....
43	2,3,6-Trimethylpyridine (2,3,6-Collidine)	176–8	......	....	....	....	....	146	250–2	....
44	Benzyl dimethyl amine	181	......	....	....	....	....	93	192	Picrolonate, 151
45	2,3,5-Trimethylpyridine (2,3,5-Collidine)	184	......	....	....	....	....	183; 179	227–8	....
46	N,N,2-Trimethylaniline (N,N-Dimethyl o-toluidine)....	185	......	1.515^{20}	0.9286^{20}_4	....	....	122; 116	....	1,3,5-Trinitrobenzene add. comp., 113
47	2,6-Dimethyl-4-ethylpyridine	186	......	....	0.916^{14}_{14}	....	....	119–20	210	....
48	2,4-Diethylpyridine..........	187–8	......	....	0.9338^0	....	....	98–100	170–1	....
49	3-Diethylaminopropyl alcohol.....	190	......	....	....	....	175	....	....	....
50	Methyl 2-pyridyl ketone	192	......	....	....	....	161	131	220	Oxime, 121; Phenylhydrazone, 155; Ethiodide, 205
51	2,3,4-Trimethylpyridine (2,3,4-Collidine)	192–3	......	....	0.912^{15}	....	....	163–4	259	....
52	N,N-Dimethylaniline	193	2–2.5	1.5582^{20}	0.9557^{20}_4	161	228; 220	163	173	p-Toluenesulfonate salt, 133; 3,5-Dinitrobenzoate salt, 115
53	2-Bromopyridine	194	......	1.657^{15}		127		....	....	
54	3-Ethyl-4-methylpyridine	195–6	......	....	0.9656^0	....	....	148–50	234; 205	....
55	3,5-Dimethyl-2-ethylpyridine	198	......	....	....	....	....	152	189	....
56	N,N,2,6-Tetramethylaniline (N,N-Dimethyl-m-2-xylidine; 2-Dimethylamino-m-xylene)	199–200	......	1.513^{20}	0.912^{20}_4	....	....	....	....	1,3,5-Trinitrobenzene add. comp., 108
57	N-Ethyl-N-methylaniline........	201	......	....	0.919^{55}_4	....	125	134–5	....	Hydrochloride, 114
58	N,N,2,5-Tetramethyl-aniline (N,N-Dimethyl-p-2-xylidine; 2-Dimethylamino-p-xylene)	204	......	....	....	....	....	158	196	....
59	N,N,2,4-Tetramethyl-aniline (N,N-Dimethyl-m-4-xylidine; 4-Dimethylamino-m-xylene)..................	205	......	1.5201^{20}	0.9164^{20}_4	....	....	123–4	219	1,3,5-Trinitrobenzene add. comp., 114
60	N,N-Diethyl-2-methylaniline (N,N-Diethyl-o-toluidine)......	206; 210				....	224	180	....	
61	2-Chloro-N,N-dimethylaniline	207	......	....	....	....	152	132	....	
62	N,N,4-Trimethylaniline (N,N-Dimethyl-p-toluidine)	210	......	1.536^{20}	0.929^{20}_4	85	219	129	....	1,3,5-Trinitrobenzene add. comp., 124, vlt.
63	3,4-Diethylpyridine............	211	......	....	....	....	....	139	221	....
64	Tri-n-butylamine..............	211; 216	......	....	0.778^{20}_{20}	....	180	106	....	β-Resorcylic acid salt, 121
65	N,N,3-Trimethylaniline (N,N-Dimethyl-m-toluidine)	212	......	1.5492^{20}	0.941^{20}_4	....	177	....	....	....
66	Methyl 4-pyridyl ketone	212–4				....	....	130	205	Oxime, 142; Phenylhydrazone, 150; HgCl₂ double salt, 183–4
67	N,N-Diethylaniline	218; 216	......	....	0.9351^{20}_4	....	102	142	....	
68	Methyl 3-pyridyl ketone	220	......	....	....	....	....	....	....	Oxime, 133; Phenylhydrazone, 137; HgCl₂ double salt, 158
69	N,N-Diethyl-4-methylaniline (N,N-Diethyl-p-toluidine)......	229	......	....	0.924^{16}	....	184	....	....	
70	2,3,4,5-Tetramethylpyridine......	232–4	......	....	....	....	....	170–2	210	....
71	Quinoline....................	239	–15.6	1.6268^{20}	1.0929^{20}_4	126	133	203	227; 218	p-Toluenesulfonate salt, 155; Ethiodide, 159; Styphnate, 207–8
72	Isoquinoline..................	243	26; 24	1.615^{20}; 1.622^{25}	1.0986^{20}_4	163	159	222	263	Ethiodide, 148

*Derivative data given in order: m.p., crystal color, solvent from which crystallized.

TABLE XVIII. ORGANIC DERIVATIVES OF AMINES

2. Tertiary amines a) Liquids 1) (Listed in order of increasing atmospheric b.p.)* (Continued)

No.	Name	Boiling point, °C	Melting point, °C	n_D	Density g/ml	Methyl p-toluene sulfonate	Methiodide	Picrate	Chloroplatinate	Miscellaneous
73	**DL-Nicotine** (DL-1-Methyl-2-(3-pyridyl)-pyrrolidine)	243	. . . . . .	. . . .	1.008_4^{20}	. . . .	219	218	abt. 280	. . . .
74	**2-Dimethylaminobenzaldehyde** . . .	244, yel.	. . . . .	. . . .	. . . .	. . . .	164	. . . .	205–6	Oxime, 87; p-Nitro-phenylhydrazone, 191
75	**Tri-isoamylamine**	245; 237	. . . . .	. . . .	0.786_4^{20}	. . . .	. . . .	125	. . . .	. . . .
76	**N,N-Dipropylaniline**.	245	. . . . .	. . . .	0.9104^{20}	. . . .	156	261	. . . .	. . . .
77	**2-Ethylquinoline** (α-Ethyl-quinoline)	245–6	. . . . .	1.598^{23}	1.050^{17}	. . . .	180	148	188	. . . .
78	**2-Methylquinoxaline**.	245–7	. . . . .	. . . .	. . . .	. . . .	215	. . . .	>250	. . . .
79	**1-Methylindole**	247; 239	. . . . .	. . . .	1.0707^0	. . . .	. . . .	150	. . . .	. . . .
80	**2-Methylquinoline** (Quinaldine) . . .	247	. . . . .	1.6126^{20}	1.0585_4^{20}	161; 134	195	191; 195	228	β-Resorcylic acid salt, 145; Ethiodide, 233
81	**8-Methylquinoline**	248	. . . . .	1.616^{20}	1.072_4^{21}	. . . .	200	. . . .	. . . .	. . . .
82	**L-Nicotine** (L-1-Methyl-2-(3-pyridyl)-pyrrolidine)	248	. . . . .	1.528^{20}	1.0097_4^{20}	. . . .	. . . .	218	275	$[\alpha]_D^{20}$: − 167–8
83	**1-Ethylisoquinoline**	250	. . . . .	. . . .	. . . .	. . . .	. . . .	207–10	200	. . . .
84	**2,4-Dimethyl-5,6,7,8-tetrahydro-quinoline**	250	20	1.5415^{20}	1.0043_4^{20}	. . . .	157	144	. . . .	Hydrochloride, 195
86	**N-Methyl-2-pyridone**	255	. . . . .	. . . .	. . . .	. . . .	145	141	. . . .	Styphnate, 162
87	**4,6-Dimethylquinoline**	255–6; 280	. . . . .	. . . .	. . . .	. . . .	236–7	238	. . . .	. . . .
88	**3-Ethylisoquinoline**	257	. . . . .	. . . .	. . . .	. . . .	171–2	180	. . . .	. . . .
89	**Tri-n-amylamine** (Tri-n-pentyl-amine).	257; 245	. . . . .	. . . .	. . . .	80	. . . .	. . . .	. . . .	. . . .
90	**3-Methylquinoline** (β-Methyl-quinoline)	257–9	16–7	1.6171^{20}	1.0673_4^{20}	. . . .	221	187	249	Ethiodide, 220
91	**6-Methylquinoline**.	258	. . . . .	1.6157^{20}	1.0654_4^{20}	154	219; 216	229	. . . .	. . . .
92	**3-Chloroquinoline** (β-Chloro-quinoline)	258–60	. . . . .	. . . .	. . . .	. . . .	276 subl.	182	>300	. . . .
93	**3-Bromo-N,N-dimethylaniline**	259	11	. . . .	. . . .	. . . .	. . . .	135	. . . .	. . . .
94	**5-Methylquinoline**.	260	. . . . .	. . . .	. . . .	. . . .	105	210–3	. . . .	. . . .
95	**4-Methylquinoline** (γ-Methyl-quinoline)	261–3	. . . . .	. . . .	1.0862^{20}	. . . .	173–4	210–1	226–30	Ethiodide, 141–3
96	**2,4-Dimethylquinoline**	264–5	. . . . .	. . . .	1.061^{15}	. . . .	263–5	193–4	229	Ethiodide, 214
97	**2-Phenylpyridine**.	268–9	. . . . .	. . . .	. . . .	. . . .	175	204	. . . .	. . . .
98	**6,8-Dimethylquinoline**	269	. . . . .	. . . .	1.066^4	. . . .	288–9	235	. . . .	. . . .
99	**N,N-Di-n-butylaniline**	271	. . . . .	. . . .	. . . .	180	. . . .	125	. . . .	. . . .
100	**4-Ethylquinoline** (γ-Ethyl-quinoline)	272–4	. . . . .	. . . .	. . . .	. . . .	149	178–80	204	. . . .
101	**N,N-Dimethyl-1-aminonaphthalene** (N,N-Dimethyl-α-naphthyl-amine)	273	. . . . .	1.624^{15}	1.0446_{15}^{15}	. . . .	. . . .	145	. . . .	1,3,5-Trinitrobenzene add. comp., 105–7
102	**5,8-Dimethylquinoline**	273–5	4–5	. . . .	1.07^{21}	. . . .	. . . .	198	234	. . . .
103	**3-Bromoquinoline** (β-Bromo-quinoline)	274–6	12	. . . .	. . . .	. . . .	. . . .	190	. . . .	Oxalate, 107
104	**3,5-Dimethyl-1-phenylpyrazole** . . .	275	. . . . .	. . . .	. . . .	. . . .	190	103	186	. . . .
105	**6-Bromoquinoline**	278	19; 24	. . . .	. . . .	. . . .	278	217	. . . .	. . . .
106	**2,4,7-Trimethylquinoline**	280–1	. . . . .	1.5973^{24}	1.0337^{20}	. . . .	322	232	272	. . . .
107	**4,7-Dimethylquinoline**	283	. . . . .	. . . .	. . . .	. . . .	. . . .	224	227	. . . .
108	**3,4-Dimethyl-1-phenylpyrazole** . . .	285	. . . . .	1.5724^{20}	1.0574^{20}	. . . .	. . . .	122.5	180	. . . .
109	**8-Chloroquinoline**	288	. . . . .	. . . .	. . . .	. . . .	165	. . . .	235	. . . .
110	**2,3'-Bipyridyl**	289; 298	. . . . .	. . . .	. . . .	. . . .	. . . .	150; di: 165–8	. . . .	. . . .
111	**8-Bromoquinoline**	302–4	. . . . .	. . . .	. . . .	. . . .	281	. . . .	230	. . . .
112	**6-Methoxyquinoline**	305d.	20; 28	. . . .	. . . .	. . . .	236	305	. . . .	. . . .
113	**N-Benzyl-N-methylaniline**	306	. . . . .	1.601^{30}	1.0422_{26}^{26}	. . . .	164	127	. . . .	. . . .

*Derivative data given in order: m.p., crystal color, solvent from which crystallized.

TABLE XVIII. ORGANIC DERIVATIVES OF AMINES
2. Tertiary amines a) Liquids (b.p. at reduced pressure only)
2) (Listed in order of increasing m.p. of the corresponding picrate derivative)*

No.	Name	Boiling point, °C	Melting point, °C	n_D	Density g/ml	Methyl p-toluene sulfonate	Methio-dide	Picrate	Chloro-platinate	Miscellaneous
1	N,N-Dimethyl-2-nitroaniline	151–3[30–3], or.-yel.	......	1.6102	......	......	......	102–3	......	Hydrochloride, 174; 1,3,5-Trinitrobenzene add. comp., 112
2	2,6-Diethylpyridine.................	71–3[17]	......	......	......	......	142	115	211–2	....
3	2-Iodopyridine (α-Iodopyridine)	93[13]	......	1.6366[20]	1.9735$_0^{20}$	......	207	120	210	....
4	1,3-Dimethyl-1,2,3,4-tetrahydroquinoline	130–2[17]	......	......	......	......	204	131	......	....
5	β,β-Diethylphenylhydrazine..........	110–2[14]	......	......	......	......	......	131	......	Reduces Fehling's and Tollen's reagent; Zn + AcOH → aniline, b.p. 184 + diethylamine, b.p. 56
6	3-Dimethylaminobenzaldehyde	138[9], yel.	......	......	......	......	185–6	147	168 r. h.	Oxime, 75–6; Semicarbazone, 229; p-Nitrophenyl-hydrazone, 188
7	4-Methyl-5,6,7,8-tetrahydroquinoline ...	122[11]	......	......	......	......	183	170	......	Hydrochloride, 203
8	3-Methyl-5,6,7,8-tetrahydroquinoline ...	126–7[17]	......	......	......	......	162	171	219	....
9	3-Ethylquinoline (β-Ethylquinoline)	135–8[12]	......	1.603[18]	1.0508$_4^{20}$	......	191	197	......	Hydrochloride, 173
10	4-Bromopyridine (γ-Bromopyridine) ...	27.5–30[0.3–0.5]	0–1	1.5679[20]	......	......	......	223	......	Decomposes to yel.-br. solid on standing

*Derivative data given in order: m.p., crystal color, solvent from which crystallized.

No.	Name	Melting point, °C	Boiling point, °C	Methyl p-toluene sulfonate	Methio-dide	Picrate	Chloro-platinate	Miscellaneous
1	**Pyrimidine**	21	123–4			156		Chloroaurate, 226
2	**4,6-Dimethylpyrimidine**	25	160			12–34	103–4	
3	**2,8-Dimethylquinoline**	27	252		221	180		n_D: 1.6022; Ethiodide, 229
4	**4-Bromoquinoline** (γ-Bromoquinoline)	29–30; 25	270d.		265–70			
5	**Quinoxaline** (Benzopyrazine)	30	230		176			D_4^{48}: 1.1334; n_D^{48}: 1.6231; Oxalate, 169; Ethiodide, 146
6	**5-Methylpyrimidine**	30.5	154			141		$HgCl_2$ double salt, 246; Chloroaurate, 209
7	**4-Chloroquinoline** (γ-Chloroquinoline)	31	263			212	278	
8	**7-Chloroquinoline**	31–2	267–8		250		253	
9	**8-Iodoquinoline**	36			200		251	
10	**1,3,5-Trimethylpyrazole**	37	170			147	187–91	n_{He}^{58}: 1.4589
11	**2-Chloroquinoline** (α-Chloroquinoline)	38	267			122		
12	**2,3-Dimethyl-5,6,7,8-tetrahydroquinoline**	38	125^{14}		117	169		
13	**7-Methylquinoline**	39	252			237	223–4	
14	**4-Bromoisoquinoline**	40	280–5		233			
15	**6-Chloroquinoline**	41	262	143	248			Ethiodide, 168–9
16	**4-(Diethylamino)benzaldehyde**	41, yel.						Oxime, 93; Semicarbazone, 214; Phenylhydrazone, 103; Anil, 108–9
17	**3-Nitropyridine**	41	216				254	Hydrochloride, 154
18	**2,4,8-Trimethylquinoline**	42	270		229	193		
19	**4-Chloro-2-methylquinoline** (4-Chloroquinaldine)	42–3			212	178		
20	**5-Chloroquinoline**	45	256		231; 172		255	
21	**2,6,8-Trimethylquinoline**	46	264–5			187–9	206–7	Hydrochloride, 207
22	**2-(Dimethylamino)naphthalene** (N,N-Dimethyl-β-naphthylamine)	47	305			206		
23	**2,4,5,8-Tetramethylquinoline**	48, pa. yel.	168–72^{12}			161		Hydrochloride, 254
24	**5-Bromoquinoline**	48; 52	280		205			Hydrochloride, 225
25	**2-Bromoquinoline** (α-Bromoquinoline)	49			210			
26	**8-Methoxyquinoline**	50	283		160	143		
27	**7-Bromoquinoline**	52; 34	290		240, yel.			Hydrochloride, 213
28	**2-Iodoquinoline** (α-Iodoquinoline)	52–3			211–2			Ethiodide, 220
29	**3-Iodopyridine** (β-Iodopyridine)	53; 50					211	Clinice cold $CHCl_3$ → chloride, 128–30, yel.
30	**4,8-Dimethylquinoline**	54–5	258–9			216–7	226	
31	**2,3,8-Trimethylquinoline**	55–6	281			242–5		Hydrochloride, 260
32	**2,6-Dimethylquinoline**	60	266	175	236–7	186; 178		Styphnate, 200
33	**N,N-Dimethyl-3-nitroaniline**	60, red			205	119		
34	**3,4'-Bipyridyl**	62	297		215			
35	**2,4,6-Trimethylquinoline**	65.5 (anh.); 39.5 (hyd.)	281–2		245–7; 225	200–1		Hydrochloride, 268–72
36	**3,3'-Bipyridyl** (β,β-Bipyridyl)	68	291–2			232		
37	**2,3-Dimethylquinoline**	68–9	263		218	230–1	230	
38	**2,2'-Bipyridyl** (α,α'-Bipyridyl)	69				158		
39	**5-Nitroquinoline**	72 (anh.)			215			Hydrochloride, 214
40	**N,N-Dibenzylaniline**	72; 70			135	131		
41	**2,2'-Bis-(dimethylamino)-biphenyl**	72–3			190–2			Hydroiodide, 256–7
42	**3,4-Dimethylquinoline**	73–4; 65	293		191	215; 205		Hydrochloride, 290
43	**8-Hydroxy-2-methylquinoline** (8-Hydroxyquinaldine)	74	266					Me. eth., 125; b.p. 282
44	**4-Dimethylaminobenzaldehyde**	74						Semicarbazone, 222; p-Nitrophenylhydrazone, 182; 2,4-Dinitrophenylhydrazone, 325; Anil, 100, grn.-yel.
45	**8-Hydroxyquinoline**	75			143	204		Benzoate, 120; 1,3,5-Trinitrobenzene add. comp., 124
46	**N,N-Dimethyl-4-aminophenol** (N,N-Dimethyl-4-hydroxyaniline)	76						Acetate, 78; Me. eth., 49; p-Toluenesulfonyl, 130
47	**8-Bromoisoquinoline**	80.5			274			Nitrate, 193

*Derivative data given in order: m.p., crystal color, solvent from which crystallized.

TABLE XVIII. ORGANIC DERIVATIVES OF AMINES
2. Tertiary amines b) Solids (Listed in order of increasing m.p.)* (Continued)

No.	Name	Melting point, °C	Boiling point, °C	Methyl p-toluene sulfonate	Methio-dide	Picrate	Chloro-platinate	Miscellaneous
48	3,4'-Biquinolyl	83–4				244		Diethiodide, 198
49	2,2'-Dipyridylamine	84; 95 after resolidification	307–8			227–8	160	
50	N,N-Dimethyl-4-nitrosoaniline (4-Nitroso-N,N-dimethylaniline)	85				140		Hydrochloride, 177
51	N,N-Dimethyl-3-aminophenol (N,N-Dimethyl-3-hydroxyaniline)	85						Benzoate, 95; Me. urethane, 87
52	2,3,6-Trimethylquinoline	86–7	285			212		
53	6,6'-Dimethyl-2,2'-bipyridyl	89–90				170–1		HgCl₂ add. comp., 238
54	4,4'-Bis-(dimethylamino)-diphenylmethane	91	390		214 (di)	185 (mono), 178 (di)		1,3,5-Trinitrobenzene add. comp., 114, vlt.
55	6-Iodoquinoline	91			>300		265	Hydrochloride, 210
56	Tribenzylamine	91			184	190		Ethiodide, 190; β-Resorcylic acid salt, 141
57	2,3,4-Trimethylquinoline	92	285		260	216	215	Hydrochloride, 274
58	4-Dimethylaminobenzophenone	92			188–90			Anil, 151; Phenylhydrazone, 105
59	N-Methyl-4-pyridone	92–4 hyg.					176 (anh.)	HgCl₂ double salt, 177–80
60	1,5-Dimethylbenzimidazole	95	300			255		Hydrochloride, 215
61	1-Phenylisoquinoline	95–6	300			165	242	Hydrochloride, 235
62	6-Bromo-2-methylquinoline (6-Bromoquinaldine)	96–7			237			Ethiodide, 218
63	4-Iodoquinoline (γ-Iodoquinoline)	97			251		185	
64	4,4'-Bis-(dimethylamino)-benzhydrol	98; 102, grn.			195 (di)			Me. eth., 71–2; 1,3,5-Trinitrobenzene add. comp., 76
65	5-Iodoquinoline	100			245		263	
66	4,4'-Bis-(dimethylamino)-triphenylmethane (Leuco-malachite green)	102, bz.; 93, al.			231 (220) (di)			1,3,5-Trinitrobenzene add. comp., 89
67	2,2'-Biquinolylmethane	103			205	239, 210 (di)		
68	2-Hydroxypyridine (α-Hydroxypyridine; α-Pyridone)	106–7	280–1					Benzoate, 42; HgCl₂ comp. with Me. eth., 200; HgCl₂ comp. with Et. eth., 141–2
69	Acridine	111			224	208		
70	3,5-Dibromopyridine	112	222	219	274			
71	1,2-Dimethylbenzimidazole	112 (anh.); 65(hyd.)	290		254	238		
72	Antipyrine (2,3-Dimethyl-1-phenyl-5-pyrazolone)	113	319			188		Salicylate, 92
73	4,4'-Bipyridyl	114 (anh.); 73 (hyd.)				257		Nitrate, 256
74	4-Dimethylaminoazobenzene	117, yel.			174, al.			Methochloride, 194
75	Triphenylamine	127	365					Hydrochloride, 214; Fuming HNO₃ in ac. a. → trinitro deriv., 280
76	3-Hydroxypyridine (β-Hydroxypyridine)	129						Oxalate, 177; Chloroplatinate of Et. eth., 192
77	Methyleneaminoacetonitrile	129	210			127,		Acid hydrolysis → glycine
78	7-Nitroquinoline	132–3			231–2			Ethiodide, 220
79	6,8'-Biquinolyl	148			126 (mono)	268		
80	4-Hydroxypyridine (γ-Hydroxypyridine)	149 (anh.)						Acetate, 140–50; Benzoate, 81; Zn → pyridine, b.p. 116
81	6-Nitroquinoline	154; 149			245			Styphnate, 190; Hydrobromide, 245
82	Quinuclidine	158 (sealed tube)				275–6	238–40	Ethiodide, 270–1
83	2,7'-Biquinolyl	160			263	240		

*Derivative data given in order: m.p., crystal color, solvent from which crystallized.

TABLE XVIII. ORGANIC DERIVATIVES OF AMINES
2. Tertiary amines b) Solids (Listed in order of increasing m.p.)* (Continued)

No.	Name	Melting point, °C	Boiling point, °C	Methyl p-toluene sulfonate	Methiodide	Picrate	Chloroplatinate	Miscellaneous
84	2,2'-Biquinolyl ketone	164				179		Oxime, 201; Phenylhydrazone, 199
85	6-Hydroxypyrimidine	164–5				190		Acetyl, 180; 215–20 after resolidification
86	7,7'-Biquinolyl	171–2			310 (mono)	300		
87	4,4'-Bis-(dimethylamino) benzophenone (Michler's ketone)	174			105	156		Oxime, 233
88	5,5'-Biquinolyl	175			272	>300		Hydrochloride, 292
89	2,3'-Biquinolyl	176			286		278	
90	6,6'-Biquinolyl	181			>290 (di)			Diethiodide, 270
91	6-Hydroxyquinoline	193				236		1,3,5-Trinitrobenzene add. comp., 193–5
92	2,2'-Biquinolyl (α,α'-Biquinolyl)	196				210; 215		
93	3-Hydroxyquinoline (β-Hydroxyquinoline)	198				240–5		
94	1,2-Di(2-pyrrolyl)ethanedione (2,2'-Bipyrroyl)	200, pa. yel.						o-Phenylenediamine → dipyrrylquinoxaline, 158; Monoxime, 147; Diphenylhydrazone, 146
95	4-Hydroxyquinoline (γ-Hydroxyquinoline)	201 (anh.)						Hydrochloride, 187 (anh.); KMnO$_4$ → kynuric acid, 200 (anh.)
96	6-Hydroxy-2-methylquinoline (6-Hydroxyquinaldine)	213						Et. eth., 71; Picrate of Et. eth., 192; Ethiodide of Et. eth., 182
97	5-Hydroxyquinoline	224			224		230	Hydrochloride, 240
98	4-Hydroxy-2-methylquinoline (4-Hydroxyquinaldine)	232 (anh.)			201 (anh.)	200	215	Me. eth., 82
99	7-Hydroxyquinoline	235			251	244–5		Benzoate, 88–9
100	5-Hydroxy-2-methylquinoline (5-Hydroxyquinaldine)	246; 232–4						Picrate of Me. eth., 217; Picrate of Et. eth., 213; 206
101	4,4'-Dipyridylamine	273–5				235; 174	>280	Hydrochloride, > 300
102	Hexamethylene tetramine	280		205	190	179		Dil. acid → formaldehyde, semicarbazone, 169

*Derivative data given in order: m.p., crystal color, solvent from which crystallized.

325

EXPLANATIONS AND REFERENCES TO TABLE XIX

Substituted phenylhydrazones. *

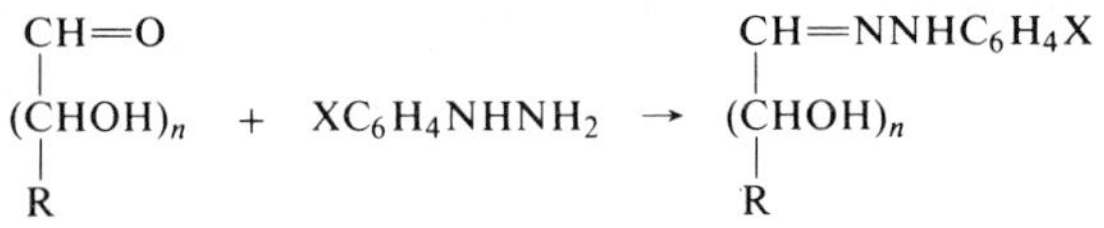

$$
\begin{array}{ccc}
\begin{array}{l} CH{=}O \\ | \\ (CHOH)_n \\ | \\ R \end{array}
& + \ XC_6H_4NHNH_2 \ \rightarrow &
\begin{array}{l} CH{=}NNHC_6H_4X \\ | \\ (CHOH)_n \\ | \\ R \end{array}
\end{array}
$$

Substituted
phenylhydrazone

From the carbohydrate and one equivalent of phenylhydrazine in aqueous acetic acid.
For directions and examples see: Cheronis, p. 520; Wild, p. 77.

From the carbohydrate, phenylhydrazine hydrochloride and sodium acetate in water.
See: Wild, p. 77.

From the carbohydrate with *p*-nitrophenylhydrazine hydrochloride and sodium acetate in methanol.
See: Cheronis, p. 523.

From the carbohydrate and *p*-nitrophenylhydrazine in alcohol.
See: Vogel, p. 456.

From the carbohydrate and benzylphenylhydrazine in aqueous alcohol.
See: Shriner, p. 77.

Phenylosazone. *

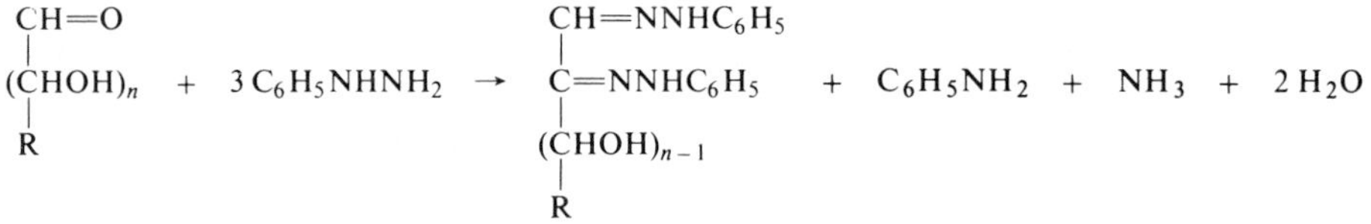

$$
\begin{array}{l} CH{=}O \\ | \\ (CHOH)_n \\ | \\ R \end{array}
\ + \ 3\,C_6H_5NHNH_2 \ \rightarrow \
\begin{array}{l} CH{=}NNHC_6H_5 \\ | \\ C{=}NNHC_6H_5 \\ | \\ (CHOH)_{n-1} \\ | \\ R \end{array}
\ + \ C_6H_5NH_2 \ + \ NH_3 \ + \ 2\,H_2O
$$

Phenylosazone

From the carbohydrate and excess of phenylhydrazine in glacial acetic acid.
For directions and examples see: Cheronis, pp. 523, 524.

From the carbohydrate, excess of phenylhydrazine hydrochloride and sodium acetate in aqueous acetic acid.
See: Cheronis, pp. 524, 525; Linstead, p. 38; Shriner, p. 132; Vogel, p. 455.

From the carbohydrate and excess of phenylhydrazine in methyl cellosolve-glacial acetic acid mixture.
See: W. T. Haskins, R. M. Hann and C. S. Hudson, *J. Amer. Chem. Soc.*, **68**, 1766 (1946).

Methylphenylosazone.

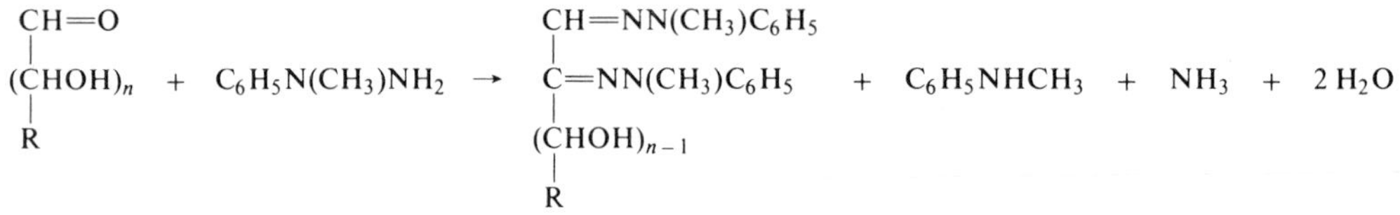

$$
\begin{array}{l} CH{=}O \\ | \\ (CHOH)_n \\ | \\ R \end{array}
\ + \ C_6H_5N(CH_3)NH_2 \ \rightarrow \
\begin{array}{l} CH{=}NN(CH_3)C_6H_5 \\ | \\ C{=}NN(CH_3)C_6H_5 \\ | \\ (CHOH)_{n-1} \\ | \\ R \end{array}
\ + \ C_6H_5NHCH_3 \ + \ NH_3 \ + \ 2\,H_2O
$$

Methylphenylosazone

From the carbohydrate and *as*-methylphenylhydrazine in aqueous alcohol.
For directions and examples see: Vogel, p. 456.

p-Phenylazobenzoate (azoate). *

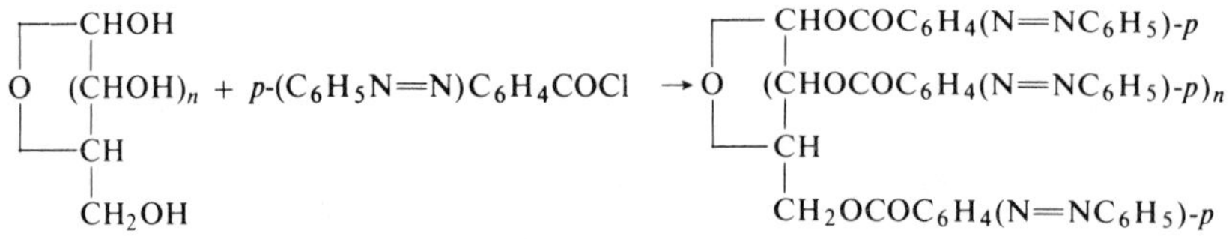

$$
O\!\!\left\langle\begin{array}{l} CHOH \\ | \\ (CHOH)_n \\ | \\ CH \\ | \\ CH_2OH \end{array}\right.
\ + \ p\text{-}(C_6H_5N{=}N)C_6H_4COCl \ \rightarrow \
O\!\!\left\langle\begin{array}{l} CHOCOC_6H_4(N{=}NC_6H_5)\text{-}p \\ | \\ (CHOCOC_6H_4(N{=}NC_6H_5)\text{-}p)_n \\ | \\ CH \\ | \\ CH_2OCOC_6H_4(N{=}NC_6H_5)\text{-}p \end{array}\right.
$$

Azoyl chloride   Carbohydrate azoate

From the carbohydrate and azoyl chloride (*p*-phenylazobenzoyl chloride) in anhydrous pyridine.
For directions and examples see: Cheronis, p. 526; G. H. Coleman, A. G. Farnham and A. Miller, *J. Amer. Chem. Soc.*, **64**, 1501 (1942); G. H. Coleman and C. M. McClosky, *J. Amer. Chem. Soc.*, **65**, 1588 (1943).

*Derivatives recommended for first trial.
WARNING: This is not an instruction manual. References should be consulted for the preparation of derivatives.

Acetate.

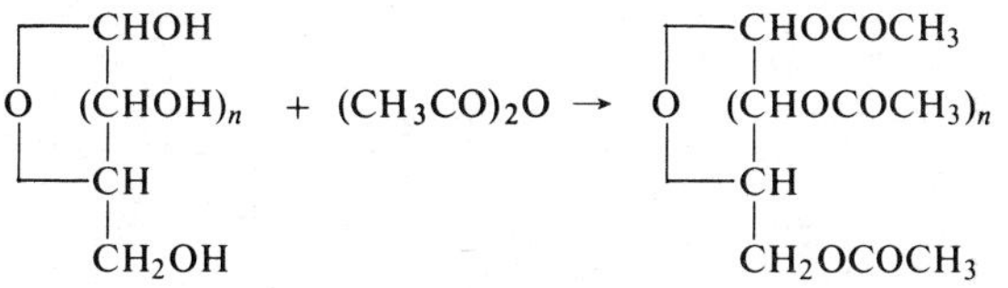

$$\begin{array}{c}\text{---CHOH}\\ \mid\\ \text{O}\quad\text{(CHOH)}_n\\ \mid\\ \text{---CH}\\ \mid\\ \text{CH}_2\text{OH}\end{array} + (\text{CH}_3\text{CO})_2\text{O} \rightarrow \begin{array}{c}\text{---CHOCOCH}_3\\ \mid\\ \text{O}\quad\text{(CHOCOCH}_3)_n\\ \mid\\ \text{---CH}\\ \mid\\ \text{CH}_2\text{OCOCH}_3\end{array}$$

Carbohydrate
acetate

From the carbohydrate, acetic anhydride and sodium acetate.
For directions and examples see: Linstead, p. 39; Shriner, p. 212; Vogel, p. 451; Wild, p. 78.

Specific rotation.

Specific rotation can be used as means for identification.
For directions and examples see: Cheronis, p. 578; Wild, p. 78.

General references:

C. A. Browne and F. W. Zerban, *Physical and Chemical Methods of Sugar Analysis.*, 3rd edition, John Wiley and Sons, New York, 1941; J. Stanek, M. Carny, J. Kocourek and J. Pacak, *The Monosaccharides*, Academic Press, New York, 1963, pp. 865–955; G. R. Pigman in *The Carbohydrates*, (Ed. W. Pigman), Academic Press, New York, 1957, pp. 602–640.

*Derivatives recommended for first trial.
WARNING: This is not an instruction manual. References should be consulted for the preparation of derivatives.

TABLE XIX. ORGANIC DERIVATIVES OF CARBOHYDRATES
a) Liquids (Listed in order of increasing m.p. of the corresponding phenylosazone derivatives)*

No.	Name	Melting point, °C	Specific rotation $[\alpha]_D$	T, °C	Conc.; Solvent	Rf-Values in n-butanol-acetic acid-water (4:1:5)	Phenylosazone M.P.	$[\alpha]_D^T$: Solvent	Azoate (p-Phenylazobenzoate) M.P.	$[\alpha]_{6438}^{25}$, chl.	p-Nitrophenyl-hydrazone	p-Bromophenyl-hydrazone	Miscellaneous
1	DL-Methyl-tetrose.........	......	......			....	140–2	....	....	....	......	....	Phenylbenzylhydrazone, 99–100
2	Apiose..........	......	+3.8; +5.6	20 15	c = 3.4; water	....	156	....	....	....	......	....	p-Bromophenyl-osazone, 210–2
3	DL-Gulose......	......	....		....	....	157–9	....	....	....	......	....	Phenylhydrazone, 143; p-Bromophenyl-osazone, 183
4	L-Erythrose......	......	+11.5 → +15.2 → +30.5	24	c = 3; water	....	164	....	....	....	......	....	Benzylphenylhydra-zone, 105, $[\alpha]_D^{20}$: +32.8, c = 5, 95% al.; Triacetyl, 134
5	D-Erythrose......	......	−14.5	20	c = 11; water	....	164; 166	0.5, pyr.-al.	....	....	......	....	Benzylphenylhydra-zone, 105–6, $[\alpha]_D^{20}$: +32.8, c = 5, 95% al.; p-Bromophenylosa-zone, 195; Phenyl-hydrazone, 116
6	DL-Erythrose.....	......	....		....	....	164; 166–8		....	....	......	....	Benzylphenylhydra-zone, 83
7	DL-Erythrulose...	......	....		....	....	164	....	....	....	......	....	Methylphenylosazone, 158–9
8	L-Erythrulose.....	......	+12	20	water	....	164	....	....	....	......	....	p-Bromophenyl-osazone, 195
9	L-Idose..........	......	+52.7	20	c = 6.2; water	....	168; 160	....	....	....	......	....	....
10	D-Gulose	......	−20.4 → +61.6	20	water	....	168; 160	+6, c = 0.4; me. al.	....	....	......	....	Phenylhydrazone, 143; p-Bromophenyl-osazone, 186; Benzylphenylhydra-zone, 124, $[\alpha]_D$: −24, c = 0.5, me. al.
11	β-Methylglycer-aldehyde	......	....			....	171	....	....	....	......	....	Benzylphenylhydra-zone, 116
12	L-Methyltetrose...	......	−30.5 → −16.5	20	c = 9.47; 96% al.	....	172–3	....	....	....	......	....	Benzylphenylhydra-zone, 96–7; Diethyl mercaptal, 109
13	D-Rhamnose (6-Desoxy-D-mannose).......,	......	−8.25	16.5	c = 10; water	....	185; 191	−95.2, pyr.	....	....	......	....	p-Bromophenylosa-zone, 225; 222–3
14	3-Amino-3-desoxy-D-glucose	......	−61 → −78	18	water	....	207	....	....	....	......	....	N-Benzoyl, 128–30
15	DL-Xylulose (DL-Xylo-ketose)	......	....			....	210–5	....	....	....	......	....	Methylphenylosazone, 173

*Derivative data given in order: m.p., crystal color, solvent from which crystallized.

TABLE XIX. ORGANIC DERIVATIVES OF CARBOHYDRATES
b) Solids (Listed in order of increasing m.p.)*

No.	Name	Melting point, °C	Specific rotation $[\alpha]_D$	T, °C	Conc.; Solvent	Rf-Values in n-butanol-acetic acid-water (4:1:5)	Phenylosazone M.P.	$[\alpha]_D^T$: Solvent	Azoate (p-Phenyl-azobenzoate) M.P.	$[\alpha]_{5438}^{25}$, chl.	p-Nitro-phenyl-hydrazone	p-Bromo-phenyl-hydrazone	Miscellaneous
1	1,3-Dihydroxy-acetone	65–71 (mono-mer); 80 (dimer)	....	.....	......	....	132	....	....	....	156; 160	....	Phenylhydrazone, 115; Me. phenylosazone, 127–130; Diacetate, 46–7; Dibenzoate, 120; Di-p-nitroben-zoate, 198; Oxime, 84
2	β-Melibiose di-hydrate (6-[α-D-Galacto-sido]-D-glu-cose)	82–5	+111.7 → +129.5	20	c = 4; water	....	176–8	+43.2; pyr.	280	+172	....	....	p-Bromophenyl-osazone, 181–2; Phenylhydrazone, 145; 160; Oxime, 186; Octaacetate, 177.5, $[\alpha]_D^{20}$: +102.5
3	D-Ribose	87; 95	−21.5; −23.7	20	c = 4; water	0.31	164; 160	....	....	....	170, ($[\alpha]_D$: −5.7, al.)	p-Bromophenyl-osazone, 180–5; β-Me. glucoside, 83–4, $[\alpha]_D^{20}$: −113.6
4	L-Ribose	87	+20.3 → +20.7	20	c = 4; water	....	166	....	....	....	....	164–5	Phenylhydrazone, 154–5
5	2-Desoxy-D-ribose	90	+2.88 → +2.13	23	water	....	....	....	....	....	....	....	Benzylphenylhydra-zone, 127–9
6	Glycollic alde-hyde (Glyco-aldehyde)	95–7	....	.....	......	....	179	....	....	....	....	....	p-Nitrophenylosazone, 311; Diphenyl-osazone, 207; Benzyl-phenylosazone, 198; Monoacetate, 157–8; Phenylhydrazone, 162
7	D-Fructose	102–4	−132.2 → −92.4	20	c = 4; water	0.23	210	....	125	−440	176, ($[\alpha]_D$: +16, pyr.-al.)	....	o-Nitrophenylhydra-zone, 156–7; α-Me. phenylosazone, 161–2; Pentaacetate: α-form, 70, $[\alpha]_D^{20}$: +34.7, chl.; β-form, 108–9, $[\alpha]_D^{20}$: −120.5, chl.
8	Lactic aldehyde (Lactaldehyde)	105	....	.....	......	....	154; 145	....	....	....	128–9	....	Phenylhydrazone, 93
9	α-L-Rhamnose (6-Desoxy-L-mannose)	105; 93–4 (hyd.)	−8.6 → +8.2	20	c = 4; water	0.37	222; 182	+94, pyr.	....	....	186; 191	....	p-Nitrophenylosa-zone, 208; 2-Naph-thylhydrazone, 170, $[\alpha]_D$: +8.4, me. al.; Semicarbazone, 183, $[\alpha]_D^{20}$: +75 → +57, w.
10	L-Altrose	105; 107–9	−32.3	20	water	....	165; 178	....	....	....	....	....	Benzylphenylhydra-zone, 148
11	D-Altrose	105	+32.6	20	c = 7.6; water	....	178	....	....	....	....	....	Benzylphenylhydra-zone, 148–50
12	2-Amino-2-desoxy-D-glu-cose (Glucosa-mine)	105–10	+48; +44		water	....	210	....	....	....	....	....	Phenylurea, 210; N-Acetyl, 190; Oxime, 127; Semicarbazone, 165
13	α-D-Lyxose	106–7; 101	+5.5 → −14.0		c = 8; water	....	164	....	....	....	172	162; 156–7	Benzylphenylhydra-zone, 116; 128; $[\alpha]_D^{20}$: +26.4, c = 4.9, abs. al.

*Derivative data given in order: m.p., crystal color, solvent from which crystallized.

TABLE XIX. ORGANIC DERIVATIVES OF CARBOHYDRATES
b) Solids (Listed in order of increasing m.p.)* (Continued)

No.	Name	Melting point, °C	Specific rotation $[\alpha]_D$	T, °C	Conc.; Solvent	Rf-Values in n-butanol-acetic acid-water (4:1:5)	Phenylosazone M.P.	$[\alpha]_D^T$ Solvent	Azoate (p-Phenyl-azobenzoate) M.P.	$[\alpha]_{6438}^{25}$, chl.	p-Nitro-phenyl-hydrazone	p-Bromo-phenyl-hydrazone	Miscellaneous
14	**Raffinose** (2-[6-(α-D-Galactosido)-α-D-glucoside]-β-D-fructose) ...	118–9, (anh.); 80 (hyd.)	+123; +105.2	 20	water c = 4; water	0.05	......	....	145	+146	....	....	Trityl, 130; α-Methylphenylhydrazone, 190; Emulsin → sucrose + galactose.
15	**β-L-Rhamnose** .	122–6	+9.1	20	water	....	222; 185	+94; pyr.	......		190–1, ($[\alpha]_D^{20}$: +21.4)	....	α-Methylphenylhydrazone, 124; p-Bromophenylosazone, 222
16	**D-Tagatose**	124	+1.0	22	c = 1; water	....	196–7; 201						Diacetone deriv., 65–6, $[\alpha]_D^{20}$: +71.8, w.
17	**D-Threose**	126–32	+29 → +19.6	22	water	....	164						Benzylphenylhydrazone, 194, bz.; Acetone deriv., 84; Triacetate, 113–4, $[\alpha]_D^{20}$: +35.5, chl.
18	**D-1-Amino-fructose**	127–8	....				210						....
19	**β-D-Allose**	128	+0.58 → +14.41	20	c = 5; water		178; 174					145, ($[\alpha]_D^{20}$: −6.7, al.)	....
20	**β-L-Allose**	128–9	−1.9	20	water		165					141–5, ($[\alpha]_D^{20}$: +6.4, al.)	....
21	**D-Talose**	128–30	+30 → +20.6	21	water	....	201; 197					205	Phenylhydrazone, 178; Benzylphenylhydrazone, 199
22	**DL-Xylose**	129–31	....				210–5						
23	**α-D-Mannose** ..	133	+29.3 → +14.2	20	c = 4; water	0.20	210	....	194–5, ($[\alpha]_D^{20}$: +56, pyr.-al. (1:1))			208–10	Phenylhydrazone, 199–200, $[\alpha]_D^{20}$: +26.3 → +33.8, pyr.; α-Methylphenylhydrazone, 178, $[\alpha]_D$: +8.6, c = 0.5, me. al.; Benzylphenylhydrazone, 165
24	**β-D-Mannose** ..	132	−17.0 → +14.2	20	c = 4; water	....	210		194–5			208–10	Phenylhydrazone, 199–200; α-Methylphenylhydrazone, 178; CaCl$_2$ add. comp., 101–2
25	**L-Mannose**	132	+14.0 → −14.0	.....	water		208						Phenylhydrazone, 195, $[\alpha]_D$: +1.2, HCl
26	**DL-Mannose** ..	132–3	....				217–9						Phenylhydrazone, 195
27	**α-D-Manno-heptose**	134–5	+85.05 → +68.64	20	c = 11; water	....	200					207–8	Phenylhydrazone, 197; Hexaacetate, 106, 50% al.; 139–40, eth.
28	**DL-Glyceralde-hyde** (*dimer*) ..	139; 142	....				132						p-Bromophenylosazone, 168; Dibenzoate, 231; Di-p-nitrobenzoate, 247; Semicarbazone, 160; 2,4-Dinitrophenylhydrazone, 166–7
29	**D-Isorhamnose** (6-Desoxy-D-glucose	139–40	+72.3 → 29.7	20	c = 10; water		185; 187–9						....

*Derivative data given in order: m.p., crystal color, solvent from which crystallized.

TABLE XIX. ORGANIC DERIVATIVES OF CARBOHYDRATES
b) Solids (Listed in order of increasing m.p.)* (Continued)

No.	Name	Melting point, °C	Specific rotation $[\alpha]_D$	T, °C	Conc.; Solvent	Rf-Values in n-butanol-acetic acid-water (4:1:5)	Phenylosazone M.P.	$[\alpha]_D^T$: Solvent	Azoate (p-Phenyl-azobenzoate) M.P.	$[\alpha]_{6438}^{25}$, chl.	p-Nitrophenyl-hydrazone	p-Bromophenyl-hydrazone	Miscellaneous
30	α-L-Glucose ...	141–3	−95.5 → −51.4	20	c = 4; water	....	208	....	....	....	....	....	Diphenylhydrazone, 162
31	L-Xylose	144; 141–3	−79.3 → −18.6	20	c = 9.94; water	....	159–61	....	....	....	....	....	Tetraacetate, 126 (β-form)
32	α-D-Xylose	145; 143	+93.6 → +18.8	20	c = 4; water	0.28	164; 155	−40.9, al.	157	+244	155	128	Me. phenylhydrazone, 108; Benzylphenyl-hydrazone, 95, $[\alpha]_D$: −33, c = 0.57, al.; 2-Naphthylhydra-zone, 124, $[\alpha]_D$: +18.6, me. al.
33	D-Fucose (6-Desoxy-D-galactose; D-Rhodeose) .	145	+89.3 → +75.7	22	water	....	177	....	....	....	....	....	Methylphenylosazone, 181; Oxime, 188–9, $[\alpha]_D$: +13.2, w.
34	L-Fucose (6-Desoxy-L-galactose; L-Rhodeose) .	145	−152.6 → −75.9	20	c = 4; water	0.27	178	....	....	....	210–1	181–4	Phenylhydrazone, 170; α-Methylphenylhy-drazone, 174, $[\alpha]_D^{20}$: −17.0, pyr.; Oxime, 188–9; α-Me. gluco-side, 158, $[\alpha]_D^{20}$: −197.5, w.; β-Me. glucoside, 119, $[\alpha]_D^{20}$: +16.04, w.
35	α-D-Glucose ...	146 (anh.); 83 (hyd.)	+112.2 → +52.7	20	c = 4; water	0.18	210	−1.5, c = 2, pyr.-al. (1:1)	266	+223	88; 196 ($[\alpha]_D$: +21.5, me. al.)	164–6, ($[\alpha]_D$: −43.6 → +18.9; c = 4, pyr.)	2-Naphthylhydrazone, 178; p-Nitrophenyl-osazone, 257; 2,4-Dinitrophenylosa-zone, 256–7; α-Penta-acetate, 112
36	2-Desoxy-D-glucose	148	+46.6	18	water	....	....	....	....	....	....	....	Benzylphenylhydra-zone, 158–9
37	β-D-Glucose ...	148–50	+18.7 → +52.7	20	c = 4; water	....	210	....	....	....	....	....	p-Nitrophenylosa-zone, 257; 2,4-Di-nitrophenylosazone, 256–7; β-Penta-acetate, 132
38	Melezitose (2-[3-(α-D-Glucosido)-D-fructosido]-α-D-glucose) ...	153–4 (+2 H₂O)	+88.2	20	c = 4; water	....	130 (sin-teres)	....	+188	....	....	....	....
39	Turanose (3-[α-D-Glucosido]-D-fructose)...	157 (anh.); 60–5 (hyd.)	+27.3 → +75.8	20	c = 4; water	....	215–20	....	....	....	....	....	Heptaacetate, 140–1, eth., $[\alpha]_D^{20}$: +37, chl.
40	β-D-Arabinose .	158–9	−175 → −105	......	c = 9.45; water	0.21	162–3; 160	....	....	....	....	....	Oxime, 136; Benzyl-phenylhydrazone, 173
41	β-L-Arabinose .	160	+190.6 → +104.5	20	water	....	166	....	262	+755	186	155	Benzylphenylhydra-zone, 174; Me. phenylhydrazone, 165; Oxime, 139; Tetraacetate, α-form: 94–6, eth., β-form: 86, w.; Tetraben-zoate, 160–1; 173; Semicarbazone, 190; 163

*Derivative data given in order: m.p., crystal color, solvent from which crystallized.

331

No.	Name	Melting point, °C	Specific rotation			Rf-Values in n-butanol-acetic acid-water (4:1:5)	Phenylosazone		Azoate (p-Phenyl-azobenzoate)		p-Nitro-phenyl-hydrazone	p-Bromo-phenyl-hydrazone	Miscellaneous
			$[\alpha]_D$	T, °C	Conc.; Solvent		M.P.	$[\alpha]_D^T$: Solvent	M.P.	$[\alpha]_{5438}^{25}$, chl.			
42	β-D-Galac-turonic acid...	160	(+55.3)	20	water	....	....	....	....	....	....	151 ($[\alpha]_D$ +11.5, me. al.)	Phenylhydrazone, 140; Brucine salt, 180
43	β-Maltose (4-[α-D-Glucosido]-β-D-glucose)...	160–5; 102–3 (+1 H_2O)	+111.7 → +130.4	20	c = 4; water	0.11	205–6	....	275	+2	....	....	Phenylhydrazone, 130; p-Nitrophenyl-osazone, 261; p-Bro-mophenylosazone, 198; 2-Naphthyl-hydrazone, 176; Octa-acetate, 160–1, $[\alpha]_D^{20}$: +62.59, chl.
44	DL-Fucose (6-Desoxy-DL-galactose; DL-Rhodeose)	161	....		....	....	187						Diacetone deriv., 41
45	DL-Sorbose...	162–3	....		.....	0.20	169–70	....	....				....
46	L-Galactose...	162–3	–120 → –73.6 →		c = 10; water	....	192–5						Phenylhydrazone, 158–60, $[\alpha]_D$: +21.6, w.
47	DL-Galactose..	163; 144	....		....	....	206						Phenylhydrazone, 158–60; α-Methyl-phenylhydrazone, 183
48	β-D-Glucuronic acid........	163	+36.3	20	water	....	....	....	225			....	Semicarbazone, 188; Cinchonine salt, 204, $[\alpha]_D^{20}$: +139.9, w.
49	DL-Arabinose.	164	....		....	....	169					160	Diphenylhydrazone, 204; p-Bromophenyl-osazone, 200–2
50	L-Sorbose.....	165; 159–61	–43.7 → –43.4	20	c = 12; water	0.20	156; 168	–6, me. al.	....				p-Bromophenylosa-zone, 181; o-Nitro-phenylosazone, 211–2; β-Me. glucoside, 120–2, $[\alpha]_D^{20}$: –88.5, w.; Pentaacetate, 97, $[\alpha]_D$: +2.9, chl.
51	D-Sorbose....	165	+42.9	20	c = 1; water	....	168; 160	....					β-Me. glucoside, 119, $[\alpha]_D$: +88.5, w.
52	3-(β-D-Galac-tosido)-D-arabinose....	166–8	–50.3 → +63.1	19	water	....	242						Benzylphenylhydra-zone, 223–5
53	α-D-Galactose.	167	+150.7 → +80.2	20	c = 5; water	0.16	196; 201	....	276	+436	154; 197	168	α-Methylphenylhydra-zone, 190–1; Benzyl-phenylhydrazone, 157; Diphenylhydra-zone, 157; α-Penta-acetate, 95
54	Stachyose (α-Galactosyl [1–6'] α-galactosyl [1'–4] α-glucosyl [1–2'] β-fructo-side)........	167–70	+148	20	water								Tetradecaacetate, 95–6, $[\alpha]_D^{22}$: +120, al.; Tetradeca-p-nitro-benzoate, 166
55	Sucrose (2-[α-D-Glucosido]-β-D-fructose).	169–70, me. al., 185, w.-al.	+66.53	20	c = 26; water	0.14	.....	....	125	+35			Nonreducing; Octa-acetate, 72; 69, $[\alpha]_D^{20}$: +59.6, chl.; Tritrityl, 128

*Derivative data given in order: m.p., crystal color, solvent from which crystallized.

TABLE XIX. ORGANIC DERIVATIVES OF CARBOHYDRATES
b) Solids (Listed in order of increasing m.p.)* (Continued)

No.	Name	Melting point, °C	Specific rotation $[\alpha]_D$	T, °C	Conc.; Solvent	Rf-Values in n-butanol-acetic acid-water (4:1:5)	Phenylosazone M.P.	$[\alpha]_D^T$ Solvent	Azoate (p-Phenyl-azobenzoate) M.P.	$[\alpha]_{6438}^{25}$, chl.	p-Nitro-phenyl-hydrazone	p-Bromo-phenyl-hydrazone	Miscellaneous
56	D-Glucoheptu-lose	171	+67.4	20	water		209–10						Hexaacetate, 112, $[\alpha]_D^{20}$: +87.0; α-Me. gluco-side, 138–40, $[\alpha]_D^{22}$: +108.5, w.
57	4-[β-D-Gluco-sido]-β-D-mannose	176 (anh.); 139–40 (hyd.)	+15.1 → +10.7	16	water		198						
58	α-L-Rhamno-hexose	180–1	−80 → −61.4	20	c = 9.67; water		200						Benzylphenylhydra-zone, 183–4
59	L-Ascorbic acid	190; 187	+49	18	me. al.	0.38					262 (di)	170 (di)	Diphenylhydrazone, 187; Di-2,4-dinitro-phenylhydrazone, 282
60	Gentiobiose (6-[β-D-Glu-cosido]-D-glu-cose)	190–5 (anh.); 86 (hyd.)	+21.4 → +8.7	20	c = 5; water		163–4; 170; 179.	−42.9²⁰, 95% al.					Octaacetate (α): 188–9, $[\alpha]_D^{20}$: +52.3, chl.; (β):192–3, $[\alpha]_D^{20}$: −5.3, chl.
61	α-D-Glucohep-tose	193	−20	20	water		194–5						β-Me. glucohepto-side, 169, $[\alpha]_D$: −75, w.; Hexaacetate, 164 (α-form), 135 (β-form)
62	α,α-Trehalose (1-[α-D-Glu-cosido]-α-D-glucose)	210; 203 (anh.); 97 (+2 H₂O)	+178.3	20	c = 7; water				134–5	+210			Nonreducing; Octani-trate, 124; Octaace-tate, 100–2, $[\alpha]_D^{20}$: +162, chl.; Hexaace-tate, 93–6, $[\alpha]_D^{19}$: +158.3, chl.
63	Primeverose (6-[β-D-Xylo-sido]-D-glu-cose)	210; 208	+24.1 → −3.3		c = 2.5; water		220						β-Heptaacetate, 216, $[\alpha]_D^{20}$: −23.5, chl.
64	Lactose (4-[β-D-Galactosido]-D-glucose)	α-form, 223 (anh.); 201 (hyd.); β-form, 252 (anh.)	+90 → +55.3 (+52.3) +35 → +55.3 (+52.3)	20 20	c = 4; water c = 4; water	0.09	200; 210–2						p-Nitrophenylosa-zone, 258; Octaace-tate, 100; Benzyl-phenylhydrazone, 128; 2-Naphthylhy-drazone, 203
65	β-Cellobiose (4-[α-D-Glu-cosido]-β-D-glucose)	225	+14.2 → +34.6	20	c = 8; water		208–10; 198	−6.5²⁰, pyr.-al. (1:1)	273	+105			Phenylhydrazone, 90; Octaacetate, (α) 229–30, $[\alpha]_D^{20}$: +42, chl.; Semicarbazone, 183–5; Octaacetate (β), 192; 202, $[\alpha]_D^{20}$: −14.5, chl.; Oxime, 123–5
66	6-[β-Cellobio-sido]-α-D-glu-cose	247–52 (anh.); 200 (hyd.)	+15.0 → +8.4		water		224						
67	6-[-β-Lacto-sido]-α-D-glucose	257	+34.7 → +22.6	24	water		233						

*Derivative data given in order: m.p., crystal color, solvent from which crystallized.

EXPLANATIONS AND REFERENCES TO TABLE XX

Formation of amine by reduction. *

$$RNO_2 \; + \; 6[H] \; \rightarrow \; RNH_2 \; + \; 2\,H_2O$$
$$\text{Amine}$$

From the nitro compound and tin in hydrochloric acid.

For directions and examples see: Cheronis, p. 625; Linstead, p. 69; Shriner, p. 262; Vogel, p. 529; Wild, p. 247.

From catalytic hydrogenation (Raney nickel, platinum oxide and palladium on charcoal) of the nitro compound in ethanol, methanol or dioxane.

See: Cheronis, p. 626; Linstead, p. 70; N. D. Cheronis and M. Koeck, *J. Chem. Ed.*, **20**, 488 (1943); K. Johnson and E. F. Degering, *J. Amer. Chem. Soc.*, **61**, 3194 (1939); S. V. Voris and P. E. Spoerri, *J. Amer. Chem. Soc.*, **60**, 935 (1938); E. R. Blout and D. C. Silverman, *J. Amer. Chem. Soc.*, **66**, 1442 (1944).

From the nitro compound and lithium aluminum hydride in ethers.

See: N. G. Gaylord, *Reduction with Complex Metal Hydrides*, Interscience, New York, 1956, pp. 762–773.

For partial reduction of polynitro compounds with sodium or ammonium polysulfide *see:* Linstead, p. 71; Vogel, p. 551.

NOTE: For directions and examples for the preparation of the derivatives of the amine formed on reduction of the nitro compounds see explanations and references to Table XVIII, p. 291, 292, 293, 294.

Polynitro derivative. *

$$ArNO_2 \; \xrightarrow{\;HNO_3\;} \; Ar(NO_2)_n \quad n > 1$$

From the aromatic nitro compound with concentrated or fuming nitric acid and sulfuric acid.

For directions and examples see: Cheronis, pp. 580, 627; Shriner, p. 249; Vogel, pp. 526, 527.

From fuming nitric acid in acetic acid or acetic anhydride.

See: Wild, pp. 24, 247, 248; J. Reilly and W. J. Hickinbottom, *J. Chem. Soc.*, **117**, 135 (1920); O. L. Brady and W. H. Gibson, *J. Chem. Soc.*, **119**, 102 (1921).

Molecular compounds of aromatic polynitro compounds with aromatic hydrocarbons.

$$Ar(NO_2)_n \; + \; Ar'H \; \rightarrow \; Ar(NO_2)_n \cdot Ar'H \quad n > 1$$
$$\text{Molecular}$$
$$\text{compound}$$

Molecular addition compounds are formed from the aromatic polynitro compound and aromatic hydrocarbons.

For directions and examples see: Table IV, p. 32, 33, 34. Wild, p. 248; T. Asahina and C. Shinomiya, *J. Chem. Soc. Japan*, **59**, 341 (1938); O. C. Dermer and R. B. Smith, *J. Amer. Chem. Soc.*, **61**, 748 (1939).

Nitroaromatic acids from side-chain oxidation.

$$Ar(NO_2)R \; \rightarrow \; Ar(NO_2)COOH$$
$$\text{Nitroaromatic}$$
$$\text{acid}$$

From the alkylaromatic nitro compound and basic aqueous potassium permanganate.

For directions and examples see: Cheronis, p. 627; Vogel, p. 629.

From the alkylaromatic nitro compound and sodium bichromate and sulfuric acid in water.

See: Cheronis, p. 628; Vogel, p. 629.

---

*Derivatives recommended for first trial.
WARNING: This is not an instruction manual. References should be consulted for the preparation of derivatives.

TABLE XX. ORGANIC DERIVATIVES OF NITRO COMPOUNDS
a) Liquids (Listed in order of increasing atmospheric b.p.)*

No.	Name	Boiling point, °C	Melting point, °C	n_D	Density g./ml.	Amine Boiling point, °C	Amine Melting point, °C	Acet-amide	Benz-amide	Benzene sulfon-amide	Picrate	Nitration product Melting point, °C	Nitration product Position of nitro groups	Miscellaneous
1	Nitroethylene	98–9	......	....	1.073^{14}	16.5	......	......	71	58	165	......	......	Polymerizes readily on contact with base
2	Nitromethane	101	−17	1.3797^{25}	1.1297^{25}_4	−6	......	28	80	30	207; 215	......	......	....
3	Nitroethane	114	......	1.392^{20}	1.0497^{20}_4	16.5	......	......	71	58	165	......	......	....
4	2-Nitropropane	120	......	1.394	1.024^0	33	......	......	26	......	......	......	......	Phenylthiourea deriv. of amine, 101; 1-Naphthyl-urea deriv. of amine, 200
5	3-Nitropropylene (3-Nitropropene)	125–30	......	....	1.051^{21}	58	......	......	......	39	140	......	......	....
6	1-Nitropropane	132	......	1.4002^{24}	1.008^{24}_4	49	......	......	84	36	135	......	......	....
7	DL-2-Nitrobutane	140	......	1.4013	0.9877^0	63	......	......	76	70	139–40	......	......	....
8	2-Methyl-1-nitro-propane	140–1	......	....	$0.987^{7.5}$	69	......	......	57	53	150	......	......	p-Toluenesulfona-mide deriv. of amine, 78
9	2-Methyl-2-nitro-butane	150	......	....	....	78	......	......	......	......	183	......	......	....
10	DL-2-Nitropentane	152–4	......	....	....	92	......	......	......	......	......	......	......	Hydrochloride of amine, 168; Oxa-late, 226; 131; Chloroaurate, 82–3
11	1-Nitrobutane	153	......	1.4103^{20}	0.9710^{20}_4	77	......	......	......	......	151	......	......	Phenylthiourea de-riv. of amine, 65; 1-Naphthylurea deriv. of amine, 149
12	1-Nitroisobutylene (1-Nitroisobutene)	154–8	......	....	1.052^0_0	69	......	......	57	53	150	......	......	....
13	3-Methyl-1-nitro-butane	164	......	....	....	96	......	......	......	......	138	......	......	Phenylthiourea de-riv. of amine, 102; 1-Naphthylurea deriv. of amine, 132
14	1-Nitropentane	173	......	1.4175	0.9525^{20}_4	104	......	......	......	......	139	......	......	Phenylthiourea de-riv. of amine, 69; 2-Naphthylthio-urea deriv. of am-ine, 114
15	2-Nitrohexane	176	......	....	0.9357^{20}_4	116–8	......	......	......	......	......	......	......	....
16	1-Nitrohexane	193	......	1.4234	0.9396^{20}_4	130	......	......	40	96	126	......	......	....
17	1-Nitroheptane	193–5	......	....	0.9476^{17}	155	......	......	......	......	121	......	......	Phenylthiourea de-riv. of amine, 75
18	2-Nitroheptane	194–8	......	....	0.9466^0	142	......	......	......	......	......	......	......	Hydrochloride of amine, 133; Oxa-late, 204–5; Chlo-roaurate, 63–4
19	Nitrocyclohexane	205–6	−34	1.4612^{19}	1.068^{19}_4	134	......	104	147	......	......	......	......	$SnCl_2 + HCl \rightarrow$ Cyclohexanone oxime, 89–90
20	1-Nitrooctane	206–10 part. d.	......	....	0.9346^{20}	180	......	......	......	......	112	......	......	1-Naphthylthio-urea deriv. of am-ine, 72
21	Nitrobenzene	210–1	......	1.553^{20}	1.2031^{20}_4	184	......	114	160	112	......	90	1,3	....
22	2-Nitrotoluene	222	......	1.5474^{20}	1.1622^{19}_{15}	200	......	110–11	146; 143	124	213	70–1	2,4	....

*Derivative data given in order: m.p., crystal color, solvent from which crystallized.

TABLE XX. ORGANIC DERIVATIVES OF NITRO COMPOUNDS

a) Liquids (Listed in order of increasing atmospheric b.p.)* (Continued)

No.	Name	Boiling point, °C	Melting point, °C	n_D	Density g./ml.	Amine Boiling point, °C	Amine Melting point, °C	Acetamide	Benzamide	Benzene sulfonamide	Picrate	Nitration product Melting point, °C	Nitration product Position of nitro groups	Miscellaneous
23	1-Ethyl-2-nitrobenzene	224	......	1.5407[19]	1.126[24]	210–11; 214	......	111–2	147	......	194–5	......	......	......
24	1,3-Dimethyl-2-nitrobenzene (2-Nitro-m-xylene) ..	226	13	......	1.112[15]	215; 218	......	177	168	......	180	182	1,3,5	p-Toluenesulfonamide deriv. of amine, 212
25	Phenylnitromethane	226d.	......	1.5323[20]	1.1598[20 over 0]	184–5	......	60	105	88	194	......	......	......
26	3-Nitrotoluene	233	16	1.5470[21]	1.1571[20 over 4]	203	......	65	125	95	200	......	......	Oxid. → 3-nitrobenzoic acid, 140
27	1,4-Dimethyl-2-nitrobenzene (2-Nitro-p-xylene) ...	241–2	......	......	1.132[15]	213–5	......	139	140	138	171	139	1,2,4	....
28	1-Ethyl-4-nitrobenzene..........	241	......	1.5458[19]	1.124[25]	216; 214	......	94	151	......	......	37	2,4,6	p-Toluenesulfonamide deriv. of amine, 104
29	1,3-Dimethyl-4-nitrobenzene (4-Nitro-m-xylene) ..	246	2	......	1.126[17.5]	217	......	133; 130	192	129–30	209	182	1,3,5	....
30	1,2-Dimethyl-3-nitrobenzene (3-Nitro-o-xylene) ...	250	15	......	......	221–2	......	135	189	......	221	82	1,2	....
31	2-Nitro-p-cymene ..	264	......	1.5309[20]	1.0744[20 over 4]	241	......	71	102	......	......	54	2,6	....
32	2-Nitroanisole	265	10	1.5620[20]	1.2540[20 over 4]	225	5–6	85; 88	60; 84	89	200	68	2,4,6	....
33	1-tert-Butyl-4-nitrobenzene......	267	......	......	......	230	17	169–70	134–6	......	......	......	......	Dil. HNO₃ → 4-nitrobenzoic acid, 240
34	2-Nitrophenetole ...	268	5–6	1.5425[20]	1.1903[15]	229	......	79	104	102	......	86	2,4	....

*Derivative data given in order: m.p., crystal color, solvent from which crystallized.

TABLE XX. ORGANIC DERIVATIVES OF NITRO COMPOUNDS
b) Solids (Listed in order of increasing m.p.)*

No.	Name	Melting point, °C	Boiling point, °C	Data for the corresponding amine obtained on reduction of all nitro groups						Nitration product		Miscellaneous
				Amine		Acet-amide	Benz-amide	Benzene sulfon-amide	Picrate	Melting point, °C	Position of nitro groups	
				Boiling point, °C	Melting point, °C							
1	2-Methyl-2-nitropropane (*tert*-Nitrobutane)	25–6	127	46	......	......	134	.....	198	......		Phenylthiourea deriv. of amine, 120
2	4-Fluoro-1-nitrobenzene	25–7	......	184–6	......	152	185	......	......			N-*p*-Nitrobenzamide deriv. of amine, 181
3	1-Nitro-2,3,6-trimethyl-benzene	30	......	235	......	186	......					....
4	3,4-Dimethyl-1-nitro-benzene (4-Nitro-*o*-xylene)	30	......	226	51; 47–8	99	......	......		82	1,2	N-Formyl deriv. of amine, 52; N-Chloroacetyl, 109
5	2-Chloro-1-nitrobenzene	32	......	209	......	87	99	129	134	50; 52	2,4	....
6	2,4-Dichloro-1-nitro-benzene	33	258	......	63	143–6	115	128	106			N-Formyl deriv. of amine, 154
7	4-Bromo-3-nitrotoluene .	33	......	136[16]	......	121; 114						
8	3-Nitrophenetole (3-Ethoxy-1-nitro-benzene)	34	......	248	......	97	103	......	158			*p*-Toluenesulfonamide deriv. of amine, 157
9	2-Nitrobiphenyl	37; 33	......	......	49–50	121	102	......				N-Formyl deriv. of amine, 75
10	6-Chloro-2-nitrotoluene .	37	238	......	245	157–9; 136	173					Oxid. → 6-chloro-2-nitro-benzoic acid, 161
11	3-Iodo-1-nitrobenzene ..	38; 35	......	......	33; 27	119	157					*p*-Toluenesulfonamide deriv. of amine, 128
12	3-Nitroanisole (3-Methoxy-1-nitro-benzene)	38	258	251	......	81	......	......	169	106	3,5	*p*-Toluenesulfonamide deriv. of amine, 68
13	4-Chloro-2-nitrotoluene .	38	240	......	21–2	139–40; 131	......					Oxid. → 4-chloro-2-nitro-benzoic acid, 142
14	5-Nitroindane	40, yel.	152[14]	250	37–8	106	137	......				....
15	2-Bromo-1-nitrobenzene	43	......	250	32	99	116	......	129	72	1,3	....
16	4-Nitroindane	44	139[10]	236	–3	126	136	......				
17	Nitromesitylene	44	......	232–3	......	216–7	204	......	189–91	86	di	*p*-Toluenesulfonamide deriv. of amine, 167
18	3-Chloro-1-nitro-benzene	45	......	230	......	72; 78	119–20	121	177			
19	2-Nitroazoxybenzene ...	49, yel.	......	......	98	156	......					
20	2-Iodo-1-nitrobenzene ..	49	......	......	61; 58	109	139	......	112			
21	4-Nitrotoluene	52	234	200	45	147	158	120	182	70	2,4	....
22	4-Chloro-1,3-dinitro-benzene (1-Chloro-2,4-dinitrobenzene).......	52	......	......	91	142 (*di*)	178 (*di*)	......		183	2,4,6	NaOH → 2,4-Dinitrophenol, 114; Hydrazine → 2,4-di-nitrophenylhydrazine, 199
23	4-Nitroanisole (4-Methoxy-1-nitro-benzene)............	53	......	240	58	130; 127	154	95	......	89	2,4	....
24	2,5-Dichloro-1-nitro-benzene	54	......	......	50, lgr.	132	120	......	......	104	1,3	....
25	4-Iodo-3-nitrotoluene ...	55	......	......	48; 38	151; 136	......	......				N-Formyl deriv. of amine, 129; NaOH → 3-Nitro-*p*-cresol, 36–7
26	3-Bromo-1-nitrobenzene	56	......	251	18	87	120; 136	......	180	59	1,2	....
27	1-Nitronaphthalene	57; 60	......	......	50	159	160	167	163; 181	......		
28	β-Nitrostyrene	58, yel.	250–60d.	......	......	......	......	......		......		Irradiation → dimer, 180–7
29	1-Methyl-2-nitro-naphthalene	58–9, yel.	......	......	51	188–9	222					....

*Derivative data given in order: m.p., crystal color, solvent from which crystallized.

No.	Name	Melting point, °C	Boiling point, °C	Amine Boiling point, °C	Amine Melting point, °C	Acetamide	Benzamide	Benzene sulfonamide	Picrate	Nitration product Melting point, °C	Nitration product Position of nitro groups	Miscellaneous
30	4-Nitrophenetole (1-Ethoxy-4-nitro-benzene)	60; 58	......	248; 254	3–4	137	173	143	69	86	2,4	....
31	3,4-Dinitrotoluene	61	......	265	89–90	4-*mono*: 131–2; *di*: 210	3-*mono*: 193–4; *di*: 263–4	178–9 (*di*)	......	......	......	Oxid. → 3,4-dinitrobenzoic acid, 165; 161
32	3-Nitrobiphenyl	61; 59	......	......	30	148	......	......	......	......	......	....
33	2,3-Dinitrotoluene	63	......	255	63–4	......	......	......	......	......	......	(NH₄)₂ S → 2-Nitro-*m*-toluidine, 108, red; HNO₃ → 2,3-Dinitrobenzoic acid, 201
34	1-Methyl-8-nitro-naphthalene	63–4	......	......	67–8	183–4	195–6	......	......	......	......	....
35	2,6-Dinitrotoluene	66	......	......	105	202–3	......	......	......	80; 82	2,4,6	Oxid. → 2,6-Dinitrobenzoic acid, 202–3
36	2,4,6-Trinitroanisole (1-Methoxy-2,4,6-trinitro-benzene)	68	......	......	......	......	......	......	......	......	......	Naphthalene adduct, 69–70; NH₃ in al. → picramide, 188
37	2,4-Dinitrotoluene	70; 72	......	292	99	224 (*di*)	224 (*di*)	2-*mono*: 138; *di*: 191	......	80; 82	2,4,6	Naphthalene adduct, 60; Oxid. → 2,4-dinitrobenzoic acid, 182–3; SnCl₂ + HCl → 4-Nitro-*o*-toluidine, 107, yel.
38	2-Nitroazobenzene	71, or.-red	......	......	59	126	122	......	......	......	......	....
39	1-Methyl-4-nitro-naphthalene	71–2, pa. yel.	......	......	51–2	166–7	238–9	......	......	......	......	Dil. HNO₃ → 4-nitro-1-naphthoic acid, 220–1
40	4-Bromo-1,3-dinitro-benzene (1-Bromo-2,4-dinitrobenzene)	75; 72	......	......	......	......	......	......	......	......	......	NaOH → 2,4-Dinitrophenol, 114; Al. NH₃ → 2,4-dinitro-aniline, 180; 188, yel.; Sn + HCl → 1,3-Diaminobenzene, 63; Hydrazine → 2,4-dinitro-phenylhydrazine, 199
41	3,5-Dimethyl-1-nitro-benzene	75	273	220–1	......	144; 140	......	......	......	......	......	N-Formyl deriv. of amine, 76–7
42	4,5-Dimethyl-1,3-dinitrobenzene	76	......	......	......	......	......	......	......	......	......	(NH₄)₂ S → 1-Amino-3,4-dimethyl-5-nitrobenzene, 75; Acetyl deriv. of this, 209–10; Benzoyl deriv. of this, 223–4
43	5-Methyl-2-nitro-naphthalene	76–7	......	......	63–4	123–4	155–6	......	......	......	......	....
44	2-Nitronaphthalene	78	......	......	112	132	162	102	195	......	......	....
45	2,4,6-Trinitrophenetole (1-Ethoxy-2,4,6-tri-nitrobenzene)	78	......	......	......	......	......	......	......	......	......	Naphthalene adduct, 39; NH₃ in al. → picramide, 188
46	2,4,6-Trinitrotoluene (T. N. T.)	80; 82	......	......	......	......	......	......	......	......	......	Naphthalene adduct, 97; CrO₃/conc. H₂SO₄ → 2,4,6-trinitrobenzoic acid, 220
47	2-Methyl-1-nitro-naphthalene	81, yel.	......	......	32, pet. eth.	188	180	......	......	......	......	....
48	4-Nitrophenanthrene	81	......	......	105	190	224	......	......	......	......	....
49	3,4-Dimethyl-1,2-dinitrobenzene	82	......	......	......	......	......	......	......	......	......	Reduct. → 1-amino-3,4-dimethyl-2-nitrobenzene, 66, red.; Acetyl deriv. of this, 115–6; Benzoyl deriv. of this, 199–200

*Derivative data given in order: m.p., crystal color, solvent from which crystallized.

TABLE XX. ORGANIC DERIVATIVES OF NITRO COMPOUNDS

b) Solids (Listed in order of increasing m.p.)* (Continued)

No.	Name	Melting point, °C	Boiling point, °C	Amine Boiling point, °C	Amine Melting point, °C	Acet-amide	Benz-amide	Benzene sulfon-amide	Picrate	Nitration product Melting point, °C	Position of nitro groups	Miscellaneous
50	5-Methyl-1-nitronaphthalene	82–3	......	......	77–8	194–5	173–4		......			......
51	Picryl chloride	83	......	......	....	208 (*tri*)	....	211		......		NaOH → Picric acid, 122; NH₃ → 2,4,6-Trinitroaniline, 188; 192–5
52	4-Chloro-1-nitrobenzene	84	......	......	72	179; 172	192	122		......		NaOH → 4-Nitrophenol, 114
53	2,5-Dibromo-1,4-dinitrobenzene	127	......	......	....	......	......					2,5-Dibromo-*p*-nitroaniline, 175, yel.
54	2,4-Dimethyl-1,3-dinitrobenzene	84; 82	......	......	65–6, lgr.	>260 (*di*)	232 (227) (*di*)					Reduct. → 1-amino-2,4-dimethyl-3-nitrobenzene, 84; N,N′-Diformyl, 219–20
55	2,4-Dinitromesitylene	85	......	......	....	....	....	......	......	232	2,4,6	Reduct. → 2-amino-4-nitromesitylene, 75; Acetyl deriv. of this, 191; Benzoyl deriv. of this, 169; Benzenesulfonyl deriv. of this, 163
56	2,5-Dibromo-1-nitrobenzene	85	......	......	51–2	171–2	......		......			......
57	4-Bromo-1-nitronaphthalene	85	......	......	102	193						N-Formyl deriv. of amine, 172
58	2,4-Dinitrophenetole 1,3-Dinitro-4-ethoxybenzene	86	......	......	67–8	193 (*di*)	....			78	2,4,6	Reduct. → 2-nitro-*p*-phenetidine, 40
59	4-Chloro-1-nitronaphthalene	87; 85	......	......	98	186		......				......
60	1,3-Dinitrobenzene	90	......	......	63	191 (*di*), 87–9 (*mono*)	240 (*di*), 125 (*mono*)	194	184	......		(NH₄)₂S → 3-Nitroaniline, 114; Naphthalene adduct, 52
61	2,3-Dimethyl-1,4-dinitrobenzene	90	......	......	116	275–6 (*di*)				......		1-amino-2,3-dimethyl-4-nitrobenzene, 114
62	3,5-Dinitrotoluene	92	283–5	....	....	235–6 (*di*)						Oxid. → 3,5-dinitro-benzoic acid, 204–5; (NH₄)₂ S → 5-Nitro-*m*-toluidine, 98
63	3,6-Dimethyl-1,2-dinitrobenzene	93	......	......	75	....						Oxid. → 2,3-dinitro-*p*-toluic acid, 249
64	4,6-Dimethyl-1,3-dinitrobenzene	93	......	......	105	1-*mono*: 165; *di*: 295	258–9 (*di*)	......		125	4,5,6	N,N′-Diformyl deriv. of amine, 182–3
65	2,4′-Dinitrobiphenyl	93	......	363	45	202 (*di*)	276–8 (*di*)					
66	8-Chloro-1-nitronaphthalene	94	......	......	88–9; 96	137						
67	2,4-Dinitroanisole 1,3-Dinitro-4-methoxybenzene	95	......	......	....	....						2,4-Dinitrophenol, 114; Naphthalene adduct, 50; Reduct. → 2-amino-4-nitroanisole, 118, or.-red; Acetyl deriv. of this, 175–6
68	3-Nitroazobenzene	96, or.	......	......	56–7	130–1	......					......
69	4-Chloro-2-nitroanisole (5-Chloro-2-methoxy-1-nitrobenzene)	98	......	......	84	104	77–8	.....	194	......		......
70	2-Nitrophenanthrene	99, pa. yel.	......	......	85, pa. yel.	225	216					CrO₃ → 2-Nitrophenanthraquinone, 260, golden yel.
71	8-Bromo-1-nitronaphthalene	99–100	......	......	90	138–9	......					......

* Derivative data given in order: m.p., crystal color, solvent from which crystallized.

No.	Name	Melting point, °C	Boiling point, °C	Amine Boiling point, °C	Amine Melting point, °C	Acet-amide	Benz-amide	Benzene sulfon-amide	Picrate	Nitration Melting point, °C	Nitration Position of nitro groups	Miscellaneous
72	3,5-Dimethyl-1,4-dinitrobenzene	101	......	......	104	....	....					Reduct. → 1-amino-2,6-dimethyl-4-nitrobenzene, 158
73	1,2-Dinitronaphthalene	102–3, br.	......	......	98	234 (di)	291 (di)	1-mono: 215	......			....
74	5-Nitroacenaphthene ...	106; 101, yel.	......	......	108	238	210; 199	......	190–200			....
75	2,β-Dinitrostyrene	106–7, yel.										Alk. KMnO₄ → 2-nitrobenzoic acid, 146–8
76	5-Chloro-1-nitronaphthalene	111			85	128						....
77	3-Nitrodurene (3-Nitro-1,2,4,5-tetramethylbenzene)	112–3	......	261–2	75	207						
78	4-Nitrobiphenyl	114		302	53	171	230					N-Formyl deriv. of amine, 172
79	9-Nitrophenanthrene ...	116–7			137–8; 104	207–8	199		190			Picrate, 78–9
80	1,2-Dinitrobenzene.....	118			102	185 (di)	301 (di)	185				(NH₄)S → 2-Nitroaniline, 71; Hot aq. NaOH → 2-nitrophenol, 45
81	2,4'-Dinitrodiphenylmethane............	118, yel.			88–9	224–5 (210) (di)						Oxid. → 2,4'-dinitrobenzophenone, 197
82	4,5-Dimethyl-1,2-dinitrobenzene	118; 115			126	227–8 (di)						Reduct. → 1-amino-4,5-dimethyl-2-nitrobenzene, 140
84	3-Methyl-2-nitro-1,4-naphthoquinone	121–2; 125, yel.										Dil. KMnO₄ → phthalic acid, 200–6; Na₂SO₃ Reduct. or Fe + ac. a. → 2-amino-3-methyl-1,4-naphthoquinone, 167, red
85	1,3,5-Trinitrobenzene...	122										Anthracene adduct, 164; Naphthalene adduct, 156; Fluorene adduct, 105
86	5-Bromo-1-nitronaphthalene.........	122			69; 63–4	215						
87	Picric acid	122										Naphthalene adduct, 149; Fluorene adduct, 84; Anthracene adduct, 138; n-Butylammonium picrate, 151
88	2,5-Dimethyl-1,3-dinitrobenzene	123–4			102–3							Reduct. → 1-amino-2,5-dimethyl-3-nitrobenzene, 98
89	1-Nitro-2,4,6-tribromobenzene	125			122	232	198					Formyl deriv. of amine, 222
90	4-Bromo-1-nitrobenzene	126			66	168	204	134	180			NaOH → 4-Nitrophenol, 114
91	2,2'-Dinitrobiphenyl....	124; 128			81	mono: 89–90; di: 161	mono: 158–60; di: 190–1					N,N'-Diformyl deriv. of amine, 137
92	1,4-Dinitronaphthalene .	131–2, yel.	......		120, yel.	303–4 (di)	280 (di)					
93	3,5-Dimethyl-1,2-dinitrobenzene	132	......	......	78	....	....					Reduct. → 1-amino-2,4-dimethyl-6-nitrobenzene, 76
94	4-Nitroazobenzene	135, or.	......	>360	126	144–6	211; 205					

*Derivative data given in order: m.p., crystal color, solvent from which crystallized.

No.	Name	Melting point, °C	Boiling point, °C	Amine Boiling point, °C	Amine Melting point, °C	Acet-amide	Benz-amide	Benzene sulfon-amide	Picrate	Nitration product Melting point, °C	Nitration product Position of nitro groups	Miscellaneous
95	4-Chloro-1,5-dinitro-naphthalene	138										NaOH → 4,8-Dinitro-1-naphthol, 235; Sn + HCl → 1,5-Diaminonaphthalene, 190; PCl$_5$ → 1,4,5-Trichloronaphthalene, 131
96	2-Nitroindene	141										Zn + Ac. a. → 2-indanone oxime, 155
97	4-Bromo-1,5-dinitro-naphthalene	143										NaOH → 4,8-Dinitro-1-naphthol, 235; HNO$_3$ At 180 → 3-nitrophthalic acid, 218
98	2,4-Dinitrostilbene	143–5, yel.			119–20, pa. yel.							
99	1,3-Dinitronaphthalene	144–5, yel.			96	263–5 (di)						
100	9-Nitroanthracene	146, yel.			145–50	273–4						CrO$_3$ Oxid. of amine → anthraquinone, 273; 286, yel.
101	4-Chloro-1,3-dinitro-naphthalene	146–7										Warm dil. NaOH → 2,4-dinitro-1-naphthol, 140
102	2,5-Dimethyl-1,4-dinitrobenzene	147; 142			150							Reduct. → 1-amino-2,5-dimethyl-4-nitrobenzene, 144–5
103	3-Nitroacenaphthene	151–2, yel.			81–2	192–3	209–10		221			
104	4-Nitroazoxybenzene	153, pa. yel.			138	151; 172						
105	3,8-Dinitroacenaphthene	155–6, br.-yel.			167–8, yel.							
106	2-Nitrofluorene	156; 154			129	191						
107	3-Nitro-1,2-naphtho-quinone	156, red										SnCl$_2$ + HCl → 3-Amino-1,2-naphthohydroquinone, 164; Dil. HNO$_3$ → phthalic acid, 200–6
108	2,2'-Dinitrodiphenyl-methane	159			160							Oxid. → 2,2'-dinitrobenzophenone, 188–9
109	1,6-Dinitronaphthalene	161–2; 166, pa. yel.			85–6; 77	257 (263) (di)	265 (di)					
110	1,8-Dinitronaphthalene	170; 173			66		311–2 (di)			218	1,3,8	
111	4-Bromo-1,8-dinitro-naphthalene	170, yel.										NaOH → 4,5-Dinitro-1-naphthol, 235; HNO$_3$ At 180 → 3-nitrophthalic acid, 218; Al. NH$_3$ → 4-amino-1,8-dinitronaphthalene, 246, red; Acetyl deriv. of this, 245
112	3-Nitrophenanthrene	170–1			87–8	200–1	213					
113	4-Iodo-1-nitrobenzene	173			67–8	184	222					
114	1,4-Dinitrobenzene	173; 171			140; 147	304 (di), 162–3 (mono)	300 (di), 128 (mono)	247 (di)				Aq. NaOH → 4-nitrophenol, 114
115	3,3'-Dinitrodiphenyl-methane	175			53–4	193 (di)						Oxid. → 3,3'-dinitrobenzophenone, 155

*Derivative data given in order: m.p., crystal color, solvent from which crystallized.

TABLE XX. ORGANIC DERIVATIVES OF NITRO COMPOUNDS

b) Solids (Listed in order of increasing m.p.)* (Continued)

No.	Name	Melting point, °C	Boiling point, °C	Amine Boiling point, °C	Amine Melting point, °C	Acetamide	Benzamide	Benzene sulfonamide	Picrate	Nitration Melting point, °C	Nitration Position of nitro groups	Miscellaneous
116	4-Nitrophenanthrenequinone	179–80, pa. yel.										CrO₃ → 6-Nitrodiphenic acid, 248–50; o-Phenylene diamine → quinoxaline deriv., 217–8; Monoxime, 169–70; Dioxime, 210; Monosemicarbazone, 210–11
117	4-Chloro-1,8-dinitronaphthalene	180, pa. yel.										NaOH → 4,5-Dinitro-1-naphthol, 235; PCl₅ → 1,4,5-Trichloronaphthalene, 131
118	9-Nitrofluorene	181–2			64; 47	262	260–1					
119	4,4'-Dinitrodiphenyl-methane	183			93	236–7 (di)						Oxid. → 4,4'-dinitrobenzophenone, 189
120	4-Chloro-3-nitro-1,2-naphthoquinone	184, red										Aniline → 2-anilino-3-nitro-1,4-naphthoquinone-4-anil, 250
121	2-Nitroanthraquinone	185			303–6	262	227–8					
122	2,2'-Dinitrostilbene	196, yel.			176, golden yel.	304 (di)			209			
123	4,β-Dinitrostyrene	199, yel.										Acid K₂Cr₂O₇ → 4-nitrobenzoic acid, 240
124	3,3'-Dinitrobiphenyl	200, yel.			94	257–8 (di)						
125	2,5-Dinitrofluorene	207, yel.			175	289 (di)						
126	2,2'-Dinitroazobenzene	209–10; 194–5, yel.			134, red	271 (di), or.						
127	1,5-Dinitronaphthalene	214; 217			190	360 (di)				154	1,4,5	
128	4,4'-Dinitroazobenzene	222–3; 216, or.-red			250–1	212 (mono)						
129	2,5-Dinitrophenanthrenequinone	228, yel.-red										Monoxime, 190–1; o-Phenylenediamine → quinoxaline deriv., 262–4
130	1-Nitroanthraquinone	230			252; 243	218	255					
131	2,7-Dinitronaphthalene	234, yel.			166; 159	261 (di)	267 (di)		210 (di)			
132	4,4'-Dinitrobiphenyl	237; 240			128	317 (di), 199 (mono)	352 (di), 203–5 (mono)					
133	1,3-Dinitroanthraquinone	240, yel.			290		>300					
134	1,6-Dinitroanthraquinone	255–7, yel.			262, red	295 (di)	275 (di)					
135	2-Nitrophenanthrenequinone	258–60, yel.			205–10, dk. vlt.							CrO₃ → 4-Nitrodiphenic acid, 217; Monoxime, 213; Monothiosemicarbazone, 234–5
136	3-Nitrophenanthrenequinone	279–80, or.										CrO₃ → 5-Nitrodiphenic acid, 268; Monoxime, 240; Dioxime, 200; Monosemicarbazone, 254, red
137	2,7-Dinitroanthraquinone (Fritzsche's reagent)	280; 262, pa. yel.			>330	>350 (di)	300 (di)					

*Derivative data given in order: m.p., crystal color, solvent from which crystallized.

TABLE XX. ORGANIC DERIVATIVES OF NITRO COMPOUNDS

b) Solids (Listed in order of increasing m.p.)* (Continued)

No.	Name	Melting point, °C	Boiling point, °C	Amine Boiling point, °C	Amine Melting point, °C	Acet-amide	Benz-amide	Benzene sulfon-amide	Picrate	Nitration product Melting point, °C	Nitration product Position of nitro groups	Miscellaneous
138	*trans*-4,4'-Dinitro-stilbene............	288, yel.	......	......	231, yel.	353 (*di*)	352 (*di*)	......	......	......	......	....
139	**9,10-Dinitroanthracene** .	294; 263, yel.	......	......	......	....	....	......	......	......	......	Oxid. → anthraquinone, 273; 286, yel.
140	**1,7-Dinitroanthra-quinone**	295	......	......	290, red	283 (*di*)	325 (*di*)	......	......	......	......	....
141	**2,7-Dinitrophenanthrene-quinone**	301–3, pa. yel.	......	......	>360, dk. vlt.	....	....	......	......	......	......	CrO₃ → 4,4'-Dinitrodiphenic acid, 257–8; Monoxime, 246–8; *o*-Phenylenediamine → quinoxaline deriv., 356; Fluorene adduct, 270, red.-yel.
142	**1,8-Dinitroanthra-quinone**	311–2	......	......	262, red	284 (*di*)	324 (*di*)	......	......	......	......	....
143	**1,5-Dinitroanthra-quinone**	384–5, pa. yel.	......	......	319, red	317 (*di*)	>350 (*di*)	......	......	......	......	Monoxime, 253

*Derivative data given in order: m.p., crystal color, solvent from which crystallized.

343

EXPLANATIONS AND REFERENCES TO TABLE XXI

Hydrolysis to the corresponding acid.*

$$RCN + 2H_2O \xrightarrow{H_2SO_4} \underset{\text{Acid}}{RCOOH} + NH_4HSO_4$$

$$RCN + H_2O \xrightarrow{NaOH} \underset{\substack{\text{Sodium salt} \\ \text{of the acid}}}{RCOONa} + NH_3$$

From the nitrile and 75% sulfuric acid, or 4:1 phosphoric acid-sulfuric acid.

For directions and examples see: Cheronis, pp. 618, 619; Linstead, p. 65; Shriner, p. 258; Vogel, p. 410; Wild, p. 250.

From the nitrile and potassium hydroxide in aqueous methanol, ethanol or benzyl alcohol.

See: Linstead, p. 65; Vogel, pp. 410, 805; Wild, p. 250; L. Palfray, S. Sabetai and S. Rovira, *Compt. Rend.*, **209**, 483 (1939).

From the nitrile and potassium hydroxide in ethylene glycol or glycerol.

See: Cheronis, pp. 618, 620; S. Rovira and L. Palfray, *Compt. Rend.*, **211**, 396 (1940).

NOTE: For the directions and examples for the preparation of the derivatives of the carboxylic acid formed on hydrolysis see the explanations and references to Tables XII, XIII and XIV, pp. 186, 187, 188, 189.

Partial Hydrolysis to the corresponding amide.*

$$RCN + H_2O \xrightarrow{H_2SO_4} \underset{\text{Amide}}{RCONH_2}$$

From the nitrile with sulfuric acid.

For directions and examples see: Cheronis, pp. 619, 620; Vogel, p. 411.

Reduction to the corresponding amine.*

$$RCN + 4[H] \rightarrow \underset{\text{Amine}}{RCH_2NH_2}$$

From the nitrile and sodium in absolute ethanol.

For directions and examples see: Cheronis, p. 621; Shriner, p. 259; Vogel, p. 411; Wild, p. 254; H. B. Cutter and M. Taras, *Ind. Eng. Chem., Anal. Ed.*, **13**, 830 (1941).

From the nitrile and lithium aluminum hydride in ether.

See: W. G. Brown in *Organic Reactions*, Vol. 6 (Ed. R. Adams), John Wiley and Sons, New York, 1951, p. 469; N. G. Gaylord, *Reduction with Complex Metal Hydrides*, Interscience, New York, 1956, pp. 731–750.

For summary of reduction methods see: V. Migrdichian, *Organic Cyanogen Compounds*, Reinhold Publishing Corp., New York, 1947, pp. 151–172.

NOTE: For directions and examples for the preparation of the derivatives of the amine formed on reduction of the nitrile see explanations and references to Table XVIII, pp. 291, 292, 293, 294.

---

*Derivatives recommended for first trial.

WARNING: This is not an instruction manual. References should be consulted for the preparation of derivatives.

TABLE XXI. ORGANIC DERIVATIVES OF NITRILES
a) Liquids 1) (Listed in order of increasing atmospheric b.p.)*

No.	Name	Boiling point, °C	Melting point, °C	n_D	Density g/ml	Acid B.P., °C	Acid M.P., °C	Amide	Anilide	S-Benzyl thiuronium chloride	Amine B.P., °C	Benzamide	Benzene sulfonamide	Phenyl thiourea	Picrate	Miscellaneous
1	Cyanoacetylene	42.5	5	1.38699^{17}	0.8159^{17}	144d.; 83–4^{50}	18	61–2	87	......	......	......	......	......	......	HgNO₃ → Wh. precipitate; Dimer, 64.5–5.0
2	Cyanoacetaldehyde	71–2	......	......	0.881^{15}	......	......	......	......	......	45–6^{15}	......	......	......	......	HNO₃ → Cyanoacetic acid, 66; 2,4-Dinitrophenylhydrazone, 170–1, aq. al.; p-Nitrophenylhydrazone, 153–4
3	Acrylonitrile	77–8	......	1.393^{20}	0.797_4^{20}	140	13	85	105	......	......	......	......	......	......	Gives solid polymer on addition of conc. NaOMe sol.
4	Fluoroacetonitrile	80	......	......	......	165	33	77	......	......	......	......	......	......	......	......
5	Acetonitrile (Cyanomethane)	81–2	......	1.3442^{20}	0.7828_4^{20}	118	......	82	114	135	16.5	71	58	106; 135	165	......
6	Trichloroacetonitrile	86; 83–4	......	......	1.439^{12}	196–7	58	141	95–7	......	......	......	......	......	......	......
7	Methacrylonitrile (α-Methylacrylonitrile)	90–1	......	1.399^{25}	0.7991^{18}	160	15–6	102–6	......	......	......	......	......	......	......	Gives solid polymer on heating with benzoyl peroxide
8	Propionitrile (Cyanoethane)	97; 103.5	......	1.3659^{25}	0.777_4^{25}	141	......	79	106	151	49	84	36	63	135	......
9	Isobutyronitrile	104; 107–8	......	......	0.773	154.3	–47	129	109–10	143	69	57	53	82	150	......
10	Trimethylacetonitrile (tert-Butylcyanide)	106	15–6	1.3792	......	......	35	153–4	127–9	......	......	......	......	......	......	......
11	2-Ethylacrylonitrile	111	......	1.4132	......	180	......	84, bz.	......	......	......	......	......	......	......	......
12	Dichloroacetonitrile	113	......	......	$1.374^{11.5}$	194	5–6	98	......	......	......	......	......	......	......	Me. ester of acid, 143–4
13	α-Chloroisobutyronitrile	116	......	1.4045^{25}; 1.435^{14}	1.064_{14}^{14}	......	31	......	69–70	......	......	......	......	......	......	......
14	n-Butyronitrile (1-Cyanopropane)	117	......	1.3812^{24}	0.796^{15}	162	......	116	96	146	77	......	......	65	151	......

*Derivative data given in order: m.p., crystal color, solvent from which crystallized.

TABLE XXI. ORGANIC DERIVATIVES OF NITRILES
a) Liquids 1) (Listed in order of increasing atmospheric b.p.)* (Continued)

No.	Name	Boiling point, °C	Melting point, °C	n_D	Density g/ml	Acid B.P., °C	Acid M.P., °C	Amide	Anilide	S-Benzyl thiuronium chloride	Amine B.P., °C	Benzamide	Benzene sulfonamide	Phenyl thiourea	Picrate	Miscellaneous
15	*trans*-Crotononitrile.......	119	......	1.4217	....	189	72	158	115	......	83–4	.....	.....	109.5–10.5	131.5–2.5	n_D^{26} of amine: 1.4263; 1-Naphthylthiourea of amine, 129–30
16	Allylcyanide ...	119	......	1.4060^{20}	0.835^{15}	169; 163	.....	73	58	75–7	......	.....	.....	123–7; 56–7	136.8–7.4	n_D^{25} of amine: 1.4191; Al. KOH → crotonic ac., 72; 1-Naphthylthiourea of amine, 109–10
17	Methoxyacetonitrile.......	120	......	1.380^{28}	0.9373^{28}_{4}	203–4	.....	96	58	......	......	.....	.....	.....	.....	....
18	2-Hydroxyisobutyronitrile (Acetone cyanohydrin) .	120d.	–19	1.3996^{20}	0.93^{20}_{4}	.....	79	98	136	......	......	.....	.....	.....	.....	....
19	3-Hydroxy-4-methoxybenzonitrile.......	124; 131.5–2.0					255–7	.....	.....	......	.....	.....	.....	.....		Acetate, 116; Acetate of acid, 206–7; Me. ester of acid, 83–4; 66–7
20	2-Methylbutyronitrile.......	125–6	......	1.380^{25}	0.8061^{0}_{4}	177	.....	112	110	......	95.5–6.0	.....	.....	.....	.....	HCl salt of amine, 176; H_2PtCl_6 salt of amine, 240
21	Chloroacetonitrile.......	127	......	....	1.193^{20}	189	63	120	137	......	......	.....	.....	.....	142–3	Addition comp. with $AlCl_3$, 38, HCl salt of amine, 144
22	Isovaleronitrile .	130	......	....	0.7884^{20}_{0}	176	.....	136	113; 109–10	153	96	.....	.....	102	138	3-Nitrohydrogen phthalate of amine, 108
23	2,4-Pentadienonitrile.......	135–8	......	1.4880	0.8444	.....	72	124	.....	.....	.....	.....	.....	.....	.....	Me. ester of acid, $50–2^{20}$

*Derivative data given in order: m.p., crystal color, solvent from which crystallized.

TABLE XXI. ORGANIC DERIVATIVES OF NITRILES
a) Liquids 1) (Listed in order of increasing atmospheric b.p.)* (Continued)

No.	Name	Boiling point, °C	Melting point, °C	n_D	Density g/ml	Acid B.P., °C	Acid M.P., °C	Amide	Anilide	S-Benzyl thiuronium chloride	Amine B.P., °C	Benz-amide	Benzene sulfon-amide	Phenyl thiourea	Picrate	Miscellaneous
24	2-Chlorocro-tononitrile	136	......	......	......	212; 85–95[10]	99	212	.....	......	128–31	.....	.....	......	191	Me. ester of acid, 161; HCl salt of amine, 220
25	Ethoxyaceto-nitrile.......	136–7[753]	......	......	......	156–7[16]	.....	80–2	.....	......	108[750]	.....	.....	......	121–3	n_D^{25} of amine: 1.4108; Et. ester of acid, b.p. 152
26	2-Methylcro-tononitrile	138	......	1.4319	0.8313	198.5	64	75–6	77	......	......	.....	.....	......	....	p-Tolui-dide of acid, 70.5–1,5
27	2-Bromoiso-butyronitrile ..	139–40	.....	1.445[25]	....	.....	48–9	148	83	......	......	.....	.....	......	....	....
28	4-Pentenonitrile	140	.....	1.4213[14]	0.848$_{15}^{14}$	188–9	.....	94	.....	91–4; 98	......	.....	.....	......	....	H_2PtCl_6 salt of amine, 166; Thiourea deriv. of amine, 43.5–4.0
29	Thiophene-2,3-dicarbonitrile (2,3-Dicyano-thiophene)....	140	......	......	......	.....	272–4	di: 228	.....	......	......	.....	.....	......	....	Di-Me, ester of acid, 32–3
30	3,3-Dimethyl-acrylonitrile ..	140–2	......	......	0.8292[14]	199	70	107–8; 65–6	126–7	......	105–8	.....	.....	105.0–5.5	139–40d.	HCl salt of amine, 193–4
31	Valeronitrile (1-Cyanobutane).	141	.....	1.3991[15]	0.8035$_4^{15}$	186	.....	106	63	......	104	.....	.....	69	139	....
32	2-Chlorobutyro-nitrile.......	143	......	......	......	189[627]	.....	75.5–6.0	74–5	......	......	.....	.....	......	142; 124	
33	Diethylaceto-nitrile.......	145	......	......	......	190	.....	107	.....	......	71.5	.....	.....	......	168–9	....
34	2-Furanecarbo-nitrile (α-Furo-nitrile; 2-Cyanofuran)..	147	......	1.4798[20]	1.0822$_4^{20}$	.....	133–4	142	124	211	145–6	.....	.....	......	150	....
35	2-Methylaceto-acetonitrile ...	147; 145–6	......	1.4239[20]	0.9794$_4^{20}$	224[34]	73	138–40	.....	......	......	.....	.....	......	.....	Semicar-bazone, 153; p-Nitro-phenyl-hydra-zone, 147
36	Cyclobutane-carbonitrile (Cyanocyclo-butane)	150	......	......	......	195	.....	152–3; 155	112.5–3.0	......	110[753]	.....	.....	......	.....	HCl salt of amine, 235.5

*Derivative data given in order: m.p., crystal color, solvent from which crystallized.

TABLE XXI. ORGANIC DERIVATIVES OF NITRILES

a) Liquids 1) (Listed in order of increasing atmospheric b.p.)* (Continued)

No.	Name	Boiling point, °C	Melting point, °C	n_D	Density g/ml	Acid B.P. °C	Acid M.P. °C	Amide	Anilide	S-Benzyl thiuronium chloride	Amine B.P., °C	Benzamide	Benzene sulfonamide	Phenyl thiourea	Picrate	Miscellaneous
37	2-Chloro-3-methylbutyro-nitrile.......	154–5	......	......	0.9922[12]	210–2; 126[32]	37–9	......	......	......	......	......	......	......	137–8; 160	Et. ester of acid, b.p. 178–9; HCl salt of amine, 190d.; H_2PtCl_6 salt of amine, 193–5
38	Isocapronitrile (4-Methyl-pentano-nitrile)......	155	......	1.4085[14]	0.807[14]	199.4	−33	119; 121	111–2	......	125	......	......	......	123–5	......
39	2,2-Dimethyl-acetoaceto-nitrile.......	163–4	......	......	1.008[13]	103[1]	......	121	......	......	......	......	......	......	......	Oxime, 99–100
40	2-Methyl-hexanonitrile..	165	......	1.4070	0.7985	210	......	72	98	......	45–54[15]	......	......	......	......	......
41	3-Methoxy-propionitrile ..	165	......	1.4032	0.9367[25]	107[10]	......	50	......	......	120	......	......	......	......	......
42	n-Capronitrile (n-Hexano-nitrile)......	165	−80.31	1.4115[20]	0.8093[20][20]	205	......	100	95	......	130	40	96	77	126	HCl salt of amine, 219
43	(Ethylamino)acetonitrile (N-Ethylglycino-nitrile)......	166–7; 81–3[29]				......	180–2	......	......	......	126–9	di: 117–8	......	......	di: 194–5	HCl salt, 141–2
44	d,l-3-Methyl-hexanonitrile..	171–2[749]	......	1.4143	0.8109	212–3[755]	......	99–100	......	......	148–9[756]	......	......	......	......	p-Tolui-dide of acid, 73–4; n_D^{20} of amine, 0.7787
45	Chlorofumaro-nitrile.......	172; 64[10]	......	1.49571[20]	1.2499[20]	......	191–2	......	di: 186	......	......	......	......	......	......	Di-Me. ester of acid, b.p. 113–4[17]
46	2-Acetoxy-propionitrile (O-Acetyl lactonitrile)...	172–3	......	......	......	167–70[78]	57–60	......	......	......	......	......	......	......	......	H_2PtCl_6 salt of amine, 207–9
47	3-Ethoxypropio-nitrile.......	173	......	1.4068	0.9189[25][4]	119[19]	......	50	......	......	138	......	......	......	......	n_D^{20} of amine, 1.4242; D_4^{20} of amine, 0.8697
48	3-Chlorobutyro-nitrile.......	176	......	......	1.0772[9]	101[13]	43–4.5	......	89–90	......	......	......	......	......	147	H_2PtCl_6 salt of amine, 212
49	3-Chloropro-pionitrile	178	......	......	1.1443[18.5]	204	41	......	119, w.	......	......	......	......	......	......	p-Tolui-dide of acid, 121; HCl salt of amine, 146–8

*Derivative data given in order: m.p., crystal color, solvent from which crystallized.

TABLE XXI. ORGANIC DERIVATIVES OF NITRILES
a) Liquids 1) (Listed in order of increasing atmospheric b.p.)* (Continued)

No.	Name	Boiling point, °C	Melting point, °C	n_D	Density g/ml	Acid B.P., °C	Acid M.P., °C	Amide	Anilide	S-Benzyl thiuronium chloride	Amine B.P., °C	Benzamide	Benzene sulfonamide	Phenyl thiourea	Picrate	Miscellaneous
50	Indole-3-carbonitrile (3-Cyanoindole) .	178, rose	. . . . .	. . . .	. . . .	208–10	. . . . .	. . . . .	. . . . .	. . . . .	. . . . .	. . . .	. . . .	. . . .	. . . .	N-Acetyl deriv., 202; Et. ester of acid, 82; M.p. of amine, 84
51	5-Methyl-hexanonitrile . .	178–80	. . . . .	. . . .	. . . .	216	. . . . .	104	75	149.5	. . . . .	. . . .	. . . .	. . . .	. . . .	Me. ester of acid, b.p. 166–7
52	Thiophene-3-carbonitrile (3-Cyanothiophene)	179; 203–5	. . . . .	1.5565^{21}	1.1956^{20}_{20}	. . . . .	138	180	. . . . .		. . . . .	. . . .	. . . .	. . . .	. . . .	
53	d,l-4-Methyl-hexanonitrile . .	180	. . . . .	1.4144	0.8141	217–8	. . . . .	98			$152–3^{750}$	. . . .	. . . .	. . . .	. . . .	n_D^{20} of amine, 1.4238; D_4^{20} of amine, 0.7802
54	d,l-Lactonitrile (Acetaldehyde cyanohydrin) .	182–4	. . . . .	1.4058^{18}	0.9877^{20}_4	122^{15}	18	79; 75–6	59	153	161–2	. . . .	. . . .	. . . .	142	. . . .
55	Glycolonitrile (Formaldehyde cyanohydrin) .	183 sl.d.	. . . . .	. . . .	. . . .	. . . .	80	120	97	141; 146–7	171	. . . .	. . . .	. . . .	160	Benzoyl deriv., 195–6
56	Heptanonitrile .	183; 187	. . . . .	. . . . .	0.8107^{20}_0	223	. . . . .	95	71		155	. . . .	. . . .	75	121	. . . .
57	4-Cyanoheptane	183–4				221–2	. . . . .	123–4			167	. . . .	. . . .	. . . .	. . . .	Et. ester of acid, b.p. 183; H_2PtCl_6 salt of amine, 211 d.
58	Benzonitrile	190	− 13	1.5289^{20}	1.0102^{15}_{15}	. . . .	122	129	162	167	184–5	105	88	156	194	. . . .
59	Thiophene-2-carbonitrile (2-Cyanothiophene)	192	. . . . .	1.5641^{15}	1.1800^{15}_4	. . . .	129–30; 192	180	140	. . . . .	58^5	. . . .	. . . .	. . . .	181–2	HCl salt of amine, 193–4
60	2-Octynonitrile .	194–6	. . . . .	. . . . .	. . . .	$148–9^{19}$	f.p. 2–5	91	44		. . . . .	. . . .	. . . .	. . . .	. . . .	. . . .
61	4-Chlorobutyronitrile	196–7	. . . . .	. . . . .	1.162^{10}	196^{22}	16	88–9	60–70, bz.-pet. eth.		. . . . .	. . . .	. . . .	. . . .	. . . .	. . . .

*Derivative data given in order: m.p., crystal color, solvent from which crystallized.

| No. | Name | Boiling point, °C | Melting point, °C | n_D | Density g/ml | Derivatives of the corresponding acid RCN → RCOOH | | | | | Derivatives of the corresponding amine RCN → RCH₂NH₂ | | | | | Miscellaneous |
| | | | | | | Acid | | Amide | Anilide | S-Benzyl thiuronium chloride | Amine | Benzamide | Benzene sulfonamide | Phenyl thiourea | Picrate | |
						B.P., °C	M.P., °C				B.P., °C					
62	Methyl cyanoacetate.......	200; 115[36]	−22.5	1.4170[25]	1.0962[25]₄	.....	......				58[15]	.....	......	......	.....	Hydrolysis → malonic acid, 135; HCl salt of amine, 102.5; H₂PtCl₆ salt of amine, 192; NH₃ → Cyanoacetamide, 118
63	Dibenzyl acetonitrile ...	200–15	......	......	......	.....	89	128–9	155	......	......	......	......	......	.....	....
64	2-Tolunitrile (2-Methylbenzonitrile).......	205	−13	1.5272[25]	0.9912[25]₂₅	.....	104	140	125	146	208	88	......	......	215	Acetyl deriv. of amine, 69
65	2,3,3-Trimethyl-1-cyclopentene-1-carbonitrile (β-Campholytonitrile).......	205; 225	......	....	0.9127[15]	255–6	135	130, al.	104, aq. al.	......	205.5–6.5	......	......	......	178	Et. ester of acid, 222–5; HCl salt of amine, 175–6
66	Caprylonitrile (Octanonitrile)	206; 199	......	1.4224[15] He	0.8172[15]₄	239	16	110	57	......	180	......	......	......	112	....
67	1,1-Dicyanopropane (Ethyl malononitrile)	206[756]; 90–1[20]	......	....	0.9515[11]	.....	111.5	di: 216	.....	......	......	......	......	......	.....	Dihydrazide of acid, 168, al.
68	Ethyl cyanoacetate.......	207	......	1.4179[20]	1.056[25]₄	.....	......	......	198–9	......	......	......	......	......	.....	NH₃ → Cyanoacetamide, 118
69	1,1-Dicyanobutane (Propyl malononitrile)	210[750]	......	....	0.9224[18]	.....	96	di: 184	di: 198	......	......	......	......	......	.....	....
70	3-Tolunitrile (3-Methylbenzonitrile)	212	−23	....	1.0316[20]	.....	113; 110	95	126	140	207	......	......	......	198; 156	Acetyl deriv. of amine, 150
71	Cyclohexylacetonitrile ...	215	......	1.457[18]	0.913[18]	244.6	33	171–2	.....	......	188–9	79–81	......	......	155–6	HCl salt of amine, 252–3; n_D^{25} of amine, 1.4625
72	4,4-Dicyano-1-butene (Allyl malononitrile)	217–8	f.p. −12	....	....	.....	105, eth.	......	......	......	......	......	......	......	.....	Di-Et. ester of acid, b.p. 222–3; p-Nitrobenzyl ester of acid, 46, al.

*Derivative data given in order: m.p., crystal color, solvent from which crystallized.

TABLE XXI. ORGANIC DERIVATIVES OF NITRILES
a) Liquids 1) (Listed in order of increasing atmospheric b.p.)* (Continued)

No.	Name	Boiling point, °C	Melting point, °C	n_D	Density g/ml	Derivatives of the corresponding acid RCN → RCOOH					Derivatives of the corresponding amine RCN → RCH$_2$NH$_2$					Miscellaneous
						Acid B.P. °C	Acid M.P. °C	Amide	Anilide	S-Benzyl thiuronium chloride	Amine B.P., °C	Benzamide	Benzene sulfonamide	Phenyl thiourea	Picrate	
73	3-Isopropylidene-1-methylcyclopentane-1-carbonitrile (β-Fencholenonitrile)	217–9	.....	......	$0.9203^{15.6}$	259	72–3	86.5–7.5	.....						.....	Me. ester of acid, 97–9
74	3-Hydroxypropionitrile ..	220	.....	......	1.059^0	d	syrup	......	.....		188	.....	.....	......	222	Sodium salt of acid, 143; P$_2$O$_5$ → acrylonitrile, b.p. 77
75	1,1-Dicyano-3-methylbutane (Isobutyl malononitrile)	222	.....	......		.....	108, bz.	di: 195–6, al.	.....							....
76	Nonanonitrile ..	224	−34.2	1.42522	0.8221^{15}_4	255	15	99	57	.....	201	49	.....	......	111	Acetyl deriv. of amine, 34.5
77	2-Phenylcrotononitrile	$224–6^{751}$	.....	1.5555	1.013	.....	136	98–9	.....						......	
78	Ethylenecyanohydrin	229.7; 220	.....	......	1.059^0	.....	.....	......	.....		$107–8^{756}$	.....	.....	......	130; 222	Me. ester of acid, b.p. 177–84; H$_2$PtCl$_6$ salt of amine, 199
79	2-Phenylpropionitrile.......	232	.....	......	......	265–8	.....	97.5	.....	210	.....	.....	.....	......	182	Me. ester of acid, 221; M.p. of amine, 85; HCl salt of amine, 123–4; H$_2$PtCl$_6$ salt of amine, 229d.
80	Phenylacetonitrile (Benzyl cyanide)......	234	.....	1.5211^{25}	1.0214^{15}_{15}	.....	76–7	157	118	163	198	116	69	135	174; 167	....
81	Phenoxyacetonitrile.......	239–40	.....	......	$1.09^{17.5}$	285	98–9	101.5	.....		$228–9^{755}$	.....	.....	......	167–8	HCl salt of amine, 215; HBr salt of amine, 192–3

*Derivative data given in order: m.p., crystal color, solvent from which crystallized.

No.	Name	Boiling point, °C	Melting point, °C	n_D	Density g/ml	Acid B.P., °C	Acid M.P., °C	Amide	Anilide	S-Benzyl thiuronium chloride	Amine B.P., °C	Benzamide	Benzene sulfonamide	Phenyl thiourea	Picrate	Miscellaneous
						acid					amine					
82	4-Hydroxy-butyronitrile	240	......	....	1.029^8	204	......	....	.....	....	205–6	......	....	......	.....	Acid forms readily butyro-lactone, b.p. 204; CrO_3 + butyro-lactone → succinic acid, 186
83	(3-Tolyl)aceto-nitrile (m-Xylylcyanide)	240–1	......	....	1.0022^{22}	$120–3^{26}$	61	141	.....	....	$214–5^{744}$	......	....	......	176	Et. ester of acid, b.p. 237–8; HCl salt of amine, 159
84	(4-Tolyl)aceto-nitrile (p-Xylylcyanide)	242–3	18	1.5153^{25}	0.9922^{22}	265–7	94	185	.....	....	214.5	95–6	....	......	155	Et. ester of acid, b.p. 240; HCl salt of amine, 216–7
85	4-Isopropyl-benzonitrile	$243–4^{734}$	......	....	....	.....	117–8, al.	153; 133	.....	....	227^{724}	......	....	......	.....	Et. ester of acid, b.p. 263–4; HCl salt of amine, 239–40
86	(2-Tolyl)aceto-nitrile (o-Xylyl cyanide)	244	......	....	1.0156^{22}	.....	88–9	161	.....	....	215.5–7.0	......	....	......	177	HCl salt of amine, 227–8
87	Decanonitrile	245	......	1.4320^{15}_{He}	0.8294^{15}_4	269	31	100; 108	70	....	....	......	....	......	.....	....
88	3-Methyl-2-phenylbutyro-nitrile	$245–9^{745}$	......	1.5038^{25}	$0.967^{15.5}$	$159–60^{14}$	61–2	111–2	132–3	....	$155–6^{24}$	......	....	......	.....	HCl salt of amine, 128
89	1,2-Dicyano-propane (Methyl suc-cinonitrile)	252–4	12	....	....	.....	115	di: 225	di: 200	....	172–3	154	....	......	.....	HCl salt of diamine, 144–5
90	1-Undecano-nitrile (1-Hendecano-nitrile)	253–4	......	....	....	164^{15}	28.2–.6	98.0–.7	71	....	....	......	....	......	.....	Hydrazide of acid, 101–2; Phenyl-hydra-zide of acid, 110
91	2-Phenylvalero-nitrile	$254–5^{750}$	......	1.5000	0.960^{15}	280	58	83–5	.....	....	90^3	......	....	......	.....	....
92	10-Undeceno-nitrile (10-Hendeceno-nitrile)	257; $129–30^{14}$	......	1.4442^{20}	0.8443^{20}_4	274	24.5	87	.....	....	....	......	....	......	.....	....

*Derivative data given in order: m.p., crystal color, solvent from which crystallized.

TABLE XXI. ORGANIC DERIVATIVES OF NITRILES

a) Liquids 1) (Listed in order of increasing atmospheric b.p.)* (Continued)

No.	Name	Boiling point, °C	Melting point, °C	n_D	Density g/ml	Acid B.P., °C	Acid M.P., °C	Amide	Anilide	S-Benzyl thiuronium chloride	Amine B.P., °C	Benzamide	Benzene sulfonamide	Phenyl thiourea	Picrate	Miscellaneous
93	3-Phenylpropionitrile.......	261	......	....	1.0014[18]	280[754]	48.5	105	98	......	221.5[755]	57–8	.....	......	152–3	HCl salt of amine, 218; H₂PtCl₆ salt of amine, 233
94	2-Cyanobenzal chloride (α,α-Dichloro-o-tolunitrile)....	261	......	....	......	.....	155, bz.	117	.....							......
95	N-Methylanilinonitrile (N-Cyano-N-methylaniline)	266	13	....	......	.....	95–100	163	.....							Picrate, 195
96	3-(2-Chlorophenyl)propionitrile.......	267–8	.....	1.5390	1.1390[20]	.....	96.5, w.	119, bz.								Me. ester of acid, 255; HCl salt of amine, 167
97	1,3-Dicyano-2-methylpropane (2-Methylglutaronitrile).......	269–71	......	....	......	205–8[12]	79	di: 175–6	.....		78[11]	.....	......			Di-p-toluidide of acid, 174–5; n_D^{25} of amine, 1.4585
98	O-Benzoyl lactonitrile (Lactonitrile benzoate).......	269–70	......	....	......	.....		112	124	.....						Di-p-nitrobenzyl ester of acid, 119.5
99	3-Cyanobenzal chloride (α,α-Dichloro-m-tolunitrile)....	272–5	......	....	......	.....		132	.....							......
100	4-Cyanobenzal chloride (α,α-Dichloro-p-tolunitrile)....	273–6[770]	......	....	......	.....		151–8	.....							Et. ester of acid, 45–6
101	Dodecanonitrile (Lauronitrile).	276.7	4	1.43595	0.8273[15]	225	44	110; 102, aq. al.	78	141	247–9	.....	.....	....		M.p. of amine, 28.3; Acetyl deriv. of amine, 68.5–9.5, bz.; p-Toluenesulfonyl deriv. of amine, 73
102	1,3-Dicyanopropane (Glutaronitrile).......	286	......	1.4365[23]	0.995[15]₄	.....	97	175	224	.....	178–80	di: 135	119	148	237	....

*Derivative data given in order: m.p., crystal color, solvent from which crystallized.

TABLE XXI. ORGANIC DERIVATIVES OF NITRILES
a) Liquids 1) (Listed in order of increasing atmospheric b.p.)* (Continued)

No.	Name	Boiling point, °C	Melting point, °C	n_D	Density g/ml	Derivatives of the corresponding acid RCN → RCOOH					Derivatives of the corresponding amine RCN → RCH$_2$NH$_2$					Miscellaneous
						Acid		Amide	Anilide	S-Benzyl thiuronium chloride	Amine	Benzamide	Benzene sulfonamide	Phenyl thiourea	Picrate	
						B.P., °C	M.P., °C				B.P., °C					
103	4-Methoxyhydrocinnamonitrile (3-(4-Methoxyphenyl)-propionitrile).	290–300	. . . . .	. . . .	. . . .	. . . . .	104–5	125	. . . . .	. . . . . .	118–20[2]	. . . . .	. . . . .	. . . . . .	. . . . .	Me. ester of acid, 38; M.p. of amine, 65.7–6.3; HCl salt of amine, 220–5
104	1,4-Dicyanobutane (Adiponitrile)	295	0–1	1.4597[20]	0.951$^{19}_{19}$	. . . . .	153; 150	220	239	. . . . . .	204–5	di: 155	di: 154	. . . . . .	220	M.p. of amine, 42

*Derivative data given in order: m.p., crystal color, solvent from which crystallized.

TABLE XXI. ORGANIC DERIVATIVES OF NITRILES
a) Liquids 2) (b.p. at reduced pressure only) (Listed in order of increasing m.p. of the corresponding acid)*

No.	Name	Boiling point, °C	Melting point, °C	n_D	Density g/ml	Derivatives of the corresponding acid RCN → RCOOH					Derivatives of the corresponding amine RCN → RCH₂NH₂					Miscellaneous
						Acid B.P., °C	Acid M.P., °C	Amide	Anilide	S-Benzyl thiuronium chloride	Amine B.P., °C	Benzamide	Benzene sulfonamide	Phenylthiourea	Picrate	
1	1,2,2,3-Tetramethyl-3-cyclopentene-1-acetonitrile (5-Methyl-α-campholenonitrile) ...	115–9[18]	.....	1.47221[18]	0.9217[18]	150–1.5[12]	35.6–7.0	99–100, lgr.	......	......	......	......	......	......	......	....
2	1-Cyanocyclohexene	81[12]	.....	1.4818[25]	0.954[25]	238–40[638]	38	127–8	......	......	55–7[12]	......	......	......	181	....
3	2-Hydroxybutyronitrile (Propanal cyanohydrin)	102–3[23]	.....	1.4150	0.9621	225–60d.	43–4	......	89–90	......	172[755]	112.1–3.1	......	......	......	N,O-Di-p-nitrobenzoyl deriv. of amine, 119.2; Acetate, 43, CS₂
4	Hydnocarponitrile .	155–6[2–3]	.....	1.4559[25]	0.8580[25]	.....	59–60	108.5d.	......	......	......	......	......	......	......	....
5	α-Chloro-α-phenylacetonitrile	131–5[13]	.....				60–1	116	......	......	......	......	......	......	......	
6	Butyl cyanoacetate .	115[15]	.....	1.4243[25]	0.998[25]	.....	66	119–20	198–9	......	......	......	......	......	......	
7	3-Bromopropionitrile	69[7]	.....	1.4789[25]	....	.....	62	......	......	......	......	......	......	......	154, yel.	Nitrile + alkali → acrylic acid, p-toluidide of which, 141
8	2,4-Diphenylbutyronitrile	152–6[1]	.....			190[1]	72–3; 76	96	......	......	......	......	......	......	......	
9	Thiophene-2-acetonitrile	115–20[22]	.....	1.5399[25]	1.153[25]	.....	76; 63–4	146–7	......	......	72–4[3]	61, bz.-lgr.	......	109.5–10	......	HCl salt of amine, 202–4; N-Acetyl deriv. of amine, 45.5–6.5
10	trans-4-Chlorocrotononitrile	61–1.4[11]	.....	1.4705	1.1207	117–8[13]	83	130–2	......	......	......	......	......	......	......	
11	2-Cyanopentanoic acid	125–30[92]	.....			.....	96	di: 184	di: 198	......	......	......	......	......	......	Amide of nitrile, 124–5; Anilide of nitrile, 88–9
12	Azelaonitrile (1,7-Dicyanoheptane)..	183[11]; 160[3]	.....	1.4426[25]	....	.....	106	di: 175	di: 186–7	......	258–9	......	......	......	......	M.p. of amine, 37
13	3-Chloro-2-hydroxy-2-methylpropionitrile (Chloroacetone cyanohydrin)	110[22]	.....	1.4520	1.2027[15]	230–5	110	......	......	......	......	......	......	......	......	....
14	1,11-Dicyanoundecane (1,11-Dicyanohendecane) .	189–90[7]	.....			.....	111–2	di: 175–6	mono: 112.5–3.0; di 160–1	......	......	......	......	......	......	M.p. of amine, 58; HCl salt of amine, 254–5, al.

*Derivative data given in order: m.p., crystal color, solvent from which crystallized.

TABLE XXI. ORGANIC DERIVATIVES OF NITRILES
a) Liquids 2) (b.p. at reduced pressure only) (Listed in order of increasing m.p. of the corresponding acid)* (Continued)

No.	Name	Boiling point, °C	Melting point, °C	n_D	Density g/ml	Acid B.P., °C	Acid M.P., °C	Amide	Anilide	S-Benzyl thiuronium chloride	Amine B.P., °C	Benzamide	Benzene sulfonamide	Phenylthiourea	Picrate	Miscellaneous
						Derivatives of the corresponding acid RCN → RCOOH					**Derivatives of the corresponding amine RCN → RCH_2NH_2**					
15	2-Cyanobutyric acid	153^{15}	.....	....		.....	111.5	212–4	......		.....	......		......	.....	Hydrazide of acid, 95–6; Amide, 113
16	2-Cyanobiphenyl...	172^{15}	.....	....		.....	114	177	......		.....	......		......	.....	Acid + conc. $H_2SO_4 →$ fluorenone, 83
17	1,12-Dicyanododecane (α,ω-Dodecane dicyanide)...	$225-8^{17}$	.....	....		.....	129, w.	di: 189	di: 191; 170–1	......					.....	Di-p-toluidide of acid, 165; M.p. of amine, 61.5; HCl salt of diamine, 309–10
18	1-Cyano-4-isopropenylcyclohexene	$116-8^{11}$	.....	1.4978	0.9439	$164-5^{10}$	132–3	164–5	......						.....	....
19	Sebaconitrile (1,8-Dicyanooctane)..	$201-3^{16}$	.....	....	......	.....	134	di: 210	di: 198						.....	M.p. of amine, 60
20	Suberonitrile (1,6-Dicyanohexane)..	180^{12}	.....	1.4448^{22}	....	.....	141	di: 216–7	di: 186–7	......	240–1; 225–6	di: 121–2	......	......	di: 180	M.p. of amine, 52
21	3-Cyanoindene (Indene-3-carbonitrile).......	$140-2^{13}$	.....	....		.....	161; 156–7	180	158	......					.....	Hydrazide of acid, 186
22	Aminoacetonitrile (Glycinonitrile)..	58^{15}, part d.	.....	....		.....	262 d.	65–6	62 (+2 H_2O)	......	116.5	di: 244	di: 168	......	di: 233–5	HCl salt, 165.5–6.5; M.p. of amine, 8.5

*Derivative data given in order: m.p., crystal color, solvent from which crystallized.

TABLE XXI. ORGANIC DERIVATIVES OF NITRILES
b) Solids (Listed in order of increasing m.p.)*

No.	Name	Melting point, °C	Boiling point, °C	Derivatives of the corresponding acid RCN → RCOOH					Derivatives of the corresponding amine RCN → RCH$_2$NH$_2$				Miscellaneous
				Acid		Amide	Anilide	S-Benzyl thiuronium chloride	Amine	Benz-amide	Benzene sulfon-amide	Picrate	
				B.P., °C	M.P., °C				B.P., °C				
1	2-Cyanodiphenyl-methane	19	313–4	.....	117	163	......	......	.....	.....	.....	......	
2	N-Piperidinoacetonitrile	19	210	.....	215–7	......	......	......	182–3; 58–61[9]	......	.....	di: 225; 220–1	Methiodide, 192–3; HCl salt of acid, 215–6; Phenylthiourea deriv. of amine, 92–3
3	3-Chloro-2-tolunitrile ..	19	107[28]	.....	159; 154	......	......	......	.....	......	.....	......	
4	Tetradecanonitrile	19–20	......	.....	54	103	82	......	.....	......	.....	......	M.p. of amine, 37
5	Cinnamonitrile........	20	255–6	.....	133	153; 109	147	175; 179	.....	......	.....	181, yel., aq. al.	HCl salt of amine, 235
6	Trichloroacrylonitrile ..	20		.....	76	97–8; 87	98	......	.....	......	.....	......	n$_D^{20.5}$: 1.5100
7	DL-Mandelonitrile ..... (Benzaldehyde cyano-hydrin)	22	......	.....	118–9	133–4	151–2; 146	......	116–7[2]	148–9	.....	153–4, al.	M.p. of amine, 56.5–8.0; O-Benzoyl deriv., 63–4; 3-Nitrobenzoyl deriv., 83–4; p-Toluidide of acid, 174; HCl salt of amine, 208–19, acet.
8	Pentadecanonitrile.....	23	322	.....	52	102	78	......	299–301	......	.....	......	M.p. of amine, 34.0–6.5; HCl salt of amine, 199; N-Acetyl deriv. of amine, 72
9	2-Methoxybenzonitrile .	24–5	255–6	.....	100–1	129	131	......	226–8	......	.....	......	H$_2$PtCl$_6$ salt of amine, 187
10	(2-Chlorophenyl) aceto-nitrile (2-Chlorobenzyl cyanide)	25	251	.....	95	175, w.	138–9	......	105–8[12]	......	.....	187, bz.	
11	1,1-Dicyanoethane (Methylmalononi-trile)	26	......	.....	135; 138 d.	di: 217	di: 182	......	.....	......	.....	di: 252 d.	Di-p-toluidide of acid, 227–8; Di-HCl salt of amine, 201
12	2-Cyanopyridine (2-Pyridinecarbonitrile; Picolinonitrile).......	26	212–5	.....	136–7	106–7	76	......	95–8[20]	......	.....	159–60 d.	HAuCl$_4$ salt, 190; N-Acetyl deriv. of amine, 59–60, bz.-pet. eth.; Oxalate salt of amine, 166–7 d.
13	4-Tolunitrile (4-Methyl-benzonitrile)........	27; 29	217	.....	179–80	155; 165	147–8	......	208	137	.....	205–15 d.	D$_{30}^{30}$: 0.9805; M.p. of amine, 13; N-Acetyl deriv. of amine, 107–8
14	D-Mandelonitrile.....	28–9	170 d.	.....	133	123	......	......	.....	......	.....	......	[α]$_{5461}^{25}$: +46.9 in bz.
15	d,l-(2-Bromophenyl) acetonitrile (2-Bromo-benzyl cyanide)	29	242 d.	.....	84	143–4; 144–8	......	......	.....	......	.....	150–1	Et. ester of acid, b.p. 150–2[13]; M.p. of amine, 163–4; HBr salt of amine, 163–4
16	(4-Chlorophenyl) aceto-nitrile (4-Chlorobenzyl cyanide)	30	265–7	.....	105	175	164–5	......	114–6[15]	......	.....	212	HCl salt of amine, 218.0–8.5; p-Toluene-sulfonate deriv. of amine, 235
17	Malononitrile (Methyl-ene cyanide)	30	218–9	.....	135	170	......	......	136	140	96	250	n$_D^{34}$: 1.4146; N-Acetyl deriv. of amine, 126
18	Hexadecanonitrile.....	31	......	.....	63	106–7	90	......	.....	......	.....	......	D$_4^{31}$: 0.8224
19	Maleonitrile (cis-1,2-Dicyanoethylene)	31	......	.....	130	181	187	163; 173	163–70	di: 178.5–9.5	di: 155, al.	di: 250 d.	......
20	2,2-Dicyanopropane (Dimethylmalono-nitrile)............	31–2	169.5, subl.	.....	192–3, subl.; part. d. >130	di: 269	203–4	159–60	153	di: 152	.....	......	Di-Me. ester of acid, b.p. 177[753]; HCl salt of amine, 256–7

* Derivative data given in order: m.p., crystal color, solvent from which crystallized.

TABLE XXI. ORGANIC DERIVATIVES OF NITRILES
b) Solids (Listed in order of increasing m.p.)* (Continued)

No.	Name	Melting point, °C	Boiling point, °C	Acid B.P., °C	Acid M.P., °C	Amide	Anilide	S-Benzyl thiuronium chloride	Amine B.P., °C	Benzamide	Benzene sulfonamide	Picrate	Miscellaneous
						Derivatives of the corresponding acid RCN → RCOOH			Derivatives of the corresponding amine RCN → RCH_2NH_2				
21	*tert*-Butylacetonitrile (Neopentyl cyanide)	32–3	138	186–8	....	132	131	......	......	......	......	......	HCl salt of amine, 158–60, subl.
22	1-Naphthylacetonitrile	33	......	....	131	154; 180	......						
23	4,4-Dicyanoheptane (Dipropylmalononitrile)	33–4 (*anh.*); 49–50 (*hyd.*)	223–4; 135–40$^{0.2}$	....	....	161	*di*: 214	*di*: 168.0–.5, me. al.	132^{25}	*di*: 154	......	......	......
24	Heptadecanonitrile	34	......	....	61	106	......	......	335–40	91	......	......	M.p. of amine, 49; N-Acetyl deriv. of amine, 62
25	1-Naphthonitrile (1-Cyanonaphthalene)	34; 37	299	....	162	202	......		155^{12}		148	223	N-Acetyl deriv. of amine, 134, lgr.; Methiodide deriv. of amine, 213, al.
26	2-Cyanopropionic acid	35	142–5^{11}	....	135; 120	*di*: 206	*di*: 182; 214	......					Amide, 105; 81; Mono-*p*-toluidide of acid, 145 d.; Di-*p*-toluidide of acid, 227–8; 245
27	4-Fluorobenzonitrile	35	189–90	....	183	154–5	......		183			203	....
28	Coumarilonitrile (Coumarin-2-carbonitrile)	36	......	310–5, sl.d.	192–3, w.	159	159	......					Et. ester of acid, 27; Ph. ester of acid, 101
29	Indole-3-acetonitrile	36.0–.5	157$^{0.2}$	....	164–5; 199	150–1	149.5–50	......		137–8	....	247 d.	Me. ester of acid, 135; M.p. of amine, 116–7; 146; HCl salt of amine, 248–9; N-Acetyl deriv. of amine, 77, pet. eth.
30	3-Bromobenzonitrile	38	225	....	155	155	......		244–5	135–6	....	205	....
31	2-(N-Anilino)-butyronitrile	39	......		141	123, w.	92, al.						Et. ester of acid, 26; b.p. 278
32	*trans-o*-Chlorocinnamonitrile	40			212, yel., al.	168	176						Me. ester of acid, b.p. 278–9
33	Octadecanonitrile	41; 43			70	109	95; 88						
34	3-Chlorobenzonitrile	41	......	....	158; 155	134	122–5	......	110–2^{17}	214	....	203	....
35	2-Chlorobenzonitrile	43	232	....	141	142	118	......	103–4^{11}	116–7	....	217	N-Acetyl deriv. of amine, 79–80
36	4-Chloromandelonitrile	43			119–22; 112–3	122–3							O-Benzoyl deriv., 57–8; Me. ester of acid, 85–8; bz.-pet. eth.
37	Nonadecanonitrile	43	......		69	109.5–.8	95.5–6.5						Me. ester of acid, 39.5–40
38	2-Bromo-4-tolunitrile	44	......		204	......							
39	3,3-Dicyanopentane (Diethylmalononitrile)	44–5	195; 92^{24}	....	125, w.	*mono*: 146; *di*: 224	......						....
40	4-Cyanobutyric acid	45	......		97–8	182–3	*di*: 221–2	......	......		105		Amide, 69–70; Et. ester of acid, b.p. 245; M.p. of amine, 157–8; HCl salt of amine, 92–4; HAuCl₄ salt of amine, 86–7; 106
41	5-Chloro-2-tolunitrile	45–6		....	169	......							Me. ester of acid, 30–1
42	(4-Aminophenyl)acetonitrile (4-Aminobenzyl cyanide)	46	312	....	199	161–2							N-Benzoyl deriv., 176–7; Picrate, 185
43	*meso*-2,3-Dimethyl-succinonitrile	46			209; 198 d.	*di*: 310–3	*di*: 235						Di-Me. ester of acid, b.p. 198–9

*Derivative data given in order: m.p., crystal color, solvent from which crystallized.

TABLE XXI. ORGANIC DERIVATIVES OF NITRILES

b) Solids (Listed in order of increasing m.p.)* (Continued)

No.	Name	Melting point, °C	Boiling point, °C	Acid B.P., °C	Acid M.P., °C	Amide	Anilide	S-Benzyl thiuronium chloride	Amine B.P., °C	Benzamide	Benzene sulfonamide	Picrate	Miscellaneous
44	**3-Bromo-4-tolunitrile**	47	......	.....	140	137	......	......	......	......	......	......	......
45	**(4-Bromophenyl)aceto-nitrile (4-Bromobenzyl cyanide)**	47	......	.....	114, subl.	192–4	......						Et. ester of acid, 30; HCl salt of amine, 240–3
46	**N-Anilinoacetonitrile**	48, lgr.	......	.....	127–8	136	113	......					......
47	**3-Cyanopropionic acid**	48–50	......	.....	235	185	mono: 157; di: 269 d.	mono: 148.5; di: 230	......	79–80	.....		Amide, 97; Me. ester of acid, b.p. 215; Di-p-toluidide of acid, 256; M.p. of amine, 193; 203 d.; HCl salt of amine, 135–6; 95–6
48	**3-Chloro-4-tolunitrile**	48–50	......	.....	200–2								Me. ester of acid, 28
49	**3,3-Diphenylacrylo-nitrile (β-Phenyl cinnamonitrile)**	49	......	.....	162; 167	......	130–1						......
50	**3-Bromo-2-hydroxy benzonitrile**	49–50	......	.....	184	165	......						......
51	**4,4-Dicyanoheptane (Dipropylmalono-nitrile)**	49–50 (hyd.); 33–4 (anh.)	......	.....	161	di: 214	di: 168.0–.5, me. al.	......	132[25]	di: 154	.....		
52	**trans-2,3-Diphenyl-acrylonitrile**	49–51	213–4[23]	.....	172	127, acet.	141, al.						......
53	**Eicosanonitrile**	49.5	......	.....	77; 75.2	108.9	92	.....	196.2[4.0]	......			M.p. of amine, 57.8; n_D^{70} of amine, 1.4341
54	**3-Cyanopyridine (Nicotinonitrile)**	50	240–5	.....	232	122	85, w.; 132, bz.-lgr.	.....	88–90[2]	132	.....	tri: 206–8	HAuCl₄ salt of amine, 196–8; Di-HCl salt of amine, 224
55	**(4-Iodophenyl)aceto-nitrile (4-Iodobenzyl cyanide)**	50–1	......	.....	135	......	......						HCl salt of amine, 294–6 d.
56	**4-Cyanodiphenyl-methane**	51	......	.....	157–8								Acid $\xrightarrow{CrO_3}$ 4-Benzoyl-benzoic acid, 197–200, al.
57	**2-(N-Anilino)-valero-nitrile**	51, pet. eth.	......	.....	147–8, al.	99, eth.-pet. eth.	......						......
58	**2-Aminobenzonitrile (Anthranilonitrile)**	51, yel., CS₂	267–8[777]	.....	147	109–11	130–1	.....	118[9]	167			Et. ester of acid, b.p., 266–8; p-Toluidide of acid, 151
59	**2-Bromobenzonitrile**	53	253	.....	150	155–6	......	.....	118[9]				HCl salt of amine, 241; 208
60	**5-Cyanothiazole**	53	......	.....	218	186							Me. ester of acid, 68–9; N-Acetyl deriv. of amine, 159–60
61	**3-Aminobenzonitrile**	53–4, aq. al.	288–90	.....	174	78–9	114						N-Acetyl deriv., 248; Me. ester of acid, 36–8; Et. ester of acid, b.p. 294; Fe/AcOH → Benzonitrile, b.p. 190 + NH₃
62	**2-Quinolinoacetonitrile**	53–4	140[1]	.....	274–5	......	......	.....	138–40[7]	......	.....	di: 202	Picrate, 176–7 d.; Me. ester of acid, 72, lgr.
63	**2-Iodobenzonitrile**	55	......	.....	162	184	......	.....		154	.....	.....	HCl salt of amine, 248–50; N-Acetyl deriv. of amine, 134–5

*Derivative data given in order: m.p., crystal color, solvent from which crystallized.

TABLE XXI. ORGANIC DERIVATIVES OF NITRILES

b) Solids (Listed in order of increasing m.p.)* (Continued)

No.	Name	Melting point, °C	Boiling point, °C	Acid B.P. °C	Acid M.P. °C	Amide	Anilide	S-Benzyl thiuronium chloride	Amine B.P. °C	Benz-amide	Benzene sulfonamide	Picrate	Miscellaneous
64	2,4,6-Trimethylbenzonitrile	55, bz.	225–30	.....	155, lgr.	187–8	......	......	98.5–100[6]	153–4	.....	......	Me. ester of acid, b.p. 241–2[718]
65	α-Aminobenzyl cyanide	55	......	.....	256	130–2, al.	......	......	135[19]	*di:* 223–5	.....	.....	N-Acetyl deriv., 130–8, et. ac.-pet. eth.; N-Benzoyl deriv., 151; Picrate, 160–1, yel.
66	Cyanoform (Tricyanomethane)	55–6, pet. eth.	......	......	......	......	......	......	......	......	......	......	NH$_4$ salt, 183
67	Succinonitrile	56.6	265–7	.....	186–8	260	230	149	159	177	.....	250–5	M.p. of amine, 27; N-Acetyl deriv. of amine, 137
68	2-Iodo-4-tolunitrile (2-Iodo-4-methylbenzonitrile)	57–8	......	.....	205–6	167							......
69	2,6-Dinitrobenzonitrile	58 (145)		.....	202–3								Me. ester of acid, 147; M.p. of amine, 88; HCl salt of amine, 185; H$_2$PtCl$_6$ salt of amine, 193
70	d,l-2,3-Dimethylsuccinonitrile	58–9	......	.....	135	*mono:* 148–9; *di:* 244	*di:* 222	......					Di-Me. ester of acid, b.p. 200
71	2-Chloro-4-tolunitrile	61–2	......	.....	155–6	182	......						......
72	4-Methoxybenzonitrile	61–2	256–7	.....	184–6	162–3	169	177; 184–5	236–7	......	.....		N-Acetyl deriv. of amine, 96; H$_2$PtCl$_6$ salt of amine, 210
73	2,4-Dichlorobenzonitrile	61–2		.....	164; 160	194	......		140[122]				Me. ester of acid, b.p. 132[15]; n_D^{25} of amine: 1.5738; HCl salt of amine, 281.2–4.3
74	4-Methoxycinnamonitrile	64		.....	170	186							Me. ester of acid, 90; Et. ester of acid, 49–50
75	3,5-Dichlorobenzonitrile	65, subl.		.....	188	......			254[6]			241	Me. ester of acid, 58; n_D^{26} of amine: 1.5690; HCl salt of amine, 300
76	cis-1,4-Dicyanocyclohexane	65		.....	168–9								Acid + conc. HCl at 180 → *trans* isomer, 300
77	Bromomalononitrile	65–6		.....	113d.	181	......						Di-p-toluidide of acid, 217
78	2-Naphthonitrile (2-Cyanonaphthalene)	66; 62	305	.....	184	192; 195	......					226	M.p. of amine, 60; N-Acetyl deriv. of amine, 126; Methiodide deriv. of amine, 168, al.
79	Cyanoacetic acid	66		......	135–6	170	226–7			120, w.	.....		Amide, 119–20; Anilide, 198–9; Warming with benzaldehyde → α-cyanocinnamic ac., 180; M.p. of amine, 200; HCl salt of amine, 123
80	2-Cyano-2-ethylbutyric acid (Diethylcyanoacetic acid)	66; 57	240–5; 164[8]	.....	125, w.	*mono:* 146; *di:* 224							NH$_3$ → Diethylcyanoacetamide, 121

*Derivative data given in order: m.p., crystal color, solvent from which crystallized.

TABLE XXI. ORGANIC DERIVATIVES OF NITRILES
b) Solids (Listed in order of increasing m.p.)* (Continued)

No.	Name	Melting point, °C	Boiling point, °C	Acid B.P., °C	Acid M.P., °C	Amide	Anilide	S-Benzyl thiuronium chloride	Amine B.P., °C	Benzamide	Benzene sulfonamide	Picrate	Miscellaneous
				Derivatives of the corresponding acid RCN → RCOOH					Derivatives of the corresponding amine RCN → RCH₂NH₂				
81	2,4-Diphenylglutaronitrile	66; 70-1	220-7[5]	.....	2 forms: (a) 164-5, w.; (b) 185-6, chl.-pet. eth.	......						.....	Di-HCl salt of amine, 319-20, iso-PrOH
82	α-Chloro-3-tolunitrile	67, al.	258-60	.....	132, w.	124	......					173	HCl salt of amine, 169; H₂PtCl₆ salt of amine, 219
83	4-Chloro-2-tolunitrile	67	......	.....	172	183							
84	1-Cyanoacenaphthene (Acenaphthene-1-carbonitrile)	68	......	.....	161							......	
85	Phenylmalononitrile (α-Cyanobenzyl cyanide)	69	......	.....	152-3	233						......	
86	6-Nitro-2-tolunitrile	69-70	......	.....	184	163							Me. ester of acid, 66
87	(4-Hydroxyphenyl)acetonitrile (4-Hydroxybenzyl cyanide)	69-70, w.	330	.....	148-50	175						200	Acetate, 49-50; M.p. of amine, 160-1
88	5-Bromo-2-tolunitrile	70	......	.....	187	180						......	
89	α-Bromo-2-tolunitrile (2-Cyanobenzyl bromide)	71-2	124[4]	.....	147	......							Me. ester of acid, 32.0-.5
90	2,2-Diphenylglutaronitrile	71-2.5	......	.....	195-7; 183	1-mono: 142-4	......						Anhydride of acid, 134; Imide of acid, 158-9
91	(2-Aminophenyl)acetonitrile (2-Aminobenzyl cyanide)	72	......	.....	119	93							N-Acetyl deriv., 120; Benzamide, 175-9; N-Acetyl deriv. of acid, 158
92	3,4-Dichlorobenzonitrile	72	......	.....	208-9	133	......		139-40[17]			228	Me. ester of acid, 46.5-7.5; HCl salt of amine, 240-2
93	1,2,2,3-Tetramethylcyclopentene-1-carbonitrile (Campholic nitrile)	73	217-9	255	106	80	91, al.	......	210	98		......	
94	Dicyanodimethyl amine (Bis(cyanomethyl)amine)	75; 77	......	.....	247.5; 225d.	di: 143	di: 140.5	......	208	tri: 166, chl.	.....	tri: 212d.	N-Nitroso deriv., 43; N-Benzoyl deriv., 131-2; Tri-HCl salt of amine, 233
95	Diphenylacetonitrile (α-Phenylbenzyl cyanide)	75	......	.....	148	168	180	145	134[2]	144-5	.....	211-2 d.	M.p. of amine, 42-3.5; HBr salt of amine, 205-7
96	4-Cyano-N,N-dimethylaniline	75.6	318	.....	242.5-3.5	206	182-3	......					Me. ester of acid, 102; HCl salt of amine, 212
97	1-Cyanoisoquinoline	78; 93	......	.....	161	......							
98	4-Cyanopyridine	78; 83	......	.....	315	155-6			120-5[12]			179-80, al.	HCl salt, 199; HAuCl₄ salt, 208-10
99	α-Chloro-4-tolunitrile (4-Cyanobenzyl chloride)	78-80, al.	263	.....	203	173						185	H₂PtCl₆ salt of amine, 226
100	2,5-Diphenylvaleronitrile	79	......	.....	80-1, lgr.	......							
101	3-Cyanobenzaldehyde (3-Formylbenzonitrile)	79-81	......	.....	175, w.	190d.							Oxime, 99-101, w.; Semicarbazone of acid, 265; Phenylhydrazone of acid, 164

*Derivative data given in order: m.p., crystal color, solvent from which crystallized.

TABLE XXI. ORGANIC DERIVATIVES OF NITRILES
b) Solids (Listed in order of increasing m.p.)* (Continued)

No.	Name	Melting point, °C	Boiling point, °C	Acid B.P., °C	Acid M.P., °C	Amide	Anilide	S-Benzyl thiuronium chloride	Amine B.P., °C	Benzamide	Benzene sulfonamide	Picrate	Miscellaneous
				Derivatives of the corresponding acid RCN → RCOOH					Derivatives of the corresponding amine RCN → RCH₂NH₂				
102	6-Nitro-3-tolunitrile....	80	......	....	219	151	......	......	......	....	......	......	Me. ester of acid, 81–2; Et. ester of acid, 55, al.
103	Benzoylacetonitrile	81	......	....	103–4d.	113	108	......	......	....	......	155d.	Carbazone of acid, 125; HCl salt of amine, 128
104	6-Chloro-2-tolunitrile ..	82–3	......	....	102	167	......	......	......	....	......	......	Me. ester of acid, 104
105	8-Cyanoquinoline......	84, 50% al.	......	....	187	171–3, w.	......	......	......	....	......	......	Et. ester of acid, 38; b.p. 175–7[7]
106	2-Nitro-3-tolunitrile....	84	......	....	223	192	......	......	......	....	......	......	Me. ester of acid, 74, me. al.
107	2,3,4,5-Tetrachloro-benzonitrile.........	84	......	....	194–5	207–8, acet.-al.	197, al.	......	......	....	......	......	M.p. of amine, 90–1
108	4-Cyanobiphenyl	85–6	......	....	228	223	......	......	195–200[15]	....	......	......	Me. ester of acid, 117–8; M.p. of amine, 128; HCl salt of amine, 308–10; 282
109	2-Naphthylacetonitrile .	86; 81	......	....	142	200	......	......	......	....	......	......	......
110	cis-2,3-Diphenylacrylo-nitrile	86	228–30[23]	....	137–8	167–8, chl.-lgr.	179, chl.-pet. eth.	......	......	....	......	......	......
111	4-Aminobenzonitrile (4-Cyanoaniline)	86	......	....	188	182.9	......	......	268–70	....	......	......	N-Acetyl deriv., 205; N-Benzoyl deriv., 170; Picrate, 150
112	1-Cyano-2-phenylacrylo-nitrile (Benzalmalono-nitrile).............	87	......	....	195–6d.	di: 189–90	......	......	......	....	......	......	Di-Me. ester of acid, 44; Di-Et. ester of acid, 32
113	5-Bromo-2,4-dimethyl-benzonitrile.........	88–90	......	....	180–1	197.5–8.5	......	......	......	....	......	......	......
114	2-Cyanotriphenyl-methane	89	270–85[20–30]	....	162	......	......	......	......	....	......	......	Me. ester of acid, 98, me. al.
115	5-Cyanoquinoline......	89 (anh.); 70 (+1.5 H₂O)	......	....	342	......	......	......	......	....	......	......	Picrate, 241; Styphnate, 241d., yel.; HCl salt, 248, yel.; Me. ester of acid, b.p. 128–32[0.2]
116	2,6-Dimethylbenzo-nitrile	90–1	......	....	116	138.5–9.0; 120–5	......	......	96–8	....	......	......	Me. ester of acid, b.p. 94[9]
117	Phenylcyanoacetic acid .	92	......	....	......	......	......	......	......	....	......	......	Amide, 147; Anilide, 136
118	2-(N-Anilino)-propio-nitrile	92, al.	......	....	162, w.	144	127, al.	......	......	....	......	......	Et. ester of acid, b.p. 272
119	2,4-Dibromobenzonitrile	92	......	....	174	198	......	......	......	....	......	......	Me. ester of acid, 33
120	β-(2-Nitrophenyl)-acrylonitrile.........	92, w.	194–6[7–8]	....	cis: 146–7, yel.; trans: 240	185	......	......	......	....	......	......	Et. ester of acid, 60
121	5-Chloro-2-nitro-4-tolunitrile	93	......	....	180–1	......	......	......	......	....	......	......	......
122	α-Bromo-3-tolunitrile (3-Cyanobenzyl bromide)..........	93, al.	......	....	155–6	......	......	......	......	....	......	......	......
123	4-Nitro-3-tolunitrile....	93–4	......	....	134	176–7	......	......	......	....	......	......	Me. ester of acid, 78–9, me. al.
124	2-(N-Anilino)-isobutyro-nitrile	93–4, al.	......	....	184–5, w.	136	155, al.	......	......	....	......	......	......
125	2-Cyanoquinoline......	94		....	157	133; 123	......	.,....					

*Derivative data given in order: m.p., crystal color, solvent from which crystallized.

TABLE XXI. ORGANIC DERIVATIVES OF NITRILES
b) Solids (Listed in order of increasing m.p.)* (Continued)

No.	Name	Melting point, °C	Boiling point, °C	Acid B.P., °C	Acid M.P., °C	Amide	Anilide	S-Benzyl thiuronium chloride	Amine B.P., °C	Benzamide	Benzene sulfonamide	Picrate	Miscellaneous
				Derivatives of the corresponding acid RCN → RCOOH					**Derivatives of the corresponding amine RCN → RCH$_2$NH$_2$**				
126	4-Cyanovaleric acid (2-Methylglutaromononitrile)	95.6, w.	......	205–8[12]	79	*di:* 175–6							Di-*p*-toluidide of acid, 74–5; H$_2$PtCl$_6$ salt of amine, 202–3
127	Fumaronitrile	96	......	....	295; 300	266	313–4	178; 182	......				
128	4-Chlorobenzonitrile	96; 94–6	223; 95[5]	....	240	179; 170	194	......	217	140		210	
129	9-Phenanthrylacetonitrile	96.5–7.0			224.5	250–2							
130	3,5-Dibromobenzonitrile	97, subl.			219–20	187							Me. ester of acid, 63
131	2-Chloro-3-nitrobenzonitrile	97–101, pa. yel.			185								Me. ester of acid, 70; Et. ester of acid, b.p. 314d.
132	2-Hydroxybenzonitrile (2-Cyanophenol)	98		211[20]	159	133	135						D$_4^{99.6}$: 1.1052; n$_\alpha^{99.6}$: 1.53716; O-Benzoyl deriv., 106, pet. eth.; M.p. of amine, 125; 129; H$_2$PtCl$_6$ salt of amine, 197d.; HI salt of amine, 184
133	4-Chloro-2-nitrobenzonitrile	98			142	172							Me. ester of acid, 42–3
134	4-Cyanotriphenylmethane	100, me. al.			165, ac. a.		196, ac. a.						
135	4-Chloro-3-nitrobenzonitrile	100–1			181–2	156	131					210d.	Me. ester of acid, 83; HCl salt of amine, 227–8
136	3-Nitro-4-tolunitrile	101			164–5	153							
137	2-Cyano-3-phenylpropionic acid	101–2; 75			120–1	225							Amide, 130; Di-Me. ester of acid, 35–6
138	3-Cyanophenanthrene	102			269	233–4; 227–8	216–7						
139	2,3,3-Triphenylpropionitrile	102, me. al.			222–3, al.	213							Me. ester of acid, 162; HCl salt of amine, 269–72
140	4-Cyanoquinoline	102			253–4	181							Methiodide, 216; HAuCl$_4$ salt, 232; Di-HCl salt of amine, 255d.
141	4-Bromo-1-naphthonitrile (1-Bromo-4-cyanonaphthalene)	102–3			212; 220								Me. ester of acid, 42
142	4-Bromo-2,5-dimethylbenzonitrile	103–4			171.5–2.5	209–10							
143	5-Nitro-3-tolunitrile	104–5			174	164–5							Me. ester of acid, 84–5
144	2,4-Dinitrobenzonitrile	104–5, al.			182–3	203–4							Me. ester of acid, 70; Hydrazide of acid, 231–3
145	4-Nitro-2-tolunitrile	105			179	173–4							Me. ester of acid, 69
146	6-Chloro-3-nitrobenzonitrile	105–6			165	178							Me. ester of acid, 73 ·
147	5-Bromo-3-nitro-2-tolunitrile	106–7			226	235							
148	2-Nitro-4-tolunitrile	107, pa. yel.			190	166							
149	9-Cyanophenanthrene	107; 111.0–.3			256.5–7.0	232–3; 226	218			167		241	M.p. of amine, 108.5; HCl salt of amine, 294; N-Acetyl deriv. of amine, 182–3
150	3-Cyanoquinoline	108			275	198–9; 195							Picrate of acid, 217–8

* Derivative data given in order: m.p., crystal color, solvent from which crystallized.

TABLE XXI. ORGANIC DERIVATIVES OF NITRILES
b) Solids (Listed in order of increasing m.p.)* (Continued)

No.	Name	Melting point, °C	Boiling point, °C	Derivatives of the corresponding acid RCN → RCOOH					Derivatives of the corresponding amine RCN → RCH$_2$NH$_2$				Miscellaneous
				Acid		Amide	Anilide	S-Benzyl thiuronium chloride	Amine	Benzamide	Benzene sulfonamide	Picrate	
				B.P., °C	M.P., °C				B.P., °C				
151	**2-Cyanophenanthrene**	109			259–60	242–3	217–8						
152	**3-Nitro-2-toluinitrile**	109–10			151–2, pa. yel.	158							Me. ester of acid, 50
153	**2-Nitrobenzonitrile**	110			146	174–6	155						HCl salt of amine, 248–50
154	**4-Chloro-1-naphtho-nitrile** (1-Chloro-4-cyanonaphthalene)	110			223–4; 210	235–6							
155	**5-Cyanoacenaphthene** (Acenaphthene-5-carbonitrile)	110–1			219, bz.	198				182–4	148–9		
156	**4-Bromobenzonitrile**	112	235–7		251	189–90			250	143		221	M.p. of amine, 20; N-Acetyl deriv. of amine, 113
157	**2,4,5-Trimethoxybenzo-nitrile**	112–4, al.		ca. 300	144, bz.-pet. eth.	184.5	154.5						Me. ester of acid, 97.5; Et. ester of acid, 72
158	**4-Hydroxybenzonitrile** (4-Cyanophenol)	113			213–4	162 (+1 H$_2$O)	196–7						M.p. of amine, 109 (anh.); 95 (+1H$_2$O); HCl salt of amine, 195; HI salt of amine, 198–200
159	**2,3-Diphenylvaleronitrile** (2 forms)	(a) 115, al.	(a) 235–40^{20} (b) 210–2^{20}			(a) 152–3 (b) 178, lgr.							
160	**α-Bromo-4-toluinitrile** (4-Cyanobenzyl bromide)	115–6, al.			229–30								Me. ester of acid, 54; Et. ester of acid, 35–6
161	**(4-Nitrophenyl)aceto-nitrile** (4-Nitrobenzyl cyanide)	116			153	197–8; 191	198						
162	**6-Bromo-3-nitrobenzo-nitrile**	117			180	197–8	166						
163	**(2-Hydroxyphenyl)acetonitrile** (2-Hydroxybenzyl cyanide)	117–9		240–3	147–9	116–8, al.		163					O-Benzoyl deriv., 50, lgr.; HCl salt of amine, 155
164	**3-Nitrobenzonitrile**	118			140	141–3	153–4						HCl salt of amine, 224
165	**4-Bromo-3-nitrobenzo-nitrile**	120			203–4	156							
166	**4-Cyanoazobenzene**	120–1, br., bz.			241, red, al.	224–5, red							Me. ester of acid, 123–4, or., me. al.; Et. ester of acid, 86–7, red, al.
167	**Dipicolinonitrile** (2,6-Dicyanopyridine)	123; 113			252d.; 228	di: 302							Di-Me. ester of acid, 124–5; Di-Et. ester of acid, 41–2
168	**2-Cyanohexanoic acid**	123–6.5			101	di: 200	di: 193						
169	**Dibromomalononitrile** (Bromodicyano-methane)	124			147d.; 136–7d.	di: 200; 206d.							Di-Me. ester, 67
170	**1-Cyanoanthracéne**	126			245, yel., al.	260							Me. ester of acid, 108, yel., ac. a.
171	**2,2,3-Triphenylpropio-nitrile**	126			162; 132	111, al.							Me. ester of acid, 127
172	**1-Cyanophenanthrene**	128			232–3	284	245						

* Derivative data given in order: m.p., crystal color, solvent from which crystallized.

TABLE XXI. ORGANIC DERIVATIVES OF NITRILES
b) Solids (Listed in order of increasing m.p.)* (Continued)

No.	Name	Melting point, °C	Boiling point, °C	Derivatives of the corresponding acid RCN → RCOOH					Derivatives of the corresponding amine RCN → RCH₂NH₂				Miscellaneous
				Acid		Amide	Anilide	S-Benzyl thiuronium chloride	Amine	Benz-amide	Benzene sulfon-amide	Picrate	
				B.P., °C	M.P., °C				B.P., °C				
173	2,3-Diphenylbutyro-nitrile (2 forms)	(a) 129–30	210–2[16]	.....	133–4	173–4	......	......	......				......
			(b) 188–92[13]		186	193							Et. ester of acid, 91–2
174	5-Bromo-3-nitro-4-tolunitrile	130	......	.....	206	171		......					......
175	2,5-Dichlorobenzonitrile	130	......	301	154	155			135[9]			230	HCl salt of amine, 151–2; n_D^{23} of amine: 1.5885
176	2,5-Dibromobenzonitrile	132			157								Me. ester of acid, 40–1; p-Nitrophenyl ester of acid, 183–4
177	5-Bromo-2-nitro-4-tolunitrile	132	......		203; 200	191		......					Et. ester of acid, 61
178	2-Hydroxy-3-nitro-benzonitrile (2-Cyano-6-nitrophenol)	132–3			148	155; 145–6	......						Me. ester of acid, 132 (94), al.
179	4-Nitro-1-naphthonitrile (1-Cyano-4-nitro-naphthalene)	133			220–1	218		......					Me. ester of acid, 107–8
180	4-Acetamidobenzo-nitrile	133, w.			185, ac. a.	177, al.	167–8, al.						N-Methylamide of acid, 172, al.
181	6-Cyanoquinoline	135, bz.-lgr.			291–2	174, bz.-al.		......	158–9[2]	......			Me. ester of acid, 90; HCl salt of amine, 239; n_D^{20} of amine: 1.6390
182	Apiolonitrile (2,5-Di-methoxy-3,4-methyl-enedioxybenzonitrile) .	135.5, al.			175, w.	......							Me. ester of acid, 71–2, al.
183	1-Nitro-2-naphthonitrile (2-Cyano-1-nitro-naphthalene)	138			239	......	......						Hot aq. Ba(OH)₂ → 1-hydroxy-2-naphthoic acid, 191–2, al.
184	3,5-Dichloro-2-hydroxy benzonitrile	139	......	.....	219.5; 223	209	......						Me. ester of acid, 147
185	trans-1,4-Dicyanocyclo-hexane	140	......	.....	300								Di-Me. ester of acid, 71
186	3,3,3-Triphenylpro-pionitrile	140, al.			177	198							Me. ester of acid, 120–1.5
187	4-Cyano-2-phenyl-quinoline (2-Phenyl-4-quinolinonitrile)	140, al.	ca. 365	.....	218	196	198						M.p. of amine, >287; HCl salt of amine, 232–5
188	Phthalonitrile (o-Di-cyanobenzene)	141	......		200–6	di: 219	di: 253	di: 151; 157		di: 184		170	N,N'-Diacetyl deriv. of diamine, 146
189	8-Nitro-2-naphthonitrile (2-Cyano-8-nitro-naphthalene)	143			295; 288	218		......					Et. ester of acid, 121, lgr.
190	5-Chloro-2-naphtho-nitrile (5-Chloro-2-cyanonaphthalene) ...	144	......		270	186–7	202						Me. ester of acid, 81; Et. ester of acid, 45
191	5-Chloro-1-naphtho-nitrile (5-Chloro-1-cyanonaphthalene) ...	145			245; 241–2	239							Et. ester of acid, 42
192	3,5-Dichloro-4-hydroxy-benzonitrile	146			269; 265	......							Acetate, 93; Me. ester of acid, 122
193	4-Nitrobenzonitrile	147; 149			241	200	211; 204	......				184–5	HCl salt of amine, 235–40
194	5-Bromo-1-naphtho-nitrile (1-Bromo-5-cyanonaphthalene) ...	147			261; 256	241		......					Et. ester of acid, 48–9

*Derivative data given in order: m.p., crystal color, solvent from which crystallized.

TABLE XXI. ORGANIC DERIVATIVES OF NITRILES

TABLE XXI. ORGANIC DERIVATIVES OF NITRILES
b) Solids (Listed in order of increasing m.p.)* (Continued)

No.	Name	Melting point, °C	Boiling point, °C	Acid B.P. °C	Acid M.P. °C	Amide	Anilide	S-Benzyl thiuronium chloride	Amine B.P. °C	Benz-amide	Benzene sulfon-amide	Picrate	Miscellaneous
195	**5-Iodo-2-naphthonitrile** (2-Cyano-5-iodo-naphthalene)	148.5	......	.....	264, al.	196	203	......	......	......	......	......	M.p. of amine, 120; HCl salt of amine, 251–3; N-Acetyl deriv. amine, 135–6
196	**3-Cyano-3-phenylpro-pionic acid**	150	215–8¹⁰	.....	166–7	......	......	......	......	......	......	.....	Di-Me. ester of acid, 55, al.; Imide of acid, 88–90; M.p. of amine, 209; HCl salt of amine, 203; N-Acetyl deriv. of amine, 65
197	**2-Cyano-2-propylvaler-amide** (Dipropyl-cyanoacetamide)	153	......	.....	161	*di:* 214	*di:* 168–8.5, me. al.	......	......	......	......	......	....
198	**2,6-Dibromobenzonitrile**	155, subl.	308–9	.....	157	208.5	......	......	......	......	......	......	Me. ester of acid, 83; Hydrazide of acid, 204
199	**3-Chloro-4-hydroxy-benzonitrile**..........	155	......	.....	169–70, w.; 164–5	180–2	......	......	......	......	......	......	Me. ester of acid, 106–7; Et. ester of acid, 77–8
200	**5-Chloro-2,4-dinitro-benzonitrile**..........	156	......	.....	182–3	212	226	......	......	......	......	......	....
201	**4-Benzamidobenzonitrile** (N-Benzoylanthra-nilonitrile)	156	......	.....	181	218–9	279	......	......	......	......	......	Me. ester of acid, 100; Et. ester of acid, 98
202	**5-Bromo-2-hydroxy-benzonitrile**..........	158–9	......	.....	165	232	222	......	......	......	......	......	O-Acetyl deriv. of acid, 60
203	*d,l-***2,3-Diphenyl-succinonitrile**	160	......	.....	183 (+1 H₂O)	......	*mono:* 173–5, al.	......	......	......	......	......	....
204	**Isophthalonitrile** (*m*-Dicyanobenzene)..	161.5–2.0	......	.....	345–7	*di:* 280	......	......	......	......	......	......	....
205	**2-Hydroxy-4-nitrobenzo-nitrile** (2-Cyano-5-nitrophenol)	160–1, yel.	......	.....	235; 226, w.	192–4	......	......	......	......	......	......	Benzoate, 122, yel., al.; Et. ester of acid, 84–5
206	*d,l-***4-Cyano-3,4-di-phenylbutyric acid** (*d,l*-2,3-Diphenyl-glutaromononitrile) ..	162–3, bz.	......	.....	208–10 (cor.), aq. ac. a.	*mono:* 200–5 d., al.	*mono:* 201–2, 50% al.	......	......	......	......	......	Et. ester, 99–100, al.; Imide of acid, 225–9, aq. al.
207	*d-***3-Carboxy-2,2,3-tri-methylcyclopentyl-acetonitrile**	164	......	.....	......	......	233, PhNO₂	......	......	......	......	......	$[\alpha]_D$: +64.41; Me. ester, 77; Et. ester, 58
208	**5-Chloro-2-hydroxy-benzonitrile** (4-Chloro-2-cyanophenol)	165–7	......	.....	172; 167.5	226	......	......	......	......	......	.....	Me. ester of acid, 84
209	**2,3-Diphenylcinnamo-nitrile** (Cyanotri-phenylethylene)	166–7; 162–3	......	.....	213	223	......	......	......	......	......	......	....
210	**1,7-Dicyanonaphthalene**	167, al.	......	.....	294–6	......	......	......	......	......	......	......	....
211	**4,4'-Dicyanodiphenyl-methane**	169	......	.....	334; 290	......	......	......	......	......	......	......	Di-Me. ester of acid, 81–2
212	**2,2'-Diphenic acid mono-nitrile** (2-Carboxy-2'-cyanobiphenyl)	170–2, bz.	......	.....	233.5	*mono:* 193; *di:* 212	*mono:* 181–3	......	......	......	......	......	Di-Me. ester of acid, 74; Me. ester, 110; Et. ester, 91–2
213	**5-Nitro-2-naphthonitrile** (2-Cyano-5-nitro-naphthalene)	172–3	......	.....	295, yel.	261–3, br.-yel.	......	......	......	......	......	......	Me. ester of acid, 112, yel.

*Derivative data given in order: m.p., crystal color, solvent from which crystallized.

No.	Name	Melting point, °C	Boiling point, °C	Derivatives of the corresponding acid RCN → RCOOH					Derivatives of the corresponding amine RCN → RCH₂NH₂				Miscellaneous
				Acid		Amide	Anilide	S-Benzyl thiuronium chloride	Amine	Benzamide	Benzene sulfonamide	Picrate	
				B.P. °C	M.P. °C				B.P. °C				
214	9-Cyanoanthracene (9-Anthracenecarbonitrile)............	175; 170–2	......	....	207, pa. yel.	......	......	......	......	....	....	....	Me. ester of acid, 111; M.p. of amine, 90
215	2,3-Dicyanopyridine ...	175–6	......	....	228–9	2-mono: 168.5 d; di: 165	......	......	......	....	....	....	....
216	1,3-Dicyanonaphthalene	179, ac. a.	......	....	267–8	......							
217	3-Cyanocoumarin	182	......	....	187 d.	236	250	......	....	....	....		Me. ester of acid, 116–7; Et. ester of acid, 64
218	2-Cyanocinnamic acid ..	183, al.	......	....	195–6 d.	di: 189–90	......	......	......	....	....	....	NH₃ → 2-Cyanocinnamamide, 123; Me. ester of nitrile, 89; Di-Me. ester of acid, 44–5
219	2-Cyanobenzoic acid	187; 192	......	....	200–6	di: 219	di: 253	di: 151; 157	......	di: 184	.....	170	NH₃ → 2-Cyanobenzamide, 173; Me. ester 51; Et. ester, 70; 65
220	1,2-Dicyanonaphthalene	190, bz.	......	....	175	di: 265	......	......	......	....	....	....	....
221	2-Hydroxy-5-nitrobenzonitrile (2-Cyano-4-nitrophenol)........	194–6, yel., w.			229–30; 227	225, al.							p-Nitrobenzoyl ester of acid, 115
222	Tetracyanoethylene	198–200, subl.	223										n_D^{25}: 1.560; D^{25}: 1.348; N,N-Dimethylaniline → N,N-dimethyl 4-tricyanovinylaniline, 173–5, ac. a.; Anthracene → adduct, 268–70, acet.
223	5-Nitro-1-naphthonitrile (1-Cyano-5-nitronaphthalene)	205	......	....	241–2; 239	235–6	......	......	......	....	....	....	Me. ester of acid, 109–10; Et. ester of acid, 93
224	1,4-Dicyanonaphthalene	208, ac. a.	......	....	309; 288								Di-Me. ester of acid, 67; Di-Et. ester of acid, 64
225	1,6-Dicyanonaphthalene	208–10	......	....	310								Di-Me. ester of acid, 99
226	1,5-Dicyanonaphthalene	211; 266–7	......	....	310, ac. a.								Di-Me. ester of acid, 114–5; Di-Et. ester of acid, 123–4
227	3-Cyanobenzoic acid ...	217	......	....	345	280	......	......	......	....	....	....	NH₃ → 3-Cyanobenzamide, >300; Me. ester, 65; Et. ester, 56
228	4-Cyanobenzoic acid ...	219	......	....	ca. 300, subl.	di: >250	di: 334–7	......	......	....	....	232 d.	NH₃ → 4-Cyanobenzamide, 223; Me. ester, 62; Et. ester, 54
229	Terephthalonitrile (p-Dicyanobenzene).....	222	......	....	ca. 300, subl.	di: >250	di: 334–7	......	......	....	....	232 d.	N,N′-Diacetyl deriv. of diamine, 225
230	1,8-Dicyanonaphthalene	232	......	....	260	di: 250–82 d.	......	......	......	....	....	....	Di-Me. ester of acid, 102–3; Di-Et. ester of acid, 58–60
231	4,4′-Dicyanobiphenyl ...	233–4	......	....	......	......	......	......	......	....	....	....	Di-Me. ester of acid, 214; 224; Di-Et. ester of acid, 112; Heating acid with lime → biphenyl, 71
232	l-2,3-Diphenylsuccinonitrile	239–40	......	....	176–7; 190, w.	......	......	......	......	....	....	....	Imide of acid, 196–8, ac. a.

*Derivative data given in order: m.p., crystal color, solvent from which crystallized.

TABLE XXI. ORGANIC DERIVATIVES OF NITRILES
b) Solids (Listed in order of increasing m.p.)* (Continued)

No.	Name	Melting point, °C	Boiling point, °C	Derivatives of the corresponding acid RCN → RCOOH					Derivatives of the corresponding amine RCN → RCH₂NH₂					Miscellaneous
				Acid		Amide	Anilide	S-Benzyl thiuronium chloride	Amine	Benz-amide	Benzene sulfon-amide	Picrate	
				B.P., °C	M.P., °C				B.P., °C				
233	**1-Cyano-9,10-anthra-quinone**	247, yel., ac. a.	. . . . . .	. . . . .	293–4, pa. yel., ac. a.	280, pa. yel., al.	288–9, pa. yel., PhNO₂	. . . . . .	. . . . . .	. . . . . .	. . . . .	. . . . . .	Me. ester of acid, 189, yel.; Et. ester of acid, 169, yel.
234	**2,3-Dicyanonaphthalene**	251, al.	. . . .	. . . . .	239–41	. . . .	. . . . . .	. . . . . .	. . . . . .	. . . . . .	. . . . .	. . . . . .	Imide of diacid, 275
235	**2,7-Dicyanonaphthalene**	267–8, ac. a.	. . . .	. . . . .	>320	di: 297–8	. . . . . .	. . . . . .	. . . . . .	. . . . . .	. . . . .	. . . . . .	Di-Me. ester of acid, 135–6
236	**2,6-Dicyanonaphthalene**	293, al.	. . . .	. . . . .	>300 d.	di: >320	. . . . . .	. . . . . .	. . . . . .	. . . . . .	. . . . .	. . . . . .	Di-Me. ester of acid, 191

*Derivative data given in order: m.p., crystal color, solvent from which crystallized.

EXPLANATIONS AND REFERENCES TO TABLES XXII AND XXIII

The main derivatives of the sulfonic acids are obtained by converting them to the corresponding sulfonyl chlorides. Hence sulfonic acids and sulfonyl chlorides are often identified through the same derivatives.

Sulfonyl chloride.

$$RSO_3Na \ + \ PCl_5 \ \rightarrow \ RSO_2Cl \ + \ POCl_3 \ + \ NaCl$$
Sulfonyl
chloride

From the sulfonic acid or its sodium or potassium salt with phosphorus pentachloride without solvent.
For directions and examples see: Cheronis, p. 638; Linstead, p. 89; Shriner, pp. 268, 270; Vogel, p. 553; Wild, p. 161.

*Sulfonamide.**

$$RSO_3Na \ \rightarrow \ RSO_2Cl \ \xrightarrow{NH_3} \ RSO_2NH_2$$
Sulfonyl   Sulfon-
chloride   amide

From the sulfonyl chloride (prepared from the acid or its salt) and concentrated aqueous ammonia solution.
For directions and examples see: Linstead, p. 49; Shriner, pp. 268, 270; Vogel, p. 553; Wild, p. 161.
From the sulfonyl chloride, aqueous ammonia and ammonium carbonate.
See: Cheronis, p. 639.

*Sulfonanilide.**

$$RSO_3Na \ \rightarrow \ RSO_2Cl \ \xrightarrow[NaOH]{C_6H_5NH_2} \ RSO_2NHC_6H_5 \ + \ NaCl$$
Sulfonyl   Sulfonanilide
chloride

From the sulfonyl chloride (prepared from the acid), aniline, and sodium hydroxide.
For directions and examples see: Linstead, p. 89; Wild, p. 161.
From the sulfonyl chloride and aniline.
See: Vogel, p. 553.

*Sulfon-p-toluidide.**

$$RSO_3Na \ \rightarrow \ RSO_2Cl \ \xrightarrow[NaOH]{p\text{-}CH_3C_6H_4NH_2} \ RSO_2NHC_6H_4CH_3\text{-}p \ + \ NaCl$$
Sulfonyl   Sulfon-p-toluidide
chloride

From the sulfonyl chloride, *p*-toluidine and aqueous sodium hydroxide.
For directions and examples see: Vogel, p. 553.

*Sulfon-1-naphthylamide.**

$$RSO_3Na \ \rightarrow \ RSO_2Cl \ \xrightarrow{1\text{-}C_{10}H_7NH_2} \ RSO_2NHC_{10}H_7\text{-}1 \ + \ NaCl$$
Sulfonyl   Sulfon-1-
chloride   naphthylamide

From the sulfonyl chloride (prepared from the acid) and 1-naphthylamine in benzene.
For directions and examples see: Cheronis, p. 639.

*S-Benzylthiuronium salt.**

$$RSO_3Na \ + \ [C_6H_5CH_2SC(NH_2)_2]^+Cl^- \ \rightarrow \ [C_6H_5CH_2SC(NH_2)_2]^+RSO_3^- \ + \ NaCl$$
S-Benzylthiuronium chloride   S-Benzylthiuronium sulfonate

From the sodium, potassium or ammonium salt of the sulfonic acid and S-benzylthiuronium chloride in water.
For directions and examples see: Cheronis, pp. 635, 636; Linstead, pp. 15, 89; Shriner, p. 269; Wild, pp. 149, 159; E. Chambers and G. W. Watt, *J. Org. Chem.*, **6**, 376 (1941); E. Campaigne and C. M. Suter, *J. Amer. Chem. Soc.*, **64**, 3040 (1942); S. Veibel, *J. Amer. Chem. Soc.*, **67**, 1867 (1945).

---

*Derivatives recommended for first trial.
WARNING: This is not an instruction manual. References should be consulted for the preparation of derivatives.

EXPLANATIONS AND REFERENCES TO TABLE XXII AND XXIII (Continued)

From sodium or potassium sulfonate and S-benzylthiuronium chloride in alcohol.
See: J. J. Donleavy, *J. Amer. Chem. Soc.*, **58**, 1004 (1936).

*Anilinium salt.**

$$RSO_3Na + C_6H_5NH_2 + HCl \rightarrow [C_6H_5NH_3]^+RSO_3^- + NaCl$$
Anilinium sulfonate

From the sodium or potassium salt of the acid, aniline and hydrochloric acid in water.
For directions and examples see: Cheronis, p. 637; O. C. Dermer and V. H. Dermer, *J. Org. Chem.*, **7**, 581 (1942).

*p-Toluidinium salt.**

$$RSO_3Na + p\text{-}CH_3C_6H_4NH_2 + HCl \rightarrow [p\text{-}CH_3C_6H_4NH_3]^+RSO_3^- + NaCl$$
p-Toluidinium sulfonate

From the sulfonic acid and *p*-toluidine.
For directions and examples see: Shriner, p. 269.
From the sodium, potassium or ammonium salt with *p*-toluidine or *p*-toluidine hydrochloride in water.
See: Shriner, p. 269; Vogel, p. 555; O. C. Dermer and V. H. Dermer, *J. Org. Chem.*, 7, 581 (1942); A. D. Barton and L. Young, *J. Amer. Chem. Soc.*, **65**, 294 (1943).
From the sodium, potassium, ferric or barium salt (after boiling with sulfuric acid) with *p*-toluidine and hydrochloric acid.
See: Cheronis, p. 637; Fieser, *J. Amer. Chem. Soc.*, **51**, 2460 (1929).

*o-Toluidinium salt.**

$$RSO_3Na + o\text{-}CH_3C_6H_4NH_2 + HCl \rightarrow [o\text{-}CH_3C_6H_4NH_3]^+RSO_3^- + NaCl$$
o-Toluidinium sulfonate

From the sodium or the potassium salt of the sulfonic acid with *o*-toluidine and hydrochloric acid in water.
For directions and examples see: Cheronis, p. 637; O. C. Dermer and V. H. Dermer, *J. Org. Chem.*, 7, 581 (1942).

*Phenylhydrazinium salt.**

$$RSO_3H + C_6H_5NHNH_2 \rightarrow [C_6H_5NHNH_3]^+RSO_3^-$$
Phenylhydrazinium
sulfonate

From the sulfonic acid with phenylhydrazine in water-alcohol.
For directions and examples see: Wild, p. 159; P. H. Latimer and R. W. Bost, *J. Amer. Chem. Soc.*, **59**, 2500 (1937).

Phenols by KOH fusion.

$$ArSO_3Na + NaOH \rightarrow ArOH + Na_2SO_3$$
Phenol

By fusion of the sodium arylsulfonate and sodium hydroxide in a nickel crucible.
For directions and examples see: Vogel, p. 552.
By fusion of the sodium arylsulfonate, potassium hydroxide and zinc dust.
See: Linstead, p. 89.

---

*Derivatives recommended for first trial.
WARNING: This is not an instruction manual. References should be consulted for the preparation of derivatives.

TABLE XXII. ORGANIC DERIVATIVES OF SULFONIC ACIDS
a) Listed in order of increasing m.p. of the corresponding sulfonamide*

No.	Name	Melting point, °C	S-Benzyl thiuronium salt	p-Toluidinium salt	Anilinium salt	o-Toluidinium salt	Sulfonyl chloride	Sulfonamide	Sulfonanilide	Sulfon-1-naphthylamide	Miscellaneous
1	Propane-1-sulfonic acid			67–8			b.p. 180	52, eth.		84	Phenylhydrazine salt, 204–5
2	Ethane sulfonic acid		115				b.p. 171	60, eth.		66	Phenylhydrazine salt, 182.8
3	Propane-2-sulfonic acid	−37					b.p. 79^{18}	60, eth.-pet. eth.		134	B.p. $159^{1.4}$; D_4^{25} = 1.1877; n_D^{20} = 1.4332; m-T uidide, 109
4	Heptane-1-sulfonic acid						16	75			Phenylhydrazine salt, 100–0.5
5	2,4,5-Trimethoxybenzene sulfonic acid						130	76	170		
6	Methane sulfonic acid	20					b.p. 60^{21}	90	100	125–6	Phenylhydrazine salt, 193.5–4d.
7	Hexadecane-1-sulfonic acid (Cetyl sulfonic acid)	54					54	97			
8	10-Bromocamphor-3-sulfonic acid						97	100.2			
9	Benzyl sulfonic acid			113, al.	102	83	92–3, eth., bz.	105, w., al.	102, al.	166 (146), yel., al.	Diethylamide, 29; Ethylamide, 65–6; Hydrazide, 131–2; Methylamide, 109–10; Phenylhydrazine salt, 173
10	3-Toluene sulfonic acid (3-Methylbenzene sulfonic acid)	oil		106		108	12	108, al.	96, al.		
11	3-Hydroxynaphthalene-2-sulfonic acid				241–2		112	110			
12	Pyridine-3-sulfonic acid	357					Hydrochloride, 141–4 d.	110–1	145		Diethylamide, 49–50; Hydrazide, 94, me. al.
13	2,6-Dimethylbenzene sulfonic acid	98					39	113 (96)			
14	2-(N-Methylamino)benzene sulfonic acid	182 d.						114.5–5.5			N-4-Toluene sulfonyl, 193
15	Indane-4-sulfonic acid						53–3.5	118–9, w.			
16	2-Phenylethane-1-sulfonic acid	91					33	122	77		
17	4-Formylbenzene sulfonic acid (Benzaldehyde-4-sulfonic acid)							122–4			Amide oxime, 185; Dimethylamide, 134–7; Di-O-acetate, 86–7.5; Chloride diacetate, 111–3
18	2-Methylnaphthalene-1-sulfonic acid						83–5	124			
19	4-Fluorobenzene sulfonic acid						36 (30)	125 (123), w.			
20	4-Chloro-3-methylbenzene sulfonic acid						65	128	92		Sulfonyl bromide, 67.5
21	3-Chloro-4-methoxybenzene sulfonic acid						82	131			
22	3-Ethoxybenzene sulfonic acid						38	131			
23	D-Camphor-10-sulfonic acid	193					67	132	121		
24	2-Fluoronaphthalene-6-sulfonic acid	105					97, chl.	133	129		
25	DL-Camphor-8-sulfonic acid	56–8					106	133–5, w.			
26	3-Methyl-4-nitrobenzene sulfonic acid						50	133.5			
27	3-Chloro-4-methylbenzene sulfonic acid						38	134	96		
28	3,5-Dimethylbenzene sulfonic acid			121–2, al.			94 (90), bz.	135, al.	119, al.		

Derivative data given in order: m.p., crystal color, solvent from which crystallized.

TABLE XXII. ORGANIC DERIVATIVES OF SULFONIC ACIDS
a) Listed in order of increasing m.p. of the corresponding sulfonamide* (Continued)

No.	Name	Melting point, °C	S-Benzylthiuronium salt	p-Toluidinium salt	Anilinium salt	o-Toluidinium salt	Sulfonyl chloride	Sulfonamide	Sulfonanilide	Sulfon-1-naphthylamide	Miscellaneous
29	Indane-5-sulfonic acid	92					46–7, eth.	135.5–6, al.	129, al.		
30	D-Camphor-8-sulfonic acid						138, ($[\alpha]_D^{14}$: +128.7 in chl.)	137, ($[\alpha]_D^{13}$: +93.6 in al.)			
31	4-Toluene sulfonic acid (4-Methylbenzene sulfonic acid)	104–5 (92)	181–2	198	238	190	71, eth.	*anh.*: 138.5–9; *dihyd.*: 105	103, eth.-al.	157, al.	Phenylhydrazide, 155, al.; o-Toluidide, 110, aq. ac. a.; Triethylammonium salt, 65
32	2,4-Dimethylbenzene sulfonic acid	62 (*hyd.*)	146				34	139 (137)	110		
33	4-Vinylbenzene sulfonic acid (4-Styrene sulfonic acid)			182–3				139–40			Dimethylamide, 62–3
34	3,4-Dichlorobenzene sulfonic acid						22.4 (19)	140 (135)			
35	2-Bromonaphthalene-1-sulfonic acid						97	140			
36	2,4,6-Trimethylbenzene sulfonic acid	77					56	142	109		
37	3-Aminobenzene sulfonic acid (Metanilic acid)		148					142			
38	D-Camphor-2-sulfonic acid			196–7			88	143	124		Me. ester, 77, ($[\alpha]_D$ +98.6 in chl.)
39	4-Methylnaphthalene-2-sulfonic acid						124–5, eth.	143–4, al.			
40	3,4-Dimethylbenzene sulfonic acid	64	208				52	144			
41	4-Chlorobenzene sulfonic acid	93	175	208–10	222–3	163–4	53	144	104	190	
42	4-Methyl-3-nitrobenzene sulfonic acid	92, hyg.		130	109	128	36	144.5		153	
43	3-Bromocamphor-8-sulfonic acid	195–6, (*anh.*)					136–7	145			
44	5-Chloro-2-methylbenzene sulfonic acid						24 (21)	145, aq. al.			
45	4-Bromo-3-methylbenzene sulfonic acid						50	146			
46	4-Methoxy-3-nitrobenzene sulfonic acid						66	146.3			
47	3-Chlorobenzene sulfonic acid			199–200	206–7			148			
48	2,5-Dimethylbenzene sulfonic acid		184				24–6	148			
49	Naphthalene-1-sulfonic acid	90	137	181	183	237	68 (66)	150	112 (152)		
50	4-Ethoxybenzene sulfonic acid						39	150			
51	4,6-Dichloro-2,5-dimethylbenzene sulfonic acid						81	150	175		
52	3-Bromo-4-methylbenzene sulfonic acid						60	151			
53	Benzene sulfonic acid	66	148	205	240	176	14.5	153 (156)	112 (110)	170–1	N-Xanthylsulfonamide, 200
54	2-Aminobenzene sulfonic acid (Orthanilic acid)		132					153			N-Benzoyl deriv. of amide, 198; Hydrochloride, 201
55	2-Iodonaphthalene-1-sulfonic acid						110	154			
56	2,6-Dichloro-4-methylbenzene sulfonic acid						56	154–5			
57	2-Chloro-5-methylbenzene sulfonic acid						56	156	229–30.5		
58	2-Toluene sulfonic acid (2-Methylbenzene sulfonic acid)	57	170	203–4	218		10	156.3	136, aq. al.		Sulfonyl bromide, b.p. 138[10]; N-Xanthylsulfonamide, 182–3.5

*Derivative data given in order: m.p., crystal color, solvent from which crystallized.

TABLE XXII. ORGANIC DERIVATIVES OF SULFONIC ACIDS
a) Listed in order of increasing m.p. of the corresponding sulfonamide* (Continued)

No.	Name	Melting point, °C	S-Benzyl thiuronium salt	p-Tolui-dinium salt	Ani-linium salt	o-Tolui-dinium salt	Sulfonyl chloride	Sulfon-amide	Sulfon-anilide	Sulfon-1-naph-thylamide	Miscellaneous
59	3-Bromocamphor-10-sulfonic acid	47.5	......	....	....	....	65	156.5	....	....	
60	2,4-Dinitrobenzene sulfonic acid	106–8 (hyd.)	......	....	....	....	102	157 (154)	....	....	Hydrazide, 110
61	2-Methyl-4-nitrobenzene sulfonic acid	130 (anh.)	......	....	....	177	106	157	....		
62	Benzophenone-3,3'-disulfonic acid	......	......	....	....	....	di: 137–8	di: 157	di: 177–8	....	Methylamide, 106–7; Dimethylamide, 118–9
63	3-Nitrobenzyl sulfonic acid	74	......	....	....	....	100, bz.	159, w.	....	....	
64	4,6-Dimethyl-2-hydroxy-1,3-benzene disulfonic acid	......	......	....	....	....	di: 89–91	di: 160–1	....	....	
65	2-Nitrodiphenylamine-4-sulfonic acid	220 d.	......	....	....	....	157	162	....	....	
66	2-Ethoxybenzene sulfonic acid	......	......	....	....	....	65–6	163	158	....	Phenylhydrazide, 132–3
67	3-Methyl-2-nitrobenzene sulfonic acid	......	......	....	....	....	58.5	163.5	....		
68	3-Amino-6-methylbenzene sulfonic acid	......	......	....	....	....	....	164	146–7	....	Chloride of N-acetyl deriv., 124; Chloride of N-chloroacetyl deriv., 87; Amide of N-acetyl deriv., 242
69	4-Chloro-2-nitrobenzene sulfonic acid	......	......	....	....	....	75	164	138	....	Anhydride, 114–5; Phenylhydrazide, 151; Ph. ester, 82
70	4-Aminobenzene sulfonic acid (Sulfanilic acid)	......	185	109	......	132	....	165	200	196	N-Acetyl, 214; Me. ester, 92
71	3,6-Dichloro-2,5-dimethylbenzene sulfonic acid	......	......	....	....	....	71	165	171	....	
72	6-Aminonaphthalene-1-sulfonic acid	......	172–4	....	....	....	....	165	....	....	Benzoylguanidine salt, 210–11
73	4-Bromobenzene sulfonic acid	102–3 (88–9)	170	215–6	237–8	182–3	76, eth.	166 (161)	119	183–4	
74	5-Bromo-2-methylbenzene sulfonic acid	......	......	....	....	....	33–5	167	....		
75	2,3-Dimethylbenzene sulfonic acid	......	......	....	....	....	47	167			
76	5-Chloro-2-methyl-3-nitrobenzene sulfonic acid	......	......	....	....	....	60	167			
77	3-Nitrobenzene sulfonic acid	48	146	222	222	193	64	167	126	166–7	
79	4,6-Dichloro-2-methylbenzene sulfonic acid	......	......	....	....	....	43	168	....		
80	4-Chloronaphthalene-2-sulfonic acid	......	......	....	....	....	106	168	....	....	Et. ester, 76–9
81	4-Bromo-2-methylbenzene sulfonic acid	......	......	....	....	....	50	168			
82	Propane-1,3-disulfonic acid	92, d. without melting	......	....	....	....	di: 45	di: 169, w.	di: 129		B.p. 157$^{1.4}$; Di-m-toluidide, 222; Di-hydrazide, 105
83	Propane-1,1-disulfonic acid	......	......	....	....	....	....	di: 169–70	151–2, al.	....	Di-(N-ethyl)anilide, 128–9, al.
84	3-Carboxybenzene sulfonic acid (3-Sulfobenzoic acid)	98 (hyd.)	163	....	224–6	....	di: 20	di: 170	....	....	
85	4-Methyl-2-nitrobenzene sulfonic acid	141 (anh.)	......	....	....	....	98–9	170	....	....	o-Anisidide, 135
86	2,4-Dimethyl-3-nitrobenzene sulfonic acid	144 (anh.)	......	....	....	....	96	172	....	....	

Derivative data given in order: m.p., crystal color, solvent from which crystallized.

No.	Name	Melting point, °C	S-Benzyl thiuronium salt	p-Toluidinium salt	Anilinium salt	o-Toluidinium salt	Sulfonyl chloride	Sulfonamide	Sulfonanilide	Sulfon-1-naphthylamide	Miscellaneous
87	2,5-Dimethyl-3-nitrobenzene sulfonic acid	128	......	135–6	......	126.5–7.5	60–1	172–3	143–4	....	....
88	4-Nitrodiphenylamine-2-sulfonic acid	......					102–4	173	164	....	....
90	3,4-Dibromobenzene sulfonic acid	66.5–7.5 (anh.)	......				34	175		....	
91	4-Chloro-3-nitrobenzene sulfonic acid	......					40–1	175–6, yel., al.			
92	3-Amino-4-methylbenzene sulfonic acid	......					....	176	....		N-Acetyl of chloride, 144; N-Benzoyl of Chloride, 196; N-Benzoyl, 203
93	7-Chloronaphthalene-2-sulfonic acid	118 (anh.); 68 (tetrahyd.)	......				86.5	176			Me. ester, 89; Et. ester, 65
94	4-Bromo-3-nitrobenzene sulfonic acid	......	......				55–7	176–7	....		
95	4-Hydroxybenzene sulfonic acid	......	169	202	170	192	....	176–7	141		
96	5-Methylnaphthalene-1-sulfonic acid	115	......					176–8	....		
97	4-Methylnaphthalene-1-sulfonic acid	......					81	177	158		
98	4-Nitrobenzene sulfonic acid	95 (109–11), hyg.	......	179–80	......		80, lgr.	180, 50% al.	136 (171), al.		
99	3-Chloro-2-methylbenzene sulfonic acid	60–72; 72 (anh.)	......				72, pet. eth.	180, w.	....		
100	2,5-Dichlorobenzene sulfonic acid	93–7 (>100)	170	247–8	262–3	250–1	38	181 (186)	160	160	....
101	2,4,5-Trimethylbenzene sulfonic acid	112	......	......	......	......	61–2	181	....		
102	8-Aminonaphthalene-2-sulfonic acid	......						181	146–7	....	N-Acetyl deriv. of amide, 213; Benzoylguanidine salt, 214–6
103	5-Chloro-4-methyl-2-nitrobenzene sulfonic acid	128	......	......			99	181	....		
104	2,4-Dichlorobenzene sulfonic acid	86		204–6		170–2	55	182	....		
105	4-Iodobenzene sulfonic acid	......					85	183	143		....
106	Quinoline-8-sulfonic acid	312					124	183–4	....		Picrate of Na salt, 226–7; Me. ester, 96; Et. ester, 73
107	6-Chloronaphthalene-2-sulfonic acid	......					110.5	183–4		....	Me. ester, 89; Et. ester, 79; Sulfonyl bromide, 124
108	5-Nitronaphthalene-2-sulfonic acid	118–9, yel.			260		125	184, yel.	....		
109	4-Chloro-2-methylbenzene sulfonic acid	......	......				54	185			
110	4-Chloro-2,5-dimethylbenzene sulfonic acid	100	......	......	......		50	185	155	....	....

*Derivative data given in order: m.p., crystal color, solvent from which crystallized.

TABLE XXII. ORGANIC DERIVATIVES OF SULFONIC ACIDS
a) Listed in order of increasing m.p. of the corresponding sulfonamide* (Continued)

No.	Name	Melting point, °C	S-Benzyl thiuro-nium salt	p-Tolui-dinium salt	Ani-linium salt	o-Tolui-dinium salt	Sulfonyl chloride	Sulfon-amide	Sulfon-anilide	Sulfon-1-naph-thylamide	Miscellaneous
111	2-Carboxy-5-methylbenzene sulfonic acid	190 (158), (anh.)	......	....	......	....	di: 59	185	....	....	Nitrile of the sulfonyl chloride, 67, lgr.; NH₄ salt of 2-amide, 186 (mono-hyd.)
112	8-Chloronaphthalene-2-sulfonic acid		......	....	......	....	94	185–6	....	....	Et. ester, 92
113	2-Chloro-5-nitrobenzene sulfonic acid	168–9 d. (hyd.)	......	....	......	....	89–90, w.	185–6, w.			
114	2-Bromobenzene sulfonic acid						51, eth.	186, w.			
115	2,3-Dichloro-6-methylbenzene sulfonic acid						49	186			
116	2-Methyl-5-nitrobenzene sulfonic acid	133.5 (di-hyd.)	......	256–7	......	256–8	46–7	186	148	......	
117	2-Chloro-4-methylbenzene sulfonic acid		......				46 (52)	186			
118	3,5-Dichloro-2-methylbenzene sulfonic acid						54	186			
119	6-Chloro-3-nitrobenzene sulfonic acid	168–9					90, eth.	186, w.			
120	2,4-Dimethyl-5-nitrobenzene sulfonic acid	132 (122), dil. HNO₃					98	187 (179)			Sulfonyl fluoride, 109–10
121	2-Methyl-5-nitrobenzene sulfonic acid	131					46–7 (44), eth.-pet. eth.	187	148	......	
122	4-Chloronaphthalene-1-sulfonic acid	130–3 d.	......		145–6	151	95	187	145–6	162	Me. ester, 83; Et. ester, 104
123	8-Iodonaphthalene-1-sulfonic acid .						115	187	140		
124	2,4-Diaminobenzene-1,5-disulfonic acid						275	187	236	......	
125	2,6-Dichloro-3-methylbenzene sulfonic acid						19.5	188			
126	4-Nitronaphthalene-1-sulfonic acid						99	188			
127	2-Chlorobenzene sulfonic acid						28.5, eth.	188	......		
128	5-Methylnaphthalene-2-sulfonic acid						120–2	188–9	248–50		
129	Phenanthrene-3-sulfonic acid.....	175–6 (anh.); 120–1 (mono-hyd.); 88–9 (di-hyd.)	......	222	......		110–1 (108)	190	......	......	Me. ester, 119–20, al.; Et. ester, 107–8
130	2,4-Dibromobenzene sulfonic acid .	110 (anh.)	......	....	......		79, eth.	190	......	......	
131	8-Nitronaphthalene-1-sulfonic acid	115 (tri-hyd.)	......				165 d.	190.5–1.5	178–8.5		
132	4-Methylbenzene-2,4-disulfonic acid	oil	......	277 d.	di: 189	170–1	54 (56), eth.	di: 190.5–1	189	......	
133	3,5-Dichloro-4-methylbenzene sulfonic acid						69	191			
134	2,5-Dimethyl-6-nitrobenzene sulfonic acid	145 (anh.)	......	158.5–9, al.	......	143–5, 50% al.	110, eth.-pet.-eth.	192, 50% al.	182, bl., al.	......	

*Derivative data given in order: m.p., crystal color, solvent from which crystallized.

No.	Name	Melting point, °C	S-Benzyl thiuronium salt	p-Toluidinium salt	Anilinium salt	o-Toluidinium salt	Sulfonyl chloride	Sulfon-amide	Sulfon-anilide	Sulfon-1-naphthylamide	Miscellaneous
135	Quinoline-6-sulfonic acid.......	>260	.....	....	......	....	91	192	.....	....	
136	3,4-Dicarboxybenzene sulfonic acid (4-sulfophthalic acid)......	138–40 (mono-hyd.)	.....	....	......	....	167–70 d., eth.	192– 200 d., w.			
137	2-Nitrobenzene sulfonic acid.....	70 (85)	.....	....	......	....	69	193	115	....	
138	Phenanthrene-9-sulfonic acid.....	174 (anh.); 134 (hyd.)	.....	235	......	....	127	193–4, al.	....	....	Me. ester, 106, me. al.; Et. ester, 108, al.
139	2-Carboxybenzene sulfonic acid (2-Sulfobenzoic acid).........	68–9 (hyd.); 134 (anh.)	206	197 (200)	165	127–8	79, pet. eth.; 40, eth.	C: 193–4 (anh.); S: 153–4, sl. htng	194–5		
140	4-Bromonaphthalene-1-sulfonic acid.................	....	.....	....	......	....	87	195	.....		
141	2,5-Dibromobenzene sulfonic acid.................	128 (anh.)	.....	....	......	....	71, pet. eth.-eth.	195, al.			
142	7-Methylnaphthalene-1-sulfonic acid.................	....	.....	....	......	....	96	195–6	162–4	....	
143	5-Fluoronaphthalene-1-sulfonic acid.................	105–6	.....	....	......	....	122–3	196–7	.....	....	Me. ester, 118, eth.
144	Acenaphthene-3-sulfonic acid....	....	.....	....	284–6	....	113–4	196–9	.....	....	Me. ester, 122–3; Et. ester, 137–9, lgr.
145	2,5-Dimethyl-4-nitrobenzene sulfonic acid	140	.....	143.5–4.5	......	143.5–4.5	75	197–8	131	....	
146	8-Chloronaphthalene-1-sulfonic acid.................	....	.....	....	......	....	101 (96–8)	197.6	.....	....	Me. ester, 70; Et. ester, 67–8
147	4-Chloro-3-methyl-5-nitrobenzene sulfonic acid.................	....	.....	....	......	....	52	201			
148	5-Amino-2-hydroxybenzene sulfonic acid (4-Aminophenol-2-sulfonic acid).............	100 (anh.)	.....	....	......	....	....	202	159	....	Pyridine salt of O,N-diacetyl deriv., 143–4
149	4-Nitrobenzyl sulfonic acid.....	71	.....	....	......	....	90	204	220	....	
150	Anthracene-1-sulfonic acid......	....	.....	....	......	....	90	205	.....	....	
151	6-Methylnaphthalene-2-sulfonic acid.................	....	.....	....	......	....	97–8	205–6	.....	....	
152	4-Iodonaphthalene-1-sulfonic acid.................	....	.....	....	......	....	124 (121)	206 (204)	136	....	
153	4-Fluoronaphthalene-1-sulfonic acid.................	100	.....	....	......	....	86, chl.	206	144	....	Et. ester, 93
154	4-Aminonaphthalene-1-sulfonic acid (Naphthionic acid).......	....	195	....	......	....		206			N-Acetyl deriv. of amide, 241; N-Acetyl deriv. of anilide, 231
155	4-Amino-2-nitrobenzene sulfonic acid.................	....	.....	....	......	....	59–60	206–7	Sulfon-	....	
156	Retene-6-sulfonic acid..........	121–3	.....	....	......	....	146–7.5	206–7.5	....	....	Me. ester, 117–9; Et. ester, 114–5
157	2,6-Dimethyl-4-hydroxybenzene-1,3-disulfonic acid.............	....	.....	....	......	....	119	208	207	....	
158	4-(N-methylamino)benzene sulfonic acid.................	244–5 d.	.....	....	......	....		210–1	....		N-4-Toluenesulfonyl, benzidine salt, 255
159	6-Chloronaphthalene-1-sulfonic acid.................	....	.....	....	......	....	70	214	....		Et. ester, 114–5
160	1-Nitronaphthalene-2-sulfonic acid.................	105, grn.	.....	....	(202)	....	120–1, pink, bz.-pet. eth.	214, 50% al.	202	....	
161	2,5-Dichlorobenzene-1,3-disulfonic acid.................	....	.....	....	......	....	114	215–7	....		

*Derivative data given in order: m.p., crystal color, solvent from which crystallized.

No.	Name	Melting point, °C	S-Benzyl thiuronium salt	p-Toluidinium salt	Anilinium salt	o-Toluidinium salt	Sulfonyl chloride	Sulfonamide	Sulfonanilide	Sulfon-1-naphthylamide	Miscellaneous
162	5-Methylbenzene-1,3-disulfonic acid	......	......	......	......	......	94, eth.	216, w.	153, al.	......	......
163	Naphthalene-2-sulfonic acid	91, hyg.	190–1	221	269	213	76 (79)	217 (213)	132	......	......
164	4-Acetamidobenzene-1-sulfonic acid	......	......	......	......	......	......	218	......	215	......
165	5-Aminonaphthalene-2-sulfonic acid	......	191	......	......	......	......	218–9 d.	127–8	......	N-Acetyl, 238–9
166	8-Nitronaphthalene-2-sulfonic acid	135–6 (hyd.)	......	......	......	......	169	223; 228	172–3	......	......
167	5-Chlorobenzene-1,3-disulfonic acid	......	......	......	......	......	106	224			
168	2-Methylbenzene-1,4-disulfonic acid	......	......	......	......	......	98, bz.-pet. eth.	di: 224	di: 178	......	......
169	4-Carboxy-3-nitrobenzene sulfonic acid	111 ($+2\frac{1}{2}$-H_2O)					di: 160	di: 226; mono-S: 192			
170	5-Chloronaphthalene-1-sulfonic acid	......					95	226	138	......	Me. ester, 89; Et. ester, 46; Sulfonyl bromide, 110
171	3,4-Di-iodobenzene sulfonic acid	122–5, after drying at 100					82, bz.-pet. eth.	227, aq. al.		......	Me. ester, 93, al.; Et. ester, 82.5, al.
172	2,4,6-Tribromobenzene sulfonic acid	64	......				64	228	220–2	......	......
173	4'-Nitrobiphenyl-4-sulfonic acid						178	228			
174	3,4-Dichloro-2-methylbenzene sulfonic acid	......					52	228			
175	Benzene-1,3-disulfonic acid	......	214				63	229	148–50	245	Alkali fusion → resorcinol, 110
176	Biphenyl-4-sulfonic acid	......					115	230	125	......	
177	2,4-Di-iodobenzene sulfonic acid	162 (anh.)					77–8	230			Me. ester, 78, al.; Et. ester, 57, al.
178	1-Ethoxybenzene-2,5-disulfonic acid	......					di: 106–8	di: 233			
179	Methane disulfonic acid (Methionic acid)	......					8; b.p. 133[10]	di: 233	di: 192–3	......	B.p. 220–70[10–15] d.
180	3,5-Dinitrobenzene sulfonic acid	......					99, chl.-lgr.	235, w., al.			
181	4-Aminobenzene-1,3-disulfonic acid (Aniline-2,4-disulfonic acid)	120 d.					......	di: 235			
182	5-Iodonaphthalene-1-sulfonic acid	......					114	236 (239)	......		
183	5-Nitronaphthalene-1-sulfonic acid	......			265		113	236	123		Me. ester, 117–8, chl.
184	4-Carboxybenzene sulfonic acid (4-Sulfobenzoic acid)	94 (hyd.), 260 (anh.)	213		......		......	di: 57	di: 236	di: 252	
185	2,3-Dichloro-4-methylbenzene sulfonic acid	......					41	237			
186	6-Hydroxynaphthalene-2-sulfonic acid	167 (anh.); 129 (hyd.)	217 (207)	248	264	208	......	237	161	......	......
188	4-Methylbenzene-1,2-disulfonic acid	......	......	......	......	......	109–11, bz.-pet. eth.	237–9	190, al.	......	......

*Derivative data given in order: m.p., crystal color, solvent from which crystallized.

TABLE XXII. ORGANIC DERIVATIVES OF SULFONIC ACIDS
a) Listed in order of increasing m.p. of the corresponding sulfonamide* (Continued)

No.	Name	Melting point, °C	S-Benzyl thiuronium salt	o-Toluidinium salt	Anilinium salt	p-Toluidinium salt	Sulfonyl chloride	Sulfonamide	Sulfonanilide	Sulfon-1-naphthylamide	Miscellaneous
189	4,5-Dimethylbenzene-1,3-disulfonic acid	......	......	......	......	......	79, yel., al.	239	200, al.	......	......
190	4-Hydroxybenzene-1,3-disulfonic acid (Phenol-2,4-disulfonic acid)	>100 d.	......	......	......	......	di: 89	di: 239	di: 205	......	......
191	4-Methoxybenzene-1,3-disulfonic acid	......	......	......	......	......	86	240	209	......	......
192	4-Acetamidonaphthalene-1-sulfonic acid (Acetylnaphthionic acid)	......	......	232–3	231–2	......	....	241	231		
193	Naphthalene-2,7-disulfonic acid	......	212	299	251–2	238	158 (162), bz.	242	......	......	......
194	5-Nitrobenzene-1,3-disulfonic acid		......	......	......	......	di: 97–8	di: 242	......	......	......
195	2,4,6-Trimethylbenzene-1,3-disulfonic acid (Mesitylene disulfonic acid)		......	......	......	......	di: 125	di: 244	di: 150–1	......	......
196	2,3,4,6-Tetrabromobenzene sulfonic acid		......	......	......	......	96.5	ca. 245 d.	......	......	......
197	4,6-Dimethylbenzene-1,3-disulfonic acid		......	......	......	......	130, pet. eth.	249, w.	196, 50% al.	......	......
198	1-Chloronaphthalene-2-sulfonic acid	130–3 d. (anh.)	......	......	......	......	84–5	250	171–2	......	Et. ester, 104
199	Azobenzene-4,4'-disulfonic acid	169 d. (anh.)	......	......	......	......	di: 222 (170)	di: 250 d.	......	......	......
200	Phenanthrene-2-sulfonic acid	150	......	291	......	......	156, ac. a.	253–4, al.	157–8	......	Me. ester, 101–2 (96–8); Et. ester, 89, yel.-br., al.
201	Benzene-1,2-disulfonic acid	......	206	......	......	......	143	254	241	......	Imide, 186
202	2-Amino-5-methylbenzene-1,4-disulfonic acid	d. at 290	......	......	......	......	di: 156, chl.	di: 257, w.	di: 196–7, aq. al.	......	......
203	5-Aminonaphthalene-1-sulfonic acid (Laurent's acid)	......	179	......	......	......	......	259–60	171	......	N-Acetyl deriv. of aniline salt, 344
204	2-Methylbenzene-1,3-disulfonic acid	......	......	......	......	......	88, bz.	>260	162, al.	......	......
205	Anthracene-2-sulfonic acid	......	......	......	......	......	122, yel., tol.	261	201	......	Me. ester, 157, yel.; Et. ester, 160; Phenylhydrazide, 210
206	Anthraquinone-2-sulfonic acid	......	211	308	309	......	197, pa. yel., bz.	261, yel., ac. a.	193, yel.-br.	......	Me. ester, 123; Et. ester, 125
207	7-Nitronaphthalene-1-sulfonic acid	......	......	......	......	......	169–70	261–2	......	......	......
208	Azoxybenzene-3,3'-disulfonic acid	126	......	......	......	......	138	273	......	......	......
209	Naphthalene-1,4-disulfonic acid	......	......	......	......	......	160 (166)	273, w., al.	179	......	......
210	4,6-Dichlorobenzene-1,3-disulfonic acid	......	......	......	......	......	123	276	......	......	......
211	Benzidine-2,2'-disulfonic acid	......	......	......	......	......	Hydro-chloride, 205	278	......	......	......
212	9,10-Dichloroanthracene-2-sulfonic acid	......	......	......	......	......	221	279	248	......	......
213	4-Nitronaphthalene-2,7-disulfonic acid	......	......	......	......	......	di: 140–1	di: 286–7	......	......	......
214	Benzene-1,4-disulfonic acid	......	......	......	......	......	131 (139)	288	249	......	......
215	Azobenzene-3,4'-disulfonic acid	......	......	......	......	......	di: 123–5	di: 288	......	......	......
216	Naphthalene-1,3-disulfonic acid	......	......	......	......	......	di: 137.5	di: 292–3	......	......	......
217	2,5-Dimethylbenzene-1,3-disulfonic acid	......	......	......	......	......	81, lgr.	295, al.	174, al.	......	......
218	Naphthalene-1,6-disulfonic acid	125 (anh.)	81	314–5	298–9	323–4	129, bz.	297–8	......	......	......

*Derivative data given in order: m.p., crystal color, solvent from which crystallized.

TABLE XXII. ORGANIC DERIVATIVES OF SULFONIC ACIDS
a) Listed in order of increasing m.p. of the corresponding sulfonamide* (Continued)

No.	Name	Melting point, °C	S-Benzyl thiuro-nium salt	p-Tolui-dinium salt	Ani-linium salt	o-Tolui-dinium salt	Sulfonyl chloride	Sulfon-amide	Sulfon-anilide	Sulfon-1-naph-thylamide	Miscellaneous
219	Naphthalene-1,7-disulfonic acid ..	......	......	......	......	......	di: 123	di: 298–300	......	......	....
220	Biphenyl-4,4′-disulfonic acid	72	171	330	......	......	203	300	......	......	....
221	Naphthalene-2,6-disulfonic acid ..	......	256	......	......	......	225, bz.	305	......	......	....
222	Azobenzene-3,3′-disulfonic acid...	......	......	......	......	......	166	305	......	......	....
223	Naphthalene-1,5-disulfonic acid ..	240–5 (anh.)	257 (251)	......	......	......	di: 183	di: 310 (>340)	di: 249	......	Di-Me. ester, 205, chl.
224	Benzene-1,3,5-trisulfonic acid	>100					tri: 187	tri: 310–5	tri: 237	......	Tri-Et. ester, 147, bz.
225	Anthracene-1,5-disulfonic acid ...	......	......				di: 249, pa. yel.	di: >330, yel.	di: 293	......	....
226	Anthracene-1,8-disulfonic acid ...	......	......				225, yel., bz.	333	224		
227	Anthraquinone-1,8-disulfonic acid	293–4	......				222–3, yel., PhNO$_2$	>340	237–8, yel., PhNO$_2$	......	
228	Anthraquinone-1,5-disulfonic acid	310–1 (hyd.), yel.	......	......	......	......	265–70, yel., PhNO$_2$	>350	269–70, red-yel., PhNO$_2$		

*Derivative data given in order: m.p., crystal color, solvent from which crystallized.

TABLE XXII. ORGANIC DERIVATIVES OF SULFONIC ACIDS

b) M.p. of the amide deriv. not available; listed in order of increasing m.p. of the corresponding anilide deriv.*

No.	Name	Melting point, °C	S-Benzyl thiuronium salt	p-Toluidinium salt	Anilinium salt	o-Toluidinium salt	Sulfonyl chloride	Sulfonamide	Sulfonanilide	Sulfon-1-naphthylamide	Miscellaneous
1	Ethane-1,2-disulfonic acid ...	104	201–2	270 d.	270, w.	......	di: 95	......	69	....	m-Toluidide, 230; Di-Et. ester, 77, eth.
2	Vinylsulfonic acid (Ethylenesulfonic acid)	....	145–6, w.	....	....	....	b.p. 118–20^{250}	......	71; b.p. 174^{1}	......	B.p. 135^{2}; n_D^{25}: 1.4505
3	Phenol-O-sulfonic acid	145 (hyd.)	....	124–5	165	......	....	....	126.5–7.5	....	
4	8-Aminonaphthalene-1-sulfonic acid (Peri acid)......	....	300	....	....	......	....	....	139–40	....	Anilinium salt of N-acetyl deriv., 273
5	3,5-Dimethyl-2-hydroxybenzene sulfonic acid	121–5 d.	....	....	....	......	....	....	142–3	......	Chloride of O-acetyl deriv., 62, pet. eth.; Methylanilide, 111–2
6	Benzophenone-2-sulfonic acid						96–7	......	143–5	....	....
7	7-Hydroxynaphthalene-1,3,6-trisulfonic acid						196	......	152–5	....	....
8	1-Hydroxynaphthalene-2-sulfonic acid..............	>250	169						O-acetyl deriv., 157–8		O-acetyl deriv. of p-toluidide, 135–6
9	2-Hydroxynaphthalene-1,6-disulfonic acid						di: 111	......	di: 191		
10	Anthraquinone-2,7-disulfonic acid						di: 186, chl. eth.		di: 192		
12	7-Hydroxynaphthalene-1,3-disulfonic acid (G-Acid)....	....	228	294	......	271	161–2	......	195		
13	7-Hydroxynaphthalene-1-sulfonic acid (Bayer Acid)....	....	218	232	240	242	....		195		
14	4-Hydroxynaphthalene-1-sulfonic acid (NW-Acid)......	170, r. h.	103	196	186–7	203–4	....		199–200	......	2-Naphthylamide, 204
15	5-Hydroxynaphthalene-1-sulfonic acid..............	110–2 d.							201	......	Chloride of O-acetyl deriv., 129
16	2-Hydroxynaphthalene-3,6-disulfonic acid (R-Acid)....	....	233	250	254	257			202		
17	Anthraquinone-1-sulfonic acid	218	191	......	284	......	216–8, yel., PhNO₂	......	214, gold-yel.		NH₃ → 1-Aminoanthraquinone, 252 (243)
18	Anthraquinone-1,6-disulfonic acid	215–7, gold, ac. a.					197–8, yel., PhNO₂	......	227–8, yel., cl.-bz.		
19	2-Hydroxynaphthalene-1,5-disulfonic acid..........	....					di: 177	......	di: 231		
20	2-Hydroxynaphthalene-1,7-disulfonic acid	....		219			169		233		
21	Anthraquinone-1,7-disulfonic acid	120 (hyd.)	......	......			231–2, br.-yel., PhNO₂	......	237–8, yel., cl.-bz.		
22	Anthraquinone-2,6-disulfonic acid	....					di: 250, yel., cl.-bz.		di: 321		....

*Derivative data given in order: m.p., crystal color, solvent from which crystallized.

TABLE XXIII. ORGANIC DERIVATIVES OF SULFONYL CHLORIDES
(Listed in order of increasing m.p.)*

No.	Name	Melting point, °C	Sulfonic acid	Amide	Anilide	1-Naphthyl amide	Salts of the corresponding acid				Miscellaneous
							S-Benzyl thiuronium	p-Toluidine	Aniline	o-Toluidine	
1	Methanedisulfonyl chloride	8	b.p.: 220–70[15–20]d.	233	192–3	......	......	......	......	......	B.p. 133[10]
2	2-Ethylbenzenesulfonyl chloride	11.7	....	100							
3	3-Methylbenzenesulfonyl chloride (Toluene-3-sulfonyl chloride)	12	....	108, al.	96	......					
4	4-Ethylbenzenesulfonyl chloride	12	....	110	....	......					N-Xanthylsulfonamide, 196
5	Benzenesulfonyl chloride	14.5	43–4 (mono-hyd.); 66 (anh.)	156; 153	....	170–1	148	205	240	176	N-Xanthylsulfonamide, 200
6	Heptane-1-sulfonyl chloride	16	....	75							
7	2,6-Dichloro-3-methylbenzenesulfonyl chloride	19.5	....	188							
8	3-Carboxybenzenesulfonyl chloride	20	98 (hyd); 148 (anh.)	di: 170	....		163	224–6	......		The m.p. is that of the 1,3-dichloride.
9	3,4-Dichlorobenzenesulfonyl chloride	22.4;19	....	140; 135							
10	5-Chloro-2-methylbenzenesulfonyl chloride	24	....	145, aq. al.							
11	2,5-Dimethylbenzenesulfonyl chloride	24–6	48 (anh.); 86 (hyd.)	148	....		184				N-Xanthylsulfonamide, 176
12	2-Chlorobenzenesulfonyl chloride	28.5	....	188	....						
13	2-Phenylethane-1-sulfonyl chloride	33	91	122	77						
14	5-Bromo-2-methylbenzenesulfonyl chloride	33–5	....	166–7							
15	2,4-Dimethylbenzenesulfonyl chloride	34	62 (hyd.)	110	....						
16	3,4-Dibromobenzenesulfonyl chloride	34	66.5–7.5 (anh.)	175	....						
17	2,4,6-Trichlorobenzenesulfonyl chloride	35–40	....	210–2d.							
18	4-Methyl-3-nitrobenzenesulfonyl chloride	36	92, hyg.	144.5	109	153	......	130–1	......	28	
19	4-Fluorobenzenesulfonyl chloride	36; 30	....	125							
20	3-Ethoxybenzenesulfonyl chloride	38	....	131							
21	3-Chloro-4-methylbenzenesulfonyl chloride	38	....	134	96						
22	2,5-Dichlorobenzenesulfonyl chloride	38	93–7	181	160	160	170	247–8	262–3	250–1	
23	2,6-Dimethylbenzenesulfonyl chloride	39	98	113; 96							
24	4-Ethoxybenzenesulfonyl chloride	39	....	150							
25	4-Chloro-3-nitrobenzenesulfonyl chloride	40–1; 60–2	....	175–6, yel., al.							
26	2,3-Dichloro-4-methylbenzenesulfonyl chloride	41	....	237							
27	4,6-Dichloro-2-methylbenzenesulfonyl chloride	43	....	168							
28	3,3-Dimethylbutane-1-sulfonyl chloride	43–4	....	96–7	....						
29	Propane-1,3-disulfonyl chloride	45	92d.	169, w.	di: 125	......					Di-m-toluidide 222
30	2-Chloro-4-methylbenzenesulfonyl chloride	46; 52	....	186	....						
31	Indane-5-sulfonyl chloride	46–7, eth.	92	135.5–6, al.	129, al.						
32	2-Methyl-5-nitrobenzenesulfonyl chloride	46–7`	133.5 (+2H₂O)	186	148	......		256–7	......	256–8	B.p. 183–5[10]
33	2,3-Dimethylbenzenesulfonyl chloride	47	....	167	....						

*Derivative data given in order: m.p., crystal color, solvent from which crystallized.

TABLE XXIII. ORGANIC DERIVATIVES OF SULFONYL CHLORIDES
(Listed in order of increasimg m.p.)* (Continued)

No.	Name	Melting point, °C	Sulfonic acid	Amide	Anilide	1-Naphthyl amide	Salts of the corresponding acid				Miscellaneous
							S-Benzyl thiuronium	p-Toluidine	Aniline	o-Toluidine	
34	4-Bromo-3-methylbenzenesulfonyl chloride	50	....	146	....	......					
35	3-Methyl-4-nitrobenzenesulfonyl chloride	50	....	133.5							
36	4-Bromo-2-methylbenzenesulfonyl chloride	50	....	168	....	......					
37	4-Chloro-2,5-dimethylbenzene-sulfonyl chloride	50	100	185	155						
38	2-Bromobenzenesulfonyl chloride .	51, eth.	....	186, w.							
39	3,4-Dichloro-2-methylbenzene-sulfonyl chloride	51–2	....	228							
40	3,4-Dimethylbenzenesulfonyl chloride	52	64; 55	144	....	......	208				
41	4-Chloro-3-methyl-5-nitrobenzene-sulfonyl chloride	52		201							
42	4-Chlorobenzenesulfonyl chloride .	53	69; 93	144	104	190	175	208–10	222–3	163–4	....
43	Indane-4-sulfonyl chloride	53–3.5	....	118–9, w.	......	......	......	......	......	......	B.p. 140–1
44	4-Methylbenzene-1,3-disulfonyl chloride (Toluene-2,4-disulfonyl chloride)	54; 46	....	di: 186–7; 191	189	......					
45	4-Chloro-2-methylbenzenesulfonyl chloride	54	....	185	....	......					
46	3,5-Dichloro-2-methylbenzene-sulfonyl chloride	54	....	186							
47	Hexadecane-1-sulfonyl chloride ..	54	54	97							
48	2,4-Dichlorobenzenesulfonyl chloride	55	86	182	....	......	......	204–6	......	170–2	
49	4-Bromo-3-nitrobenzenesulfonyl chloride	55–7	....	176–7							
50	2,4,6-Trimethylbenzenesulfonyl chloride	56	78	142	109						
51	2-Chloro-5-methylbenzenesulfonyl chloride	56	....	156	229–30.5						
52	2,6-Dichloro-4-methylbenzene-sulfonyl chloride	56	....	154–5							
53	4-Carboxybenzenesulfonyl chloride	57	94 (hyd.); 260 (anh.)	di: 236	di: 252	......					The m.p. is that of the 1,4-dichloride
54	Tetralin-6-sulfonyl chloride	58	....	135	155–6						
55	3-Methyl-2-nitrobenzenesulfonyl chloride	58.5	....	163.5	....						
56	5-Carboxy-2-methylbenzenesul-fonyl chloride (4-Toluic acid-2-sulfonyl chloride)	59	190; 158 (anh.)	185							The m.p. is that of the 1,5-dichloride
57	4-Amino-3-nitrobenzenesulfonyl chloride (2-Nitroaniline-4-sulfonyl chloride)	59–60	....	206–7							
58	5-Chloro-2-methyl-3-nitrobenzene-sulfonyl chloride	60	....	167							
59	3-Bromo-4-methylbenzenesulfonyl chloride	60	....	151	....						
60	2,5-Dimethyl-3-nitrobenzene-sulfonyl chloride	61	128; 200	173	143–4	......	136				
61	2,4,5-Trimethylbenzenesulfonyl chloride	61	112	181	....						
62	Benzene-1,3-disulfonyl chloride...	63	....	229	148–50	245	214	......	......	.....	N-Xanthylsulfon-amide, 170; Alk. fusion → resor-cinol, 110
63	4-Chloro-3-methylbenzenesulfonyl chloride	63	....	128	92	......					....

*Derivative data given in order: m.p., crystal color, solvent from which crystallized.

TABLE XXIII. ORGANIC DERIVATIVES OF SULFONYL CHLORIDES
(Listed in order of increasing m.p.)* (Continued)

| No. | Name | Melting point, °C | Sulfonic acid | Amide | Anilide | 1-Naphthyl amide | Salts of the corresponding acid | | | | Miscellaneous |
							S-Benzyl thiuronium	p-Toluidine	Aniline	o-Toluidine	
64	7-Methylnaphthalene-2-sulfonyl chloride	63–4	....	163–4	....	......	......	......	......	......	....
65	2,4,6-Tribromobenzenesulfonyl chloride	64	64	228	220–2d.	......					
66	3-Nitrobenzenesulfonyl chloride ..	64	48	167	126	166.5	146	222	126.5–7.5	193	....
67	2,3,4-Trichlorobenzenesulfonyl chloride	64–5	....	227–30	....	......					
68	3-Bromocamphor-10-sulfonyl chloride	65	47.5	156							
70	2-Ethoxybenzenesulfonyl chloride .	65–6	....	163	158	......					
71	4-Methoxy-3-nitrobenzenesulfonyl chloride	66	....	146.3							
72	D-Camphor-10-sulfonyl chloride ..	67	193	132	121; 88	......					
73	2-Methylbenzenesulfonyl chloride (Toluene-2-sulfonyl chloride) ...	68	57	156.3	136	......	170	203–4	218	......	N-Xanthylsulfon-amide, 182–3.5
74	Naphthalene-1-sulfonyl chloride ..	68; 66	90	150	112; 152	......	137	181	183	237	....
75	4-Methylbenzenesulfonyl chloride (Toluene-4-sulfonyl chloride) ...	69	104–5	138.5–9 (anh.); 105 (+2H₂O)	103	157	181–2	198	238	190	....
76	2-Nitrobenzenesulfonyl chloride ..	69	70; 85	193	115						
77	3,5-Dichloro-4-methylbenzene-sulfonyl chloride	69	....	191	....						
78	2,4-Dimethoxybenzenesulfonyl chloride	70	....	167							
79	6-Chloronaphthalene-1-sulfonyl chloride	70	....	214							
80	3,4-Dimethyl-5-nitrobenzenesul-fonyl chloride	70	....	180							
81	2,5-Dibromobenzenesulfonyl chloride	71	128 (anh.)	195							
82	3,6-Dichloro-2,5-dimethylbenzene-sulfonyl chloride	71	....	165	171						
83	3-Chloro-2-methylbenzenesulfonyl chloride	72, pet. eth.	60–72	180, w.	....						
84	4-Methoxynaphthalene-2-sulfonyl chloride	75	....	157	145						
85	2-Chloronaphthalene-1-sulfonyl chloride	75	....	153							
86	4-Chloro-2-nitrobenzenesulfonyl chloride	75	82	164; 237	138	......			......	Phenylhydrazide 151; Ph. ester, 82
87	2,5-Dimethyl-4-nitrobenzene-sulfonyl chloride	75	140	197–8	131	......		143.5–4.5	......	143.5–4.5	....
88	4-Bromobenzenesulfonyl chloride .	76, eth.	88–90	166; 161	119	183.5	170	215–6	237–8	182–3	....
89	Naphthalene-2-sulfonyl chloride ..	76; 79	91, hyg.; 122	217; 213	132	......	190–1	221	269	213	
90	6-Bromonaphthalene-1-sulfonyl chloride	77	....	217							
91	2,4-Di-iodobenzenesulfonyl chloride	77–8	167 (anh.)	230	....	......			......	Me. ester, 78, al.; Et. ester, 57, al.
92	2-Carboxybenzenesulfonyl chloride	79, pet. eth.	68–9 (hyd.); 134 (anh.)	....	194–5	......	206	196; 200	165	127–8	The m.p. is that of the 1,2-dichloride
93	2,4-Dibromobenzenesulfonyl chloride	79, eth.	110 (anh.)	190	....						
94	4,5-Dimethylbenzene-1,3-disul-fonyl chloride	79, yel.	....	239	200, al.	......					

*Derivative data given in order: m.p., crystal color, solvent from which crystallized.

No.	Name	Melting point, °C	Sulfonic acid	Amide	Anilide	1-Naphthyl amide	Salts of the corresponding acid				Miscellaneous
							S-Benzyl thiu-ronium	p-Tolui-dine	Aniline	o-Tolui-dine	
95	4-Nitrobenzenesulfonyl chloride ..	80, lgr.	109–11; 95 hyg.	180, 50% al.	171; 136 d.	......	......	179–80	......	......	
96	6-Methoxynaphthalene-1-sulfonyl chloride	80.5	....	149.5	177.5	......			......		
97	2,5-Dimethylbenzene-1,3-disul-fonyl chloride	81, lgr.	....	295, al.	174, al.	......			......		
98	4,6-Dichloro-2,5-dimethylbenzene-sulfonyl chloride	81	....	150	175						
99	4-Methylnaphthalene-1-sulfonyl chloride	81, eth.	....	174; 177 al.	158	......			......		
100	3,4-Di-iodobenzenesulfonyl chloride	82, bz.-pet. eth.	122–5	227, aq. al.	....						
101	3-Chloro-4-methoxybenzenesul-fonyl chloride	82	....	131							
102	7-Methoxynaphthalene-2-sulfonyl chloride	83	....	220	121						
103	2-Methylnaphthalene-1-sulfonyl chloride	83–5	....	124							
104	1-Chloronaphthalene-2-sulfonyl chloride	84–5	130–3d. (anh.)	250	171–2						
105	4-Iodobenzenesulfonyl chloride	85	....	183	143	......					
106	4-Ethoxynaphthalene-2-sulfonyl chloride	85	....	183	143.5	......					
107	4,5-Dichloro-3-methylbenzene-sulfonyl chloride	85–8	....	183–5							
108	4-Methoxybenzene-1,3-disulfonyl chloride	86	....	240	209						
109	4-Fluoronaphthalene-1-sulfonyl chloride	86	100 (hyd.)	206	144	......			......		Et. ester, 93
110	7-Chloronaphthalene-2-sulfonyl chloride	87	68 (tetra-hyd.); 118 (anh.)	76							Me. ester, 89; Et. ester, 65
111	4-Bromonaphthalene-1-sulfonyl chloride	87	....	195		......					
112	D-Camphor-3-sulfonyl chloride ...	88	77	143	124	......		196–7	......		Me. ester, 77
113	2-Methylbenzene-1,3-disulfonyl chloride	88	....	260	162	......					
114	8-Methylnaphthalene-2-sulfonyl chloride	88	.........*	116		......					
115	4-Hydroxybenzene-1,3-disulfonyl chloride (Phenol-2,4-disulfonyl chloride)	89	>100d.	239							
116	3,5-Dimethyl-2-hydroxybenzene-1,3-disulfonyl chloride	89–91	....	160–1	....						
117	Anthracene-1-sulfonyl chloride ...	90	....	205							
118	2-Chloro-5-nitrobenzenesulfonyl chloride	90, w.	168–9d. (hyd.)	185–6							
119	Quinoline-6-sulfonyl chloride.....	91	>260	192							
120	Benzylsulfonyl chloride	92–3, eth., lgr.	....	105, w., al.	102	166; 146	......	113, al.	102	83	Hydrazide, 131–2, Phenylhydrazide, 173
121	6-Iodonaphthalene-1-sulfonyl chloride	92.5	....	213	....						
122	6-Methoxynaphthalene-2-sulfonyl chloride	93	....	189	120	......					
123	5-Chloro-2-nitrobenzenesulfonyl chloride	93	....	159	....						
124	1-Bromonaphthalene-2-sulfonyl chloride	93	....	271	....						

*Derivative data given in order: m.p., crystal color, solvent from which crystallized.

No.	Name	Melting point, °C	Sulfonic acid	Amide	Anilide	1-Naph-thyl amide	Salts of the corresponding acid				Miscellaneous
							S-Benzyl thiu-ronium	p-Tolui-dine	Aniline	o-Tolui-dine	
125	5-Methylbenzene-1,3-disulfonyl chloride	94, eth.	....	216, w.	153, al.	......	......	......	......	......	
126	3,5-Dimethylbenzenesulfonyl chloride	94; 90, bz.	....	135, al.	129, al.	......	......	121–2, al.			
127	8-Chloronaphthalene-2-sulfonyl chloride	94	....	185							
128	1-Iodonaphthalene-2-sulfonyl chloride	94	....	247							
129	4-Chloronaphthalene-1-sulfonyl chloride	94–5	130–3d.	187		162				151	o-Toluidide, 145–6; Me. ester, 83; Et. ester, 104
130	Ethane-1,2-disulfonyl chloride	95	....	69	....	......	201–2	270d.	270, w.	......	Di-Et ester, 77, eth.; m-Toluidide, 230
131	5-Bromonaphthalene-1-sulfonyl chloride	95	....	232–3							....
132	5-Chloronaphthalene-1-sulfonyl chloride	95	....	226	138	......	......	......	......	......	Sulfonyl bromide, 110; Me. ester, 89; Et. ester, 46
133	7-Methylnaphthalene-1-sulfonyl chloride	96	....	195–6	162–4	......	......	......	......	......	
134	2,4-Dimethyl-3-nitrobenzene-sulfonyl chloride	96	144 (anh.)	172							
135	5-Bromonaphthalene-2-sulfonyl chloride	96	....	220							
136	2-Benzoylbenzenesulfonyl chloride (Benzophenone-2-sulfonyl chloride)	96–7	....	....	143–5						
137	2,3,4,6-Tetrabromobenzenesulfonyl chloride	96.5	....	245d.							
138	10-Bromocamphor-3-sulfonyl chloride	97	....	100–2							
139	6-Fluoronaphthalene-2-sulfonyl chloride	97, chl.	105 (hyd.)	133	129	......	190–1	221	269	213	
140	2-Bromonaphthalene-1-sulfonyl chloride	97	....	140							......
141	6-Methylnaphthalene-2-sulfonyl chloride	97–8	....	205–6							
142	5-Nitrobenzene-1,3-disulfonyl chloride	97–8	....	242							
143	2,4-Dimethyl-5-nitrobenzene-sulfonyl chloride	98	132; 122, dil. HNO₃	187; 179	....						Sulfonyl fluoride, 109–10
144	2-Methylbenzene-1,4-disulfonyl chloride	98	....	224	di: 178						
145	4-Methyl-3-nitrobenzenesulfonyl chloride	98–9	....	170							o-Anisidide, 135
146	4-Methoxynaphthalene-1-sulfonyl chloride	98.5	....	226	147.5	......	......	......	......	......	
147	5-Chloro-4-methyl-2-nitro-benzenesulfonyl chloride	99	128	181							
148	3,5-Dinitrobenzenesulfonyl chloride	99, chl., lgr.	....	235							
149	4-Nitronaphthalene-1-sulfonyl chloride	99	....	188	....						
150	7-Iodonaphthalene-2-sulfonyl chloride	100	....	210	....						
151	7-Bromonaphthalene-2-sulfonyl chloride	100	....	218							

*Derivative data given in order: m.p., crystal color, solvent from which crystallized.

No.	Name	Melting point, °C	Sulfonic acid	Amide	Anilide	1-Naphthyl amide	Salts of the corresponding acid				Miscellaneous
							S-Benzyl thiuronium	p-Toluidine	Aniline	o-Toluidine	
152	3-Nitrobenzylsulfonyl chloride....	100, bz.	74 (hyd.)	159 d., w.	....	.....	....	....	....	.....	Me. amide, 106–7; Dimethylamide, 118–9
153	8-Chloronaphthalene-1-sulfonyl chloride	101	....	197	....						....
154	2,4-Dinitrobenzenesulfonyl chloride	102	106–8 (hyd.); 130 (anh.)	157; 154	....						Hydrazide, 110
155	4-Nitrodiphenylamine-2-sulfonyl chloride	102–4	....	173	164						
156	4-Ethoxynaphthalene-1-sulfonyl chloride	103	....	170	180						
157	7-Ethoxynaphthalene-2-sulfonyl chloride	103	....	142	153						
158	3,7-Diethylnaphthalene-1-sulfonyl chloride	105–7	....	207							
159	5,6-Dichloronaphthalene-1-sulfonyl chloride	106	....	223							
160	4-Chloronaphthalene-2-sulfonyl chloride	106	....	168							Et. ester, 76–9
161	2-Methyl-4-nitrobenzene-sulfonyl chloride	106	....	157							
162	DL-Camphor-8-sulfonyl chloride..	106	56–8	133–5, w.							
163	5-Chlorobenzene-1,3-disulfonyl chloride	106	....	224							
164	3-Ethoxybenzene-1,4-disulfonyl chloride	106–8	....	di: 233							
165	6-Ethoxynaphthalene-2-sulfonyl chloride	107.5	....	183	153						
166	4-Methylbenzene-1,2-disulfonyl chloride	109–11	....	237–9	190	.....					
167	2,5-Dimethyl-6-nitrobenzene-sulfonyl chloride	110	145 (anh.)	192	182	.....		158.5–9	.....	143–5, 50% al.	
168	2-Iodonaphthalene-1-sulfonyl chloride	110	....	154							
169	Phenanthrene-3-sulfonyl chloride .	110–1	175–6 (anh.); 120–1 (+1 H$_2$O); 88d. (+2H$_2$O)	190	....	.....	222			.....	Me. ester, 119–20, al.; Et. ester, 107–8
170	6-Chloronaphthalene-2-sulfonyl chloride	110.5	....	184							
171	2-Hydroxynaphthalene-1,6-disulfonyl chloride	111	....	di: 191							
172	Acenaphthene-5-sulfonyl chloride	111	....	223	178						
173	3-Hydroxynaphthalene-2-sulfonyl chloride	112	....	110	....				241–2		
174	5-Nitronaphthalene-1-sulfonyl chloride	113	....	236	123						
175	Acenaphthene-3-sulfonyl chloride	113–4	87–9	199	284–6	.....			.....		Me. ester, 122–3; Et. ester, 137–9, lgr.
176	2,5-Dichlorobenzene-1,3-disulfonyl chloride	114	....	215–7							
177	5-Chloronaphthalene-2-sulfonyl chloride	115	....	216							
178	Biphenyl-4-sulfonyl chloride (4-Phenylbenzenesulfonyl chloride .	115	....	230	125						

*Derivative data given in order: m.p., crystal color, solvent from which crystallized.

No.	Name	Melting point, °C	Sulfonic acid	Amide	Anilide	1-Naphthyl amide	Salts of the corresponding acid				Miscellaneous
							S-Benzyl thiuronium	p-Toluidine	Aniline	o-Toluidine	
179	2-Ethoxynaphthalene-1-sulfonyl chloride	116	....	158	187	......					
180	4,5-Dichloronaphthalene-1-sulfonyl chloride	117	....	229	....						
181	2,6-Dimethyl-4-hydroxybenzene-1,3-disulfonyl chloride	117–8	....	206–8	205–7	......					
182	6-Ethoxynaphthalene-1-sulfonyl chloride	118	....	154	194.5						
183	4,6-Dichloronaphthalene-1-sulfonyl chloride	119	....	226							
184	5-Methoxynaphthalene-1-sulfonyl chloride	119.5	....	194.5	157						
185	5-Methylnaphthalene-2-sulfonyl chloride	120–2	....	188–9	248–50; 133–4	......					
186	8-Bromonaphthalene-2-sulfonyl chloride	121	....	187	....						
187	5-Ethoxynaphthalene-1-sulfonyl chloride	121	....	182.5	130						
188	2-Methoxynaphthalene-1-sulfonyl chloride	121	....	159	196.5	......					
189	1-Nitronaphthalene-2-sulfonyl chloride	121, pink, bz.-pet. eth.	105, lgr.	214	202						
190	6,8-Dichloronaphthalene-2-sulfonyl chloride	121	....	228	....						
191	2-Chloro-5-methyl-6-nitro-benzene sulfonyl chloride	122	....	177	....						
192	Anthracene-2-sulfonyl chloride ...	122	....	261	201						Me. ester, 157; Et. ester, 160
193	5-Fluoronaphthalene-1-sulfonyl chloride	122–3	105 (hyd.)	196–7	....						Me. ester, 118, eth.
194	4,6-Dichlorobenzene-1,3-disulfonyl chloride	123	....	276	....						
195	Azobenzene-3,4'-disulfonyl chloride	123–5	....	di: 288	....						
196	7,8-Dichloronaphthalene-2-sulfonyl chloride	124	....	227	....						
197	4-Iodonaphthalene-1-sulfonyl chloride	124; 121	....	206; 204	136	....					
198	6-Bromonaphthalene-2-sulfonyl chloride	124	....	127							
199	Quinoline-8-sulfonyl chloride.....	124	312	183–4	....					......	Me. ester, 96; Et. ester, 73
200	4-Methylnaphthalene-2-sulfonyl chloride	124–5	....	143–4	....					....	
201	1,5-Dichloronaphthalene-1-sulfonyl chloride	125	....	282							
202	2,4,6-Trimethylbenzene-1,3-disulfonyl chloride	125	....	di: 244	di: 150–1						
203	5-Nitronaphthalene-2-sulfonyl chloride	125	118–9, yel.	184	....						
204	6-Nitronaphthalene-1-sulfonyl chloride	127	....	223–4	....						
205	Phenanthrene-9-sulfonyl chloride .	127	174 (anh.)	193–4	....	......		235	......	....	Me. ester, 106, me. al.; Et. ester, 108 al.
206	Naphthalene-1,6-disulfonyl chloride	129	125 (anh.)	298	....	......	81; 235	314–5	298–9	323–4	....
207	7-Chloronaphthalene-1-sulfonyl chloride	129	....	235	....	......					

*Derivative data given in order: m.p., crystal color, solvent from which crystallized.

TABLE XXIII. ORGANIC DERIVATIVES OF SULFONYL CHLORIDES
(Listed in order of increasing m.p.)* (Continued)

No.	Name	Melting point, °C	Sulfonic acid	Amide	Anilide	1-Naphthyl amide	S-Benzyl thiuronium	p-Toluidine	Aniline	o-Toluidine	Miscellaneous
208	4,6-Dichloronaphthalene-2-sulfonyl chloride	130	....	218	....	....	....	....	...,.	....	
209	2,4,5-Trimethoxybenzenesulfonyl chloride	130	....	76	170	....	....	....	....	....	
210	4,6-Dimethylbenzene-1,3-di-sulfonyl chloride	130, pet. eth.	....	249, w.	196, 50% al.	....	....	....	....	....	
211	5,6,7-Trichloronaphthalene-1-sulfonyl chloride	131	....	249	....	....	....	....	....	....	
212	Benzene-1,4-disulfonyl chloride...	131; 139	....	288	249	....	....	....	....	....	
213	5,8-Dichloronaphthalene-2-sulfonyl chloride	134	....	244	....	....	....	....	....	....	
214	3,7-Dichloronaphthalene-1-sulfonyl chloride	136	....	269	....	....	....	....	....	....	
215	3-Bromocamphor-8-sulfonyl chloride	136–7									
216	8-Chloro-7-methoxynaphthalene-1-sulfonyl chloride	137	....	153	196	....	....	....	....	....	
217	Benzophenone-3,3'-disulfonyl chloride	137–8	....	di: 157	di: 177–8	....	....	....	....	....	N-Xanthylsulfon-amide, 197
218	Naphthalene-1,3-disulfonyl chloride	137.5	....	di: 292–3	....	....	....	....	....	....	....
219	7,8-Dichloronaphthalene-1-sulfonyl chloride	138	....	221	....	....	....	....	....	....	
220	Azoxybenzene-3,3'-disulfonyl chloride	138	126	di: 273	....	....	....	....	....	....	....
221	D-Camphor-8-sulfonyl chloride...	138	....	137	....	....	....	....	....	....	
222	3-Methoxynaphthalene-2-sulfonyl chloride	138	....	113	174	....	....	....	....	....	
223	8-Cyanonaphthalene-1-sulfonyl chloride	139	....	333.4	....	....	....	....	....	....	
224	4-Nitronaphthalene-2-sulfonyl chloride	139.5	....	225	....	....	....	....	....	....	
225	6-Iodonaphthalene-2-sulfonyl chloride	140	....	222	....	....	....	....	....	....	
226	8-Iodonaphthalene-1-sulfonyl chloride	140	115	187	....	....	....	....	....	....	
227	4-Nitronaphthalene-2,7-disulfonyl chloride	140–1	....	286–7	....	....	....	....	....	....	
228	4-Hydroxybenzenesulfonyl chloride	141	....	176–7	....	......	169	202	170	192	
229	4,8-Dichloronaphthalene-2-sulfonyl chloride	141	....	205	....	....	....	....	....	....	
230	5-Iodonaphthalene-1-sulfonyl chloride	141	....	239	....	....	....	....	....	....	
231	Pyridine-3-sulfonyl chloride......	Hydro-chloride, 141–4	357	110–1	145	....	....	....	....	......	Hydrazide, 94
232	6,7-Dichloronaphthalene-1-sulfonyl chloride	142	....	268	....	....	....	....	....	....	
233	Benzene-1,2-disulfonyl chloride...	143	....	254	241	....	....	....	....	....	
234	6-Ethoxy-1-nitronaphthalene-2-sulfonyl chloride	146	....	218	....	....	....	....	....	....	
235	Retene-6-sulfonyl chloride	146–7.5, yel.-br.	121–3	206–7.5	....	....	....	....	....	....	
236	7-Bromonaphthalene-1-sulfonyl chloride	147	....	209	....	....	....	....	....	....	....
237	4-Acetamidobenzenesulfonyl chloride	149	....	219	214	215	....	....	....	....	
238	5,7-Dichloronaphthalene-1-sulfonyl chloride	149	....	272	....	....	....	....	....	....	

*Derivative data given in order: m.p., crystal color, solvent from which crystallized.

TABLE XXIII. ORGANIC DERIVATIVES OF SULFONYL CHLORIDES
(Listed in order of increasing m.p.)* (Continued)

No.	Name	Melting point, °C	Sulfonic acid	Amide	Anilide	1-Naphthyl amide	Salts of the corresponding acid				Miscellaneous
							S-Benzyl thiuronium	p-Toluidine	Aniline	o-Toluidine	
239	4,5-Dibenzylnaphthalene-1-sulfonyl chloride	151	....	168	....	......	......	......	......	......	
240	4,7-Dichloronaphthalene-1-sulfonyl chloride	151	....	217	....	......	......	......	......	......	
241	7-Ethoxy-8-nitronaphthalene-1-sulfonyl chloride	155	....	173.4	....	......	......	......	......	......	
242	2-Amino-5-methylbenzene-1,3-disulfonyl chloride	156, chl.	....	di: 257, w.	di: 196–7, aq. al.	......	......	......	......	......	
243	Phenanthrene-2-sulfonyl chloride	156	150	253–4	157–8	......	......	291	......	......	Me. ester, 101–2; Et. ester, 89, yel.-br.
244	4,7-Dichloronaphthalene-2-sulfonyl chloride	156	....	196	....	......	......	......	......	......	
245	6,7,8-Trichloronaphthalene-2-sulfonyl chloride	157	....	245	....	......	......	......	......	......	
246	4,5-Dichloronaphthalene-2-sulfonyl chloride	158	....	197	....	......	......	......	......	......	
247	5,6,8-Trichloronaphthalene-2-sulfonyl chloride	158	....	235	....	......	......	......	......	......	
248	Naphthalene-2,7-disulfonyl chloride	158; 162	....	242	....	......	212	299	251–2	238	
249	Naphthalene-1,4-disulfonyl chloride	160; 166	....	273, w., or al.	179	......	......	......	......	......	
250	4-Carboxy-3-nitrobenzene-sulfonyl chloride	di: 160	111 (+2½ H_2O)	192	....	......	......	......	......	......	The m.p. is that of the 1,4-dichloride; Diamide, 226
251	7-Hydroxynaphthalene-1,3-disulfonyl chloride	161–2	....	....	195	......	228	294	......	271	
252	Fluorene-2-sulfonyl chloride	164	155 (hyd.)	213d.	....	......	......	......	......	......	
253	2,5-Dimethylbenzene-1,4-disulfonyl chloride	164	....	310	223	......	......	......	......	......	
254	8-Nitronaphthalene-1-sulfonyl chloride	165 d.	115 d. (+3H_2O)	191	178–8.5	......	......	......	......	......	
255	7-Iodonaphthalene-1-sulfonyl chloride	165	....	240	....	......	......	......	......	......	
256	Azobenzene-3,3'-disulfonyl chloride	166	....	di: 305	....	......	......	......	......	......	
257	3,6-Dichloronaphthalene-2-sulfonyl chloride	166	....	218	....	......	......	......	......	......	
258	5,6-Dichloronaphthalene-2-sulfonyl chloride	167	....	192	....	......	......	......	......	......	
259	Phthalic acid-4-sulfonyl chloride (3,4-Dicarboxybenzenesulfonyl chloride)	167–70, d. eth.	138–40 (+1H_2O)	192–200 d., w.	....	......	......	......	......	......	
260	8-Nitronaphthalene-2-sulfonyl chloride	169	135–6 (+1½ H_2O)	223	172–3	......	......	......	......	......	
261	4-Amino-2-hydroxybenzene-sulfonyl chloride	169	....	155	....	......	......	......	......	......	
262	2-Hydroxynaphthalene-1,7-disulfonyl chloride	169	....	233	....	......	......	......	......	......	
263	7-Nitronaphthalene-1-sulfonyl chloride	169–70	....	261–2	....	......	......	......	......	......	
264	4,6,7,8-Tetrachloronaphthalene-2-sulfonyl chloride	176	....	235	....	......	......	......	......	......	
265	4'-Nitrobiphenyl-4-sulfonyl chloride	178	....	228	....	......	......	......	......	......	
266	Naphthalene-1,5-disulfonyl chloride	183	245 (anh.)	310; 340	249	......	257; 251 d.	332	......	......	Di-Me. ester, 205, chl.
267	Anthraquinone-2,7-disulfonyl chloride	186, chl.	....	....	di: 192	......	......	......	......	......	

*Derivative data given in order: m.p., crystal color, solvent from which crystallized.

TABLE XXIII. ORGANIC DERIVATIVES OF SULFONYL CHLORIDES
(Listed in order of increasing m.p.)* (Continued)

No.	Name	Melting point, °C	Sulfonic acid	Amide	Anilide	1-Naphthyl amide	S-Benzyl thiuronium	p-Toluidine	Aniline	o-Toluidine	Miscellaneous
268	Benzene-1,3,5-trisulfonyl chloride .	187	....	tri: 310–5	tri: 237	......	......	......	......	......	Tri-Et. ester, 147, bz.
269	7-Hydroxynaphthalene-1,3,6-trisulfonyl chloride	196	....	....	152–5	......	......	......	......	......	....
270	Anthraquinone-2-sulfonyl chloride	197	....	261	193	......	211	308	309	......	Me. ester, 123; Et. ester, 125
271	Anthraquinone-1,6-disulfonyl chloride	197–8,yel., PhNO₂	215–7, gold	....	227–8, yel.	......	......	......	......	......	
272	Biphenyl-4,4'-disulfonyl chloride	203	....	300	....	......	171	330 d.	......	......	
273	Benzidine-2,2'-disulfonyl chloride .	Hydro- chloride, 205	....	di: 278	....	......	......	......	......	......	
274	Anthraquinone-1-sulfonyl chloride	216–8, yel., PhNO₂	218	....	214	......	191	......	284	......	NH₃ → 1-Amino- anthraquinone, 252; 243
275	9,10-Dichloroanthracene-2-sulfonyl chloride	221	....	279	248	......	......	......	......	......	....
276	Azobenzene-4,4'-disulfonyl chloride	222	169d. (anh.)	250 d.	....	......	......	......	......	......	Et. ester, 104
277	Anthraquinone-1,8-disulfonyl chloride	222–3, yel, PhNO₂	293–4	>340	237–8, yel., PhNO₂	......	......	......	......	......	
278	Anthracene-1,8-disulfonyl chloride	225	....	333	224	......	......	......	......	......	
279	Naphthalene-2,6-disulfonyl chloride	225	....	305	....	......	256	......	360	......	
280	2-Hydroxynaphthalene-1,5-disulfonyl chloride	231	....	....	231	......	......	......	......	......	
281	Anthracene-1,5-disulfonyl chloride	240	....	di: 330	di: 293	......	......	......	......	......	
282	Anthraquinone-2,6-disulfonyl chloride	250, yel. cl. bz.	....	....	di: 321	......	......	......	......	......	
283	Anthraquinone-1,3-disulfonyl chloride	265–70, yel.	310–1 (hyd.)	>350	269–70, yel.-red	......	......	......	......	......	
284	2,4-Diaminobenzene-1,5-disulfonyl chloride	275	....	187	236	......	......	......	......	......	
285	Anthraquinone-1,7-disulfonyl chloride	301–2, br.- yel, PhNO₂	120 (hyd.)	....	237–8, yel. cl. bz.	......	......	......	......	......	

*Derivative data given in order: m.p., crystal color, solvent from which crystallized.

*Formation of sulfonic acid by hydrolysis.**

$$RSO_2NH_2 \ + \ H_2O \ \xrightarrow{\ HCl\ } \ RSO_3H \ + \ NH_4Cl$$

Sulfonic
acid

From hydrolysis of the sulfonamide by 25% hydrochloric acid.

For directions and examples see: Cheronis, p. 631; Shriner, pp. 104, 267; R. S. Schreiber and R. L. Shriner, *J. Amer. Chem. Soc.*, **56**, 1618 (1934).

From hydrolysis of the sulfonamide in a mixture of concentrated sulfuric acid and 85% phosphoric acid.

See: Cheronis, p. 633.

Formation of the amine by hydrolysis.

$$2\,ArSO_2NHR \ + \ 5\,HBr \ + \ 5\,C_6H_5OH \ \rightarrow \ ArSSAr \ + \ 2\,RNH_2 \ + \ 5\,p\text{-}BrC_6H_4OH \ + \ 4\,H_2O$$

Amine

From hydrolysis of the sulfonamide in 48% hydrobromic acid and phenol.

For directions and examples see: Shriner, pp. 105, 267; H. R. Snyder and R. E. Heckert, *J. Amer. Chem. Soc.*, **74**, 2006 (1952); H. R. Snyder and H. C. Geller, *J. Amer. Chem. Soc.*, **74**, 4864 (1952).

NOTE: For directions and explanations for the preparation of derivatives of the sulfonic acid formed by the hydrolysis of the sulfonamide see explanations and references to Table XXII, pp. 369, 370.

For directions and explanations for the preparation of the amine formed by the hydrolysis of the sulfonamide see explanations and references to Table XVIII, pp. 291, 292, 293.

*N-Xanthylsulfonamide.**

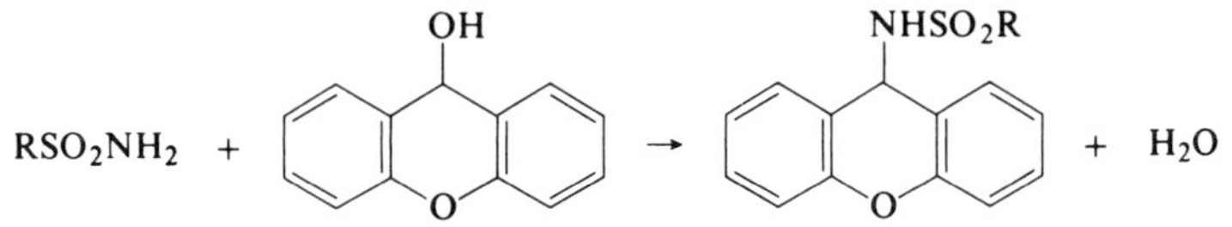

$$RSO_2NH_2 \ + \ \text{Xanthydrol} \ \rightarrow \ \text{N-Xanthylsulfonamide} \ + \ H_2O$$

From the sulfonamide and xanthydrol in glacial acetic acid.

For directions and examples see: Cheronis, p. 634; Linstead, p. 95; Shriner, p. 267; R. F. Phillips and V. S. Frank, *J. Org. Chem.*, **9**, 9 (1944).

N-Acetylsulfonamide.

$$RSO_2NH_2 \ + \ (CH_3CO)_2O \ \rightarrow \ RSO_2NHCOCH_3 \ + \ CH_3COOH$$

$$RSO_2NHR' \ + \ CH_3COCl \ \rightarrow \ RSO_2NR'COCH_3 \ + \ HCl$$

N-Acetyl-
sulfonamide

From the sulfonamide and acetic anhydride.

For directions and examples see: Cheronis, p. 634.

From the sulfonamide and acetyl chloride in acetic acid.

See: Linstead, p. 95; Vogel, p. 555.

*Derivatives recommended for first trial.
WARNING: This is not an instruction manual. References should be consulted for the preparation of derivatives.

TABLE XXIV. ORGANIC DERIVATIVES OF SULFONAMIDES AND SULFONANILIDES
(Listed in order of increasing m.p.)*

No.	Name	Melting point, °C	Sulfonic acid	Derivatives of the corresponding acid				Salts of the corresponding acid				Miscellaneous derivatives of the acid
				Chloride	Amide	Anilide	1-Naph-thyl-amide	S-Benzyl thiu-ronium	p-Tolui-dinium	Ani-linium	o-Tolui-dinium	
1	2-Methylpropane-1-sulfonamide..........	14–6	....	b.p. 80[13]	....	38	107	......	......	......	......	
2	N,N-Diethylbenzylsul-fonamide............	29	....	92	105, w., al.	102	166 (146)	......	113, al.	102	83	....
3	2-Methylpropane-1-sulfonanilide..........	38	....	b.p. 80[13]	14–6	....	107	......	......	......	......	
4	3-Methylbutane-1-sulfonanilide..........	42	....	b.p. 98[13]	3	....	90–1	......	......	......	......	
5	Butane-1-sulfonamide ...	45	−15	b.p. 75[10]	....	10–5	60.5	......	......	......	......	Phenylhydrazi-nium salt, 114–5
6	Propane-1-sulfonamide ..	52, eth.	7.5	b.p. 78[15]	....	10	84	......	67–8		......	Phenylhydrazi-nium salt, 204.5d.
7	Ethane sulfonamide	58, eth.	−17	b.p. 178	....	58	66	115	......		......	Phenylhydrazi-nium salt, 182.8
8	Ethane sulfonanilide.....	58	−17	b.p. 178	58	....	66	115	......		......	Phenylhydrazi-nium salt, 182.8
9	N,N-Diethyltoluene-4-sulfonamide..........	60	104–5	69	105 (dihyd.); 138.5–9 (anh.)	103	157	181–2	198	238	190	m-Toluidide, 109
10	Propane-2-sulfonamide ..	60, eth.-pet. eth.	−37	b.p. 61[9]	....	84	134	......				
11	N-Ethyltoluene-4-sulfonamide..........	64	104–5	69	105 (dihyd.); 138.5–9 (anh.)	103	157	181–2	198	238	190	....
12	N-Ethylbenzyl sulfon-amide...............	65–6, eth., lgr.	....	92	105, w., al.	102	166 (146)	......	113, al.	102	83	
13	N-(1-Naphthyl)ethane sulfonamide..........	66	−17	b.p. 178	58	58	....	115	......			
14	Ethane-1,2-disulfon-anilide..............	69	....	di: 95, eth.	....	....	....	201–2	d. at 270	270, w.	......	Di-Et. ester, 77, eth.; m-Tolui-dide, 230
15	N-Methyltoluene-2-sul-fonamide............	74–5	57	68	....	136	....	170	203–4	218	......	....
16	Heptane-1-sulfonamide ..	75	....	16	....	....	....	......				
17	2,4,5-Trimethoxybenzene sulfonamide..........	76	....	130	....	170	....	......				
18	2-Phenylethane-1-sulfon-anilide..............	77	91	33	122							
19	N-Methyltoluene-4-sulfonamide..........	78–9	104–5	69	105 (dihyd.); 138.5–9 (anh.)	103	157	181–2	198	238	190	....
20	Propane-2-sulfonanilide..	84	....	b.p. 61[9]	60	....	134	......				
21	N-(1-Naphthyl)propane-1-sulfonamide..........	84	7.5	b.p. 78[15]	52	10	....	......	67–8			
22	N,N-Dimethyltoluene-4-sulfonamide..........	86–7	104–5	69	105 (dihyd.); 138.5–9 (anh.)	103	157	181–2	198	238	190	....
23	Methane sulfonamide....	90	20	b.p. 60[21]	....	100.5	125.5	......				Phenylhydrazi-nium salt, 193.5–4d.
24	N-(1-Naphthyl)-3-methyl-butane-1-sulfonamide...	90–1	....	b.p. 98[13]	3	42	....	......				....
25	4-Chloro-3-methylbenzene sulfonanilide.........	92	....	63	128							
26	Toluene-3-sulfonanilide ..	96	....	12	108	....	......		106		......	

*Derivative data given in order: m.p., crystal color, solvent from which crystallized.

No.	Name	Melting point, °C	Sulfonic acid	Derivatives of the corresponding acid				Salts of the corresponding acid				Miscellaneous derivatives of the acid
				Chloride	Amide	Anilide	1-Naphthylamide	S-Benzyl thiuronium	p-Toluidinium	Anilinium	o-Toluidinium	
27	3-Chloro-4-methylbenzene sulfonanilide	96	. . . .	38	134	. . . .		. . . . .		. . . . .		. . . .
28	3,3-Dimethylbutane sulfonamide	96–7	. . . .	43–4								
29	Hexadecane-1-sulfonamide.	97	54	54								
30	2-Ethylbenzene sulfonamide.	100	. . . .	11.7								
31	10-Bromocamphor-3-sulfonamide	100–2	. . . .	97								. . . .
32	Methane sulfonanilide . . .	100.5	20	b.p. 60[21]	90	. . . .	125.5	. . . . . .		. . . . .		Phenylhydrazinium salt, 193.5–4d.
33	N,N-Dimethylbenzyl sulfonamide.	101	. . . .	92–3, eth., lgr.	105, w., al.	102, al.	166 (146)	. . . . .	113, al.	102	83	Hydrazide, 131–2; Phenylhydrazide, 173
34	Benzyl sulfonanilide (Toluene-α-sulfonanilide)	102, al.	. . . .	92	105, w., al.	. . . .	166 (146)	. . . . .	113, al.	102	83	Hydrazide, 131–2; Phenylhydrazide, 173
35	Toluene-4-sulfonanilide . .	103	92 (104–5)	69	137	. . . .	. . . . . .	181–2	198	238	190	N-Xanthylsulfonamide, 197
36	4-Chlorobenzene sulfonanilide	104	69 (93)	53	144		190	175	208–10	222–3	163–4	. . . .
37	Benzyl sulfonamide (Toluene-α-sulfonamide)	105, w., al.	. . . .	92–3, eth., lgr.	. . . .	102, al.	166 (146)	. . . . .	113, al.	102	83	Hydrazide, 131–2; Phenylhydrazide, 173
38	N-(1-Naphthyl)-2-methylpropane sulfonamide . . .	107	. . . .	b.p. 80[13]	14–6	38						
39	Toluene-3-sulfonamide. . .	108, al.	. . . .	12	. . . .	96	. . . . . .		106	. . . . .	108	. . . .
40	2,4-Dimethyl-6-nitrobenzene sulfonamide. . . .	108	97	. . . .								. . . .
41	2,4,6-Trimethylbenzene sulfonanilide	109	78	56	142	. . . .	. . . . . .			. . . . .		N-Xanthylsulfonamide, 203
42	N-Methylbenzyl sulfonamide.	109–10	. . . .	92–3, eth., bz.	105, w., al.	102, al.	166 (146)	. . . . .	113, al.	102	83	N-Xanthylsulfonamide, 188; Hydrazide, 131–2; Phenylhydrazide, 173
43	2,4-Dimethylbenzene sulfonanilide	110	62 (hyd.)	34	. . . .							. . . .
44	4-Ethylbenzene sulfonamide.	110	. . . .	12								N-Xanthylsulfonamide, 196
45	3-Hydroxynaphthalene-2-sulfonamide.	110	. . . .	112						241–2		. . . .
46	Pyridine-3-sulfonamide . .	110–1	357	Hydrochloride, 141–4 d.	. . . .	145	. . . . . .		. . . . .		. . . . .	Hydrazide, 94
47	Naphthalene-1-sulfonanilide	112 (152)	90	68; 66	150	. . . .	. . . . . .	137	181	183	237	. . . .
48	Benzene sulfonanilide. . . .	112	66 (anh.)	14.5	156	. . . .	170–1	148	205	240	176	N-Xanthylsulfonamide, 200
49	3-Methoxynaphthalene-2-sulfonamide.	113	. . . .	138	. . . .	174						
50	2,6-Dimethylbenzene sulfonamide.	113 (96)	98	39	. . . .							
51	2-(N-Methylamino)benzene sulfonamide	114.5–5.5	182 d.	. . . .								2-N-p-Toluenesulfonyl deriv. of sulfonamide, 193

*Derivative data given in order: m.p., crystal color, solvent from which crystallized.

TABLE XXIV. ORGANIC DERIVATIVES OF SULFONAMIDES AND SULFONANILIDES
(Listed in order of increasing m.p.)* (Continued)

No.	Name	Melting point, °C	Sulfonic acid	Derivatives of the corresponding acid				Salts of the corresponding acid				Miscellaneous derivatives of the acid
				Chloride	Amide	Anilide	1-Naphthylamide	S-Benzyl thiuronium	p-Toluidinium	Anilinium	o-Toluidinium	
52	2-Nitrobenzene sulfonanilide	115	70 (85)	69	193	....	....	....	....	....	....	
53	N-Benzyltoluene-4-sulfonamide	115–6	92	69	137	....	......	181–2	198	238	190	N-Xanthylsulfonamide, 197
54	8-Methylnaphthalene-2-sulfonamide	116	....	88								
55	Indane-4-sulfonamide	118–9, w.	....	53–3.5; b.p. 140–1^5	....							
56	4-Bromobenzene sulfonanilide	119	88–90	76, eth.	166; 161	....	183.5	170	215–6	237–8	182–3	....
57	3,5-Dimethylbenzene sulfonanilide	119	....	94; 90	135	....	....		121–2 d.			
58	6-Methoxynaphthalene-2-sulfonanilide	120	....	93	189							
59	D-Camphor-10-sulfonanilide	121	193	67	132							
60	7-Methoxynaphthalene-2-sulfonanilide	121	....	83	220							
61	2-Phenylethane-1-sulfonamide	122	91	33	....	77						
62	4-Formylbenzene sulfonamide	122–4										Oxime, 158; O,O-Diacetate of sulfonamide, 86–7.5
63	5-Nitronaphthalene-1-sulfonanilide	123		113	236	....						Me. ester, 117–8, chl.
64	2-Methylnaphthalene-1-sulfonamide	124		83–5	....							....
65	D-Camphor-3-sulfonanilide	124	77	88	143	....			196–7	......	......	Me. ester, 77, $[\alpha]_D$ +98.6 in chl.
66	Biphenyl-4-sulfonanilide	125		115	230	....						
67	4-Fluorobenzene sulfonamide	125		36; 30	....							
68	N-(1-Naphthyl)methane sulfonamide	125.5	20	b.p. 60^{21}	90	100.5						
69	3-Nitrobenzene sulfonanilide	126	48	64	167	....	166.5	146	222	222	193	....
70	Phenol-O-sulfonanilide	126.5–7.5	145 (mono-hyd.)	....	....	....	......	......	124–5	126.5–7.5		
71	5-Aminonaphthalene-2-sulfonanilide	127–8	....	....	219	....		191		......		5-N-Acetyl deriv. of sulfonamide, 247
73	4-Chloro-3-methylbenzene sulfonamide	128	....	63	....	92						Sulfonyl bromide, 67.5
74	6-Fluoronaphthalene-2-sulfonanilide	129	105 (hyd.)	97	133	....						
75	5-Ethoxynaphthalene-1-sulfonanilide	130	....	121	182.5							
76	3-Chloro-4-methoxybenzene sulfonamide	131	....	82	....							
77	4-Aminonaphthalene-2-sulfonamide	131 (hyd.)	....									4-N-Acetyl deriv. of sulfonamide, 220
78	2,5-Dimethyl-4-nitrobenzene sulfonanilide	131	140	75	197–8	....			143.5–4.5	......	143.5–4.5	....

*Derivative data given in order: m.p., crystal color, solvent from which crystallized.

No.	Name	Melting point, °C	Sulfonic acid	Derivatives of the corresponding acid				Salts of the corresponding acid				Miscellaneous derivatives of the acid
				Chloride	Amide	Anilide	1-Naphthylamide	S-Benzyl thiuronium	p-Toluidinium	Anilinium	o-Toluidinium	
79	3-Ethoxybenzene sulfonamide	131	....	38	....	....	......	......	......	......	......	....
80	D-Camphor-10-sulfonamide	132	193	67	....	121 (88)	......	......	......	......	......	....
81	Naphthalene-2-sulfonanilide	132	91 (hyg.)	76; 79	217; 213	....	......	......	......	......	......	....
82	6-Fluoronaphthalene-2-sulfonamide	133	105 (hyd.)	97, chl.	....	129	......	190–1	221	269	213	....
83	DL-Camphor-8-sulfonamide	133–5, w.	56–8	106	....	....	......	......	......	......	......	....
84	3-Methyl-4-nitrobenzene sulfonamide	133.5	....	50	....	....	......	......	......	......	......	....
85	3-Chloro-4-methylbenzene sulfonamide	134	....	38	....	96	......	......	......	......	......	....
86	N-(1-Naphthyl)propane-2-sulfonamide	134	−37	b.p. 61⁹	60	84	......	......	......	......	......	m-Toluidide, 109
87	3,5-Dimethylbenzene sulfonamide	135, al.	....	94; 90, bz.	....	129, al.	......	......	121–2, al.	......	......	....
88	Tetraline-6-sulfonamide	135	....	58	....	155–6	......	......	......	......	......	....
89	Indane-5-sulfonamide	135.5–6, al.	92	46–7, eth.; b.p. 148–9⁴	....	129, al.	......	......	......	......	......	....
90	Toluene-2-sulfonanilide	136	57	68	156	....	......	170	203–4	218	......	N-Xanthylsulfonamide, 183
91	4-Iodonaphthalene-1-sulfonanilide	136	....	124; 121	206; 204	....	......	......	......	......	......	....
92	D-Camphor-8-sulfonamide	137	....	138	....	....	......	......	......	......	......	[α]ᴅ¹³: +93.6, in al.
93	4-Chloro-2-nitrobenzene sulfonanilide	138	82	75	237	....	......	......	......	......	......	....
94	2,4-Dimethylbenzene sulfonamide	138	62 (hyd.)	34	....	110	......	146	......	......	......	N-Xanthylsulfonamide, 188
95	Toluene-4-sulfonamide	138.5–9.0 (anh.); 105 (dihyd.)	104–5	69	....	103	157	181–2	198	238	190	N-Xanthylsulfonamide, 197
96	8-Aminonaphthalene-1-sulfonanilide	139–40	....	....	....	....	......	300	......	......	......	....
97	4-Vinylbenzene sulfonamide	139–40	....	....	....	....	......	......	182–3	......	......	Dimethylamide, 62–3
98	2-Bromonaphthalene-1-sulfonamide	140	....	97	....	....	......	......	......	......	......	....
99	3,4-Dichlorobenzene sulfonamide	140; 135	....	22.4; 19	....	....	......	......	......	......	......	....
100	2,4,6-Trimethylbenzene sulfonamide	142	78	56	....	109	......	......	......	......	......	N-Xanthylsulfonamide, 203
101	7-Ethoxynaphthalene-2-sulfonamide	142	....	103	....	153	......	......	......	......	......	....
102	3-Aminobenzene sulfonamide	142	....	....	....	....	......	148	......	......	......	....
103	3,5-Dimethyl-2-hydroxybenzene sulfonanilide	142–3	121–5	....	....	....	......	......	......	......	......	2-O-Acetyl deriv. of sulfonyl chloride, 62, pet. eth.
104	D-Camphor-3-sulfonamide	143	77	88	....	124	......	......	196–7	......	......	N-Methylanilide, 111–2; Me. ester, 77, [α]ᴅ: +98.6 in chl.
105	5-Chloro-2-methylbenzene sulfonamide	143	21; 24	....	....	....	......	......	......	......	......	....

* Derivative data given in order: m.p., crystal color, solvent from which crystallized.

TABLE XXIV. ORGANIC DERIVATIVES OF SULFONAMIDES AND SULFONANILIDES
(Listed in order of increasing m.p.)* (Continued)

No.	Name	Melting point, °C	Sulfonic acid	Derivatives of the corresponding acid				Salts of the corresponding acid				Miscellaneous derivatives of the acid
				Chloride	Amide	Anilide	1-Naphthyl-amide	S-Benzyl thiu-ronium	p-Tolui-dinium	Ani-linium	o-Tolui-dinium	
106	4-Iodobenzene sulfonanilide	143	....	85	183	....	....	....	....	......	....	....
107	2,5-Dimethyl-3-nitrobenzene sulfonanilide	143–4	128 (200)	61	173	....	....	....	135–6	......	126.5–7.5	....
108	4-Methylnaphthalene-2-sulfonamide	143–4	....	124–5	....	....	....	....	....	......	....	....
109	Benzophenone-2-sulfonanilide	143–5	....	96–7	....	....	....	....	....	......	....	....
110	4-Ethoxynaphthalene-2-sulfonanilide	143.5	....	85	183	....	....	....	....	......	....	....
111	3,4-Dimethylbenzene sulfonamide	144	64 (55)	52	....	....	....	208	....	......	....	....
112	4-Fluoronaphthalene-1-sulfonanilide	144	100 (hyd.)	86	206	....	....	....	....	......	....	Et. ester, 93
113	4-Chlorobenzene sulfonamide	144	69 (93)	53	....	104	190	175	208–10	222–3	163–4	....
114	4-Methyl-3-nitrobenzene sulfonamide	144.5	92 (hyg.)	36	....	109	153	....	130–1	......	28	....
115	3-Chloro-6-methylbenzene sulfonamide	145, aq. al.	....	24	....	....	....	....	....	......	....	....
116	Pyridine-3-sulfonanilide	145	....	....	110–1	....	....	....	....	......	....	....
117	3-Bromocamphor-8-sulfonamide	145	195–6 (anh.)	136–7	....	....	....	....	....	......	....	....
118	4-Chloronaphthalene-1-sulfonanilide	145–6	130–3	94–5	187	....	162	....	....	145–6	151	Me. ester, 83; Et. ester, 104
119	4-Bromo-3-methylbenzene sulfonamide	146	....	50	....	....	....	....	....	......	....	....
120	4-Methoxy-3-nitrobenzene sulfonamide	146.3	....	66	....	....	....	....	....	......	....	....
121	1-Aminonaphthalene-7-sulfonanilide	147	....	....	181 (hyd.)	....	....	....	....	......	....	Benzoylguanidine salt, 214–6
122	4-Methoxynaphthalene-2-sulfonanilide	147.5	....	98.5	226	....	....	....	....	......	....	....
123	3-Chlorobenzene sulfonamide	148	....	....	....	....	....	....	199–200	206–7	....	....
124	2-Methyl-5-nitrobenzene sulfonanilide	148	133.5	46–7	186	....	....	....	256–7	......	256–8	....
125	2,5-Dimethylbenzene sulfonamide	148	48 (anhyd.); 86 (hyd.)	24–6	....	....	....	184	....	......	....	N-Xanthylsulfonamide, 176
126	Benzene-1,3-disulfonanilide	148–50	....	63	229	......	245	214	....	......	....	N-Xanthylsulfonamide, 170
127	6-Methoxynaphthalene-1-sulfonamide	149.5	....	80.5	....	177.5	....	....	....	......	....	....
128	Naphthalene-1-sulfonamide	150	90	68; 66	....	112 (152)	......	137	181	183	237	....
129	4-Ethoxybenzene sulfonamide	150	....	39	....	....	....	....	....	......	....	....
130	4,6-Dichloro-2,5-dimethylbenzene sulfonamide	150	....	81	....	175	....	....	....	......	....	....
131	3-Bromo-4-methylbenzene sulfonamide	151	....	60	....	....	....	....	....	......	....	....
132	2-Hydroxynaphthalene-3,6,8-trisulfonanilide	152–5	....	tri: 196	....	....	....	....	....	......	....	....
133	7-Ethoxynaphthalene-2-sulfonanilide	153	....	103	142	....	....	....	....	......	....	....
134	2-Chloronaphthalene-1-sulfonamide	153	....	75	....	....	....	....	....	......	....	....

* Derivative data given in order: m.p., crystal color, solvent from which crystallized.

TABLE XXIV. ORGANIC DERIVATIVES OF SULFONAMIDES AND SULFONANILIDES
TABLE XXIV. ORGANIC DERIVATIVES OF SULFONAMIDES AND SULFONANILIDES
(Listed in order of increasing m.p.)* (Continued)

No.	Name	Melting point, °C	Sulfonic acid	Derivatives of the corresponding acid				Salts of the corresponding acid				Miscellaneous derivatives of the acid
				Chloride	Amide	Anilide	1-Naphthyl-amide	S-Benzyl thiuronium	p-Tolui-dinium	Ani-linium	o-Tolui-dinium	
135	8-Chloro-7-methoxynaphthalene-1-sulfonamide	153	....	137	....	196	......					
136	2-Aminobenzene sulfonamide	153	....				......	132	......	......	......	2-N-Benzoyl deriv. of sulfonamide, 198; Hydrochloride, 201
137	5-Methylbenzene-1,3-disulfonanilide	153, al.	....	94, eth.	216, w.		....					
138	6-Ethoxynaphthalene-2-sulfonanilide	153	....	107.5	183		....					
139	2-Iodonaphthalene-1-sulfonamide	154	....	110								
140	6-Ethoxynaphthalene-1-sulfonamide	154	....	118	....	194.5	....					
141	2,6-Dichloro-4-methylbenzene sulfonamide	154–5	....	56								
142	4-Chloro-2,5-dimethylbenzene sulfonanilide	155	100	50	185		....					
143	4-Amino-2-hydroxybenzene sulfonamide	155	....	169								
144	Tetraline-6-sulfonanilide	155–6	....	58	135		....					
145	Benzene sulfonamide	156 (153)	43–4 (mono-hyd); 66 (anh.)	14.5	....	......	170–1	148	205	240	176	N-Xanthylsulfonamide, 200
146	2-Chloro-5-methylbenzene sulfonamide	156	....	56	....	229–30.5						
147	D-3-Bromocamphor-10-sulfonamide	156	47.5	65								
148	Toluene-2-sulfonamide	156.3	57	68	....	136	......	170	203–4	218	......	N-Xanthylsulfonamide, 182–3,5
149	4-Methoxynaphthalene-2-sulfonamide	157	....	75.5	145							
150	5-Methoxynaphthalene-1-sulfonanilide	157	....	119.5	194.5		....					
151	N-(1-Naphthyl)toluene-4-sulfonamide	157	104–5	69	105 (dihyd.); 138.5–9 (anh.)	103	......	181–2	198	238	190	....
152	Benzophenone-3,3'-disulfonamide	157	....	di: 137–8	....	di: 177–8						N-Xanthylsulfonamide, 197
153	2,4-Dinitrobenzene sulfonamide	157; 154	106–8 (hyd.); 130 (anh.)	102	....							Hydrazide, 110
154	2-Methyl-4-nitrobenzene sulfonamide	157	....	106								
155	2-Nitrodiphenylamine-4-sulfonanilide	157	220d.	....	162							
156	Phenanthrene-2-sulfonanilide	157–8	150	156	253–4	......			291	......	......	Me. ester, 101–2; Et. ester, 89, yel.-br.
157	4-Methylnaphthalene-1-sulfonanilide	158	....	81	174; 177		....					....
158	2-Ethoxynaphthalene-1-sulfonamide	158	....	116	....	187						
159	4-Methoxynaphthalene-1-sulfonanilide	158	....	81	177		......					

Derivative data given in order: m.p., crystal color, solvent from which crystallized.

TABLE XXIV. ORGANIC DERIVATIVES OF SULFONAMIDES AND SULFONANILIDES
(Listed in order of increasing m.p.)* (Continued)

No.	Name	Melting point, °C	Sulfonic acid	Derivatives of the corresponding acid				Salts of the corresponding acid				Miscellaneous derivatives of the acid
				Chloride	Amide	Anilide	1-Naph-thyl-amide	S-Benzyl thiu-ronium	p-Tolui-dinium	Ani-linium	o-Tolui-dinium	
160	2-Ethoxybenzene sulfon-anilide	158	....	65–6	163	....	....	....	....	....	.....	Phenylhydrazide, 132–3
161	2-Methoxynaphthalene-1-sulfonamide	159	....	121	....	196.5	....					....
162	3-Nitrobenzylsulfonamide (3-Nitrotoluene-α-sul-fonamide)	159d., w.	74 (hyd.)	100, bz.	....							Me. amide, 106–7; Dimethyl-amide, 118–9
163	5-Amino-2-hydroxyben-zene sulfonanilide	159 (98)	100 (anh.)	....	202 d.							....
164	5-Chloro-2-nitrobenzene sulfonamide	159	....	93								....
165	2,5-Dichlorobenzene sul-fonanilide	160	93–7	38	181			170				
166	3,5-Dimethyl-2-hydroxy-benzene-1,4-disulfon-amide	160–1	....	di: 89–91								
167	6-Hydroxynaphthalene-2-sulfonanilide	161	129 (hyd.); 167 (anhyd.)	....	238			217 (207)	247	264	208	
168	2-Methylbenzene-1,3-disulfonanilide	162	....	88	260							
169	N-(1-Naphthyl)-4-chloro-naphthalene-1-sulfon-amide	162	130–3	94–5	187	145–6	....			145–6	151	Me. ester, 83; Et. ester, 104
170	2-Nitrodiphenylamine-4-sulfonamide	162	220 d.	....	....	157						
171	7-Methylnaphthalene-1-sulfonanilide	162–4	....	96	197							
172	2-Ethoxybenzene sulfon-amide	163	....	65–6	....	158						Phenylhydrazide, 132–3
173	7-Methylnaphthalene-2-sulfonamide	163–4	....	63–4								
174	3-Methyl-2-nitrobenzene sulfonamide	163.5	....	58.5								
175	5-Amino-2-methylbenzene sulfonamide	164	....	....	....	146–7						5-N-Acetyl deriv. of sulfonamide, 242
176	4-Chloro-2-nitrobenzene sulfonamide	164	....	75	....	138						Phenylhydrazide, 151; Ph. ester, 82
177	4-Nitrodiphenylamine-2-sulfonanilide	164	....	102–4	174							
178	4-Aminobenzene sulfon-amide	165	....	....	....	200	196	185				
179	3,6-Dichloro-2,5-di-methylbenzene sulfon-amide	165	....	71	....	171						
180	6-Aminonaphthalene-1-sulfonamide	165	....	....	....	....		172–4				Benzoylguanidine salt, 210–1
181	N-(1-Naphthyl)benzyl sulfonamide	166 (146)	....	92–3, eth., bz.	105, w., al.	102, al.			113, al.	102	83	Hydrazide, 131–2; Phenyl-hydrazide, 173
182	4-Bromobenzene sulfon-amide	166; 161	88–90	76, eth.	....	119	183.5	170	215–6	237–8	182–3	....
183	5-Bromo-2-methylbenzene sulfonamide	166–7	....	33–5								
184	N-(1-Naphthyl)-3-nitro-benzene sulfonamide	166.5	48	64.	167	126		146	222	126.5-7.5	193	....
185	3-Nitrobenzene sulfon-amide	167	48	64	....	126	166.5	146	222	126.5-7.5	193	....

*Derivative data given in order: m.p., crystal color, solvent from which crystallized.

No.	Name	Melting point, °C	Sulfonic acid	Derivatives of the corresponding acid				Salts of the corresponding acid				Miscellaneous derivatives of the acid
				Chloride	Amide	Anilide	1-Naphthyl-amide	S-Benzyl thiu-ronium	p-Tolui-dinium	Ani-linium	o-Tolui-dinium	
186	2,4-Dimethoxybenzene sulfonamide..........	167	....	70	....							
187	5-Chloro-2-methyl-3-nitrobenzene sulfonamide	167		60								
188	4-Hydroxynaphthalene-1-sulfonamide..........	167				199–200						
189	2,3-Dimethylbenzene sulfonamide..........	167	....	47								
190	4,6-Dichloro-2-methyl-benzene sulfonamide....	168		43								
191	4-Chloronaphthalene-2-sulfonamide..........	168	....	106							......	Et. ester, 76–9
192	4,5-Dibenzylnaphthalene-1-sulfonamide........	168		151								
193	4-Bromo-2-methylbenzene sulfonamide..........	168		50								
194	Propane-1,3-disulfon-amide..............	169, w.	92 d.	di: 45	....	di: 125						Di-m-toluidide, 222; Di-hydrazide, 105
195	Propane-1,1-disulfon-amide..............	169–70		....	....	151–2, al.						
196	2,4,5-Trimethoxybenzene sulfonanilide.........	170	....	130	76	....						
197	4-Acetamidonaphthalene-1-sulfonanilide	170	....		241	....						
198	4-Ethoxynaphthalene-1-sulfonamide..........	170	....	103	....	180						
199	3-Carboxybenzene sulfon-amide..............	di: 170	98 (hyd.); 148 (anh.)	di: 20	....	....		163	224–6	......		
200	4-Methyl-2-nitrobenzene sulfonamide..........	170	....	98–9	....						......	o-Anisidide, 135
201	N-(1-Naphthyl)benzene sulfonamide..........	170–1	43–4 (mono-hyd.); 66 (anh.)	14.5	156	112	148	148	205	240	176	Me. ester, b.p. 150[15]; N-Xan-thylsulfon-amide, 200
202	3,6-Dichloro-2,5-di-methylbenzene sulfon-anilide	171	....	71	165	....		....				
203	5-Aminonaphthalene-1-sulfonanilide..........	171	....	....	260			179				
204	4-Nitrobenzene sulfon-anilide	171 (136)	109–11 (95)	80, lgr.	180	....			179–80	......		
205	2,4-Dimethyl-3-nitroben-zene sulfonamide	172	144 (anh.)	96	....							
206	8-Nitronaphthalene-2-sulfonanilide..........	172–3	135.6 (hyd.)	169	223							
207	2,5-Dimethyl-3-nitroben-zene sulfonamide	173	128 (200)	61	....	143–4	......	......	136			
208	4-Nitrodiphenylamine-2-sulfonamide..........	173	....	102–4	....	164						
209	7-Ethoxy-8-nitro-naphthalene-1-sulfon-amide..............	173.4	....	155								
210	2,5-Dimethylbenzene-1,3-disulfonanilide	174, al.	....	81, lgr.	295, al.	....						
211	3-Methoxynaphthalene-2-sulfonanilide..........	174	....	138	113							
212	3,4-Dibromobenzene sul-fonamide.............	175	66.5–7.5 (anhyd.)	34	....	....		....				....

*Derivative data given in order: m.p., crystal color, solvent from which crystallized.

No.	Name	Melting point, °C	Sulfonic acid	Derivatives of the corresponding acid				Salts of the corresponding acid				Miscellaneous derivatives of the acid
				Chloride	Amide	Anilide	1-Naphthyl-amide	S-Benzyl thiu-ronium	p-Tolui-dinium	Ani-linium	o-Tolui-dinium	
213	4,6-Dichloro-2,5-di-methylbenzene sulfon-anilide	175	....	81	150	....						
214	4-Chloro-3-nitrobenzene sulfonamide	175–6, yel., al.	....	40–1 (60–2)								
215	7-Chloronaphthalene-2-sulfonamide	176	68 (tetra-hyd.); 118 (anhyd.)	87	....							Me. ester, 89; Et. ester, 65
216	3-Amino-4-methylbenzene sulfonamide	176	....	....								3-N-Benzoyl deriv. of sulfon-amide, 203; 3-N-Acetyl deriv. of sulfonyl chloride, 144; 3-N-Benzoyl deriv. of sulfonyl chloride, 196
217	4-Bromo-3-nitrobenzene sulfonamide	176–7	....	55–7	....							
218	4-Hydroxybenzene sulfonamide	176–7	....	141				169	202	170	192	
219	5-Methylnaphthalene-1-sulfonamide	176–8	115	....								
220	4-Methylnaphthalene-1-sulfonamide	177 d.; 174	....	81, lgr.		158						
221	2-Chloro-5-methyl-6-nitrobenzene sulfonamide	177	....	122								
222	Benzophenone-3,3'-disul-fonanilide	177–8	....	di: 137–8	di: 157							
223	6-Methoxynaphthalene-1-sulfonanilide	177.5		80.5	149.5							
224	Acenaphthene-5-sulfon-anilide	178	....	111	223							
225	2-Methylbenzene-1,4-di-sulfonamide	di: 178	....	98	224							
226	8-Nitronaphthalene-1-sulfonanilide	178–8.5	115 (trihyd.)	165 d.	190.5–1.5							
227	Naphthalene-1,4-disulfon-anilide	179		160 (166)	273, w., al.							
228	4-Nitrobenzene sulfon-amide	180, 50% al.	109–11 (95) (hyg.)	80, lgr.	....	171 (136) d.			179–80			
229	4-Ethoxynaphthalene-1-sulfonanilide	180	....	103	170							
230	3,4-Dimethyl-5-nitroben-zene sulfonamide	180	....	70								
231	3-Chloro-2-methylbenzene sulfonamide	180, w.	60–72	72, pet. eth.	....							
232	2-Aminonaphthalene-8-sulfonamide	181 (hyd.)	....	....	147							1-N-Acetyl deriv. of sulfonamide, 213; Benzoyl-guanidine salt, 214–6
233	2,5-Dichlorobenzene sul-fonamide	181	93–7 (>1.00)	38	....	160	160	170	247–8	262–3	250–1	....
234	2,4,5-Trimethylbenzene sulfonamide	181	112	61								

*Derivative data given in order: m.p., crystal color, solvent from which crystallized.

TABLE XXIV. ORGANIC DERIVATIVES OF SULFONAMIDES AND SULFONANILIDES
(Listed in order of increasing m.p.)* (Continued)

No.	Name	Melting point, °C	Sulfonic acid	Derivatives of the corresponding acid				Salts of the corresponding acid				Miscellaneous derivatives of the acid
				Chloride	Amide	Anilide	1-Naph-thyl-amide	S-Benzyl thiu-ronium	p-Tolui-dinium	Ani-linium	o-Tolui-dinium	
235	5-Chloro-4-methyl-2-nitrobenzene sulfonamide..............	181	128	99	....	....	....					
236	2,4-Dichlorobenzene sulfonamide..........	182	86	55	....	....	....		204–6	......	170–2	
237	2,5-Dimethyl-6-nitrobenzene sulfonanilide	182	145 (anh.)	110	192	....	....		158.5–9, al.	......	143–5, 50% al.	
238	5-Ethoxynaphthalene-1-sulfonamide..........	182.5	....	121	....	130	....					
239	6-Ethoxynaphthalene-2-sulfonamide..........	183	....	107.5	....	153	....					
240	4-Iodobenzene sulfonamide..............	183	....	85	....	143	....					
241	4-Ethoxynaphthalene-2-sulfonamide..........	183	....	85	....	143.5	....					
242	Quinoline-8-sulfonamide .	183–4	312	124	....	....	....				......	Me. ester, 96; Et. ester, 73; Picrate of Na salt, 226–7
243	4,5-Dichloro-3-methylbenzene sulfonamide....	183–5	....	85–8	....	....	....					
244	N-(1-Naphthyl)-4-bromobenzene sulfonamide....	183.5	88–90	76, eth.	166; 161	119	....	170	215–6	237–8	182–3	....
245	6-Chloronaphthalene-2-sulfonamide..........	184	....	110.5	....	....	....					
246	5-Nitronaphthalene-2-sulfonamide..........	184	118–9, yel.	125	....	....	....			260 d.		
247	8-Chloronaphthalene-2-sulfonamide..........	185	....	94	....	....	....					
248	4-Chloro-2-methylbenzene sulfonamide..........	185	....	54	....	....	....					
249	4-Chloro-2,5-dimethylbenzene sulfonamide....	185	100	50	....	155	....					
250	2-Carboxy-5-methylbenzene sulfonamide (4-Toluic acid-2-sulfonamide).............	185	190 (158) (anh.)	di: 59	....	....	....					
251	2-Chloro-5-nitrobenzene sulfonamide..........	185–6	168–9 d. (hyd.)	90, w.	....	....	....					
252	2-Bromobenzene sulfonamide..............	186, w.	....	51, eth.	....	....	....					
253	3,5-Dichloro-2-methylbenzene sulfonamide....	186	....	54	....	....	....					
254	2-Methyl-5-nitrobenzene sulfonamide..........	186	133.5 (dihyd.)	46–7; b.p. 183–5[10]	....	148	....		256–7	......	256–8	
255	2-Chloro-4-methylbenzene sulfonamide....	186	....	46; 52	....	....	....		......	......	....	....
256	4-Methylbenzene-1,3-disulfonamide (Toluene-2,4-disulfonamide).....	186–7 (191)	....	54; 46	....	189	....	......	277 d.	di: 189	di: 170–1	Di-m-toluidide, 138
257	2-Ethoxynaphthalene-1-sulfonanilide	187	....	116	158	....	....					
258	4-Chloronaphthalene-1-sulfonamide..........	187	130–3 d.	94–5	....	145–6	162			145–6	151	Me. ester, 83; Et. ester, 104
259	8-Iodonaphthalene-1-sulfonamide..........	187	115	140	....	....	....				....	....

*Derivative data given in order: m.p., crystal color, solvent from which crystallized.

TABLE XXIV. ORGANIC DERIVATIVES OF SULFONAMIDES AND SULFONANILIDES
(Listed in order of increasing m.p.)* (Continued)

No.	Name	Melting point, °C	Sulfonic acid	Derivatives of the corresponding acid				Salts of the corresponding acid				Miscellaneous derivatives of the acid
				Chloride	Amide	Anilide	1-Naphthylamide	S-Benzylthiouronium	p-Toluidinium	Anilinium	o-Toluidinium	
260	2,4-Dimethyl-5-nitrobenzene sulfonamide....	187 (179)	132 (122) dil. HNO₃	98	....	....	......	......	......	......	......	Sulfonyl chloride, 109–10
261	2,4-Diaminobenzene-1,5-disulfonamide.........	187	....	275	....	236	......	......				
262	8-Bromonaphthalene-2-sulfonamide..........	187	....	121								
263	2-Chlorobenzene sulfonamide...............	188	....	28.5								
264	4-Nitronaphthalene-1-sulfonamide..........	188	....	99								
265	2,6-Dichloro-3-methylbenzene sulfonamide....	188	....	19.5								
266	5-Methylnaphthalene-2-sulfonamide..........	188–9	....	120–2	....	248–50 (133–4)	......					
267	4-Methylbenzene-1,3-disulfonanilide	189		54; 46	186–7	....						
268	6-Methoxynaphthalene-2-sulfonamide..........	189	....	93	....	120						
269	Sulfanilylguanidine (Sulfaguanidine)	189–90 (anh.); 143 (hyd.)	....	....	....	....	......	......	......	......		4-N-Acetyl deriv. of sulfonamide, 262–6 (248–51); Hydrochloride, 205–6
270	N-(1-Naphthyl)-4-chlorobenzene sulfonamide....	190	69 (93)	53	144	104	.....	175	208–10	222–3	163–4	....
271	2,4-Dibromobenzene sulfonamide..........	190	110 (anh.)	79, eth.								
272	4-Methylbenzene-1,2-disulfonanilide (Toluene-3,4-disulfonanilide)....	190	....	109–11	237–9							
273	Phenanthrene-3-sulfonamide...............	190	175–6 (anh.); 120–1 (monohyd.); 88 d. (dihyd.)	110–1	....	......	......	222				Me. ester, 119–20, al.; Et. ester, 107–8
274	4-Aminonaphthalene-1-sulfonanilide	190	....	....								4-N-Acetyl deriv. of sulfonamide, 247
275	8-Nitronaphthalene-1-sulfonamide..........	191	115 d. (trihyd.)	165 d.	....	178–8.5						
276	2-Hydroxynaphthalene-1,6-disulfonanilide	191	....	di: 111								
277	3,5-Dichloro-4-methylbenzene sulfonamide....	191	....	69								
278	Sulfapyridine	191–2										4-N-Acetyl deriv. of sulfonamide, 226–7, acet.
279	Anthraquinone-2,7-disulfonanilide	192	....	di: 186, chl.								
280	4-Carboxy-3-nitrobenzene sulfonamide (2-Nitrobenzoic acid-4-sulfonamide).............	192	111 (+2½ H₂O)	di: 160	....							Diamide, 226
281	Quinoline-6-sulfonamide .	192	>260	91	....							
282	5,6-Dichloronaphthalene-2-sulfonamide	192	....	167	....							

*Derivative data given in order: m.p., crystal color, solvent from which crystallized.

TABLE XXIV. ORGANIC DERIVATIVES OF SULFONAMIDES AND SULFONANILIDES
(Listed in order of increasing m.p.)* (Continued)

No.	Name	Melting point, °C	Sulfonic acid	Derivatives of the corresponding acid				Salts of the corresponding acid				Miscellaneous derivatives of the acid
				Chloride	Amide	Anilide	1-Naphthylamide	S-Benzylthiuronium	p-Toluidinium	Anilinium	o-Toluidinium	
283	Anthraquinone-2,7-disulfonamide	192	....	186	....	....	....	....	....	....	....	....
284	2,5-Dimethyl-6-nitrobenzene sulfonamide	192	145 (anh.)	110	....	182	....	....	158.5-9	....	143-5, 50% al.	....
285	2-Amino-5-methylbenzene-1,3-disulfonanilide	192	....	156	257	....	....	....	....	....	....	....
286	Methane disulfonanilide	192-3	220-7^{15} d.	8; b.p. 133^{10}	di: 233	....	....	....	....	....	....	....
287	3,4-Dicarboxybenzene sulfonamide (Phthalic acid-4-sulfonamide)	192-200 d., w.	138-40 (mono-hyd.)	167-70 d., eth.	....	....	....	....	....	....	....	....
288	2-Nitrobenzene sulfonamide	193	70 (85)	69	....	115	....	....	....	....	....	....
289	Anthraquinone-2-sulfonanilide	193	....	197	261	....	....	211	308	309	......	Me. ester, 123; Et. ester, 125
290	Phenanthrene-9-sulfonamide	193-4	174 (anh.)	127	....	....	....	....	235	....	....	Me. ester, 106, me. al.; Et. ester, 108, al.
291	2-Carboxybenzene sulfonanilide	194-5	68-9 (hyd.); 134 (anh.)	79, pet. eth.	....	....	....	206	196 (200)	165	127-8	....
292	5-Methoxynaphthalene-1-sulfonamide	194.5	....	119.5	....	157	....	....	....	....	....	....
293	6-Ethoxynaphthalene-1-sulfonanilide	194.5	....	118	154	....	....	....	....	....	....	....
294	7-Hydroxynaphthalene-1,3-disulfonanilide	195	....	161-2	....	....	....	228	294	......	271	....
295	4-Bromonaphthalene-1-sulfonamide	195	....	87	....	....	....	....	....	....	....	....
296	7-Hydroxynaphthalene-1-sulfonanilide	195	....	....	....	....	....	218	232	240	242	....
297	2,5-Dibromobenzene sulfonamide	195	128 (anh.)	71	....	....	....	....	....	....	....	....
298	7-Methylnaphthalene-1-sulfonamide	195-6	....	96	....	162-4	....	....	....	....	....	....
299	4,6-Dimethylbenzene-1,3-disulfonanilide	196, 50% al.	....	130, pet. eth.	249, w.	....	....	....	....	....	....	....
300	4,7-Dichloronaphthalene-2-sulfonamide	196	....	156	....	....	....	....	....	....	....	....
301	N-(1-Naphthyl)-4-aminobenzene sulfonamide	196	....	....	165	200	....	185	....	....	....	....
302	8-Chloro-7-methoxynaphthalene-1-sulfonanilide	196	....	137	153	....	....	....	....	....	....	....
303	5-Fluoronaphthalene-1-sulfonamide	196-7	105 (hyd.)	122-3	....	....	....	....	....	....	....	Me. ester, 118, eth.
304	2-Methoxynaphthalene-1-sulfonanilide	196.5	....	121	159	....	....	....	....	....	....	....
305	8-Chloronaphthalene-1-sulfonamide	197	....	101	....	....	....	....	....	....	....	....
306	4,5-Dichloronaphthalene-2-sulfonamide	197	....	158	....	....	....	....	....	....	....	....
307	2,5-Dimethyl-4-nitrobenzene sulfonamide	197-8	140	75	....	131	....	....	143.5-4.5	......	143.5-4.5	....

*Derivative data given in order: m.p., crystal color, solvent from which crystallized.

TABLE XXIV. ORGANIC DERIVATIVES OF SULFONAMIDES AND SULFONANILIDES
(Listed in order of increasing m.p.)* (Continued)

No.	Name	Melting point, °C	Sulfonic acid	Derivatives of the corresponding acid				Salts of the corresponding acid				Miscellaneous derivatives of the acid
				Chloride	Amide	Anilide	1-Naphthylamide	S-Benzyl thiuronium	p-Toluidinium	Anilinium	o-Toluidinium	
308	**Sulfamethazine** (Sulfadimethylpyrimidine).....	198–9 ($+\frac{1}{2}$ H₂O), pa. yel.	....	....	....	....	....	....	....	....	....	4-N-Acetyl deriv. of sulfonamide, 249–50
309	**Acenaphthene-3-sulfonamide**.............	199	87–9	113–4	....	284–6	....	....	....	....	....	Me. ester, 122–3; Et. ester, 137–9, lgr.
310	**4-Hydroxynaphthalene-1-sulfonanilide**.........	199–200	170	....	167	....	....	103	196	186–7	203–4	2-Naphthyl sulfonamide, 204
311	**4-Aminobenzene sulfonanilide**.............	200	....	....	165	....	196	185	109	....	132	4-N-Acetyl deriv. of sulfonanilide, 214; Me. ester, 92
312	**4,5-Dimethylbenzene-1,3-disulfonanilide**........	200, al.	....	79, yel.	239	....	....	....	....	....	....	....
313	**4-Chloro-3-methyl-5-nitrobenzene sulfonamide**...............	201	....	52	....	....	....	....	....	....	....	....
314	**Anthracene-2-sulfonanilide**.............	201	....	122	261	....	....	....	....	....	....	Phenylhydrazide, 210; Me. ester, 157; Et. ester, 160
315	**5-Hydroxynaphthalene-1-sulfonanilide**.........	201	110–2 d.	....	....	....	....	....	....	....	....	O-Acetyl deriv. of sulfonyl chloride, 129
316	**2-Hydroxynaphthalene-3,6-disulfonanilide**.....	202	....	....	....	....	....	233	250	254	257	....
317	**5-Amino-2-hydroxybenzene sulfonamide**......	202 d.	100 (anh.)	....	....	159 (98)	....	....	....	....	....	....
318	**1-Nitronaphthalene-2-sulfonanilide**.........	202	105, grn.	121, bz.-pet. eth.	214	....	....	....	....	....	....	....
319	**Sulfathiazole**..........	202.5	....	....	....	....	....	....	....	....	....	4-N-Acetyl deriv. of sulfonamide, 256–7
320	**4-Nitrobenzyl sulfonamide** (4-Nitrotoluene-α-sulfonamide).........	204	71	90	220 d.	....	....	....	....	....	....	....
321	**Anthracene-1-sulfonamide**...............	205	....	90	....	....	....	....	....	....	....	....
322	**4,8-Dichloronaphthalene-2-sulfonamide**.........	205	....	141	....	....	....	....	....	....	....	....
323	**3-Amino-4-hydroxybenzene sulfonamide**......	205 (170)	155–6 d.	....	....	....	....	....	....	....	....	....
324	**4-Hydroxybenzene-1,3-disulfonanilide** (Phenol-2,4-disulfonanilide)....	205	>100 d.	89	239	....	....	....	....	....	....	....
325	**6-Methylnaphthalene-2-sulfonamide**........	205–6	....	97–8	....	....	....	....	....	....	....	....
326	**2,6-Dimethyl-4-hydroxybenzene-1,3-disulfonanilide**...............	205–7	....	117–8	206–8	....	....	....	....	....	....	....
327	**4-Iodonaphthalene-1-sulfonamide**...........	206; 204	....	124; 121	....	136	....	....	....	....	....	....
328	**4-Fluoronaphthalene-1-sulfonamide**...........	206	100 (hyd.)	86	....	144	....	....	....	....	....	Et. ester, 93
330	**4-Amino-3-nitrobenzene sulfonamide**..........	206–7	....	59–60	....	....	....	....	....	....	....	....

* Derivative data given in order: m.p., crystal color, solvent from which crystallized.

No.	Name	Melting point, °C	Sulfonic acid	Derivatives of the corresponding acid				Salts of the corresponding acid				Miscellaneous derivatives of the acid
				Chloride	Amide	Anilide	1-Naphthyl-amide	S-Benzyl thiuronium	p-Tolui-dinium	Ani-linium	o-Tolui-dinium	
331	Retene-6-sulfonamide	206–7.5	121–3	146–7.5, yel.-br.	....	....	......	......	......	......	......	Me. ester, 117–9; Et. ester, 114–5
332	2,6-Dimethyl-4-hydroxy-benzene-1,3-disulfon-amide	206–8	....	117–8	....	205–7	......	......	......	......	......	
333	3,7-Diethylnaphthalene-1-sulfonamide	207		105–7								
334	6-Bromonaphthalene-2-sulfonamide	207		124								
335	7-Bromonaphthalene-1-sulfonamide	209		147								
336	4-Methoxybenzene-1,3-disulfonanilide	209		86	240							
337	7-Iodonaphthalene-2-sulfonamide	210		100								
338	4-(N-Methylamino)ben-zene sulfonamide	210–11	244–5 d.	....								p-Toluenesul-fonate, benzi-dine salt, 255
339	2,4,6-Trichlorobenzene sulfonamide	210–2 d.	....	35–40								
340	4-Aminonaphthalene-1-sulfonamide	212	....	....		190						
341	6-Iodonaphthalene-1-sulfonamide	213	....	92.5								
342	Fluorene-2-sulfonamide	213 d.	155 (hyd.)	164								
343	Anthraquinone-1-sul-fonanilide	214	218	216–8, yel., PhNO₂	....	....	......	191		284	......	NH₃ → 1-Amino-anthraquinone, 252 (243)
344	1-Nitronaphthalene-2-sulfonamide	214	105, grn.	121, pink, bz.-pet. eth.	....	202						
345	4-Acetamidobenzene sulfonanilide	214	....	149	219	......	215					
346	6-Chloronaphthalene-1-sulfonamide	214	....	70							......	Et. ester, 114–5
347	N-(1-Naphthyl)-4-acetamidobenzene sul-fonamide	215		149	219	214						
348	2,5-Dichlorobenzene-1,3-disulfonamide	215–7		114								
349	5-Methylbenzene-1,3-disulfonamide	216, w.	....	94, eth.	....	153, al.						
350	5-Chloronaphthalene-2-sulfonamide	216		115								
351	Naphthalene-2-sulfon-amide	217; 213	91 (hyg.); (122)	76; 79	....	132	......	190–1	221	269	213	....
352	4,7-Dichloronaphthalene-1-sulfonamide	217		151								
353	6-Bromonaphthalene-1-sulfonamide	217		77								
354	4,6-Dichloronaphthalene-2-sulfonamide	218	....	136								
355	7-Bromonaphthalene-2-sulfonamide	218	....	100								
356	3,6-Dichloronaphthalene-2-sulfonamide	218	....	166								

* Derivative data given in order: m.p., crystal color, solvent from which crystallized.

TABLE XXIV. ORGANIC DERIVATIVES OF SULFONAMIDES AND SULFONANILIDES
(Listed in order of increasing m.p.)* (Continued)

No.	Name	Melting point, °C	Sulfonic acid	Derivatives of the corresponding acid				Salts of the corresponding acid				Miscellaneous derivatives of the acid
				Chloride	Amide	Anilide	1-Naphthyl-amide	S-Benzyl thiuronium	p-Tolui-dinium	Ani-linium	o-Tolui-dinium	
357	6-Ethoxy-1-nitro-naphthalene-2-sulfonamide	218	....	146	....	....						....
358	5-Aminonaphthalene-2-sulfonamide	218–9 d.		....		127–8	....					1-N-Acetyl deriv. of sulfonamide, 238–9
359	4-Acetamidobenzene sulfonamide	219	....	149	....	214	215					
360	5-Bromonaphthalene-2-sulfonamide	220	....	96								
361	4-Nitrobenzyl sulfonanilide (4-Nitrotoluene-α-sulfonanilide)	220 d.	71	90	204	......						
362	7-Methoxynaphthalene-2-sulfonamide	220	....	83	....	121						
363	2,4,6-Tribromobenzene sulfonanilide	220–2 d.	64	64	228		...					
364	7,8-Dichloronaphthalene-1-sulfonamide	221	....	138			...					
365	8-Hydroxynaphthalene-1-sulfonamide	222 d.	107 (hyg.)	....								
366	6-Iodonaphthalene-2-sulfonamide	222	....	140								
367	2,5-Dimethylbenzene-1,4-disulfonanilide	223	....	164	310							
368	5,6-Dichloronaphthalene-1-sulfonamide	223	....	106								
369	Acenaphthene-5-sulfonamide	223	....	111	....	178						
370	6-Nitronaphthalene-1-sulfonamide	223–4	....	127	....							
371	2-Methylbenzene-1,4-disulfonamide	224	....	98		di: 178						
372	5-Chlorobenzene-1,3-disulfonamide	224	....	106								
373	Anthracene-1,8-disulfonanilide	224	....	225	333	....						
374	4-Nitronaphthalene-2-sulfonamide	225	....	139.5								
375	4,6-Dichloronaphthalene-1-sulfonamide	226	....	119								
376	4-Carboxamido-3-nitrobenzene sulfonamide (2-Nitrobenzamide-4-sulfonamide)	226	111 (+2.5 H₂O)	di: 160	192	....						
377	5-Chloronaphthalene-1-sulfonamide	226	....	95	....	138	....					Sulfonyl bromide, 110; Me. ester, 89; Et. ester, 46
378	4-Methoxynaphthalene-1-sulfonamide	226	....	98.5	....	147.5						....
379	3,4-Di-iodobenzene sulfonamide	227, aq. al.	122–5	82, bz.-pet. eth.								
380	7,8-Dichloronaphthalene-2-sulfonamide	227	....	124								
381	Anthraquinone-1,6-disulfonanilide	227–8, yel.	215–7, gold	197–8, yel. PhNO₂	....							
382	2,3,4-Trichlorobenzene sulfonamide	227–30	....	64–5	....							

*Derivative data given in order: m.p., crystal color, solvent from which crystallized.

TABLE XXIV. ORGANIC DERIVATIVES OF SULFONAMIDES AND SULFONANILIDES
(Listed in order of increasing m.p.)* (Continued)

No.	Name	Melting point, °C	Sulfonic acid	Derivatives of the corresponding acid				Salts of the corresponding acid				Miscellaneous derivatives of the acid
				Chloride	Amide	Anilide	1-Naphthylamide	S-Benzyl thiuronium	p-Toluidinium	Anilinium	o-Toluidinium	
383	3,4-Dichloro-2-methylbenzene sulfonamide	228	. . . .	51–2	. . . .	. . . .	. . . .	. . . .	. . . .	. . . .	. . . .	. . . .
384	4′-Nitrobiphenyl-4-sulfonamide.	228	. . . .	178	. . . .	. . . .	. . . .	. . . .	. . . .	. . . .	. . . .	. . . .
385	2,4,6-Tribromobenzene sulfonamide	228	64	64	. . . .	220–2 d.	. . . .	. . . .	. . . .	. . . .	. . . .	. . . .
386	8-Nitronaphthalene-2-sulfonamide	228; 223	135–6 (+1.5 H₂O)	169	. . . .	172–3	. . . .	. . . .	. . . .	. . . .	. . . .	. . . .
387	6,8-Dichloronaphthalene-2-sulfonamide	228	. . . .	121	. . . .	. . . .	. . . .	. . . .	. . . .	. . . .	. . . .	. . . .
388	Benzene-1,3-disulfonamide.	229	. . . .	63	. . . .	148–50	245	214	. . . .	. . . .	. . . .	N-Xanthylsulfonamide, 170; Alk. fusion → resorcinol, 110.
389	4,5-Dichloronaphthalene-1-sulfonamide	229	. . . .	117	. . . .	. . . .	. . . .	. . . .	. . . .	. . . .	. . . .	. . . .
390	Biphenyl-4-sulfonamide . .	230	. . . .	115	. . . .	125	. . . .	. . . .	. . . .	. . . .	. . . .	. . . .
391	2,4-Di-iodobenzene sulfonanilide	230	167 (anh.)	77–8	. . . .	. . . .	. . . .	. . . .	. . . .	. . . .	. . . .	Me. ester, 78, al; Et. ester, 52, al.
392	2-Hydroxynaphthalene-1,5-disulfonanilide	231	. . . .	di: 231	. . . .	. . . .	. . . .	. . . .	. . . .	. . . .	. . . .	. . . .
393	5-Bromonaphthalene-1-sulfonamide.	232–3	. . . .	95	. . . .	. . . .	. . . .	. . . .	. . . .	. . . .	. . . .	. . . .
394	2-Hydroxynaphthalene-1,7-disulfonanilide	233	. . . .	169	. . . .	. . . .	. . . .	. . . .	. . . .	. . . .	. . . .	. . . .
395	Methane disulfonamide . .	233	b.p. 220–70¹⁵⁻²⁰ d.	di: 8; b.p. 133¹⁰	. . . .	di: 192–3	. . . .	. . . .	. . . .	. . . .	. . . .	. . . .
396	2-Ethoxybenzene-1,4-disulfonamide	233	. . . .	di: 106–8	. . . .	. . . .	. . . .	. . . .	. . . .	. . . .	. . . .	. . . .
397	3,5-Dinitrobenzene sulfonamide.	235	. . . .	99, chl., lgr.	. . . .	. . . .	. . . .	. . . .	. . . .	. . . .	. . . .	. . . .
398	4-Aminobenzene-1,3-disulfonamide (Aniline-2,4-disulfonamide).	235, w.	120 d.	. . . .	. . . .	. . . .	. . . .	. . . .	. . . .	. . . .	. . . .	. . . .
399	7-Chloronaphthalene-1-sulfonamide.	235	. . . .	129	. . . .	. . . .	. . . .	. . . .	. . . .	. . . .	. . . .	. . . .
400	5,6,8-Trichloronaphthalene-2-sulfonamide . .	235	. . . .	158	. . . .	. . . .	. . . .	. . . .	. . . .	. . . .	. . . .	. . . .
401	4,6,7,8-Tetrachloronaphthalene-2-sulfonamide . .	235	. . . .	176	. . . .	. . . .	. . . .	. . . .	. . . .	. . . .	. . . .	. . . .
402	5-Nitronaphthalene-1-sulfonamide.	236	. . . .	113	. . . .	123	. . . .	. . . .	. . . .	. . . .	. . . .	. . . .
403	4-Carboxamidobenzene sulfonamide.	236	94 (hyd.); 260 (anh.)	di: 57	. . . .	di: 252	. . . .	. . . .	. . . .	. . . .	. . . .	. . . .
404	2,4-Diaminobenzene-1,5-disulfonanilide	236	. . . .	275	187	. . . .	. . . .	. . . .	. . . .	. . . .	. . . .	. . . .
405	2,3-Dichloro-4-methylbenzene sulfonamide	237	. . . .	41	. . . .	. . . .	. . . .	. . . .	. . . .	. . . .	. . . .	. . . .
406	Benzene-1,3,5-trisulfonanilide	tri: 237	>100	tri: 187	tri: 310–5	. . . .	. . . .	. . . .	. . . .	. . . .	. . . .	Tri-Et. ester, 147, bz.
408	Anthraquinone-1,7-disulfonanilide	237–8, yel., cl. bz.	120 (hyd.)	321–2, br.-yel., PhNO₂	. . . .	. . . .	. . . .	. . . .	. . . .	. . . .	. . . .	. . . .
409	Anthraquinone-1,8-disulfonanilide	237–8, yel., PhNO₂	293–4	222–3	>340	. . . .	. . . .	. . . .	. . . .	. . . .	. . . .	. . . .
410	4-Methylbenzene-1,2-disulfonamide.	237–9	. . . .	109–11	. . . .	190	. . . .	. . . .	. . . .	. . . .	. . . .	. . . .

*Derivative data given in order: m.p., crystal color, solvent from which crystallized.

No.	Name	Melting point, °C	Sulfonic acid	Derivatives of the corresponding acid				Salts of the corresponding acid				Miscellaneous derivatives of the acid
				Chloride	Amide	Anilide	1-Naphthyl-amide	S-Benzyl thiouronium	p-Tolui-dinium	Ani-linium	o-Tolui-dinium	
411	6-Hydroxynaphthalene-2-sulfonamide	238	129 (hyd.); 167 (anh.)	....	....	161	......	217 (207)	248	264	208	....
412	4-Hydroxybenzene-1,3-disulfonamide (Phenol-2,4-disulfonamide)	239	>100 d.	89	....	205	......	......		......		....
413	5-Iodonaphthalene-1-sulfonamide	239	....	141								
414	4,5-Dimethylbenzene-1,3-disulfonamide	239	....	79, yel.		200, al.						
415	7-Iodonaphthalene-1-sulfonamide	240	....	165								
416	4-Methoxybenzene-1,3-disulfonamide	240	....	86		209						
417	4-Acetamidonaphthalene-1-sulfonamide	241				170	......					
418	Benzene-1,2-disulfonanilide	241	....	143	254	....		206				
419	5-Nitrobenzene-1,3-disulfonamide	242	....	di: 97–8								
420	Naphthalene-2,7-disulfonamide	242	....	158; 162	....			212	299	251–2	238	....
421	2,4,6-Trimethylbenzene-1,3-disulfonamide	244	....	di: 125		di: 150–1						
422	5,8-Dichloronaphthalene-2-sulfonamide	244	....	134								
423	2,3,4,6-Tetrabromobenzene sulfonamide	245 d.	....	96.5								
424	6,7,8-Trichloronaphthalene-2-sulfonamide	245	....	157								
425	N,N'-Di(1-naphthyl)benzene-1,3-disulfonamide	245	....	63	229	148–50	......	214				Alk. fusion → resorcinol, 110; N-Xanthylsulfonamide, 170
426	Anthraquinone-1,5-disulfonamide	246 (350)	310 d.	265–70	....	270 d.						
427	1-Iodonaphthalene-2-sulfonamide	247	....	94								
428	9,10-Dichloroanthracene-2-sulfonanilide	248	....	221	279							
430	5-Methylnaphthalene-2-sulfonanilide	248–50	....	120–2	188–9							
431	4,6-Dimethylbenzene-1,3-disulfonamide	249, w.	....	130, pet. eth.	....	196, 50% al.						
432	Benzene-1,4-disulfonanilide	249	....	131 (139)	288	....						
433	5,6,7-Trichloronaphthalene-1-sulfonamide	249	....	131								
434	Naphthalene-1,5-disulfonanilide	249	245 (anh.)	di: 183	310 (340)	......		257; 251d.	332			Di-Me. ester, 205, chl.
435	1-Chloronaphthalene-2-sulfonamide	250	130–3d. (anh.)	84–5	....	171–2						
436	Azobenzene-4,4'-disulfonamide	250 d.	169d. (anh.)	di: 222	....	......					......	Et. ester, 104
437	4-Carboxanilidobenzene sulfonanilide	252	94 (hyd.); 260 (anh.)	di: 57	di: 236	......						

*Derivative data given in order: m.p., crystal color, solvent from which crystallized.

No.	Name	Melting point, °C	Sulfonic acid	Derivatives of the corresponding acid				Salts of the corresponding acid				Miscellaneous derivatives of the acid
				Chloride	Amide	Anilide	1-Naphthyl-amide	S-Benzyl thiuronium	p-Tolui-dinium	Ani-linium	o-Tolui-dinium	
438	Phenanthrene-2-sulfonamide	253–4	150	156	....	157–8	......	......	291	......	......	Me. ester, 101–2; Et. ester, 89, yel.-br.
439	Benzene-1,2-disulfonamide	254	....	143	....	241	......	206	......	......	......	....
440	2-Amino-5-methylbenzene-1,3-disulfonamide	257	....	156	....	192						
441	2-Amino-5-methylbenzene-1,4-disulfonamide	257, w.	290	di: 156, chl.	....	di: 196-7, aq. al.						....
442	5-Aminonaphthalene-1-sulfonamide	260	....	....	....	171	......	179	......	......	......	5-N-Acetyl deriv. of sulfonamide, 231–2
443	2-Methylbenzene-1,3-disulfonamide	260	....	88	....	162						....
444	Anthracene-2-sulfonamide	261	....	122	....	201						Me. ester, 157; Et. ester, 160
445	Anthraquinone-2-sulfonamide	261	....	197	....	193	......	211	308	309	......	Me. ester, 123; Et. ester, 125
446	7-Nitronaphthalene-1-sulfonamide	261–2	....	169–70								
447	6,7-Dichloronaphthalene-1-sulfonamide	268	....	142								
448	3,7-Dichloronaphthalene-1-sulfonamide	269	....	136								
449	Anthraquinone-2,6-disulfonanilide	269–70	310–11 (hyd.)	265–70, yel.	> 350	......						
450	Anthraquinone-1,5-disulfonanilide	270 d.	310 d.	265–70	246 (350)	......						
451	1-Bromonaphthalene-2-sulfonamide	271	....	93								
452	5,7-Dichloronaphthalene-1-sulfonamide	272	....	149								
453	Naphthalene-1,4-disulfonamide	273, w., al.	....	160 (166)	....	179						
454	Azoxybenzene-3,3'-disulfonamide	273	126	di: 138	....							
455	4,6-Dichlorobenzene-1,3-disulfonamide	276	....	123								
456	Benzidine-2,2'-disulfonamide	278	....	Hydro-chlo-ride, 205								
457	9,10-Dichloroanthracene-2-sulfonamide	279	....	221	....	248						
458	1,5-Dichloronaphthalene-2-sulfonamide	282	....	125								
459	4-Nitronaphthalene-2,7-disulfonamide	286–7	....	140–1								
460	Azobenzene-3,4'-disulfonamide	288	....	123–5								
461	Benzene-1,4-disulfonamide	288	....	131 (139)	....	249						
462	Naphthalene-1,3-disulfonamide	292–3	....	di: 137.5								
463	Anthracene-1,5-disulfonanilide	293	....	di: 240	di: > 330							
464	2,5-Dimethylbenzene-1,3-disulfonamide	295, al.	....	81, lgr.	....	174, al.	......					

*Derivative data given in order: m.p., crystal color, solvent from which crystallized.

No.	Name	Melting point, °C	Sulfonic acid	Derivatives of the corresponding acid				Salts of the corresponding acid				Miscellaneous derivatives of the acid
				Chloride	Amide	Anilide	1-Naphthylamide	S-Benzyl thiuronium	p-Toluidinium	Anilinium	o-Toluidinium	
465	Naphthalene-1,6-disulfonamide	298	125 (anh.)	129	....	....	......	81 (235)	314–5	298–9	323–4	....
466	Naphthalene-1,7-disulfonamide	298–300	123	....								
467	Biphenyl-4,4'-disulfonamide	300	72	203	....	....	......	171	330 d.	......	......	
468	Naphthalene-2,6-disulfonamide	305	....	225				256	......	360	......	
469	Azobenzene-3,3'-disulfonamide	305	....	di: 166								
470	2,5-Dimethylbenzene-1,4-disulfonamide	310		164	....	223						
471	Naphthalene-1,5-disulfonamide	310 (340)	245 (anh.)	183	....	249	......	257; 251 d.	332	......	......	Di-Me. ester, 205, chl.
472	Benzene-1,3,5-trisulfonamide	310–5	>100	tri: 187	....	tri: 237						Tri-Et. ester, 147, bz.
473	Anthraquinone-2,6-disulfonanilide	321	....	di: 250, yel., cl. bz.	....	....						....
474	Anthracene-1,5-disulfonamide	330	....	di: 240	....	di: 293	......					
475	Anthracene-1,8-disulfonamide	333	....	225	....	224						
476	8-Cyanonaphthalene-1-sulfonamide	333–4		139	....	....						
477	Anthraquinone-1,8-disulfonamide	>340	293–4	222–3, yel., PhNO₂	....	237–8, yel., PhNO₂						....
478	Anthraquinone-1,3-disulfonamide	>350	310–11 (hyd.)	265–70, yel.	....	269–70, red-yel.	......					....

*Derivative data given in order: m.p., crystal color, solvent from which crystallized.

EXPLANATIONS AND REFERENCES TO TABLE XXV

2,4-Dinitrophenyl thioether (2,4-Dinitrophenyl sulfide). *

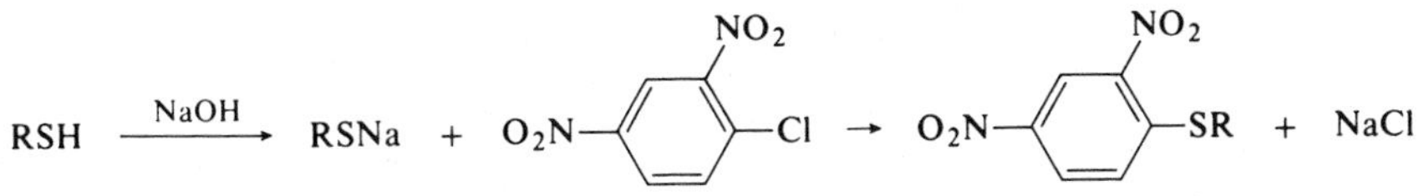

$$RSH \xrightarrow{\text{NaOH}} RSNa + O_2N-\text{C}_6H_3(NO_2)-Cl \rightarrow O_2N-\text{C}_6H_3(NO_2)-SR + NaCl$$

2,4-Dinitrophenyl
thioether

From the sodium thiolate (prepared from the thiol and sodium hydroxide) and 2,4-dinitrochlorobenzene in methanol.

For directions and examples see: Cheronis, p. 642.

From the sodium thiolate and 2,4-dinitrochlorobenzene in aqueous or absolute alcohol.

See: Linstead, p. 86; Shriner, p. 255; Vogel, p. 500; Wild, p. 91; R. W. Bost, J. O. Turner and R. D. Norton, *J. Amer. Chem. Soc.*, **54**, 1985 (1932); R. W. Bost, J. O. Turner and M. W. Conn, *J. Amer. Chem. Soc.*, **55**, 4956 (1933).

2,4-Dinitrophenyl sulfone. *

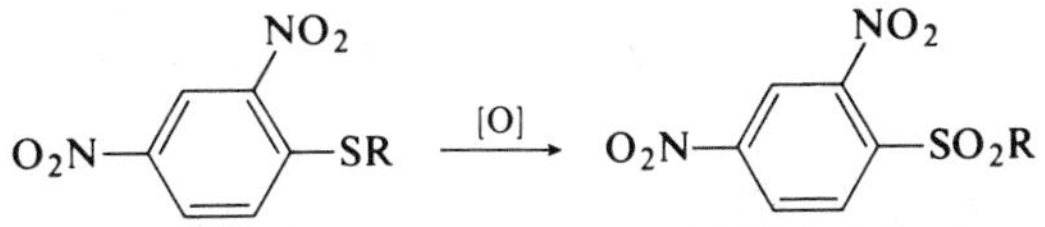

$$O_2N-\text{C}_6H_3(NO_2)-SR \xrightarrow{[O]} O_2N-\text{C}_6H_3(NO_2)-SO_2R$$

2,4-Dinitrophenyl
sulfone

From the thioether (prepared from the thiol and 2,4-dinitrochlorobenzene) and potassium permanganate in aqueous or glacial acetic acid.

For directions and examples see: Cheronis, p. 641; Linstead, p. 87; Vogel, p. 501; Wild, pp. 91–2.

From the thioether with hydrogen peroxide, ammonium molybdate and perchloric acid in water.

See: Cheronis, p. 642.

Hg salt.

$$2\,RSH + Hg(CN)_2 \rightarrow (RS)_2Hg + 2\,HCN$$

Mercuric
thiolate

From the thiol and aqueous mercuric cyanide in ethanol.

For directions and examples see: Linstead, p. 86; Wild, pp. 90–91; E. Wertheim, *J. Amer. Chem. Soc.*, **51**, 3661 (1929).

3,5-Dinitrothiobenzoate. *

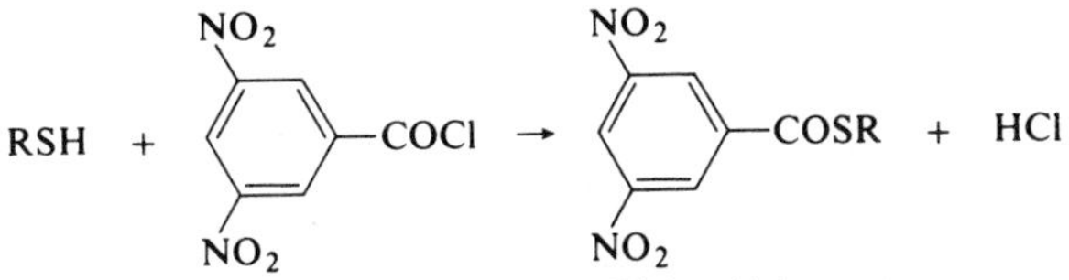

$$RSH + (NO_2)_2\text{C}_6H_3-COCl \rightarrow (NO_2)_2\text{C}_6H_3-COSR + HCl$$

3,5-Dinitrothiobenzoate

From the thiol, 3,5-dinitrobenzoyl chloride and pyridine.

For directions and examples see: Cheronis, p. 643; Shriner, p. 255; Vogel, p. 501; Wild, p. 92; E. Wertheim, *J. Amer. Chem. Soc.*, **51**, 3661 (1929).

3-Nitrothiophthalate. *

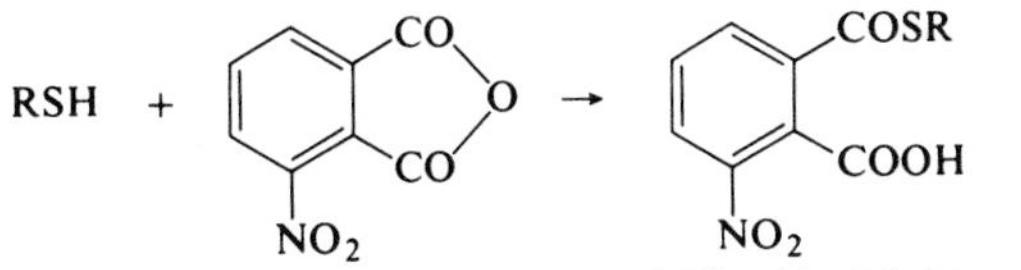

$$RSH + \text{3-nitrophthalic anhydride} \rightarrow \text{(COSR, COOH, NO}_2\text{ product)}$$

3-Nitrothiophthalate

From 3-Nitrophthalic anhydride and the thiol.

For directions and examples see: Wild, p. 93; E. Wertheim, *J. Amer. Chem. Soc.*, **51**, 3661 (1929).

*Derivatives recommended for first trial.
WARNING: This is not an instruction manual. References should be consulted for the preparation of derivatives.

411

S-Alkylmercaptosuccinic acid (Alkylthiosuccinic acid).*

$$RSH \; + \; \begin{array}{l} CHCOONa \\ \| \\ CHCOONa \end{array} \xrightarrow{\; H^+ \;} \begin{array}{l} RSCHCOOH \\ | \\ CH_2COOH \end{array}$$

S-Alkylmercap-
tosuccinic acid

From the thiol and disodium maleate in ethanol.

For directions and examples see: J. G. Hendrickson and L. F. Hatch, *J. Org. Chem.*, **25**, 1747 (1960).

1-Anthraquinonyl thioether.

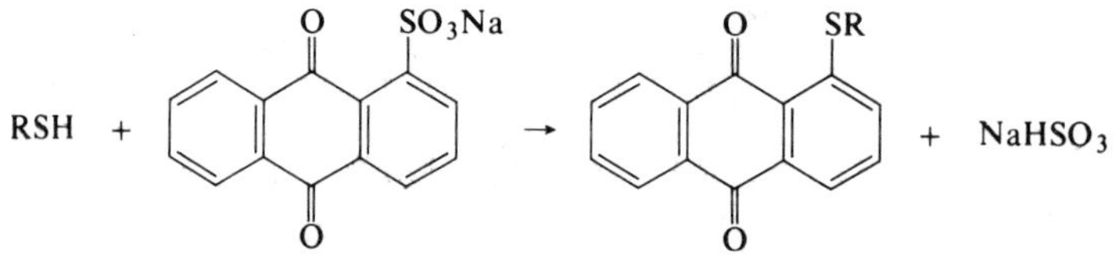

1-Anthraquinonyl
thioether

From the thioether with sodium anthraquinone-1-sulfonate and sodium hydroxide in water.

For directions and examples see: E. E. Reid, C. M. MacKall and G. E. Miller, *J. Amer. Chem. Soc.*, **43**, 2104 (1921); W. S. Hoffman and E. E. Reid, *J. Amer. Chem. Soc.*, **45**, 1831 (1923); L. M. Ellis and E. E. Reid, *J. Amer. Chem. Soc.*, **54**, 1674 (1932).

Acetate.

$$RSH \; + \; CH_3COCl \; \rightarrow \; CH_3COSR \; + \; HCl$$

$$RSH \; + \; (CH_3CO)_2O \; \rightarrow \; CH_3COSR \; + \; CH_3COOH$$

$$RSH \; + \; CH_2{=}C{=}O \; \rightarrow \; CH_3COSR$$

Acetate

From the thiol with acetyl chloride, or from the thiol with acetic anhydride and aqueous sodium hydroxide, or from the thiol with ketene.

For directions and examples see: A. Schöberl and A. Wagner in *Methoden der Organischen Chemie (Houben-Weyl)*, Vol. 9 (Ed. E. Muller), Georg Thieme Verlag, Stuttgart, 1955, pp. 753–756.

Methyl thioether.

$$RSNa \; + \; CH_3I \; \rightarrow \; RSCH_3 \; + \; NaI$$

Methyl
thioether

From the sodium thiolate with alkyl halide.

For directions and examples see: E. E. Reid, *The Chemistry of Bivalent Sulfur*, Vol. 2, Chemical Publishing Co., New York, 1960, p. 25.

Disulfide.

$$2\,RSH \; \xrightarrow{\; [O] \;} \; RSSR \; + \; H_2O$$

Disulfide

From the thiol (or thiophenol) with ferric chloride in aqueous acetic acid.

For directions and examples see: Linstead, p. 87; Wild, p. 95; T. Zincke and W. Frohneberg, *Chem. Ber.*, **43**, 840 (1910).

From the thiol with chlorine, bromine or iodine in hydrocarbon solvent.

See: E. E. Reid, *The Chemistry of Bivalent Sulfur*, Vol. 1, Chemical Publishing Co., New York, 1958, p. 124.

---

*Derivatives recommended for first trial.
WARNING: This is not an instruction manual. References should be consulted for the preparation of derivatives.

TABLE XXV. ORGANIC DERIVATIVES OF THIOLS (MERCAPTANS)
a) Liquids (Listed in order of increasing atmospheric b.p.)*

No.	Name	Boiling point, °C	Melting point, °C	n_D^{20}	D_4^{20}	2,4-Dinitrophenyl thio-ether	2,4-Dinitrophenyl sulfone	3,5-Dinitrothiobenzoate	3-Nitrothiophthalate	1-Anthraquinonyl thio-ether	Mercury salt	S-Alkyl mercaptosuccinic acid	Disulfide	Miscellaneous
1	Methanethiol (Methyl mercaptan)	5.96	−123	....	0.8599[25]	127–8	189.5	......	......	221	176	133	−84.72; b.p. 116–8	....
2	Ethanethiol (Ethyl mercaptan)	36	−144.4	1.4318	0.83147[25]	114–5	160	62	149	184	85	119.5	−101.4; b.p. 153.5	Acetyl, b.p. 114
3	2-Propanethiol (sec-Propyl mercaptan)	56	−130.7	1.4256	0.8142	93.5–5.0	140.5	84	145	134	63	......	−69; b.p. 176	....
4	2-Methyl-2-propanethiol (tert-Butyl mercaptan)	64.2	1.26	1.4230	0.7999	109–11	......	......	......	......	......	164	b.p. 200–1	Benzoyl, b.p. 110[28]
5	1-Propanethiol (n-Propyl mercaptan)	67.5	−113.8	1.4348	0.8047	85–6.5	127.5	52	137	151	72	118–9	−85.59; b.p. 195–6	....
6	2-Butanethiol (sec-Butyl mercaptan)	84–5	−165.0	1.4367	0.8294	65.6–6.0	120	......	......	......	189	135	b.p. 95–7[14]	Benzoyl, b.p. 150[20]
7	2-Methyl-1-propanethiol (Isobutyl mercaptan)	88.72	....	1.4386	0.8357	74.5–5.0	105.5	63–4	136	144	95	120.9–1.4	b.p. 220	....
8	2-Propene-1-thiol (Allyl mercaptan)	90; 67–9	....	1.4680	0.93044	72	105	52					b.p. 174d	....
9	1-Butanethiol (n-Butyl mercaptan)	98–100	−119 to −115	1.44402	0.8337	66	92	49	144	112.5	86	103.7–4; 144.5	b.p. 226	Benzoyl, b.p. 160[23]
10	2-Pentanethiol (sec-Amyl mercaptan)	112.9	−169	1.4386[25]	0.82815[25]	....							b.p. 122–3[10]	
11	2-Methoxyethanethiol	113	....	1.4488[23]	....	90						....		Acetyl, b.p. 110[110]
12	D,L-3-Methylbutanethiol (Isoamyl mercaptan)	117; 118–20	....	1.44118	0.83475	....						115.6–6.0	b.p. 250	....
13	1-Phenylethane-1-thiol	119–20	....	1.557	1.022	....	161						58	Acetyl, b.p. 123–5[13]
14	D-2-Methyl-1-butanethiol	119–21	....		0.8403[25]	78–9					60	122.3–.6	b.p. 122–3	$[\alpha]_D^{23}$: +3.21
15	2-Chloropropanethiol	125	....	1.4844	1.1062	76–7							b.p. 113–20[20]	Acetyl, b.p. 70–1[9]
16	2-Chloroethanethiol	125–6		1.5289	1.203	95–7							b.p. 170–80	Acetyl, b.p. 51[4]
17	2-Ethoxyethanethiol	125–6		1.5795	0.9462	66							b.p. 150–2[15]	....
18	1-Pentanethiol (n-Amyl mercaptan)	126.64	−75.83	1.44366	0.8390	79.5–80.5	83	40	132	114	75	107–8	b.p. 140–5[17]	PdCl₂ deriv., 41
19	2-Hexanethiol (sec-Hexyl mercaptan)	142; 138–9	−147.0	1.4426[25]	0.83050[25]	....								....
20	1,2-Ethanedithiol (Ethylene dithioglycol)	147; 46–7[16]	−41.0	1.5550	1.1185[25]	248								Di-Me. eth., b.p. 183
21	1-Hexanethiol (n-Hexyl mercaptan)	151–2	−81.03	1.4490	0.8526	73.5–5.0	97	......		129	58	96.0–.5	....	....
22	2-Hydroxyethanethiol	158; 54[12]	....	1.4443	1.1143	100–2	......				123	.....	28	Diacetyl, b.p. 98–9; Dibenzoyl, 39; S-Phenylurethane, 59–60, bz.

*Derivative data given in order: m.p., crystal color, solvent from which crystallized.

TABLE XXV. ORGANIC DERIVATIVES OF THIOLS (MERCAPTANS)
a) Liquids (Listed in order of increasing m.p.)* (Continued)

No.	Name	Boiling point, °C	Melting point, °C	n_D^{20}	D_4^{20}	2,4-Dinitrophenyl thioether	2,4-Dinitrophenyl sulfone	3,5-Dinitrothiobenzoate	3-Nitrothiophthalate	1-Anthraquinonyl thioether	Mercury salt	S-Alkyl mercaptosuccinic acid	Disulfide	Miscellaneous
23	**Cyclohexanethiol** (Cyclohexyl mercaptan).........	158–60	....	1.4933	0.9782	148	172	......	......	......	78	150.5–1.5	b.p. 288	....
24	*cis*-**3-Methyl-1-cyclohexanethiol**	165		1.4647[25]	0.916[25]	......	......						......	$[\alpha]_{546}$: −2.24
25	**Thiophenol** (Mercaptobenzene)..	169.5; 172.5	−14.9	1.5888	1.0780	121	161	149	131	......	......	......	66.5; 61	4-Nitrothiobenzoyl, 115.7–.8; Phenylurethane, 128
26	*trans*-**3-Methyl-1-cyclohexanethiol**	171	....	1.4663[25]	0.914[25]	......	......						......	$[\alpha]_{546}$: + 5.50
27	**2-Thiophenethiol**	171.1; 166	....	1.6021	1.168[19.5]	119	143						56	Acetyl, b.p. 230–2
28	**1,3-Propanedithiol**	172.9	−79.0	1.5371[25]	1.0775[25]	194	......						b.p. 198[20]	Dibenzoyl, 56.3
29	**1-Heptanethiol** (*n*-Heptyl mercaptan) ..	176–7	−43.4	1.4498[25]	0.83891[25]	81–2	101	53	96	132	77	105.8–6.2	b.p. 164[6]	....
30	**2-Octanethiol** (*sec*-Octyl mercaptan) ...	186.4	−79.0	1.4481[25]	0.83293[25]	......							b.p. 161–71[6]	....
31	**2-Thiocresol** (2-Toluenethiol)	194.3	15	......	......	101; 98–9.5	155				170.3	......	......	S-4-Nitrobenzoyl, 90–1
32	**Phenylmethanethiol** (Benzyl mercaptan)..	194–5	....		1.058	130; 128.5–9.5	182.5	120	137			189–90; 192	74; 70	Ph. eth., 42
33	**3-Thiocresol** (3-Toluenethiol)	195.4		......	......	100–1.5; 91	145					......	......	S-4-Nitrobenzoyl, 95–6; Ag. deriv., 126–7
34	**1,4-Butanedithiol**	195.6	− 53.9	1.5265[25]	1.0395[25]	......							......	Dibenzoyl, 49.5
35	**2-Phenylethanethiol**...	199		1.5643[19]	1.0318[18]	93–4.5	133					......	b.p. 168–80[1.5]	Acetyl, b.p. 134–5[13]
36	**1-Octanethiol** (*n*-Octyl mercaptan) ...	199.1	−49.2	1.4519[25]	0.83956[25]	78; 76–7.5	98			95	71	96.1–.6	b.p. 178–83[5]	....
37	**2-Chlorothiophenol** ...	205			1.2752[19.5]	138	......					......	90	....
38	**2,5-Dimethylthiophenol** (*p*-Xylene-2-thiol)...	211–2		......	......	......							......	S-Ph. eth., b.p. 172.5[11]; S-4-Tolyl, eth., b.p. 188[11]
39	**2,4-Dimethylthiophenol** (*m*-Xylene-4-thiol) ..	214		....	....	......							......	S-Ph. eth., b.p. 171[11]; S-Benzyl eth., 35

* Derivative data given in order: m.p., crystal color, solvent from which crystallized.

TABLE XXV. ORGANIC DERIVATIVES OF THIOLS (MERCAPTANS)
a)Liquids (Listed in order of increasing m.p.)* (Continued)

No.	Name	Boiling point, °C	Melting point, °C	n_D^{20}	D_4^{20}	2,4-Dinitro-phenyl thio-ether	2,4-Dinitro-phenyl sulfone	3,5-Dinitro-thioben-zoate	3-Nitro-thio-phthal-ate	1-Anthra-quinonyl thio-ether	Mercury salt	S-Alkyl mer-capto-succinic acid	Disulfide	Miscellaneous
40	2-Hydroxythiophenol (Thiocatechol)......	216–7[751]; 88–90[8]	5–6	......	1.2373[0]	......	......			......	......	......	......	O-Me. eth., b.p. 218–9; Di-Me. eth., b.p. 237
41	Bis(2-mercaptoethyl) ether	217	−80	1.5339	1.1648[25]	......						......	......	4-Nitro-benzoyl, 106.5
42	1,5-Pentanedithiol	217.3; 110[16]	−72.5	......	......									Dibenzoyl, 45
43	1-Nonanethiol (n-Nonyl mercaptan)...	220.2	−20.1	1.45197	0.83714	86; 84–5	92			117.5	......	105–6	b.p. 211–2[6]	....
44	2-Isopropyl-5-methyl-benzenethiol (Thiothymol).......	230–1	......	......	......						......	78, al.	......	....
45	5-Isopropyl-2-methyl-benzenethiol (Thiocarvacrol).....	235–6	......	......	0.9975[17.5]	......	......				109	......		S-Me. eth., b.p. 244
46	1,6-Hexanedithiol	237.1	−21.0	1.5077[25]	0.9886[25]	......							......	Dibenzoyl, 57
47	1-Naphthalenethiol (α-Mercapto naphthalene).......	285; 114.8[10.5]	......	1.6802	1.1607	......						......	......	Benzyl eth., 78–80; 4-Nitro-benzoyl, 121–30; Acetyl, b.p., 200–3[25]

*Derivative data given in order: m.p., crystal color, solvent from which crystallized.

TABLE XXV. ORGANIC DERIVATIVES OF THIOLS (MERCAPTANS)
b) Solids (Listed in order of increasing m.p.)*

No.	Name	Melting point, °C	Boiling point, °C	2,4-Dinitrophenyl thioether	2,4-Dinitrophenyl sulfone	Acetate	Benzoate	S-Alkyl mercaptosuccinic acid	Disulfide	Methyl thioether	Miscellaneous
1	3-Hydroxythiophenol (Thioresorcinol).	17	168[35]	......	......	....	di: 78	......	95	15; b.p. 224 sl. d.	Di-Me. eth., b.p. 224–5
2	Hexadecanethiol (Cetyl mercaptan)	19	123–8[0.5]	91; 96	105	....		105–6.5	55.5	......	
3	1,10-Decanedithiol	20; 17.8	297.1	......			di: 57; 55			......	D_4^{25}: 0.9432; n_D^{25}: 1.4940
4	Benzoylmethanethiol (2-Mercaptoacetophenone)	23–4	116–22[4]						81	......	D_4^{20}: 1.1753; Oxime, 70; Phenylhydrazone, 90–1
5	2-Aminothiophenol	26	234; 125–7[6]	152	......	di: 135			93	b.p. 234 sl. d.	....
6	3-Phenylenedithiol (Dithioresorcinol).	27.1	245; 123[17]	......						......	Trinitrobenzene add. comp., 76–7
7	4-Dimethylaminothiophenol	28.5	259–60	176	......				118		
8	2-Phenylenedithiol (Dithiocatechol)	29	238–9			di: 88.5	di: 74–5; 94–5				
9	4-Hydroxybenzenethiol (Thiohydroquinone)	30	144–6[20]			S-: 85–6, bz.-lgr.; di: 66	75; di: 161			84–5	S-Et. eth., 39–41
10	threo-Dithiothreitol (threo-2,3-Dihydroxy-1,4-dithiobutane)	42–3	123–5[2]			tetra: 73, me. al.					Oxid. → trans-4,5-dihydroxy-o-dithiane, 132; Diisopropylidene deriv., 78, me. al.
11	4-Thiocresol (4-Toluenethiol)	43–4	195	102.5–4	189.5	di: 66	di: 161				S-4-Tolyl deriv., 57; S-4-Nitrobenzoyl 114–5; S-Chloroacetyl, 40
12	4-Chloro-1-naphthalenethiol	43–4, al.							122	......	
13	4-Aminothiophenol	46	140–5[16]		......	N-mono: 154; 163, yel.			82	b.p. 272–3	
14	1,4-(Dimethylthio)-benzene (p-Xylylene dimercaptan)	46–7	156[12]	......			di: 135				
15	5-Amino-2-methylthiophenol	47, bz.-pet. eth.	......			di: 125, yel., bz.-pet. eth.				147, lgr.	Tri-Me. eth., b.p. 159[17]
16	4-Chlorothiophenol	54	123					163–4	73	......	
17	2-Hydroxy-1-naphthalenethiol	55, pet. eth.				O-: 120, CCl₄; di: 57					
18	1,3,5-Benzenetrithiol	57–60	......			tri: 73–4, al.			......	tri: 66–8, al.	
19	2-Nitrothiophenol	58–61	......	131–3	......				199	64–5	
20	1-Amino-2-propanethiol	63–5	......						Dihydrochloride, 214	......	Hydrochloride, 87–8, al.; Picrate, 143–4 d.
21	4-Bromothiophenol	75	231	142	190	51–2, me. al.			94.5	38, al.	Benzenesulfonate, 75
22	3-Nitrothiophenol	77	......						84	......	
23	4-Nitrothiophenol	77	......	160	......		123–7, 50% ac. a.		184	72	......
24	4-Nitro-1-naphthalenethiol	77–9	......	193	......				189		
25	2-Naphthalenethiol (2-Mercaptonaphthalene).	81	286	145	228	53.5	....		139	......	4-Nitrobenzoyl, 183–4; 2-Tolyl eth., b.p. 229.5[11]

*Derivative data given in order: m.p., crystal color, solvent from which crystallized.

TABLE XXV. ORGANIC DERIVATIVES OF THIOLS (MERCAPTANS)
b) Solids (Listed in order of increasing m.p.)* (Continued)

No.	Name	Melting point, °C	Boiling point, °C	2,4-Dinitro-phenyl thio-ether	2,4-Dinitro-phenyl sulfone	Acetate	Benzoate	S-Alkyl mer-capto-succinic acid	Disulfide	Methyl thio-ether	Miscellaneous
26	DL-*erythro*-Dithiothreitol (*erythro*-2,3-Dihydroxy-1,4-dithiobutane).	82–3, lt. pet. eth.				*tetra*: 126, bz.					Oxid. → *cis*-4,5-di-hydroxy-*o*-di-thiane, 132; Di-iso-propylidene deriv., 145, pet. eth.
27	4-Iodothiophenol	85–6, al.		140.5					124	45; 38	CrO₃/AcOH → sulfone, 83
28	4-Amino-1-naphthalenethiol	91–3				N-*mono*: 173			168		
29	2-(3-Aminopropionamido)ethane-thiol (Aletheine)	93–6; subl.$_{-5}$ 140^{10}				*di*: 139–40, et. ac.					Hydrochloride, 214–7, 95% al.; Oxalate, 121–2, al.
30	Bis(4-mercaptophenyl)ether	98				*di*: 68					
31	4-Phenylenedithiol (1,4-Dimer-captobenzene).	98, aq. al.				*di*: 126, pet. eth.				*di*: 85, me. al.	Di-Et. eth., 81.5, al.
32	2-Aminoethanethiol	98–100	130	94.5		*di*: 30			Dihydro-chlo-ride, 216		Hydrochloride, 70–2, al.; Picrate, 126; S-Acetyl hy-drochloride, 137
33	Triphenylmethanethiol	107		190		139–41	185		*ca.* 155		
34	4-Bromo-2-nitrothiophenol	110		142							
35	4-Biphenylthiol	111–2, al.		146	170				150	107–8, al.	
36	3-Amino-1-propanethiol	112–3							Dihydro-chlo-ride, 219		Hydrochloride, 69, dil. al.
37	1,8-Naphthalenedithiol	113–4, al.								*di*: 84	
38	4-Hydroxy-1-naphthalenethiol	114				*di*: 77					
39	Bis(4-thiophenyl)sulfide	114				*di*: 65			190		
40	2,4,6-Trinitrothiophenol	114		217							
41	1,5-Naphthalenedithiol	118–21, yel.				*di*: 187–9	*di*: 232			*di*: 250	
42	4-Chloro-2-nitrothiophenol	120–2		141							
43	2,4-Dinitrothiophenol	131–2		193–7	240–1	107	113		280		
44	6-Hydroxy-2-naphthalenethiol	137, al.				*di*: 110; 107, al.			221		
45	2-Benzothiazolthiol	177–9, aq. me. al.								52, aq. al.	Et. eth., 26, al.
46	2,7-Naphthalenedithiol	181; 174, al.				*di*: 110	*di*: 152–3				
47	1-Anthraquinonethiol	187, yel., ac. a.							>350	218, yel., al.	Et. eth., 183, yel., al.; Benzyl eth., 241, yel., ac. a.
48	2-Anthraquinonethiol	206, yel., ac. a.							257	162, yel., ac. a.	Et. eth., 138, yel., al.; Benzyl eth., 138, yel., al.
49	2-Benzimidazolthiol	298; 296–7, dil. al.									Benzyl eth., 186–7; 1-N-Me. eth., 190–2

*Derivative data given in order: m.p., crystal color, solvent from which crystallized.

*Sulfone.**

$$RSR' \xrightarrow{[O]} RSO_2R'$$
Sulfone

From the thioether in glacial acetic acid with dilute aqueous potassium permanganate.

For directions and examples see: Cheronis, p. 641; Linstead, p. 87; G. W. Fenton and C. K. Ingold, *J. Chem. Soc.*, 2338 (1929); H. Rheinboldt and E. Giesbrecht, *J. Amer. Chem. Soc.*, **68**, 973 (1946).

From the thioether with aqueous hydrogen peroxide in acetone or in acetic acid.

See: O. Hinsberg, *J. prakt. Chem.*, **90**, 350 (1914); H. Rheinboldt and E. Giesbrecht, *J. Amer. Chem. Soc.*, **68**, 973 (1946); C. G. Overberger, S. P. Lighthelm and E. A. Swire, *J. Amer. Chem. Soc.*, **72**, 2856 (1950).

From the thioether with hydrogen peroxide and ammonium molybdate in aqueous perchloric acid.

See: Cheronis, p. 641.

From the thioether with potassium bichromate and sulfuric acid in water.

See: G. Raiziss, L. W. Clemence, M. Severac and J. C. Moetsch, *J. Amer. Chem. Soc.*, **61**, 2763 (1939).

From the thioether with chromic anhydride in glacial acetic acid.

See: C. G. Overberger, S. P. Lightelm and E. A. Swire, *J. Amer. Chem. Soc.*, **72**, 2856 (1950).

For extensive lists of references for the oxidation of thioethers to the corresponding sulfones *see:* A. Schöberl and A. Wagner in *Methoden der Organischen Chemie (Houben-Weyl)*, Vol. 9, (Ed. E. Miller), Georg Thieme Verlag, Stuttgart, 1955, pp. 227-231; E. E. Reid, *Organic Chemistry of Bivalent Sulfur*, Vol. 2, Chemical Publishing Co., New York, 1960, pp. 64-65.

Sulfoxide.

$$RSR' \xrightarrow{[O]} RSOR'$$
Sulfoxide

From the thioether with hydrogen peroxide in acetic acid, in acetone or in alcohol-acetic acid mixture.

For directions and examples see: O. Hinsberg, *Chem. Ber.*, **43**, 289 (1910); R. L. Shriner, H. C. Struck and W. V. Jorison, *J. Amer. Chem. Soc.*, **52**, 2060 (1930); P. Karrer, N. J. Antia and R. Schwyzer, *Helv. chim. Acta*, **34**, 1392 (1951).

From the thioether with perphthalic acid in ether.

See: H. Böhme, *Chem. Ber.*, **70**, 378 (1937).

From the thioether with chromic anhydride in aqueous acetic acid.

See: R. Knoll, *J. prakt. Chem.*, **113**, 40 (1926).

For additional references for the oxidation of thioethers to the corresponding sulfoxides *see:* A. Schöberl and A. Wagner in *Modern Methoden der Organischen Chemie (Houben-Weyl)*, Georg Thieme Verlag, Stuttgart, 1955, pp. 211-215.

Mercuric halide addition compound.

$$R_2S + HgX_2 \rightarrow R_2S \cdot HgX_2$$
Mercuric halide
addition compound

From the thioether with mercuric chloride, bromide or iodide in ethanol, acetone or aqueous solution.

For directions and examples see: W. F. Faragher, J. C. Morrell and S. Comay, *J. Amer. Chem. Soc.*, **51**, 2774 (1929); E. E. Reid, *Organic Chemistry of Bivalent Sulfur*, Vol. 2, Chemical Publishing Co., New York, 1960.

Mercuric halide or mercuric acetate derivative.

2-Mercuric chloride derivative   2,5-Bis(mercuric chloride) derivative

2,5-Bis(mercuric acetate) derivative

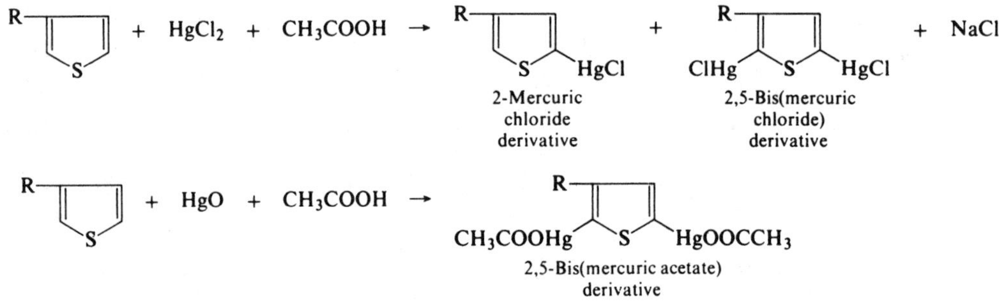

*Derivatives recommended for first trial.
WARNING: This is not an instruction manual. References should be consulted for the preparation of derivatives.

418

EXPLANATIONS AND REFERENCES TO TABLE XXVI (Continued)

These derivatives are for substituted thiophenes only.

From the substituted thiophene with the mercuric salt (with or without sodium acetate) in ethanol or in acetic acid.

For directions and examples see: H. D. Hartough, *Thiophene and its Derivatives,* (*The Chemistry of Heterocyclic Compounds,* Vol. 3), Interscience, London, 1952, pp. 444–453.

For various additional compounds of thioethers with metal salts *see*: E. E. Reid, *Organic Chemistry of Bivalent Sulfur,* Vol. 2, Chemical Publishing Co., New York, 1960, pp. 52–60.

*Derivatives recommended for first trial.
WARNING: This is not an instruction manual. References should be consulted for the preparation of derivatives.

TABLE XXVI. ORGANIC DERIVATIVES OF THIOETHERS (SULFIDES)

a) Liquids 1) Noncyclic (Listed in order of increasing b.p.)*

No.	Name	Boiling point, °C	Melting point, °C	n_D^{20}	D_4^{20}	Sulfone	Sulfoxide	Disulfide	Miscellaneous
1	Dimethyl sulfide (Methyl sulfide)	37.3	−98.27	1.4356	0.8458[21]	109; b.p. 238	18.45; b.p. 189	109.7	Tri-HgCl₂ add. comp., 158; 150–1, rapid htng.; HgI₂ add. comp., 75; SnBr₄ add. comp., 85–7; PtCl₂ add. comp., 159; PtBr₂ add. comp., 159; PdCl₂ add. comp., 130; PdBr₂ add. comp., 125; AgNO₃ add. comp., 126
2	Ethyl methyl sulfide..........	66.9	−104.8	1.4353	0.8483	36, eth.	....	130	HgCl₂ add. comp., 128; HgI₂ add. comp., 59; PdCl₂ add. comp., 67
3	Methyl vinyl sulfide..........	67.3	......	1.4845	0.9026	....	....	....	....
4	Divinyl sulfide (Vinyl sulfide)...	84	......	......	0.9174	....	......:	....	....
5	Allyl methyl sulfide	91–3	......	1.4712	0.8767	....	....	....	....
6	Diethyl sulfide (Ethyl sulfide)...	92	−102.05	1.44233	0.8368	73–4; b.p. 248	4–6; 15; b.p. 88–9[15]	152.6; 154	Disulfoxide, 123–4[11]; Mono-HgCl₂ add. comp., 90; Di-HgCl₂ add. comp., 119.5; HgI₂ add. comp., 110; AgNO₃ add. comp., 122; SnCl₄ add. comp., 102; SnBr₄ add. comp., 84; PtCl₂ add. comp., 106; PtBr₂ add. comp., 118; PtI₂ add. comp., 136; PdCl₂ add. comp., 83; PdBr₂ add. comp., 100
7	Ethyl vinyl sulfide	92	......	1.4756	0.8756	....	....	....	....
8	Isopropyl methyl sulfide	93.5; 85	−101.48	1.4392	0.8291	....	....	....	....
9	Methyl n-propyl sulfide	95.5	−112.98	1.4442	0.8438	....	....	....	....
10	tert-Butyl methyl sulfide	99	......	1.4402	0.8257	....	....	....	....
11	Chloromethyl methyl sulfide	107.1	......	1.4967	....	....	....	....	....
12	Ethyl isopropyl sulfide.........	107.3	−122.19	1.4407	0.8246	....	....	....	....
13	Isobutyl methyl sulfide	112.5	......	1.4433	0.8335	....	....	....	....
14	Methyl 2-methylallyl sulfide ...	113.0–.2	......	1.4712		....	....	....	....
15	Allyl ethyl sulfide.............	115–6	......	......	0.8676	....	....	....	....
16	Ethyl n-propyl sulfide..........	118.5; 110–2	−117.04	1.4461	0.84448	25; b.p. 142[23]	....	173.7	....
17	Di-isopropyl sulfide (Isopropyl sulfide).................	120.7	−78.08	1.4381	0.8135	36	68.5	176	PtCl₂ add. comp., 163; PtBr₂ add. comp., 174; PtI₂ add. comp., 176; PtI₄ add. comp., 139
18	n-Butyl methyl sulfide	122.5	......	1.4477	0.8427	....	....	....	HgCl₂ add comp., 116.5
19	Chloromethyl ethyl sulfide	128	......	......	....	33	....	b.p. 78–80[16]	
20	Isopropyl propyl sulfide........	132	......	1.4440	0.8269	....	....	....	....
21	sec-Butyl ethyl sulfide	133.6	......	1.4477	0.8353	....	....	....	....
22	Ethyl isobutyl sulfide	134.2	......	1.4452	0.8306	....	....	....	HgCl₂ add. comp., 108
23	Diallyl sulfide (Allyl sulfide)...	139[758]; 35[5–7]	−83	1.4877[27]	0.88765[27]	b.p. 109[3]	b.p. 107–9[7–8]	b.p. 78–80[16]	....
24	Methyl 2-methylbutyl sulfide ...	139–40	......	......	0.8410[19]	....	....	....	....
25	2-Chloroethyl methyl sulfide....	140; 44[20]	......	1.4908	1.1155	....	....	....	....
26	Di-n-propyl sulfide (n-Propyl sulfide).................	141–2	−101.9	1.4481	0.8358	30	14.5–5.0; b.p. 82[15]	194	Mono-HgCl₂ add. comp., 88–9; Di-HgCl₂ add. comp., 122; 127.5; Chloroamine-T → sulfilimide, 110–1.5
27	n-Butyl ethyl sulfide	144.2	−95.13	1.4491	0.8376	50	....	193	....
28	Methyl pentyl sulfide (Amyl methyl sulfide).............	144.5–5.5	......	1.448	0.843	....	....	....	HgCl₂ add. comp., 127
29	Di-tert-butyl sulfide (tert-Butyl sulfide).............	150	......	1.4505	....	....	....	201; b.p. 88[21]	....
30	Diacetyl sulfide (Thioacetic anhydride).................	155–8d.; 63[20]	......	1.4810[21]	1.124	....	....	....	Reduction → acetaldehyde, b.p. 20.2, 2,4-dinitrophenylhydrazone, 168
31	Dichloromethyl sulfide (Chloromethyl sulfide).............	156.5	−54; −37	1.5313	1.4065	70.5–2.0	....	....	....
32	2-Chloroethyl ethyl sulfide	157; 63–5[47]	......	1.06644	1.4878	....	....	....	....

*Derivative data given in order: m.p., crystal color, solvent from which crystallized.

TABLE XXVI. ORGANIC DERIVATIVES OF THIOETHERS (SULFIDES)
a) Liquids 1) Noncyclic (Listed in order of increasing b.p.)* (Continued)

No.	Name	Boiling point, °C	Melting point, °C	n_D^{20}	D_4^{20}	Sulfone	Sulfoxide	Disulfide	Miscellaneous
33	2-Aminoisopropyl methyl sulfide	158	......	......	......	b.p. 140[4]	......	......	......
34	Ethyl 3-methylbutyl sulfide	160	......	1.4495	0.8349	13.5	......	......	$HgCl_2$ add. comp., 87
35	4-Chlorophenyl methyl sulfide	170	......	1.6023^{25}	1.1224^{25}	57–8	......	......	......
36	3-Aminopropyl methyl sulfide	170	......	......	......	44; b.p. 165–8[6]	197	......	Hydrochloride, 136; Picrate, 127; Oxalate, 207d.
37	Di-isobutyl sulfide (Isobutyl sulfide)	172–3	......	1.4463	0.8262	17; b.p. 265	68.5	215	Mono-$HgCl_2$ add. comp., 116; Di-$HgCl_2$ add. comp., 131; $PtCl_2$ add. comp., (i) 83; (ii) 139; $PtBr_2$ add. comp., 143–4; PtI_2 add. comp., 187; $PtCl_4$ add. comp., 162; $AuCl_3$ add. comp., 87; $PdCl_2$ add. comp., 95; $PdBr_2$ add. comp., 140; PdI_2 add. comp., 145
38	Di-2-methylallyl sulfide (2-Methylallyl sulfide)	173	......	......	......	......	......	......	......
39	Chloromethyl dichloromethyl sulfide	177.2^{753}	......	1.5395	1.5258	......	......	......	......
40	Dicrotyl sulfide (Crotyl sulfide)	186.5; 88.9^{20}	......	1.495^{25}	0.9032^0	......	......	......	......
41	Di-n-butyl sulfide (n-Butyl sulfide)	188.0; 182; $109–15^{15}$	−79.7	1.45405	0.8386	44, resolidifies at 32.5	33	231	Di-$HgCl_2$ add. comp., 113; 110.5; $PtBr_2$ add. comp., 65; PtI_2 add. comp., 67; $PdCl_2$ add. comp., 32; $Pd(NO_3)_2$ add. comp., 166; $AgNO_3$ add. comp., 98
42	4-Aminobutyl methyl sulfide	188–90	......	......	......	42; b.p. 165[4]	......	......	Hydrochloride, 153–4, acet.; Picrate, 116–8
43	Di-dichloromethyl sulfide (Di-chloromethyl sulfide)	189; $62–4^{10}$	......	1.5464	1.6273	......	......	......	......
44	Methyl phenyl sulfide	194–6; $78–9^{13}$	......	1.5870	1.0533^{25}	88, w.	......	......	......
45	2-Chloroethyl chloromethyl sulfide	194.5; 77^{10}	......	1.5311	1.338	......	......	......	......
46	Benzyl methyl sulfide	195–8	......	1.5550^{25}	......	127, w.	......	......	......
47	Ethyl phenyl sulfide	205; 200–2	......	1.5701^{15}	1.024^{15}	42; b.p. 160[12]	......	......	$PdCl_2$ add. comp., 140
48	Isopropyl phenyl sulfide	208	......	1.5468	0.9855	......	......	......	$PdCl_2$ add. comp., 162
49	Methyl 4-tolyl sulfide	211–2; $104–5^{20}$	......	1.57537^{16}	1.0302^{16}	89, bz.-pet. eth.	50–4	......	......
50	Di-(3-methylbutyl)sulfide (Isoamyl sulfide)	215.3	......	1.4471	0.8285	31; b.p. 295	......	250	$PdCl_2$ add. comp., 95; $PdBr_2$ add. comp., 133; PdI_2 add. comp., 143; $SnCl_4$ add. comp., 64; $SnBr_4$ add. comp., 45–6
51	Allyl phenyl sulfide	215–8; $104–6^{25}$	......	1.5760	1.0275	......	......	......	......
52	Di-2-chloroethyl sulfide (2-Chloroethyl sulfide; Mustard gas; Yperite)	217	14.4	1.53125	1.2741	56; b.p. 183[20]	109–11	......	......
53	Phenyl propyl sulfide	219–20	......	1.5571	0.9995	44	......	......	$PdCl_2$ add. comp., 91
54	Ethyl 4-tolyl sulfide	220–1	......	1.5568	$1.0016^{17.5}$	55–6	......	......	......
55	Benzyl ethyl sulfide	222–3; $98–9^{13}$	......	......	......	84	......	......	Mono-$HgCl_2$ add. comp., 84; Di-$HgCl_2$ add. comp., 142
56	3-Methylbutyl phenyl sulfide	240–2	......	1.5380	0.9681	36	......	......	$PdCl_2$ add. comp., 97
57	2-Chloroethyl phenyl sulfide	245; 121^{15}	......	1.5838	1.1799	45	......	......	......
58	2-Chloroethyl 4-tolyl sulfide	255–7; $150–2^{20}$	......	......	......	78	......	......	......
59	2-Hydroxyethyl 4-tolyl sulfide	282–3; $119–20^1$	......	......	......	55	......	......	......
60	Di-(3-tolyl)sulfide (m-Tolyl sulfide)	290; 174^{12}	......	......	......	94	b.p. 215^{15}	b.p. 150 d.	......

Derivative data given in order: m.p., crystal color, solvent from which crystallized.

No.	Name	Boiling point, °C	Melting point, °C	n_D^{20}	D_4^{20}	Sulfone	Sulfoxide	Disulfide	Miscellaneous
61	Diphenyl sulfide (Phenyl sulfide)	296; 157–8[16.5]	−21.5	1.6312	1.1160	128–9; b.p. 379	70.5	61	Disulfone of the disulfide, 193–4
62	Di-n-heptyl sulfide (n-Heptyl sulfide)	298	. . . . . .	. . . .	. . . .	80	. . . .	. . . .	. . . .
63	4-Chlorophenyl phenyl sulfide. . .	305–15d.; 167–8[10]	. . . . . .		. . . .	34; 90	. . . .	. . . .	. . . .
64	Phenyl 3-tolyl sulfide	309.5; 164.5[11]	−6.5	. . . .	1.0937[16]	. . . .	. . . .	. . . .	. . . .
65	Phenyl 2-tolyl sulfide	309.9; 300.5; 160.5[11]	. . . . . .	. . . .	1.1012[15]	81, al.	. . . .	. . . .	. . . .
66	Phenyl 4-tolyl sulfide	311.5; 167.5[11]	15.7	. . . .	1.0900[15.7]	127–8, al.	. . . .	. . . .	. . . .

*Derivative data given in order: m.p., crystal color, solvent from which crystallized.

TABLE XXVI. ORGANIC DERIVATIVES OF THIOETHERS (SULFIDES)
a) Liquids 2) Cyclic (Listed in order of increasing b.p.)*

No.	Name	Boiling point, °C	Melting point, °C	n_D^{20}	D_4^{20}	Sulfone	HgCl$_2$ addition compound	Miscellaneous
1	Ethylene sulfide (Thiirane; Thiacyclopropane)	55–6	....	1.4914	1.0046	....	....	Polymerizes rapidly
2	2-Methylethylene sulfide (2-Methylthiirane)	76	....	1.4731[19]	0.946[18]	....	....	....
3	Thiophene	84.12	−38.30	1.5287	1.0644	....	....	2-HgCl deriv., 182–3; 2-HgBr deriv., 169–70; 2-HgI deriv., 116–7
4	2,2-Dimethylethylene sulfide (2,2-Dimethylthiirane)	87; 84–6	....	1.4641	....	....	....	....
5	Trimethylene sulfide (Thietane; Thiacyclobutane)	95	−73.25	1.506[23]	1.0200	76, w.	93–5 d.	Methiodide, 98.5–9.0
6	2-Ethylethylene sulfide (2-Ethylthiirane)	104	....	1.475[19]	0.930[18]	....	....	....
7	2-Methyltrimethylene sulfide (2-Methylthietane)	106	....	1.4831	0.9571	b.p. 251.5–3.5	....	....
8	2-Methylthiophene	112.4	....	1.52042	1.02183	....	....	5-HgCl deriv., 204; 5-HgBr deriv., 179–80; 5-HgI deriv., 111–2
9	2,4-Dimethyltrimethylene sulfide (2,4-Dimethylthietane)	113–4	....	1.4502[18]	0.8710[18]	b.p. 255.0–5.5	....	....
10	3-Methylthiophene	115.4	....	1.52042	1.02183	....	....	2-HgCl deriv., 128–9; 2,5-Di-HgOOCCH$_3$ deriv., >240d.
11	2,2-Dimethyltrimethylenesulfide (2,2-Dimethylthietane)	120	....	1.4739[18]	....	55	118	....
12	Tetramethylene sulfide (Thiolane; Thiophane; Thiacyclopentane)	121.2; 120.2–.5	−96.17	1.5047	0.99869	28.36	128	Sulfoxide, 105–7[12]
13	2-Methyltetramethylene sulfide (2-Methylthiolane)	132.5[750]	−100.71	1.4922	0.9555	b.p. 279–80[758]	di: 162	....
14	2-Ethylthiophene	132.5–4.0	....	1.5127	0.990[24]	....	....	5-HgCl deriv., 147–8; 5-HgI deriv., 96–7
15	3-Ethylthiophene	135–6	....	1.5146	0.9980	....	....	2-HgCl deriv., 67–8; 2,5-Di-HgCl deriv., 295–7d.
16	2,5-Dimethylthiophene	135.5–6.0	−62.57	1.5126	0.98587	....	....	3-HgCl deriv., 156–7; 3-HgI deriv., 175
17	2,4-Dimethylthiophene	137–8; 140	....	1.5130	0.9956	....	....	5-HgCl deriv., 138–9; 5-HgI deriv., 137–9
18	3-Methyltetramethylene sulfide (3-Methylthiolane)	138.2	−81.10	1.4924	0.9634	1; 0.5	83	....
19	2,3-Dimethylthiophene	140.2–1.2	−49.1 to −48.9	1.5188	1.0021	....	....	5-HgCl deriv., 218.5–9.5; 5-HgI deriv., 184.0–4.5; 4,5-Di-HgOOCCH$_3$, 237–40
20	trans-2,5-Dimethyltetramethylene sulfide (trans-2,5-Dimethylthiolane)	142	−76.35	1.4766	0.9188	3; b.p. 278	111	....
21	cis-2,5-Dimethyltetramethylene sulfide (cis-2,5-Dimethylthiolane)	142.3	....	1.4799	0.9222	4; b.p. 278	di: 180	....
22	3,4-Dimethylthiophene	144–6	....	1.5212	1.008[23][21]	....	....	2-HgCl deriv., 139–40.5; 2-HgBr deriv., 152; 2-HgI deriv., 142
23	1,4-Thioxane	148.9	−17	1.5070	1.1177	130; 105.5	171	Sulfoxide, 25; 45
24	2-Methylpentamethylene sulfide (2-Methylthiane)	151	−58.14	1.4905	0.9428	68.5	102	....
25	2-Isopropylthiophene	152	....	1.5037	0.9673	....	....	....
26	3-Isopropylthiophene	155.7; 157	....	1.5052	0.9733	....	....	5-HgCl deriv., 137
27	3-Methylpentamethylene sulfide (3-Methylthiane)	157–8	−60.17	1.4922	0.9473	83	136	....
28	2-Ethyltetramethylene sulfide (2-Ethylthiolane)	157–8	....	1.4896	0.9451	....	mono: 100; di: 146–8	....
29	2-Propylthiophene	157.5–9.5	....	1.5048	0.9683	....	....	5-HgCl deriv., 155; 5-HgSCN deriv., 169.0–9.5
30	4-Methylpentamethylene sulfide (4-Methylthiane)	158.6	−28.11	1.5049	0.9687	121.5	136	....
31	2-Ethyl-5-methylthiophene	159.8–60.4	−68.6	1.5073	0.9663	....	....	....

*Derivative data given in order: m.p., crystal color, solvent from which crystallized.

No.	Name	Boiling point, °C	Melting point, °C	n_D^{20}	D_4^{20}	Sulfone	HgCl₂ addition compound	Miscellaneous
32	**2-Ethyl-3-methylthiophene**	160.0–1.5	....	1.5092[22.5]	0.9792[22.5]	....	....	5-HgCl deriv., 172–3; 5-HgI deriv., 156–7
33	**2,6-Dimethyl-1,4-thioxane**	160–1	....	1.4733	....	105.5	....	....
34	**3-Propylthiophene**	160–2	....	1.5057	0.9716		....	....
35	**2-Allylthiophene**........................	161; 158.5–9.0	....	1.5281[20.5]	1.0175[20.5]	....	....	....
36	**3,5-Dimethyl-1,4-thioxane**	162	....	....	....	102	....	....
37	**3-Ethyl-5-methylthiophene**	162–4	−60 to −59	1.5098	0.9742	....	....	....
38	**2,3,5-Trimethylthiophene**	163–5; 164.5[746]	....	1.5131	0.9753	....	....	x-HgCl deriv., 160–1
39	**2-tert-Butylthiophene**	163.9	−59.2	1.49788	0.9514	....	....	....
40	**3-tert-Butylthiophene**	168.9	−54.8	1.50149	0.9574	....	....	....
41	**Hexamethylene sulfide** (Thiepane; Thiacyclo-heptane)	170; 173–4	....	1.5125	0.9883	71	149	Methiodide, 141.5–2.0
42	**2,3,4-Trimethylthiophene**	172.7; 160–3	....	1.5208	0.995	....	....	....
43	**1,3-Dithiolane**(1,3-Dithiacyclopentane)	175	....	1.5975[15]	1.259[17]	*di*: 205	117; 126	Methiodide, 96; Disulf-oxide, 134
44	**2-Methyl-5-propylthiophene**	179.5–80.5	....	1.5026	....	....	....	....
45	**Cyclohexene sulfide** (7-Thiabicycloheptane)...	180; 71.5–3.5[21]	....	1.5309	0.9274	....	....	....
46	**2-n-Butylthiophene**........................	181–2[740]	....	1.50896	0.9537	....	....	....
47	**2,5-Diethylthiophene**	181–2; 63–6[14]	....	1.5036	0.962[14]	....	....	....
48	**3-n-Butylthiophene**	181–3	....	1.51005	0.9570	....	....	....
49	**2,3,4,5-Tetramethylthiophene**...............	182–4; 187–9	....	1.5196	0.9442[21/21]	....	....	....
50	**3,4-Diethylthiophene**	185–7	....	1.5157[17]	....	....	....	2-HgCl deriv., 118
51	**2-Ethyl-1,3-dithiolane**	191–2	....	....	....	124	....	....
52	**2,4-Dimethyltetramethylene sulfide** (2,4-Dimethylthiolane)	197–8[742]	....	1.4818	0.9265	b.p. 123.3[5]	89	....
53	**3-Ethyl-2,4,5-trimethylthiophene**	204–6[748]	....	1.5132	0.9609	....	....	....
54	**2-n-Octylthiophene**	257–9; 106–8[1]	....	1.4824	0.920	....	....	....

* Derivative data given in order: m.p., crystal color, solvent from which crystallized.

TABLE XXVI. ORGANIC DERIVATIVES OF THIOETHERS (SULFIDES)
b) Solids (Listed in order of increasing m.p.)*

No.	Name	Melting point, °C	Boiling point, °C	Sulfone	Sulfoxide	Disulfide	Miscellaneous
1	Ethyl 2-naphthyl sulfide..................	16	170.5[15]	43–5			
2	Ethyl hexadecyl sulfide	19	....	88			
3	Thiane (Pentamethylene sulfide; Thiacyclo-hexane)................................	19.07; 13	142	98.5–9.0			D_4^{20}: 0.9849; n_D^{20}: 1.5067; $HgCl_2$ add. comp., 137.5, al.; Meth-iodide, 192 subl.
4	2-Methyl-1,4-dithiane	20	....	di: 304			
5	Didecyl sulfide (Decyl sulfide).............	27	217–8[8]	206–7			
6	3-Tolyl 4-tolyl sulfide	27.8, al.	179[11]	116, ac. a.	72		
7	Diundecyl sulfide (Undecyl sulfide; n-Hendecyl sulfide)	34.8					
8	Di-(2-bromoethyl) sulfide (Bromoethyl sulfide).	35	....	111–2			
9	4-Iodophenyl phenyl sulfide (4-Iododiphenyl sulfide)	35	....	141			
10	2-Aminophenyl phenyl sulfide (2-Amino-diphenyl sulfide).........................	35–6, al.	212[25]	122, dil. al.			N-Benzenesulfonyl, 225–6
11	4-Bromophenyl methyl sulfide	37.5	....	56–7			
12	Didodecyl sulfide (n-Dodecyl sulfide)	40.5	....			34.5	
13	Benzyl phenyl sulfide	41; 44.5, al.	197[27]	146–6.5; 148, al.	123		
14	Di-(4-chlorobenzyl)sulfide (4-Chlorobenzyl sulfide)	41	....			59	
15	1-Naphthyl phenyl sulfide	41.8, aq. al.	220–5[11]	99.5–100.5, al.			
16	3-Nitrophenyl phenyl sulfide (3-Nitrodiphenyl sulfide)	42.5	....	80.5–81			
17	Ethyl 4-nitrophenyl sulfide	44	....	138.5			
18	Isopropyl 4-nitrophenyl sulfide	44.5	....	115.3			
19	Di-(4-methoxyphenyl)sulfide (4-Methoxyphenyl sulfide)	46	....	130	96	45	Disulfone, 221, r.h.; 210–2, s.h.
20	Dibenzoyl sulfide (Thiobenzoic anhydride; Benzoyl sulfide)	48, al.	....			133; 128	
21	Dibenzyl sulfide (Benzyl sulfide)	50, chl.	....	151.7, al.-bz.	134.8	73	D_{50}^{50}: 1.0712; $HgCl_2$ add. comp., 131; HgI_2 add. comp., 37–8; $FeCl_3$ add. comp., 94; $PtCl_2$ add. comp., 159; $PtCl_4$ add. comp., 172d.; $PtBr_2$ add. comp., 139; PtI_2 add. comp., 129
22	2,5-Dichlorophenyl methyl sulfide	51	....	88			
23	Methyl 2-naphthyl sulfide..................	51.8; 64	226[11]	115–6, al.	67.5	....	Disulfone, 166
24	Bis(phenylthio)methane	52	....	di: 120–1			
25	Ditetradecyl sulfide (n-Tetradecyl sulfide)	53.8	....			46	
26	Di-(2-diphenoxyethyl)sulfide...............	54, al.	....	108, pink, al.		97	
27	1,3-Dithiane	54	....	di: 330; 308			
28	Bis(benzylthio)methane	55	....	di: 216			
29	4-Nitrophenyl phenyl sulfide (4-Nitrodiphenyl sulfide)	55, yel., lgr.	240[25]	142, aq. al.			
30	Di-n-octyl sulfide (n-Octyl sulfide)	57	....	76	....	b.p. 178–83[5]	
31	Di-(4-tolyl)sulfide (4-Tolyl sulfide)	57.3	179[11]	158, bz.	95, pet. eth.	48, al.	
32	2-Aminophenyl 4-aminophenyl sulfide (2,4′-Diaminodiphenyl sulfide)	61; 62.5, aq. al.	....	124–6	....		Diacetyl, 208
33	Dihexadecyl sulfide (n-Hexadecyl sulfide)......	61.3	....	103.4	99.8		
34	2-Chloroethyl 4-nitrophenyl sulfide	62	....	128			
35	Di-(2-tolyl)sulfide (2-Tolyl sulfide)..........	64, al.	285; 174[15]	134–5, al.	121, pet. eth.	38–9, al.	
36	Methyl 2-nitrophenyl sulfide	64.5, yel., al.	....	106	....		$D_4^{78.2}$: 1.2626; $n_D^{78.2}$: 1.62458; $AgNO_3$ add. comp., 122, yel., al.
37	Benzyl 4-bromophenyl sulfide	65	....	159	....		

*Derivative data given in order: m.p., crystal color, solvent from which crystallized.

No.	Name	Melting point, °C	Boiling point, °C	Sulfone	Sulfoxide	Disulfide	Miscellaneous
38	1,2-Bis(phenylthio)ethane	69–70	....	*di*: 180	....	....	....
39	Dioctadecyl sulfide (Octadecyl sulfide)........	71; 64.5	....	....	....	62.5	....
40	2-Phenyl-1,3-dithiane.	71–2	....	*di*: 265	....	....	....
41	Methyl 4-nitrophenyl sulfide	72	....	141	....	....	$D_4^{80.1}$: 1.2391; $n_D^{80.1}$: 1.64008
42	Di-(3-phenylpropyl)sulfide (3-Phenylpropyl sulfide)	73	....	117	....	b.p. 165–6$^{0.03}$	....
43	Di-(2-methoxyphenyl)sulfide (2-Anisyl sulfide) .	73, al.	252–3[10]	157–8, bz.	....	120	....
44	Di-(4-methylbenzyl)sulfide (4-Methylbenzyl sulfide)	76	....	197	....	....	....
45	Cinnamyl phenyl sulfide	78	....	111–2	90–1	....	....
46	4-(Methylthio)benzaldehyde	78, yel., lgr.	273; 153[17]	....	....	....	Phenylhydrazone, 138; Thiosemicarbazone, 177–9, yel.
47	4-Nitrobenzyl phenyl sulfide	79	....	209.5	....	....	....
48	1,2-Bis(tolylthio)ethane	80–1	....	*di*: 200–1	....	....	....
49	2-Nitrophenyl phenyl sulfide (2-Nitrodiphenyl sulfide)	80.2; 77, yel.	210[15]	147.5, al.	....	....	....
50	Di-(2-aminobenzyl) sulfide.............	81	....	....	....	90–1, lgr.-et. ac.	N,N′-Diformyl, 163; N,N′-Diacetyl, 209; Picrate, 203–4d.
51	4-Nitrophenyl 4-tolyl sulfide	81.5, yel.	....	170–1, yel.	....	....	....
52	Di-(2-aminophenyl) sulfide	87	....	146–7	....	93, al.	N,N′-Diacetyl, 164–5; N,N′-Dibenzoyl, 162–3
53	Di-(2-phenylethyl) sulfide (2-Phenylethyl sulfide)	92.5; 90	....	100.6	69	b.p. 172–5$^{0.8}$	....
54	4-Aminophenyl phenyl sulfide (4-Aminodiphenyl sulfide)...................	96, lgr.	243[29]	176, al.	152, w.	....	Hydrochloride, 197–8d.; N-4-Toluenesulfonyl, 73
55	Di-(4-amino-3-methylphenyl) sulfide.....	96, 25% al.	....	....	....	....	Dihydrochloride, 248–9, dil. HCl; N,N′-Diacetyl, 22, al.; N,N′-Dibenzoyl, 233, me. al.; Dipicrate, 186, w.
56	Di-(4-chlorophenyl)sulfide (4-Chlorophenyl sulfide)	98; 88–90	212[18]	148–9, subl.	143	73	....
57	3-Aminophenyl 4-nitrophenyl sulfide (3-Amino-4′-nitrodiphenyl sulfide)	99–100	....	....	....	....	N-Acetyl, 115–6; N,N-Dimethyl, 83–4
58	Di-(2-amino-5-methylphenyl) sulfide.........	103–4, al.	....	....	....	98	Dihydrochloride, 100d., al.; N,N′-Diacetyl, 165, al.; N,N′-Dibenzoyl, 185–6, al.; Diurethane, 113, bz.-pet. eth.; Dipicrate, 179, bz.
59	Di-(4-aminobenzyl) sulfide................	104–5	....	....	....	96–8, al.	N,N′-Diacetyl, 188; N,N′-Dibenzoyl, 224
60	Di-(4-aminophenyl)sulfide................	108–9, w.	....	178, me. al.	175d., al.	85; 106, al.	N,N′-Diacetyl, 220–1; N,N′-Dibenzoyl, 234
61	Di-(1-naphthyl)sulfide	110, al.	290[15]	187, al.	166, al.	91	....
62	1,4-Dithiane	111–2	199–200	*mono*: 200; *di*: >330	....	....	Methiodide, 73–4; Sulfoxide-sulfone, 279
63	Di-(4-bromophenyl)sulfide (4-Bromophenyl sulfide)	112, al.	243[20]	172	153, al.	94.5	....
64	2,4-Dinitrophenyl phenyl sulfide (2,4-Dinitrodiphenyl sulfide)...................	121; 117, bz.	....	161	....	....	....
65	Di-(2-nitrophenyl)sulfide (2-Nitrophenyl sulfide)	122–3, yel., al.-ac. a.	....	164	....	198–9, ac. a. or bz.	....
66	Benzyl 4-nitrophenyl sulfide	123	....	172	....	....	....
67	Di-(3-hydroxyphenyl)sulfide (3-Hydroxyphenyl sulfide)	130	....	190–1; 186–7	94–5, pet. eth.	95	Acetyl, 87
68	1,3-Bis(2-nitrophenylthio)propane	140	....	*di*: 156–7	....	....	....
69	Di-(2-hydroxyphenyl)sulfide (2-Hydroxyphenyl sulfide)	142, bz.	....	179; 164–5, bz.	....	b.p. >200 d.	Diacetyl, 95–6, al. Di-Me. eth., 73
70	1-Naphthyl 2-naphthyl sulfide (1,2′-Dinaphthyl sulfide)	151	291–2[15]	123	....	....	....
71	Di-(2-naphthyl)sulfide (2-Naphthyl sulfide) ...	151	296[15]	177, al.	137.5–8.5	139	....

*Derivative data given in order: m.p., crystal color, solvent from which crystallized.

No.	Name	Melting point, °C	Boiling point, °C	Sulfone	Sulfoxide	Disulfide	Miscellaneous
72	**Di-(4-hydroxyphenyl) sulfide** (4-Hydroxyphenyl sulfide)	151, al.	....	240–1, w.	195, acet.	150–1	Diacetyl, 94; 55; Di-Me. eth., 46; Di-Et. eth. 55
73	**Di-(4-nitrobenzyl) sulfide** (4-Nitrobenzyl sulfide)	158–9	....	260.5	212	126.5	....
74	**Di-(4-nitrophenyl) sulfide** (4-Nitrophenyl sulfide)	174–5; 156–7, or., or pa. yel.	....	282; 245; 225	....	182, ac. a.	....
75	**Di-(3-nitrophenyl) sulfide** (3-Nitrophenyl sulfide)	193	....	201	....	84, al.	....
76	**Di-(2,4-dinitrophenyl) sulfide** (2,4-Dinitro-phenyl sulfide)	193–7; 193–4, yel., ac. a.	....	240–1	....	280	....
77	**Di-(2,4,6-trinitrophenyl) sulfide** (Dipicryl sulfide)	230–1; 226, yel.	....	307	....		....

*Derivative data given in order: m.p., crystal color, solvent from which crystallized.

EXPLANATIONS AND REFERENCES TO TABLES XXVII, XXVIII AND XXIX

The dissociation of an organic carboxylic acid, phenol or the conjugate acid of an amine in aqueous solution is expressed by the equation

$$HA_{aq}^{\pm n} + H_2O \rightleftharpoons A_{aq}^{\pm n-1} + H_3O_{aq}^+$$

and the corresponding equilibrium constant K_e is given by

$$K_e = \frac{(a_{A_{aq}^{\pm n-1}})(a_{H_3O^+})}{(a_{HA_{aq}^{\pm n}})(a_{H_2O})}$$

where the a's are the activities of the species. At low electrolyte concentrations a_{H_2O} is virtually constant, and a second constant, K_a, the thermodynamic dissociation constant, is defined as

$$K_a = K_e(a_{H_2O}) = \frac{(a_{A_{aq}^{\pm n-1}})(a_{H_3O^+})}{(a_{HA_{aq}^{\pm n}})} = \frac{(c_{A_{aq}^{\pm n-1}})(c_{H_3O^+})}{(c_{HA_{aq}^{\pm n}})} \cdot \frac{f_\pm^2}{f_{HA_{aq}^{\pm n}}}$$

where the c's are the concentrations and the f's are the activity coefficients of the species.

The dissociation constants in the Tables are given in the more convenient pK_a notation, where

$$pKa = -\log K_a$$

For Table XXVII, $HA^{\pm n} = RCOOH$, and $A^{\pm n-1} = RCOO^-$.

For Table XXVIII, $HA^{\pm n} = ArOH$, and $A^{\pm n-1} = ArO^-$.

For Table XXIX, $HA^{\pm n} = RR'R''NH^+$ and $A^{\pm n-1} = RR'R''N$ (R, R' and R'' may be alkyl or aryl groups or a hydrogen atom).

For monobasic acids $pK_a = pK_1$.

For dicarboxylic acids both pK_1 and pK_2 are given; pK_1 is defined above, and pK_2 is the analogous dissociation constant of the monoanion, $A^{\pm n-1}$, obtained on the first dissociation. For other dibasic acids such as the conjugate acids of amino acids, or diamines, pK_1 is as defined above, and pK_2 is the dissociation constant for the species obtained after the first protonation.

For a comprehensive compilation of the dissociation constants of organic acids (including phenols) in aqueous solution, as well as summary of the methods for pK determinations, see: G. Kortüm, W. Vogel and K. Andrussow, in *Pure and Applied Chemistry*, Vol. 1, Butterworths, London, 1961, pp. 190–536.

For a comprehensive compilation of the dissociation constants of organic bases (especially amines) in aqueous solution see: D. D. Perrin, *Dissociation Constants of Organic Bases in Aqueous Solution*, Butterworths, London, 1965.

For general references including methods of determination of pK, and data for both acids and bases see: J. F. King, in *Elucidation of Structures by Physical and Chemical Methods*, Vol 1 (Ed. K. W. Bently) (*Technique of Organic Chemistry*, Vol. 9), Interscience, New York, 1963, Chapter 6, pp. 318–401; H. C. Brown, D. H. McDaniel and O. Hafliger in *Determination of Organic Structures by Physical Methods*, Vol. 1 (Ed. E. A. Braude and F. C. Nachod), Academic Press, New York, 1955, Chapter 14, p. 567.

---

*Derivatives recommended for first trial.
WARNING: This is not an instruction manual. References should be consulted for the preparation of derivatives.

TABLE XXVII. ACID DISSOCIATION CONSTANTS OF ORGANIC ACIDS IN AQUEOUS SOLUTION
(Listed in order of increasing pKa)

No.	Name	T,°C	pK_1	pK_2	No.	Name	T,°C	pK_1	pK_2
1	Heptafluoro-*n*-butyric acid	25	0.17	....	51	1,3,5-Benzenetricarboxylic acid	25	2.12	3.89
2	Trifluoroacetic acid	25	0.23	....					$pK_3 = 4.70$
3	Trichloroacetic acid	25	0.63	....	52	D,L-3,5-Di-iodotyrosine	25	2.12	6.48
4	2,4,6-Trinitrobenzoic acid	25	0.65		53	3-(2-Fluorophenyl)alanine	24	2.12	9.01
5	Tribromoacetic acid	25	0.66	....	54	3-(4-Fluorophenyl)alanine	24	2.13	9.05
6	3-Chlorophenylglycine	25	1.05	3.93	55	D,L-β-Phenylalanine	25	2.16	9.15
7	2,6-Dinitrobenzoic acid	25	1.14		56	*l*-Lysine	25	2.16	9.18
8	Trichloroacrylic acid	25	1.15						$pK_3 = 10.79$
9	Difluoroacetic acid	25	1.24		57	*d*-Lysine	25	2.16	9.16
10	Oxalic acid	25	1.27	4.28					$pK_3 = 10.81$
11	Dichloroacetic acid	25	1.29	....	58	Glutamic acid	20	2.16	4.324
12	2-Chloro-6-nitrobenzoic acid	25	1.34						$pK_3 = 9.96$
13	2-Bromo-6-nitrobenzoic acid	25	1.37		59	6,6,6-Trifluoronorleucine	25	2.164	9.463
14	Benzenehexacarboxylic acid	25	1.40	2.19	60	3-(3-Chlorophenyl)alanine	25	2.17	8.91
				$pK_3 = 3.31$	61	2-Chloro-5-nitrobenzoic acid	25	2.17	....
				$pK_4 = 4.78$	62	D,L-Serine	25	2.21	4.15
				$pK_5 = 5.89$	63	Diethylmalonic acid	25	2.21	7.29
				$pK_6 = 6.96$	64	N-Propylalanine	25	2.21	10.19
15	*d,l*-2,3-Dibromosuccinic acid	20	1.42	3.24	65	N-Ethylalanine	25	2.22	10.22
16	2,4-Dinitrobenzoic acid	25	1.42		66	D,L-N-Methylalanine	25	2.22	10.19
17	*erythro*-2-Bromo-3-chlorosuccinic acid	19	1.43	2.60	67	2-Nitrobenzoic acid	25	2.22	
18	*d,l*-2,3-Dichlorosuccinic acid	20	1.46	2.86	68	3-(2-Chlorophenyl)alanine	25	2.23	8.94
19	*threo*-2-Bromo-3-chlorosuccinic acid	20	1.46	2.77	69	6-Bromo-2-methylolbenzoic acid	20	2.25	
20	*meso*-2,3-Dibromosuccinic acid	20	1.51	2.71	70	6-Chloro-2-methylolbenzoic acid	20	2.26	
21	*meso*-2,3-Dichlorosuccinic acid	20	1.52	2.94	71	*d,l*-Valine	25	2.286	9.744
22	4,4,4-Trifluorovaline	25	1.537	8.098	72	*d,l*-2-Aminobutyric acid	25	2.29	9.83
23	4,4,4-Trifluorothreonine	25	1.554	7.822	73	2-Benzyl-2-cyanopropionic acid	25	2.29	
24	2-Amino-4,4,4-trifluoro-*n*-butyric acid	25	1.600	8.169	74	2-Amino-*n*-pentanoic acid	25	2.318	9.808
25	2,5-Dinitrobenzoic acid	25	1.62		75	(3,4-Dihydroxyphenyl)alanine	25	2.32	8.68
26	Nitroacetic acid	25	1.68						$pK_3 = 9.88$
27	Trifluoroacrylic acid	25	1.79		76	2-Chloro-3-hydroxysuccinic acid	25	2.32	
28	Benzenepentacarboxylic acid	25	1.80	2.73	77	3-Hydroxyglutaric acid	25	2.32	4.24
				$pK_3 = 3.97$					$pK_3 = 9.56$
				$pK_4 = 5.25$	78	*d,l*-Leucine	25	2.32	9.74
				$pK_5 = 6.46$	79	*d,l*-Norleucine	25	2.335	9.83
29	Cyclopropane-1,1-dicarboxylic acid	25	1.82	5.43	80	*d,l*-Alanine	25	2.34	9.87
30	Hydroxyproline	25	1.82	9.66	81	*cis*-Caronic acid (*cis*-1,1-Dimethyl-2,3-cyclopropanedicarboxylic acid)	25	2.34	8.31
31	DL-Histidine	25	1.82	6.04	82	N-Ethylglycine	25	2.34	10.23
				$pK_3 = 9.12$	83	Glycine	25	2.35	9.78
32	2,3-Dinitrobenzoic acid	25	1.85	....	84	N-Methylglycine	25	2.35	10.18
33	2-Methyl-4-nitrobenzoic acid	25	1.86		85	N-Propylglycine	25	2.35	10.19
34	1,2,4,5-Benzenetetracarboxylic acid	25	1.92	2.87	86	N-*n*-Butylglycine	25	2.35	10.25
				$pK_3 = 4.49$	87	N-Isobutylglycine	25	2.35	10.12
				$pK_4 = 5.63$	88	2-Aminoisobutyric acid	25	2.36	10.25
35	*trans*-Ethylene oxide-1,2-dicarboxylic acid	19	1.93	3.25	89	(Methylsulfonyl)acetic acid	25	2.36	
36	*cis*-Ethylene oxide-1,2-dicarboxylic acid	18	1.94	3.92	90	2-Cyano-2-cyclohexylacetic acid	25	2.37	
37	Maleic acid	25	1.94	6.23	91	2-Cyanopropionic acid	25	2.37	....
38	Ornithine	25	1.94	8.65	92	D,L-Tryptophane	25	2.38	9.39
39	2-Chloro-4-nitrobenzoic acid	25	1.96		93	1,2,3,5-Benzenetetracarboxylic acid	25	2.38	3.51
40	4-Aminosalicylic acid (4-Amino-2-hydroxybenzoic acid)	25	1.99	3.92					$pK_2 = 4.44$
41	2-Chloro-3-nitrobenzoic acid	25	2.02	....					$pK_3 = 5.81$
42	Asparagine	25	2.05	3.87	94	4-Aminobenzoic acid	25	2.38	....
43	Anthranilic acid	25	2.05	4.95	95	2-Cyanoisobutyric acid	25	2.42	
44	5,5,5-Trifluoroleucine	25	2.05	8.92	96	Cyanoacetic acid	25	2.46	
45	S-Ethylcysteine	25	2.05	8.60	97	Pyruvic acid	25	2.49	
46	1,2,3,4-Benzenetetracarboxylic acid	25	2.06	3.25	98	O-Acetylcitric acid	25	2.49	
				$pK_3 = 4.73$	99	3-Pentenoic acid	25	2.51	
				$pK_4 = 6.21$	100	1,2,4-Benzenetricarboxylic acid	25	2.52	3.84
47	Di-*n*-propylmalonic acid	25	2.07	7.51					$pK_3 = 5.20$
48	3-(4-Chlorophenyl)alanine	25	2.08	8.96	101	(2-Chlorovinyl)acetic acid	25	2.54	....
49	3-(3-Fluorophenyl)alanine	24	2.10	8.98	102	Oxaloacetic acid	25	2.55	4.37
50	Arginine	25	2.10	9.07	103	Fluoroacetic acid	25	2.58	....

TABLE XXVII. ACID DISSOCIATION CONSTANTS OF
ORGANIC ACIDS IN AQUEOUS SOLUTION
(Listed in order of increasing pKa) (Continued)

No.	Name	T,°C	pK₁	pK₂	No.	Name	T,°C	pK₁	pK₂
104	Phenylmalonic acid	25	2.58	5.03	162	(2,4-Dichloro-6-methylphenoxy)acetic acid	20	3.13	. . . .
105	2-Chloro-3-hydroxybutyric acid	25	2.58	. . . .	163	2-Cyanobenzoic acid	25	3.14	. . . .
106	2-Fluoroacrylic acid	25	2.58	. . . .	164	(3-Methoxyphenoxy)acetic acid	25	3.14	
107	2,4-Dioxo-n-pentanoic acid	25	2.58	. . . .	165	Ethyl-n-propylmalonic acid	25	3.15	7.43
108	2-Chloro-3-hydroxy-3-phenyl propionic acid	25	2.61		166	4,4,4-Trifluorocrotonic acid	25	3.15	
109	2-Chloro-6-hydroxybenzoic acid	25	2.63	. . . .	167	(4-Iodophenoxy)acetic acid	25	3.16	
110	2-Butynoic acid (Tetrolic acid).	25	2.65	. . . .	168	(2-Iodophenoxy)acetic acid	25	3.17	
111	1-Aminocyclohexanecarboxylic acid	25	2.66		169	3,3-Difluoroacrylic acid	25	3.17	. . . .
112	Bromosuccinic acid	50	2.69	4.69	170	Dimethylmalonic acid	25	3.17	6.06
113	3-Hydroxy-2-naphthoic acid	25	2.71		171	Phenoxyacetic acid	25	3.17	
114	5-Aminosalicylic acid (5-Amino-2-hydroxybenzoic acid	25	2.74	5.84	172	Iodoacetic acid	25	3.18	
115	Triethylsuccinic acid	25	2.74		173	3-Iodo-2-methylolbenzoic acid	20	3.18	
116	Salicylic acid (2-Hydroxybenzoic acid) . .	30	2.75; (3.00)	12.38	174	2-Hydroxy-3-chloroisobutyric acid	25	3.20	
117	2,4-Dichlorobenzoic acid	25	2.76	. . . .	175	(4-Methoxyphenoxy)acetic acid	25	3.21	
118	1,2,3-Benzenetricarboxylic acid	25	2.80	4.20 pK₃ = 5.87	176	(4-Methylphenoxy)acetic acid	25	3.22	
					177	meso-Tartaric acid	25	3.22	4.82
119	3,4-Dinitrobenzoic acid.	25	2.82		178	2-Chlorocrotonic acid	25	3.22	
120	3,5-Dinitrobenzoic acid.	25	2.82		179	2,4-Dihydroxybenzoic acid.	30	3.22	
121	Guanidinoacetic acid	25	2.82		180	(2-Methoxyphenoxy)acetic acid	25	3.23	
122	2-Bromobenzoic acid	25	2.85		181	(2-Methylphenoxy)acetic acid	25	3.23	
123	Malonic acid	25	2.86	5.65	182	Cyclopentane-1,1-dicarboxylic acid	25	3.23	4.08
124	Chloroacetic acid	25	2.86		183	3-Bromomandelic acid	25	3.23	
125	Ethylmethylmalonic acid	25	2.86	6.43	184	3-Chloromandelic acid	25	3.24	
126	2-Iodobenzoic acid	25	2.86		185	2,6-Dimethylbenzoic acid	25	3.25	
127	2-Chloropropionic acid	18	2.88		186	3-Iodomandelic acid	25	3.26	
128	(4-Nitrophenoxy)acetic acid	25	2.89		187	2-Fluorobenzoic acid	25	3.27	
129	Bromoacetic acid	25	2.90		188	3-Chloro-2-methylbenzoic acid	25	3.27	
130	2-Chlorobenzoic acid	25	2.92		189	(4-Chloro-2-methylphenoxy)acetic acid . .	25	3.28	
131	(4-Cyanophenoxy)acetic acid	25	2.93		190	3-Bromo-2-methylbenzoic acid	20	3.28	
132	Isopropylmalonic acid.	25	2.94	5.38	191	cis-3-Chloroacrylic acid	18	3.32	
133	(3-Nitrophenoxy)acetic acid	25	2.95		192	cis-Cyclopropane-1,2-dicarboxylic acid . .	24	3.33	6.47
134	Phthalic acid	25	2.95	5.41	193	(2,6-Dimethylphenoxy)acetic acid	25	3.36	
135	2-Chloroisobutyric acid	18	2.97		194	Anthraquinone-1-carboxylic acid	20	3.37	
136	3,5-Dinitro-4-methylbenzoic acid	25	2.97		195	3-Hydroxy-3-phenylpropionic acid	18	3.40	
137	(2-Cyanophenoxy)acetic acid	25	2.97		196	d,l-Mandelic acid	25	3.41	
138	2-Bromopropionic acid	18	2.97		197	Anthraquinone-2-carboxylic acid	20	3.42	
139	Ethylmalonic acid	25	2.99	5.83	198	N-Formylglycine.	19	3.43	
140	n-Propylmalonic acid	25	2.996	5.84	199	2,4,6-Trimethylbenzoic acid	25	3.44	
141	d-Tartaric acid	25	3.00	4.34	200	3-Nitrobenzoic acid.	25	3.44	
142	Fumaric acid	25	3.02	4.38	201	Cyclohexane-1,1-diacetic acid	25	3.45	7.08
143	(3-Cyanophenoxy)acetic acid	25	3.03		202	2-Phenylbenzoic acid (Biphenyl-2-carboxylic acid)	25	3.46	
144	Benzilic acid	18	3.05		203	meso-2,3-Diphenylsuccinic acid	25	3.48	
					204	(2,4-Dinitrophenyl)acetic acid	25	3.50	. . . .
146	Methylmalonic acid	25	3.05	5.76	205	Tetramethylsuccinic acid	25	3.50	7.28
147	(2-Chlorophenoxy)acetic acid	25	3.05		206	d,l-2,3-Diethylsuccinic acid	25	3.51	6.60
148	3,3,3-Trifluoropropionic acid	25	3.06		207	2-Phenoxybenzoic acid	20	3.53	
149	3-Aminobenzoic acid	25	3.07	4.73	208	2-Hydroxy-2-phenylpropionic acid	18	3.53	
150	(3-Chlorophenoxy)acetic acid	25	3.07		210	Terephthalic acid.	25	3.54	4.46
151	2-Hydroxy-3-chlorobutyric acid	25	3.08		211	2-tert-Butylbenzoic acid	25	3.54	
152	(3-Fluorophenoxy)acetic acid	25	3.09		212	3-Aminopropionic acid	25	3.55	
153	(2-Fluorophenoxy)acetic acid	25	3.09		213	4-Cyanobenzoic acid	25	3.55	
154	(4-Chlorophenoxy)acetic acid	25	3.10		214	3-(Methylamino)benzoic acid	25	3.55	
155	(4-Bromophenoxy)acetic acid	25	3.10		215	3-Methoxy-2-methylolbenzoic acid	20	3.58	
156	α-Iodopropionic acid	18	3.11		216	d,l-2,3-Diphenylsuccinic acid	25	3.58	
157	(2-Bromophenoxy)acetic acid	25	3.12		217	2-Aminocyclohexanecarboxylic acid	25	3.59	10.21
158	3-Chlorolactic acid	25	3.12		218	3-Cyanobenzoic acid	25	3.60	
159	(4-Bromophenoxy)acetic acid	25	3.13		219	3-Ethoxy-2-methylolbenzoic acid	20	3.62	
160	(3-Iodophenoxy)acetic acid	25	3.13		220	Isophthalic acid.	25	3.62	4.60
161	(4-Fluorophenoxy)acetic acid	25	3.13	. . . .	221	3-Ethyl-3-methylglutaric acid	25	3.62	6.70

No.	Name	T,°C	pK_1	pK_2	No.	Name	T,°C	pK_1	pK_2
222	3,3-Diethylglutaric acid	25	3.62	7.12	279	cis-Tetrahydronaphthalene-2,3-dicarboxylic acid..............	20	3.98	6.47
223	meso-2,3-Diethylsuccinic acid	25	3.63	6.46	280	4-Chlorobenzoic acid	25	3.98	....
224	(2,4-Dichlorophenoxy)acetic acid	20	3.64	....	281	2,5-Dimethylbenzoic acid	25	3.98	
225	Malonamic acid (Malonic acid mono-amide) ..	25	3.64	....	282	3-Bromopropionic acid	18	3.99	
226	2-Isopropylbenzoic acid	25	3.64		283	2-Hydroxy-2-methylbutyric acid ...	18	3.99	
227	Decahydronaphthyloxyacetic acid	25	3.64		284	3-Cyanopropionic acid	25	3.99	
228	2-Cyclohexyloxypropionic acid	25	3.64		285	3-Chloropropionic acid	25	3.996	....
229	trans-3-Chloroacrylic acid	18	3.65		286	trans-Tetrahydronaphthalene-2,3-dicarboxylic acid..............	20	4.00	5.70
230	9-Anthracenecarboxylic acid	20	3.65		287	(2-Nitrophenyl)acetic acid	25	4.004	
231	Ethoxyacetic acid	18	3.65		288	3-Aminopentanoic acid	25	4.02	10.40
232	trans-Cyclopropane-1,2-dicarboxylic acid ..	24	3.65	5.13	289	4-Aminobutyric acid	25	4.03	
233	N-Acetylglycine	25	3.67		290	cis-Cyclobutane-1,3-dicarboxylic acid ...	25	4.03	5.31
234	1-Anthracenecarboxylic acid	20	3.68		291	2-Hydroxyisobutyric acid	18	4.04	
235	2-Benzyl-2-phenylsuccinic acid	20	3.69	6.49	292	(2-Iodophenyl)acetic acid	25	4.04	
236	3,3-Di-n-propylglutaric acid..........	25	3.69	7.31	293	trans-4-Nitrocinnamic acid	25	4.05	
237	Cyclopentyloxyacetic acid	25	3.70	....	294	(2-Bromophenyl)acetic acid	25	4.05	
238	3,3-Dimethylglutaric acid	25	3.70	6.29	295	(2-Chlorophenyl)acetic acid	25	4.07	
239	cis-3-Aminocyclohexanecarboxylic acid ..	15	3.70		296	2-Methoxybenzoic acid	20	4.08	
240	d,l-N-Acetylalanine.................	25	3.72		297	3-Methoxybenzoic acid	25	4.09	
241	N-Propionylglycine.................	25	3.72		298	3-Iodopropionic acid	18	4.09	
242	2,3-Dimethylbenzoic acid	25	3.74		299	cis-Cyclohexane-1,3-dicarboxylic acid ...	16	4.10	5.46
243	Formic acid	25	3.74		300	Ethylsuccinic acid	25	4.00	.....;
244	Phthalamic acid (Phthalic acid mono-amide) ..	25	3.75		301	Iminodipropionic acid	30	4.11	9.61
245	Glutaconic acid	25	3.77	5.08	302	Benzylsuccinic acid	20	4.11	5.65
246	meso-2,3-Dimethylsuccinic acid	25	3.77	5.94	303	trans-3-Nitrocinnamic acid	25	4.12	
247	2-Ethylbenzoic acid................	25	3.79	....	304	4-Fluorobenzoic acid	25	4.14	
248	3-Methylcyclopentyl-1,1-diacetic acid....	25	3.79	6.74	305	(3-Chlorophenyl)acetic acid	25	4.14	
249	trans-Cyclobutane-1,2-dicarboxylic acid..	20	3.79	5.61	306	4,4,4-Trifluorobutyric acid	25	4.15	
250	Cyclohexyloxyacetic acid	25	3.80	....	307	trans-2-Nitrocinnamic acid	25	4.15	
251	2-Hydroxybutyric acid	18	3.80		308	3-Isopropoxybenzoic acid	20	4.15	
252	Cyclopentyl-1,1-diacetic acid	25	3.80	6.77	309	2-Naphthoic acid	25	4.16	
253	3-Bromobenzoic acid...............	25	3.81		310	Succinic acid	25	4.16	5.61
254	3-Chlorobenzoic acid	25	3.82		311	(3-Iodophenyl)acetic acid	25	4.16	
255	trans-Caronic acid (trans-1,1-Dimethyl-2,3-cyclopropanedicarboxylic acid)	25	3.82	5.32	312	3-Ethoxybenzoic acid	20	4.17	
256	2,2-Diethylsuccinic acid	25	3.84		313	2,2-Diphenyladipic acid	20	4.17	5.80
257	trans-3-Aminocyclohexanecarboxylic acid	15	3.85	....	314	(4-Iodophenyl)acetic acid	25	4.18	
258	3-Iodobenzoic acid	25	3.85		315	2-Anthracenecarboxylic acid	20	4.18	
259	(4-Nitrophenyl)acetic acid	25	3.85		316	trans-Cyclohexane-1,4-dicarboxylic acid ..	16	4.18	
260	cis-3-Methylcyclohexyloxyacetic acid....	25	3.85		317	4,4,5,5,6,6,6-Heptafluorohexanoic acid...	25	4.18	
261	Lactic acid	25	3.86		318	2,4-Dimethylbenzoic acid	25	4.18	
262	cis-Cinnamic acid	25	3.88		319	trans-Cyclohexane-1,2-dicarboxylic acid .	19	4.18	5.93
263	Hydroxyacetic acid................	25	3.89		320	(4-Bromophenyl)acetic acid	25	4.19	
264	1,2,3-Cyclohexanetricarboxylic acid	23	3.89	4.85 pK_3 = 8.83	321	(4-Chlorophenyl)acetic acid	25	4.19	
265	cis-Cyclobutane-1,2-dicarboxylic acid....	19	3.90	5.89	322	Mesaconic acid	18	4.20	
266	2-Methylcyclohexyloxyacetic acid......	25	3.90	....	323	Benzoic acid....................	25	4.20	
267	3-Fluorobenzoic acid...............	25	3.90		324	3-Propoxybenzoic acid	20	4.20	
268	3-Hydroxybenzoic acid.............	30	3.90	9.78	325	2-Ethoxybenzoic acid	20	4.21	
269	2-Methylbenzoic acid..............	25	3.91		326	3,4-Diphenyladipic acid	25	4.22	5.19
270	2,2-Diphenylglutaric acid	20	3.91	5.38	327	trans-2-Chlorocinnamic acid	25	4.23	
271	3-Phenoxybenzoic acid.............	20	3.91		328	3-Fluoromandelic acid	25	4.24	
272	2-(Bromomethyl)butyric acid	18	3.92		329	(1-Naphthyl)acetic acid	25	4.24	
273	Diphenylacetic acid	25	3.94		330	(2-Isopropoxy)benzoic acid	20	4.24	
274	2,2-Dibenzylsuccinic acid	20	3.96	6.66	331	2-Propoxybenzoic acid	20	4.24	
275	Triphenylacetic acid	25	3.96		332	3-Butoxybenzoic acid	20	4.25	
276	trans-Cyclopentane-1,2-dicarboxylic acid.	25	3.96	5.85	333	(4-Fluorophenyl)acetic acid	25	4.25	
277	4-Bromobenzoic acid..............	25	3.97		334	Acrylic acid	25	4.25	
278	(3-Nitrophenyl)acetic acid	25	3.97	....	335	3-Methylglutaric acid	25	4.25	5.41
					336	(2-Naphthyl)acetic acid	25	4.26	
					337	cis-Cyclopentane-1,3-dicarboxylic acid...	25	4.26	5.51
					338	5-Aminopentanoic acid	25	4.27	....

No.	Name	T, °C	pK$_1$	pK$_2$	No.	Name	T, °C	pK$_1$	pK$_2$
339	3-Methylbenzoic acid	25	4.27	....	401	3-(2-Chlorophenyl)propionic acid	25	4.58	....
340	2,2-Diphenylpimelic acid	25	4.28	5.39	402	3-(3-Chlorophenyl)propionic acid	25	4.58	
341	3-Ethylglutaric acid	25	4.29	5.33	403	4-Methyl-3-pentenoic acid	25	4.60	
342	Angelic acid	18	4.29	....	404	cis-3-Hydroxycyclohexanecarboxylic acid	25	4.60	
343	trans-3-Chlorocinnamic acid	25	4.29		405	2-Hydroxycinnamic acid	25	4.61	
344	3,5-Dimethylbenzoic acid	25	4.30		406	4-Hydroxybenzoic acid	27.8	4.61	9.31
345	3-Isopropylglutaric acid	25	4.30	5.51	407	Levulinic acid	18	4.64	
346	3-n-Propylglutaric acid	25	4.31	5.31	408	2-Methyl-3-hydroxybutyric acid	18	4.65	
347	Phenylacetic acid	25	4.31		409	3-(3-Methoxyphenyl)propionic acid	25	4.65	
348	trans-Cyclohexane-1,3-dicarboxylic acid	19	4.31	5.73	410	4-Acetylbutyric acid	18	4.66	
349	2,2-Diphenylsuberic acid	20	4.31	5.39	411	2-Methylacrylic acid	18	4.66	
350	trans-Cyclopentane-1,3-dicarboxylic acid	25	4.32	5.42	412	3-Phenylpropionic acid	25	4.66	
351	2,2-Diphenylazelaic acid	20	4.33	5.38	413	3-(2-Methylphenyl)propionic acid	25	4.66	
352	(3,4-Dimethoxyphenyl)acetic acid	25	4.33		414	4-Pentenoic acid	25	4.67	
353	3,4,5-Trihydroxybenzoic acid	30	4.33		415	4-Ureidobutyric acid	25	4.68	
354	cis-Cyclohexane-1,2-dicarboxylic acid	20	4.34	6.77	416	3-(3-Methylphenyl)propionic acid	25	4.68	
355	4-Isopropylbenzoic acid	25	4.35		417	3-(4-Methylphenyl)propionic acid	25	4.68	
356	Vinylacetic acid	25	4.35		418	(4-Isopropoxy)benzoic acid	20	4.68	
357	Glutaric acid	25	4.35	5.42	419	trans-2-Hydroxycyclohexanecarboxylic acid	25	4.68	
358	4-Ethylbenzoic acid	25	4.35	....					
359	(4-Methoxyphenyl)acetic acid	25	4.36		420	trans-4-Hydroxycyclohexanecarboxylic acid	25	4.68	
360	4-Methylbenzoic acid	25	4.37						
361	(4-Methylphenyl)acetic acid	25	4.37		421	3-(4-Methoxyphenyl)propionic acid	25	4.69	
362	(4-Ethylphenyl)acetic acid	25	4.37		422	2-Pentenoic acid	25	4.69	
363	4-Methoxycinnamic acid	25	4.38		423	trans-Crotonic acid	25	4.69	
364	trans-Cyclohexane-1,2-dicarboxylic acid	20	4.38	5.42	424	4-Hydroxypentanoic acid	18	4.69	
365	(4-Isopropylphenyl)acetic acid	25	4.39		425	2-Hexenoic acid	25	4.70	
366	trans-4-Aminocyclohexanecarboxylic acid	25	4.39	10.55	426	4-Hexenoic acid	25	4.72	
367	3-Hydroxycinnamic acid	25	4.40		427	5-Hexenoic acid	25	4.72	
368	4-tert-Butylbenzoic acid	25	4.40		428	2-Ethylpentanoic acid	18	4.72	
369	cis-Crotonic acid	18	4.41		429	2-Ethylbutyric acid	25	4.75	
370	3,4-Dimethylbenzoic acid	25	4.41		430	4-Phenylbutyric acid	25	4.76	
371	trans-2-Bromocinnamic acid	25	4.41		431	Acetic acid	25	4.76	
372	cis-Cyclohexane-1,2-diacetic acid	20	4.42	5.45	432	Isovaleric acid	25	4.78	
373	cis-Cyclopentane-1,2-diacetic acid	20	4.42	5.42	433	4-Propoxybenzoic acid	20	4.78	
374	(4-tert-Butylphenyl)acetic acid	25	4.42		434	4,4-Dimethylpentanoic acid	18	4.79	
375	trans-Cyclopentane-1,2-diacetic acid	20	4.43	5.43	435	Cyclobutanecarboxylic acid	25	4.79	
376	trans-4-Chlorocinnamic acid	25	4.43		436	d,l-2-Methylpentanoic acid	18	4.79	
377	Adipic acid	25	4.43	5.42	437	5-Methyl-4-hexenoic acid	25	4.80	
378	4-Cyanobutyric acid	25	4.44		438	4-Methyl-2-pentenoic acid	25	4.80	
379	trans-Cinnamic acid	25	4.44		439	3-(2-Methoxyphenyl)propionic acid	25	4.80	
380	trans-3-Methylcinnamic acid	25	4.44		440	2-Methylbutyric acid	18	4.81	
381	3-(Acetylamino)propionic acid	25	4.45		441	n-Butyric acid (n-Butanoic acid)	25	4.82	
382	2-Ureidoisobutyric acid	25	4.46		442	trans-3-Hydroxycyclohexanecarboxylic acid	25	4.82	
383	4-Methoxybenzoic acid	25	4.47						
384	3-(4-Nitrophenyl)propionic acid	25	4.47		443	Nicotinic acid	25	4.82	11.98
385	Pimelic acid	25	4.48	5.42	444	Cyclopropanecarboxylic acid	25	4.83	
386	3-(1-Naphthoyl)propionic acid	20	4.48		445	cis-4-Aminocyclohexanecarboxylic acid	25	4.83	10.62
387	3-Methylcyclohexyl-1,1-diacetic acid	25	4.49	6.08	446	d,l-3-Methylpentanoic acid	18	4.84	
388	4-Methylcyclohexyl-1,1-diacetic acid	25	4.49	6.10	447	Isonicotinic acid	20	4.84	12.23
389	5,5,5-Trifluoropentanoic acid	25	4.49		448	cis-4-Hydroxycyclohexanecarboxylic acid	25	4.84	
390	3-Ureidopropionic acid	25	4.49		449	Isocaproic acid	18	4.84	
391	3-(2-Nitrophenyl)propionic acid	25	4.50		450	n-Pentanoic acid (n-Valeric acid)	25	4.84	
392	trans-2-Methylcinnamic acid	25	4.50		451	Isobutyric acid	25	4.86	
393	2-Methoxycinnamic acid	25	4.50		452	Propionic acid (Propanoic acid)	25	4.87	
394	3-Hexenoic acid	25	4.52		453	4-Hydroxyisocaproic acid	18	4.87	
395	Suberic acid	25	4.52	5.40	454	n-Hexanoic acid (n-Caproic acid)	25	4.88	
396	Azelaic acid	25	4.53	5.40	455	n-Heptanoic acid	25	4.89	
					456	n-Octanoic acid (n-Caprylic acid)	25	4.89	
398	Succinamic acid (Succinic acid mono-amide)	25	4.54		457	Cyclohexylpropionic acid	25	4.91	
					458	d,l-2,3-Dimethylsuccinic acid	25	4.94	6.20
399	trans-4-Methylcinnamic acid	20	4.56	....	459	n-Nonanoic acid (Pelargonic acid)	25	4.94	
400	O-Acetylsalicylic acid (Aspirin)	17	4.57	....	460	Cyclohexylbutyric acid	25	4.95	....

No.	Name	T,°C	pK_1	pK_2	No.	Name	T,°C	pK_1	pK_2
461	Tiglic acid	18	4.96	....	468	cis-3-Methyl-2-pentenoic acid	25	5.15	....
462	3-(2-Naphthoyl)propionic acid	20	4.96	....	469	Itaconic acid	18	5.54	....
463	Cyclopentanecarboxylic acid	25	4.99	....	470	Citraconic acid	18	6.17	....
464	2,2-Dimethylbutyric acid	18	5.03	....	471	Ethylenediamine-N,N,N′,N′-tetra-acetic acid	25	6.27	10.95
465	Trimethylacetic acid	25	5.05	....	472	Ethylenediamine-N,N′-diacetic acid	30	6.42	9.46
466	3,3-Dimethylacrylic acid (3-Methyl-crotonic acid)	25	5.12	....	473	Ethylenediamine-N,N′-dipropionic acid	30	6.87	9.60
467	trans-3-Methyl-2-pentenoic acid	25	5.13	....					

TABLE XXVIII. ACID DISSOCIATION CONSTANTS OF PHENOLS IN AQUEOUS SOLUTION

(Listed in order of increasing pKa)

No.	Name	T,°C	pK_1	pK_2
1	Picric acid (2,4,6-Trinitrophenol)	25	0.29; (0.71)	....
2	4-Hydroxypyrimidine	20	1.85	8.59
3	2-Hydroxypyrimidine	20	2.24	9.17
4	4-Chloro-2,6-dinitrophenol	25	2.97	....
5	2,6-Dinitrophenol	25	3.71	
6	2,4-Dinitrophenol	25	4.09	....
7	2,6-Dinitrohydroquinone	21	4.42	9.14
8	8-Hydroxyquinoline	20	5.017	9.813
9	2,5-Dinitrophenol	25	5.04	
10	3,4-Dinitrophenol	25	5.42	
11	8-Hydroxyquinaldine	25	5.55	10.31
12	4-Methyl-8-hydroxyquinoline	25	5.58	10.00
13	3,4-Dimethyl-8-hydroxyquinoline	25	5.80	10.05
14	5-Formyl-2-nitrophenol (3-Hydroxy-4-nitrobenzaldehyde)	25	6.00	....
15	5-Chloro-2-nitrophenol	25	6.05	....
16	5-Carboethoxy-2-nitrophenol (Ethyl 3-hydroxy-4-nitrobenzoate)	25	6.11	....
17	5-Carbomethoxy-2-nitrophenol (Methyl 3-hydroxy-4-nitrobenzoate)	25	6.15	....
18	2,4-Dimethyl-8-hydroxyquinoline	25	6.20	10.60
19	3-Nitrocatechol	25	6.68	....
20	2-Nitro-5-phenylphenol (3-Hydroxy-4-nitrobiphenyl)	25	6.74	....
21	2-Formylphenol .(Salicylaldehyde)	25	6.79	
22	5-Methoxy-2-nitrophenol	25	7.09	
23	4-Nitrophenol	25	7.16	
24	2-Nitrophenol	25	7.21	
25	2,6-Dimethyl-4-nitrophenol	25	7.22	
26	5-Methyl-2-nitrophenol	25	7.25	
27	2,6-Dichlorohydroquinone	25	7.30	9.99
28	Vanillin (4-Formyl-2-methoxyphenol; 4-Hydroxy-3-methoxybenzaldehyde)	25	7.396	....
29	2-Nitrohydroquinone	21	7.63	10.06
30	4-Formylphenol (4-Hydroxybenzaldehyde)	25	7.66	
31	4-(Methylsulfonyl)phenol	25	7.83	
33	o-Vanillin (2-Formyl-6-methoxyphenol; 2-Hydroxy-3-methoxybenzaldehyde)	25	7.91	
34	4-Cyanophenol (4-Hydroxybenzonitrile)	25	7.95	
35	3-Formylphenol (3-Hydroxybenzaldehyde)	25	8.00	
36	4-Acetylphenol	25	8.05	
37	3,5-Dimethyl-4-(methylsulfonyl)phenol	25	8.13	
38	4-Cyano-3,5-dimethylphenol (2,6-Dimethyl-4-hydroxybenzonitrile)	25	8.21	
39	3,5-Dimethyl-4-nitrophenol	25	8.25	
40	4-Cyano-2,6-dimethylphenol (3,5-Dimethyl-4-hydroxybenzonitrile)	25	8.27	
41	2-Carboxamidophenol (Salicylamide)	20	8.37	
42	3-Nitrophenol	25	8.38	
43	4-Carbobenzyloxyphenol (Benzyl 4-hydroxybenzoate)	25	8.41	
44	2-Bromophenol	25	8.42	....
45	Phloroglucinol (1,3,5-Trihydroxybenzene)	25	8.45 (7.0)	8.88
46	2-Iodophenol	25	8.46	
47	4-Carbobutoxyphenol (n-Butyl 4-hydroxybenzoate)	25	8.47	
48	4-Carbomethoxyphenol (Methyl 4-hydroxybenzoate)	25	8.47	
49	2-Chlorophenol	25	8.48	
50	4-Carboethoxyphenol (Ethyl 4-hydroxybenzoate)	25	8.50	

No.	Name	T,°C	pK_1	pK_2
51	2-Hydroxy-3-methoxybenzylamine	25	8.70	10.52
52	3-Hydroxypyridine	20	8.72	....
53	2-Fluorophenol	25	8.82	
54	Isovanillin (5-Formyl-2-methoxyphenol; 3-Hydroxy-4-methoxybenzaldehyde)	25	8.889	....
55	3-Hydroxy-2-methoxybenzylamine	25	8.89	10.52
56	4-Hydroxy-3-methoxybenzylamine	25	8.94	10.42
57	3-Nitro-2,4,6-trimethylphenol	25	8.98	
58	Sodium 4-hydroxybenzenesulfonate	25	9.01	
59	Pyrogallol (1,2,3-Trihydroxybenzene)	25	9.01	11.64
60	2-Methylhydroquinone (Toluhydroquinone)	25	9.05	11.62
61	3-Chlorophenol	25	9.08; (9.02)	....
62	3-Bromophenol	25	9.11	....
63	3-Iodophenol	25	9.17	
64	3-Acetylphenol	25	9.19	
65	4-Iodophenol	25	9.20	
66	Sodium 3-hydroxybenzenesulfonate	25	9.29	
67	1-Naphthol	20.5	9.30; (9.85)	....
68	3-(Methylsulfonyl)phenol	25	9.33	
69	4-Bromophenol	25	9.34	
70	3-Fluorophenol	25	9.36; (9.28)	
71	1,4-Naphthohydroquinone (1,4-Dihydroxynaphthalene)	26.5	9.37	10.93
72	4-Chlorophenol	25	9.38; (9.42)	....
73	Sodium 4-hydroxybenzoate	20	9.39	
74	Resorcinol (1,3-Dihydroxybenzene)	30	9.44	11.32
75	Catechol (1,2-Dihydroxybenzene)	30	9.48	12.08
76	4-Phenylphenol (4-Hydroxybiphenyl)	22.5	9.51	
77	3-(Methylthio)phenol	25	9.53	
78	4-(Methylthio)phenol	25	9.53	
79	2,4,6-Trimethylphenol	25	9.56	
80	2-Naphthol	19.5	9.57; (9.93)	
81	3-Phenylphenol (3-Hydroxybiphenyl)	22.5	9.64	
82	3-Methoxyphenol	25	9.65	
83	2,6-Dimethylolphenol	25	9.66	
84	2-Aminophenol (2-Hydroxyaniline)	28	9.71	
85	2,4-Dimethylphenol	25	9.77	
86	4-Methylolphenol	25	9.82	
87	3-Methylolphenol	25	9.83	
88	3-Aminophenol (3-Hydroxyaniline)	21.5	9.87	9.92
89	Sodium 4-hydroxybenzenephosphonate	25	9.90	
90	3-Ethylphenol	28	9.90	
91	2,6-Dimethylol-4-methylphenol	25	9.92	
92	2-Methylolphenol	25	9.92	
93	4-Fluorophenol	25	9.92	
94	Sodium 3-hydroxybenzoate	20	9.94	
95	2-Methoxyphenol	25	9.98	
96	Phenol	25	9.99; (9.95)	
97	Eugenol (4-Allyl-2-methoxyphenol)	25	10.00	
98	2-Phenylphenol (2-Hydroxybiphenyl)	22.5	10.01	....
99	4-Ethylphenol	28	10.01	
100	3-Methylphenol (m-Cresol)	25	10.09	
101	3-Ethyl-5-methylphenol	28	10.10	
102	3,5-Dimethylphenol	25	10.15	
103	4-Methyl-2-methylolphenol	25	10.15	
104	4-Methoxyphenol	25	10.20	
105	Sodium 3-hydroxybenzenephosphonate	25	10.2	

TABLE XXVIII. ACID DISSOCIATION CONSTANTS OF PHENOLS
IN AQUEOUS SOLUTION
(Listed in order of increasing pKa)

No.	Name	T, °C	pK_1	pK_2	No	Name	T, °C	pK_1	pK_2
106	2-Ethylphenol	28	10.2	. . . .	115	2,6-Dimethylphenol	25	10.59	. . . .
107	2,5-Dimethylphenol	24	10.22	. . . .	116	2,4,6-Trimethylphenol	25	10.88;	. . . .
108	4-Methylphenol (*p*-Cresol)	25	10.26	. . . .				(10.99)	
109	2-Methylphenol (*o*-Cresol)	25	10.28	. . . .	117	Hydroquinone (1,4-Dihydroxybenzene)	25	10.85;	11.39
110	4-Indanol	25	10.32	. . . .				(9.96)	
111	3,4-Dimethylphenol	25	10.32	. . . .	118	4-Hydroxypyridine	20	11.09	. . . .
112	2,4,5-Trimethylphenol	25	10.45	. . . .	119	Tetramethylhydroquinone (Durohydro-	25	11.51	. . . .
113	2,4-Dimethylphenol	25	10.45	. . . .		quinone)			
114	2,3-Dimethylphenol	25	10.50	. . . .	120	2-Hydroxypyridine	20	11.62	. . . .

TABLE XXIX. DISSOCIATION CONSTANTS OF ORGANIC BASES IN AQUEOUS SOLUTION

(Listed in order of increasing pKa)

(The data relates to the acid dissociation constant (Ka) of the conjugate acid (BH$^+$) of the listed base (B))

No.	Name	T,°C	pK$_1$	pK$_2$	No.	Name	T,°C	pK$_1$	pK$_2$
1	2,4,6-Trinitroaniline	25	−9.41		59	2-Amino-6-nitronaphthalene	25	2.62	
2	2,4-Dinitroaniline	25	−4.53		60	N,N-Dimethyl-3-nitroaniline	25	2.63	
3	2,6-Dinitro-4-methylaniline	25	−3.96		61	2-Chloroaniline	25	2.65	
4	2,4-Dichloro-6-nitroaniline	25	−3.61		62	2-Amino-4-cyanonaphthalene	25	2.66	
5	2,6-Dichloro-4-nitroaniline	20	−2.55		63	1-Amino-3-bromonaphthalene	25	2.67	
6	N-Cyanodiethylamine	25	−2.0		64	3-Bromoquinoline	25	2.69	
7	1-Amino-2-nitronaphthalene	25	−1.74		65	1-Amino-3-chloronaphthalene	25	2.69	
8	4-Chloro-2-nitroaniline	25	−1.02		66	5-Nitroquinoline	20	2.69	
9	Dicyanomethyl ethyl amine	25	−0.6		67	6-Nitroquinoline	20	2.72	
10	2-Fluoropyridine	25	−0.44		68	1-Amino-5-nitronaphthalene	25	2.73	
11	Pyrrole	25	−0.27; (−3.8)		69	2-Amino-8-nitronaphthalene	25	2.73	
					70	3-Cyanoaniline	25	2.748	
12	2-Nitroaniline	25	−0.26		71	1-Amino-8-nitronaphthalene	23	2.79	
13	Bis (cyanomethyl) amine	25	0.2		72	1-Amino-3-iodonaphthalene	25	2.82	
14	2-Chloropyridine	25	0.49; (0.72)		73	3-Bromopyridine	25	2.84	
					74	3-Chloropyridine	25	2.84	
15	1-Amino-4-nitronaphthalene	20	0.54		75	1-Amino-6-nitronaphthalene	25	2.89	
16	Quinoxaline	20	0.56		76	4-Methyl-3-nitroaniline	25	2.96	
17	N,N-Dimethyl-4-nitroaniline	25	0.607		77	3-Cyano-N,N-dimethylaniline	25	2.969	
18	Pyrazine	27	0.65	−5.78	78	3-Fluoropyridine	25	2.97	
19	Diphenylamine	25	0.79		79	2-Amino-5-nitronaphthalene	25	3.01	
20	3-Nitropyridine	25	0.81		80	2-Methoxypyridine	25	3.06	
21	2-Bromopyridine	25	0.90		81	2-Amino-7-nitronaphthalene	25	3.10	
22	2-Cyanoaniline	25	0.95		82	1,3-Dimethylpyrazole	25	3.11	
23	2,6-Dimethyl-4-nitroaniline	22	0.98		83	2-Methoxyquinoline	20	3.16	
24	4-Nitroaniline	25	1.00		84	3-Acetylpyridine	25	3.18	
25	N-Chlorodiethylamine	25	1.02		85	4-Hydroxypyridine	20	3.20	11.12
26	2-Methyl-4-nitroaniline	24.5	1.04		86	2-Fluoroaniline	25	3.20	
27	2-Bromoquinoline	25	1.05		87	1-Amino-4-bromonaphthalene	25	3.21	
28	Tris (2-cyanoethyl) amine	25	1.1		88	3-Iodopyridine	25	3.25	
29	Pyrimidine	20	1.23; (1.31)		89	1-Amino-3-methoxynaphthalene	25	3.26	
					90	1-Amino-3-hydroxynaphthalene	25	3.30	
30	2-Amino-3-nitronaphthalene	25	1.48		91	1-Amino-5-chloronaphthalene	25	3.34	
31	3-Methyl-4-nitroaniline	25	1.50		92	2-Amino-4-chloronaphthalene	25	3.38	
32	2,5-Dichloroaniline	22	1.57		93	4-Aminofluorene	25	3.39	
33	4-Cyanoaniline	25	1.74		94	2-Amino-4-bromonaphthalene	25	3.40	
34	4-Cyano-N,N-dimethyl aniline	25	1.78		95	2-Amino-4-iodonaphthalene	25	3.41	
35	2-Iodopyridine	25	1.82		96	1-Amino-7-chloronaphthalene	25	3.48	
36	2,3-Dimethyl-4-nitroaniline	25	1.96		97	1-Amino-6-chloronaphthalene	25	3.48	
37	2,4-Dichloroaniline	25	2.00		98	Quinazoline	20	3.49	
38	1-Amino-3-nitronaphthalene	25	2.07		99	3-Chloroaniline	25	3.52	
39	3-Methoxy-5-nitroaniline	25	2.11		100	4-Bromo-2,6-dimethylaniline	25	3.54	
40	4-Aminobenzophenone (4-Benzoylaniline)	25	2.17		101	3-Methylpyrazole	25	3.56	
41	4-Aminoacetophenone (4-Acetylaniline)	25	2.19; (2.75)		102	3-Aminoacetophenone (3-Acetylaniline)	25	3.56	
					103	3-Fluoroaniline	25	3.57	
42	2-Aminoacetophenone (2-Acetylaniline)	25	2.22		104	4-Bromo-2-methylaniline	25	3.58	
43	Pyridazine	20	2.24		105	3-Bromoaniline	25	3.58	
44	1-Amino-3-cyanonaphthalene	25	2.26		106	2-(Methylthio)pyridine	20	3.59	
45	Cinnoline	20	2.27		107	3-Iodoaniline	25	3.61	
46	1,2,4-Triazole	25	2.30		108	5-Bromoquinoline	25	3.62	
47	2,4-Dibromoaniline	15	2.3		109	1-(Methylamino)naphthalene	27	3.67	
48	2,6-Dibromoaniline	25	2.34		110	2-Amino-7-chloronaphthalene	23	3.71	
49	7-Nitroquinoline	20	2.40		111	4-Bromopyridine	20	3.78	
50	2-Amino-4-nitronaphthalene	25	2.43		112	4-Iodoaniline	25	3.78	
51	Thiazole	20	2.44		113	2-Aminobiphenyl	22	3.82	
52	3-Nitroaniline	25	2.466		114	3,5-Dimethoxyaniline	25	3.82	
53	Pyrazole	25	2.48		115	4-Chloropyridine	20	3.84	
54	2-Bromoaniline	25	2.53		116	4-Bromoaniline	25	3.86	
55	1-Amino-7-nitronaphthalene	25	2.55		117	1-Aminofluorene	25	3.87	
56	8-Nitroquinoline	20	2.55		118	1-Amino-6-methoxynaphthalene	25	3.90	
57	3,5-Dimethyl-4-nitroaniline	25	2.59		119	1-Aminonaphthalene	25	3.92	
58	2-Iodoaniline	25	2.60		120	2,6-Dimethylaniline	25	3.95	

TABLE XXIX. DISSOCIATION CONSTANTS OF ORGANIC BASES IN AQUEOUS SOLUTION
(Listed in order of increasing pKa) (Continued)
(The data relates to the acid dissociation constant (Ka) of the conjugate acid (BH$^+$) of the listed base (B))

No.	Name	T,°C	pK$_1$	pK$_2$	No.	Name	T,°C	pK$_1$	pK$_2$
121	8-Aminoquinoline	20	3.95	......	183	8-Methylquinoline	25	4.91	......
122	1-Amino-5-hydroxynaphthalene	25	3.96	......	184	3,5-Dimethylaniline	25	4.91	
123	1-Amino-3-methylnaphthalene	25	3.96	......	185	3-Aminoquinoline	20	4.91	......
124	1-Amino-6-hydroxynaphthalene	25	3.97	......	186	3-Methoxypyridine	25	4.91	
125	4-Chloroaniline	25	4.00	......	187	4-tert-Butylaniline	25	4.95	
126	3-(Methylthio)aniline	25	4.00	......	188	3,5-Di-tert-butylaniline	25	4.97	
127	4-Iodopyridine	20	4.02	......	189	2-Vinylpyridine	25	4.98	
128	2-Amino-4-methoxynaphthalene	25	4.05	......	190	1,3-Diaminobenzene (m-Phenylenediamine)	25	4.98	2.41
129	2-Amino-5-hydroxynaphthalene	25	4.07	......	191	N,3-Dimethylaniline	21	5.00	
130	1-Amino-7-methoxynaphthalene	25	4.07	.	192	2-tert-Butylaniline	25	5.03	
131	3-Bromo-4-methoxyaniline	23	4.08	......	193	6-Methoxyquinoline	20	5.03	
132	1-Aminoanthracene	25	4.1	......	194	N,N-Dimethylaniline	25	5.068	
133	2-Aminonaphthalene	25	4.16	......	195	4-Methylaniline	25	5.08	......
134	N-Allylaniline	25	4.17	......	196	3-Hydroxypyridine	25	5.10	8.60
135	3-Ethoxyaniline	25	4.17	......	197	N-Ethylaniline	24	5.12	
136	1-(Ethylamino)naphthalene	25	4.18	......	198	N-n-Butylaniline	25	5.12	
137	2-Amino-7-methoxynaphthalene	25	4.19	......	199	3,5,6-Trimethylaniline	20.5	5.12	......
138	1-Amino-7-hydroxynaphthalene	25	4.20	......	200	2-Benzylpyridine	25	5.13	
139	3-Methoxyaniline	25	4.20	......	201	6-Methylquinoline	25	5.15	
140	4-Aminobiphenyl	29	4.22	......	202	2-Amino-5,6,7,8-tetrahydronaphthalene	17	5.17	
141	4-Bromo-N,N-dimethylaniline	25	4.232	......	203	3,4-Dimethylaniline	25	5.17	
142	3-Aminobiphenyl	17	4.25	......	204	4-Ethoxyaniline	28	5.20	
143	2-Amino-7-hydroxynaphthalene	25	4.25	......	205	4-Methylquinoline	25	5.20; (5.59)	
144	2,3,5,6-Tetramethylaniline	25	4.30	......					
145	2-Bromo-N,N-dimethylaniline	25	4.31		206	Pyridine	25	5.25	
146	4-(Methylthio)aniline	25	4.35	......	207	Bis(2-cyanoethyl)amine	25	5.26	
147	2-n-Propylaniline	25	4.36	......	208	7-Methylquinoline	25	5.29	
148	Tris(2-chloroethyl)amine	25	4.37	......	209	N-Cyclopentylaniline	25	5.30	
149	2-Ethylaniline	25	4.37	......	210	5-Aminoindane	16	5.31	
150	2,4,6-Trimethylaniline	25	4.37	......	211	4-Methoxyaniline	25	5.31	
151	3,5-Dimethylpyrazole	25	4.38	......	212	Aminoacetonitrile	25	5.34	
152	3-(Methylthio)pyridine	20	4.42	......	213	N,N,3-Trimethylaniline	25	5.344	
153	2-Isopropylaniline	25	4.42	......	214	N,4-Dimethylaniline	23	5.36	
154	N-Isobutylaniline	25	4.43	......	215	Isoquinoline	20	5.42	
155	2-Ethoxyaniline	28	4.43	......	216	N-n-Hexylaniline	19	5.42	
156	2-Methylaniline	25	4.44	......	217	5-Aminoquinoline	20	5.42	
157	1-Amino-5,6,7,8-tetrahydronaphthalene	16	4.47	......	218	2,2,2-Trichloroethylamine	20	5.47	
158	Phenanthridine	20	4.48	......	219	Benzimidazole	25	5.53	
159	2-Methoxyaniline	25	4.52	......	220	1-Methylbenzimidazole	25	5.54	
160	2,5-Dimethylaniline	25	4.53	......	221	3-Ethylpyridine	25	5.56	
161	1-(Cyanomethyl)pyridine	25	4.55	......	222	1-Ethylbenzimidazole	25	5.59	
162	(Cyanomethyl) diethyl amine	25	4.55	......	223	6-Aminoquinoline	20	5.59	
163	Bis(2-Cyanoethyl) ethyl amine	25	4.55	......	224	N-Cyclohexylaniline	25	5.60	
164	2-(Dimethylamino)naphthalene	25	4.566	......	225	4-Vinylpyridine	25	5.62	
165	Aniline	25	4.603	......	226	N,N,4-Trimethylaniline	25	5.627	
166	5-Methylquinoline	25	4.62	......	227	3-Methylpyridine	25	5.63	
167	N,2-Dimethylaniline	23	4.62	......	228	4-Methylbenzimidazole	25	5.65	
168	2-Aminofluorene	25	4.64	......	229	N,N-Di-n-propylaniline	23	5.68	
169	2-Amino-6-methoxynaphthalene	25	4.64	......	230	2-Methylquinoline	25	5.69	
170	4-Fluoroaniline	25	4.65	......	231	1-Isopropylbenzimidazole	25	5.71	
171	3-tert-Butylaniline	25	4.66	......	232	3-Isopropylpyridine	25	5.72	
172	3-Isopropylaniline	25	4.67	......	233	2-tert-Butylpyridine	25	5.76	
173	2,3-Dimethylaniline	25	4.70	......	234	N-Isopropylaniline	25	5.77	
174	3-Methylaniline	25	4.70	......	235	5-Methylbenzimidazole	25	5.78	
175	3-Ethylaniline	25	4.70	......	236	3-tert-Butylpyridine	25	5.82	
176	1,2-Diaminobenzene (o-Phenylenediamine)	25	4.74	0.6	237	2-Isopropylpyridine	25	5.83	
177	N-n-Propylaniline	25	4.79	......	238	4-Ethylpyridine	25	5.87	
178	Quinoline	25	4.81	......	239	2-Ethylpyridine	25	5.89	
179	3-Aminofluorene	25	4.82	......	240	2-Methylpyridine	25	5.94	
180	1-(Dimethylamino)naphthalene	28	4.83	......	241	4-(Methylthio)pyridine	20	5.94	
181	N-Methylaniline	25	4.848	......	242	2-Hexylpyridine	25	5.95	......
182	2,4-Dimethylaniline	25	4.89	......					

**TABLE XXIX. DISSOCIATION CONSTANTS OF
ORGANIC BASES IN AQUEOUS SOLUTION**
(Listed in order of increasing pKa) (Continued)
(The data relates to the acid dissociation constant (Ka) of the conjugate acid (BH$^+$) of the listed base (B))

No.	Name	T,°C	pK$_1$	pK$_2$	No.	Name	T,°C	pK$_1$	pK$_2$
243	2-*n*-Propylpyridine	25	5.97	......	305	Bis(2-hydroxyethyl)amine	25	8.88	......
244	4-*tert*-Butylpyridine	25	5.99	......	306	N,N-Dimethylbenzylamine	25	8.91	......
245	2-Pentylpyridine	25	6.00	......	307	3-Bromopropylamine	21	8.93	......
246	4-Isopropylpyridine	25	6.02	......	308	1,2-Bis(dimethylamino)ethane	30	8.97	5.85
247	4-Methylpyridine	25	6.03	......	309	1-(Aminoethyl)benzene	25	9.08	......
248	3-Aminopyridine	25	6.03	......	310	2-Acetoxyethylamine	25	9.1	......
249	N,N,2-Trimethylaniline	25	6.11	......	311	4-Aminopyridine	25	9.114	......
250	N,2,6-Trimethylaniline	25	6.12	......	312	4-Aminoquinoline	20	9.13	......
					313	(3-Cyanopropyl) diethyl amine	25	9.13	......
252	2-Methylbenzimidazole	25	6.19	......	314	3-Methoxybenzylamine	25	9.15	......
253	1,4-Diaminobenzene (*p*-Phenylenediamine)	25	6.2	2.67	315	2-Methylbenzylamine	25	9.19	......
254	2-Isopropylbenzimidazole	25	6.21	......	316	1-Aminoindane	22.5	9.21	......
255	2-Ethylbenzimidazole	25	6.27	......	317	*trans*-1-Amino-2-hydroxycyclopentane	25	9.28	......
256	N,N-Di-*n*-butylaniline	19	6.30	......	318	Diallylamine	25	9.29	......
257	2,5-Dimethylpyridine	25	6.40	......	319	Dimethyl (2-hydroxyethyl) amine	20	9.31	......
258	2-Phenylimidazole	25	6.40	......	320	3-Methylbenzylamine	25	9.33	......
259	Bis(2-chloroethyl)methyl amine	25	6.43	......	321	Benzylamine	25	9.35; (9.62)	......
260	4-Methoxyquinoline	30	6.45	......					
261	3,4-Dimethylpyridine	25	6.46	......	322	4-Methylbenzylamine	25	9.36	......
262	4-Methoxypyridine	25	6.47	......	323	3,4-Dimethoxybenzylamine	25	9.39	......
263	Bis(2-chloroethyl) ethyl amine	25	6.55	......	324	2,3-Dimethoxybenzylamine	25	9.41	......
264	1,2-Dimethylbenzimidazole	25	6.55	......	325	2-(Phenethylamino)ethylamine	25	9.44	6.59
265	2,3-Dimethylpyridine	25	6.57	......	326	N,N-Diethylbenzylamine	25	9.44	......
266	2,6-Dimethylpyridine	25	6.60	......	327	4-Methoxybenzylamine	25	9.47	......
267	7-Aminoquinoline	20	6.61	......	328	2-(Methylthio)ethylamine	20	9.49	......
268	N,N-Diethylaniline	22	6.61	......	329	2-Hydroxyethylamine	25	9.498	......
269	4-Ethoxypyridine	20	6.67	......	330	2-(Dimethylamino)ethylamine	25	9.53	6.63
270	2-Aminopyridine	25	6.71	......	331	N-Methylbenzylamine	25	9.54	......
271	2,4-Dimethylpyridine	25	6.77	......	332	1,2-Bis(diethylamino)ethane	25	9.55	6.18
272	Imidazole	25	6.95	......	333	2-Aminoindane	21.5	9.57	......
273	N-*tert*-Butylaniline	25	7.00	......	334	N-Propylbenzylamine	25	9.58	......
274	(2-Cyanoethyl) dimethyl amine	29	7.0	......	335	1,2,3-Triaminopropane	20	9.59	7.95; pK$_3$= 3.72
275	2-Benzyl-2-pyrroline	25	7.06	......					
276	2-Aminoquinoline	20	7.30	......					
277	N,N-Di-isopropylaniline	25	7.37	......	336	2-Methoxyethylamine	20	9.61	......
278	N-Methylmorpholine	25	7.38	......	337	5-Bromo-*n*-pentylamine	21	9.62	......
279	2,4,6-Trimethylpyridine	25	7.43	......	338	*trans*-1-Amino-2-hydroxycyclohexane	25	9.63	......
280	4-Methylimidazole	25	7.518	......	339	1-Amino-1,2,3,4-tetrahydronaphthalene	20	9.63	......
281	1,3-Triazine	25	7.6	......	340	N-Ethylbenzylamine	25	9.64	......
282	N-Ethylmorpholine	25	7.67	......	341	3,3,3-Trichloro-*n*-propylamine	20	9.65	......
283	2-Cyanoethylamine	29	7.7	......	342	1-Allylpiperidine	25	9.65	......
284	Tris(2-hydroxyethyl)amine	25	7.762	......	343	2-Methoxybenzylamine	25	9.70	......
285	2-Methylimidazole	25	7.85	......	344	*cis*-1-Amino-2-hydroxycyclopentane	25	9.70	......
286	N-Methylaziridine	25	7.86	......	345	2-(Furfurylamino)ethylamine	20	9.72	6.20
287	N-*n*-Butylaziridine	25	7.86	......	346	*cis*-1-Amino-2-hydroxycyclohexane	25	9.72	......
288	2-Ethyl-2-pyrroline	25	7.87	......	347	Piperazine	25	9.81	5.55
289	2,3,5,6-Tetramethylpyridine	20	7.90	......	348	Trimethylamine	25	9.81	......
290	2-Cyclohexyl-2-pyrroline	25	7.91	......	349	Phenethylamine ((2-Aminoethyl)benzene)	25	9.84	......
291	N-Ethylaziridine	24	7.93	......	350	Diethyl 2-hydroxyethyl amine	20	9.87	......
292	Aziridine	25	8.01	......	351	1-Methyl-3-pyrroline	25	9.88	......
293	N,2-Diethylaziridine	25	8.18	......	352	1,2-Diaminoethane (1,2-Ethylenediamine)	25	9.928	6.848
294	2-Ethylaziridine	25	8.29	......	353	*cis*-1,2-Diaminocyclohexane	20	9.93	6.13
295	Triallylamine	25	8.31	......	354	2-Amino-1,2,3,4-tetrahydronaphthalene	17	9.93	......
296	Morpholine	25	8.33	......	355	Tri-*n*-butylamine	25	9.93	......
297	2,4-Dimethylimidazole	25	8.36	......	356	4,4,4-Trichloro-*n*-butylamine	20	9.93	......
298	2-Bromoethylamine	24	8.49	......	357	*trans*-1,2-Diaminocyclohexane	20	9.94	6.47
299	Bis(2-hydroxyethyl) methyl amine	25	8.52	......	358	3-Hydroxypropylamine	25	9.96	......
300	1,2-Bis(furfurylamino)ethane	20	8.61	5.74	359	*meso*-2,3-Diaminobutane	25	9.97	6.92
301	2,2-Dimethylaziridine	25	8.64	......	360	*d,l*-2,3-Diaminobutane	25	10.00	6.91
302	*trans*-2,3-Dimethylaziridine	24	8.69	......	361	1,2-Diaminopropane (1,2-Propylenediamine)	25	10.00	7.13
303	*cis*-2,3-Dimethylaziridine	23	8.72	......					
304	(2-Chloroethyl) diethyl amine	25	8.80	......	362	*cis*-Neobornylamine	25	10.01	......

TABLE XXIX. DISSOCIATION CONSTANTS OF
ORGANIC BASES IN AQUEOUS SOLUTION
(Listed in order of increasing pKa) (Continued)
(The data relates to the acid dissociation constant (Ka) of the conjugate acid (BH⁺) of the listed base (B))

No.	Name	T,°C	pK$_1$	pK$_2$	No.	Name	T,°C	pK$_1$	pK$_2$
363	2-(Diethylamino)ethylamine	25	10.02	7.07	417	n-Pentylamine	25	10.63	......
364	1-Methylpiperidine	25	10.08	......	418	n-Undecylamine (n-Hendecylamine)	25	10.63	......
365	Dimethyl isobutyl amine	20	10.08	......	419	n-Nonylamine	25	10.64	
366	5,5,5-Trichloro-n-pentylamine	20	10.12	......	420	Acridine	25	10.65	
367	2-(Methylamino)ethylamine	20	10.15	6.86	421	n-Octylamine	25	10.65	
368	cis-1,2,6-Trimethylpiperidine	30	10.15	......	422	Methylamine	25	10.657	
369	2,2-Dimethyl-n-propylamine	25	10.15	......	423	1-Ethyl-2-methylpiperidine	25	10.66	
370	1,2-Bis(methylamino)ethane	25	10.16	7.40	424	n-Heptylamine	25	10.66	
371	Dimethyl propyl amine	20	10.16	......	425	2-Aminoheptane	25	10.67	
372	Dimethyl ethyl amine	20	10.16	......	426	Cyclohexylamine	25	10.68	
373	trans-Bornylamine	25	10.17	......	427	tert-Butylamine	25	10.68	
374	1,2,2,4-Tetramethylpiperidine	30	10.18	......	428	tert-Butyl dimethyl amine	20	10.69	
375	n-Butyl dimethyl amine	25	10.19	......	429	n-Propylamine	25	10.69	
376	1,2-Dimethylpyrrolidine	26	10.20	......	430	1,2,2,6-Tetramethylpiperidine	30	10.70	
377	1,2-Dimethylpiperidine	25	10.22	......	431	Ethylamine	25	10.70	
378	1,5-Diaminopentane	25	10.25	9.13	432	Isobutylamine	25	10.72	
379	Tri-n-propylamine	25	10.26	......	433	3-(Trimethylsilyl)-n-propylamine	25	10.73	
380	1,2-Bis(propylamino)ethane	25	10.27	7.53	434	Dimethylamine	25	10.73	
381	2-(Butylamino)ethylamine	25	10.30	7.53	435	Triethylamine	25	10.75	
382	1,3-Diaminopropane (1,3-Propylenediamine)	25	10.30	8.29	436	2-Cyclohexylpyrrolidine	25	10.76	
					437	1,4-Diaminobutane	20	10.80	9.35
383	2-Benzylpyrrolidine	25	10.31		438	Di-isobutylamine	21	10.91	
384	Tri-isobutylamine	25	10.32		439	1,6-Diaminohexane	25	10.93	9.83
385	N-Methylpyrrolidine	25	10.32	......	440	Di-isoamylamine	27.8	10.94	
386	2-(propylamino)ethylamine	25	10.34	7.54	441	Quinuclidine	25	10.95	
387	4-Hydroxy-n-butylamine	20	10.35		442	2-Methylpiperidine	25	10.95	
388	1,2-Bis(isopropylamino)ethane	25	10.40	7.59	443	2-(Trimethylsilyl)ethylamine	25	10.97	
389	1-n-Propylpiperidine	26.5	10.41		444	Di-n-tridecylamine	25	11.00	
390	3-Aminopentane	25	10.42		445	Di-n-octadecylamine	25	11.00	
391	3-Aminocyclohexene	25	10.42		446	1,8-Diaminooctane	20	11.00	10.1
392	1-n-Butylpiperidine	26	10.43		447	Di-n-propylamine	25	11.00	
393	1-Ethylpiperidine	23	10.45		448	Di-n-dodecylamine	25	11.00	
394	1,2-Bis(ethylamino)ethane	25	10.46	7.70	449	Di-n-pentadecylamine	25	11.00	
395	Diethyl methyl amine	20	10.46		450	Di-n-hexylamine	25	11.01	
396	5-Hydroxy-1-pentylamine	23	10.46		451	Di-n-octylamine	25	11.01	
397	Dimethyl isopropyl amine	20	10.47		452	2,2,4-Trimethylpiperidine	30	11.04	
398	trans-1-Amino-4-methylcyclohexane	25	10.48		453	Cyclohexyl methyl amine	25	11.04	
399	(Aminomethyl)cyclohexane	25	10.49		454	Diethylamine	25	11.04	
400	cis-1-Amino-2-methylcyclohexane	25	10.49	......	455	2,2,6,6-Tetramethylpiperidine	25	11.07	
401	2-Aminooctane	25	10.49		456	Azepine	25	11.07	
402	trans-1-Amino-2-methylcyclohexane	25	10.51		457	cis-2,6-Dimethylpiperidine	25	11.07	
403	cis-1-Amino-3-methylcyclohexane	25	10.56	......	458	3-Methylpiperidine	25	11.07	
404	2-(Ethylamino)ethylamine	25	10.56	7.63	459	Piperidine	25	11.123	
405	Dimethyl sec-butyl amine	20	10.57	......	460	Di-isopropylamine	21	11.13	
406	6-Hydroxy-n-hexylamine	21	10.60		461	Di-n-pentylamine	26	11.16	
407	6-Bromo-n-hexylamine	21	10.60		462	2,2,6-Trimethylpiperidine	30	11.21	
408	1-Aminoheptadecane	25	10.60		463	(tert-Butylamino)cyclohexane	25	11.23	
409	1-Aminodocosane	25	10.60		464	Di-n-butylamine	25	11.25	
410	1-Aminooctadecane	25	10.60		465	1,2,2,4,4-Pentamethylpiperidine	25	11.25	
411	1-Aminopentadecane	25	10.61		466	Pyrrolidine	25	11.27	
412	n-Butylamine	25	10.61		467	Azetidine	25	11.29	
413	1-Aminohexadecane	25	10.61	......	468	Isopropylamine	25	11.54; (10.63)	
414	trans-1-Amino-3-methylcyclohexane	25	10.61	......					
415	1-Aminotetradecane	25	10.62	......	469	1,2-Dimethyl-2-pyrroline	25	11.90	......
416	2-(Isopropylamino)ethylamine	25	10.62	7.70	470	Acetamidine (Methylamidine)	25	12.40	......

INFRARED CORRELATION CHART No. 1

Prepared from information supplied by Beckman Instruments

WAVENUMBERS IN KAYSERS

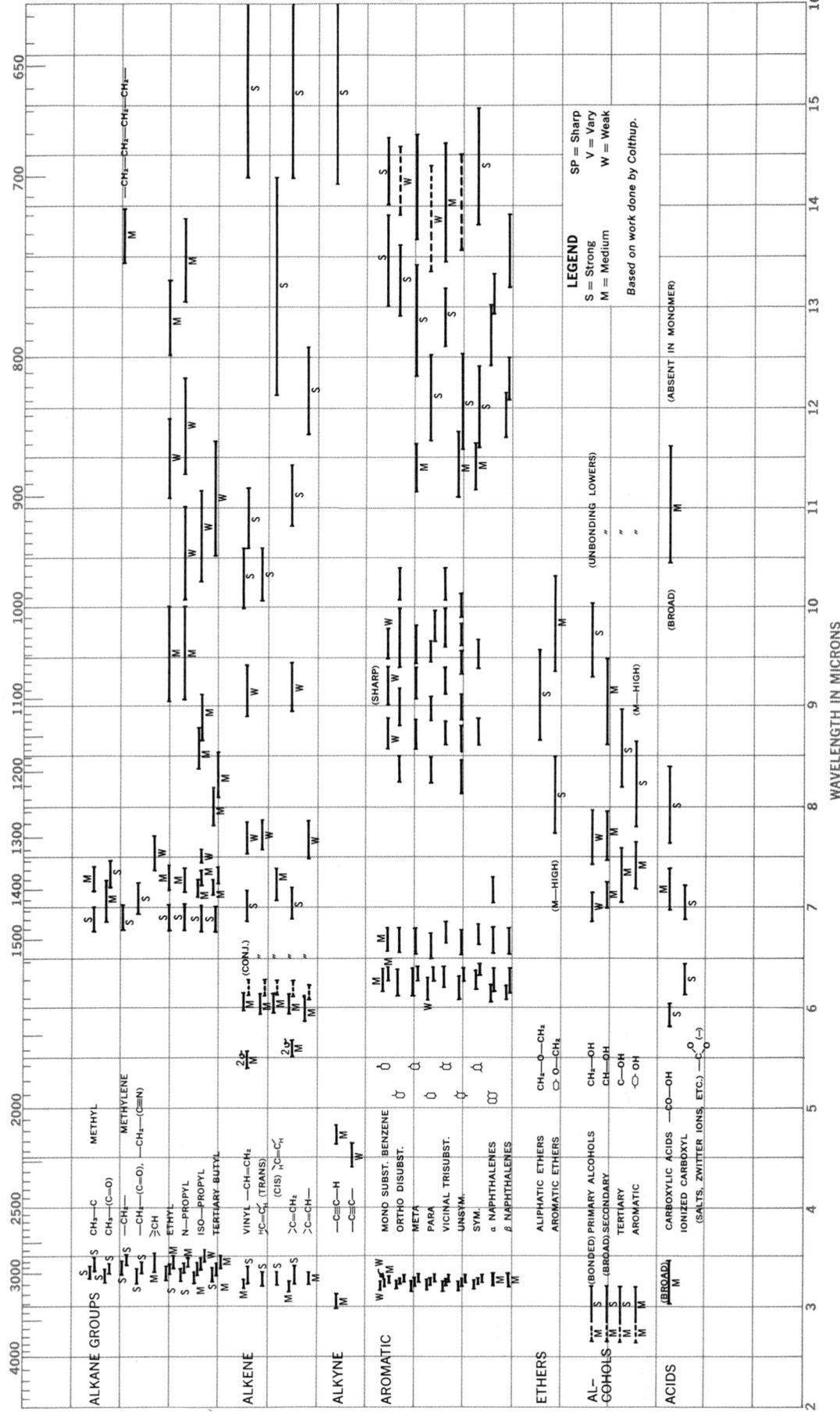

WAVELENGTH IN MICRONS

441

INFRARED CORRELATION CHART No. 1 (Con't.)

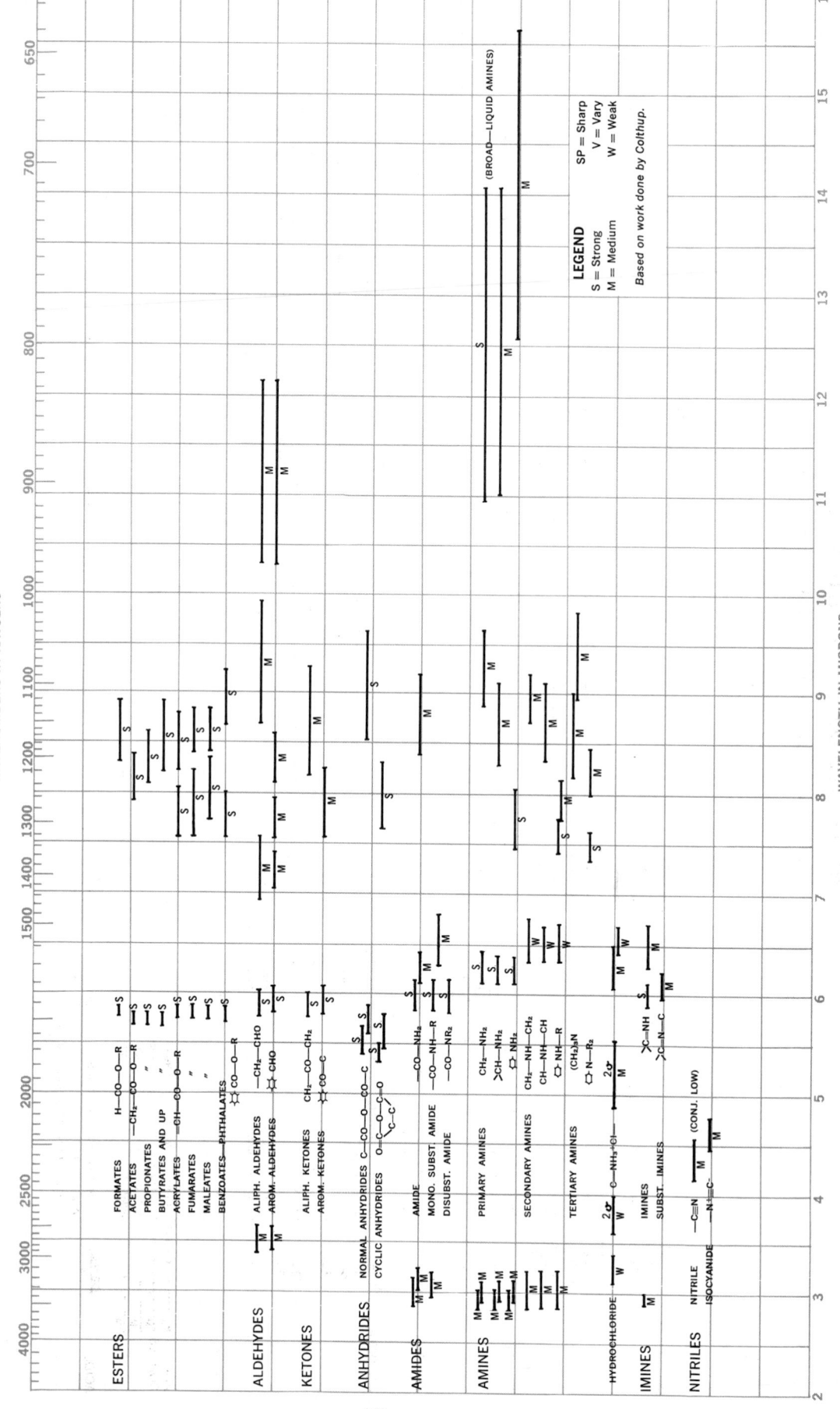

WAVENUMBERS IN KAYSERS

WAVELENGTH IN MICRONS

LEGEND

S = Strong   V = Vary
M = Medium   W = Weak
SP = Sharp

Based on work done by Colthup.

INFRARED CORRELATION CHART No. 1 (Con't.)

WAVENUMBERS IN KAYSERS

WAVELENGTH IN MICRONS

MISCELLANEOUS

X=C=X (ISOCYANATES)
1—2 DIENOID ETC.)
STRAINED RING C=O (β LACTAMS)
CHLOROCARBONATE C=O
ACID CHLORIDE C=O
EPOXY RING
O C—C
CH₂—O—(SI, P, OR S)
CH₂—S—CH₂

SULFUR GROUPS — SH, PH
PHOSPHORUS — PH
SILICON — SiH, Si—CH₃, Si—C
FLUORINE — CF₂ AND CF₃, C—F (UNSAT.), C—F (SAT.)
CHLORINE — CCl₂ AND CCl₃, CCl (ALIPH.)
BROMINE — CBr₂ AND CBr₃, CBr (ALIPH.)

C=S
P=O
Si ; CH₃
P₂S

SULFUR OXYGEN COMPOUNDS

IONIC SULFATE (SO₄)⁻⁻
IONIC SULFONATE R—SO₃
SULFONIC ACID R—SO₃H
COVALENT SULFATE R—O—SO₂—O—R
COVALENT SULFONATE R—O—SO₂—R
SULFONAMIDE R—SO₂—NH₂
SULFONE R—SO₂—R
SULFOXIDE R—SO—R

PHOSPHORUS OXYGEN

IONIC PHOSPHATE (PO₄)⁻⁻⁻
COVALENT PHOSPHATE ≡(R—O)₃P→O
CH₂—O—P→O
O=O—P→O

CARBON OXYGEN

IONIC CARBONATE (CO₃)⁻⁻
COVALENT CARBONATE O=C (O—R)₂
IMINO CARBONATE HN=C (O—R)₂

NITROGEN OXYGEN

IONIC NITRATE (NO₃)
COVALENT NITRATE R—O—NO₂
NITRO R—NO₂ UNCONJ.
CONJ.
COVALENT NITRITE R—O—NO
NITROSO R—NO
AMMONIUM (NH₄)⁺

OH AND NH STR.
CH STR.
C≡X STR.
C=O STR.
C=N STR.
C=C STR.
NH BEND
CH BEND
OH BEND
C—O STR.
C—N STR.
C—C STR.
CH ROCK
NH ROCK

ASSIGNMENTS

LEGEND

SP = Sharp
V = Vary
W = Weak
S = Strong
M = Medium

Based on work done by Colthup.

443

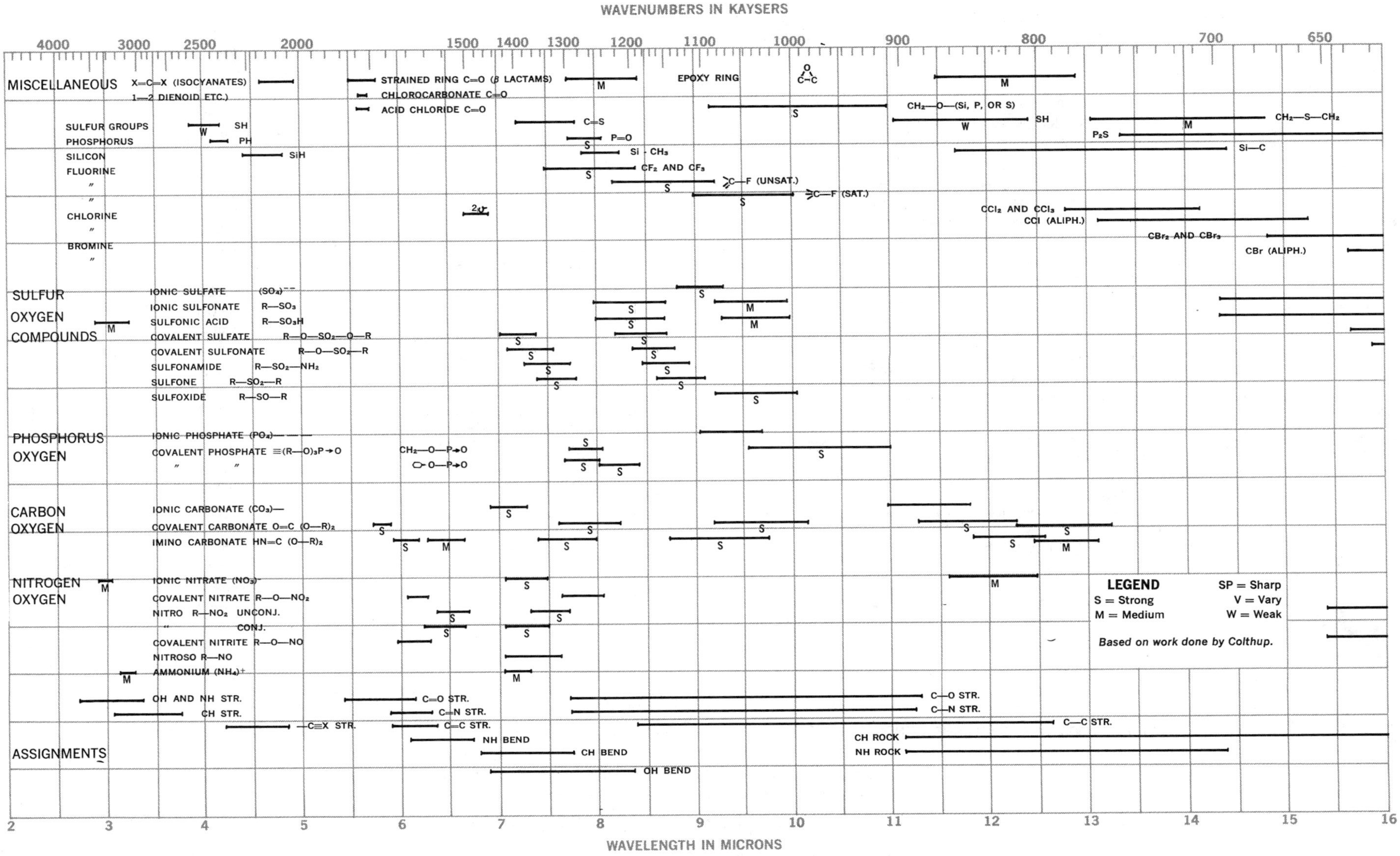

INFRARED CORRELATION CHART No. 2

Prepared from information supplied by Beckman Instruments

This chart presents some information regarding structure, double-bond vibrations, hydrogen stretching and triple-bond vibrations.

HYDROGEN STRETCHING AND TRIPLE-BOND VIBRATIONS, 3750-2000 CM.⁻¹

DOUBLE-BOND VIBRATIONS, ETC. 2000-1500 CM.⁻¹

WAVENUMBERS IN KAYSERS

WAVELENGTH IN MICRONS

LEGEND
S = Strong
M = Medium
SP = Sharp
V = Vary
W = Weak

Based on work done by Bellamy.

Left-axis group labels (hydrogen stretching/triple-bond region):
CH_3, CH_2, CH, $CH=CH$, $CH=CH_2$, $>C=CH_2$, $CH\equiv CR$, $RC\equiv CR^1$, CHO, AROMATIC CH, $COOH$, OH, $C=O$, NH_2, NH, $=NH$, $CONH_2$, $CONH$ (open-chain), $CONH$ (lactams), NH_3^+RCOOH, $NH_4^+CLRCOOH$, $NH_2\,RCOO—X^+$, PYRIDINE, Etc., AZIDES, $C\equiv N$, $\alpha\beta$—UNSATURATED NITRILES, $C\equiv N—$, $P—H$, NH_4^+, CN^-, OCN^-, Etc., $P—OH$, SH, $Si\,H$

Right-axis group labels (double-bond region):
ALLENES, $—CH=CH_2$, $>C=C$, $C=C$, $C=C$ (conjugated), AROMATIC (all types of substitution except as below), AROMATIC (para- and unsymmetrical trisubstitution), AROMATIC (vicinal trisubstitution), AROMATIC (conjugated), AROMATIC (all types), $C=N$, $C=N$ (conjugated or cyclic), $N=N$, $CONH_2$ (see chart 4 for CO), $CONH$ (open-chains only), NH_2, NH, NH_3^+, NH_4^+CLR, COO^-, $C—NO_2$, $O—NO_2$, $N—NO_2$, $—O—N=O$, PYRIDINE, Etc., PYRIMIDINE, Etc., TROPOLONES

Inline annotations: CHARACTERISTIC BAND PATTERNS FOR VARIOUS TYPES OF SUBSTITUTION; 2 BANDS(S); (MONOMER); POLYMERS (S); FREE (V); SINGLE BRIDGE (V); W/BROAD DIMER; CHELATES (W); FREE 2 BANDS (M); BONDED (S); FREE BAND (M); FREE (M); BONDED (S); BONDED (M); FREE (S); BONDED (M); FREE; (W) MOST; M CONTINUOUS BAND SERIES; BONDED (S); FREE (S); FREE (S); BONDED (S)

444

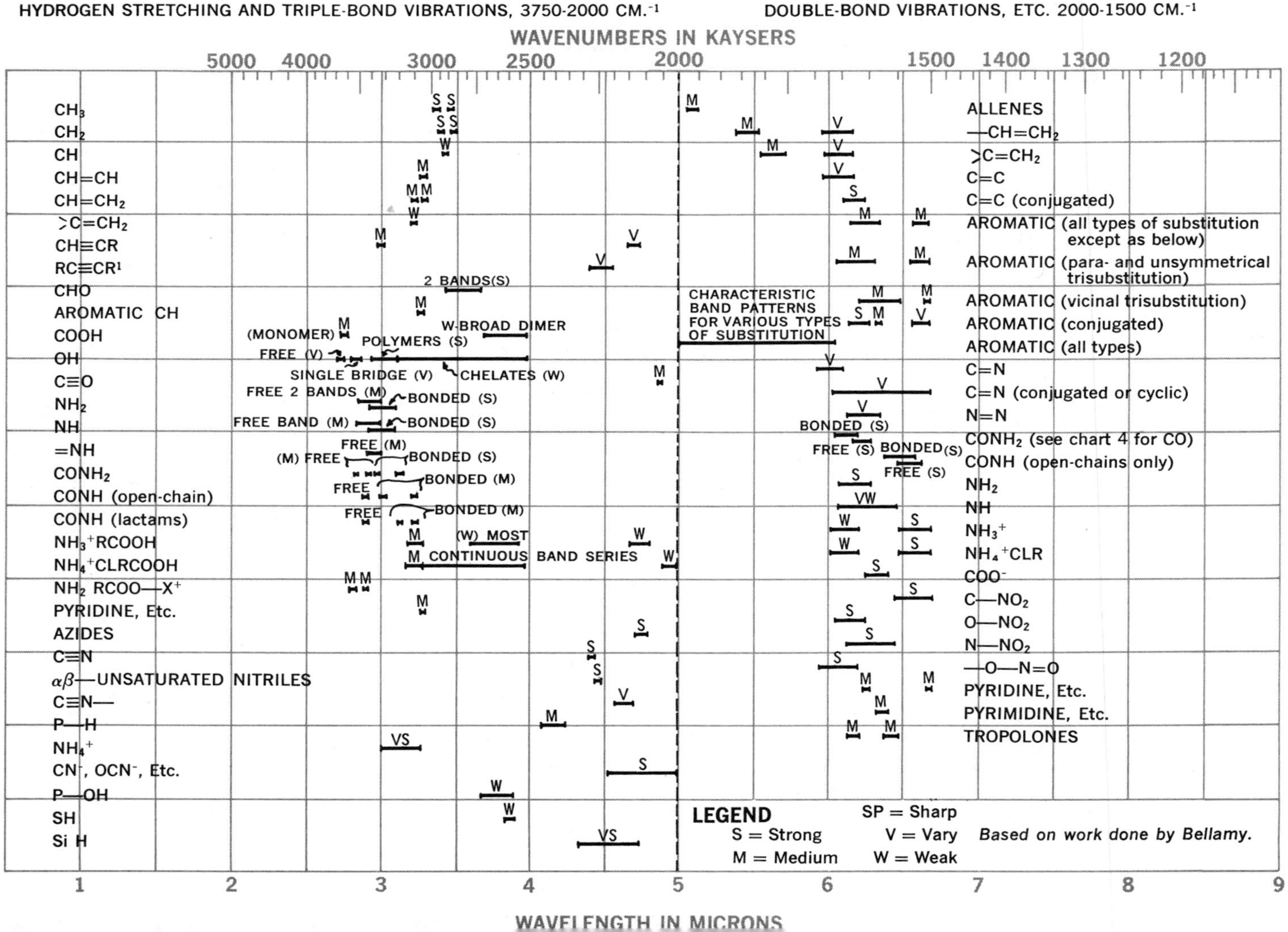

INFRARED CORRELATION CHART No. 3

Prepared from information supplied by Beckman Instruments

This chart presents some correlations between structure and the carbonyl vibrations of some classes of organic compounds. In all cases the absorption bands are strong and fall within the range of 1900-1500 cm^{-1}.

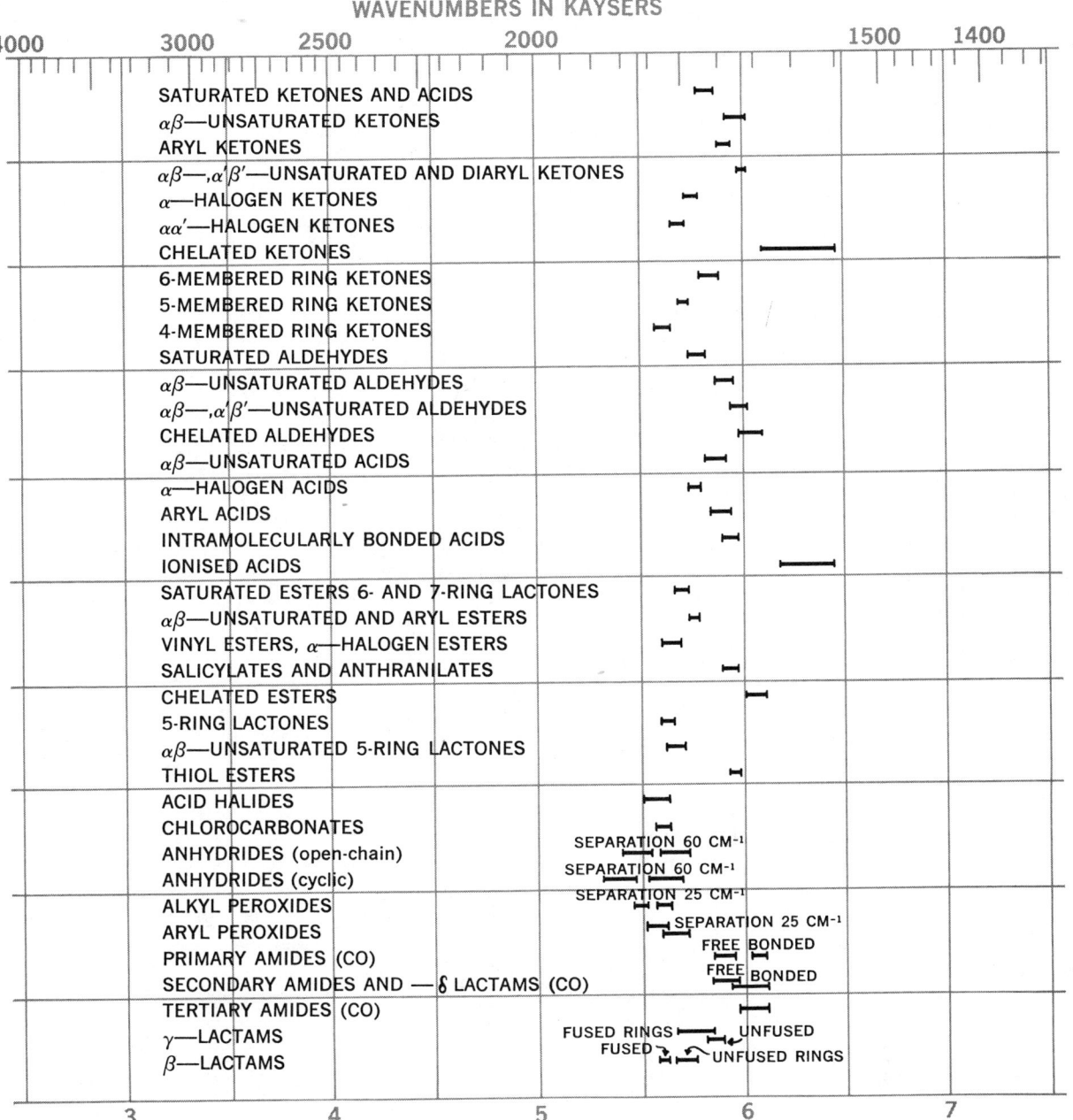

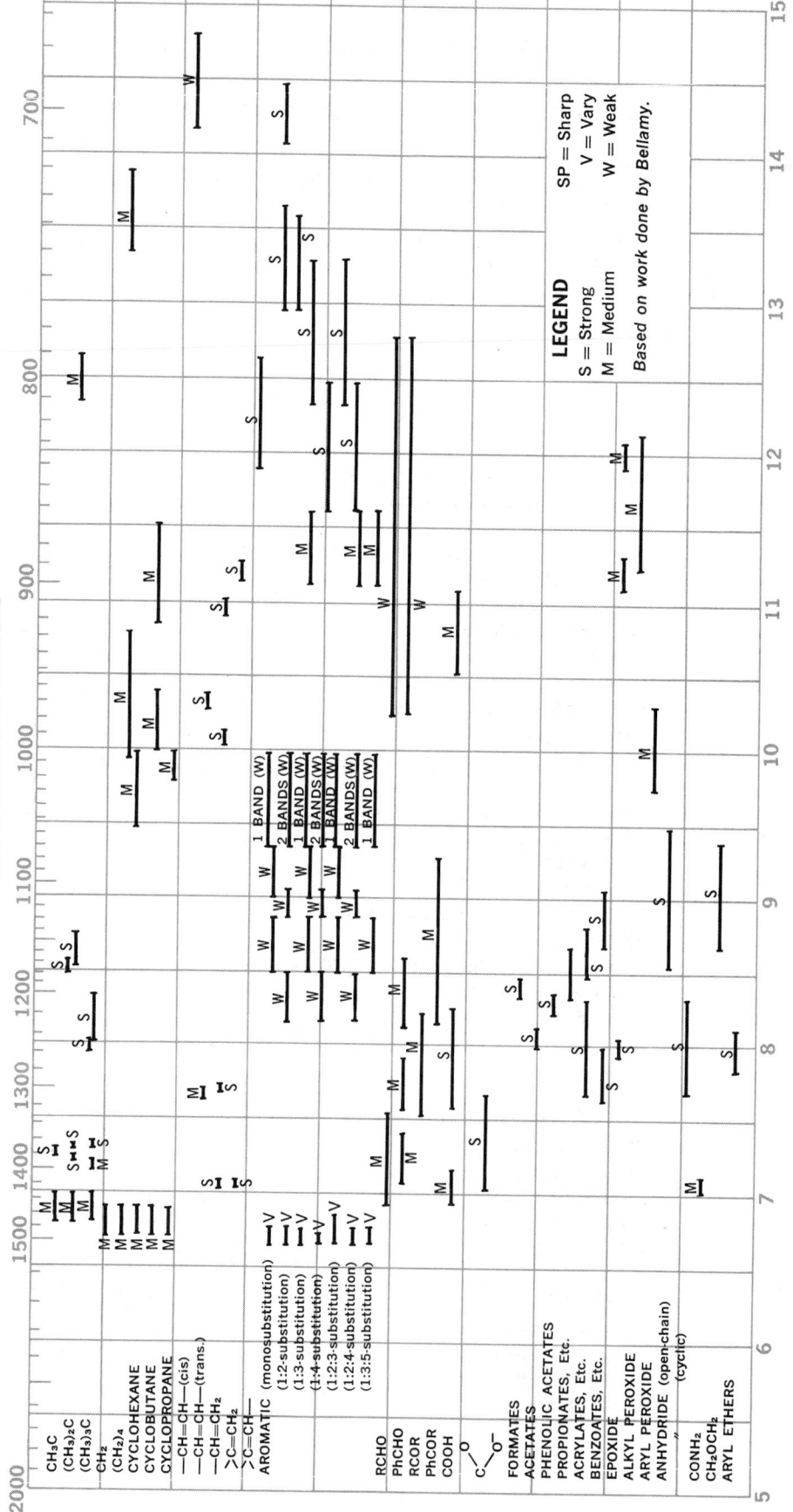

INFRARED CORRELATION CHART No. 4

Prepared from information supplied by Beckman Instruments

This chart presents some correlations between structure and single-bond vibrations for a number of classes of compounds having absorption between 1500-650 cm⁻¹.

LEGEND

SP = Sharp   V = Vary
S = Strong   W = Weak
M = Medium

Based on work done by Bellamy.

WAVENUMBERS IN KAYSERS

WAVELENGTH IN MICRONS

INFRARED CORRELATION CHART No. 4 (Con't.)

Prepared from information supplied by Beckman Instruments

This chart presents some correlations between structure and single-bond vibrations for a number of classes of compounds having absorption between 1500-650 cm⁻¹.

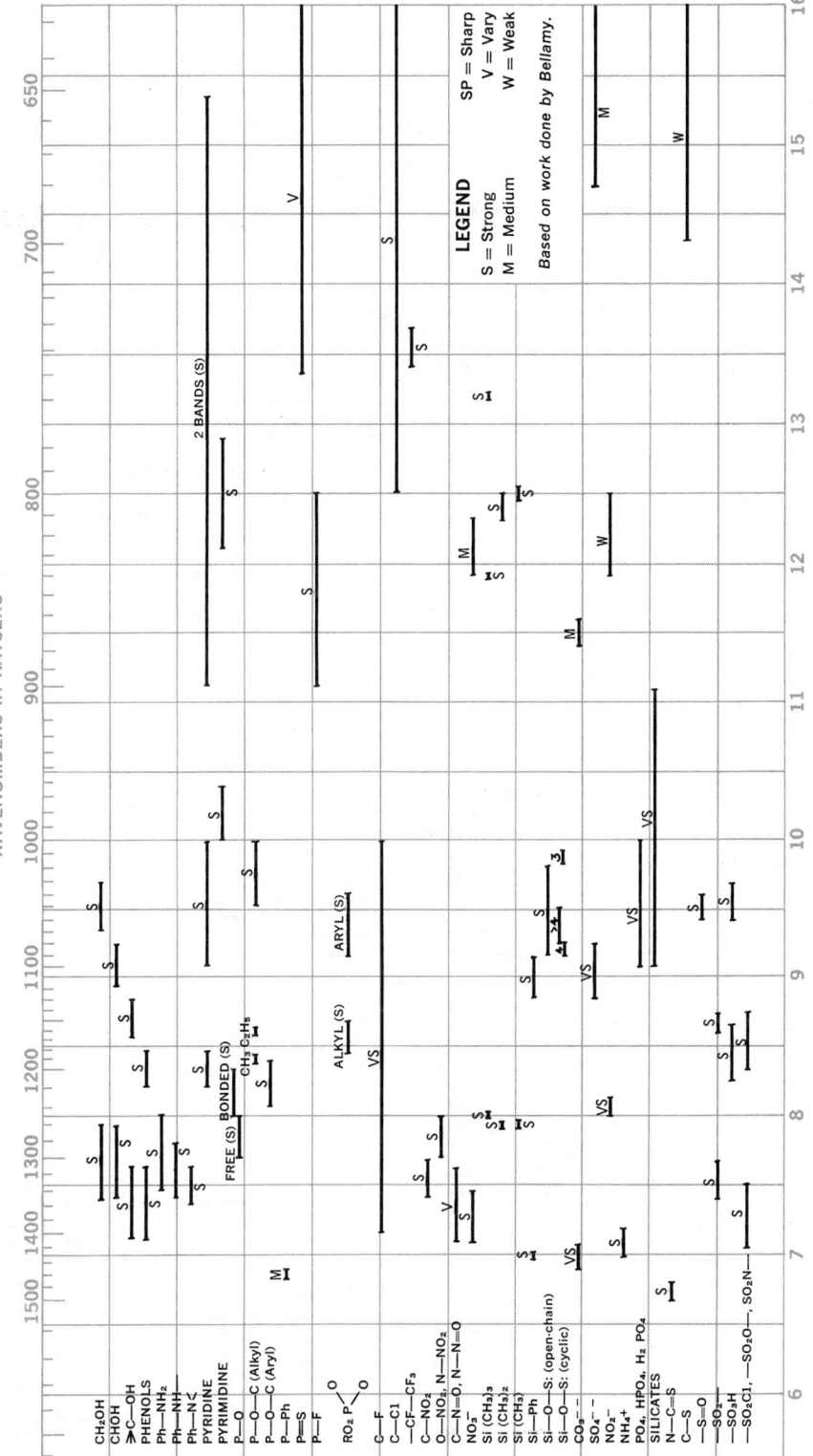

FAR INFRARED VIBRATIONAL FREQUENCY CORRELATION CHART

Based on evidence compiled by James E. Stewart of Beckman Instruments.
This chart shows the vibrational frequency correlation in the far infrared region.
Because research is continuing in the far infrared region, this chart is not
all-inclusive.

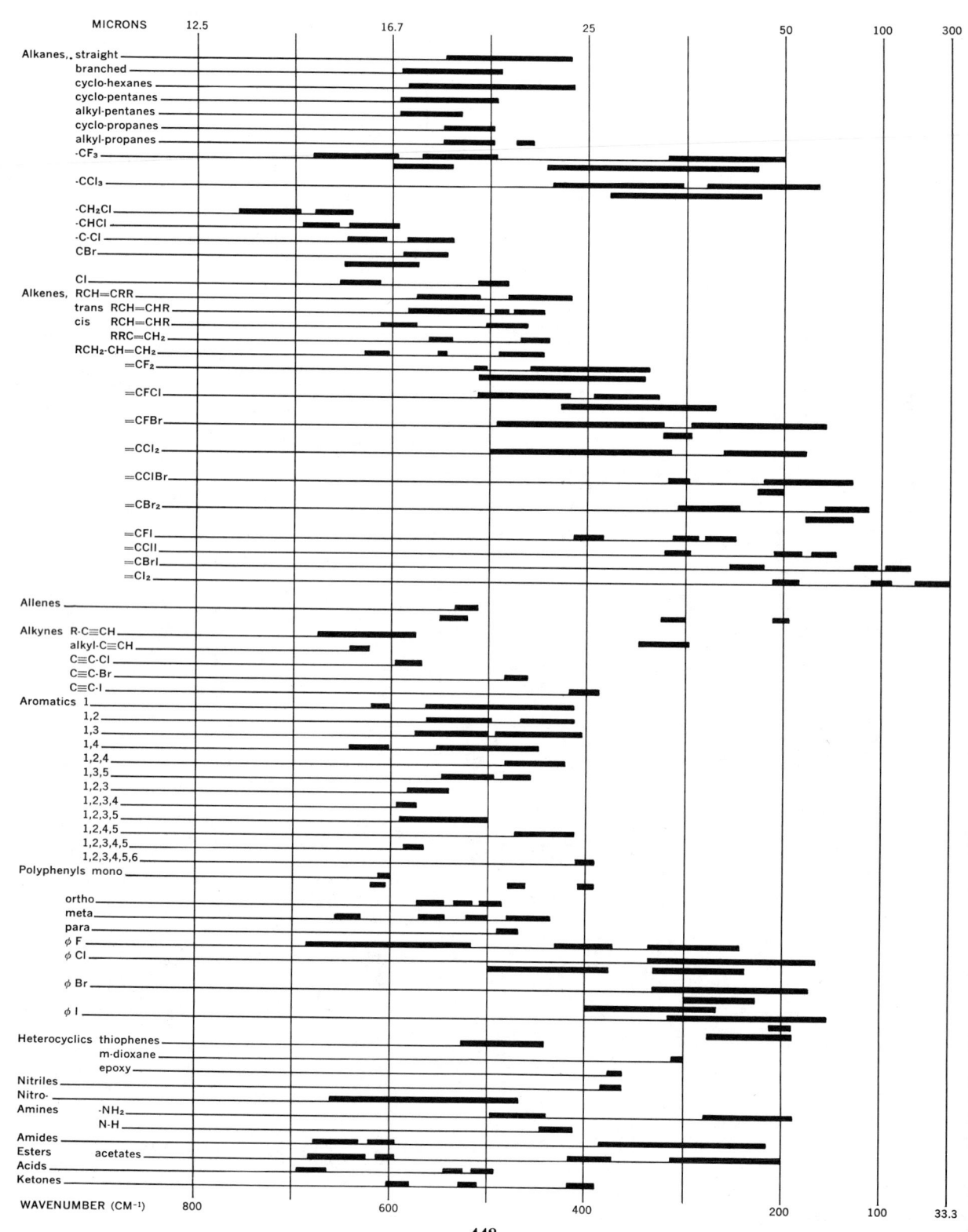

FAR INFRARED VIBRATIONAL FREQUENCY CORRELATION CHART (Con't.)

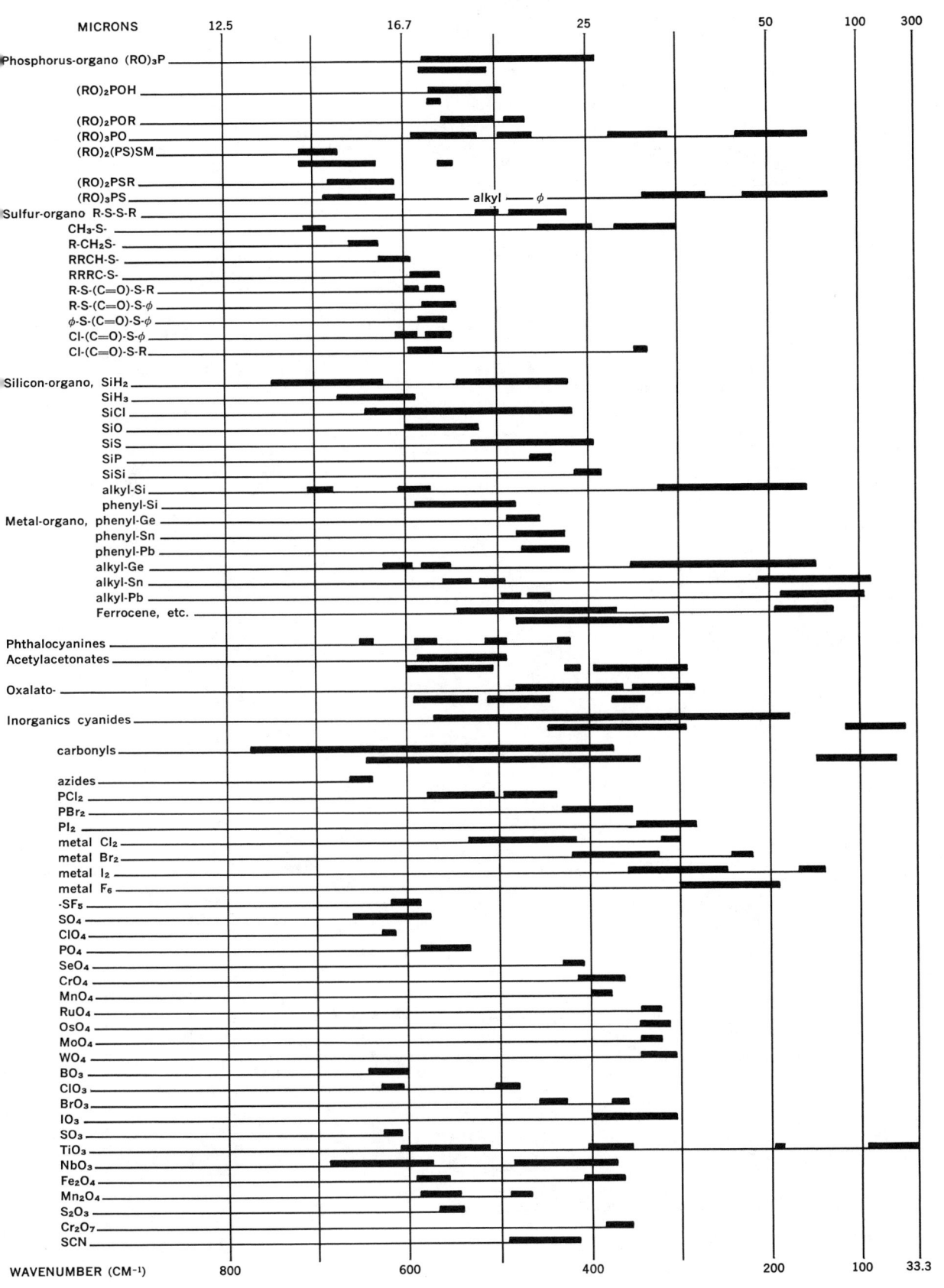

CHARACTERISTIC NMR SPECTRAL POSITIONS
FOR HYDROGEN IN ORGANIC STRUCTURES

By permission from Erno Mohacsi, J. of Chemical Education, 41, 38 (1964)

This table is useful for quick qualitative determination of proton spectrum lines by providing a tabulation of line positions obtained using tetramethylsilane as an internal reference. The listing has been kept as simple as possible for this purpose. The proton spectrum lines are arranged according to the chemical shift relative to tetramethylsilane and are given in values of τ and σ. The purpose of this table is to supplement tables available in standard references and to summarize information available in the literature.

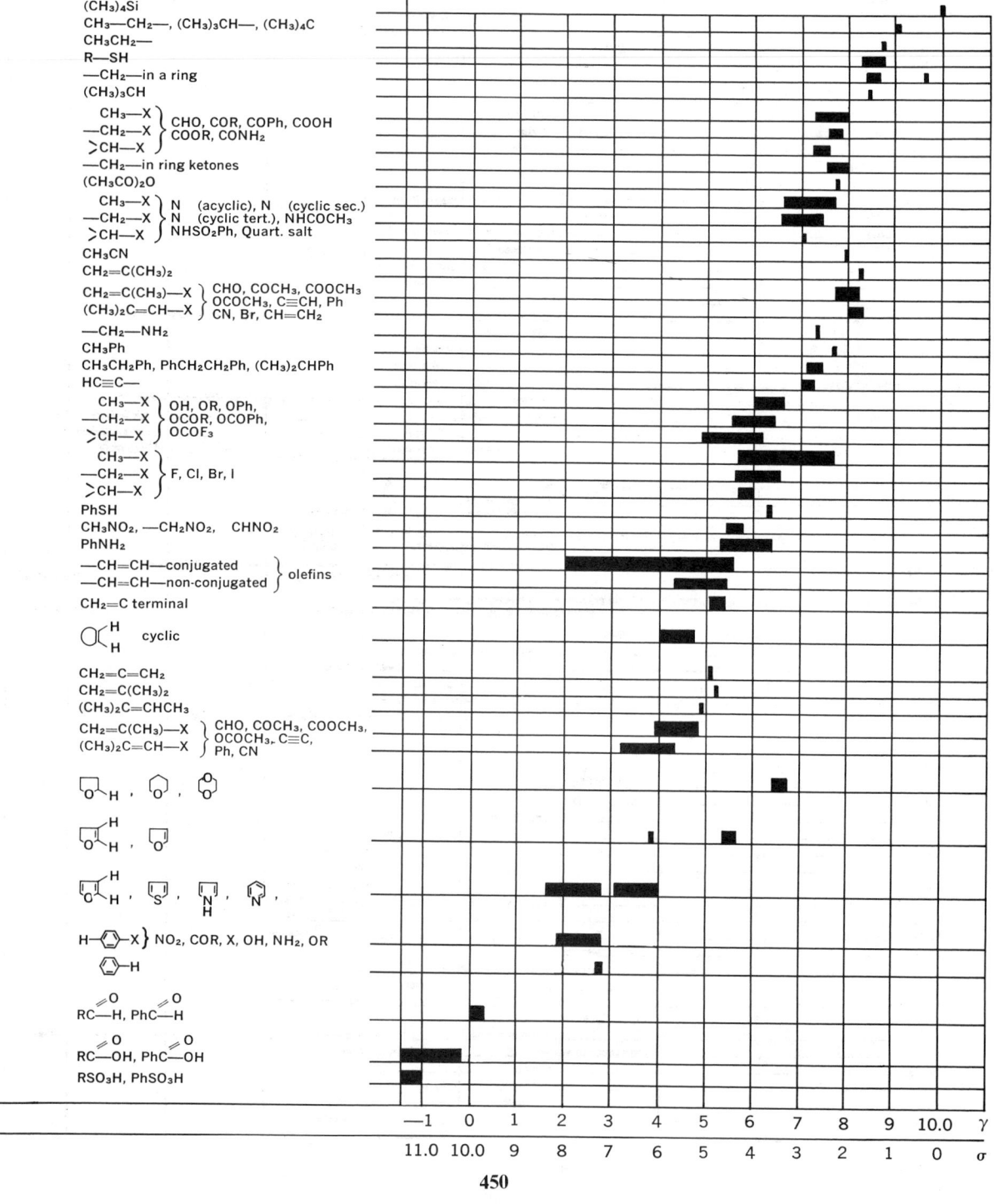

MISCIBILITY OF ORGANIC SOLVENT PAIRS

Table A

Doctor J. S. Drury

Industrial and Engineering Chemistry   Vol. 44, No. 11, Nov. 1952

(Reprinted by permission)

The classifications were made by shaking together 5 ml. of each of the solvents listed in a test tube for 1 minute, then allowing the mixture to settle. If no interfacial meniscus was observed, the solvent pair was considered miscible. If such a meniscus was present, the solvent pair was regarded as immiscible. The classification of immiscible is a qualitative one since solvent pairs may exhibit some degree of partial miscibility while existing as separate phases. Solvent pairs possessing a pronounced degree of partial miscibility are designated by the symbol Is.

#	Compounds	Acetone	Acetyl acetone	2-Amino-2-methyl-1-propanol	Aniline	Benzaldehyde	Benzene	Benzin	Benzyl alcohol	Butyl acetate	Butyl alcohol	n-Butyl ether	Capryl alcohol	Carbon tetrachloride	Diacetone alcohol	Diethanolamine	Diethyl cellosolve	Diethyl ether	Dimethylaniline	Ethyl alcohol	Ethyl benzoate	Ethylene glycol	2-Ethylhexanol	Formamide	Furfuryl alcohol	Glycerol	Hydroxyethyl-ethylenediamine	Isoamyl alcohol	Methyl isobutyl ketone	Nitromethane	Dibutoxytetra-ethylene glycol	Pyridine	Triethanolamine	Trimethylene glycol
1	Acetone	..	M	M	..	M	M	M	M	M	M	M	M	M	M	M	M	M	M	M	M	M	M	M	M	I	M	M	M	M	M	M	..	M
2	Acetyl acetone	M	..	R	..	M	M	M	M	M	M	M	M	M	M	M	R	M	M	M	M	M	M	M	M	I	M	R	M	M	M	M	..	M
3	Adiponitrile	M	M	M	M	..	M	..	M	M	M	I	..	I	M	..	M	I	M	M	M	I	I	M	M	I	M	I	M	..	M	M	M	M
4	2-Amino-2-methyl-1-propanol	M	R	..	..	M	M	I	M	M	M	Is	M	M	R	M	M	M	M	M	M	M	M	M	M	M	M	M	M	M	M	M	..	M
5	Benzaldehyde	M	M	M	..	..	M	M	M	M	M	M	M	M	M	I	M	M	M	M	M	Is	M	M	M	Is	R	M	M	M	M	M	..	M
6	Benzene	M	M	M	..	M	..	M	M	M	M	M	M	M	M	Is	I	M	M	M	M	M	M	I	M	I	I	M	M	M	I	M	..	I
7	Benzin	M	M	I	..	M	M	..	I	M	M	M	M	M	I	I	M	M	M	M	M	M	I	I	M	I	Is	M	M	M	M	M	..	I
8	Benzonitrile	M	M	M	M	..	M	..	M	M	M	M	M	M	M	M	M	M	M	M	M	I	M	I	M	I	M	M	M	M	M	M	M	I
9	Benzothiazole	M	M	M	M	..	M	..	M	M	M	M	M	M	M	M	M	M	M	M	M	I	M	I	M	I	M	M	M	M	M	M	M	M
10	Benzyl alcohol	M	M	M	M	M	M	I	..	M	M	M	M	M	M	M	M	M	M	M	M	M	M	M	M	M	M	M	M	M	M	M	..	M
11	Benzyl mercaptan	M	M	I	M	..	M	..	M	M	M	M	M	M	M	M	M	M	M	M	I	M	I	M	I	M	M	M	..	M	M	M	R	I
12	Butyl acetate	M	M	M	..	M	M	M	M	..	M	M	M	M	M	I	M	M	M	M	M	Is	M	I	M	I	I	M	M	M	M	M	..	Is
13	Butyl alcohol	M	M	M	..	M	M	M	M	M	..	M	M	M	M	M	M	M	M	M	M	M	M	M	M	M	M	M	M	M	M	M	..	M
14	n-Butyl ether	M	M	Is	..	M	M	M	M	M	M	..	M	M	M	I	M	M	M	M	M	I	M	I	M	I	M	M	M	I	M	M	..	I
15	Capryl alcohol	M	M	M	..	M	M	M	M	M	M	M	..	M	M	M	M	M	M	M	M	M	M	I	M	I	M	M	M	Is	M	M	..	M
16	Carbon tetrachloride	M	M	M	..	M	M	M	M	M	M	M	M	..	Is	I	M	M	M	M	M	I	M	I	M	I	M	M	M	M	M	M	..	I
17	Diacetone alcohol	M	M	R	..	M	Is	I	M	M	M	M	M	Is	..	M	M	M	M	M	M	M	M	M	M	M	R	M	M	M	M	M	..	M
18	Diethanolamine	M	R	M	..	I	I	I	M	I	M	I	M	I	M	..	I	I	Is	M	M	M	M	M	M	M	M	I	I	I	M	M	..	M
19	Diethyl Cellosolve	M	M	M	..	M	M	M	M	M	M	M	M	M	M	I	..	M	M	M	M	M	M	M	M	I	M	M	M	M	M	M	..	M
20	Diethyl ether	M	M	M	..	M	M	M	M	M	M	I	M	M	M	I	M	..	M	M	M	I	M	I	M	I	M	M	M	M	M	M	..	I
21	Dimethylaniline	M	M	M	..	M	M	M	M	M	M	M	M	M	M	Is	M	M	..	M	M	I	M	I	M	I	M	M	M	M	M	M	..	I
22	Di-N-propylaniline	M	M	I	M	..	M	..	M	I	M	M	M	M	I	..	M	M	M	I	M	I	M	I	M	I	M	M	..	M	M	M	I	I
23	Ethyl alcohol	M	M	M	..	M	M	M	M	M	M	M	M	M	M	M	M	M	M	..	M	M	M	M	M	M	M	M	M	M	M	M	..	M
24	Ethyl benzoate	M	M	M	..	M	M	M	M	M	M	M	M	M	M	..	M	M	M	M	..	I	M	I	M	I	M	M	M	M	M	M	..	M
25	Ethyl isothiocyanate	M	M	R	M	..	M	..	M	M	M	M	M	M	M	M	M	M	M	M	M	I	M	I	M	I	R	M	M	..	M	M	M	M
26	Ethyl thiocyanate	M	M	M	M	..	M	..	M	M	M	M	M	M	M	M	M	M	M	M	M	I	M	I	M	I	M	M	M	..	M	M	M	I
27	Ethylene glycol	M	M	M	..	Is	I	M	M	Is	M	I	M	I	M	M	M	I	I	M	I	..	M	M	M	M	M	M	I	I	M	M	..	M
28	2-Ethylhexanol	M	M	M	..	M	M	M	M	M	M	M	M	M	M	M	M	M	M	M	M	..	..	I	M	I	M	M	M	I	M	M	..	M
29	Formamide	M	M	M	..	M	I	I	M	I	M	I	I	I	M	M	M	M	I	I	M	I	M	..	M	M	M	M	Is	M	M	M	..	M
30	Furfuryl alcohol	M	M	M	..	M	M	I	M	M	M	M	M	M	M	M	M	M	M	M	M	M	M	M	..	M	M	M	M	M	M	M	..	M
31	Glycerol	I	I	M	..	I	I	I	M	I	M	I	I	I	I	M	I	I	I	M	I	M	I	M	M	..	M	I	I	I	I	M	..	..
32	Hydroxyethyl-ethylenediamine	M	R	M	..	R	I	Is	M	I	M	I	M	I	R	M	I	I	I	M	M	M	M	M	M	M	..	M	M	M	M	M	..	M
33	Isoamyl alcohol	M	M	M	..	M	M	M	M	M	M	M	M	M	M	M	M	M	M	M	M	M	M	M	M	I	M	..	M	M	M	M	..	M
34	Isoamyl sulfide	M	M	I	M	..	M	..	M	M	M	..	M	I	..	M	M	M	M	I	M	I	I	I	M	M	..	M	M	..	M	M	I	I
35	Isobutyl mercaptan	M	M	M	M	..	M	..	M	M	M	..	M	I	..	M	M	M	M	M	M	I	M	I	I	I	M	M	..	M	M	R	R
36	Methyl disulfide	M	M	M	M	..	M	..	M	M	M	M	M	M	M	M	M	M	M	I	M	I	M	I	M	M	..	M	M	..	M	M	I	R
37	Methyl isobutyl ketone	M	M	M	..	M	M	M	M	M	M	M	M	M	M	I	M	M	M	M	M	I	M	Is	M	I	M	..	M	M	M	M	..	I
38	Nitromethane	M	M	M	..	M	I	M	M	M	M	M	I	Is	M	M	I	M	M	M	M	M	I	I	M	M	M	M	..	..	M	M	..	I
39	Dibutoxytetra-ethylene glycol	M	M	M	..	M	M	M	M	M	M	M	M	M	M	I	M	M	M	M	M	I	M	I	M	I	M	M	M	M	..	M	..	M
40	Pyridine	M	M	M	..	M	M	M	M	M	M	M	M	M	M	M	M	M	M	M	M	M	M	M	M	M	M	M	M	M	M	..	..	M
41	Tri-n-butylamine	M	M	I	I	..	M	..	M	M	M	M	..	M	I	..	M	M	M	M	I	M	I	M	I	M	M	M	..	M	M	M	I	I
42	Trimethylene glycol	M	M	M	..	M	I	I	M	Is	M	I	M	I	M	M	M	I	I	M	Is	M	M	M	M	M	M	M	M	I	I	M	M	..

451

MISCIBILITY OF ORGANIC SOLVENT PAIRS (Continued)

Tables B and C

W. M. Jackson and J. S. Drury

Reprinted from Vol. 51 pp. 1491 to 1493, December 1959.
Copyright 1959 by the American Chemical Society and reprinted by permission of the copyright owner.

> The classifications were made at 20°C in the following manner. One-milliliter portions of each solvent comprising a pair were shaken together for approximately a minute. If no interfacial meniscus was observed after the contents of the tube were allowed to settle, the solvent pair was considered to be miscible, M. If a meniscus was observed without apparent change in the volume of either solvent, the pair was regarded as immiscible, I. This classification is a qualitative one, since solvent pairs may exhibit various degrees of partial miscibility while existing as separate phases. If an obvious change occurred in the volume of each solvent, but a meniscus was present, the pair was classified as partially miscible, S. The designation R indicates that the two solvents reacted.

Table B

#	Compounds	Acetone	Isoamyl acetate	n-Amyl cyanide	Benzene	Benzyl ether	2-Bromoethyl acetate	Chloroform	Cinnamaldehyde	Di-n-amylamine	Di-n-butyl carbonate	Diethylacetic acid	Diethylenetriamine	Diethyl formamide	Diisobutyl ketone	Diisopropylamine	Di-n-propyl aniline	Ethyl alcohol	Ethyl benzoate	Ethyl ether	Ethyl phenylacetate	Heptadecanol[a]	3-Heptanol	n-Heptyl acetate	n-Hexyl ether	Methyl isopropyl ketone	4-Methyl-n-valeric acid	o-Phenetidine	Sulfuric acid (concd.)	Tetradecanol[a]	Tri-n-butyl phosphate	Triethylene glycol	Triethylenetetramine	2,6,8-Trimethyl 4-nonanone	#
1	Acetone	..	M	M	M	M	M	M	M	M	M	M	M	M	M	M	M	M	M	M	M	M	M	M	M	M	M	M	R	M	M	M	M	M	1
2	Isoamyl acetate	M	..	M	M	M	M	M	M	M	M	M	M	M	M	M	M	M	M	M	M	M	M	M	M	M	M	M	R	M	M	M	M	M	2
3	n-Amyl cyanide	M	M	..	M	M	M	M	M	M	M	M	M	M	M	M	M	M	M	M	M	M	M	M	M	M	M	M	I	M	M	M	M	M	3
4	Benzene	M	M	M	..	M	S	M	M	M	M	M	M	M	M	M	M	M	M	M	M	M	M	M	M	M	M	M	I	M	S	I	M	M	4
5	Benzyl ether	M	M	M	M	..	M	S	M	M	M	M	M	M	M	M	M	M	M	M	M	M	M	M	M	M	M	M	I	M	M	M	M	M	5
6	2-Bromoethyl acetate	M	M	M	S	M	..	M	M	M	M	R	M	M	R	M	R	M	M	M	M	M	M	M	S	M	M	R	M	M	M	R	M	M	6
7	Chloroform	M	M	M	M	M	M	..	M	M	M	M	M	R	M	M	M	M	M	M	M	M	M	M	M	M	M	I	M	M	M	R	M	M	7
8	Cinnamaldehyde	M	M	M	M	M	M	M	..	M	M	M	R	M	R	M	M	M	M	M	M	M	M	M	M	R	R	M	M	M	R	M	M	M	8
9	Di-n-amylamine	M	M	M	M	M	R	R	M	..	M	M	R	I	M	M	M	M	M	M	M	M	M	M	M	M	R	M	M	I	I	M	M	M	9
10	Di-n-butyl carbonate	M	M	M	M	M	M	M	M	M	..	M	M	M	M	M	M	M	M	M	M	M	M	M	M	M	M	R	M	M	I	I	M	M	10
11	Diethylacetic acid	M	M	M	M	M	R	M	R	M	M	..	M	R	M	M	M	M	M	M	M	M	M	M	M	M	R	M	M	I	R	M	M	M	11
12	Diethylenetriamine	M	M	M	M	M	R	M	R	R	M	M	..	R	M	M	M	M	M	M	R	I	R	R	M	R	R	M	I	R	M	M	M	M	12
13	Diethyl formamide	M	M	M	M	M	R	M	M	M	M	M	M	..	M	R	M	M	M	M	M	I	M	M	M	R	R	M	I	R	M	M	M	M	13
14	Diisobutyl ketone	M	M	M	M	M	R	M	R	M	M	M	M	M	..	M	M	M	M	M	M	M	M	M	M	R	R	M	I	R	M	M	M	M	14
15	Diisopropylamine	M	M	M	M	M	M	M	M	M	M	M	R	M	M	..	M	M	M	M	M	M	M	M	M	R	R	M	I	M	M	M	M	M	15
16	Di-n-propylaniline	M	M	M	M	M	M	M	M	M	M	M	M	M	M	M	..	M	M	M	M	M	M	M	M	M	R	M	M	M	M	M	M	M	16
17	Ethyl alcohol	M	M	M	M	M	M	M	M	M	M	M	M	M	M	M	M	..	M	M	M	M	M	M	M	M	M	M	M	M	M	M	M	M	17
18	Ethyl benzoate	M	M	M	M	M	M	M	M	M	M	M	M	M	M	M	M	M	..	M	M	M	M	M	M	M	M	M	M	M	M	M	M	M	18
19	Ethyl ether	M	M	M	M	M	M	M	M	M	M	M	M	M	M	M	M	M	M	..	M	M	M	M	M	M	M	M	R	M	M	M	I	M	19
20	Ethyl phenylacetate	M	M	M	M	M	M	M	M	M	M	M	M	M	M	M	M	M	M	M	..	M	M	M	M	M	M	M	M	M	M	M	M	M	20
21	Heptadecanol[a]	M	M	M	M	M	M	M	M	M	M	M	M	M	M	M	M	M	M	M	M	..	M	M	M	M	M	M	R	M	M	I	M	M	21
22	3-Heptanol	M	M	M	M	M	M	M	M	M	M	M	M	M	M	M	M	M	M	M	M	M	..	M	M	M	M	R	M	M	I	M	M	M	22
23	n-Heptyl acetate	M	M	M	M	M	M	M	M	M	M	M	R	M	M	M	M	M	M	M	M	M	M	..	M	M	M	M	R	M	M	I	M	M	23
24	n-Hexyl ether	M	M	M	M	S	M	M	M	M	I	I	M	M	M	M	M	M	M	M	M	M	M	M	..	M	M	M	R	M	M	I	R	M	24
25	Methyl isopropyl ketone	M	M	M	M	M	M	M	M	M	M	M	R	M	M	M	M	M	M	M	M	M	M	M	M	..	M	M	R	M	M	I	M	M	25
26	4-Methyl-n-valeric acid	M	M	M	M	M	M	M	M	R	M	M	R	M	M	M	R	M	M	M	M	M	M	M	M	M	..	M	R	M	M	M	R	M	26
27	o-Phenetidine	M	M	M	M	M	M	M	M	M	M	M	M	M	M	M	M	M	M	M	M	M	M	M	M	M	M	..	R	M	M	M	M	M	27
28	Sulfuric acid (concd.)	R	R	I	R	R	I	R	R	R	R	R	R	R	R	R	R	R	M	M	R	M	R	R	R	R	R	R	..	R	R	R	R	M	28
29	Tetradecanol[a]	M	M	M	M	M	M	M	M	M	M	M	M	M	M	M	M	M	M	M	M	M	M	M	M	M	M	R	M	..	M	M	M	M	29
30	Tri-n-butyl phosphate	M	M	M	M	M	M	M	M	M	M	M	M	M	M	M	M	M	M	M	M	M	M	M	M	M	M	R	M	M	..	M	M	M	30
31	Triethylene glycol	M	I	M	S	I	M	M	M	S	I	M	M	M	I	I	M	I	M	M	I	M	M	I	I	M	M	M	R	I	M	..	M	I	31
32	Triethylenetetramine	M	M	M	M	M	M	R	M	M	R	I	I	R	M	M	M	I	M	M	I	M	M	M	I	R	M	R	R	M	M	M	..	I	32
33	2,6,8-Trimethyl 4-nonanone	M	M	M	M	M	M	M	M	M	M	M	M	M	M	M	M	M	M	M	M	M	M	M	M	M	M	M	M	M	M	I	I	..	33

[a] Union Carbide name.

Compound number	Compounds	Acetone	Isoamyl acetate	n-Amyl cyanide	Anisaldehyde	Benzene	Benzyl ether	Chloroform	o-Cresol	Diisobutyl ketone	Diethylacetic acid	Diethyl formamide	Di-n-propyl aniline	Ethyl alcohol	Ethyl ether	3-Heptanol	n-Heptyl acetate	n-Hexyl ether	α-Methylbenzylamine	α-Methylbenzyldiethanolamine	α-Methylbenzyldimethylamine	α-Methylbenzylethanolamine	2-Methyl-5-ethylpyridine	Methyl isopropyl ketone	4-Methyl-n-valeric acid	o-Phenetidine	2-Phenylethylamine	Isopropanolamine	Pyridine	Salicylaldehyde	Tetradecanol[a]	Tri-n-butyl phosphate	Triethylenetetramine	2,6,8-Trimethyl 4-nonanone
1	1,3-Butylene glycol	M	I	M	I	I	I	M	M	I	M	M	M	M	S	M	I	I	M	M	M	M	M	M	M	M	M	M	M	M	M	M	M	I
2	2,3-Butylene glycol	M	M	M	M	S	I	M	M	M	M	M	M	M	M	M	M	I	M	M	M	M	M	M	M	M	M	M	M	M	M	M	M	I
3	2-Chloroethanol	M	M	M	M	M	M	M	M	M	M	M	M	M	M	M	M	M	R	M	M	M	R	M	M	R	R	M	M	M	M	M	M	M
4	3-Chloro-1,2-propanediol	M	M	M	M	I	M	M	M	M	M	M	I	M	M	M	M	I	R	M	M	M	M	M	M	M	R	R	M	M	S	M	R	S
5	Dibutyl hydrogen phosphite	M	M	M	M	M	M	M	M	M	M	M	M	M	M	M	M	M	M	M	M	M	M	M	M	M	M	M	M	M	M	M	M	M
6	Diethylene glycol dibutyl ether	M	M	M	M	M	M	M	M	M	M	M	M	M	M	M	M	R	M	S	M	M	M	M	M	R	R	M	M	M	M	M	R	M
7	Diethylene glycol diethyl ether	M	M	M	M	M	M	M	M	M	M	M	M	M	M	M	M	M	M	M	M	M	M	M	M	M	M	M	M	M	M	M	M	M
8	Diethylene glycol monobutyl ether	M	M	M	M	M	M	M	M	M	M	M	M	M	M	M	M	M	M	M	M	M	M	M	M	M	M	M	M	M	M	M	M	M
9	Diethylene glycol monoethyl ether	M	M	M	M	M	M	M	M	M	M	M	M	M	M	M	M	I	M	M	M	M	M	M	M	M	M	M	M	M	M	M	M	M
10	Diethylene glycol monomethyl ether	M	M	M	M	M	M	M	M	M	M	M	M	M	M	M	M	I	M	M	M	M	M	M	M	M	M	M	M	M	M	M	M	M
11	Dipropylene glycol	M	M	M	M	M	M	M	M	M	M	M	M	M	M	M	M	I	M	M	M	M	M	M	M	M	M	M	M	M	M	M	M	M
12	Ethylene diacetate	M	M	M	M	M	M	M	M	M	M	M	M	M	M	M	M	I	M	M	M	M	M	M	M	M	M	M	M	M	M	M	M	M
13	Ethylene glycol	M	I	I	I	I	I	S	M	I	M	M	M	M	I	M	I	I	M	M	M	M	I	M	M	M	M	M	M	I	I	S	M	I
14	Ethyl glycol ethylbutyl ether	M	M	M	M	M	M	M	M	M	M	M	M	M	M	M	M	M	M	M	M	M	M	M	M	M	M	M	M	M	M	M	M	M
15	Ethylene glycol monobutyl ether	M	M	M	M	M	M	M	M	M	M	M	M	M	M	M	M	M	M	M	M	M	M	M	M	M	M	M	M	M	M	M	M	M
16	Ethylene glycol monoethyl ether	M	M	M	M	M	M	M	M	M	M	M	M	M	M	M	M	M	M	M	M	M	M	M	M	M	M	M	M	M	M	M	M	M
17	Ethylene glycol monomethyl ether	M	M	M	M	M	M	M	M	M	M	M	M	M	M	M	M	M	M	M	M	M	M	M	M	M	M	M	M	M	M	M	M	M
18	Ethylene glycol monophenyl ether	M	M	M	M	M	M	M	I	M	M	I	M	I	M	M	M	M	M	M	M	M	M	M	M	M	M	M	M	M	I	I	I	M
19	Glycerol	I	I	I	I	I	I	I	M	I	I	M	I	M	I	I	M	I	M	M	I	I	M	M	I	I	M	M	M	I	I	I	M	I
20	1,2-Propanediol	M	M	M	M	I	M	M	I	M	M	M	I	M	S	M	I	I	M	M	M	M	M	M	M	M	M	M	M	M	I	S	M	I
21	1,3-Propanediol	M	I	I	I	I	I	M	M	I	M	M	I	M	I	M	I	I	M	M	M	M	M	M	M	M	M	M	M	M	I	S	M	I
22	Triethylene glycol	M	I	M	M	S	I	M	M	I	M	M	I	M	I	M	I	I	M	M	M	M	M	M	M	M	M	M	M	M	M	M	M	I
23	Triethyl phosphate	M	M	M	M	M	M	M	M	M	M	M	M	M	M	M	M	M	M	M	M	M	M	M	M	M	M	M	M	M	M	M	M	M
24	Trimethylene chlorohydrin	M	M	M	M	M	M	M	M	M	M	M	M	M	M	M	M	M	R	M	M	M	R	M	M	R	R	M	M	M	M	M	R	M

a Union Carbide name.

EMERGENT STEM CORRECTION FOR
LIQUID-IN-GLASS THERMOMETERS

Accurate thermometers are calibrated with the entire stem immersed in the bath which determines the temperature of the thermometer bulb. However, for reasons of convenience it is common practice when using a thermometer to permit its stem to extend out of the apparatus. Under these conditions both the stem and the mercury in the exposed stem are at a temperature different from that of the bulb. This introduces an error into the observed temperature. Since the coefficient of thermal expansion of glass is less than that of mercury, the observed temperature will be less than the true temperature if the bulb is hotter than the stem and greater than the true temperature, providing the thermal gradient is reversed. For exact work the magnitude of this error can only be determined by experiment. However, for most purposes it is sufficiently accurate to apply the following equation which takes into account the difference of the thermal expansion of glass and mercury:

$$T_c = T_o + F \times L(T_o - T_m)$$

Where
- T_c = corrected temperature
- T_o = observed temperature
- T_m = mean temperature of exposed stem. The mean temperature of the exposed stem may be determined by fastening the bulb of a second thermometer against the midpoint of the exposed liquid column.
- L = the length of the exposed column in degrees above the surface of the substance whose temperature is being determined.
- F = correction factor. For approximate work and when the liquid in the thermometer is mercury a value for F of 0.00016 is generally used. For more accurate work with mercury filled thermometers values as given in the following table are used. For thermometers filled with organic liquids it is customary to use 0.001 for the value of F.

			Values of F for various glasses		
$T_m°C.$	Corning 0041	Corning 8800	Corning 8810	Jena 16 III	Jena 59 III
50	0.000157	0.000166	0.000156	0.000158	0.000164
150	0.000159	0.000167	0.000157	0.000158	0.000165
250	0.000163	0.000168	0.000161	0.000161	0.000170
350	0.000168	0.000173	0.000166	. . . .	0.000177

CORRECTION OF BOILING POINTS TO STANDARD PRESSURE

By H. B. Hass and R. F. Newton

This correction may be made by using the equation:

$$\Delta t = \frac{(273.1 + t)(2.8808 - \log p)}{\phi + .15(2.8808 - \log p)} \tag{1}$$

where Δt = degrees C to be added to the observed boiling point.

t = the observed boiling point.

$\log p$ = the logarithm of the observed pressure in millimeters of mercury.

ϕ = the entropy of vaporization at 760 mm.

The value of ϕ may be estimated from the graph and the table. Substances not included in the table may be classified by grouping them with compounds which bear a close physical or structural resemblance to them.

Example 1. Benzene boils at 20°C. at 75 mm pressure. What is its normal boiling point? We do not find benzene in the table but we find hydrocarbons in group 2, and a group 2 compound with a boiling point of 20° has a ϕ of 4.6.

Substituting in the equation

$$\Delta t = \frac{(273.1 + 20)(2.8808 - 1.8751)}{4.60 + .15(2.8808 - 1.8751)} = 62°$$

Adding this to 20° gives 82° as a first approximation.

The graph shows that the ϕ for a compound of group 2 boiling at 82° is 4.72 instead of 4.60 which we originally used. Since ϕ is in the denominator, this increase will lower our Δt by the ratio, 4.60/4.72, or the corrected Δt is 62 × 4.60/4.72 = 60.4. Adding Δt to t, gives 80.4° as a second approximation.

The formula can best be used in a slightly different form when the reverse calculation is desired, i.e., when one calculates the vapor pressure at a given temperature, lower than the normal boiling point.

$$2.8808 - \log p = \frac{\phi \Delta t}{273.1 + t - .15\Delta t} \tag{2}$$

Example 2. Alcohol boils at 78.4°C. What is its vapor pressure at 20°C.? Substituting in equation 2:

$$2.8808 - \log p = \frac{6.06 \times 58.4}{293.1 - (.15 \times 58.4)} = 1.245$$

$$\log p = 2.8808 - 1.245 = 1.6358$$

$$p = 43.2 \text{ mm.}$$

Here no second approximation is necessary, since the correct value of ϕ was taken immediately, the normal boiling point having been known.

Compound	Group	Compound	Group
Acetaldehyde	3	Benzyl alcohol	5
Acetic acid	4	Butylethylene	1
Acetic anhydride...........	6	Butyric acid	7
Acetone	3	Camphor....................	2
Acetophenone	4	Carbon monoxide	1
Amines.....................	3	Carbon oxysulfide	2
n-Amyl alcohol	8	Carbon suboxide	2
Anthracene	1	Carbon sulfoselenide	2
Anthraquinone	1	m.p. Chloroanilines..........	3
Benzaldehyde	2	Chlorinated derivatives......	Same group as though Cl was H
Benzoic acid	5		
Benzonitrile.................	2	o.m.p. Cresols...............	4
Benzophenone	2	Cyanogen	4

455

Compound	Group	Compound	Group
Cyanogen chloride..........	3	Methyl benzoate.............	3
Dibenzyl ketone	2	Methyl ether	3
Dimethyl amine	4	Methyl ethyl ether	3
Dimethyl oxalate	4	Methyl ethyl ketone	2
Dimethyl silicane	2	Methyl fluoride	3
Esters	3	Methyl formate	4
Ethanol	8	Methyl salicylate	2
Ethers.......................	2	Methyl silicane	1
Ethylamine	4	α,β Naphthols	3
Ethylene glycol	7	Nitrobenzene	3
Ethylene oxide	3	Nitromethane...............	3
Formic acid.................	3	o.m.p. Nitrotoluenes	2
Glycol diacetate	4	o.m.p. Nitrotoluidines......	2
Halogen derivatives	Same group as though halogen were hydrogen.	Phenanthrene	1
		Phenol	5
		Phosgene....................	2
Heptylic acid................	7	Phthalic anhydride	2
Hydrocarbons	2	Propionic acid	5
Hydrogen cyanide	3	n-Propyl alcohol............	8
Isoamyl alcohol.............	7	Quinoline	2
Isobutyl alcohol	8	Sulfides.....................	2
Isobutyric acid..............	6	Tetranitromethane	3
Isocaproic acid	7	Trichloroethylene............	1
Methane	1	Valeric acid	7
Methanol	7	Water	6
Methyl amine...............	5		

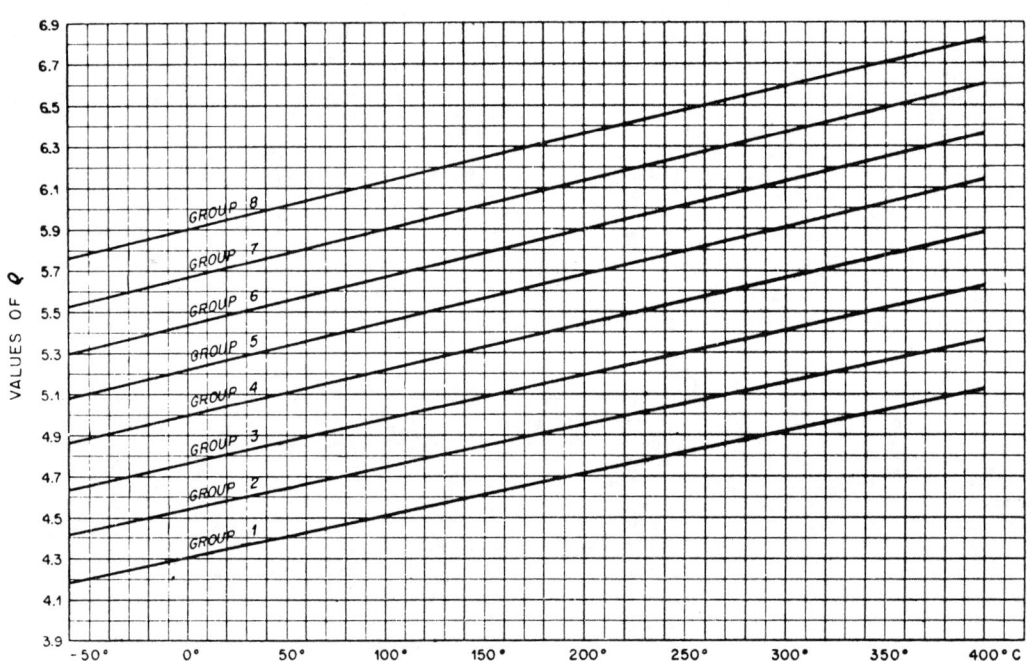

MOLECULAR ELEVATION OF THE BOILING POINT

(Most values from Hoyt, C.S. and Fink, C.K., Journal of Physical Chemistry, Vol. 41, No. 3., March, 1937.)
Molecular elevation of the boiling point showing the elevation of the boiling point in degrees C due to the addition of one gram molecular weight of the dissolved substance to 1000 grams of any one of the solvents below. The correction in the last column gives the number of degrees to be subtracted for each mm. of difference between the barometric reading and 760 mm.

Solvent	K_B	Barometric Correction per mm.
Acetic acid	3.07	0.0008
Acetone	1.71	0.0004
Aniline	3.52	0.0009
Benzene	2.53	0.0007
Bromobenzene	6.26	0.0016
Carbon bisulfide	2.34	0.0006
Carbon tetrachloride	5.03	0.0013
Chloroform	3.63	0.0009
Cyclohexane	2.79	0.0007
Ethanol (ethyl alcohol)	1.22	0.0003
Ethyl acetate	2.77	0.0007
Ethyl ether	2.02	0.0005
n-Hexane	2.75	0.0007
Methanol (methyl alcohol)	0.83	0.0002
Methyl acetate	2.15	0.0005
Nitrobenzene	5.24	0.0013
n-Octane	4.02	0.0010
Phenol	3.56	0.0009
Toluene	3.33	0.0008
Water	0.512	0.0001

MOLECULAR DEPRESSION OF THE FREEZING POINT

Showing the depression of the freezing point due to the addition of one gram molecular weight of solute, for various solvents.

Solvent	Depression for one gram molecular weight dissolved in 1000 grams, °C	Solvent	Depression for one gram molecular weight dissolved in 1000 grams, °C
Acetic acid	3.90	Diphenyamine	8.60
Acetophenone	5.65	Diphenyl ether	8.00
Aniline	5.87	Ethylene dibromide	11.80
Anthracene	14.65	Ethyl ether	11.79
Anthraquinone	14.80	Formic acid	2.77–2.80
Benzene	4.90–5.23	Hexachlorobenzene	20.75
Benzoic acid	7.85–8.79	Menthol	12.4
Benzophenone	9.88	Naphthalene	6.90–7.10
d-Bromocamphor	11.87	β-Naphthol	11.25
Bromoform	14.25	Nitrobenzene	6.89–7.10
tert-Butyl alcohol	12.80	Phenenthrene	12.0
Camphor	49.80	Phenol	7.20–7.50
Carbazole	12.30	Phenylhydrazine	5.86
Carbon disulfide	3.83	Pyridine	4.97
Carbon tetrachloride	29.8–34.8	Stearic acid	4.50
Chloroform	4.67–4.90	Triphenylmethane	12.45
Cyclohexane	20.0–20.30	Urethane	5.00–5.14
Dicyclohexyl	14.50	Water	1.85–1.87
m-Dinitrobenzene	10.60	p-Xylene	4.30
Diphenyl	8.00–8.35		

These data for carbohydrates were compiled originally for the Biology Data Book by M. L. Wolfram, G. G. Maher and R. G. Pagnucco (1964). Data are reproduced here by permission of the copyright owners of the above publication, the Federation of American Societies for Experimental Biology, Washington, D.C. pp. 351–359.

All data are for crystalline substances, unless otherwise specified. Selection of substances was restricted to natural carbohydrates found free (or in chemical combination and released on hydrolysis) and to biological oxidation products of the natural carbohydrates. The nomenclature conforms with that of the British-American report as published in the *Journal of Organic Chemistry*, 28:281 (1963). Substances have been arranged alphabetically under the name of the parent sugar within groups formulated according to increasing carbon content (excluding carbon in substituents), with synonymous common names in parentheses. **Melting Point:** b.p. = boiling point; d. = decomposes; s. = sinters. **Specific Rotation** was determined in water at concentrations of 1–5 g per 100 ml. of solution and at 20°–25°C, unless otherwise specified; other temperatures or wavelengths are shown in brackets; c = grams solute per 100 ml of solution.

Part I. NATURAL MONOSACCHARIDES: ALDOSES AND KETOSES

Substance (Synonym)	Chemical Formula	Melting Point °C	Specific Rotation $[\alpha]_D$
(A)	(B)	(C)	(D)
Aldoses			
1 D-Glyceraldehyde	$C_3H_6O_3$	....	+13.5 ± 0.5 (syrup)
2 D-Glyceraldehyde, 3-deoxy-3,3-*C*-bis-(hydroxymethyl)- (Cordycepose)	$C_5H_{10}O_4$	....	−26 (c 0.6, C_2H_5OH
3 D-Glyceraldehyde, 3,3-bis(*C*-hydroxymethyl)- (Apiose)	$C_5H_{10}O_5$	....	+5.6 (c 10) [15°] syrup
4 β-D-Arabinose	$C_5H_{10}O_5$	155	−175 → −103
5 D-Arabinose, 2-*O*-methyl-	$C_6H_{12}O_5$	Syrup	−102
6 α-L-Arabinose	$C_5H_{10}O_5$	158 amorphous	+55.4 → +105
7 β-L-Arabinose	$C_5H_{10}O_5$	160	+190.6 → +104.5
8 DL-Arabinose	$C_5H_{10}O_5$	163.5–164.5	None
9 α-L-Lyxose	$C_5H_{10}O_5$	105	+5.8 → +13.5
10 L-Lyxose, 5-deoxy-3-*C*-formyl- (Streptose)	$C_6H_{10}O_5$	....	....
11 L-Lyxose, 3-*C*-formyl- (Hydroxy-streptose)	$C_6H_{10}O_6$	....	....
12 Pentose, 4,5-anhydro-5-deoxy-D-*erythro*-	$C_5H_8O_3$	....	....
13 Pentose, 2-deoxy-D-*erythro*-	$C_5H_{10}O_4$	96–98	−91 → −58
14 D-Ribose	$C_5H_{10}O_5$	87	−23.1 → −23.7
15 D-Ribose, 2-*C*-hydroxymethyl- (Hamamelose)	$C_6H_{12}O_6$	....	−7.1 [λ578]
16 α-D-Xylose	$C_5H_{10}O_5$	145	+93.6 → +18.8
17 D-Xylose, 5-deoxy-	$C_5H_{10}O_4$	....	+16
18 β-D-Xylose, 2-*O*-methyl-	$C_6H_{12}O_5$	137–138	−21 → +34
19 α-D-Xylose, 3-*O*-methyl-	$C_6H_{12}O_5$	95	+45 → +19
20 D-Allose, 6-deoxy-	$C_6H_{12}O_5$	140–143 146–148	+1.6 [18°] (c 0.6) −4.7 → 0
21 D-Allose, 6-deoxy-2,3-di-*O*-methyl- (Mycinose)	$C_8H_{16}O_5$	102–106	−46 → −29
22 Amicetose (a trideoxy hexose)	$C_6H_{12}O_3$	Oil, b.p. 65–70	+28.6 ($CHCl_3$)
23 Antiarose	$C_6H_{12}O_5$	....	Levo
24 α-D-Galactose	$C_6H_{12}O_6$	167	+150.7 → +80.2
25 β-D-Galactose	$C_6H_{12}O_6$	143–145	+52.8 → +80.2
26 D-Galactose, 3,6-anhydro-	$C_6H_{10}O_5$	....	+21.3 [10°]
27 α-D-Galactose, 6-deoxy- (D-Fucose; Rhodeose)	$C_6H_{12}O_5$	140–145	+127 → +76.3 (c 10)
28 D-Galactose, 6-deoxy-3-*O*-methyl- (Digitalose)	$C_7H_{14}O_5$	106[1], 119[2]	+106

	Substance (Synonym)	Chemical Formula	Melting Point °C	Specific Rotation $[\alpha]_D$
	(A)	(B)	(C)	(D)
	Aldoses (Con't)			
29	D-Galactose, 6-deoxy-4-O-methyl-	$C_7H_{14}O_5$	131–132	+82
30	D-Galactose, 6-deoxy-2,3-di-O-methyl-	$C_8H_{16}O_5$	....	+73
31	α-D-Galactose, 3-O-methyl-	$C_7H_{14}O_6$	144–147	$+150.6 \rightarrow +108.6$
32	α-D-Galactose, 6-O-methyl-	$C_7H_{14}O_6$	122–123	$+117 \rightarrow +77.3$
33	L-Galactose	$C_6H_{12}O_6$	....	See D-Galactose
34	α-L-Galactose, 3,6-anhydro-	$C_6H_{10}O_5$	....	$-39.4 \rightarrow -25.2$
35	α-L-Galactose, 6-deoxy- (L-Fucose)	$C_6H_{12}O_5$	145	$-124.1 \rightarrow -76.4$
36	L-Galactose, 6-deoxy-2-O-methyl-	$C_7H_{14}O_5$	149–150	-75 ± 4 (c 0.5)
37	L-Galactose, 6-sulfate	$C_6H_{12}O_9S$	....	-47 (c 0.2) (Na salt)
38	DL-Galactose	$C_6H_{12}O_6$	143–144, 163	None (racemic)
39	α-D-Glucose	$C_6H_{12}O_6$	146, 83 (H_2O)	$+112 \rightarrow +52.7$
40	β-D-Glucose	$C_6H_{12}O_6$	148–150	$+18.7 \rightarrow +52.7$
41	D-Glucose, 6-acetate	$C_7H_{14}O_7$	135	+48
42	D-Glucose, 2,3-di-O-methyl-	$C_8H_{16}O_6$	85–86, 121	+50
43	D-Glucose, 6-O-benzoyl- (Vaccinin)	$C_{13}H_{16}O_7$	Amorphous	+48 (C_2H_5OH)
44	α-D-Glucose, 6-deoxy- (Chinovose; Epirhamnose; Glucomethylose; Isorhamnose; Isorhodeose; Quinovose)	$C_6H_{12}O_5$	139–140	$+73.3 \rightarrow +29.7$ (c 8)
45	α-D-Glucose, 6-deoxy-3-O-methyl- (D-Thevetose)	$C_7H_{14}O_5$	116	$+84 \rightarrow +33$
46	D-Glucose, 6-sulfonic acid, 6-deoxy- (6-Sulfoquinovose)	$C_6H_{12}O_8S$	173–174	$+87^3$
47	D-Glucose, 3-O-methyl-	$C_7H_{14}O_6$	162–167	$+98 \rightarrow +59.5$
48	α-L-Glucose	$C_6H_{12}O_6$	141–143	$-95.5 \rightarrow -51.4$
49	L-Glucose, 6-deoxy-3-O-methyl- (L-Thevetose)	$C_7H_{14}O_5$	126–129	-36.9 ± 2
50	D-Gulose, 6-deoxy-	$C_6H_{12}O_5$	....	....
51	Hexose, 2-deoxy-D-arabino-[4]	$C_6H_{12}O_5$	148	+46.6 [18°]
52	Hexose, 2,6-dideoxy-3-O-methyl-D-arabino- (D-Oleandrose)	$C_7H_{14}O_4$	....	-11
53	Hexose, 3,6-dideoxy-D-arabino- (Tyvelose)	$C_6H_{12}O_4$	....	$+24 \pm 2$
54	Hexose, 2,6-dideoxy-3-O-methyl-L-arabino- (L-Oleandrose)	$C_7H_{14}O_4$	62–63	$+11.9 \pm 2.5$
55	Hexose, 3,6-dideoxy-L-arabino- (Ascarylose)	$C_6H_{12}O_4$	....	-24 ± 2
56	Hexose, 2,6-dideoxy-3-O-methyl-D-lyxo- (Diginose)	$C_7H_{14}O_4$	90–92	$+56 \pm 4$
57	Hexose, 2,6-dideoxy-L-lyxo- (L-Fucose, 2-deoxy-)	$C_6H_{12}O_4$	103–106	-61.6
58	Hexose, 2,6-dideoxy-3-O-methyl-L-lyxo-	$C_7H_{14}O_4$	78–85	-65
59	Hexose, 2,6-dideoxy-D-ribo- (Digitoxose; D-Altrose, 2,6-dideoxy-)	$C_6H_{12}O_4$	110	$+46.4$
60	Hexose, 2,6-dideoxy-3-O-methyl-D-ribo- (Cymarose)	$C_7H_{14}O_4$	93	$+52$
61	Hexose, 3,6-dideoxy-D-ribo- (Paratose)	$C_6H_{12}O_4$	....	$+10 \pm 2$ (c 0.9)
62	Hexose, 4,6-dideoxy-3-O-methyl-D-ribo- (D-Gulose, 4,6-dideoxy-3-O-methyl-; Chalcose)	$C_7H_{14}O_4$	96–99	$+120 \rightarrow +76$
63	Hexose, 2,6-dideoxy-D-xylo- (Boivinose)	$C_6H_{12}O_4$	96–98	$-3.9 \rightarrow +3.9$

	Substance (Synonym)	Chemical Formula	Melting Point °C	Specific Rotation $[\alpha]_D$
	(A)	(B)	(C)	(D)
	Aldoses (Con't)			
64	Hexose, 2,6-dideoxy-3-O-methyl-D-xylo- (Sarmentose)	$C_7H_{14}O_4$	78–79	$+12 \rightarrow +15.8$
65	Hexose, 3,6-dideoxy-D-xylo- (Abequose)	$C_6H_{12}O_4$	....	-3.2 ± 0.6
66	Hexose, 2,6-dideoxy-3-C-methyl-L-xylo- (Mycarose)	$C_7H_{14}O_4$	129–129	-31.1
67	Hexose, 2,6-dideoxy-3-C-methyl-3-O-methyl-L-xylo-(Cladinose)	$C_8H_{16}O_4$	oil, b.p. 120–132 (0.25 mm)	-23.1
68	Hexose, 3,6-dideoxy-L-xylo- (Colitose)	$C_6H_{12}O_4$	....	$+4$ (H_2O); -51 ± 2 (CH_3OH)
69	D-Idose[5]	$C_6H_{12}O_6$	....	....
70	L-Idose, 1,6-anhydro-	$C_6H_{10}O_5$	....	....
71	α-D-Mannose	$C_6H_{12}O_6$	133	$+29.3 \rightarrow +14.5$
72	β-D-Mannose	$C_6H_{12}O_6$	132	$-16.3 \rightarrow +14.5$
73	D-Mannose, 6-deoxy- (D-Rhamnose)	$C_6H_{12}O_5$	86–90	-7.0
74	α-L-Mannose, 6-deoxy-monohydrate (L-Rhamnose)	$C_6H_{14}O_6$	93–94	$-8.6 \rightarrow +8.2$
75	β-L-Mannose, 6-deoxy-	$C_6H_{12}O_5$	123–125	$+38.4 \rightarrow +8.9$
76	L-Mannose, 6-deoxy-2-O-methyl-	$C_7H_{14}O_5$	....	....
77	L-Mannose, 6-deoxy-3-O-methyl- (L-Acofriose)	$C_7H_{14}O_5$	114–115	$+30$ [18°]
78	L-Mannose, 6-deoxy-2,4-di-O-methyl-	$C_8H_{16}O_5$	82	-19 [16°]
79	L-Mannose, 6-deoxy-5-C-methyl-4-O-methyl-(Noviose)	$C_8H_{16}O_5$	128–130	$+19.9$ (50% C_2H_5OH)
80	Rhodinose (a 2,3,6-trideoxyhexose)	$C_6H_{12}O_3$	....	-11 ± 1.6
81	D-Talose	$C_6H_{12}O_6$	128–132	$+16.9$
82	D-Talose, 6-deoxy- (D-Talomethylose)	$C_6H_{12}O_5$	129–131	$+20.6$
83	L-Talose, 6-deoxy- (L-Talomethylose)	$C_6H_{12}O_5$	116–118	-19.5 ± 2 [18°]
84	L-Talose, 6-deoxy-2-O-methyl- (L-Acovenose)	$C_7H_{14}O_5$	....	-19.4
85	Heptose, D-glycero-D-galacto-	$C_7H_{14}O_7$	139–140	$+47 \rightarrow +64$ (c 0.5)
86	Heptose, D-glycero-D-manno-	$C_7H_{14}O_7$	....	....
87	Heptose, D-glycero-L-manno-	$C_7H_{14}O_7$	....	....
	Ketoses			
88	Dihydroxyacetone	$C_3H_6O_3$	80 (dimer)	None
89	Tetrulose, L-glycero-[8] (L-Erythrulose; Ketoerythritol; L-Threulose)	$C_4H_8O_4$	Syrup	$+12$
90	Pentulose, D-erythro- (Adonose; D-Ribulose)	$C_5H_{10}O_5$	Syrup	$+16.6$ [27°]
91	Pentulose, L-erythro- (L-Ribulose)	$C_5H_{10}O_5$	....	-16.6
92	Pentulose, D-threo- (D-Xylulose)	$C_5H_{10}O_5$	....	-33
93	Pentulose, 5-deoxy-D-threo-	$C_5H_{10}O_4$	....	-5 ± 1 (CH_3OH)
94	Pentulose, L-threo- (L-Xylulose; L-Lyxulose; Xyloketose)	$C_5H_{10}O_5$	Syrup	$+33.1$
95	Hexulose, β-D-arabino-(β-D-Fructose; Levulose)	$C_6H_{12}O_6$	102–104[7]	$-133.5 \rightarrow -92$
96	Hexulose, 6-deoxy-D-arabino- (D-Rhamnulose)	$C_6H_{12}O_5$	....	-13 ± 2
97	Hexulose, D-lyxo- (D-Tagatose)	$C_6H_{12}O_6$	131–132	$+2.7 \rightarrow -4, -5$
98	5-Hexulose, D-lyxo	$C_6H_{12}O_6$	158	-86.6
99	Hexulose, 6-deoxy-L-lyxo- (L-Fuculose)	$C_6H_{12}O_5$	....	....

Substance (Synonym)	Chemical Formula	Melting Point °C	Specific Rotation $[\alpha]_D$
(A)	(B)	(C)	(D)
Ketoses (Con't)			
100 Hexulose, D-*ribo*- (D-Psicose)	$C_6H_{12}O_6$	Amorphous	+4.7
101 Hexulose, L-*xylo*- (L-Sorbose)	$C_6H_{12}O_6$	159–161	−43.1
102 Hexulose, 6-deoxy-L-*xylo*-	$C_6H_{12}O_5$	88	-25 ± 2 (c 0.7)
103 Heptulose, D-*altro*- (Sedoheptulose; Sedoheptose)	$C_7H_{14}O_7$	Amorphous	+2.5 (c 10)
104 Heptulose·hemihydrate, L-*galacto*- (Perseulose)	$C_7H_{14}O_7\cdot\frac{1}{2}H_2O$	110–115	$-90 \to -80$
105 Heptulose, L-*gulo*-	$C_7H_{14}O_7$	....	−28
106 Heptulose, D-*ido*-	$C_7H_{14}O_7$	172	-34 ± 8 (c 0.3)
107 Heptulose, D-*manno*- (Mannoketoheptose; D-Mannotagatoheptose)	$C_7H_{14}O_7$	152	+29.4
108 Heptulose, D-*talo*-	$C_7H_{14}O_7$	....	....
109 Octulose, D-*glycero*-L-*galacto*-	$C_8H_{16}O_8$	....	$-57, -43.4 \to -13.4$
110 Octulose, D-*glycero*-D-*manno*-	$C_8H_{16}O_8$	....	+20 (CH_3OH)

[1] Original melting point. [2] Melting point after four-months' storage. [3] As a methyl glycoside cyclohexylamine salt. [4] Included because of speculations concerning it in biological processes. [5] Either D-idose or L-altrose is in the polysaccharide varianose. [6] Early literature refers to this as D-erythrose. [7] The $\cdot\frac{1}{2}H_2O$ and $\cdot 2H_2O$ forms also exist.

Part II. NATURAL MONOSACCHARIDES: AMINO SUGARS

Substance (Synonym)	Chemical Formula	Melting Point °C	Specific Rotation $[\alpha]_D$
(A)	(B)	(C)	(D)
Aldosamines			
1 D-Ribose, 3-amino-3-deoxy-	$C_5H_{11}NO_4$	158–158.5 d.	−24.6 (hydrochloride)
2 D-Galactose, 2-amino-2-deoxy- (Galactosamine; Chondrosamine)	$C_6H_{13}NO_5$	185	$+121 \to +80$ (hydrochloride)
3 α-L-Galactose, 2-amino-2,6-dideoxy- (L-Fucosamine)	$C_6H_{13}NO_4$	192–193 d.	$-119 \to -92$ [27°] (hydrochloride)
4 α-D-Glucose, 2-amino-2-deoxy- (Glucosamine; Chitosamine)	$C_6H_{13}NO_5$	88	$+100 \to +47.5$
5 β-D-Glucose, 2-amino-2-deoxy-	$C_6H_{13}NO_5$	110–111	$+28 \to +47.5$
6 D-Glucose, 3-amino-3-deoxy- (Kanosamine)	$C_6H_{13}NO_5$	128 d.	+19 [14°]
7 D-Glucose, 6-amino-6-deoxy-	$C_6H_{13}NO_5$	161–162 d.	$+23 \to +50.1$ (hydrochloride)
8 D-Glucose, 2,6-diamino-2,6-dideoxy- (Neosamine C)	$C_6H_{14}N_2O_4$	>230	+61.5 (dihydrochloride)
9 D-Glucose, 3,6-dideoxy-3-dimethylamino- (Mycaminose)	$C_8H_{17}NO_4$	115–116	+31 (hydrochloride)
10 D-Glucose, 4,6-dideoxy-4-dimethylamino-	$C_8H_{17}NO_4$	192–193	+45.5 (hydrochloride)
11 L-Glucose, 2-deoxy-2-methylamino-	$C_7H_{15}NO_5$	130–132	−64
12 D-Gulose, 2-amino-1,6-anhydro-2-deoxy-	$C_6H_{11}NO_4$	250–260 d.	$+41 \pm 2$ (hydrochloride)
13 D-Gulose, 2-amino-2-deoxy-	$C_6H_{13}NO_5$	152–162 d.	$+5.6 \to -18.7$ (hydrochloride)

461

Substance (Synonym)	Chemical Formula	Melting Point °C	Specific Rotation [α]$_D$
(A)	(B)	(C)	(D)
Aldosamines (Con't)			
14 Hexose, 3,4,6-trideoxy-3-dimethyl-amino-D-*xylo*- (Desosamine; Picrocine)	$C_8H_{17}NO_3$	189–191 d.	+49.5 (*c* 10).(hydrochloride)
15 Hexose, a 4-acetamido-2-amino-2,4,6-trideoxy-	$C_8H_{16}N_2O_4$	216–219	+115 → +94 [26°] (*c* 0.05)
16 Hexose, an amino-deoxy-3-*O*-carboxy-ethyl-	$C_9H_{17}NO_7$	. . . .	. . . .
17 Hexose, a 2,6-diamino-2,6-dideoxy- (Neosamine B; Paramose)	$C_6H_{14}N_2O_4$	135–150 d.	+17.5 (*c* 0.9 (hydrochloride)
18 Hexose, a 3-dimethylamino-2,3,6-trideoxy- (Rhodosamine)	$C_8H_{17}NO_3$	. . . .	. . . .
19 D-Mannose, 2-amino-2-deoxy- (Mannosamine)	$C_6H_{13}NO_5$	142 d.	−4.3 (*c* 9) (hydrochloride)
20 D-Mannose, 3-amino-3,6-dideoxy- (Mycosamine)	$C_6H_{13}NO_4$	162	−11.5 (hydrochloride)
21 D-Talose, 2-amino-2-deoxy- (Talosamine)	$C_6H_{13}NO_5$	151–153	+3.4 → −5.7 (*c* 0.9) (hydrochloride)
22 L-Talose, 2-amino-2,6-dideoxy- (Pneumosamine)	$C_6H_{13}NO_4$	162–163	+6.9 → +10.4 (hydrochloride)
Ketosamines			
23 Pentulose, 1-(*o*-carboxyanilino)-1-deoxy-D-*erythro*-	$C_{12}H_{14}NO_6$	. . . .	. . . .
24 Hexulose, 1-(*o*-carboxyanilino)-1-deoxy-D-*arabino*-	$C_{13}H_{16}NO_7$	. . . .	. . . .
25 Hexulose, 5-amino-5-deoxy-L-*xylo*-	$C_6H_{13}NO_5$	174–176	−62
26 Hexulose, 6-deoxy-6-(*N*-methyl-acetamido)-L-*xylo*-	$C_9H_{17}NO_6$	. . . .	. . . .

Part III. NATURAL ALDITOLS AND INOSITOLS (with Inososes and Inosamines)

Substance (Synonym)	Chemical Formula	Melting Point °C	Specific Rotation [α]$_D$
(A)	(B)	(C)	(D)
Alditols			
1 Glycerol	$C_3H_8O_3$	20	None
2 Glycerol, 1-deoxy- (1,2-Propane-diol)[1]	$C_3H_8O_2$	Oil, b.p. 188–189	None (racemic)
3 Erythritol	$C_4H_{10}O_4$	118–120	None (meso)
4 Erythritol, 1,4-dideoxy- (2,3-Butyleneglycol)	$C_4H_{10}O_2$	25, 34	None (meso)
5 D-Threitol, 1,4-dideoxy-	$C_4H_{10}O_2$	19	−13.0
6 L-Threitol, 1,4-dideoxy-	$C_4H_{10}O_2$	. . . .	+10.2
7 DL-Threitol, 1,4-dideoxy-	$C_4H_{10}O_2$	7.6	None (racemic)
8 D-Arabinitol	$C_5H_{12}O_5$	103	+7.82 (*c* 8, borax solution)
9 L-Arabinitol	$C_5H_{12}O_5$	101–102	−32 (*c* 0.4, 5% molybdate)
10 Ribitol (Adomitol)	$C_5H_{12}O_5$	102	None (meso)
11 Galactitol (Dulcitol)	$C_6H_{14}O_6$	186–188	None (meso)

Substance (Synonym)	Chemical Formula	Melting Point °C	Specific Rotation $[\alpha]_D$
(A)	(B)	(C)	(D)
Alditols (Con't)			
12 D-Glucitol (Sorbitol)	$C_6H_{14}O_6$	112	−1.8 [15°]
13 D-Glucitol, 1,5-anhydro- (Polygalitol)	$C_6H_{12}O_5$	140–141	+42.4
14 L-Iditol	$C_6H_{14}O_6$	73.5	−3.5 (c 10)
15 D-Mannitol	$C_6H_{14}O_6$	166	−0.21
16 D-Mannitol, 1,5-anhydro- (Styracitol)	$C_6H_{12}O_5$	157	−49.9
17 Heptitol, D-*glycero*-D-*galacto*- (Heptitol, L-*glycero*-D-*manno*-; Perseitol)	$C_7H_{16}O_7$	183–185, 188	−1.1
18 Heptitol, D-*glycero*-D-*gluco*- (Heptitol, L-*glycero*-D-*talo*-; β-Sedoheptitol)	$C_7H_{16}O_7$	131–132	+46 (5% NH₄ molybdate)
19 Heptitol, D-*glycero*-D-*manno*- (Heptitol, D-*glycero*-D-*talo*-; Volemitol)	$C_7H_{16}O_7$	153	+2.65
20 Octitol, D-*erythro*-D-*galacto*-	$C_8H_{18}O_8 \cdot H_2O$	169–170	−11 (5% NH₄ molybdate)
Inositols			
21 Betitol (a dideoxy inositol)	$C_6H_{12}O_4$	224	....
22 Bioinosose (*scyllo*-Inosose; *myo*-Inosose-2; a deoxy keto inositol)	$C_6H_{10}O_6$	198–200	None (meso)
23 *h*-Bornesitol (a *myo*-inositol monomethyl ether)	$C_7H_{14}O_6$	200	+31.6
24 *l*-Bornesitol (a *myo*-inositol monomethyl ether)	$C_7H_{14}O_6$	205–206	−32.1
25 Conduritol (a 2,3-dehydro-2,3-dideoxyinositol)	$C_6H_{10}O_4$	142–143	None (meso)
26 Cordycepic acid (a tetrahydroxycyclohexanecarboxylic acid)[2]	$C_7H_{12}O_6$	....	....
27 Dambonitol (a *myo*-inositol dimethyl ether)	$C_8H_{16}O_6$	206	None (meso)
28 DL-Inositol	$C_6H_{12}O_6$	253	None (racemic)
29 *d*-Inositol	$C_6H_{12}O_6$	....	+60
30 *l*-Inositol	$C_6H_{12}O_6$	240	−65
31 Laminitol (a *C*-methyl *myo*-inositol)	$C_7H_{14}O_6$	266–269	−3
32 Liriodendritol (a *myo*-inositol dimethyl ether)	$C_8H_{16}O_6$	224	−25
33 *muco*-Inositol monomethyl ether	$C_7H_{14}O_6$	322–325	....
34 *myo*-Inositol (*meso*-Inositol)	$C_6H_{12}O_6$	217–218	None (meso)
35 *d-myo*-Inosose-1 (a deoxy keto inositol)	$C_6H_{10}O_6$	138–139	+19.6
36 Mytilitol (a *C*-methyl *scyllo*-inositol)	$C_7H_{14}O_6$	259	None (meso)
37 *neo*-Inosamine-2 (a deoxy amino inositol)	$C_6H_{13}O_5N$	239–241 d.	None (meso)
38 *d*-Ononitol (a *myo*-inositol monomethyl ether)	$C_7H_{14}O_6$	172	+6.6
39 *h*-Pinitol (a *dextro*-inositol monomethyl ether)	$C_7H_{14}O_6$	186	+65.5
40 *l*-Pinitol (a *levo*-inositol monomethyl ether)	$C_7H_{14}O_6$	186	−65
41 *l*-Quebrachitol (a *levo*-inositol monomethyl ether)	$C_7H_{14}O_6$	190–191	−80.2 [28°]
42 *d*-Quercitol (a deoxy *dextro*-inositol)	$C_6H_{12}O_5$	235	+24.2
43 *d*-Quinic acid (a trideoxy carboxy *dextro*-inositol)	$C_7H_{12}O_6$	164	+44 (c 10)

	Substance (Synonym)	Chemical Formula	Melting Point °C	Specific Rotation $[\alpha]_D$
	(A)	(B)	(C)	(D)
	Inositols (Con't)			
44	*l*-Quinic acid (a trideoxy carboxy *levo*-inositol)	$C_7H_{12}O_6$	162	−42.1
45	Quinic acid, 5-dehydro-	$C_7H_{10}O_6$	140–142 (138 s.)	−82.4 [28°]
46	Scyllitol (*scyllo*-Inositol; Cocositol)	$C_6H_{12}O_6$	352–353	None (meso)
47	Sequoyitol (a *myo*-inositol monomethyl ether)	$C_7H_{14}O_6$	234–235	None (meso)
48	Shikimic acid (a 3,4-anhydro-quinic acid)	$C_7H_{10}O_5$	183–184	−200 [16°]
49	Shikimic acid, 5-dehydro-	$C_7H_8O_5$	150–152	−57.5 [28°] (EtOH)
50	Streptamine (2,4-diaminodideoxy-scyllitol)	$C_6H_{14}O_4N_2$	88, 210–250 d.	None (meso)
51	Streptamine, 2-deoxy-	$C_6H_{14}O_3N_2$	. . . .	None (meso)
52	Streptadine (1,3-Dideoxy-1,3-diguanidino-scyllitol)	$C_8H_{18}N_6O_4$	. . . .	None (meso)
53	Viburnitol (a deoxy *levo*-inositol)[3]	$C_6H_{12}O_5$	174	−73.9

[1] The 1-phosphate ester of this diol is said to occur in brain tissue and sea-urchin eggs. [2] Strong evidence that cordycepic acid is really D-mannitol. [3] Not an enantiomorph of *d*-quercitol; other isomeric relationship is involved.

Part IV. NATURAL ALDONIC, URONIC, AND ALDARIC ACIDS

	Substance (Synonym)	Chemical Formula	Melting Point °C	Specific Rotation $[\alpha]_D$
	(A)	(B)	(C)	(D)
	Aldonic Acids			
1	D-Glyceric acid	$C_3H_6O_4$	Gum	Dextro
2	L-Glyceric acid	$C_3H_6O_4$	Gum	Levo
3	D-Arabinonic acid	$C_5H_{10}O_6$	114–116	+10.5 (*c* 6)
4	L-Arabinonic acid	$C_5H_{10}O_6$	118–119	−9.6 → −41.7[1]
5	L-Arabinonic-1,4-lactone	$C_5H_8O_5$	97–99	−72
6	D-Ribonic acid	$C_5H_{10}O_6$	112–113	−17.0
7	D-Xylonic acid	$C_5H_{10}O_6$	. . . .	−2.9 → +20.1[1]
8	L-Xylonic acid	$C_5H_{10}O_6$	. . . .	−91.8[1]
9	D-Altronic acid	$C_6H_{12}O_7$	. . . .	+11.5 → +24.8[1] (Ca salt, N HCl)
10	D-Galactonic acid	$C_6H_{12}O_7$	122	−11.2 → +57.6[1]
11	D-Gluconic acid	$C_6H_{12}O_7$	130–132 (110–112 s.)	−6.7 → +11.9[1]
12	L-Gulonic acid	$C_6H_{12}O_7$	Exists only in soln.	[ca. 0°]
13	Hexsonic acid, 2-deoxy-D-*arabino*-	$C_6H_{12}O_6$	93–95	+68 (lactone)
14	2-Hexulosonic acid, D-*arabino*-	$C_6H_{10}O_7$	. . . .	−81.7 (Na salt)
15	2-Hexulosonic acid, 3-deoxy-D-*erythro*-	$C_6H_{10}O_6$	. . . .	−29.2 (*c* 6, Ca salt)
16	2-Hexulosonic acid, D-*lyxo*-	$C_6H_{10}O_7$	169	−5
17	5-Hexulosonic acid, D-*arabino*-	$C_6H_{10}O_7$	108–109	. . . .
18	5-Hexulosonic acid, D-*xylo*-	$C_6H_{10}O_7$	. . . .	−14.5
19	D-Mannonic acid	$C_6H_{12}O_7$	. . . .	−15.6
20	D-Gluconic acid, *O*-β-D-galactopyranosyl- (1 → 4)- (Lactobionic acid)	$C_{12}H_{22}O_{12}$	. . . .	+25.1 (Ca salt)

	Substance (Synonym)	Chemical Formula	Melting Point °C	Specific Rotation $[\alpha]_D$
	(A)	(B)	(C)	(D)
	Uronic Acids			
21	L-Lyxuronic acid	$C_5H_8O_6$	. . . .	. . . .
22	β-D-Galacturonic acid	$C_6H_{10}O_7$	160	$+27 \rightarrow +55.6$
23	α-D-Galacturonic acid·monohydrate	$C_6H_{12}O_8$	159–160 (110–115 s.)	$+97.9 \rightarrow +50.9$
24	D-Galacturonic acid, 2-amino-2-deoxy-	$C_6H_{11}O_6N$	160 d.	$+84.5$ (pH 2 HCl)
25	β-D-Glucuronic acid	$C_6H_{10}O_7$	156	$+11.7 \rightarrow +36.3$
26	D-Glucuronic acid, 2-amino-2-deoxy-	$C_6H_{11}O_6N$	120–172 d.	$+55$
27	D-Glucuronic acid, 3-O-methyl-	$C_7H_{12}O_7$	Syrup	$+6$
28	L-Guluronic acid	$C_6H_{10}O_7$	. . . .	. . . .
29	L-Iduronic acid	$C_6H_{10}O_7$	. . . .	$+30$
30	β-D-Mannuronic acid	$C_6H_{10}O_7$	165–167	$-47.9 \rightarrow -23.9$
31	α-D-Mannuronic acid·monohydrate	$C_6H_{12}O_8$	110 s., 120–130 d.	$+16 \rightarrow -6.1$ (c 6.8)
	Aldaric Acids			
32	D-Tartaric acid	$C_4H_6O_6$	170	-15
33	L-Tartaric acid	$C_4H_6O_6$	170	$+15$ [15°]
34	L-Malic acid	$C_4H_6O_5$	100	-2.3 (c 8.4)

[1] Equilibrates with the lactone.

FATS AND OILS

These data for fats and oils were compiled originally for the Biology Data Book by H. J. Harwood, and R. P. Geyer. 1964. Data are reproduced here by permission of the copyright owners of the above publication, the Federation of American Societies for Experimental Biology, Washington, D.C. pp. 380–382.

Values are typical rather than average, and frequently were derived from specific analyses for particular samples (especially the constituent fatty acids). Extreme variations may occur, depending on a number

	Fat or Oil	Source	Constants				
			Melting (or Solidi-fication) Point, °C	Specific Gravity (or Density)	Refractive Index $n_{D}^{40°}$	Iodine Value	Saponification Value
	(A)	(B)	(C)	(D)	(E)	(F)	(G)
	Land Animals						
1	Butterfat	*Bos taurus*	32.2	0.911[40°/15°]	1.4548	36.1	227
2	Depot fat	*Homo sapiens*	(15)	0.918[15°]	1.4602	67.6	196.2
3	Lard oil	*Sus scrofa*	(30.5)	0.919[15°]	1.4615	58.6	194.6
4	Neat's-foot oil	*B. taurus*	....	0.910[25°]	1.464[25°]	69–76	190–199
5	Tallow, beef	*B. taurus*	....	....	....	49.5	197
6	Tallow, mutton	*Ovis aries*	(42.0)	0.945[15°]	1.4565	40	194
	Marine Animals						
7	Cod-liver oil	*Gadus morhua*	....	0.925[25]	1.481[25°]	165	186
8	Herring oil	*Clupea harengus*	....	0.900[60°]	1.4610[60°]	140	192
9	Menhaden oil	*Brevoortia tyrannus*	....	0.903[60°]	1.4645[60°]	170	191
10	Sardine oil	*Sardinops caerulea*	....	0.905[60°]	1.4660[60°]	185	191
11	Sperm oil, body	*Physeter macrocephalus*	....	....	....	76–88	122–130
12	Sperm oil, head	*P. macrocephalus*	....	....	....	70	140–144
13	Whale oil	*Balaena mysticetus*	....	0.892[60]	1.460[60°]	120	195
	Plants						
14	Babassu oil	*Attalea funifera*	22–26	(0.893[60°])	1.443[60°]	15.5	247
15	Castor oil	*Ricinus communis*	(−18.0)	0.961[15°]	1.4770	85.5	180.3
16	Cocoa butter	*Theobroma cacao*	34.1	0.964[15°]	1.4568	36.5	193.8
17	Coconut oil	*Cocos nucifera*	25.1	0.924[15°]	1.4493	10.4	268
18	Corn oil	*Zea mays*	(−20.0)	0.922[15°]	1.4734	122.6	192.0
19	Cotton seed oil	*Gossypium hirsutum*	(−1.0)	0.917[25°]	1.4735	105.7	194.3
20	Linseed oil	*Linum usitatissimum*	(−24.0)	0.938[15°]	1.478[25°]	178.7	190.3
21	Mustard oil	*Brassica hirta*	....	0.9145[15°]	1.475	102	174
22	Neem oil	*Melia azadirachta*	−3	0.917[15°]	1.4615	71	194.5
23	Niger-seed oil	*Guizotia abyssinica*	....	0.925[15°]	1.471	128.5	190
24	Oiticica oil	*Licania rigida*	....	0.974[25°]	....	140–180	....
25	Olive oil	*Olea europaea sativa*	(−6.0)	0.918[15°]	1.4679	81.1	189.7
26	Palm oil	*Elaeis guineensis*	35.0	0.915[15°]	1.4578	54.2	199.1
27	Palm-kernel oil	*E. guineensis*	24.1	0.923[15°]	1.4569	37.0	219.9
28	Peanut oil	*Arachis hypogaea*	(3.0)	0.914[15°]	1.4691	93.4	192.1
29	Perilla oil	*Perilla frutescens*	....	(0.935[15°])	1.481[25°]	195	192
30	Poppy-seed oil	*Papaver somniferum*	(−15)	0.925[15°]	1.4685	135	·194
31	Rapeseed oil	*Brassica campestris*	(−10)	0.915[15°]	1.4706	98.6	174.7
32	Safflower oil	*Carthamus tinctorius*	....	(0.900[60°])	1.462[60°]	145	192
33	Sesame oil	*Sesamum indicum*	(−6.0)	0.919[25°]	1.4646	106.6	187.9
34	Soybean oil	*Glycine soja*	(−16.0)	0.927[15°]	1.4729	130.0	190.6
35	Sunflower-seed oil	*Helianthus annuus*	(−17.0)	0.923[15°]	1.4694	125.5	188.7
36	Tung oil	*Aleurites fordi*	(−2.5)	0.934[15°]	1.5174[25°]	168.2	193.1
37	Wheat-germ oil	*Triticum aestivum*	....	....	....	125	....

[1] Caproic. [2] Capryli. [3] Capric. [4] Butyric. [5] Decenoic. [6] C_{12} monoethenoic. [7] C_{14} monoethenoic. [8] Gadoleic plus erucic. [9] C_{12} n-pentadecanoic. [10] C_{17} margaric. [11] 12-Methyl tetradecanoic. [12] C_{20} polyethenoic.

of variables such as source, treatment, and age of a fat or oil. **Specific Gravity** (column D) was calculated at the specified temperature (degrees centigrade) and referred to water at the same temperature, unless otherwise specified. **Density,** shown in parentheses (column D), was measured at the specified temperature (degrees centigrade). **Refractive Index** (column E) was measured at 50°C, unless otherwise specified.

Constituent Fatty Acids, g/100 g total fatty acids

	Saturated						Unsaturated				
	Lauric	Myristic	Palmitic	Stearic	Arachidic	Other	Palmitoleic	Oleic	Linoleic	Linolenic	Other
	(H)	(I)	(J)	(K)	(L)	(M)	(N)	(O)	(P)	(Q)	(R)
1	2.5	11.1	29.0	9.2	2.4	2.0[1]; 0.5[2]; 2.3[3]	4.6	26.7	3.6	....	3.6[4]; 0.1[5]; 0.1[6]; 0.9[7] 1.4[8]; 1.0[9]; 1.0[10]; 0.4[11]
2	....	2.7	24.0	8.4	....	....	5	46.9	10.2	....	2.5[8]
3	....	1.3	28.3	11.9	....	....	2.7	47.5	6	....	0.2[7]; 2.1[8]
4	....	....	17–18	2–3	....	....	....	74–76	....	....	....
5	....	6.3	27.4	14.1	....	....	....	49.6	2.5	....	....
6	....	4.6	24.6	30.5	....	....	....	36.0	4.3	....	....
7	....	5.8	8.4	0.6	....	....	20.0	←——29.1——→		....	25.4[12]; 9.6[13]
8	....	7.3	13.0	Trace	....	....	4.9	....	....	20.7	30.1[12]; 23.2[13]
9	....	5.9	16.3	0.6	0.6	....	15.5	....	....	29.6	19.0[12]; 11.7[13]; 0.8[14]
10	....	5.1	14.6	3.2	....	....	11.8	←——17.8——→		....	18.1[12]; 14.0[13]; trace[7]; 15.4[15]
11	1	5	6.5	....	....	....	26.5	37	19	....	1[13]; 4[7]; 19[16]
12	16	14	8	2	....	3.5[3]	15	17	6.5	....	4[6]; 14[7]; 6.5[16]
13	0.2	9.3	15.6	2.8	....	....	14.4	35.2	....	....	13.6[12]; 5.9[13]; 2.5[7]; 0.2[17]
14	44.1	15.4	8.5	2.7	0.2	0.2[1]; 4.8[2]; 6.6[3]	....	16.1	1.4	....	....
15	←————2.4————→						....	7.4	3.1	....	87[18]
16	....	....	24.4	35.4	....	....	....	38.1	2.1	....	....
17	45.4	18.0	10.5	2.3	0.4[19]	0.8[1]; 5.4[2]; 8.4[3]	0.4	7.5	Trace	....	....
18	....	1.4	10.2	3.0	....	....	1.5	49.6	34.3	....	....
19	....	1.4	23.4	1.1	1.3	....	2.0	22.9	47.8	....	....
20	....	....	6.3	2.5	0.5	....	....	19.0	24.1	47.4	0.2[11]
21	....	1.3[20]	....	....	....	....	....	27.2[20]	16.6[20]	1.8[20]	1.1[11]; 1.0[21]; 51.0[22]
22	....	2.6[20]	14.1[20]	24.0[20]	0.8[20]	....	....	58.5[20]	....	....	....
23	....	3.3[20]	8.2[20]	4.8[20]	0.5[20]	....	....	30.3[20]	57.3[20]	....	....
24	←————11.3[23]————→						....	6.2	....	....	82.5[21]
25	....	Trace	6.9	2.3	0.1	....	....	84.4	4.6	....	....
26	....	1.4	40.1	5.5	....	....	....	42.7	10.3	....	....
27	46.9	14.1	8.8	1.3	....	2.7[2]; 7.0[3]	....	18.5	0.7	....	....
28	....	....	8.3	3.1	2.4	....	....	56.0	26.0	....	3.1[11]; 1.1[21]
29	←————9.6[23]————→						....	17.8	....	17.5	....
30	....	....	4.8[20]	2.9[20]	....	....	....	30.1[20]	62.2[20]	....	....
31	....	....	1	....	....	....	....	32	15	1	50[22]
32	←————6.8[23]————→						....	18.6	70.1	3.4	....
33	....	....	9.1	4.3	0.8	....	....	45.4	40.4	....	....
34	0.2	0.1	9.8	2.4	0.9	....	0.4	28.9	50.7	6.5	0.1[7]
35	....	....	5.6	2.2	0.9	....	....	25.1	66.2	....	....
36	←————4.6[23]————→						....	4.1	0.6	....	90.7[25]
37	←————16.0[23]————→						....	28.1	52.3	3.6	....

[13] C_{22} polyethenoic. [14] Behenic. [15] C_{14} polyethenoic. [16] Gadoleic. [17] C_{24} polyethenoic. [18] Ricinoleic. [19] Includes behenic and lignoceric. [20] Percent by weight. [21] Lignoceric. [22] Erucic. [23] Includes behenic. [24] Licanic. [25] Eleostearic.

WAXES

These data for waxes were compiled originally for the Biology Data Book by A. H. Warth. Data are reproduced here by permission of the copyright owners of the above publication, the American Societies for Experimental Biology, Washington, D.C. p. 382.

Specific Gravity (column C) was calculated at the specified temperature, degrees centigrade, and referred to water at the same temperature. **Density.** shown in parentheses (column C), and **Refractive Index** (column D) were measured at the specified temperature, degrees centigrade.

Wax	Melting Point °C	Specific Gravity or (Density)	Refractive Index $n\frac{°C}{D}$	Iodine Value	Acid Value	Saponification Value
(A)	(B)	(C)	(D)	(E)	(F)	(G)
1 Bamboo leaf	79–80	$(0.961^{25°})$	....	7.8^1	14.5	43.4
2 Bayberry (myrtle)	46.7–48.8	$(0.985^{15°})$	$1.436^{80°}$	2.9^2–3.9^3	3.5	20.5–21.7
3 Beeswax, crude	62–66	$(0.927–0.970^{15°})$	$1.439–1.483^{40°}$	$6.8–16.4^2$	16.8–35.8	89.3–149.0
4 Beeswax, white, U.S.P.	61–69	$(0.959–0.975^{15°})$	$1.447–1.465^{65°}$	$7–11^3$	17–24	90–96
5 Beeswax, yellow	62–65	$(0.960–0.964^{15°})$	$1.443–1.449^{65°}$	6–11	18–24	90–97
6 Candelilla, refined	67–69	$(0.982–0.986^{15°})$	$1.454–1.463^{85°}$	14.4–20.4	12.7–18.1	35–86
7 Cape berry[4]	40.5–45.0	$(1.004–1.007^{15°})$	$1.450^{45°}$	0.6–2.4	2.5–3.7	211–215
8 Carandá	79.7–84.5	$(0.990^{25°})$	...·........	8.0–8.9	5.0–9.5	64.5–78.5
9 Carnauba	83–86	$0.990–1.001^{15°}$	$1.467–1.472^{40°}$	7.2–13.5	2.9–9.7	78–95
10 Castor oil, hydrogenated	83–88	$(0.980–0.990^{20°})$	....	2.5–8.5	1.0–5.0	177–181
11 Chinese insect	81.5–84.0	$0.950–0.970^{15°}$	$1.457^{40°}$	1.4	0.2–1.5	73–93
12 Cotton	68–71	$0.959^{15°}$	....	24.5	32	70.6
13 Cranberry	207–218	$(0.970–0.975^{15°})$	....	$44.2–53.2^2$	42.2–59.1	131–134
14 Douglas-fir bark	59.0–72.8	$(1.030^{25°})$	$1.468^{80°}$	25.8–62.5	58.6–80.7	112–200
15 Esparto	67.5–78.1	$0.988^{15°}$	....	22–23	22.7–23.9	69.8–79.3
16 Flax	61.5–69.8	$0.908–0.985^{15°}$	....	21.6–28.8	17.5–48.3	77.5–101.5
17 Ghedda, E. Indian beeswax	60.5–66.4	$0.956–0.973^{15°}$	$1.440^{50°}$	5.6–12.6	5.8–7.9	84.5–118.3
18 Indian corn	80–81	....	....	4.2^2	1.9	120.3
19 Japan wax	48–53	$0.975–0.993^{15°}$	....	4.5–12.5	6–20	206.5–237.5
20 Jojoba	11.2–11.8	$0.864–0.899^{25°}$	$1.465^{25°}$	$81.7–88.4^2$	0.2–0.6	92.2–95.0
21 Madagascar	88	....	....	3.2–5.3	17.7–28.0	140.0–159.6
22 Microcrystalline, amber	64–91	$0.913–0.943^{15°}$	$1.424–1.452^{80°}$	0	0	0
23 Microcrystalline, white	71–89	$0.928–0.941^{15°}$	$1.441^{80°}$	0	0	0
24 Montan, crude	76–86	$(1.010–1.020^{25°})$	....	13.9–17.6	22.7–31.0	59.4–92.0
25 Montan, refined	77–84	$(1.010–1.030^{25°})$	....	10–14	24–43	72–103
26 Orange peel	44.0–46.5	$0.985^{15°}$	$1.502^{20°}$	115.7^2	48.3	120.9
27 Ouricury, refined	79.0–83.8	$1.053^{15°}$	....	$6.9–7.8^2$	3.4–21.1	61.8–85.8
28 Ozocerite, refined	74.4–75.0	$0.907–0.920^{15°}$	....	0	0	0
29 Palm	74–86	$(0.991–1.045^{15°})$	....	$8.9–16.9^2$	5.0–10.6	64.5–104.0
30 Paraffin, American	49–63	$0.896–0.925^{15°}$	$1.442–1.448^{80°}$	0	0	0
31 Peat wax, natural	73–76	$0.980^{15°}$	....	16–40	60.0–73.3	73.9–136.0
32 Rice bran, refined	75.3–79.9	....	$1.469^{30°}$	11.1–19.4	15–17	56.9–104.4
33 Shellac wax	79–82	$0.971–0.980^{15°}$	....	$6.0–8.8^3$	12.1–24.3	63.8–83.0
34 Sisal hemp	74–81	$1.007–1.010^{15°}$	....	$28–29^2$	$16–19^2$	56–58
35 Sorghum grain	77–82	....	....	15.7–20.9	10.1–16.2	16–44
36 Spanish moss	79–80	....	....	33.0	25.0	120.4
37 Spermaceti	42–50	$0.905–0.945^{15°}$	$1.440^{70°}$	4.8–5.9	2.0–5.2	108–134
38 Sugarcane, crude	52–67	$0.988–0.998^{25°}$	....	32–84	24–57	128–177
39 Sugarcane, double-refined	77–82	$0.961–0.979^{25°}$	$1.510^{25°}$	13–29	8–23	55–95
40 Wool wax, refined	36–43	$0.932–0.945^{15°}$	$1.478–1.482^{40°}$	15.0–46.9	5.6–22.0	80–127

[1] Wijs test. [2] Hanus test. [3] Hubl test. [4] *Myrica cordifolia.*

DIAMAGNETIC SUSCEPTIBILITIES OF ORGANIC COMPOUNDS

Compiled by George W. Smith

The following table contains values for the molar susceptibility of χ_M, specific susceptibility χ, and volumetric susceptibility K. The cgs Gaussian system of units is employed. In the Gaussian units the relation between magnetic induction B and magnetic field strength H is

$$B = H + 4\pi I \qquad (1)$$

where I is the magnetization or magnetic moment per unit volume. Actually, the quantities involved in equation (1) are all vectors, but one may assume that all three are collinear, a reasonable assumption for organic diamagnetic substances in the liquid state. For crystals I may vary with crystal orientation. Equation (1) may be rewritten

$$B = H + 4\pi KH = (I + 4\pi K)H \qquad (2)$$

Here, K is the magnetic susceptibility, often called the volumetric susceptibility and is a unitless quantity.

Other susceptibilities of use to chemists and physicists are the specific or mass susceptibility which is defined

$$\chi = K/\rho \qquad (3)$$

where ρ is the density of the sample in grams per cc., and the molar susceptibility which is defined as

$$\chi_M = M\chi = MK/\rho \qquad (4)$$

where M is the molecular weight of the substance in grams.

Temperatures, when listed, are enclosed in parentheses and are listed in degrees C.

Literature references for values contained in this table may be found in General Motors Research Laboratories publication GMR-317.

Compound	$-\chi_M \times 10^6$	$-\chi \times 10^6$	$-K \times 10^6$	Compound	$-\chi_M \times 10^6$	$-\chi \times 10^6$	$-K \times 10^6$
Acenaphthanthracene	184	.73		Amyl iodide	(118.7)	.5996 (18°)	(.910) (20°)
Acenaphthene	109.3	(.709)	(.726) (99°)	n-Amyl methyl ketone	80.50	.705$_3$	(.580) (15°)
Acetal	81.39	.688 (32°)	(.568) (32°)	Amyl nitrate	(76.4)	.574	
Acetaldehyde	22.70	.515$_3$	(.403) (18°)	iso-Amyl propionate	101.73	705$_1$ (25°)	(.609) (25°)
Acetamide	34.1	.577	(.618) (20°)	n-Amyl valerate	124.55	.723$_9$	(.638) (0°)
Acetic acid	31.54	.525 (32°)	(.551) (32°)	Anethole	(96.0)	.648	(.644) (15°)
Acetic anhydride	(52.8)	.517	(.562) (15°)	Aniline	62.95	(.676)	(.691) (20°)
Acetoaminofluorene	(141)	.63		Anisidine	(80.5)	.654	(.72) (20°)
Acetone	33.7$_8$	.581$_1$	(.460) (20°)	Anisole	72.79	(.673)	(.672) (15°)
Acetonitrile	28.0	(.682)	(.534) (20°)	Anthanthrene	204.2	.739	
Acetonylacetone	62.51	.547$_6$	(.531) (20°)	Anthanthrone	178.1	.581	
Acetophenone	72.05	.599$_8$	(.615) (20°)	Anthracene	(130)	.731	(.914) (27°)
Acetophenone oxime	79.9$_0$	.592$_0$		Anthracenedinitrile	154.6	(.678)	
Acetophenone oxime-O-methyl ether	92.3$_1$	.618$_5$		Anthracenonitrile	142.1	(.700)	
Acetoxime	44.4$_2$	607$_6$	(.480) (20°)	Anthraquinone	(119.6)	.575	(.825) (20°)
Acetoxime-O-benzyl ether	104.8$_9$	.642$_7$		Anthrazine	245.7	.646	
Acetoxime-O-methyl-ether	54.8$_7$	.629$_8$		Arabinose	85.70	.571	(.905) (20°)
Acetylacetone	54.88	.548$_1$	(.535) (20°)	Arbutoside	158.0	(.589)	
Acetyl chloride	38.9	.496	(.548) (20°)	Asarone	131.4	(.631)	(.735) (18°)
Acetylene	12.5	(.480)		Asparagine	69.5	(.526)	(.812) (15°)
Acetylphenylacetylene	86.9	.508		Aspartic acid	64.2 ± .4	(.482)	(.800) (12°)
Acetylthiophene	71.7	(.568)		Aurin	(161.4)	.556	
Acridine	(123.3)	.688	(.757) (20°)	p-Azoanisole	(147.7)	.610	
Adonitol	91.30	.600		Azobenzene	(106.8)	.586	.611 (70.5°)
Alanine	50.5	(.567)		p-Azophenetole	(171.7)	.635	
Allyl acetate	(56.7)	.566	(.525) (20°)	m-Azotoluene	(127.8)	.608	.643 (58°)
Allyl alcohol	36.70	.632	.540 (20°)	Azulene	98.5	.768	
1-Allylpyrrole	73.80	.685 (20°)		Barbituric acid (Anh.)	53.8	(.420)	
Aminoazobenzene	(118.3)	.600		Barbituric acid (.2H$_2$O)	78.6	(4.79)	
Aminoazotoluene	(142.2)	.631		Benzalazine	(123.7)	.594	
o-Aminoazotoluene	(138)	.61 ± .02		Benzaldehyde	60.78	(.573)	(.602) (15°)
α-Aminobutyric acid	62.1	(.602)		Benzaldoxime	(69.8)	.576	(.639) (20°)
Aminomethyldiethyldiazine	114.8	(.696)		Benzamide	(72.3)	.597	(.801) (4°)
4-Aminostilbene	122.5	(.628)		Benzanthrone	142.9	.620	
2-Aminothiazole	56.0	.564		Benzene	54.84	.702 (32°)	.611
n-Amyl acetate	89.06	684$_5$	.5979 (20.7°)	Benzidine	110.9	.603	(.754) (20°)
iso-Amyl acetate	89.40	.687	.599 (20°)	Benzil	(118.6)	.564	.616 (100°)
n-Amyl alcohol	(67.5)	.766	(.624) (20°)	Benzoic acid	70.28	(.575)	(.728) (15°)
γ-Amyl alcohol	(71.0)	.8060	(.655) (25°)	Benzoic anhydride	(124.9)	.552	(.662) (15°)
Inactive Amyl alcohol	69.06	783$_1$ (25°)		Benzonitrile	65.19	(.632)	(.638) (15°)
iso-Amyl alcohol	68.96	.782$_3$ (25°)	(.64) (15°)	Benzophenone	109.60	.601$_3$	(.66) (50°)
sec-Amyl alcohol	69.1	.785	(.635) (20°)	3,4-Benzopyrene	135.7	.538	
tert-Amyl alcohol	(70.9)	.804	(.654) (15°)	Benzopyrene	194.0		
n-Amylamine	69.4	(.796)	(.606) (20°)	Benzoyl acetone	(95.0)	.586	(.639) (60°)
iso-Amylamine	71.6	(.821)	(.616) (20°)	Benzoyl chloride	(75.8)	.539 (20°)	(.657) 15°)
n-Amylbenzene	112.55	(.759)	(.652) (20°)	Benzyl acetate	93.18	(.620)	(.655) (16°)
iso-Amyl bromide	(88.7)	.587	(.706) (20°)	Benzyl alcohol	71.83	(.664)	(.697) (15°)
iso-Amyl-n-butyrate	113.52	.717$_1$ (25°)	(.616) (25°)	Benzylamine	75.26	(.702)	(.690) (19°)
iso-Amyl chloride	(79.0)	.741	(.662) (20°)	Benzyl chloride	81.98	(.647)	(.713) (18°)
iso-Amyl cyanide	73.4	(.755)	(.609) (20°)	Benzyl formate	81.43	(.598)	(.646) (20°)
iso-Amylene	53.7	.766		Benzylideneaniline	(100.4)	.554	
Amylene bromide	(114.5)	.498		Benzylidene chloride	(97.9)	.608	(.763) (14°)
Amylene chloride	(95.2)	.675		Benzylidenemethylamine	(73.1)	.613	
iso-Amyl ether	(129)	.813	(.635) (15°)	Benzyl methyl ketone	83.44	.621$_9$	(.624) (20°)
iso-Amyl formate	78.38	.674$_5$	(.591) (25°)	Bibenzyl	(126.8)	.696	.671 (54.5°)
Amylidene chloride	(93.4)	.662					

Compound	$-\chi_M \times 10^6$	$-\chi \times 10^6$	$-K \times 10^6$
3,4-Bis-(p-hydroxyphenyl)-2,4-hexadiene	(157)	.59	
m,m'-Bitolyl	(127.4)	.6993 (27.4°)	(.699) (16°)
m,m'-Bitolyl sulfide	(140.0)	.6530 (27.4°)	
Borneol	126.0	(.817)	(.826) (20°)
Bromobenzene	78.92	.5030 (20°)	(.753) (20°)
Bromobenzenediazocyanide	86.88	(.414)	
Bromochloromethane	55.0 ± .6	(.425)	(.846) (19°)
Bromodichloromethane	66.3 ± .3	(.405)	(.812) (15°)
Bromoform	82.60	.327	.948 (20°)
Bromonaphthalene	(123.8)	.598	
α-Bromonaphthalene	115.90	.560	.840 (20°)
m-Bromotoluene	(93.4)	.546	(.770) (20°)
Bromotrichloromethane	73.1 ± .7	(.369)	(.758) (0°)
Butane	57.4	(.988)	
iso-Butane	51.7	(.890)	
1,4-Butanediol	61.5	(.682)	(.696) (20°)
2-Butene (cis)	42.6	(.759)	
2-Butene (trans)	43.3	(.772)	
1-Butene-3,4-diacetate	95.5	(.555)	
2-Butene-1,4-diacetate (cis)	95.2	(.553)	
2-Butene-1,4-diacetate (trans)	95.1	(.552)	
2-Butene-1,4-diol (cis)	54.3	(.616)	
2-Butene-1,4-diol (trans)	53.5	(.607)	
n-Butyl acetate	77.47	.666$_9$ (25°)	(.583) (25°)
iso-Butyl acetate	78.52	.676$_9$	(.584) (25°)
n-Butyl alcohol	56.536 (20°)	(.7627)	(.6176) (20°)
iso-Butyl alcohol	57.704 (20°)	(.7785)	(.624) (20°)
sec-Butyl alcohol	57.683 (20°)	(.7782)	(.629) (20°)
tert-Butyl alcohol	57.42	.774$_7$ (25°)	(.611) (20°)
n-Butylamine	58.9	(.805)	(.596) (20°)
iso-Butylamine	59.8	(.818)	(.599) (20°)
9-Butyl anthracene	176.0	(.751)	
n-Butylbenzene	100.79	(.751)	(.646) (20°)
iso-Butylbenzene	101.81	(.759)	(.648) (20°)
tert-Butylbenzene	102.5	(.764)	(.662) (20°)
n-Butyl benzoate	116.69	.654$_8$	(.656) (25°)
Butyl bromide	77.14	.563 (20°)	(.730) (20°)
iso-Butyl bromide	79.88	.583 (20°)	(.737) (20°)
1-n-Butyl chloride	67.10	.725	.642 (20°)
2-n-Butyl chloride	67.40	.728	.635 (20°)
n-Butyl cyanide	(62.8)	.7558 (27.4°)	(.606) (20°)
tert-Butyl cyclohexane	115.09	.8205	(.6670) (20°)
1,4-Butyl diacetate	103.4	(.594)	
n-Butyl ethyl ketone	80.73	.707$_2$	(.579) (20°)
n-Butyl formate	65.83	.644$_6$	(.571) (25°)
iso-Butyl formate	66.79	.654$_9$	(.574) (25°)
iso-Butylideneazine	(95.8)	.683	
Butyl iodide	(93.6)	.5086 (18°)	(.822) (20°)
iso-Butyl methyl ketone	70.05	.699$_5$	(.561) (20°)
tert-Butyl methyl ketone	69.86	.697$_9$	(.558) (16°)
n-Butyl perfluor-n-butyrate	126.7	(.469)	
p-tert-Butylphenol	108.0	(.719)	(.653) (114°)
Butyl sulfide	(113.7)	.7774 (27.4°)	(.652) (16°)
Butyl thiocyanate	(79.38)	.6891 (27.4°)	(.659) (25°)
2-Butyne-1,4-diacetate	95.9	(.564)	
2-Butyne-1,4-dibenzoate	169.0	(.561)	
2-Butyne-1,4-diol	50.3	(.584)	
n-Butyraldehyde	46.08	.639$_4$	(.522) (20°)
iso-Butyraldehyde	46.38	.643$_6$	(.511) (20°)
iso-Butyraldoxime	56.1$_2$	.644$_3$	(.576) (20°)
n-Butyric acid	55.10	.625	.598 (20°)
iso-Butyric acid	56.06	.636$_3$ (25°)	(.601) (25°)
Butyronitrile	49.4	(.715)	(.569) (15°)
Butyrylphenylacetylene	(106.4)	.618	
Cacodyl	(99.9)	.476	(.689) (15°)
Cacodylic acid	(79.9)	.579	
Camphor	(103)	.68	(.67) (25°)
Camphoric acid	129.0	(.644)	(.791) (20°)
Camphoric anhydride	(113)	.620	(.740) (20°)
n-Caproic acid	78.55	.676$_2$	(.624) (25°)
Caproylphenylacetylene	(130 4)	.651	
n-Caprylic acid	101.60	.7053	(.642) (20°)
Carbanilide	134.05	(.6316)	(.783) (20°)
Carbazole	117.4	(.702)	
Carbon disulfide	42.2	554	(.699) (22°)
Carbon tetrabromide	93.73	.2826 (20°)	(.966)
Carbon tetrachloride	66.60	433	.691 (20°)
Carbon tetraiodide	(136)	.261	(1.13) (20°)
Carvacrol	(109.1)	.726	(.709) (20°)
Carvone	(92.2)	.614	(.590) (20°)
Cetyl alcohol	(183.5)	.757 (17.5°)	(.619) (50°)
Cetyl mercaptan	390.4	(1.510)	
Chloral	(67.7)	.459	(.694) (20°)
Chloranil	(112.6)	.458	
Chloracetic acid	48.1	(.509)	(.804) (20°)
Chloroacetone	(50.9)	.550	(.633) (20°)
Chloroacetylchloride	53.7	(.475)	(.710) (0°)
p-Chloranisole	89.1	(.625)	
Chlorobenzene	69.97	(.6216)	(.688) (20°)
Chlorobenzene diazocyanide	65.02	(.393)	
Chlorodibromomethane	75.1 ± .4	(.361)	(.883) (15°)
Chlorodifluoromethane	38.6	.446	
1-Chloro-2,3-dihydroxypropane	(77.9)	.604	
Chlorodiphenylmethane	⌐131.9	(.651)	
Chloroethylene	35.9	.574	(.528) (liq., 15°)
Chloroform	59.30	.497	.740 (20°)
Chlorofumaric acid	67.02	(.445)	
p-Chloroiodo benzene	99.42	(.417)	(.786) (57°)
Chloromaleic acid	67.36	(.448)	
Chloromethylstilbene	144.8	(.633)	
α-Chloronaphthalene	107.60	.661	.789 (20°)
m-Chloronitrobenzene	(74.8)	.475	.638 (48°)
o-Chlorophenol	77.4	(.602)	(.747) (18°)
p-Chlorophenol	77.6	(.604)	(.789) (20°)
1-(o-Chlorophenylazo)-2-naphthol	161.0	(.570)	
1-(p-Chlorophenylazo)-2-napthol	161.4	(.571)	
Chlorotrifluoroethylene	49.1	.422	
Chlorotrifluoromethane	45.3 ± 1.5	(.434)	
Cholesterol	(284.2)	.735	(.784) (20°)
Chrysene	166.67	.731	
Chrysoidine	(126.3)	.595	
Cinnamic acid	78.36	.529	(.660) (4°)
Cinnamic acid (α-trans)	78.2	(.528)	
Cinnamic acid (β-trans)	79.0	(.533)	
Cinnamic acid (cis-MP 68°)	77.6	(.524)	
Cinnamic acid (cis-MP 58°)	77.9	(.526)	
Cinnamic acid (cis-MP 42°)	83.2	(.562)	
Cinnamic aldehyde	(74.8)	.566	(.629) (15°)
Cinnamyl alcohol	(87.2)	.650	(.679) (20°)
Cinnamylideneaniline	123.2	(.595)	
Citral	(98.9)	.650	(.577) (20°)
Coronene	(243.3)	.810	
Coumarin	82.5	(.565)	(.528) (20°)
o-Cresol	72.90	.675	(.706) (20°)
m-Cresol	72.02	.667 (26°)	(.690) (20°)
p-Cresol	72.1	.667 (26°)	(.690) (20°)
o-Cresylmethyl ether	81.94	.671 (40°)	(.661) (15°)
m-Cresylmethyl ether	77.91	.638 (40°)	(.623) (15°)
p-Cresylmethyl ether	79.13	.648 (40°)	(.629) (19°)
Cumene	89.53	.744$_9$	(.642)
Cyamelide	(56.1)	.435	(.490) (15°)
Cyameluric acid	101.1 (10°)	(.457)	
9-Cyanoanthracene	142.1	(.699)	
Cyanogen	(21.6)	.415	(.359) (liq., 17°)
Cyanuric acid	61.5	.476	(.842) (0°)
Cyclobutanecarboxylic acid	58.16	.5816 (30°)	(.613) (30°)
1,3-Cyclohexadiene	48.6	(.607)	(.510) (20°)
1,4-Cyclohexadiene	48.7	(.608)	(.515) (20°)
Cyclohexane	68.13	.8100 (27.5°)	(.627) (20°)
Cyclohexanecarboxylic acid	83.24	.6499 (30°)	(.668) (30°)
Cyclohexanol	73.40	.732	.694 (20°)
Cyclohexanone	62.0$_1$	.632$_3$	(.599) (20°)
Cyclohexanone oxime	71.5$_2$	.632$_1$	
Cyclohexanoneoxime-O-methyl ether	82.9$_6$	.652$_3$	
Cyclohexene	57.5	(.700)	(.567) (20°)
Cyclohexenol	64.1	(.653)	
Cyclooctane	91.4	(.815)	(.684) (20°)
Cyclooctene	84.6	(.769)	(.654) (20°)
Cyclooctatetraene	(53.9)	.518	
Cyclopentane	59.18	.8439	.6290 (20°)
Cyclopentanecarboxylic acid	73.48	.6446 (30°)	(.677) (30°)
Cyclopentanone	51.63	.6141 (30°)	(.582) (30°)
Cyclopropane	39.9	(.948)	(.683) (−79°)
Cyclopropanecarboxylic acid	45.33	.5271 (30°)	(.569) (30°)
p-Cymene	102.8	.766 (20°)	(.656) (20°)
Decalin	106.70	.7718	.6814 (20°)
cis-Decalin	(107.0)	.774	(.686) (35°)
trans-Decalin	(107.7)	.779	(.670) (35°)
n-Decane	119.74	(.8416)	(.6143) (20°)
1-Deuterio pyrrole	48.75	.716 (20°)	
Deuteroindene	80.88	.690	(.692) (13°)
Diacetal	(153.8)	.668	
Di-iso-amylamine	(133.1)	.846	(.649) (21°)
Diazoacetic ester	57	(.50)	(.54) (24°)
Dibenzocoronene	289.4	.778	
1,2,5,6-Dibenzofluorene	184	.69 ± .03	
3,4,5,6-Dibenzophenanthrene	(203)	.73 ± .03	
Dibenzphenanthrone	200.5	.716	
Dibenzpyrene	213.6	.706	
Dibenzpyrenequinone	183.1	.551	
iso-Dibenzpyrenequinone	194.6	.586	
Dibenzyl ketone	131.70	.626$_2$	
p-Dibromobenzene	(101.4)	.430	.786 (100°)
2,3-Dibromo-2-butene-1,4-diol	94.2	(.383)	
Dibromodichloromethane	81.1 ± .4	(.334)	(.808) (25°)
1,2-Dibromodiiodoethylene	(140.1)	.320	
1,2-Dibromoethylene	(71.7)	.386	(.877) (17.5°)
1,2-Dibromo-2-fluoroethane	(78.0)	.379	(.855) (17°)
Dibromo-4-nitrophenol	(167.5)	.564	
1,2-Dibromotetrachloroethane	(126.0)	.387	(1.049)
Di-n-butylamine	103.7	(.802)	(.767) (20°)
Di-iso-butylamine	105.7	(.817)	(.609) (20°)
Di-sec-butylamine	105.9	(.819)	(.641) (0°)
Di-iso-butyl ketone	104.30	.733$_8$	(.591) (20°)
Di-tert-butyl ketone	104.06	.732$_1$	
2,6-Di-tert-butyl-4-methyl phenol	165.3	(.750)	
2,4-Di-tert-butyl phenol	155.6	(.754)	
Dibutyl phthalate	175.1	(.629)	(.657) (21°)
Di-iso-butyralacetylene	(125.6)	.738	
Dicetyl sulfide	401.7	(.832)	
Dichloroacetic acid	58.2	(.451)	(.705) (20°)
Dichloroacetyl chloride	69.0	(.468)	

Compound	$-\chi_M \times 10^6$	$-\chi \times 10^6$	$-K \times 10^6$	Compound	$-\chi_M \times 10^6$	$-\chi \times 10^6$	$-K \times 10^6$
o-Dichlorobenzene	84.26	.5734	(.748) (20°)	2,5-Dimethylpyrrole	71.92	.756 (20°)	(.707) (20°)
m-Dichlorobenzene	83.19	.5661	(.729) (20°)	α,ω-Dimethyl styrene	(90.7)	.686	
p-Dichlorobenzene	82.93	.5644	(.823) (20.5°)	Dimethyl succinate	81.50	.5581	(.625) (18°)
1,4-Dichloro-2-butyne	74.2	(.603)		Dimethyl sulfate	(62.2)	.493	(.657) (20°)
1,2-Dichloro-1,2-dibromoethane	(108.6)	.423		Dimethyl sulfide	(44.9)	.723	(.612) (21°)
1,1-Dichloro-difluoroethylene	60.0	.451		Dimethyltrichloromethylcarbinol	(105)	.59	
Dichlorodifluoromethane	52.2	.432	(.642) (−30°)	N,N-Dimethyl urea	55.1	(.625)	(.784)
1,1-Dichloroethylene	49.2	.508	(.635) (15°)	N,N′-Dimethyl urea	56.3	(.639)	(.730)
cis-1,2-Dichloroethylene	51.0	.526	(.679) (15°)	o-Dinitrobenzene	65.98	.3921	(.614) (17°)
trans-1,2-Dichloroethylene	48.9	.504	(.638) (15°)	m-Dinitrobenzene	70.53	.4197	(.659) (0°)
1,3-Dichloro-2-hydroxypropane	(80.1)	.621		p-Dinitrobenzene	68.30	.4064	(.660) (30°)
Dicyandiamide	44.55	(.530)	(.742) (14°)	2,4-Dinitrophenol	(73.1)	.397	(.668) (24°)
Dicyclohexanol acetylene	(151.6)	.682		Dinitroresorcinol	(62.4)	.312	
Dicyclohexyl	129.31	.7776	.6889 (20°)	1,4-Dioxane	52.16	.592 (32°)	(.606) (32°)
1,1-Dicyclohexylnonane	231.98	.7930	.7001 (20°)	Diphenyl	103.25	.6695	(.664) (73°)
Diethanolacetylene	(75.3)	.660		1,1-Diphenylallyl-3-chloride	146.1	(.639)	
Diethyl acetaldehyde	70.71	.705₉	(.576) (20°)	1,3-Diphenylallyl-3-chloride	140.7	(.615)	
Diethylallylacetophenone	146.2	(.676)	(.663) (16°)	Diphenylamine	(109.7)	.648	.686 (55.5°)
Diethyl allylmalonate	118.8	(.593)	(.602) (14°)	Diphenyl-bis-diazo cyanide	85.03	(.327)	
Diethylamine	56.8	(.777)	(.552) (18°)	Diphenylbutadiene	129.6	(.629)	
Diethylcyclohexylamine	(124.5)	.802	(.699) (0°)	Diphenylchloroarsine	(145.5)	.550	(.871) (40°)
Diethyl ethylmalonate	115.2	(.612)	(.614) (20°)	Diphenyldecapentaene	180.5	(.635)	
Diethyl ketone	58.1₄	.675₁	(.551) (19°)	Diphenyldiacetylene	(134.6)	.640	
Diethyl ketoxime	68.3₁	.6754		Diphenyldiazomethane	115	(.592)	
Diethyl malonate	(92.6)	.5782	(.611) (20°)	Diphenyldihydrotetrazine	129.9	(.545)	
Diethyl-3-(1-methyl butane) ethyl-malonate	175	(.677)		1,1-Diphenylethylene	(118.0)	.655	(.680) (14°)
Diethyl oxalate	81.71	.5595	(.603) (15°)	1,6-Diphenylhexane	171.81	.7208	.6877 (20°)
Diethyl phthalate	127.5	(.574)	(.645) (25°)	Diphenylhexatriene	146.9	(.632)	
Diethyl sebacate	(177.0)	.685	(.661) (20°)	Diphenylmethane	(115.7)	.688	.684 (35.5°)
Diethylstilbestrol	172.0	(.547)		Diphenylmethanol	119.1	.647	
Diethylstilbestrol dipropionate	265.2	.720		1,1-Diphenylnonane	206.32	.7357	.6935 (20°)
Diethyl succinate	105.07	.6035	(.628) (20°)	Diphenyloctatetraene	164.3	(.636)	
Diethyl sulfate	(86.8)	.563	(.667) (15°)	Diphenylphenoxyarsine	(225.2)	.567	
Diethyl sulfide	(67.9)	.753	(.630) (20°)	N,N-Diphenyl urea	126.3	(.595)	(.759)
Diethyl tartrate	(113.4)	.550	(.662) (20°)	N,N′-Diphenyl urea	127.5	(.600)	(.743) (20°)
Difluoroacetamide	(41.2)	.433		Di-n-propyl ketone	80.45	.705₀	(.576) (20°)
1-Difluoro-2-dibromoethane	(85.5)	.382	(.883) (20°)	Di-iso-propyl ketone	81.14	.711₀	(.573) (20°)
1,1-Difluoro-2,2-dichloroethyl amyl ether	129.84	(.587)	(.694) (20°)	Dipropyl oxalate	105.27	.6046	(.628) (0°)
1,1-Difluoro-2,2-dichloroethyl butyl ether	119.48	(.577)	(.703) (20°)	Di-iso-propyl oxalate	106.02	.6089	
1,1-Difluoro-2,2-dichloroethyl ethyl ether	96.13	(.537)	(.723) (20°)	Dodecyl alcohol	147.70	.7849 (20.7°)	(.652) (24°)
1,1-Difluoro-2,2-dichloroethyl methyl ether	80.68	(.489)	(.696) (20°)	Dulcitol	112.40	.617	(.905) (15°)
				Elaidic acid	204.8	(.725)	(.619) (79°)
				Erythritol	73.80	.604	(.876) (20°)
1,1-Difluoro-2,2-dichloroethyl propyl ether	107.19	(.555)	(.701) (20°)	Ethane	27.3₇	(.910)	(.511) (−100°)
Difluoroethanol	(41.3)	.503		4-Ethoxy-3-methoxybenzyl acetate	138.5	.619	
Di-n-heptylamine	171.5	(.805)		4-Ethoxy-3-methoxybenzyl benzoate	177.3	.620	
Di-n-hexylamine	148.9	(.803)		1-Ethoxynaphthalene	119.9	(.696)	(.738) (20°)
Dihydronaphthalene	(85.1)	.654	(.652) (12°)	2-Ethoxynaphthalene	119.2	(.692)	(.734) (25°)
o-Dimethoxybenzene	87.39	.6329	(.686) (25°)	Ethyl acetate	54.10	.614	.554 (20°)
m-Dimethoxybenzene	87.21	.6316	(.682) (0°)	Ethyl acetoacetate	71.67	.550₈	(.565) (20°)
p-Dimethoxybenzene	86.65	.6275	(.661) (55°)	Ethylacetophenone	95.5	(.644)	(.639) (16°)
o-(2,5-Dimethoxybenzoyl)-benzoic acid	161.0	(.562)		Ethyl alcohol	33.60	.728	.575 (20°)
Dimethoxymethane	(47.3)	.621	(.532)	Ethylally acetophenone	122.5	(.651)	(.634) (16°)
Dimethylacetophenone	96.8	(.653)	(.645) (16°)	Ethyl amylpropiolate	(112.7)	.670	
Dimethylallylacetophenone	122.4	(.650)	(.635) (16°)	Ethylaniline	89.30	(.737)	(.709)
Dimethylaniline	89.66	(.740)		9-Ethyl anthracene	153.0	(.741)	(.771) (99°)
2,2-Dimethylbutane	76.24	.8848	.5744 (20°)	Ethylbenzene	77.20	.7272	.6341 (20°)
2,3-Dimethylbutane	76.22	.8845	.5853 (20°)	Ethyl benzoate	93.32	.621₁	(.648) (25°)
2,3-Dimethyl-2-butene	65.9	(.783)	(.557)	Ethyl benzoylacetate	(115.3)	.600	(.673) (20°)
Dimethylcyclohexanone	(84.8)	.672		Ethyl benzylidenecyanoacetate	(116.3)	.578	
1,2 and 1,3 Dimethylcyclopentanes	81.31	.8281	.6224 (20°)	Ethyl benzylmalonate	(154.5)	.6172	(.663) (20°)
2,5-Dimethyl-2,5-dibromo-3-hexine	(135.6)	.506		Ethyl bromide	54.70	.502	.719 (20°)
Dimethyl diethylketo tetrahydro-furfurane	(116.2)	.753		Ethyl bromoacetate	(82.8)	.496	(.747) (20°)
2,5-Dimethyl-1-ethylpyrrole	94.61	.768 (20°)		Ethyl-1-isobutylacetoacetate	(121.4)	.652	
2,5-Dimethyl-3-ethylpyrrole	93.87	.762 (20°)		Ethyl butylmalonate	139.3	.644₂	(.629) (20°)
2,5-Dimethylfuran	66.37	.687 (20°)	(.620) (18°)	Ethyl-n-butyrate	(77.7)	.6693	(.585) (25°)
Dimethyl furazan	57.27	.584		Ethyl-iso-butyrate	78.32	.674₃	(.583) (25°)
2,5-Dimethyl-4-heptene	100.6	(.797)		Ethyl chloroacetate	(72.3)	.590	(.684) (20°)
2,4-Dimethyl-2,4-hexadiene	(78.7)	.714		Ethyl cinnamate	(107.5)	.610	(.640) (20°)
2,3-Dimethylhexane	98.77	.8648	.6164 (20°)	Ethyl-iso-cyanate	(45.6)	.642	(.582) (16°)
2,5-Dimethylhexane	98.15	.8593	.5969 (20°)	Ethyl cyanoacetate	(67.3)	.595	(.632) (20°)
3,4-Dimethylhexane	99.06	.8673	.6240 (20°)	Ethylcyclohexane	91.09	.8118	.6324 (20°)
2,6-Dimethyl-4 hexanol	116.9	.812		Ethyldiallylacetophenone	147.4	(.646)	(.636) (16°)
2,5-Dimethyl-3-hexine-2,5-diol	(103.0)	.724		Ethyl dibromocinnamate	174.5	.519	
Dimethyl isoxazole	59.7	(.615)		Ethyl dichloroacetate	85.2	(.543)	(.696) (20°)
Dimethylketo tetrahydrofurfurane	(68.5)	.600		Ethyl diethylacetoacetate	(117.9)	.6328	(.615) (20°)
Dimethyl malonate	69.69	.5277	(.609) (20°)	Ethyl diethylmalonate	(140.4)	.6492	(.641) (20°)
1,6-Dimethylnaphthalene	113.3	(.725)		Ethyl dithiolacetate	(71.0)	.5904 (27.4°)	
2,4-Dimethylnonane	134.68	(.862)	(.636) (20°)	Ethylene	12.0	(.428)	(.242) (−102°)
3,4-Dimethylnonane	134.70	(.862)	(.647) (20°)	Ethylene	15.30	.546 (32°)	(.309) (−102°)
4,5-Dimethylnonane	134.52	(.861)	(.647) (20°)	Ethylene bromide	78.80	.419	.915 (20°)
2,6-Dimethyl-2,6,8-nonatriene	(108.8)	.724		Ethylene chloride	59.62	.602 (32°)	(.757) (20°)
Dimethyl-2,4-nonatriene	(148.8)	.990		Ethylenediamine	46.26	.771 (32°)	(.686) (20°)
2,6-Dimethyloctane	122.54	(.861)	(.627) (20°)	Ethylene iodide	104.7	.371 (32°)	(.791) (10°)
3,4-Dimethyloxadiazole	57.17	.583		Ethylene oxide	30.7	(.697)	(.618) (7°)
Dimethyl oxalate	(55.7)	.472	(.542) (54°)	Ethyl ether	55.10	.743	.534 (20°)
Dimethyl oxamide	(63.2)	.544		Ethyl ethylacetoacetate	93.9	.5937	(.582) (20°)
2,2-Dimethylpentane	86.97	.8680	.5849 (20°)	Ethyl ethylbutylmalonate	(163.3)	.6683	(.650) (20°)
2,3-Dimethylpentane	87.51	.8733	.6070 (20°)	Ethyl ethylpropylmalonate	(152.4)	.6619	(.648) (20°)
2,4-Dimethylpentane	87.48	.8732	.5876 (20°)	Ethyl formate	43.00	.580	.531 (20°)
2,2-Dimethylpropane	63.1	(.875)	(.536) (0°)	Ethyl hexylpropiolate	129.9	.713	
2,5-Dimethyl-3-propyl-pyrrole	106.07	.773 (20°)		Ethyl hydroxylamine	(43.0)	.704	
2,4-Dimethylpyrrole	69.64	.732 (20°)	(.679) (14°)	Ethylidene chloride	(57.4)	.580	(.681) (20°)
				Ethyl iodide	(69.7)	.4470 (17.5°)	(.864) (20°)
				Ethyl iodoacetate	(97.6)	.456	(.829) (13°)
				Ethyl lactate	(72.6)	.615	(.633) (25°)

Compound	$-\chi_M \times 10^6$	$-\chi \times 10^6$	$-K \times 10^6$	Compound	$-\chi_M \times 10^6$	$-\chi \times 10^6$	$-K \times 10^6$
Ethyl methylacetoacetate	(81.9)	.5684	(.569) (20°)	Indene (natural)	84.79	(.730)	(.723) 25°)
Ethyl methyl ketoxime	57.3₂	.658₀	(.530) (20°)	Indene (synthetic)	80.89	(.696)	(.690) (25°)
Ethyl-1-methyl-2-oxocyclohexane-carboxylate	112.1	(.608)		Indole	85.0	(.726)	
Ethyl methylphenylmalonate	(153.2)	.6121	(.658) (20°)	Iodobenzene	92.00	.451	.826 (20°)
Ethyl nitrophenylpropiolate	114.6	.523		Iodoform (in sol'n)	117.1	.2974 (20°)	(1.192) (17°)
Ethyl oxamate	62.0	(.529)	(.427) (19°)	1-Iodo-2-phenylacetylene	(110.1)	.483	
Ethyl perfluor-n-butyrate	103.5	(.427)		o-Iodotoluene	(112.2)	.5145 (30°)	(.874) (20°)
Ethyl phenylacetate	104.27	(.635)	(.656) (20°)	m-Iodotoluene	(112.3)	.5152 (30°)	(.875) (20°)
Ethyl phenylmalonate	(142.2)	.6017	(.659) (20°)	p-Iodotoluene	101.31	(.465)	(.780) (40°)
Ethyl phenylpropiolate	(104.2)	.598	(.636) (13°)	Leucine	84.9	(.647)	
Ethyl phosphate	(98.2)	.539	(.576) (25°)	iso-Leucine	84.9	(.647)	
Ethyl propionate	(66.5)	.6514	(.584) (15°)	Maleic acid	49.71	(.428)	(.681) (20°)
Ethyl propylacetoacetate	(105.7)	.6135	(.593) (20°)	Maleic anhydride	(35.8)	.365	(.341) (20°)
Ethyl-n-propyl ketone	69.03	.689₁	(.560) (22°)	Malonic acid	(46.3)	.4453	(.726) (15°)
Ethyl succinimide	(72.0)	.566		Mannitol	111.20	.610	(.908) (20°)
Ethyl sulfine	(67.0)	.631		Mannose	102.90	.571	(.879)
Ethyl sulfite	(75.4)	.546	(.604) (0°)	Mesitylene	92.32	.7682 (20°)	(.665) (20°)
Ethylsulfone ethyl ether	(81.8)	.592		Methane	12.2₇	(.765)	
Ethyl thiocyanate	(55.7)	.6392 (27.4°)	(.637) (25°)	Methione	91.0	(.610)	
Ethyl isothiocyanate	(59.0)	.6772 (27.4°)	(.680) (15°)	p-Methoxyazobenzene	(118.9)	.560	
Ethyl thiolacetate	(62.7)	.6019 (27.4°)	(.586) (25°)	o-Methoxybenzaldehyde	76.0	(.558)	(.632) (20°)
Ethyl thionacetate	(63.5)	.6098 (27.4°)		p-Methoxybenzaldehyde	78.0	(.572)	(.642) (20°)
Ethyl tribromoacetate	(119.5)	.368	(.821) (20°)	o-Methoxybenzyl alcohol	87.9	.637	(.664) (25°)
Ethyl trichloroacetate	(99.6)	.520	(.719) (20°)	1-Methoxynaphthalene	107.0	(.676)	(.741) (14°)
N-Ethyl urea	55.5	(.630)	(.764) (18°)	2-Methoxynaphthalene	107.6	(.680)	
Ethyl iso-valerate	(91.1)	.700	(.607) (20°)	1-(o-Methoxyphenylazo)-2-naphthol	163.6	(.588)	
Eucalyptol	(116.3)	.754	(.699) (20°)	Methoxysaligenin acetate	110.3	.613	
Eugenol and iso-eugenol	(102.1)	.622	(.663) (20°)	Methyl acetate	42.60	.575	.537 (20°)
Flavanthrone	241.0	.590		Methyl acetoacetate	59.60	.513₂	(.553) (20°)
Fluorene	110.5	.655		Methylacetylacetone	(65.0)	.569	
Fluorenone	99.4	.552	(.623) (100°)	Methyl alcohol	21.40	.668	.530 (20°)
Fluorobenzene	(58.4)	.608	(.623) (20°)	Methylallylaketone	111.9	(1.330)	
Fluorobromoacetic acid	(59.5)	.379		Methylamine	(27.0)	.870	(.608) (−11°)
Fluorodichloromethane	48.8	.474	(.676) (0°)	N-Methylaniline	82.74	(.773)	(.762) ,20°)
p-Fluorophenetole	(88.0)	.628		9-Methylanthracene	146.5	.762	(.812) (99°)
Fluoro trichloroethylene	72.5	.485	(.742) (25°)	Methyl benzoate	81.59	.599₃	(.651) (25°)
Fluorotrichloromethane	58.7	.427	(.638) (17°)	Methyl-o-benzoylbenzoate	139.4	(.580)	(.690) (19°)
Formaldehyde	(18.6)	.62	(.51) (−20°)	Methylbenzylaniline	(132.2)	.670	
Formamide	(21.9)	.486	(.551) (20°)	Methyl bromide	42.8	.451	(1.044) (0°)
Formic acid	19.90	.432	.527 (20°)	2-Methylbutane	64.40	.8925	.5531 (20°)
β-Formylpropionic acid	55.3	(.542)		2-Methyl-2-butene	54.14	(.772)	(.516) (13°)
Fructose	102.60	.570		Methyl butyl ketone	(69.1)	.690	(.563) (15°)
Fulvene (Benzene χ_M measured to be 49)	42.9	(.549)	(.452) (20°)	Methyl iso-butyl ketone	(69.3)	.692	(.554) (20°)
Fulvene $\left(\chi \dfrac{54.8}{49}\right)$	48.0	(.614)	(.505) (20°)	Methyl tert-butyl ketone	(70.4)	.703	(.562) (16°)
Fumaric acid	49.11	(.423)	(.692) (20°)	p-Methyl-o-tert-butylphenol	120.3	(.732)	
Furan	43.09	.633 (20°)	(.598) (15°)	Methyl butyrate	(66.4)	.6498	(.588) (16°)
Furfural	47.1	(.490)	(.568) (20°)	o-Methylcarbanilide	154.0	(.681)	
Galactose	103.00	.572		Methyl chloride	(32.0)	.633	
Gallic acid	90.0	(.529)	(.896) (4°)	Methyl chloroacetate	58.1	(.535)	(.661) (20°)
Geraniol formate	(119.9)	.658	(.610) (20°)	Methylcholanthrene	(182)	.68 ± .04	
Glucose	102.60	.570		3-Methylcholanthrene	194.0	(.723)	
D-Glucose	101.5	(.563)	(.869) (25°)	Methylcyclohexane	78.91	.8038	.6181 (20°)
Glutamic acid	78.5	(.533)		2-Methylcyclohexanone	(74.0)	.660	(.610) (18°)
Glycerol	57.06	.619	.779 (20°)	3-Methylcyclohexanone	(74.8)	.667	(.610) (20°)
Glycine	40.3	(.537)	(.846) (50°)	4-Methylcyclohexanone	(63.5)	.566	(.516) (24°)
Glycol	38.80	.624	.698 (20°)	Methylcyclopentane	70.17	.8338	.6245 (20°)
Guaiacol	(79.2)	.638	(.720) (21°)	4-Methyl-2,6-di-tert-butylphenol	167.6	(.761)	
Helianthrone	189.9	.497		Methyl dichloroacetate	73.1	(.511)	
1,2-Heptadiene	73.5	(.764)		Methyldiphenoxyphosphine oxide	(152.9)	.616	
2,3-Heptadiene	72.1	(.749)		Methyldiphenyltriazine	(155.1)	.627	
Heptaldehyde	81.02	.709₆	(.603) (20°)	Methylene bromide	65.10	.375	.935 (20°)
n-Heptane	85.24	.8507	.5817 (20°)	Methylene chloride	(46.6)	.549	(.733) (20°)
4-Heptanol	91.5	.789	(.647) (20°)	Methylene iodide	93.10	.348	1.156 (20°)
n-Heptanoic acid	88.60	.680	.626 (20°)	Methylene succinic acid	57.57	(.443)	(.723)
n-Heptyl amine	93.1	(.808)	(.628) (20°)	Methyl ether	26.3	.571	
n-Heptyl benzene	134.41	.7625	.6528 (20°)	Methylethylallylacetophenone	133.3	(.659)	(.643) (16°)
Heptyl cyclohexane	147.40	.8084	.6559 (20°)	Methyl ethyl ketone	45.5₈	.632₂	(.509) (20°)
n-Heptylic acid	89.74	.6900	(.630) (25°)	Methyl formate	(32.0)	.5327	(.519) (20°)
1-Heptyne	77.0	(.801)	(.584) (25°)	Methylfumaric acid	56.98	(.438)	(.642)
2-Heptyne	79.5	(.826)	(.615) (25°)	3-Methylheptane	97.99	.8580	.6056 (20°)
Hexabromoethane	(148.0)	.294	(1.124) (20°)	2-Methyl-4-heptene	88.0	(.784)	
Hexachlorobenzene	(147.5)	.518	(1.059) (24°)	5-Methyl-1,2-hexadiene	73.6	(.765)	(.553) (19°)
Hexachloroethane	(112.7)	.476	(.995) (20°)	2-Methylhexane	86.24	.8607	.5841 (20°)
Hexachlorohexatrione	(145.0)	.433		Methyl hexyl ketone	(93.3)	.728	(.596) (20°)
n-Hexadecane	187.63	.8286	.6421 (20°)	Methyl-m-hydroxybenzoate	88.4	(.581)	
1,5-Hexadiene	(55.1)	.671	(.462) (20°)	Methyl-p-hydroxybenzoate	88.7	(.583)	
2,3-Hexadiene	60.9	(.741)		Methyl iodide	(57.2)	.403	(.918) (20°)
n-Hexaldehyde	69.40	.693₀		Methylmaleic acid	57.84	(.446)	(.721)
2,2,4,7,9,9-Hexamethyldecane	191.52	.8458	.6596 (20°)	9-Methyl-10-methoxyanthracene	158.1	(.711)	
Hexamethyl disiloxane	118.9	.7324		Methyl-o-methoxybenzoate	95.6	(.575)	(.665) (19°)
Hexamethylene glycol	84.30	.713		Methyl-p-methoxybenzoate	98.6	(.593)	
n-Hexane	(74.6)	.8654 (27.4°)	(.565)	Methyl-α-methoxy-isobutyrate	(81.9)	.620	
Hexene	65.7	(.781)		1-Methylnaphthalene	102.8	(.723)	(.741) (14°)
Hexestrol	(165)	.61 ± .02		2-Methylnaphthalene	102.6	(.722)	(.743) (20°)
n-Hexyl alcohol	79.20	.774	.637 (20°)	4-Methylnonane	121.39	(.853)	(.625) (20°)
n-Hexyl benzene	124.23 (20°)	(.767)	(.658) (20°)	5-Methyl-5-nonene	111.6	(.796)	
n-Hexyl methyl ketone	91.4₇	.713₁	(.583)	4-Methyloctane	109.63	(.855)	(.618) (20°)
n-Hexyl methyl ketoxime	102.5₈	.716₂	(.634) (20°)	Methylol urea	48.3	(.493)	
Hexylpropiolamide	(103.7)	.677		2-Methylpentane	75.26	.8734	.5705 (20°)
Hydrindene	(78.5)	.664	(.639) (16°)	3-Methylpentane	75.52	.8764	.5823 (20°)
Hydroquinone	64.63	.587	(.797) (20°)	4-Methyl-2-pentanol	80.4	.788	(.641) (20°)
Hydroxyazobenzene	(99.7)	.503		Methyl perfluor-n-butyrate	92.5	(.406)	
p-Hydroxybenzaldehyde	66.8	(.547)	(.618) (130°)	Methyl phenylacetate	92.73	(.618)	(.645) (16°)
4-Hydroxy-2-butanone	48.5	.55	(.573) (14°)	Methyl phenylpropiolate	95.6	.597	
				2-Methylpropene	44.4	(.791)	
				Methyl propionate	(55.0)	.6240	(.571) (20°)

Compound	$-\chi_M \times 10^6$	$-\chi \times 10^6$	$-K \times 10^6$	Compound	$-\chi_M \times 10^6$	$-\chi \times 10^6$	$-K \times 10^6$
Methyl-n-propyl ketone	57.41	.666₄	(.541) (15°)	Pentachloroethane	(99.1)	.490	(.819) (25°)
Methyl-iso-propyl ketone	58.45	.679₀	(.545) (20°)	Pentachlorohexadione	(129.5)	.452	
1-Methylpyrrole	58.56	.722 (20°)	(.664) (10°)	2,3-Pentadiene	49.1	(.721)	(.501) (20°)
2-Methylpyrrole	60.10	.741 (20°)	(.700)	n-Pentane	63.05	.8739	.5472 (20°)
Methyl salicylate	86.30	.567	.668 (20°)	2,4-Pentanediol	70.4	.677	
Methyl silicone	(172.7)	.730		Perfluoroacetic acid	43.3	(.380)	
α-Methyl styrene	(80.1)	.678	(.620) (20°)	Perfluoro-n-butyric acid	81.0	(.378)	
2-Methylthiazole	59.56	.601 (20°)		Perfluorobutyric anhydride	149.4	(.387)	
2-Methylthiophene	66.35	.676 (20°)	(.689) (20°)	Perfluorocyclooctane oxide	157.6	(.379)	
Methyl trichloroacetate	84.2	(.475)	(.707) (19°)	Perfluoropropionic acid	61.0	(.372)	
N-Methyl urea	44.6	(.602)	(.725)	Perhydroanthracene	146.01	.7592	.7178 (20°)
Morpholine	55.0	(.631)	(.631)	Perylene	166 8	.662	
Myleran	169.7	.69	(.834) (4°)	Phenanthrene	(127.9)	.718	(.763) (100°)
Myristic acid	176.0	(.771)	(.661) (60°)	Phenanthrenequinone	104.5	.502	(.698)
Naphthalaldehydic acid	117.6	(.588)		Phenanthrenonitrile	139.0	(.685)	
Naphthalene	(91.9)	.717	(.821) (20°)	o-Phenetidine	(101.7)	.741 (25°)	
Naphthalene picrate	185.9	(.523)		p-Phenetidine	(96.8)	.706 (25°)	(.749) (15°)
2-Naphthalenesulfonylamine	127.6	(.616)		Phenetole	(84.5)	.692	(.689) (20°)
2-Naphthalenesulfonyl chloride	121.91	(.538)		Phenol	60.21	(.640)	(.675) (45°)
meso-Naphthodianthrene	214.6	.612		Phenothiazine	114.8	(.576)	
meso-Naphthodianthrone	221.8	.583		Phenylacetaldehyde	72.01	.599₄	(.614) (20°)
1-Naphthol	98.2	.681	(.834) (4°)	Phenyl acetate	82.04	(.603)	(.647) (25°)
	97.0	.673	(.819) (4°)	Phenylacetic acid	82.72	(.608)	(.657) (80°)
2-Naphthol	98.25	(.682)	(.829) (4°)	Phenylacetylene	72.01	(.705)	(.655) (20°)
α-Naphthonitrile	103.3	(.674)	(.753) (5°)	1-Phenylazo-2-naphthol	137.6	(.554)	
β-Naphthonitrile	101.0	(.659)	(.721) (60°)	2-Phenylbenzofuran	130.5	(.672)	
α-Naphthoquinone	73.5	(.465)	(.661)	1-Phenyl-4-benzoyl-1,3-butadiene	(140.3)	.599	
β-Naphthoquinone	67.9	(.429)		Phenylbutadiene	(85.7)	.658	
N-1-Naphthylacetamide	117.8	(.636)		4-Phenyl-1-butene	93.49	(.7077)	(.6239) (20°)
N-2-Naphthylacetamide	117.8	(.636)		Phenylbutyl acetate	134.5	.653	
1-Naphthylamine	98.8	.690	.757 (54°)	Phenyl n-butyrate	105.46	(.643)	
2-Naphthylamine	98.00	(.684)	(.726) (98°)	Phenyl iso-cyanate	(72.7)	.610	(.699) (20°)
1-Naphthylamine hydrochloride	(127.6)	.710		o-Phenylenediamine	71.98	.6662	
Nicotine	113.328	(.699)	(.705) (20°)	m-Phenylenediamine	70.53	.6529	(.723) (58°)
o-Nitroaniline	66.47	(.481)	(.694) (15°)	p-Phenylenediamine	70.28	.6503	
m-Nitroaniline	70.09	(.507)	(.725) (20°)	Phenyl ether	(108.1)	.635	(.681) (20°)
p-Nitroaniline	66.43	(.481)	(.691) (14°)	Phenylethyl sulfide	(94.4)	.6826 (27.4°)	
o-Nitrobenzaldehyde	68.23	.4517		Phenylfluoroform	(77.3)	.529	
m-Nitrobenzaldehyde	68.55	.4538		Phenylhydrazine	67.82	(.627)	(.688) (23°)
p-Nitrobenzaldehyde	66.57	.4407	(.507) (0°)	Phenylhydroxylamine	(68.2)	.625	
Nitrobenzene	61.80	.502	.604 (20°)	Phenyl mercaptan	(70.8)	.6425 (27.4°)	(.693) (20°)
Nitrobenzene diazo cyanide	59.22	(.336)		1-Phenyl-2-Methylbutane	113.53	(.766)	(.660) (20°)
o-Nitrobenzoic acid	76.11	.4556	(.718) (4°)	Phenylmethyl sulfide	(83.2)	.6695 (27.4°)	
m-Nitrobenzoic acid	80.22	.4802	(.717) (4°)	Phenylpropiolamide	(83.3)	.574	
p-Nitrobenzoic acid	78.81	.4718	(.731) (32°)	Phenyl propionate	93.79	(.625)	(.654) (25°)
o-Nitrobromobenzene	87.3	(.432)	(.700) (80°)	Phenylsulfone	(129.0)	.591	(.740) (20°)
m-Nitrobromobenzene	89.5	(.443)	(.755) (20°)	Phenyl thiocyanate	(81.5)	.6027 (27.4°)	(.677) (24°)
p-Nitrobromobenzene	89.6	(.444)	(.859) (22°)	Phenyl isothiocyanate	(86.0)	.6365 (27.4°)	(.719) (24°)
m-Nitro carbanilide	148.1	(.576)		1-Phenyl-4,6,6-trimethylheptane	173.90	(.796)	(.682) (20°)
Nitroethane	(35.4)	.472	(.497) (20°)	N-Phenyl urea	82.1	(.603)	(.785)
Nitromethane	21.1	.3457	(.391) (25°)	Phloroglucinol	(73.4)	.582	
1-Nitronaphthalene	98.47	(.569)	(.696) (62°)	Phthalamide	(91.3)	.556	
o-Nitrophenol	73.3	.527 (24°)	(.873) (20°)	Phthalic acid	83.61	.5035	(.802) (20°)
m-Nitrophenol	70.8	.509 (25°)	(.756)	iso-Phthalic acid	84.64	.5097	
p-Nitrophenol	69.5	.500 (22°)	(.740)	tere-Phthalic acid	83.51	.5029	(.759)
1-(m-Nitrophenylazo)-2-naphthol	142.0	(.484)		Phthalic anhydride	67.31	(.454)	(.694) (4°)
1-(p-Nitrophenylazo)-2-naphthol	141.7	(.483)		Phthalimide	(78.4)	.533	
Nitrophenylfluoroform	(84.1)	.440		Picric acid	84.38	(.368)	(.649)
2-Nitropropane	45.73	.5135	(.509) (20°)	Piperazine	56.8	(.659)	
Nitrosobenzene	59.1	(.552)		Piperidine	64.2	(.754)	(.650) (20°)
N-Nitrosodiethylamine	59.3	(.580)	(.546) (20°)	Propane	40.5	(.919)	(.538) (−45°)
p-Nitrosodiethylaniline	92.6	(.520)	(.644) (15°)	Propene	31.5	(.749)	(.456) (−47°)
p-Nitrosodimethylaniline	73.3	(.488)		Propionaldehyde	34.32	.591₀	(.477) (20°)
N-Nitrosodiphenylamine	110.7	(.558)		Propionic acid	43.50	.586	.582 (20°)
1-Nitroso-2-naphthol	83.9	(.485)		Propionitrile	38.5	(.699)	(.547) (21°)
2-Nitroso-1-naphthol	82.7	(.478)		Propionylphenylacetylene	(95.1)	.601	
4-Nitroso-1-naphthol	91.8	(.530)		Propiophenone	83.73	.624₀	(.631) (20°)
m-Nitrosonitrobenzene	66.0	(.433)		n-Propyl acetate	65.91	.645₃ (25°)	(.569) (25°)
p-Nitrosonitrobenzene	65.8	(.433)		iso-Propyl acetate	67.04	.656₄	(.566) (25°)
p-Nitrosophenol	50.7	(.412)		n-Propyl alcohol	45.176 (20°)	(.7518)	(.6047) (20°)
Nitrosopiperidine	(63.4)	.555	(.590) (20°)	iso-Propyl alcohol	45.794 (20°)	(.7621)	(.5985) (20°)
p-Nitrosotoluene	70.4	(.581)		9-Propylanthracene	164.0	(.744)	
o-Nitrotoluene	72.28	.5272	(.613) (20°)	n-Propylbenzene	89.24	(.742)	(.640) (20°)
m-Nitrotoluene	72.71	.5304	(.614) (20°)	n-Propyl benzoate	105.00	(.640)	(.646) (25°)
p-Nitrotoluene (in sol'n)	72.06	.5257	(.676) (20°)	n-Propyl bromide	(65.6)	.533	(.721) (20°)
n-Nonane	108.13	.8431	.6057 (20°)	iso-Propyl bromide	(65.1)	.529	(.693) (20°)
1,2-Octadiene	83.6	(.759)		Propyl butyrate	(89.4)	.6867	(.604) (15°)
n-Octane	96.63	.8460	.5949 (20°)	prim-Propyl chloride	56.10	.715	.633 (20°)
Octanonoxime	(102.7)	.717		iso-Propylcyclohexane	102.65	.8131	.6528 (20°)
Octyl alcohol	102.65	.7766 (20°)	(.640) (20°)	Propylenediamine	(58.1)	.784	(.688) (15°)
Octyl chloride	(114.9)	.773	(.676) (20°)	Propylene oxide	42.5	(.732)	(.629) (0°)
Octylcyclohexane	158.09	.8051	.6578 (20°)	Propyl formate	(55.0)	.6248	(.563) (20°)
Octylene	(89.5)	.798	(.576) (17°)	Propyl hexylpropiolate	(136.8)	.697	
Octylene bromide	(150.4)	.553		Propyl iodide	(84.3)	.4958 (30°)	(.864) (20°)
n-Octyl mercaptan	(115.1)	.7866 (27.4°)		Propyl propionate (extrap.)	(77.95)	.6711	(.593) (20°)
Oenanthylidene chloride	(116.5)	.689		Propyl sulfide	(92.1)	.7787 (27.4°)	(.634) (17°)
Oleic acid	208.5	(.738)	(.661) (18°)	N-Propyl urea	67.4	(.660)	
Opianic acid	111.5	(.530)		Pseudocumene	(101.6)	.815 (20°)	(.740) (20°)
Ovalene	353.8	.888		Pyramidone	149.0	(.645)	
Oxalic acid (anh.)	33.8	(.375)		Pyranthrene	266.9	.709	
Oxalic acid	60.05	(.4763)	(.787)	Pyranthrone	250.3	.616	
Oxamide	(39.0)	.443	(.738)	Pyrazine	37.6	(.469)	(.484) (61°)
Palmitic acid	198.6	(.775)	(.661) (62°)	Pyrene	147.9	.731	(.933) (0°)
Paraldehyde	(86.2)	.652	(.648) (20°)	Pyridine	49.21	(.622)	(.611) (20°)
Pentabromophenol	(194.0)	.397		Pyrocatechol	68.76	.6248	(.857) (15°)
Pentacene	(205.4)	.738		Pyrrole	47.6	(.709)	(.688) (20°)

DIAMAGNETIC SUSCEPTIBILITIES OF ORGANIC COMPOUNDS (Continued)

Compound	$-\chi_M \times 10^6$	$-\chi \times 10^6$	$-K \times 10^6$	Compound	$-\chi_M \times 10^6$	$-\chi \times 10^6$	$-K \times 10^6$
Pyrrolidine	54.8	(.771)		Trianilinophosphine oxide	(201.7)	.624	
Quinoline	86.0	(.666)	(.729) (20°)	1,2,3-Tribromopropane	(117.9)	.420	(1.023) (23°)
Quinone	38.4	(.355)	(.468) (20°)	Tri-iso-butylamine	(156.8)	.846	(.646) (25°)
Quinonoxime	(50.4)	.409		Trichloroacetic acid (in sol'n)	73.0	(.44)	(.723) (46°)
Resorcinol	67.26	.6112	(.785) (15°)	Trichlorobenzene	(106.5)	.587	
Rhamnose	99.20	.605	(.890) (20°)	Trichloro-tert-butyl alcohol (in sol'n)	98.01	.552	
Safrol and iso-Safro	(97.5)	.601	(.66) (20°)	Trichloroethylene	65.8	.501	(.734) (20°)
Salicylaldehyde	64.4	(.527)	(.615) (20°)	Trichloronitromethane	(75.3)	.458	(.756) (20°)
Salicylic acid	72.23	.523	(.755) (20°)	Triethvlamine	81.4	(.804)	(.586) (20°)
Saligenin	76.9	.620	(.720) (25°)	Triethyl citrate	(161.9)	.586	(.666) (20°)
Salol	(123.2)	.575	.678 (45°)	Triethyl phosphate	(125.3)	.688	(.735) (20°)
Salvarsan dihydrochloride	(246.1)	.518		Triethylphosphine	(90.0)	.762	(.610) (15°)
Selenophene	66.82	.510		Triethylphosphine oxide	(91.6)	.683	
iso-Selenophene	110–111	.84–.85		Triethyl phosphite	(104.8)	.631	(.611) (20°)
trans-Selenophene	70–77	.53–.59		Triethyl triazinetricarbonate	(164.1)	.552	
Sorbitol	107.80	.592		Trifluorocresol	(83.8)	.517	
Stearic acid	220.8	(.776)	(.657) (69°)	Tri-n-heptylamine	251.3	(.806)	
Stilbene	(120.0)	.666	(.646) (125°)	Tri-n-hexylamine	221.7	(.823)	
Stilbestrol	(130)	.62, .63		Trimethylacetophenone	108.2	(.667)	(.648) (16°)
Styrene	(68.2)	.655	(.594) (20°)	2,2,3-Trimethylbutane	88.36	.8818	.6086 (20°)
Succinic acid	(57.9)	.4902	(.767) (15°)	2,2,3-Trimethylpentane	99.86	.8743	.6261 (20°)
Succinic anhydride	(47.5)	.475	(.524)	2,2,4-Trimethylpentane	98.34	.8610	.5958 (20°)
Succinimide	(47.3)	.477	(.674) (16°)	2,3,5-Trimethylpyrrole	82.31	.754 (20°)	
Sulfamide	44.4	(.462)	(.832)	1,3,5-Trinitrobenzene	74.55	(.350)	(.591) (20°)
p-Sulfanilamide	80 15	(.465)		Triperfluorobutylamine	253.0	(.377)	
Terpineol	111.9	(.725)	(.678) (room temp.)	Triphenoxyarsine	(195.2)	.551	
				Triphenylarsine	(177.0)	.578	
Tetrabenzylmonosilane	266.2	(.678)		Triphenylarsine dihydroxide	(270.5)	.795	
1,1,2,2-Tetrabromoethane	(123.4)	.357	(1.058) (20°)	Triphenylarsine oxide	(199.1)	.618	
Tetrabromoethylene	(114.8)	.334		Triphenylbismuthine	(196.8)	.447	(.708) (20°)
Tetracene	(168.0)	.736		Triphenylbismuthine dinitrate	(254.5)	.451	
1,1,2,2-Tetrachloroethane	(89.8)	.535	(.856) (20°)	Triphenylcarbinol	(175.7)	.675	(.802) (20°)
Tetrachloroethylene	81.6	.492	(.802) (15°)	Triphenylmethane	(165.6)	.678	.686 (100°)
Tetrahydroquinoline	(89.0)	.668	(.715) (4°)	Triphenylphosphine	(166.8)	.636	(.759)
Tetraiodoethylene	(164.3)	.309	(.922) (20°)	Triphenyl phosphite	(183.7)	.592	(.701) (18°)
Tetraiodopyrrole	(188.9)	.331		Triphenylstibine	(182.2)	.516	
Tetramethylketotetrahydrofurfurane	(104.7)	.736		Triphenylstibine dihydroxide	(238.5)	.616	
Tetranitromethane	43.02	.2195	(.360) (20°)	N,N',N-Triphenyl urea	176.5	(.613)	
Tetraphenylbutadiene	228.0	(.636)		Triquinoyl	(133.0)	.426	
Tetraphenyldecapentaene	280.8	(.643)		Tropolone	61	.50	
Tetraphenylhexatriene	246.4	(.641)		Tryptophan	132.0	(.646)	
Tetraphenyloctatetraene	264.1	(.643)		Tyrosine	105.3	(.581)	
Tetraphenylrubene	344.0	(.646)		Undecane	131.84	(.8435)	(.6247) (20°)
Tetra-p-tolylmonosilane	276.4	(.704)		Urea	33.4	(.556)	(.742) (20°)
Tetrolic acetal	(97.8)	.688		Urethan	(57)	.64	(.63) (21°)
Tetronic acid	(52.5)	.525		iso-Valeraldehyde	(57.5)	.668	(.536) (17°)
Thiacoumerin	93.6	.577		n-Valeric acid	66.85	.6548	(.617) (20°)
Thiazole	50.55	.595 (20°)	(714) (17°)	iso-Valeric acid	(67.7)	.663	(.621) (15°)
Thiobarbituric acid	72.9	(.506)		Valerylphenylacetylene	(119.0)	.639	
Thiophene	57.38	.682 (20°)	(.726) (20°)	Valine	74.3	(.634)	
Tolane	(118.9)	.667	(.644) (100°)	Violanthrene	273.5	.641	
Toluene	66.11	.7176	.6179 (20°)	Violanthrone	204.8	.449	
o-Toluidine	76.0	.710 (24°)	(.709) (20°)	iso-Violanthrone	215.9	.473	
m-Toluidine	74.6	.697 (25°)	(.689) (20°)	Water	(13.00)	.7218 (20°)	(.7205) (20°)
p-Toluidine	72.1	.673 (25°)	(.704) (20°)	Water (value usually used as standard)	(12.97)	.720 (20°)	(.719) (20°)
α-Tolunitrile	76.87	(.656)	(.666) (18°)	Xanthone	(108.1)	.551	
1-(o-Tolylazo)-2-naphthol	148.7	(.567)		o-Xylene	77.78	.7327	.6440 (20°)
1-(p-Tolylazo)-2-naphthol	157.6	(.601)		m-Xylene	76.56	.7212	.6235 (20°)
Triallylacetophenone	152.5	(.634)		p-Xylene	76.78	.7232	.6226 (20°)
Tri-iso-amylamine	192	.845	(.647) (25°)	Xylose	84.80	.565	(.862) (20°)

N	0	1	2	3	4	5	6	7	8	9	Proportional Parts 1 2 3 4 5 6 7 8 9
10	0000	0043	0086	0128	0170	0212	0253	0294	0334	0374	*4 8 12 17 21 25 29 33 37
11	0414	0453	0492	0531	0569	0607	0645	0682	0719	0755	4 8 11 15 19 23 26 30 34
12	0792	0828	0864	0899	0934	0969	1004	1038	1072	1106	3 7 10 14 17 21 24 28 31
13	1139	1173	1206	1239	1271	1303	1335	1367	1399	1430	3 6 10 13 16 19 23 26 29
14	1461	1492	1523	1553	1584	1614	1644	1673	1703	1732	3 6 9 12 15 18 21 24 27
15	1761	1790	1818	1847	1875	1903	1931	1959	1987	2014	*3 6 8 11 14 17 20 22 25
16	2041	2068	2095	2122	2148	2175	2201	2227	2253	2279	3 5 8 11 13 16 18 21 24
17	2304	2330	2355	2380	2405	2430	2455	2480	2504	2529	2 5 7 10 12 15 17 20 22
18	2553	2577	2601	2625	2648	2672	2695	2718	2742	2765	2 5 7 9 12 14 16 19 21
19	2788	2810	2833	2856	2878	2900	2923	2945	2967	2989	2 4 7 9 11 13 16 18 20
20	3010	3032	3054	3075	3096	3118	3139	3160	3181	3201	2 4 6 8 11 13 15 17 19
21	3222	3243	3263	3284	3304	3324	3345	3365	3385	3404	2 4 6 8 10 12 14 16 18
22	3424	3444	3464	3483	3502	3522	3541	3560	3579	3598	2 4 6 8 10 12 14 15 17
23	3617	3636	3655	3674	3692	3711	3729	3747	3766	3784	2 4 6 7 9 11 13 15 17
24	3802	3820	3838	3856	3874	3892	3909	3927	3945	3962	2 4 5 7 9 11 12 14 16
25	3979	3997	4014	4031	4048	4065	4082	4099	4116	4133	2 3 5 7 9 10 12 14 15
26	4150	4166	4183	4200	4216	4232	4249	4265	4281	4298	2 3 5 7 8 10 11 13 15
27	4314	4330	4346	4362	4378	4393	4409	4425	4440	4456	2 3 5 6 8 9 11 13 14
28	4472	4487	4502	4518	4533	4548	4564	4579	4594	4609	2 3 5 6 8 9 11 12 14
29	4624	4639	4654	4669	4683	4698	4713	4728	4742	4757	1 3 4 6 7 9 10 12 13
30	4771	4786	4800	4814	4829	4843	4857	4871	4886	4900	1 3 4 6 7 9 10 11 13
31	4914	4928	4942	4955	4969	4983	4997	5011	5024	5038	1 3 4 6 7 8 10 11 12
32	5051	5065	5079	5092	5105	5119	5132	5145	5159	5172	1 3 4 5 7 8 9 11 12
33	5185	5198	5211	5224	5237	5250	5263	5276	5289	5302	1 3 4 5 6 8 9 10 12
34	5315	5328	5340	5353	5366	5378	5391	5403	5416	5428	1 3 4 5 6 8 9 10 11
35	5441	5453	5465	5478	5490	5502	5514	5527	5539	5551	1 2 4 5 6 7 9 10 11
36	5563	5575	5587	5599	5611	5623	5635	5647	5658	5670	1 2 4 5 6 7 8 10 11
37	5682	5694	5705	5717	5729	5740	5752	5763	5775	5786	1 2 3 5 6 7 8 9 10
38	5798	5809	5821	5832	5843	5855	5866	5877	5888	5899	1 2 3 5 6 7 8 9 10
39	5911	5922	5933	5944	5955	5966	5977	5988	5999	6010	1 2 3 4 5 7 8 9 10
40	6021	6031	6042	6053	6064	6075	6085	6096	6107	6117	1 2 3 4 5 6 8 9 10
41	6128	6138	6149	6160	6170	6180	6191	6201	6212	6222	1 2 3 4 5 6 7 8 9
42	6232	6243	6253	6263	6274	6284	6294	6304	6314	6325	1 2 3 4 5 6 7 8 9
43	6335	6345	6355	6365	6375	6385	6395	6405	6415	6425	1 2 3 4 5 6 7 8 9
44	6435	6444	6454	6464	6474	6484	6493	6503	6513	6522	1 2 3 4 5 6 7 8 9
45	6532	6542	6551	6561	6571	6580	6590	6599	6609	6618	1 2 3 4 5 6 7 8 9
46	6628	6637	6646	6656	6665	6675	6684	6693	6702	6712	1 2 3 4 5 6 7 7 8
47	6721	6730	6739	6749	6758	6767	6776	6785	6794	6803	1 2 3 4 5 5 6 7 8
48	6812	6821	6830	6839	6848	6857	6866	6875	6884	6893	1 2 3 4 4 5 6 7 8
49	6902	6911	6920	6928	6937	6946	6955	6964	6972	6981	1 2 3 4 4 5 6 7 8
50	6990	6998	7007	7016	7024	7033	7042	7050	7059	7067	1 2 3 3 4 5 6 7 8
51	7076	7084	7093	7101	7110	7118	7126	7135	7143	7152	1 2 3 3 4 5 6 7 8
52	7160	7168	7177	7185	7193	7202	7210	7218	7226	7235	1 2 2 3 4 5 6 7 7
53	7243	7251	7259	7267	7275	7284	7292	7300	7308	7316	1 2 2 3 4 5 6 6 7
54	7324	7332	7340	7348	7356	7364	7372	7380	7388	7396	1 2 2 3 4 5 6 6 7
N	0	1	2	3	4	5	6	7	8	9	1 2 3 4 5 6 7 8 9

* Interpolation in this section of the table is inaccurate.

N	0	1	2	3	4	5	6	7	8	9	Proportional Parts								
											1	2	3	4	5	6	7	8	9
55	7404	7412	7419	7427	7435	7443	7451	7459	7466	7474	1	2	2	3	4	5	5	6	7
56	7482	7490	7497	7505	7513	7520	7528	7536	7543	7551	1	2	2	3	4	5	5	6	7
57	7559	7566	7574	7582	7589	7597	7604	7612	7619	7627	1	2	2	3	4	5	5	6	7
58	7634	7642	7649	7657	7664	7672	7679	7686	7694	7701	1	1	2	3	4	4	5	6	7
59	7709	7716	7723	7731	7738	7745	7752	7760	7767	7774	1	1	2	3	4	4	5	6	7
60	7782	7789	7796	7803	7810	7818	7825	7832	7839	7846	1	1	2	3	4	4	5	6	6
61	7853	7860	7868	7875	7882	7889	7896	7903	7910	7917	1	1	2	3	4	4	5	6	6
62	7924	7931	7938	7945	7952	7959	7966	7973	7980	7987	1	1	2	3	3	4	5	6	6
63	7993	8000	8007	8014	8021	8028	8035	8041	8048	8055	1	1	2	3	3	4	5	5	6
64	8062	8069	8075	8082	8089	8096	8102	8109	8116	8122	1	1	2	3	3	4	5	5	6
65	8129	8136	8142	8149	8156	8162	8169	8176	8182	8189	1	1	2	3	3	4	5	5	6
66	8195	8202	8209	8215	8222	8228	8235	8241	8248	8254	1	1	2	3	3	4	5	5	6
67	8261	8267	8274	8280	8287	8293	8299	8306	8312	8319	1	1	2	3	3	4	5	5	6
68	8325	8331	8338	8344	8351	8357	8363	8370	8376	8382	1	1	2	3	3	4	4	5	6
69	8388	8395	8401	8407	8414	8420	8426	8432	8439	8445	1	1	2	2	3	4	4	5	6
70	8451	8457	8463	8470	8476	8482	8488	8494	8500	8506	1	1	2	2	3	4	4	5	6
71	8513	8519	8525	8531	8537	8543	8549	8555	8561	8567	1	1	2	2	3	4	4	5	5
72	8573	8579	8585	8591	8597	8603	8609	8615	8621	8627	1	1	2	2	3	4	4	5	5
73	8633	8639	8645	8651	8657	8663	8669	8675	8681	8686	1	1	2	2	3	4	4	5	5
74	8692	8698	8704	8710	8716	8722	8727	8733	8739	8745	1	1	2	2	3	4	4	5	5
75	8751	8756	8762	8768	8774	8779	8785	8791	8797	8802	1	1	2	2	3	3	4	5	5
76	8808	8814	8820	8825	8831	8837	8842	8848	8854	8859	1	1	2	2	3	3	4	5	5
77	8865	8871	8876	8882	8887	8893	8899	8904	8910	8915	1	1	2	2	3	3	4	4	5
78	8921	8927	8932	8938	8943	8949	8954	8960	8965	8971	1	1	2	2	3	3	4	4	5
79	8976	8982	8987	8993	8998	9004	9009	9015	9020	9025	1	1	2	2	3	3	4	4	5
80	9031	9036	9042	9047	9053	9058	9063	9069	9074	9079	1	1	2	2	3	3	4	4	5
81	9085	9090	9096	9101	9106	9112	9117	9122	9128	9133	1	1	2	2	3	3	4	4	5
82	9138	9143	9149	9154	9159	9165	9170	9175	9180	9186	1	1	2	2	3	3	4	4	5
83	9191	9196	9201	9206	9212	9217	9222	9227	9232	9238	1	1	2	2	3	3	4	4	5
84	9243	9248	9253	9258	9263	9269	9274	9279	9284	9289	1	1	2	2	3	3	4	4	5
85	9294	9299	9304	9309	9315	9320	9325	9330	9335	9340	1	1	2	2	3	3	4	4	5
86	9345	9350	9355	9360	9365	9370	9375	9380	9385	9390	1	1	2	2	3	3	4	4	5
87	9395	9400	9405	9410	9415	9420	9425	9430	9435	9440	0	1	1	2	2	3	3	4	4
88	9445	9450	9455	9460	9465	9469	9474	9479	9484	9489	0	1	1	2	2	3	3	4	4
89	9494	9499	9504	9509	9513	9518	9523	9528	9533	9538	0	1	1	2	2	3	3	4	4
90	9542	9547	9552	9557	9562	9566	9571	9576	9581	9586	0	1	1	2	2	3	3	4	4
91	9590	9595	9600	9605	9609	9614	9619	9624	9628	9633	0	1	1	2	2	3	3	4	4
92	9638	9643	9647	9652	9657	9661	9666	9671	9675	9680	0	1	1	2	2	3	3	4	4
93	9685	9689	9694	9699	9703	9708	9713	9717	9722	9727	0	1	1	2	2	3	3	4	4
94	9731	9736	9741	9745	9750	9754	9759	9763	9768	9773	0	1	1	2	2	3	3	4	4
95	9777	9782	9786	9791	9795	9800	9805	9809	9814	9818	0	1	1	2	2	3	3	4	4
96	9823	9827	9832	9836	9841	9845	9850	9854	9859	9863	0	1	1	2	2	3	3	4	4
97	9868	9872	9877	9881	9886	9890	9894	9899	9903	9908	0	1	1	2	2	3	3	4	4
98	9912	9917	9921	9926	9930	9934	9939	9943	9948	9952	0	1	1	2	2	3	3	4	4
99	9956	9961	9965	9969	9974	9978	9983	9987	9991	9996	0	1	1	2	2	3	3	3	4
N	0	1	2	3	4	5	6	7	8	9	1	2	3	4	5	6	7	8	9

PERIODIC TABLE OF THE ELEMENTS

KEY TO CHART

- Atomic Number → **50**
- Symbol → **Sn** (Oxidation States → +2 +4)
- Atomic Weight → 118.69
- Electron Configuration → -18-18-4

Group	1a	2a	3b	4b	5b	6b	7b	8			1b	2b	3a	4a	5a	6a	7a	0	Orbit
	1 H +1 -1 1.00797 1																	2 He 0 4.0026 2	K
	3 Li +1 6.939 2-1	4 Be +2 9.0122 2-2											5 B +3 10.811 2-3	6 C +2 +4 -4 12.01115 2-4	7 N +1 +2 +3 +4 +5 -2 -3 14.0067 2-5	8 O -2 15.9994 2-6	9 F -1 18.9984 2-7	10 Ne 0 20.183 2-8	K-L
	11 Na +1 22.9898 2-8-1	12 Mg +2 24.312 2-8-2											13 Al +3 26.9815 2-8-3	14 Si +4 -4 28.086 2-8-4	15 P +3 +5 -3 30.9738 2-8-5	16 S +4 +6 -2 32.064 2-8-6	17 Cl +1 +5 +7 -1 35.453 2-8-7	18 Ar 0 39.948 2-8-8	K-L-M
	19 K +1 39.102 -8-8-1	20 Ca +2 40.08 -8-8-2	21 Sc +3 44.956 -8-9-2	22 Ti +2 +3 +4 47.90 -8-10-2	23 V +2 +3 +4 +5 50.942 -8-11-2	24 Cr +2 +3 +6 51.996 -8-13-1	25 Mn +2 +3 +4 +6 +7 54.9380 -8-13-2	26 Fe +2 +3 55.847 -8-14-2	27 Co +2 +3 58.9332 -8-15-2	28 Ni +2 +3 58.71 -8-16-2	29 Cu +1 +2 63.54 -8-18-1	30 Zn +2 65.37 -8-18-2	31 Ga +3 69.72 -8-18-3	32 Ge +2 +4 72.59 -8-18-4	33 As +3 +5 -3 74.9216 -8-18-5	34 Se +4 +6 -2 78.96 -8-18-6	35 Br +1 +5 -1 79.909 -8-18-7	36 Kr 0 83.80 -8-18-8	-L-M-N
	37 Rb +1 85.47 -18-8-1	38 Sr +2 87.62 -18-8-2	39 Y +3 88.905 -18-9-2	40 Zr +4 91.22 -18-10-2	41 Nb +3 +5 92.906 -18-12-1	42 Mo +6 95.94 -18-13-1	43 Tc +4 +6 +7 (99) -18-13-2	44 Ru +3 101.07 -18-15-1	45 Rh +3 102.905 -18-16-1	46 Pd +2 +4 106.4 -18-18-0	47 Ag +1 107.870 -18-18-1	48 Cd +2 112.40 -18-18-2	49 In +3 114.82 -18-18-3	50 Sn +2 +4 118.69 -18-18-4	51 Sb +3 +5 -3 121.75 -18-18-5	52 Te +4 +6 -2 127.60 -18-18-6	53 I +1 +5 +7 -1 126.9044 -18-18-7	54 Xe 0 131.30 -18-18-8	-M-N-O
	55 Cs +1 132.905 -18-8-1	56 Ba +2 137.34 -18-8-2	57* La +3 138.91 -18-9-2	72 Hf +4 178.49 -32-10-2	73 Ta +5 180.948 -32-11-2	74 W +6 183.85 -32-12-2	75 Re +4 +6 +7 186.2 -32-13-2	76 Os +3 +4 190.2 -32-14-2	77 Ir +3 +4 192.2 -32-15-2	78 Pt +2 +4 195.09 -32-16-2	79 Au +1 +3 196.967 -32-18-1	80 Hg +1 +2 200.59 -32-18-2	81 Tl +1 +3 204.37 -32-18-3	82 Pb +2 +4 207.19 -32-18-4	83 Bi +3 +5 208.980 -32-18-5	84 Po +2 +4 (210) -32-18-6	85 At (210) -32-18-7	86 Rn (222) -32-18-8	-N-O-P
	87 Fr +1 (223) -18-8-1	88 Ra +2 (226) -18-8-2	89** Ac +3 (227) -18-9-2																-O-P-Q

Group 8: 26 Fe, 27 Co, 28 Ni, 44 Ru, 45 Rh, 46 Pd, 76 Os, 77 Ir, 78 Pt

Transition Elements

*Lanthanides

3b	4b	5b	6b	7b	8			1b	2b	3a	4a	5a	6a	Orbit
58 Ce +3 +4 140.12 -19-9-2	59 Pr +3 +4 140.907 -20-9-2	60 Nd +3 144.24 -22-8-2	61 Pm +3 (145) -23-8-2	62 Sm +2 +3 150.35 -24-8-2	63 Eu +2 +3 151.96 -25-8-2	64 Gd +3 157.25 -25-9-2	65 Tb +3 158.924 -26-9-2	66 Dy +3 162.50 -28-8-2	67 Ho +3 164.930 -29-8-2	68 Er +3 167.26 -30-8-2	69 Tm +3 168.934 -31-8-2	70 Yb +2 +3 173.04 -32-8-2	71 Lu +3 174.97 -32-9-2	-N-O-P

**Actinides

3b	4b	5b	6b	7b	8			1b	2b	3a	4a	5a	6a	Orbit
90 Th +4 232.038 -19-9-2	91 Pa +4 +5 (231) -20-9-2	92 U +3 +4 +5 +6 238.03 -21-9-2	93 Np +3 +4 +5 +6 (237) -22-9-2	94 Pu +3 +4 +5 +6 (242) -23-9-2	95 Am +3 +4 +5 +6 (243) -24-9-2	96 Cm +3 +4 (247) -25-9-2	97 Bk +3 +4 (249) -26-9-2	98 Cf +3 (251) -28-8-2	99 Es (254) -29-8-2	100 Fm (252) -30-8-2	101 Md (256) -31-8-2	102 (254) -32-8-2	103 Lw	-O-P-Q

Numbers in parentheses are mass numbers of most stable isotope of that element.

ATOMIC WEIGHTS

For the sake of completeness all known elements are included in the list. Several of those more recently discovered are represented only by the unstable isotopes. The value in parenthesis in the atomic weight column is, in each case, the mass number of the most stable isotope.**

Name	Symbol	At. No.	International atomic weight 1961	1959	Valence
Actinium	Ac	89	....	(227)	....
Aluminum	Al	13	26.9815	26.98	3
Americium	Am	95	....	(243)	3, 4, 5, 6
Antimony, stibium	Sb	51	121.75	121.76	3, 5
Argon	Ar	18	39.948	39.944	0
Arsenic	As	33	74.9216	74.92	3, 5
Astatine	At	85	....	(210)	1, 3, 5, 7
Barium	Ba	56	137.34	137.36	2
Berkelium	Bk	97	....	(217)	3, 4
Beryllium	Be	4	9.0122	9.013	2
Bismuth	Bi	83	208.980	208.99	3, 5
Boron	B	5	10.811	10.82	3
Bromine	Br	35	79.909	79.916	1, 3, 5, 7
Cadmium	Cd	48	112.40	112.41	2
Calcium	Ca	20	40.08	40.08	2
Californium	Cf	98	....	(251)	....
Carbon	C	6	12.01115	12.011	2, 4
Cerium	Ce	58	140.12	140.13	3, 4
Cesium	Cs	55	132.905	132.91	1
Chlorine	Cl	17	35.453	35.457	1, 3, 5, 7
Chromium	Cr	24	51.996	52.01	2, 3, 6
Cobalt	Co	27	58.9332	58.94	2, 3
Columbium, see *Niobium*					
Copper	Cu	29	63.54	63.54	1, 2
Curium	Cm	96	....	(247)	3
Dysprosium	Dy	66	162.50	162.51	3
Einsteinium	Es	99	....	(254)	....
Erbium	Er	68	167.26	167.27	3
Europium	Eu	63	151.96	152.0	2, 3
Fermium	Fm	100	....	(257)	....
Fluorine	F	9	18.9984	19.00	1
Francium	Fr	87	....	(223)	1
Gadolinium	Gd	64	157.25	157.26	3
Gallium	Ga	31	69.72	69.72	2, 3
Germanium	Ge	32	72.59	72.60	4
Gold, aurum	Au	79	196.967	197.0	1, 3
Hafnium	Hf	72	178.49	178.50	4
Helium	He	2	4.0026	4.003	0
Holmium	Ho	67	164.930	164.94	3
Hydrogen	H	1	1.00797	1.0080	1
Indium	In	49	114.82	114.82	3
Iodine	I	53	126.9044	126.91	1, 3, 5, 7
Iridium	Ir	77	192.2	192.2	3, 4
Iron, ferrum	Fe	26	55.847	55.85	2, 3
Krypton	Kr	36	83.80	83.80	0
Lanthanum	La	57	138.91	138.92	3
Lead, plumbum	Pb	82	207.19	207.21	2, 4
Lithium	Li	3	6.939	6.940	1
Lutetium	Lu	71	174.97	174.99	3
Magnesium	Mg	12	24.312	24.32	2
Manganese	Mn	25	54.9380	54.94	2, 3, 4, 6, 7
Mendelevium	Md	101	....	(256)	....
Mercury, hydrargyrum	Hg	80	200.59	200.61	1, 2
Molybdenum	Mo	42	95.94	95.95	3, 4, 6
Neodymium	Nd	60	144.24	144.27	3
Neon	Ne	10	20.183	20.183	0
Neptunium	Np	93	....	(237)	4, 5, 6
Nickel	Ni	28	58.71	58.71	2, 3
Niobium (columbium)	Nb	41	92.906	92.91	3, 5
Nitrogen	N	7	14.0067	14.008	3, 5
Nobelium	No	102	....	(254)	
Osmium	Os	76	190.2	190.2	2, 3, 4, 8
Oxygen	O	8	15.9994	16.000	2
Palladium	Pd	46	106.4	106.4	2, 4, 6
Phosphorus	P	15	30.9738	30.975	3, 5
Platinum	Pt	78	195.09	195.09	2, 4
Plutonium	Pu	94	....	(244)	3, 4, 5, 6
Polonium	Po	84	....	(209)	....
Potassium, kalium	K	19	39.102	39.100	1
Praseodymium	Pr	59	140.907	140.92	3
Promethium	Pm	61	....	(145)	3
Protactinium	Pa	91	....	(231)	....
Radium	Ra	88	....	(226)	2
Radon	Rn	86	....	(222)	0
Rhenium	Re	75	186.2	186.22	....
Rhodium	Rh	45	102.905	102.91	3
Rubidium	Rb	37	85.47	85.48	1
Ruthenium	Ru	44	101.07	101.1	3, 4, 6, 8
Samarium	Sm	62	150.35	150.35	2, 3
Scandium	Sc	21	44.956	44.96	3
Selenium	Se	34	78.96	78.96	2, 4, 6
Silicon	Si	14	28.086	28.09	4
Silver, argentum	Ag	47	107.870	107.873	1
Sodium, natrium	Na	11	22.9898	22.991	1
Strontium	Sr	38	87.62	87.63	2
Sulfur	S	16	32.064	32.066*	2, 4, 6
Tantalum	Ta	73	180.948	180.95	5
Technetium	Tc	43	....	(97)	6, 7
Tellurium	Te	52	127.60	127.61	2, 4, 6
Terbium	Tb	65	158.924	158.93	3
Thallium	Tl	81	204.37	204.39	1, 3
Thorium	Th	90	232.038	(232)	4
Thulium	Tm	69	168.934	168.94	3
Tin, stannum	Sn	50	118.69	118.70	2, 4
Titanium	Ti	22	47.90	47.90	3, 4
Tungsten (wolfram)	W	74	183.85	183.86	6
Uranium	U	92	238.03	238.07	4, 6
Vanadium	V	23	50.942	50.95	3, 5
Xenon	Xe	54	131.30	131.30	0
Ytterbium	Yb	70	173.04	173.04	2, 3
Yttrium	Y	39	88.905	88.91	3
Zinc	Zn	30	65.37	65.38	2
Zirconium	Zr	40	91.22	91.22	4

* Because of natural variations in the relative abundances of the isotopes of sulfur the atomic weight of this element has a range of ± 0.003.

** The 1959 atomic weights are based on O = 16.000 whereas those of 1961 are based on the isotope C¹².

ORGANIC COMPOUND INDEX

A

Compound	Compound No.	Page No.
Aceanthrenequinone	48	184
Acenaphthene	99	47
Acenaphthene-1-carbonitrile	84	361
Acenaphthene-5-carbonitrile	155	364
Acenaphthene-5-carboxylic acid	347	207
5,6-Acenaphthenediol	497	122
Acenaphthenequinone	44	183
Acenaphthene-3-sulfonamide	309	404
Acenaphthene-5-sulfonamide	369	406
Acenaphthene-5-sulfonanilide	224	400
Acenaphthene-3-sulfonic acid	144	376
Acenaphthene-3-sulfonyl chloride	175	386
Acenaphthene-5-sulfonyl chloride	172	386
1-Acenaphthenol	353	114
7-Acenaphthenone	151	177
1-Acenaphthoic acid	387	209
cis-Acenaphthylene glycol	527	124
Acepleiadene	123	48
Acetaldehyde	3	144
Acetaldehyde cyanohydrin	54	349
Acetaldehyde trimer	34	145
Acetamide	56	232
Acetamidine	470	439
4-Acetamidobenzene sulfonamide	359	406
4-Acetamidobenzene sulfonanilide	345	405
4-Acetamidobenzene-1-sulfonic acid	164	377
4-Acetamidobenzenesulfonyl chloride	237	388
4-Acetamidobenzonitrile	180	365
4-Acetamidonaphthalene-1-sulfonamide	417	408
4-Acetamidonaphthalene-1-sulfonanilide	197	399
4-Acetamidonaphthalene-1-sulfonic acid	192	378
2-Acetamidophenol	511	123
3-Acetamidophenol	375	115
Acetanilide	174	235
Acetic acid	3	190
" "	431	432
Acetic anhydride	4	223
Acetoacetamide	15	231
Acetoacetanilide	64	232
4-Acetobutanoic acid	51	191
γ-Acetobutyric acid	51	191
d, l-Acetoin	39	82
" "	54	162
Acetol	40	82
Acetone	1	161
Acetone cyanohydrin	18	346
Acetone-1,3-dicarboxylic acid	189	200
Acetonitrile	5	345
Acetonylacetone	148	165
2-Acetophenelidide	51	232
3-Acetophenelidide	100	233
4-Acetophenelidide	242	237
Acetophenone	160	166
Acetophenone-o-carboxylic acid	144	198
Aceto-N-piperidide	7	230
Acetopyruvic acid	117	197
2-Acetotoluidide	171	235
3 Acetotoluidide	31	231
4-Acetotoluidide	290	239
Acetoveratron	55	174
Acetoxyacetic acid	62	195
Acetoxyacetone	120	165
w-Acetoxyacetophenone	60	276
2-Acetoxyethylamine	310	438
3-Acetoxyphenol	29	95
α-Acetoxypropionic acid	48	194
2-Acetoxypropionitrile	46	348
α-Acetoxypropionyl chloride	30	212
Acetylacetone	44	162

Compound	Compound No.	Page No.
d, l-N-Acetylalanine	240	431
3-(Acetylamino) propionic acid	381	432
2-Acetylaniline	42	436
3-Acetylaniline	102	436
4-Acetylaniline	41	436
2-Acetylanisole	199	167
3-Acetylanisole	200	167
Acetylanthranilamide	365	241
Acetylanthranilanilide	332	240
Acetylanthranilic acid	293	205
2-Acetylbenzoic acid	144	198
Acetylbenzoylketone	178	167
2-Acetylbiphenyl	80	170
3-Acetylbiphenyl	94	170
4-Acetylbiphenyl	152	177
Acetyl bromide	2	218
4-Acetylbutyric acid	410	432
Acetyl carbinol	40	82
3-Acetylcatechol	188	106
" "	123	176
4-Acetylcatechol	266	110
" "	148	177
Acetyl chloride	1	212
0-Acetylcitric acid	98	429
1-Acetylcyclohexene	157	166
1-Aceylcyclopentene	88	170
2-Acetyl-p-cymene	202	167
2-Acetyldibenzothiophene	144	177
Acetyl diethyl carbinol	47	162
1-Acetyl 2,2-dimethylcyclohexene	82	170
Acetylene	1	28
Acetylenedicarboxylic acid	277	204
Acetylenedicarboxylic acid diamide	505	246
9-Acetylfluorene	101	175
Acetyl fluoride	1	211
2-Acetylfuran	15	172
3-Acetylfuran	33	169
β-Acetylglutaric acid	49	194
N-Acetylglycine	233	431
Acetylglycyl chloride	17	212
8-Acetylguanine	39	173
2-Acetylhydroquinone	512	123
1-Acetyl-2-hydroxynaphthalene	83	174
2-Acetyl-1-hydroxynaphthalene	130	176
3-Acetylindazole	193	178
N-Acetylindole	313	239
3-Acetylindole	195	179
Acetyl iodide	1	221
0-Acetyllactic acid	48	194
0-Acetyl lactonitrile	46	348
0-Acetyl lactoyl chloride	30	212
Acetyl methyl carbinol	39	82
1-Acetyl-2-methylcyclohexanone	155	166
d,l-1-Acetyl-3-methylcyclohexanone	50	169
1-Acetyl-4-methylcyclohexanone	151	166
2-Acetyl-5-methylfuran	69	170
2-Acetyl-5-methylthiophene	89	170
α-Acetyl-β-methylurea	376	242
1-Acetylnaphthalene	215	168
2-Acetylnaphthalene	57	174
Acetylnaphthionic acid	192	378
1-Acetyl-2-naphthol	73	100
2-Acetyl-1-naphthol	196	106
δ-Acetylpentanoic acid	23	193
2-Acetylphenanthrene	172	178
3-Acetylphenanthrene	95	175
9-Acetylphenanthrene	99	175
N-Acetyl-o-phenetidine	51	232
2-Acetylphenol	10	96

Compound	Compound No.	Page No.
3-Acetylphenol	180	105
" "	64	434
4-Acetylphenol	223	107
" "	141	176
" "	36	434
Acetyl phenyl carbinol	74	170
Acetyl β-phenylhydrazine	203	236
Acetylphloroglucinol	545	125
N-Acetylpiperidine	7	230
β-Acetylpropionic Acid	16	172
" " "	18	193
2-Acetylpyridine	144	165
3-Acetylpyridine	84	436
2-Acetylquinoline	42	173
4-Acetylresorcinol	362	115
5-Acetylresorcinol	368	115
Acetylsalicylaldehyde diacetate	141	280
Acetyl salicylanilide	246	237
0-Acetylsalicylic acid	186	200
" " "	400	432
6-Acetyltetralin	99	170
3-Acetylthianaphthene	102	171
2-Acetylthiophene	169	166
N-Acetyl-m-toluidine	31	231
N-Acetyl-p-toluidine	290	239
Acetylurea	461	244
5-Acetyl-n-valeric acid	23	193
cis-Aconitanilide	344	240
trans-Aconitic acid	314	206
Aconitic acid dianilide	400	242
cis-Aconitic acid dianilide	427	243
trans-Aconityl trichloride	37	215
Acraldehyde	6	144
Acridine	69	324
"	420	439
2,3-Acridinediol	564	126
2,8-Acridinediol	588	127
3,7-Acridinediol	590	127
Acrolein	6	144
Acrylamide	62	232
Acrylanilide	139	234
Acrylic acid	5	190
" "	334	431
Acrylonitrile	3	345
Acrylyl chloride	4	212
Adamantane	80	9
Adipic acid	232	202
" "	377	432
Adipic acid dianilide	490	245
Adipic acid monoamide	204	236
Adipic acid monoanilide	286	238
Adipic aldehyde	26	151
Adiponitrile	104	354
Adipyl dichloride	34	215
β-Alanine	32	285
DL-Alanine	103	289
" "	80	429
L-Alanine	105	289
3-Aldehydobenzoic acid	271	204
Aldol	23	150
Aletheine	29	417
Aleuritic acid	121	197
Alizarin	54	184
Allene	4	12
L-(+)-Alloisoleucine	128	290
Allomucic acid	274	204
β-D-Allose	19	330
β-L-Allose	20	330
DL-Allothreonine	65	287
Alloxanic acid	249	203
Allyl acetate	32	251
Allylacetic acid	22	190
Allylacetone	28	162
Allylacetyl chloride	23	212
Allylacetylene	10	28
Allyl alcohol	5	80
Allylamine	12	295
N-Allylaniline	134	437
Allyl benzoate	238	263
Allyl bromide	6	58
Allyl n-butyrate	74	253
Allyl chloride	6	55
Allyl cyanide	16	346
Allylcyclohexane	281	21
4-Allyl-1,2-dimethoxybenzene	99	136
Allyl ethyl ether	17	131
Allyl ethyl sulfide	15	420
Allyl formate	15	250
Allyl iodide	6	60
Allyl isobutyrate	66	252
Allylmalonic acid	126	198
Allyl malononitrile	72	350
Allyl mercaptan	8	413
2-Allyl-6-methoxyphenol	22	95
4-Allyl-2-methoxyphenol	23	95
" "	97	434
4-Allyl-1,2-methylenedioxybenzene	96	135
Allyl methyl ether	8	131
Allyl methyl sulfide	5	420
1-Allylnaphthalene	157	41
2-Allylphenol	10	94
4-Allylphenol	15	94
Allyl phenyl sulfide	51	421
1-Allylpiperidine	342	438
Allyl propionate	51	252
Allylsulfide	23	420
2-Allylthiophene	35	424
Allylurea	63	232
D-Altrose	11	329
L-Altrose	10	329
2-Aminoacenaphthene	305	312
3-Aminoacenaphthene	135	307
4-Aminoacenaphthene	152	307
5-Aminoacenaphthene	220	309
Aminoacetamide	95	305
3-Aminoacetanilide	153	307
4-Aminoacetanilide	321	240
"	364	314
Aminoacetonitrile	22	356
"	212	437
2-Aminoacetophenone	1	172
"	166	300
"	42	436
3-Aminoacetophenone	125	176
"	193	308
"	102	436
4-Aminoacetophenone	137	176
"	213	309
"	41	436
3-Aminoalizarin	56	184
1-Aminoanthracene	132	437
1-Amino-9,10-anthraquinone	41	183
" "	459	318
2-Amino-9,10-anthraquinone	59	185
" "	472	318
2-Amino-9,10-anthraquinone-1-carboxylic acid	381	209
4-Aminoantipyrine	222	309
2-Aminoazobenzene	71	305
4-Aminoazobenzene	277	311
4-Aminobenzalacetophenone	337	313
2-Aminobenzaldehyde	16	153
4-Aminobenzaldehyde	62	156

ORGANIC COMPOUND INDEX

Compound	Compound No.	Page No.
2-Aminobenzamide	224	309
3-Aminobenzamide	167	235
4-Aminobenzamide	173	235
"	395	315
3-Aminobenzanilide	254	237
4-Aminobenzene-1,3-disulfonamide	398	407
4-Aminobenzene-1,3-disulfonic acid	181	377
2-Aminobenzene sulfonamide	136	397
3-Aminobenzene sulfonamide	102	395
4-Aminobenzene sulfonamide	178	398
4-Aminobenzene sulfonanilide	311	404
2-Aminobenzene sulfonic acid	54	372
3-Aminobenzene sulfonic acid	37	372
4-Aminobenzene sulfonic acid	70	373
4-Aminobenzhydrol	260	311
2-Aminobenzoic acid	9	284
3-Aminobenzoic acid	15	284
"	149	430
4-Aminobenzoic acid	26	285
"	94	429
2-Aminobenzonitrile	58	359
3-Aminobenzonitrile	61	359
4-Aminobenzonitrile	150	307
"	111	362
2-Aminobenzophenone	135	176
"	209	309
4-Aminobenzophenone	155	177
"	272	311
"	40	436
3-Aminobenzopyrazole	345	314
6-Aminobenzopyrazole	433	317
2-Aminobenzothiazole	297	312
2-Aminobenzyl alcohol	136	307
3-Aminobenzyl alcohol	186	308
4-Aminobenzyl alcohol	92	305
DL-α-Aminobenzyl cyanide	63	304
α-Aminobenzl cyanide	65	360
2-Aminobenzyl cyanide	91	361
4-Aminobenzyl cyanide	36	304
"	42	358
"	46	304
2-Aminobiphenyl	113	436
" "	142	437
3-Aminobiphenyl	60	304
4-Aminobiphenyl	140	437
1-Amino-3-bromoanthraquinone	455	317
2-Amino-3-bromoanthraquinone	473	318
2-Amino-4-bromobenzaldehyde	146	307
2-Amino-5-bromobiphenyl	66	304
4-Amino-4'-bromobiphenyl	326	313
3-Amino-5-bromo-p-cresol	165	308
2-Amino-1-bromo-3-methylanthraquinone	422	316
1-Amino-3-bromonaphthalene	107	306
"	63	436
1-Amino-4-bromonaphthalene	197	309
" "	87	436
2-Amino-1-bromonaphthalene	83	305
2-Amino-4-bromonaphthalene	94	436
2-Amino-5-bromonaphthalene	23	303
2-Amino-6-bromonaphthalene	282	311
2-Amino-3-bromophenol	333	113
2-Amino-4-bromophenol	283	311
2-Amino-5-bromophenol	380	115
4-Amino-2-bromophenol	432	118
5-Amino-2-bromophenol	379	115
3-Amino-5-bromopyridine	100	306
2-Aminobutane	13	295
α-Amino-n-butylbenzene	131	299
γ-Amino-n-butylbenzene	133	299
4-Aminobutyl methyl sulfide	42	421
DL-α-Aminobutyric acid	115	290
" " "	72	429
L-(+)-α-Aminobutyric acid	100	289
DL-β-Aminobutyric acid	29	285
γ-Aminobutyric acid	35	285
"	289	431
4-Aminobutyrophenone	143	307
3-Aminocamphor	228	310
γ-Amino-n-caproic acid	37	285
w-Amino-n-caproic acid	34	285
3-Aminocarbazole	460	318
4-Aminochalcone	337	313
1-Amino-4-chloroanthraquinone	392	315
1-Amino-5-chloroanthraquinone	443	317
4-Amino-2-chlorobenzaldehyde	329	313
2-Amino-5-chloro-p-cresol	213	107
2-Amino-6-chloro-p-cresol	157	104
4-Amino-6-chloro-m-cresol	551	125
6-Amino-4-chloro-o-cresol	218	107
2-Amino-4-chloro-N,N-dimethylaniline	178	300
2-Amino-5-chloro-4-methylphenol	213	107
2-Amino-6-chloro-4-methylphenol	157	104
4-Amino-6-chloro-3-methylphenol	551	125
6-Amino-4-chloro-2-methylphenol	218	107
1-Amino-2-chloronaphthalene	72	305
1-Amino-3-chloronaphthalene	80	305
" "	65	436
1-Amino-4-chloronaphthalene	190	308
1-Amino-5-chloronaphthalene	91	436
1-Amino-6-chloronaphthalene	97	436
1-Amino-7-chloronaphthalene	96	436
2-Amino-1-chloronaphthalene	70	305
2-Amino-4-chloronaphthalene	92	436
2-Amino-7-chloronaphthalene	110	436
2-Amino-4-chloro-5-nitrophenol	554	126
2-Amino-4-chloro-6-nitrophenol	382	115
4-Amino-2-chloro-3-nitrophenol	434	118
4-Amino-2-chloro-5-nitrophenol	445	119
2-Amino-3-chlorophenol	278	110
2-Amino-4-chlorophenol	334	113
2-Amino-5-chlorophenol	387	116
4-Amino-2-chlorophenol	385	116
4-Amino-3-chlorophenol	402	117
2-Amino-4-chloropyridine	294	312
8-Amino-6-chloroquinoline	113	306
trans-2-Aminocinnamic acid	12	284
trans-3-Aminocinnamic acid	20	285
trans-4-Aminocinnamic acid	16	284
3-Aminocoumarin	292	312
6-Aminocoumarin	379	315
2-Amino-p-cresol	323	113
3-Amino-o-cresol	309	112
"	156	307
3-Amino-p-cresol	350	114
4-Amino-m-cresol	466	120
4-Amino-o-cresol	458	120
5-Amino-o-cresol	410	117
6-Amino-m-cresol	416	117
6-Amino-o-cresol	154	104
1-Amino-3-cyanonaphthalene	44	436
2-Amino-4-cyanonaphthalene	62	436
1-Aminocyclohexanecarboxylic acid	125	290
"	111	430
2-Aminocyclohexanecarboxylic acid	217	430
cis-3-Aminocyclohexanecarboxylic acid	239	431
trans-3-Aminocyclohexanecarboxylic acid	257	431
cis-4-Aminocyclohexanecarboxylic acid	445	432
trans-4-Aminocyclohexanecarboxylic acid	366	432
4-Aminocyclohexanecarboxylic acid	114	289
2-Aminocyclohexanol	96	305
3-Aminocyclohexene	391	439

Compound	Compound No.	Page No.
2-Amino-p-cymene	158	300
cis-9-Aminodecalin	140	299
trans-9-Aminodecalin	132	299
2-Amino-2-desoxy-D-glucose	12	329
3-Amino-3-desoxy-D-glucose	14	328
2-Aminodibenzfuran	172	308
2-Aminodibenzyl	12	303
4-Aminodibenzyl	42	304
3-Amino-4,6-dichlorophenol	324	113
4-Amino-N,N-diethylaniline	173	300
2-Aminodiethylether	35	296
4-Amino-2,6-diethyl-5-methylpyrimidine	404	316
2-Amino-1,4-dihydroxy-9,10-anthraquinone	474	318
3-Amino-1,2-dihydroxy-9,10-anthraquinone	56	184
5-Amino-1,4-dihydroxy-9,10-anthraquinone	437	317
1-Amino-2,4-dihydroxynaphthalene	417	117
2-Amino-3,4-dihydroxynaphthalene	424	118
3-Amino-4-(dimethylamino)toluene	148	299
2-Amino-N,N-dimethylaniline	120	299
4-Amino-N,N-dimethylaniline	29	303
2-Amino-5,4′-dimethylazobenzene	251	310
4-Amino-3,2′-dimethylazobenzene	195	309
2-Amino-4,4′-dimethylbiphenyl	81	305
3-Amino-4,4′-dimethylbiphenyl	204	309
6-Amino-3,4′-dimethylbiphenyl	6	302
1-Amino-2,6-dimethylnaphthalene	161	307
2-Amino-1,4-dimethylnaphthalene	119	306
2-Amino-3,6-dimethylnaphthalene	313	312
2-Amino-3,7-dimethylnaphthalene	286	311
2-Amino-3,5-dimethylphenol	420	118
2-Amino-3,6-dimethylphenol	376	115
2-Amino-4,6-dimethylphenol	318	112
3-Amino-4,5-dimethylphenol	429	118
4-Amino-2,5-dimethylphenol	569	126
4-Amino-2,6-dimethylphenol	332	113
4-Amino-3,5-dimethylphenol	472	121
4-Amino-2,3-dimethyl-1-phenylpyrazolone-5	222	309
4-Amino-2,6-dimethylpiperidine	89	298
4-Amino-2,6-dimethylpyridine	400	316
2-Amino-4,5-dimethylpyrimidine	438	317
2-Amino-4,6-dimethylpyrimidine	413	316
4-Amino-2,6-dimethylpyrimidine	394	315
6-Amino-5,7-dimethylquinoline	385	315
7-Amino-2,4-dimethylquinoline	174	308
1-Amino-4,5-dimethyl-1,2,3-triazole	178	308
4-Amino-2,6-dinitro-m-cresol	436	119
6-Amino-2,4-dinitro-m-cresol	403	117
4-Amino-2,6-dinitro-3-methylphenol	436	119
6-Amino-2,4-dinitro-3-methylphenol	403	117
1-Amino-2,4-dinitronaphthalene	453	317
2-Amino-1,5-dinitronaphthalene	409	316
2-Amino-1,8-dinitronaphthalene	450	317
2-Amino-3,5-dinitrophenol	542	125
2-Amino-4,6-dinitrophenol	444	119
4-Amino-2,5-dinitrophenol	435	119
4-Amino-2,6-dinitrophenol	448	119
2-Aminodiphenylamine	129	306
4-Aminodiphenylamine	115	306
2-Aminodiphenyl ether	28	138
3-Aminodiphenyl ether	20	137
4-Aminodephenyl ether	49	139
α-Aminodiphenylmethane	185	301
2-Aminodiphenylmethane	58	304
2-Aminodiphenyl sulfide	10	425
4-Aminodiphenyl sulfide	54	426
1-Aminodocosane	409	439
2-Aminoethanethiol	32	417
2-Aminoethyl alcohol	62	83
〃 〃 〃	71	297
α-Aminoethylbenzene	83	297
α-Aminoethylbenzene	309	438
β-Aminoethylbenzene	93	298
〃 〃	349	438
1-(2-Aminoethyl)-4-hydroxybenzene	368	314
2-(w-Aminoethyl)-indole	256	310
4-(2-Aminoethyl)-phenol	368	314
1-Aminoethyl phenyl ketone	233	310
1-Aminofluorene	117	436
2-Aminofluorene	168	437
3-Aminofluorene	179	437
4-Aminofluorene	93	436
9-Aminofluorene	90	305
2-Aminofluorenone	367	314
1-Amino-4-fluoronaphthalene	41	304
D-1-Aminofructose	18	330
4-Aminoguaiacol	275	311
1-Aminohendecane	155	300
2-Aminohendecane	151	299
1-Aminoheptadecane	408	439
2-Amino-n-heptane	55	296
〃	425	439
4-Amino-n-heptane	53	296
DL-α-Aminoheptanoic acid	86	288
1-Aminohexadecane	413	439
3-Amino-n-hexane	47	296
4-Aminohexanoic acid	37	285
6-Aminohexanoic acid	34	285
D-α-Aminohydratropic acid	104	289
DL-α-Aminohydratropic acid	76	287
1-Aminohydrocinnamic acid	1	284
4-Aminohydrocinnamic acid	7	284
D-β-Aminohydrocinnamic acid	62	287
DL-β-Aminohydrocinnamic acid	60	287
L-β-Aminohydrocinnamic acid	63	287
3-Amino-6-hydroxyacetophenone	261	311
4-Amino-2-hydroxyacetophenone	265	311
1-Amino-2-hydroxyanthraquinone	458	317
4-Amino-4′-hydroxyazobenzene	401	316
3-Amino-4-hydroxybenzenesulfonamide	323	404
4-Amino-2-hydroxybenzenesulfonamide	143	397
5-Amino-2-hydroxybenzenesulfonamide	317	404
5-Amino-2-hydroxybenzenesulfonamide	163	398
5-Amino-2-hydroxybenzene sulfonic acid	148	376
4-Amino-2-hydroxybenzenesulfonyl chloride	261	389
4-Amino-2-hydroxybenzoic acid	40	429
5-Amino-2-hydroxybenzoic acid	114	430
trans-1-Amino-2-hydroxycyclohexane	338	438
cis-1-Amino-2-hydroxycyclohexane	346	438
cis-1-Amino-2-hydroxycyclopentane	344	438
trans-1-Amino-2-hydroxycyclopentane	317	438
2-Amino-4-hydroxy-3-methoxybenzaldehyde	285	311
1-Amino-3-hydroxynaphthalene	397	316
〃	90	436
1-Amino-5-hydroxynaphthalene	122	437
1-Amino-6-hydroxynaphthalene	406	316
〃 〃	124	437
1-Amino-7-hydroxynaphthalene	430	317
〃 〃	138	437
1-Amino-8-hydroxynaphthalene	179	308
2-Amino-3-hydroxynaphthalene	452	317
2-Amino-5-hydroxynaphthalene	129	437
2-Amino-6-hydroxynaphthalene	436	317
2-Amino-7-hydroxynaphthalene	421	316
〃 〃	143	437
1-Amino-4-hydroxy-3-nitronaphthalene	359	314
2-Amino-3-hydroxypentane	74	297
3-Amino-2-hydroxypentane	72	297
2-Amino-10-hydroxyphenanthrene	446	317
5-Amino-8-hydroxyquinoline	321	313
7-Amino-8-hydroxyquinoline	273	311
5-Amino-2-hydroxy-m-xylene	310	312

Compound No.	Page No.		Compound No.	Page No.
1-Aminoindane............................	125	299	4-Amino-2-methyl-6-nitrophenol	258	109
"	316	438	4-Amino-3-methyl-2-nitrophenol	509	123
2-Aminoindane............................	143	299	DL-α-Amino-α-methylpentanoic acid...........	102	289
"	333	438	2-Amino-4-methylphenol	323	113
4-Aminoindane............................	150	299	2-Amino-5-methylphenol......................	416	117
5-Aminoindane............................	21	303	2-Amino-6-methylphenol	154	104
"	210	437	3-Amino-2-methylphenol	309	112
4-Amino-4'-iodobiphenyl	372	315	3-Amino-4-methylphenol	350	114
1-Amino-3-iodonaphthalene	142	307	4-Amino-2-methylphenol	458	120
" "	72	436	4-Amino-3-methylphenol	466	120
2-Amino-4-iodonaphthalene	95	436	5-Amino-2-methylphenol	410	117
2-Amino-5-iodophenol........................	341	113	4-Amino-3-methyl-1-phenylpyrazole	155	307
α-Aminoisobutyric acid......................	85	288	5-Amino-3-methyl-1-phenylpyrazole	241	310
" "	88	429	2-Amino-3-methylpyridine	4	303
4-Amino-6-isopropyl-m-cresol	464	120	2-Amino-4-methylpyridine	188	308
4-Amino-5-isopropyl-2-methyl-6-nitrophenol	319	112	2-Amino-6-methylpyridine	28	303
4-Amino-6-isopropyl-3-methylphenol	464	120	3-Amino-4-methylpyridine	211	309
(2-Aminoisopropyl) methyl sulfide.............	33	421	5-Amino-2-methylpyridine	180	308
4-Amino-5-isopropyl-6-nitro-o-cresol	319	112	2-Amino-4-methylquinoline	298	312
1-Aminoisoquinoline.........................	266	311	3-Amino-2-methylquinoline	358	314
5-Aminoisoquinoline.........................	284	311	4-Amino-2-methylquinoline	378	315
β-Aminoisovaleric acid......................	48	286	5-Amino-2-methylquinoline	246	310
Aminomalonic acid	3	284	6-Amino-2-methylquinoline	403	316
1-Amino-4-mercaptonaphthalene	162	307	7-Amino-2-methylquinoline	333	313
1-Amino-3-methoxynaphthalene	89	436	8-Amino-6-methylquinoline	82	305
1-Amino-6-methoxynaphthalene	118	436	α-Amino-α-methyl-succinic acid	61	287
1-Amino-7-methoxynaphthalene	130	437	5-Amino-2-methylthiophenol	15	416
2-Amino-4-methoxynaphthalene	128	437	1-Amino-5-methyl-1,2,3-triazole	108	306
2-Amino-6-methoxynaphthalene	169	437	5-Amino-3-methyl-1,2,4-triazole	331	313
2-Amino-7-methoxynaphthalene	137	437	DL-α-Amino-α-methylvaleric acid...............	102	289
4-Amino-N-methylaniline	17	303	1-Aminonaphthalene	51	304
1-Amino-2-methylanthraquinone	428	317	"	119	436
2-Amino-5-methylbenzene-1,3-disulfonalinide	285	403	2-Aminonaphthalene	232	310
2-Amino-5-methylbenzene-1,3-disulfonamide	440	409	"	133	437
2-Amino-5-methylbenzene-1,4-disulfonamide	441	409	2-Aminonaphthalene-8-sulfonamide	232	400
2-Amino-5-methylbenzene-1,4-disulfonic acid	202	378	4-Aminonaphthalene-1-sulfonamide	340	405
2-Amino-5-methylbenzene-1,3-disulfonyl chloride .	242	389	4-Aminonaphthalene-2-sulfonamide	77	394
3-Amino-4-methylbenzenesulfonamide	216	400	5-Aminonaphthalene-1-sulfonamide	442	409
5-Amino-2-methylbenzenesulfonamide	175	398	5-Aminonaphthalene-2-sulfonamide	358	406
3-Amino-4-methylbenzene sulfonic acid	92	374	6-Aminonaphthalene-1-sulfonamide	180	398
3-Amino-6-methylbenzene sulfonic acid	68	373	1-Aminonaphthalene-7-sulfonanilide	121	396
2-Amino-5-methylbenzophenone	97	305	4-Aminonaphthalene-1-sulfonanilide	274	402
3-Amino-4-methylbenzophenone	223	309	5-Aminonaphthalene-1-sulfonanilide	203	399
3'-Amino-4-methylbenzophenone...............	230	310	5-Aminonaphthalene-2-sulfonanilide	71	394
4-Amino-3-methylbenzophenone	231	310	8-Aminonaphthalene-1-sulfonanilide	96	395
4-Amino-4'-methylbenzophenone...............	402	316	4-Aminonaphthalene-1-sulfonic acid	154	376
4-Amino-3-methylbiphenyl....................	32	303	5-Aminonaphthalene-1-sulfonic acid	203	378
4-Amino-4'-methylbiphenyl	192	308	5-Aminonaphthalene-2-sulfonic acid	165	377
2-Amino-2-methylbutane......................	18	295	6-Aminonaphthalene-2-sulfonic acid	72	373
(Aminomethyl) cyclohexane	399	439	8-Aminonaphthalene-1-sulfonic acid	4	380
cis-1-Amino-2-methylcyclohexane	400	439	8-Aminonaphthalene-2-sulfonic acid	102	374
trans-1-Amino-2-methylcyclohexane	402	439	4-Amino-1-naphthalenethiol....................	28	417
cis-1-Amino-3-methylcyclohexane	403	439	1-Amino-2-naphthoic acid	39	285
trans-1-Amino-3-methylcyclohexane	414	439	2-Amino-1-naphthoic acid	5	284
trans-1-Amino-4-methylcyclohexane	398	439	3-Amino-1-naphthoic acid	21	285
2-Amino-4-methyldiphenylamine................	317	313	4-Amino-1-naphthoic acid	19	285
4-Amino-2-methyldiphenylamine	47	304	5-Amino-1-naphthoic acid	42	286
α-Amino-α-methylmalonic acid	72	287	6-Amino-1-naphthoic acid	38	285
1-Amino-2-methylnaphthalene	9	303	7-Amino-1-naphthoic acid	52	286
1-Amino-3-methylnaphthalene	54	304	1-Amino-2-naphthol..........................	580	127
" "	123	437	3-Amino-2-naphthol..........................	562	126
1-Amino-4-methylnaphthalene	55	304	4-Amino-2-naphthol..........................	479	121
1-Amino-5-methylnaphthalene	124	306	5-Amino-1-naphthol..........................	491	122
1-Amino-7-methylnaphthalene	68	304	5-Amino-2-naphthol..........................	489	122
1-Amino-8-methylnaphthalene	101	306	6-Amino-1-naphthol..........................	502	122
2-Amino-1-methylnaphthalene	53	304	6-Amino-2-naphthol..........................	529	124
2-Amino-3-methylnaphthalene	304	312	7-Amino-2-naphthol..........................	508	123
2-Amino-4-methylnaphthalene	105	306	8-Amino-1-naphthol	179	308
2-Amino-5-methylnaphthalene	86	305	8-Amino-2-naphthol	520	123
2-Amino-4-methyl-5-nitrophenol	501	122	2-Amino-1,4-naphthoquinone	427	317
2-Amino-4-methyl-6-nitrophenol	263	109	5-Amino-4-nitroacenaphthene	447	317

	Compound No.	Page No.
1-Amino-4-nitroanthraquinone	470	318
1-Amino-5-nitroanthraquinone	469	318
4-Amino-4′-nitroazobenzene	440	317
2-Amino-4-nitrobenzaldehyde	271	311
2-Amino-5-nitrobenzaldehyde	420	316
4-Amino-3-nitrobenzaldehyde	408	316
4-Amino-3-nitrobenzenesulfonamide	330	404
4-Amino-2-nitrobenzenesulfonic acid	155	376
4-Amino-3-nitrobenzenesulfonyl chloride	57	382
4-Amino-3-nitrobenzophenone	316	312
2-Amino-4′-nitrobiphenyl	357	314
2-Amino-5-nitrobiphenyl	274	311
4-Amino-3-nitrobiphenyl	373	315
4-Amino-4′-nitrobiphenyl	419	316
4-Amino-6-nitrocatechol	557	126
2-Amino-5-nitro-p-cresol	501	122
2-Amino-6-nitro-p-cresol	263	109
4-Amino-2-nitro-m-cresol	509	123
4-Amino-6-nitro-o-cresol	258	109
3-Amino-4-nitrodiphenyl sulfide	57	426
2-Amino-3-nitrofluorenone	464	318
2-Amino-5-nitrohydroquinone	346	314
2-Amino-4-nitromesitylene	120	306
1-Amino-2-nitronaphthalene	322	313
" "	7	436
1-Amino-3-nitronaphthalene	38	436
1-Amino-4-nitronaphthalene	412	316
" "	15	436
1-Amino-5-nitronaphthalene	252	310
" "	68	436
1-Amino-6-nitronaphthalene	75	436
1-Amino-7-nitronaphthalene	55	436
1-Amino-8-nitronaphthalene	185	308
" "	71	436
2-Amino-1-nitronaphthalene	276	311
2-Amino-3-nitronaphthalene	30	436
2-Amino-4-nitronaphthalene	50	436
2-Amino-5-nitronaphthalene	323	313
" "	79	436
2-Amino-6-nitronaphthalene	59	436
2-Amino-7-nitronaphthalene	81	436
2-Amino-8-nitronaphthalene	201	309
" "	69	436
2-Amino-3-nitrophenol	538	124
2-Amino-4-nitrophenol	344	114
2-Amino-5-nitrophenol	522	123
2-Amino-6-nitrophenol	233	108
3-Amino-4-nitrophenol	481	121
4-Amino-2-nitrophenol	312	112
4-Amino-3-nitrophenol	390	116
5-Amino-2-nitrophenol	414	117
5-Amino-3-nitrophenol	427	118
3-Amino-4′-nitrophenyl sulfide	57	426
4-Amino-3-nitropyridine	418	316
2-Amino-8-nitroquinoline	355	314
6-Amino-5-nitroquinoline	388	315
8-Amino-6-nitroquinoline	411	316
2-Amino-4-nitroresorcinol	473	121
4-Amino-6-nitroresorcinol	405	117
2-Amino-4-nitrostilbene	320	313
4-Amino-2-nitrostilbene	227	310
DL-α-Aminononanoic acid	82	288
1-Aminooctadecane	410	439
2-Aminooctane	401	439
DL-α-Aminooctanoic acid	80	288
1-Aminooxindole	50	304
DL-2-Amino-n-pentane	27	295
1-Aminopentadecane	411	439
3-Aminopentane	390	439
2-Amino-n-pentanoic acid	74	429
DL-3-Aminopentanoic acid	13	284
" "	288	431
DL-4-Aminopentanoic acid	46	286
5-Aminopentanoic acid	11	284
" "	338	431
5-Amino-1-pentene	26	295
2-Aminophenacyl alcohol	189	308
4-Aminophenacyl alcohol	369	315
2-Amino-9,10-phenanthraquinone	55	184
3-Amino-9,10-phenanthraquinone	42	183
1-Aminophenanthrene	327	313
2-Aminophenanthrene	149	307
3-Aminophenanthrene	154	307
4-Aminophenanthrene	208	309
9-Aminophenanthrene	308	312
2-Aminophenol	456	120
"	84	434
3-Aminophenol	277	110
"	263	311
"	88	434
4-Aminophenol	476	121
"	396	316
4-Aminophenol-2-sulfonic acid	148	376
(2-Aminophenyl)acetic acid	4	284
DL-α-Aminophenylacetic acid	74	287
L-α-Aminophenylacetic acid	118	290
(3-Aminophenyl) acetic acid	10	284
(4-Aminophenyl)acetic acid	31	285
(2-Aminophenyl)acetonitrile	91	361
(4-Aminophenyl)acetonitrile	42	358
3-Amino-5-phenylacridine	426	317
2-Aminophenyl-4-aminophenyl sulfide	32	425
2-(2-Aminophenyl)ethyl alcohol	5	302
2-(4-Aminophenyl)ethyl alcohol	219	309
β-Amino-β-phenylisobutyric acid	68	287
3-Aminophenyl-4-nitrophenyl sulfide	57	426
2-Aminophenyl phenyl sulfide	10	425
4-Aminophenyl phenyl sulfide	54	426
3-Amino-6-phenylpyridine	210	309
2-(4-Aminophenyl)-quinoline	312	312
3-Amino-2-phenylquinoline	240	310
4-Amino-2-phenylquinoline	376	315
3-Amino-1-phenyl-1,2,4-triazole	336	313
5-Amino-1-phenyl-1,2,4-triazole	351	314
4-Aminophenylurethane	114	306
2-Amino-β-picoline	4	303
2-Amino-γ-picoline	188	308
3-Amino-γ-picoline	211	309
5-Amino-α-picoline	180	308
6-Amino-α-picoline	28	303
1-Amino-2-propanethiol	20	416
3-Amino-1-propanethiol	36	417
2-(3-Aminopropionamido)ethanethiol	29	417
3-Aminopropionic acid	212	430
2-Aminopropiophenone	46	173
"	37	304
3-Aminopropiophenone	34	173
4-Aminopropiophenone	170	178
"	315	312
β-Aminopropiophenone	233	310
2-Aminopropyl alcohol	73	297
3-Aminopropyl alcohol	87	297
DL-3-Aminopropyleneglycol	179	300
3-Aminopropyl methyl sulfide	36	421
4-Aminopyrazole	131	306
2-Aminopyridine	74	305
"	270	438
3-Aminopyridine	89	305
"	248	438
4-Aminopyridine	353	314
"	311	438

Compound		Compound	
	No.	Page No.	

	Compound No.	Page No.		Compound No.	Page No.
γ-Aminopyridine	353	314	w-Amino-n-valeric acid	11	284
2-Aminopyrimidine	281	311	2-Aminovanillin	285	311
4-Aminopyrimidine	338	313	4-Aminoveratrol	148	307
5-Aminoquinaldine	246	310	4-Amino-o-xylene	43	304
2-Aminoquinizarin	474	318	n-Amyl acetate	94	254
5-Aminoquinizarin	437	317	sec.-Amyl(2) acetate	65	252
γ-Aminoquinoline	344	314	sec.-Amyl(3) acetate	64	252
2-Aminoquinoline	289	312	tert.-Amyl acetate	50	252
"	276	438	active-Amylacetic acid	42	191
3-Aminoquinoline	173	308	active-Amyl alcohol	23	81
"	185	437	n-Amyl alcohol	32	81
4-Aminoquinoline	344	314	sec-Amyl alcohol	15	80
"	312	438	sym-sec-Amyl alcohol	13	80
5-Aminoquinoline	226	310	tert-Amyl alcohol	8	80
"	217	437	active-Amylamine	29	296
6-Aminoquinoline	238	310	n-Amylamine	32	296
"	223	437	sec-n-Amylamine	27	295
7-Aminoquinoline	171	308	n-Amyl bromide	17	58
"	267	438	tert-Amyl bromide	12	58
8-Aminoquinoline	109	306	n-Amyl-n-butyrate	146	257
"	121	437	n-Amyl-n-caproate	226	262
5-Aminoresorcinol	361	114	n-Amyl-n-caprylate	302	267
"	328	313	tert-Amyl carbinol	31	81
3-Aminosalicylic acid	64	287	n-Amyl chloride	22	55
4-Aminosalicylic acid	40	429	tert-Amyl chloride	15	55
5-Aminosalicylic acid	88	288	2-n-Amylcinnamaldehyde	9	150
" "	114	430	n-Amyl n-enanthate	277	265
2-Aminostyrene	9	302	n-Amyl ether	72	134
3-Aminostyrene	1	302	n-Amyl ethyl ether	43	133
4-Aminostyrene	2	303	tert-Amyl ethyl ether	35	132
1-Aminotetradecane	415	439	tert-Amyl ethyl ketone	68	163
1-Amino-1,2,3,4-tetrahydronaphthalene	339	438	n-Amyl formate	63	252
2-Amino-1,2,3,4-tetrahydronaphthalene	354	438	n-Amyl iodide	15	60
1-Amino-5,6,7,8-tetrahydronaphthalene	175	300	tert-Amyl iodide	10	60
" "	157	437	n-Amyl levulinate	292	266
2-Amino-5,6,7,8-tetrahydronaphthalene	22	303	sec-n-Amylmalonic acid	50	194
" "	202	437	n-Amyl mercaptan	18	413
3-Amino-1,2,4,5-tetramethylbenzene	117	306	sec-Amyl mercaptan	10	413
4-Amino-1,2,3,5-tetramethylbenzene	170	300	n-Amyl methyl carbinol	50	82
4-Amino-2-thiocresol	39	304	n-Amyl methyl ether	34	132
4-Amino-1-thionaphthol	162	307	tert-Amyl methyl ether	28	132
2-Aminothiophene	10	302	n-Amyl methyl ketone	69	163
2-Aminothiophenol	5	303	tert-Amyl methyl ketone	26	162
"	5	416	Amyl methyl sulfide	28	420
3-Aminothiophenol	3	302	n-Amyl 2-naphthyl ether	5	137
4-Aminothiophenol	35	303	4-n-Amylphenol	4	96
"	13	416	4-tert-Amylphenol	166	104
3-Aminothioxanthone	445	317	n-Amyl phenyl ketone	3	172
6-Aminothymol	389	315	n-Amylpropiolic acid	16	190
2-Amino-4-toluic acid	14	284	n-Amylpropionate	120	255
3-Amino-4-toluic acid	18	284	n-Amyl stearate	25	274
1-Amino-1,2,3-triazole	52	304	n-Amyl n-valerate	183	259
1-Amino-1,3,4-triazole	137	307	Anethole	3	137
3-Amino-1,2,4-triazole	356	314	Angelamide	212	236
3-Amino-2,4,6-tribromophenol	264	109	Angelanilide	207	236
3-Amino-2,4,6-trichlorophenol	179	105	Angelic acid	34	194
4-Amino-2,3,6-trichlorophenol	401	117	" "	342	432
w-Aminotridecylic acid	17	284	Anhalamine	58	140
2-Amino-4,4,4-trifluoro-n-butyric acid	24	429	Anhydrocamphoronic acid	201	201
1-Amino-2,4,5-trimethylbenzene	103	306	Aniline	80	297
1-Amino-3,4,5-trimethylbenzene	118	306	"	165	437
4-Amino-2,3,6-trinitrophenol	354	114	Aniline-2,4-disulfonamide	398	407
2-Aminotriphenyl carbinol	262	311	Aniline-2,4,-disulfonic acid	181	377
3-Aminotriphenyl carbinol	348	314	N-Anilinoacetonitrile	46	359
4-Aminotriphenyl carbinol	242	310	2-Anilinobutyric acid	209	201
2-Aminotriphenylmethane	288	311	2-(N-Anilino)-butyronitrile	31	358
4-Aminotriphenylmethane	141	307	2-Anilinoisobutanoic acid	290	204
1-Aminoundecane	155	300	2-Anilinoisobutyric acid	290	204
2-Aminoundecane	151	299	2-(N-Anilino)-isobutyronitrile	124	362
DL-β-Amino-n-valeric acid	13	284	Anilinoisovaleric acid	188	200
DL-γ-Amino-n-valeric acid	46	286	Anilinomalonic acid	155	199

ORGANIC COMPOUND INDEX

Compound	Compound No.	Page No.
2-Anilinopentanoic acid	221	202
2-Anilinopropionic acid	247	203
2-(N-Anilino)-propionitrile	118	362
2-Anilinovaleric acid	221	202
2-(N-Anilino)-valeronitrile	57	359
Anisalacetone	96	175
Anisalacetophenone	104	175
α-Anisal-α-cinnamalacetone	169	178
o-Anisaldehyde	15	153
3-Anisaldehyde	85	149
4-Anisaldehyde	89	149
4-Anisamide	330	240
4-Anisanilide	341	240
2-Anisic acid	114	197
4-Anisic acid	291	204
Anisic anhydride	43	226
2-Anisidine	94	135
3-Anisidine	102	136
4-Anisidine	35	138
m-Anisidine	168	300
o-Anisidine	134	299
p-Anisidine	67	304
Anisil	56	140
"	161	177
Anisoin	54	139
"	146	177
Anisole	56	133
Anisoyl chloride	58	213
4-Anisyl alcohol	5	88
"	4	137
4-Anisyl methyl carbinol	133	87
2-Anisyl phenyl ketone	29	173
4-Anisyl phenyl ketone	80	174
2-Anisyl sulfide	43	426
Anthracene	189	51
9-Anthracenecarbonitrile	214	367
Anthracene-1-carboxylic acid	380	209
" "	234	431
Anthracene-2-carboxylic acid	397	209
" "	315	431
Anthracene-9-carboxylic acid	331	207
" "	230	431
1,2-Anthracenediol	408	117
1,5-Anthracenediol	578	127
1,8-Anthracenediol	556	126
2,3-Anthracenediol	582	127
2,6-Anthracenediol	586	127
2,7-Anthracenediol	581	127
9,10-Anthracenediol	468	120
Anthracene-1,5-disulfonamide	474	410
Anthracene-1,8-disulfonamide	475	410
Anthracene-1,5-disulfonanilide	463	409
Anthracene-1,8-disulfonanilide	373	406
Anthracene-1,5-disulfonic acid	225	379
Anthracene-1,8-disulfonic acid	226	379
Anthracene-1,5-disulfonyl chloride	281	390
Anthracene-1,8-disulfonyl chloride	278	390
Anthracene-1-sulfonamide	321	404
Anthracene-2-sulfonamide	444	409
Anthracene-2-sulfonanilide	314	404
Anthracene-1-sulfonic acid	150	376
Anthracene-2-sulfonic acid	205	378
Anthracene-1-sulfonyl chloride	117	384
Anthracene-2-sulfonyl chloride	192	387
1,8,9-Anthracentriol	465	120
9,10-Anthradiol	468	120
Anthragallol	61	185
9-Anthraldehyde	91	158
Anthralin	465	120
Anthranilamide	155	235
Anthranilanilide	223	237
Anthranilic acid	9	284

Compound	Compound No.	Page No.
Anthranilic acid	43	429
Anthranilonitrile	58	359
9-Anthraphenone	179	178
Anthrapurpurin	63	185
1,2-Anthraquinone	28	182
1,4-Anthraquinone	35	183
9,10-Anthraquinone	52	184
9,10-Anthraquinone-1-carboxylic acid	402	210
" " "	194	430
9,10-Anthraquinone-2-carboxylic acid	401	210
" " "	197	430
9,10-Anthraquinone-2-carboxylic acid chloride	32	217
9,10-Anthraquinone-2,3-dicarboxylic acid	407	210
9,10-Anthraquinone-1,4-dicarboxylic acid dichloride	35	217
9,10-Anthraquinone-1,5-dicarboxylic acid dichloride	36	217
9,10-Anthraquinone-2,6-dicarboxylic acid dichloride	34	217
9,10-Anthraquinone-1,3-disulfonamide	478	410
9,10-Anthraquinone-1,5-disulfonamide	426	408
9,10-Anthraquinone-1,8-disulfonamide	477	410
9,10-Anthraquinone-2,7-disulfonamide	283	403
9,10-Anthraquinone-1,5-disulfonanilide	450	409
9,10-Anthraquinone-1,6-disulfonanilide	381	406
9,10-Anthraquinone-1,7-disulfonanilide	408	407
9,10-Anthraquinone-1,8-disulfonanilide	409	407
9,10-Anthraquinone-2,6-disulfonanilide	449	409
9,10-Anthraquinone-2,6-disulfonanilide	473	410
9,10-Anthraquinone-2,7-disulfonanilide	279	402
9,10-Anthraquinone-1,5-disulfonic acid	228	379
9,10-Anthraquinone-1,6-disulfonic acid	18	380
9,10-Anthraquinone-1,7-disulfonic acid	21	380
9,10-Anthraquinone-1,8-disulfonic acid	227	379
9,10-Anthraquinone-2,6-disulfonic acid	22	380
9,10-Anthraquinone-2,7-disulfonic acid	10	380
9,10-Anthraquinone-1,3-disulfonyl chloride	283	390
9,10-Anthraquinone-1,6-disulfonyl chloride	271	390
9,10-Anthraquinone-1,7-disulfonyl chloride	285	390
9,10-Anthraquinone-1,8-disulfonyl chloride	277	390
9,10-Anthraquinone-2,6-disulfonyl chloride	282	390
9,10-Anthraquinone-2,7-disulfonyl chloride	267	389
9,10-Anthraquinone-2-sulfonamide	445	409
9,10-Anthraquinone-1-sulfonanilide	343	405
9,10-Anthraquinone-2-sulfonanilide	289	403
9,10-Anthraquinone-1-sulfonic acid	17	380
9,10-Anthraquinone-2-sulfonic acid	206	378
9,10-Anthraquinone-2-sulfonyl chloride	274	390
9,10-Anthraquinone-2-sulfonyl chloride	270	390
1-Anthraquinonethiol	47	417
2-Anthraquinonethiol	48	417
Anthrarufin	50	184
α-Anthroic acid	380	209
β-Anthroic acid	397	209
meso-Anthroic acid	331	207
1-Anthrol	381	115
Anthrone	183	178
Anthroxanic acid	307	205
Antiarol	57	140
Antipyrine	72	324
Apiolic acid	273	204
Apiolonitrile	182	365
Apionol	413	117
Apiose	2	328
cis-Apocamphoric acid	327	206
β-D-Arabinose	40	331
DL-Arabinose	49	332
β-L-Arabinose	41	331
l-Arabonic acid	152	199
Arachidamide	154	234
Arachidanilide	79	233
Arachidic acid	78	196

Compound	Compound No.	Page No.	Compound	Compound No.	Page No.
Arachidic anhydride	32	225	Benzalmalononitrile	112	362
DL-Arginine	66	287	Benzamide	220	236
"	50	429	2-Benzamidobenzoic acid	282	204
L-Arginine	40	286	4-Benzamidobenzonitrile	201	366
Asaronic acid	216	202	2-Benzamidophenol	437	119
l-Ascorbic acid	309	205	3-Benzamidophenol	457	120
L-Ascorbic acid	59	333	4-Benzamidophenol	539	124
DL-Asparagine	42	429	Benzanilide	323	240
α-L-Asparagine	45	286	1,2-Benzanthracene	158	49
β-L-Asparagine	54	286	Benzanthrone	186	178
DL-Aspartic acid	79	288	Benzene	1	35
L-Aspartic acid	81	288	Benzeneazocatechol	433	118
Aspirin	186	200	2-Benzeneazophenol	134	103
Atropic acid	130	198	4-Benzeneazophenol	383	116
" "	400	432	Benzeneazoresorcinol	449	119
Azelaic acid	132	198	Benzene-1,2-dicarboxylic acid	321	206
" "	396	432	Benzene-1,3-dicarboxylic acid	408	210
Azelaic acid diamide	351	241	Benzene-1,4-dicarboxylic acid	404	210
Azelaic acid monoamide	91	233	Benzene-1,2-disulfonamide	439	409
Azelaic acid monoanilide	153	234	Benzene-1,3-disulfonamide	388	407
Azelayl chloride	26	214	Benzene-1,4-disulfonamide	461	409
Azelonitrile	12	355	Benzene-1,2-disulfonanilide	418	408
Azepine	456	439	Benzene-1,3-disulfonanilide	126	396
Azetidine	15	295	Benzene-1,4-disulfonanilide	432	408
"	467	439	Benzene-1,2-disulfonic acid	201	378
Azidoacetyl chloride	2	214	Benzene-1,3-disulfonic acid	175	377
d,l-Azidobutyric acid	3	193	Benzene-1,4-disulfonic acid	214	378
α-Azidoisobutyric acid	9	193	Benzene-1,2-disulfonyl chloride	233	388
2-Azidoisovaleric acid	6	192	Benzene-1,3-disulfonyl chloride	62	382
d,l-2-Azidopropionic acid	8	192	Benzene-1,4-disulfonyl chloride	212	388
Aziridine	292	438	Benzenehexacarboxylic acid	14	429
Azobenzene-3-carboxylic acid	264	204	α-Benzene hexachloride	4	64
Azobenzene-4-carboxylic acid	374	208	β-Benzene hexachloride	6	64
Azobenzene-2,2′-dicarboxylic acid	379	209	γ-Benzene hexachloride	3	64
Azobenzene-4,4′-dicarboxylic acid dichloride	31	217	Benzenepentacarboxylic acid	28	429
Azobenzene-3,3′-disulfonamide	469	410	Benzene sulfonamide	145	397
Azobenzene-3,4′-disulfonamide	460	409	Benzene sulfonanilide	48	393
Azobenzene-4,4′-disulfonamide	436	408	Benzene sulfonic acid	53	372
Azobenzene-3,3′-disulfonic acid	222	379	Benzenesulfonyl chloride	5	381
Azobenzene-3,4′-disulfonic acid	215	378	1,2,3,4-Benzenetetracarboxylic acid	46	429
Azobenzene-4,4′-disulfonic acid	199	378	1,2,3,5-Benzenetetracarboxylic acid	93	429
Azobenzene-3,3′-disulfonyl chloride	256	389	1,2,4,5-Benzenetetracarboxylic acid	34	429
Azobenzene-3,4′-disulfonyl chloride	195	387	1,2,3-Benzenetricarboxylic acid	118	430
Azobenzene-4,4′-disulfonyl chloride	276	390	1,2,4-Benzenetricarboxylic acid	100	429
2-Azophenol	453	120	Benzenetricarboxylic-1,3,5-acid	410	210
3-Azophenol	521	123	" "	51	429
4-Azophenol	537	124	Benzene-1,3,5-trisulfonamide	472	410
Azoxybenzene-3,3′-disulfonamide	454	409	Benzene-1,3,5-trisulfonanilide	406	407
Azoxybenzene-3,3′-disulfonic acid	208	378	Benzene-1,3,5-trisulfonic acid	224	379
Azoxybenzene-3,3′-disulfonyl chloride	220	388	Benzene-1,3,5-trisulfonyl chloride	268	390
2-Azoxyphenol	391	116	1,3,5-Benzenetrithiol	18	416
3-Azoxyphenol	474	121	10,11-Benzfluoranthene	166	50
4-Azoxyphenol	552	125	11,12-Benzfluoranthene	190	51
Azulene	102	47	1,2-Benzfluorene	179	50
			2,3-Benzfluorene	187	51
			3,4-Benzfluorene	132	48
B			Benzhydrol	39	89
			Benzhydrylamine	185	301
Bayer acid	13	380	Benzidine	280	311
Behenic acid	83	196	Benzidine-2,2′-disulfonamide	456	409
Benz[bc]aceanthrylene	127	48	Benzidine-2,2′-disulfonic acid	211	378
3,4-Benzacenaphthenequinone	48	184	Benzidine-2,2′-disulfonyl chloride	273	390
Benzalacenaphthenone	138	176	Benzil	119	176
Benzalacetone	31	173	Benzilamide	295	239
Benzalacetophenone	66	174	Benzilic acid	227	202
Benzal chloride	22	63	" "	144	430
Benzaldehyde	66	147	Benzilic acid chloride	23	214
Benzaldehyde-2-carboxylic acid	86	157	Benzimidazole	219	437
Benzaldehyde-3-carboxylic acid	122	159	2-Benzimidazolthiol	49	417
Benzaldehyde-4-carboxylic acid	130	160	4,5-Benzindane	171	42
Benzaldehyde cyanohydrin	7	357	5,6-Benzindane	97	47
Benzaldehyde-4-sulfonic acid	17	371	Benzindole	34	231
Benzalfluorene	67	45			

	Compound No.	Page No.
Benzo-4-bromoanilide	438	243
Benzo-4-chloroanilide	410	243
Benzo-2-ethylanilide	275	238
Benzo-4-fluoroanilide	388	242
2-Benzofurylacetic acid	111	197
2-Benzofurylacetic acid	251	203
2-Benzofuryl methyl ketone	102	175
Benzoic acid	164	199
" "	323	431
Benzoic anhydride	13	225
d,l-Benzoin	56	90
" "	162	177
Benzoin acetate	122	279
Benzo-2-iodovanilide	251	237
Benzonitrile	58	349
2-Benzophenetidide	132	234
3-Benzophenetidide	129	234
4-Benzophenetidide	354	241
Benzophenone	47	173
Benzophenone-2-carboxylic acid	171	200
Benzophenone-2,4-dicarboxylic acid	365	208
Benzophenone-4,4′-dicarboxylic acid	409	210
Benzophenone-3,3′-disulfonamide	152	397
Benzophenone-3,3′-disulfonanilide	222	400
Benzophenone-3,3′-disulfonic acid	62	373
Benzophenone-3,3′-disulfonyl chloride	217	388
Benzophenone-2-sulfonanilide	109	396
Benzophenone-2-sulfonic acid	6	380
Benzophenone-2-sulfonyl chloride	136	385
Benzopyrazine	5	323
1,4-Benzoquinone	12	181
Benzotetronic acid	516	123
2-Benzothiazolthiol	45	417
2-Benzotoluidide	266	238
3-Benzotoluidide	201	236
4-Benzotoluidide	315	239
Benzotrichloride	23	63
Benzo-2,4,6-trichloroanilide	350	241
Benzoylacetone	75	174
Benzoylacetonitrile	103	362
w-Benzoylacetophenone	111	175
2-Benzoylacrylic acid	126	176
4-Benzoylaniline	40	436
9-Benzoylanthracene	179	178
N-Benzoylanthranilic acid	282	204
N-Benzoylanthranilonitrile	201	366
2-Benzoylbenzamide	327	240
2-Benzoylbenzanilide	416	243
2-Benzoylbenzenesulfonyl chloride	136	385
2-Benzoylbenzoic acid	171	200
Benzoyl bromide	14	218
N-Benzoyl-p-bromoaniline	438	243
α-Benzoylbutanoic acid	90	196
α-Benzoylbutyric acid	90	196
Benzoyl carbinol	47	89
" "	150	177
Benzoylcarbinyl acetate	60	276
4-Benzoylcatechol	355	114
Benzoyl chloride	41	213
N-Benzoyl-p-chloroaniline	410	243
N-Benzoyl-o-ethylaniline	275	238
Benzoylformic acid	85	174
" "	61	195
Benzoylformyl chloride	7	214
2-Benzoylfuran	40	173
Benzoyl-2-furoylmethane	90	175
Benzoyl iodide	3	222
O-Benzoyllactic acid	138	198
O-Benzoyllactonitrile	98	353
Benzoylmethanethiol	4	416
Benzoylnitromethane	136	176
N-Benzoyl-m-phenetidine	129	234
N-Benzoyl-o-phenetidine	132	234
N-Benzoyl-p-phenetidine	354	241
2-Benzoylphenol	26	97
3-Benzoylphenol	251	109
4-Benzoylphenol	322	112
" "	165	177
Benzoyl phenyl carbinol	56	90
α-Benzoyl-β-phenylhydrazine	334	240
2-Benzoylphloroglucinol	426	118
N-Benzoylpiperidine	3	231
β-Benzoylpropionanilide	283	238
α-Benzoylpropionic acid	114	175
" "	85	196
β-Benzoylpropionic acid	147	199
2-Benzoylpyridine	217	168
Benzoylpyruvic acid	241	202
4-Benzoylresorcinol	352	114
Benzoyl sulfide	20	425
Benzoyl-2-thenoylmethane	107	175
2-Benzoylthiophene	60	174
N-Benzoyl-m-toluidine	201	236
N-Benzoyl-o-toluidine	266	238
N-Benzoyl-p-toluidine	315	239
N-Benzoyl-2,4,6-trichloroaniline	350	241
1,2-Benzphenanthrene	195	51
3,4-Benzphenanthrene	56	45
N-Benzpiperidide	3	231
1,2-Benzpyrene	172	50
4,5-Benzpyrene	174	50
3,4-Benzpyrene-5-carboxaldehyde	127	160
N-Benzylacetamide	26	231
N-Benzylacetanilide	20	231
Benzyl acetate	209	261
Benzylacetic acid	123	198
Benzylacetophenone	97	175
N-Benzylacrylamide	487	245
Benzyl alcohol	104	85
Benzylamine	81	297
"	321	438
Benzylanilide	355	241
N-Benzylaniline	19	303
N-Benzylbenzamide	143	234
N-Benzylbenzanilide	148	234
Benzyl benzoate	8	273
Benzyl bromide	23	58
Benzyl 4-bromophenyl sulfide	37	425
Benzyl n-butyl ether	91	135
Benzyl n-butyrate	258	264
N-Benzyl-n-caproamide	9	231
Benzyl chloride	38	56
Benzyl chloroacetate	89	254
Benzyl 3-chlorophenyl ketone	79	174
Benzyl 4-chlorophenyl ketone	139	176
α-Benzylcinnamic anhydride	47	226
Benzyl cinnamate	44	275
N-Benzylcrotonamide	175	235
Benzyl cyanide	80	351
Benzylcyanoacetic acid	119	197
2-Benzyl-2-cyanopropionic acid	73	429
Benzyl dimethyl amine	44	320
Benzyl dimethyl carbinol	4	88
Benzyl ether	109	136
Benzyl ethyl amine	93	298
Benzyl ethyl ether	71	134
Benzyl ethyl ketone	182	167
Benzyl ethyl sulfide	55	421
N-Benzylformamide	25	231
Benzyl formate	176	259
Benzylglycolaldehyde	26	154
Benzyl 4-hydroxybenzoate	43	434
N-Benzylidene-2-aminophenol	155	104
N-Benzylidene-4-aminophenol	475	121

	Compound No.	Page No.
Benzylidenemalonic acid	315	206
Benzylidene malonyldichloride	17	216
Benzyl iodide	2	61
Benzyl isobutyl ether	85	135
N-Benzylisovaleramide	14	231
Benzylmalonic acid	148	199
Benzylmalonic acid diamide	475	245
Benzylmalonic acid dianilide	459	244
Benzylmalonyl dichloride	35	215
Benzyl mercaptan	32	414
Benzyl-4-methoxyphenyl ketone	103	164
Benzyl methyl amine	79	297
N-Benzyl-N-methylaniline	113	321
Benzyl methyl ether	61	134
Benzyl methyl ketone	5	172
Benzyl methyl sulfide	46	421
1-Benzylnaphthalene	40	44
2-Benzylnaphthalene	39	44
4-Benzyl-1-naphthol	296	111
Benzyl-1-naphthyl ether	47	139
Benzyl-2-naphthyl ether	53	139
Benzyl 4-nitrophenyl sulfide	66	426
Benzyl oxalate	120	279
3-Benzyloxybenzaldehyde	35	154
2-Benzyloxyethanol	128	86
N-Benzylpalmitamide	90	233
2-Benzylphenol	1	96
4-Benzylphenol	137	103
d,l-Benzylphenyl carbinol	37	89
Benzyl phenyl ketone	73	174
Benzyl phenyl sulfide	13	425
2-Benzyl-2-phenylsuccinic acid	235	431
N-Benzylphthalamide	371	241
α-Benzylpropionic acid	20	193
2-Benzylpyridine	200	437
2-Benzylpyrrolidine	383	439
2-Benzyl-2-pyrroline	275	438
Benzylpyruvic acid	40	194
Benzyl salicylate	354	271
N-Benzylstearamide	101	233
Benzyl succinate	47	275
Benzylsuccinic acid	302	431
DL-Benzylsuccinic anhydride	44	226
Benzyl sulfide	21	425
Benzyl sulfonamide	37	393
Benzyl sulfonanilide	34	393
Benzyl sulfonic acid	9	371
Benzyl sulfonyl chloride	120	384
N-Benzyltoluene-4-sulfonamide	53	394
Benzylurea	281	238
Betaine	28	285
Betol	178	105
Biacetyl	5	161
Biallyl	38	13
Bibenzyl	27	44
Bicetyl	55	9
Bicyclo[2,2,1]heptane	77	9
Bicyclohexyl	262	7
Bicyclo[12,2,2]octadeca-14,16,18-triene	1	25
cis-Bicyclo[3,3,0]octane	103	4
Bicyclo[4,2,0]octane	102	4
Bicyclo[2,2,2]oct-2-ene	24	26
Bicyclo[4,2,0]oct-2-ene	241	19
Bicyclo[4,2,0]oct-3-ene	2	25
Bicyclo[4,2,0]oct-7-ene	230	19
Bicyclo[4,7]pentadiene	3	26
Bifluorenylidene	182	50
2,2′-Bifuroyl	185	178
Bi-α-naphthol	587	127
1,1′-Binaphthyl	160	49
1,2′-Binaphthyl	71	46
2,2′-Binaphthyl	178	50
2,2′-Biphenol	227	108
2,4′-Biphenol	418	118
3,3′-Biphenol	290	111
4,4′-Biphenol	579	127
Biphenyl	60	45
Biphenyl-2-carboxylic acid	202	430
Biphenyl-2,2′-dicarboxylic acid	354	208
Biphenyl-4,4′-disulfonamide	467	410
Biphenyl-4,4′-disulfonic acid	220	379
Biphenyl-4,4′-disulfonyl chloride	272	390
Biphenyleneethylene	28	44
Biphenylene oxide	50	139
2-Biphenyl ethyl ether	14	137
3-Biphenyl ethyl ether	16	137
4-Biphenyl ethyl ether	46	139
2-Biphenyl methyl ether	10	137
4-Biphenyl methyl ether	51	139
Biphenyl-4-sulfonamide	390	407
Biphenyl-4-sulfonanilide	66	394
Biphenyl-4-sulfonic acid	176	377
Biphenyl-4-sulfonyl chloride	178	386
4-Biphenylthiol	35	417
α,α′-Bipyridyl	38	323
β,β′-Bipyridyl	36	323
2,2′-Bipyridyl	38	323
2,3′-Bipyridyl	110	321
3,3′-Bipyridyl	36	323
3,4′-Bipyridyl	34	323
4,4′-Bipyridyl	73	324
2,2′-Bipyrroyl	94	325
α,α′-Biquinolyl	92	325
2,2′-Biquinolyl	92	325
2,3′-Biquinolyl	89	325
2,7′-Biquinolyl	83	324
3,4′-Biquinolyl	48	324
5,5′-Biquinolyl	88	325
6,6′-Biquinolyl	90	325
6,8′-Biquinolyl	79	324
7,7′-Biquinolyl	86	325
2,2′-Biquinolyl ketone	84	325
2,2′-Biquinolylmethane	67	324
Bis(benzylthio)methane	28	425
Bis(2-cyanoethyl)amine	207	437
Bis(2-cyanoethyl)ethylamine	163	437
Bis(cyanomethyl)amine	94	361
"	13	436
Bis(2-chloroethyl)ethylamine	263	438
Bis(2-chloroethyl)methylamine	259	438
1,2-Bis(chloromethyl)-benzene	6	57
1,3-Bis(chloromethyl)-benzene	1	57
1,4-Bis-(chloromethyl)-benzene	8	57
1,1-Bis-(4-chlorophenyl)-2,2,2-trichloroethane	3	57
2,2-Bis-(4-chlorophenyl)-1,1-trichloroethane	2	64
1,2-Bis(diethylamino)ethane	332	438
4,4′-Bis-(dimethylamino)-benzhydrol	64	324
4,4′-Bis-(dimethylamino)-benzophenone	87	325
" "	190	178
2,2′-Bis-(dimethylamino)-biphenyl	41	323
1,4-Bis(dimethylamino)butane	35	319
4,4′-Bis(dimethylamino)diphenylmethane	54	324
1,2-Bis(dimethylamino)ethane	308	438
4,4′-Bis(dimethylamino)triphenylmethane	66	324
1,2-Bis(ethylamino)ethane	394	439
1,2-Bis(furfurylamino)ethane	300	438
Bis(2-hydroxyethyl)amine	305	438
Bis(2-hydroxyethyl)-methyl amine	299	438
2,2-Bis(hydroxymethyl)-1,3-propanediol	69	90
1,2-Bis(isopropylamino)ethane	388	439
Bis(2-mercaptoethyl)ether	41	415
Bis(4-mercaptophenyl)ether	30	417
1,2-Bis(methylamino)ethane	370	439
1,4-Bis(methylamino)ethane	61	304

	Compound No.	Page No.
1,3-Bis(2-nitrophenylthio)propane	68	426
1,2-Bis(phenylthio)ethane	38	426
Bis(phenylthio)methane	24	425
1,2-Bis(propylamino)ethane	380	439
Bis(4-thiophenyl)sulfide	39	417
1,2-Bis(tolylthio)ethane	48	426
Biuret	411	243
Bornane	79	9
"Borneo camphor"	66	90
d-Borneol	66	90
d-Bornyl acetate	21	274
trans-Bornylamine	373	439
trans-Brassidic acid	51	195
Brometone	63	90
Bromoacetamide	76	233
Bromoacetanilide	224	237
2-Bromoacetanilide	111	233
3-Bromoacetanilide	69	232
4-Bromoacetanilide	331	240
Bromoacetic acid	41	194
"	129	430
Bromoactic anhydride	11	225
Bromoacetone	33	162
2-Bromoacetophenone	52	169
3-Bromoacetophenone	97	170
4-Bromoacetophenone	52	173
w-Bromoacetophenone	51	173
Bromoacetyl bromide	8	218
Bromoacetyl chloride	24	212
cis-2-Bromoallocinnamic acid	158	199
2-Bromoaniline	10	303
"	54	436
3-Bromoaniline	167	300
"	105	436
4-Bromoaniline	98	305
"	116	436
3-Bromo-o-anisidine	93	305
4-Bromo-o-anisidine	187	308
2-Bromo-p-anisidine	87	305
2-Bromoanisole	84	135
4-Bromoanisole	86	135
3-Bromobenzaldehyde	86	149
4-Bromobenzaldehyde	43	155
2-Bromobenzamide	298	239
3-Bromobenzamide	301	239
4-Bromobenzamide	402	242
2-Bromobenzanilide	258	238
3-Bromobenzanilide	247	237
Bromobenzene	1	73
2-Bromobenzene-1,4-dicarboxylic acid	403	210
2-Bromobenzenesulfonamide	252	401
4-Bromobenzenesulfonamide	182	398
4-Bromobenzenesulfonanilide	56	394
2-Bromobenzenesulfonic acid	114	375
4-Bromobenzenesulfonic acid	73	373
2-Bromobenzenesulfonyl chloride	38	382
4-Bromobenzenesulfonyl chloride	88	383
2-Bromobenzoic acid	226	202
" "	122	430
3-Bromobenzoic acid	235	202
" "	253	431
4-Bromobenzoic acid	382	209
" "	277	431
2-Bromobenzoic anhydride	31	225
3-Bromobenzoic anhydride	66	226
4-Bromobenzoic anhydride	80	227
2-Bromobenzonitrile	59	359
3-Bromobenzonitrile	30	358
4-Bromobenzonitrile	156	364
2-Bromobenzophenone	35	173
4-Bromobenzophenone	113	175
2-Bromo-1,4-benzoquinone	4	181
2-Bromobenzoyl bromide	6	219
4-Bromobenzoyl bromide	11	219
2-Bromobenzoyl chloride	56	213
3-Bromobenzoyl chloride	55	213
4-Bromobenzoyl chloride	5	216
2-Bromobenzoylformic acid	99	197
2-Bromobenzyl bromide	2	59
3-Bromobenzyl bromide	3	59
4-Bromobenzyl bromide	7	59
4-Bromobenzyl chloride	2	57
2-Bromobenzyl cyanide	15	357
4-Bromobenzyl cyanide	45	359
2-Bromobiphenyl	17	73
4-Bromobiphenyl	6	74
α-Bromobutanoic acid	39	191
β-Bromobutanoic acid	1	193
1-Bromo-2-butanone	75	163
α-Bromobutyric acid	39	191
β-Bromobutyric acid	1	193
DL-α-Bromobutyric anhydride	1	224
DL-α-Bromobutyryl bromide	11	218
DL-α-Bromobutyryl chloride	31	212
3-Bromocamphor-8-sulfonamide	117	396
D-3-Bromocamphor-10-sulfonamide	147	397
10-Bromocamphor-3-sulfonamide	31	393
3-Bromocamphor-8-sulfonic acid	43	372
3-Bromocamphor-10-sulfonic acid	59	373
10-Bromocamphor-3-sulfonic acid	8	371
3-Bromocamphor-8-sulfonyl chloride	215	388
3-Bromocamphor-10-sulfonyl chloride	68	383
10-Bromocamphor-3-sulfonyl chloride	138	385
4-Bromocatechol	151	104
Bromochloroacetic acid	14	193
1-Bromo-4-chlorobenzene	19	71
1-Bromo-2-chloroethane	12	63
2-Bromo-4-chlorophenol	16	96
1-Bromo-3-chloropropane	18	63
erythro-2-Bromo-3-chlorosuccinic acid	17	429
threo-2-Bromo-3-chlorosuccinic acid	19	429
trans-α-Bromocinnamic acid	178	200
trans-2-Bromocinnamic acid	371	432
4-Bromocinnamic acid	384	209
4-Bromo-m-cresol	71	100
5-Bromo-m-cresol	55	99
4-Bromo-o-cresol	72	100
5-Bromo-o-cresol	124	103
2-Bromo-p-cresol	54	99
3-Bromo-p-cresol	50	99
γ-Bromocrotonic acid	72	196
1-Bromo-4-cyanonaphthalene	141	363
1-Bromo-5-cyanonaphthalene	194	365
2-Bromocymene	13	73
2-Bromo-1,4-diaminobenzene	122	306
2-Bromo-4,6-dichlorophenol	89	101
3-Bromo-2,6-dichlorophenol	113	102
Bromodicyanomethane	169	364
2-Bromo-1,4-dihydroxy-9,10-anthraquinone	46	184
6-Bromo-1,4-dihydroxy-9,10-anthraquinone	27	182
4-Bromo-1,2-dihydroxybenzene	151	104
3-Bromo-1,2-dihydroxynaphthalene	253	109
4-Bromo-1,5-dihydroxynaphthalene	250	109
2-Bromo-4,6-dimethylaniline	48	304
4-Bromo-2,6-dimethylaniline	100	436
2-Bromo-N,N-dimethylaniline	145	437
3-Bromo-N,N-dimethylaniline	93	321
4-Bromo-N,N-dimethylaniline	141	437
1-Bromo-2,3-dimethylbenzene	9	73
4-Bromo-2,5-dimethylbenzoic acid	267	204
5-Bromo-2,4-dimethylbenzoic acid	280	204
4-Bromo-2,5-dimethylbenzonitrile	142	363
5-Bromo-2,4-dimethylbenzonitrile	113	362
2-Bromo-4,5-dimethylphenol	123	102

Compound	Compound No.	Page No.
1-Bromo-2,4-dinitrobenzene	40	338
4-Bromo-1,3-dinitrobenzene	40	338
2-Bromo-3,5-dinitrobenzoic acid	339	207
4-Bromo-3,5-dinitrobenzoic acid	305	205
4-Bromo-1,5-dinitronaphthalene	97	341
4-Bromo-1,8-dinitronaphthalene	111	341
2-Bromo-4,6-dinitrophenol	261	109
2-Bromo-2,6-dinitrophenol	118	102
2-Bromoethanol	43	82
2-Bromo-4-ethoxyaniline	2	302
3-Bromo-4-ethoxyaniline	38	304
5-Bromo-2-ethoxyaniline	65	304
β-Bromoethyl acetate	112	255
2-Bromoethylamine	298	438
1-Bromo-2-ethylbenzene	4	73
1-Bromo-4-ethylbenzene	5	73
2-Bromo-4-ethylphenol	8	94
β-Bromoethyl phenyl ether	11	137
Bromoethyl sulfide	8	425
Bromoform	7	65
4-Bromofuran-2-carboxylic acid	174	200
5-Bromofuran-2-carboxylic acid	295	205
2-Bromohexane	19	58
6-Bromo-2-hexanone	171	166
6-Bromo-n-hexylamine	407	439
Bromohydroquinone	228	108
3-Bromo-4-hydroxyaniline	370	315
4-Bromo-3-hydroxyaniline	335	313
5-Bromo-2-hydroxyaniline	283	311
3-Bromo-2-hydroxybenzoic acid	289	204
5-Bromo-2-hydroxybenzoic acid	255	203
3-Bromo-2-hydroxybenzonitrile	50	359
5-Bromo-2-hydroxybenzonitrile	202	366
3-Bromo-4-hydroxybiphenyl	174	105
3-Bromo-2-hydroxy-5-methylaniline	165	308
5-Bromo-2-hydroxy-3-methylaniline	236	310
5-Bromo-2-hydroxy-4-methylaniline	243	310
5-Bromo-4-hydroxy-2-methylaniline	439	317
5-Bromo-2-hydroxy-3-methylbenzoic acid	366	208
4-Bromo-3-hydroxy-2-naphthoic acid	362	208
5-Bromo-2-hydroxy-3-toluic acid	366	208
α-Bromoisobutanoic acid	36	194
2-Bromoisobutyraldehyde	28	145
α-Bromoisobutyramide	277	238
α-Bromoisobutyranilide	59	232
α-Bromoisobutyric acid	36	194
α-Bromoisobutyric anhydride	23	225
2-Bromoisobutyronitrile	27	347
α-Bromoisobutyryl bromide	10	218
α-Bromoisobutyryl chloride	21	214
1-Bromo-2-isopropylbenzene	7	73
1-Bromo-4-isopropylbenzene	11	73
4-Bromo-5-isopropyl-2-methylphenol	37	98
4-Bromoisoquinoline	14	323
8-Bromoisoquinoline	47	323
α-Bromoisovaleramide	237	237
α-Bromoisovaleranilide	181	235
d,l-α-Bromoisovaleric acid	30	194
DL-α-Bromoisovaleryl bromide	13	218
DL-α-Bromoisovaleryl chloride	19	214
Bromol	171	105
Bromomalonic acid	142	198
Bromomalononitrile	77	360
3-Bromomandelic acid	183	430
2-Bromo-6-methoxyaniline	93	305
3-Bromo-4-methoxyaniline	87	305
" "	131	437
5-Bromo-2-methoxyaniline	187	308
2-Bromo-4-methylaniline	154	300
3-Bromo-2-methylaniline	169	300
3-Bromo-4-methylaniline	3	303
3-Bromo-5-methylaniline	16	303
4-Bromo-2-methylaniline	69	304
" "	104	436
4-Bromo-3-methylaniline	132	307
5-Bromo-2-methylaniline	8	303
3-Bromo-4-methylbenzenesulfonamide	131	396
4-Bromo-2-methylbenzenesulfonamide	193	399
4-Bromo-3-methylbenzenesulfonamide	119	396
5-Bromo-2-methylbenzenesulfonamide	183	398
3-Bromo-4-methylbenzenesulfonic acid	52	372
4-Bromo-2-methylbenzenesulfonic acid	81	373
4-Bromo-3-methylbenzenesulfonic acid	45	372
5-Bromo-2-methylbenzenesulfonic acid	74	373
3-Bromo-4-methylbenzenesulfonyl chloride	59	382
4-Bromo-2-methylbenzenesulfonyl chloride	36	382
4-Bromo-3-methylbenzenesulfonyl chloride	34	382
5-Bromo-2-methylbenzenesulfonyl chloride	14	381
2-Bromo-4-methylbenzoic acid	205	201
4-Bromo-2-methylbenzoic acid	298	205
(2-Bromomethyl)butyric acid	272	431
Bromomethyl ethyl ketone	75	163
2-Bromo-4-methyl-5-nitrobenzoic acid	324	206
4-Bromo-4-methyl-6-nitrobenzoic acid	352	208
6-Bromo-4-methyl-2-nitrobenzoic acid	330	207
2-Bromo-4-methyl-3-nitrophenol	115	102
2-Bromo-4-methyl-5-nitrophenol	203	106
2-Bromo-4-methyl-6-nitrophenol	90	101
6-Bromo-2-methyl-4-nitrophenol	270	110
3-Bromo-2-methylolbenzoic acid	190	430
6-Bromo-2-methylolbenzoic acid	69	429
2-Bromo-4-methylphenol	54	99
3-Bromo-4-methylphenol	50	99
4-Bromo-2-methylphenol	72	100
4-Bromo-3-methylphenol	71	100
5-Bromo-2-methylphenol	124	103
5-Bromo-3-methylphenol	55	99
6-Bromo-2-methylquinoline	62	324
1-Bromo-2-naphthaldehyde	105	159
1-Bromonaphthalene	16	73
2-Bromonaphthalene	2	74
3-Bromo-1,2-naphthalenediol	253	109
4-Bromo-1,5-naphthalenediol	250	109
1-Bromonaphthalene-2-sulfonamide	451	409
2-Bromonaphthalene-1-sulfonamide	98	395
4-Bromonaphthalene-1-sulfonamide	295	403
5-Bromonaphthalene-1-sulfonamide	393	407
5-Bromonaphthalene-2-sulfonamide	360	406
6-Bromonaphthalene-1-sulfonamide	353	405
6-Bromonaphthalene-2-sulfonamide	334	405
7-Bromonaphthalene-1-sulfonamide	335	405
7-Bromonaphthalene-2-sulfonamide	355	405
8-Bromonaphthalene-2-sulfonamide	262	402
2-Bromonaphthalene-1-sulfonic acid	35	372
4-Bromonaphthalene-1-sulfonic acid	140	376
1-Bromonaphthalene-2-sulfonyl chloride	124	384
2-Bromonaphthalene-1-sulfonyl chloride	140	385
4-Bromonaphthalene-1-sulfonyl chloride	111	384
5-Bromonaphthalene-1-sulfonyl chloride	131	385
5-Bromonaphthalene-2-sulfonyl chloride	135	385
6-Bromonaphthalene-1-sulfonyl chloride	90	383
6-Bromonaphthalene-2-sulfonyl chloride	198	387
7-Bromonaphthalene-1-sulfonyl chloride	236	388
7-Bromonaphthalene-2-sulfonyl chloride	151	385
8-Bromonaphthalene-2-sulfonyl chloride	186	387
5-Bromo-1-naphthoic acid	390	209
7-Bromo-1-naphthoic acid	369	208
8-Bromo-1-naphthoic acid	275	204
1-Bromo-2-naphthol	141	103
3-Bromo-2-naphthol	142	103
4-Bromo-1-naphthol	305	112
4-Bromo-2-naphthol	279	110

Compound	Compound No.	Page No.
6-Bromo-2-naphthol	310	112
4-Bromo-1-naphthonitrile	141	363
5-Bromo-1-naphthonitrile	194	365
1-Bromo-2-naphthylamine	83	305
3-Bromo-1-naphthylamine	107	306
4-Bromo-1-naphthylamine	197	309
5-Bromo-1-naphthylamine	23	303
6-Bromo-2-naphthylamine	282	311
2-Bromo-4-nitroacetanilide	219	236
2-Bromo-4-nitroaniline	206	309
4-Bromo-2-nitroaniline	229	310
4-Bromo-3-nitroaniline	296	312
5-Bromo-2-nitroaniline	341	313
2-Bromo-1-nitrobenzene	15	337
3-Bromo-1-nitrobenzene	26	337
4-Bromo-1-nitrobenzene	90	340
4-Bromo-3-nitrobenzenesulfonamide	217	400
4-Bromo-3-nitrobenzenesulfonic acid	94	374
4-Bromo-3-nitrobenzenesulfonyl chloride	49	382
2-Bromo-6-nitrobenzoic acid	13	429
4-Bromo-3-nitrobenzoic acid	325	206
6-Bromo-3-nitrobenzoic acid	279	204
4-Bromo-3-nitrobenzonitrile	165	364
6-Bromo-3-nitrobenzonitrile	162	364
2-Bromo-3-nitro-p-cresol	115	102
2-Bromo-5-nitro-p-cresol	203	106
2-Bromo-6-nitro-p-cresol	90	101
6-Bromo-4-nitro-o-cresol	270	110
4-Bromo-1-nitronaphthalene	57	339
5-Bromo-1-nitronaphthalene	86	340
8-Bromo-1-nitronaphthalene	71	339
2-Bromo-4-nitrophenol	241	108
2-Bromo-5-nitrophenol	262	109
2-Bromo-6-nitrophenol	83	100
3-Bromo-2-nitrophenol	77	100
3-Bromo-5-nitrophenol	356	114
4-Bromo-2-nitrophenol	165	104
5-Bromo-2-nitrophenol	32	98
4-Bromo-2-nitrothiophenol	34	417
4-Bromo-3-nitrotoluene	7	337
5-Bromo-2-nitro-4-toluic acid	324	206
5-Bromo-3-nitro-2-toluic acid	352	208
5-Bromo-3-nitro-4-toluic acid	330	207
5-Bromo-2-nitro-4-tolunitrile	177	365
5-Bromo-3-nitro-2-tolunitrile	147	363
5-Bromo-3-nitro-4-tolunitrile	174	365
5-Bromo-n-pentylamine	337	438
3-Bromophenacyl bromide	53	173
4-Bromophenacyl bromide	140	176
1-Bromo-9,10-phenanthraquinone	38	183
2-Bromo-9,10-phenanthraquinone	37	183
3-Bromo-9,10-phenanthraquinone	47	184
4-Bromo-o-phenetidine	65	304
2-Bromo-p-phenetidine	38	304
3-Bromo-p-phenetidine	2	302
2-Bromophenetole	90	135
4-Bromophenetole	97	135
2-Bromophenol	2	94
"	44	434
3-Bromophenol	14	96
"	62	434
4-Bromophenol	74	100
"	69	434
(2-Bromophenoxy)acetic acid	214	201
" " "	157	430
(3-Bromophenoxy)acetic acid	155	430
(4-Bromophenoxy)acetic acid	233	202
" " "	159	430
d,l-α-Bromophenylacetic acid	91	196
(2-Bromophenyl)acetic acid	122	198
" " "	294	431
(4-Bromophenyl)acetic acid	143	198
(4-Bromophenyl)acetic acid	320	431
d,l-(2-Bromophenyl)acetonitrile	15	357
(4-Bromophenyl)acetonitrile	45	359
2-Bromo-p-phenylenediamine	122	306
2-Bromophenyl ethyl ketone	57	170
4-Bromophenylhydrazine	212	309
4-Bromophenyl methyl sulfide	11	425
4-Bromophenyl sulfide	63	426
3-Bromophthalic acid	304	205
3-Bromophthalic anhydride	57	226
4-Bromophthalic anhydride	49	226
1-Bromopropane	4	58
3-Bromo-1-propanol	73	84
1-Bromo-2-propanone	33	162
α-Bromopropionamide	196	236
β-Bromopropionamide	166	235
α-Bromopropionanilide	114	233
d,l-α-Bromopropionic acid	6	193
" " "	138	430
β-Bromopropionic acid	57	195
" "	282	431
DL-α-Bromopropionic anhydride	2	224
3-Bromopropionitrile	7	355
DL-α-Bromopropionyl bromide	9	218
2-Bromopropiophenone	57	170
3-Bromopropiophenone	30	173
4-Bromopropiophenone	44	173
α-Bromopropiophenone	203	168
3-Bromopropylamine	307	438
2-Bromopyridine	21	436
"	53	320
3-Bromopyridine	39	319
"	73	436
4-Bromopyridine	10	322
"	111	436
γ-Bromopyridine	10	322
4-Bromopyromucic acid	174	200
5-Bromopyromucic acid	295	205
6-Bromoquinaldine	62	324
2-Bromoquinizarin	46	184
6-Bromoquinizarin	27	182
2-Bromoquinoline	25	323
"	27	436
3-Bromoquinoline	103	321
"	64	436
4-Bromoquinoline	4	323
5-Bromoquinoline	24	323
"	108	436
6-Bromoquinoline	105	321
7-Bromoquinoline	27	323
8-Bromoquinoline	111	321
α-Bromoquinoline	25	323
β-Bromoquinoline	103	321
γ-Bromoquinoline	4	323
3-Bromosalicylic acid	289	204
5-Bromosalicylic acid	255	203
β-Bromostyrene	28	58
o-Bromostyrene	6	73
p-Bromostyrene	8	73
Bromosuccinic acid	112	430
Bromoterephthalic acid	403	210
4-Bromothiophenol	21	416
2-Bromotoluene	2	73
3-Bromotoluene	3	73
4-Bromotoluene	1	74
3-Bromo-4-toluic acid	205	201
5-Bromo-2-toluic acid	298	205
5-Bromo-m-toluidine	16	303
6-Bromo-m-toluidine	132	307
4-Bromo-o-toluidine	8	303
5-Bromo-o-toluidine	69	304
6-Bromo-o-toluidine	169	300

Compound	Compound No.	Page No.
2-Bromo-*p*-toluidine	3	303
3-Bromo-*p*-toluidine	154	300
α-Bromo-2-tolunitrile	89	361
α-Bromo-3-tolunitrile	122	362
α-Bromo-4-tolunitrile	160	364
2-Bromo-4-tolunitrile	38	358
3-Bromo-4-tolunitrile	44	359
5-Bromo-2-tolunitrile	88	361
1-Bromo-2-vinylbenzene	6	73
1-Bromo-4-vinylbenzene	8	73
5-Bromo-*m*-4-xylidine	48	304
1,2-Butadiene	11	12
1,3-Butadiene	7	12
1,3-Butadiyne	4	28
Butanal	11	144
n-Butane	6	3
Butanedial	59	147
Butane-1,4-dicarboxylic acid	232	202
Butanedioic acid	294	205
d,l-1,3-Butanediol	105	85
1,4-Butanediol	119	86
2,3-Butanediol	83	84
2,3-Butanedione	5	161
1,4-Butanedithiol	34	414
Butane-1-sulfonamide	5	392
Butane-1,2,3,4-tetracarboxylic acid (low melting form)	306	205
Butane-1,2,3,4-tetracarboxylic acid (high melting form)	368	208
1-Butanethiol	9	413
2-Butanethiol	6	413
Butane-1,1,4-tricarboxylic acid	202	201
n-Butanoic acid	10	190
" "	441	432
1-Butanol	14	80
2-Butanol	7	80
2-Butanone	3	161
2-Butenal	22	145
1-Butene	6	12
cis-2-Butene	10	12
trans-2-Butene	8	8
cis-Butenedioic acid	175	200
trans-Butenedioic acid	399	209
cis-2-Butenoic acid	13	190
trans-2-Butenoic acid	71	195
3-Butenoic acid	12	190
3-Buten-1-ol	11	80
d,l-3-Buten-2-ol	4	80
3-Buten-2-one	2	161
n-Butoxybenzene	82	135
2-Butoxybenzoic acid	332	431
2-*n*-Butoxyethanol	61	83
2-*sec*-Butoxyethanol	52	82
2-(2-*n*-Butoxyethoxy)ethanol	118	86
β-*n*-Butoxyethyl benzoate	4	272
2-Butoxytoluene	93	135
tert-Butylacetaldehyde	21	145
4-*n*-Butylacetanilide	138	234
n-Butyl acetate	54	252
sec-Butyl acetate	40	251
tert-Butyl acetate	22	250
tert-Butylacetic acid	17	190
tert-Butylacetonitrile	21	358
4-*n*-Butylacetophenone	62	170
n-Butyl acetylene	17	28
sec-Butyl acetylene	16	28
tert-Butyl acetylene	8	28
n-Butyl alcohol	14	80
d,l-*sec*-Butyl alcohol	7	80
tert-Butyl alcohol	7	88
n-Butylallene	135	16
tert-Butylallene	77	14
n-Butylamine	17	295
" "	412	439
Di-*n*-butylamine	65	297
DL-*sec*-*n*-Butylamine	13	295
tert-Butylamine	6	295
" "	427	439
(*tert*-Butylamino)cyclohexane	463	439
2-(Butylamino)ethylamine	381	439
N-*n*-Butylaniline	159	300
" "	198	437
2-*n*-Butylaniline	4	302
4-*n*-Butylaniline	174	300
N-*tert*-Butylaniline	273	438
2-*tert*-Butylaniline	147	299
" "	192	437
3-*tert*-Butylaniline	171	437
4-*tert*-Butylaniline	139	299
" "	187	437
N-*n*-Butylaziridine	287	438
n-Butylbenzene	26	36
sec-Butylbenzene	16	35
tert-Butylbenzene	13	35
n-Butyl benzoate	288	266
2-*tert*-Butylbenzoic acid	211	430
4-*tert*-Butylbenzoic acid	368	432
n-Butyl bromide	11	58
sec-Butyl bromide	10	58
tert-Butyl bromide	8	58
n-Butyl *tert*-butyl ketone	100	164
n-Butyl *n*-butyrate	115	255
tert-Butyl *n*-butyrate	84	253
n-Butyl *n*-caproate	185	259
n-Butyl *n*-caprylate	275	265
n-Butyl carbamate	13	231
d-*sec*-Butyl carbinol	23	81
tert-Butyl carbinol	28	89
n-Butyl carbonate	184	259
n-Butyl cellosolve	60	134
Butyl cellosolve benzoate	4	272
n-Butylchloral	58	147
n-Butyl chloride	13	55
sec-Butyl chloride	10	55
tert-Butyl chloride	8	55
n-Butyl chloroacetate	129	256
n-Butyl chloroformate	80	253
4-Butyl-*o*-cresol	289	111
6-*tert*-Butyl-*m*-cresol	38	98
tert-Butylcyanide	10	345
Butyl cyanoacetate	6	355
n-Butylcyclohexane	243	7
n-Butyl cyclooctatetraene	3	25
n-Butylcyclopentane	164	5
n-Butyl dimethyl amine	375	439
tert-Butyl dimethyl amine	428	439
4-*tert*-Butyl-1,3-dimethylbenzene	94	38
n-Butyl *n*-enanthate	227	262
d,l-1,3-Butylene glycol	105	85
2,3-Butylene glycol	83	84
α-Butylene oxide	13	131
n-Butyl ether	54	133
Butyl ethyl acetylene	44	29
1-*tert*-Butyl-4-ethylbenzene	97	39
n-Butyl ethyl ether	31	132
sec-Butyl ethyl ether	24	132
tert-Butyl ethyl ether	20	132
n-Butyl ethyl ketone	64	163
sec-Butyl ethyl ketone	35	162
tert-Butyl ethyl ketone	23	161
n-Butyl ethyl sulfide	27	420
sec-Butyl ethyl sulfide	21	420
n-Butyl formate	35	251
sec-Butyl formate	21	250

Compound No.	Page No.

tert-Butyl formate 14 250
N-*n*-Butylglycine........................ 86 429
n-Butyl *n*-heptanoate 227 262
n-Butyl *n*-hexanoate....................... 185 259
n-Butyl 2-hydroxybenzoate 27 95
n-Butyl 4-hydroxybenzoate 47 434
n-Butyl iodide.......................... 11 60
sec-Butyl iodide.......................... 8 60
tert-Butyl iodide......................... 7 60
n-Butyl isobutyl ketone 102 164
sec-Butyl isobutyl ketone 103 175
tert-Butyl isobutyl ketone 85 163
tert-Butyl isobutyrate...................... 56 252
3-*tert*-Butyl-1-isopropylbenzene 118 39
4-*tert*-Butyl-1-isopropylbenzene 120 39
n-Butyl isopropyl ether 40 133
sec-Butyl isopropyl ketone 53 162
tert-Butyl isopropyl ketone 30 162
n-Butyl lactate 151 257
n-Butyl levulinate.......................... 257 264
sec-Butyl levulinate......................... 222 262
n-Butylmalonic acid........................ 118 197
sec-Butylmalonic acid....................... 76 196
n-Butyl mercaptan.......................... 9 413
sec-Butyl mercaptan......................... 6 413
tert-Butyl mercaptan......................... 4 413
n-Butyl methoxymethyl ketone 106 164
sec-Butyl methoxymethyl ketone 95 164
tert-Butyl methoxymethyl ketone 87 163
n-Butyl methylacetic acid 37 191
n-Butyl methyl acetylene.................... 39 29
tert-Butyl methyl acetylene.................. 23 29
n-Butyl methyl amine...................... 25 295
DL-*sec*-Butyl methyl amine................... 19 295
2-Butyl-1-methylbenzene 67 38
3-Butyl-1-methylbenzene 79 38
4-Butyl-1-methylbenzene 86 38
2-*sec*-Butyl-1-methylbenzene 54 37
3-*sec*-Butyl-1-methylbenzene 48 37
4-*sec*-Butyl-1-methylbenzene 64 37
2-*tert*-Butyl-1-methylbenzene 65 38
3-*tert*-Butyl-1-methylbenzene 38 36
4-*tert*-Butyl-1-methylbenzene 44 37
n-Butyl methyl carbinol 33 81
sec-Butyl methyl carbinol 28 81
tert-Butyl methyl carbinol 16 80
n-Butyl methyl ether....................... 19 132
tert-Butyl methyl ether...................... 10 131
n-Butylmethylglycol aldehyde 11 150
n-Butyl-methylglucolic acid 16 193
n-Butyl methyl ketone 24 161
sec-Butyl methyl ketone 16 161
tert-Butyl methyl ketone 12 161
4-Butyl-2-methylphenol 289 111
2-*tert*-Butyl-5-methylphenol.................. 38 98
n-Butyl methyl sulfide 18 420
tert-Butyl methyl sulfide 10 420
1-Butylnaphthalene.......................... 167 41
2-Butylnaphthalene.......................... 170 41
1-*tert*-Butylnaphthalene 169 41
2-*tert*-Butylnaphthalene 168 41
n-Butyl-2-naphthyl ether 18 137
sec-Butyl-2-naphthyl ether 15 137
n-Butyl nitrate............................ 69 253
Butyl nitrite................................ 8 250
1-*tert*-Butyl-4-nitrobenzene 33 336
n-Butyl *n*-octanoate 275 265
n-Butyl oxalate 266 264
n-Butyl oxamate 71 232
n-Butyl *n*-pentanoate 147 257
sec-Butyl *n*-pentanoate 127 256
2-*n*-Butylphenol............................ 17 94

4-*n*-Butylphenol............................ 3 96
2-*sec*-Butylphenol.......................... 14 94
4-*tert*-Butylphenol......................... 192 106
n-Butyl phenylacetate 293 266
(4-*tert*-Butylphenyl) acetic acid.............. 374 432
n-Butyl phenyl ether....................... 82 135
n-Butyl phenyl ketone 207 168
tert-Butyl phenyl ketone 181 167
n-Butyl phthalate 355 271
1-*n*-Butylpiperidine......................... 392 439
tert-Butylpropiolic acid 37 194
n-Butyl propionate......................... 88 254
n-Butyl propyl acetylene.................... 46 29
n-Butyl *n*-propyl ketone 107 164
tert-Butyl *n*-propyl ketone 57 163
2-*tert*-Butylpyridine........................ 233 437
3-*tert*-Butylpyridine........................ 236 437
4-*tert*-Butylpyridine........................ 244 438
n-Butyl salicylate 27 95
    "   " 318 268
n-Butyl stearate 19 274
n-Butyl styryl ketone 25 172
n-Butyl sulfide 41 421
tert-Butyl sulfide 29 420
2-*n*-Butylthiophene 46 424
3-*n*-Butylthiophene 48 424
2-*tert*-Butylthiophene 39 424
3-*tert*-Butylthiophene 40 424
2-Butyltoluene............................. 67 38
3-Butyltoluene............................. 79 38
4-Butyltoluene............................. 86 38
2-*sec*-Butyltoluene......................... 54 37
3-*sec*-Butyltoluene......................... 48 37
4-*sec*-Butyltoluene......................... 64 37
2-*tert*-Butyltoluene......................... 65 38
3-*tert*-Butyltoluene......................... 38 36
4-*tert*-Butyltoluene......................... 44 37
n-Butyl 2-tolyl ether....................... 93 135
n-Butyl urethane.......................... 13 231
n-Butyl *n*-valerate........................ 147 257
sec-Butyl *n*-valerate....................... 127 256
2-Butynal................................. 26 145
1-Butyne................................. 3 28
2-Butyne................................. 6 28
2-Butynoic acid............................ 110 430
3-Butyn-2-one............................. 4 161
n-Butyraldehyde........................... 11 144
n-Butyramide............................. 179 235
N-Butyranilide............................. 89 233
n-Butyric acid............................ 10 190
    "   " 441 432
n-Butyric anhydride 9 223
Butyroin................................. 132 165
Butyronitrile.............................. 14 345
n-Butyrophenone.......................... 191 167
n-Butyrylacetone.......................... 121 165
n-Butyryl bromide 5 218
n-Butyryl chloride 9 212
n-Butyryl fluoride 5 211
n-Butyryl iodide 3 221

C

Cadaverine................................ 76 297
Camphane................................. 79 9
d,l-Camphene............................. 10 26
l-Camphene 11 26
d,l-cis-Camphenecamphoric acid.............. 190 200
d,l-trans-Camphenecamphoric acid............. 163 199
d,l-cis-Camphenic acid 190 200
d,l-trans-Camphenic acid.................... 163 199
Campholene............................... 237 19

	Compound No.	Page No.
β-Campholenic acid	45	194
d-Campholic acid	129	198
l-Campholic acid	131	198
Campholic nitrile	93	361
d,l-α-Campholytic acid	24	193
β-Campholytic acid	187	200
Δ⁵-Campholytic acid	97	196
β-Campholytonitrile	65	350
d,l-Camphor	192	178
d-Camphoric acid	303	205
D-Camphoric acid diamide	413	243
D-Camphoric acid dianilide	479	245
D-Camphoric acid monoamide	367	241
D-Camphoric acid monoanilide	437	243
D-Camphoric anhydride	81	227
Camphorquinone	31	183
D-Camphor-3-sulfonamide	104	395
D-Camphor-8-sulfonamide	92	395
DL-Camphor-8-sulfonamide	83	395
D-Camphor-10-sulfonamide	80	395
D-Camphor-3-sulfonanilide	65	394
D-Camphor-10-sulfonanilide	59	394
D-Camphor-2-sulfonic acid	38	372
D-Camphor-8-sulfonic acid	30	372
DL-Camphor-8-sulfonic acid	25	371
D-Camphor-10-sulfonic acid	23	371
D-Camphor-3-sulfonyl chloride	112	384
D-Camphor-8-sulfonyl chloride	221	388
DL-Camphor-8-sulfonyl chloride	162	386
D-Camphor-10-sulfonyl chloride	72	383
DL-Canaline	27	285
L-Canaline	47	286
L-Canavanine	23	285
Capramide	110	233
Capranilide	36	231
Capric acid	13	193
Capric anhydride	2	225
Caproaldehyde	38	146
n-Caproamide	115	233
n-Caproanilide	88	233
n-Caproic acid	33	191
" "	454	432
Caproic anhydride	15	223
n-Capronitrile	42	348
n-Caprophenone	3	172
Caproylacetic acid	73	196
n-Caproyl bromide	12	218
n-Caproyl chloride	32	212
4-n-Caproylresorcinol	53	99
" "	62	174
Caprylaldehyde	60	147
n-Caprylamide	161	235
Caprylanilide	18	231
n-Caprylic acid	47	191
" "	456	432
Caprylic anhydride	18	223
Caprylonitrile	66	350
n-Capryloyl chloride	40	213
Carballylic acid triamide	441	244
Carballylic acid trianilide	495	246
Carbamide	233	237
Carbanilide	488	245
Carbazole	457	317
4-Carbenzyloxyphenol	43	434
N-Carbethoxy-1,4-diaminobenzene	114	306
5-Carbethoxy-2-nitrophenol	16	434
4-Carbethoxyphenol	50	434
4-Carbobutoxyphenol	47	434
5-Carbomethoxy-2-nitrophenol	17	434
4-Carbomethoxyphenol	48	434
Carbon tetrabromide	2	66
Carbon tetrachloride	8	63
	Compound No.	Page No.
4-Carboxamidobenzen sulfonamide	403	407
1-Carboxamido-3-mitro-benzene sulfonanilide	376	406
2-Carboxamidophenol	41	434
4-Carboxanilidobenzenesulfonanilide	437	408
3-Carboxybenzenesulfonamide	199	399
2-Carboxybenzenesulfonanilide	291	403
2-Carboxybenzenesulfonic acid	139	376
3-Carboxybenzenesulfonic acid	84	373
4-Carboxybenzenesulfonic acid	184	377
2-Carboxybenzenesulfonyl chloride	92	383
3-Carboxybenzenesulfonyl chloride	8	381
4-Carboxybenzenesulfonyl chloride	53	382
2-Carboxy-2′-cyanobiphenyl	212	366
2-Carboxy-5-methylbenzenesulfonamide	250	401
2-Carboxy-5-methylbenzenesulfonic acid	111	375
5-Carboxy-2-methylbenzenesulfonyl chloride	56	382
4-Carboxy-3-nitrobenzenesulfonamide	280	402
4-Carboxy-3-nitrobenzenesulfonic acid	169	377
4-Carboxy-3-nitrobenzenesulfonyl chloride	250	389
(2-Carboxyphenyl)acetic acid	292	205
(2-Carboxyphenyl)acetic anhydride	64	226
d-3-Carboxy-2,2,3-trimethylcyclopentylaceto-nitrile	207	366
l-3-Carene	297	22
l-4-Carene	294	22
cis-Caronic acid	81	429
trans-Caronic acid	255	431
Carvacrol	19	94
Carvacryl acetate	273	265
d-Carvone	192	167
Catechol	207	107
"	75	434
Catechol dibenzoate	124	279
Catechol dibenzyl ether	39	138
Catechol diethyl ether	24	138
Catechol monomethyl ether	9	137
Cedrene	333	24
β-Cellobiose	65	333
6-(β-Cellobiosido)-α-D-glucose	66	333
Cellosolve	51	133
Cellosolve benzoate	300	267
Cetane	2	8
Cetyl acetate	13	273
Cetyl alcohol	25	88
Cetyl bromide	1	59
Cetyl chloride	54	56
Cetyl iodide	1	61
Cetylmalonic acid	161	199
Cetyl mercaptan	2	416
Cetyl palmitate	66	276
Cetyl stearate	79	277
Cetyl sulfonic acid	7	371
Chalcone	66	174
D-Chaulmoogramide	144	234
D-Chaulmoogranilide	73	232
d-Chaulmoogric acid	69	195
Chelidonic acid	391	209
Chloral	19	145
Chloranil	53	184
Chloranilic acid	51	184
Chloroacetaldehyde	13	144
Chloroacetamide	191	236
Chloroacetanilide	243	237
2-Chloroacetanilide	72	232
4-Chloroacetanilide	372	241
Chloroacetic acid	56	195
" "	124	430
Chloroacetic anhydride	16	225
Chloroacetone	17	161
Chloroacetone cyanohydrin	13	355
Chloroacetonitrile	21	346
2-Chloroacetophenone	188	167

ORGANIC COMPOUND INDEX

Compound	Compound No.	Page No.
3-Chloroacetophenone	184	167
4-Chloroacetophenone	194	167
w-Chloroacetophenone	68	174
Chloroacetyl bromide	4	218
Chloroacetyl chloride	12	212
Chloroacetyl fluoride	6	211
Chloroacetyl iodide	2	222
2-Chloroacrolein	1	152
cis-3-Chloroacrylic acid	191	430
trans-3-Chloroacrylic acid	229	431
cis-2-Chloroallocinnamic acid	170	200
cis-α-Chloroallocinnamic acid	135	198
cis-β-Chloroallocinnamic acid	180	200
2-Chloroaniline	102	298
" "	61	436
3-Chloroaniline	144	299
" "	99	436
4-Chloroaniline	112	306
" "	125	437
3-Chloroanisic acid	342	207
2-Chloro-p-anisidine	79	305
3-Chloro-o-anisidine	162	300
3-Chloro-p-anisidine	8	302
4-Chloro-o-anisidine	140	307
5-Chloro-o-anisidine	56	304
2-Chloroanisole	79	135
3-Chloroanisole	78	135
4-Chloroanisole	81	135
1-Chloroanthraquinone-2-carboxylic acid	396	209
2-Chlorobenzal chloride	24	63
3-Chlorobenzal chloride	25	63
4-Chlorobenzal chloride	26	63
2-Chlorobenzaldehyde	78	148
3-Chlorobenzaldehyde	1	153
4-Chlorobenzaldehyde	24	154
2-Chlorobenzamide	260	238
3-Chlorobenzamide	241	237
2-Chlorobenzanilide	177	235
3-Chlorobenzanilide	195	236
4-Chlorobenzanilide	414	243
Chlorobenzene	1	70
2-Chlorobenzene-1,4-dicarboxylic acid	405	210
5-Chlorobenzene-1,3-disulfonamide	372	406
5-Chlorobenzene-1,3-disulfonic acid	167	377
5-Chlorobenzene-1,3-disulfonyl chloride	163	386
2-Chlorobenzenesulfonamide	263	402
3-Chlorobenzenesulfonamide	123	396
4-Chlorobenzenesulfonamide	113	396
4-Chlorobenzenesulfonanilide	36	393
2-Chlorobenzenesulfonic acid	127	375
3-Chlorobenzenesulfonic acid	47	372
4-Chlorobenzenesulfonic acid	41	372
2-Chlorobenzenesulfonyl chloride	12	381
4-Chlorobenzenesulfonyl chloride	42	382
2-Chlorobenzoic acid	213	201
" "	130	430
3-Chlorobenzoic acid	243	203
" "	254	431
4-Chlorobenzoic acid	375	208
" "	280	431
2-Chlorobenzoic anhydride	33	225
3-Chlorobenzoic anhydride	39	225
4-Chlorobenzoic anhydride	74	226
2-Chlorobenzonitrile	35	358
3-Chlorobenzonitrile	34	358
4-Chlorobenzonitrile	128	363
4-Chlorobenzophenone	108	175
2-Chloro-1,4-benzoquinone	5	181
4-Chloro-1,2-benzoquinone	8	181
2-Chlorobenzotrichloride	1	71
2-Chlorobenzoyl bromide	4	219
3-Chlorobenzoyl bromide	3	219
4-Chlorobenzoyl bromide	10	219
2-Chlorobenzoyl chloride	53	213
3-Chlorobenzoyl chloride	49	213
4-Chlorobenzoyl chloride	47	213
4-Chlorobenzyl bromide	5	59
2-Chlorobenzyl chloride	47	56
3-Chlorobenzyl chloride	49	56
4-Chlorobenzyl chloride	48	56
2-Chlorobenzyl cyanide	10	357
4-Chlorobenzyl cyanide	16	357
2-Chlorobenzylidene chloride	24	63
3-Chlorobenzylidene chloride	25	63
4-Chlorobenzylidene chloride	26	63
2-Chlorobenzyl phenyl ketone	93	175
3-Chlorobenzyl phenyl ketone	38	173
4-Chlorobenzyl phenyl ketone	168	177
4-Chlorobenzyl sulfide	14	425
2-Chlorobiphenyl	3	71
3-Chlorobiphenyl	29	70
4-Chlorobiphenyl	22	71
1-Chloro-2-butanone	40	162
2-Chloro-2-butenal	49	146
3-Chloro-n-butyraldehyde (trimer)	83	157
4-Chloro-n-butyraldehyde	2	152
2-Chlorobutyronitrile	32	347
3-Chlorobutyronitrile	48	348
4-Chlorobutyronitrile	61	349
3-Chlorocatechol	39	98
4-Chlorocatechol	160	104
3-Chlorocinchonic acid	377	208
trans-α-Chlorocinnamic acid	195	201
trans-β-Chlorocinnamic acid	212	201
trans-2-Chlorocinnamic acid	337	207
" " "	327	431
trans-3-Chlorocinnamic acid	343	432
trans-4-Chlorocinnamic acid	376	432
trans-o-Chlorocinnamonitrile	32	358
1-Chlorocyclohexanecarboxylic acid	21	192
2-Chloro-m-cresol	47	99
2-Chloro-p-cresol	3	94
4-Chloro-m-cresol	78	100
6-Chloro-m-cresol	36	98
4-Chloro-o-cresol	42	98
5-Chloro-o-cresol	103	102
2-Chlorocrotonaldehyde	49	146
α-Chlorocrotonic acid	178	430
β-Chlorocrotonic acid	100	197
γ-Chlorocrotonic acid	87	196
2-Chlorocrotononitrile	24	347
trans-4-Chlorocrotononitrile	10	355
p-Chlorocumene	17	70
1-Chloro-4-cyanonaphthalene	154	364
5-Chloro-1-cyanonaphthalene	191	365
5-Chloro-2-cyanonaphthalene	190	365
4-Chloro-2-cyanophenol	208	366
2-Chloro-p-cymene	26	70
3-Chloro-p-cymene	27	70
4-Chloro-1,3-diaminobenzene	160	307
2-Chloro-4,6-dibromoaniline	176	308
4-Chloro-2,6-dibromoaniline	166	308
4-Chloro-2,6-dibromophenol	164	104
4-Chloro-3,5-dibromophenol	275	110
N-Chlorodiethylamine	25	436
3-Chloro-1,2-dihydroxybenzene	39	98
4-Chloro-1,2-dihydroxybenzene	160	104
4-Chloro-1,3-dihydroxybenzene	214	107
2-Chloro-N,N-dimethylaniline	61	320
2-Chloro-4,6-dimethylaniline	27	303
1-Chloro-2,3-dimethylbenzene	12	70
1-Chloro-2,4-dimethylbenzene	14	70
1-Chloro-3,4-dimethylbenzene	16	70
2-Chloro-1,4-dimethylbenzene	10	70

Compound	Compound No.	Page No.
4-Chloro-2,5-dimethylbenzenesulfonamide	249	401
4-Chloro-2,5-dimethylbenzenesulfonanilide	142	397
4-Chloro-2,5-dimethylbenzenesulfonic acid	110	374
4-Chloro-2,5-dimethylbenzenesulfonyl chloride	37	382
2-Chloro-2,3-dimethylbutane	29	55
3-Chloro-2,2-dimethylbutane	25	55
4-Chloro-2,2-dimethylbutane	27	55
5-Chloro-2,3-dimethylpentane	36	56
2-Chloro-3,4-dimethylphenol	6	96
2-Chloro-4,5-dimethylphenol	101	101
2-Chloro-4,6-dimethylphenol	11	94
3-Chloro-4,5-dimethylphenol	190	106
3-Chloro-4,6-dimethylphenol	161	104
4-Chloro-2,3-dimethylphenol	138	103
4-Chloro-3,5-dimethylphenol	248	109
5-Chloro-2,3-dimethylphenol	130	103
1-Chloro-2,4-dinitrobenzene	22	337
4-Chloro-1,3-dinitrobenzene	22	337
5-Chloro-2,4-dinitrobenzonitrile	200	366
4-Chloro-1,3-dinitronaphthalene	101	341
4-Chloro-1,5-dinitronaphthalene	95	341
4-Chloro-1,8-dinitronaphthalene	117	342
2-Chloro-4,6-dinitrophenol	239	108
4-Chloro-2,6-dinitrophenol	128	103
" "	4	434
5-Chloro-2,4-dinitrophenol	163	104
α-Chlorodiphenylacetic acid	153	199
2-Chloroethanethiol	16	413
2-Chloroethanol	24	81
1-Chloro-4-ethoxybenzene	1	137
β-Chloroethyl acetate	81	253
1-Chloro-2-ethylbenzene	6	70
1-Chloro-3-ethylbenzene	8	70
1-Chloro-4-ethylbenzene	9	70
1-Chloro-2-ethylbutane	32	56
2-Chloroethyl chloromethyl sulfide	45	421
(2-Chloroethyl)diethylamine	304	438
α-Chloroethyl ethyl ether	33	132
β-Chloroethyl ethyl ether	39	133
2-Chloroethyl ethyl sulfide	32	420
1-Chloro-2-ethyl-3-hexane	18	169
2-Chloroethyl methyl sulfide	25	420
2-Chloroethyl 4-nitrophenyl sulfide	34	425
2-Chloroethyl phenyl sulfide	57	421
2-Chloroethyl sulfide	52	421
2-Chloroethyl 4-tolyl sulfide	58	421
Chloroform	5	63
Chlorofumaranilide	393	242
Chlorofumaric acid	310	206
Chlorofumaronitrile	45	348
2-Chlorohexane	31	56
3-Chlorohexane	30	55
o-Chlorohydrocinnamic acid	104	197
Chlorohydroquinone	210	107
3-Chloro-4-hydroxyaniline	342	313
4-Chloro-2-hydroxyaniline	343	314
2-Chloro-3-hydroxybenzaldehyde	112	159
2-Chloro-4-hydroxybenzaldehyde	366	115
" " " "	116	159
2-Chloro-5-hydroxybenzaldehyde	95	158
3-Chloro-2-hydroxybenzaldehyde	36	155
3-Chloro-4-hydroxybenzaldehyde	111	159
4-Chloro-2-hydroxybenzaldehyde	28	154
4-Chloro-3-hydroxybenzaldehyde	106	159
5-Chloro-2-hydroxybenzaldehyde	88	158
2-Chloro-6-hydroxybenzoic acid	109	430
3-Chloro-4-hydroxybenzoic acid	262	203
5-Chloro-2-hydroxybenzoic acid	268	204
3-Chloro-4-hydroxybenzonitrile	199	366
5-Chloro-2-hydroxybenzonitrile	208	366
3-Chloro-4-hydroxybiphenyl	112	102
5-Chloro-2-hydroxybiphenyl	35	98
2-Chloro-2-hydroxybutyric acid	105	430
5-Chloro-4-hydroxy-2-methylaniline	448	317
6-Chloro-3-hydroxy-4-methylbenzoic acid	519	123
3-Chloro-2-hydroxy-2-methylpropionitrile	13	355
4-Chloro-1-hydroxy-2-naphthoic acid	363	208
4-Chloro-3-hydroxy-2-naphthoic acid	359	208
5-Chloro-2-hydroxy-3-nitrobenzoic acid	250	203
2-Chloro-3-hydroxy-3-phenylpropionic acid	108	430
2-Chloro-3-hydroxysuccinic acid	76	429
4-Chloro-2-iodophenol	119	102
α-Chloroisobutanoic acid	8	193
α-Chloroisobutyric acid	8	193
" "	135	430
α-Chloroisobutyronitrile	13	345
β-Chloroisocrotonic acid	54	195
1-Chloro-2-isopropylbenzene	13	70
1-Chloro-4-isopropylbenzene	17	70
2-Chloro-4-isopropyl-1-methylbenzene	26	70
3-Chloro-4-isopropyl-1-methylbenzene	27	70
α-Chloroisopropyl methyl ketone	59	163
α-Chloroisovaleric acid	38	191
3-Chlorolactic acid	158	430
Chloromalonic acid	182	200
3-Chloromandelic acid	184	430
4-Chloromandelic acid	156	199
4-Chloromandelonitrile	36	358
2-Chloro-4-methoxyaniline	8	302
2-Chloro-6-methoxyaniline	162	300
3-Chloro-4-methoxyaniline	79	305
4-Chloro-2-methoxyaniline	56	304
5-Chloro-2-methoxyaniline	140	307
3-Chloro-4-methoxybenzenesulfonamide	76	394
3-Chloro-4-methoxybenzenesulfonic acid	21	371
3-Chloro-4-methoxybenzenesulfonyl chloride	101	384
3-Chloro-4-methoxybenzoic acid	342	207
8-Chloro-7-methoxynaphthalene-1-sulfonamide	135	397
8-Chloro-7-methoxynaphthalene-1-sulfonanilide	302	403
8-Chloro-7-methoxynaphthalene-1-sulfonyl chloride	216	388
5-Chloro-2-methoxy-1-nitrobenzene	69	339
Chloromethyl acetate	38	251
2-Chloro-N-methylaniline	118	298
4-Chloro-N-methylaniline	156	300
2-Chloro-4-methylaniline	130	299
4-Chloro-2 methylaniline	157	300
4-Chloro-3-methylaniline	138	307
2-Chloro-4-methylbenzenesulfonamide	255	401
2-Chloro-5-methylbenzenesulfonamide	146	397
3-Chloro-2-methylbenzenesulfonamide	231	400
3-Chloro-4-methylbenzenesulfonamide	85	395
3-Chloro-6-methylbenzenesulfonamide	115	396
4-Chloro-2-methylbenzenesulfonamide	248	401
4-Chloro-3-methylbenzenesulfonamide	73	394
5-Chloro-2-methylbenzenesulfonamide	105	395
3-Chloro-4-methylbenzenesulfonanilide	27	393
4-Chloro-3-methylbenzenesulfonanilide	25	392
2-Chloro-4-methylbenzenesulfonic acid	117	375
2-Chloro-5-methylbenzenesulfonic acid	57	372
3-Chloro-2-methylbenzenesulfonic acid	99	374
3-Chloro-4-methylbenzenesulfonic acid	27	371
4-Chloro-2-methylbenzenesulfonic acid	109	374
4-Chloro-3-methylbenzenesulfonic acid	20	371
5-Chloro-2-methylbenzenesulfonic acid	44	372
2-Chloro-4-methylbenzenesulfonyl chloride	30	381
2-Chloro-5-methylbenzenesulfonyl chloride	51	382
3-Chloro-2-methylbenzenesulfonyl chloride	83	383
3-Chloro-4-methylbenzenesulfonyl chloride	21	381
4-Chloro-2-methylbenzenesulfonyl chloride	45	382
4-Chloro-3-methylbenzenesulfonyl chloride	63	382
5-Chloro-2-methylbenzenesulfonyl chloride	10	381
2-Chloro-4-methylbenzoic acid	237	202
2-Chloro-6-methylbenzoic acid	120	197

	Compound No.	Page No.
(3-Chloromethyl)benzoic acid	199	201
(4-Chloromethyl)benzoic acid	323	206
3-Chloro-2-methylbenzoic acid	188	430
4-Chloro-2-methylbenzoic acid	269	204
6-Chloro-2-methylbenzoic acid	70	429
4-Chloro-3-methyl-2-butanone	59	163
DL-3-Chloro-2-methyl-1-butene	17	55
3-Chloro-2-methyl-2-butene	18	55
2-Chloro-3-methylbutyronitrile	37	348
Chloromethyl dichloromethyl sulfide	39	421
Chloromethyl ethyl ether	22	132
Chloromethyl ethyl ketone	40	162
Chloromethyl ethyl sulfide	19	420
Chloromethyl methyl ether	12	131
Chloromethyl methyl sulfide	11	420
2-Chloro-5-methyl-6-nitrobenzenesulfonamide	221	400
4-Chloro-3-methyl-5-nitrobenzenesulfonamide	313	404
5-Chloro-2-methyl-3-nitrobenzenesulfonamide	187	399
5-Chloro-4-methyl-2-nitrobenzenesulfonamide	235	401
4-Chloro-3-methyl-5-nitrobenzenesulfonic acid	147	376
5-Chloro-2-methyl-3-nitrobenzenesulfonic acid	76	373
5-Chloro-4-methyl-2-nitrobenzenesulfonic acid	103	374
2-Chloro-5-methyl-6-nitrobenzenesulfonyl chloride	191	387
4-Chloro-3-methyl-5-nitrobenzenesulfonyl chloride	41	382
5-Chloro-2-methyl-3-nitrobenzenesulfonyl chloride	58	382
5-Chloro-4-methyl-2-nitrobenzenesulfonyl chloride	147	385
2-Chloro-2-methylpentane	24	55
3-Chloro-3-methylpentane	28	55
1-Chloro-2-methyl-3-pentanone	1	169
2-Chloro-3-methylphenol	47	99
2-Chloro-5-methylphenol	36	98
3-Chloro-4-methylphenol	3	94
4-Chloro-2-methylphenol	42	98
4-Chloro-3-methylphenol	78	100
5-Chloro-2-methylphenol	103	102
(4-Chloro-2-methylphenoxy)acetic acid	189	430
Chloromethyl 2-phenylethyl ketone	58	174
3-Chloro-2-methyl-1-propene	12	55
4-Chloro-2-methylquinoline	19	323
Chloromethyl sulfide	31	420
1-Chloronaphthalene	28	70
2-Chloronaphthalene	14	71
1-Chloronaphthalene-2-sulfonamide	435	408
2-Chloronaphthalene-1-sulfonamide	134	396
4-Chloronaphthalene-1-sulfonamide	258	401
4-Chloronaphthalene-2-sulfonamide	191	399
5-Chloronaphthalene-1-sulfonamide	377	406
5-Chloronaphthalene-2-sulfonamide	350	405
6-Chloronaphthalene-1-sulfonamide	346	405
6-Chloronaphthalene-2-sulfonamide	245	401
7-Chloronaphthalene-1-sulfonamide	399	407
7-Chloronaphthalene-2-sulfonamide	215	400
8-Chloronaphthalene-1-sulfonamide	305	403
8-Chloronaphthalene-2-sulfonamide	247	401
4-Chloronaphthalene-1-sulfonanilide	118	396
1-Chloronaphthalene-2-sulfonic acid	198	378
4-Chloronaphthalene-1-sulfonic acid	122	375
4-Chloronaphthalene-2-sulfonic acid	80	373
5-Chloronaphthalene-1-sulfonic acid	170	377
6-Chloronaphthalene-1-sulfonic acid	159	376
6-Chloronaphthalene-2-sulfonic acid	107	374
7-Chloronaphthalene-2-sulfonic acid	93	374
8-Chloronaphthalene-1-sulfonic acid	146	376
8-Chloronaphthalene-2-sulfonic acid	112	375
1-Chloronaphthalene-2-sulfonyl chloride	104	384
2-Chloronaphthalene-1-sulfonyl chloride	85	383
4-Chloronaphthalene-1-sulfonyl chloride	129	385
4-Chloronaphthalene-2-sulfonyl chloride	160	386
5-Chloronaphthalene-1-sulfonyl chloride	132	385
5-Chloronaphthalene-2-sulfonyl chloride	177	386
6-Chloronaphthalene-1-sulfonyl chloride	79	383
6-Chloronaphthalene-2-sulfonyl chloride	170	386
7-Chloronaphthalene-1-sulfonyl chloride	207	387
7-Chloronaphthalene-2-sulfonyl chloride	110	384
8-Chloronaphthalene-1-sulfonyl chloride	153	386
8-Chloronaphthalene-2-sulfonyl chloride	127	385
4-Chloro-1-naphthalenethiol	12	416
3-Chloro-2-naphthoic acid	345	207
4-Chloro-1-naphthoic acid	378	209
5-Chloro-2-naphthoic acid	393	209
7-Chloro-2-naphthoic acid	376	208
8-Chloro-1-naphthoic acid	266	204
1-Chloro-2-naphthol	94	101
3-Chloro-1-naphthol	346	114
4-Chloro-1-naphthol	273	110
4-Chloro-2-naphthol	201	106
5-Chloro-1-naphthol	315	112
6-Chloro-1-naphthol	173	105
7-Chloro-2-naphthol	299	111
4-Chloro-1-naphthonitrile	154	364
5-Chloro-1-naphthonitrile	191	365
5-Chloro-2-naphthonitrile	190	365
2-Chloro-1,4-naphthoquinone	13	181
3-Chloro-1,2-naphthoquinone	22	182
4-Chloro-1,2-naphthoquinone	29	182
1-Chloro-2-naphthylamine	70	305
2-Chloro-1-naphthylamine	72	305
3-Chloro-1-naphthylamine	80	305
4-Chloro-1-naphthylamine	190	308
4-Chloro-2-nitroaniline	244	310
" "	8	436
5-Chloro-2-nitroaniline	279	311
4-Chloro-2-nitroanisole	69	339
2-Chloro-1-nitrobenzene	5	337
3-Chloro-1-nitrobenzene	18	337
4-Chloro-1-nitrobenzene	52	339
2-Chloro-5-nitrobenzenesulfonamide	251	401
4-Chloro-2-nitrobenzenesulfonamide	176	398
4-Chloro-3-nitrobenzenesulfonamide	214	400
4-Chloro-2-nitrobenzenesulfonanilide	93	395
5-Chloro-2-nitrobenzenesulfonanilide	164	398
2-Chloro-5-nitrobenzenesulfonic acid	113	375
4-Chloro-2-nitrobenzenesulfonic acid	69	373
4-Chloro-3-nitrobenzenesulfonic acid	91	374
6-Chloro-3-nitrobenzenesulfonic acid	119	375
2-Chloro-5-nitrobenzenesulfonyl chloride	118	384
4-Chloro-2-nitrobenzenesulfonyl chloride	86	383
4-Chloro-3-nitrobenzenesulfonyl chloride	25	381
5-Chloro-2-nitrobenzenesulfonyl chloride	123	384
2-Chloro-3-nitrobenzoic acid	41	429
2-Chloro-4-nitrobenzoic acid	206	201
" " "	39	429
2-Chloro-5-nitrobenzoic acid	61	429
2-Chloro-6-nitrobenzoic acid	12	429
3-Chloro-2-nitrobenzoic acid	364	208
4-Chloro-2-nitrobenzoic acid	211	201
4-Chloro-3-nitrobenzoic acid	283	204
5-Chloro-2-nitrobenzoic acid	200	201
6-Chloro-3-nitrobenzoic acid	256	203
2-Chloro-3-nitrobenzonitrile	131	363
4-Chloro-2-nitrobenzonitrile	133	363
4-Chloro-3-nitrobenzonitrile	135	363
6-Chloro-3-nitrobenzonitrile	146	363
4-Chloro-1-nitronaphthalene	59	339
5-Chloro-1-nitronaphthalene	76	340
8-Chloro-1-nitronaphthalene	66	339
4-Chloro-3-nitro-1,2-naphthequinone	120	342
2-Chloro-3-nitrophenol	274	110
2-Chloro-4-nitrophenol	232	108
2-Chloro-5-nitrophenol	267	110

Compound	Compound No.	Page No.
3-Chloro-5-nitrophenol	363	115
4-Chloro-2-nitrophenol	149	104
4-Chloro-3-nitrophenol	300	111
4-Chloro-4-nitrophenol	42	417
5-Chloro-2-nitrophenol	15	434
5-Choro-3-nitrosalicylic acid	250	203
4-Chloro-2-nitrotoluene	13	337
6-Chloro-2-nitrotoluene	10	337
5-Chloro-2-nitro-4-tolunitrile	121	362
DL-2-Chloropentane	19	55
3-Chloropentane	20	55
2-Chloropentanoic acid	43	191
1-Chloro-2-pentene	23	55
3-Chloro-1-pentene	16	55
4-Chlorophenacyl bromide	122	176
2-Chlorophenetole	83	135
4-Chlorophenetole	1	137
2-Chlorophenol	1	94
"	49	434
3-Chlorophenol	15	96
"	61	434
4-Chlorophenol	20	97
"	72	434
2-Chlorophenoxyacetamide	285	238
4-Chlorophenoxyacetamide	238	237
2-Chlorophenoxyacetanilide	192	236
4-Chlorophenoxyacetanilide	200	236
(2-Chlorophenoxy)acetic acid	217	202
" " "	147	430
(3-Chlorophenoxy)acetic acid	150	430
(4-Chlorophenoxy)acetic acid	238	202
" " "	154	430
(2-Chlorophenyl)acetic acid	102	197
" " "	295	431
(3-Chlorophenyl)acetic acid	305	431
(4-Chlorophenyl)acetic acid	127	198
" " "	321	431
(2-Chlorophenyl)acetonitrile	10	357
(4-Chlorophenyl)acetonitrile	16	357
α-Chloro-α-phenylacetonitrile	5	355
3-(2-Chlorophenyl)alanine	68	429
3-(3-Chlorophenyl)alanine	60	429
3-(4-Chlorophenyl)alanine	48	429
2-Chlorophenyl ethyl ketone	49	169
3-Chlorophenylglycine	6	429
4-Chlorophenyl methyl sulfide	35	421
4-Chlorophenyl phenyl sulfide	63	422
3-(2-Chlorophenyl)propionic acid	401	432
3-(3-Chlorophenyl)propionic acid	402	432
3-(2-Chlorophenyl)propionitrile	96	353
4-Chlorophenyl sulfide	56	426
3-Chlorophthalic anhydride	54	226
4-Chlorophthalic anhydride	42	226
4-Chloropicolinic acid	286	204
Chloroprene	9	55
2-Chloropropanethiol	15	413
1-Chloro-2-propanol	22	81
d,l-2-Chloro-1-propanol	27	81
3-Chloro-1-propanol	54	83
1-Chloro-2-propanone	17	161
1-Chloropropene	5	55
2-Chloropropionaldehyde	14	144
3-Chloropropionaldehyde	37	146
d,l-α-Chloropropionamide	54	232
d,l-α-Chloropropionanilide	80	233
d,l-2-Chloropropionic acid	18	190
" "	127	430
3-Chloropropionic acid	25	193
" "	285	431
3-Chloropropionitrile	49	348
DL-α-Chloropropionyl chloride	13	212
β-Chloropropionyl chloride	28	212
2-Chloropropiophenone	49	169
3-Chloropropiophenone	43	173
4-Chloropropiophenone	21	172
3-Chloropropyl alcohol	54	83
2-Chloropyridine	38	319
"	14	436
3-Chloropyridine	23	319
"	74	436
4-Chloropyridine	21	319
"	115	436
4-Chloropyridine-2-carboxylic acid	286	204
4-Chloroquinaldine	19	323
2-Chloroquinoline-3-carboxylic acid	372	208
2-Chloroquinoline-4-carboxylic acid	377	208
2-Chloroquinoline	11	323
3-Chloroquinoline	92	321
4-Chloroquinoline	7	323
5-Chloroquinoline	20	323
6-Chloroquinoline	15	323
7-Chloroquinoline	8	323
8-Chloroquinoline	109	321
α-Chloroquinoline	11	323
β-Chloroquinoline	92	321
γ-Chloroquinoline	7	323
4-Chlororesorcinol	214	107
3-Chlorosalicylaldehyde	36	155
5-Chlorosalicylic acid	268	204
β-Chlorostyrene	45	56
o-Chlorostyrene	11	70
p-Chlorostyrene	15	70
Chloroterephthalic acid	405	210
4-Chloro-2,3,5,6-Tetrabromophenol	534	124
1-Chloro-2,2,3,3-tetramethylbutane	5	57
2-Chlorothiophenol	37	414
4-Chlorothiophenol	16	416
2-Chlorotoluene	2	70
3-Chlorotoluene	3	70
4-Chlorotoluene	4	70
5-Chloro-o-toluidine	157	300
6-Chloro-m-toluidine	138	307
2-Chloro-5-tolunitrile	71	360
3-Chloro-2-tolunitrile	3	357
3-Chloro-4-tolunitrile	48	359
4-Chloro-2-tolunitrile	83	361
5-Chloro-2-tolunitrile	41	358
6-Chloro-2-tolunitrile	104	362
α-Chloro-3-tolunitrile	82	361
α-Chloro-4-tolunitrile	99	361
4-Chloro-1,2,3-trihydroxy-9,10-anthraquinone	36	183
2-Chloro-1,3,5-trimethylbenzene	22	70
3-Chloro-2,2,3-trimethylbutane	33	56
2-Chloro-n-valeric acid	43	191
(2-Chlorovinyl)acetic acid	101	429
1-Chloro-2-vinylbenzene	11	70
1-Chloro-4-vinylbenzene	15	70
5-Chloro-m-xylidine	27	303
Cholanic acid	252	203
Cholanthrene	169	50
Cholesterol	59	90
Cholesteryl acetate	147	280
Chrysazol	556	126
Chrysene	195	51
Chrysenequinone	39	183
α-Chrysenic acid	308	205
Chrysodiphenic acid	318	206
Chrysofluorene	179	50
Chrysoquinone	39	183
Cinchol	57	90
Cinchomeronic acid	389	209
Cinchoninic acid	385	209
Cinnamalacetone	89	175

Compound	Compound No.	Page No.
Cinnamacetophenone	131	176
Cinnamaldehyde	90	149
Cinnamalfluorene	156	49
Cinnamamide	278	238
Cinnamanilide	287	238
cis-Cinnamic acid	262	431
trans-Cinnamic acid	181	200
" " "	379	432
Cinnamic anhydride	62	226
Cinnamonitrile	5	357
trans-Cinnamoyl bromide	2	220
Cinnamyl alcohol	13	88
trans-Cinnamyl chloride	4	216
Cinnamyl cinnamate	50	275
Cinnamyl phenyl sulfide	45	426
Cinnoline	45	436
Citraconic acid	98	196
" "	470	433
Citraconic acid diamide	392	242
Citraconic acid dianilide	359	241
Citraconic anhydride	10	223
Citral a.	82	148
Citral b.	83	148
d,l-Citramalic acid	154	199
Citric acid	113	197
" "	228	202
Citric acid triamide	448	244
Citric acid trianilide	425	243
d-Citronellal	76	148
d-Citronellic acid	49	191
l-Citronellic acid	12	192
Citronellol	115	86
L-Citrulline	51	286
2,3,4-Collidine	51	320
2,3,5-Collidine	45	320
2,3,6-Collidine	43	320
2,4,5-Collidine	34	319
2,4,6-Collidine	41	320
Coronene	204	51
2-Coumaramide	445	244
4-Coumaramide	415	243
trans-2-Coumaric acid	333	207
p-Coumaric acid	336	207
Coumarilic acid	312	206
Coumarilonitrile	28	358
Coumarin-2-carbonitrile	28	358
Coumarin-3-carboxylic acid	302	205
Coumarone-2-carboxylic acid	312	206
Creatine	113	289
Creatinine	117	290
Creosol	92	135
m-Cresol	5	94
"	100	434
o-Cresol	11	96
"	109	435
p-Cresol	17	97
"	108	435
m-Cresyl acetate	200	260
o-Cresyl acetate	187	259
p-Cresyl acetate	203	260
m-Cresyl benzoate	76	277
o-Cresyl benzoate	350	271
p-Cresyl benzoate	101	278
2-Cresyl ethyl ether	70	134
3-Cresyl ethyl ether	76	135
4-Cresyl ethyl ether	75	135
2-Cresyl methyl ether	62	134
3-Cresyl methyl ether	66	134
4-Cresyl methyl ether	65	134
Crotonaldehyde	22	145
Crotonamide	318	240
trans-α-Crotonanilide	182	235
Crotonchloral	58	147
α-Crotonic acid	109	197
cis-Crotonic acid	13	190
" " "	369	432
trans-Crotonic acid	71	195
" " "	423	432
Crotonic anhydride	14	223
trans-Crotononitrile	15	346
trans-Crotonyl chloride	21	212
Crotylacetone	71	163
Crotyl sulfide	40	421
Cumaldehyde	87	149
Cumene	7	35
p-Cumidine	135	299
Cuminic acid	150	199
Cyanethine	404	316
1-Cyanoacenaphthene	84	361
5-Cyanoacenaphthene	155	364
Cyanoacetaldehyde	2	345
Cyanoacetamide	188	235
Cyanoacetanilide	424	243
Cyanoacetic acid	60	195
" "	79	360
" "	96	429
Cyanoacetyl chloride	15	214
Cyanoacetylene	1	345
2-Cyanoaniline	22	436
3-Cyanoaniline	70	436
4-Cyanoaniline	111	362
"	33	436
1-Cyanoanthracene	170	364
9-Cyanoanthracene	214	367
1-Cyano-9,10-anthraquinone	233	368
4-Cyanoazobenzene	166	364
2-Cyanobenzal chloride	94	353
3-Cyanobenzal chloride	99	353
4-Cyanobenzal chloride	100	353
3-Cyanobenzaldehyde	101	361
4-Cyanobenzamide	470	245
4-Cyanobenzanilide	370	241
2-Cyanobenzoic acid	299	205
" "	219	367
" "	163	430
3-Cyanobenzoic acid	227	367
" "	218	430
4-Cyanobenzoic acid	348	207
" "	228	367
" "	213	430
2-Cyanobenzyl bromide	89	361
3-Cyanobenzyl bromide	122	362
4-Cyanobenzyl bromide	160	364
3-Cyanobenzyl chloride	82	361
4-Cyanobenzyl chloride	99	361
α-Cyanobenzyl cyanide	85	361
2-Cyanobiphenyl	16	356
4-Cyanobiphenyl	108	362
1-Cyanobutane	31	347
2-Cyanobutanoic acid	22	192
4-Cyanobutanoic acid	32	194
2-Cyanobutyric acid	15	356
4-Cyanobutyric acid	32	194
" "	40	358
" "	378	432
2-Cyanocinnamic acid	218	367
3-Cyanocoumarin	217	367
Cyanocyclobutane	36	347
1-Cyanocyclohexane	2	355
2-Cyano-2-cyclohexylacetic acid	90	429
N-Cyanodiethylamine	6	436
3-Cyano-N,N-dimethylaniline	77	436
4-Cyano-N,N-dimethylaniline	96	361
" "	34	436

Compound	Compound No.	Page No.
4-Cyano-2,6-dimethylphenol	40	434
4-Cyano-3,5-dimethylphenol	38	434
d,l-4-Cyano-3,4-diphenylbutyric acid	206	366
2-Cyanodiphenylmethane	1	357
4-Cyanodiphenylmethane	56	359
Cyanoethane	8	345
2-Cyanoethylamine	283	438
2-Cyano-2-ethylbutyric acid	80	360
(2-Cyanoethyl)dimethylamine	274	438
Cyanoform	66	360
2-Cyanofuran	34	347
4-Cyanoheptane	57	349
2-Cyanohexanoic acid	168	364
α-Cyanohydrocinnamic acid	119	197
3-Cyanoindene	21	356
3-Cyanoindole	50	349
2-Cyano-5-iodonaphthalene	195	366
2-Cyanoisobutyric acid	95	429
1-Cyano-4-isopropenylcyclohexane	18	356
1-Cyanoisoquinoline	97	361
Cyanomethane	5	345
N-Cyano-N-methylaniline	95	353
(Cyanomethyl)diethylamine	162	437
1-(Cyanomethyl)pyridine	161	437
1-Cyanonaphthalene	25	358
2-Cyanonaphthalene	78	360
8-Cyanonaphthalene-1-sulfonamide	476	410
8-Cyanonaphthalene-1-sulfonyl chloride	223	388
1-Cyano-4-nitronaphthalene	179	365
1-Cyano-5-nitronaphthalene	223	367
2-Cyano-1-nitronaphthalene	183	365
2-Cyano-5-nitronaphthalene	213	366
2-Cyano-8-nitronaphthalene	189	365
2-Cyano-4-nitrophenol	221	367
2-Cyano-5-nitrophenol	205	366
2-Cyano-6-nitrophenol	178	365
2-Cyanopentanoic acid	11	355
1-Cyanophenanthrene	172	364
2-Cyanophenanthrene	151	364
3-Cyanophenanthrene	138	363
9-Cyanophenanthrene	149	363
2-Cyanophenol	189	106
"	132	363
4-Cyanophenol	158	364
"	34	434
(2-Cyanophenoxy)acetic acid	137	430
(3-Cyanophenoxy)acetic acid	143	430
(4-Cyanophenoxy)acetic acid	131	430
1-Cyano-2-phenylacrylonitrile	112	362
2-Cyano-3-phenylpropionic acid	137	363
3-Cyano-3-phenylpropionic acid	196	366
4-Cyano-2-phenylquinoline	187	365
1-Cyanopropane	14	345
2-Cyanopropionic acid	19	192
" "	26	358
" "	91	429
3-Cyanopropionic acid	39	194
" "	47	359
" "	284	431
(3-Cyanopropyl) diethyl amine	313	438
2-Cyano-2-propylvaleramide	197	366
2-Cyanopyridine	12	357
3-Cyanopyridine	54	359
4-Cyanopyridine	98	361
2-Cyanoquinoline	125	362
3-Cyanoquinoline	150	363
4-Cyanoquinoline	140	363
5-Cyanoquinoline	115	362
6-Cyanoquinoline	181	365
8-Cyanoquinoline	105	362
5-Cyanothiazole	60	359
2-Cyanothiophene	59	349
3-Cyanothiophene	52	349
Cyanotriphenylethylene	209	366
2-Cyanotriphenylmethane	114	362
4-Cyanotriphenylmethane	134	363
4-Cyanovaleric acid	126	363
Cyclobutane	8	3
Cyclobutanecarbonitrile	36	347
Cyclobutanecarbonyl chloride	25	212
Cyclobutanecarboxylic acid	26	191
" "	435	432
Cyclobutane-1,1-dicarboxylic acid	242	203
cis-Cyclobutane-1,2-dicarboxylic acid	197	201
" " "	265	431
cis-Cyclobutane-1,3-dicarboxylic acid	290	431
trans-Cyclobutane-1,2-dicarboxylic acid	249	431
Cyclobutanone	8	161
Cyclobutene	9	12
Cyclobutylamine	22	295
Cyclobutyl methyl ketone	37	162
Cyclodecane	248	7
cis-Cyclodecene	329	24
trans-Cyclodecene	4	25
Cyclodecyne	58	30
Δ²-Cyclogeranic acid	128	198
α-Cyclogeraniolene	246	20
Cyclohept(fg)acenaphthene	123	48
1,3-Cycloheptadiene	211	19
Cycloheptane	68	4
Cycloheptanone	134	165
1,3,5-Cycloheptatriene	191	18
Cycloheptene	187	18
Cyclohexadecane	35	8
1,5,9,13-Cyclohexadecatetraene	5	25
1,3-Cyclohexadiene	78	14
Cyclohexane	29	3
Cyclohexencarboxaldehyde	30	151
Cyclohexancarboxanilide	272	238
Cyclohexanecarboxylic acid	10	193
Cyclohexanecarboxylic acid chloride	36	212
Cyclohexane-1,1-diacetic acid	201	430
cis-Cyclohexane-1,2-diacetic acid	372	432
trans-Cyclohexane-1,2-diacetic acid	364	432
cis-Cyclohexane-1,2-dicarboxylic acid	354	432
trans-Cyclohexane-1,2-dicarboxylic acid	319	431
cis-Cyclohexane-1,3-dicarboxylic acid	299	431
trans-Cyclohexane-1,3-dicarboxylic acid	348	432
trans-Cyclohexane-1,4-dicarboxylic acid	316	431
1,3-Cyclohexanedione	132	176
1,4-Cyclohexanedione	109	175
Cyclohexanethiol	23	414
Cyclohexane-1,2,3-tricarboxylic acid	264	431
Cyclohexanol	6	88
Cyclohexanone	79	163
Cyclohexene	81	15
Cyclohexene sulfide	45	424
2-Cyclohexenone	48	169
1-Cyclohexenylcarboxylic acid	21	193
Cyclohexylacetaldhyde	15	150
N-Cyclohexylacetamide	134	234
Cyclohexyl acetate	128	256
Cyclohexylacetic acid	46	191
Cyclohexylacetonitrile	71	350
Cyclohexylacetyl chloride	25	214
β-Cyclohexylacrylic acid	53	195
Cyclohexylamine	49	296
"	426	439
N-Cyclohexylaniline	224	437
N-Cyclohexylbenzamide	280	238
Cyclohexylbenzene	133	40
Cyclohexyl bromide	21	58

Compound	Compound No.	Page No.
Cyclohexyl n-butyrate	199	260
Cyclohexylbutyric acid	460	432
Cyclohexyl carbinol	84	84
Cyclohexyl chloride	35	56
4-Cyclohexylcyclohexanone	12	172
Cyclohexyl ethyl amine	69	297
Cyclohexyl ethyl ether	55	133
Cyclohexyl formate	111	255
Cyclohexyl iodide	17	60
Cyclohexyl isobutyrate	178	259
Cyclohexyl mercaptan	23	414
Cyclohexyl methoxymethyl ketone	9	169
Cyclohexyl methyl amine	58	297
" " "	453	439
Cyclohexyl methyl ether	50	133
Cyclohexyl methyl ketone	130	165
Cyclohexyloxyacetic acid	250	431
2-Cyclohexyloxypropionic acid	228	431
2-Cyclohexylphenol	56	99
4-Cyclohexylphenol	316	112
4-Cyclohexylphenyl methyl ether	36	138
3-Cyclohexylpropene	281	21
Cyclohexyl propionate	159	257
Cyclohexylpropionic acid	457	432
2-Cyclohexylpyrrolidine	436	439
2-Cyclohexyl-2-pyrroline	290	438
Cyclononane	235	7
cis-1,4,7-Cyclononatriene	9	26
trans-Cyclononene	6	25
Cyclononyne	55	30
1,3-Cyclooctadecadiene	7	25
1,3-Cyclooctadiene	8	25
cis-1,4-Cyclooctadiene	263	21
1,5-Cyclooctadiene	270	21
Cyclooctane	142	5
Cyclooctatetraene	252	20
1,3,5-Cyclooctatriene	264	21
1,3,6-Cyclooctatriene	9	25
Cyclooctene	258	20
Cyclooctyne	49	30
Cyclopentadecane	40	8
Cyclopentadiene	23	12
Cyclopentadienophenanthrene	165	50
Cyclopentane	17	3
Cyclopentanecarboxylic acid	29	192
" " "	463	433
cis-Cyclopentane-1,2-diacetic acid	373	432
trans-Cyclopentane-1,2-diacetic acid	375	432
Cyclopentane-1,1-dicarboxylic acid	182	430
trans-Cyclopentane-1,2-dicarboxylic acid	276	431
cis-1,3-Cyclopentane-1,3-dicarboxylic acid	162	199
" " " " "	337	431
trans-Cyclopentane-1,3-dicarboxylic acid	350	432
Cyclopentanol	35	81
1,2-Cyclopentanonaphthalene	171	42
Cyclopentanone	27	162
4-H-Cyclopenta(def)phenanthrene	118	47
Cyclopentene	28	13
2,3-Cyclopentenonaphthalene	97	47
1-Cyclopentenylcarboxylic acid	160	199
1-Cyclopentylacetone	35	169
Cyclopentylaldehyde	42	146
N-Cyclopentylaniline	209	437
Cyclopentyl bromide	18	58
3-Cyclopentyl-2-butanone	7	169
Cyclopentyl chloride	26	55
Cyclopentyl-1,1-diacetic acid	252	431
Cyclopentyl ethyl ether	45	133
1-Cyclopentylformaldehyde	47	146
Cyclopentyl iodide	16	60
α-Cyclopentyl-α-methylacetone	7	169
Cyclopentyl methyl ether	38	133
Cyclopentyl methyl ketone	76	163
Cyclopentyloxyacetic acid	237	431
5-Cyclopentylpentanoic acid	25	192
1-Cyclopentyl-2-propanone	35	169
d-w-Cyclopentyltridecanoic acid	69	195
α-Cyclopentylundecylamide	172	235
Cyclopropane	4	3
Cyclopropanecarbonyl chloride	19	212
Cyclopropanecarboxylic acid	19	190
" " "	444	432
Cyclopropane-1,1-dicarboxylic acid	29	429
cis-Cyclopropane-1,2-dicarboxylic acid	192	430
trans-Cyclopropane-1,2-dicarboxylic acid	232	431
Cyclopropene	3	12
Cyclopropylacetic acid	23	190
Cyclopropylamine	9	295
1,3-Cyclotetradecadiene	10	25
m-Cymene	17	35
o-Cymene	22	36
p-Cymene	21	36
p-Cymidine	158	300
L-Cysteic acid	96	288
L-Cysteine	130	290
L-Cystine	75	287

D

Compound	Compound No.	Page No.
DDT	2	64
cis-Decahydronaphthalene	246	7
trans-Decahydronaphthalene	244	7
Decahydronaphthyloxyacetic acid	227	431
cis-Decalin	246	7
trans-Decalin	244	7
cis-1-Decalone	93	170
Decamethylene glycol	44	89
Decanal	77	148
n-Decanamide	110	233
n-Decaneanilide	36	231
n-Decane	239	7
1,10-Decanedicarboxylic acid	172	200
Decanedioic acid	183	200
Decanedioic anhydride (dimer)	27	225
1,10-Decanediol	44	89
1,10-Decanedithiol	3	416
n-Decanoic acid	13	193
n-Decanoic anhydride	2	225
1-Decanol	121	86
d,l-2-Decanol	106	85
2-Decanone	166	166
4-Decanone	163	166
Decanonitrile	87	352
n-Decanoyl chloride	52	213
1-Decene	305	22
4-Decene	306	22
5-Decene	299	22
Decyl acetylene	60	30
n-Decyl alcohol	121	86
n-Decylbenzene	172	42
n-Decyl chloride	50	56
n-Decylcyclohexane	272	7
n-Decylcyclopentane	269	7
n-Decyl methyl ketone	2	172
1-Decylnaphthalene	186	42
Decyl sulfide	5	425
1-Decyne	52	30
3-Decyne	53	30
5-Decyne	54	30
Dehydroacetic acid monoanilide	178	235
Desoxybenzoin	73	174
6-Desoxy-D-galactose	33	331
6-Desoxy-DL-galactose	44	332

Compound	Compound No.	Page No.
6-Desoxy-L-galactose	34	331
2-Desoxy-D-glucose	36	331
6-Desoxy-D-glucose	29	330
6-Desoxy-L-mannose	9	329
6-Desoxy-D-mannose	13	328
2-Desoxy-D-ribose	5	329
1,1-Diacenaphthene	130	48
Diacetone alcohol	59	83
" "	98	164
β,β'-Diacetoxydiethyl ether	274	265
1,3-Diacetoxypropane	193	260
Diacetylacetone	48	173
1,4-Diacetylbenzene	147	177
N,N'-Diacetylbenzidine	509	246
Diacetylene	4	28
N,N'-Diacetylethylenediamine	349	241
N,N'-Diacetyl-m-phenylenediamine	405	242
N,N'-Diacetyl-o-phenylenediamine	389	242
N,N'-Diacetyl-p-phenylenediamine	507	246
4,6-Diacetylpyrogallol	486	122
4,6-Diacetylresorcinol	478	121
Diacetylsulfide	30	420
N,N-Diacetyltetramethylenediamine	250	237
1,5-Diacetyl-2,3,4-trihydroxybenzene	486	122
N,N'-Diacetyltrimethylenediamine	120	234
Diallylamine	318	438
Diallyl sulfide	23	420
2,8-Diaminoacridone	478	318
1,4-Diaminoanthraquinone	463	318
1,5-Diaminoanthraquinone	475	318
1,6-Diaminoanthraquinone	468	318
1,7-Diaminoanthraquinone	466	318
1,8-Diaminoanthraquinone	462	318
2,7-Diaminoanthraquinone	476	318
2,2'-Diaminoazobenzene	300	312
3,3'-Diaminoazobenzene	350	314
4,4'-Diaminoazoxybenzene	407	316
1,2-Diaminobenzene	196	309
"	176	437
1,3-Diaminobenzene	84	305
"	190	437
1,4-Diaminobenzene	318	313
"	253	438
2,4-Diaminobenzene-1,5-disulfonamide	261	402
2,4-Diaminobenzene-1,5-disulfonanilide	404	407
2,4-Diaminobenzene-1,5-disulfonic acid	124	375
2,4-Diaminobenzene-1,5-disulfonyl chloride	284	390
3,5-Diaminobenzoic acid	55	286
2,2'-Diaminobenzophenone	299	312
3,3'-Diaminobenzophenone	381	315
4,4'-Diaminobenzophenone	456	317
Di-(2-aminobenzyl)sulfide	50	426
Di-4-aminobenzyl sulfide	59	426
2,2'-Diaminobiphenyl	134	307
2,4'-Diaminobiphenyl	33	303
3,4-Diaminobiphenyl	199	309
1,3-Diaminobutane	54	296
1,4-Diaminobutane	66	297
" "	437	439
d,l-2,3-Diaminobutane	360	438
meso-2,3-Diaminobutane	359	438
2,7-Diaminocarbazole	461	318
cis-1,2-Diaminocyclohexane	353	438
trans-1,2-Diaminocyclohexane	357	438
1,2-Diaminocyclohexane	84	297
2,2'-Diaminodibenzyl	104	306
2,2'-Diaminodibenzyl disulfide	159	307
2,2'-Diaminodibenzyl sulfide	133	307
4,4'-Diaminodibenzyl sulfide	203	309
2,2'-Diaminodiethyl sulfide	145	299
1,4-Diamino-5,8-dihydroxy-9,10-anthraquinone	57	184
1,4-Diamino-5,8-dihydroxy-9,10-anthraquinone	471	318
4,4'-Diamino-3,3'-dimethoxybiphenyl	309	312
4,4'-Diamino-2,2'-dimethylazoxybenzene	334	313
α,α'-Diamino-1,4-dimethylbenzene	15	303
1,3-Diamino-2,6-dimethylbenzene	94	305
1,3-Diamino-4,6-dimethylbenzene	205	309
2,2'-Diamino-4,4'-dimethylbiphenyl	257	310
DL-2,2'-Diamino-6,6'-dimethylbiphenyl	306	312
2,2'-Diamino-6,6'-dimethylbiphenyl	349	314
4,4'-Diamino-2,2'-dimethylbiphenyl	221	309
4,4'-Diamino-2,3'-dimethylbiphenyl	11	302
4,4'-Diamino-3,3'-dimethylbiphenyl	290	312
4,4'-Diamino-2,2'-dimethyldiphenylmethane	268	311
4,4'-Diamino-3,3'-dimethyldiphenylmethane	354	314
6,6'-Diamino-3,3'-dimethyldiphenylmethane	182	308
4,4'-Diamino-3,3'-dimethyldiphenyl sulfide	181	308
6,6'-Diamino-3,3'-dimethyldiphenyl sulfide	200	309
6,6'-Diamino-3,3'-dimethyltriphenylmethane	399	316
1,1'-Diamino-2,2'-dinaphthyl	465	318
DL-2,2'-Diamino-1,1'-dinaphthyl	410	316
4,4'-Diamino-1,1'-dinaphthyl	423	317
2,4-Diaminodiphenylamine	291	312
2,2'-Diaminodiphenyl disulfide	167	308
4,4'-Diaminodiphenyl disulfide	144	307
4,4'-Diaminodiphenylmethane	169	308
2,2'-Diaminodiphenyl sulfide	147	307
2,4'-Diaminodiphenyl sulfide	32	425
4,4'-Diaminodiphenyl sulfone	386	315
1,2-Diaminoethane	352	438
1,9-Diaminofluorene	255	310
2,7-Diaminofluorenone	467	318
1,6-Diaminohexane	30	303
" "	439	439
1,2-Diamino-4-hydroxybenzene	374	315
1,4-Diamino-2-iodobenzene	225	309
4,6-Diaminoisophthaldehyde	431	317
1,3-Diaminoisopropyl alcohol	31	303
2,6-Diamino-4-methylphenol	360	114
Di-(2-amino-5-methylphenyl) sulfide	58	426
Di-(4-amino-3-methylphenyl) sulfide	55	426
4,6-Diamino-2-methylquinoline	414	316
1,2-Diaminonaphthalene	191	308
1,4-Diaminonaphthalene	254	310
1,8-Diaminonaphthalene	99	305
2,7-Diaminonaphthalene	371	315
1,2-Diamino-4-nitrobenzene	415	316
1,3-Diamino-4-nitrobenzene	361	314
1,4-Diamino-2-nitrobenzene	307	312
1,18-Diaminooctadecane	168	308
1,8-Diaminooctane	446	439
1,5-Diaminopentane	76	297
" "	378	439
2,4-Diaminophenol	120	102
" "	130	306
3,4-Diaminophenol	450	119
" "	374	315
3,5-Diaminophenol	442	119
Di-(2-aminophenyl) sulfide	52	426
Di-(4-aminophenyl) sulfide	60	426
1,2-Diaminopropane	361	438
D,L-1,2-Diaminopropane	42	296
1,3-Diaminopropane	50	296
" "	382	439
2,4-Diaminopyridine	214	309
2,5-Diaminopyridine	217	309
2,6-Diaminopyridine	258	311
3,4-Diaminopyridine	442	317
5,8-Diaminoquinizarin	57	184
"	471	318
cis-2,2'-Diaminostilbene	269	311
trans-2,2'-Diaminostilbene	387	315

Compound	Compound No.	Page No.
cis-4,4′-Diaminostilbene	259	311
trans-4,4′-Diaminostilbene	451	317
DL-2,3-Diaminosuccinic acid	126	290
meso-2,3-Diaminosuccinic acid	119	290
4,4′-Diamino-2,5,2′,5′-tetramethyltriphenylmethane	434	317
2,4-Diaminotoluene	194	308
2,5-Diaminotoluene	88	305
3,4-Diaminotoluene	157	307
4,4′-Diaminotriphenylcarbinol	375	315
3,4-Diaminotriphenylmethane	111	306
α,α′-Diamino-m-xylene	161	300
2,4-Diamino-m-xylene	94	305
4,6-Diamino-m-xylene	205	309
Di-n-amylamine	97	298
Di-n-amyl ether	72	134
Di-n-amyl ketone	183	167
2,5-Dianilino-1,4-benzoquinone	477	318
Dianisalacetone	160	177
Dianisidine	309	312
9,9′-Dianthranyl-10,10′-quinone	58	184
Dianthraquinone	58	184
Dibenzalacetone	145	177
1,2,3,4-Dibenzanthracene	185	51
1,2,5,6-Dibenzanthracene	200	51
1,2,7,8-Dibenzanthracene	183	50
Dibenzofuran	50	139
Dibenzofuran-2-carboxaldehyde	57	156
4-Dibenzofurylacetic acid	340	207
Dibenzoyl	119	176
1,2-Dibenzoylbenzene	178	178
1,3-Dibenzoylbenzene	129	176
1,4-Dibenzoylbenzene	184	178
N,N′-Dibenzoylbenzidine	511	247
1,2-Dibenzoylethane	174	178
cis-1,2-Dibenzoylethylene	163	177
trans-1,2-Dibenzoylethylene	142	176
N,N′-Dibenzoylethylenediamine	491	246
Dibenzoylmethane	111	175
1,3-Dibenzoylpropane	87	175
Dibenzoyl sulfide	20	425
N,N-Dibenzoyltetramethylenediamine	364	241
4-Dibenzothienylacetic acid	248	203
N,N′-Dibenzoyltrimethylenediamine	274	238
1,2,7,8-Dibenzphenanthrene	202	51
2,3,5,6-Dibenzphenanthrene	199	51
2,3,6,7-Dibenzphenanthrene	198	51
Dibenzylacetanilide	299	239
Dibenzylacetic acid	94	196
Dibenzylacetone	221	168
Dibenzyl acetonitrile	63	350
Dibenzylacetyl chloride	17	214
Dibenzylamine	184	301
N,N-Dibenzylaniline	40	323
Dibenzyl ether	109	136
Dibenzyl ketone	19	172
4,5-Dibenzylnaphthalene-1-sulfonamide	192	399
4,5-Dibenzylnaphthalene-1-sulfonyl chloride	239	389
1,2-Dibenzyloxybenzene	39	138
1,4-Dibenzyloxybenzene	55	139
Dibenzyl phthalate	48	275
Dibenzyl succinate	64	276
2,2-Dibenzylsuccinic acid	274	431
Dibenzyl sulfide	21	425
Dibenzyl d-tartarate	63	276
Dibiphenyleneethylene	182	50
Dibromoacetamide	308	239
3,5-Dibromoacetic acid	35	194
3,5-Dibromoacetophenone	88	175
2,4-Dibromoaniline	127	306
" "	47	436
2,6-Dibromoaniline	139	307
" "	48	136
1,2-Dibromobenzene	12	73
1,3-Dibromobenzene	10	73
1,4-Dibromobenzene	5	74
2,4-Dibromobenzenesulfonamide	271	402
2,5-Dibromobenzenesulfonamide	297	403
3,4-Dibromobenzenesulfonamide	212	399
2,4-Dibromobenzenesulfonic acid	130	375
2,5-Dibromobenzenesulfonic acid	141	376
3,4-Dibromobenzenesulfonic acid	90	374
2,4-Dibromobenzenesulfonyl chloride	93	383
2,5-Dibromobenzenesulfonyl chloride	81	383
3,4-Dibromobenzenesulfonyl chloride	16	381
2,4-Dibromobenzoic acid	270	204
2,4-Dibromobenzonitrile	119	362
2,5-Dibromobenzonitrile	176	365
2,6-Dibromobenzonitrile	198	366
3,5-Dibromobenzonitrile	130	363
2,4′-Dibromobenzophenone	77	174
3,3′-Dibromobenzophenone	171	178
4,4′-Dibromobenzophenone	191	178
2,6-Dibromo-1,4-benzoquinone	16	182
4,4′-Dibromobiphenyl	8	74
1,2-Dibromobutane	11	65
1,3-Dibromobutane	14	65
1,4-Dibromobutane	16	65
2,3-Dibromobutane	10	65
1,2-Dibromo-1-butene	6	65
1,3-Dibromo-2-butene	13	65
3,5-Dibromo-o-cresol	194	106
4,6-Dibromo-o-cresol	58	99
2,6-Dibromo-p-cresol	43	98
1,3-Dibromo-2,4-dihydroxynaphthalene	303	111
1,5-Dibromo-4,8-dihydroxynaphthalene	367	115
2,6-Dibromo-1,5-dihydroxynaphthalene	589	127
2,3-Dibromo-5,6-dimethylphenol	186	106
2,5-Dibromo-1,4-dinitrobenzene	53	339
1,1-Dibromoethane	2	65
1,2-Dibromoethane	3	65
2,2-Dibromoethanol	81	84
Di-(2-bromoethyl)sulfide	8	425
Dibromogallic acid	336	113
1,7-Dibromoheptane	1	66
2,6-Dibromohydroquinone	425	118
Dibromomalononitrile	169	364
Dibromomethane	1	65
2,6-Dibromo-4-methylphenol	43	98
3,5-Dibromo-2-methylphenol	194	106
4,6-Dibromo-2-methylphenol	58	99
1,2-Dibromo-2-methylpropane	5	65
1,1-Dibromo-2-methylpropene	9	65
1,2-Dibromonaphthalene	3	74
1,4-Dibromonaphthalene	4	74
1,3-Dibromo-2,4-naphthalenediol	303	111
1,3-Dibromo-2-naphthol	107	102
1,6-Dibromo-2-naphthol	212	107
2,4-Dibromo-1-naphthol	208	107
4,6-Dibromo-2-naphthol	320	112
2,5-Dibromo-1-nitrobenzene	56	339
2,4-Dibromo-6-nitrophenol	259	109
2,6-Dibromo-4-nitrophenol	348	114
1,9-Dibromononane	21	65
1,8-Dibromooctane	20	65
1,5-Dibromopentane	18	65
2,4-Dibromophenol	18	97
2,5-Dibromophenol	104	102
2,6-Dibromophenol	52	99
3,5-Dibromophenol	127	103
Di-(4-bromophenyl)sulfide	63	426
DL-1,2-Dibromopropane	4	65
1,3-Dibromopropane	12	65

	Compound No.	Page No.
2,3-Dibromo-1-propanol	112	85
1,3-Dibromopropene	8	65
α,β-Dibromopropionamide	222	237
2,3-Dibromopropionic acid	63	195
3,5-Dibromopyridine	70	324
d,l-2,3-Dibromosuccinic acid	15	429
meso-2,3-Dibromosuccinic acid	20	429
2,5-Dibromotoluene	14	73
3,4-Dibromotoluene	15	73
2,6-Dibromo-3,4,5-trihydroxybenzoic acid	336	113
Di-(β-n-butoxyethyl)carbonate	344	270
Dibutyl acetylene	54	30
Di-tert-butyl acetylene	40	29
Di-n-butylamine	65	297
" "	464	439
N,N-Di-n-butylaniline	99	321
" " "	256	438
3,5-Di-tert-butylaniline	188	437
1,4-Di-sec-butylbenzene	139	40
Di-n-butyl carbonate	184	259
Di-n-butyl ether	54	133
Di-sec-butyl ether	44	133
Di-n-butyl ketone	140	165
Di-sec-butyl ketone	93	164
Di-tert-butyl ketone	74	163
Di-n-butyl oxalate	266	264
Di-n-butyl phthalate	355	271
Di-n-butyl racemate	353	271
Di-n-butyl sebacate	356	271
Di-n-butyl succinate	320	268
Di-n-butyl sulfide	41	421
Di-tert-butyl sulfide	29	420
Di-n-butyl d-tartarate	9	273
Di-n-butyl d,l-tartarate	353	271
3,4-Dicarboxybenzenesulfonamide	287	403
3,4-Dicarboxybenzenesulfonic acid	136	376
3,4-Dicarboxybenzenesulfonyl chloride	259	389
Dichloroacetaldehyde	15	144
Dichloroacetamide	106	233
Dichloroacetanilide	184	235
2,4-Dichloroacetanilide	268	238
2,5-Dichloroacetanilide	231	237
Dichloroacetic acid	25	190
" "	11	429
Dichloroacetic anhydride	12	223
1,1-Dichloroacetone	19	161
1,3-Dichloroacetone	41	173
Dichloroacetonitrile	12	345
2,4-Dichloroacetophenone	33	173
2,5-Dichloroacetophenone	209	168
3,5-Dichloroacetophenone	4	172
Dichloroacetyl chloride	11	212
Dichloroacetyl iodide	1	222
2,4-Dichloroaniline	85	305
" "	37	436
2,5-Dichloroaniline	49	304
" "	32	436
9,10-Dichloroanthracene	34	72
meso-Dichloroanthracene	34	72
9,10-Dichloroanthracene-2-sulfonamide	457	409
9,10-Dichloroanthracene-2-sulfonanilide	428	408
9,10-Dichloroanthracene-2-sulfonic acid	212	378
9,10-Dichloroanthracene-2-sulfonyl chloride	275	390
2,3-Dichlorobenzaldehyde	55	155
2,4-Dichlorobenzaldehyde	63	156
2,5-Dichlorobenzaldehyde	45	155
2,6-Dichlorobenzaldehyde	58	156
3,4-Dichlorobenzaldehyde	19	154
3,5-Dichlorobenzaldehyde	54	155
2,5-Dichlorobenzamide	297	239
1,2-Dichlorobenzene	7	70
1,3-Dichlorobenzene	5	70
1,4-Dichlorobenzene	12	71
2,5-Dichlorobenzene-1,3-disulfonamide	348	405
4,6-Dichlorobenzene-1,3-disulfonamide	455	409
2,5-Dichlorobenzene-1,3-disulfonic acid	161	376
4,6-Dichlorobenzene-1,3-disulfonic acid	210	378
2,5-Dichlorobenzene-1,3-disulfonyl chloride	176	386
4,6-Dichlorobenzene-1,3-disulfonyl chloride	194	387
2,4-Dichlorobenzenesulfonamide	236	401
2,5-Dichlorobenzenesulfonamide	233	400
3,4-Dichlorobenzenesulfonamide	99	395
2,5-Dichlorobenzenesulfonanilide	165	398
2,4-Dichlorobenzenesulfonic acid	104	374
2,5-Dichlorobenzenesulfonic acid	100	374
3,4-Dichlorobenzenesulfonic acid	34	372
2,4-Dichlorobenzensulfonyl chloride	48	382
2,5-Dichlorobenzenesulfonyl chloride	22	381
3,4-Dichlorobenzenesulfonyl chloride	9	381
2,4-Dichlorobenzoic acid	117	430
2,5-Dichlorobenzoic acid	230	202
3,4-Dichlorobenzoic acid	322	206
2,2'-Dichlorobenzoin	63	174
2,4-Dichlorobenzonitrile	73	360
2,5-Dichlorobenzonitrile	175	365
3,4-Dichlorobenzonitrile	92	361
3,5-Dichlorobenzonitrile	75	360
4,4'-Dichlorobenzophenone	177	178
3,4-Dichlorobenzotrichloride	1	64
2,6-Dichlorobenzyl chloride	4	57
Di-(4-chlorobenzyl)sulfide	14	425
2,2'-Dichlorobiphenyl	15	71
4,4'-Dichlorobiphenyl	30	71
1,2-Dichlorobutane	15	63
2,3-Dichloro-n-butyraldehyde	4	150
2,4-Dichloro-m-cresol	7	96
2,6-Dichloro-m-cresol	61	99
2,6-Dichloro-p-cresol	23	97
4,6-Dichloro-m-cresol	99	101
2,5-Dichloro-3,6-dihydroxy-1,4-benzoquinone	51	184
1,5-Dichloro-4,8-dihydroxynaphthalene	495	122
2,3-Dichloro-1,4-dihydroxynaphthalene	321	112
2,5-Dichloro-1,4-dimethylbenzene	21	71
3,6-Dichloro-2,5-dimethylbenzenesulfonamide	179	398
4,6-Dichloro-2,5-dimethylbenzenesulfonamide	130	396
3,6-Dichloro-2,5-dimethylbenzenesulfonanilide	202	399
4,6-Dichloro-2,5-dimethylbenzenesulfonanilide	213	400
3,6-Dichloro-2,5-dimethylbenzenesulfonic acid	71	373
4,6-Dichloro-2,5-dimethylbenzenesulfonic acid	51	372
3,6-Dichloro-2,5-dimethylbenzenesulfonyl chloride	82	383
4,6-Dichloro-2,5-dimethylbenzenesulfonyl chloride	98	384
4,4'-Dichlorodiphenylmethane	13	71
1,1-Dichloroethane	3	63
1,2-Dichloroethane	9	63
cis-1,2-Dichloroethylene	4	63
trans-1,2-Dichloroethylene	2	63
2,2'-Dichloroethyl ether	42	133
β,β'-Dichloroethyl ether	69	134
Di-2-chloroethyl sulfide	52	421
2,6-Dichlorohydroquinone	27	434
2,4-Dichloro-3-hydroxybenzaldehyde	114	159
2,6-Dichloro-3-hydroxybenzaldehyde	113	159
3,5-Dichloro-2-hydroxybenzaldehyde	81	157
3,5-Dichloro-4-hydroxybenzaldehyde	119	159
4,6-Dichloro-3-hydroxybenzaldehyde	109	159
3,5-Dichloro-2-hydroxybenzonitrile	184	365
3,5-Dichloro-4-hydroxybenzonitrile	192	365
Dichloromethane	1	63
2,3-Dichloro-4-methylbenzenesulfonamide	405	407
2,6-Dichloro-3-methylbenzenesulfonamide	265	402

	Compound No.	Page No.
2,6-Dichloro-4-methylbenzenesulfonamide	141	397
3,4-Dichloro-2-methylbenzenesulfonamide	383	407
3,5-Dichloro-2-methylbenzenesulfonamide	253	401
3,5-Dichloro-4-methylbenzenesulfonamide	277	402
4,5-Dichloro-3-methylbenzenesulfonamide	243	401
4,6-Dichloro-2-methylbenzenesulfonamide	190	399
2,3-Dichloro-4-methylbenzenesulfonic acid	185	377
2,3-Dichloro-6-methylbenzenesulfonic acid	115	375
2,6-Dichloro-3-methylbenzenesulfonic acid	125	375
2,6-Dichloro-4-methylbenzenesulfonic acid	56	372
3,4-Dichloro-2-methylbenzenesulfonic acid	174	377
3,5-Dichloro-2-methylbenzenesulfonic acid	118	375
3,5-Dichloro-4-methylbenzenesulfonic acid	133	375
4,6-Dichloro-2-methylbenzenesulfonic acid	79	373
2,3-Dichloro-4-methylbenzenesulfonyl chloride	26	381
2,6-Dichloro-3-methylbenzenesulfonyl chloride	7	381
2,6-Dichloro-4-methylbenzenesulfonyl chloride	52	382
3,4-Dichloro-2-methylbenzenesulfonyl chloride	39	382
3,5-Dichloro-2-methylbenzenesulfonyl chloride	46	382
3,5-Dichloro-4-methylbenzenesulfonyl chloride	77	383
4,5-Dichloro-3-methylbenzenesulfonyl chloride	107	384
4,6-Dichloro-2-methylbenzenesulfonyl chloride	27	381
2',4-Dichloro-3-methylphenol	7	96
2,6-Dichloro-3-methylphenol	61	99
2,6-Dichloro-4-methylphenol	23	97
4,6-Dichloro-3-methylphenol	99	101
(2,4-Dichloro-6-methylphenoxy)acetic acid	162	430
1,3-Dichloro-2-methylpropane	17	63
Dichloromethyl sulfide	31	420
Dichloromethyl sulfide	43	421
1,2-Dichloronaphthalene	4	71
1,3-Dichloronaphthalene	16	71
1,4-Dichloronaphthalene	20	71
1,5-Dichloronaphthalene	26	71
1,6-Dichloronaphthalene	9	71
1,7-Dichloronaphthalene	18	71
1,8-Dichloronaphthalene	25	71
2,6-Dichloronaphthalene	28	71
2,7-Dichloronaphthalene	27	71
1,5-Dichloronaphthalene-2-sulfonamide	458	409
3,6-Dichloronaphthalene-2-sulfonamide	356	405
3,7-Dichloronaphthalene-1-sulfonamide	448	409
4,5-Dichloronaphthalene-1-sulfonamide	389	407
4,5-Dichloronaphthalene-2-sulfonamide	306	403
4,6-Dichloronaphthalene-1-sulfonamide	375	406
4,6-Dichloronaphthalene-2-sulfonamide	354	405
4,7-Dichloronaphthalene-1-sulfonamide	352	405
4,7-Dichloronaphthalene-2-sulfonamide	300	403
4,8-Dichloronaphthalene-2-sulfonamide	322	404
5,6-Dichloronaphthalene-1-sulfonamide	368	406
5,6-Dichloronaphthalene-2-sulfonamide	282	402
5,7-Dichloronaphthalene-1-sulfonamide	452	409
5,8-Dichloronaphthalene-2-sulfonamide	422	408
6,7-Dichloronaphthalene-1-sulfonamide	447	409
6,8-Dichloronaphthalene-2-sulfonamide	387	407
7,8-Dichloronaphthalene-1-sulfonamide	364	406
7,8-Dichloronaphthalene-2-sulfonamide	380	406
1,5-Dichloronaphthalene-1-sulfonyl chloride	201	387
3,6-Dichloronaphthalene-1-sulfonyl chloride	257	389
3,7-Dichloronaphthalene-1-sulfonyl chloride	214	388
4,5-Dichloronaphthalene-1-sulfonyl chloride	180	387
4,5-Dichloronaphthalene-2-sulfonyl chloride	246	389
4,6-Dichloronaphthalene-1-sulfonyl chloride	183	387
4,6-Dichloronaphthalene-1-sulfonyl chloride	208	388
4,7-Dichloronaphthalene-1-sulfonyl chloride	240	389
4,7-Dichloronaphthalene-2-sulfonyl chloride	244	389
4,8-Dichloronaphthalene-2-sulfonyl chloride	229	388
5,6-Dichloronaphthalene-1-sulfonyl chloride	159	386
5,6-Dichloronaphthalene-2-sulfonyl chloride	258	389
5,7-Dichloronaphthalene-1-sulfonyl chloride	238	388
5,8-Dichloronaphthalene-2-sulfonyl chloride	213	388

	Compound No.	Page No.
6,7-Dichloronaphthalene-1-sulfonyl chloride	232	388
6,8-Dichloronaphthalene-2-sulfonyl chloride	190	387
7,8-Dichloronaphthalene-1-sulfonyl chloride	219	388
7,8-Dichloronaphthalene-2-sulfonyl chloride	196	387
2,4-Dichloro-1-naphthol	219	107
1,4-Dichloro-2-naphthol	288	111
2,4-Dichloro-6-nitroaniline	4	436
2,6-Dichloro-4-nitroaniline	5	436
2,4-Dichloro-1-nitrobenzene	6	337
2,5-Dichloro-1-nitrobenzene	24	337
2,4-Dichloro-5-nitrophenol	209	107
2,4-Dichloro-6-nitrophenol	283	111
2,3-Dichlorophenol	57	99
2,4-Dichlorophenol	33	98
" "	32	434
2,5-Dichlorophenol	62	99
2,6-Dichlorophenol	81	100
3,5-Dichlorophenol	85	101
(2,4-Dichlorophenoxy)acetic acid	224	431
2,5-Dichlorophenyl methyl sulfide	22	425
Di-(4-chlorophenyl)sulfide	56	426
3,5-Dichlorophthalic anhydride	37	225
3,6-Dichlorophthalic anhydride	75	226
4,5-Dichlorophthalic anhydride	72	226
1,2-Dichloropropane	11	63
1,3-Dichloropropane	16	63
2,2-Dichloropropane	6	63
1,3-Dichloro-2-propanol	72	83
2,3-Dichloropropanol	85	84
1,1-Dichloro-2-propanone	19	161
1,3-Dichloro-2-propanone	41	173
d,l-2,3-Dichloropropionaldehyde	3	152
3,5-Dichlorosalicylaldehyde	81	157
d,l-2,3-Dichlorosuccinic acid	18	429
meso-2,3-Dichlorosuccinic acid	21	429
2,3-Dichlorotoluene	23	70
2,4-Dichlorotoluene	20	70
2,5-Dichlorotoluene	19	70
2,6-Dichlorotoluene	18	70
3,4-Dichlorotoluene	24	70
3,5-Dichlorotoluene	21	70
α,α-Dichloro-m-tolunitrile	99	353
α,α-Dichloro-o-tolunitrile	94	353
α,α-Dichloro-p-tolunitrile	100	353
2,6-Dichloro-3,4,5-tribromophenol	524	123
2,5-Dichloro-p-xylene	21	71
Dicinnamalacetone	173	178
Di-m-cresyl carbonate	61	276
Di-o-cresyl carbonate	85	277
Di-p-cresyl carbonate	145	280
Di-m-cresyl oxalate	138	280
Di-o-cresyl oxalate	128	279
Di-p-cresyl oxalate	164	281
Di-p-cresyl succinate	153	281
Dicrotyl sulfide	40	421
m-Dicyanobenzene	204	366
o-Dicyanobenzene	188	365
p-Dicyanobenzene	229	367
4,4'-Dicyanobiphenyl	231	367
1,1-Dicyanobutane	69	350
1,4-Dicyanobutane	104	354
4,4-Dicyano-1-butene	72	350
cis-1,4-Dicyanocyclohexane	76	360
trans-1,4-Dicyanocyclohexane	185	365
Dicyanodimethylamine	94	361
4,4'-Dicyanodiphenylmethane	211	366
1,12-Dicyanododecane	17	356
1,1-Dicyanoethane	11	357
cis-1,2-Dicyanoethylene	19	357
1,11-Dicyanohendecane	14	355
1,7-Dicyanoheptane	12	355

	Compound No.	Page No.
4,4-Dicyanoheptane	23	358
" "	51	359
1,6-Dicyanohexane	20	356
1,1-Dicyano-3-methylbutane	75	351
Dicyanomethyl ethyl amine	9	436
1,3-Dicyano-2-methylpropane	97	353
1,2-Dicyanonaphthalene	220	367
1,3-Dicyanonaphthalene	216	367
1,4-Dicyanonaphthalene	224	367
1,5-Dicyanonaphthalene	226	367
1,6-Dicyanonaphthalene	225	367
1,7-Dicyanonaphthalene	210	366
1,8-Dicyanonaphthalene	230	367
2,3-Dicyanonaphthalene	234	368
2,6-Dicyanonaphthalene	236	368
2,7-Dicyanonaphthalene	235	368
1,8-Dicyanonoctane	19	356
3,3-Dicyanopentane	39	358
1,1-Dicyanopropane	67	350
1,2-Dicyanopropane	89	352
1,3-Dicyanopropane	102	353
2,2-Dicyanopropane	20	357
2,3-Dicyanopyridine	215	367
2,6-Dicyanopyridine	167	364
2,3-Dicyanothiophene	29	347
1,11-Dicyanoundecane	14	355
N,N-Dicyclohexylacetamide	130	234
Dicyclohexylamine	171	300
Dicyclohexyl oxalate	55	275
Dicyclohexyl phthalate	90	277
Dicyclopentyl ketone	43	169
Di-n-decyl ketone	84	174
Didecyl sulfide	5	425
Di-dichloromethyl sulfide	43	421
Di-(2,4-dinitrophenyl)sulfide	76	427
Di-(2-diphenoxyethyl)sulfide	26	420
Di-n-dodecylamine	448	439
Didodecyl ether	13	137
Didodecyl sulfide	12	425
Diethanolamine	10	88
"	180	301
sym-Diethoxyacetone	5	169
3,4-Diethoxybenzaldehyde	92	149
1,2-Diethoxybenzene	24	138
1,4-Diethoxybenzene	43	139
Di-(β-ethoxyethyl)carbonate	279	265
Di-(β-ethoxyethyl)phthalate	32	274
1,3-Diethoxy-2-propanone	5	169
Diethylacetaldehyde	29	145
Diethylacetic acid	27	191
Diethyl acetonedicarboxylate	287	266
Diethylacetonitrile	33	347
Diethylacetyl chloride	27	212
Diethylacetylene	22	28
Diethyl adipate	270	265
Diethylamine	11	295
"	454	439
4-(Diethylamino)benzaldehyde	17	153
"	16	323
β-Diethylaminoethyl alcohol	29	319
2-Diethylaminoethyl alcohol	29	319
2-(Diethylamino)ethylamine	363	439
1-Diethylaminoisopropyl alcohol	37	319
1-(N,N-Diethylamino)naphthalene	182	301
2-(N,N-Diethylamino)naphthalene	189	301
3-(Diethylamino)phenol	116	102
3-Diethylaminopropyl alcohol	49	320
N,N-Diethylaniline	67	320
"		268	438
Diethyl azelate	339	270
N,2-Diethylaziridine	293	438
1,2-Diethylbenzene	28	36
1,3-Diethylbenzene	23	36
1,4-Diethylbenzene	29	36
N,N-Diethylbenzylamine	326	438
Diethyl benzyl malonate	345	270
N,N-Diethylbenzylsulfonamide	2	392
Diethyl bromomalonate	252	263
Diethyl butylmalonate	254	264
Diethyl d-camphorate	334	269
Diethyl carbinol	13	80
Diethylcarbinyl acetate	64	252
Diethyl carbonate	55	252
Diethyl citraconate	241	263
Diethylcyanoacetic acid	80	360
Diethylene glycol	126	86
Diethylene glycol diacetate	274	265
Diethylene glycol diethyl ether	73	134
Diethylene glycol dimethyl ether	58	133
Diethylene glycol mono-n-butyl ether	118	86
Diethylene glycol mono-n-butyl ether acetate	280	265
Diethylene glycol monoethyl ether	100	85
Diethylene glycol monoethyl ether acetate	211	261
Diethylene glycol monomethyl ether	96	85
" " " "	77	135
Diethyl ether	4	131
sym-Diethylethylenediamine	61	297
unsym-Diethylethylenediamine	56	296
N,N-Diethylformamide	2	230
Diethyl fumarate	212	261
Diethyl glutarate	247	263
3,3-Diethylglutaric acid	222	431
3,3-Diethylhexane	217	6
3,4-Diethylhexane	190	6
Diethyl (2-hydroxyethyl) amine	350	438
Diethyl isophthalate	333	269
Diethyl itaconate	235	262
Diethyl ketone	9	161
Diethyl l-malate	291	266
Diethyl maleate	219	262
Diethyl malonate	173	258
Diethylmalonic acid	167	199
"	63	429
Diethylmalonic acid diamide	471	245
Diethylmalonic acid monoamide	271	238
Diethylmalononitrile	39	358
Diethyl malonyl dichloride	42	213
Diethyl mesaconate	236	263
Diethyl meso-tartrate	75	276
Diethyl methylacetic acid	32	191
Diethyl methyl amine	395	439
N,N-Diethyl-2-methylaniline	60	320
N,N-Diethyl-4-methylaniline	69	320
1,2-Diethyl-3-methylbenzene	85	38
1,2-Diethyl-4-methylbenzene	74	38
2,4-Diethyl-1-methylbenzene	80	38
2,5-Diethyl-1-methylbenzene	87	38
2,6-Diethyl-1-methylbenzene	90	38
3,5-Diethyl-1-methylbenzene	66	38
Diethyl methylmalonate	167	258
Diethyl mucate	166	281
Diethyl naphthalate	83	277
3,7-Diethylnaphthalene-1-sulfonamide	333	405
3,7-Diethylnaphthalene-1-sulfonyl chloride	158	386
Diethyl 3-nitrophthalate	53	275
Diethyl 4-nitrophthalate	36	274
Diethyl oxalate	145	257
3,3-Diethylpentane	133	5
2,4-Diethylphenol	21	97
β,β-Diethylphenylhydrazine	5	322
Diethyl phthalate	336	270
Diethyl pimelate	296	266

Compound	Compound No.	Page No.
1,3-Diethyl-2-propylbenzene	89	38
2,4-Diethylpyridine	48	320
2,6-Diethylpyridine	2	322
3,4-Diethylpyridine	63	320
Diethyl sebacate	349	271
Diethyl suberate	330	269
Diethyl succinate	210	261
2,2-Diethylsuccinic acid	256	431
d,l-2,3-Diethylsuccinic acid	206	430
meso-2,3-Diethylsuccinic acid	223	431
Diethyl sulfate	188	259
Diethyl sulfide	6	420
Diethyl d-tartarate	4	273
Diethyl-d,l-tartronate	218	261
Diethyl terephthalate	49	275
2,5-Diethylthiophene	47	424
3,4-Diethylthiophene	50	424
2,4-Diethyltoluene	80	38
2,5-Diethyltoluene	87	38
2,6-Diethyltoluene	90	38
3,5-Diethyltoluene	66	38
N,N-Diethyltoluene-4-sulfonamide	9	392
N,N-Diethyl-o-toluidine	60	320
N,N-Diethyl-p-toluidine	69	320
Di-2-fluorenylmethane	186	51
Difluoroacetamide	8	231
Difluoroacetic acid	4	190
" "	9	429
3,3-Difluoroacrylic acid	169	430
1,2-Difluorobenzene	4	69
1,3-Difluorobenzene	1	69
1,4-Difluorobenzene	3	69
1,3-Difluoropropane	2	62
Difurfuralacetone	72	174
Diglycolic acid	223	202
Diguaiacol carbonate	126	279
Di-n-hendecyl ketone	91	175
Di-n-heptadecyl ketone	117	176
Di-n-heptylamine	1	303
Di-n-heptyl ether	103	136
Di-n-heptyl ketone	36	173
Di-n-heptyl sulfide	62	422
Dihexadecyl sulfide	33	425
Di-n-hexylamine	450	439
Di-n-hexyl ether	95	135
Di-n-hexyl ketone	114	177
Dihydroacetic acid monoanilide	178	235
Dihydropyran	27	132
cis-1,2-Dihydroxyacenaphthene	527	124
5,6-Dihydroxyacenaphthene	497	122
1,3-Dihydroxyacetone	42	89
" "	94	175
" "	1	329
2,3-Dihydroxyacetophenone	188	106
" "	123	176
2,4-Dihydroxyacetophenone	362	115
" "	175	178
2,5-Dihydroxyacetophenone	512	123
" "	197	179
3,4-Dihydroxyacetophenone	266	110
" "	148	177
3,5-Dihydroxyacetophenone	368	115
" "	176	178
2,3-Dihydroxyacridine	564	126
2,8-Dihydroxyacridine	588	127
3,7-Dihydroxyacridine	590	127
3,9-Dihydroxyacridine	487	122
1,3-Dihydroxyacridone	592	128
3,5-Dihydroxyaniline	361	114
" "	328	313
1,2-Dihydroxyanthracene	408	117
1,5-Dihydroxyanthracene	578	127
1,8-Dihydroxyanthracene	556	126
2,3-Dihydroxyanthracene	582	127
2,6-Dihydroxyanthracene	586	127
2,7-Dihydroxyanthracene	581	127
9,10-Dihydroxyanthracene	468	120
1,2-Dihydroxy-9,10-anthraquinone	54	184
1,4-Dihydroxy-9,10-anthraquinone	33	183
1,5-Dihydroxy-9,10-anthraquinone	50	184
1,7-Dihydroxyanthrone	567	126
2,2'-Dihydroxyazobenzene	453	120
2,4-Dihydroxyazobenzene	449	119
3,3'-Dihydroxyazobenzene	521	123
3,4-Dihydroxyazobenzene	433	118
4,4'-Dihydroxyazobenzene	537	124
2,2'-Dihydroxyazoxybenzene	391	116
3,3'-Dihydroxyazoxybenzene	474	121
4,4'-Dihydroxyazoxybenzene	552	125
2,3-Dihydroxybenzaldehyde	94	158
2,4-Dihydroxybenzaldehyde	110	159
3,4-Dihydroxybenzaldehyde	117	159
3,5-Dihydroxybenzaldehyde	118	159
2,4-Dihydroxybenzanilide	208	236
1,2-Dihydroxybenzene	207	107
" "	75	434
1,3-Dihydroxybenzene	226	108
" "	74	434
1,4-Dihydroxybenzene	452	119
" "	117	435
2,4-Dihydroxybenzoic acid	530	124
" "	338	207
" "	179	430
3,4-Dihydroxybenzoic acid	319	206
2,4-Dihydroxybenzophenone	352	114
2,4'-Dihydroxybenzophenone	378	115
3,3'-Dihydroxybenzophenone	446	119
3,4-Dihydroxybenzophenone	355	114
4,4'-Dihydroxybenzophenone	526	124
1,1'-Dihydroxy-2,2'-binaphthyl	546	125
2,2'-Dihydroxy-1,1'-binaphthyl	543	125
4,4'-Dihydroxy-1,1'-binaphthyl	587	127
2,2'-Dihydroxybiphenyl	227	108
2,4'-Dihydroxybiphenyl	418	118
3,3'-Dihydroxybiphenyl	290	111
3,4-Dihydroxybiphenyl	357	114
4,4'-Dihydroxybiphenyl	579	127
2,4-Dihydroxy-n-caprylbenzene	62	174
2,2'-Dihydroxychalcone	404	117
2,4'-Dihydroxychalcone	359	114
β,β'-Dihydroxydiethylamine	10	88
β,β'-Dihydroxydiethyl ether	126	86
2,2'-Dihydroxy-3,3'-dimethylbiphenyl	237	108
2,2'-Dihydroxy-4,4'-dimethylbiphenyl	271	110
2,2'-Dihydroxy-5,5'-dimethylbiphenyl	386	116
2,2'-Dihydroxy-6,6'-dimethylbiphenyl	423	118
4,4'-Dihydroxy-2,2'-dimethylbiphenyl	240	108
4,4'-Dihydroxy-3,3'-dimethylbiphenyl	406	117
2,2'-Dihydroxy-1,1'-dinaphthylmethane	504	123
1,2-Dihydroxy-3,5-dinitrobenzene	422	118
2,4'-Dihydroxydiphenylmethane	256	109
3,3'-Dihydroxydiphenylmethane	200	106
4,4'-Dihydroxydiphenylmethane	398	116
erythro-2,3-Dihydroxy-1,4-dithiobutane	26	417
threa-2,3-Dihydroxy-1,4-dithiobutane	10	416
2,5-Dihydroxy-3-n-dodecyl-1,4-benzoquinone	18	182
Di-(2-hydroxyethyl)amine	180	301
2,2'-Dihydroxyhydrazobenzene	332	313
1,4-Dihydroxyisoquinoline	573	126
1,3-Dihydroxy-4-methyl-9,10-anthraquinone	45	184
1,8-Dihydroxy-2-methyl-9,10-anthraquinone	23	182
1,3-Dihydroxy-2-methylbenzene	265	110

ORGANIC COMPOUND INDEX

Compound	No.	Page No.
1,3-Dihydroxy-2-methylnaphthalene	339	113
1,4-Dihydroxy-2-methylnaphthalene	447	119
3,5-Dihydroxy-2-methyl-1,4-naphthoquinone	25	182
2,7-Dihydroxy-4-methylquinoline	585	127
1,2-Dihydroxynaphthalene	64	99
1,3-Dihydroxynaphthalene	291	111
1,4-Dihydroxynaphthalene	461	120
" "	71	434
1,5-Dihydroxynaphthalene	575	127
1,6-Dihydroxynaphthalene	330	113
1,7-Dihydroxynaphthalene	462	120
1,8-Dihydroxynaphthalene	343	114
2,3-Dihydroxynaphthalene	407	117
2,6-Dihydroxynaphthalene	535	124
2,7-Dihydroxynaphthalene	484	121
2,4-Dihydroxy-5-nitroaniline	360	314
2,5-Dihydroxy-4-nitroaniline	346	314
1,2-Dihydroxy-3-nitrobenzene	148	104
1,3-Dihydroxy-2-nitrobenzene	143	103
1,3-Dihydroxy-4-nitrobenzene	280	110
1,2-Dihydroxyphenanthrene	463	120
1,9-Dihydroxyphenanthrene	477	121
2,5-Dihydroxyphenanthrene	467	120
2,6-Dihydroxyphenanthrene	561	126
2,7-Dihydroxyphenanthrene	576	127
2,8-Dihydroxyphenanthrene	513	123
3,4-Dihydroxyphenanthrene	345	114
3,6-Dihydroxyphenanthrene	548	125
3,8-Dihydroxyphenanthrene	570	126
9,10-Dihydroxyphenanthrene	371	115
(3,4-Dihydroxyphenyl)alanine	75	429
1,5-Di-(2-hydroxyphenyl)-1,4-pentadiene-3-one	441	119
Di-(2-hydroxyphenyl)sulfide	69	426
Di-(3-hydroxyphenyl)sulfide	67	426
Di-(4-hydroxyphenyl)sulfide	72	427
1,3-Dihydroxy-2-propanone	42	89
2,3-Dihydroxypropylamine	179	300
1,2-Dihydroxy-3-propylbenzene	97	101
1,2-Dihydroxy-4-propylbenzene	63	99
1,3-Dihydroxy-5-propylbenzene	135	103
1,4-Dihydroxy-2-propylbenzene	159	104
1,6-Dihydroxypyrene	568	126
3,5-Dihydroxypyrene	547	125
3,8-Dihydroxypyrene	591	127
2,5-Dihydroxypyridine	572	126
2,3-Dihydroxyquinoline	574	126
2,2′-Dihydroxystilbene	172	105
3,5-Dihydroxystilbene	392	116
4,4′-Dihydroxystilbene	583	127
α,4-Dihydroxytoluene	294	111
2,3-Dihydroxytoluene	84	101
2,5-Dihydroxytoluene	293	111
2,6-Dihydroxytoluene	265	110
3,4-Dihydroxytoluene	76	100
3,5-Dihydroxytoluene	217	107
1,3-Dihydroxy-2,4,6-tribromobenzene	234	108
4,4′-Dihydroxytriphenylmethane	411	117
2,6-Dihydroxy-p-xylene	419	118
3,6-Dihydroxy-o-xylene	549	125
4,5-Dihydroxy-m-xylene	102	101
4,6-Dihydroxy-m-xylene	295	111
2,4-Di-iodoaniline	184	308
1,3-Di-iodobenzene	3	76
1,4-Di-iodobenzene	6	76
2,4-Di-iodobenzenesulfonamide	391	407
3,4-Di-iodobenzenesulfonamide	379	406
2,4-Di-iodobenzenesulfonic acid	177	377
3,4-Di-iodobenzenesulfonic acid	171	377
2,4-Di-iodobenzenesulfonyl chloride	91	383
3,4-Di-iodobenzenesulfonyl chloride	100	384
1,2-Di-iodoethane	1	68

Compound	No.	Page No.
Di-iodomethane	1	67
2,4-Di-iodophenol	100	101
2,5-Di-iodophenol	191	106
2,6-Di-iodophenol	86	101
3,4-Di-iodophenol	133	103
3,5-Di-iodophenol	202	106
3,4-Di-iodophthalic anhydride	76	226
1,3-Di-iodopropane	2	67
DL-3,5-Di-iodotyrosine	52	429
L-3,5-Di-iodotyrosine	44	286
Di-isoamylamine	86	297
"	440	439
Di-isoamyl carbonate	246	263
Di-isoamyl ether	64	134
Di-isoamyl oxalate	312	268
Di-isoamyl phthalate	357	271
Di-isobutylamine	52	296
"	438	439
Di-isobutyl carbinol	63	83
Di-isobutyl carbonate	154	257
Di-isobutyl ether	46	133
Di-isobutyl ketone	101	164
Di-isobutyl oxalate	237	263
Di-isobutyl sulfide	37	421
Di-isobutyl d-tartarate	106	278
Di-isobutyl d,l-tartarate	80	277
Di-isobutyryl	63	163
Di-isopropylamine	23	295
" "	460	439
N,N-Di-isopropylaniline	277	438
1,2-Di-isopropylbenzene	75	38
1,3-Di-isopropylbenzene	73	38
1,4-Di-isopropylbenzene	95	38
Di-isopropyl carbonate	90	254
Di-isopropyl ether	18	131
Di-isopropylideneacetone	8	172
Di-isopropyl ketone	20	161
2,4-Di-isopropyl-1-methylbenzene	122	39
2,6-Di-isopropyl-1-methylbenzene	126	39
Di-isopropyl oxalate	160	258
Di-isopropyl phthalate	347	270
Di-isopropyl sulfide	17	420
Di-isopropyl d-tartarate	323	268
Di-isopropyl d,l-tartarate	34	274
2,4-Di-isopropyltoluene	122	39
2,6-Di-isopropyltoluene	126	39
2,4-Diketo-n-valeric acid	117	197
Dilauryl ether	13	137
Dimedone	181	178
1,4-Dimercaptobenzene	31	417
sym-Dimethoxyacetone	19	169
3,4-Dimethoxyacetophenone	55	174
3,4-Dimethoxyaniline	148	307
3,5-Dimethoxyaniline	114	436
2,3-Dimethoxybenzaldehyde	32	154
2,4-Dimethoxybenzaldehyde	59	156
3,4-Dimethoxybenzaldehyde	23	134
1,2-Dimethoxybenzene	2	137
1,3-Dimethoxybenzene	89	135
1,4-Dimethoxybenzene	34	138
2,4-Dimethoxybenzenesulfonamide	186	399
2,4-Dimethoxybenzenesulfonyl chloride	78	383
4,4′-Dimethoxybenzil	56	140
" "	161	177
3,4-Dimethoxybenzoic acid	281	204
4,4′-Dimethoxybenzoin	54	139
" "	146	177
3,4-Dimethoxybenzyl alcohol	132	86
2,3-Dimethoxybenzylamine	324	438
3,4-Dimethoxybenzylamine	323	438
Di-(β-methoxyethyl)carbonate	245	263

Compound	Compound No.	Page No.
3,5-Dimethoxy-4-hydroxybenzoic acid	515	123
6,7-Dimethoxy-8-hydroxy-1,2,3,4-tetra-hydroisoquinoline	58	140
2,5-Dimethoxy-3,4-methylenedioxybenzoic acid	273	204
2,5-Dimethoxy-3,4-methylenedioxybenzonitrile	182	365
(3,4-Dimethoxyphenyl)acetic acid	352	432
Di-(2-methoxyphenyl)carbonate	126	279
Di-(2-methoxyphenyl)sulfide	43	426
Di-(4-methoxyphenyl)sulfide	19	425
1,3-Dimethoxy-2-propanone	19	169
3,5'-Dimethoxy-5,7,4'-trihydroxyflavonol	584	127
Dimethylethylacetaldehyde	23	145
2,3-Dimethylacetanilide	244	237
2,4-Dimethylacetanilide	234	237
2,5-Dimethylacetanilide	253	237
2,6-Dimethylacetanilide	366	241
3,5-Dimethylacetanilide	265	238
2,2-Dimethylacetoacetonitrile	39	348
2,4-Dimethylacetophenone	187	167
2,5-Dimethylacetophenone	193	167
3,4-Dimethylacetophenone	204	168
3,5-Dimethylacetophenone	198	167
Dimethyl acetylene	6	28
3,3-Dimethylacrylic acid	64	195
" " "	466	433
3,3-Dimethylacrylonitrile	30	347
Dimethyl adipate	3	272
1,1-Dimethylallene	22	12
1,3-Dimethylallene	31	13
Di-2-methylallyl sulfide	38	421
Dimethylamine	2	295
"	434	439
Dimethylaminoacetone	10	319
3-(Dimethylamino)aniline	181	301
4-Dimethylaminoazobenzene	74	324
2-Dimethylaminobenzaldehyde	74	321
3-Dimethylaminobenzaldehyde	6	322
4-Dimethylaminobenzaldehyde	64	156
"	44	323
4-Dimethylaminobenzoic acid	67	287
4-Dimethylaminobenzophenone	58	324
2-Dimethylaminodiethyl ether	9	319
2-(Dimethylamino)ethylamine	330	438
2-(Dimethylamino)ethyl alcohol	14	319
N,N-1-(Dimethylamino)naphthalene	101	321
" " "	180	437
2-(Dimethylamino)naphthalene	22	323
" "	164	437
3-(Dimethylamino)phenol	144	103
N,N-Dimethyl-3-aminophenol	51	324
4-(Dimethylamino)phenol	111	102
N,N-Dimethyl-4-aminophenol	46	323
4-Dimethylaminothiophenol	7	416
2-Dimethylamino-m-xylene	56	320
4-Dimethylamino-m-xylene	59	320
2-Dimethylamino-p-xylene	58	320
N,N-Dimethylaniline	52	320
"	194	437
N,2-Dimethylaniline	103	298
"	167	437
N,3-Dimethylaniline	191	437
N,4-Dimethylaniline	109	298
"	214	437
2,3-Dimethylaniline	124	299
"	173	437
2,4-Dimethylaniline	117	298
"	182	437
2,5-Dimethylaniline	112	298
"	160	437
2,6-Dimethylaniline	115	298
"	120	436
3,4-Dimethylaniline	43	304
"	203	437
3,5-Dimethylaniline	121	299
"	184	437
1,3-Dimethylanthracene	83	46
1,4-Dimethylanthracene	64	45
1,5-Dimethylanthracene	140	48
9,10-Dimethylanthracene	175	50
Dimethyl azelate	2	272
2,2-Dimethylaziridine	301	438
cis-2,3-Dimethylaziridine	303	438
trans-2,3-Dimethylaziridine	302	438
1,2-Dimethylazulene	41	44
1,3-Dimethylazulene	30	44
4,8-Dimethylazulene	59	45
2,6-Dimethylbenzaldehyde	84	149
3,5-Dimethylbenzaldehyde	80	148
2,4-Dimethylbenzamide	374	242
3,4-Dimethylbenzamide	221	237
3,5-Dimethylbenzamide	235	237
3,4-Dimethylbenzanilide	131	234
2,6-Dimethyl-1,2-benzanthracene	164	50
3,9-Dimethyl-1,2-benzanthracene	96	47
4,9-Dimethyl-1,2-benzanthracene	66	45
4,10-Dimethyl-1,2-benzanthracene	117	47
5,6-Dimethyl-1,2-benzanthracene	177	50
5,8-Dimethyl-1,2-benzanthracene	135	48
5,10-Dimethyl-1,2-benzanthracene	173	50
6,7-Dimethyl-1,2-benzanthracene	171	50
8,10-Dimethyl-1,2-benzanthracene	146	49
9,10-Dimethyl-1,2-benzanthracene	126	48
2,5-Dimethylbenzene-1,3-disulfonamide	464	409
2,5-Dimethylbenzene-1,4-disulfonamide	470	410
4,5-Dimethylbenzene-1,3-disulfonamide	414	408
4,6-Dimethylbenzene-1,3-disulfonamide	431	408
2,5-Dimethylbenzene-1,3-disulfonanilide	210	399
2,5-Dimethylbenzene-1,4-disulfonanilide	367	406
4,5-Dimethylbenzene-1,3-disulfonanilide	312	404
4,6-Dimethylbenzene-1,3-disulfonanilide	299	403
2,5-Dimethylbenzene-1,3-disulfonic acid	217	378
4,5-Dimethylbenzene-1,3-disulfonic acid	189	378
4,6-Dimethylbenzene-1,3-disulfonic acid	197	378
2,5-Dimethylbenzene-1,3-disulfonyl chloride	97	384
2,5-Dimethylbenzene-1,4-disulfonyl chloride	253	389
4,5-Dimethylbenzene-1,3-disulfonyl chloride	94	383
4,6-Dimethylbenzene-1,3-disulfonyl chloride	210	388
2,3-Dimethylbenzenesulfonamide	189	399
2,4-Dimethylbenzenesulfonamide	94	395
2,5-Dimethylbenzenesulfonamide	125	396
2,6-Dimethylbenzenesulfonamide	50	393
3,4-Dimethylbenzenesulfonamide	111	396
3,5-Dimethylbenzenesulfonamide	87	395
2,4-Dimethylbenzenesulfonanilide	43	393
3,5-Dimethylbenzene sulfonanilide	57	394
2,3-Dimethylbenzene sulfonic acid	75	373
2,4-Dimethylbenzene sulfonic acid	32	372
2,5-Dimethylbenzene sulfonic acid	48	372
2,6-Dimethylbenzene sulfonic acid	13	371
3,4-Dimethylbenzene sulfonic acid	40	372
3,5-Dimethylbenzene sulfonic acid	28	371
2,3-Dimethylbenzenesulfonyl chloride	33	381
2,4-Dimethylbenzenesulfonyl chloride	15	381
2,5-Dimethylbenzenesulfonyl chloride	11	381
2,6-Dimethylbenzenesulfonyl chloride	23	381
3,4-Dimethylbenzenesulfonyl chloride	40	382
3,5-Dimethylbenzenesulfonyl chloride	126	385
4,4'-Dimethylbenzhydrol	38	89
4,4'-Dimethylbenzil	133	176
1,2-Dimethylbenzimidazole	71	324
1,2-Dimethylbenzimidazole	264	438
1,5-Dimethylbenzimidazole	60	324

Compound	Compound No.	Page No.
2,3-Dimethylbenzoic acid	242	431
2,4-Dimethylbenzoic acid	169	200
" "	318	431
2,5-Dimethylbenzoic acid	179	200
" " "	281	431
2,6-Dimethylbenzoic acid	185	430
3,4-Dimethylbenzoic acid	257	203
" "	370	432
3,5-Dimethylbenzoic acid	259	203
" " "	344	432
2,6-Dimethylbenzonitrile	116	362
2,2′-Dimethylbenzophenone	86	174
4,4′-Dimethylbenzophenone	118	176
2,3-Dimethyl-1,4-benzoquinone	3	181
2,5-Dimethyl-1,4-benzoquinone	15	182
2,6-Dimethyl-1,4-benzoquinone	7	181
α,4-Dimethylbenzyl alcohol	110	85
N,N-Dimethylbenzylamine	306	438
2,4-Dimethylbenzylamine	119	298
3,5-Dimethylbenzylamine	123	299
Di-(4-methylbenzyl)sulfide	44	426
N,N-Dimethylbenzyl sulfonamide	33	393
7,7-Dimethylbicyclo(3,1,1)heptane	140	5
2,3-Dimethylbicyclo(2,1,1)hept-2-ene	250	20
6,6′-Dimethyl-2,2′-bipyridyl	53	324
2,3-Dimethyl-1,3-butadiene	52	13
2,2-Dimethylbutanamide	227	237
d,l-2,3-Dimethylbutanamide	228	237
3,3-Dimethylbutanamide	230	237
2,2-Dimethylbutananilide	81	233
d,l-2,3-Dimethylbutananilide	47	232
3,3-Dimethylbutananilide	229	237
2,2-Dimethylbutane	18	3
2,3-Dimethylbutane	20	3
3,3-Dimethylbutanesulfonamide	28	393
3,3-Dimethylbutane-1-sulfonyl chloride	28	381
2,2-Dimethylbutanoic acid	21	190
d,l-2,3-Dimethylbutanoic acid	24	190
3,3-Dimethylbutanoic acid	17	190
2,2-Dimethyl-1-butanol	31	81
2,3-Dimethyl-1-butanol	38	82
2,3-Dimethyl-2-butanol	17	80
3,3-Dimethyl-2-butanol	16	80
3,3-Dimethyl-2-butanone	12	161
2,3-Dimethyl-1-butene	34	13
2,3-Dimethyl-2-butene	60	14
3,3-Dimethyl-1-butene	25	12
Dimethyl sec-butyl amine	405	439
1,3-Dimethyl-5-tert-butylbenzene	83	38
Di-(3-methylbutyl)sulfide	50	421
3,3-Dimethyl-1-butyne	8	28
2,2-Dimethylbutyric acid	464	433
Dimethyl d-camphorate	308	267
2,6-Dimethylcarbazole	449	317
3,6-Dimethylcarbazol	444	317
Dimethylethylcarbinyl acetate	50	252
Dimethyl carbonate	18	250
3,5-Dimethylcatechol	102	101
1,4-Dimethylchrysene	144	49
2,8-Dimethylchrysene	192	51
6,12-Dimethylchrysene	194	51
Dimethyl citraconate	196	260
1,1-Dimethylcyclohexane	70	4
cis-1,2-Dimethylcyclohexane	86	4
trans-1,2-Dimethylcyclohexane	78	4
cis-1,3-Dimethylcyclohexane	71	4
trans-1,3-Dimethylcyclohexane	81	4
cis-1,4-Dimethylcyclohexane	80	4
trans-1,4-Dimethylcyclohexane	69	4
5,5-Dimethyl-1,3-cyclohexanedione	181	178
2,2-Dimethylcyclohexanol	77	84
cis-2,4-Dimethylcyclohexanol	74	84
trans-2,4-Dimethylcyclohexanol	71	83
cis-2,5-Dimethylcyclohexanol	70	83
trans-2,5-Dimethylcyclohexanol	76	84
2,6-Dimethylcyclohexanol	69	83
3,3-Dimethylcyclohexanol	88	84
3,4-Dimethylcyclohexanol	92	84
cis-3,5-Dimethylcyclohexanol	89	84
trans-3,5-Dimethylcyclohexanol	90	84
2,2-Dimethylcyclohexanone	109	164
2,3-Dimethylcyclohexanone	125	165
cis-2,4-Dimethylcyclohexanone	123	165
trans-2,4-Dimethylcyclohexanone	111	164
d-2,5-Dimethylcyclohexanone	116	164
d,l-2,5-Dimethylcyclohexanone	113	164
2,6-Dimethylcyclohexanone	118	165
3,3-Dimethylcyclohexanone	127	165
3,4-Dimethylcyclohexanone	141	165
cis-3,5-Dimethylcyclohexanone	136	165
trans-3,5-Dimethylcyclohexanone	131	165
4,4-Dimethylcyclohexanone	26	173
1,2-Dimethylcyclohexene	239	19
1,3-Dimethylcyclohexene	240	19
1,4-Dimethylcyclohexene	223	19
1,5-Dimethylcyclohexene	224	19
1,6-Dimethylcyclohexene	232	19
3,3-Dimethylcyclohexene	201	18
4,4-Dimethylcyclohexene	197	18
1,2-Dimethylcyclohexen-3-one	95	170
1,5-Dimethylcyclohexen-3-one	164	166
1,5-Dimethylcyclohexen-4-one	147	165
2,5-Dimethylcyclohexen-3-one	143	165
5,5-Dimethylcyclohexen-3-one	79	170
1,2-Dimethylcyclooctatetraene	11	25
1,1-Dimethylcyclopentane	32	3
cis-1,2-Dimethylcyclopentane	43	3
trans-1,2-Dimethylcyclopentane	38	3
cis-1,3-Dimethylcyclopentane	36	3
trans-1,3-Dimethylcyclopentane	35	3
3,3-Dimethylcyclopentanone	73	163
1,2-Dimethylcyclopentene	136	16
1,3-Dimethylcyclopentene	99	15
1,4-Dimethylcyclopentene	102	15
3,3-Dimethylcyclopentene	94	15
2,3-Dimethyl-2-cyclopentenone	100	170
1,1-Dimethylcyclopropane	9	3
cis-1,2-Dimethylcyclopropane	15	3
trans-1,2-Dimethylcyclopropane	11	3
cis-1,1-Dimethyl-2,3-cyclopropanedicarboxylic acid	81	429
trans-1,1-Dimethyl-2,3-cyclopropanedicarboxylic acid	255	431
1,10-Dimethyl-cis-decahydronaphthalene	255	7
1,10-Dimethyl-trans-decahydronaphthalene	252	7
1′,10-Dimethyl-1,2-dibenzanthracene	125	48
4,4′-Dimethyldibenzylamine	11	303
2,3-Dimethyl-1,4-dinitrobenzene	61	339
2,4-Dimethyl-1,3-dinitrobenzene	54	339
2,5-Dimethyl-1,3-dinitrobenzene	88	340
2,5-Dimethyl-1,4-dinitrobenzene	102	341
3,4-Dimethyl-1,2-dinitrobenzene	49	338
3,5-Dimethyl-1,2-dinitrobenzene	93	340
3,5-Dimethyl-1,4-dinitrobenzene	72	340
3,6-Dimethyl-1,2-dinitrobenzene	63	339
4,5-Dimethyl-1,2-dinitrobenzene	82	340
4,5-Dimethyl-1,3-dinitrobenzene	42	338
4,6-Dimethyl-1,3-dinitrobenzene	64	339
3,5-Dimethyl-2,4-dinitrophenol	249	109
Dimethylethyl acetic acid	21	190
Dimethyl ethyl amine	2	319
" "	372	439

Compound	Compound No.	Page No.
1,2-Dimethyl-3-ethylbenzene	47	37
1,2-Dimethyl-4-ethylbenzene	39	36
1,3-Dimethyl-2-ethylbenzene	40	36
1,3-Dimethyl-4-ethylbenzene	37	36
1,3-Dimethyl-5-ethylbenzene	30	36
1,4-Dimethyl-2-ethylbenzene	33	33
3,3-Dimethyl-2-ethyl-1-butene	155	17
2,2-Dimethylethylene sulfide	4	423
2,2-Dimethyl-3-ethylhexane	172	5
2,2-Dimethyl-4-ethylhexane	134	5
2,3-Dimethyl-3-ethylhexane	231	6
2,3-Dimethyl-4-ethylhexane	203	6
2,4-Dimethyl-3-ethylhexane	199	6
2,4-Dimethyl-4-ethylhexane	171	5
2,5-Dimethyl-3-ethylhexane	167	5
3,3-Dimethyl-4-ethylhexane	208	6
3,4-Dimethyl-3-ethylhexane	233	7
2,2-Dimethyl-3-ethylpentane	97	4
2,3-Dimethyl-3-ethylpentane	122	5
2,4-Dimethyl-3-ethylpentane	104	4
2,6-Dimethyl-4-ethylpyridine	47	320
3,5-Dimethyl-2-ethylpyridine	55	320
1,7-Dimethylfluorene	111	47
N,N-Dimethylformamide	1	230
Dimethyl fumarate	135	280
2,5-Dimethylfuran	32	132
Dimethyl glutarate	206	260
3,3-Dimethylglutaric acid	238	431
Dimethylglycolic acid	80	196
2,6-Dimethyl-1,3-heptadiene	251	20
2,6-Dimethyl-1,5-heptadiene	254	20
2,6-Dimethyl-2,4-heptadiene	247	20
2,6-Dimethyl-2,5-heptadiene	279	21
3,5-Dimethyl-2,4-heptadiene	233	19
3,6-Dimethyl-2,4-heptadiene	259	20
2,2-Dimethylheptane	91	4
2,3-Dimethylheptane	117	4
2,4-Dimethylheptane	93	4
2,5-Dimethylheptane	100	4
2,6-Dimethylheptane	99	4
3,3-Dimethylheptane	106	4
3,4-Dimethylheptane	118	5
3,5-Dimethylheptane	101	4
4,4-Dimethylheptane	98	4
2,6-Dimethyl-4-heptanol	63	83
2,2-Dimethyl-3-heptanone	100	164
2,5-Dimethyl-4-heptanone	103	164
2,6-Dimethyl-4-heptanone	101	164
3,5-Dimethyl-4-heptanone	93	164
2,6-Dimethyl-2-heptene	225	19
2,3-Dimethyl-2-hepten-6-one	45	169
2,4-Dimethyl-2,4-hexadiene	184	18
2,5-Dimethyl-1,3-hexadiene	195	18
2,5-Dimethyl-1,5-hexadiene	203	18
2,5-Dimethyl-2,4-hexadiene	235	19
3,3-Dimethyl-1,5-hexadiene	121	16
3,4-Dimethyl-1,5-hexadiene	122	16
3,4-Dimethyl-2,4-hexadiene	228	19
3,5-Dimethylhexahydrobenzaldehyde	19	150
2,2-Dimethylhexane	47	3
2,3-Dimethylhexane	57	3
2,4-Dimethylhexane	50	3
2,5-Dimethylhexane	48	3
3,3-Dimethylhexane	53	3
3,4-Dimethylhexane	63	4
2,5-Dimethyl-3,4-hexanedione	63	163
2,2-Dimethylhexanoic acid	41	191
2,2-Dimethyl-3-hexanone	57	163
2,4-Dimethyl-3-hexanone	53	162
3,4-Dimethyl-2-hexanone	83	163
4,4-Dimethyl-3-hexanone	68	163
2,2-Dimethyl-cis-3-hexene	134	16
2,2-Dimethyl-trans-3-hexene	118	16
2,3-Dimethyl-1-hexene	161	17
2,3-Dimethyl-2-hexene	214	19
2,3-Dimethyl-3-hexene	182	18
2,4-Dimethyl-1-hexene	164	17
2,4-Dimethyl-2-hexene	162	17
2,4-Dimethyl-cis-3-hexene	154	17
2,4-Dimethyl-trans-3-hexene	147	17
2,5-Dimethyl-1-hexene	165	17
2,5-Dimethyl-2-hexene	172	17
2,5-Dimethyl-3-hexene	123	16
3,3-Dimethyl-1-hexene	130	16
3,4-Dimethyl-1-hexene	167	17
3,4-Dimethyl-2-hexene	192	18
3,4-Dimethyl-trans-3-hexene	215	19
3,5-Dimethyl-1-hexene	129	16
3,5-Dimethyl-2-hexene	169	17
4,4-Dimethyl-1-hexene	144	17
4,4-Dimethyl-2-hexene	138	16
4,5-Dimethyl-1-hexene	152	17
4,5-Dimethyl-2-hexene	157	17
5,5-Dimethyl-1-hexene	124	16
5,5-Dimethyl-cis-2-hexene	140	16
5,5-Dimethyl-trans-2-hexene	131	16
3,4-Dimethyl-3-hexen-2-one	84	163
4,5-Dimethyl-4-hexen-3-one	99	164
4,5-Dimethyl-5-hexen-3-one	92	164
5,5-Dimethyl-3-hexen-2-one	55	170
sym-Dimethylhydrazine	21	295
unsym-Dimethylhydrazine	14	295
2,3-Dimethylhydroquinone	549	125
2,5-Dimethylhydroquinone	540	124
2,6-Dimethylhydroquinone	377	115
N,N-Dimethyl-3-hydroxyaniline	51	324
N,N-Dimethyl-4-hydroxyaniline	46	323
2,4-Dimethyl-6-hydroxyaniline	366	314
2,5-Dimethyl-4-hydroxyaniline	454	317
2,6-Dimethyl-4-hydroxyaniline	393	315
3,5-Dimethyl-2-hydroxyaniline	301	312
3,5-Dimethyl-4-hydroxyaniline	310	312
4,5-Dimethyl-2-hydroxyaniline	382	315
4,5'-Dimethyl-2-hydroxyazobenzene	235	108
2,6-Dimethyl-4-hydroxybenzene-1,3-disulfonamide	332	405
3,5-Dimethyl-2-hydroxybenzene-1,4-disulfonamide	166	398
2,6-Dimethyl-4-hydroxybenzene-1,3-disulfonanilide	326	404
2,6-Dimethyl-4-hydroxybenzene-1,3-disulfonic acid	157	376
4,6-Dimethyl-2-hydroxy benzene-1,3-disulfonic acid	64	373
3,5-Dimethyl-2-hydroxybenzene-1,3-disulfonyl chloride	116	384
3,5-Dimethyl-2-hydroxybenzenesulfonanilide	103	395
3,5-Dimethyl-2-hydroxybenzenesulfonic acid	5	380
2,6-Dimethyl-4-hydroxybenzenesulfonyl chloride	181	387
2,6-Dimethyl-4-hydroxybenzonitrile	38	434
3,5-Dimethyl-4-hydroxybenzonitrile	40	434
Dimethyl (2-hydroxyethyl) amine	319	438
2,2-Dimethyl-3-hydroxypropionaldehyde	84	157
2,4-Dimethyl-8-hydroxyquinoline	18	434
3,4-Dimethyl-8-hydroxyquinoline	13	434
2,4-Dimethylimidazole	297	438
3,5-Dimethylindole	116	306
Dimethyl isobutyl amine	365	439
Dimethyl isophthalate	93	277
Dimethyl isopropyl amine	397	439
1,4-Dimethyl-7-isopropylazulene	6	43

	Compound No.	Page No.
1,2-Dimethyl-3-isopropylbenzene	71	38
1,2-Dimethyl-4-isopropylbenzene	69	38
1,3-Dimethyl-2-isopropylbenzene	61	37
1,3-Dimethyl-4-isopropylbenzene	62	37
1,3-Dimethyl-5-isopropylbenzene	51	37
1,4-Dimethyl-2-isopropylbenzene	56	37
Dimethyl isopropyl carbinol	17	80
1,7-Dimethyl-4-isopropylnaphthalene	43	44
2,4-Dimethyl-3-isopropylpentane	169	5
Dimethyl itaconate	41	275
Dimethyl l-malate	263	264
Dimethyl maleate	179	259
Dimethyl malonate	142	257
Dimethylmalonic acid	313	206
" "	170	430
Dimethylmalononitrile	20	357
Dimethyl mesaconate	175	259
Dimethyl mesotartarate	144	280
3,3-Dimethyl-1-methoxy-2-butanone	87	163
2,4-Dimethyl-2-methoxyphenol	286	111
2,2-Dimethyl-3-methylene-bicyclo(2,2,1)heptene	10	26
2,2-Dimethyl-5-methylene-bicyclo(2,2,1)heptane	278	21
3,5-Dimethyl-4-(methylsulfonyl)phenol	37	434
Dimethyl mucate	168	281
Dimethyl naphthalate	137	280
1,3-Dimethylnaphthalene	155	41
1,4-Dimethylnaphthalene	159	41
1,5-Dimethylnaphthalene	75	46
1,6-Dimethylnaphthalene	154	41
1,7-Dimethylnaphthalene	153	41
1,8-Dimethylnaphthalene	53	45
2,3-Dimethylnaphthalene	109	47
2,6-Dimethylnaphthalene	114	47
2,7-Dimethylnaphthalene	100	47
1,3-Dimethyl-2-naphthol	158	104
1,4-Dimethyl-2-naphthol	326	113
2,4-Dimethyl-1-naphthol	139	103
3,4-Dimethyl-1-naphthol	276	110
2,5-Dimethyl-1,4-naphthoquinone	9	181
3,7-Dimethyl-1,2-naphthoquinone	20	182
1,4-Dimethyl-2-naphthylamine	119	306
N,N-Dimethyl-α-naphthylamine	101	321
N,N-Dimethyl-β-naphthylamine	22	323
2,6-Dimethyl-1-naphthylamine	161	307
3,6-Dimethyl-2-naphthylamine	313	312
3,7-Dimethyl-2-naphthylamine	286	311
Dimethylneopentylacetic acid	33	194
N,N-Dimethyl-2-nitroaniline	1	322
N,N-Dimethyl-3-nitroaniline	33	323
" "	60	436
N,N-Dimethyl-4-nitroaniline	17	436
2,3-Dimethyl-4-nitroaniline	36	436
2,4-Dimethyl-5-nitroaniline	270	311
2,5-Dimethyl-4-nitroaniline	325	313
2,6-Dimethyl-4-nitroaniline	23	436
3,5-Dimethyl-4-nitroaniline	57	436
4,6-Dimethyl-1-nitroaniline	123	306
1,2-Dimethyl-3-nitrobenzene	30	336
1,3-Dimethyl-2-nitrobenzene	24	336
1,3-Dimethyl-4-nitrobenzene	29	336
1,4-Dimethyl-2-nitrobenzene	27	336
3,4-Dimethyl-1-nitrobenzene	4	337
3,5-Dimethyl-1-nitrobenzene	41	338
2,4-Dimethyl-3-nitrobenzenesulfonamide	205	399
2,4-Dimethyl-5-nitrobenzenesulfonamide	260	402
2,4-Dimethyl-6-nitrobenzenesulfonamide	40	393
2,5-Dimethyl-3-nitrobenzenesulfonamide	207	399
2,5-Dimethyl-4-nitrobenzenesulfonamide	307	403
2,5-Dimethyl-6-nitrobenzenesulfonamide	284	403
3,4-Dimethyl-5-nitrobenzenesulfonamide	230	400
2,5-Dimethyl-3-nitrobenzenesufonanilide	107	396
2,5-Dimethyl-4-nitrobenzenesulfonanilide	78	394
2,5-Dimethyl-6-nitrobenzenesulfonanilide	237	401
2,4-Dimethyl-3-nitrobenzenesulfonic acid	86	373
2,4-Dimethyl-5-nitrobenzenesulfonic acid	120	375
2,5-Dimethyl-3-nitrobenzenesulfonic acid	87	374
2,5-Dimethyl-4-nitrobenzenesulfonic acid	145	376
2,5-Dimethyl-6-nitrobenzenesulfonic acid	134	375
2,4-Dimethyl-3-nitrobenzenesulfonyl chloride	134	385
2,4-Dimethyl-5-nitrobenzenesulfonyl chloride	143	385
2,5-Dimethyl-3-nitrobenzenesulfonyl chloride	60	382
2,5-Dimethyl-4-nitrobenzenesulfonyl chloride	87	383
2,5-Dimethyl-6-nitrobenzenesulfonyl chloride	167	386
3,4-Dimethyl-5-nitrobenzenesulfonyl chloride	80	383
2,6-Dimethyl-4-nitrophenol	25	434
3,5-Dimethyl-4-nitrophenol	39	434
Dimethyl-3-nitrophthalate	96	278
Dimethyl-4-nitrophthalate	89	277
N,N-Dimethyl-4-nitrosoaniline	50	324
2,3-Dimethyl-2-norbornene	250	20
2,4-Dimethyl-2,4-octadiene	289	22
2,6-Dimethyl-2,5-octadiene	12	25
2,6-Dimethyl-2,6-octadiene	296	22
2,6-Dimethyl-2,7-octadiene	286	21
2,7-Dimethyl-2,6-octadiene	292	22
3,6-Dimethyl-2,6-octadiene	285	21
3,7-Dimethyl-2,4-octadiene	293	22
4,4-Dimethyl-1,7-octadiene	265	21
4,5-Dimethyl-2,6-octadiene	283	21
2,2-Dimethyloctane	156	5
2,3-Dimethyloctane	205	6
2,4-Dimethyloctane	160	5
d,l-2,5-Dimethyloctane	162	5
2,6-Dimethyloctane	182	6
2,7-Dimethyloctane	176	5
3,3-Dimethyloctane	188	6
3,4-Dimethyloctane	212	6
3,5-Dimethyloctane	178	6
3,6-Dimethyloctane	177	6
4,4-Dimethyloctane	185	6
4,5-Dimethyloctane	189	6
2,6-Dimethyl-1-octene-8-carboxylic acid	49	191
2,6-Dimethyl-1-octen-8-carboxylic acid	12	192
2,6-Dimethylol-4-methylphenol	91	434
2,4-Dimethylolphenol	85	434
2,6-Dimethylolphenol	83	434
Dimethyl oxalate	73	276
2,4-Dimethyl-1,3-pentadiene	101	15
2,4-Dimethyl-2,3-pentadiene	89	15
4,4-Dimethyl-1,2-pentadiene	77	14
3,3-Dimethylpentanal	40	146
2,2-Dimethylpentane	27	3
2,3-Dimethylpentane	33	3
2,4-Dimethylpentane	28	3
3,3-Dimethylpentane	31	3
2,3-Dimethylpentanoic acid	18	192
4,4-Dimethylpentanoic acid	434	432
2,4-Dimethyl-1-pentanol	53	82
2,4-Dimethyl-3-pentanol	34	81
3,3-Dimethyl-1-pentanol	40	146
2,2-Dimethyl-3-pentanone	23	161
2,4-Dimethyl-3-pentanone	20	161
3,3-Dimethyl-2-pentanone	26	162
3,4-Dimethyl-2-pentanone	42	162
2,3-Dimethyl-1-pentene	84	15
2,3-Dimethyl-2-pentene	113	16
2,4-Dimethyl-2-pentene	82	15
3,3-Dimethyl-1-pentene	69	14
3,4-Dimethyl-1-pentene	80	15
3,4-Dimethyl-2-pentene	91	15
4,4-Dimethyl-1-pentene	58	14
4,4-Dimethyl-cis-2-pentene	79	14

Compound	Compound No.	Page No.
4,4-Dimethyl-*trans*-2-pentene	68	14
3,4-Dimethyl-3-penten-2-one	61	163
3,4-Dimethyl-4-penten-2-one	51	162
4,4-Dimethyl-1-pentyne	19	28
4,4-Dimethyl-2-pentyne	23	29
1,2-Dimethylphenanthrene	145	49
1,3-Dimethylphenanthrene	68	45
1,4-Dimethylphenanthrene	26	44
1,5-Dimethylphenanthrene	38	44
1,6-Dimethylphenanthrene	88	46
1,7-Dimethylphenanthrene	86	46
1,8-Dimethylphenanthrene	180	50
1,9-Dimethylphenanthrene	89	46
2,3-Dimethylphenanthrene	72	46
2,4-Dimethylphenanthrene	115	47
2,5-Dimethylphenanthrene	20	43
2,6-Dimethylphenanthrene	7	43
2,7-Dimethylphenanthrene	105	47
2,9-Dimethylphenanthrene	37	44
3,4-Dimethylphenanthrene	45	44
3,5-Dimethylphenanthrene	29	44
3,6-Dimethylphenanthrene	142	49
2,3-Dimethylphenol	109	103
'' ''	114	435
2,4-Dimethylphenol	8	96
'' ''	113	435
2,5-Dimethylphenol	108	102
'' ''	107	435
2,6-Dimethylphenol	115	435
3,4-Dimethylphenol	69	100
'' ''	111	435
3,5-Dimethylphenol	82	100
'' ''	102	434
(2,6-Dimethylphenoxy)acetic acid	193	430
2,4-Dimethylphenyl acetate	224	262
2,5-Dimethylphenyl acetate	256	264
2,6-Dimethylphenyl acetate	207	261
3,4-Dimethylphenyl acetate	10	273
3,5-Dimethylphenyl acetate	1	272
1,1-Dimethyl-2-phenylelhanol	4	88
sym-Dimethyl-*p*-phenylenediamine	61	304
2,2-Dimethyl-1-phenylpropane	32	36
3,4-Dimethyl-1-phenylpyrazole	108	321
3,5-Dimethyl-1-phenylpyrazole	104	321
2,3-Dimethyl-1-phenyl-5-pyrazolone	72	324
Dimethyl phthalate	332	269
3,4-Dimethylphthalic anhydride	56	226
4,5-Dimethylphthalic anhydride	78	226
Dimethyl pimelate	9	272
cis-2,5-Dimethylpiperazine	239	310
trans-2,5-Dimethylpiperazine	247	310
1,2-Dimethylpiperidine	377	439
cis-2,6-Dimethylpiperidine	457	439
2,6-Dimethylpiperidine	45	296
2,2-Dimethylpropane	7	3
2,2-Dimethyl-1-propanol	28	89
Dimethylpropyl amine	371	439
2,2-Dimethyl-*n*-propylamine	369	439
1,2-Dimethyl-3-propylbenzene	96	38
1,2-Dimethyl-4-propylbenzene	91	38
1,3-Dimethyl-4-propylbenzene	84	38
1,3-Dimethyl-5-propylbenzene	92	38
1,4-Dimethyl-2-propylbenzene	76	38
Dimethyl *n*-propyl carbinol	19	81
1,3-Dimethylpyrazole	15	319
'' ''	82	436
3,5-Dimethylpyrazole	151	437
2,3-Dimethylpyridine	31	319
'' ''	265	438
2,4-Dimethylpyridine	25	319
'' ''	271	438
2,5-Dimethylpyridine	26	319
'' ''	257	438
2,6-Dimethylpyridine	18	319
'' ''	266	438
3,4-Dimethylpyridine	32	319
'' ''	261	438
3,5-Dimethylpyridine	40	319
4,6-Dimethylpyrimidine	2	323
1,2-Dimethylpyrrolidine	5	319
'' ''	376	439
1,3-Dimethylpyrrolidine	6	319
2,4-Dimethylpyrrolidine	38	296
2,5-Dimethylpyrrolidine	37	296
1,2-Dimethyl-2-pyrroline	469	439
2,3-Dimethylquinoline	37	323
2,4-Dimethylquinoline	96	321
2,6-Dimethylquinoline	32	323
2,8-Dimethylquinoline	3	323
3,4-Dimethylquinoline	42	323
4,6-Dimethylquinoline	87	321
4,7-Dimethylquinoline	107	321
4,8-Dimethylquinoline	30	323
5,8-Dimethylquinoline	102	321
6,8-Dimethylquinoline	98	321
2,5-Dimethylresorcinol	419	118
4,6-Dimethylresorcinol	295	111
Dimethyl sebacate	17	273
2,2′-Dimethylstilbene	84	46
3,3′-Dimethylstilbene	34	44
Dimethyl suberate	313	268
Dimethyl succinate	1	273
d,*l*-2,3-Dimethylsuccinic acid	458	432
meso-2,3-Dimethylsuccinic acid	246	431
α,α-Dimethylsuccinic anhydride	5	225
unsym-Dimethylsuccinic anhydride	5	225
cis-α,β-Dimethylsuccinic anhydride	36	225
trans-α,β-Dimethylsuccinic anhydride	14	225
d,*l*-2,3-Dimethylsuccinonitrile	70	360
meso-2,3-Dimethylsuccinonitrile	43	358
Dimethyl sulfate	150	257
Dimethyl sulfide	1	420
Dimethyl *d*-tartarate	87	277
Dimethyl *d*,*l*-tartarate	127	279
Dimethyl tartaronate	72	276
Dimethyl terephthalate	161	281
1,1-Dimethyl-1,2,3,4-tetrahydronaphthalene	117	39
1,5-Dimethyl-1,2,3,4-tetrahydronaphthalene	140	40
2,2-Dimethyl-1,2,3,4-tetrahydronaphthalene	129	40
2,5-Dimethyl-1,2,3,4-tetrahydronaphthalene	135	40
2,6-Dimethyl-1,2,3,4-tetrahydronaphthalene	138	40
2,7-Dimethyl-1,2,3,4-tetrahydronaphthalene	137	40
2,8-Dimethyl-1,2,3,4-tetrahydronaphthalene	136	40
5,6-Dimethyl-1,2,3,4-tetrahydronaphthalene	147	40
5,7-Dimethyl-1,2,3,4-tetrahydronaphthalene	149	40
5,8-Dimethyl-1,2,3,4-tetrahydronaphthalene	150	41
6,7-Dimethyl-1,2,3,4-tetrahydronaphthalene	148	40
1,3-Dimethyl-1,2,3,4-tetrahydroquinoline	4	322
2,3-Dimethyl-1,2,3,4-tetrahydroquinoline	25	303
2,3-Dimethyl-5,6,7,8-tetrahydroquinoline	12	323
2,4-Dimethyl-5,6,7,8-tetrahydroquinoline	84	321
DL-2,6-Dimethyl-1,2,3,4-tetrahydroquinoline	7	303
1,1-Dimethyltetralin	117	39
1,5-Dimethyltetralin	140	40
2,2-Dimethyltetralin	129	40
2,5-Dimethyltetralin	135	40
2,6-Dimethyltetralin	138	40
2,7-Dimethyltetralin	137	40
2,8-Dimethyltetralin	136	40
5,6-Dimethyltetralin	147	40
5,7-Dimethyltetralin	149	40
5,8-Dimethyltetralin	150	41

Compound	Compound No.	Page No.
6,7-Dimethyltetralin	148	40
1,1-Dimethyl-2-tetralone	83	170
2,4-Dimethyltetramethylene sulfide	52	424
cis-2,5-Dimethyltetramethylene sulfide	21	423
trans-2,5-Dimethyltetramethylene-sulfide	20	423
2,2-Dimethylthietane	11	423
2,4-Dimethylthietane	9	423
2,2-Dimethylthiirane	4	423
1,4-(Dimethylthio)benzene	14	416
2,4-Dimethylthiolane	52	424
cis-2,5-Dimethylthiolane	21	423
trans-2,5-Dimethylthiolane	20	423
2,3-Dimethylthiophene	19	423
2,4-Dimethylthiophene	17	423
2,5-Dimethylthiophene	16	423
3,4-Dimethylthiophene	22	423
2,4-Dimethylthiophenol	39	414
2,5-Dimethylthiophenol	38	414
2,6-Dimethyl-1,4-thioxane	33	424
3,5-Dimethyl-1,4-thioxane	36	424
N,N-Dimethyltoluene-4-sulfonamide	22	392
N,N-Dimethyl-m-toluidine	65	320
N,N-Dimethyl-o-toluidine	46	320
N,N-Dimethyl-p-toluidine	62	320
2,2-Dimethyltrimethylene sulfide	11	423
2,4-Dimethyltrimethylene sulfide	9	423
sym-Dimethylurea	145	234
unsym-Dimethylurea	381	242
N,N-Dimethyl-m-2-xylidine	56	320
N,N-Dimethyl-m-4-xylidine	59	320
N,N-Dimethyl-p-2-xylidine	58	320
Di-1-naphthastilbene	161	49
Di-2-naphthastilbene	196	51
Di-(1-naphthyl)acetyl chloride	33	217
N,N'-Di-(1-naphthyl)benzene-1,3-disulfonamide	425	408
Di-1-naphthylethylene	161	49
Di-2-naphthylethylene	196	51
Di-1-naphthyl ketone	128	176
1,1'-Dinaphthylmethane	112	47
Di-(1-naphthyl) sulfide	61	426
1,2'-Dinaphthyl sulfide	70	426
Di-(2-naphthyl) sulfide	71	426
Dineopentylacetic acid	93	196
3,8-Dinitroacenaphthene	105	341
2,4-Dinitroaniline	391	315
" "	2	436
2,6-Dinitroaniline	311	312
2,4-Dinitroanisole	67	339
9,10-Dinitroanthracene	139	343
1,3-Dinitroanthraquinone	133	342
1,5-Dinitroanthraquinone	143	343
1,6-Dinitroanthraquinone	134	342
1,7-Dinitroanthraquinone	140	343
1,8-Dinitroanthraquinone	142	343
2,7-Dinitroanthraquinone	137	342
2,2'-Dinitroazobenzene	126	342
4,4'-Dinitroazobenzene	128	342
2,4-Dinitrobenzamide	436	243
3,5-Dinitrobenzamide	385	242
3,5-Dinitrobenzanilide	483	245
1,2-Dinitrobenzene	80	340
1,3-Dinitrobenzene	60	339
1,4-Dinitrobenzene	114	341
2,4-Dinitrobenzenesulfonamide	153	397
3,5-Dinitrobenzenesulfonamide	397	407
2,4-Dinitrobenzenesulfonic acid	60	373
3,5-Dinitrobenzenesulfonic acid	180	377
2,4-Dinitrobenzenesulfonyl chloride	154	386
3,5-Dinitrobenzenesulfonyl chloride	148	385
2,3-Dinitrobenzoic acid	32	429
2,4-Dinitrobenzoic acid	287	204
" " "	16	429
2,5-Dinitrobenzoic acid	25	429
2,6-Dinitrobenzoic acid	7	429
3,4-Dinitrobenzoic acid	119	430
3,5-Dinitrobenzoic acid	328	206
" " "	120	430
2,4-Dinitrobenzoic anhydride	68	226
3,5-Dinitrobenzoic anhydride	48	226
2,4-Dinitrobenzonitrile	144	363
2,6-Dinitrobenzonitrile	69	360
3,5-Dinitrobenzoyl bromide	4	220
2,4-Dinitrobenzoyl chloride	8	216
3,5-Dinitrobenzoyl chloride	14	216
Di-(4-nitrobenzyl) sulfide	73	427
2,2'-Dinitrobiphenyl	91	340
2,4'-Dinitrobiphenyl	65	339
3,3'-Dinitrobiphenyl	124	342
4,4'-Dinitrobiphenyl	132	342
3,5-Dinitrocatechol	422	118
4,6-Dinitro-m-cresol	195	106
4,6-Dinitro-o-cresol	147	104
2,2'-Dinitrodiphenylmethane	108	341
2,4'-Dinitrodiphenylmethane	81	340
3,3'-Dinitrodiphenylmethane	115	341
4,4'-Dinitrodiphenylmethane	119	342
2,4-Dinitrodiphenyl sulfide	64	426
1,3-Dinitro-4-ethoxybenzene	58	339
2,5-Dinitrofluorene	125	342
2,6-Dinitrohydroquinone	325	113
" "	7	434
3,5-Dinitro-2-hydroxyaniline	380	315
3,5-Dinitro-2-hydroxybenzoic acid	284	204
2,6-Dinitro-4-iodophenol	238	108
4,6-Dinitro-2-iodophenol	211	107
2,4-Dinitromesitylene	55	339
1,3-Dinitro-4-methoxybenzene	67	339
2,6-Dinitro-4-methylaniline	377	315
" "	3	436
3,5-Dinitro-4-methylbenzoic acid	136	430
4,6-Dinitro-2-methylphenol	147	104
4,6-Dinitro-3-methylphenol	195	106
1,2-Dinitronaphthalene	73	340
1,3-Dinitronaphthalene	99	341
1,4-Dinitronaphthalene	92	340
1,5-Dinitronaphthalene	127	342
1,6-Dinitronaphthalene	109	341
1,8-Dinitronaphthalene	110	341
2,7-Dinitronaphthalene	131	342
1,6-Dinitro-2-naphthol	496	122
2,4-Dinitro-1-naphthol	331	113
1,5-Dinitro-2-naphthylamine	409	316
1,8-Dinitro-2-naphthylamine	450	317
2,4-Dinitro-1-naphthylamine	453	317
2,5-Dinitrophenanthrenequinone	129	342
2,7-Dinitrophenanthrenequinone	141	343
2,4-Dinitrophenetole	58	339
2,4-Dinitrophenol	243	108
" "	6	434
2,5-Dinitrophenol	9	434
2,6-Dinitrophenol	70	100
" "	5	434
3,4-Dinitrophenol	10	434
3,5-Dinitrophenol	298	111
(2,4-Dinitrophenyl)acetic acid	204	430
2,4-Dinitrophenyl hydrazine	416	316
2,4-Dinitrophenyl phenyl sulfide	64	426
Di-(2-nitrophenyl) sulfide	65	426
Di-(3-nitrophenyl) sulfide	75	427
Di-(4-nitrophenyl) sulfide	74	427
2,4-Dinitrophenyl sulfide	76	427
3,5-Dinitrophthalic anhydride	71	226
2,4-Dinitroresorcinol	369	115
4,6-Dinitroresorcinol	533	124

Compound	Compound No.	Page No.
3,5-Dinitrosalicylamide	378	242
3,5-Dinitrosalicylic acid	284	204
" " "	455	120
2,2′-Dinitrostilbene	122	342
2,4-Dinitrostilbene	98	341
trans-4,4′-Dinitrostilbene	138	343
2,β-Dinitrostyrene	75	340
4,β-Dinitrostyrene	123	342
2,4-Dinitrothiophenol	43	417
2,3-Dinitrotoluene	33	338
2,4-Dinitrotoluene	37	338
2,6-Dinitrotoluene	35	338
3,4-Dinitrotoluene	31	338
3,5-Dinitrotoluene	62	339
Di-n-octadecylamine	445	439
Dioctadecyl sulfide	39	426
Di-n-octylamine	451	439
Di-n-octyl ketone	50	173
Di-n-octyl sulfide	30	425
1,4-Dioxane	36	132
2,4-Dioxopentanoic acid	117	197
" "	107	430
Di-n-pentadecylamine	449	439
Dipentene	319	23
Dipentylacetylene	59	30
Di-n-pentylamine	461	439
2,2′-Diphenanilide	481	245
2,2′-Diphenic acid	354	208
2,2′-Diphenic acid diamide	452	244
2,2′-Diphenic acid dichloride	22	216
2,2′-Diphenic acid mononitrile	212	366
2,2′-Diphenic anhydride	79	227
2,4′-Diphenol	418	118
1,2-Diphenoxyethane	52	139
1,3-Diphenoxypropane	38	138
Diphenylacetaldehyde	93	149
Diphenylacetamide	333	240
N,N-Diphenylacetamide	118	234
Diphenylacetanilide	377	242
Diphenylacetic acid	222	202
" "	273	431
Diphenylacetic anhydride	41	225
Diphenylacetoin	56	174
1,1-Diphenylacetone	76	174
1,3-Diphenylacetone	19	172
Diphenylacetonitrile	95	361
Diphenylacetyl chloride	12	216
Sym-Diphenylacetylene	47	44
cis-2,3-Diphenylacrylonitrile	110	362
trans-2,3-Diphenylacrylonitrile	52	359
3,3-Diphenylacrylonitrile	49	359
Diphenyl adipate	140	280
2,2-Diphenyladipic acid	313	431
3,4-Diphenyladipic acid	326	431
Diphenylamine	62	304
"	19	436
Diphenylaminoformyl chloride	21	216
2,2-Diphenylazelaic acid	351	432
2,6-Diphenyl-1,4-benzoquinone	17	182
trans-trans-1,4-Diphenyl-1,3-butadiene	155	49
2,3-Diphenylbutyronitrile	173	365
2,4-Diphenylbutyronitrile	8	355
Diphenylcarbamyl chloride	21	216
Diphenyl carbinol	39	89
Diphenyl carbonate	115	279
Diphenyl-4-carbonyl chloride	26	216
2,3-Diphenylcinnamonitrile	209	366
Diphenylclycolaldehyde	121	159
Diphenylene ketone oxide	189	178
1,1-Diphenylethane	160	41
1,2-Diphenylethane	27	44
1,2-Diphenylethanol	37	89
Diphenyl ether	7	137
1,2-Diphenylethylamine	186	301
2,2-Diphenylethylamine	24	303
Di-(2-phenylethyl)sulfide	53	426
N,N-Diphenylformamide	39	232
2,2-Diphenylglutaric acid	270	431
d,l-2,3-Diphenylglutaromononitrile	206	366
2,2-Diphenylglutaronitrile	90	361
2,4-Diphenylglutaronitrile	81	361
unsym-Diphenylhydrazine	14	303
Diphenylmethane	2	43
1,6-Diphenylnaphthalene	87	46
1,5-Diphenyl-3-pentadienone	145	177
1,5-Diphenyl-1,5-pentanedione	87	175
1,5-Diphenyl-3-pentanone	221	168
Diphenyl phthalate	108	278
2,2-Diphenylpimelic acid	340	432
1,3-Diphenyl-2-propanone	19	172
2,3-Diphenylpropionaldehyde	34	154
2,2-Diphenylpropionic acid	265	204
α,α-Diphenylpropionyl chloride	23	216
β,β-Diphenylpropiophenone	120	176
2,2-Diphenylpropylamine	7	302
2,3-Diphenylpropylamine	187	301
Di-(3-phenylpropyl)sulfide	42	426
2,2-Diphenylsuberic acid	349	432
Diphenyl succinate	152	281
d,l-2,3-Diphenylsuccinic acid	216	430
meso-2,3-Diphenylsuccinic acid	203	430
d,l-2,3-Diphenylsuccinonitrile	203	366
l-2,3-Diphenylsuccinonitrile	232	367
Diphenyl sulfide	61	422
Diphenyl triketone	92	175
sym-Diphenylurea	488	245
unsym-Diphenylurea	399	242
2,3-Diphenylvaleronitrile	159	364
2,5-Diphenylvaleronitrile	100	361
Dipicolinonitrile	167	364
Dipicryl sulfide	77	427
Dipropargyl	29	29
Di-n-propylacetaldehyde	54	146
Di-n-propyl acetylene	43	29
Di-n-propyladipate	7	272
Di-n-propylamine	36	296
" "	447	439
N,N-Di-n-propylaniline	76	321
" "	229	437
1,3-Dipropylbenzene	106	39
1,4-Dipropylbenzene	114	39
Di-n-propyl carbinol	48	82
Di-n-propyl carbonate	119	255
Dipropylcyanoacetamide	197	366
Di-n-propyl ether	30	132
3,3-Di-n-propylglutaric acid	236	431
Di-n-propyl ketone	52	162
Di-n-propyl maleate	10	272
Di-n-propylmalonic acid	47	429
Dipropylmalononitrile	23	358
"	51	359
Di-n-propyl oxalate	204	260
Di-n-propyl succinate	283	265
Di-n-propyl sulfide	26	420
Di-n-propyl d-tartarate	342	270
Di-n-propyl d,l-tartarate	16	273
2,2′-Dipyridylamine	49	324
4,4′-Dipyridylamine	101	325
1,2-Di-(2-pyrrolyl)ethanedione	94	325
trans-trans-Distyryl	155	49
Ditetradecyl sulfide	25	425
Di-2-thenoylmethane	127	176
1,3-Dithiacyclopentane	43	424

	Compound No.	Page No.
1,3-Dithiane	27	425
1,4-Dithiane	62	426
Dithiocatechol	8	416
1,3-Dithiolane	43	424
Dithioresorcinol	6	416
DL-*erythro*-Dithiothreitol	26	417
threo-Dithiothreitol	10	416
2,2′-Ditolylamine	59	304
Di-*o*-tolylamine	59	304
Di-4-tolyl carbinol	38	89
Di-2-tolyl carbonate	85	277
Di-3-tolyl carbonate	61	276
Di-4-tolyl carbonate	145	280
sym-Di-*m*-tolylethylene	34	44
sym-Di-*o*-tolylethylene	84	46
Di-2-tolyl ketone	86	174
Di-4-tolyl ketone	118	176
Di-2-tolyl oxalate	128	279
Di-3-tolyl oxalate	138	280
Di-4-tolyl oxalate	164	281
Di-4-tolyl succinate	153	281
Di-(2-tolyl) sulfide	35	425
Di-(3-tolyl) sulfide	60	421
Di-(4-tolyl) sulfide	31	425
sym-Di-2-tolylurea	494	246
sym-Di-3-tolylurea	463	244
sym-Di-4-tolylurea	502	246
Di-*n*-tridecylamine	444	439
Di-*n*-tridecyl ketone	110	175
Di-(2,4,6-trinitrophenyl) sulfide	77	427
Di-*n*-undecyl ketone	91	175
Diundecyl sulfide	7	425
Divarinol	135	103
Divinyl acetylene	26	29
Divinyl sulfide	4	420
p-Dixylylamine	11	303
L-Djenkolic acid	108	289
n-Docosane	20	8
n-Docosanoic acid	83	196
1-Docosene	5	26
cis-13-Docosenoic acid	17	193
Docosylcyclohexane	31	8
Docosylcyclopentane	21	8
1-Docosyne	9	31
Dodecanal	18	153
Dodecanamide	160	235
Dodecananilide	46	232
Dodecane	254	7
α,ω-Dodecanedicyanide	17	356
n-Dodecanoic acid	29	194
n-Dodecanoic anhydride	12	225
1-Dodecanol	3	88
2-Dodecanone	2	172
3-Dodecanone	4	169
Dodecanonitrile	101	353
Dodecanoyl chloride	13	214
1-Dodecene	330	24
Dodecyl acetylene	62	30
n-Dodecyl alcohol	3	88
n-Dodecylamine	6	303
n-Dodecylbenzene	178	42
n-Dodecyl bromide	29	58
n-Dodecyl chloride	53	56
n-Dodecylcylohexane	277	7
n-Dodecylcylopentane	274	7
n-Dodecyl ether	13	137
n-Dodecyl methyl ketone	17	172
n-Dodecyl sulfide	12	425
1-Dodecyne	60	30
6-Dodecyne	59	30
n-Dotriacontane	55	9
1-Dotriacontene	15	26
Dotriacontylcyclohexane	66	9
Dotriacontylcyclopentane	54	9
1-Dotriacontyne	19	31
Droserone	25	182
Dulcin	353	241
Dulcitol	64	90
Dunnione	10	181
Durene	73	46
Durenol	257	109
Duridine	117	306
Durohydroquinone	119	435
Dypnone	220	168

E

	Compound No.	Page No.
n-Eicosane	14	8
Eicosanoic acid	78	196
n-Eicosanoic anhydride	32	225
Eicosanol	35	89
Eicosanonitrile	53	359
Eicosene	2	26
Eicosylcyclohexane	24	8
Eicosylcyclopentane	16	8
1-Eicosyne	6	31
Elaidamide	84	233
Elaidic acid	31	194
Elaidyl alcohol	15	88
Embelin	18	182
Enanthaldehyde	52	146
n-Enanthophenone	213	168
Enanthoyl chloride	35	212
α-Epichlorohydrin	41	133
1,2-Epoxybutane	13	131
2,3-Epoxybutane	11	131
Epoxyethane	1	131
1,2-Epoxypropane	5	131
Ergosterol	61	90
Ericanilide	30	231
Erucic acid	17	193
Erythritol	41	89
meso-Erythritol	54	90
D-Erythrose	5	328
DL-Erythrose	6	328
L-Erythrose	4	328
DL-Erythrulose	7	328
L-Erythrulose	8	328
Ethanal	3	144
Ethane	2	3
Ethane-1,2-dicarboxylic acid	294	205
1,2-Ethanediol	101	85
Ethane-1,2-disulfonanilide	14	392
Ethane-1,2-disulfonic acid	1	380
Ethane-1,2-disulfonyl chloride	130	385
1,2-Ethanedithiol	20	413
Ethanesulfonamide	7	392
Ethanesulfonanilide	8	392
Ethanesulfonic acid	2	371
Ethanethiol	2	413
Ethanoic acid	3	190
Ethanol	2	80
Ethanolamine	62	83
"	71	297
Ethene	1	12
Ethoxyacetaldehyde	24	145
Ethoxyacetamide	57	232
Ethoxyacetic acid	34	191
" "	231	431
Ethoxyacetone	6	169
Ethoxyacetonitrile	25	347
α-Ethoxyacetophenone	23	169

Compound	Compound No.	Page No.
Ethoxyacetyl chloride	20	212
2-Ethoxyaniline	141	299
" "	155	437
3-Ethoxyaniline	163	300
" "	135	437
4-Ethoxyaniline	164	300
" "	204	437
2-Ethoxybenzaldehyde	3	153
3-Ethoxybenzaldehyde	88	149
4-Ethoxybenzaldehyde	91	149
3-Ethoxybenzamide	252	237
4-Ethoxybenzamide	435	243
4-Ethoxybenzanilide	347	241
2-Ethoxybenzene-2,5-disulfonamide	396	407
1-Ethoxybenzene-2,5-disulfonic acid	178	377
2-Ethoxybenzene-1,4-disulfonyl chloride	164	386
2-Ethoxybenzenesulfonamide	172	398
3-Ethoxybenzenesulfonamide	79	395
4-Ethoxybenzenesulfonamide	129	396
2-Ethoxybenzenesulfonanilide	160	398
2-Ethoxybenzenesulfonic acid	66	373
3-Ethoxybenzenesulfonic acid	22	371
4-Ethoxybenzenesulfonic acid	50	372
2-Ethoxybenzenesulfonyl chloride	70	383
3-Ethoxybenzenesulfonyl chloride	20	381
4-Ethoxybenzenesulfonyl chloride	24	381
2-Ethoxybenzoic acid	5	193
" "	325	431
3-Ethoxybenzoic acid	193	201
" " "	312	431
4-Ethoxybenzoic acid	316	206
4-Ethoxybenzoic anhydride	46	226
3-Ethoxybenzoyl chloride	20	214
4-Ethoxybenzoyl chloride	32	214
2-Ethoxybiphenyl	14	137
3-Ethoxybiphenyl	16	137
4-Ethoxybiphenyl	46	139
1-Ethoxy-2-butanone	62	163
2-Ethoxyethanethiol	17	413
2-Ethoxyethanol	29	81
" "	51	133
2-(2-Ethoxyethoxy)ethanol	100	85
β-Ethoxyethyl acetate	104	255
β-Ethoxyethyl benzoate	300	267
1-Ethoxy-4-iodobenzene	6	137
4-Ethoxy-3-methoxybenzaldehyde	53	155
1-Ethoxy-2-methoxyethane	37	133
3-Ethoxy-2-methyl-2-butanol	52	133
Ethoxymethyl ethyl ketone	62	163
3-Ethoxy-2-methylolbenzoic acid	219	430
2-Ethoxy-1-naphthaldehyde	100	158
4-Ethoxy-1-naphthaldehyde	61	156
1-Ethoxynaphthalene	108	136
2-Ethoxynaphthalene-1-sulfonamide	158	397
4-Ethoxynaphthalene-1-sulfonamide	198	399
4-Ethoxynaphthalene-2-sulfonamide	241	401
5-Ethoxynaphthalene-1-sulfonamide	238	401
6-Ethoxynaphthalene-1-sulfonamide	140	397
6-Ethoxynaphthalene-2-sulfonamide	239	401
7-Ethoxynaphthalene-2-sulfonamide	101	395
2-Ethoxynaphthalene-1-sulfonanilide	257	401
4-Ethoxynaphthalene-1-sulfonanilide	229	400
4-Ethoxynaphthalene-2-sulfonanilide	110	396
5-Ethoxynaphthalene-1-sulfonanilide	75	394
6-Ethoxynaphthalene-1-sulfonanilide	293	403
6-Ethoxynaphthalene-2-sulfonanilide	138	397
7-Ethoxynaphthalene-2-sulfonanilide	133	396
2-Ethoxynaphthalene-1-sulfonyl chloride	179	387
4-Ethoxynaphthalene-1-sulfonyl chloride	156	386
4-Ethoxynaphthalene-2-sulfonyl chloride	106	384
5-Ethoxynaphthalene-1-sulfonyl chloride	187	387
6-Ethoxynaphthalene-1-sulfonyl chloride	182	387
6-Ethoxynaphthalene-2-sulfonyl chloride	165	386
7-Ethoxynaphthalene-2-sulfonyl chloride	157	386
2-Ethoxy-6-nitroaniline	44	304
4-Ethoxy-2-nitroaniline	235	310
1-Ethoxy-4-nitrobenzene	30	338
3-Ethoxy-1-nitrobenzene	8	337
6-Ethoxy-1-nitronaphthalene-2-sulfonamide	357	406
7-Ethoxy-8-nitronaphthalene-1-sulfonamide	209	399
6-Ethoxy-1-nitronaphthalene-2-sulfonyl chloride	234	388
7-Ethoxy-8-nitronaphthalene-1-sulfonyl chloride	241	389
2-Ethoxyphenol	9	96
3-Ethoxyphenol	24	97
1-Ethoxy-2-propanone	6	169
3-Ethoxypropionic acid	1	192
3-Ethoxypropionitrile	47	348
β-Ethoxypropionyl chloride	1	214
4-Ethoxypyridine	269	438
2-Ethoxytoluene	70	134
3-Ethoxytoluene	76	135
4-Ethoxytoluene	75	135
1-Ethoxy-2,4,6-trinitrobenzene	45	338
N-Ethylacetanilide	12	231
2-Ethylacetanilide	165	235
Ethyl acetate	10	250
Ethyl acetoacetate	133	165
" "	140	256
2-Ethylacetophenone	59	170
Ethyl acetopyruvate	201	260
Ethyl acetylene	3	28
Ethyl acetylglycolate	136	256
Ethyl acrylate	28	251
2-Ethylacrylonitrile	11	345
N-Ethylalanine	65	429
Ethyl alcohol	2	80
Ethylallene	29	13
Ethyl allylacetoacetate	182	259
Ethylamine	3	295
"	431	439
(Ethylamino)acetonitrile	43	348
Ethyl 2-aminobenzoate	15	273
2-(Ethylamino)ethylamine	404	439
1-(Ethylamino)naphthalene	136	437
Ethyl-n-amyl alcohol	56	83
N-Ethylaniline	99	298
"	197	437
2-Ethylaniline	110	298
"	149	437
3-Ethylaniline	175	437
4-Ethylaniline	116	298
Ethyl anisate	315	268
Ethyl anthranilate	15	273
2-Ethylaziridine	294	438
N-Ethylaziridine	291	438
2-Ethylazulene	19	43
N-Ethylbenzanilide	23	231
Ethylbenzene	3	35
2-Ethylbenzenesulfonamide	30	393
4-Ethylbenzenesulfonamide	44	393
2-Ethylbenzenesulfonyl chloride	2	381
4-Ethylbenzenesulfonyl chloride	4	381
Ethyl benzilate	37	274
1-Ethylbenzimidazole	222	437
2-Ethylbenzimidazole	255	438
Ethyl benzoate	202	260
2-Ethylbenzoic acid	247	431
4-Ethylbenzoic acid	358	432
Ethyl benzoylacetate	212	168
" "	311	268
Ethyl 2-benzoylbenzoate	82	277
Ethyl benzoylformate	297	266

	Compound No.	Page No.
N-Ethylbenzylamine	340	438
N-Ethylbenzylsulfonamide	12	392
Ethyl bromide	3	58
Ethyl bromoacetate	107	255
Ethyl-α-bromopropionate	110	255
Ethyl β-bromopropionate	135	256
2-Ethyl-1,3-butadiene	61	14
2-Ethylbutanamide	168	235
2-Ethylbutananilide	209	236
2-Ethylbutane-1,1-dicarboxylic acid	50	194
2-Ethylbutanoic acid	27	194
Ethylbutanol	42	82
2-Ethyl-1-butene	41	13
Ethyl n-butoxyethyl carbonate	221	262
2-Ethylbutyraldehyde	29	145
Ethyl n-butyrate	48	252
2-Ethylbutyric acid	429	432
Ethyl n-caprate	269	264
Ethyl n-caproate	116	255
α-Ethylcaproic acid	45	191
Ethyl n-caprylate	190	259
Ethyl carbamate	4	231
Ethyl carbonate	55	252
Ethyl cellosolve acetate	104	255
Ethyl chloride	3	55
Ethyl chloroacetate	83	253
Ethyl chloroformate	20	250
Ethyl α-chloropropionate	87	253
Ethyl cinnamate	319	268
Ethyl crotonate	73	253
Ethyl cyanoacetate	68	350
Ethylcyclobutane	25	3
Ethylcyclohexane	90	4
Ethyl cyclohexanecarboxylate	166	258
2-Ethylcyclohexanone	44	169
3-Ethylcyclohexanone	146	165
1-Ethylcyclohexene	238	19
3-Ethylcyclohexene	236	19
4-Ethylcyclohexene	231	19
3-Ethyl-2-cyclohexenone	64	170
2-Ethylcyclohexylamine	70	297
Ethylcyclooctatetraene	13	25
Ethylcyclopentane	45	3
2-Ethylcyclopentanone	90	164
1-Ethylcyclopentene	139	16
3-Ethylcyclopentene	115	16
4-Ethylcyclopentene	137	16
Ethylcyclopropane	12	3
S-Ethylcysteine	45	429
9-Ethyl-cis-decahydronaphthalene	259	7
9-Ethyl-trans-decahydronaphthalene	258	7
Ethyl n-decanoate	269	264
Ethyl dichloroacetate	106	255
Ethyl difluoroacetate	25	250
Ethyl 3,5-dinitrobenzoate	129	279
Ethyl 3,5-dinitrosalicylate	134	280
Ethyl diphenylacetate	81	277
2-Ethyl-1,3-dithiolane	51	424
Ethyl enanthate	152	257
Ethylene	1	12
Ethylene bromohydrin	43	82
Ethylene chlorohydrin	24	81
Ethylene cyanohydrin	78	351
1,2-Ethylenediamine	39	296
"	352	438
Ethylenediamine N,N′-diacetic acid	472	433
Ethylenediamine-N,N′-dipropionic acid	473	433
Ethylenediamine-N,N,N′,N′-tetraacetic acid	471	433
Ethylene dichloride	9	63
Ethylene dithioglycol	20	413
Ethylene glycol diacetate	156	257
Ethylene glycol dibenzoate	104	278
Ethylene glycol di-n-butyrate	253	264
Ethylene glycol diformate	126	256
Ethylene glycol di-(β-hydroxyethyl)ether	130	86
Ethylene glycol dilaurate	69	276
Ethylene glycol dimethyl ether	26	132
Ethylene glycol dimyristate	88	277
Ethylene glycol dipalmitate	98	278
Ethylene glycol diphenyl ether	52	139
Ethylene glycol dipropionate	198	260
Ethylene glycol di-n-stearate	111	278
Ethylene glycol monoacetate	148	257
Ethylene glycol monobenzyl ether	128	86
Ethylene glycol mono-n-butyl ether	61	83
" " " "	60	134
Ethylene glycol mono-sec-butyl ether	52	82
Ethylene glycol monoethyl ether	29	81
" " " "	51	133
Ethylene glycol monoethyl ether acetate	104	255
Ethylene glycol monoethyl monomethyl ether	37	133
Ethylene glycol monoformate	138	256
Ethylene glycol monoisobutyl ether	51	82
Ethylene glycol monoisopropyl ether	36	81
Ethylene glycol monomethyl ether	20	81
Ethylene glycol monomethyl ether acetate	78	253
Ethylene glycol monomethyl mono-n-propyl ether	47	133
Ethylene glycol monophenyl ether	122	86
Ethylene glycol mono-n-propyl ether	44	82
Ethyleneimine	10	295
Ethylene oxide	1	131
cis-Ethylene oxide-1,2-dicarboxylic acid	36	429
trans-Ethylene oxide-1,2-dicarboxylic acid	35	429
Ethylene sulfide	1	423
Ethylenesulfonic acid	2	380
Ethyl ether	4	131
Ethyl ethoxyacetate	62	252
Ethyl 4-ethoxybenzoate	324	268
Ethyl β-ethoxyethyl carbonate	161	258
Ethyl ethylacetoacetate	169	258
2-Ethylethylene sulfide	6	423
N-Ethylformamide	5	230
N-Ethylformanilide	9	230
Ethyl formate	3	250
Ethyl 2-furoate	35	274
Ethyl furoylacetate	5	272
Ethyl β-(2-furyl)acrylate	244	263
Ethyl 2-furyl ketone	11	172
N-Ethylglycine	22	285
" "	82	429
N-Ethylglycinonitrile	43	348
Ethyl glycolate	108	255
3-Ethylglutaric acid	341	432
Ethyl hendecylenate	309	267
3-Ethylheptane	127	5
4-Ethylheptane	120	5
Ethyl n-heptanoate	152	257
5-Ethyl-3-heptanone	115	164
5-Ethyl-4-hepten-3-one	128	165
Ethyl hexadecyl sulfide	2	425
Ethyl hexahydrobenzoate	166	258
2-Ethylhexanal-1	57	147
3-Ethylhexane	66	4
Ethyl n-hexanoate	116	255
2-Ethylhexanoic acid	45	191
2-Ethyl-1-hexanol	87	84
2-Ethyl-2-hexenal	13	150
2-Ethyl-3-hexenal	14	150
2-Ethyl-1-hexene	204	18
3-Ethyl-1-hexene	160	17
3-Ethyl-2-hexene	207	18
3-Ethyl-3-hexene	193	18
4-Ethyl-1-hexene	177	18

	Compound No.	Page No.
2-Ethyl-2-hexene	178	18
2-Ethyl-3-hexenoic acid	9	192
2-Ethyl-1-hexen-3-one	86	163
3-Ethyl-5-hexen-2-one	70	163
2-Ethyl-1-hexylacetate	171	258
Ethyl hexyl acetylene	53	30
Ethyl *n*-hexyl ketone	142	165
Ethyl hydrocinnamate	282	265
2-Ethyl-4-hydroxybenzaldehyde	30	154
Ethyl 2-hydroxybenzoate	16	94
Ethyl 3-hydroxybenzoate	106	102
" "	107	278
Ethyl 4-hydroxybenzoate	247	109
" "	148	280
" "	50	434
4-Ethyl-4-hydroxy-3-hexanone	124	165
Ethyl α-hydroxyisobutyrate	95	254
Ethyl 3-hydroxy-2-naphthoate	125	279
Ethyl 3-hydroxy-4-nitrobenzoate	16	434
Ethylidene diacetate	121	255
Ethylidenefluorene	110	47
1-Ethylidene-2-methylcyclohexane	288	22
1-Ethylidene-3-methylcyclohexane	284	21
1-Ethylidene-4-methylcyclohexane	282	21
Ethyl iodide	3	60
Ethyl isoamyl ketone	94	164
Ethyl isobutylacetaldehyde	53	146
Ethyl isobutylacetic acid	13	192
Ethyl isobutyl ether	23	132
Ethyl isobutyl ketone	31	162
Ethyl isobutyl sulfide	22	420
Ethyl isobutyrate	36	251
Ethyl isocrotonate	53	252
Ethyl isopropyl ketone	18	161
Ethyl isopropylacetaldehyde	39	146
1-Ethyl-2-isopropylbenzene	45	37
1-Ethyl-3-isopropylbenzene	42	37
1-Ethyl-4-isopropylbenzene	57	37
Ethyl isopropyl carbinol	21	81
Ethyl isopropyl ether	9	131
3-Ethyl-4-isopropyl-1-methylbenzene	100	39
Ethyl isopropyl sulfide	12	420
1-Ethylisoquinoline	83	321
3-Ethylisoquinoline	88	321
Ethyl isovalerate	67	252
Ethyl *d,l*-lactate	100	254
Ethyl laurate	316	268
Ethyl levulinate	162	166
" "	181	259
Ethyl malonate	173	258
Ethylmalonic acid	136	198
" "	139	430
Ethyl malonic acid diamide	454	244
Ethylmalonic acid monoanilide	284	238
Ethyl malononitrile	67	350
d,l-Ethyl mandelate	40	275
Ethyl margarate	6	273
Ethyl mercaptan	2	413
Ethyl methacrylate	45	251
Ethyl methoxyacetate	60	252
Ethyl 2-methoxybenzoate	305	267
Ethyl 3-methoxybenzoate	299	267
Ethyl 4-methoxybenzoate	315	268
Ethyl β-methoxyethyl carbonate	143	257
Ethyl 1-methoxyethyl ketone	36	162
Ethyl methoxymethyl ketone	29	162
Ethylmethylacetic acid	14	190
Ethyl methylacetoacetate	139	256
D,L-Ethylmethylacetyl chloride	15	212
Ethyl methyl acetylene	13	28
3-Ethyl-2-methylacrolein	43	146
trans-β-Ethyl-α-methylacrylic acid	2	193
Ethyl methyl amine	5	295
N-Ethyl-N-methylaniline	57	320
N-Ethyl-3-methylaniline	122	299
1-Ethyl-2-methylbenzene	12	35
1-Ethyl-3-methylbenzene	9	35
1-Ethyl-4-methylbenzene	10	35
Ethyl 2-methylbenzoate	230	262
2-Ethyl-2-methylbutanoic acid	32	191
2-Ethyl-3-methylbutanoic acid	24	192
4-Ethyl-4-methylbutanoic acid	42	191
2-Ethyl-3-methyl-1-butene	95	15
Ethyl 3-methylbutyl sulfide	34	421
Ethyl methyl carbinol	7	80
1-Ethyl-2-methylcyclohexane	143	5
d,l-cis-1-Ethyl-3-methylcyclohexane	161	5
cis-1-Ethyl-4-methylcyclohexane	148	5
trans-1-Ethyl-4-methylcyclohexane	146	5
1-Ethyl-2-methyl-3-cyclohexenone	101	171
2-Ethyl-4-methylcyclohexanone	161	166
d-2-Ethyl-5-methylcyclohexanone	39	169
d,l-2-Ethyl-5-methylcyclohexanone	154	166
1-Ethyl-3-methylcyclohexene	274	21
1-Ethyl-4-methylcyclohexene	273	21
1-Ethyl-5-methylcyclohexene	277	21
5-Ethyl-1-methylcyclohexene-3-one	180	167
1-Ethyl-1-methylcyclopentane	75	4
cis-1-Ethyl-2-methylcyclopentane	85	4
trans-1-Ethyl-2-methylcyclopentane	73	4
cis-1-Ethyl-3-methylcyclopentane	74	4
trans-1-Ethyl-3-methylcyclopentane	72	4
3-Ethyl-2-methylcyclopentanone	117	164
3-Ethyl-4-methylcyclopentanone	129	165
Ethyl methyl ether	2	131
3-Ethyl-3-methylglutaric acid	221	430
β-Ethyl-β-methylglutaric anhydride	3	225
3-Ethyl-2-methylheptane	211	6
3-Ethyl-3-methylheptane	198	6
3-Ethyl-4-methylheptane	220	6
4-Ethyl-2-methylheptane	181	6
4-Ethyl-3-methylheptane	221	6
4-Ethyl-4-methylheptane	222	6
5-Ethyl-2-methylheptane	175	5
5-Ethyl-3-methylheptane	187	6
3-Ethyl-2-methylhexane	109	5
3-Ethyl-3-methylhexane	119	5
4-Ethyl-2-methylhexane	96	4
4-Ethyl-3-methylhexane	115	4
Ethyl methyl ketone	3	161
Ethylmethylmalonic acid	125	430
1-Ethyl-5-methylnaphthalene	13	43
2-Ethyl-6-methylnaphthalene	16	43
3-Ethyl-2-methylpentane	58	3
3-Ethyl-3-methylpentane	65	4
2-Ethyl-4-methylpentanoic acid	13	192
2-Ethyl-3-methyl-1-pentene	174	17
2-Ethyl-4-methyl-1-pentene	159	17
3-Ethyl-2-methyl-1-pentene	156	17
3-Ethyl-3-methyl-1-pentene	166	17
3-Ethyl-4-methyl-1-pentene	145	17
3-Ethyl-4-methyl-*cis*-2-pentene	189	18
3-Ethyl-4-methyl-*trans*-2-pentene	186	18
1-Ethyl-2-methylphenanthrene	74	46
1-Ethyl-7-methylphenanthrene	79	46
3-Ethyl-6-methylphenanthrene	22	43
7-Ethyl-1-methylphenanthrene	92	46
3-Ethyl-5-methylphenol	101	434
1-Ethyl-2-methylpiperidine	423	439
3-Ethyl-4-methylpyridine	54	320
Ethyl-2-methylstyryl ketone	56	170
Ethyl methyl-sulfide	2	420

Compound	Compound No.	Page No.
2-Ethyl-3-methylthiophene	32	424
2-Ethyl-5-methylthiophene	31	423
3-Ethyl-5-methylthiophene	37	424
N-Ethylmorpholine	282	438
Ethyl myristate	341	270
1-Ethylnaphthalene	152	41
2-Ethylnaphthalene	151	41
Ethyl-1-naphthoate	351	271
Ethyl-2-naphthoate	28	274
Ethyl-1-naphthyl ether	108	136
Ethyl-2-naphthyl ether	19	137
Ethyl-1-naphthyl ketone	216	168
Ethyl-2-naphthyl ketone	71	174
Ethyl-2-naphthyl sulfide	1	425
Ethyl nitrate	16	250
Ethyl nitrite	1	250
N-Ethyl-4-nitroacetanilide	185	235
N-Ethyl-4-nitroaniline	183	308
1-Ethyl-2-nitrobenzene	23	336
1-Ethyl-4-nitrobenzene	28	336
Ethyl-2-nitrobenzoate	26	274
"	321	268
Ethyl-3-nitrobenzoate	54	275
Ethyl-4-nitrobenzoate	77	277
Ethyl-2-nitrocinnamate	51	275
Ethyl-3-nitrocinnamate	117	279
Ethyl-4-nitrocinnamate	160	281
Ethyl-4-nitrophenyl-sulfide	17	425
Ethyl-3-nitrosalicylate	150	280
Ethyl-5-nitrosalicylate	136	280
3-Ethyloctane	225	6
4-Ethyloctane	226	6
Ethyl-n-octanoate	190	259
Ethyl orthoformate	85	253
Ethyl orthosilicate	113	255
Ethyl oxalate	145	257
Ethyl oxamate	176	235
" "	146	280
Ethyl oxanilate	33	231
" "	92	277
Ethyl palmitate	14	273
Ethyl pelargonate	231	262
2-Ethylpentanamide	136	234
2-Ethylpentananilide	86	233
3-Ethylpentane	39	3
Ethyl n-pentanoate	86	253
2-Ethyl pentanoic acid	36	191
" "	428	432
2-Ethyl-1-pentanol	56	83
3-Ethyl-3-pentanol	37	82
3-Ethyl-2-pentanone	46	162
2-Ethyl-1-pentene	106	16
3-Ethyl-1-pentene	85	15
3-Ethyl-2-pentene	112	16
3-Ethyl-3-penten-1-yne	33	29
3-Ethyl-1-pentyne	28	29
1-Ethylphenanthrene	46	44
2-Ethylphenanthrene	55	45
9-Ethylphenanthrene	48	44
2-Ethylphenol	6	94
" "	106	435
3-Ethylphenol	9	94
" "	90	434
4-Ethylphenol	40	98
" "	99	434
Ethyl phenoxyacetate	289	266
Ethyl phenoxymethyl ketone	10	169
d,l-α-Ethylphenylacetamide	65	232
Ethyl phenylacetate	233	262
d,l-α-Ethylphenylacetic acid	26	193
(4-Ethylphenyl)acetic acid	362	432
Ethylphenylglycolaldehyde	22	150
3-Ethyl-3-phenylhexane	141	40
sym-Ethylphenylhydrazine	153	300
unsym-Ethylphenylhydrazine	152	299
Ethyl-β-phenylpropionate	282	265
Ethyl phenyl sulfide	47	424
Ethyl phthalate	336	270
α-Ethylpimelic acid	27	191
N-Ethylpiperidine	12	319
"	393	439
2-Ethylpiperidine	59	297
3-Ethylpiperidine	63	297
4-Ethylpiperidine	62	297
α-Ethylpiperidine	59	297
β-Ethylpiperidine	63	297
γ-Ethylpiperidine	62	297
Ethyl piperonylate	3	273
Ethyl pivalate	44	251
Ethylpropiolic acid	42	194
Ethyl propionate	27	251
Ethyl-n-propylacetic acid	36	191
Ethyl propyl acetylene	38	29
2-Ethyl-3-n-propylacrolein	61	147
Ethyl propyl amine	20	295
1-Ethyl-2-propylbenzene	72	38
1-Ethyl-3-propylbenzene	68	38
1-Ethyl-4-propylbenzene	78	38
Ethyl n-propyl carbinol	30	81
Ethyl-n-propyl ether	14	131
Ethyl n-propyl ketone	22	161
Ethyl n-propylmalonic acid	165	430
Ethyl 3-propylphenyl ketone	24	169
Ethyl n-propylsulfide	16	420
2-Ethylpyridine	22	319
"	239	437
3-Ethylpyridine	28	319
"	221	437
4-Ethylpyridine	33	319
"	238	437
Ethyl pyromucate	35	274
2-Ethyl-2-pyrroline	288	438
Ethyl pyruvate	101	254
2-Ethylquinoline	77	321
3-Ethylquinoline	9	322
4-Ethylquinoline	100	321
α-Ethylquinoline	77	321
γ-Ethylquinoline	9	322
β-Ethylquinoline	100	321
Ethyl salicylate	249	263
" "	16	94
Ethyl stearate	33	274
Ethyl succinic acid	300	431
Ethyl sulfate	188	259
Ethyl sulfide	6	420
1-Ethyl-1,2,3,4-tetrahydronaphthalene	134	40
2-Ethyl-1,2,3,4-tetrahydronaphthalene	132	40
5-Ethyl-1,2,3,4-tetrahydronaphthalene	144	40
6-Ethyl-1,2,3,4-tetrahydronaphthalene	142	40
1-Ethyltetralin	134	40
2-Ethyltetralin	132	40
5-Ethyltetralin	144	40
6-Ethyltetralin	142	40
2-Ethyltetramethylene sulfide	28	423
Ethyl α-thienyl ketone	185	167
2-Ethylthiirane	6	423
2-Ethylthiolane	28	423
2-Ethylthiophene	14	423
3-Ethylthiophene	15	423
Ethyl 2-toluate	230	262
Ethyl 3-toluate	248	263
Ethyl 4-toluate	251	263

	Compound No.	Page No.
m-Ethyltoluene	9	35
o-Ethyltoluene	12	35
p-Ethyltoluene	10	35
N-Ethyltoluene-4-sulfonamide	11	392
N-Ethyl-m-toluidine	122	299
Ethyl 2-(-4-toluyl) benzoate	94	277
Ethyl 2-tolyl ether	70	134
Ethyl 3-tolyl ether	76	135
Ethyl 4-tolyl ether	75	135
Ethyl 4-tolyl sulfide	54	421
Ethyl trichloroacetate	117	255
Ethyl trifluoroacetate	5	250
Ethyl trimethylacetate	44	251
2-Ethyl-1,3,5-trimethylbenzene	99	39
3-Ethyl-1,2,4-trimethylbenzene	108	39
4-Ethyl-1,2,3-trimethylbenzene	113	39
5-Ethyl-1,2,3-trimethylbenzene	107	39
5-Ethyl-1,2,4-trimethylbenzene	101	39
6-Ethyl-1,2,4-trimethylbenzene	102	39
3-Ethyl-2,2,3-trimethylpentane	227	6
3-Ethyl-2,2,4-trimethylpentane	157	5
3-Ethyl-2,4,5-trimethylthiophene	53	424
Ethyl undecylenate	309	267
Ethylurethane	4	231
Ethyl n-valerate	86	253
Ethyl vinyl accetylene	25	29
Ethyl vinyl ether	6	131
Ethyl vinyl ketone	11	161
Ethyl vinyl sulfide	7	420
Ethyne	1	28
Ethynyl methyl ketone	4	161
Ethynyl phenyl ketone	54	173
Eucarvane	192	6
Eucarvene	290	22
Eugenol	23	95
"	97	434
Eugenol acetate	23	274
Eugenol methyl ether	99	136
Euxanthone	567	126

F

Fenchane	145	5
α-Fenchane	197	6
β-Fenchene	278	21
δ-Fenchene	245	20
ξ-Fenchene	266	21
β-Fencholenonitrile	73	351
d-Fenchone	149	166
d, l-Fenchyl alcohol	16	88
Flavol	586	127
Flavopurpurin	62	185
Fluoranthrene	113	47
Fluorene	116	47
Fluorene-2-acetic acid	301	205
Fluorene-9-carboxylic acid	353	208
Fluorene-9-carboxylic acid chloride	18	216
Fluorene-2-sulfonamide	342	405
Fluorene-2-sulfonyl chloride	252	389
Fluorenone	115	176
Fluorenone-1-carboxylic acid chloride	30	216
Fluorenone-4-carboxylic acid chloride	29	216
Di-2-fluorenylmethane	186	51
Fluoroacetamide	151	234
4-Fluoroacetanilide	288	239
Fluoroacetic acid	12	193
" "	103	429
Fluoroacetonitrile	4	345
4-Fluoroacetophenone	153	166
Fluoroacetyl chloride	3	212
Fluoroacetyl fluoride	3	211

	Compound No.	Page No.
2-Fluoroacrylic acid	106	430
2-Fluoroaniline	75	297
" "	86	436
3-Fluoroaniline	85	297
" "	103	436
4-Fluoroaniline	82	297
" "	170	437
2-Fluorobenzaldehyde	65	147
3-Fluorobenzaldehyde	62	147
4-Fluorobenzaldehyde	64	147
4-Fluorobenzamide	293	239
Fluorobenzene	2	69
4-Fluorobenzenesulfonamide	67	394
4-Fluorobenzenesulfonic acid	19	371
4-Fluorobenzenesulfonyl chloride	19	381
2-Fluorobenzoic acid	187	430
3-Fluorobenzoic acid	267	431
4-Fluorobenzoic acid	285	204
" "	304	431
4-Fluorobenzonitrile	27	358
2-Fluorobenzoyl chloride	43	213
3-Fluorobenzoyl chloride	37	212
4-Fluorobenzoyl chloride	39	212
2-Fluoroethanol	9	80
3-Fluoromandelic acid	328	431
1-Fluoronaphthalene	8	69
2-Fluoronaphthalene	9	69
4-Fluoronaphthalene-1-sulfonamide	328	404
5-Fluoronaphthalene-1-sulfonamide	303	403
6-Fluoronaphthalene-2-sulfonamide	82	395
4-Fluoronaphthalene-1-sulfonanilide	112	396
6-Fluoronaphthalene-2-sulfonanilide	74	394
2-Fluoronaphthalene-6-sulfonic acid	24	371
4-Fluoronaphthalene-1-sulfonic acid	153	376
5-Fluoronaphthalene-1-sulfonic acid	143	376
4-Fluoronaphthalene-1-sulfonyl chloride	109	384
5-Fluoronaphthalene-1-sulfonyl chloride	193	387
6-Fluoronaphthalene-2-sulfonyl chloride	139	385
4-Fluoro-1-naphthylamine	41	304
4-Fluoro-1-nitrobenzene	2	337
3-Fluoro-2-nitrophenol	22	97
3-Fluoro-4-nitrophenol	29	97
5-Fluoro-2-nitrophenol	13	96
2-Fluorophenol	53	434
3-Fluorophenol	70	434
4-Fluorophenol	93	434
(2-Fluorophenoxy) acetic acid	153	430
(3-Fluorophenoxy) acetic acid	152	430
(4-Fluorophenoxy) acetic acid	161	430
(4-Fluorophenyl) acetic acid	333	431
3-(2-Fluorophenyl) alanine	53	429
3-(3-Fluorophenyl) alanine	49	429
3-(4-Fluorophenyl) alanine	54	429
2-Fluoropropionic acid	5	192
2-Fluoropyridine	10	436
3-Fluoropyridine	78	436
2-Fluorotoluene	5	69
3-Fluorotoluene	6	69
4-Fluorotoluene	7	69
Formaldehyde	1	144
Formaldehyde cyanohydrine	55	349
Formamide	4	230
Formanilide	5	231
Formic acid	2	190
" "	243	431
Form-N-piperidide	6	230
4-Formylbenzenesulfonamide	62	394
4-Formylbenzenesulfonic acid	17	371
2-Formylbenzoic acid	86	157
3-Formylbenzoic acid	122	159
" "	271	204
4-Formylbenzoic acid	130	160

Compound	Compound No.	Page No.
3-Formylbenzonitrile	101	361
3-Formylbenzoyl chloride	31	214
4-Formylbenzoyl chloride	9	216
N-Formylglycine	198	430
2-Formyl-6-methoxyphenol	33	434
4-Formyl-2-methoxyphenol	28	434
5-Formyl-2-methoxyphenol	54	434
5-Formyl-2-nitrophenol	14	434
2-Formylphenol	21	434
3-Formylphenol	204	106
" "	35	434
4-Formylphenol	30	434
N-Formylpiperidine,	6	230
3-Formylsalicylic acid	123	160
5-Formylsalicylic acid	129	160
4-Formylstyrene	18	152
Fritzsches reagent	137	342
D-Fructose	7	329
D-Fucose	33	331
DL-Fucose	44	332
L-Fucose	34	331
Fumaric acid	399	209
" "	142	430
Fumaric acid diamide	501	246
Fumaric acid dianilide	508	246
Fumaronitrile	127	363
Fumaryl dichloride	38	215
3-Furaldehyde	46	146
Furamide	262	238
Furan	3	131
2-Furanacrylamide	335	240
Furanacrylic acid	208	201
Furancarbinol	59	134
2-Furancarbonitrile	34	347
2-Furancarboxaldehyde	55	147
2-Furancarboxylic acid	184	200
3-Furancarboxylic acid	159	199
Furanilide	197	236
Furfural	55	147
Furfuralacetone	23	172
Furfuralacetophenone	10	172
Furfural diacetate	27	154
Furfuryl acetate	130	256
Furfuryl alcohol	64	83
" "	59	134
" "	57	296
Furfurylamine	345	438
2-(Furfurylamino) ethylamine	68	276
Furfuryl diacetate	185	178
Furil	184	200
2-Furoic acid	159	199
3-Furoic acid	58	90
Furoin	166	177
" "	34	347
α-Furonitrile	34	212
Furoyl chloride	58	90
Furoyl furyl carbinol	65	195
2-Furylacetic acid	31	154
2-(2-Furyl)acrolein	208	201
β-(2-Furyl) acrylic acid	335	240
" " "	64	83
2-Furylcarbinol	57	296
α-Furylmethylamine	15	172
2-Furyl methyl ketone	40	173
2-Furyl phenyl ketone	2	150
3-(2-Furyl) propionaldehyde	59	174
2-Furyl-2-thenoylmethane		

G

G-Acid	12	380
α-D-Galactose	53	332
β-D-Galacturonic acid	42	332

Compound	Compound No.	Page No.
DL-Galactose	47	332
L-Galactose	46	332
3-(β-D-Galactosido)-D-arabinose	52	332
4-(β-D-Galactosido)-D-glucose	64	333
6-(α-D-Galactosido)-D-glucose	2	329
2-[6-(α-D-Galactosido]α-D-glucoside)-β-D-fruc-tose	14	330
Gallacetophenone	454	120
" "	188	178
Gallaldehyde	128	160
Gallaldehyde trimethyl ether	67	156
Gallamide	492	246
Gallanilide	442	244
Gallic acid	386	209
Gallodiacetophenone	486	122
Gammexane, 666	3	64
Gentiobiose	60	333
Geranial	82	148
Geraniol	120	86
Geraniolene	254	20
Geranyl acetate	264	264
Gibberene	111	47
α-D-Glucoheptose	61	333
D-Glucoheptulose	56	333
Glucosamine	12	329
α-D-Glucose	35	331
α-L-Glucose	30	331
β-D-Glucose	37	331
2-(α-D-Glucosido)-β-D-fructose	55	332
3-[α-D-Glucosido]-D-fructose	39	331
2-[3-(α-D-Glucosido)-D-fructosido]-α-D-glucose	38	331
1-(α-D-Glucosido)-α-D-glucose	62	333
4-(α-D-Glucosido)-β-D-glucose	43	332
4-(α-D-Glucosido)-β-D-glucose	65	333
6-(β-D-Glucosido)-D-glucose	60	333
4-(β-D-Glucosido)-β-D-mannose	57	333
β-D-Glucuronic acid	48	332
trans-Glutaconic acid	192	201
" "	245	431
L-Glutamamide	328	240
DL-Glutamic acid	30	285
" "	58	429
L-Glutamic acid	41	286
L-Glutamine	24	285
Glutaraldehyde	69	147
Glutaric acid	107	197
" "	357	432
Glutaric acid diamide	360	241
Glutaric acid dianilide	472	245
Glutaric anhydride	20	225
Glutaronitrile	102	353
Glutaryl dichloride	46	213
DL-Glyceraldehyde (dimer)	115	159
" "	28	330
Glycerol	131	86
α-Glyceryl phenyl ether	40	89
" " " "	41	138
Glyceryl triacetate	298	267
Glyceryl tribenzoate	102	278
Glyceryl tributyrate	352	271
Glyceryl tripropionate	335	269
Glyceryl tristearate	100	278
Glycine	57	286
" "	83	429
Glycineamide	95	305
Glycinonitrile	22	356
Glycol	101	85
Glycolaldehyde	82	157
" "	6	329
Glycolaldehyde phenyl ether	79	148
Glycolanilide	102	233

Compound	No.	Page No.		Compound	No.	Page No.
Glycolamide	189	236		1-Heptadecanol	30	89
Glycolic acid	79	196		"	11	153
Glycolic acid ethyl ether	34	191		9-Heptadecanone	50	173
Glycolic acid methyl ether	31	191		Heptadecanonitrile	24	358
Glycollic aldehyde	6	329		1-Heptadecene	336	24
Glycolonitrile	55	349		Heptadecylamine	45	304
Glycylglycine	77	288		Heptadecylcyclohexane	15	8
Glyoxal	5	144		Heptadecylcyclopentane	7	8
Guaiacol	12	96		n-Heptadecyl methyl ketone	61	174
"	9	137		1-Heptadecyne	1	31
Guaiacol acetate	261	264		1,2-Heptadiene	135	16
Guaiacol carbonate	126	279		1,4-Heptadiene	119	16
δ-Guaiazulene	6	43		2,4-Heptadiene	146	17
Guaiene	109	47		1,6-Heptadiyne	41	29
Guanidinoacetic acid	121	430		Heptafluoro-n-butyric acid	1	429
D-Gulose	10	328		4,4,5,5,6,6,6-Heptafluorohexanoic acid	317	431
DL-Gulose	3	330		Heptanal	52	146
				Heptanamide	95	233
				Heptananilide	35	231
H				n-Heptane	41	3
				Heptane-1,7-dicarboxylic acid	132	198
Hemimellitene	18	36		2,4-Heptanedione	121	165
Hendecanal	6	150		Heptane-1-sulfonamide	16	392
n-Hendecanamide	128	234		Heptane-1-sulfonic acid	4	371
n-Hendecananilide	37	231		Heptane-1-sulfonyl chloride	6	381
1-Hendecane	247	7		1-Heptanethiol	29	414
n-Hendecanoic acid	7	193		2,4,6-Heptanetrione	48	173
n-Hendecanoic anhydride	8	225		n-Heptanoic acid	44	191
1-Hendecanol	125	86		"	455	432
d, l-2-Hendecanol	117	86		n-Heptanoic anhydride	16	223
2-Hendecanone	186	167		1-Heptanol	75	84
6-Hendecanone	183	167		d, l-2-Heptanol	50	82
1-Hendecanonitrile	90	352		d, l-4-Heptanol	48	82
10-Hendecanonitrile	92	352		2-Heptanone	69	163
1-Hendecene	328	24		3-Heptanone	64	163
2-Hendecene	326	24		4-Heptanone	52	162
5-Hendecene	327	24		Heptanonitrile	56	349
10-Hendecen-1-oic acid	4	193		n-Heptanoyl chloride	35	212
w-Hendecenoyl chloride	6	214		Heptatriacontane	70	9
n-Hendecylamine	155	300		1-Heptatriacontene	20	26
"	418	439		1-Heptatriacontyne	24	31
n-Hendecylbenzene	175	42		2-Heptenal	17	150
n-Hendecyl chloride	52	56		1-Heptene	104	15
n-Hendecylcyclohexane	275	7		cis-2-Heptene	116	16
n-Hendecylcyclopentane	271	7		trans-2-Heptene	114	16
α-Hendecyleneamide	67	232		cis-3-Heptene	110	16
n-Hendecyl methyl ketone	9	172		trans-3-Heptene	109	16
n-Hendecyl phenyl ketone	45	173		1-Hepten-4-one	60	163
n-Hendecyl sulfide	7	425		1-Hepten-5-one	3	169
1-Hendecyne	57	30		1-Hepten-6-one	12	169
n-Heneicosane	17	8		2-Hepten-4-one	81	163
11-Heneicosanone	84	174		cis-3-Hepten-2-one	37	169
1-Heneicosene	4	26		trans-3-Hepten-2-one	21	169
Heneicosylcyclohexane	28	8		3-Hepten-6-one	15	169
Heneicosylcyclopentane	19	8		5-Hepten-2-one	71	163
1-Heneicosyne	8	31		n-Heptoic acid	44	191
2-Heneicosyne	7	31		n-Heptyl acetate	158	257
n-Hentriacontane	52	9		n-Heptyl acetylene	47	30
1-Hentriacontene	14	26		n-Heptyl alcohol	75	84
Hentriacontylcyclohexane	63	9		sec-Heptyl alcohol	50	82
Hentriacontylcyclopentane	51	9		n-Heptylamine	64	297
1-Hentriacontyne	18	31		"	424	439
n-Heptacosane	38	8		n-Heptylbenzene	145	40
1-Heptacosene	12	26		n-Heptyl bromide	22	58
Heptacosylcyclohexane	50	9		n-Heptyl n-butyrate	223	262
Heptacosylcyclopentane	37	8		n-Heptyl n-caproate	304	267
1-Heptacosyne	14	31		n-Heptyl n-caprylate	337	270
Heptadecanamide	149	234		n-Heptyl chloride	37	56
n-Heptadecane	5	8		n-Heptylcyclohexane	264	7
Heptadecane-1,1-dicarboxylic acid	161	199		n-Hepthylcyclopentane	257	7
n-Heptadecanoic acid	55	195		n-Heptyl n-enanthate	327	269
n-Heptadecanoic anhydride	25	225		n-Heptyl ether	103	136

Compound	Compound No.	Page No.		Compound	Compound No.	Page No.
n-Heptyl formate	132	256		*trans*-Hexahydro-*p*-cresol	67	83
n-Heptyl *n*-heptanoate	327	269		1,2,3,4,5,6-Hexahydrocyclohexane	67	90
n-Heptyl *n*-hexanoate	304	267		*cis*-Hexahydroisophthalic anhydride	19	223
n-Heptyl iodide	19	60		Hexahydrophenylacetyl chloride	25	214
n-Heptyl mercaptan	29	414		*cis*-Hexahydrophthalic anhydride	6	225
n-Heptyl methyl ketone	150	166		*trans*-DL-Hexahydrophthalic acid anhydride	63	226
1-Heptylnaphthalene	179	42		L-Hexahydrotyrosine	93	288
2-Heptylnaphthalene	180	42		Hexahydroxybenzene	503	122
n-Heptyl *n*-octanoate	337	270		Hexamethylbenzene	167	50
n-Heptyl *n*-pentanoate	276	265		Hexamethylenediamine	30	303
n-Heptyl propionate	195	260		Hexamethylene sulfide	41	424
n-Heptyl sulfide	62	422		Hexamethylene tetramine	102	325
n-Heptyl *n*-valerate	276	265		Hexanal	38	146
2-Heptynal	4	152		*n*-Hexanamide	115	233
1-Heptyne	35	29		*n*-Hexananilide	88	233
2-Heptyne	39	29		*n*-Hexane	24	3
3-Heptyne	38	29		1,6-Hexanedial	26	151
Heptyne-1-carboxylic acid	16	190		Hexane-1,6-dicarboxylic acid	215	202
2-Heptynoic acid	3	192		2,3-Hexanedione	82	163
3-Heptynoic acid	2	192		2,5-Hexanedione	148	165
Hexabenzobenzene	204	51		1,6-Hexanedithiol	46	415
Hexachlorobenzene	35	72		1,2,3,4,5,6-Hexanehexol	64	90
Hexachloroethane	5	64		1-Hexanethiol	21	413
n-Hexacosane	34	8		2-Hexanethiol	19	413
Hexacosylcyclohexane	46	9		*n*-Hexanoic acid	33	191
Hexacosylcyclopentane	32	8		" "	454	432
1-Hexacosyne	13	31		*n*-Hexanoic anhydride	15	223
n-Hexadecanal	6	153		1-Hexanol	49	82
n-Hexadecane	2	8		*d, l*-2-Hexanol	33	81
Hexadecane-1-sulfonamide	29	393		3-Hexanol	30	81
Hexadecane-1-sulfonic acid	7	371		2-Hexanone	24	161
Hexadecane-1-sulfonyl chloride	47	382		3-Hexanone	22	161
Hexadecanethiol	2	416		*n*-Hexanonitrile	42	348
n-Hexadecanoic acid	58	195		*n*-Hexanoyl bromide	12	218
n-Hexadecanoic anhydride	24	225		*n*-Hexanoyl chloride	32	212
1-Hexadecanol	25	88		Hexatriacontane	68	9
Hexadecanonitrile	18	357		1-Hexatriacontene	19	26
1-Hexadecene	335	24		Hexatriacontylcyclohexane	75	9
n-Hexadecylacetate	13	273		Hexatriacontylcyclopentane	67	9
n-Hexadecyl bromide	1	59		1-Hexatriacontyne	23	31
Hexadecyl chloride	54	56		*cis*-1,3,5-Hexatriene	73	14
Hexadecylcyclohexane	12	8		*trans*-1,3,5-Hexatriene	72	14
Hexadecylcyclopentane	4	8		2-Hexenal	50	146
n-Hexadecyl iodide	1	61		3-Hexenal	51	146
n-Hexadecyl palmitate	66	276		1-Hexene	40	13
n-Hexadecyl stearate	79	277		*cis*-2-Hexene	53	13
n-Hexadecyl sulfide	33	425		*trans*-2-Hexene	49	13
1-Hexadecyne	64	30		2-Hexene (*cis,trans* mixture)	50	13
2,4-Hexadienal	5	152		*cis*-3-Hexene	44	13
1,2-Hexadiene	65	14		*trans*-3-Hexene	46	13
trans-1,3-Hexadiene	42	13		3-Hexene (*cis, trans* mixture)	45	13
1,(*cis*-or *trans*)3-Hexadiene	59	14		2-Hexenoic acid	15	193
1,4-Hexadiene	57	14		" "	425	432
1,5-Hexadiene	38	13		3-Hexenoic acid	394	432
2,3-Hexadiene	51	13		4-Hexenoic acid	426	432
2,4-Hexadiene	75	14		5-Hexenoic acid	427	432
2,4-Hexadien-1-ol	11	88		4-Hexen-3-one	43	162
1,5-Hexadien-3-yne	26	29		5-Hexen-2-one	28	162
1,4-Hexadiyne	21	28		1-Hexen-3-yne	25	29
1,5-Hexadiyne	29	29		1-Hexen-4-yne	27	29
Hexahydrobenzaldehyde	56	147		2-Hexen-4-yne	30	29
Hexahydrobenzanilide	272	238		*n*-Hexyl acetate	133	256
Hexahydrobenzoic acid	10	193		*n*-Hexyl acetylene	42	29
Hexahydrobenzoic anhydride	4	225		*n*-Hexyl alcohol	49	82
Hexahydrobenzoyl chloride	36	212		*n*-Hexylamine	46	296
Hexahydrobenzyl alcohol	84	84		N-*n*-Hexylaniline	216	437
Hexahydrobenzylamine	68	297		*n*-Hexylbenzene	124	39
cis-Hexahydro-*m*-cresol	65	83		*n*-Hexyl bromide	20	58
trans-Hexahydro-*m*-cresol	68	83		2-Hexyl bromide	19	58
cis-Hexahydro-*o*-cresol	58	83		*n*-Hexyl *n*-butyrate	186	259
trans-Hexahydro-*o*-cresol	60	83		*n*-Hexyl *n*-caproate	278	265
cis-Hexahydro-*p*-cresol	66	83		*n*-Hexyl *n*-caprylate	328	269

Compound	Compound No.	Page No.
n-Hexyl chloride	34	56
2-Hexyl chloride	31	56
3-Hexyl chloride	30	56
n-Hexylcyclohexane	256	7
n-Hexylcyclopentane	250	7
n-Hexyl n-decanoate	278	265
n-Hexyl n-enanthate	303	267
n-Hexyl ether	95	135
n-Hexyl ethyl ether	53	133
n-Hexyl formate	103	254
n-Hexyl n-heptanoate	303	267
n-Hexyl iodide	18	60
n-Hexyl mercaptan	21	413
sec-Hexyl mercaptan	19	413
n-Hexyl methyl acetylene	45	29
n-Hexyl methyl ether	48	133
n-Hexylmethylglycolic acid	22	193
n-Hexyl methyl ketone	114	164
1-Hexylnaphthalene	176	42
2-Hexylnaphthalene	177	42
n-Hexyl n-octanoate	328	269
n-Hexyl n-pentanoate	228	262
n-Hexyl phenyl carbinol	129	86
n-Hexyl phenyl ketone	213	168
n-Hexyl propionate	155	257
2-Hexylpyridine	242	437
n-Hexyl n-valerate	228	262
1-Hexyne	17	28
2-Hexyne	24	29
3-Hexyne	22	28
Hippuramide	383	242
Hippuranilide	444	244
Hippuric acid	296	205
" "	398	242
" "	25	285
DL-Histidine	31	429
L-Histidine	95	288
Homoanthranilic acid	18	284
DL-Homoaspartic acid	61	287
Homophenetole	71	134
Homophthalic acid	292	205
Homophthalic anhydride	64	226
Homopimanthrene	79	46
d-Hydnocarpamide	172	235
Hydnocarponitrile	4	355
Hydrazobenzene	278	311
1-Hydrindamine	125	299
2-Hydrindamine	143	299
4-Hydrindamine	150	299
5-Hydrindamine	21	303
α-Hydrindone	32	173
Hydrocinnamaldehyde	81	148
Hydrocinnamamide	140	234
Hydrocinnamanilide	104	233
Hydrocinnamic acid	38	194
Hydrocinnamoyl chloride	48	213
Hydrocinnamyl alcohol	123	86
Hydrophlorone	540	124
Hydroquinone	452	119
" "	117	435
Hydroquinone diacetate	154	281
Hydroquinone dibenzoate	170	281
Hydroquinone dibenzyl ether	55	139
Hydroquinone diethyl ether	43	139
Hydroquinone dimethyl ether	34	138
Hydroquinone monomethyl ether	48	99
" " "	32	138
Hydroquinone triacetate	133	279
Hydrotropilidene	211	19
1-Hydroxyacenaphthene	353	114
Hydroxyacetaldehyde	82	157
2-Hydroxyacetanilide	511	123
3-Hydroxyacetanilide	375	115
4'-Hydroxyacetanilide	443	119
" "	340	240
Hydroxyacetic acid	79	196
" "	263	431
α-Hydroxyacetophenone	47	89
" "	150	177
2-Hydroxyacetophenone	10	96
" "	7	172
3-Hydroxyacetophenone	180	105
" "	121	176
4-Hydroxyacetophenone	223	107
" "	141	176
2-Hydroxyaniline	456	120
" "	383	315
" "	88	434
3-Hydroxyaniline	277	110
" "	263	311
" "	88	434
4-Hydroxyaniline	476	121
" "	396	316
1-Hydroxyanthracene	381	115
9-Hydroxyanthracene	269	110
1-Hydroxy-9,10-anthraquinone	32	183
2-Hydroxy-9,10-anthraquinone	60	185
Hydroxyanthrarufin	49	184
2-Hydroxyazobenzene	134	103
4-Hydroxyazobenzene	383	116
2-Hydroxybenzaldehyde	4	94
" "	71	148
3-Hydroxybenzaldehyde	204	106
" "	90	158
" "	35	434
4-Hydroxybenzaldehyde	102	158
" "	252	109
" "	30	434
2-Hydroxybenzaldehyde-3-carboxylic acid	123	160
4-Hydroxybenzaldehyde-3-carboxylic acid	129	160
3-Hydroxybenzamide	342	240
4-Hydroxybenzamide	322	240
2-Hydroxybenzanilide	437	119
3-Hydroxybenzanilide	457	120
" "	311	239
4-Hydroxybenzanilide	539	124
" "	421	243
4-Hydroxybenzene-1,3-disulfonamide	412	408
4-Hydroxybenzene-1,3-disulfonanilide	324	404
4-Hydroxybenzene-1,3-disulfonic acid	190	378
4-Hydroxybenzene-1,3-disulfonyl chloride	115	384
4-Hydroxybenzenesulfonamide	218	400
4-Hydroxybenzenesulfonic acid	95	374
4-Hydroxybenzenesulfonyl chloride	228	388
4-Hydroxybenzenethiol	9	416
2-Hydroxybenzoic acid	400	117
" " "	244	203
" " "	116	430
3-Hydroxybenzoic acid	505	123
" " "	320	206
" " "	268	431
4-Hydroxybenzoic acid	532	124
" " "	343	207
" " "	406	432
2-Hydroxybenzonitrile	132	363
4-Hydroxybenzonitrile	158	364
" "	34	434
2-Hydroxybenzophenone	26	97
" "	28	173
3-Hydroxybenzophenone	251	109
" "	149	177
4-Hydroxybenzophenone	322	112
" "	165	177
7-Hydroxybenzo[a]pyrene	544	125

Compound	Compound No.	Page No.
2-Hydroxybenzyl alcohol	49	89
" " "	146	103
4-Hydroxybenzyl alcohol	294	111
4-Hydroxybenzylamine	177	308
2-Hydroxybenzylaniline	234	310
N-4-Hydroxybenzylaniline	432	317
2-Hydroxybenzyl cyanide	163	364
4-Hydroxybenzyl cyanide	87	361
2-Hydroxybiphenyl	51	99
" "	98	434
3-Hydroxybiphenyl	117	102
" "	81	434
4-Hydroxybiphenyl	431	118
" "	76	434
6-Hydroxybiphenyl-2-carboxylic acid	389	116
3-Hydroxy-2-butanone	39	82
" " "	54	162
4-Hydroxy-n-butylamine	387	439
2-Hydroxybutyraldehyde	23	150
4-Hydroxybutyraldehyde	6	152
2-Hydroxybutyric acid	251	431
2-Hydroxybutyronitrile	3	355
4-Hydroxybutyronitrile	82	352
3-Hydroxycarbostyril	574	126
2-Hydroxy-3-chlorobutyric acid	151	430
2-Hydroxy-3-chloroisobutyric acid	174	430
2-Hydroxychrysazin	40	183
trans-2-Hydroxycinnamic acid	333	207
" " "	405	432
3-Hydroxycinnamic acid	367	432
4-Hydroxycinnamic acid	336	207
4-Hydroxycoumarin	516	123
5-Hydroxycoumarin	558	126
trans-2-Hydroxycyclohexanecarboxylic acid	419	432
cis-3-Hydroxycyclohexanecarboxylic acid	404	432
trans-3-Hydroxycyclohexanecarboxylic acid	442	432
cis-4-Hydroxycyclohexanecarboxylic acid	448	432
trans-4-Hydroxycyclohexanecarboxylic acid	420	432
2-(1'-Hydroxycyclopentyl)cyclopetanone	13	172
3-Hydroxydimethylaniline	144	103
α-Hydroxydiphenylacetic acid	227	202
α-Hydroxydiphenylacetyl chloride	23	214
2-Hydroxydiphenyl ether	216	107
2-Hydroxydiphenylmethane	46	99
4-Hydroxydiphenylmethane	137	103
2-Hydroxyethanethiol	22	413
β-Hydroxyethyl acetate	148	257
Di-2-Hydroxyethylamine	180	301
2-Hydroxyethylamine	329	438
4-Hydroxyethylbenzene	40	98
β-Hydroxyethyl formate	138	256
2-Hydroxyethyl 4-tolyl sulfide	59	421
1-Hydroxyfluorenone	246	109
3-Hydroxyglutaric acid	77	429
6-Hydroxy-n-hexylamine	406	439
4-Hydroxy-3-iodo-5-methoxybezaldehyde	470	121
2-Hydroxyisobutanoic acid	80	196
α-Hydroxyisobutyramide	105	233
α-Hydroxyisobutyranilide	249	237
α-Hydroxyisobutyric acid	80	196
" " "	291	431
2-Hydroxyisobutyronitrile	18	346
4-Hydroxyisocaproic acid	453	432
3-Hydroxy-2-isopropylpropionaldehyde	7	152
Hydroxymalonic acid	240	202
2-Hydroxy-4-methoxyacetophenone	45	99
3-Hydroxy-4-methoxyaniline	275	311
2-Hydroxy-3-methoxybenzaldehyde	33	434
3-Hydroxy-4-methoxybenzaldehyde	54	434
4-Hydroxy-3-methoxybenzaldehyde	126	103
" " " "	70	156
" " " "	28	434
4-Hydroxy-3-methoxybenzoic acid	525	124
3-Hydroxy-4-methoxybenzonitrile	19	346
4-Hydroxy-3-methoxybenzyl alcohol	245	109
2-Hydroxy-3-methoxybenzylamine	51	434
3-Hydroxy-2-methoxybenzylamine	55	434
4-Hydroxy-3-methoxybenzylamine	56	434
4-Hydroxy-3-methoxystyryl methyl ketone	159	177
2-Hydroxy-3-methylaniline	156	307
2-Hydroxy-4-methylaniline	363	314
2-Hydroxy-5-methylaniline	302	312
3-Hydroxy-4-methylaniline	362	314
4-Hydroxy-2-methylaniline	390	315
4-Hydroxy-3-methylaniline	384	315
5-Hydroxy-2-methylaniline	324	313
2-Hydroxy-5-methylbenzaldehyde	41	155
3-Hydroxy-3-methyl-2-butanone	47	162
3-Hydroxy-3-methylbutyraldehyde	8	152
2-Hydroxy-2-methylbutyric acid	283	431
7-Hydroxy-4-methylcoumarin	480	121
5-Hydroxymethylfurfural	10	153
3-Hydroxy-3-methyl-2-heptanone	38	169
5-Hydroxy-5-methyl-3-heptanone	22	169
2-Hydroxy-2-methylhexanal	11	150
2-Hydroxy-5-methyl-4-nitroaniline	417	316
3-Hydroxy-3-methyl-2-pentanone	34	169
4-Hydroxy-4-methyl-2-pentanone	59	83
" " " "	98	164
4-Hydroxy-2-methylquinoline	98	325
5-Hydroxy-2-methylquinoline	100	325
6-Hydroxy-2-methylquinoline	96	325
8-Hydroxy-2-methylquinoline	43	323
8-Hydroxy-4-methylquinoline	12	434
d,l-2-Hydroxy-2-methylsuccinic acid	154	199
1-Hydroxy-2-naphthaldehyde	49	155
2-Hydroxy-1-naphthaldehyde	132	103
" " "	71	156
3-Hydroxy-2-naphthaldehyde	87	157
4-Hydroxy-1-naphthaldehyde	124	160
7-Hydroxynaphthalene-1,3-disulfonamide	294	403
2-Hydroxynaphthalene-1,5-disulfonanilide	392	407
2-Hydroxynaphthalene-1,6-disulfonanilide	276	402
2-Hydroxynaphthalene-1,7-disulfonanilide	394	407
2-Hydroxynaphthalene-3,6-disulfonanilide	316	404
2-Hydroxynaphthalene-1,5-disulfonic acid	19	380
2-Hydroxynaphthalene-1,6-disulfonic acid	9	380
2-Hydroxynaphthalene-1,7-disulfonic acid	20	380
2-Hydroxynaphthalene-3,6-disulfonic acid	16	380
7-Hydroxynaphthalene-1,3-disulfonic acid	12	380
2-Hydroxynaphthalene-1,5-disulfonyl chloride	280	390
2-Hydroxynaphthalene-1,6-disulfonyl chloride	171	386
2-Hydroxynaphthalene-1,7-disulfonyl chloride	262	389
7-Hydroxynaphthalene-1,3-disulfonyl chloride	251	389
3-Hydroxynaphthalene-2-sulfonamide	45	393
4-Hydroxynaphthalene-1-sulfonamide	188	399
6-Hydroxynaphthalene-2-sulfonamide	411	408
8-Hydroxynaphthalene-1-sulfonamide	365	406
4-Hydroxynaphthalene-1-sulfonanilide	310	404
5-Hydroxynaphthalene-1-sulfonanilide	315	404
6-Hydroxynaphthalene-2-sulfonanilide	167	398
7-Hydroxynaphthalene-1-sulfonanilide	296	403
1-Hydroxynaphthalene-2-sulfonic acid	8	380
3-Hydroxynaphthalene-2-sulfonic acid	11	371
4-Hydroxynaphthalene-1-sulfonic acid	14	380
5-Hydroxynaphthalene-1-sulfonic acid	15	380
6-Hydroxynaphthalene-2-sulfonic acid	186	377
7-Hydroxynaphthalene-1-sulfonic acid	13	380
3-Hydroxynaphthalene-2-sulfonyl chloride	173	386
2-Hydroxy-1-naphthalenethiol	17	416
4-Hydroxy-1-naphthalenethiol	38	417
6-Hydroxy-2-naphthalenethiol	44	417
2-Hydroxynaphthalene-3,6,8-trisulfonanilide	132	396
7-Hydroxynaphthalene-1,3,6-trisulfonic acid	7	380

Compound	No.	Page No.	Compound	No.	Page No.
7-Hydronaphthalene-1,3,6-trisulfonyl chloride	269	390	4-Hydroxyphenyl sulfide	72	427
3-Hydroxy-2-naphthamide	462	244	3-Hydroxyphthalic acid	412	117
4-Hydroxy-2-naphthamide	460	244	DL-Hydroxyproline	30	429
3-Hydroxy-2-naphthanilide.................	493	246	L-Hydroxyproline	83	288
3-Hydroxy-1-naphthoic acid.................	571	126	2-Hydroxypropane-1,2,3-tricarboxylic acid.......	228	202
3-Hydroxy-2-naphthoic acid.................	550	125	1-Hydroxy-2-propanone	40	82
" " "	350	207	" " "	58	163
" " "	113	430	2-Hydroxypropionaldehyde	8	150
4-Hydroxy-2-naphthoic acid.................	351	208	3-Hydroxypropionitrile.....................	74	351
5-Hydroxy-1-naphthoic acid.................	563	126	4-Hydroxypropiophenone	370	115
5-Hydroxy-1,4-naphthoquinone	21	182	" "	180	178
3-Hydroxy-1-naphthylamine	397	316	DL-2-Hydroxy-n-propylamine.............	67	297
3-Hydroxy-2-naphthylamine	452	317	3-Hydroxypropylamine	358	438
6-Hydroxy-1-naphthylamine	406	316	4-Hydroxypyrazole	250	310
6-Hydroxy-2-naphthylamine	436	317	α-Hydroxypyridine	68	324
7-Hydroxy-1-naphthylamine	430	317	β-Hydroxypyridine	76	324
7-Hydroxy-2-naphthylamine	421	316	γ-Hydroxypyridine	80	324
8-Hydroxy-1-naphthylamine	179	308	2-Hydroxypyridine	215	107
2-Hydroxy-5-nitroaniline	319	313	"	68	324
2-Hydroxy-6-nitroaniline	441	317	"	120	435
3-Hydroxy-4-nitroaniline	365	314	3-Hydroxypyridine	307	112
4-Hydroxy-2-nitroaniline	347	314	"	76	324
4-Hydroxy-3-nitroaniline	295	312	"	52	434
5-Hydroxy-2-nitroaniline	398	316	"	196	437
4-Hydroxy-4'-nitroazobenzene...............	528	124	4-Hydroxypyridine	373	115
3-Hydroxy-4-nitrobenzaldehyde..............	14	434	"	80	324
4-Hydroxy-3-nitrobenzene	308	112	"	118	435
2-Hydroxy-5-nitrobenzoic acid	559	126	"	85	436
2-Hydroxy-3-nitrobenzonitrile	178	365	2-Hydroxypyrimidine......................	3	434
2-Hydroxy-4-nitrobenzonitrile	205	366	4-Hydroxypyrimidine......................	2	434
2-Hydroxy-5-nitrobenzonitrile	221	367	6-Hydroxypyrimidine......................	85	325
2-Hydroxy-2'-nitrobiphenyl	337	113	Hydroxypyruvic aldehyde...................	120	159
3-Hydroxy-2-nitrobiphenyl	145	103	4-Hydroxyquinaldine......................	98	325
3-Hydroxy-4-nitrobiphenyl	206	107	5-Hydroxyquinaldine......................	100	325
3-Hydroxy-6-nitrobiphenyl	59	99	6-Hydroxyquinaldine......................	96	325
4-Hydroxy-2-nitrobiphenyl	342	114	8-Hydroxyquinaldine......................	43	323
4-Hydroxy-2'-nitrobiphenyl	229	108	"	11	434
4-Hydroxy-3-nitrobiphenyl	79	100	β-Hydroxyquinoline	93	325
4-Hydroxy-4'-nitrobiphenyl	514	123	γ-Hydroxyquinoline	95	325
4-Hydroxy-3-nitro-1-naphthylamine...........	359	314	2-Hydroxyquinoline	500	122
9-Hydroxynonanal	33	154	3-Hydroxyquinoline	507	123
DL-β-Hydroxynorvaline.....................	59	286	"	93	325
5-Hydroxy-4-octanone......................	132	165	"	95	325
4-Hydroxypentanoic acid...................	424	432	4-Hydroxyquinoline	553	126
5-Hydroxy-2-pentanone	40	169	5-Hydroxyquinoline	97	325
5-Hydroxy-1-pentylamine	396	439	"	493	122
1-Hydroxyphenanthrene	393	116	6-Hydroxyquinoline	91	325
2-Hydroxyphenanthrene	440	119	7-Hydroxyquinoline	566	126
3-Hydroxyphenanthrene	282	111	"	99	325
4-Hydroxyphenanthrene	236	108	8-Hydroxyquinoline	110	102
9-Hydroxyphenanthrene	396	116	" "	45	323
4-Hydroxyphenylacetamide	358	241	" "	8	434
α-Hydroxyphenylacetic acid................	151	199	w-Hydroxyresacetopenone	483	121
(2-Hydroxyphenyl)acetic acid...............	220	202	6-Hydroxyretene........................	409	117
(4-Hydroxyphenyl)acetic acid...............	225	202	Hydroxysuccinanilide......................	420	243
(2-Hydroxyphenyl)acetonitrile..............	163	364	Hydroxysuccinic acid......................	115	197
(4-Hydroxyphenyl)acetonitrile..............	87	361	2-Hydroxythiophenol	40	415
2-Hydroxy-2-phenylbutyraldehyde...........	22	150	3-Hydroxythiophenol	1	416
4-Hydroxy-m-phenylenediamine..............	120	102	4-Hydroxy-3,5-toluenediamine..............	360	114
" " "	130	306	2-Hydroxy-p-toluidine....................	362	314
4-Hydroxyphenylethyl alcohol..............	167	104	3-Hydroxy-p-toluidine....................	363	314
4-Hydroxyphenylglycine...................	71	287	4-Hydroxy-m-toluidine	302	312
3-Hydroxyphenyl methyl sulfide	77	434	4-Hydroxy-o-toluidine...................	324	313
4-Hydroxyphenyl methyl sulfide	78	434	5-Hydroxy-o-toluidine...................	390	315
2-Hydroxy-2-phenylpropionaldehyde...........	20	150	1-Hydroxytriacontane	48	89
2-Hydroxy-2-phenylpropionic acid	208	430	5-Hydroxy-2,4,6-tribromoaniline...........	253	310
2-Hydroxy-3-phenylpropionic acid	105	197	3-Hydroxy-2,4,6-trichlorobenzaldehyde..........	99	158
3-Hydroxy-2-phenylpropionic acid	149	199	5-Hydroxy-1,2,3-trimethoxybenzene.............	57	140
3-Hydroxy-3-phenylpropionic acid	195	430	β-Hydroxyvaline.........................	49	286
2-Hydroxyphenyl sulfide	69	426	9-Hydroxyxanthene......................	154	177
3-Hydroxyphenyl sulfide	67	426			

Compound	Compound No.	Page No.
I		
L-Idose	9	328
Imidazole	272	438
Iminodipropionic acid	301	431
Indane	20	36
Indane-4-sulfonamide	55	394
Indane-5-sulfonamide	89	395
Indane-4-sulfonic acid	15	371
Indane-5-sulfonic acid	29	372
Indane-4-sulfonyl chloride	43	382
Indane-5-sulfonyl chloride	31	381
4-Indanol	110	435
1-Indanone	32	173
2-Indanone	67	174
Indene	25	36
Indene-3-carbonitrile	21	356
Indole	57	304
Indole-3-acetonitrile	29	358
Indole-3-carbonitrile	50	349
Indole-3-carboxaldehyde	125	160
2-(2-Indolyl)ethylamine	256	310
3-(3-Indolyl)propylamine	75	305
3-Indolylpyruvic acid	435	317
meso-Inositol	67	90
Iodoacetamide	93	233
Iodoacetanilide	264	238
2-Iodoacetanilide	156	235
3-Iodoacetanilide	187	235
4-Iodoacetanilide	387	242
Iodoacetic acid	86	196
" "	172	430
Iodoacetic anhydride	15	225
4-Iodoacetophenone	116	176
Iodoacetyl chloride	8	214
2-Iodoaniline	77	305
" "	58	436
3-Iodoaniline	13	303
" "	107	436
4-Iodoaniline	102	306
" "	112	436
4-Iodoanisole	30	138
2-Iodobenzaldehyde	13	153
3-Iodobenzaldehyde	42	155
4-Iodobenzaldehyde	68	156
2-Iodobenzamide	386	242
3-Iodobenzamide	394	242
4-Iodobenzamide	458	244
2-Iodobenzanilide	257	238
4-Iodobenzanilide	447	244
Iodobenzene	1	75
4-Iodobenzenesulfonamide	240	401
4-Iodobenzenesulfonanilide	106	396
4-Iodobenzenesulfonic acid	105	374
4-Iodobenzenesulfonyl chloride	105	384
2-Iodobenzoic acid	246	203
" "	126	430
3-Iodobenzoic acid	297	205
" "	258	431
4-Iodobenzoic acid	392	209
" "	126	430
3-Iodobenzoic anhydride	59	226
4-Iodobenzoic anhydride	82	227
2-Iodobenzonitrile	63	359
4-Iodobenzoyl bromide	3	220
2-Iodobenzoyl chloride	3	216
4-Iodobenzoyl chloride	20	216
4-Iodobenzyl cyanide	55	359
4-Iodobiphenyl	5	76
4-Iododiphenyl sulfide	9	425
Iodoform	2	68
1-Iodo-4-isopropylbenzene	4	75

Compound	Compound No.	Page No.
3-Iodomandelic acid	186	430
2-Iodo-4-methylaniline	26	303
2-Iodo-5-methylaniline	40	304
3-Iodo-4-methylaniline	20	303
4-Iodo-2-methylaniline	151	307
2-Iodo-4-methylbenzonitrile	68	360
3-Iodo-2-methylolbenzoic acid	173	430
1-Iodonaphthalene	5	75
2-Iodonaphthalene	4	76
1-Iodonaphthalene-2-sulfonamide	427	408
2-Iodonaphthalene-1-sulfonamide	139	397
4-Iodonaphthalene-1-sulfonamide	327	404
5-Iodonaphthalene-1-sulfonamide	413	408
6-Iodonaphthalene-1-sulfonamide	341	405
6-Iodonaphthalene-2-sulfonamide	366	406
7-Iodonaphthalene-1-sulfonamide	415	408
7-Iodonaphthalene-2-sulfonamide	337	405
8-Iodonaphthalene-1-sulfonamide	259	401
4-Iodonaphthalene-1-sulfonanilide	91	395
2-Iodonaphthalene-1-sulfonic acid	55	372
4-Iodonaphthalene-1-sulfonic acid	152	376
5-Iodonaphthalene-1-sulfonic acid	182	377
8-Iodonaphthalene-1-sulfonic acid	123	375
1-Iodonaphthalene-2-sulfonyl chloride	128	385
2-Iodonaphthalene-1-sulfonyl chloride	168	386
4-Iodonaphthalene-1-sulfonyl chloride	197	387
5-Iodonaphthalene-1-sulfonyl chloride	230	388
6-Iodonaphthalene-1-sulfonyl chloride	121	384
6-Iodonaphthalene-2-sulfonyl chloride	225	388
7-Iodonaphthalene-1-sulfonyl chloride	255	389
7-Iodonaphthalene-2-sulfonyl chloride	150	385
8-Iodonaphthalene-1-sulfonyl chloride	226	388
4-Iodo-2-naphthol	304	112
5-Iodo-2-naphthonitrile	195	366
3-Iodo-1-naphthylamine	142	307
2-Iodo-1-nitrobenzene	20	337
3-Iodo-1-nitrobenzene	11	337
4-Iodo-1-nitrobenzene	113	341
2-Iodo-4-nitrophenol	170	105
2-Iodo-6-nitrophenol	225	108
3-Iodo-2-nitrophenol	105	102
3-Iodo-4-nitrophenol	292	111
4-Iodo-2-nitrophenol	125	103
4-Iodo-3-nitrophenol	394	116
5-Iodo-2-nitrophenol	181	105
5-Iodo-3-nitrophenol	328	113
4-Iodo-3-nitrotoluene	25	337
2-Iodophenetole	100	136
4-Iodophenetole	6	137
2-Iodophenol	31	98
" "	46	434
3-Iodophenol	25	97
" "	63	434
4-Iodophenol	168	105
" "	65	434
(2-Iodophenoxy)acetic acid	168	430
(3-Iodophenoxy)acetic acid	160	430
(4-Iodophenoxy)acetic acid	167	430
(2-Iodophenyl)acetic acid	292	431
(3-Iodophenyl)acetic acid	311	431
(4-Iodophenyl)acetic acid	314	431
(4-Iodophenyl)acetonitrile	55	359
2-Iodo-p-phenylenediamine	225	309
4-Iodophenyl phenyl sulfide	9	425
3-Iodophthalic anhydride	67	226
4-Iodophthalic anhydride	55	226
β-Iodopropionamide	119	234
α-Iodopropionic acid	156	430
β-Iodopropionic acid	84	196
" "	298	431
β-Iodopropionyl chloride	9	214

Compound	Compound No.	Page No.
α-Iodopyridine	3	322
"	35	436
β-Iodopyridine	29	323
"	88	436
2-Iodopyridine	3	322
"	35	436
3-Iodopyridine	29	323
"	88	436
4-Iodopyridine	127	437
α-Iodoquinoline	28	323
γ-Iodoquinoline	63	324
2-Iodoquinoline	28	323
4-Iodoquinoline	63	324
5-Iodoquinoline	65	324
6-Iodoquinoline	55	324
8-Iodoquinoline	9	323
4-Iodothiophenol	27	417
2-Iodotoluene	3	75
3-Iodotoluene	2	75
4-Iodotoluene	1	76
4-Iodo-m-toluidine	40	304
5-Iodo-o-toluidine	151	307
2-Iodo-p-toluidine	20	303
3-Iodo-p-toluidine	26	303
2-Iodo-4-tolunitrile	68	360
1-Iodo-2,4,5-trimethylbenzene	2	76
5-Iodovanillin	470	121
Isatin	431	243
"	425	317
Isoamyl acetate	75	253
prim-Isoamyl alcohol	26	81
sec-Isoamyl alcohol	12	80
Isoamylamine	28	295
Isoamylbenzoate	307	267
Isoamyl bromide	16	58
Isoamyl n-butyrate	134	256
Isoamyl carbamate	29	231
Isoamyl carbinol	47	82
Isoamyl chloride	21	55
Isoamyl chloroformate	99	254
Isoamylcyclohexane	245	7
Isoamyl ether	64	134
N-Isoamylformanilide	12	230
Isoamyl formate	52	252
Isoamyl iodide	14	60
Isoamyl isobutyrate	122	255
Isoamyl isovalerate	157	257
Isoamyl levulinate	286	266
Isoamyl mercaptan	12	413
Isoamyl-1-naphthyl ether	110	136
Isoamyl 2-naphthyl ether	8	137
Isoamyl nitrite	24	250
Isoamyl oxalate	312	268
Isoamyl phthalate	357	271
Isoamyl propionate	109	255
Isoamyl salicylate	28	95
" "	326	269
Isoamyl stearate	12	273
Isoamyl succinate	343	270
Isoamyl sulfide	50	421
Isoamyl urethane	29	231
Isoaspartic acid	72	287
d,l-Isobornane	48	9
Isobutane	5	3
Isobutanoic acid	8	190
2-Isobutoxyethanol	51	82
Isobutylacetaldehyde	33	145
Isobutyl acetate	43	251
Isobutylacetic acid	30	191
Isobutyl acetylene	15	28
Isobutyl alcohol	10	80
Isobutylallene	111	16
Isobutylamine	16	295
"	432	439
N-Isobutylaniline	137	299
"	154	437
4-Isobutylaniline	149	299
Isobutylbenzene	15	35
Isobutyl benzoate	265	264
Isobutyl bromide	9	58
Isobutyl n-butyrate	105	255
Isobutyl carbamate	16	231
Isobutyl carbonate	154	257
Isobutyl chloride	11	55
Isobutyl chloroformate	59	252
Isobutyl enanthate	191	260
Isobutylene glycol	79	84
Isobutyl ether	46	133
N-Isobutylformanilide	11	230
Isobutyl formate	23	250
N-Isobutylglycine	87	429
Isobutyl n-heptanoate	191	260
Isobutyl 2-hydroxybenzoate	24	95
Isobutyl iodide	9	60
Isobutyl isobutyrate	93	254
Isobutyl isovalerate	124	255
Isobutyl levulinate	239	263
Isobutylmalononitrile	75	351
Isobutyl mercaptan	7	413
Isobutyl methoxymethyl ketone	96	164
2-Isobutyl-1-methylbenzene	55	37
3-Isobutyl-1-methylbenzene	49	37
4-Isobutyl-1-methylbenzene	53	37
Isobutyl methyl carbinol	25	81
Isobutyl methyl ketone	15	161
Isobutyl methyl sulfide	13	420
Isobutyl 2-naphthyl ether	12	137
Isobutyl oxalate	237	263
Isobutyl n-pentanoate	137	256
4-Isobutylphenol	18	94
Isobutyl phenylacetate	281	265
Isobutyl phenyl ketone	195	167
Isobutyl propionate	72	253
Isobutyl n-propyl ketone	67	163
Isobutyl salicylate	301	267
"	24	429
Isobutyl stearate	11	273
Isobutyl succinate	310	267
Isobutyl sulfide	37	421
m-Isobutyltoluene	49	37
p-Isobutyltoluene	53	37
o-Isobutyltoluene	55	37
Isobutyl urethane	16	231
Isobutyl n-valerate	137	256
Isobutyraldehyde	9	144
Isobutyramide	217	236
Isobutyranilide	137	234
Isobutyric acid	8	190
" "	451	432
Isobutyric anhydride	7	223
Isobutyronitrile	9	345
Isobutyrophenone	177	167
Isobutyryl chloride	6	212
Isocaproaldehyde	33	145
Isocaproamide	190	236
Isocaproanilide	169	235
Isocaproic acid	30	191
" "	449	432
Isocapronitrile	38	348
Isocaproyl chloride	29	212
Isocarvestrene	317	23
Isocrotonamide	126	234
Isocrotonanilide	127	234
Isocrotonic acid	13	190

Compound	Compound No.	Page No.		Compound	Compound No.	Page No.
Isodurene	59	37		Isopropyl benzoate	213	261
Isodurenol	122	102		2-Isopropylbenzoic acid	226	431
Isoduridine	170	300		4-Isopropylbenzoic acid	150	199
Isoeugenol	26	95		" "	355	432
Isoeugenol methyl ether	104	136		4-Isopropylbenzonitrile	85	352
Isoeugenyl acetate	118	279		4-Isopropylbenzoyl chloride	22	214
Isogeraniolene	251	20		α-Isopropylbenzyl alcohol	116	86
Isohexanamide	190	236		4-Isopropylbenzyl chloride	51	56
Isohexyl alcohol	47	82		Isopropyl bromide	5	58
Isohexylamine	43	296		2-Isopropylbutanoic acid	24	192
Isohexyl methyl ketone	110	164		Isopropyl n-butyrate	58	252
Isolaurolene	153	17		Isopropyl carbamate	78	233
D-(-)-Isoleucine	94	288		Isopropyl carbonate	90	254
DL-Isoleucine	99	289		Isopropyl chloride	4	55
L-(+)Isoleucine	91	288		Isopropylcyclohexane	153	5
Isonicotinic acid	406	210		5-Isopropyl-1,3-cyclohexanedione	78	174
" "	447	432		3-Isopropylcyclohexanone	76	170
Isopentane-1,1-dicarboxylic acid	76	196		4-Isopropylcyclohexanone	67	170
Isopentylbenzene	60	37		3-Isopropyl-2-cyclohexenone	58	170
Isopentyl salicylate	28	95		l-6-Isopropyl-3-cyclohexenone	61	170
Isophorone	172	166		Isopropylcyclopentane	83	4
Isophthalaldehyde	75	157		2-Isopropylcyclopentanone	119	165
Isophthalic acid	408	210		Isopropyl ether	18	131
" "	220	430		Isopropyl formate	7	250
Isophthalic acid diamide	504	246		3-Isopropylglutaric acid	345	432
Isophthalic acid monoamide	503	246		4-Isopropylheptane	179	6
Isophthalonitrile	204	366		Isopropyl 2-hydroxybenzoate	20	94
Isophthaloyl dichloride	6	216		Isopropylideneacetone	25	162
Isopicramic acid	448	119		5-Isopropylidene-2-methyl-1,3-cyclohexadiene	321	23
Isoprene	17	12		1-Isopropylidene-2-methylcyclohexane	309	23
Isopropenyl acetylene	7	28		1-Isopropylidene-3-methylcyclohexane	310	23
Isopropanolamine	67	297		1-Isopropylidene-4-methylcyclohexane	307	23
1-Isopropenyl-1,4-cyclohexadiene	308	23		4-Isopropylidene-1-methylcyclohexane	325	24
l-5-Isopropyl-2-methyl-1,3-cyclohexadiene	313	23		3-Isopropylidene-6-methylcyclohexane	324	24
d-1-Isopropenyl-3-methylcyclohexane	301	22		3-Isopropylidene-1-methylcyclopentane-1-		
l-1-Isopropenyl-3-methylcyclohexane	302	22		carbonitrile	73	351
1-Isopropenyl-4-methylcyclohexane	300	22		Isopropyl iodide	4	60
cis-3-Isopropenyl-4-methylcyclohexene	298	22		Isopropyl isobutyrate	47	252
1-Isopropenyl-3-methylcyclohexene	323	23		Isopropyl isovalerate	76	253
1-Isopropenyl-5-methylcyclohexene	320	23		Isopropyl d,l-lactate	118	255
l-3-Isopropenyl-1-methylcyclohexene	318	23		Isopropyl levulinate	192	260
3-Isopropenyl-2-methylcyclohexene	303	22		Isopropylmalonic acid	132	430
4-Isopropenyl-3-methylcyclohexene	304	22		Isopropyl methoxymethyl ketone	56	162
5-Isopropenyl-1-methylcyclohexene	317	23		Isopropylmethylacetic acid	24	190
Isopropenyl methyl ketone	7	161		5-Isopropyl-2-methylacetophenone	202	167
(2-Isopropoxy)benzoic acid	330	431		Isopropyl methyl acetylene	18	28
(3-Isopropoxy)benzoic acid	308	431		Isopropyl methylamine	8	295
(4-Isopropoxy)benzoic acid	418	432		2-Isopropyl-1-methylbenzene	22	36
2-Isopropoxyethanol	36	81		3-Isopropyl-1-methylbenzene	17	35
1-Isopropoxy-3-methyl-2-butanone	88	163		4-Isopropyl-1-methylbenzene	21	36
Isopropoxymethyl methyl ketone	49	162		2-Isopropyl-5-methylbenzenethiol	44	415
Isopropyl acetate	17	250		5-Isopropyl-2-methylbenzenethiol	45	415
Isopropyl acetylene	5	28		2-Isopropyl-5-methyl-1,4-benzoquinone	2	181
4-Isopropylacetophenone	210	168		2-Isopropyl-3-methyl-1-butene	127	16
2-Isopropylacrolein	25	145		d,l-Isopropyl methyl carbinol	12	80
Isopropyl alcohol	3	80		1-Isopropyl-4-methyl-1,3-cyclohexadiene	311	23
1-Isopropylallene	54	13		1-Isopropyl-4-methyl-1,4-cyclohexadiene	322	23
Isopropylamine	4	295		2-Isopropyl-5-methyl-1,3-cyclohexadiene	312	23
"	468	439		1-5-Isopropyl-2-methyl-1,3-cyclohexadiene	313	23
2-(Isopropylamino)ethylamine	416	439		5-Isopropyl-2-methyl-1,3-cyclohexadiene	314	23
N-Isopropylaniline	234	437		1-Isopropyl-2-methylcyclohexane	236	7
2-Isopropylaniline	153	437		l-1-Isopropyl-3-methylcyclohexane	228	6
3-Isopropylaniline	172	437		d,l-1-Isopropyl-3-methylcyclohexane	214	6
4-Isopropylaniline	135	299		d-1-Isopropyl-3-methylcyclohexane	219	6
2-Isopropylazulene	5	43		cis-1-Isopropyl-4-methylcyclohexane	229	6
5-Isopropylazulene	10	43		trans-1-Isopropyl-4-methylcyclohexane	232	7
6-Isopropylazulene	15	43		trans-1-Isopropyl-4-methylcyclohexane	184	6
4-Isopropylbenzaldehyde	87	149		1-Isopropyl-2-methylcyclopentane	126	5
4-Isopropylbenzamide	236	237		1-Isopropyl-3-methylcyclopentane	124	5
Isopropylbenzene	7	35		7-Isopropyl-1-methylfluorene	101	47
1-Isopropylbenzimidazole	231	437		3-Isopropyl-2-methylhexane	196	6
2-Isopropylbenzimidazole	254	438		2-Isopropyl-5-methylhydroquinone	347	114

Compound	Compound No.	Page No.
Isopropyl methyl ketone	6	161
1-Isopropyl-7-methylnaphthalene	165	41
7-Isopropyl-1-methyl-9,10-phenanthraquinone	30	182
1-Isopropyl-7-methylphenanthrene	81	46
3-Isopropyl-1-methylphenanthrene	70	45
4-Isopropyl-1-methylphenanthrene	58	45
6-Isopropyl-1-methylphenanthrene	18	43
7-Isopropyl-1-methylphenanthrene	103	47
2-Isopropyl-5-methylphenol	44	98
5-Isopropyl-2-methylphenol	19	94
Isopropyl methyl sulfide	8	420
5-Isopropylnaphthanthracene	95	46
8-Isopropylnaphthanthracene	136	48
9-Isopropylnaphthanthracene	133	48
12-Isopropylnaphthanthracene	98	47
1-Isopropylnaphthalene	158	41
2-Isopropylnaphthalene	161	41
Isopropyl 2-naphthyl ether	22	138
Isopropyl 4-nitrophenyl sulfide	18	425
Isopropyl oxalate	160	258
Isopropyl n-pentanoate	98	254
2-Isopropyl-1-pentene	179	18
2-Isopropylphenanthrene	17	43
9-Isopropylphenanthrene	14	43
2-Isopropylphenol	7	94
4-Isopropylphenol	65	99
(4-Isopropylphenyl)acetic acid	365	432
d,l-Isopropyl phenyl carbinol	116	86
Isopropyl phenyl ketone	177	167
Isopropyl phenyl sulfide	48	421
Isopropyl phthalate	347	270
Isopropyl propionate	39	251
4-Isopropyl-1-propylbenzene	105	39
Isopropyl n-propyl ether	25	132
Isopropyl n-propyl ketone	34	162
Isopropyl propyl sulfide	20	420
2-Isopropylpyridine	237	437
3-Isopropylpyridine	232	437
4-Isopropylpyridine	246	438
Isopropyl salicylate	20	94
" "	262	264
Isopropyl sulfide	17	420
2-Isopropylthiophene	25	423
3-Isopropylthiophene	26	423
2-Isopropyltoluene	22	36
3-Isopropyltoluene	17	35
4-Isopropyltoluene	21	36
Isopropyl urethane	78	233
Isopropyl n-valerate	98	254
Isoquinoline	72	320
"	215	437
1-Isoquinoline-1-carboxaldehyde	38	155
1,4-Isoquinolinediol	573	126
D-Isorhamnose	29	330
trans(β)-Isosafrole	101	136
DL-Isoserine	69	287
L-Isoserine	33	285
Isovaleraldehyde	17	145
Isovaleramide	245	237
Isovaleranilide	162	235
Isovaleric acid	15	190
" "	432	432
Isovaleric anhydride	11	223
Isovalerone	101	164
Isovaleronitrile	22	346
Isovalerophenone	197	167
Isovaleryl bromide	6	218
Isovaleryl chloride	16	212
Isovanilline	54	434
Itaconic acid	254	203
" "	469	433

Compound	Compound No.	Page No.
Itaconic acid diamide	406	242
Itaconic acid dianilide	403	242
Itaconic anhydride	26	225

J

Compound	Compound No.	Page No.
Jasminaldehyde	9	150
Juglone	21	182

K

Compound	Compound No.	Page No.
α-Ketobutyric acid	11	193
2-Ketocyclohexanecarboxylic acid	82	196
3-Ketocyclohexanecarboxylic acid	74	196
4-Ketocyclohexanecarboxylic acid	67	195
γ-Ketovaleric acid	18	193

L

Compound	Compound No.	Page No.
Lactaldehyde	8	329
d,l-Lactamide	49	232
d,l-Lactanilide	22	231
d,l-Lactic acid	7	192
Lactic acid benzoate	138	198
" " "	261	431
Lactic aldehyde	8	329
d,l-Lactonitrile	54	349
Lactonitrile benzoate	98	353
Lactose	64	333
6-[β-Lactosido]-α-D-glucose	67	333
DL-Lanthionine	89	288
meso-Lanthionine	116	290
Lauraldehyde	18	153
Lauramide	160	235
Lauranilide	46	232
Laurent's acid	203	378
Lauric acid	29	194
Lauric anhydride	12	225
Laurolene	205	18
Laurone	91	175
Lauronitrile	101	353
Laurophenone	45	173
Lauroyl chloride	13	214
Lauryl alcohol	3	88
Lauryl bromide	29	58
Lauryl chloride	53	56
DL-Leucine	101	289
" "	78	429
L-Leucine	123	290
Leuco-malachite green	66	324
Levulinamide	152	234
Levulinanilide	124	234
Levulinic acid	16	172
" "	18	193
" "	407	432
Lignoceric acid	88	196
d-Limonene	316	23
d,l-Limonene	319	23
l-Linalool	103	85
l-Linalyl acetate	215	261
l-Linalyl alcohol	103	85
2,3-Lutidine	31	319
2,4-Lutidine	25	319
2,5-Lutidine	26	319
2,6-Lutidine	18	319
3,4-Lutidine	32	319
3,5-Lutidine	40	319
D-Lysine	57	429

	Compound No.	Page No.
DL-Lysine	129	290
L-Lysine	53	286
" "	208	429
α-D-Lyxose	13	329

M

	Compound No.	Page No.
L-Malamide	309	239
L-Malanilide	420	243
l-Malic acid	115	197
Maleamic acid	352	241
Maleic acid	175	200
" "	37	429
Maleic acid diamide	380	242
Maleic acid dianilide	396	242
Maleic acid monoamide	352	241
Maleic acid monoanilide	397	242
Maleic anhydride	21	225
Maleimide	83	233
Maleonitrile	19	357
Malonamic acid	6	231
" "	225	431
Malondiamide	345	241
Malonic acid	185	200
" "	123	430
Malonic acid diamide	345	241
Malonic acid dianilide	477	245
Malonic acid monoamide	6	231
" "	225	431
Malonic acid monoanilide	232	237
Malononitrile	17	357
β-Maltose	43	332
d,l-Mandelamide	239	237
d,l-Mandelanilide	289	239
d,l-Mandelic acid	151	199
" "	196	430
D-Mandelonitrile	14	357
DL- Mandelonitrile	7	357
d-Mannitol	62	90
α-D-Mannoheptose	27	330
α-D-Mannose	23	330
β-D-Mannose	24	330
DL-Mannose	26	330
L-Mannose	25	330
Margaramide	149	234
Margaric acid	55	195
Margaric aldehyde	11	153
Margaric anhydride	25	225
Melezitose	38	331
β-Melibiose dihydrate	2	329
Menogene	324	24
Menogerene	321	23
1,5-p-Menthadiene	314	23
2,4-p-Menthadiene	312	23
d,l-2,8-m-Menthadiene	323	23
3,8-m-Menthadiene	320	23
3,8-o-Menthadiene	298	22
5,8-o-Menthadiene	304	22
6,8-o-Menthadiene	303	22
d-m-Menthane	219	6
l-m-Menthane	228	6
o-Menthane	236	7
p-Menthane	232	7
cis-p-Menthane	229	6
trans-p-Menthane	184	6
m-Menthene	310	23
o-Menthene	309	23
p-Menthene	307	23
d-m-8-Menthene	301	22
l-m-8-Menthene	302	22
p-8-Menthene	300	22

	Compound No.	Page No.
4(8)-p-Menthen-3-one	180	167
l-Menthol	20	88
l-Menthone	165	166
l-Menthyl acetate	232	262
L-Menthylamine	111	298
2-Mercaptoacetophenone	4	416
Mercaptobenzene	25	414
1-Mercaptonaphthalene	47	415
2-Mercaptonaphthalene	25	416
Mesaconic acid	329	207
" "	322	431
Mesaconic acid diamide	368	241
Mesaconic acid dianilide	391	242
Mesaconyl dichloride	28	214
Mesidine	142	299
Mesityl acetate	255	269
Mesitylacetic acid	261	203
Mesitylacetone	70	174
Mesitylene	11	35
Mesityleneamide	235	237
Mesitylene disulfonic acid	195	378
Mesitylenic acid	259	203
Mesityl oxide	25	162
Mesityl phenyl ketone	167	177
Metaldehyde	101	158
Metanilic acid	37	372
Methacrolein	10	144
Methacrylamide	180	235
Methacrylic acid	9	190
Methacrylonitrile	7	345
Methacrylyl chloride	7	212
Methallylacetone	66	163
Methallyl chloride	12	55
Methanal	1	144
Methane	1	3
Methane disulfonamide	395	407
Methane disulfonanilide	286	403
Methane disulfonic acid	179	377
Methane disulfonyl chloride	1	381
Methanesulfonamide	23	392
Methanesulfonanilide	32	393
Methanesulfonic acid	6	371
Methanethiol	1	413
Methanol	1	80
Methionic acid	179	377
DL-Methionine	87	288
L-Methionine	90	288
Methone	181	178
Methoxyacetaldehyde	16	145
Methoxyacetamide	103	233
Methoxyacetanilide	21	231
4-Methoxyacetanilide	211	236
Methoxyacetic acid	31	191
Methoxyacetonitrile	17	346
α-Methoxyacetophenone	25	169
2-Methoxyacetophenone	199	167
3-Methoxyacetophenone	200	167
4-Methoxyacetophenone	24	172
Methoxyacetyl chloride	14	212
2-Methoxyaniline	94	135
" "	134	299
" "	159	437
3-Methoxyaniline	102	136
" "	168	300
" "	139	437
4-Methoxyaniline	35	138
" "	67	304
" "	211	437
4-Methoxybenzalacetone	96	175
4-Methoxybenzalacetophenone	104	175
2-Methoxybenzaldehyde	15	153
3-Methoxybenzaldehyde	85	149

	Compound No.	Page No.
4-Methoxybenzaldehyde	89	149
2-Methoxybenzamide	218	236
2-Methoxybenzanilide	225	237
Methoxybenzene	56	133
4-Methoxybenzene-1,3-disulfonamide	416	408
4-Methoxybenzene-1,3-disulfonanilide	336	405
4-Methoxybenzene-1,3-disulfonic acid	191	378
4-Methoxybenzene-1,3-disulfonyl chloride	108	384
4-Methoxybenzil	82	174
2-Methoxybenzoic acid	114	197
" " "	296	431
3-Methoxybenzoic acid	297	431
4-Methoxybenzoic acid	291	204
" " "	383	432
4-Methoxybenzoic anhydride	43	226
2-Methoxybenzonitrile	9	357
4-Methoxybenzonitrile	72	360
2-Methoxybenzophenone	29	173
4-Methoxybenzophenone	80	174
4-Methoxybenzoyl bromide	9	219
2-Methoxybenzoyl chloride	57	213
3-Methoxybenzoyl chloride	54	213
4-Methoxybenzoyl chloride	58	213
2-Methoxybenzyl alcohol	127	86
4-Methoxybenzyl alcohol	5	88
" " "	4	137
2-Methoxybenzylamine	343	438
3-Methoxybenzylamine	314	438
4-Methoxybenzylamine	327	438
2-Methoxybiphenyl	10	137
4-Methoxybiphenyl	51	139
1-Methoxy-2-butanone	29	162
1-Methoxy-3-butanone	14	161
4-Methoxychalcone	104	175
2-Methoxycinnamic acid	393	432
4-Methoxycinnamic acid	363	432
4-Methoxycinnamonitrile	74	360
4-Methoxycyclohexanone	54	170
2-Methoxyethanethiol	11	413
2-Methoxyethanol	20	81
2-(2-Methoxyethoxy)ethanol	96	85
β-Methoxyethyl acetate	78	253
2-Methoxyethylamine	336	438
β-Methoxyethyl benzoate	294	266
Methoxyethyl methyl ketone	14	161
1-Methoxyethyl n-propyl ketone	78	163
1-Methoxy-2-hexanone	106	164
1-Methoxy-3-hexanone	78	163
4-Methoxyhydrocinnamonitrile	103	354
3-Methoxyisobutyraldehyde	36	145
1-Methoxy-3-methyl-2-butanone	56	162
4-Methoxy-2-methyl-n-butyraldehyde	9	152
Methoxymethyl methyl ketone	13	161
3-Methoxy-2-methylolbenzoic acid	215	430
1-Methoxy-4-methyl-2-pentanone	96	164
2-Methoxy-3-methyl-2-pentanone	95	164
3-Methoxy-2-methyl-1-propanol	57	133
Methoxymethyl n-propyl ketone	72	163
2-Methoxy-1-naphthaldehyde	73	156
3-Methoxy-1-naphthaldehyde	52	155
4-Methoxy-1-naphthaldehyde	9	153
5-Methoxy-1-naphthaldehyde	56	155
1-Methoxynaphthalene	106	136
2-Methoxynaphthalene-1-sulfonamide	161	398
3-Methoxynaphthalene-2-sulfonamide	49	393
4-Methoxynaphthalene-1-sulfonamide	378	406
4-Methoxynaphthalene-2-sulfonamide	149	397
5-Methoxynaphthalene-1-sulfonamide	292	403
6-Methoxynaphthalene-1-sulfonamide	127	396
6-Methoxynaphthalene-2-sulfonamide	268	402
7-Methoxynaphthalene-2-sulfonamide	362	406
2-Methoxynaphthalene-1-sulfonanilide	304	403
3-Methoxynaphthalene-2-sulfonanilide	211	399
4-Methoxynaphthalene-1-sulfonanilide	159	397
4-Methoxynaphthalene-3-sulfonanilide	122	396
5-Methoxynaphthalene-1-sulfonanilide	150	397
6-Methoxynaphthalene-1-sulfonanilide	223	400
6-Methoxynaphthalene-2-sulfonanilide	58	394
7-Methoxynaphthalene-2-sulfonanilide	60	394
2-Methoxynaphthalene-1-sulfonyl chloride	188	387
3-Methoxynaphthalene-2-sulfonyl chloride	222	388
4-Methoxynaphthalene-1-sulfonyl chloride	146	385
4-Methoxynaphthalene-2-sulfonyl chloride	84	383
5-Methoxynaphthalene-1-sulfonyl chloride	184	387
6-Methoxynaphthalene-1-sulfonyl chloride	96	384
6-Methoxynaphthalene-2-sulfonyl chloride	122	384
7-Methoxynaphthalene-2-sulfonyl chloride	102	384
2-Methoxy-2-nitroaniline	121	306
2-Methoxy-4-nitroaniline	314	312
2-Methoxy-5-nitroaniline	249	310
3-Methoxy-5-nitroaniline	39	436
4-Methoxy-2-nitroaniline	287	311
4-Methoxy-3-nitroaniline	64	304
3-Methoxy-1-nitrobenzene	12	337
4-Methoxy-1-nitrobenzene	23	337
4-Methoxy-3-nitrobenzenesulfonamide	120	396
4-Methoxy-3-nitrobenzenesulfonic acid	46	372
4-Methoxy-3-nitrobenzenesulfonyl chloride	71	383
4-Methoxy-3-nitrobenzoic acid	300	205
2-Methoxy-4-nitrophenol	428	118
5-Methoxy-2-nitrophenol	22	434
1-Methoxy-2-pentanone	72	163
2-Methoxy-3-pentanone	36	162
2-Methoxyphenol	12	96
" "	9	137
" "	95	434
3-Methoxyphenol	21	95
" "	82	434
4-Methoxyphenol	48	99
" "	104	434
(2-Methoxyphenoxy)acetic acid	180	430
(3-Methoxyphenoxy)acetic acid	164	430
(4-Methoxyphenoxy)acetic acid	175	430
2-Methoxyphenyl acetate	261	264
(4-Methoxyphenyl)acetic acid	92	196
" " " "	359	432
4-Methoxyphenyl methyl carbinol	133	87
3-(2-Methoxyphenyl)propionic acid	439	432
3-(3-Methoxyphenyl)propionic acid	409	432
3-(4-Methoxyphenyl)propionic acid	421	432
3-(4-Methoxyphenyl)propionitrile	103	354
4-Methoxyphenyl sulfide	19	425
2-Methoxy-1-propanol	49	133
1-Methoxy-2-propanone	13	161
1-Methoxy-4-propenylbenzene	3	137
2-Methoxy-4-propenylphenol	26	95
2-Methoxy-5-propenylphenol	182	105
2-Methoxypropionic acid	11	192
3-Methoxypropionitrile	41	348
β-Methoxypropionyl chloride	26	212
1-Methoxy-2-n-propoxyethane	47	133
2-Methoxypyridine	80	436
3-Methoxypyridine	186	437
4-Methoxypyridine	262	438
2-Methoxyquinoline	83	436
4-Methoxyquinoline	260	438
6-Methoxyquinoline	112	321
"	193	437
7-Methoxyquinoline	59	140
8-Methoxyquinoline	27	138
"	26	323
2-Methoxytoluene	62	134

Compound	Compound No.	Page No.
3-Methoxytoluene	66	134
4-Methoxytoluene	65	134
1-Methoxy-2,4,6-trinitrobenzene	36	338
N-Methyl acetanilide	123	234
Methyl acetate	4	250
Methylacetoacetate	108	164
"	123	255
2-Methylacetoacetonitrile	35	347
2-Methylacetophenone	170	166
3-Methylacetophenone	176	166
4-Methylacetophenone	6	172
N-Methyl-2-acetotoluidide	17	231
N-Methyl-3-acetotoluidide	32	231
N-Methyl-4-acetotoluidide	58	232
N-Methylacetylanthranilic acid	311	206
Methyl acetylene	2	28
Methyl acrylate	12	250
2-Methylacrylic acid	411	432
α-Methylacrylonitrile	7	345
β-Methyladipic acid	89	196
D-N-Methyl-α-alanine	109	289
DL-N-Methyl-α-alanine	98	289
"	66	429
N-Methyl-β-alanine	2	284
Methyl alcohol	1	80
Methylallene	11	12
2-Methylallyl sulfide	38	421
Methylamidine	470	439
Methylamine	1	295
"	422	439
L-Methylamine	111	298
2-(N-Methylamino)benzenesulfonamide	51	393
4-(N-Methylamino)benzenesulfonamide	338	405
2-(N-Methylamino)benzenesulfonic acid	14	371
4-(N-Methylamino)benzenesulfonic acid	158	376
Methyl-2-aminobenzoate	15	273
3-(Methylamino)benzoic acid	214	430
(2-Methylamino)ethylamine	367	439
1-(Methylamino)naphthalene	183	301
"	109	436
2-(Methylamino)naphthalene	188	301
2-(Methylamino)-n-pentane	33	296
2-(Methylamino)phenol	184	105
4-(Methylamino)phenol	152	104
Methyl-n-amyl alcohol	41	82
N-Methylaniline	90	298
"	181	437
2-Methylaniline	94	298
"	156	437
3-Methylaniline	96	298
"	174	437
4-Methylaniline	34	303
"	195	437
N-Methylanilinonitrile	95	353
Methyl anisate	59	276
2-Methylanisole	62	134
3-Methylanisole	66	134
4-Methylanisole	65	134
1-Methylanthracene	85	46
9-Methylanthracene	80	46
Methylanthranilate	15	273
2-Methyl-9,10-anthraquinone	24	182
N-Methylaziridine	286	438
2-Methylazulene	23	43
6-Methylazulene	82	46
2-Methylbenzaldehyde	74	148
3-Methylbenzaldehyde	73	148
4-Methylbenzaldehyde	75	148
2-Methylbenzene-1,3-disulfonamide	443	409
2-Methylbenzene-1,4-disulfonamide	371	406
4-Methylbenzene-1,2-disulfonamide	410	407
4-Methylbenzene-1,3-disulfonamide	256	401
5-Methylbenzene-1,3-disulfonamide	349	405
2-Methylbenzene-1,3-disulfonanilide	168	398
2-Methylbenzene-1,4-disulfonanilide	225	400
4-Methylbenzene-1,2-disulfonanilide	272	402
4-Methylbenzene-1,3-disulfonanilide	267	402
5-Methylbenzene-1,3-disulfonanilide	137	397
2-Methylbenzene-1,3-disulfonic acid	204	378
2-Methylbenzene-1,4-disulfonic acid	168	377
4-Methylbenzene-1,2-disulfonic acid	188	377
4-Methylbenzene-2,4-disulfonic acid	132	375
5-Methylbenzene-1,3-disulfonic acid	162	377
2-Methylbenzene-1,3-disulfonyl chloride	113	384
2-Methylbenzene-1,4-disulfonyl chloride	144	385
4-Methylbenzene-1,2-disulfonyl chloride	166	386
4-Methylbenzene-1,3-disulfonyl chloride	44	382
5-Methylbenzene-1,3-disulfonyl chloride	125	385
2-Methylbenzenesulfonic acid	58	372
3-Methylbenzenesulfonic acid	10	371
4-Methylbenzenesulfonic acid	31	372
2-Methylbenzenesulfonyl chloride	73	383
3-Methylbenzenesulfonyl chloride	3	381
4-Methylbenzenesulfonyl chloride	75	383
4-Methylbenzhydrol	29	89
Methyl benzilate	110	278
1-Methylbenzimidazole	220	437
2-Methylbenzimidazole	252	438
4-Methylbenzimidazole	228	437
5-Methylbenzimidazole	235	437
Methyl benzoate	172	258
2-Methylbenzoic acid	125	198
"	269	431
3-Methylbenzoic acid	137	198
"	339	432
4-Methylbenzoic acid	278	204
"	360	432
2-Methylbenzoic anhydride	9	225
3-Methylbenzoic anhydride	29	225
4-Methylbenzoic anhydride	40	225
2-Methylbenzonitrile	64	350
3-Methylbenzonitrile	70	350
4-Methylbenzonitrile	13	357
4-Methylbenzophenone	74	174
2-Methyl-1,4-benzoquinone	6	181
Methyl 2-benzoylbenzoate	119	279
2-Methylbenzoyl bromide	5	219
3-Methylbenzoyl bromide	1	219
4-Methylbenzoyl bromide	8	219
4-Methylbenzoyl chloride	51	213
1-Methyl-3,4-benzphenanthrene	69	45
2-Methyl-3,4-benzphenanthrene	61	45
5-Methyl-3,4-benzphenanthrene	143	49
6-Methyl-3,4-benzphenanthrene	76	46
7-Methyl-3,4-benzphenanthrene	31	44
8-Methyl-3,4-benzphenanthrene	54	45
2-Methyl-1',2'-benzpyrene	139	48
3-Methyl-1',2'-benzpyrene	148	49
4-Methyl-1',2'-benzpyrene	191	51
5-Methyl-1',2'-benzpyrene	188	51
6-Methyl-1',2'-benzpyrene	170	50
8-Methyl-1',2'-benzpyrene	181	50
9-Methyl-1',2'-benzpyrene	149	49
2-Methylbenzyl alcohol	14	88
3-Methylbenzyl alcohol	109	85
4-Methylbenzyl alcohol	32	89
N-Methylbenzylamine	331	438
2-Methylbenzylamine	104	298
"	315	438
3-Methylbenzylamine	101	298
"	320	438

Compound	Compound No.	Page No.
4-Methylbenzylamine	105	298
"	322	438
2-Methylbenzyl chloride	44	56
3-Methylbenzyl chloride	43	56
4-Methylbenzyl chloride	41	56
4-Methylbenzyl sulfide	44	426
N Methylbenzyl sulfonamide	42	393
Methyl bromide	1	58
Methyl bromoacetate	79	253
Methyl 2-bromobenzoate	267	264
Methyl 3-bromobenzoate	29	274
Methyl 4-bromobenzoate	121	279
2-Methyl-1,3-butadiene	17	12
3-Methyl-1,2-butadiene	22	12
3-Methylbutanal	17	145
d,l-2-Methylbutanamide	170	235
d,l-2-Methylbutananilide	163	235
2-Methylbutane	10	3
D,L-3-Methylbutane-1-sulfonanilide	4	392
D-2-Methyl-1-butanethiol	14	413
3-Methylbutanethiol	12	413
d,l-2-Methylbutanoic acid	14	190
3-Methylbutanoic acid	15	190
2-Methyl-1-butanol	23	81
2-Methyl-2-butanol	8	80
3-Methyl-1-butanol	26	81
d,l-3-Methyl-2-butanol	12	80
2-Methyl-3-butanone	6	161
2-Methyl-1-butenal	18	145
2-Methyl-2-butenal	31	145
3-Methyl-2-butenal	41	146
2-Methyl-1-butene	15	12
2-Methyl-2-butene	21	12
3-Methyl-1-butene	12	12
cis-2-Methyl-2-butenoic acid	59	195
trans-2-Methyl-2-butenoic acid	34	194
2-Methyl-1-buten-3-one	7	161
2-Methyl-3-butenyl phenyl ketone	53	170
3-Methyl-3-buten-1-yne	7	28
3-Methylbutyl acetate	75	253
D-2-Methyl-n-butylamine	29	296
3-Methylbutyl phenyl sulfide	56	421
3-Methyl-1-butyne	5	28
α-Methylbutyraldehyde	18	145
Methyl n-butyrate	31	251
2-Methylbutyric acid	440	432
2-Methylbutyronitrile	20	346
5-Methyl α-campholenonitrile	1	355
Methyl n-caprate	225	262
Methyl n-caproate	97	254
Methyl n-caprylate	163	258
Methyl carbamate	11	231
Methyl carbonate	18	250
3-Methylcatechol	84	101
4-Methylcatechol	76	100
2-Methylcatechol 2-methyl ether	92	135
Methyl cellosolve	20	81
Methyl cellosolve acetate	78	253
Methyl cellosolve benzoate	294	266
Methyl chloride	1	55
Methyl chloroacetate	61	252
Methyl 2-chlorobenzoate	250	263
Methyl 3-chlorobenzoate	242	263
" "	7	273
Methyl 2-chlorocinnamate	52	275
Methyl chloroformate	9	250
2-Methylchrysazin	23	182
1-Methylchrysene	197	51
2-Methylchrysene	193	51
3-Methylchrysene	168	50
4-Methylchrysene	154	49
5-Methylchrysene	121	48
6-Methylchrysene	162	50
2-Methylcinnamaldehyde	27	151
Methyl cinnamate	39	275
α-Methylcinnamic acid	81	196
trans-2-Methylcinnamic acid	392	432
trans-3-Methylcinnamic acid	380	432
trans-4-Methylcinnamic acid	399	432
3-Methycrotonaldehyde	41	146
Methyl crotonate	46	252
3-Methylcrotonic acid	466	433
2-Methylcrotononitrile	26	347
Methyl cyanoacetate	174	258
"	62	350
Methylcyclobutane	14	3
1-Methylcyclobutene	20	12
3-Methylcyclobutene	16	16
Methylcycloheptane	94	4
1-Methylcycloheptene	243	20
5-Methylcycloheptene	14	25
1-Methyl-1,4-cyclohexadiene	188	18
5-Methyl-1,3-cyclohexadiene	117	16
Methylcyclohexane	44	3
Methyl cyclohexanecarboxylate	144	257
cis-4-Methylcyclohexanecarboxylic acid	28	192
trans-4-Methylcyclohexanecarboxylic acid	134	198
cis-3-Methyl-1-cyclohexanethiol	24	414
trans-3-Methyl-1-cyclohexanethiol	26	414
cis-2-Methylcyclohexanol	58	83
trans-2-Methylcyclohexanol	60	83
trans-2-Methylcyclohexanol	2	88
cis-3-Methylcyclohexanol	65	83
trans-3-Methylcyclohexanol	68	83
cis-4-Methylcyclohexanol	66	83
trans-4-Methylcyclohexanol	67	83
2-Methylcyclohexanone	97	164
d-3-Methylcyclohexanone	104	164
d,l-3-Methylcyclohexanone	105	164
4-Methylcyclohexanone	112	164
1-Methylcyclohexene	158	17
3-Methylcyclohexene	126	16
4-Methylcyclohexene	125	16
3-Methylcyclohexyl-1,1-diacetic acid	81	170
3-Methylcyclohexyl-1,1-diacetic acid	387	432
4-Methylcyclohexyl-1,1-diacetic acid	388	432
2-Methylcyclohexyloxyacetic acid	266	431
cis-3-Methylcyclohexyloxy acetic acid	260	431
Methylcyclooctatetraene	15	25
Methylcyclopentane	26	3
2-Methylcyclopentanecarboxylic acid	26	192
2-Methylcyclopentanone	45	162
d-3-Methylcyclopentanone	50	162
d,l-3-Methylcyclopentanone	55	162
1-Methylcyclopentene	63	14
3-Methylcyclopentene	43	13
4-Methylcyclopentene	62	14
2-Methyl-3-cyclopentenone	90	163
3-Methylcyclopentyl-1,1-diacetic acid	248	431
1-Methyl-trans-decahydronaphthalene	260	7
9-Methyl-cis-decahydronaphthalene	253	7
9-Methyl-trans-decahydronaphthalene	251	7
Methyl n-decanoate	225	262
Methyl dibenzylacetate	45	275
Methyl 3,5-dinitrobenzoate	143	280
Methyl 3,5-dinitrosalicylate	156	281
Methyl diphenylacetate	84	277
2-Methyl-1,4-dithiane	4	425
Methyl enanthate	125	256
Methyleneaminoacetonitrile	77	324
Methylene bromide	1	65
Methylene chloride	1	63
Methylene cyanide	17	357

Compound	Compound No.	Page No.
Methylene-di-2-naphthol	504	123
3,4-Methylenedioxybenzaldehyde	12	153
3,4-Methylenedioxychalcone	153	177
3,4-Methylenedioxypropiophenone	27	173
1,2-Methylenedioxy-4-propylbenzene	101	136
Methylenefluorene	28	44
Methylene iodide	1	67
Methylenesuccinic acid	254	203
Methylenesuccinic anhydride	26	225
endo-4,7-Methylene-4,7,8,9-tetrahydroindene	3	26
Methyl ethoxyacetate	92	254
Methyl ethylacetoacetate	153	257
2-Methylethylene sulfide	2	423
4-Methylfluorene	50	45
N-Methylformamide	3	230
N-Methylformanilide	8	230
Methyl formate	2	250
Methylfumaric acid	329	207
Methylfumaryl dichloride	28	214
2-Methylfuran	15	131
5-Methylfurfural	68	147
Methyl furoate	141	256
Methyl furoylacetate	6	272
Methyl β-(2-furyl)acrylate	18	274
Methyl gallate	506	123
" "	169	281
cis-α-Methylglutaconic anhydride	34	225
cis-β-Methylglutaconic anhydride	35	225
3-Methylglutaric acid	335	431
α-Methylglutaric anhydride	17	223
β-Methylglutaric anhydride	10	225
2-Methylglutaromononitrile	126	363
2-Methylglutaronitrile	97	353
β-Methylglyceraldehyde	11	328
N-Methylglycine	84	429
Methyl glycolate	96	254
Methyl glyoxal	103	171
α-Methylguanidoacetic acid	113	289
Methyl hendecylenate	285	265
2-Methylheptadecane	273	7
2-Methyl-2-heptadecene	337	24
2-Methyl-1,3-heptadiene	222	19
3-Methyl-1,5-heptadiene	173	17
3-Methyl-2,4-heptadiene	229	19
4-Methyl-2,4-heptadiene	227	19
6-Methyl-1,3-heptadiene	194	18
6-Methyl-2,4-heptadiene	185	18
2-Methylheptane	61	4
3-Methylheptane	67	4
4-Methylheptane	62	4
Methyl n-heptanoate	125	256
4-Methyl-1-heptanol	86	84
2-Methyl-4-heptanone	67	163
3-Methyl-2-heptanone	91	164
6-Methyl-2-heptanone	110	164
6-Methyl-3-heptanone	94	164
2-Methyl-1-heptene	202	18
2-Methyl-2-heptene	219	19
2-Methyl-3-heptene	170	17
3-Methyl-1-heptene	163	17
3-Methyl-2-heptene	212	19
3-Methyl-3-heptene	208	18
4-Methyl-1-heptene	175	18
4-Methyl-2-heptene	181	18
4-Methyl-3-heptene	206	18
5-Methyl-1-heptene	180	18
5-Methyl-2-heptene	200	18
5-Methyl-3-heptene	171	17
6-Methyl-1-heptene	216	19
6-Methyl-2-heptene	198	18
6-Methyl-3-heptene	176	18
3-Methyl-3-hepten-2-one	122	165
3-Methyl-3-hepten-5-one	17	169
4-Methyl-6-hepten-3-one	80	163
5-Methyl-3-heptyne	34	29
2-Methyl-1,3-hexadiene	142	17
2-Methyl-1,5-hexadiene	100	15
2-Methyl-2,4-hexadiene	141	17
3-Methyl-1,5-hexadiene	76	14
3-Methyl-2,4-hexadiene	143	17
4-Methyl-1,3-hexadiene	150	17
5-Methyl-1,2-hexadiene	111	16
5-Methyl-1,4-hexadiene	97	15
Methyl hexahydrobenzoate	144	257
5-Methylhexanal	45	146
2-Methylhexanamide	38	232
4-Methylhexanamide	108	233
2-Methylhexananilide	107	233
4-Methylhexananilide	45	232
2-Methylhexane	34	3
3-Methylhexane	37	3
3-Methyl-2,4-hexanedione	139	165
3-Methyl-2,5-hexanedione	92	170
Methyl n-hexanoate	97	254
2-Methylhexanoic acid	37	191
3-Methylhexanoic acid	17	192
4-Methylhexanoic acid	40	191
5-Methylhexanoic acid	35	191
2-Methyl-1-hexanol	55	83
d,l-4-Methyl-1-hexanol	57	83
2-Methyl-3-hexanone	34	162
3-Methyl-2-hexanone	38	162
4-Methyl-2-hexanone	48	162
4-Methyl-3-hexanone	35	162
5-Methyl-2-hexanone	31	162
2-Methylhexanonitrile	40	348
d,l-3-Methylhexanonitrile	44	348
d,l-4-Methylhexanonitrile	53	349
5-Methylhexanonitrile	51	349
2-Methyl-1-hexene	98	15
2-Methyl-2-hexene	108	16
2-Methyl-3-hexene	88	15
3-Methyl-1-hexene	83	15
3-Methyl-trans-2-hexene	105	16
3-Methyl-cis-3-hexene	107	16
3-Methyl-trans-3-hexene	103	15
4-Methyl-1-hexene	90	15
4-Methyl-cis-2-hexene	92	15
4-Methyl-trans-2-hexene	93	15
5-Methyl-1-hexene	86	15
5-Methyl-cis-2-hexene	96	15
5-Methyl-trans-2-hexene	87	15
5-Methyl-4-hexenoic acid	437	432
3-Methyl-1-hexen-5-one	39	162
5-Methyl-4-hexen-3-one	65	163
5-Methyl-5-hexen-2-one	66	163
Methyl hexyl acetylene	50	30
2-Methyl-3-hexyne	31	29
4-Methyl-2-hexyne	32	29
5-Methyl-2-hexyne	36	29
α-Methylhydrocinnamamide	159	235
Methyl hydrocinnamate	259	264
d,l-α-Methylhydrocinnamic acid	20	193
2-Methylhydroquinone	293	111
"	60	434
Methyl 2-hydroxybenzoate	12	94
Methyl 3-hydroxybenzoate	95	101
" "	97	278
Methyl 4-hydroxybenzoate	313	112
" "	157	281
" "	48	434
2-Methyl-3-hydroxybutyric acid	408	432
Methyl α-hydroxyisobutyrate	71	253

Compound	Compound No.	Page No.
Methyl 3-hydroxy-2-naphthoate	109	278
Methyl 3-hydroxy-4-nitrobenzoate	17	434
4-Methyl-8 hydroxyquinoline	12	434
2-Methylimidazole	285	438
4-Methylimidazole	280	438
Methyliminodiacetic acid	355	208
Methyliminodiacetic acid diamide	339	240
Methyliminodiacetic acid monoamide	338	240
1-Methylindane	36	36
2-Methylindane	34	36
3-Methylindene	63	37
1-Methylindole	79	321
2-Methylindole	76	305
3-Methylindole	175	308
4-Methylindole	177	300
5-Methylindole	73	305
7-Methylindole	145	307
Methyl iodide	1	60
Methyl isobutylcarbinyl acetate	91	254
Methyl isobutyrate	19	250
Methyl isocrotonate	34	251
Methylisopropylacetaldehyde	27	145
Methyl isovalerate	42	251
Methyl d,l-lactate	82	253
Methyl laurate	314	268
Methyl levulinate	152	166
" "	165	258
Methylmaleic acid	98	196
Methylmaleic anhydride	10	223
Methyl malonate	142	257
Methylmalonic acid	194	201
" "	146	430
Methylmalonic acid diamide	457	244
Methylmalononitrile	11	357
Methyl d,l-mandelate	71	276
Methyl margarate	22	274
Methyl mercaptan	1	413
Methyl methacrylate	26	251
Methyl methoxyacetate	60	252
Methyl 2-methoxybenzoate	284	265
Methyl 3-methoxybenzoate	290	266
Methyl 4-methoxybenzoate	59	276
Methyl methylacetoacetate	131	256
Methyl 2-methylallyl sulfide	14	420
Methyl 2-methylbenzoate	208	261
Methyl 3-methylbenzoate	216	261
Methyl 2-methylbutyl sulfide	24	420
2-Methyl-6-methylene-2,7-octadiene	295	22
4-Methyl-2-methylolphenol	103	434
N-Methylmorpholine	278	438
Methyl myristate	2	273
1-Methylnaphthalene	146	40
2-Methylnaphthalene	12	43
2-Methyl-1,3-naphthalenediol	339	113
2-Methylnaphthalene-1-sulfonamide	64	394
4-Methylnaphthalene-1-sulfonamide	220	400
4-Methylnaphthalene-2-sulfonamide	108	396
5-Methylnaphthalene-1-sulfonamide	219	400
5-Methylnaphthalene-2-sulfonamide	266	402
6-Methylnaphthalene-2-sulfonamide	325	404
7-Methylnaphthalene-1-sulfonamide	298	403
7-Methylnaphthalene-2-sulfonamide	173	398
8-Methylnaphthalene-2-sulfonamide	54	394
4-Methylnaphthalene-1-sulfonanilide	157	397
5-Methylnaphthalene-2-sulfonanilide	430	408
7-Methylnaphthalene-1-sulfonanilide	171	398
2-Methylnaphthalene-1-sulfonic acid	18	371
4-Methylnaphthalene-1-sulfonic acid	97	374
4-Methylnaphthalene-2-sulfonic acid	39	372
5-Methylnaphthalene-1-sulfonic acid	96	374
5-Methylnaphthalene-2-sulfonic acid	128	375
6-Methylnaphthalene-2-sulfonic acid	151	376
7-Methylnaphthalene-1-sulfonic acid	142	376
2-Methylnaphthalene-1-sulfonyl chloride	103	384
4-Methylnaphthalene-1-sulfonyl chloride	99	384
4-Methylnaphthalene-2-sulfonyl chloride	200	387
5-Methylnaphthalene-2-sulfonyl chloride	185	387
6-Methylnaphthalene-2-sulfonyl chloride	141	385
7-Methylnaphthalene-1-sulfonyl chloride	133	385
7-Methylnaphthalene-2-sulfonyl chloride	64	383
8-Methylnaphthalene-2-sulfonyl chloride	114	384
2-Methylnaphthanthracene	151	49
3-Methylnaphthanthracene	163	50
4-Methylnaphthanthracene	184	51
5-Methylnaphthanthracene	157	49
6-Methylnaphthanthracene	134	48
7-Methylnaphthanthracene	141	48
8-Methylnaphthanthracene	159	49
9-Methylnaphthanthracene	153	49
10-Methylnaphthanthracene	176	50
11-Methylnaphthanthracene	120	47
12-Methylnaphthanthracene	138	48
Methyl 2-naphthoate	113	278
2-Methyl-1,4-naphthohydroquinone	447	119
1-Methyl-2-naphthol	231	108
2-Methyl-1-naphthol	75	100
4-Methyl-1-naphthol	140	103
4-Methyl-2-naphthol	129	103
5-Methyl-1-naphthol	187	106
5-Methyl-2-naphthol	205	106
7-Methyl-1-naphthol	230	108
2-Methyl-1,4-naphthoquinone	11	181
N-Methyl-N-(1-naphthyl)acetamide	87	233
1-Methyl-2-naphthylamine	53	304
2-Methyl-1-naphthylamine	9	303
3-Methyl-1-naphthylamine	54	304
3-Methyl-2-naphthylamine	304	312
4-Methyl-1-naphthylamine	55	304
4-Methyl-2-naphthylamine	105	306
5-Methyl-1-naphthylamine	124	306
5-Methyl-2-naphthylamine	86	305
7-Methyl-1-naphthylamine	68	304
8-Methyl-1-naphthylamine	101	306
d,l-Methyl 1-naphthyl carbinol	36	89
Methyl-1-naphthyl ether	106	136
Methyl 2-naphthyl ether	44	139
Methyl 1-naphthyl ketone	215	168
" " "	18	172
Methyl 2-naphthyl ketone	57	174
Methyl 2-naphthyl sulfide	23	425
Methylneopentylacetic acid	23	192
Methylneopentylglycolic acid	133	198
Methyl neopentyl ketone	21	161
Methyl nitrate	6	250
N-Methyl-4-nitroacetanilide	291	239
2-Methyl-4-nitroacetanilide	419	243
4-Methyl-2-nitroacetanilide	109	233
N-Methyl-3-nitroaniline	106	306
N-Methyl-4-nitroaniline	340	313
2-Methyl-3-nitroaniline	163	307
2-Methyl-4-nitroaniline	293	312
" " "	26	436
2-Methyl-5-nitroaniline	216	309
2-Methyl-6-nitroaniline	164	307
3-Methyl-4-nitroaniline	303	312
" " "	31	436
4-Methyl-2-nitroaniline	245	310
4-Methyl-3-nitroaniline	126	306
" " "	76	436
5-Methyl-2-nitrobenzene	43	338
2-Methyl-4-nitrobenzenesulfonamide	154	397
2-Methyl-5-nitrobenzenesulfonamide	254	401
3-Methyl-4-nitrobenzenesulfonamide	174	398
3-Methyl-4-nitrobenzenesulfonamide	84	395

Compound	Compound No.	Page No.
4-Methyl-2-nitrobenzenesulfonamide	200	399
4-Methyl-3-nitrobenzenesulfonamide	114	396
2-Methyl-5-nitrobenzenesulfonanilide	124	396
2-Methyl-4-nitrobenzenesulfonic acid	61	373
2-Methyl-5-nitrobenzenesulfonic acid	121	375
2-Methyl-5-nitrobenzenesulfonic acid	116	375
3-Methyl-2-nitrobenzenesulfonic acid	67	373
3-Methyl-4-nitrobenzenesulfonic acid	26	371
4-Methyl-2-nitrobenzenesulfonic acid	85	373
4-Methyl-3-nitrobenzenesulfonic acid	42	372
2-Methyl-3-nitrobenzenesulfonyl chloride	145	385
2-Methyl-4-nitrobenzenesulfonyl chloride	161	386
2-Methyl-5-nitrobenzenesulfonyl chloride	32	381
3-Methyl-2-nitrobenzenesulfonyl chloride	55	382
3-Methly-4-nitrobenzenesulfonyl chloride	35	382
4-Methyl-3-nitrobenzenesulfonyl chloride	18	381
Methyl 3-nitrobenzoate	116	279
Methyl 4-nitrobenzoate	131	279
2-Methyl-4-nitrobenzoic acid	33	429
2-Methyl-3-nitrobenzonitrile	153	364
2-Methyl-2-nitrobutane	9	335
3-Methyl-1-nitrobutane	13	335
Methyl 2-nitrocinnamate	105	278
Methyl 3-nitrocinnamate	155	281
Methyl 4-nitrocinnamate	165	281
1-Methyl-2-nitronaphthalene	29	337
1-Methyl-4-nitronaphthalene	39	338
1-Methyl-8-nitronaphthalene	34	338
2-Methyl-1-nitronaphthalene	47	338
5-Methyl-1-nitronaphthalene	50	339
3-Methyl-2-nitro-1,4-naphthoquinone	84	340
2-Methy-3-nitrophenol	365	115
2-Methyl-4-nitrophenol	183	105
2-Methyl-5-nitrophenol	260	109
2-Methyl-6-nitrophenol	96	101
3-Methyl-2-nitrophenol	27	97
3-Methyl-4-nitrophenol	306	112
3-Methyl-6-nitrophenol	49	99
4-Methyl-2-nitrophenol	19	97
4-Methyl-3-nitrophenol	121	102
5-Methyl-2-nitrophenol	26	434
Methyl 2-nitrophenyl sulfide	36	425
Methyl 4-nitrophenyl sulfide	41	426
2-Methyl-1-nitropropane	8	335
2-Methyl-2-nitropropane	1	337
Methyl 3-nitrosalicylate	158	281
Methyl 5-nitrosalicylate	151	280
28-Methylnonacosane	62	9
2-Methylnonane	218	6
3-Methylnonane	224	6
4-Methylnonane	210	6
5-Methylnonane	209	6
Methyl n-nonyl carbinol	117	86
Methyl n-nonyl ketone	186	167
8-Methyl-4-nonyne	37	29
4-Methyl-3,5-octadiene	271	21
7-Methyl-2,4-octadiene	272	21
2-Methyloctanal	3	150
7-Methyloctanal	1	150
2-Methyloctane	128	5
3-Methyloctane	130	5
4-Methyloctane	125	5
Methyl n-octanoate	163	258
6-Methyloctanoic acid	15	192
7-Methyloctanoic acid	20	192
2-Methyl-4-octanone	102	164
4-Methyl-4-octanone	2	169
2-Methyl-4-octene	244	20
3-Methyl-2-octene	257	20
4-Methyl-2-octene	183	18
7-Methyl-3-octene	253	20
2-Methyl-3-octen-5-one	66	170
2-Methyl-3-octen-6-one	28	169
7-Methyl-1-octen-5-one	8	169
Methyl n-octyl carbinol	106	85
Methyl n-octyl ketone	166	166
2-Methylolphenol	92	434
3-Methylolphenol	87	434
4-Methylolphenol	86	434
Methyl orthoformate	33	251
Methyl palmitate	24	274
Methyl pelargonate	205	260
2-Methyl-1,3-pentadiene	64	14
2-Methyl-1,3-pentadiene	66	14
2-Methyl-1,4-pentadiene	35	13
2-Methyl-2,3-pentadiene	56	14
3-Methyl-1,2-pentadiene	74	14
3-Methyl-1,3-pentadiene	71	14
3-Methyl-1,4-pentadiene	33	13
4-Methyl-1,2-pentadiene	54	13
4-Methyl-1,3-pentadiene	67	14
2-Methylpentamethylene sulfide	24	423
3-Methylpentamethylene sulfide	27	423
4-Methylpentamethylene sulfide	30	423
d, l-2-Methylpentanamide	52	232
3-Methylpentanamide	199	236
d, l-2-Methylpentananilide	92	233
d, l-3-Methylpentananilide	70	232
4-Methylpentananilide	169	235
2-Methylpentane	21	3
3-Methylpentane	22	3
2-Methyl-2,4-pentanediol	99	85
Methyl n-pentanoate	57	252
d, l-2-Methylpentanoic acid	28	191
" " "	436	432
d, l-3-Methylpentanoic acid	29	191
" " "	446	432
4-Methylpentanoic acid	30	191
2-Methyl-1-pentanol	41	82
2-Methyl-2-pentanol	19	81
2-Methyl-3-pentanol	21	81
3-Methyl-1-pentanol	46	82
3-Methyl-2-pentanol	28	81
3-Methyl-3-pentanol	18	80
4-Methyl-1-pentanol	47	82
d, l-4-Methyl-2-pentanol	25	81
3-Methyl-2-pentanone	16	161
4-Methyl-2-pentanone	15	161
4-Methylpentanonitrile	38	348
4-Methylpentanoyl chloride	29	212
2-Methyl-2-pentenal-1	43	146
2-Methyl-1-pentene	39	13
2-Methyl-2-pentene	47	13
3-Methyl-1-pentene	32	13
3-Methyl-cis-2-pentene	55	13
3-Methyl-trans-2-pentene	48	13
4-Methyl-1-pentene	31	13
4-Methyl-cis-2-pentene	36	13
4-Methyl-trans-2-pentene	37	13
cis-3-Methyl-2-pentenoic acid	468	433
trans-3-Methyl-2-pentenoic acid	467	433
4-Methyl-2-pentenoic acid	438	432
4-Methyl-3-pentenoic acid	403	432
2-Methyl-1-penten-3-one	18	161
2-Methyl-1-penten-4-one	32	162
3-Methyl-1-penten-4-one	41	162
4-Methyl-3-penten-2-one	25	162
Methyl pentyl acetylene	48	30
1-Methyl-3-pentylbenzene	119	39
Methyl pentyl sulfide	28	420
3-Methyl-1-pentyne	16	28
4-Methyl-1-pentyne	15	28
4-Methyl-2-pentyne	18	28
Methyl phenacyl ketone	75	174

	Compound No.	Page No.
1-Methylphenanthrene	128	48
3-Methylphenanthrene	44	44
4-Methylphenanthrene	25	43
9-Methylphenanthrene	90	46
2-Methylphenol	11	96
"	109	435
3-Methylphenol	5	94
"	100	434
4-Methylphenol	17	97
"	108	435
Methyl phenoxyacetate	271	265
(2-Methylphenoxy) acetic acid	181	430
(4-Methylphenoxy) acetic acid	176	430
Methyl 3-phenoxypropyl ketone	49	173
Methyl phenylacetate	214	261
(4-Methylphenyl) acetic acid	361	432
d-2-Methyl-1-phenylbutane	50	37
d,l-2-Methyl-1-phenylbutane	58	37
2-Methyl-2-phenylbutane	43	37
3-Methyl-1-phenylbutane	60	37
3-Methyl-2-phenylbutane	35	36
3-Methyl-3-phenyl-2-butanone	65	170
3-Methyl-4-phenyl-2-butanone	16	169
2-Methyl-1-phenyl-1-butene	143	40
Methyl α-phenyl n-butyrate	114	278
3-Methyl-2-phenylbutyronitrile	88	352
d, l-Methyl phenyl carbinol	1	88
Methyl 4-phenylcyclohexyl ketone	72	170
Methyl phenyl ether	56	133
Methyl β-phenylethyl ketone	196	167
Methylphenylglycol aldehyde	20	150
2-Methyl-2-phenylhexane	121	39
3-Methyl-3-phenylhexane	123	39
α-Methyl-α-phenylhydrazine	138	299
4-Methylphenlhydrazine	91	305
Methyl phenyl ketone	160	166
2-Methyl-1-phenylpentane	104	39
2-Methyl-3-phenylpentane	93	38
3-Methyl-1-phenylpentane	115	39
3-Methyl-3-phenylpentane	82	38
2-Methyl-3-phenylpropionaldehyde	10	150
Methyl β-phenylpropionate	259	264
3-(2-Methylphenyl) propionic acid	413	432
3-(3-Methylphenyl) propionic acid	416	432
3-(4-Methylphenyl) propionic acid	417	432
Methyl phenyl sulfide	44	423
Methylphloroglucinol	531	124
Methyl phthalate	332	269
3-Methylphthalic anydride	50	226
4-Methylphthalic anhydride	38	225
1-Methylpiperidine	364	439
DL-3-Methylpiperidine	44	296
" "	458	439
DL-2-Methylpiperidine	41	296
" "	442	439
L-2-Methylpiperidine	40	296
Methyl piperonylate	65	276
Methyl pivalate	29	251
2-Methylpropane	5	3
2-Methylpropane-1-sulfonamide	1	392
2-Methylpropane-1-sulfonanilide	3	392
2-Methyl-1,2-prepanediol	79	84
2-Methyl-1-propanethiol	7	413
2-Methyl-2-propanethiol	4	413
2-Methyl-1-propanol	10	80
2-Methyl-2-propenal	10	144
2-Methylpropene	5	12
Methylpropionate	11	250
Methyl n-propylacetaldehyde	30	145
Methyl-n-propylacetic acid	28	191
Methyl propyl acetylene	24	29
1-Methyl-2-propylbenzene	31	36
1-Methyl-3-propylbenzene	24	36
1-Methyl-4-propylbenzene	27	36
Methyl n-propylcarbinyl acetate	65	252
1-Methyl-1-propylcyclohexane	238	7
1-Methyl-2-propylcyclohexane	241	7
1-Methyl-3-propylcyclohexane	204	6
1-Methyl-4-propylcyclohexane	240	7
3-Methyl-2-n-propylcyclopentanone	86	170
Methyl n-propyl ether	7	131
Methyl n-propyl ketone	10	161
Methyl n-propyl sulfide	9	420
2-Methyl-5-propylthiophene	44	424
4-Methylpurpuroxanthin	45	184
2-Methylpyrazine	16	319
1-Methylpyrazole	11	319
3-Methylpyrazole	106	298
"	101	436
4-Methylpyrazole	100	298
5-Methyl-2-pyrazoline	78	297
3-Methylpyrene	62	45
3-Methylpyridazine	114	298
2-Methylpyridine	13	319
"	240	437
3-Methylpyridine	19	319
"	227	437
4-Methylpyridine	20	319
"	247	438
N-Methyl-2-pyridone	86	321
N-Methyl-4-pyridone	59	324
Methyl 2-pyridyl ketone	144	165
" " "	50	320
Methyl 3-pyridyl ketone	68	320
" " "	175	166
Methyl-4-pyridyl ketone	167	166
" " "	66	320
DL-1-Methyl-2-(3-pyridyl)pyrrolidine	73	321
L-1-Methyl-2-(3-pyridyl)pyrrolidine	82	321
4-Methylpyrimidine	17	319
5-Methylpyrimidine	6	323
5-Methylpyrogallol	268	110
Methyl pyromucate	141	256
N-Methylpyrrolidine	3	319
"	385	439
2-Methylpyrrolidine	30	296
3-Methylpyrrolidine	31	296
1-Methyl-3-pyrroline	351	438
Methyl pyruvate	70	253
β-Methylquinoline	90	321
γ-Methylquinoline	95	321
2-Methylquinoline	80	321
"	230	437
3-Methylquinoline	90	321
4-Methylquinoline	95	321
"	205	437
5-Methylquinoline	94	321
"	166	437
6-Methylquinoline	91	321
"	201	437
7-Methylquinoline	13	323
"	208	437
8-Methylquinoline	81	321
"	183	437
2-Methylquinoxaline	78	321
2-Methylresorcinol	265	110
5-Methylresorcinol	217	107
5-Methylsalicylaldehyde	41	155
Methyl salicylate	12	94
"	220	262
Methyl sebacate	42	275
Methyl stearate	43	275
Methyl styryl ketone	37	173

	Compound No.	Page No.
α-Methylstyryl phenyl ketone	220	168
Methylsuccinic acid	145	199
Methylsuccinic acid diamide	476	245
Methylsuccinic acid dianilide	428	243
DL-Methylsuccinic anhydride	7	225
Methyl succinonitrile	89	352
Methyl sulfate	150	257
Methyl sulfide	1	420
(Methylsulfonyl)acetic acid	89	429
3-(Methylsulfonyl)phenol	68	434
4-(Methylsulfonyl)phenol	31	434
DL-Methylterose	1	328
L-Methylterose	12	328
2-Methyl-3,4,5,6-tetrabromophenol	523	123
3-Methyl-2,4,5,6-tetrabromophenol	494	122
4-Methyl-2,3,5,6-tetrabromophenol	499	122
2-Methyl-4,5,6,7-tetrahydroindole	127	299
6-Methyl-1,2,3,4-tetrahydroisoquinoline	172	300
1-Methyl-1,2,3,4-tetrahydronaphthalene	112	39
2-Methyl-1,2,3,4-tetrahydronaphthalene	111	39
5-Methyl-1,2,3,4-tetrahydronaphthalene	131	40
6-Methyl-1,2,3,4-tetrahydronaphthalene	128	40
4-Methyl-5,6,7,8-tetrahydro-2-naphthol	222	107
3-Methyl-5,6,7,8-tetrahydroquinoline	8	322
4-Methyl-5,6,7,8-tetrahydroquinoline	7	322
7-Methyl-1,2,3,4-tetrahydroquinoline	176	300
1-Methyltetralin	112	39
2-Methyltetralin	111	39
5-Methyltetralin	131	40
6-Methyltetralin	128	40
2-Methyl-1-tetralone	84	170
4-Methyl-1-tetralone	87	170
2-Methyltetramethylene sulfide	13	423
3-Methyltetramethylene sulfide	18	423
2-Methylthiane	24	423
3-Methylthiane	27	423
4-Methylthiane	30	423
4-Methylthiazole-5-carboxaldehyde	65	156
Methyl 2-thienyl ketone	169	166
2-Methylthietane	7	423
2-Methylthiirane	2	423
3-(Methylthio)aniline	126	437
4-(Methylthio)aniline	146	437
4-(Methylthio)benzaldehyde	46	426
2-(Methylthio)ethylamine	328	438
2-Methylthiolane	13	423
3-Methylthiolane	18	423
2-Methylthiophene	8	423
3-Methylthiophene	10	423
3-Methyl-2-thiophenecarboxaldehyde	11	152
5-Methyl-2-thiophenecarboxaldehyde	10	152
3-(Methylthio)phenol	77	434
4-(Methylthio)phenol	78	434
2-(Methylthio)pyridine	106	436
3-(Methylthio)pyridine	152	437
4-(Methylthio)pyridine	241	437
Methyl thymyl ether	87	135
Methyl 2-toluate	208	261
Methyl 3-toluate	216	261
Methyl 4-toluate	30	274
N-Methyltoluene-2-sulfonamide	15	392
N-Methyltoluene-4-sulfonamide	19	392
N-Methyl-o-toluidine	103	298
N-Methyl-p-toluidine	109	298
Methyl 2-(4-toluyl)benzoate	86	277
α-Methyl-4-tolyl alcohol	110	85
Methyl 2-tolyl ether	62	134
Methyl 3-tolyl ether	66	134
Methyl 4-tolyl ether	65	134
Methyl 2-tolyl ketone	170	166
Methyl 3-tolyl ketone	176	166
Methyl 4-tolyl sulfide	49	421
2-Methyl-3,4,5-tribromophenol	153	104
2-Methyl-3,4,6-tribromophenol	162	104
3-Methyl-2,4,6-tribromophenol	131	103
4-Methyl-2,3,6-tribromophenol	198	106
2-Methyl-3,5,6-trichlorophenol	68	100
2-Methyl-4,5,6-trichlorophenol	114	102
3-Methyl-2,4,6-trichlorophenol	41	98
4-Methyl-2,3,5-trichlorophenol	80	100
1-Methyl-3,4,5-trihydroxybenzene	268	110
5-Methyl-1,2,3-trihydroxybenzene	268	110
5-Methyl-1,2,4-trihydroxybenzene	314	112
Methyl 3,4,5-trihydroxybenzoate	506	123
Methyl trimethylacetate	29	251
2-Methyltrimethylene sulfide	7	423
2-Methyl-3,4,5-trinitrophenol	197	106
3-Methyl-2,4,6-trinitrophenol	224	107
4-Methylumbelliferone	480	121
Methyl undecylenate	285	265
Methyl n-undecyl ketone	9	172
Methylurea	121	234
Methyl urethane	11	231
Methyl n-valerate	57	252
Methyl vinyl acetylene	14	28
Methyl vinyl carbinol	4	80
Methyl vinyl ketone	2	161
Methyl vinyl sulfide	3	420
Methyl 4-xenyl ketone	152	177
Michler's ketone	87	325
Morphol	345	114
Morpholine	48	296
"	296	438
Mucic acid	341	207
Mucic acid diamide	465	244
Mucic acid monoamide	407	242
Mucic acid monoanilide	506	246
Muconic acid	400	209
Muconic acid diamide	489	245
Mustard gas	52	421
Myrcene	295	22
Myricyl alcohol	46	89
Myristaldehyde	4	153
Myristamide	147	234
Myristanilide	61	232
Myristic acid	46	194
Myristic anhydride	19	225
Myristone	110	175
Myristoyl chloride	11	214
Myristyl alcohol	17	88

N

1-Naphthaldehyde	7	153
2-Naphthaldehyde	50	155
Naphthalene	77	46
1-Naphthaleneacetic acid	177	200
2-Naphthaleneacetic acid	210	201
1,8-Naphthalenedicarboxylic anhydride	84	227
1,2-Naphthalenediol	64	99
1,3-Naphthalenediol	291	111
1,4-Naphthalenediol	490	122
1,5-Naphthalenediol	575	127
1,6-Naphthalenediol	330	113
1,7-Naphthalenediol	462	120
1,8-Naphthalenediol	343	114
2,3-Naphthalenediol	407	117
2,6-Naphthalenediol	535	124
2,7-Naphthalenediol	484	121
Naphthalene-1,3-disulfonamide	462	409
Naphthalene-1,4-disulfonamide	453	409
Naphthalene-1,5-disulfonamide	471	410

ORGANIC COMPOUND INDEX

	Compound No.	Page No.
Naphthalene-1,6-disulfonamide	465	410
Naphthalene-1,7-disulfonamide	466	410
Naphthalene-2,6-disulfonamide	468	410
Naphthalene-2,7-disulfonamide	420	408
Naphthalene-1,4-disulfonanilide	227	400
Naphthalene-1,5-disulfonanilide	434	408
Naphthalene-1,3-disulfonic acid	216	378
Naphthalene-1,4-disulfonic acid	209	378
Naphthalene-1,5-disulfonic acid	223	379
Naphthalene-1,6-disulfonic acid	218	378
Naphthalene-1,7-disulfonic acid	219	379
Naphthalene-2,6-disulfonic acid	221	379
Naphthalene-2,7-disulfonic acid	193	378
Naphthalene-1,3-disulfonyl chloride	218	388
Naphthalene-1,4-disulfonyl chloride	249	389
Naphthalene-1,5-disulfonyl chloride	266	389
Naphthalene-1,6-disulfonyl chloride	206	387
Naphthalene-2,6-disulfonyl chloride	279	390
Naphthalene-2,7-disulfonyl chloride	248	389
1,5-Naphthalenedithiol	41	417
1,8-Naphthalenedithiol	37	417
2,7-Naphthalenedithiol	46	417
Naphthalene-1-sulfonamide	128	396
Naphthalene-2-sulfonamide	351	405
Naphthalene-1-sulfonanilide	47	393
Naphthalene-2-sulfonanilide	81	395
Naphthalene-1-sulfonic acid	49	372
Naphthalene-2-sulfonic acid	163	377
Naphthalene-1-sulfonyl chloride	74	383
Naphthalene-2-sulfonyl chloride	89	383
1,2,3,4-Naphthalenetetrol	555	126
1,2,5,8-Naphthalenetetrol	469	121
1,4,5,8-Naphthalenetetrol	485	121
1-Naphthalenethiol	47	415
2-Naphthalenethiol	25	416
1,2,4-Naphthalenetriol	388	116
1,2,6-Naphthalenetriol	482	121
1,2,7-Naphthalenetriol	498	122
1,3,6-Naphthalenetriol	175	105
1,4,6-Naphthalenetriol	335	113
1,6,7-Naphthalenetriol	459	120
2-Naphthamide	412	243
1-Naphthanilide	324	240
Naphthionic acid	154	376
1-Naphthoamide	434	243
β-Naphthohydroquinone	220	107
1,4-Naphthohydroquinone	461	120
"	71	434
1-Naphthoic acid	245	203
2-Naphthoic acid	288	204
" "	309	431
α-Naphthoic anhydride	65	226
β-Naphthoic anhydride	60	226
1-Naphthoic anhydride	65	226
2-Naphthoic anhydride	60	226
2-Naphthoin	157	177
α-Naphthol	169	105
" "	67	434
β-Naphthol	284	411
" "	80	434
1-Naphthol	169	105
"	67	434
2-Naphthol	284	111
"	80	434
1-Naphthonitrile	25	358
2-Naphthonitrile	78	360
α-Naphthoquinone	14	181
β-Naphthoquinone	19	182
1,2-Naphthoquinone	19	182
1,4-Naphthoquinone	14	181
1-Naphthoxyacetone	11	169
2-Naphthoxyacetone	105	175
1-Naphthoxyacetyl chloride	24	214
2-Naphthoxyacetyl chloride	11	216
1-Naphthoyl chloride	61	213
2-Naphthoyl chloride	7	216
3-(1-Naphthoyl)propionic acid	386	432
3-(2-Naphthoyl)propionic acid	462	433
1-Naphthylacetamide	379	242
N-(1-Naphthyl)acetamide	316	239
2-Naphthylacetamide	430	243
N-(2-Naphthyl)acetamide	240	237
N-(1-Naphthyl)-4-acetamidobenzenesulfonamide	347	405
N-(1-Naphthyl)acetanilide	302	239
1-Naphthyl acetate	58	276
2-Naphthyl acetate	99	278
(1-Naphthyl)acetic acid	177	200
" " "	329	431
(2-Naphthyl)acetic acid	210	201
" " "	336	431
1-Naphthylacetonitrile	22	358
2-Naphthylacetonitrile	109	362
1-Naphthylacetyl chloride	29	214
1-Naphthylamine	51	304
2-Naphthylamine	232	310
N-(1-Naphthyl)-4-aminobenzenesulfonamide	301	403
2-Naphthylanilide	348	241
N-(1-Naphthyl)benzamide	319	240
N-(2-Naphthyl)benzamide	320	240
N-(1-Naphthyl)benzenesulfonamide	201	399
1-Naphthyl benzoate	78	277
2-Naphthyl benzoate	142	280
o-2-Naphthylbenzoic acid	308	205
N-(1-Naphthyl)benzylsulfonamide	181	398
N-(1-Naphthyl)-4-bromobenzenesulfonamide	244	401
N-(1-Naphthyl)-4-chlorobenzenesulfonamide	270	402
N-(1-Naphthyl)-4-chloronaphthalene-1-sulfonamide	169	398
N-(1-Naphthyl)ethanesulfonamide	13	392
1-(1-Naphthyl)ethanol	36	89
(1-Naphthyl)glycolic acid	112	197
2-Naphthylglyoxal	124	176
(1-Naphthyl)glyoxylic acid	141	198
N-(1-Naphthyl)methanesulfonamide	68	394
N-(1-Naphthyl)-3-methylbutane-1-sulfonamide	24	392
N-(1-Naphthyl)-2-methylpropanesulfonamide	38	393
1-Naphthyl 2-naphthyl sulfide	70	426
N-(1-Naphthyl)-3-nitrobenzenesulfonamide	184	398
1-Naphthyl phenyl amine	78	305
2-Naphthyl phenyl amine	218	309
1-Naphthyl phenyl ketone	100	175
1-Naphthyl phenyl sulfide	15	425
N-(1-Naphthyl)propane-1-sulfonamide	21	392
N-(1-Naphthyl)propane-2-sulfonamide	86	395
2-(1-Naphthyl)propionaldehyde	25	150
3-(1-Naphthyl)propionic acid	239	202
2-Naphthyl n-propyl ether	23	138
2-Naphthyl salicylate	178	105
" "	130	279
2-Naphthyl sulfide	71	426
N-(1-Naphthyl)toluene-4-sulfonamide	151	397
1-Naphthyl 4-tolyl amine	128	306
2-Naphthyl 4-tolyl amine	198	309
cis-Neobornylamine	362	438
Neonerolin	19	137
Neopentane	7	3
Neopentyl alcohol	28	89
Neopentyl benzene	32	36
Neopentyl bromide	13	58
Neopentyl chloride	14	55
Neopentyl cyanide	21	358
Neopentyl phenyl ketone	91	170
Neral	83	148
Nerolin	44	139

Compound	Compound No.	Page No.
Nicotinaldehyde	15	152
Nicotinamide	214	236
Nicotinanilide	226	237
DL-Nicotine	73	321
L-Nicotine	82	321
Nicotinic acid	443	432
Nicotinic anhydride	53	226
Nicotinonitrile	54	359
Nicotinyl chloride	16	214
Ninhydrin	203	179
3-Nitroacenaphthene	103	341
5-Nitroacenaphthene	74	340
2-Nitroacetanilide	82	233
3-Nitroacetanilide	304	239
4-Nitroacetanilide	453	244
Nitroacetic acid	26	429
3-Nitroacetophenone	112	175
2-Nitroaniline	110	306
"	12	436
3-Nitroaniline	237	310
"	52	436
4-Nitroaniline	330	313
"	24	436
2-Nitroaniline-4-sulfonyl chloride	57	382
3-Nitroanisic acid	300	205
2-Nitro-p-anisidine	64	304
3-Nitro-o-anisidine	121	306
3-Nitro-p-anisidine	287	311
4-Nitro-o-anisidine	249	310
5-Nitro-o-anisidine	314	312
2-Nitroanisole	107	136
"	32	336
3-Nitroanisole	21	138
"	12	337
4-Nitroanisole	34	138
"	23	337
9-Nitroanthracene	100	341
1-Nitroanthraquinone	130	342
2-Nitroanthraquinone	121	342
2-Nitroazobenzene	38	338
3-Nitroazobenzene	68	339
4-Nitroazobenzene	94	340
2-Nitroazoxybenzene	19	337
4-Nitroazoxybenzene	104	341
2-Nitrobenzaldehyde	22	154
3-Nitrobenzaldehyde	44	155
4-Nitrobenzaldehyde	92	158
2-Nitrobenzamide	362	241
3-Nitrobenzamide	263	238
4-Nitrobenzamide	429	243
2-Nitrobenzamide-4-sulfonamide	376	406
2-Nitrobenzanilide	300	239
3-Nitrobenzanilide	294	239
4-Nitrobenzanilide	449	244
Nitrobenzene	21	335
5-Nitrobenzene-1,3-disulfonamide	419	408
5-Nitrobenzene-1,3-disulfonic acid	194	378
5-Nitrobenzene-1,3-disulfonyl chloride	142	385
2-Nitrobenzenesulfonamide	288	403
3-Nitrobenzenesulfonamide	185	398
4-Nitrobenzenesulfonamide	228	400
2-Nitrobenzenesulfonanilide	52	394
3-Nitrobenzenesulfonanilide	69	394
4-Nitrobenzenesulfonanilide	204	399
2-Nitrobenzenesulfonic acid	137	376
3-Nitrobenzenesulfonic acid	77	373
4-Nitrobenzenesulfonic acid	98	374
2-Nitrobenzenesulfonyl chloride	76	383
3-Nitrobenzenesulfonyl chloride	66	383
4-Nitrobenzenesulfonyl chloride	95	384
2-Nitrobenzoic acid	218	202
" "	67	429
3-Nitrobenzoic acid	204	201
" "	200	430
4-Nitrobenzoic acid	373	208
2-Nitrobenzoic acid-4-sulfonamide	280	402
2-Nitrobenzoic anhydride	61	226
3-Nitrobenzoic anhydride	69	226
4-Nitrobenzoic anhydride	73	226
2-Nitrobenzonitrile	153	364
3-Nitrobenzonitrile	164	364
4-Nitrobenzonitrile	193	365
3-Nitrobenzoyl bromide	1	220
4-Nitrobenzoyl bromide	5	220
3-Nitrobenzoyl carbinol	51	89
4-Nitrobenzoyl carbinol	55	90
2-Nitrobenzoyl chloride	27	214
3-Nitrobenzoyl chloride	60	213
4-Nitrobenzoyl chloride	15	216
2-Nitrobenzyl alcohol	43	89
3-Nitrobenzyl alcohol	8	88
4-Nitrobenzyl alcohol	52	90
2-Nitrobenzyl bromide	4	59
3-Nitrobenzyl bromide	6	59
4-Nitrobenzyl bromide	8	59
4-Nitrobenzyl chloride	7	57
4-Nitrobenzyl cyanide	161	364
4-Nitrobenzyl phenyl sulfide	47	426
4-Nitrobenzyl sulfide	73	427
3-Nitrobenzylsulfonamide	162	398
4-Nitrobenzylsulfonamide	320	404
4-Nitrobenzylsulfonanilide	361	406
3-Nitrobenzylsulfonic acid	63	373
4-Nitrobenzylsulfonic acid	149	376
3-Nitrobenzylsulfonyl chloride	152	386
2-Nitrobiphenyl	9	337
3-Nitrobiphenyl	32	338
4-Nitrobiphenyl	78	340
4'-Nitrobiphenyl-4-sulfonamide	384	407
4'-Nitrobiphenyl-4-sulfonic acid	173	377
4'-Nitrobiphenyl-4-sulfonyl chloride	265	389
1-Nitrobutane	11	335
DL-2-Nitrobutane	7	335
tert-Nitrobutane	1	337
3-Nitrocatechol	148	104
"	19	434
4-Nitrocatechol	460	120
2-Nitrocinnamamide	390	242
3-Nitrocinnamamide	417	243
4-Nitrocinnamamide	439	243
trans-2-Nitrocinnamic acid	371	208
" " "	307	431
trans-3-Nitrocinnamic acid	317	206
" " "	303	431
trans-4-Nitrocinnamic acid	398	209
" " "	293	431
trans-2-Nitrocinnamoyl chloride	13	216
trans-4-Nitrocinnamoyl chloride	28	216
2-Nitro-m-cresol	27	97
2-Nitro-p-cresol	19	97
3-Nitro-o-cresol	365	115
3-Nitro-p-cresol	121	102
4-Nitro-m-cresol	306	112
4-Nitro-o-cresol	183	105
5-Nitro-o-cresol	260	109
6-Nitro-m-cresol	49	99
6-Nitro-o-cresol	96	101
Nitrocyclohexane	19	335
2-Nitro-p-cymene	31	336
2-Nitrodiphenyl sulfide	49	426
3-Nitrodiphenyl sulfide	16	425
4-Nitrodiphenyl sulfide	29	425
2-Nitrodiphenylamine-4-sulfonamide	170	398
4-Nitrodiphenylamine-2-sulfonamide	208	399

Compound	Compound No.	Page No.
2-Nitrodiphenylamine-4-sulfonanilide	155	397
4-Nitrodiphenylamine-2-sulfonanilide	177	398
4-Nitrodiphenylamine-2-sulfonic acid	88	374
2-Nitrodiphenylamine-4-sulfonic acid	65	373
4-Nitrodiphenylamine-2-sulfonyl chloride	155	386
3-Nitrodurene	77	340
Nitroethane	3	335
Nitroethylene	1	335
2-Nitrofluorene	106	341
9-Nitrofluorene	118	342
1-Nitroheptane	17	335
2-Nitroheptane	18	335
1-Nitrohexane	16	335
2-Nitrohexane	15	335
2-Nitrohydroquinone	29	434
4-Nitroindane	16	337
5-Nitroindane	14	337
2-Nitroindane	96	341
1-Nitroisobutene	12	335
1-Nitroisobutylene	12	335
4-Nitromesidine	120	306
Nitromesitylene	17	337
Nitromethane	2	335
3-Nitro-N-methylaniline	106	306
1-Nitronaphthalene	27	337
2-Nitronaphthalene	44	338
4-Nitronaphthalene-2,7-disulfonamide	459	409
4-Nitronaphthalene-2,7-disulfonic acid	213	378
4-Nitronaphthalene-2,7-disulfonyl chloride	227	388
1-Nitronaphthalene-2-sulfonamide	344	405
4-Nitronaphthalene-1-sulfonamide	264	402
4-Nitronaphthalene-2-sulfonamide	374	406
5-Nitronaphthalene-1-sulfonamide	402	407
5-Nitronaphthalene-2-sulfonamide	246	401
6-Nitronaphthalene-1-sulfonamide	370	406
7-Nitronaphthalene-1-sulfonamide	446	409
8-Nitronaphthalene-1-sulfonamide	275	402
8-Nitronaphthalene-2-sulfonamide	386	407
1-Nitronaphthalene-2-sulfonanilide	318	404
5-Nitronaphthalene-1-sulfonanilide	63	394
8-Nitronaphthalene-1-sulfonanilide	226	400
8-Nitronaphthalene-2-sulfonanilide	206	399
1-Nitronaphthalene-2-sulfonic acid	160	376
4-Nitronaphthalene-1-sulfonic acid	126	375
5-Nitronaphthalene-1-sulfonic acid	183	377
5-Nitronaphthalene-2-sulfonic acid	108	374
7-Nitronaphthalene-1-sulfonic acid	207	378
8-Nitronaphthalene-1-sulfonic acid	131	375
8-Nitronaphthalene-2-sulfonic acid	166	377
1-Nitronaphthalene-2-sulfonyl chloride	189	387
4-Nitronaphthalene-1-sulfonyl chloride	149	385
4-Nitronaphthalene-2-sulfonyl chloride	224	388
5-Nitronaphthalene-1-sulfonyl chloride	174	386
5-Nitronaphthalene-2-sulfonyl chloride	203	387
6-Nitronaphthalene-1-sulfonyl chloride	204	387
7-Nitronaphthalene-1-sulfonyl chloride	263	389
8-Nitronaphthalene-1-sulfonyl chloride	254	389
8-Nitronaphthalene-2-sulfonyl chloride	260	389
4-Nitro-1-naphthalenethiol	24	416
1-Nitro-2-naphthol	199	106
2-Nitro-1-naphthol	302	111
4-Nitro-1-naphthol	421	118
4-Nitro-2-naphthol	272	110
5-Nitro-1-naphthol	451	119
5-Nitro-2-naphthol	364	115
6-Nitro-1-naphthol	471	121
8-Nitro-1-naphthol	311	112
8-Nitro-2-naphthol	351	114
1-Nitro-2-naphthonitrile	183	365
4-Nitro-1-naphthonitrile	179	365
5-Nitro-1-naphthonitrile	223	367
5-Nitro-2-naphthonitrile	213	366
8-Nitro-2-naphthonitrile	189	365
3-Nitro-1,2-naphthoquinone	107	341
1-Nitro-2-naphthylamine	276	311
2-Nitro-1-naphthylamine	322	313
4-Nitro-1-naphthylamine	412	316
5-Nitro-1-naphthylamine	252	310
5-Nitro-2-naphthylamine	323	313
8-Nitro-1-naphthylamine	185	308
8-Nitro-2-naphthylamine	201	309
1-Nitrooctane	20	335
1-Nitropentane	14	335
DL-2-Nitropentane	10	335
3-Nitrophenacyl alcohol	51	89
4-Nitrophenacyl alcohol	55	90
2-Nitrophenanthrene	70	339
3-Nitrophenanthrene	112	341
4-Nitrophenanthrene	48	338
9-Nitrophenanthrene	79	340
2-Nitrophenanthrenequinone	135	342
3-Nitrophenanthrenequinone	136	342
4-Nitrophenanthrenequinone	116	342
3-Nitro-p-phenetidine	235	310
2-Nitrophenetole	34	336
3-Nitrophenetole	17	137
"	8	337
4-Nitrophenetole	30	338
2-Nitrophenol	34	98
"	24	434
3-Nitrophenol	185	105
"	42	434
4-Nitrophenol	242	108
"	23	434
(3-Nitrophenoxy)acetic acid	133	430
(4-Nitrophenoxy)acetic acid	128	430
3-Nitrophenylacetamide	164	235
4-Nitrophenylacetamide	422	243
4-Nitrophenylacetanilide	423	243
(2-Nitrophenyl)acetic acid	207	201
" " "	287	431
(3-Nitrophenyl)acetic acid	157	199
" " "	278	431
(4-Nitrophenyl)acetic acid	229	202
" " "	259	431
(4-Nitrophenyl)acetonitrile	161	364
4-Nitrophenylacetyl chloride	10	216
β-(2-Nitrophenyl)acrylonitrile	120	362
2-Nitro-p-phenylenediamine	307	312
4-Nitro-m-phenylenediamine	361	314
4-Nitro-o-phenylenediamine	415	316
2-Nitrophenylhydrazine	158	307
3-Nitrophenylhydrazine	170	308
4-Nitrophenylhydrazine	352	314
2-Nitro-5-phenylphenol	20	434
2-Nitrophenyl phenylsulfide	49	426
3-Nitrophenyl phenyl sulfide	16	425
4-Nitrophenyl phenyl sulfide	29	425
3-(2-Nitrophenyl)propionic acid	391	432
3-(4-Nitrophenyl)propionic acid	384	432
2-Nitrophenyl sulfide	65	426
3-Nitrophenyl sulfide	75	427
4-Nitrophenyl 4-tolyl sulfide	51	426
4-Nitrophthalamide	426	243
4-Nitrophthalanilide	408	243
3-Nitrophthalic acid	346	207
4-Nitrophthalic acid	253	203
3-Nitrophthalic acid diamide	432	243
3-Nitrophthalic acid dianilide	484	245
3-Nitrophthalic anhydride	70	226
4-Nitrophthalic anhydride	51	226

	Compound No.	Page No.
3-Nitrophthalimide	455	244
3-Nitrophthaloyl chloride	16	216
1-Nitropropane	6	335
2-Nitropropane	4	335
3-Nitropropene	5	335
3-Nitropropylene	5	335
3-Nitropyridine	17	323
"	20	436
5-Nitroquinoline	39	323
"	66	436
6-Nitroquinoline	81	324
"	67	436
7-Nitroquinoline	78	324
"	49	436
8-Nitroquinoline	56	436
2-Nitroresorcinol	143	103
4-Nitroresorcinol	280	110
5-Nitroresorcinol	397	116
3-Nitrosalicylamide	296	239
3-Nitrosalicylamide (hydrate)	267	238
5-Nitrosalicylamide	474	245
5-Nitrosalicylanilide	473	245
3-Nitrosalicylic acid	374	115
" "	166	199
5-Nitrosalicylic acid	358	208
4-Nitroso-N,N-dimethylaniline	50	324
β-Nitrostyrene	28	337
3-Nitro-1,2,4,5-tetramethylbenzene	77	340
2-Nitrothiophenol	19	416
3-Nitrothiophenol	22	416
4-Nitrothiophenol	23	416
2-Nitrotoluene	22	335
3-Nitrotoluene	26	336
4-Nitrotoluene	21	337
3-Nitrotoluene-α-sulfonamide	162	398
4-Nitrotoluene-α-sulfonamide	320	404
4-Nitrotoluene-α-sulfonanilide	361	406
2-Nitro-p-toluidine	126	306
3-Nitro-o-toluidine	164	307
3-Nitro-p-toluidine	245	310
4-Nitro-o-toluidine	216	309
5-Nitro-o-toluidine	293	312
6-Nitro-m-toluidine	303	312
6-Nitro-o-toluidine	163	307
2-Nitro-3-tolunitrile	106	362
2-Nitro-4-tolunitrile	148	363
3-Nitro-2-tolunitrile	152	364
3-Nitro-4-tolunitrile	136	363
4-Nitro-2-tolunitrile	145	363
4-Nitro-3-tolunitrile	123	362
5-Nitro-3-tolunitrile	143	363
6-Nitro-2-tolunitrile	86	361
6-Nitro-3-tolunitrile	102	362
1-Nitro-2,4,6-tribromobenzene	89	340
2-Nitro-3,4,6-tribromophenol	301	111
2-Nitro-4,5,6-tribromophenol	287	111
3-Nitro-2,4,6-tribromophenol	156	104
1-Nitro-2,3,6-trimethylbenzene	3	337
3-Nitro-2,4,6-trimethylphenol	57	434
2-Nitro-m-xylene	24	336
2-Nitro-p-xylene	27	336
3-Nitro-o-xylene	30	336
4-Nitro-m-xylene	29	336
4-Nitro-o-xylene	4	337
5-Nitro-m-4-xylidine	123	306
5-Nitro-p-2-xylidine	325	313
n-Nonacosane	45	8
Nonacosylcyclohexane	56	9
Nonacosylcyclopentane	44	8
1-Nonacosyne	16	31
n-Nonadecane	11	8
1-Nonadecanol	34	89
10-Nonadecanol	46	89
2-Nonadecanone	61	174
Nonadecanonitrile	37	358
1-Nonadecene	1	26
Nonadecylcyclohexane	22	8
Nonadecylcyclopentane	13	8
1-Nonadecyne	5	31
1,8-Nonadiene	255	20
2,7-Nonadiene	280	21
1,8-Nonadiyne	51	30
2,7-Nonadiyne	56	30
Nonanal	67	147
n-Nonane	144	5
1-Nonanethiol	43	415
n-Nonanoic acid	48	191
" "	459	432
1-Nonanol	107	85
d,l-2-Nonanol	102	85
5-Nonanol	97	85
2-Nonanone	150	166
3-Nonanone	142	165
5-Nonanone	140	165
Nonanonitrile	76	351
n-Nonanoyl chloride	45	213
Nonantriacontane	74	9
1-Nonatriacontene	22	26
1-Nonatriacontyne	26	31
2-Nonenal	16	150
1-Nonene	268	21
2-Nonene	275	21
3-Nonene	269	21
4-Nonene	260	20
n-Nonylacetylene	57	30
n-Nonyl alcohol	107	85
n-Nonylamine	95	298
"	419	439
n-Nonylbenzene	166	41
n-Nonyl bromide	27	58
n-Nonyl chloride	46	56
n-Nonylcyclohexane	270	7
n-Nonylcyclopentane	266	7
n-Nonyl iodide	20	60
n-Nonyl mercaptan	43	415
1-Nonylnaphthalene	184	42
2-Nonylnaphthalene	185	42
1-Nonyne	47	30
2-Nonyne	50	30
3-Nonyne	48	30
4-Nonyne	46	29
Nopinane	140	5
Nopinene	291	22
Norbornane	77	9
Norbricyclene	33	8
DL-Norleucine	106	289
"	79	429
L-(+)-Norleucine	111	289
DL-Norvaline	112	289
L-(+)-Norvaline	120	290
NW-acid	14	380

O

	Compound No.	Page No.
Octachloronaphthalene	33	72
n-Octacosane	42	8
Octacosylcyclohexane	53	9
Octacosylcyclopentane	41	8
1-Octacosyne	15	31
Octadecanal	14	153
n-Octadecane	8	8
n-Octadecanoic acid	70	195
n-Octadecanoic anhydride	28	225

	Compound No.	Page No.
1-Octadecanol	33	89
Octadecanonitrile	33	358
1-Octadecene	338	24
cis-Octa-9-decen-1-ol	134	87
trans-Octa-9-decen-ol	15	88
cis-Octadecenyl alcohol	134	87
n-Octadecyl acetate	27	274
Octadecylcyclohexane	18	8
Octadecylcyclopentane	10	8
Octadecyl sulfide	39	426
1-Octadecyne	3	31
2-Octadecyne	4	31
2,4-Octadiene	234	19
2,6-Octadiyne	2	31
Octanal	60	147
n-Octanamide	161	235
n-Octane	82	4
Octane-1,8-dicarboxylic acid	183	200
Octanedioic acid	215	202
Octanedioic anhydride (dimer)	22	225
1-Octanethiol	36	414
2-Octanethiol	30	414
n-Octanoic acid	47	191
" "	456	432
n-Octanoic anhydride	18	223
1-Octanol	98	85
d,l-2-Octanol	80	84
2-Octanone	114	164
4-Octanone	107	164
Octanonitrile	66	350
n-Octanoyl chloride	40	213
Octatriacontane	72	9
1-Octatriacontene	21	26
1-Octatriacontyne	25	31
Octatrienal	37	155
4-Octenal	12	152
1-Octene	209	18
cis-2-Octene	221	19
trans-2-Octene	220	19
cis-3-Octene	217	19
trans-3-Octene	218	19
cis-4-Octene	213	19
trans-4-Octene	210	19
n-Octyl acetate	194	260
sec-Octyl acetate	162	258
n-Octylacetylene	52	30
n-Octyl alcohol	98	85
n-Octylamine	77	297
"	421	439
n-Octylbenzene	156	41
n-Octyl bromide	24	58
n-Octyl *n*-butyrate	268	264
n-Octyl *n*-caproate	325	269
n-Octyl *n*-caprylate	348	270
n-Octyl chloride	39	56
n-Octylcyclohexane	267	7
n-Octylcylopentane	263	7
n-Octyl *n*-enanthate	338	270
n-Octyl formate	170	258
n-Octyl *n*-heptanoate	338	270
n-Octyl *n*-hexanoate	325	269
n-Octyl iodide	21	60
n-Octyl mercaptan	36	414
sec-Octyl mercaptan	30	414
1-Octylnaphthalene	182	42
2-Octylnaphthalene	183	42
n-Octyl *n*-octanoate	348	270
n-Octyl *n*-pentanoate	306	267
n-Octyl propionate	234	262
n-Octyl sulfide	30	425
2-*n*-Octylthiophene	54	424
n-Octyl *n*-valerate	306	267
1-Octyne	42	29
2-Octyne	45	29
3-Octyne	44	29
4-Octyne	43	29
2-Octynoic acid	14	192
3-Octyn-2-one	14	169
2-Octynonitrile	60	349
Oleamide	41	232
Oleanilide	2	231
trans-Oleic acid	31	194
" " "	4	192
Oleic anhydride	1	225
Oleyl alcohol	134	87
Oleyl chloride	3	214
Orcinol	217	107
DL-Ornithine	127	290
" "	38	429
L-Ornithine	8	284
Orthanilic acid	54	372
Oxalic acid	116	197
" "	10	429
Oxalic acid diamide	513	247
Oxalic acid dianilide	497	246
Oxaloacetic acid	102	429
Oxalyl chloride	2	212
Oxalyl dibromide	1	218
Oxamic acid	334	207
Oxanilic acid	224	202
Oxine	110	102
2-Oxobutanoic acid	11	193
2-Oxocyclohexane-carboxylic acid	82	196
3-Oxocyclohexane-carboxylic acid	74	196
4-Oxocyclohexane-carboxylic acid	67	195
2-Oxopropionaldehyde	103	171
α-Oxopropionic acid	11	190

P

	Compound No.	Page No.
Palmitaldehyde	6	153
Palmitamide	142	234
Palmitanilide	74	232
Palmitic acid	58	195
Palmitic anhydride	24	225
Palmitoyl chloride	10	214
Paraisobutyraldehyde	48	155
Paraldehyde	34	145
Pararosaniline	429	317
Pararosaniline base	65	90
Pelargonaldehyde	67	147
Pelargonamide	113	233
Pelargonanilide	19	231
Pelargonic acid	48	191
" "	459	432
Pelargonyl chloride	45	213
Pentabromophenol	560	126
Pentachlorobenzaldehyde	126	160
Pentachlorobenzene	24	71
Pentachloroethane	21	63
Pentachlorophenol	488	122
n-Pentacosane	29	8
1-Pentacosene	8	26
Pentacosylcyclohexane	43	8
Pentacosylcyclopentane	30	8
1-Pentacosyne	12	31
Pentadecanal	5	153
Pentadecanamide	125	234
Pentadecananilide	48	232
n-Pentadecane	268	7
n-Pentadecanoic acid	44	194
1-Pentadecanol	21	88
8-Pentadecanone	36	173
Pentadecanonitrile	8	357

Compound	Compound No.	Page No.
1-Pentadecene	334	24
7-Pentadecene	16	25
Pentadecylamine	18	303
Pentadecylcyclohexane	9	8
Pentadecylcyclopentane	1	8
n-Pentadecylic acid	44	194
1-Pentadecyne	63	30
1,2-Pentadiene	29	13
1-cis-3-Pentadiene	27	12
1-trans-3-Pentadiene	26	12
1,3-Pentadiene	24	12
1,4-Pentadiene	13	12
2,3-Pentadiene	30	13
2,4-Pentadienonitrile	23	346
1,3-Pentadiyne	20	28
Pentaerythritol	69	90
Pentaerythritol tetraacetate	123	279
Pentahydrocyclohexane	68	90
Pentamethylaniline	339	313
Pentamethylbenzene	32	44
Pentamethylbenzoic acid	335	207
Pentamethylenediamine	76	297
Pentamethyleneglycol	124	86
2,2,3,3,4-Pentamethylpentane	215	6
2,2,3,4,4-Pentamethylpentane	173	5
Pentamethylphenol	297	111
1,2,2,4,4-Pentamethylpiperidine	465	439
Pentamethylene sulfide	3	425
Pentanal	20	145
n-Pentanamide	146	234
n-Pentananilide	28	231
n-Pentane	13	3
Pentane-1,1-dicarboxylic acid	118	197
1,5-Pentanediol	124	86
2,4-Pentanedione	44	162
1,5-Pentanedithiol	42	415
1-Pentanethiol	18	413
2-Pentanethiol	10	413
n-Pentanoic acid	20	190
" " "	450	432
n-Pentanoic anhydride	13	223
1-Pentanol	32	81
d,l-2-Pentanol	15	80
3-Pentanol	13	80
2-Pentanone	10	161
3-Pentanone	9	161
n-Pentanoyl bromide	2	219
n-Pentanoyl chloride	22	212
Pentatriacontane	65	9
1-Pentatriacontene	18	26
Pentatriacontylcyclohexane	73	9
Pentatriacontylcyclopentane	64	9
1-Pentatriacontyne	22	31
2-Pentenal	35	145
1-Pentene	14	12
cis-2-Pentene	19	12
trans-2-Pentene	18	12
2-Pentenoic acid	422	432
3-Pentenoic acid	99	429
4-Pentenoic acid	22	190
" " "	414	432
4-Pentenonitrile	28	347
1-Penten-3-one	11	161
1-Penten-3-yne	14	28
1-Penten-4-yne	10	28
cis-3-Penten-1-yne	11	28
trans-3-Penten-1-yne	12	28
n-Pentyl acetate	94	254
n-Pentylacetylene	35	29
n-Pentylamine	417	439
n-Pentylbenzene	81	38
n-Pentyl bromide	17	58
DL-2-Pentyl bromide	14	58
3-Pentyl bromide	15	58
n-Pentyl n-butyrate	146	257
3-Pentyl chloride	20	55
n-Pentyl chloride	22	55
2-n-Pentylcinnamaldehyde	9	150
n-Pentylcyclohexane	249	7
n-Pentylcyclopentane	242	7
n-Pentyl n-heptanoate	277	265
n-Pentyl n-hexanoate	226	262
n-Pentyl iodide	15	60
2-Pentyl iodide	12	60
3-Pentyl iodide	13	60
sec-n-Pentylmalonic acid	50	194
1-Pentylnaphthalene	173	42
2-Pentylnaphthalene	174	42
n-Pentyl n-octanoate	302	267
n-Pentyl n-Pentanoate	183	259
4-n-Pentylphenol	4	96
n-Pentyl phenyl ketone	3	172
n-Pentyl propionate	120	255
2-Pentylpyridine	245	438
n-Pentylstearate	25	274
3-Pentyltoluene	119	39
1-Pentyne	9	28
2-Pentyne	13	28
2-Pentynoic acid	42	194
Peonol	45	99
Perchloroethylene	14	63
Perfluoro-n-butyric anhydride	3	223
Perfluoro-n-caproic anhydride	6	223
Perfluorocyclohexane	3	62
Perfluorocyclopentane	1	62
Perfluoro-n-decane	8	62
Perfluoro-n-hendecane	9	62
Perfluoro-n-heptane	6	62
Perfluoro-n-hexane	4	62
Perfluoro-n-hexanoic anhydride	6	223
Perfluoro-2-methylpentane	5	62
Perfluoro-n-nonane	7	62
Perfluoropropionic anhydride	2	223
Perfluoro-n-undecane	9	62
Peri acid	4	380
Perylene	201	51
Phenacetin	242	237
Phenacridone	204	179
Phenacyl acetate	60	276
Phenacyl alcohol	47	89
" "	150	177
Phenacyl bromide	51	173
Phenacyl chloride	68	174
Phenacyl phenyl ketone	111	175
1-Phenanthraldehyde	96	158
2-Phenanthraldehyde	47	155
3-Phenanthraldehyde	69	156
9-Phenanthraldehyde	89	158
9,10-Phenanthraquinone	34	183
Phenanthrene	104	47
Phenanthrene-2-carboxylic acid chloride	24	216
Phenanthrene-3-carboxylic acid chloride	27	216
Phenanthrene-9-carboxylic acid chloride	25	216
1,2-Phenanthrenediol	463	120
1,9-Phenanthrenediol	477	121
2,5-Phenanthrenediol	467	120
2,6-Phenanthrenediol	561	126
2,7-Phenanthrenediol	576	127
2,8-Phenanthrenediol	513	123
3,6-Phenanthrenediol	548	125
3,8-Phenanthrenediol	570	126
9,10-Phenanthrenediol	371	115
Phenanthrene-2-sulfonamide	438	409
Phenanthrene-3-sulfonamide	273	402

Compound	Compound No.	Page No.
Phenanthrene-9-sulfonamide	290	403
Phenanthrene-2-sulfonanilide	156	397
Phenanthrene-2-sulfonic acid	200	378
Phenanthrene-3-sulfonic acid	129	375
Phenanthrene-9-sulfonic acid	138	376
Phenanthrene-2-sulfonyl chloride	243	389
Phenanthrene-3-sulfonyl chloride	169	386
Phenanthrene-9-sulfonyl chloride	205	387
3,4,5-Phenanthrenetriol	372	115
Phenanthridine	158	437
Phenanthrindene	118	47
1-Phenanthroic acid	360	208
2-Phenanthroic acid	388	209
3-Phenanthroic acid	394	209
9-Phenanthroic acid	383	209
1-Phenanthrol	393	116
2-Phenanthrol	440	119
3-Phenanthrol	282	111
4-Phenanthrol	236	108
9-Phenanthrol	396	116
3-Phenanthrylacetic acid	276	204
9-Phenanthrylacetonitrile	129	363
9-Phenanthrylmethylamine	215	309
2-Phenethyl alcohol	113	85
Phenethylamine	349	438
2-(Phenethylamino)ethylamine	325	438
o-Phenetidine	141	299
m-Phenetidine	163	300
p-Phenetidine	164	300
Phenetole	63	134
Phenetrol	413	117
4-Phenetylurea	353	241
Phenol	28	97
"	'96	434
Phenol-2,4-disulfonamide	412	408
Phenol-2,4-disulfonanilide	324	404
Phenol-2,4-disulfonic acid	190	378
Phenol-2,4-disulfonyl chloride	115	384
Phenolphthalein	577	127
Phenol-o-sulfonanilide	70	394
Phenol-o-sulfonic acid	3	380
Phenoxyacetaldehyde	79	148
Phenoxyacetamide	122	234
Phenoxyacetanilide	112	233
Phenoxyacetic acid	110	197
" "	171	430
Phenoxyacetone	189	167
Phenoxyacetonitrile	81	351
α-Phenoxyacetophenone	98	175
Phenoxyacetyl chloride	50	213
2-Phenoxybenzaldehyde	31	151
2-Phenoxybenzoic acid	207	430
3-Phenoxybenzoic acid	271	431
4-Phenoxybutanoic acid	10	192
1-Phenoxy-2-butanone	10	169
α-Phenoxybutyryl chloride	14	214
γ-Phenoxybutyryl chloride	4	214
Phenoxyethanal	79	148
2-Phenoxyethanol	122	86
Phenoxymethyl phenyl ketone	98	175
Phenoxymethyl n-propyl ketone	13	169
1-Phenoxy-2-pentanone	13	169
5-Phenoxy-2-pentanone	49	173
2-Phenoxyphenol	216	107
1-Phenoxy-2-propanone	189	167
2-Phenoxypropionic acid	146	199
3-Phenoxypropionic acid	108	197
α-Phenoxypropionyl chloride	18	214
Phenylacetaldehyde	70	147
"	8	153
Phenylacetamide	310	239
Phenylacetanilide	183	235
Phenyl acetate	168	258
Phenylacetic acid	77	196
" "	347	432
Phenylacetic anhydride	30	225
Phenylacetone	5	172
Phenylacetonitrile	80	351
4-Phenylacetophenone	152	177
Phenylacetyl bromide	7	219
Phenylacetyl chloride	44	213
Phenylacetyl fluoride	8	211
1-Phenylacrylic acid	130	198
DL-α-Phenylalanine	76	287
DL-β-Phenylalanine	78	288
" "	55	429
L-Phenylalanine	92	288
4-Phenylanisole	51	139
4-Phenylbenzaldehyde	51	155
4-Phenylbenzenesulfonyl chloride	178	386
Phenyl benzoate	95	277
2-Phenylbenzoic acid	140	198
" "	202	430
4-Phenylbenzoic acid	349	207
4-Phenylbenzoyl chloride	26	216
α-Phenylbenzyl cyanide	95	361
1-Phenyl-1,3-butanedione	75	174
d,l-α-Phenylbutanoic acid	26	193
1-Phenyl-2-butanone	182	167
3-Phenyl-2-butanone	41	169
4-Phenyl-2-butanone	196	167
2-Phenyl-cis-2-butene	52	37
4-Phenyl-3-buten-2-one	37	173
α-Phenylbutyramide	65	232
γ-Phenylbutyramide	60	232
4-Phenylbutyric acid	430	432
γ-Phenylbutyric acid	43	194
γ-Phenylbutyryl chloride	5	214
Phenyl carbamate	10	231
4-Phenylcatechol	357	114
2-Phenylcinnamaldehyde	80	157
3-Phenylcinnamaldehyde	20	154
Phenyl cinnamate	103	278
β-Phenylcinnamonitrile	49	359
2-Phenylcrotononitrile	77	351
Phenylcyanoacetic acid	117	362
2-Phenylcyclohexanone	81	174
3-Phenylcyclohexanone	214	168
4-Phenylcyclohexanone	106	175
2-Phenylcyclopentanone	22	172
1-Phenyldecane	172	42
ω-Phenyldibenzofulvene	67	45
2-Phenyl-1,3-dithiane	40	426
1-Phenyldodecane	178	42
m-Phenylenediamine	84	305
"	190	437
o-Phenylenediamine	196	309
"	176	437
p-Phenylenediamine	318	313
2-Phenylenedithiol	8	416
3-Phenylenedithiol	6	416
4-Phenylenedithiol	31	417
Phenylethanal	70	147
2-Phenylethane-1-sulfonamide	61	394
2-Phenylethane-1-sulfonanilide	18	392
2-Phenylethane-1-sulfonic acid	16	371
2-Phenylethane-1-sulfonyl chloride	13	381
1-Phenylethanethiol	13	413
2-Phenylethanethiol	35	414
2-Phenylethanol	113	85
Phenyl ether	7	137
β-Phenylethyl acetate	243	263

Compound	Compound No.	Page No.	Compound	Compound No.	Page No.
1-Phenylethyl alcohol	1	88	3-Phenylpropanol	123	86
DL-α-Phenylethylamine	83	297	Phenylpropargyl aldehyde	13	152
β-Phenylethylamine	92	298	3-Phenylpropenal	90	149
α-Phenylethyl bromide	25	58	Phenylpropiolamide	117	233
β-Phenylethyl bromide	26	58	Phenylpropiolic acid	191	200
d,l-Phenylethyl carbinol	111	85	2-Phenylpropionaldehyde	14	152
α-Phenylethyl chloride	42	56	α-Phenylpropionamide	77	233
β-Phenylethyl chloride	40	56	β-Phenylpropionamide	140	234
2-Phenylethyl cinnamate	56	275	Phenylpropiolanilide	213	236
Phenyl ethyl ether	63	134	β-Phenylpropionanilide	104	233
2-Phenylethyl sulfide	53	426	Phenyl propionate	5	273
9-Phenylfluorene	150	49	β-Phenylpropionic acid	38	194
β-Phenylglutaconic anhydride	77	226	2-Phenylpropionic acid	50	191
β-Phenylglutaric anhydride	45	226	3-Phenylpropionic acid	412	432
N-Phenylglycine	6	284	2-Phenylpropionitrile	79	351
Phenylglyoxal	28	151	3-Phenylpropionitrile	93	353
Phenylglyoxal hydrate	77	157	Phenylpropiolyl chloride	12	214
Phenylglyoxylic acid	85	174	3-Phenylpropionaldehyde	81	148
2-Phenylheptane	130	40	β-Phenylpropionyl chloride	48	213
3-Phenylheptane	125	39	1-Phenyl-n-propyl alcohol	111	85
1-Phenyl-1-hepten-3-one	25	172	1-Phenylpropylamine	107	298
4-Phenylhexahydroacetophenone	72	170	2-Phenylpropylamine	108	298
2-Phenylhexane	103	39	3-Phenylpropylamine	126	299
d,l-3-Phenylhexane	98	39	Phenyl-n-propyl ether	74	134
1-Phenyl-1-hexen-5-one	29	169	Phenyl-n-propylglycolic acid	101	197
1-Phenyl-4-hexen-1-one	26	169	Phenyl n-propyl ketone	191	167
3-Phenyl-1-hexen-5-one	30	169	Phenyl propyl sulfide	53	421
3-Phenyl-2-hexen-5-one	63	170	3-Phenylpropyl sulfide	42	426
3-Phenyl-3-hexen-5-one	42	169	2-Phenylpyridine	97	321
Phenylhydrazine	160	300	Phenyl 2-pyridyl ketone	217	168
Phenyl 2-hydroxybenzoate	30	97	Phenylpyruvic acid	234	202
2-Phenylimidazole	258	438	2-Phenyl-4-quinolinonitrile	187	365
1-Phenylisobutylamine	113	298	Phenyl salicylate	30	97
1-Phenylisopropylamine	91	298	"	46	275
DL-2-Phenylisopropylamine	98	298	Phenyl stearate	70	276
Phenyl isopropyl ether	68	134	Phenyl styryl ketone	66	174
1-Phenylisoquinoline	61	324	α-d,l-Phenylsuccinamide	314	239
d,l-2-Phenyllactic acid	68	195	β-d,l-Phenylsuccinamide	269	238
N-Phenylmaleimide	75	232	d,l-Phenylsuccinic acid	260	203
Phenylmalonic acid	231	202	d,l-Phenylsuccinic acid diamide	450	244
" "	104	430	d,l-Phenylsuccinic acid dianilide	468	244
Phenylmalononitrile	85	361	α-d,l-Phenylsuccinic acid monoanilide	357	241
Phenylmalonyl dichloride	36	215	β-d,l-Phenylsuccinic acid monoanilide	346	241
Phenylmethanethiol	32	414	DL-Phenylsuccinic anhydride	18	225
2-Phenylnaphthalene	107	47	N-Phenylsuccinimide	307	239
2-Phenylnaphthalene-1,2'-dicarboxylic acid	318	206	Phenyl sulfide	61	422
4-Phenyl-1-naphthol	338	113	Phenyl 2-thienyl ketone	60	174
N-Phenyl-1-naphthylamine	78	305	Phenyl 4-tolyl carbinol	29	89
Phenylnitromethane	25	336	Phenyl 4-tolyl ketone	74	174
1-Phenylnonane	166	41	Phenyl 2-tolyl sulfide	65	422
1-Phenyloctane	156	41	Phenyl 3-tolyl sulfide	64	422
2-Phenylpentane	46	37	Phenyl 4-tolyl sulfide	67	422
3-Phenylpentane	41	37	1-Phenyltridecane	181	42
3-Phenyl-2,4-pentanedione	69	174	1-Phenylundecane	175	42
5-Phenylpentanoic acid	52	195	Phenyl n-undecyl ketone	45	173
3-Phenyl-2-pentanone	71	170	Phenylurea	273	238
4-Phenyl-2-pentanone	31	169	Phenyl urethane	10	231
5-Phenyl-2-pentanone	27	169	2-Phenylvaleronitrile	91	352
5-Phenyl-3-pentanone	201	167	Phloracetophenone	200	179
4-Phenylphenacyl bromide	156	177	"	545	125
4-Phenylphenacyl chloride	158	177	Phloroglucinaldehyde	17	152
2-Phenylphenol	51	99	Phloroglucinol	255	109
"	98	434	"	541	125
3-Phenylphenol	117	102	"	45	434
"	81	434	Phloroglucinol triacetate	139	280
4-Phenylphenol	431	118	Phloroglucinol trimethyl ether	31	138
"	76	434	Phorone	8	172
2-Phenyl-2-phenoxyethanol	48	139	Phthalaldehyde	39	155
1-Phenyl-2-phenoxymethanol	40	138	Phthalaldehydic acid	86	157
N-Phenylphthalimide	440	243	Phthalamic acid	282	238
Phenyl phthalate	108	278	" "	244	431

Compound	Compound No.	Page No.
Phthalic acid	321	206
" "	134	430
Phthalic acid diamide	464	244
Phthalic acid dianilide	496	246
Phthalic acid monoamide	282	238
" " "	244	431
Phthalic acid mononitrile	299	205
Phthalic acid 4-sulfonamide	287	403
Phthalic acid 4-sulfonyl chloride	259	389
Phthalic anhydride	58	226
Phthalimide	482	245
α-Phthalonamide	373	241
β-Phthalonamide	303	239
Phthalonic acid	219	202
Phthalonic acid dianilide	443	244
Phthalonic acid monoanilide	363	241
Phthalonitrile	188	365
Phthaloyl dibromide	6	220
Phthaloyl dichloride	59	213
Phthaloyl difluoride	7	211
Picene	202	51
α-Picolene	13	319
β-Picolene	19	319
γ-Picolene	20	319
Picolinic acid	198	201
Picolinonitrile	12	357
Picramic acid	444	119
" "	380	315
Picramide	404	242
"	405	316
Picric acid	281	110
" "	87	340
" "	1	434
Picryl chloride	51	339
Pimelic acid	124	198
" "	385	432
Pimelic acid diamide	356	241
Pimelic acid dianilide	305	239
Pimelic acid monoanilide	157	235
Pinacol	19	88
Pinacolone	12	161
d,l-Pinacolyl alcohol	16	80
Pinacolyl chloride	25	55
d,α-Pinane	213	6
d,l-Pinane	206	6
l-β-Pinane	223	6
α-Pinene	287	22
β-Pinene	291	22
Pinosylvyn	392	116
Piperazine	202	309
"	347	438
Piperic acid	344	207
Piperidine	34	296
"	459	439
N-Piperidinoacetonitrile	2	357
Piperonal	12	153
Piperonalacetone	143	176
Piperonalacetophenone	153	177
Piperonyl alcohol	31	89
Piperonylamide	336	240
Piperonylic acid	357	208
Piperylene	24	12
Pivalaldehyde	12	144
Pivalamide	306	239
Pivalanilide	215	236
Pivalic acid	19	193
Pivalic anhydride	8	223
Pivalophenone	181	167
Pivalyl chloride	10	212
Prehnilene	77	38
Primeverose	63	333
DL-Proline	36	285
L-Proline	50	286
Propanal	4	144
Propanal cyanohydrin	3	355
Propane	3	3
1,3-Propanedicarboxylic acid	107	197
Propanedioic acid	185	200
d,l-1,2-Propanediol	91	84
1,3-Propanediol	108	85
Propane-1,1-disulfonamide	195	399
Propane-1,3-disulfonamide	194	399
Propane-1,1-disulfonic acid	83	373
Propane-1,3-disulfonic acid	82	373
Propane-1,3-disulfonyl chloride	29	381
1,3-Propanedithiol	28	414
Propane-1-sulfonamide	6	392
Propane-2-sulfonamide	10	392
Propane-2-sulfonanilide	20	392
Propane-1-sulfonic acid	1	371
Propane-2-sulfonic acid	3	371
1-Propanethiol	5	413
2-Propanethiol	3	413
Propane-1,2,3-tricarboxylic acid	258	203
Propanoic acid	6	190
" "	452	432
1-Propanol	6	80
2-Propanol	3	80
2-Propanone	1	161
Propargyl aldehyde	7	144
Propene	2	12
2-Propene-1-thiol	8	413
2-Propen-1-ol	5	80
cis-Propenyl acetylene	11	28
trans-Propenyl acetylene	12	28
trans-Propenylbenzene	19	36
Propiolamide	27	231
Propiolanilide	68	232
Propiolic acid	7	190
Propionaldehyde	4	144
Propionamide	55	232
Propionanilide	141	234
Propionic acid	6	190
" "	452	432
n-Propionic anhydride	5	223
Propionitrile	8	345
Propionyl bromide	3	218
Propionyl chloride	5	212
1-Propionylcyclohexene	77	170
Propionyl fluoride	2	211
N-Propionylglycine	241	431
Propionyl iodide	2	221
2-Propionylphenanthrene	134	176
3-Propionylphenanthrene	64	174
9-Propionylphenanthrene	65	174
4-Propionylphenol	370	115
6-Propionyltetralin	85	170
Propiophenone	174	166
Propiopiperone	27	173
n-Propoxyacetaldehyde	32	145
n-Propoxybenzene	74	134
2-Propoxybenzoic acid	331	431
3-Propoxybenzoic acid	324	431
4-Propoxybenzoic acid	433	432
2-n-Propoxyethanol	44	82
1-Propoxy-2-propanone	49	162
N-Propylacetanilide	7	231
2-n-Propylacetanilide	135	234
4-n-Propylacetanilide	66	232
n-Propyl acetate	30	251
Propyl acetylene	9	28
N-Propylalanine	64	429

	Compound No.	Page No.
n-Propyl alcohol	6	80
n-Propylallene	65	14
n-Propylamine	7	295
"	429	439
2-(Propylamino) ethylamine	386	439
N-*n*-Propylaniline	128	299
" " "	177	437
2-*n*-Propylaniline	129	299
" " "	147	437
4-*n*-Propylaniline	136	299
n-Propylbenzene	8	35
n-Propyl benzoate	240	263
d,l- -Propylbenzyl alcohol	23	88
N-Propylbenzylamine	334	438
n-Propyl bromide	7	58
n-Propyl *n*-butyrate	77	253
n-Propyl *n*-caproate	149	257
n-Propyl *n*-caprylate	229	262
n-Propyl carbamate	24	231
n-Propyl carbonate	119	255
3-Propylcatechol	97	101
4-Propylcatechol	63	99
n-Propyl chloride	7	55
n-Propyl chloroformate	41	251
n-Propylcyclohexane	165	5
2-*n*-Propylcyclohexanone	156	166
3-*n*-Propylcyclohexanone	46	169
4-*n*-Propylcyclohexanone	168	166
3-*n*-Propyl-2-cyclohexenone	51	169
Propylcyclooctatetraene	17	25
n-Propylcyclopentane	88	4
2-Propylcyclopentanone	137	165
3-Propylcyclopentanone	145	165
n-Propyl *n*-enanthate	189	259
Propylene	2	12
1,2-Propylenediamine	361	438
1,3-Propylenediamine	50	296
" " "	382	439
α-Propylene glycol	91	84
Propylene oxide	5	131
1,2,3-Propylenetricarboxylic acid trichloride	37	215
n-Propyl ether	30	132
N-Propylformanilide	10	230
n-Propyl formate	13	250
n-Propyl 2-furoate	197	260
3-*n*-Propylglutaric acid	346	432
N-Propylglycine	85	429
4-Propylheptane	191	6
n-Propyl *n*-heptanoate	189	259
n-Propyl *n*-hexanoate	149	257
3-Propyl-3-hexen-2-one	32	169
2-Prophylhydroquinone	159	104
n-Propyl 4-hydroxybenzoate	132	279
n-Propyl iodide	5	60
n-Propyl isobutyrate	68	252
n-Propyl isovalerate	102	254
n-Propyl levulinate	217	261
n-Propylmalonic acid	140	430
n-Propylmalononitrile	69	350
n-Propyl mercaptan	5	413
sec-Propyl mercaptan	3	413
1-Propylnaphthalene	163	41
2-Propylnaphthalene	162	41
n-Propyl nitrate	37	251
n-Propyl *n*-octanoate	229	262
n-Propyl oxalate	204	260
n-Propyl *n*-pentanoate	114	255
2-*n*-Propyl-1-pentene	199	18
1-Propyl phenanthrene	9	43
2-Propyl phenanthrene	11	43
9-Propyl phenanthrene	42	44
2-*n*-Propylphenol	13	94

	Compound No.	Page No.
3-*n*-Propylphenol	5	96
4-*n*-Propylphenol	2	96
1-*n*-Propylpiperidine	389	439
n-Propyl propionate	49	252
3-Propylpropiophenone	24	169
2-*n*-Propylpyridine	243	438
n-Propyl 2-pyridyl ketone	173	166
n-Propyl 3-pyridyl ketone	206	168
n-Propyl 4-pyridyl ketone	190	167
n-Propyl pyromucate	197	260
5-Propylresorcinol	135	103
n-Propyl salicylate	260	264
n-Propyl sulfide	26	420
2-Propylthiophene	29	423
3-Propylthiophene	34	424
m-Propyltoluene	24	36
o-Propyltoluene	31	36
p-Propyltoluene	27	36
2-Propyl-1,3,5-trimethylbenzene	116	39
5-Propyl-1,2,4-trimethylbenzene	40	127
n-Propylurethane	24	231
n-Propyl *n*-valerate	114	255
Propynal	7	144
Propyne	2	28
Protocatechualdehyde	117	159
Protocatechualdehyde diethyl ether	92	149
Protocatechuamide	451	244
Protocatechuanilide	329	240
Protocatechuic acid	319	206
Pseudocumene	14	35
Pseudocumenol	98	101
Pseudocumenyl acetate	38	275
Pseudocumidine	103	306
Pseudopinene	291	22
Pulegan	124	5
Pulegone	180	167
Pulenene	248	20
Putrescine	66	297
Pyrazine	18	436
Pyrazole	53	436
Pyrene	152	49
Pyrene-3-carboxadlehyde	107	159
1,6-Pyrenediol	568	126
3,8-Pyrenediol	591	127
Pyridazine	43	436
Pyridine	7	319
"	206	437
2-Pyridinecarbonitrile	12	357
3-Pyridinecarboxaldehyde	15	152
2-Pyridinecarboxylic acid	198	201
3-Pyridinecarboxylic acid	370	208
4-Pyridinecarboxylic acid	406	210
3-Pyridinecarboxylic anhydride	53	226
Pyridine-3,4-dicarboxylic acid	389	209
2, 5-Pyridinediol	572	126
Pyridine-3-sulfonamide	46	393
Pyridine-3-sulfonanilide	116	396
Pyridine-3-sulfonic acid	12	371
Pyridine-3-sulfonyl chloride	231	388
α-Pyridone	215	107
"	68	324
β-Pyridone	112	307
γ-Pyridone	115	373
3-Pyridylacetone	60	170
3-Pyridylacrylic acid	361	208
Pyrimidine	1	323
"	29	436
Pyrocatechol	207	107
Pyrocatechol dibenzoate	124	279
Pyrocatechol monomethyl ether	9	137
Pyrogallol	317	112
"	59	434

Compound No.	Page No.

Compound	Compound No.	Page No.
Pyrogallol triacetate	167	281
Pyrogallol trimethyl ether	29	138
Pyromucic acid	184	200
γ-Pyrone-2,6-dicarboxylic acid	391	209
Pyrotartaric acid	145	199
Pyrrole	11	436
2-Pyrrolecarboxaldehyde	25	154
Pyrrolidine	24	295
"	466	439
Pyruvamide	198	236
Pyruvanilide	133	234
Pyruvic acid	11	190
" "	97	429
Pyruvic aldehyde	103	171

Q

Compound	Compound No.	Page No.
Quadricycleane	40	3
Quadricyclo (2.2.1.0²·⁶.0³·⁵) heptane	40	3
d-Quercitol	68	90
Quinacetophenone	197	179
Quinaldine	80	321
Quinazoline	98	436
Quinhydrone	187	178
Quinizarin	33	183
Quinoline	71	320
"	178	437
Quinoline-2-carboxaldehyde	60	156
Quinoline-4-carboxaldehyde	29	154
Quinoline-6-carboxaldehyde	66	156
Quinoline-8-carboxaldehyde	79	157
Quinoline-3-carboxylic acid	395	209
Quinoline-4-carboxylic acid	385	209
2,3-Quinolinediol	574	126
Quinoline-6-sulfonamide	281	402
Quinoline-8-sulfonamide	242	401
Quinoline-6-sulfonic acid	135	376
Quinoline-8-sulfonic acid	106	374
Quinoline-6-sulfonyl chloride	119	384
Quinoline-8-sulfonyl chloride	199	387
2-Quinolinoacetonitrile	62	359
Quinone	12	181
Quinoxaline	5	323
"	16	436
Quinuclidine	82	324
"	441	439

R

Compound	Compound No.	Page No.
R-Acid	16	380
Rafinose	14	330
Resacetophenone	362	115
"	175	178
Resodiacetophenone	478	121
Resorcinol	226	108
"	74	434
Resorcinol diacetate	329	269
Resorcinol dibenzoate	149	280
Resorcinol diethyl ether	98	136
Resorcinol dimethyl ether	89	135
Resorcinol monoacetate	29	95
"	331	269
Resorcinol monoethyl ether	24	97
Resorcinol monomethyl ether	21	95
β-Resorcylaldehyde	110	159
β-Resorcylaldehyde dimethyl ether	59	156
β-Resorcylamide	469	245
β-Resorcylanilide	208	236
β-Resorcylic acid	338	207
α-Resorcylic aldehyde	118	159
Retene	103	47

Compound	Compound No.	Page No.
Retenequinone	30	182
Retene-6-sulfonamide	331	405
Retene-6-sulfonic acid	156	376
Retene-6-sulfonyl chloride	235	388
6-Retenol	409	117
α-L-Rhamnohexose	58	333
D-Rhamnose	13	328
α-L-Rhamnose	9	329
β-L-Rhamnose	15	330
D-Rhodeose	33	331
DL-Rhodeose	44	332
L-Rhodeose	34	331
d-Rhodinal	76	148
D-Ribose	3	329
L-Ribose	4	331
Rufol	578	127

S

Compound	Compound No.	Page No.
Saccharin	466	244
Salicylaldehyde	4	94
"	71	148
"	21	434
Salicylaldehyde ethyl ether	3	153
Salicylaldehyde methyl ether	15	153
Salicylaldehyde triacetate	141	280
Salicylamide	261	238
"	41	434
Salicylanilide	248	237
Salicylic acid	400	117
" "	244	203
" "	116	430
Salicylic acid ethyl ether	5	193
Salicylic acid methyl ether	114	197
Salicyloyl chloride	1	216
Saligenin	49	89
"	146	103
Saligenin-2-methyl ether	127	86
Salol	30	97
"	46	275
Santene	250	20
Sarcosine	43	286
Sebacic acid	183	200
Sebacic acid diamide	446	244
Sebacic acid dianilide	433	243
Sebacic acid monoamide	343	240
Sebacic acid monoanilide	194	236
Sebacic anhydride (dimer)	27	225
Sebaconitrile	19	356
Sebacoyl dichloride	33	214
Semicarbazide	97	233
DL-Serine	70	287
"	62	429
L-Serine	56	286
d-Silversterene	315	23
β-Sitosterol	57	90
Skatole	175	308
Sodium 3-hydroxybenzenephosphonate	105	434
Sodium 4-hydroxybenzenephosphonate	89	434
Sodium 3-hydroxybenzenesulfonate	66	434
Sodium 4-hydroxybenzenesulfonate	58	434
Sodium 3-hydroxybenzoate	94	434
Sodium 4-hydroxybenzoate	73	434
Sorbaldehyde	5	152
d-Sorbitol	50	89
D-Sorbose	51	332
DL-Sorbose	45	332
L-Sorbose	50	332
Spiropentane	16	3
Stachyose	54	332
Stearaldehyde	14	153
Stearamide	158	235

	Compound No.	Page No.
Stearanilide	94	233
Stearic acid	70	195
Stearic anhydride	28	225
Stearoyl chloride	2	216
Stearyl alcohol	33	89
trans-Stilbene	131	48
Stilbene-2-carboxaldehyde	72	156
2,2-Stilbenediol	172	105
3,5-Stilbenediol	392	116
4,4' Stilbenediol	583	127
4-Styrenesulfonic acid	33	372
Suberane	68	4
Suberene	187	18
Suberic acid	215	202
" "	395	432
Suberic anhydride (dimer)	22	225
Suberic acid diamide	456	244
Suberic acid dianilide	395	242
Suberic acid monoamide	210	236
Suberic acid monoanilide	216	236
Suberonitrile	20	356
Succinaldehyde	59	147
Succinamic acid	312	239
" "	398	432
Succinic acid	294	205
" "	310	431
Succinic acid diamide	498	246
Succinic acid dianilide	480	245
Succinic acid monoamide	312	239
" "	398	432
Succinic acid monoanilide	279	238
Succinic anhydride	52	226
Succinimide	205	236
Succinonitrile	67	360
Succinyl dibromide	12	219
Succinyl dichloride	38	212
Sucrose	55	332
Sulfadimethylpyrimidine	308	404
Sulfaguanidine	269	402
Sulfamethazine	308	404
Sulfanilic acid	70	373
Sulfanilylguanidine	269	402
Sulfapyridine	278	402
Sulfathiazole	319	404
2-Sulfobenzoic acid	139	376
3-Sulfobenzoic acid	84	373
4-Sulfobenzoic acid	184	377
4-Sulfophthalic acid	136	376
Sylvan	15	131
Syringetin	584	127

T

D-Tagatose	16	330
D-Talose	21	330
d-Tartaric acid	263	203
" "	141	430
d, l-Tartaric acid	326	206
meso-Tartaric acid	203	201
" " "	177	430
D-Tartaric acid diamide	418	243
meso-Tartaric acid diamide	401	242
D-Tartaric acid dianilide	500	246
D, L-Tartaric acid monoamide	478	245
D-Tartaric acid monoanilide	375	242
D, L-Tartaric acid monoanilide	486	245
Tartronic acid	240	202
Taurine	110	289
Terephthalaldehyde	103	158
Terephthalic acid	404	210
" "	210	430
Terephthalic acid dianilide	510	247
Terephthalic acid monoanilide	485	245
Terephthalic acid mononitrile	348	207
Terephthalonitrile	229	367
Terephthaloyl dichloride	19	216
α-Terpinene	311	23
γ-Terpinene	322	23
d, l-α-Terpineol	114	85
Terpinolene	325	24
1,2,3,4,5,6,7,8-Tetrabenzanthracene	203	51
2,3,4,6-Tetrabromoaniline	248	310
1,2,4,5-Tetrabromobenzene	9	74
2,3,4,6-Tetrabromobenzenesulfonamide	423	408
2,3,4,6-Tetrabromobenzenesulfonic acid	196	378
2,3,4,6-Tetrabromobenzenesulfonyl chloride	137	385
Tetrabromocatechol	492	122
2,4,5,6-Tetrabromo-m-cresol	494	122
3,4,5,6-Tetrabromo-o-cresol	523	123
2,3,5,6-Tetrabromo-p-cresol	499	122
1,1,2,2-Tetrabromoethane	19	65
2,3,4,5-Tetrabromophenol	285	111
Tetrabromophthalic anhydride	85	227
2,3,4,5-Tetrachlorobenzaldehyde	93	158
2,3,4,6-Tetrachlorobenzaldehyde	85	157
1,2,3,4-Tetrachlorobenzene	6	71
1,2,3,5-Tetrachlorobenzene	10	71
1,2,4,5-Tetrachlorobenzene	29	71
2,3,4,5-Tetrachlorobenzonitrile	107	362
2,3,5,6-Tetrachloro-1,4-benzoquinone	53	184
1,1,2,2-Tetrachloroethane	19	63
1,1,2,2-Tetrachloroethylene	14	63
2,3,5,6-Tetrachlorohydroquinone	565	126
4,6,7,8-Tetrachloronaphthalene-2-sulfonamide	401	407
4,6,7,8-Tetrachloronaphthalene-2-sulfonyl chloride	264	389
2,3,4,6-Tetrachlorophenol	93	101
Tetrachlorophthalic anhydride	83	227
1,2,3,4-Tetrachlorotetralin	32	72
5,6,7,8-Tetrachlorotetralin	31	72
Tetracontane	76	9
1-Tetracontene	23	26
1-Tetracontyne	27	31
n-Tetracosane	26	8
Tetracosanoic acid	88	196
1-Tetracosene	7	26
Tetracosylcyclohexane	39	8
Tetracosylcyclopentane	27	8
1-Tetracosyne	11	31
Tetracyanoethylene	222	367
Tetradecanal	4	153
n-Tetradecane	265	7
1,14-Tetradecanedicarboxylic acid	168	199
n-Tetradecanoic acid	46	194
n-Tetradecanoic anhydride	19	225
1-Tetradecanol	17	88
2-Tetradecanone	17	172
Tetradecanonitrile	4	357
1-Tetradecene	332	24
n-Tetradecylacetylene	64	30
n-Tetradecyl bromide	30	58
n-Tetradecylcyclohexane	6	8
n-Tetradecylcyclopentane	278	7
n-Tetradecyl sulfide	25	425
1-Tetradecyne	62	30
Tetraethylene glycol dimethyl ether	105	136
Tetraethyl pyromellitate	74	276
Tetraethyl silicate	113	255
1,2,3,4-Tetrahydroanthracene	108	47
1,2,3,4-Tetrahydroanthranol	221	107
Tetrahydrofuran	16	131
Tetrahydrofurfural	44	146
α-Tetrahydrofurfuryl acetate	164	258

Compound	Compound No.	Page No.
Tetrahydrofurfuryl alcohol	78	84
" "	67	134
α-Tetrahydrofurfuryl benzoate	346	270
α-Tetrahydrofurfuryl propionate	180	259
1,2,3,4-Tetrahydroisoquinoline	146	299
Tetrahydromesitylene	256	20
bz-Tetrahydro-6-methoxyquinoline	25	138
1,2,3,4-Tetrahydro-2-naphthaldehyde	24	150
1,2,3,4-Tetrahydronaphthalene	88	38
cis-Tetrahydronaphthalene-2,3-dicarboxylic acid	279	431
trans-Tetrahydronaphthalene-2,3-dicarboxylic acid	286	431
1,2,3,4-Tetrahydro-2-naphthoic acid	106	197
d,l-1,2,3,4-Tetrahydro-2-naphthol	25	95
5,6,7,8-Tetrahydro-1-naphthol	88	101
5,6,7,8-Tetrahydro-2-naphthol	66	100
5,6,7,8-Tetrahydro-2-naphthylamine	22	303
1,2,3,4-Tetrahydrophenanthrene-9-carboxaldehyde	108	159
Tetrahydropyran	29	132
1,2,3,4-Tetrahydroquinoline	165	300
Tetrahydrosylvan	21	132
2,3,4,5-Tetrahydroxyadipic acid	274	204
1,2,3,4-Tetrahydroxybenzene	413	117
1,2,3,5-Tetrahydroxybenzene	430	118
1,2,4,5-Tetrahydroxybenzene	536	124
2,3,4,2′-Tetrahydroxybenzophenone	182	178
2,5,2′,6′-Tetrahydroxybenzophenone	196	179
3,4,3′,4′-Tetrahydroxybenzophenone	202	179
d,l-1,2,3,4-Tetrahydroxybutane	41	89
1,2,3,4-Tetrahydroxynaphthalene	555	126
1,2,5,8-Tetrahydroxynaphthalene	469	121
1,4,5,8-Tetrahydroxynaphthalene	485	121
Tetraiodophthalic anhydride	86	227
Tetralin	88	38
Tetralin-6-sulfonamide	88	395
Tetralin-6-sulfonanilide	144	397
Tetralin-6-sulfonyl chloride	54	382
1-Tetralone	90	170
2-Tetralone	75	170
Tetramethylallene	89	15
N,N,2,4-Tetramethylaniline	59	320
N,N,2,5-Tetramethylaniline	58	320
N,N,2,6-Tetramethylaniline	56	320
2,3,5,6-Tetramethylaniline	144	437
2,3,5,6-Tetramethylbenzaldehyde	2	153
1,2,3,4-Tetramethylbenzene	77	38
1,2,3,5-Tetramethylbenzene	59	37
1,2,4,5-Tetramethylbenzene	73	46
2,2,3,3-Tetramethylbutane	78	9
1,1,3,3-Tetramethylcyclohexane	154	5
1,1,3,4-Tetramethylcyclohexane	174	5
cis-1,1,3,5-Tetramethylcyclohexane	147	5
trans-1,1,3,5-Tetramethylcyclohexane	163	5
cis-1,2,3,5-Tetramethylcyclohexane	230	6
trans-1,2,3,5-Tetramethylcyclohexane	193	6
cis-1,2,4,5-Tetramethylcyclohexane	237	7
trans-1,2,4,5-Tetramethylcyclohexane	216	6
2,2,6,6-Tetramethylcyclohexanone	138	165
d,1,2,2,3-Tetramethylcyclopentane-1-carboxylic acid	129	198
2,2,5,5-Tetramethylcyclopentanone	77	163
1,2,3,3-Tetramethylcyclopentene	237	19
1,2,2,3-Tetramethyl-3-cyclopentene-1-acetonitrile	1	355
1,2,2,3-Tetramethylcyclopentene-1-carbonitrile	93	361
Tetramethylenediamine	66	297
Tetramethylene glycol	119	86
Tetramethylene sulfide	12	423
Tetramethylethylene glycol	19	88
2,2,3,3-Tetramethylhexane	183	6
2,2,3,4-Tetramethylhexane	155	5
2,2,3,5-Tetramethylhexane	139	5
2,2,4,4-Tetramethylhexane	152	5
2,2,4,5-Tetramethylhexane	136	5
2,2,5,5-Tetramethylhexane	107	4
2,3,3,4-Tetramethylhexane	207	6
2,3,3,5-Tetramethylhexane	151	5
2,3,4,4-Tetramethylhexane	194	6
2,3,4,5-Tetramethylhexane	186	6
3,4,4,4-Tetramethylhexane	234	7
2,2,5,5-Tetramethyl-3-hexyne	40	29
Tetramethylhydroquinone	119	435
1,2,4,8-Tetramethylnaphthalene	36	44
1,2,5,6-Tetramethylnaphthalene	122	48
1,3,6,8-Tetramethylnaphthalene	78	46
1,4,5,7-Tetramethylnaphthalene	35	44
1,4,5,8-Tetramethylnaphthalene	137	48
1,4,6,7-Tetramethylnaphthalene	51	45
2,2,3,3-Tetramethylpentane	114	4
2,2,3,4-Tetramethylpentane	92	4
2,2,4,4-Tetramethylpentane	76	4
2,3,3,4-Tetramethylpentane	121	5
3,3,4,4-Tetramethylpentanoic acid	66	195
2,2,4,4-Tetramethyl-3-pentanone	74	163
2,3,4,6-Tetramethylphenol	122	102
2,3,5,6-Tetramethylphenol	257	109
1,2,2,4-Tetramethylpiperidine	374	439
1,2,2,6-Tetramethylpiperidine	430	439
2,2,6,6-Tetramethylpiperidine	455	439
2,2,6,6-Tetramethyl-4-piperidone	20	172
2,3,4,5-Tetramethylpyridine	70	320
2,3,5,6-Tetramethylpyridine	289	438
Tetramethyl pyromellitate	162	281
2,4,5,8-Tetramethylquinoline	23	323
Tetramethylsuccinic acid	205	430
2,3,4,5-Tetramethylthiophene	49	424
Tetratriacontane	61	9
1-Tetratriacontene	17	26
Tetratriacontylcyclohexane	71	9
Tetratriacontylcyclopentane	60	9
1-Tetratriacontyne	21	31
Tetrolic acid	110	430
2-Thenoic acid	173	200
3-Thenoic acid	196	201
7-Thiabicycloheptane	45	424
Thiacyclobutane	5	423
Thiacycloheptane	41	424
Thiacyclohexane	3	425
Thiacyclopentane	12	423
Thiacyclopropane	1	423
2-Thianaphthenecarboxylic acid	367	208
3-Thianaphthenecarboxylic acid	272	204
Thiane	3	425
Thiazole	51	436
2-Thienylacetic acid	75	196
α-Thienylacetone	78	170
α-Thienylglyoxylic acid	96	196
1-(α-Thienyl)-2-propanone	78	170
Thiepane	41	424
Thietane	5	423
Thiirane	1	423
Thioacetamide	150	234
Thioacetanilide	42	232
Thioacetic acid	1	190
Thioacetic anhydride	30	420
Thiobenzoic anhydride	20	425
Thiocarvacrol	45	415
Thiocatechol	40	415
2-Thiocresol	31	414
3-Thiocresol	33	414

Compound	Compound No.	Page No.
4-Thiocresol	11	416
Thiohydroquinone	9	416
Thiolane	12	423
Thiophane	12	423
Thiophene	3	423
Thiophene-2-acetonitrile	9	355
Thiophene-2-carbonitrile	59	349
Thiophene-3-carbonitrile	52	349
2-Thiophenecarboxaldehyde	72	148
2-Thiophenecarboxylic acid	173	200
3-Thiophenecarboxylic acid	196	201
Thiophene-2,3-dicarbonitrile	29	347
2-Thiophenethiol	27	414
Thiophenol	25	414
Thioresorcinol	1	416
Thiothymol	44	415
Thiourea	382	242
1,4-Thioxane	23	423
DL-Threonine	73	287
D-Threose	17	330
B-Thujone	159	166
Thymol	44	98
Thymoquinol	347	114
Thymoquinone	2	181
Thymyl acetate	272	265
Thymyl benzoate	31	274
DL-Thyroxine	58	286
Tiglamide	43	232
Tiglanilide	44	232
Tiglic acid	59	195
" "	461	433
T.N.T.	46	338
Tolane	47	44
m-Tolidine	221	309
o-Tolidine	290	312
4-Tolil	133	176
2-Tolualdehyde	74	148
3-Tolualdehyde	73	148
4-Tolualdehyde	75	148
2-Toluamide	255	238
3-Toluamide	98	233
4-Toluamide	317	240
2-Toluanilide	202	236
3-Toluanilide	206	236
4-Toluanilide	270	238
Toluene	2	35
2-p-Tolueneazo-p-cresol	235	108
Toluene-2,4-disulfonamide	256	401
Toluene-3,4-disulfonamide	272	402
Toluene-2,4-disulfonyl chloride	44	382
Toluene-α-sulfonamide	37	393
Toluene-2-sulfonamide	148	397
Toluene-3-sulfonamide	39	393
Toluene-4-sulfonamide	95	395
Toluene-α-sulfonanilide	34	393
Toluene-2-sulfonanilide	90	395
Toluene-3-sulfonanilide	26	392
Toluene-4-sulfonanilide	35	393
Toluene-2-sulfonic acid	58	372
Toluene-3-sulfonic acid	10	371
Toluene-4-sulfonic acid	31	372
Toluene-2-sulfonyl chloride	73	383
Toluene-3-sulfonyl chloride	3	381
Toluene-4-sulfonyl chloride	75	383
2-Toluenethiol	31	414
3-Toluenethiol	33	414
4-Toluenethiol	11	416
Toluhydroquinone	60	434
2-Toluic acid	125	198
3-Toluic acid	137	198
4-Toluic acid	278	204
4-Toluic acid-2-sulfonamide	250	401
4-Toluic acid-2-sulfonyl chloride	56	382
m-Toluic anhydride	29	225
o-Toluic anhydride	9	225
p-Toluic anhydride	40	225
m-Toluidine	96	298
o-Toluidine	94	298
p-Toluidine	34	303
2-Tolunitrile	64	350
3-Tolunitrile	70	350
4-Tolunitrile	13	357
p-Toluquinone	6	181
2-(4-Toluyl)benzamide	361	241
4-Toluyl chloride	51	213
2-Tolyl acetate	187	259
3-Tolyl acetate	200	260
4-Tolyl acetate	203	260
(2-Tolyl)acetic acid	95	196
(4-Tolyl)acetic acid	103	197
(2-Tolyl)acetonitrile	86	352
(3-Tolyl)acetonitrile	83	352
(4-Tolyl)acetonitrile	84	352
2-Tolyl benzoate	350	271
3-Tolyl benzoate	76	277
4-Tolyl benzoate	101	278
2-Tolyl carbinol	14	88
3-Tolyl carbinol	109	85
4-Tolyl carbinol	32	89
p-Tolylhydrazine	91	305
2-Tolyl sulfide	35	425
3-Tolyl sulfide	60	421
4-Tolyl sulfide	31	425
3-Tolyl 4-tolyl sulfide	6	425
2-Tolylurea	409	243
3-Tolylurea	259	238
4-Tolylurea	384	242
Transbrassidamide	85	233
Transbrassidanilide	50	232
α,α-Trehalose	62	333
1,2,4-Triacetoxybenzene	133	279
1,3,5-Triacetoxybenzene	139	280
n-Triacontane	49	3
Triacontanol	48	89
1-Triacontene	13	26
Triacontylcyclohexane	59	9
Triacontylcyclopentane	47	9
1-Triacontyne	17	31
Triallylamine	295	438
1,2,3-Triaminopropane	88	298
" "	335	438
4,4',4''-Triaminotriphenyl carbinol	65	90
" " "	429	317
4,4',4''-Triaminotriphenylmethane	424	317
Tri-n-amylamine	89	321
1,3-Triazine	281	438
d,l,α-Triazobutyric acid	3	193
α-Triazoisobutyric acid	9	193
2-Triazoisovaleric acid	6	192
1,2,4-Triazole	46	436
d,l-2-Triazopropionic acid	8	192
Tribenzylamine	56	324
Tribromoacetamide	193	236
Tribromoacetic acid	176	200
" "	5	429
1,1,1-Tribromoacetone	211	168
3,4,5-Tribromoacetophenone	164	177
2,4,6-Tribromoaniline	264	311
3,4,5-Tribromoaniline	267	311
1,3,5-Tribromobenzene	7	74
2,4,6-Tribromobenzenesulfonamide	385	407
2,4,6-Tribromobenzenesulfonanilide	363	406
2,4,6-Tribromobenzenesulfonic acid	172	377
2,4,6-Tribromobenzenesulfonyl chloride	65	383

Compound	Compound No.	Page No.
1,1,1-Tribromo-*tert*-butyl alcohol	63	90
3,4,5-Tribromocatechol	349	114
2,3,6-Tribomo-*p*-cresol	198	106
2,4,6-Tribromo-*m*-cresol	131	103
3,4,5-Tribromo-*o*-cresol	153	104
3,4,6-Tribromo-*o*-cresol	162	104
2,2,2-Tribromoethanal	63	147
1,1,2-Tribromoethane	15	65
2,2,2-Tribromoethanol	45	89
2,4,5-Tribromophenetol	45	139
2,4,6-Tribromophenetol	42	138
2,3,4-Tribromophenol	176	105
2,4,5-Tribromophenol	150	104
2,4,6-Tribromophenol	171	105
Tribromophloroglucinol	384	116
1,2,3-Tribromopropane	17	65
2,4,6-Tribromoresorcinol	234	108
Tri-*n*-butylamine	64	320
" "	355	438
Tricarballylic acid	258	203
Trichloroacetaldehyde	19	145
Trichloroacetamide	256	238
Trichloroacetanilide	99	233
2,4,6-Trichloroacetanilide	276	238
Trichloroacetic acid	47	194
" "	3	429
Trichloroacetonitrile	6	345
Trichloroacetyl bromide	7	218
Trichloroacetyl chloride	18	212
Trichloroacetyl fluoride	4	211
Trichloroacetyl iodide	4	222
Trichloroacrylic acid	8	429
Trichloroacrylonitrile	6	357
2,4,6-Trichloroaniline	125	306
2,4,6-Trichloroanisole	37	138
2,3,4-Trichlorobenzaldehyde	78	157
2,3,5-Trichlorobenzaldehyde	40	155
2,3,6-Trichlorobenzaldehyde	74	157
2,4,5-Trichlorobenzaldehyde	98	159
2,4,6-Trichlorobenzaldehyde	46	155
3,4,5-Trichlorobenzaldehyde	76	157
1,2,3-Trichlorobenzene	11	71
1,2,4-Trichlorobenzene	25	70
1,3,5-Trichlorobenzene	17	71
2,3,4-Trichlorobenzenesulfonamide	382	406
2,4,6-Trichlorobenzenesulfonamide	339	405
2,3,4-Trichlorobenzenesulfonyl chloride	67	381
2,4,6-Trichlorobenzenesulfonyl chloride	17	381
4,4,4-Trichloro-*n*-butylamine	356	438
2,2,3-Trichloro-*n*-butyraldehyde	58	147
2,2,4-Trichloro-*n*-butyraldehyde	16	152
2,3,5-Trichloro-*p*-cresol	80	100
2,4,6-Trichloro-*m*-cresol	41	98
3,5,6-Trichloro-*o*-cresol	68	100
4,5,6-Trichloro-*o*-cresol	114	102
Trichloroethanal	19	145
1,1,1-Trichloroethane	7	63
1,1,2-Trichloroethane	13	63
Trichloroethanol	45	82
2,2,2-Trichloroethylamine	218	437
1,1,2-Trichloroethylene	10	63
Trichlorohydroquinone	327	113
1,1,1-Trichloroisopropanol	24	88
3,3,3-Trichlorolactic acid	165	199
Trichlorolactamide	96	233
Trichlorolactanilide	325	240
5,6,7-Trichloronaphthalene-1-sulfonamide	433	408
5,6,8-Trichloronaphthalene-2-sulfonamide	400	407
6,7,8-Trichloronaphthalene-2-sulfonamide	424	408
5,6,7-Trichloronaphthalene-1-sulfonyl chloride	211	388
5,6,8-Trichloronaphthalene-2-sulfonyl chloride	247	389
6,7,8-Trichloronaphthalene-2-sulfonyl chloride	245	389
1,3,4-Trichloro-2-naphthol	415	117
1,4,5-Trichloro-2-naphthol	395	116
2,3,4-Trichloro-1-naphthol	439	119
5,5,5-Trichloro-*n*-pentylamine	366	439
2,4,6-Trichlorophenetole	26	138
2,3,4-Trichlorophenol	136	103
2,3,5-Trichlorophenol	67	100
2,3,6-Trichlorophenol	60	99
2,4,5-Trichlorophenol	87	101
2,4,6-Trichlorophenol	91	101
3,4,5-Trichlorophenol	193	106
Trichlorophloroglucinol	329	113
1,2,3-Trichloropropane	20	63
3,3,3-Trichloro-*n*-propylamine	341	438
2,3,4-Trichlorotoluene	5	71
2,3,5-Trichlorotoluene	8	71
2,4,5-Trichlorotoluene	23	71
2,4,6-Trichlorotoluene	2	71
3,4,5-Trichlorotoluene	7	71
n-Tricosane	23	8
12-Tricosanone	91	175
1-Tricosene	6	26
Tricosylcyclohexane	36	8
Tricosylcyclopentane	25	8
1-Tricosyne	10	31
Tricresyl phosphate	358	271
Tricyanomethane	66	360
Tricyclo(2,2,1,0^{2,6})heptane	33	8
Tridecanal	7	150
Tridecanamide	116	233
Tridecananilide	53	232
n-Tridecane	261	7
n-Tridecanonic acid	28	194
n-Tridecanoic anhydride	17	225
1-Tridecanol	12	88
2-Tridecanone	9	172
7-Tridecanone	14	172
1-Tridecene	331	24
Tridecyl acetylene	63	30
n-Tridecyl aldehyde	7	150
Tridecylbenzene	181	42
Tridecylcyclohexane	3	8
Tridecylcyclopentane	276	7
Tridecylic acid	28	194
1-Tridecyne	61	30
Triethyl aconitate	322	268
Triethylamine	4	319
" "	435	439
1,2,4-Triethylbenzene	109	39
1,3,5-Triethylbenzene	110	39
2,4,6-Triethylbenzoic acid	139	198
Triethyl carbinol	37	82
Triethyl citrate	340	270
Triethylene bromohydrin	73	84
Triethylene glycol	130	86
Triethylene glycol diacetate	193	260
Triethylene glycol dimethyl ether	88	135
Triethyl orthoformate	85	253
1,4,5-Triethylpyrazole	42	320
2,4,5-Triethylpyridine	34	319
Triethylsuccinic acid	115	430
Triethyl trimesate	159	281
Trifluoroacetaldehyde	2	144
Trifluoroacetic acid	2	429
Trifluoroacetic anhydride	1	223
Trifluoroacrylic acid	27	429
4,4,4-Trifluorobutyric acid	306	431
4,4,4-Trifluorocrotonic acid	166	430
5,5,5-Trifluoroleucine	44	429
3-(Trifluoromethyl)acetophenone	158	166
6,6,6-Trifluoronorleucine	59	429

	Compound No.	Page No.		Compound No.	Page No.
5,5,5-Trifluoropentanoic acid	389	432	Trimethyl aconitate	317	268
2,2,2-Trifluoropropionaldehyde	8	144	Trimethylallene	56	14
3,3,3-Trifluoropropionic acid	148	430	Trimethylamine	1	319
4,4,4-Trifluorothreonine	23	429	"	348	438
4,4,-Trifluorovaline	22	429	N,N,2-Trimethylaniline	46	320
α-2,4-Trihydroxyacetophenone	483	121	"	249	438
2,3,4-Trihydroxyacetophenone	454	120	N,N,3-Trimethylaniline	65	320
" "	189	178	" "	213	437
2,3,5-Trihydroxyacetophenone	517	123	N,N,4-Trimethylaniline	62	320
" "	200	179	" "	226	437
2,4,5-Trihydroxyacetophenone	518	123	N,2,6-Trimethylaniline	250	438
" "	199	179	2,4,6-Trimethylaniline	142	299
2,4,6-Trihydroxyacetophenone	545	125	" "	150	437
" "	201	179	3,5,6-Trimethylaniline	199	437
3,4,5-Trihydroxyacetophenone	195	178	1,2,10-Trimethylanthracene	91	46
1,8,9-Trihydroxyanthracene	465	120	2,3,6-Trimethylbenzaldehyde	18	150
1,2,3-Trihydroxy-9,10-anthraquinone	61	185	2,4,5-Trimethylbenzaldehyde	21	154
1,2,4-Trihydroxy-9,10-anthraquinone	43	183	2,4,6-Trimethylbenzaldehyde	21	150
1,2,5-Trihydroxy-9,10-anthraquinone	49	184	1,2,3-Trimethylbenzene	18	36
1,2,6-Trihydroxy-9,10-anthraquinone	62	185	1,2,4-Trimethylbenzene	14	35
1,2,7-Trihydroxy-9,10-anthraquinone	63	185	1,3,5-Trimethylbenzene	11	35
1,2,8-Trihydroxy-9,10-anthraquinone	40	183	2,4,6-Trimethylbenzene-1,3-disulfonamide	421	408
2,4,6-Trihydroxybenzaldehyde	17	152	2,4,6-Trimethylbenzene-1,3-disulfonic acid	195	378
3,4,5-Trihydroxybenzaldehyde	128	160	2,4,6-Trimethylbenzene-1,3-disulfonyl chloride	202	387
1,2,3-Trihydroxybenzene	317	112	2,4,5-Trimethylbenzenesulfonamide	234	400
" "	59	434	2,4,6-Trimethylbenzenesulfonamide	100	395
1,2,4-Trihydroxybenzene	340	113	2,4,6-Trimethylbenzenesulfonanilide	41	393
1,3,5-Trihydroxybenzene	255	109	2,4,5-Trimethylbenzenesulfonic acid	101	374
" "	45	434	2,4,6-Trimethylbenzenesulfonic acid	36	372
3,4,5-Trihydroxybenzoic acid	386	209	2,4,5-Trimethylbenzenesulfonyl chloride	61	382
" "	353	432	2,4,6-Trimethylbenzenesulfonyl chloride	50	382
2,4,6-Trihydroxybenzophenone	426	118	2,4,6-Trimethylbenzoic acid	236	202
1,2,4-Trihydroxynaphthalene	388	116	" " "	199	430
1,2,6-Trihydroxynaphthalene	482	121	2,4,6-Trimethylbenzonitrile	64	360
1,2,7-Trihydroxynaphthalene	498	122	2,4,4'-Trimethylbenzophenone	219	168
1,3,6-Trihydroxynaphthalene	175	105	2,4,5-Trimethylbenzophenone	218	168
1,4,6-Trihydroxynaphthalene	335	113	2,3,5-Trimethyl-1,4-benzoquinone	1	181
1,6,7-Trihydroxynaphthalene	459	120	2,4,6-Trimethylbenzoyl chloride	30	214
9,10,16-Trihydroxypalmitic acid	121	197	2,4,6-Trimethylbenzyl chloride	3	57
2,3,4-Trihydroxy-9,10-phenanthraquinone	26	182	1,3,3-Trimethylbicyclo(2.2.1)heptane	145	5
3,4,5-Trihydroxyphenanthrene	372	115	2,2,3-Trimethylbicyclo(2.2.1)heptane	48	9
1,2,3-Trihydroxypropane	131	86	2,2,7-Trimethylbicyclo(2.2.1)heptane	197	6
1,2,3-Trihydroxytoluene	268	110	1,5,5-Trimethylbicyclo(2.2.1)hept-2-ene	245	20
2,4,5-Trihydroxytoluene	314	112	2,7,7-Trimethylbicyclo(2.2.1)hept-2-ene	266	21
2,4,6-Trihydroxytoluene	531	124	3,7,7-Trimethylbicyclo(2.2.1)hept-2-ene	294	22
3,4,5-Trihydroxytoluene	268	110	3,7,7-Trimethylbicyclo(2.2.1)hept-3-ene	297	22
2,3,5-Tri-iodophenol	244	108	2,2,3-Trimethylbutane	30	3
2,4,6-Tri-iodophenol	399	116	2,3,3-Trimethyl-1-butene	70	14
2,4,6-Tri-iodoresorcinol	358	114	Trimethyl carbinol	7	88
Tri-isoamylamine	75	321	Trimethyl citrate	112	278
Tri-isobutylamine	384	439	1,1,4-Trimethylcycloheptane	192	6
2,4,6-Tri-isopropyl-1,3,5-trioxan	48	155	2,6,6-Trimethylcycloheptanone	73	170
Triketohydrindene hydrate	203	179	1,4,4-Trimethylcycloheptene	290	22
Trimesic acid	410	210	1,1,3-Trimethylcyclohexane	105	4
Trimesic acid triamide	512	247	cis-1,2,3-Trimethylcyclohexane	129	5
Trimesic acid trianilide	186	235	trans-1,2,3-Trimethylcyclohexane	123	5
2,4,6-Trimethoxybenzaldehyde	104	158	cis-1,2,4-Trimethylcyclohexane	132	5
3,4,5-Trimethoxybenzaldehyde	67	156	trans-1,2,4-Trimethylcyclohexane	113	4
1,2,3-Trimethoxybenzene	29	138	cis-1,3,5-Trimethylcyclohexane	112	4
1,3,5-Trimethoxybenzene	31	138	Trans-1,3,5-Trimethylcyclohexane	110	4
2,4,5-Trimethoxybenzene sulfonamide	17	392	1,2,2-Trimethylcyclohexanol	18	88
2,4,5-Trimethoxybenzene sulfonanilide	196	399	1,3,5-Trimethylcyclohexanol	82	84
2,4,5-Trimethoxybenzene sulfonic acid	5	371	2,2,6-Trimethylcyclohexanol	26	88
2,4,5-Trimethoxybenzene sulfonyl chloride	209	388	2,3,3-Trimethylcyclohexanol	9	88
2,4,5-Trimethoxybenzoic acid	216	202	2,3,6-Trimethylcyclohexanol	95	84
2,4,5-Trimethoxybenzonitrile	157	364	2,4,5-Trimethylcyclohexanol	93	84
Trimethylacetaldehyde	12	144	3,3,5-Trimethylcyclohexanol	27	89
Trimethylacetic acid	19	193	2,2,6-Trimethylcyclohexanone	126	165
" "	465	433	1,2,3-Trimethylcyclohexene	276	21
Trimethylacetic anhydride	8	223	1,3,5-Trimethylcyclohexene	256	20
Trimethylacetonitrile	10	345	1,4,4-Trimethylcyclohexene	248	20
2,4,5-Trimethylacetophenone	205	168	1,4,5-Trimethylcyclohexene	261	20

	Compound No.	Page No.		Compound No.	Page No.
1,5,5-Trimethylcyclohexene	246	20	2,3,3-Trimethylpentane	56	3
1,5,6-Trimethylcyclohexene	249	20	2,3,4-Trimethylpentane	54	3
1,6,6-Trimethylcyclohexene	267	21	3,4,4-Trimethylpentanoic acid	27	192
1,5,5-Trimethylcyclohexene-6-carboxylic acid	128	198	2,2,4-Trimethyl-3-pentanone	30	162
1,3,3-Trimethylcyclohexen-6-ol	94	84	2,3,3-Trimethyl-1-pentene	151	17
1,3,5-Trimethyl-1-cyclohexen-3-ol	22	88	2,3,4-Trimethyl-1-pentene	149	17
1,5,5-Trimethylcyclohexen-3-one	172	166	2,3,4-Trimethyl-2-pentene	196	18
4,4,6-Trimethyl-2-cyclohexen-1-one	68	170	2,4,4-Trimethyl-1-pentene	120	16
1,1,2-Trimethylcyclopentane	55	3	2,4,4-Trimethyl-2-pentene	132	16
1,1,3-Trimethylcyclopentane	46	3	3,3,4-Trimethyl-1-pentene	133	16
1,cis-2,cis-3-Trimethylcyclopentane	77	4	3,4,4-Trimethyl-1-pentene	128	16
1,cis-2,trans-3-Trimethylcyclopentane	60	4	3,4,4-Trimethyl-2-pentene	168	17
1,trans-2,cis-3-Trimethylcyclopentane	52	3	1,2,3-Trimethylphenanthrene	52	45
1,cis-2,cis-4-Trimethylcyclopentane	64	4	1,2,7-Trimethylphenanthrene	124	48
1,cis-2,trans-4-Trimethylcyclopentane	59	4	1,2,8-Trimethylphenanthrene	147	49
1,trans-2,cis-4-Trimethylcyclopentane	49	9	1,3,7-Trimethylphenanthrene	57	45
1,2,2-Trimethylcyclopentane-1,3-dicarboxylic acid	303	205	1,3,8-Trimethylphenanthrene	119	47
d,l-1,2,3-Trimethylcyclopentene	205	18	1,4,7-Trimethylphenanthrene	63	45
1,5,5-Trimethylcyclopentene	153	17	1,6,7-Trimethylphenanthrene	129	48
2,3,3-Trimethyl-1-cyclopentene-1-carbonitrile	65	350	2,3,5-Trimethylphenol	177	105
1,5,5-Trimethylcyclopentene-2-carboxylic acid	187	200	2,4,5-Trimethylphenol	98	101
1,5,5-Trimethylcyclopentene-4-carboxylic acid	24	193	" "	112	435
4,5,5-Trimethylcyclopentene-1-carboxylic acid	97	196	2,4,6-Trimethylphenol	92	101
1,1,2-Trimethylcyclopropane	19	3	" "	116	435
1,2,3-Trimethylcyclopropane	23	3	2,4,5-Trimethylphenyl acetate	38	275
Trimethylenediamine	50	296	2,4,6-Trimethylphenyl acetate	255	264
Trimethylene glycol	108	85	cis-1,2,6-Trimethylpiperidine	368	439
Trimethylene glycol diacetate	193	260	2,2,4-Trimethylpiperidine	60	297
Trimethylene glycol diphenyl ether	38	138	" "	452	439
Trimethyleneimine	15	295	2,2,6-Trimethylpiperidine	51	296
Trimethylene sulfide	5	423	" "	462	439
Trimethylglycine	28	285	1,3,4-Trimethylpyrazole	27	319
2,2,3-Trimethylheptane	170	5	1,3,5-Trimethylpyrazole	10	323
2,2,4-Trimethylheptane	135	5	2,3,4-Trimethylpyridine	51	320
2,2,5-Trimethylheptane	137	5	2,3,5-Trimethylpyridine	45	320
2,2,6-Trimethylheptane	138	5	2,3,6-Trimethylpyridine	43	320
2,3,3-Trimethylheptane	180	6	2,4,6-Trimethylpyridine	41	320
2,3,4-Trimethylheptane	195	6	" "	279	438
2,3,5-Trimethylheptane	166	5	1,2,5-Trimethylpyrrolidine	8	319
2,3,6-Trimethylheptane	159	5	2,3,4-Trimethylquinoline	57	324
2,4,4-Trimethylheptane	150	5	2,3,6-Trimethylquinoline	52	324
2,4,5-Trimethylheptane	168	5	2,3,8-Trimethylquinoline	31	323
2,4,6-Trimethylheptane	131	5	2,4,6-Trimethylquinoline	35	323
2,5,5-Trimethylheptane	149	5	2,4,7-Trimethylquinoline	106	321
3,3,4-Trimethylheptane	201	6	2,4,8-Trimethylquinoline	18	323
3,3,5-Trimethylheptane	158	5	2,6,8-Trimethylquinoline	21	323
3,4,4-Trimethylheptane	200	6	2-(Trimethylsilyl)ethylamine	443	439
3,4,5-Trimethylheptane	202	6	3-(Trimethylsilyl)-n-propylamine	433	439
2,2,3-Trimethylhexane	95	4	2,3,4-Trimethylthiophene	42	424
2,2,4-Trimethylhexane	84	4	2,3,5-Trimethylthiophene	38	424
2,2,5-Trimethylhexane	79	4	Trimethyl trimesate	163	281
2,3,3-Trimethylhexane	108	4	4,4',4''-Trimethyltriphenyl carbinol	53	90
2,3,4-Trimethylhexane	111	4	Trimethylvinyl chloride	18	55
2,3,5-Trimethylhexane	89	4	2,4,6-Trinitroaniline	405	316
2,4,4-Trimethylhexane	87	4	" "	1	436
3,3,4-Trimethylhexane	116	4	2,4,6-Trinitroanisole	36	338
2,2,4-Trimethyl-3-hexanone	85	163	2,4,6-Trinitrobenzamide	499	246
1,2,3-Trimethylnaphthalene	3	43	1,3,5-Trinitrobenzene	85	340
1,2,4-Trimethylnaphthalene	33	44	2,4,6-Trinitrobenzoic acid	356	208
1,2,5-Trimethylnaphthalene	8	43	" " "	4	429
1,2,6-Trimethylnaphthalene	1	43	2,4,6-Trinitro-m-cresol	224	107
1,3,5-Trimethylnaphthalene	21	43	3,4,5-Trinitro-o-cresol	197	106
1,3,7-Trimethylnaphthalene	164	41	2,4,6-Trinitrophenetole	45	338
1,3,8-Trimethylnaphthalene	24	43	2,4,6-Trinitrophenol	281	110
1,4,5-Trimethylnaphthalene	49	45	" "	1	434
1,6,7-Trimethylnaphthalene	4	43	2,4,6-Trinitrothiophenol	40	417
2,3,6-Trimethylnaphthalene	106	47	2,4,6-Trinitrotoluene	46	338
2,4,6-Trimethylolphenol	79	434	Tri-n-pentylamine	89	321
Trimethyl orthoformate	33	251	Triphenylacetic acid	275	431
2,2,3-Trimethylpentane	51	3	Triphenylamine	75	324
2,2,4-Trimethylpentane	42	3	Triphenylcarbinol	60	90

Compound	No.	Page No.
2,3,-(3′,4′,5′-Triphenylcyclopentadieno)-indone ..	201	179
Triphenylmethane	93	46
Triphenylmethanethiol	33	417
Triphenylmethanol	60	90
Triphenylmethylamine	207	309
Triphenylmethyl chloride	9	57
Triphenyl phosphate	62	276
2,2,3-Triphenylpropionitrile	171	364
2,3,3-Triphenylpropionitrile	139	363
3,3,3-Triphenylpropionitrile	186	365
Tri-n-propylamine	24	319
" "	379	439
Tris(2-chloroethyl)amine	148	437
Tris(2-cyanoethyl)amine	28	436
Tris(2-hydroxyethyl)amine	284	438
Tris-p-tolylcarbinol	53	90
Tritriacontane	58	9
1-Tritriacontene	16	26
Tritriacontylcyclohexane	69	9
Tritriacontylcyclopentane	57	9
1-Tritriacontyne	20	31
Trityl chloride	9	57
d,l-Tropamide	337	240
Tropane	36	319
2-Tropene	30	319
d,l-Tropic acid	149	199
Tropidine	30	319
Tropilidene	191	18
DL-Tryptophane	84	288
"	92	429
L-Tryptophane	97	289
Turanose	39	331
Tyramine	368	314
DL-Tyrosine	124	290
L-Tyrosine	121	290
Tyrosol	167	104

U

Compound	No.	Page No.
Undecanal	6	150
n-Undecanamide	128	234
n-Undecananilide	37	231
n-Undecane	247	7
n-Undecanoic acid	7	193
n-Undecanoic anhydride	8	225
1-Undecanol	125	86
d,l-2-Undecanol	117	86
2-Undecanone	186	167
6-Undecanone	183	167
1-Undecanonitrile	90	352
1-Undecene	328	24
2-Undecene	326	24
5-Undecene	327	24
10-Undecen-1-oic acid	4	193
10-Undecenonitrile	92	352
w-Undecenoyl chloride	6	214
n-Undecyl acetylene	61	30
n-Undecyl alcohol	125	86
n-Undecylamine	155	300
"	418	439
sec-n-Undecylamine	151	299
n-Undecylbenzene	175	42
n-Undecyl chloride	52	56
n-Undecylcyclohexane	275	7
n-Undecylcyclopentane	271	7
α-Undecylenamide	67	232
Undecylenic acid	4	193
w-Undecylenoyl chloride	6	214
n-Undecylic acid	7	193
n-Undecyl sulfide	7	425
1-Undecyne	57	30
Urea	233	237

Compound	No.	Page No.
4-Ureidobutyric acid	415	432
2-Ureidoisobutyric acid	382	432
3-Ureidopropionic acid	390	432

V

Compound	No.	Page No.
Valeraldehyde	20	145
n-Valeramide	146	234
n-Valeranilide	28	231
n-Valeric acid	20	190
" "	450	432
Valeric anhydride	13	223
Valeronitrile	31	347
n-Valerophenone	207	168
n-Valeryl bromide	2	219
n-Valeryl chloride	22	212
DL-Valine	107	289
" "	71	429
L-(+)-Valine	122	290
Vanillalacetone	159	177
Vanillic acid	525	124
" "	332	207
Vanillin	126	103
"	70	156
"	28	434
o-Vanillin	33	434
Veratraldehyde	23	154
o-Veratraldehyde	32	154
Veratramide	326	240
Veratranilide	292	239
Veratric acid	281	204
Veratrole	2	137
Veratryl alcohol	132	86
Vinylacetanilide	40	232
Vinylacetic acid	12	190
" "	356	432
Vinylacetyl chloride	8	212
4-Vinylbenzaldehyde	18	152
4-Vinylbenzensulfonamide	97	395
4-Vinylbenzenesulfonic acid	33	372
Vinyl benzoate	177	259
Vinyl bromide	2	58
Vinyl chloride	2	55
1-Vinylcyclohexene	262	21
4-Vinylcyclohexene	226	19
Vinyl iodide	2	60
2-Vinylpyridine	189	437
4-Vinylpyridine	225	437
Vinyl sulfide	4	420
Vinylsulfonic acid	2	380

X

Compound	No.	Page No.
Xanthone	189	178
Xanthydrol	154	177
1,2-Xylene	6	35
1,3-Xylene	5	35
1,4-Xylene	4	35
m-Xylene-4-thiol	39	414
p-Xylene-2-thiol	38	414
p-2-Xylenol	108	102
o-3-Xylenol	109	102
m-4-Xylenol	8	96
o-4-Xylenol	69	100
m-5-Xylenol	82	100
p-Xylenyl acetate	256	264
sym-m-Xylenyl acetate	1	272
unsym-m-Xylenyl acetate	224	262
vic-m-Xylenyl acetate	207	261
m-2-Xylidine	115	298
p-2-Xylidine	112	298

Compound	Compound No.	Page No.
o-3-Xylidine	124	299
m-4-Xylidine	117	298
o-4-Xylidine	43	304
m-5-Xylidine	121	299
DL-Xyloketose	15	328
m-Xylo-p-quinone	7	181
o-Xylo-p-quinone	3	181
p-Xylo-p-quinone	15	182
α-D-Xylose	32	331
DL-Xylose	22	330
L-Xylose	31	331
6-[β-D-Xylosido]-D-glucose	63	333
DL-Xylulose	15	328
m-Xylylamine	101	298
o-Xylylamine	104	298
p-Xylylamine	105	298
o-Xylyl cyanide	86	352
m-Xylyl cyanide	83	352
p-Xylyl cyanide	84	352
m-Xylylenediamine	161	300
p-Xylylenediamine	15	303
m-Xylylene dichloride	1	57
o-Xylylene dichloride	6	57
p-Xylylene dichloride	8	57
p-Xylylene dimercaptan	14	416

Y

Compound	Compound No.	Page No.
Yperite	52	421

Index

This Index lists the names of all tables and major subjects contained in this book in alphabetical order.
The Index listing the individual compounds for which derivatives are cited begins on page 479.

	Table No.	Page No.
A		
Abbreviations	—	IX
Acetylenes	III	27–31
Acid anhydrides	XIV	223–227
Acids, amino	XVII	282–290
Acids, carboxylic	XII	186–210
Acids, natural, aldonic, uronic, aldaric, physical characteristics of	—	464
Acids, organic, dissociation constants of, in aqueous solution	XXVII	428–433
Acids, phenols, dissociation constants of, in aqueous solution	XXVIII	428; 434–435
Acids, sulfonic	XXII	369–380
Acyl halides	XIII	211–222
Alcohols	VI	77–90
Aldaric acids, natural, physical characteristics of	—	464
Aldehydes	IX	141–160
Alditols and inositols, physical characteristics of	—	462
Aldonic acids, natural, physical characteristics of	—	464
Aldosamines, physical characteristics of	—	462
Aldoses and ketoses, physical characteristics of	—	458
Alkanes and Cycloalkanes	I	2–9
Alkenes, Cycloalkenes, Dienes and Polyenes	II	10–26
Alkyl halides	V	55–61
Alkynes (acetylenes)	III	27–31
Amides and Imides	XV	228–247
Amines, primary, secondary, tertiary	XVIII	291–325
Amino acids	XVII	282–290
Amino sugars, physical characteristics of	—	461
Anhydrides, acid	XIV	223–227
Animal fats and oils, physical and chemical characteristics of	—	466–467
Aromatic Hydrocarbons	IV	32–51
Aryl halides	V	69–76
Atomic weights, table	—	478
B		
Bases, organic, dissociation constants of, in aqueous solution	XXIX	428; 436–439
Boiling point, molecular elevation of	—	457
Boiling points, correction to standard pressure	—	455–456
Bromides, Acyl Halides	XIII	218–220
Bromides, Alkyl and Cycloalkyl Halides	V	58–59
Bromides, Aryl Halides	V	73–74
Bromides, Dihalides and Polyhalides (non-aromatic)	V	65–66
C		
Carbohydrates	XIX	326–333
Carbohydrates, physical characteristics of	—	458–465
Carboxylic acids	XII	186–210
Chlorides, Acyl Halides	XIII	212–217
Chlorides, Alkyl and Cycloalkyl Halides	V	55–57
Chlorides, Aryl Halides	V	70–72
Chlorides, Dihalides and Polyhalides (non-aromatic)	V	63–64
Chlorides, Sulfonyl	XXIII	369–370; 381–390
Compounds, diamagnetic susceptibilities of organic		469–474
Compounds, organic, index		479
Cycloalkanes	I	2–9
Cycloalkenes	II	10–26
Cycloalkyl Halides	V	55–61

	Table No.	Page No.

D

Diamagnetic susceptibilities of organic compounds	—	469–474
Dienes	II	10–26
Dihalides	V	62–68
Dissociation constants of organic acids in aqueous solution	XXVII	428–433
Dissociation constants of organic bases in aqueous solution	XXIX	428; 436–439
Dissociation constants of phenols in aqueous solution	XXVIII	428; 434–435

E

Elements, periodic table of the	—	477
Emergent stem correction for liquid-in-glass thermometers	—	454
Esters	XVI	248–281
Ethers	VIII	129–140
Explanations and references to the Tables		1

F

Far I.R. charts— Far Infra-red vibrational frequency correlation charts	—	448–449
Fats, plant and animals, physical and chemical characteristics of	—	466–467
Fatty acids, concentration of, in various fats and oils	—	466–467
Fluorides, Acyl Halides	XIII	211
Fluorides, Aryl Halides	V	69
Fluorides, Dihalides and Polyhalides (non-aromatic)	V	62
Four-place Logarithms	—	475–476
Freezing point, molecular depression of	—	457

H

Halides, Acyl	XIII	211–222
Halides, Alkyl, Aryl, Cycloalkyl, Dihalides and Polyhalides	V	52–76
Hydrocarbons, aromatic	IV	32–51

I

Imides and amides	XV	228–247
Index of organic compounds	—	479
Inositols and alditols, physical characteristics of	—	462
Iodides, Acyl Halides	XIII	211–222
Iodides, Alkyl and Cycloalkyl Halides	V	60–61
Iodides, Aryl Halides	V	75–76
Iodides, Dihalides and Polyhalides (non-aromatic)	V	67–68
I.R. charts—Infra-red correlation charts	—	441–447

K

Ketones	X	161–179
Ketosamines, physical characteristics of	—	462
Ketoses and aldoses, physical characteristics of	—	458

L

Logarithms, four-place	—	475–476

M

Magnetic susceptibilities of organic compounds	—	470–475
Mercaptans (Thiols)	XXV	411–417
Miscibility of organic solvent pairs	—	451–453

562

	Table No.	Page No.

Molecular elevation of the boiling point . . . . — . . 457
Molecular depression of the freezing point . . . . — . . 457
Monosaccharides, natural, physical characteristics of . . . . — . . 458

N

Natural aldonic, aldaric and uronic acids, physical characteristics of . . . . — . . . 465
Nitriles . . . . XXI . . . 344–368
Nitro compounds . . . . XX . . . 334–343
NMR Charts—characteristic NMR (nuclear magnetic resonance) spectral positions for
   hydrogen in organic structures . . . . — 450

O

Oils, plant and animal, physical and chemical characteristics of . . . . — . . . 466–467
Optical rotation of certain sugars . . . . — 458
Organic acids, dissociation constants of, in aqueous solutions . . . . XXVII . . . 428–433
Organic compounds, diamagnetic susceptibilities . . . . — . . . 469–474
Organic compounds index . . . . — 479
Organic solvent pairs, miscibility of . . . . — . . . 451–453

P

Periodic table of the elements . . . . — 477
Phenols . . . . VII . . . 91–128
Phenols, dissociation constants of, in aqueous solution . . . . XXVIII 428; 434–435
Polyenes . . . . II 10–26
Polyhalides . . . . V 62–68

Q

Quinones . . . . XI . . . 180–185

S

Solvent, miscibility of organic . . . . — . . . 451–453
Specific rotation of sugars . . . . — 458
Stem correction, thermometer . . . . — 454
Sugars, amino, physical characteristics of . . . . — 461
Sulfides (Thioethers) . . . . XXVI . . . 418–427
Sulfonamides . . . . XXIV . . . 391–410
Sulfonanilides . . . . XXIV . . . 391–410
Sulfonic acids . . . . XXII . . . 369–380
Sulfonyl chlorides . . . . XXIII . . . 369–370; 381–390
Susceptibilities, diagmagnetic, of organic compounds . . . . — . . . 469–474

T

Thermometer correction stem . . . . — 454
Thioethers (Sulfides) . . . . XXVI . . . 418–427
Thiols (Mercaptans) . . . . XXV . . . 411–417

U

Uronic acids, natural, physical characteristics of . . . . — 464

W

Waxes, physical and chemical characteristics . . . . — 468

SELECTED CRC HANDBOOK SERIES

CRC HANDBOOK OF BIOCHEMISTRY AND MOLECULAR BIOLOGY
Gerald D. Fasman
Brandeis University

CRC HANDBOOK SERIES IN CLINICAL LABORATORY SCIENCE
David Seligson
Yale University

CRC HANDBOOK OF ELECTROPHORESIS
Lena A. Lewis
Cleveland Clinic
J. J. Opplt
Metropolitan General Hospital

CRC HANDBOOK SERIES IN ENGINEERING IN MEDICINE AND BIOLOGY
David G. Fleming
Case Western Reserve University
Barry N. Feinberg
Purdue University

CRC HANDBOOK OF ENVIRONMENTAL CONTROL
Richard G. Bond
Univeristy of Minnesota
Conrad P. Straub
University of Minnesota

CRC FENAROLI'S HANDBOOK OF FLAVOR INGREDIENTS
Nicolo Bellanca
CIBA-GEIGY Corp.
Giovanni Fenaroli
University of Milano, Italy
Thomas E. Furia
Dynapol

CRC HANDBOOK OF MARINE SCIENCE
F. G. Walton Smith
International Oceanographic Foundation
Frederick A. Kalber
Hydrobiological Services
Joseph T. Baker
Vreni Murphy
Roche Institute of Marine Pharmacology, Australia

CRC HANDBOOK OF MATERIALS SCIENCE
C. T. Lynch
Wright-Patterson Air Force Base

CRC STANDARD MATH TABLES
William H. Beyer
University of Akron

CRC HANDBOOK OF MICROBIOLOGY
Allen I. Laskin
Esso Research and Engineering Co.
Hubert Lechevalier
Rutgers University

CRC HANDBOOK SERIES IN NUTRITION AND FOOD
Miloslav Rechcigl, Jr.
Agency for International Development

CRC ATLAS OF SCINTIMAGING FOR CLINICAL NUCLEAR MEDICINE
Henry N. Wellman
Indiana University School of Medicine

CRC ATLAS OF SPECTRAL DATA AND PHYSICAL CONSTANTS FOR ORGANIC COMPOUNDS
Jeanette Grasselli
Standard Oil Company (Ohio)
William M. Ritchey
Case Western Reserve University

CRC HANDBOOK SERIES IN ZOONOSES
James H. Steele
University of Texas